TH
7011
H8
200

D1196315

LIBRARY
NSCC, WATERFRONT CAMPUS
80 MAWIO'MI PLACE
DARTMOUTH, NS B2Y 0A5 CANADA

THE HVAC
HANDBOOK

THE HVAC HANDBOOK

Robert C. Rosaler, P.E.   Editor-in-Chief

NSCC-INSTITUTE OF TECHNOLOGY LIBRARY
P.O. BOX 2210 (5685 LEEDS STREET)
HALIFAX, NOVA SCOTIA B3J 3C4

McGRAW-HILL

New York   Chicago   San Francisco   Lisbon   London   Madrid
Mexico City   Milan   New Delhi   San Juan   Seoul
Singapore   Sydney   Toronto

The **McGraw·Hill** *Companies*

Library of Congress Cataloging-in-Publication Data

The HVAC handbook / Robert C. Rosaler, editor in chief.
    p.   cm.
    Includes index.
    ISBN 0-07-140202-0
    1. Heating—Handbooks, manuals. etc.   2. Air conditioning—Handbook, manuals, etc.   3. Ventilation—Handbooks, manuals, etc.
    I. Rosaler, Robert C.

    TH7011.H8   2004
    697—dc22   2004040272

Copyright © 2004 by The McGraw-Hill Companies, Inc. All rights reserved. Printed in the United States of America. Except as permitted under the United States Copyright Act of 1976, no part of this publication may be reproduced or distributed in any form or by any means, or stored in a data base or retrieval system, without the prior written permission of the publisher.

1 2 3 4 5 6 7 8 9 0   DOC/DOC   0 1 0 9 8 7 6 5 4

ISBN 0-07-140202-0

The sponsoring editor for this book was Larry S. Hager and the production supervisor was Sherri Souffrance. It was set in Times Roman by International Typesetting and Composition. The art director for the cover was Anthony Landi.

Printed and bound by RR Donnelley.

This book was printed on recycled, acid-free paper containing a minimum of 50% recycled, de-inked fiber.

McGraw-Hill books are available at special quantity discounts to use as premiums and sales promotions, or for use in corporate training programs. For more information, please write to the Director of Special Sales, Professional Publishing, McGraw-Hill, Two Penn Plaza, New York, NY 10121-2298. Or contact your local bookstore.

Information contained in this work has been obtained by The McGraw-Hill Companies, Inc. ("McGraw-Hill") from sources believed to be reliable. However, neither McGraw-Hill nor its authors guarantee the accuracy or completeness of any information published herein, and neither McGraw-Hill nor its authors shall be responsible for any errors, omissions, or damages arising out of use of this information. This work is published with the understanding that McGraw-Hill and its authors are supplying information but are not attempting to render engineering or other professional services. If such services are required, the assistance of an appropriate professional should be sought.

This handbook is dedicated to all the practitioners of the HVAC industry: entrepreneurs, engineers, maintenance technicians, marketers, manufacturers, and editors who created and developed this wonderful technology that literally added new horizons to the realm of human comfort.

CONTENTS

CONTRIBUTORS

W.P. Bahnfleth   *The Pennsylvania State University, Department of Architectural Engineering, University Park, Pennsylvania (CHAP. 22)*

Donald W. Batz   *Senior Vice-President, Comm Air Mechanical Service, Oakland, California (SEC. 18.1)*

William N. Berryman   *Berryman Consulting, Gilbert, Arizona (CHAP. 13)*

Michael S. Bilecky   *Von Otto & Bilecky, Washington, D.C. (CHAP. 6)*

Gary M. Bireta   *Project Engineer, Mechanical Engineering, Arcadis-Giffels, Inc., Southfield, Michigan (SEC. 19.1)*

Richard T. Blake   *The Metro Group, Inc., Long Island City, New York (CHAP. 8)*

E.D. Cancilla   *Ajax Boiler Inc., Santa Ana, California (SECS. 16.1, 16.2)*

Nick S. Cassimatis   *Gas Energy, Inc., Brooklyn, New York (SEC. 18.3)*

Cleaver-Brooks   *Division of Aqua-Chem, Inc., Milwaukee, Wisconsin (SEC. 25.3)*

Keiron O'Connell   *AAF International, Louisville, Kentucky (CHAP. 20)*

Edward Di Donato   *Nordstrom Valves, Inc., Sulphur Springs, Texas (SEC. 25.5)*

Peter M. Fairbanks   *Bluestone Energy Services, Inc., Braintree, Massachusetts (CHAP. 9)*

Bayley Fan   *LAU Industries, Lebanon, Indiana (SEC. 23.2)*

Colin Frayne   *The Metro Group, Inc., Long Island City, New York (CHAP. 8)*

Joe Gosmano   *Gulf Coast General Manager, Marley Cooling Technologies, Houston, Texas (SEC. 19.4)*

Ernest H. Graf   *Assistant Director, Mechanical Engineering, Arcadis-Giffels, Inc., Southfield, Michigan (CHAP. 4; SECS. 19.1, 19.2)*

Nils R. Grimm   *Consulting Engineer, Hastings-on-Hudson, New York (CHAP. 3; SECS. 25.2, 26.1)*

Susan Perdue Hammock   *Alfa Laval Inc., Richmond, Virginia (SEC. 17.8)*

Lew Harriman   *Mason-Grant Company Portsmouth, New Hampshire (SEC. 18.9)*

Ben Harstine   *Supervisor of Field Service, Joy/Green Fan Division, New Philadelphia Fan Co., New Philadelphia, Ohio (SEC. 23.3)*

John C. Hensley   *Marketing Services Manager, The Marley Cooling Tower Company, Overland Park, Kansas (SEC. 19.4)*

M.B. Herbert   *Willow Grove, Pennsylvania (CHAP. 2)*

Hudy C. Hewitt   *Chairman, Department of Mechanical Engineering, University of New Orleans (SEC. 18.8)*

Martin Hirschorn   *President, Industrial Acoustics Company, Bronx, New York (CHAP. 10)*

Terry Hoffmann   *Johnson Controls, Inc., Milwaukee, Wisconsin (CHAP. 7)*

James E. Hope   *Director of Technical Services, ITT Bell & Gossett, Morton Grove, Illinois* (SEC. 25.4)

H. Michael Hughes   *Engineering Consultant, Lavonia, Georgia* (SEC. 18.1)

Robert Jorgensen   *Retired Vice-President-Engineering, Buffalo Forge Company, Buffalo, New York* (SEC. 23.1)

Jay Kohler   *York International Corp., York, Pennsylvania* (SEC. 18.3)

Ronald A. Kondrat   *Product Manager, Heating Division, Modine Manufacturing Co., Racine, Wisconsin* (SECS. 17.6, 17.7)

Douglas Kosar   *Senior Project Manager, Gas Research Institute, Chicago, Illinois* (SEC. 18.9)

W.J. Kowalski   *The Pennsylvania State University, Department of Architectural Engineering, University Park, Pennsylvania* (CHAP. 22)

Kenton J. Kuffner   *WaterFurnace International, Fort Wayne, Indiana* (SEC. 18.6)

Billy C. Langley   *Consulting Engineer, Azle, Texas* (SECS. 18.6, 18.7)

Melvin S. Lee   *Senior Project Designer, Arcadis-Giffels, Inc., Southfield, Michigan* (SEC. 19.2)

Lehr Associates   *New York, New York* (SECS. 17.1, 17.2, 17.3, 17.4, 17.5, 17.11, 17.12)

Robert L. Linstroth   *Product Manager, Heating Division, Modine Manufacturing Co., Racine, Wisconsin* (SEC. 17.6)

William S. Lytle   *Project Engineer, Mechanical Engineering, Giffels-Arcadis, Inc., Southfield, Michigan* (CHAP. 4)

Chan Madan   *Continental Products, Inc., Indianapolis, Indiana* (SEC. 18.2)

Howard J. McKew   *RDK Engineers, Andover, Maine* (CHAPS. 1, 14, 15)

Ravi K. Malhotra   *President, Heatrons Corp., Fenton, Missouri* (SEC. 18.7)

The Mechanical Contracting Foundation   *Rockville, Maryland* (SEC. 25.1)

Simo Milosevic   *Project Engineer, Mechanical Engineering, Arcadis-Giffels Inc., Southfield, Michigan* (SEC. 19.3)

Ron Moore   *The RM Group, Knoxville, Tennessee* (CHAPS. 11, 12)

George Player   *RDK Engineers, Andover, Maine* (CHAP. 14)

Michael J. Palazzolo   *General Manager, Safety King, Inc., Utica, Michigan* (SEC. 26.2)

Michael S. Palazzolo   *President, Safety King, Inc., Utica, Michigan* (SEC. 26.2)

Kenneth Puetzer   *Chief Engineer, Sullair Refrigeration, Subsidiary of Sundstrand Corp., Michigan City, Indiana* (SEC. 18.5)

Radiant Panel Association   *Loveland, Colorado* (SEC. 17.10)

James A. Reese   *York International Corp., York, Pennsylvania* (CHAP. 24)

Mike O'Rovrke   *Business Unit Manager, Sterling Hydronics Westfield, Massachusetts* (SEC. 17.9)

Willis Schroader   *Robur Corp., Evansville, Indiana* (SEC. 18.3)

John M. Schultz   *Retired Chief Engineer, Centrifugal Systems, York International Corp., York, Pennsylvania* (SEC. 18.4)

Aparajita Sengupta   *Kellogg Brown & Root, Inc., Houston, Texas* (CHAP. 5)

Hal M. Sindelar   *Product Manager, Heating Division, Modine Manufacturing Co., Racine, Wisconsin* (SECS. 17.6, 17.7)

Alan J. Smith   *Kellogg Brown & Root, Inc., Houston, Texas (CHAP. 5)*

Mark W. Spatz   *Honeywell International, Inc., Buffalo, New York (SEC. 18.1)*

C. Curtis Trent   *President Emeritus, Trent Technologies, Inc., Tyler, Texas (SEC. 18.8)*

Warren C. Trent   *CEO, Trent Technologies, Inc., Tyler, Texas (SEC. 18.8)*

William A. Turner   *President, Turner Building Science, LLC, The H.L. Turner Group Inc., Concord, New Hampshire (CHAP. 21)*

Webster Engineering & Manufacturing Co., Inc.   *Winfield, Kansas (SEC. 16.3)*

George White   *Aggreko, Inc., Benicia, California (SEC. 18.4)*

PREFACE

This *HVAC Handbook,* in one comprehensive volume, combines the essential contents of the previously published *HVAC Systems and Components Handbook* and the *HVAC Maintenance and Operations Handbook.*

New topics have been added and current topics expanded to reflect current practice. These include: IAQ Management, Aerobiological Congrols, Condensate Control, Commissioning, Heat Pumps, and more.

An introductory chapter includes a perspective on industry trends, and should be read as a guide prior to the handbook topics.

The editor wishes to acknowledge the advice and guidance of Larry Hager of McGraw-Hill on publication matters and Howard J. McKew, P.E. on technical guidelines.

ROBERT C. ROSALER, P.E.

THE HVAC HANDBOOK

P · A · R · T · A

INTRODUCTION

CHAPTER 1
HISTORY, PERSPECTIVE AND TRENDS IN HVAC DESIGN AND USAGE

Howard J. McKew, P.E., CPE
Andover, Massachusetts

1.1 INTRODUCTION

This volume provides information and expert advice on the operation and maintenance of heating, ventilating, and air-conditioning systems. It addresses some of the basics of HVAC equipment, automatic temperature controls, and system distribution. In compiling the data for this book it is important to reflect back on the HVAC industry as it relates to how buildings have been designed, built, operated, and maintained. Equally important is to look forward to how buildings will be designed, built, operated, and maintained in the twenty-first century. In reflecting back, old habits are hard to break. Going forward, we need to revisit *all* of our processes and integrate computer technology into our everyday facility management lives.

The future of the HVAC industry will involve this merging of the building industry with the computer age as design engineers, builders and facility operators reluctantly embrace today's technology. It is this computer technology that offers the largest challenge to all these participants, simply because old habits are hard to change. The agenda is complicated by an industry where many a participant is of the precomputer age: Engineers, contractors, and operators, who honed their skills prior to the energy crisis and oil embargo of the early 1970s, fill many of the leadership roles today based on experience and seniority.

While these old timers are trying to cope with change, a more computer-literate generation of professionals has joined the building industry workforce, skilled and comfortable with computers, software, and the Internet. The challenge for this new generation is to retrain their mentors to more efficient ways of doing business. We have seen how drafting boards and straight edges have been displaced by computer-aided drawings (CAD). This obvious change was just the tip of the iceberg and a good starting point to where we were, where we are, and where we are going.

1.2 WHERE WE WERE: CONSTRUCTION DOCUMENTS

If you go back in time to the end of the eighteenth and nineteenth centuries and pick up a set of construction drawings, you will find that the drawings were completed on a linen material and the drafting was done with the use of ink drafting tools. This approach to drafting was still being taught in schools in the early 1960s. By the end of that decade the drafting community was changing. Drawings were being completed on plastic and the drafting pencil, which once held lead, was also changed to plastic.

The design community reached a plateau in the 1970s when it came to producing construction drawings but began to change again in the early 1980s as it embraced computer drafting. Many a design firm began to enter the computer age with cumbersome software to product the contract documents. Perceived to be a drafting process, design engineers quickly began to realize that they could not apply CAD without integrating one more "D". CADD (computer-aided design drafting), showed that by designing the project as it was being produced via computer software, it could now begin to be cost-effective.

The design community was also beginning to dabble with computerized calculations, pushing aside the manual engineering calculations of the late 1960s by inputting the data first onto punch cards that were then inputted into the computer software. The computer calculation process proved to be more efficient and cost-effective several years before the CADD process.

As engineers began to dabble with the computer age, the construction industry was several years behind the design community when it came to the application of technology. What *did* change for the construction industry was how buildings would be built, as *fast track* became the buzzword for the 1970s. One can observe the progression of the builder from general contractor to construction manager to design builder in less than 20 years. Like their engineer counterpart, contractors went about their business in the 1960s the same way they went about it 50 years earlier. The 1970s signalled the beginning of the change for processes associated with design and construction documents. It also signalled a change as to how buildings would now be built.

All this transformation was then emphasized with "time is money" as the energy crisis would impact the cost of living and, for building programs, the cost of construction. If you follow the architectural community's production of AIA documents, you can also follow the demise of architect leadership. Once the standard of the building industry, the AIA Owner-Architect agreement and Owner-Contractor agreement began to be displaced with the AIA Construction Manager document in the 1970s. Contract documents were dissected and issued for bid in packages to firm up "fixed costs" as soon as possible. This fragmentation of drawings and specifications led to problematic issues that the facility operators were left to resolve—incomplete contract documents.

General contractors were quick to change hats and become construction managers where they could pick up more control of a building program and more of the "soft cost" fee associated with preconstruction services. By the end of the 1980s, there was another AIA Design-Build document but it was too little, too late. Construction managers now had a say in the design phase, as they were the perceived experts for job cost control. Surveys showed contractors were managing the majority of design–build projects and they would soon be looking to introduce program management as the next generation of design–build and single-source solutions in the building industry. With each transformation from design–bid–build to construction management to design–build to program management, construction documents were also reshaped to keep pace with the builder.

While the construction community leapfrogged over their counterparts in the design community, these builders were not as quick to embrace computer technology. For the construction industry, it would take the 1990s to push them into the computer age with estimating and project management software. The construction community is only now in the twenty-first century considering issues associated with CAD, such as CAD layering standards.

1.3   WHERE WE WERE: FACILITY MANAGEMENT

For the building operators, technology was more often dumped on them, as they became those responsible for operating and maintaining the buildings designed and built by others. Probably the most obvious change for facility management has been in name only. Up until the 1960s, you would find a *boiler operator* and a *chief engineer*. Titles changed to *facility engineer* to *facility manager* to *vice president of engineering* and eventually *vice president of support services*.

Facility management was also impacted by the energy crisis as building systems became more complex and energy management systems displaced the much simpler building systems of the previous 50 years. Utilities became more costly to operate, and strategic planning and purchasing of fuel made dual-fuel boilers (oil and gas) displace no. 6 fuel oil with its negative environmental impact. Perceived as a necessary evil, building operators still lack the respect for their design and construction counterparts. Building owners and developers may be quick to see an iceberg but fail to look below the surface of the waterline. A rule of thumb has documented that somewhere between 10 percent and 25 percent of the cost to build, operate, and maintain a building is an *first cost*. Using this benchmark, approximately 80 percent of the cost of a building will be invested *after the building is built*. Today, recognition for efficient facility management remains in the minority.

It has been this misunderstanding of what operation and maintenance does and the values associated with this profession that has compromised the success of peak building performance. A lack of respect for the job at hand continues to hold back facility management and, in doing so, holds back the ability to invest in computer technology. At the same time, facility management has hampered their own progress by not embracing today's technology as they continue to manage as they did 30 to 40 years ago—with little use of computers.

Over the years, computerization of the operation and maintenance of systems has been handed to building operators by designer engineers who know little about managing a facility. Instead of providing computer software and CAD record drawings that can facilitate operation and maintenance, the documents are a by-product of building design. How the design community came to the point where they have anointed themselves as experts in specifying what needs to go into an operation and maintenance manual is a mystery. While building technicians are knowledgeable of tours, tasking, frequency, and special instructions, it has been the design community that has set the standard for operation and maintenance documents.

As a result, facility managers must reinvest time and effort to compile data collection that can then be inputted into their preventive maintenance work order system. It has been these "manual" work order systems that have been the cornerstone to facility management for the past 100 years. With the twentieth century closing out, computerized maintenance management software (CMMS) systems began to be considered by facility management. Still time was the "fly in the ointment" as seldom was there time to transfer equipment data into these computer software systems. To compound this dilemma, few in the building management profession were computer literate and/or had the computer hardware and software to support this effort. At the same time, very few consulting engineers understood what CMMS was, so as to begin to change their standardized specifications to require these operation and maintenance tools. The end product of a building program would remain unchanged until only now.

A similar problem exists today when addressing the record drawing process. Until around 1990, construction drawings were transferred to the building owner on paper or plastic. This process had not changed from buildings constructed at the start of the 1900s. Here again facility managers were receiving documentation based on the needs and understanding of the designer and *not* the operator. Standing on the sideline, building operators had little to say and/or offered little to the design team as to facility management needs. As a result, with architects and engineers embracing CAD, the facility manager would soon be receiving computer-generated documents that operation and maintenance could not conveniently use. With the twentieth century coming to a close, building operators were not trained in the use of CAD nor did they have the necessary software to bring the drawing up on their computer or to print out this CAD document. To make matters worse, the design community had established standard layers for managing the information on their construction drawings. These layers had little to do with operation, maintenance, security, regulatory compliance, life safety, etc., that facility management faced on a daily basis. The idea of CAD layered documents that could be an integral part of a computer-aided facility management (CAFM) system was foreign to the design community who were the creators of these drawings.

1.4   WHERE WE ARE AND WHERE WE ARE GOING

"Where we were" has a direct impact on "where we are" and "where we need to go" in facility management. In reflecting back over the years, it is important to emphasize that technology and how the building industry applied this technology may make it appear that much had occurred between 1900 and 2000. Unfortunately, techniques have not been as on progressive as one may think. Today the design community continues to fail to educate itself about their facility management counterpart. The same can be said for the construction industry. Both appear to operate in a vacuum when it comes to delivering a finished product that the facility management staff can seamlessly embrace and take control of. At the same time, there is plenty of blame to go around and facility management needs to share in this blame as they too have operated within their own business world, seldom participating in joint venture discussions, training sessions, and planning that could contribute to that elusive, seamless process.

In response to facilities that are now very complex and embrace computerized technology, building owners are finding that these sophisticated mechanical, electrical, and data communication systems frequently do not operate as intended. In troubleshooting the problems, finger pointing by the designers and the builders provides no solution. Building owners are now demanding that building systems be commissioned before owner acceptance. This directive is reinforced by certification programs, such as LEED (leadership in engineering and environmental design) that make commissioning a prerequisite of this program. Similar high-performance certification and grants also make commissioning a prerequisite. It is this mandate that reinforces the need for the designer and the builder to deliver a finished product because facility management is that 80 percent cost that makes up the life cycle cost of owning and operating a building—the "necessary evil" below the waterline.

To this mandate, the designers are now jumping on the commissioning mandate with a "yes, we can do that" to hopefully grab another opportunity to sell their services. The construction community is also advocating the additional cost for commissioning because many recognize this process as a means to close the project out quicker by having documents that provide a comprehensive means to demonstrate/verify that the system functionally performs per design. With more sophisticated computerized equipment and systems, commissioning is one way for facility managers to receive a better product. At the same time, it is this continued emphasis on technology that continues to make the transition to owner occupancy a difficult task.

Today, facility management is applying all that computer technology from the design and construction industry, and facility management is slowly striving to make better use of that technology. What holds back many a building technician is the ability to take owner-ship for a facility if it was not turned over "running on all 8 cylinders." While engineers continue to design building systems and construction professionals strive to take more control of the building process, facility staffs are being downsized, reengineered, and outsourced. This places added pressure on the operation and maintenance group to suc-cessfully run the building, maintain space comfort, and consume energy at a minimum operating cost.

With their performance under a microscope, operation and maintenance groups need to figure out a means to win respect from their boss. Facility operators need to "take a page" from their leaders. In a survey of hospitals in New England, all the healthcare operations queried stated that this facility had changed logo, company letterhead, etc., based on a mar-keting consultant's advice to change and to be more competitive. Building management needs to also market their services today so that they can get the required financial support to con-tinue to be successful. Just the term "support services" implies this service is not essential. Facility management should be cataloged with other business development budgets because, without an efficient and cost-effective building, the core business will be compromised.

Embracing computer technology and leapfrogging over the design and construction industry, building operators need to aggressively use the business tools provided by the construction project. In unison with this effort, operation needs to be proactive and pro-gressive in other ways to expedite their jobs. Facility management must turn the table on the building industry and take a leadership rather than a receivership role. While the cur-rent core of design engineers have grown up on computers and software (the same for the contractors) the next generation of operation and maintenance should be setting the pace for the building industry.

To *have* respect requires *earning* respect. To date, operation and maintenance have taken a backseat to their counterparts. Facility managers must educate their counterparts long before the building is turned over to them. The word needs to get out as to what is really required to efficiently operate and maintain a facility. These requirements must find their way into antiquated construction specifications. Many of the issues, topics and con-cepts within this book should become central guideposts for construction documents instead of operation and maintenance criteria from the past.

While change has affected the design of HVAC systems with the advancement of equip-ment, automatic controls, and efficiency of these systems, operation and maintenance con-tinues to lag behind. What is needed for a jump-start is the turnover of the project to the owner and the application of handheld computers. Facility requirements should be drafted by the facility management industry and then shared with the design and construction so that the correct operation and maintenance requirements can get published; no more designer interpretation of what is needed. Building operators must get involved if they are going to be ready to receive the end product. Accepting ownership and immediate respon-sibility of the building will be a prerequisite of facility groups.

1.4.1 Handheld Computer Applications

So far, computer software programs have not facilitated good communication between the design engineer and the facility engineer. Transfer of information between the builder and the owner has progressively gotten worse as technology has gotten better. The problem can be directly attributed to a lack of communication. The second part of the problem is that many computer programs go unused or receive limited use due to an ill-prepared facility group. The solution is literally in the palm of the operator's hand with the successful advancement of handheld computers.

In the late 1990s the use of personal digital assistants (PDAs) grew dramatically by people of all ages. The use of these handheld devices started out as computer games conveniently packaged for kids. Adults quickly embraced this concept as software programs were packaged for those who wanted their contact list, telephone numbers, daily calendar, etc. with them at all times. Basically, a person's daily planner went hi-tech with the introduction and success of this minicomputerized device. This unique technological breakthrough opened the door to more advanced computer capabilities that can be held in someone's hand. Finally, the vision of having access to a CAD drawing, equipment specification, preventive maintenance work order, and so on downloaded into these new handheld computers was now a reality.

Also, bar coding is now a viable tool for the facility manager who may have been using this identification process for inventory control. Clearly, the application of bar coding equipment, panels, valves, and devices in the building operation and maintenance community has been a quarter of a century behind the times. Today, bar codes can be found on just about everything *but* HVAC equipment. Now, handheld computers can displace the cell phone and/or two-way radio as the vehicle for data collection with the power of the computerized database. The bar code now becomes the link between the equipment or device and all that operation and maintenance data stored somewhere back in a document file room. Instead of calling in a request for information to the facility operations office, a bar coded piece of equipment can quickly and efficiently bring the needed information to the technician's handheld computer screen.

For many people in the operation and maintenance of buildings, including HVAC, familiarity with PDAs will streamline the education process necessary to get up to speed with more advanced handheld programs. Also ready for application are all those CAD documents turned over to building owners at the time of project occupancy. Inputting the information is now a reality rather than a vision. What will be required now will be computer specialists who can program and transfer these business management tools into a user-friendly format that can be brought to the computer screen in a timely manner. A solution to this obstacle may fall under the umbrella of commissioning of facility management. With building owners requesting that commissioning be included as an integral part of a building program, commissioning of the operation and maintenance will be an extension of this quality process. A new generation of facility management is now on the horizon.

Through time management, operation and maintenance can now be a seamless process if specified to occur early in the construction phase. Operation and maintenance data collection being specified to begin being processed immediately after shop drawings are approved will streamline and replace an antiquated request for operation and maintenance manuals at the end of a project. The building industry needs to only look at the automotive industry and how computerization of cars has reshaped our expectations when buying a vehicle. As the car or truck rolls off the assembly line, it has been tuned up (commissioned) and the operation and maintenance manual is conveniently located in the vehicle. No person purchasing an automobile would ever expect the owner's manual to show up weeks or months later and, upon arrival, be incomplete.

As handheld technology becomes a tool of the trade for service technicians, wireless technology will further enhance this computerization process. Just as people can call a remote site to have their car unlocked because they left the keys in the vehicle, wireless control management is here, ready to assist facility operations.

SYSTEM DESIGN AND APPLICATIONS

CHAPTER 2
BASICS

M. B. Herbert, P.E.
Willow Grove, Pennsylvania

2.1 INTRODUCTION

Heating, ventilating, and air-conditioning (HVAC) systems are designed to provide control of space temperature, humidity, air contaminants, differential pressurization, and air motion. Usually an upper limit is placed on the noise level that is acceptable within the occupied spaces. To be successful, the systems must satisfactorily perform the tasks intended.

Most heating, ventilating, and air-conditioning systems are designed for human comfort. Human comfort is discussed at length in Ref. 1. This reference should be studied until it is understood because it is the objective of HVAC design.

Many industrial applications have objectives other than human comfort. If human comfort can be achieved while the demands of industry are satisfied, the design will be that much better.

Heating, ventilating, and air-conditioning systems require the solution of energy–mass balance equations to define the parameters for the selection of appropriate equipment. The solution of these equations requires the understanding of that branch of thermodynamics called "psychrometrics." Ref. 2 should be studied.

Automatic control of the HVAC system is required to maintain desired environmental conditions. The method of control is dictated by the requirements of the space. The selection and the arrangement of the system components are determined by the method of control. Controls are necessary because of varying weather conditions and internal loads. These variations must be understood before the system is designed. Control principles are discussed in Chap. 7 and in Ref. 3.

The proliferation of affordable computers has made it possible for most offices to automate their design efforts. Each office should evaluate its needs, choose from the available computer programs on the market, and then purchase a compatible computer and its peripherals.

No one office can afford the time to develop all its own programs. Time is also required to become proficient with any new program, including those developed "in-house."

Purchased programs are not always written to give the information required, thus they should be amenable to in-house modification. Documentation of purchased programs should describe operation in detail so that modification can be achieved with a minimum of effort.

2.2 CONCEPT PHASE

The *concept phase* of the project is the feasibility stage; here the quality of the project and the amount of money to be spent are decided. This information should be gathered and summarized on a form similar to Fig. 2.1.

2.2.1 Site Location and Orientation of Structure

The considerations involved in the selection of the site for a facility are economic:

1. Nearby raw materials
2. Nearby finished-goods markets
3. Cheap transportation of materials and finished goods
4. Adequate utilities and low-cost energy sources for manufacturing
5. Available labor pool
6. Suitable land
7. Weather

These factors can be evaluated by following the analysis given in the *Handbook of Industrial Engineering and Management* (See Bibliography). It is prudent to carefully evaluate several alternative sites for each project.

The orientation of the structure is dictated by considering existing transportation routes, obstructions to construction, flow of materials and products through the plant, personnel accessibility and security from intrusion, and weather.

2.2.2 Codes, Rules, and Regulations

Laws are made to establish minimum standards to protect the public and the environment from accidents and disasters. Federal, state, and local governments are involved in these formulations. Insurance underwriters may also impose restraints on the design and operation of a facility. It is incumbent upon the design team to understand the applicable restraints *before the design is begun*. Among the applicable documents that should be studied are

1. Occupational Safety and Health Act (OSHA)
2. Environmental Protection Agency (EPA) requirements
3. National Fire Protection Association (NFPA), Fire Code (referenced in OSHA)
4. Local building codes
5. Local energy conservation laws, which usually follow the American Society of Heating, Refrigeration, and Air-Conditioning Engineers (ASHRAE) Standard 90.1A

2.2.3 Concept Design Procedures

The conceptual phase requires the preparation of a definitive scope of work. Describe the project in words. Break it down to its components. Itemize all unique requirements, what is required, why, and when. Budgeting restraints on capital costs and labor hours should be included. A convenient form is shown in Fig. 2.2. This from is a starting tool for gathering data.

2.5

COMPANY _____ PROJECT _____ P.O. NO. _____ DATE _____
LOCATION _____ SPACE NAME _____ SIZE _____ SHEET NO. _____

ACTIVITY

	M	T	W	T	F	S	S
DAY OF WEEK							
NO. PEOPLE							
HOURS/DAY							

BUILDING CONSTRUCTION

FLOOR _____
WALLS _____
WINDOW GLASS _____ FRAME _____
SHADING _____
CEILING _____
ROOF _____
DOORS _____
PARTITIONS _____

CODES, BUILDING _____
MECHANICAL _____
PLUMBING _____
ELECTRICAL _____
FIRE _____

ENVIRONMENT

TEMPERATURE ± ___ dB, ± ___ WB, ± ___ RH
VENTILATION ___ CFM, ___ ACHR, ___ %OA
AIR FILTERS ___ TYPE, ___ %EFF.
AIR PRESSURE ___ IN.W.W. (+), (-)

LIGHTING TYPE _____ , ___ WATTS
ELECTRICAL CLASS _____ POWER _____
EMERGENCY LIGHTING _____
TYPE CONTROL _____

TELEPHONE _____ INTERCOM _____
CCTV _____ COMPUTER _____
WORD PROCESSOR _____

HAZARDS & SAFETY

FIRE CLASS _____
HAZARDOUS MATERIALS & QUANTITIES _____

TYPE OF FIRE PROTECTION _____

REASONS _____

TYPE OF FIRE ALARM _____
SAFETY SHOWER & EYEWASH _____ STRETCHER _____
FIRE BLANKET _____

NOTES

EQUIPMENT LIST

ITEM	QUANTITY	SIZE			COLD WATER			HOT WATER			WASTE		GAS		AIR		VACUUM		STEAM		PROCESS VENTILATION				ELECTRICAL		NOTES
		LENGTH	WIDTH	HEIGHT	PRESSURE	PRESS. LOSS	GPM	TEMP./PRESS	PRESS. LOSS	GPM	SANITARY, GPM	PROCESS, GPM	PRESSURE	SCFM/SCFH	PRESSURE	SCFM	PRESSURE	SCFM	TEMP./PRESS	LB/HR	EXH. TEMP/ΔP	EXH. SCFM	SUP. TEMP/ΔP	SUP. SCFM	VOLTS/PHASE	KVA/KW	

FIGURE 2.1 Design information form.

```
COMPANY _____  PO NO. _____  DATE _____

LOCATION _____  SHEET NO. _____

SUBJECT _____

                          PROJECT BRIEF

COMPUTED BY _____  CHECKED BY _____

TYPE OF PROJECT _____

☐  HEATING _____
☐  VENTILATING, Comfort, Process, _____
☐  AIR CONDITIONING, Comfort, Process, _____
☐  PLUMBING, Sewage Treatment _____
☐  FIRE PROTECTION _____
☐  PROCESS PIPING _____
☐  ELECTRICAL, Power, Lighting, Control _____
☐  STRUCTURAL, Civil _____
☐  ARCHITECTURAL _____
☐  _____

DUE DATES:

Preliminaries _____ , Cost Estimates _____ , Final Documents _____ ,

SCOPE OF WORK
_____
_____
_____
_____
_____
_____
_____
_____
_____
_____
_____
_____
_____
_____
_____
_____

PROJECT ASSIGNMENTS:  Proj. Mgr. _____          Proj. Engr. _____
Discipline Engrs. _____

CONTACTS
| Name & Title | Firm Name | Address | Telephone |
```

FIGURE 2.2 Project brief form.

It will suffice for many projects. For a major project, a more formal written document should be prepared and approved by the client. This approval should be obtained before proceeding with the design.

The method of design is influenced by the client's imposed schedule. Fasttracking methods will identify long delivery items that might require early purchase. Multiple construction packages are not uncommon, since they appreciably reduce the length of construction time. Usually, more engineering effort is required to divide the work into separate bid packages. Points of termination of each contract must be shown on the drawings and reflected in the scope of work in the specifications. Great care in the preparation of these documents is required to prevent omission of some work from all contracts and inclusion of some work in more than one contract.

Some drawings and some sections of the specifications will be issued in more than one bid package. To prevent problems, the bid packages should be planned in the concept stage and carried through to completion of the project. All changes must be defined clearly for everyone involved in the project.

Every step of the design effort should be documented in written form. When changes are made that are beyond the scope of work, the written documents help recover costs necessitated by these changes. Also, any litigation that may be instituted will usually result in decisions favorable to those with the proper documentation.

After the scope of work has been accurately documented and approved, assemble the data necessary to accomplish the work:

1. Applicable building codes
2. Local laws and ordinances
3. Names, titles, addresses, and telephone numbers of local officials
4. Names, titles, addresses, and telephone numbers of client contacts
5. Client's standards

If the project is similar to previous designs, review what was done before and how well the previous design fulfilled its intended function.

Use check figures from this project to make an educated guess of the sizes and capacities of the present project. Use Figs. 2.3 and 2.4 to record past projects.

Every project has monetary constraints. It is incumbent upon the consultant to live within the monies committed to the facility. Use Figs. 2.5 and 2.6 to estimate the capacities and costs of the systems. Do not forget to increase the costs from the year that the dollars were taken to the year that the construction is to take place.

Justification for the selection of types of heating, ventilating, and cooling systems is usually required. Some clients require a detailed economic analysis based on life cycle costs. Others may require only a reasonable payback time. If a system cannot be justified on a reasonable payback basis, then it is unreasonable to expect the more detailed analysis of life cycle costs to reverse the negative results. A simple comparison between two payback alternatives can be made as follows:

$$\text{Payback years } N = \frac{\$ \text{ first cost}}{\$ \text{ savings, first year}} \tag{2.1}$$

This simple payback can be refined by considering the cost of money, interest rate i (decimal), and escalation rate e (decimal). The escalation rate is the expected rate of costs of fuel, power, or services. The actual number of years for payback n is given by

$$n = \frac{\log\left[1 + N(R-1)/R\right]}{\log R} \tag{2.2}$$

JOB NAME SPACE NAME YEAR OF DESIGN TYPE OF SYSTEM	OUTSIDE DESIGN CONSIDERATIONS DB °F/°C	WB °F/°C	INSIDE DESIGN CONSIDERATIONS DB °F/°C	WB °F/°C	FLOOR AREA SQ FT (SQ M)	CFM/SQ FT (CMS/SQ M)	% O A	BTU/HR-SQ FT (W/HR-SQ M) ROOM SENS	GRAND TOTAL	LIGHT & POWER WATTS/SQ FT (WATTS/SQ M)	SQ FT/PERSON (SQ M/PERSON)	SQ FT/TON (SQ M/KW)	IND APP DEW POINT °F (°C)

FIGURE 2.3 Air-conditioning check figures.

JOB NAME SPACE NAME YEAR OF DESIGN TYPE OF SYSTEM	DESIGN CONSIDERATIONS		FLOOR AREA SQ. FT. (SQ. M)	VENTILATION		INFILTRATION		HEATING LOAD BTU/HR-SQ FT (W/HR-SQ M)	NOTES
	OUTSIDE °F (°C)	INSIDE °F (°C)		$\frac{CFM}{SQ\ FT}\left(\frac{CMS}{SQ\ M}\right)$	AC/HR	$\frac{CFM}{SQ\ FT}\left(\frac{CMS}{SQ\ M}\right)$	AC/HR		

FIGURE 2.4 Heating check figures.

COMPANY _____ PO NO. _____ DATE _____

LOCATION _____ SHEET NO. _____

SUBJECT _____

COMPUTED BY _____ CHECKED BY _____

ESTIMATED COST											
REFRIG. TONS (KW)											

AIR QUALITY	EXHAUST CFM (CMS)											
	SUPPLY CFM (CMS)											
	AC/HR											
	CFM/SQ FT (CMS/SQ M)											

ROOM VOLUME CU FT (CU M)											
FLOOR AREA SQ FT (SQ M)											
ROOM NAME & SIZE TYPE OF SYSTEM											

FIGURE 2.5   Conceptual design estimate form for cooling.

COMPANY _____	PO NO. _____	DATE _____
LOCATION _____		SHEET NO. _____
SUBJECT _____		

COMPUTED BY _____   CHECKED BY _____

ESTIMATED COST										
HEAT LOAD BTU/HR (KW)										
HEAT REQUIRED — BTU/SQ FT (W/CMS)										
HEAT REQUIRED — BTU/SQ FT (W/CU M)										
HEAT REQUIRED — BTU/SQ FT (W/SQ M)										
ROOM VOLUME CU FT (CU M)										
FLOOR AREA SQ FT (SQ M)										
ROOM NAME & SIZE TYPE OF SYSTEM										

FIGURE 2.6 Conceptual design estimate form for heating.

where

$$R = \frac{1+e}{1+i}$$

and N is defined by Eq. (2.1). This formula is easily programmed on a hand-held computer. A nomographic solution is provided in Ref. 4.

There are many other economic models that a client or an engineering staff can use for economic analysis. There is much literature available on this subject.

2.3   PRELIMINARY DESIGN PHASE

The *preliminary design phase is* the verification phase of the project. Review the concept phase documents, especially if a time lapse has occurred between phases. Verify that the assumptions are correct and complete. If changes have been made, even minor ones, document these in writing to all individuals involved.

2.3.1   Calculation Book

The calculations are the heart of decision making and equipment selection. The calculation book should be organized so that the calculations for each area or system are together. Prepare a table of contents so anyone may find the appropriate calculations for a given system. Use divider sheets between sections to expedite retrieval. All calculations should be kept in one place. Whenever calculations are required elsewhere, make the necessary reproductions and promptly return the originals to their proper place in the calculation book.

2.3.2   Calculations

The calculations reflect on the design team. The calculations should be neat, orderly, and complete, to aid checking procedures. Most industrial clients require that the calculations be submitted for their review. Also when revisions are required, much less time will be spent making the necessary recalculations. All calculations made during this phase should be considered accurate, final calculations.

Many routine calculations can now be done more rapidly and more accurately with the aid of a computer. The computer permits rapid evaluation of alternatives and changes. If a computer program is not available for a routine calculation, the calculation should be done and documented on a suitable form. If a form does not exist, develop one.

All calculations should be dated and signed by the designer and checker. Each sheet should be assigned an appropriate number. When a calculation sheet is revised, a revision date should be added. When a calculation sheet is superseded, the sheet should be marked "void." Do not dispose of superseded calculations until the project is built satisfactorily and functioning properly.

List all design criteria on sheets such as Fig 2.7, referencing sources where applicable. List all references used in the design at appropriate points in the calculations.

When you are doing calculations, especially where forms do not exist, always follow a number with its units, such as feet per second (meters per second), British thermal units (watts, foot-pounds, newton-meters), etc. This habit will help to prevent the most common blunders committed by engineers.

To avoid loose ends and errors of omission, *always try to complete one part or section of the work before beginning the next.* If this is impossible, keep a "things to do" list, and list these open ends.

COMPANY _____ P.O. NO. _____ DATE _____

LOCATION _____ SHEET NO. _____

SUBJECT _____
_____

COMPUTED BY _____ CHECKED BY _____

OUTSIDE DESIGN DATA

Data for _____ Elevation above mean sea level _____

Latitude _____ ° _____ ' _____   Latitude _____ ° _____ ' _____

Item	Winter		Summer	
Temperature, DB/WB/DP†	/	/	/	/
Pressure, Total/Vapor	/		/	
Humid. Ratio/%RH/Enthalpy	/	/	/	/
Specific Volume				
Mean Daily Temp Range				
Wind Velocity				
Hours Exceed Design, %				

Summer Design Day Temperatures

Hour	1	2	3	4	5	6	7	8	9	10	11	12	13	14	15	16	17	18	19	20	21	22	23	24
DB																								

Month	Cooling Out. Design DB	WB	CLTD Corrections To	N	NNE NNW	NE NW	ENE WNW	E W	ESE WSW	SE SW	SSE SSW	S	Horiz.
JAN													
FEB													
MAR													
APR													
MAY													
JUN													
JUL													
AUG													
SEP													
OCT													
NOV													
DEC													

Heating Degree Days

Month	JAN	FEB	MAR	APR	MAY	JUN	JUL	AUG	SEP	OCT	NOV	DEC	YEAR
D.D.													

FIGURE 2.7   Outside design data form.

2.3.3 Equipment Selection

From the calculations and the method of control, the capacity and operating conditions may be determined for each component of the system. Manufacturers' catalogs give extensive tables and sometimes performance curves for their equipment. All equipment that moves or is moved vibrates and generates noise. *In most HVAC systems, noise is of utmost importance* *to the designer (*See Chap. 10). Beware of the manufacturer that is vague or ignorant about the noise and vibration of its equipment or is reluctant to produce certified test data.

Many equipment test codes have been written by ASHRAE, American Refrigeration Institute (ARI), Air Moving and Conditioning Association (AMCA), and other societies and manufacturer groups. A comprehensive list of these codes is contained in ASHRAE handbooks. Manufacturer's catalogs usually contain references to codes by which their equipment has been rated. *Designers are warned to remember that the manufacturer's representative is awarded for sales of equipment, not for disseminating advice.* Designers should make their own selections of equipment and should write their own specifications, based on past experience.

2.3.4 Equipment Location

Mechanical and electrical equipment must be serviced periodically and eventually replaced when its useful life has expired. To achieve this end, every piece of equipment must be accessible and have a planned means of replacement.

The roof and ceiling spaces are not adequate equipment rooms. Placing equipment on the roof subjects the roof to heavy traffic, usually enough to void its guarantee. The roof location also subjects maintenance personnel to the vagaries of the weather. In severe weather, the roof may be too dangerous for maintenance personnel.

Ceiling spaces should not be used for locating equipment. Servicing equipment in the ceiling entails erecting a ladder at the proper point and removing a ceiling tile or opening an access door to gain access to the equipment. Crawling over the ceiling is dangerous and probably violates OSHA regulations. No matter how careful the maintenance personnel are, eventually the ceiling will become dirty, the tiles will be broken, and if water is involved, the ceiling will be stained.

Also, the equipment will suffer from lack of proper maintenance, because no one on a ladder can work efficiently. This work in the occupied space is disruptive to the normal activities of that space.

Equipment should be located in spaces specifically designed to house them. Sufficient space should be provided so that workers can walk around pieces of equipment, swing a wrench, rig a hoist, or replace an electric motor, fan shaft, or fan belts. Do not forget to provide space for the necessary electrical conduits, piping, and air ducts associated with this equipment. Boilers and other heat exchangers require space for replacing tubes. Valves in piping should be located so that they may be operated without resorting to a ladder or crawling through a tight space. If equipment is easily reached, it will be maintained. Adequate space also provides for good housekeeping, which is a safety feature.

Provision of adequate space in the planning stage can be made only after the types and sizes of systems have been estimated. Select equipment based on the estimated loads. Lay out each piece to a suitable scale. Arrange the equipment room with cutout copies of the equipment. Allow for air ducts, piping, electrical equipment, access aisles, and maintenance workspace. Cutouts permit several arrangements to be prepared for study.

When you are locating the equipment rooms, be sure each piece of equipment can be brought into and removed from the premises at any time during the construction. A strike may delay the delivery of a piece of equipment beyond its scheduled delivery date. This delay should not force construction to be halted, as it would if the chiller or boiler had to be set in place before the roof or walls were constructed.

2.3.5 Distribution Systems

HVAC distribution systems are of two kinds: air ducts and piping. Air ducts are used to convey air to and from desired locations. Air ducts include supply air, return-relief air, exhaust air, and air-conveying systems. Piping is used to convey steam and condensate, heating hot water, chilled water, brine, cooling tower water, refrigerants, and other heat-transfer fluids. Energy is required to force the fluids through these systems. This energy should be considered when systems are evaluated or compared.

System Layouts. Locate the air diffusers and heat exchangers on the prints of the architectural drawings. Note the air-flow rates for diffusers and the required capacities for the heat exchangers. Draw tentative single-line air ducts from the air apparatus to the air diffusers. Mark on these lines the flow rates from the most remote device to the fan. With these air quantities, the air ducts may be sized. Use Section 26.1, or *ASHRAE Handbook, Fundamentals*, Chap. 32, or the *Industrial Ventilation Manual* to size these ducts. Record these sizes on a form similar to those shown there.

A similar method is used to size the piping systems; see Chap. 25. Remember, steam, condensate, and refrigerant piping must be pitched properly for the systems to function correctly. Water systems should also be pitched to facilitate draining and elimination of air.

Piping systems are briefly described in Chap. 25 of this book and in the *ASHRAE Handbook, Fundamentals*. A more substantial treatment is contained in *Piping Handbook* (see Bibliography).

REFERENCES

1. *2003 ASHRAE Handbook, Fundamentals*, ASHRAE, Atlanta, GA, 2003, chap. 8, "Physiological Principles and Thermal Comfort."
2. *2003 ASHRAE Handbook, Fundamentals*, chap. 6, "Psychrometrics."
3. John E. Hains, *Automatic Control of Heating and Air Conditioning*, McGraw-Hill, New York, 1953.
4. John Molnar, *Nomographs—What They Are and How to Use Them*, Ann Arbor Science Publishers, Ann Arbor, MI, 1981.

BIBLIOGRAPHY

ASHRAE: *Cooling and Heating Load Calculation Manual*, 2d ed. American Society of Heating, Refrigeration, and Air-Conditioning Engineers, Atlanta, 1992.

————: Energy Conservation in Existing Buildings—High Rise Residential ASHRAE ANSI/ASHRAE/IES 100.2-1991

————: Energy Conservation in Existing Buildings—Commercial ASHRAE ANSI/ASHRAE/IES 100.3-1995

————: Energy Conservation in Existing Facilities—Industrial ASHRAE ANSI/ASHRAE/IES 100.4-1984

————: Energy Conservation in Existing Buildings—Institutional ASHRAE ANSI/ASHRAE/IES 100.5-1991

————: Energy Conservation in Existing Buildings—Public Assembly ASHRAE ANSI/ASHRAE/IES 100.6-1991

————: Energy Conservation in New Building Design—Residential only ASHRAE ANSI/ASHRAE/IES 90A-1980

————: Energy Efficient Design of New Buildings Except Low Rise Residential Buildings ASHRAE ASHRAE/IES 90.1-1989

————: Psychrometrics Theory & Practice, ASHRAE, Atlanta, 1996.

————: *Simplified Energy Analysis Using the Modified Bin Method*, ASHRAE, Atlanta, 1984.

————: 1995 ASHRAE Handbook, HVAC Applications

————: 1994 ASHRAE Handbook, Refrigeration

————: 1997 ASHRAE Handbook, Fundamentals

————: 1996 ASHRAE Handbook, HVAC Systems & Equipment

Baldwin, John L.: *Climates of the United States*, Government Printing Office, Washington, DC, 1974.

Fan Engineering, Buffalo Forge Co., Buffalo, NY.

Handbook of Industrial Engineering and Management, 2d ed., Prentice-Hall, Englewood Cliffs, NJ, 1971.

Hartman, Thomas B.: *Direct digital control for HVAC Systems*, McGraw-Hill, New York, 1993.

Hydraulic Institute: *Pipe Friction Manual*, Hydraulic Institute, Cleveland, 1975.

Industrial Ventilation, A Manual of Recommended Practice, 22d ed., American Conference of Governmental Industrial Hygienists, Lansing, MI, 1994.

Kusuda, T.: *Algorithms for Psychrometric Calculations*, National Bureau of Standards, Government Printing Office, Washington, DC, 1970.

Molnar, John: *Facilities Management Handbook*, Van Nostrand Reinhold, New York, 1983.

————: *Nomographs—What They Are and How to Use Them*, Ann Arbor Science Publishers, Ann Arbor, MI, 1981.

Naggar, Mohinder L.: *Piping Handbook*, 5th ed., McGraw-Hill, New York, 1992.

NFPA: *National Fire Codes*, National Fire Protection Association, Batterymarch Park, Quincy, MA, 1995.

CHAPTER 3
DESIGN SOFTWARE

Nils R. Grimm, P.E.*

Consulting Engineer, Hastings-on-Hudson, New York

3.1 INTRODUCTION

One of the cardinal rules for a good, economical, energy-efficient design is *not* to design the total system (be it heating, ventilating, air conditioning, exhaust, humidification, dehumidification, etc.) to meet the most critical requirements of just a small (or minor) portion of the total area served. That critical area should be isolated and treated separately.

The designer today has the option of using either a manual method or a computer program to calculate heating and cooling loads, select equipment, and size piping and ductwork. For large or complex projects, computer programs are generally the most cost effective and *should be used*. On projects where life cycle costs and/or annual energy budgets are required, *computer programs should be used*.

Where one or more of the following items will probably be modified during the design phase of a project, *computer programs should be used:*

- Building orientation
- Wall or roof construction (overall U value)
- Percentage of glazing
- Building or room sizes

However, for small projects a manual method should be seriously considered before one assumes automatically that computer design is the most cost-effective for all projects.

In the next section, heating and cooling loads are treated together since the criteria and the computer programs are similar.

3.2 HEATING AND COOLING LOADS

The first step in calculating the heating and cooling loads is to establish the project's heating design criteria:

---

*Software by Trane Corp., St. Paul, MN and Carrier Corp., Syracuse, NY.

- Ambient dry-bulb or wet-bulb temperature (or relative humidity), wind direction and speed
- Site elevation above sea level, latitude
- Space dry-bulb or wet-bulb temperature (or relative humidity), ventilation air
- Internal or process heating or cooling and exhaust air requirements
- Hours of operation of the areas or spaces to be heated or cooled (day, night, weekday, weekends, and holidays)

Even when the owner or user has established the project design criteria, the designer should determine that they are reasonable.

The winter outdoor design temperature should be based preferably on a minimum temperature that will not be exceeded for 99 percent of the total hours in the months of December, January, and February (a total of 2160 h) in the northern hemisphere and the months of June, July, and August in the southern hemisphere (a total of 2208 h). However, for energy conservation considerations, some government agencies and the American Society of Heating, Refrigeration, and Air-Conditioning Engineers (ASHRAE) Standard 90-75, *Energy Conservation in New Building Design*, require the outdoor winter design temperature to be based on a temperature that will not be exceeded 97.5 percent of the same total heating hours.

Similarly, the summer outdoor design dry-bulb temperature should be based on the lowest dry-bulb temperature that will not be exceeded 2.5 percent of the total hours in June through September (a total of 2928 h) in the northern hemisphere and in December through March in the southern hemisphere (a total of 2904 h). For energy conservation reasons, some government agencies require the outdoor summer design temperature to be based on a dry-bulb temperature that will not be exceeded 5 percent of the same total cooling hours.

More detailed or current weather data (including elevation above sea level and latitude) are sometimes required for specific site locations in this country and around the world than are included in standard design handbooks such as Refs. 1 and 2 or computer programs such as Refs. 3 and 4 or from Ref. 5.

It is generally accepted that the effect of altitude on systems installed at 2000 ft (610 m) or less is negligible and can be safely omitted. However, systems designed for installations at or above 2500 ft (760 m) must be corrected for the effects of high altitude. Appropriate correction factors and the effects of altitudes at and above 2500 ft (760 m) are discussed in App. A of this book.

To avoid overdesigning the heating, ventilating, and air-conditioning system so as to conserve energy and to minimize construction costs, each space or area should be analyzed separately to determine the minimum and maximum temperatures that can be maintained and whether humidity control is required or desirable.

The U.S. government has set 68°F (20°C) as the maximum design indoor temperature for personnel comfort during the heating season in areas where employees work. In manufacturing areas the *process requirements* govern the actual temperature. From an energy conservation point of view, if a process requires a space temperature greater than 5°F (2.8°C) above or below 68°F (20°C), the space should, if possible, be treated separately and operate independently from the general personnel comfort areas. The staff members working in such areas should be provided with supplementary spot (localized) heating, ventilating, and air-conditioning systems as the conditions require, in order to maintain personnel comfort.

The space's dry-bulb temperature, relative humidity, number of people, and ventilation air requirements can be established (once the activity to be performed in each space is known) from standard design handbook sources such as Refs. 2, 6 to 8, 10, and 22 for heating and Refs. 1, 6 to 22, 27, and 40 for cooling.

The normal internal loads generally produce a heat gain and therefore usually are not considered in the space heating load calculations but must be included in cooling load calculations. These internal loads, including process loads, are listed in standard design handbook sources such as Refs. 23 and 24.

The process engineering department or quality control group should determine the manufacturing process space temperature, humidity, and heating requirements. The manufacturer of the particular process equipment is an alternative source for the recommended space and process requirements.

The air temperature at the ceiling may exceed the comfort range and should be considered in calculating the overall heat transmission to or from the outdoors. A normal 0.75°F (0.42°C) increase in air temperature per 1 ft (0.3 m) of elevation above the breathing level [5 ft (1.5 m) above finish floor] is expected in normal applications, with approximately 75°F (24°C) temperature difference between indoors and outdoors.

There is limited information on process heating requirements in standard handbooks, such as Refs. 25 to 35, and on cooling requirements, such as Refs. 25, 27, and 29 to 35.

Usually the owner and/or user establishes the hours of operation. If the design engineer is not given the hours of operation for the basis of the design, she or he must jointly establish them with the owner and/or user.

The method of calculating the heating or cooling loads (manual or computer) should be determined next.

3.2.1 Manual Method

If the manual method is selected, the project heating loads should be calculated by following one of the accepted procedures found in standard design sources such as Refs. 21, 22, and 36 to 39. For cooling loads, see Refs. 21 to 24, 37 to 39, 41, and 42.

3.2.2 Computer Method

If the computer method has been chosen to calculate the project heating or cooling loads, one must then select a program to use among the several available. Two of the most widely used for heating and cooling are Trane's TRACE and other Customer Direct Service (CDS) Network diskettes and Carrier's E20-II programs.

Regardless of the program used, its specific input and operating instructions must be strictly followed. It is common to trace erroneous or misleading computer output data to mistakes in inputting the design data into the computer. *It cannot be overstressed that to get meaningful output results, the input data must be correctly entered and checked after entry before the program is run.* It is also a good policy, if not a mandatory one, to independently check the computer results the first time you run a new or modified computer program, to ensure the results are valid.

If the computer program used does not correct the computer output for the effects of altitude when the elevation of the project is equal to or greater than 2500 ft (760 m) above sea level, the computer output must be manually corrected by using the appropriate correction factors, listed in App. A of this book.

We outline the computer programs available with TRACE and other CDS diskettes and E-20-II in the remainder of this chapter. However, *this is not to imply that these are the only available sources of programs for the HVAC fields.* Space restraints and similarities to other programs are the reasons for describing programs from only two sources. *Programs are changing rapidly, and you should keep up-to-date on these continually.*

3.3 TRANE PROGRAMS*

Software can dramatically aid the system selection process by simulating various alternatives accurately and quickly.

Programs are available that perform accurate energy and load analyses which can then be translated into dollars and cents by modeling a particular utility's rates. Still other computerized design tools predict acoustical performance and simplify HVAC equipment selection, air and water distribution, life-cycle costing, and system comparisons.

The following summary describes programs available. (Ref. 43)

3.3.1 System Analyzer

System Analyzer helps estimate loads and perform energy and economic analyses, so as to quickly evaluate virtually any building, system, and equipment combination. It is a scoping tool to decide what systems might be appropriate for an initial design or to get a general idea of how one system-and-equipment combination performs when compared with another. The entry categories are as follows:

- Weather information
- Building type
- System
- Equipment
- Utility rates
- Equipment costs

Weather Information.   Select a weather profile from various locations around the world. It includes the option of overriding the provided weather information. The selected weather location determines the default utility provider and rate structure, as well as the currency.

Building Type.   This section allows selecting a building template and then defining its geometry (dimensions, physical orientation), construction, internal loads, and operating schedules.

The template provides default values for many of these entries. Edit them or create a new template and add it to the Building Loads template library.

System.   Select up to two airside systems for each building. After selecting the system, further define it by describing fan characteristics, static pressures, and temperatures. System Analyzer provides six constant volume systems:

- Fan coil
- Heat pump
- Packaged terminal air conditioner (PTAC)
- Variable temperature constant volume
- Terminal reheat
- Multizone

---

*This section courtesy of the Trane Corporation, St. Paul, MN.

And six variable volume systems:

- Parallel fan-powered VAV
- Series fan-powered VAV
- VAV with reheat (VRH)
- Two-fan dual-duct VAV
- Bypass VAV
- Shutoff VAV

Equipment.   Select up to two equipment types for the refrigeration side of the HVAC system. Additional entries can define pump characteristics, energy rates, and operating schedules. System Analyzer provides six equipment types:

- Chiller plant
- Self-Contained air conditioners
- Rooftop units
- Air-to-Air heat pumps
- Packaged terminal air conditioners
- Water-source heat pumps

Utility Rates.   Select from several provided utility rate structures or create one. Enter rates for electricity, gas, oil, district-chilled water, and other utilities. The weather location selected when completing the regional criteria entries automatically determines the default utility rate structure.

Equipment Costs.   Define the financial factors System Analyzer uses to generate cost analyses and comparisons. Rely on the default values, or improve the accuracy of these analyses by changing the entries to more closely reflect the project.

Reports (Output).   Once you make the necessary entries in each entry category, "tell" System Analyzer to calculate your results. Once the calculation is complete, view the results. System Analyzer has functions that enable exporting of any of the spreadsheet-style reports as Dbase tables or tab-delimited ASCII text. In addition, System Analyzer has the ability to produce output reports compatible with the GLHEPRO program which was developed to aid in the design of ground loop heat exchangers used in geothermal heat pump systems.

Job Summary Report.   Recaps the airside system and equipment definitions and capacities, as well as the outcomes of the life-cycle-cost and environmental-impact analyses.

Design Report.   Breaks out design cooling-load values and design heating loads in separate columns. Both sections flag "user over sizing" if you defined an existing building and entered load values that differ from the loads calculated by the program.

System Analyzer's design calculations are intended only as a quick—therefore rough—estimate of the maximum loads in the building. Do *not* use them to actually size equipment!

System Report.   Shows the system load profiles for each alternative (i.e., operating hours in 5 percent increments of loads and airflows), as well as the monthly coil loads (including the effect of any EMS/BAS Controls).

Energy Reports.   Shows monthly energy consumption by equipment type (heating, cooling, air-moving, miscellaneous, domestic hot water) and by utility (electric, gas, etc.). In addition the electric demand is displayed for each month and each rate type.

Economics Report.   Provides comprehensive output that allows direct comparisons between alternatives. Costs are broken down for each alternative and a yearly cash flow is determined for each year of the study. The cumulative costs are displayed in both future value and present value and a cost difference is calculated.

Graphs.   Select between several preformatted graphs (or you can create one of your own) to plot one aspect of one or more alternatives.

Profiles Reports.   Shows all of the hourly demand information (building loads, outside air temperature, thermal storage profiles, etc.) calculated by the program for the current job. Select between several preformatted reports or create one of your own.

Methodology.   System Analyzer generalizes building load characteristics. All buildings are calculated as rectangular with two zones—an interior and a perimeter—and evenly distributed internal loads, ventilation, and glass. The perimeter depth is defined as the distance between the exterior walls and the interior zone. Each zone is calculated independently with no thermal interaction.

Weather profiles are available for 24-h periods for each month. The profiles contain dry-bulb temperature, wet-bulb temperature, enthalpy, air pressure, solar radiation, and cloud cover. There are two different types of weather data used: design weather and typical weather.

Design weather is the worst-case weather and is used to determine the design building load. Equipment sizing is determined from the design building load. Design weather is also used to calculate the design-day peak-energy demand for each month. System Analyzer uses ASHRAE 2.5 percent weather for summer design and ASHRAE 97.5 percent weather for winter design.

Typical weather is an average weather profile that has been calculated based on existing weather data. It is used to calculate the weekday and weekend average energy consumption for each month.

3.3.2 TRACE 700

In 1973 the first version of Trane Air Conditioning Economics, TRACE, became the first computer program of its type. It is a complete load, system, energy, and economic analysis program that compares the energy and economic impact of such building alternatives as architectural features, HVAC systems, HVAC equipment, building utilization or scheduling, and economic options.

The TRACE program is, in essence, an analytic tool for building system designers. It enables them to optimize the building, system, and equipment designs on the basis of energy utilization and life cycle cost. Using the TRACE program in the early stages of building planning allows the building owner and building design team to receive the maximum benefit of a detailed analysis.

TRACE can also be invaluable for assessing the energy and economic impact of building renovation or system retrofit projects. To fully appreciate the capabilities of the Trane Air Conditioning Economics program and to obtain maximum value from its use, the following is a thorough explanation of the program.

Program Organization.   The TRACE program incorporates five major phases, each with specific tasks or functions that must be performed to provide a complete energy and economic analysis. The names of these phases are *load, design, system simulation, equipment simulation* and *economic analysis.*

The building heating/cooling load calculations, used in the load phase of the program for annual energy consumption analysis, are of sufficient detail to permit the evaluation of the effect of building data such as orientation, size, shape and mass, heat transfer characteristics of air and moisture, as well as hourly climatic data.

Beyond this, the calculations used to simulate the operation of the building and its service systems through a full-year operating period, are of sufficient detail to permit the evaluation of the effect of system design, climatic factors, operational characteristics, and mechanical equipment operating characteristics on annual energy usage. Manufacturers' data is used in the program for the simulation of all systems and equipment.

The calculation procedures used in TRACE are based upon 8,760 h of operation of the building and its service system. These procedures use techniques recommended in the appropriate ASHRAE publications or produce results which are consistent with such recommended techniques.

This document shows the calculation procedures used in the TRACE program. The calculations explicitly cover the following items:

• Climatic data, including coincident hourly data for temperature, solar radiation, wind, and humidity of typical days of the year, representing seasonal variations. In total, the TRACE program calculates building heat gains and losses, by zone, for 1152 h of the year, representing seasonal variations. TRACE has the ability to import hourly climatic data and calculate building heat gains and losses for all 8760 h in a year.

• Building and orientation, size, shape, mass, and heat transfer characteristics of air and moisture.

• Building operational characteristics, accounting for temperature, humidity, ventilation, illumination, and control modes for occupied and unoccupied hours.

• Mechanical operational characteristics, which take into account design capacity, part load performance and ambient dry-bulb and wet-bulb depression effects on equipment performance and energy consumption.

• Internal heat generation from illumination, equipment, and the number of people in occupied spaces during both the occupied and unoccupied hours.

Load Phase. In the *load* phase of the program, conventional load data describing the building construction, orientation and location are required entries. In addition, the utilization profile of the building, including lighting schedules, occupancy schedules, and miscellaneous load schedules, are required.

The program obtains weather data from its library for the city designated by the program user. Building loads are then actually calculated by zone and by hour, from information provided from the weather library (in the case of weather-dependent loads). It takes into account the coincident loading scheduled by the program user for items such as lights, people, and miscellaneous loads.

Beyond this, the program accounts for energy consumed by systems that do not contribute loads to the air conditioning system. These energy consumptions have an effect upon the overall energy demand of the building and the associated energy costs.

Design Phase. The second major phase of the program is the *design* phase. The purpose of this phase is to establish the building load model at design conditions. Entries required by the design phase include the type of mechanical system, as well as the percentage of wall, lighting, and miscellaneous loads assigned to the return air. In addition, the design outside air quantities is required.

The program then determines design-cooling load, heating load, outside air quantity, total air quantity, and the supply air dry-bulb temperature. The air quantities and supply air dry bulb in the cooling mode are determined psychrometrically using standard procedures outlined in the ASHRAE Handbook of Fundamentals. Design loads determined in this phase are based on 100 percent of design-entered values, even though the coincident design values of weather-affected loads may not actually occur during the weather year. The aforementioned design values are determined for both the perimeter and interior system from entries by the user.

Airside System Simulation. The next major phase of the program is the *airside system simulation* phase. Its key function in the program is to translate building heat gains and losses into equipment loads by system and by hour, utilizing all of the building variables that affect the system operation. In this phase, the program tracks an air particle around the complete airside system loop, picking up loads and canceling simultaneous gains and losses along the airflow path of each system.

The final output from the system simulation phase is the equipment load by system and by hour. This consists of air-moving loads, heating loads, cooling loads, and humidification loads where applicable. This is perhaps the most complicated phase of the program. Complication arises from the fact that each major system or system combinations and hybrids thereof, must utilize separate individual system programming subroutines to reflect the actual operation and control of that system.

The program contains system simulation programming subroutines for 32 different system types. These can be combined to form innumerable variations for the building under study.

Equipment Simulation. The equipment loads, by system and by hour, are then provided to the *equipment simulation* phase, along with a description of the equipment to be used in the system.

The previously described weather information is also input into this phase. Regardless of whether the equipment has air-cooled or water-cooled condensing, the weather affects the overall part load efficiencies.

The essential function of the equipment and simulation phase is to translate equipment loads, by system and hour, into energy consumption by source. The loads are translated into kilowatt-hours of electricity, therms of gas, oil, district hot water or chilled water, even to the extent of calculating the total gallons of makeup water required by a cooling tower or the energy consumed by the crankcase heaters of a reciprocating compressor. The entry requirements of this phase consist only of the equipment types for heating, cooling and air-moving as well as pumping heads and pump motor efficiency for each system where hydronic pumping is involved.

This data is utilized within the program to call for the equipment library, which is the performance information for the various pieces of equipment. This information is used to convert system loads into energy consumption for subsequent processing to the economic analysis phase.

It is important to note it is not necessary for the user to enter the part load performance of equipment accessories into the program. They are already contained in the equipment library and are accessed when called for by the user.

Economic Phase. The next and final major phase of the program is the *economic analysis* phase. This phase utilizes user entries, such as the utility rates and system installed cost data, along with other economic information such as mortgage life, cost of capital, etc., to compute annual owning and operating costs. It also calculates the various financial measurements of an investment such as cash flow effect, profit and loss effect, payout period, and return on additional investment between alternatives.

In very simple terms, the program determines how much it costs to operate one system compared with another. It then computes the present worth of the savings and the incremental return on the additional investment. It is keyed to provide information the owner needs to make his or her final economic decision, including monthly and yearly utility costs over the life of the HVAC system.

3.3.3 TRACE 700 Load Design

Load Phase. The *load* phase of the program computes the peak sensible and latent zone loads as well as the block sensible and latent loads for the building. In addition, the hourly sensible and latent loads, including weather-dependent loads, are calculated for each zone, based on the weather library.

Loads defined in the calculation are:

* External loads
    Wall load
    Glass load
    Roof load
    Floor load
* Internal loads
    Lighting
    People
    Miscellaneous

The specific entries required to facilitate these calculations are:

* External loads
    * Weather
        Weather library (geographic location)
        Outdoor design (winter, summer)
    * Solar
        Latitude and longitude
        Time zone
        Clearness number
        Design month
        Building orientation
    * Construction
        Roof heat transfer properties
        Roof area
        Wall heat transfer properties
        Glass U-value
        Percent glass
        Glass shading coefficient
* Internal loads
    * Room design temperature (set point)
        Summer, winter
    * Design values
        Lighting
        People sensible and latent heat
        People density
        Miscellaneous sensible and latent heat

- Utilization schedules
  - Lighting
  - People
  - Miscellaneous

Day Calculations

- Weather data for one day
- Sunrise and sunset time, solar declination angle, degrees equation of time, hour constants

Hour Calculations

- Direction cosines of sun, direct normal solar intensity, sky brightness

Zone Calculations

- Ground brightness, angle of incidence, direct solar radiation, sky diffuse radiation, ground diffuse radiation, total solar radiation
- Wall load
- Transmission and absorption factors of glass
- Glass film coefficient
- Solar heat gain through glass
- Glass load
- Ground brightness, angle of incidence, direct solar radiation, sky diffuse radiation, ground diffuse radiation, total solar radiation
- Roof load
- Floor load
- Partition load
- Internal loads, people, lights, miscellaneous

Design Phase. The design phase of the TRACE program calculates the design supply air temperatures, heating and cooling capacities, and supply air quantities given the peak load files generated by the load phase. For applications where the building design parameters are known, you can override the calculation of these values using optional entries to the system phase. This gives you the ability to simulate existing buildings with installed equipment that may not be sized according to the loads calculated in the load phase.

The entries required for these calculations are as follows:

- Output from load phase
  - Zone peak loads
  - Building block loads
- Room design conditions
- Outside air conditions
- Outside air percentage
- Percentage of internal and wall loads to return air
- System type
- System fan static pressure

- Fan motor efficiency
- Infiltration
- Reheat minimum airflow

The design phase will first assign the peak or block load calculated by the load phase to the systems providing cooling. Once the loads have been assigned, the latent and sensible components of the loads are totaled and the sensible heat ratio (SHR) for each system is determined.

Knowing the design room conditions, design outside air conditions, percent of outside air used for ventilation, and fan heat, the supply air dry-bulb temperatures for each system are psychrometrically established. Applying this supply air dry-bulb temperature to each zone, the required peak airflow for each individual zone is determined. For peak-air systems, the system airflow is determined by totaling the zone peak airflows. For block-air systems, the system airflow is based on the block load of the system. From the earlier psychrometric simulation, the coil entering and leaving enthalpy conditions can be determined. The program then modifies the enthalpy difference to correct for the design barometric pressure. The modified enthalpy difference is subsequently used to calculate the design cooling capacity. The design heating capacity for primary systems is calculated by summing the wall, glass, floor, and roof loads plus the ventilation and infiltration loads at the winter design temperature. In addition, mixing and reheat systems will include a reheat load. All internal and solar loads should conservatively be scheduled unavailable when arriving at the final design heating capacity.

Assignment of Loads.   The first step in the design phase is to assign zone sensible loads to the system that will ultimately handle that particular load. The assignment of loads to the return air also takes place. The percentage of lights, wall, and roof loads assigned to return air will reduce the zone sensible load. To determine whether the system space cooling loads are based on the block system load or the sum of the zone peak loads:

If no skin system is specified, all of the heating and cooling loads are assigned to the primary system.

$$QSYSc = QLITEs \times (1 - PCLRA) + QPEOPs + QMISCs$$

$$+ \, QWALLc \times (1 - PCWRA) + QGLASSc$$

$$+ \, QROOFc \times (1 - PCRRA) + QFLOORc + QINFc$$

$$QSYSh = QWALLh \times (1 - PCWRA) + QGLASSh$$

$$+ \, QROOFh \times (1 - PCRRA) + QFLOORh + QINFh$$

where

$$QINFc = K \times CFMINF \times (SDDB - RMDBc)$$

$$QINFh = K \times CFMINF \times (SDDB - RMDBh)$$

If a heating-only skin system is specified, only the wall, glass, and floor heating loads are assigned to the skin system, and the remaining heating loads are assigned to the primary system.

If a heating/cooling skin system is specified, both the heating and cooling wall, glass, and roof loads are assigned to the skin system, while the remaining loads are assigned to the primary system.

Only the primary system may handle latent loads.

System Cooling Supply Air Dry Bulb. Once the system loads have been assigned, the psychrometric iteration to find SADBc is performed using the following procedure.

- Step A. Calculate the system SHR and the temperature increase due to the supply and return fans.
- Step B.
   1. Assume a value for coil leaving the dry bulb, and then
   2. Determine the cooling supply air dry bulb, SADBc.

Given the SHR line and SADBc, the wet bulb can be taken from the psych chart.

- Step C. Calculate cooling coil airflow. The coil airflow may be a sum of the peaks airflow, or a block airflow.
- Step D. Determine the temperature increase due to the return air loads and return air fan. Use only block return air loads.
- Step E. Determine the return air temperature.
- Step F. Determine the return/outside air mixture condition. Note that RACFM = Coil CFM − OACFM.
- Step G. Determine the coil-entering condition. For draw-through fan configurations, the coil-entering condition is the same as the return/outside air mix. For blow-through fan configurations, the supply fan heat must be added first.
- Step H. Determine the new coil-leaving condition.

Follow the coil line from the coil-entering condition down to the previous assumption for CLDB.

If the difference between the humidity ratios is less than one percent, the iteration stops. Otherwise, follow the coil line down to where it intersects the SHR line (for draw through) or to where it intersects the SHR line minus fan heat (for blow-through).

Restart the iteration at Step B2.

When you enter cooling supply air dry bulb, the psychrometric iteration will calculate the room humidity ratio at which the psychrometric iteration of the TRACE program converges.

Zone Airflows and Heating Supply Air Dry Bulb. After the cooling supply air dry-bulb temperature has been determined for the system, the program calculates the cooling airflow for each zone served by that system. This airflow calculation is based on the peak sensible loads of the zone, as previously calculated and assigned.

Once the zone cooling airflows have been determined for both the primary and skin system, the heating supply air dry-bulb temperature (SADBh) can be calculated.

When system types FC or VTCV are used as skin systems, skin SADBh is initially calculated using the total skin winter design load (QSKINh) and the sum of the skin cooling zone airflows.

If, however, the skin SADBh calculated is greater than the skin SADBh input by the user, or is greater than 125°F, the final SADBh value will be reset to the lower value between the 125°F and user input skin SADBh. If the calculated value of skin SADBh is overridden by the user, the design skin system SADBh must be recalculated.

The zone skin cooling airflows are then recalculated.

For primary systems, the heating supply air temperature is initially based on the worst-case zone (assuming 20 percent over-design) and the zone cooling airflow.

If, however, the calculated value of SADBh is greater than the system SADBh you entered or is above 125°F, the final value will be reset to the lower value, either 125°F or your SADBh entry.

Once the system heating supply air dry bulb is known, temporary values of zone heating airflows are calculated.

At this point, both a "heating" and "cooling" airflow is known for all zones. Zones which are heating only have their final zone airflow set to the heating airflow value. Zones which are heating/cooling have their final zone airflow set to the larger of the cooling and heating airflow value.

Note: All airflows are calculated on a basis of standard airflow unless specified otherwise in the Load Parameters dialog box.

System Airflows. For peak-air systems, the fan airflow is the sum of the airflows of the zones served by that system.

For block-air systems, the fan airflow is based on the system block sensible load.

System Cooling Capacity. Next, the design capacity is determined from the coil-entering and the coil-leaving condition.

Once the humidity ratios are adjusted for altitude corrections, the return and outdoor air mixture enthalpy and coil-leaving enthalpy are calculated.

The cooling capacity is then calculated.

For the double-duct, multizone and bypass multizone systems, the coil airflow is increased by five percent to account for damper leakage.

The ventilation load is included in the overall cooling capacity.

System Heating Capacity. The heating capacity requirement is calculated for skin systems.

Repeat this process for primary systems.

Design Calculation Summary

- Assign loads
- Read building and zone peak loads
- Assign loads to zones
- Assign zone loads to system
- Determine system cooling SADB
- Read input file
- System sensible heat ratio
- Fan heat
- Cooling supply air dry bulb and supply air humidity ratio
- Determine zone airflows
- Zone cooling airflow
- Supply air dry bulb heating
- Zone heating airflow
- Zone airflow, system airflow, outside airflow
- Zone output
- Determine system capacity
- Supply air dry bulb if blow-through or coil-leaving
- Dry bulb if draw-through
- Return air dry bulb and return air humidity ratio
- Return and outside air; coil-leaving enthalpy

- Design cooling capacity and design heating capacity
- System output

3.3.4 TOPSS

This program provides a single interface for calculating performance on a wide variety of Trane products. Before TOPSS, each Trane product had its "own" selection program requiring users to learn many different user interfaces, methods for calculation, etc. TOPSS lets you select any of the 42 products with the same user interface.

Unlike most selection programs that use table lookup techniques TOPSS uses fundamental models for establishing performance of equipment. Many of these models are rated and certified in accordance with ARI 410 standards. TOPSS is a database driven program so it easily keeps up with ever-changing product lines.

Enter a set of conditions and desired performances into TOPSS and the program will determine product configurations that meet or come close to those conditions and performances. After performing these calculations, TOPSS provides an interface for reviewing, printing, graphing, selecting, exporting schedules to Microsoft Excel, and emailing your equipment selections to Trane.

Product Families.   Select from the following Trane product families in TOPSS:

- **Air Handlers**
  - Air Heating Coils (HRCL)
  - Modular Climate Changer Air Handlers (MCCB)
  - Packaged Fresh Air Unit (FAXA)
  - Cooling Coil (CLCL)
  - Heating Coil (HTCL)
  - Indoor Climate Changer Air Handler (MCCX)
  - Packaged Climate Changer (LPCA) (LPC)
  - Outdoor Climate Changer Air Handler (TSCX)
  - Vane Axial Fan (QFAN) (QFAN1)
- **Air Terminal Devices and Heating Products**
  - Basic Horizontal Fan Coil Unit (Basic)
  - Blower Coil Air Handler (BCXB)
  - Fan Coil Air Conditioners (Unitrane) (Fancoil)
  - Horizontal Unit Ventilator (HUVA)
  - Low Height Vertical Fan Coil (Lowboy)
  - Unit Cabinet Heater (ForceFlo)
  - Variable Air Volume Double Duct Terminal Units (VDDE)
  - Variable Air Volume Dual Duct Terminal Units (DUAL)
  - Variable Air Volume Fan Powered Terminal Units (FP)
  - Variable Air Volume Fan Powered Terminal Units (VFXE)
  - Variable Air Volume Fan Powered Terminal Units—Low Height (LFXE)
  - Variable Air Volume Fan Powered Terminal Units—Low Height (LH)
  - Variable Air Volume Single Duct Terminal Units (SINGLE)
  - Variable Air Volume Single Duct Terminal Units (VCXE)
  - Vertical Unit Ventilator (VUVA)
- **Refrigeration**
  - Air-Cooled Rotary Liquid Chillers Series R Chiller Model RTAC—140 to 500 Tons (RTAC)

- 20–60 Ton Packaged Air-Cooled Chiller (CGAF)
- 70–125 Ton Water-Cooled Chiller—Series R (RTWA)
- **Unitary**
  - Trane Odyssey Split System Family 7.5 through 20 Tons (SSC2 TWE-A,B)
  - 2.5, 3, 3.5, 4, 5, Ton TWE-C—Convertible Air Handlers (SSC2 TWE-C)
  - 2.5, 3, 3.5, 4, 5, 6 Ton TTA—Split System Cooling (SSC2 TTA)
  - Trane Air-Cooled Condensing Units 20–120 Tons (RAUC)
  - 3–20 Ton Packaged Unitary Heat Pump Rooftop—Dedicated (WCD)
  - 3–25 Ton Packaged Unitary Cooling Rooftop—Dedicated (TCD)
  - 3–25 Ton Packaged Unitary Gas/Electric Rooftop—Dedicated (YCD)
  - 3–5 Ton Unitary Cooling Rooftop Unit (T_C)
  - 3–5 Ton Unitary Gas/Electric Rooftop Unit (Y_C)
  - 3–5 Ton Packaged Heat Pump (W_C)
  - 20–75 Ton Packaged Industrial Rooftop (S_HF)
  - 27.5–50 Ton Packaged Commercial Rooftop (Voyager 3)
  - 90–130 Ton Packaged Industrial Rooftop (S_HG)
  - Commercial Self Contained—Signature Series (SCXF)
  - Commercial Self Contained—Modular Series (SCXG)
  - Water-Source Comfort Systems (GEHV)

Quick Select Wizard. The TOPSS Quick Select Wizard guides through the steps required to obtain a list of product selections that meet or exceed the performance requirements for a particular performance criterion. Instructions at the top of the Quick Select Wizard dialog box tell how to complete the current step.

Customizable Environment. There are many features to customize the TOPSS environment including:

Preferences

Defaults, tolerances, and units of measure

Measurement scheme

Arrange fields

Hide columns/rows

Templates

- Preferences let you customize the TOPSS environment to work habits and to make it more efficient by avoiding redundant entries. Preferences are divided into four categories; each is represented by a tab in the Preferences dialog box.
  - The View tab lets you skip unwanted messages and choose scrolling options.
  - The Defaults tab provides an "autosave" option and lets you pick a starting view so that each new job or existing job automatically opens in your favorite view.
  - The Configurable Product Reports tab can save time if you often select products like the Outdoor Climate Changer Air Handler with flexible arrangements. Use the entries on this tab to predefine the content and organization of product reports.
  - The Dimension Drawing Options tab lets you preselect the drawings you want to print from the choices available for each product. (Dimension drawings are not available for all products.)
- Defaults, tolerances and units of measure presently in effect for the selected product of a current job are displayed on the DTU View. This information is editable and is presented on three tabs.

- The Units Information tab displays the unit of measure currently in effect for each performance criterion.
- The Defaults Information tab lists the performance criteria for which default values can be set and displays the default value currently in effect, if any.
- The Tolerance Information tab displays the minimum and maximum limits that TOPSS uses to determine whether a potential selection meets the requirement you entered for a performance criterion.

- *Measurement Scheme* lets you apply a different measurement scheme—that is, a set of units-of-measure—to the current product in the current job. Each measurement scheme defines the unit-of-measure for every input and output within a specific product family. Two schemes have been predefined for most products, English I-P and Metric SI.

- *Arrange Fields* allows a change in sequence in which the performance criteria appear in the Inputs, Outputs, and Schedule views. You can also rename and establish the content of user-defined tabs.

- *Hide Columns/Rows* conceals from view selected fields (columns in a horizontal schedule, rows in a vertical schedule) on the User-Defined tab in the Schedule view so you can display only the information of particular interest.

- *Templates* allows you to create a file (.pst) that tells TOPSS how you want to view fields, defaults, tolerances, and units. Use a template to ensure that you apply any preferred parameters consistently. After you define a product template you have the option to designate it as the product default template.

On-Screen Features

- **Multiple Views for Entry and Review**
  Different TOPSS views are available to help you define and review your selections. To see only one set of performance criteria at a time, use the Inputs view. Or, use the Schedule view if you prefer to work with several sets of criteria at the same time. Similarly, the Outputs view shows you the calculated performance for a single result, while the Schedule view lets you review all of the results from the run at once.

  Tabs within each view provide an additional "filter", letting you display as much or as little information as you choose. For example, performance characteristics or "criteria" such as unit voltage that are (or can be) user-defined, appear in the Inputs view. Within this view, you would find unit voltage on the All Fields tab and on the Electrical Information tab, but not on the Main Cooling Coil tab.

- **On-screen Configuration**
  For products like the Outdoor Climate Changer Air Handler (TSCX) TOPSS allows you to configure the functionality of a unit by picking and placing modules in a desired sequence. You can add, delete, or modify modules in the unit.

- **Color Coded Fields**
  Many performance criteria are interrelated: entering a value for one criterion may make others (un)available or change them from "optional" to "required" or vice versa. Some criteria require information from you; others can only be calculated, and there are other criteria that can be entered or calculated. TOPSS uses color to flag these distinctions:

  - A *red* field identifies a required entry; without this information, TOPSS cannot calculate performance or identify possible selections.
  - A *gray* field indicates that the characteristic is irrelevant based on the entries made so far.
  - A *yellow* field "warns" you that the value for this performance characteristic is close to an operating limit or otherwise requires your attention. TOPSS can still use this

characteristic to perform the run. A result with one or more yellow fields is still considered valid and can be added to the job folder.

- A *blue* field identifies information required to generate dimension drawings.
- A performance criterion labeled with *blue* text can be either an input (entered by you) or an output (calculated by TOPSS).
- A performance criterion labeled with *black* text can only be an output; that is, it is calculated by TOPSS during a run.

- **Copy and Paste from Other Applications**
  Take advantage of the clipboard and the copy/paste functions of the Windows environment and TOPSS to move relevant data from other applications (e.g., Word, Excel, etc.) into TOPSS and vice versa. For example, if the equipment schedule exists as an electronic spreadsheet, copy the desired cells and paste them into the Schedule view of TOPSS. If TOPSS does not recognize a cell, it will prompt you to select a valid value.

- **Find and Replace Functions**
  This feature can be used to make entering and maintaining data faster and easier. For example, you can use this feature to:

  - Find all occurrences of a value in any field
  - Change all fields to a specific value
  - Change all fields to a specific value based on the value of another field

Graphic Features

- **Dimension Drawings**
  Dimension drawings are now available for all products. Since they represent the overall unit, dimension drawings for configurable products can only be printed and previewed when a result for each module is marked as In Unit.

  You can specify which of the following views to include in the dimension drawing:

  - Overall plan and elevation views
  - Detailed plan view
  - Detailed elevation view
  - Detailed end views
  - Curb view

- **Fan Curves**
  The fan curve feature displays fan curves for most products containing fans. The kinds of information displayed (when available) vary from product to product, as do the options to print and/or save this information in another form.

Results

- **Reports**
  A product report documents the performance criteria—inputs, outputs, and warranty information—for a result that TOPSS identifies as a potential selection. The report quantifies the specifications of the unit. It also includes job information, the version of the program that generated the result, and the date that the report was printed. Each result you select yields a single page of performance data.

- **Direct email**
  The export option lets you specify a format and destination for the information contained in the product report. Select Microsoft Mail (MAPI) as the destination to email the exported product report to a client.

Help Features

- **Product and Feature Help Files**
  For procedures, tips, and explanations about TOPSS windows, menus, and commands click Contents in the TOPSS Help menu. For explanations of the performance criteria associated with a particular Trane product, choose the appropriate product from the Product Help submenu.

- **Context Sensitive Help and F1**
  Click on the standard toolbar and then on a field or click an entry and press F1 to get a brief explanation about a particular object and how to use it.

- **Tutorial videos**
  The TOPSS Help system includes brief online videos. The What's New video identifies and demonstrates program features that were added to or updated for this version of TOPSS. It plays automatically when TOPSS is initially installed or updated on your PC. TOPSS tutorial videos show you how to use a particular TOPSS feature whereas product videos show you how to select a particular product.

3.3.5   VariTrane Duct Designer

The VariTrane Duct Designer duct design program was written to help optimize design while obtaining a minimum pressure system. The program is based on engineering data and procedures outlined in the ASHRAE Fundamentals Handbook. It includes tested data from the ASHRAE Fitting Database and from United McGill to provide the most accurate modeling possible.

   The program consists of three subprograms or applets: Duct Configurator, Ductulator and Fitting Loss Calculator. Each of these applets can be used individually or combined to provide a complete duct analysis.

Duct Configurator.   Duct Configurator sizes and analyzes supply duct systems. Quickly model the conditions for any duct system and specifications for each section, terminal unit, and diffuser. The section-by-section approach easily incorporates existing ductwork and fittings.

   Design Methodologies
   EQUAL FRICTION.   This design methodology sizes the supply duct system for a constant pressure loss per unit length. This is the most widely used method of sizing lower pressure, lower velocity duct systems. The main disadvantage of this method is that there is no equalization of pressure drops in duct branches unless the system has a symmetrical layout.

   Note: Set the optimal design friction rate per length of duct for the entire duct system in Duct Configurator by calculating the friction rate in the first trunk section of the system (the section connected directly to the fan). If the friction rate calculated for the trunk section is undesirable for the entire system, override the friction rate by selecting "No" to "Use calculated friction rate?" Set it directly in the field that appears.
   STATIC REGAIN.   This design methodology sizes the supply duct system to obtain uniform static pressure at all branches and outlets. Much more complex than equal friction, static regain can be used to design systems of any pressure or velocity. Duct velocities are systematically reduced, which allows the velocity pressure to convert to static pressure, offsetting friction losses in the succeeding section of duct. Systems designed using static regain require little or no balancing. One disadvantage of this methodology is that oversized ducts can occur at the ends of long branches. However, VariTrane Duct Designer lets you limit this problem by specifying a minimum velocity constraint.

Note: Set the minimum velocity constraint directly on the main tab. Choose from one of the three values listed or specify your own velocity constraint.

Fitting Entry Methodologies.   Two new fitting entry methods in Duct Configurator add increased flexibility to the data entry process. They include:

The *Lowest Friction* method reduces entry time by limiting the amount of operations required for fitting selections. When checked, the fitting chosen is based on the fitting type and shape. Fitting types include transition, 45-degree wye, 90-degree tee, bullhead tee, and cross; along with fan connection, entry, and exit, which are available under certain circumstances. When a fitting type is selected, a fitting is automatically chosen based on lowest frictional loss, which eliminates a lengthy search through the available fittings.

The *Quick Pick Fitting* method allows a user to customize the fitting types and available fittings within those types. When this method is selected the user selects a fitting type and shape. From these choices a customized list of fittings is presented based on the user's preference of fittings. The selections for a given fitting type must be set prior to using this method, but after the initial setup, the list of customized fittings will be available for any duct system modeled in the future.

The *ASHRAE* method is the traditional method of inputting fittings and it provides the full assortment of ASHRAE fittings. When checked, every ASHRAE database fitting from a given fitting type is available for selection. This method offers the most accurate, but also the most tedious entry method.

These methodologies may be used in conjunction with one another at any time in the entry process. For example, halfway through entering a duct system with the **Lowest Friction** method the user discovers a fitting that is not a close enough match to any of the fitting types. At this point the user can switch to the **ASHRAE** or the **Quick Pick Fittings** method of fitting selection and then back again if necessary.

Trane VAV Box Selections.   VariTrane Duct Designer now offers the ability to make Trane VAV box selections directly from the program through a seamless integration with the TOPSS program. This eliminates the tedious and time-consuming process of transferring product selection data into Duct Configurator. It also provides additional sizing criteria and increased accuracy in the duct design procedure. The ability to export the VAV box selections to the TOPSS software gives the user the ability to further analyze the box selections.

Insert Feature.   Modifying an existing duct system layout is easier with the new insert feature. When a selection is inserted between two sections, it is connected to the downstream section by a transition of the same shape and is connected to the upstream section by the fitting that existed prior to the insertion. The transition can then be changed to a junction to create additional takeoffs in the system.

Modified Duct System

Features

AUTO-CORRECT FOR LEAKAGE.   When this feature is checked, Duct Configurator automatically recalculates the airflow required at the supply fan to overcome leakage losses in the system. The amount of leakage is based on the leakage class you select. Leakage class is an ASHRAE term that denotes the leakage airflow per unit of duct surface area at a specific static pressure. For more information on leakage, see Chapter 32 in the 1997 *ASHRAE Fundamentals Handbook*.

AUTO-CORRECT FOR THERMAL LOSSES/GAINS.   During the journey from the supply fan to the conditioned spaces, the air stream may undergo various thermal changes. When this feature is checked, Duct Configurator automatically recalculates the airflow required at the supply fan to adjust for heat loss or gain based on the thermal insulation specified for each section and the temperatures specified inside and outside of the duct sections.

AUTO-BALANCE THE SYSTEM.   By checking this advanced feature, Duct Configurator will automatically add dampers in noncritical paths, where required, to balance the pressures in the system. Whenever the static overpressurization exceeds 1.0 in Hg in a given section, the application inserts an orifice rather than a damper to balance that section. This is done for acoustic optimization because orifices are less noisy than dampers when high-pressure drops are required to balance the system.

When this feature is not checked, Duct Configurator identifies the critical path and determines the required duct sizes and supply fan airflow accordingly, but it does not correct the oversized ductwork or reduce overpressurization with dampers in the noncritical paths.

Resizing Noncritical Paths.   Reduce your duct material and installation costs 40 percent or more without affecting annual energy costs by using the Resize Noncritical Paths feature in Duct Configurator. Duct Configurator can optimize the noncritical path sizes prior to adding any dampers or orifices. All noncritical paths sized using either Equal Friction or Static Regain have excess static pressure. By reducing the sizes of the duct sections in the noncritical paths, the majority of the excess static pressure is converted to velocity pressure. The remaining excess static pressure can then be reduced using dampers as described above.

Note: By enabling auto balancing with resizing noncritical paths, the calculation time increases dramatically. This is due to the tremendous amount of iterations that can accumulate for each path. For this reason, it is suggested to run this feature after the initial design has been optimized.

Adjust for Elevation Differences.   Duct Configurator does system calculations both at the elevation you specify and at sea level. You can specify elevation changes for individual sections of duct if there are large elevation differences from the inlet to the outlet of a given section.

Constant/Variable Volume Systems.   Duct Configurator can model both constant and variable volume systems. Constant volume systems are systems in which the *block airflow* equals that of the *peak airflow*. If these values are not the same, it is said that there is *diversity* in the system. Diversity is the decimal value that describes the ratio of block airflow and peak airflow.

$$\text{Diversity} = \frac{\text{Block Airflow}}{\text{Peak Airflow}}$$

If the diversity is something other than 1, the system is a variable volume system. This can be input in one of two ways. Diversity factor or block airflow can be input on the fan tab, but not both. The program then calculates the remaining variable. Peak airflow is always known since it is input on the VAV or diffuser tab for each terminal device. The diversity is then equally distributed throughout the system. For example, if the diversity factor is 0.8 at the fan and 1.0 at each VAV box, the root section diversity factor would be slightly more than 0.8 and the section immediately preceding the VAV box would have a diversity factor slightly less than 1.0.

Note: Modeling downstream of VAV boxes cannot currently be done, therefore VAV boxes and diffusers cannot be used in the same modeled duct system. To account for the losses downstream of a VAV box use the Ductulator in conjunction with the Fitting Loss Calculator to determine pressure loss in each section of the longest run. Then add all of the losses together and enter that value in Duct Configurator.

Existing Building or Design Constraints.   Another feature of the Duct Configurator applet is the ability to easily meet design constraints or model existing ductwork. To do this, set maximum or fixed section size constraints.

Note: An unlimited amount of take-offs can be connected to a section defined as a plenum.

Spreadsheet. Make design changes and check for entry errors with ease through the spreadsheet view in Duct Configurator. Once in the spreadsheet view, double-click on a field category to get further information. Use copy/paste functionality to make global changes.

Ductulator. This applet transforms Trane's popular manual duct sizing and layout calculator into a convenient PC-based tool. Use this applet to quickly size system components and determine the appropriate nominal duct size for equal friction applications.

Fitting Loss Calculator. This applet lets you quickly identify the optimal fittings and sizes for each duct section by comparing their efficiency and cost.

Note: Return and exhaust systems cannot yet be modeled using Duct Configurator. To account for these losses, use the Ductulator in conjunction with the Fitting Loss Calculator to obtain the static pressure losses through the return and exhaust ductwork.

These applets, combined with an integral database of accurate performance data for hundreds of ASHRAE and United McGill fittings, allow you to confidently model new or existing supply duct systems, whether round, rectangular, or flat oval. Once you complete your design, print reports with detailed information about all aspects of the duct system, including bills of material.

3.3.6 TRACE 700 Load Express

TRACE 700 Load Express is exactly what its name implies: a quick and easy software application that calculates cooling loads, heating loads, and airflow capacities using ASHRAE-approved algorithms. Combine the intuitive Microsoft Windows-based interface with entries geared specifically for small to medium-sized light commercial buildings and the result is an incredibly short learning curve. TRACE 700 Load Express lets "rookies" and experienced users alike perform accurate load calculations in minutes with just five simple steps:

1. Select a weather profile.
2. Define templates.
3. Enter room parameters.
4. Describe the air-handling system.
5. Assign rooms to systems.

Why the Slightly Different Name? We changed the program's name from *Load Express* to *TRACE 700 Load Express* because we have integrated *Load Express* with *TRACE 700* to give you the same look and feel, and extended capabilities as *TRACE 700* and *TRACE 700 Load Design*. All programs will be essentially the same now, think of them as different operating modes.

What are the Differences between Them? TRACE 700 Load Express allows a maximum of 20 rooms and 20 zones where *TRACE 700* and *TRACE 700 Load Design* do not have room and zone limitations. The majority of the differences occur in the Create Airside section. Some of the options and system types have also been limited in *TRACE 700 Load Express.*

What is the Benefit of Doing All This? There are many benefits. The most important one is that most of the added features and capabilities (applicable to load design) in *TRACE 700* will now be available to *TRACE 700 Load Express* users. Also, same calculation engines, same file extensions and libraries will enable users of all program modes to transfer archived files back and forth without any additional steps needed.

3.3.7 Trane Acoustics Program

This powerful Microsoft Windows-based tool helps designers accurately model the sound level reaching a tenant's ears. Evaluating the total effect of sound in an enclosed space entails many complex equations. Solving them manually consumes hours of precious design time and is prone to errors. TAP acoustical analysis software automates these calculations to provide fast, accurate results. Then, you can quickly compare the sound characteristics of several system alternatives and choose the one that best satisfies your client's design criteria.

How does the Trane Acoustics Program work? Put simply, it "builds" and analyzes sound paths by allowing the user to choose specific equipment and building components that generate, attenuate, reduce, or regenerate sound. Dialog-box entries let you further refine component attributes.

As components are added, moved, or deleted, the program dynamically recalculates the resulting sound power levels. Multiple paths? The TAP program not only determines the overall sound level at the receiver, but also how much sound each path contributes.

Once the analysis is complete, view the presentation-quality reports on-screen or print them. Choose detailed tables, NC or RC charts, or a combination of these formats.

Other program features include:

- Point-and-click, pictorial modeling of equipment (fans, diffusers, etc.) and building components (ceilings, walls, ductwork, etc.) in each sound path.
- The most complete, up-to-date library of sound data available for Trane products. Add a "custom element" to model equipment not found in the library.
- "On-the-fly" calculation and display of sound path summations.
- Multiple-path analysis—e.g. discharge airborne, discharge breakout and unit-radiated sound (a timesaving feature that lets you focus on other aspects of project design).
- Comprehensive, professional reports including NC and RC graphs that you can print or display.
- Industry-wide acceptance of your results. That's because the Trane acoustics program uses the latest available ASHRAE algorithms for modeling acoustical elements such as duct lagging, outdoor barriers, and duct silencers.
- A quick comparison of calculated sound levels with the desired NC value, an invaluable troubleshooting tool to isolate potential problems in an existing system.
- On-line Help with detailed explanations of ASHRAE algorithms.

3.3.8 Trane Engineering Toolbox

Power Factor Correction. This handy tool calculates the trigonometric relationships between power factor, kW, kVs and amps for electric motors. It also helps you quickly find the nominal size capacitor needed to correct your AC motor's power to a desired power factor, and generates the corrected power factor for up to nine part-load data points.

Timothy Zurick's 1977 *Air Conditioning, Heating & Refrigeration Dictionary* defines power factor as "the ratio of the power consumed (watts) in a circuit to the power drawn (volt-amperes). This ratio describes the difference between the amount of power actually used and the amount of power apparently used."

The distinction between actual and apparent power use is a crucial one. A low power factor (a ratio less than 1.0) doesn't increase a motor's power consumption, but it does require more current. The utility company must generate and distribute surplus current. Larger-than-necessary distribution components, such as wires and switching gears, are sized to provide the apparent current rather than the actual. Utility companies often impose an economic penalty on customers with low power factors or high apparent power consumption (kVA).

If enough information is known about a primarily inductive circuit, possible to calculate the required amount of capacitance for power factor correction. The characteristics of inductive and capacitive elements offset each other when applied in the same circuit. This interaction means that adding capacitive elements, or capacitors, to a primarily inductive circuit will increase its power factor and reduce the resulting current draw.

The entries for power factor correction are very versatile. Based on the information you have about the motor, you can choose one of five sets of values:

Uncorrected Power Factor. Prior to adding a capacitor, the ratio of real power (kW) to apparent power (kVA) at any given point in time in an electrical circuit. This value is generally available from the motor manufacturer.

Rated Load Amps. Rated Load Amps (RLA) are the amount of current that a rotation machine draws when operating at a rated voltage, speed (RPM), and torque.

kW Input. A kilowatt is a measure of power defined as 1000 J/S of energy flow. kW should be entered for the full load (rated) condition.

Brake Horsepower. Brake Horsepower is the actual power delivered to or by a shaft.

Motor Efficiency. Motor efficiency is entered as a percent.

You must also enter either capacitance or Corrected Power Factor. Then, the Engineering Toolbox will solve for the remaining aspects.

Corrected Power Factor. Corrected power factor is the power factor with installed capacitors. If you are using the tool to calculate the amount of capacitance required to attain a desired power factor, then the corrected power factor is the desired power factor.

Installed Capacitance. The amount of capacitance that must be added to the motor to attain the corrected power factor and line current.

Engineering Toolbox returns the line current and required capacitance, as well as corrected power factor.

Required capacitance is the amount of capacitance (in Kvar) that must be added to the motor to attain the corrected power factor and line current.

Capacitors used for power factor correction are generally offered in predetermined sizes that seldom match the required capacitance. Typically, designers pick the closest-size capacitor that is smaller than the required capacitance to avoid overcorrecting.

Standard capacitors are also cataloged at a specific voltage and frequency. If they are used at a different voltage or frequency, they must be re-rated to the new condition using the following equation:

$$Kvar\ actual = Kvar\ rated \times (V\ actual\ /\ V\ rated)^2 \times (f\ actual\ /\ f\ rated)$$

Properties of Air. Properties of air is an electronic version of the Trane psychrometric chart. You enter any two properties of a known quantity of air and the program calculates the five other properties.

Using one start point, this applet calculates all properties of moist or dry air at normal HVAC conditions. It's an electronic version of the Trane psychrometric chart, and it uses the same calculating routines as TRACE 600 and TRACE Load 700.

Enter any two values and the program will instantly calculate the other five properties (if pychrometrically possible).

Engineer's Toolbox takes any combination of:

- Altitude
- Dry bulb
- Wet bulb
- Specific humidity
- Enthalpy
- Dew point
- Relative humidity
- Specific volume

and returns the remaining properties:

- Altitude
- Barometric pressure
- Air pressure
- Dry bulb
- Wet bulb
- Specific humidity
- Enthalpy
- Dew point
- Relative humidity
- Specific volume
- Humidity ratio

Try the web enabled Properties of Air calculator.

Mixed Air Properties.   Enter psychrometric information about two known quantities of air and the program returns properties for the mixture of the two quantities. The mixed properties of air applet is very similar to the properties of air applet.

Engineering Toolbox requires values for both air quantities, including:

- Quantity
- Dry Bulb
- Wet Bulb
- Altitude

With these values entered, Engineering Toolbox calculates values for each air quantity, as well as a combination of the two. Values calculated include:

- Altitude
- Barometric pressure
- Air quantity
- Air pressure
- Dry bulb
- Wet bulb

- Humidity ratio
- Humidity
- Specific humidity
- Enthalpy
- Specific volume
- Dew point
- Relative humidity

Try the web-enabled version of the Mixed Air Properties calculator for yourself!

Fluid Properties.   Calculates physical properties of the five different fluids used in HVAC applications including water, ethylene glycol, propylene glycol, calcium chloride, and methanol brine.

Engineering toolbox requires values for fluid type, temperature, and concentration, to calculate:

- Temperature concentration
- Viscosity
- Freezing point
- Specific gravity
- Thermal conductivity
- Specific heat

Refrigerant Properties.   Calculates several properties of ten commonly used refrigerants.

- CFC-11 (R-11)
- CFC-12 (R-12)
- HCFC-22 (R-22)
- HFC-32 (R-32)
- CFC-113 (R-113)
- CFC-114 (R-114)
- HCFC-123 (R-123)
- HFC-134a (R-134a)
- CFC-500 (R-500)
- CFC-502 (R-502)

Engineering Toolbox requires entries for refrigerant type, temperature and concentration. In return, Engineering Toolbox calculates:

- Temperature concentration
- Liquid enthalpy
- Liquid viscosity
- Liquid thermal conductivity
- Specific heat

- Density
- Vapor enthalpy
- Vapor viscosity
- Vapor thermal conductivity
- Vapor entropy
- Sonic velocity
- Specific heat
- Specific volume

Refrigerant Line Sizing.   This calculator combines refrigerant properties and piping design fundamentals to compute line sizes for suction, discharge, and liquid lines of split systems using R-12 or R-22.

System values required are:

- Refrigerant type
- Saturated suction temperature
- Condensing temperature
- Evaporator tonnage
- Superheat
- Subcooling
- Compressor superheat

The discharge line connects the outlet of the compressor to the condenser. In a "condensing unit," the discharge line is an integral part of the packaged piece of equipment and need not be sized. Engineering Toolbox selects a discharge line to maintain the velocity of the refrigerant vapor below 3,500 fpm and the total pressure drop in the discharge line below 6 psi.

When selecting a line size, be sure that the minimum tons reported for the line are lower than the minimum capacity of the refrigeration circuit. If it is not, reduce the line size of vertical risers only. If the minimum tons for a 30-ton circuit are reported as 11.6 ton for a 1-5/8 in (4 cm) discharge line, this means that the refrigerant can carry oil in this size line down to 11.6 ton. As long as the minimum capacity of the circuit is greater than 11.6 ton, the line size is acceptable. If it is not, the vertical riser sections of the line should be reduced one size to 1-3/8 in (3.5 cm).

It is good practice to maintain refrigerant velocity above 500 fpm.

Discharge line values required include:

- Line length
- Ball valve count
- Angle valve count
- Short elbow count
- Long elbow count
- Tee "thru" count
- Tee "branch" count

The liquid line connects the condenser to the expansion device, which is at the inlet of the evaporator. Engineering Toolbox selects a liquid line to maintain the velocity of the liquid refrigerant below 600 fpm (180 mpm) and to maintain subcooling greater than 5°F (9°C) at the inlet to the expansion device. The system needs to have enough subcooling so that the refrigerant will not flash to vapor before it reaches the expansion valve. It ensures this by selecting a line size that will have at least 5°F (9°C) of subcooling remaining at the inlet of the expansion valve.

For R-22 . . . When vertical upward flow exists in the liquid line (i.e., when the condenser is located below the evaporator), loss of head pressure due to the column effect of the liquid must be taken into consideration. The loss of head pressure is most likely to occur when the liquid is very dense (liquid temperature is low). To avoid loss of head pressure in vertical upward lines, an adjustment factor is applied to the subcooling before the expansion valve, so that the program does not choose a line that will not perform when the outdoor conditions are not at design. The adjustment factor applied is based on the properties of refrigerant R-22 at 50°F (10°C). For every 10 ft (3m) of upward flow 3.2°F (5°C) is subtracted from the subcooling before the expansion valve; 5°F(10°C) before the expansion must still be maintained even with the adjustment factor.

Liquid line values required include:

- Line length
- Ball valve quantity
- Angle valve quantity
- Short elbow quantity
- Long elbow quantity
- Tee "thru" quantity
- Tee "branch" quantity
- Filter drier
- Solenoid valve
- Elevation change

The suction line connects the evaporator to the compressor inlet.

Engineering Toolbox selects a suction line to maintain the velocity of the refrigerant vapor below 4,000 fpm and the total pressure drop in the suction line below 6 psi.

When selecting a line size, be sure that the minimum tons reported for the line is lower than the minimum capacity of the refrigeration circuit. If not, reduce the line size of vertical risers only. For example, if the minimum tons for a 30-ton circuit is reported as 11.7 ton for a 2-1/8 in suction line, this means that the refrigerant can carry oil in this size line down to 11.7 ton. As long as the minimum capacity of the circuit is greater than 11.7 ton, the line size is acceptable. If it is not, the vertical riser sections of the line should be reduced one size to 1-5/8 in.

It is a good practice to maintain refrigerant velocity above 500 fpm.

Suction line values required include:

- Line length
- Ball valve quantity
- Angle valve quantity
- Short elbow quantity
- Long elbow quantity
- Tee "thru" quantity

- Tee "branch" quantity
- Suction filter quantity

3.4 CARRIER PROGRAMS*

Carrier's E20-II programs are available to assist HVAC engineers in the layout and design of commercial air conditioning systems. This section summarizes the features and capabilities of each E20-II program.

3.4.1 Hourly Analysis Program v4.x

Provides versatile features for designing HVAC systems in commercial buildings. Also offers powerful 8760 hour-by-hour energy analysis capabilities. All of this is packaged in a single, easy to use software program which allows input data for system design to be reused in energy studies. Key features:

- Uses a system-based approach to design calculations which tailors sizing procedures and reports to the specific type of system being designed.
- Provides data for sizing fans, ducts, cooling and heating coils, humidifiers, air terminals, terminal fan coils and heat pumps, chillers and boilers.
- Models a wide range of system types associated with central station air handlers, packaged rooftop units, vertical units, split systems, DX fan coils, hydronic fan coils, and water source heat pumps.
- Models a variety of air terminal types including CAV, CAV with reheat, VAV, VAV with reheat, series fan powered mixing box, parallel fan powered mixing box, mixing box, and induction.
- Performs outdoor ventilation airflow sizing calculations per ASHRAE Standard 62-2001 or by conventional procedures.
- Models CO_2 sensors and demand-controlled ventilation.
- Uses the ASHRAE Transfer Function method and heat extraction equations for load calculations.
- Summarizes key sizing data in two simple pages of reports.
- Provides additional reports detailing component loads, hourly load profiles, detailed hourly system performance, and psychrometric charts.
- Performs a true hour-by-hour energy analysis using measured weather data for all 8,760 hours of the year to calculate building loads, air system performance, plant performance, total energy use and cost.
- Models performance of packaged DX, split DX, packaged heat pump, and split heat pump equipment as well as chilled water, hot water and steam plants.
- Models twelve different types of air-and water-cooled chillers, including absorption.
- Models chiller networks and chiller sequencing controls.
- Offers features for analyzing primary/secondary and primary/variable speed pumping.

---

*This section by Carrier Corp., Syracuse, NY. See Ref. 44.

- Analyzes complex electric and fuel utility rates, including demand charges.
- Provides weather data for over 700 cities worldwide.
- Permits hourly scheduling of occupants, internal loads, and thermostat operation.
- Permits up to 2,500 spaces, 250 air systems, 100 plants and 100 building design alternatives per project. Storage for schedules, walls, roofs, windows, doors, shades, chillers, cooling towers boilers, and utility rates is unlimited.

3.4.2 System Design Load Program v4.x

Offers all system design features available in the Hourly Analysis Program. Designed for those needing only system design capabilities, but not energy analysis features. Eliminating energy analysis makes the program simpler, and less expensive to license.

3.4.3 Block Load v3.x

Provides powerful load estimating capabilities in a streamlined, easy to use package. Key features:

- Graphical interface makes it easy to visualize project data.
- Streamlined inputs allow projects to be completed quickly.
- Uses ASHRAE Transfer Function Method for load calculations.
- Handles projects large and small, from one-zone rooftop systems to 150-zone central air handler systems.
- Provides sizing information for fans, ducts, cooling and heating coils, and air terminals.
- Offers reports listing sizing data, component loads, and handy rule of thumb check figures.

3.4.3 Engineering Economic Analysis v3.x

Provides features for comparing the life-cycle economic performance of building and system design alternatives. Key features:

- Considers investment costs and operating costs for each design alternative. Costs can be itemized.
- Considers financing for investment costs, income taxes, and depreciation.
- Tailors calculation methods and reports to typical decision-making criteria used by private owners and public (government) organizations. Also tailors methods and reports to situations where alternatives are mutually exclusive or independent.
- Calculates cash flow, total present worth, net present worth, internal rate of return, savings to investment ratio, and payback period
- Offers detailed life-cycle analysis, simple payback, and simple cash flow as calculation options.
- Provides a concise summary report to aid in making economic decisions.
- Provides detailed reports to help users understand results.
- Reports combine graphics and text for maximum impact.
- Allows up to 20 alternatives to be compared in a single study.

3.4.5   Refrigerant Piping Design v4.x

Performs line sizing and pressure loss calculations for refrigerant piping in split DX cooling systems, condenserless chiller systems, and single-evaporator refrigeration systems. Key features:

- Sizes suction, discharge, and liquid lines.
- Offers R-22, R-134a, R-502, and R-717 as refrigerant options.
- Uses sizing procedures outlined in the ASHRAE Refrigeration Handbook.
- Sizes horizontal lines, vertical risers, and double risers as required.
- Determines minimum capacity for oil entrainment in vertical risers.
- Determines minimum subcooling required in liquid lines.
- Permits users to manually adjust line sizes and immediately examine resulting pressure drops.

REFERENCES

1. *1995 ASHRAE Handbook, Fundamentals*, ASHRAE, Atlanta, GA, 1995, chap. 24, "Weather Data."
2. Carrier Corporation, *Handbook of Air Conditioning System Design*, McGraw-Hill, New York, 1965, part 1, chap. 2.
3. Loads Design Weather Region diskettes from the Trane Company, La Crosse, WI.
4. E20-II diskettes from Carrier Corp., Syracuse, NY.
5. National Climatic Data Center, Nashville, NC.
6. *1993 ASHRAE Handbook, Fundamentals*, ASHRAE, Atlanta, GA, 1993, chap. 8, "Physiological Principles and Thermal Comfort."
7. Ibid., chap. 23, "Infiltration and Ventilation."
8. *Ventilation Standard*, ANSI/ASHRAE document 61-1981R, ASHRAE, Atlanta, GA.
9. *1995 ASHRAE Handbook, HVAC Applications*, ASHRAE, Atlanta, GA, 1995, chap. 2, "Retail Facilities."
10. Ibid., chap. 3, "Commercial and Public Buildings."
11. Ibid., chap. 4, "Places of Assembly."
12. Ibid., chap. 5, "Domiciliary Facilities."
13. Ibid., chap. 6, "Educational Facilities."
14. Ibid., chap. 7, "Health Care Facilities."
15. Ibid., chap. 9, "Aircraft."
16. Ibid., chap. 10, "Ships."
18. Ibid., chap. 13, "Laboratory Systems."
19. Ibid., chap. 15, "Clean Spaces."
20. Ibid., chap. 16, "Data Processing System Areas."
21. Carrier Corp., *Handbook of Air Conditioning System Design*, part 1, chap. 1, McGraw-Hill, New York, 1965.
22. Ibid., chap. 6.
23. *1993 ASHRAE Handbook, Fundamentals*, ASHRAE, Atlanta, GA, chap. 25, "Residential Cooling and Heating Load Calculations." Chap. 26, "Non residential Cooling and Heating Load Calculations."
24. Carrier Corp., *Handbook of Air Conditioning System Design*, part 1, chap. 7, McGraw-Hill, New York, 1965.

25. *1993 ASHRAE Handbook, Fundamentals*, chap. 9, "Environmental Control of Animals and Plants."

26. Ibid., chap. 10, "Physiological Factors in Drying and Storing Farm Crops."

27. *1995 ASHRAE Handbook, Applications*, ASHRAE, Atlanta, GA, chap. 11, "Industrial Air Conditioning."

28. Ibid., chap. 14, "Engine Test Facilities."

29. Ibid., chap. 17, "Printing Plants."

30. Ibid., chap. 18, "Textile Processing."

31. Ibid., chap. 19, "Photographic Materials."

32. Ibid., chap. 20, "Environment Control for Animals and Plants."

33. Ibid., chap. 22, "Air Conditioning of Wood and Paper Products Facilities."

34. Ibid., chap. 23, "Nuclear Facilities."

35. Ibid., chap. 25, "Mine Air Conditioning and Ventilation."

36. *1993 ASHRAE Handbook, Fundamentals*, ASHRAE, Atlanta, GA, chap. 25, "Residential Cooling and Heating Load Calculations." chap. 26, "Non Residential Cooling & Heating Load Calculations."

37. Ibid., chap. 3, "Heat Transfer."

38. Ibid., chap. 27, "Fenestration."

39. *1965 Handbook of Air Conditioning System Design*, Part 1, chap. 5, McGraw-Hill, New York,

40. 1995 *ASHRAE Handbook, Fundamentals*, chap. 12, "Enclosed Vehicular Facilities," ASHRAE, 1791 Tullie Circle N. E. Atlanta, GA, 30329.

41. *1965 Handbook of Air Conditioning Systems Design*, Part 1, chap. 3, McGraw-Hill, New York,

42. Ibid., chap. 4.

43. Trane Software Programs for HVAC. Trane Corp., CDS Dept., La Crosse, WI.

44. Carrier Software Programs for HVAC, Carrier Corp., Syracuse, NY.

CHAPTER 4
HVAC APPLICATIONS*

Ernest H. Graf, P.E.

Assistant Director, Mechanical Engineering
Arcadis-Giffels, Inc., Southfield, Michigan

William S. Lytle, P.E.

Project Engineer, Mechanical Engineering,
Arcadis-Giffels, Inc., Southfield, Michigan

4.1 GENERAL CONSIDERATIONS

As a system design develops from concept to final contract documents, the following subjects should be considered throughout the heating, ventilating, and air-conditioning (HVAC) design period. These subjects are of a general nature inasmuch as they are applicable to all HVAC designs, and they may become specific requirements inasmuch as codes are continually updated.

4.1.1 Cooling Towers and Legionnaires' Disease

Since the 1976 outbreak of pneumonia in Philadelphia, cooling towers have frequently been linked with the *Legionella pneumophila* bacteria, or Legionnaires' disease. Several precautions should be taken:

1. Keep basins and sumps free of mud, silt, and organic debris.
2. Use chemical and/or biological inhibitors as recommended by water-treatment specialists. Do not overfeed, because high concentrations of some inhibitors are nutrients for microbes.
3. Do not permit the water to stagnate. The water should be circulated throughout the system for at least 1 h each day regardless of the water temperature at the tower. The water temperature within the indoor piping will probably be 60°F (15.6°C) or warmer, and one purpose of circulating the water is to disperse active inhibitors throughout the system.
4. Minimize leaks from processes to cooling water, especially at food plants. Again, the processes may contain nutrients for microbes.

---

*Updated for *HVAC Handbook* by James R. Fox, P.E., David H. Helwig, William C. Newman, Wayne M. Lawton, Joseph D. Robinson, P.E., James Runski, P.E., and Paul D. Bohn of Arcadis-Giffels.

4.1.2 Elevator Machine Rooms

These spaces are of primary importance to the safe and reliable operation of elevators. In the United States, all ductwork or piping in these rooms must be for the sole purpose of serving equipment in these rooms unless the designer obtains permission from the authorities in charge of administering American National Standards Institute/American Society of Mechanical Engineers (ANSI/ASME) Standard 17.1, *Safety Code for Elevators and Escalators*. If architectural or structural features tend to cause an infringement of this rule, the duct or pipe must be enclosed in an approved manner.

4.1.3 Energy Conservation

A consequence of the 1973 increase in world oil prices is legislation governing the design of buildings and their HVAC systems. Numerous U.S. states and municipalities include an energy code or invoke a particular issue of American Society of Heating, Refrigeration, and Air Conditioning Engineers (ASHRAE) Standard 90 as a part of their building code. Standard 90.1 establishes indoor and outdoor design conditions, limits the overall U-factor for walls and roofs, limits reheat systems, requires the economizer cycle on certain fan systems, limits fan motor power, requires minimum duct and pipe insulation, requires minimum efficiencies for heating and cooling equipment, and so on.

In the interest of freedom of design, the energy codes permit tradeoffs between specified criteria as long as the annual consumption of depletable energy does not exceed that of a system built in strict conformance with the standard. Certain municipalities require that the drawings submitted for building-permit purposes include a statement to the effect that the design complies with the municipality's energy code. Some states issue their own preprinted forms that must be completed to show compliance with the state's energy code.

This energy code is being adopted by many states as a required minimum.

4.1.4 Equipment Maintenance

The adage "out of sight, out of mind" applies to maintenance. Equipment that a designer knows should be periodically checked and maintained may get neither when access is difficult. Maintenance instructions are available from equipment manufacturers; the system designer should be acquainted with these instructions, and the design should include reasonable access, including walk space and headroom, for ease of maintenance. Some features for ease of maintenance will increase project costs, and the client should be included in the decision to accept or reject these features.

Penthouse and rooftop equipment should be serviceable via stairs or elevators and via roof walkways (to protect the roofing). Ship's ladders are inadequate when tools, parts, chemicals, and so on are to be carried. Rooftop air handlers, especially those used in cold climates, should have enclosed service corridors (where project budgets allow). If heavy rooftop replacement parts, filters, or equipment are expected to be skidded or rolled across a roof, the architect must be advised of the loading to permit proper roof system design.

Truss-mounted air handlers, unit heaters, valves, exhaust fans, and so on should be over aisles (for servicing from mechanized lifts and rolling platforms) when catwalks are impractical. Locate isolation valves and traps within reach of building columns and trusses to provide a degree of stability for service personnel on ladders.

It is important that access to ceiling spaces by coordinated with the architect. Lay-in ceilings provide unlimited access to the space above, except possibly at lights, speakers, sprinklers, and the like. When possible, locate valves, dampers, VAV boxes, coils, and the like above corridors and janitor closets so as to disturb the client's operations the least.

Piping system diagrams are important and should be provided by the construction documents. Piping should be labeled with service and flow direction arrows, and valves should be numbered as to an identification type number. The valves normally used in modern-day piping systems fall into two categories, multiturn and quarter-turn. Quarter-turn valves would include ball valves, butterfly valves, and plug valves. The multiturn variety includes the traditional gate and globe type. Quarter-turn valves are becoming more popular and usually can be provided at less cost, depending on service, but care should be used in their application. Valves in pipe sizes up to 2 in (5 cm) are normally manufactured with threaded bodies, though some types are available with flanged ends. Valves in these sizes are ball, plug, gate, or globe type in design. For pipe sizes 2.5 in (6.3 cm) and larger, threaded ends are rarely used or available. The multiturn designs would normally be provided with flanged ends. Plug valves would also be flanged. The butterfly valve is furnished with either flanged ends, though these are rarely used in aboveground work, lug body, or wafer style. The wafer style would be the least expensive but is not a good choice because there is no way to disconnect the service downstream of the valve without a short flanged spool bolted up to the valve (see Section 25.5). The best choice for the butterfly valve is the lug body because it is the best of both worlds. It is installed like a flanged-body valve without the flanged-body cost.

Most experienced piping designers shy away from quarter-turn valves for steam service at all temperatures and pressures. This is, for the most part, out of consideration for safety, especially for the smaller pipe sizes where lever-type operators would be expected. For larger pipe sizes that carry steam, high-performance butterfly valves with gear operators are gaining popularity. The gear operator provides a measure of safety where slow operation is desired. Warm-up bypasses should always be considered for all steam isolation valves without regard to whether they are gate or butterfly type.

Pressure gauges are normally provided to monitor changes in pressure loss across pumps, heat exchangers, and strainers. Figure 4.1 suggests a piping scheme that allows one gauge, which can be used to increase the accuracy of gauge reading where the piping configuration permits. Gauge valves should be either small globe [$^3/_8$ in (10 mm)] or ball valves. These types of valves are not as likely to seize with disuse, and they allow throttling to steady a bouncing gauge needle.

The observation of steam-trap operation can be facilitated by having a $^3/_8$-in (10-mm) test valve at the trap discharge pipe (Fig. 4.2). With valve V-1 closed, trap leakage and cycling may be observed at an open test valve. The test valve can be used to monitor reverse-flow leaks at check valves.

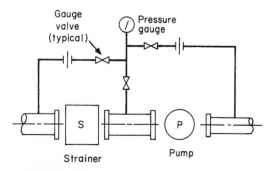

FIGURE 4.1 Multiple-point pressure gauge.

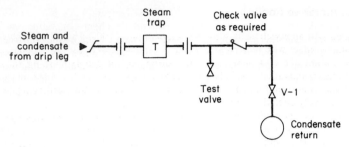

FIGURE 4.2   Test valve at steam trap.

4.1.5 Equipment Noise and Vibration

Noise and vibration can reach unacceptable levels in manufacturing plants as well as in offices, auditoriums, and the like. Once an unacceptable level is built in, it is very costly to correct. The noise control recommendations in Chap. 10 and in the latest edition of the *ASHRAE Handbook, HVAC Applications*, should be followed. Sound and vibration specialists should be consulted for HVAC systems serving auditoriums and other sensitive areas. Fans, variable-air-volume boxes, dampers, diffusers, pumps, valves, ducts, and pipes which have sudden size changes or interior protrusions or which are undersized can be sources of unwelcome noise.

Fans are the quietest when operating near maximum efficiency, yet even then they may require sound attenuation at the inlet and outlet. Silencers and/or a sufficient length of acoustically lined ductwork are commonly used to "protect" room air grilles nearest the fan. Noise through duct and fan sides must also be considered. In the United States, do not use acoustic duct lining in hospitals except as permitted by the U.S. Department of Health and Human Services (DHHS) Publication HRS-M-HF 84-1.

Dampers with abrupt edges and those used for balancing or throttling air flows cause turbulence in the air stream, which in turn is a potential noise source. Like dampers, diffusers (as well as registers, grilles, and slots) are potential noise sources because of their abrupt edges and integral balancing dampers. Diffuser selection, however, is more advanced in that sound criteria are readily available in the manufacturers' catalogs. Note, however, that a background noise (white noise) is preferable in office spaces because it imparts a degree of privacy to conversation. Diffusers can provide this.

Pumps are also the quietest when operating near maximum efficiency. Flexible connectors will help to dampen vibration transmission to the pipe wall but will not stop water- or liquid-borne noise.

Valves for water, steam, and compressed-air service can be a noise source or even a source of damaging vibration, depending on the valve pattern and on the degree of throttling or pressure reduction. Here again, the findings of manufacturers' research are available for the designer's use.

Equipment rooms with large fans, pumps, boilers, chillers, compressors, and cooling towers should not be located adjacent to sound- or vibration-sensitive spaces. General office, commercial, and institutional occupancies usually require that this equipment be mounted on springs or vibration isolation pads (with or without inertia bases) to mitigate the transfer of vibration to the building's structure. Spring-mounted equipment requires spring pipe hangers and flexible duct and conduit connections. Air-mixing boxes and variable-volume boxes are best located above corridors, toilet rooms, and public spaces. Roof fans, exhaust pipes from diesel-driven generators, louvers, and the like should be designed and located to minimize noise levels, especially when near residential areas.

4.1.6 Evaporative Cooling

An airstream will approach at its wet-bulb temperature a 100-percent saturated condition after intimate contact with recirculated water. Evaporative cooling can provide considerable relief without the cost of refrigeration equipment for people working in otherwise unbearably hot commercial and industrial surroundings, such as laundries, boiler rooms, and foundries. Motors and transformers have been cooled (and their efficiency increased) by an evaporatively cooled airstream.

Figure 4.3 shows the equipment and psychrometric elements of a *direct* evaporative cooler. Its greatest application is in hot, arid climates. For example, the 100°F (38°C), 15-percent relative humidity (RH) outdoor air in Arizona could be cooled to 70°F (21°C), 82-percent RH with an 88-percent efficient unit. Efficiency is the quotient of the dry-bulb conditions shown at (2), (3), and (4) in Fig. 4.3b. Note that the discharge air from a direct evaporative cooler is near 100-percent humidity and that condensation will result if the air is in contact with surfaces below its dew point. The discharge dew point in the above example is 64°F (18°C).

Figure 4.4 schematically shows an *indirect* evaporative cooler. Whereas a direct evaporative cooler increases the airstream's moisture, an indirect evaporative cooler does not; that is, there is sensible cooling only at (1) to (2) in Fig. 4.4b. Air is expelled externally at (5). When an indirect cooler's discharge (2) is ducted to a direct cooler's inlet, the final discharge (3) will be somewhat cooler and include less moisture that that of a direct cooler only. Various combinations of direct and indirect equipment have been used as stand-alone equipment or to augment refrigeration equipment for reduced overall operating costs. Refer to the *2000 ASHRAE Handbook, Systems and Equipment*, and the *2003 ASHRAE Handbook, HVAC Applications*.

Some evaporative cooling equipment operates with an atomizing water spray only, with any overspray going to the drain. Some additional air cooling is available when the water temperature is less than the air wet-bulb temperature. Evaporative cooling involves 100 percent outdoor air, and there must be provisions to exhaust the air. Evaporative cooling has also been applied to roof cooling; a roof is wetted by fine sprays, and the water evaporation causes cooler temperatures at the roof's upper and lower surfaces. The water supply for all applications must be analyzed for suitability and, as needed, treated to control scale, algae, and bacteria.

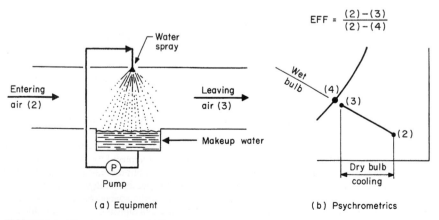

$$EFF = \frac{(2)-(3)}{(2)-(4)}$$

(a) Equipment   (b) Psychrometrics

FIGURE 4.3 Direct evaporative cooling.

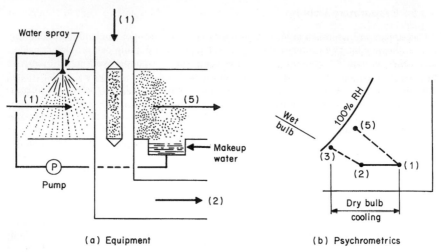

(a) Equipment   (b) Psychrometrics

FIGURE 4.4 Indirect evaporative cooling.

4.1.7 Fire and Smoke Control Dampers

Wherever practical and/or necessary, building walls and floors are made of fire-resistant material to hinder the spread of fire. Frequently, HVAC ducts must penetrate walls and floors. In order to maintain the fire resistance of a penetrated wall, fire dampers or equal protection must be provided whenever a fire-resistance-rated wall, floor, or ceiling is penetrated by ducts or grilles. Fire dampers are approved devices (approved by administrators of the building code, fire marshal, and/or insurance underwriter) that automatically close in the presence of higher-than-normal temperatures to restrict the passage of air and flame. Smoke dampers are approved devices that automatically close on a control signal to restrict the passage of smoke.

The following are general applications for fire or smoke dampers per the National Fire Protection Association Standard NFPA-90A, 2002 edition:

• Provide 3-h fire dampers in ducts that penetrate walls and partitions which require a 3-h or higher resistance rating, provide 1.5-h dampers in ducts that penetrate those requiring a rating of 2 h or higher but less than 3 h, and provide 1.5-h dampers in ducts that penetrate shaft walls requiring a rating of 1 to 2 h.

• Provide fire dampers in all nonducted air-transfer openings that penetrate partitions if they require a fire-resistance rating.

• Provide smoke dampers and associated detectors at air-handling equipment whose supply capacity exceeds 2000 ft^3/min (944 L/s). Provide supply and return smoke detectors and associated dampers for units whose return capacity exceeds 15,000 ft^3/min (7080 L/s). The dampers shall isolate the equipment (including filters) from the remainder of the system except that the smoke dampers may be omitted (subject to approval by the authority having jurisdiction) when the entire air-handling system is within the space served or when rooftop air handlers serve ducts in large open spaces directly below the air handler.

Exceptions to the preceding are allowed when the facility design includes an engineered smoke control system. Note that schools, hospitals, nursing homes, and jails may have more stringent requirements. Kitchen hood exhaust systems must be coordinated with local jurisdictions.

Dampers that snap closed have often incurred sufficient vacuum on the downstream side to collapse the duct (see Ref. 1). Smoke and other control dampers that close normally and restrict the total air flow of a rotating fan can cause pressure (or vacuum) within the duct equal to fan shutoff pressure. A fan might require a full minute after the motor is deenergized before coasting to a safe speed (pressure). Provide adequate duct construction, relief doors, or delayed damper closure (as approved by the authority having jurisdiction).

Refer to the building codes, local fire marshal rules, insurance underwriter's rules, and NFPA-90A for criteria regarding fire and smoke dampers.

For specialized systems refer to the applicable sections of the NFPA. For example, NFPA 45 does not allow the use of fire dampers for certain types of laboratory exhaust streams.

4.1.8 Outdoor Air

This is needed to make up for air removed by exhaust fans; to pressurize buildings so as to reduce the infiltration of unwanted hot, cold, moist, or dirty outdoor air; to dilute exhaled carbon dioxide, off-gassing of plastic materials, and body odors; and to replenish oxygen.

Outdoor air must be supplied at a minimum rate to provide an indoor air quality for human comfort and minimize adverse health effects. Building pressurization should be less that 0.15 in (4 mm) water gauge (WG) on ground floors that have doors to the outside so that doors do not hang open from outflow of air. The building's roof and walls must be basically airtight to attain pressurization. If there are numerous cracks, poor construction joints, and other air leaks throughout the walls, it is impractical to pressurize the building—and worse, the wind will merely blow in through the leaks on one side of the building and out through the leaks on the other side. Variable-air-volume (VAV) systems require special attention regarding outdoor air because as the supply fan's air flow is reduced, the outdoor and return air entering this fan tend to reduce proportionately. Provide minimum outdoor air at the minimum unit airflow for VAV systems.

The National Fire Protection Association (NFPA) standards recommend minimum outdoor air quantities for hazardous occupancies. NFPA standards are a requirement insofar as building codes have adopted them by reference. Building codes frequently specify minimum outdoor air requirements for numerous hazardous and nonhazardous occupancies. ASHRAE Standard 62-2002 and mechanical codes recommend minimum quantities of outdoor air for numerous activities. ASHRAE 62-2002 typically requires at least 20 ft³/min (9.5 L/s) per person. Refer to the ASHRAE standards for additional requirements.

4.1.9 Perimeter Heating

The heat loss through outside walls, whether solid or with windows, must be analyzed for occupant comfort. The floor temperature should be no less than 65°F (18°C), especially for sedentary activities. In order to have comfortable floor temperatures, it is important that perimeter insulation be continuous from the wall through the floor slab and continue below per Refs. 2 and 3.

Walls with less than 250 Btu/h · lin ft (240 W/lin m) loss may generally be heated by ceiling diffusers that provide airflow down the window—unless the occupants would be especially sensitive to cold, such as in hospitals, nursing homes, day-care centers, and swimming pools. Walls with 250 to 450 Btu/h · lin ft (240 to 433 W/lin m) of loss can be heated by warm air flowing down from air slots in the ceiling; the air supply should be approximately 85 to 110°F (29 to 43°C). Walls with more than 450 Btu/h · lin ft (433 W/lin m) of loss should be heated by under-window air supply or radiation. See Ref. 4 for additional discussion. The radiant effect of cold surfaces may be determined from the procedures in ANSI/ASHRAE Standard 55-1992.

Curtain-wall construction, custom-designed wall-to-roof closures, and architectural details at transitions between differing materials have, at times, been poorly constructed and sealed, with the result that cold winter air is admitted to the ceiling plenum and/or occupied spaces. Considering that the infiltration rates published by curtain-wall manufacturers are frequently exceeded because of poor construction practices, it is prudent to provide overcapacity in lieu of undercapacity in heating equipment. The design of finned radiation systems should provide for a *continuous* finned element along the wall requiring heat. Do not design short lengths of finned element connected by bare pipe all within a continuous enclosure. Cold downdrafts can occur in the area of bare pipe. Reduce the heating-water supply temperature and then the finned-element size as required to provide the needed heat output and water velocity.

The surface temperatures of glass, window frames, ceiling plenums, structural steel, vapor barriers, and so on should be analyzed for potential condensation, especially when humidifiers or wet processes are installed.

4.1.10 Process Loads

Heat release from manufacturing processes is frequently a major portion of an industrial air-conditioning load. Motors, transformers, hot tanks, ovens, etc. form the process load. If all motors and other equipment in large plants are assumed to be fully loaded and to be operating continuously, then invariably the air-conditioning system will be greatly oversized. The designer and client should mutually establish diversity factors that consider actual motor loads and operating periods, large equipment with motors near the roof (here the motor heat may be directly exhausted and not affect the air-conditioned zone), amount of motor input energy carried off by coolants, etc. Diversity factors could be as much as 0.5 or even 0.3 for research and development shops containing numerous machines that are used only occasionally by the few operators assigned to the shop.

4.1.11 Room Air Motion

Ideally, occupied portions [or the lower 6 ft (2 m)] of air-conditioned spaces for sedentary activities would have 20- to 40-ft/min (0.1- to 0.2-m/s) velocity of air movement, with the air being within 2°F (1°C) of a set point. It is impractical to expect this velocity throughout an entire area at all times, inasmuch as air would have to be supplied at approximately a 2-ft^3/min · ft^2 (10.2-L/s · m^2) rate or higher. This rate is easily incurred by the design load of perimeter offices, laboratories, and computer rooms but would only occur in an interior office when there is considerable heat-releasing equipment. The supply air temperature should be selected such that, at design conditions, a flow rate of at least 0.8 ft^3/min · ft^2 (4.1 L/s · m^2), but never less than 0.5 ft^3/min · ft^2 (2.5 L/s · m^2), is provided.

People doing moderate levels of work in non-air-conditioned industrial plants might require as much as a 250-ft/min (1.3-m/s) velocity of air movement in order to be able to continue working as the air temperature approaches 90°F (32°C). This would not necessarily provide a full-comfort condition, but it would provide acceptable relief. Loose paper, hair, and other light objects may start to be blown about at air movements of 160 ft/min (0.8 m/s); see Ref. 5. Workers influenced by high ambient temperatures and radiant heat may need as much as a 4000-ft/min (20-m/s) velocity of a 90°F (32°C) airstream to increase their convective and evaporative heat loss. These high velocities would be in the form of spot cooling or of a relief station that the worker could enter and exit at will. Air movement can only compensate for, but not stop, low levels of radiant heat. Only effective shielding will stop radiant energy. Continuous air movement of approximately 300 ft/min (1.5 m/s) and higher can be disturbing to workers.

Situations involving these higher air movements and temperatures should be analyzed by the methods in Refs. 6 to 9.

4.2 OCCUPANCIES

4.2.1 Clean Rooms

For some manufacturing facilities, an interior room that is conditioned by a unitary air conditioner with 2-in- (5-cm-) thick throwaway filters might be called a *clean room*; that is, it is clean relative to the atmosphere of the surrounding plant. Generally, however, clean rooms are spaces associated with the microchip, laser optics, medical, and other industries where air-borne particles as small as 0.5 μm and less are removed. One micrometer equals one-millionth of a meter, or 0.000039 in (0.000001 m).

Clean rooms are identified by the maximum permissible number of 0.5-μm particles per cubic foot. For example, a ISO class 3 clean room will have no more than 100 of these particles per cubic foot, a class 10 clean room no more than 10, and so on. This degree of cleanliness can be attained by passing the air through a high-efficiency particulate air (HEPA) filter installed in the plane of the clean-room ceiling, after which the air continues in a downward vertical laminar flow (VLF) to return grilles located in the floor or in the walls at the floor. Horizontal laminar flow (HLF) rooms are also built wherein the HEPA filters are in one wall and the return grilles are in the opposite wall. A disadvantage with an HLF room is that downstream activities may receive contaminants from upstream activities.

An alternative to an entire space being ultraclean is to provide ultraclean chambers within a clean room (e.g., ISO class chambers in a ISO class 7 room). This is feasible when a product requires the ISO class conditions for only a few operations along the entire assembly line.

The air-conditioning system frequently includes a three-fan configuration (primary, secondary, and makeup) similar to that shown in Fig. 4.5. The primary fan maintains the high air change through the room and through the final HEPA filters. The secondary fan maintains a side-stream (to the primary circuit) airflow through chilled-water or -brine cooling coils, humidifiers, and heating coils. The makeup fan injects conditioned outdoor air into the secondary circuit, thereby providing clean-room pressurization and makeup for exhaust fans. Clean-room air changes are high, such that the total room air might be replaced every 7 s, and this generally results in the fan energy being the major portion of the internal heat gain. Whenever space permits, locate filters downstream of fans so as to intercept contaminants from the lubrication and wear of drive belts, couplings, bearings, etc.

For additional discussions, refer to the *2003 ASHRAE HVAC Handbook, Applications.* ISO Standard 14 644-1 is the international standard for clean rooms.

4.2.2 Computer Rooms

These rooms are required to house computer equipment that is sensitive to swings in temperature and humidity. Equipment of this type normally requires controlled conditions 24 h per day, 7 days per week. Computer equipment can be classified as (1) data processing, (2) computer-aided design and drafting (CADD), and (3) microcomputer. Microcomputers are generally similar to standard office equipment and require no special treatment. Some CADD equipment is also microcomputer-based and falls into the same category. Data processing and larger CADD systems fall into the realm of specialized computer rooms, and these are discussed below.

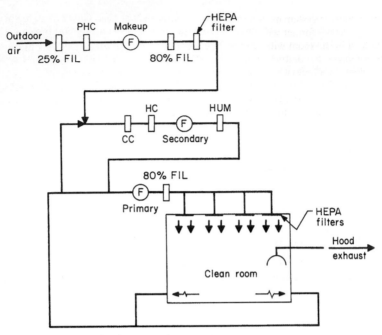

FIGURE 4.5   Three-fan clean-room air system.

Data processing systems operate on a multiple-shift basis, requiring air-conditioning during other than normal working hours. Humidity stability is of prime importance with data processing equipment and CADD plotters. The equipment is inherently sensitive to rapid changes in moisture content and temperature.

To provide for the air-conditioning requirements of computer equipment, two components are necessary: a space to house the equipment and a system to provide cooling and humidity control. Fundamental to space construction is a high-quality vapor barrier and complete sealing of all space penetrations, such as piping, ductwork, and cables. To control moisture penetration into the space effectively, it is necessary to extend the vapor barrier up over the ceiling in the form of a plenum enclosure. Vapor-sealing the ceiling itself is not generally adequate due to the nature of its construction and to penetration from lighting and other devices.

A straightforward approach to providing conditioning to computer spaces is to use packaged, self-contained computer-room units specifically designed for the service. Controls for these units have the necessary accuracy and response to provide the required room conditions. An added advantage to packaged computer-room units is flexibility. As the needs of the computer room change and as the equipment and heat loads move around, the air-conditioning units can be relocated to suit the new configuration. The units can be purchased either with chilled-water or direct-expansion coils, as desired. Remote condensers or liquid coolers can also be provided. Large installations lend themselves quite well to heat recovery; therefore, the designer should be aware of possible potential uses for the energy.

Centrally located air-handling units external to the computer space offer benefits on large installations. More options are available with regard to introduction of ventilation air, energy recovery, and control systems. Maintenance is also more convenient where systems are centrally located. There are obvious additional benefits with noise and vibration control. Use of a centrally located system must be carefully evaluated with regard to first cost and to potential savings, as the former will carry a heavy impact.

The load in the room will be primarily sensible. This will require a fairly high airflow rate as compared to comfort applications. High airflow rates require a high degree of care with air distribution devices in order to avoid drafts. One way to alleviate this problem is to utilize underfloor distribution where a raised floor is provided for computer cable access. A typical computer-room arrangement is shown in Fig. 4.6. Major obstructions to air flow below the floor must be minimized so as to avoid dead spots.

In summary, important points to remember are the following:

1. Completely surround the room with an effective vapor barrier.

2. Provide well-sealed wall penetrations where ductwork and piping pass into the computer space.

3. Provide high-quality humidity and temperature controls capable of holding close tolerances: $\pm 1°F$ (0.6°C) for temperature and ± 5 percent for relative humidity.

4. Pay close attention to air distribution, avoiding major obstructions under floors where underfloor distribution is used.

5. Be alert to opportunities for energy recovery.

6. Make sure that the chosen control parameters and design temperatures and conditions satisfy the equipment manufacturer's specifications.

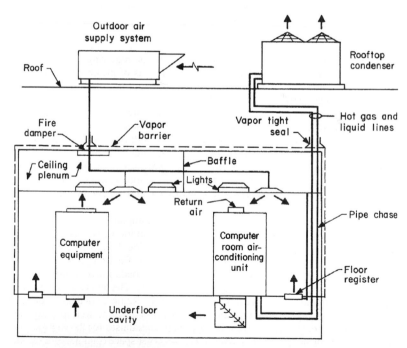

1. Locate floor registers so as to be in nontraffic areas and free from obstruction
2. Ceiling plenum baffles located where and as directed by local codes and insurance underwriters

FIGURE 4.6 Typical computer-room layout.

7. Be attentive to operating-noise levels within the computer space.

8. If chilled water or cooling water is piped to computer-room units within the computer-room space, provide a looped- or grid-type distribution system with extra valved outlets for flexibility.

9. Provide minimum outside air per ASHRAE Standard 62-2001.

4.2.3 Offices

Cooling and heating systems for office buildings and spaces are usually designed with an emphasis on the occupants' comfort and well-being. The designer should remain aware that not only the mechanical systems but also the architectural features of the space affect the comfort of the occupants. And the designer will do well to remember that the mechanical system should in all respects be invisible to the casual observer.

The application of system design is divided into three parts: the method of energy transfer, the method of energy distribution, and the method of control. Controls are discussed in Chap. 7 and will therefore not be discussed here.

To properly apply a mechanical system to control the office environment, it is necessary to completely understand the nature of the load involved. This load will have a different character depending on the part of the office that is being served. Perimeter zones will have relatively large load swings due to solar loading and heat loss because of thermal conduction. The loading from the occupants will be relatively minor. Core zones, on the other hand, will impart more loading from building occupants and installed equipment.

For the office environment, the common system used today is the variable-air-volume (VAV) system. This approach was originally developed as a cooling system, but with proper application of control it will serve equally well for heating. In climates where there is need for extensive heating, perimeter treatment is required to replace the skin loss of the building structure. An old but reliable method is fin-tube radiation supplied with hot water to replace the skin loss. A system that is being seen with more regularity is in the form of perimeter air supply. Care should be taken with the application of perimeter air systems to ensure that wall U-values are at least to the level of ASHRAE Standard 90.1. If this is not done, interior surface temperatures will be too low and the occupants in the vicinity will feel cold.

Avoid striking the surface of exterior windows with conditioned air, as this will probably cool even double-pane glass to below the dew point of the outdoor air in the summer. The result will be fogged windows and a less-than-happy client.

In the interest of economy from a final cost and operating basis, it is best to return the bulk of the air circulated to the supply fan unit. Only enough outdoor air should be made up to the building space to provide ventilation air (ASHRAE 62-2001), replace toilet exhaust, and pressurize the building. For large office systems, it is generally more practical to return spent air to the central unit or units through a ceiling plenum. If the plenum volume is excessively large, a better approach would be to duct the return air directly back to the unit. The ceiling plenum will be warmer during the cooling season when the return air is ducted, and this will require a somewhat greater room air supply because more heat will be transmitted to the room space from the ceiling rather than directly back to the coil through the return air.

Terminal devices require special attention when applied to VAV systems. At low flow rates, the diffuser will tend to dump unless care is taken in the selection to maintain adequate throw. Slot-type diffusers tend to perform well in this application, but there are other diffuser designs, such as the perforated type, that are more economical and will have adequate performance.

The air-handling, refrigeration, and heating equipment could be located either within an enclosed mechanical-equipment room or on the building roof in the form of unitary self-contained equipment. For larger systems, of 200 tons (703 kW) of refrigeration or more, the mechanical-equipment room offers distinct advantages from the standpoint of maintenance; however, the impact on building cost must be evaluated carefully. An alternate approach to the enclosed equipment room is a custom-designed, factory-fabricated equipment room. These are shipped to the jobsite in preassembled, bolted-together, ready-to-run modules. For small offices and retail stores, the most appropriate approach would be roof-mounted, packaged, self-contained, unitary equipment. It will probably be found that this is the lowest in first cost, but it will not fare well in a life-cycle analysis because of increased maintenance costs after 5 to 10 yrs of service.

4.2.4 Test Cells

The cooling and heating of test cells poses many problems.

Within the automotive industry, test cells are used for the following purposes:

• Endurance testing of transmissions and engines
• Hot and cold testing of engines
• Barometric testing and production testing
• Emissions testing

The treatment of production test cells would be very similar to the treatment of noisy areas in other parts of an industrial environment. These areas are generally a little more open in design, with localized protection to contain the scattering of loose pieces in the event of a mechanical failure of the equipment being tested. Hot and cold rooms and barometric cells are usually better left to a package purchase from a manufacturer engaged in that work as a specialty.

Endurance cells, on the other hand, are generally done as a part of the building package (Fig 4.7). It will be found that these spaces are air conditioned for personnel comfort during setup only. The cell would be ventilated while a test is under way. Heat gains for the nontest air-conditioned mode would be from the normal sources: ambient surroundings, lights, people, and so on. It should be remembered, however, that sufficient outdoor air will be needed to make up for trench and floor exhaust while maintaining the cell at a negative-pressure condition relative to other areas. Consult local building codes to ensure compliance with regulations concerning exhaust requirements in areas of this nature.

During testing, as stated previously, the cell would only be ventilated. Outdoor air would be provided at a rate of 100 percent in sufficient quantity to maintain reasonable conditions within the cell. Temperatures within the cell could often be in excess of 140°F (60°C) during a test. Internal-combustion engines are generally liquid-cooled, but even so, the frame losses are substantial and large amounts of outdoor air will be required in order to maintain space conditions to even these high temperature limits. In cold climates, it is necessary to temper ventilation air to something above freezing; 50°F (10°C) is usually appropriate, but each situation needs to be evaluated on its own merit. The engine losses are best obtained from the manufacturer, but in the absence of these data there is information in the *2003 ASHRAE Handbook, HVAC Applications*, that will aid in completing an adequate heat balance. The dynamometer is most often air-cooled and can be thought of as similar to an electric motor. The engine horsepower (wattage) output will be converted to electricity, which is usually fed into the building's electrical system; therefore, the dynamometer losses to the cell will be on the order of 15 to 20 percent of the engine shaft output.

The engine test cell will require a two-stage exhaust system for cooling. The first stage would be to provide low-level floor and trench exhaust to remove heavy fuel vapors and to maintain negative conditions in the cell at all times. The second stage would be interlocked with the ventilation system and would come on during testing and would exhaust at a rate about 5 to 10 percent greater than the supply rate to maintain negative-pressure conditions.

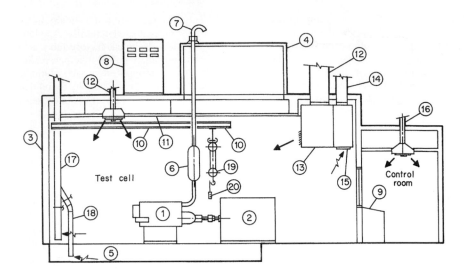

1. Engine
2. Dynamometer
3. Blast wall
4. Blast cupola
5. Fuel and service trench
6. Muffler
7. Engine exhaust
8. Dynamometer
9. Control panel
10. Crane
11. Suspended ceiling
12. Supply air (conditioned, unconditioned)
13. Supply air plenum
14. Cell exhaust
15. Exhaust plenum
16. Control room supply (conditioned)
17. Exhaust duct
18. Trench exhaust duct
19. Electric hoist
20. Hoist electric control

FIGURE 4.7   Typical test-cell layout.

The second stage would also be activated in the event of a fuel spill to purge the cell as quickly as possible. Activation of the purge should be by automatic control in the event that excessive fumes are detected. An emergency manual override for the automatic purge should be provided. Shutdown of the purge should be manual. Consult local codes for explicit requirements.

Depending on the extent of the engine exhaust system, a booster fan may be required to preclude excessive backpressure on the engine. Where more than one cell is involved, one fan would probably serve multiple cells. Controls would need to be provided to hold the back pressure constant at the engine (Fig. 4.8).

Air conditioning for the test cell could be via either direct-expansion or chilled-water coils. During a test, the cell conditioning would be shut down in all areas except the control room. Depending on equipment size, it usually is an advantage to have a separate system cooling the control room. One approach to heating and cooling an endurance-type test cell is shown schematically in Fig. 4.9. Local building codes and the latest volumes of NFPA should be reviewed to ensure that local requirements are being met. Fuel vapors within the cell should be continually monitored. The cell should purge automatically in the event that dangerous concentrations are approached.

The following is suggested as the sequence of events for the control cycle of the test cell depicted in Fig. 4.9.

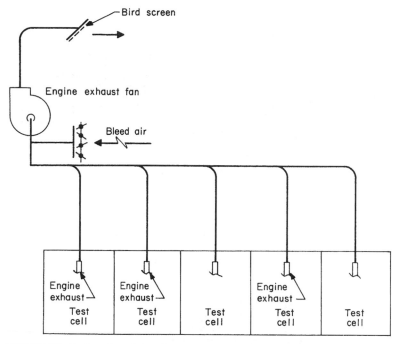

FIGURE 4.8 Engine exhaust helper fan.

Setup Mode

1. AC-1 and RF-1 are running. Outdoor-air and relief-air dampers are modulated in an economizer arrangement.
2. EF-2 is controlled manually and runs at all times, maintaining negative conditions in the cell and the control room.
3. EF-1 is off and D-1 is shut.

Emergency Ventilation Mode

1. If vapors are detected, D-2 shuts and D-1 opens.
2. EF-1 starts and AC-1 changes to high-volume delivery with cooling coil shut down and outdoor-air damper open.
3. HV-1 starts and its outdoor-air damper opens.
4. System should be returned to normal manually.

Test Mode

1. AC-1 cooling coil shuts down.
2. AC-1 changes to high-volume delivery with outdoor-air damper fully open. D-2 closes and D-3 opens.
3. HV-1 starts and EF-1 starts.

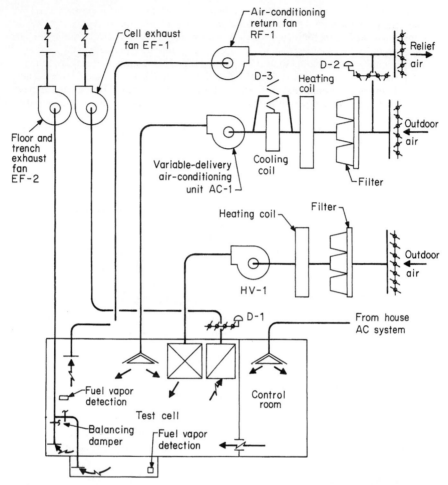

FIGURE 4.9 Test-cell heating, ventilating, and cooling.

4.3 EXHAUST SYSTEMS

One of the early considerations in the design of an exhaust (or ventilation) system should be the ultimate discharge point into the atmosphere. Most of the emissions from ventilation systems are nontoxic or inert and thus will not require a permit for installation or building operation. But should the exhaust air stream contain any of the criteria pollutants—those pollutants for which emissions and ambient concentration criteria have been established, such as CO, NO_x, SO_2, lead, particulate matter (PM), and hydrocarbons (HCs)—it is likely that a permit to install the system will be required.

Once it is determined that a permit will be necessary, an emissions estimate must be made to determine estimates of both uncontrolled (before a pollution control device) and controlled emissions. The emissions estimate may be obtained from either the supplier of the equipment being contemplated for installation or from the Environmental Protection Agency (EPA) Publication AP-42, *Compilation of Air Pollutant Emission Factors*. AP-42 contains emission factors for many common industrial processes, which, when applied to process weight figures, yield emission rates in pounds (kilograms) per hour or tons per year, depending on process operating time. The permit to install an application may be obtained from the state agency responsible for enforcing the federal Clean Air Act. In most states, the Department of Environmental Protection or Department of Natural Resources will have jurisdiction. In general, the permit-to-install application requires the information and data listed in chart A.

When designing an area containing a process exhaust system and a control system for the exhaust, it would be well to keep in mind that federal and local air-quality regulations may govern the type of emission control equipment installed and the maximum allowable emissions. The factors dictating what regulations apply include the type of process or equipment being exhausted, the type and quantity of emissions, the maximum emission rate, and the geographic location of the exhausted process. In order to determine what specific rules and regulations apply, the requirements of the U.S. Code of Federal Regulations Title 40 (40 CFR) should be understood early in the project stages so that all applicable rules may be accommodated. Should the design office lack the necessary expertise in this area, a qualified consultant should be engaged. The federal government has issued a list entitled "Major Stationary Sources." The exhaust system's designer should be acquainted with this list, for it identifies the pollutant sources governed by special requirements. Several of the more common sources are listed in chart B (40 CFR should be consulted for the complete listing). One of the major sets of rules included in 40 CFR is the Prevention of Significant Deterioration (PSD) rules, which establish the extent of pollution control necessary for the major stationary sources.

If a source is determined to be major for any pollutant, the PSD rules may require that the installation include the best available control technology (BACT). The BACT is dependent on the energy impact, environmental impact, economic impact, and other incidental costs associated with the equipment. In addition, the following items are prerequisites to the issue of a permit for pollutants from a major source:

1. Applicant's name and address.
2. Person to contact and telephone number.
3. Proposed facility location.
4. Standard Industrial Classification (SIC) Code.
5. Amount of each air contaminant from each source in pounds per hour (pph) and tons per year (tpy) at maximum and average.
6. What federal requirement will apply to the source?

   • National Emission Standards for Hazardous Air Pollutants (NESHAPs)
   • New Source Performance Standards (NSPS)
   • Prevention of Significant Deterioration (PSD)
   • Emission Offset Policy (EOP)

7. Will best available control technology (BACT) be used?
8. Will the new source cause significant degradation of air quality?
9. How will the new source affect the ambient air quality standard?
10. What monitoring will be installed to monitor the process, exhaust, or control device?
11. What is the construction schedule and the estimated cost of the pollution abatement devices?

CHART A   Commonly requested information for air quality permit applications.

1. Fossil-fuel-fired generating plants greater than 250-million-Btu/h (73-MW) input
2. Kraft pulp mills
3. Portland cement plants
4. Iron-and steel-mill plants
5. Municipal incinerators greater than 250-ton/day charging
6. Petroleum refineries
7. Fuel-conversion plants
8. Chemical process plants
9. Fossil-fuel boilers, or combination thereof totaling more than 250-million-Btu/h (73-MW) input
10. Petroleum storage and transfer units exceeding 300,000-barrel storage
11. Glass-fiber processing plants

CHART B   Major stationary sources—partial list.

1. Review and compliance of control technology with the following:

    a. State Air-Quality Implementation Plan (SIP)
    b. New Source Performance Standards (NSPS) (see chart C)
    c. National Emissions Standards for Hazardous Air Pollutants (NESHAPs)
    d. BACT

2. Evidence that the source's allowable emissions will not cause or contribute to the deterioration of the National Ambient Air-Quality Standard (NAAQS) or the increment over baseline, which is the amount the source is allowed to increase the background concentration of the particular pollutant.

1. Fossil-fuel-fired steam generators
2. Electric utility steam generators
3. Incinerators
4. Portland cement plants
5. Sulfuric acid plants
6. Asphalt concrete plants
7. Petroleum refineries
8. Petroleum liquid storage vessels
9. Petroleum gaseous storage vessels
10. Sewage treatment plants
11. Phosphate fertilizer industry—wet process phosphoric acid plants
12. Steel plants—electric arc furnaces
13. Steel plants—electric arc furnaces and argon decarburization vessels
14. Kraft pulp mills
15. Grain elevators
16. Surface coating of metal furniture
17. Stationary gas turbines
18. Automobile and light-duty truck painting
19. Graphic arts industry—rotogravure printing
20. Pressure-sensitive tape and label surface coating operations
21. Industrial surface coating: large appliance
22. Asphalt processing and asphalt roofing manufacture
23. Bulk gasoline terminals
24. Petroleum dry cleaners

CHART C   New Source Performance Standards (NSPS)—partial list.

3. The results of an approved computerized air-quality model that demonstrates the acceptability of emissions in terms of health-related criteria.

4. The monitoring of any existing NAAQS pollutant for up to 1 yr or for such time as is approved.

5. Documentation of the existing (if any) source's impact and growth since August 7, 1977, in the affected area.

6. A report of the projected impact on visibility, soils, and vegetation.

7. A report of the projected impact on residential, industrial, commercial, and other growth associated with the area.

8. Promulgation of the proposed major source to allow for public comment. Normally, the agency processing the permit application will provide for public notice.

One of the first steps regarding potential pollutant sources is to determine the applicable regulations. For this, an emissions estimate must be made, and the *in-attainment* or *nonattainment* classification of the area in which the source is to be located must be determined. The EPA has classified all areas throughout the United States, including all U.S. possessions and territories. The area is classified as either in-attainment (air quality is better than federal standards) or nonattainment (air quality is worse than federal standards).

If the source is to be located in a nonattainment area, the PSD rules and regulations do not apply, but all sources that contribute to the violation of the NAAQS are subject to the Emissions Offset Policy (EOP). The following items must be considered when reviewing a source that is to be located in a nonattainment area:

1. The lowest achievable emission rate (LAER), which is defined as the most stringent emission limit that can be achieved in practice

2. The emission limitation compliance with the SIP, NSPS, and NESHAPs

3. The contribution of the source to the violation of the NAAQS

4. The impact on the nonattainment area of the fugitive dust sources accompanying the major source.

In general, the EOP requires that for a source locating in a nonattainment area, more than equivalent offsetting emission reductions must be obtained from existing emissions prior to approval of the new major source or major modification. The *bubble concept*, wherein the total emissions from the entire facility with the new source do not exceed the emissions prior to addition of the new source, may be used to determine the emission rate. If there were emission reductions at *existing* sources, they would offset the contributions from the new source, or *offset* the new emissions. This same bubble concept may be used for sources that qualify for in-attainment or PSD review.

In the design of a polluting or pollution control facility, stack design should be considered. A stackhead rain-protection device (Figs. 4.10 and 4.14) should be used in lieu of the weather cap found on many older installations, since this cap does not allow for adequate dispersion of the exhaust gas. When specifying or designing stack heights, it should be noted that the EPA has promulgated rules governing the minimum stack height; these rules are known as *good engineering practice* (GEP). A GEP stack has sufficient height to ensure that emissions from the stack do not result in excessive concentrations of any air pollutant in the vicinity of the source as a result of atmospheric downwash, eddy currents, or wakes caused by the building itself or by nearby structures (Figs. 4.11 and 4.12). For uninfluenced stacks, the GEP height is 98 ft (30 m). For stacks on or near structures, the GEP height is (1) 1.5 times the lesser of the height or width of the structure, plus the height of

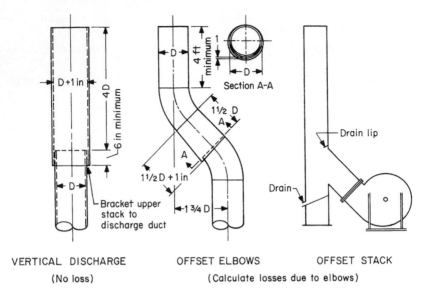

VERTICAL DISCHARGE   OFFSET ELBOWS   OFFSET STACK

(No loss)   (Calculate losses due to elbows)

1. Rain protection characteristics of these caps are superior to a deflecting cap located 0.75D from top of stack.

2. The length of upper stack is related to rain protection. Excessive additional distance may cause "blowout" of effluent at the gap between upper and lower sections.

FIGURE 4.10   Typical rain-protection devices. (*From Industrial Ventilation—A Manual of Recommended Practice, 21st ed., Committee on Industrial Ventilation, American Conference of Governmental Industrial Hygienists, copyright 1992, p. 5–53.*)

the structure, or (2) such height that the owner of the building can show is necessary for proper dispersion. In addition to GEP stack height, stack exit velocity must be maintained for proper dispersion characteristics.

Figures 4.13 and 4.14 illustrate the relationship between velocity at discharge and the velocity at various distances for the stackhead. Maintaining an adequate exit velocity ensures that the exhaust gases will not reenter the building through open windows, doors, or mechanical ventilation equipment. Depending on normal ambient atmospheric conditions, the exit velocities may range from 2700 to 5400 ft/min (14 to 28 m/s). In practice, it has been found that 3500 ft/min (18 m/s) is a good average figure for stack exit velocity, giving adequate plume rise yet maintaining an acceptable noise level within the vicinity of the stack.

Care must be taken when designing exhaust systems handling pollutants for which no specific federal emission limit exists (noncriteria pollutants). All pollutants not included in the criteria pollutant category or the NESHAPS category are considered noncriteria pollutants. When establishing or attempting to determine acceptable concentration levels for noncriteria pollutants, the local authority responsible for regulating air pollution should be consulted since policy varies from district to district. In general, however, noncriteria pollutants' allowable emission rates are based on the American Conference of Governmental Industrial Hygienists (ACGIH) time-weighted average acceptable exposure levels.

A hazardous air pollutant is one for which no ambient air-quality standard is applicable, but which may cause or contribute to increased mortality or illness in the general population.

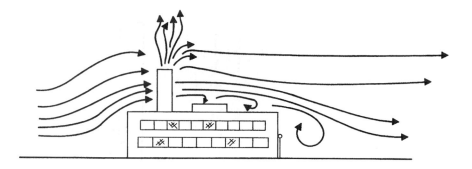

GEP stack height minimizes re-entrainment of exhaust gases into air which might enter building ventilation system.

FIGURE 4.11   GEP stack.

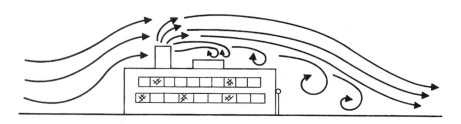

Non-GEP stack allows exhaust gases to be entrained in building wakes and eddy currents.

FIGURE 4.12   Non-GEP stack.

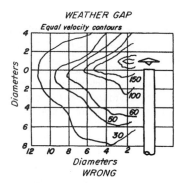

Deflecting weather cap discharges downward.

FIGURE 4.13   Weather-cap dispersion characteristics.   (*From Industrial Ventilation—A Manual of Recommended Practice, 21st ed., Committee on Industrial Ventilation, American Conference of Governmental Industrial Hygienists, copyright 1992, p. 5.62.*)

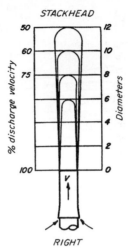

Vertical discharge cap throws upward where dilution will take place.

FIGURE 4.14 Stackhead dispersion characteristics. (*From Industrial Ventilation—A Manual of Recommended Practice, 21st ed., Committee on Industrial Ventilation, American Conference of Governmental Industrial Hygienists, copyright 1992, p. 5.62.*)

Emission standards for such pollutants are required to be set at levels that protect the public health. These allowable pollutants' emission levels are known as NESHAPS and include levels for radon-222, beryllium, mercury, vinyl chloride, radionuclides, benzene, asbestos, arsenic, and fugitive organic leaks from equipment.

An exhaust stream that includes numerous pollutants, with some being noncriteria pollutants, can be quickly reviewed by assuming that all the exhaust consists of the most toxic pollutant compound. If the emission levels are acceptable for that review, they will be acceptable for all other compounds.

REFERENCES

1. United McGill Corporation, *Engineering Bulletin*, vol. 2, no. 9, copyright 1990.

2. *Energy Conservation in New Building Design*, ASHRAE Standard 90A-1980, ASHRAE, Atlanta, GA, 1980, p. 18, para. 4.4.2.4.

3. *2001 ASHRAE Handbook, Fundamentals*, ASHRAE, Atlanta, GA, 1997.

4. Tom Zych, "Overhead Heating of Perimeter Zones in VAV Systems," *Contracting Business*, August 1985, pp. 75–78.

5. *Thermal Environmental Conditions for Human Occupancy*, ANSI/ASHRAE Standard 55–192, ASHRAE, Atlanta, GA, p. 4, para 5.1.4.

6. Knowlton J. Caplan, "Heat Stress Measurements," *Heating/Piping/Air Conditioning*, February 1980, pp. 55–62.

7. *Industrial Ventilation—A Manual of Recommended Practice*, 24th ed., Committee on Industrial Ventilation, American Conference of Governmental Industrial Hygienists, Lansing, Mich., 2001.

8. *2003 ASHRAE Handbook, HVAC Applications*, ASHRAE, Atlanta, GA.

9. W. C. L. Hemeon, *Plant and Process Ventilation*, 2d ed., Industrial Press, New York, 1963 chap. 13, pp. 325–334.

CHAPTER 5
COGENERATION SYSTEMS

Alan J. Smith, P.E.

Kellogg Brown & Root, Inc., Houston, Texas

Aparajita Sengupta, P.E.

Kellogg Brown & Root, Inc., Houston, Texas

5.1 INTRODUCTION

A *cogeneration facility* consists of equipment that produces both electric energy and forms of useful thermal energy (such as heat or steam) for industrial, commercial, heating, or cooling purposes. This integrated production of electric energy and useful thermal energy is also designated either as a *topping-cycle* or *bottoming-cycle* facility. Topping-cycle facilities first transform fuel into useful electric power output; the rejected heat from power production is then used to provide useful thermal energy. In contrast, bottoming-cycle facilities first apply input energy to a useful thermal process, and the rejected heat emerging from the process is then used for power production. Either of these cycles can efficiently apply thermal energy to meet process or comfort heating, ventilating, and air-conditioning (HVAC) by generating steam, hot water or chilled water.

This chapter describes the various methods of applying thermal energy from a cogeneration system to HVAC systems.

5.2 HVAC APPLICATIONS FOR THERMAL ENERGY

Typical cogeneration systems include prime movers such as combustion turbine or internal combustion engine generators with heat recovery systems applied to their hot exhaust streams. Feasible methods for applying thermal energy to meet process, HVAC or comfort (hereafter referred to as "utility") requirements include steam-turbine-driven mechanical chillers, steam absorption chillers and exhaust-gas-driven absorption chillers.

5.2.1 Steam-Driven Mechanical Chillers, Steam/Hot-Water Absorption Chiller Units

Steam generation and mechanical drive and/or absorption chillers (Fig. 5.1) are cost-effective in cases where the cogeneration-system-produced steam supplements a facility's

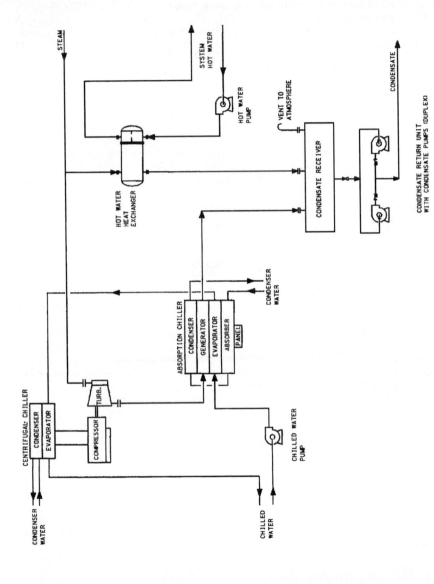

FIGURE 5.1 Piggy-back system utilizing steam for HVAC processes.

5.2

existing steam requirements. A hot-water absorption chiller system (Fig. 5.2) rather than a steam chiller system should be considered for facilities with requirements only for hot water, as equipment required to transfer energy from steam to hot water is not required. The steam or hot-water generation system design should include a standby energy source to ensure that utility requirements are met if the cogeneration system operates at a reduced electrical generation level or suffers an unplanned outage. If desired, the exhaust system for the heat-recovery steam generator (HRSG) units could be designed to operate with fresh air natural gas firing while maintenance is being conducted on the system's prime mover. In this case, a system of exhaust guillotine dampers and seal air blowers would be utilized.

The HRSG typically has a minimum exhaust temperature of approximately 250°F (121°C) when natural gas fuel is used to fire the prime mover. Maintaining this temperature protects the HRSG from water-vapor condensation and acid formation that occurs when the exhaust temperature drops below the dew point. Although an HRSG can be designed for lower exhaust temperatures, the corrosion-resistant design is not always economically feasible and should be evaluated.

The HRSG normally imposes a back pressure of 8 to 12 in · WG (inches water gauge) (2000 to 3000 Pa) on the prime mover exhaust. Specific vendor calculations should be used to estimate the impact of system backpressure on combustion turbine performance. This back pressure results in a horsepower penalty of approximately 0.35 percent/in · WG (1.9 W/Pa) for combustion gas turbines. Within the limitations specified by internal-combustion-engine vendors, exhaust-gas backpressure does not appreciably reduce the mechanical power output of the engine.

5.2.2 Steam or Hot-Water Generation Control

In facilities that can use a limited amount of thermal energy from a cogeneration system, the system or hot-water production rate can be controlled by regulating the throttle of the prime mover or by bypassing the exhaust-gas heat around the heat-recovery unit and sending it up the stack. Excess steam or hot water can also be diverted to other heat rejection systems.

Using steam turbines, combined-cycle cogeneration systems use thermal energy not required for utility service to generate additional electricity.* One patented cycle varies its steam production rate by reinjecting high-pressure steam into the power turbine section of combustion gas turbine. This procedure reduces the amount of steam that must be used by a facility and increases the electric power output of the unit.

Factors that affect the selection of an absorption chiller in a cogeneration system include:

• Available steam or hot-water pressure and temperature
• Steam consumption rate
• Physical size
• Machine performance under partial-load conditions

Steam and hot-water requirements for typical chiller units are summarized in Table 5.1. The steam consumption rate of the two-stage machine is approximately 40 percent less than that of the single-stage machines. Heat rejection requirements are also reduced more than 20 percent compared to the requirements of similar capacity single-stage absorption chillers. The rates get even lower for combination centrifugal-absorption chiller units, often

---

*Combined-cycle systems simultaneously produce power using a fossil-fueled prime mover and a steam turbine generator unit.

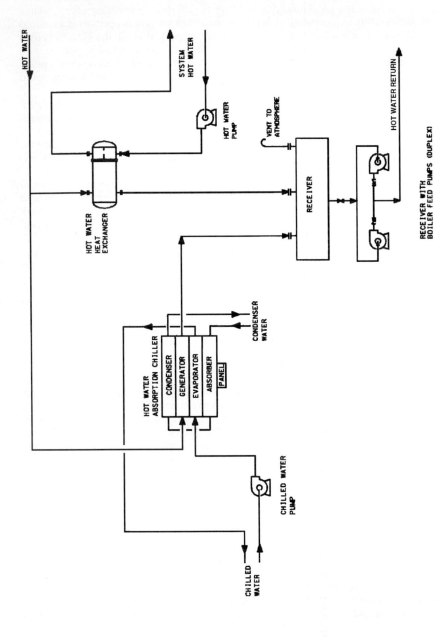

FIGURE 5.2 Hot water system for HVAC processes.

5.4

TABLE 5.1 Thermal Energy Requirements of Chillers

Chiller type	Steam supply conditions		Hot-water supply conditions		Nominal steam consumption rate	
	psig	kPag	°F	°C	lb/(h · ton)	kg/(s · W)
Single-stage absorption, small	—	—	160	71	—	—
Single-stage absorption	8–15	55–103	270	132	17.5–20	8.1–9.3
Two-stage absorption	100–120	690–830	300–400	150–204	9.9–12	4.6–5.5
Combination centrifugal-absorption	600 & higher	4135 & higher	—	—	8.0 & lower	3.7 & lower
Ammonia absorption, single-stage*	40–160	275–1100	—	—	30.6–57.8	14–27
Ammonia absorption, two-stage*	1–24	7–165	—	—	47.5–67.5	22–31

*Used primarily in low-temperature applications.

called "piggy-back systems," as described later in this section. Two-stage absorption chillers are designed with absorbent streams using parallel or series flow. The configuration of the parallel-flow machine results in reduced height in all machine sizes and reduced width in larger machine sizes. Either type of machine can be installed fully assembled in capacities up to 750 ton (2640 kW). In larger sizes, the series-flow machine must be partially assembled at the installation site, while the parallel-flow machine can be transported and installed as a single unit for capacities up to 2500 ton.

The steam utilization characteristics of absorption chillers affect their sizing in cogeneration systems. The single-stage absorption machine's electricity and steam consumption rate per unit of chilled-water production decreases with reduced load to approximately 30 percent of design capacity. At this point, consumption rises unless some other cycle enhancement is added. Steam-consumption curves decrease slightly at reduced-load conditions for series-flow two-stage machines. Two-stage machines using parallel flow maintain flat steam consumption curves over the entire load range. Occasionally, ammonia-absorption machines are used in low temperature applications for cold storage or freezer storage warehouse use.

Combination Centrifugal-Absorption Chiller Units. Noncondensing (backpressure) steam turbines driving mechanical chillers can be used in series with conventional absorption chillers by matching steam flow rates and exhaust pressure from the steam turbine (Fig. 5.1). This type of system (piggy-back system) must always run when there is another use for the exhaust steam. The traditional distribution of chiller capacity is one-third in the mechanical-drive chiller and two-thirds in the absorption chiller. At higher steam pressures, the capacity distribution may approach 50 percent for each portion of the system. Noncondensing steam turbines enhance the energy efficiency of a cogeneration cycle, because exhaust steam can be used for other heating or absorbing processes. This arrangement allows the absorption chiller to operate at a higher chilled water supply temperature, thus causing less operational problems associated with lower evaporator temperatures. Typical steam inlet pressure for noncondensing steam turbines is at least 400 lb/in^2 (2750 kPag), while the exhaust pressure is approximately 8 lb/in^2 (55 kPag). Figure 5.3

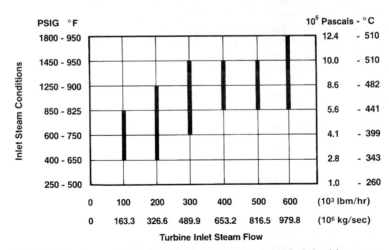

FIGURE 5.3 Range of initial steam conditions normally selected for industrial steam turbines. (*From J. M. Kovacik*, Industrial Plant Objectives and Cogeneration System Development, *General Electric Co.*)

illustrates the range of inlet steam pressures and flows commonly used with noncondensing steam turbines.

5.2.3 Modular Chiller Buildings

Inlet air cooling by using a chilled water coil increases the mass flow rate through a combustion turbine resulting in increased electrical output. A popular method for producing chilled water is to install a self-contained modular chiller building and connect it to chilled water mains in a primary loop. A separate building containing secondary (distribution) pumps will then supply chilled water to the combustion turbine by means of supply and return chilled-water headers. This will allow a variable flow rate of chilled water in the secondary (distribution) loop. Variable frequency drives and multiple pumps in the secondary chilled water loop will maintain the desired flow as needed by the inlet air coil cooling load. Each modular chiller building will generally be rated at 2000 ton (7040 kW) and will be self-sufficient with its chilled-water pumps, condenser water pumps, cooling tower (multicell), motor control centers, and wiring. These packages will arrive at a project site pretested and ready to be connected into the chilled-water mains, electrical hook-up, and controls hook-up. An integrated control system will monitor the flow and temperature of chilled water in the secondary loop and start/stop chiller modules in response to thermal load on the system, operating the chiller packages at close to full load and maintaining high overall efficiency.

5.3 OPERATIONAL CRITERIA

Electricity demand and process energy demand (chilled water, hot water, and steam) establish sizing and operating criteria for a cogeneration system. These data must be examined over various periods of time (seasonally, weekly, daily, and even hourly in some cases) to establish a specific cyclic pattern for the energy use.

The specific components and sources of the demand must be known. Careful consideration should be given to the decrease in a facility's electricity requirements if electric-driven centrifugal chillers are to be replaced by steam absorption units as part of the cogeneration system.

Typical operational criteria that could result from process data are:

• Efficient use of thermal energy produced by the prime movers
• Supply the base electric load
• Interchange sales of electrical energy (power sales to a public utility company.)

The decision to engage in interchange sales of electricity will likely increase capital cost because the facility must adopt additional electric energy metering requirements that are specified by the public utility companies.

Typical ranges for the electric power generation capacity of industrial, institutional, residential, and commercial cogeneration systems are summarized in Table 5.2. Industrial and institutional facilities can achieve significant economic benefit from cogeneration systems due to their balanced requirements for electric and thermal energy.

TABLE 5.2   Typical Cogeneration System Electric Power
Generation Capacities

Application	Electrical output (kW)
One and two family homes	5–15
Multifamily dwellings	20–5,000
Office-buildings	2,000–10,000
Local shopping centers	100–250
Distribution centers	250–2,500
Regional shopping centers	5,000–15,000
Industrial institution facilities	Site dependent

Source: Richard Stone, "Stand Alone Cogeneration By Large Building
Complexes," Energy Economics, Policy and Management (Fairmont Press,
Atlanta), vol. 62, Summer 1982.

5.4   PRIME MOVERS

Combustion gas turbines and internal-combustion engines are the typical prime movers used
in topping cycles. Each has high-temperature exhaust gas. Internal-combustion engines also
produce thermal energy in their jacket water that may be used by the heat recovery system.

5.4.1   Combustion Gas Turbine Generators

Combustion turbine generator (CTG) units exhibit the following characteristics in a cogeneration system:

* High temperature of exhaust gas
* High quantity of exhaust gas

With thermal energy recovery, the overall cycle energy efficiency of a CTG unit typically exceeds 60 percent. Common types of heat-recovery equipment used in CTG cogeneration systems are:

* Heat-recovery steam generator (HRSG)
* Hot-water heater
* Exhaust-gas chillers

Combustion turbines typically produce sufficient thermal energy to produce approximately 10 lb/h (4.54 kg/h) of 15 to 150-psig (100 to 1030-kPag) steam per horsepower
(0.7457 kW) of electrical output. Because of the volume of excess air contained in the CTG
exhaust, it is possible to supplement the heat contained in the turbine exhaust to gain additional steam-generating capacity or cooling capacity by burning additional fuel. This supplemental gas firing typically has an efficiency of 90 percent. In all cases, net efficiency of the
CTG and power generation can be increased by increasing the density of the air to the turbine.
Inlet air cooling is a method to cool this air, increasing the mass flow rate through the CTG.
In plants where chilled water is produced, inlet air cooling can be accomplished by supplying
chilled water to a cooling coil located at the inlet to the turbine compressor. A control
system, consisting of a modulating chilled-water-flow control valve that controls based on
feedback from a temperature sensor in the leaving air, is required to maintain an optimum

air temperature. This control concept yields maximum efficiency without any possibility of condensation and/or ice formation—conditions that could damage the turbine blades. Sufficient generator cooling should be confirmed due to the increased generator electrical output.

Heat Balance. Mechanical energy makes up approximately 30 percent of a CTG unit's heat balance under full-load conditions. Exhaust gas contains essentially the remainder of the energy, with small portions allocated to lube oil and other losses. This exhaust-gas thermal energy can be directly applied to driving an HRSG or an exhaust-gas chiller-heater. The lube oil temperature is low and the quantity of heat is small, and thus, in most cases, it is not economical to recover heat from this source.

Load Control. Single- and two-shaft combustion turbines are available. The multishaft units are designed with separate shafts for the compressor section and the power turbine section. Separate shafts permit the rotating speed of the compressor section to be controlled by the requirements of the power turbine, rather than by the rotating speed of the generator.

Representative partial-load operating efficiencies between the single- and two-shaft types of combustion gas turbines are illustrated in Fig. 5.4. The two-shaft units are generally able to maintain higher exhaust temperatures, and therefore greater operating efficiency,

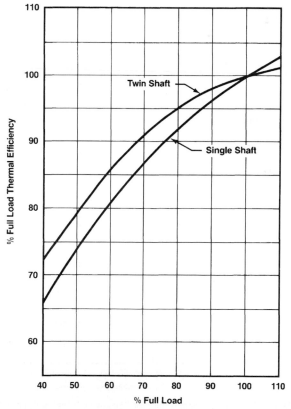

FIGURE 5.4 Part load efficiency—single and twin shaft turbines. (*Courtesy of Ruston Gas Turbines Limited*)

under partial-load conditions. The two-shaft units, however, will have higher heat rates (increased fuel consumption) at full-load conditions. If partial-load operation of a combustion turbine is required because of cogeneration system operating criteria, consideration should be given to a two-shaft combustion turbine.

5.4.2 Internal-Combustion Engines

Internal-combustion engines exhibit the following characteristics in cogeneration systems:

- High mechanical efficiency
- More efficient operation at partial load (Fig. 5.5)
- High-temperature exhaust gases (below that generated by combustion turbines)
- Readily available maintenance services

Heat-recovery equipment used in cogeneration systems using internal-combustion engines include:

- Heat-recovery steam generators
- Coil-type hot-water heaters
- Steam separators for use with high-temperature cooling of engine jackets
- Exhaust-gas-driven chillers

Internal-combustion engines typically generate 3 lb/h (1.36 kg/h) of 15- to 150-psig (100- to 1030-kPag) steam per horsepower output. Due to the lack of oxygen in the exhaust gas, supplemental heating of the exhaust gas thermal energy is not usually economically feasible.

Jacket-Water Heat Recovery.   Cogeneration heat-recovery systems that use engine jacket-water thermal energy take four forms:

1. The heated jacket water may be routed to process needs. Engine cooling is dependent on the leak-tight integrity of this system.

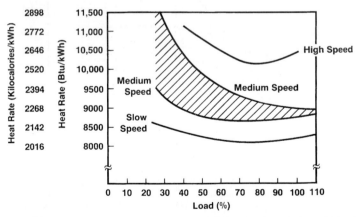

FIGURE 5.5 Typical Variation of internal combustion engine heat rate with load. (*From* Handbook of Industrial Cogeneration, *Publication DE 82 009604, Department of Energy, Washington, DC, October 1981, Fig. 16-1.*)

2. The jacket cooling-water circuit for each engine transfers heat to an overall utilization circuit serving facility process needs. The overall utilization circuit may also be heated by the engine exhaust. This configuration minimizes connections to the jacket cooling-water system.

3. Hot water in the jacket cooling-water system is flashed to steam in an attached steam flash tank. Water enters of engine at 235°F (113°C) and exits at 250°F (121°C). Steam is produced at 235°F (113°C), 8 psig (55 kPag). Flow must be restricted at the entrance to the steam flash chamber to maintain sufficient back pressure on the liquid coolant in the engine chambers.

4. Some engines use natural-convection ebullient cooling. A steam-and-water mixture rises through the engine to a separating tank, where the steam is released and the water is recirculated. A rapid coolant flow through the engine is required due to a small rise in the temperature of the fluid. Moreover, backpressure must be controlled, for the steam bubbles in the engine could rapidly expand, causing the engine to overheat. This system produces 15-psig (100 kPag) steam at 250°F (120°C).

The relatively low temperature and pressure of these jacket-water heat-recovery systems make them suitable for single-stage absorption chiller application (Table 5.1).

Heat Balance. A typical heat balance for an internal-combustion engine is illustrated in Fig. 5.6. The exhaust heat makes up the largest portion of the energy. The jacket cooling-water component of thermal energy from an internal-combustion engine contains 15 percent of the heat input (Fig. 5.6). Jacket cooling-water temperatures are summarized in Table 5.3. Some internal-combustion-engine manufacturers discourage

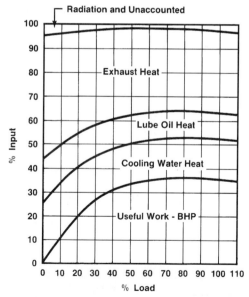

FIGURE 5.6 Heat balance for 8 cylinder diesel engine. (*From* Cogeneration, *Fairbanks-Morse Engine Division, Colt Industries.*)

TABLE 5.3   Typical Waste-Heat Temperatures

Thermal energy source	Gas turbine		Internal-combustion engine	
	°F	°C	°F	°C
Exhaust gas	900–1000	480–540	1000–1200	540–650
Lube oil	165 (max.)	74	160–200	70–95
Jacket water	—	—	180–250	80–120

operating with high jacket-water temperatures, because special gasket and seal designs are required.

The lubricating oil system also contains usable heat (Fig. 5.6). The normal operating temperature for the system is 165°F (74°C). The lube oil cooling fluid may also be routed through the exhaust heat recovery unit if process requirements specify heat at a higher temperature. If the lube oil coolant temperature is elevated above a temperature range between 180°F (82°C) and 200°F (93°C), special lubricants may be required to ensure an adequate useful life of the oil.

Load Control.   The heat rate of an internal-combustion engine remains almost constant above approximately 50 percent load (Fig. 5.6). From the engine heat balance, energy normally converted to mechanical energy is transferred to thermal energy below 50 percent power. Cogeneration systems are suited to using a large portion of this thermal energy.

HVAC OPERATIONS

CHAPTER 6
MECHANICAL SYSTEMS COMMISSIONING

Michael S. Bilecky
Von Otto & Bilecky, Washington, D.C.

6.1 INTRODUCTION TO COMMISSIONING

6.1.1 What Are the Objectives of Commissioning?

Commissioning is, quite simply, assuring that mechanical systems are installed and will function in accordance with design intent. Equally as important to properly functioning mechanical systems is assuring that those who will operate and maintain them are fully trained. Achieving these assurances is anything but simple. It involves rigorous efforts to exercise mechanical systems through their full range of operations in attempts to verify that the systems will perform as intended by the designer and as expected by the owner and providing a comprehensive training regimen for operating and maintenance personnel. *Commissioning is the final stage in any quality assurance program; and though it can be an add-on to an already completed project, commissioning provides the best results when it has been integrated into the project from the earliest stages of design.*

Commissioning as described in this manual is presented as being performed by an independent commissioning agent; i.e., a person not responsible for the design, installation, or operating and maintenance of the mechanical systems. Employing the services of a commissioner acting as an independent agent for the owner provides the commissioner autonomy, allowing for objective and unbiased evaluations and assessments in the commissioning process. The services of an independent agent, though providing the optimum, are not absolutely required. Commissioning can be performed by the design architect/engineer (A/E), by the installing contractor, by the owner, or by the owner's operating and maintenance personnel. Commissioning by each of these entities brings the potential for individual bias and self-interest to skew the commissioning process. The A/E may attempt to mask design deficiencies and avoid embarrassing change orders by accepting system performance at levels less than had been intended and expected by the owner, but not suitably provided by the contract documents. The installing contractor may assert that systems are functioning up to that level provided in contract documents; or more accurately, his assessment that the installation meets specified performance. The installing contractor has the greatest potential for exercising bias in commissioning his installations, this being the proverbial "fox guarding the chickens." The bias that can be exercised by the owner or the owner's operating and maintenance personnel can be one of high expectations not substantiated by contract documents. In this case the contractual work can get bogged down in claims and counterclaims among the design team, installing contractor, and owner. Though not a spelled out contract service of the independent commissioning agent, his or her independence should provide all parties with an unbiased source of dispute resolution; telling the A/E, the contractor, or the owner that in his or her opinion the work in dispute is or is not provided in the contract documents. Remember, the commissioning is an assurance that what has been purchased by the owner has in fact been provided. Getting the owner what he or she wants starts with the design process, and the design documents are the basis on which all subsequent assessments are made.

The initial role of the A/E is to provide the owner with mechanical systems that will serve the owner's needs. These systems are defined by the contract documents that normally consist of plans and specifications. It is wise to have the commissioning agent as a team member as early as the design phase of the project. Commissioning is *not* intended to provide a design review; it is assumed that the design of mechanical systems will be competently provided by the design engineer. The role of the commissioning agent through the design phase is primarily one of an *observer*—to gain insight into design intent and expectations and to be cognizant of design decisions and trade-offs which ultimately will affect systems' performance and operations. The commissioning agent also serves as an *advisor*, offering guidance in the preparation of documents to assure that all necessary system components and test procedures are provided to facilitate the commissioning.

6.1.2 Why Perform Commissioning?

Full commissioning is not accomplished in the conventional services offered by the A/E design team or by the installing contractor. The A/E creates their design, does shop drawing review during construction, periodically visits the site during construction to perform an often cursory review of project progress or to resolve questions and conflicts between design documents and field conditions, and performs punch list and final inspections which usually entail very little actual operation of the mechanical systems through their full range of operational sequences.

The installing mechanical contractor is looking to get off of the job as soon as possible, as doing this generally will represent his largest profit. This does not necessarily mean that the contractor has short-changed the job as far as the quality of materials and workmanship put into assembling the pieces which will comprise the total mechanical installation. But what *is* often short-changed is the testing of systems to assure proper performance and long-term operation—not just operation that appears to be proper for the few minutes during which a system is demonstrated to the owner or the A/E. Further, today's mechanical systems are becoming more and more complex, particularly with computer-based control systems being integral to nearly all equipment and systems. It is not uncommon for mechanical contractors to rely on their subcontractors for testing of systems for which they were responsible and accepting at face value the statements from the subcontractors that their work is complete and functioning properly. This may well be true when a subsystem is viewed as its own distinct entity. However, in the reality of the building mechanical systems as a whole, what may have *appeared* to be proper operation of a subsystem may in fact be incomplete or even improper when the subsystem is integrated with other interactive mechanical operations. The mechanical contractor's self-testing of the mechanical systems as a whole is seldom performed.

The owner or the owner's operating and maintenance personnel are usually in the weakest position to commission a building's mechanical system. They are normally handed a completed product, given a few hours or days of instructions on equipment operations and maintenance, and left to fend for themselves over the warranty period and eventually through the life of the system. Operating and maintenance personnel perform a hands-on type of commissioning by responding to system failures and complaints from building occupants. During the warranty period, the failures and complaints are usually passed back to the contractor who returns and "fixes" the problem; sometimes these fixes are made by totally subverting the original design intent, taking the shortest path to silencing complaints. *What often results is the masking of design flaws or installation shortcomings which will be costly over the life of the building.*

Further, at the time the building is turned over from the contractor to the owner, the limited exposure given to operating and maintenance personnel is seldom sufficient for them to understand the underlying design concepts and intricacies of the systems installed. They are provided operating and maintenance manuals which provide little insight as to how interactive subsystems are to function to create the whole of the mechanical system. They are relegated to attempting to decipher manufacturer's printed literature for equipment and extrapolate on how systems are supposed to operate based upon information on pieces of equipment. What often results is that operating and maintenance personnel, primarily from lack of adequate training and instruction from the contractor, resort to jumping controls out or making modifications to equipment and systems to get them to function in a way that they understand; which is not necessarily the way the designer intended.

The real world process of designing, installing, operating, and maintaining mechanical systems has created the need for commissioning. Someone has to be responsible for a holistic assessment of the building's mechanical systems. The assessment of the whole, completed project is most effectively achieved by the independent commissioning agent, who by the project's end probably possesses the greatest knowledge of the component systems and the interaction of these components that constitute the whole of mechanical systems.

6.1.3 Effects of Commissioning

To architects, engineers, and contractors who have not been through a mechanical commissioning process, the independent commissioning agent is often initially viewed as a threat. The perception is that the commissioner is the agent of the owner, policing the performance of the design team and the contractors serving to put light on errors and omissions. Though the perception is largely correct, viewing the commissioning agent as a threat normally subsides over the life of the construction project; and usually results in resounding support of the commissioning process. After all, all parties are, in theory, working toward the best interests of the owner; and the theory should be supported by the reality that the rewards gleaned from a good project are far greater than any short-term gain (and often losses) from a poorly designed or constructed project. Once it is realized that the commissioning agent shares the common goal of a good project which satisfies the owner and from which all parties walk away with their fair profits, support for the commissioning process grows.

By being involved with a project from the earliest design phases, the commissioning agent may identify design errors or omissions which would subsequently become requests for information from the contractors and subsequent change orders, proving both costly in man-hours and possibly damaging, or at least embarrassing, to the design team. By assuring that the work is installed by the contractors as intended by the contract documents, the commissioning agent helps assure that a properly working building is turned over to the owner. This minimizes the amount of contractor callbacks to correct deficiencies which eat into contractors' profits. *Having the commissioning agent participate in the*

demonstrations and training sessions provided to the owner's operating and mainte-
nance personnel results in a better and more comprehensive understanding of systems
operations by those personnel.

Thus, the commissioning process should be a win-win-win proposition for the design team, the contractors, and the owner.

6.2 OVERVIEW

When should mechanical systems commissioning start? Actually, the process of commissioning starts with the outset of design when building and mechanical system types are being discussed in the early planning phase. Decisions made by owners, architects, and engineers in the early stages of planning will determine the mechanical systems to be designed, installed, and ultimately commissioned. Ideally, the commissioning agent would be a participant in the design process. By being involved with project design, the commissioning agent gleans an understanding of the decisions that were made to establish the type of mechanical system to be installed. This understanding will assist in making on-site observations during construction and, more importantly, during observations of actual systems performance. The commissioning agent can also be an active contributor during the early design process by bringing mechanical engineering expertise to the table in addition to and complementary of the project A/E. During design, the commissioning agent offers a second set of eyes to review contract documents for completeness and accuracy, helping assure that all of the necessary pieces and procedures are in place by design and specification to properly commission the mechanical systems.

Perhaps the most important contribution of the commissioning agent during the
design process is assisting in the incorporation of a written commissioning plan in the
contract specifications. This is an essential component to achieve successful commissioning of the building. The commissioning process requires time and manpower of the A/E, the mechanical contractor and subcontractors, and of the owner's operating and maintenance personnel which extends beyond that normally provided when commissioning is not performed. Particularly, contractors must be placed on notice, via the commissioning specification, that they will be required to demonstrate various tests and systems operations to the commissioning agent. Without this notice being given in the contract specifications, the contractor can rightfully become less than cooperative when asked to demonstrate (and redemonstrate) mechanical tests and operations because the contractor did not allocate the necessary man-hours in his bid. Demonstrations and instructions to owner's operating and maintenance personnel in a commissioned building also extend beyond those normally provided by the contractor; and as such, the owner's personnel must also be on notice that more time will be required of them to accept the building.

The commissioning plan should be a distinct section of the mechanical specifications (normally Section 15995). The specification should delineate the responsibilities of the various participants during construction through the completion of the project and acceptance of the building by the owner. As with any good specification, the commissioning plan should be written specifically applicable to the mechanical systems being installed on the project; it does little good to enumerate the procedures for commissioning an air-cooled chiller when a water-cooled machine is designed for the project.

The involvement of the commissioning agent during design, though important to the commissioning process, does not require an extensive amount of time, and should represent about 15 percent of the total man-hours and fee for commissioning.

6.2.1 Construction Phase

Construction Schedules and the Commissioning Plan.   After a set of construction documents has been prepared, bid, and a contract for construction awarded, the commissioning agent starts to play a more active role. Usually, a preconstruction meeting is held at which time the general contractor presents a construction schedule. This schedule will contain milestones for the general construction of the project and should identify major mechanical milestones such as scheduled delivery and installation of major pieces of equipment (air handling units, boilers, chillers, etc.). The mechanical contractor should also present his construction schedule that more specifically identifies the timing of the mechanical systems (distribution systems installation, receipt and setting of equipment, insulation application, equipment start-up, testing, adjusting, balancing, demonstrations, etc.). The mechanical contractor will normally possess a construction schedule for personal use, but may not be in the habit of producing this schedule as a record document. The project commissioning specifications should delineate milestones necessary for the commissioning process, and put the mechanical contractor on notice that he needs to present a formal construction schedule responsive to the specified milestones. The construction schedules are essential to allow the commissioning agent to keep abreast of the project's progress and the installation of mechanical systems. The commissioning agent needs to periodically observe the installation of piping and duct distribution systems, witness pressure tests, observe the setting of major pieces of equipment and witness start-ups, etc. Utilizing the construction schedules, the commissioning agent should prepare and distribute to all appropriate personnel a written commissioning plan either reinforcing or amplifying the plan contained in the project specifications, integrated with the construction schedule. The commissioning plan should give brief and concise descriptions of the work to be performed by the commissioning agent that in turn provides the contractors with knowledge of whom to have present for the various acts of commissioning. This restating or reinforcing of the commissioning plan is then used to record achievement of the mechanical milestones throughout the project's construction.

For example, projects involving water piping will normally contain specifications for flushing and cleaning the piping. This is an event that should be witnessed by the commissioning agent and identified as such in the project commissioning specification section. The mechanical contractor is placed on notice that the project schedule should identify flushing and cleaning as a milestone. Keeping current with updated progress schedules, the commissioning agent and contractors are aware of an impending act that requires their coordination. The reality is that, even with progress schedules updated as often as biweekly, the actual day of occurrence often cannot be clearly defined and communication from the contractor to the commissioning agent is necessary. Per the specifications and commissioning plan, the contractor should be given the responsibility of notifying the commissioning agent at least 48 h in advance of items which can only be observed when they occur, such as the flushing of piping.

Construction Documents and Correspondence.   Via the contract specifications, and reinforced at the preconstruction meeting, the commissioning agent is to be placed on the distribution list for the receipt of equipment submittals, shop drawings, and correspondence related to the mechanical systems being commissioned: i.e., if the commissioning is only for the heating, ventilating, and air conditioning systems, the commissioning agent does not need to be furnished documentation of plumbing systems, fire protection, etc. The equipment submittals and shop drawings sent to the commissioning agent should be those that bear the A/E's review stamp as the review and approval of shop drawings and submittals for compliance with contract documents is the responsibility of the design A/E. The review of submittals by the commissioning agent is complementary to that of the A/E and is necessary for the commissioning agent to be fully aware of the actual equipment that is to be furnished on the project (substitutions, design alternates, etc.). Should the commissioning agent note some discrepancies between submittals and contract requirements not cited by the A/E, his duty is to notify the A/E of the discrepancies. Ultimate resolution of acceptance or rejection of submittals resides with the A/E.

The commissioning agent should also be copied on all correspondence of a mechanical nature throughout project construction, most notably contractor generated requests for information or clarification, the consulting engineer's responses, and mechanical change orders. The commissioning agent plays no role in evaluating change orders, but must be aware of any changes from the contract documents in the mechanical work.

The commissioning agent is required to generate correspondence throughout the commissioning process, documenting observations and milestones passed. This correspondence should be copied to all concerned parties, i.e., the owner, architect, mechanical engineer, general contractor and mechanical contractor. It is the mechanical contractor's responsibility to further copy any concerned subcontractors.

Construction—Through Approximately 90 percent Completion. As construction progresses, the commissioning agent makes periodic site visits to observe the quality of work being installed and to stay abreast of work progress and any contractual changes made. The frequency and duration of the site visits intensify as the project nears completion. The commissioning agent should record his visits and observations in a log and issue brief summary reports for record. Any discrepancies noted during construction should be brought to the attention of the A/E as the A/E bears the responsibility of enforcing the tenets of the contract documents.

During the early phases of construction, the mechanical work will largely consist of the installation of piping and duct distribution systems, and the observations made by the commissioning agent would primarily be to ascertain that materials, valves, fittings, supports, and joining materials and methods are in accordance with project specifications. Prior to the application of piping and duct insulation installation, various pressure and leak tests are to be performed (per specifications), and the commissioning agent should witness most, if not all, tests.

As the project progresses, equipment will start to arrive on site and be installed by the contractor. The commissioning agent cannot observe the actual installation of all equipment, but should observe the installation of a representative sampling. Observing an installation does not mean that the commissioning agent needs to supervise a mechanic from uncrating through mounting and fitting, but rather through periodic site visits get an assessment of the workmanship. In the case of terminal equipment such as unit ventilators, fan coil units, variable volume air terminals, etc., the observations should be made as the first few of each type of terminal device are installed so that any problems or deficiencies noted can be brought to the attention of the engineer and contractor before the problems or deficiencies are repeated throughout the project. In the case of main air handling units and central plant equipment, such as chillers, boilers, pumps, and cooling towers, the commissioning agent should observe the site prior to setting of the equipment to ascertain if specified supports, service clearances, and vibration isolation are in place, and again after equipment has been set and leveled. Further site visits and observations should be made as piping and distribution systems are connected. The commissioning agent needs to be present during *all* equipment start-ups performed by factory representatives (such as boilers, chillers, variable frequency drives, etc.).

From the start of the project through the final connections of equipment and distribution systems, the commissioning agent will have made numerous unscheduled independent visits to the project observing the work as well as several scheduled visits coordinated with the contractor for specific milestone observations. After systems have been flushed and cleaned, balancing of air and water commences along with the calibration and tuning of controls systems.

This marks a major milestone in the commissioning process. Up to this point, the commissioning agent has primarily been an observer during construction. Once balancing and control systems operations commence, the commissioning agent becomes more proactive with the greatest demand of time and effort lying ahead.

The involvement of the commissioning agent during construction, up to the point where balancing and controls calibration commence, should represent about 35 percent of the total man-hours and fee for commissioning. Thus, the commissioning process is now 50 percent complete (including the 15 percent during design).

Construction Completion. Herein resides the irony that has created the need for mechanical systems commissioning. By this time in the project, the commissioning agent has 50 percent of his work ahead of him, but the mechanical contractor normally will have assessed his work to be 90 to 95 percent complete and invoiced accordingly. Yet systems that have not been properly balanced and/or control systems that do not function properly are probably the cause of 90 percent of comfort complaints, inefficient operation, and even systems failures. In many instances, the failures of balancing and control systems are not due to the poor performance or lack of diligence of the design engineer or of the contractors, but more due to the lack of the presence of a single overseer able to devote the time to exercise systems through their full range of control sequences. The mechanical contractor relies mostly on the subcontractors, accepting their assertions that they have properly completed their work and systems are functioning in accordance with design. The mechanical contractor seldom performs his own comprehensive testing of operations to validate the assertions of his subcontractors. The individual subcontractors may feel that their work is completed and checked out, but it is only done so to the extent of the work for which they were responsible.

To illustrate, the following scenario is all too common: The controls subcontractor determines from the contract plans that an air handling unit is to provide minimum outdoor air at 20 percent of the total volume of the unit. He sets his outdoor air damper control signal to be 20 percent of the full range from closed to open and contends that, by moving the damper 20 percent, he has provided 20 percent outdoor air. The balancing subcontractor comes along and disconnects control damper linkages to set damper positions for minimum outdoor air based upon measured air volumes, marks the damper position, and then reconnects the linkage; he has completed his job. However, unless the controls subcontractor revisits the air handling unit to verify that his 20 percent control signal in fact moves the damper to the position which the balancing subcontractor marked, the air handling unit will introduce outdoor air in some indeterminate quantity based upon where the control signal has placed the damper. It is not unusual to see a 20 percent control signal not open the damper at all or to open the damper to a position which admits far more than the 20 percent outdoor air minimum. In the former case, minimum ventilation rates are not met. In the latter case, admittance of excessive outdoor air may exceed the heating or cooling capacity of the air handling equipment resulting in spatial discomfort and, even if discomfort is not experienced, energy is wasted.

In the illustration, both controls and balancing subcontractors would have reported to the mechanical contractor that they have completed their work, and most certainly, the mechanical contractor did not himself go back and exercise the controls to view damper operation.

The consulting engineer may perform this test, but most likely will not. The consulting engineer cannot be relied upon to be the omniscient overseer. The engineer performs a punch out, but this does not approach the depth that independent commissioning does. The crass reality behind this is that the fee paid for engineering services does not cover the costs of the man-hours that are required to exercise all systems' controls through their full range of sequences and observe the results. However, even if the engineer were given an adequate fee to encompass commissioning, he may be willing to accept systems deficiencies because they are the result of his own design errors or omissions, and hope that the deficiencies do not come to the owner's attention.

Once balancing and controls calibrations commence, the commissioning agent becomes a more visible presence on site. Initially, he is there to physically observe the performance of work of the balancing and controls subcontractors. This literally involves following the workmen around for a few hours at a time. The purpose is as much to observe their work as it is to enforce in their minds that a serious look at the results of their work is impending.

Balancing, controls calibration, and making controls sequences operational can take contractors a few days or several weeks, depending on the size of the project. In the case of balancing, on-site work is normally completed weeks or months before a testing and balancing report is submitted. Often, the building is actually turned over to the owner and occupied before the balancing report is submitted. This occurs because balancing is one of the last processes performed; and, more often than not, work on the building continues up to the completion date unless the building is delivered ahead of schedule. This can get into contractual gray areas unless it has been clearly written in the contract documents that the owner may take beneficial occupancy of the building without accepting mechanical systems as complete until the commissioning process has recommended acceptance of the mechanical systems. Commissioning can proceed ahead of submission of the final balance report. Upon the assertion from the balancing subcontractor that balancing has been satisfactorily completed, credence being added to the subcontractor's assertion by the commissioning agent having viewed the balancing process, the balancing contractor can provide his handwritten field balancing report and notes.

By the time the controls subcontractor has completed calibration and performed his own test of sequences of operation, the mechanical contractor asserts that the installation is complete and ready for checkout. This is when the commissioning agent assumes a dominant role and exerts his most intensive efforts. Up to this point, all previous work of the commissioning agent required the efforts of a single person (though not necessarily the same person). Now, a team of people comprise the commissioning agent. At a minimum, at least two people are required equipped with radios. This is due to the fact that it is necessary to simultaneously view operations of interactive mechanical components that are remote from each other; a simple example being a room-mounted thermostat and a rooftop air handling unit. It takes one person to exercise the thermostat while another observes the correlating operation of the rooftop unit. This action/reaction testing is the foundation of the commissioning process and is required to be performed on all major mechanical systems (central plant equipment and air handling units). It is normally performed on only a representative sampling of terminal equipment such as variable air volume boxes, fan coil units, cabinet heaters, etc. Control and operations of terminal units are repetitive, and any endemic problems will be discovered provided that a suitable number of each type of typical terminal unit is observed; usually a quarter to a third of each of all types of units installed is sufficient.

In addition to the team representing the commissioning agent, the presence of contractors is required during this stage in the commissioning process, and coordination is required. The general contractor may wish to be present, but may delegate his authority to the mechanical contractor. A representative of the mechanical contractor, usually the project manager or superintendent, is required to be with the commissioning agent throughout this stage. The controls subcontractor is needed through most of this stage as exercising and verifying operation of controls is the predominant activity taking place. Other subcontractors or equipment suppliers are necessary predicated on the nature of the work: for example, a factory representative is needed when equipment with factory furnished controls is exercised. The presence of representatives of the A/E and the owner are subject to their discretion.

The time involved with verifying the operation of controls and systems is tremendously indeterminate, based entirely upon the expertise of and diligence exercised by the contractors during installation. A system that is initially found to perform properly in accordance with design intent obviously takes less time to evaluate than one that has malfunctions. No amount of overseeing by the commissioning agent during construction can assure a flawless operational test. The reality is that mistakes will have been made, through oversight, misinterpretation of contract documents, errors, omissions, etc. It is the mission of the commissioning agent to unearth the flaws and have them corrected. Further, it is often necessary to revisit a subsystem when some other interrelated component is evaluated and found unsatisfactory. A simple illustration would be the case of a variable air volume terminal unit not being able to satisfy its space loads. The VAV terminal may be acting properly, but the air handling unit may not be delivering air at a sufficient temperature or quantity. An initial review of the air handling unit may have established that the unit appeared to work properly in response to the temperature reset or static pressure signals that it received from components elsewhere in the system. The apparent deficiency of the VAV terminal would necessitate a reevaluation of the air handling unit. Another more dramatic illustration of the difficulty in estimating time associated with tasks would be if the commissioning agent went to site to perform observations of systems operations in the cooling mode, expecting chiller operations to be producing 45°F (8°C) water, and finding that the chiller is generating 52°F (12°C) water due to chiller malfunctions. The commissioning could be terminated until proper chiller operations were established or commissioning could proceed working around the insufficient chilled water temperature; but in any case, another visit to the site would be required once chiller problems have been rectified.

Buildings furnished with computer-based direct digital controls and energy management systems are considerably easier to commission than those not so equipped. The historical recording capability of automation systems permits substantial diagnosis of operations prior to actually visiting the site for inspection. Once on site, use of the computer allows simulation of wide ranges of operational sequences. A common failure in commissioning of the automated control systems is reliance upon what the computer says the system is doing; successful commissioning dictates visual observation of equipment to assure that the signal sent from or received by the computer correlates to the actual operation of the field device. The use of an energy management automation system, though a valuable tool in commissioning the mechanical systems, creates another whole layer in the commissioning process that requires a very detailed specification and poses unique demands upon the commissioning agent. The automation system essentially becomes the brains of the mechanical systems and warrants such extensive checking that Chap. 9, addressing energy management systems is included in this handbook.

It is incumbent upon the commissioning agent to have all mechanical systems demonstrated to function in accordance with design intent as provided by the contract documents. During the demonstrations, problems found resultant from faulty installation must be rectified by the contractor, and the system reinspected. Problems that exist after all installation work checks out properly need to be brought to the attention of the A/E for rectification.

To arrive at verification that systems perform in accordance with contract documents and design intent represents approximately 38 percent of the total commissioning man-hours and fee; thus by this time, commissioning is 88 percent complete. The commissioning agent will have issued documentation of the results of the site visits made during this check-out process, citing systems that were found to perform acceptably as well as deficiencies.

6.2.2 Operation and Maintenance (O&M) Manuals and Training

Operation and Maintenance manuals are often delivered near the end of the construction phase. It is far more beneficial to have "proposed" manuals submitted well in advance of this, perhaps 60 to 90 days after all shop drawings have been approved. The A/E and commissioning agent review the O&Ms for accuracy, pertinence, and completeness. The submitted, but not yet approved, manuals should be used and further reviewed during maintenance personnel training.

Most training of maintenance personnel takes the form of having factory technicians or contractor personnel give demonstrations and instructions to maintenance personnel on servicing equipment near the end of the project. Seldom are maintenance personnel provided sufficient instructions on how the equipment forms systems and how the systems are intended to function. This can be improved by having multiple training sessions.

The initial training session should require that the design engineer or the commissioner provide maintenance personnel with an overview of systems design intent, illuminating how the individual pieces of equipment or subsystems are intended to function to create the building system. This meeting also is used to provide maintenance personnel the opportunity to review the proposed O&M manuals and to have input into the form and content of subsequent training sessions.

Subsequent training sessions then generally occur when equipment has been installed, pre-commissioned, and started. It is always beneficial to have the owner's maintenance personnel present during start-up of equipment; however, formal equipment servicing training should not be during start-ups. Equipment and systems should have had some successful runtime prior to performing service training. During the equipment service training, the proposed O&M manuals should be used to familiarize maintenance personnel with their content and to validate their usefulness. Upon completion of this phase, the final O&M manuals can be prepared for submittal by the contractor.

The final training session occurs by having maintenance personnel observe as the commissioning agent goes through the building verifying systems performance and operation.

6.2.3 Close Out

The commissioning agent reviews all close-out documents required of the contractor, including As-Built Drawings, warranty statements, and all other certifications required for submission by project specifications. As-Built drawings, including control drawings, should be furnished and used by the commissioning agent during the field check-out process. This assures their accuracy.

The review of close-out documents by the commissioning agent is complementary to those performed by the A/E.

Upon satisfactory completion of all of the foregoing, the commissioning agent provides a letter stating that the mechanical systems are ready for acceptance by the owner. It is at this point that warranty periods should begin. This phase represents approximately 2 percent of the commissioning man-hours and fee; commissioning is now 90 percent complete.

6.2.4 Follow-Up

As the acceptance of the mechanical systems usually occurs either in summer or winter, i.e., in the cooling or heating mode, a subsequent season inspection is required. Satisfactory operation of systems in the cooling mode is no guarantee that systems will perform properly in the heating mode and vice versa. Therefore, trends generated by the automation system are again reviewed (if the building is so equipped) and the commissioning team needs to revisit the site to exercise systems in the remaining mode of operation. This process constitutes approximately 10 percent of the commissioning man-hours and fee; and when satisfactorily completed, the commissioning process is 100 percent complete.

6.2.5 Practical Commissioning

Though the benefits of mechanical systems commissioning are most fully realized when the commissioning agent is brought on board as a team member in the earliest stages of design; commissioning can commence at any phase in the project, with commensurate reductions in what commissioning achieves.

For example, perhaps the commissioning agent is retained at about the time that the contractor has concluded his work; all systems being operational, balancing and controls calibration and setup having been completed. The commissioning agent essentially performs a test and verification of systems operations. Before the commissioning agent visits the site, catch-up work must be performed to understand the mechanical systems design and intent. Copies of contract drawings and specifications, construction correspondence (requests for information, clarifications, change orders, etc.), mechanical equipment and controls shop drawings, test and balance reports, and As-Built drawings for review must be provided. If the building has a computer-based automation system, trends need to be provided for review. Once the commissioning agent is satisfied with an understanding of mechanical systems design intent, and if trends indicate apparent proper systems operation when an automation system is involved, the commissioning team is ready to visit the site to perform tests and observations. Critical components of the commissioning process are omitted by this late stage commissioning; particularly, the commissioning agent will not have witnessed systems flushing and cleaning, start-up of major pieces of equipment, testing and balancing procedures, and controls setup. Many deficiencies manifested by errors or inadequacies in systems installation and setup may still be discovered by the commissioning agent, though certainly not to the same extent had the commissioning agent been involved with the project through construction; and more importantly, having been involved during construction, deficiencies that are discovered and require remedial action by the contractor may have been avoided completely by early detection.

To illustrate, suppose that in checking out an air handling system, the commissioning agent has found that the air handling unit appears to be operating properly. The filters and coils are clean, the test and balance report indicates that air and water flows are in accordance with design, controls appear to function properly, chilled water is being delivered at proper temperature, yet the space which is served by the air handler cannot maintain comfort cooling conditions. Initial perceptions may be that the system is inadequate for the load imposed upon it; maybe the coil or fan are too small, i.e. a design error. Before asserting a design error (usually the first response from contractors), the commissioning agent must perform further diagnostics. In doing so, he may unearth installation or setup failures by the contractors. It may be that when the system was balanced, full water flow was documented at the cooling coil; but during subsequent operation, some slag lodged in the control valve, preventing it from opening fully. All outward appearances would indicate that the valve has stroked fully, but checking water and air temperatures across the coil would indicate that full flow was not being realized. Diagnostics may have also revealed that the controls contractor's signal to open the air handling unit's outdoor air damper to its minimum position in fact opens the damper well beyond the minimum setpoint, introducing hot, humid outdoor air in excess of the design cooling coil's capacity. Due to the contractor's inadequate flushing of the water system or lack of coordination between balancing and controls subcontractors, latent problems arose which then require remedial work (possibly draining systems, pipe fitters, controls work, balancing work, etc.). The system then must be retested by the commissioning agent.

Some of the problems of the foregoing illustration may be avoided by bringing the commissioning agent into the project at earlier construction stages; but there is no optimal time other than at commencement of design. Fees for commissioning are much more difficult to establish when the commissioning agent is not a part of the team from the beginning.

The commissioning agent may offer a fixed fee based upon his assessment of the complexity of the mechanical systems and the time that he will have to devote, depending upon when he becomes involved with the project. However, his actual time to complete the commissioning is greatly governed by the quality of work performed by the contractor. As indicated by the illustration of the air handling unit that did not maintain space temperature, the commissioning agent was required to spend additional time performing diagnostics and reinspection beyond the time that he may have initially allocated to check a properly operating air handling unit. Therefore, if the commissioning agent is to be locked into a fixed fee, it may be inflated to protect from losing money on a poorly installed project. If the fixed fee proposed by the commissioning agent is too low to accommodate the poorly installed project, the agent has the choice of taking a loss, to provide less than the full service intended, or to solicit extra compensation from the owner. For these reasons, a fee proposal that offers compensation for hourly services with perhaps an upset limit is more practical, fair, and equitable for both owner and commissioning agent. Also, a means by which the owner may extract compensation from the contractor for extra time spent by the commissioning agent for a poorly installed project may instill greater diligence by the contractor in performing the installation.

6.3 THE COMMISSIONING SPECIFICATION

6.3.1 General

The commissioning specification is critical to the success of the commissioning process. It serves multiple purposes, both objectively and subjectively. It objectively defines the scope of work to be performed by all parties involved with the building: the design team, the builders, the owner's operating and maintenance personnel, and the commissioning agent. Via the overt statements of responsibilities and expectations, it places everyone involved on notice that an independent oversight of design and installation will be performed, instilling a reality that business will not be conducted as usual. Going through the commissioning process entails demonstration and verification of mechanical systems operations and performance not normally performed by conventional design and construction processes. This realization should result in a greater dedication to detail and workmanship by the designers and contractors.

6.3.2 Related Documents

The commissioning specification does not stand alone. It must be written as an integral part of the complete project specifications; most particularly to be in sync with Division 1, General Requirements, and Division 15, Mechanical. Division 1 will contain the general dictates and tenets and Division 15 will contain the specific dictates and tenets for submittals, shop drawings, as-builts, record documents, warranties, etc. Division 15 will further provide the specifics of the mechanical equipment and systems. It should be assured that items such as warranties, certificates, materials testing, start-up procedures, As-Built drawings, inspections, demonstrations, and close-out submittals are clearly spelled out and covered. These issues are not provided in the commissioning section of the specification, but are enforced and verified by the commissioning section. The following offer several examples:

1. The requirements for flushing, cleaning, and testing of piping systems is to be specified in Division 15, but not in the commissioning section. The commissioning section specifies that the commissioning agent will observe the flushing, cleaning, and testing.

2. Testing, adjusting, and balancing (TAB) is a complete section of Division 15 and spells out procedures for the complete testing, adjusting, and balancing of the mechanical systems. The commissioning specification cites that the commissioning agent will observe TAB procedures and allows for limited amounts of retesting to be performed to demonstrate that TAB has been performed correctly. The TAB and commissioning sections of the specification have to be coordinated so that what is expected by the commissioning agent is provided in the TAB section.

3. Demonstrations and instructions for operating and maintenance personnel should be included in the General Requirements section of Division 15; spelling out procedure and giving a minimum number of hours that will be required of the contractor to perform demonstrations and instructions, and during the construction process when the training will occur. The commissioning section of the specification will assert that the commissioning agent will be present during the demonstrations and instructions.

The foregoing are but a few examples illustrating that much of what the commissioning agent performs is observance of compliance with requirements contained elsewhere in the project specifications. The presence of a commissioning authority does not relieve the A/E from their obligations to prepare comprehensive specifications for the project.

6.3.3   Specification and Check Lists

The following sample specification and check lists are offered as a guide. They should be prepared by the independent commissioning agent to be incorporated with the project specifications prior to the project being issued for bids. The specification and check lists could also be prepared by the project mechanical consulting engineer if a commissioning agent has not been involved with the project through design. As with any good contract document, the specification and check lists should be written tailored to the specific requirements of the project to which they will be applied. It is impossible to provide guide specifications and check lists within the confines of this book that would encompass all mechanical system types and the work required to properly commission them. The sample specification and check lists presented should offer a basis on which project specific documents may be written. To facilitate this process, explanations and amplifications are given to selective paragraphs in italic type in the specifications.

SAMPLE SPECIFICATION

PART 1—GENERAL

1.01 INTENT: The intended result of the HVAC Commissioning process is to assure the owner that the HVAC systems are installed and operate in accordance with contract drawings and specifications prior to the owner's acceptance of the building.

1.02 SCOPE:

    A. Work Included: The HVAC commissioning shall provide substantial verification that systems and equipment are installed and performing in accordance with the contract documents and design intent. This independent commissioning shall be complementary to the construction period services performed by the architect and mechanical consulting engineer.

    B. Work Not Included: It shall not be incumbent upon the commissioning agent to verify adequacy of HVAC systems to accommodate the heating, cooling, and/or ventilating loads imposed upon them, i.e. to evaluate design. Systems installed and performing in accordance with plans and specifications that do not achieve and/or maintain spatial conditions in accordance with design intent will be so noted when observed. Commissioning of plumbing, fire protection sprinkler, and electrical systems are excluded from the HVAC commissioning process except as may be incidental to the operations of the HVAC system; i.e. note would be made if condensate drainage was observed to be inadequate, if a fan turned backwards due to improper power connections, etc.

1.03 RELATED DOCUMENTS: The contract drawings and requirements of DIVISION 1, GENERAL REQUIREMENTS, DIVISION 15, MECHANICAL, and DIVISION 16, ELECTRICAL apply to this Section of the Specifications.

This is important as the commissioning section of the specification essentially only provides the procedural methods that will be employed to verify that the requirements of drawings and specifications have been met.

1.04 COMMISSIONING AGENT AUTHORITY: Throughout the commissioning process, the commissioning agent's role is primarily one of an observer/witness; monitoring the installation, start-up, and operation of the mechanical heating, ventilating, and air conditioning (HVAC) systems. The commissioning agent shall have no authority to alter design or installation procedures. If acceptable performance cannot be achieved, it will be the commissioning agent's responsibility to apprise the owner, design engineer, and/or contractor of the deficiency. Corrective actions shall be the responsibility of the owner, design engineer, and/or contractor; and not that of the commissioning agent. The commissioning agent shall have the authority to require tests and demonstrations to verify proper performance.

1.05 ARCHITECT/ENGINEER RESPONSIBILITY:

    A. In addition to their normal performance of construction period services, the A/E will furnish to the commissioning agent one copy of all approved HVAC systems shop drawings and submittals and place the commissioning agent on the mailing list for all communications regarding the HVAC systems.

    B. The A/E shall respond in writing to deficiencies cited in correspondence issued by the commissioning agent.

This has been found to be a necessary means of documenting that issues raised by the commissioning agent have been addressed to minimize the amount of time required of the commissioning agent to reverify systems.

1.06 CONTRACTOR'S RESPONSIBILITY:

A. The general contractor shall be responsible for assuring that the commissioning agent is provided with all relevant correspondence, submittals, notifications, and assistance as may be required to satisfactorily complete the commissioning process using whatever personnel, time, and resources that are required. This Section provides minimum commissioning requirements; however, the contractor shall exceed those requirements whenever necessary to achieve the intent of HVAC commissioning.

B. The general contractor shall include in his bid the cost of furnishing the material requested and manpower necessary for the verification of proper HVAC system installation and operation as specified in this section.

C. The contractor shall respond in writing to deficiencies cited in correspondence issued by the commissioning agent.

This has been found to be a necessary means of documenting that remedial work has been performed, hopefully properly, to minimize the amount of time required of the commissioning agent to reverify systems. Enforcing contractor documentation assures that issues raised by the commissioning agent are not ignored or swept under the carpet.

PART 2—PRODUCTS

2.01 NOT APPLICABLE TO THIS SPECIFICATION

PART 3—EXECUTION

3.01 COMMISSIONING TEAM: At a minimum, the following are members of the commissioning team:

A. The owner's authorized representatives

This should include operating and maintenance personnel that will ultimately be responsible for the building.

B. The architect
C. The consulting mechanical engineer
D. The general contractor

Though it is primarily the mechanical HVAC systems that are being commissioned, the general contractor bears ultimate responsibility for the successful completion of the project and needs to stay abreast of the work of the subcontractors. In addition, it is not unusual to unearth construction related problems when commissioning the HVAC systems. For example, it may be found that an HVAC system is installed and performing in accordance with design intent, but it cannot maintain proper space comfort conditions. While diagnosing the problem, it was found that an opening in an exterior wall was not properly sealed, allowing leakage into the building. This becomes an issue for the general contractor to rectify.

E. The electrical subcontractor

The involvement of the electrical subcontractor can become necessary if power connections to HVAC equipment or control wiring interlocks with starters have been found to be missing or improperly connected. Also, in many installations involving computer-based automation systems, much of the wiring installation may have been subcontracted to the project electrical subcontractor by the controls subcontractor. A common example of a failure in coordination between mechanical/electrical work is when the mechanical has specified a two speed fan and the electrical has provided a single speed starter.

F. The mechanical subcontractor (and all of his subcontractors performing HVAC work with particularly critical involvement of the testing, adjusting, and balancing subcontractor and temperature controls subcontractor)

At one point or another in the project, the commissioning agent will have been in contact with nearly all of the mechanical contractor's subcontractors. The mechanical contractor may perform only the piping components of the project with in-house personnel and have subbed out the sheet metal, insulation, controls, testing, and balancing, etc. Additionally, all major pieces of equipment usually are specified to be initially started by factory authorized technicians. The commissioning agent will observe the work of all of these subcontractors and factory technicians. The mechanical contractor is also responsible for placing all of his subcontractors and equipment suppliers on notice to respond to the authority of the commissioning agent.

G. The commissioning agent

The following paragraph is necessary to place all concerned parties on notice that time and effort will be required of them to accommodate HVAC commissioning.

3.02 RESPONSIBILITIES: Each member of the commissioning team has responsibilities for the successful completion of the commissioning process as follows:

A. The Owner's Representatives shall perform their normal construction contract administration functions.

Depending upon who the owner is, this work may be delegated to the A/E if it is a small project with an owner that does not have a construction management team. On government or institutional projects, the owner may have their own contract administrator, field construction manager, or teams of people representing different divisions within the owning agency such as a boiler maintenance shop, an HVAC shop, an energy management division, etc. Whoever is involved representing the owner, their duties are essentially unaltered by that of the commissioning agent.

B. The architect shall provide adequate support to the consulting mechanical engineer as related to his duties in the commissioning process. It shall also be the architect's responsibility, either directly or through his assignee, to assure that the commissioning agent is:

1. Provided copies of approved shop drawings as they are returned to the contractor;
2. Notified of time, date, and place of all regularly scheduled progress meetings, and of any special meeting that may be called regarding HVAC systems;
3. Copied on all correspondence pertinent to the HVAC systems including but not limited to minutes of progress meetings, responses to contractor requests for information, change order documentation.

As the work being performed is HVAC commissioning, the architect's duties are only moderately affected by the commissioning process. The Architect needs to stay abreast of the commissioning process, as it is normally the architect who bears ultimate responsibility for the design team and construction management. All correspondence for the project generally flows through the architect and he needs to assure that the commissioning agent is in the loop.

C. The consulting mechanical engineer shall perform his normal construction contract administration functions.

The normal contract administration functions include attendance at progress meetings, shop drawing review, answering questions, issuing clarifications or change orders, site visits to monitor project progress or address conflicts, punchout, review of As-Built drawings, review of operating and maintenance manuals and other closeout submittals. The presence of a commissioning agent does not relieve the mechanical consulting engineer of any of these duties. Any responses from the consulting engineer necessitated by observations of the commissioning agent fall within the normal contractual obligations of the consulting engineer. The architect may delegate some of the correspondence and document transfer between the A/E team and the commissioning agent to the consulting engineer; i.e., directly providing shop drawings and permitting direct correspondence with record copies being furnished to the architect.

D. The general contractor shall, in addition to his normal responsibilities for construction of the project, assure that his subcontractors recognize the authority of the commissioning agent and perform responsive to the requirements of the commissioning process. He shall assure that proper notification, at least 48 h in advance, of the milestones of the mechanical systems installation is provided to the commissioning agent, at a minimum as follows:

1. Pressure testing of piping systems
2. Flushing and cleaning of piping systems
3. Factory start-up of central plant equipment
4. Factory start-up of rooftop equipment
5. Calibration of automatic temperature controls
6. Start date of air and water balancing
7. Date of punch-out inspections
8. Date of instructions to owner's operating personnel regarding operations of the HVAC system

The foregoing establishes that the general contractor, who has the ultimate responsibility for the construction sequencing of the project, must also take

responsibility for proper notifications of the commissioning agent. He may dele-
gate this responsibility to the mechanical subcontractor at his discretion. The mile-
stones listed are minimum requirements, and there may be others specific to the
needs of the project which should be identified here and again restated in the com-
missioning plan presented at an early progress meeting.

E. The electrical subcontractor shall perform his normal contract obligations and be
responsive to the authority of the commissioning agent.

This places the electrical subcontractor on notice that he may be required to per-
form demonstrations and in other ways be responsive to the commissioning
process.

F. The mechanical subcontractor shall, in addition to his normal responsibilities
for construction of the project, assure that proper notification of the milestones
of the mechanical systems installation as cited in Paragraph D are provided to
the general contractor. The mechanical subcontractor shall assure that his subcontractors recognize the authority of the commissioning agent and perform
responsive to the requirements of the commissioning process, particularly the
testing, adjusting and balancing (TAB) and temperature controls and/or
automation system subcontractors.

This places the mechanical subcontractor on notice that he, and all of his sub-
contractors, will be required to perform demonstrations and in other ways be
responsive to the commissioning process.

G. The commissioning agent will follow the procedures as set forth in paragraph 3.03
to execute his responsibility for:

1. Verifying that the mechanical systems are installed and operating in accordance with contract documents and specifications;
2. Assuring that owner's operating and maintenance personnel are fully trained
   on systems operation and maintenance;
3. Assuring that close-out documentation is properly provided to the owner.

3.03 COMMISSIONING AGENT PROCEDURES: The commissioning agent will perform systems commissioning following the procedures listed herein, and all members
of the commissioning team shall cooperate fully with the execution of these procedures. Initial HVAC commissioning shall be performed while the central plant is in
operation, either in the heating or mechanical cooling mode. Initial commissioning
will not take place during "swing" seasons when neither boilers nor chillers are operating. If commissioning occurs during the mechanical cooling operation, subsequent
commissioning shall be performed during the next "swing" season and at the
changeover to the heating season; and similarly, if commissioning occurs in the heating season, subsequent commissioning shall be performed during next "swing" season and during the next cooling season. The subsequent seasonal systems
commissioning shall consist of observing the central plant equipment start-up,
reviewing complete sets of automation system operational histories *(on projects with*
automation systems), and follow-up site visits to spot check automatic temperature
controls systems.

The foregoing paragraph may be modified if such complete service is not retained by
the owner or if the mechanical systems are not designed for such distinct modes of

operation. The full and complete commissioning of HVAC systems necessitates observations of systems in all modes of operation—meaning mechanical cooling, heating, and natural cooling modes. Some HVAC systems designs, particularly two-pipe heating and cooling systems, have a "swing" season of operation where neither mechanical cooling nor heating is provided; and the systems designed operate primarily in an economizer mode. An example of this could be a school that shuts down the chiller plant in late September and does not start the boiler plant until late October. There could be a 3 to 4 week period where the mechanical systems only operate in the natural ventilation and cooling mode. On the other hand, the school could be designed as a four-pipe system to operate without economizer cycles. In this case, only the heating and mechanical cooling modes need verification. In all cases, it is essential that the HVAC systems be commissioned in both heating and mechanical cooling modes, as operation in the heating mode may not reveal problems that exist in the cooling mode and vice versa.

To achieve HVAC systems commissioning, the commissioning agent shall:

A. Attend periodic construction progress meetings and perform unscheduled walks through the building to observe and keep abreast of mechanical systems installation progress, means, and methods. Commissioning agent's presence at meetings and in the building will be for his benefit in preparing to commission the building and shall in no way be construed as superseding the authority of the project architect/engineers.

This reinforces that the role of the commissioning agent is primarily one of observer and that he has no authority to alter design.

B. Perform a complementary review of HVAC shop drawings after approval of the project consulting mechanical engineer. The purpose of this review is primarily one of familiarization with equipment to be furnished on the project for on site verification by the commissioning agent and in no way relieves the consulting mechanical engineer of his duties for shop drawing review.

This reinforces that the consulting mechanical engineer is responsible for approving what gets installed on the project, and that the role of the commissioning agent is primarily one of verification. There can be benefits by having the commissioning agent perform submittals reviews concurrent with the consulting engineer. However, this will increase the time and cost of the commissioning agent, particularly if there are a lot of rejections of initial submittals followed by resubmittals. Collating and coordinating the design engineer's and commissioning agent's review comments can also be cumbersome and slow the submittal review process.

C. Issue commissioning checklists relevant to the project for contractor monitoring and verification of installation progress (see sample checklists). Establish milestones in the HVAC system installation at which time interim commissioning status reports will be prepared and issued by the commissioning agent.

The commissioning checklists are essential not only to document completion of milestones, but also to provide an amplification of the scope of work which will be required of various parties in the commissioning process. It is recommended that a complete set of commissioning checklists be included with this specification section, specific to the project, to assist contractors in allotting time for commissioning.

D. Observe piping system pressure tests, flushing, and cleaning.

The wording of this paragraph leaves open to the discretion of the commissioning agent as to how many pressure tests, flushing, and cleaning he will observe. On a small project, he may observe all such operations. On large projects, particularly those done in phases, he may opt to observe only selected systems once he is satisfied that work is being performed properly.

E. Observe representative sampling of duct systems, plenums, coils, and filters for cleanliness, damage, or leakage.

The extent of a representative sampling is left up to the discretion of the commissioning agent, but places contractors on notice that such systems will be inspected.

F. Be present to observe the start-up of central plant equipment; i.e., chillers, cooling towers, and boilers. This shall be the start-up that is supervised and certified by the equipment manufacturer's authorized agent. Commissioning agent will observe the start-up of a representative number of self-contained refrigeration units (rooftops, split-systems) and verify that the operations have been certified by manufacturers' representatives.

This paragraph should be further amplified specific to the project to include any other types of equipment which are specified to be started or certified by factory authorized representatives, such as variable frequency drives, packaged pump sets, ice storage systems, energy recovery equipment, etc.

G. Visit the project periodically during testing, adjusting, and balancing of the air and water systems to observe that actual work being performed, and review the certified reports submitted by the TAB agency.

Again, the review of the TAB report by the commissioning agent is complementary to that of the mechanical consulting engineer. It is the consulting engineer's duty to address discrepancies or deficiencies and it is the commissioning agent's responsibility to assure that this work is performed. It is the commissioning agent's duty to apprise the engineer of discrepancies or deficiencies that may have been missed by the consulting engineer in his review.

H. Periodically observe the installation of the automatic temperature controls system: observe the operation of the main air compressor (running time, pressure, moisture), verify calibration of a representative sampling of sensing devices, observe the operation of all air handling units and exercise their respective control sequences including safeties, exercise the controls and observe operation of a representative sampling of incremental units, and observe the operation and controlling sequence of all central system piping isolation, changeover, and modulating control valves.

I. Verify the commissioning of the energy management automation system by reviewing point histories and by being present during hardware and software punch outs at the facility. The operational and monitoring capabilities of the EMS will be employed in the commissioning of the mechanical systems.

Paragraphs H. and I. need to be tailored to be harmonious with one another and specific to the project. Though most modern day projects are being designed with computer-based control systems, not all are done using full direct digital controls and electronic/electric actuating devices. There still exist some projects with only pneumatic/electric controls and often some hybrid form of controls exist where pneumatics are used for actuators while the rest of the control system is computer based. Commissioning of computer-based automation controls systems requires substantial elaboration and unique processes which are amplified further in the Chapter on this subject.

J. Be present during mechanical contractor instructions to the owner's operating personnel regarding operations of the HVAC system.

K. Review contractor-prepared Operating and Maintenance manuals, As-Built drawings, and all certifications and warranties required for submission by project specifications.

This review is complementary to that of the consulting mechanical engineer. Certifications can include such items as equipment compliance with specified standards such as ASME and ARI, water treatment, etc. Warranties can be those for manufactured equipments such as compressors, variable frequency drives, etc., as well as the contractors' warranties.

L. Furnish a written report and recommend acceptance of the HVAC system upon satisfactorily completing the commissioning process. Recommendations for approval, when appropriate, will be forwarded to the project A/E for inclusion in their final submission of project close-out documentation to the Owner.

". . . recommend acceptance of the HVAC system upon satisfactorily completing the commissioning process" is a key phrase that needs to be defined in other procedural sections of the project specifications. Normally, acceptance of the project has occurred at what has been viewed as project completion; that is, systems have been inspected and found to be operating satisfactorily, and the contractors start their warranty periods. Problems that subsequently arise are viewed as warranty issues. The fallacy in this is that the mechanical systems are normally accepted after having been viewed in only one mode of operation, either heating or cooling, depending on what time of year the general contractor has completed the work; and, as stated previously, the mechanical systems need to be inspected in all modes of operation to satisfactorily pass the commissioning process. A mechanism may be inserted in the contract documents to afford the contractors their due in allowing a building to be accepted and warranty periods to commence after this single season inspection with a proviso that acceptance of the mechanical systems is given conditionally in anticipation that the systems will perform acceptably in the next seasonal changeover. Should the systems not perform acceptably, the initial acceptance is rescinded until such time as the subsequent seasonal operations are accepted. The warranty of the mechanical systems then commences at the new date of acceptance. For example, suppose that the building was accepted while the mechanical systems were operating in the cooling mode; the date of acceptance and commencement of the warranty period was September 1. The subsequent inspection of systems in the heating mode revealed failures of the controls system to properly operate or an inability to establish proper water flow in a heating loop. The contractor should not be let off the hook for providing a full year's warranty [or two year warranty if so specified] for the mechanical systems; i.e., the owner is purchasing a timed warranty of the mechanical systems. If successful commissioning of the mechanical systems in the heating mode did not occur until December 1, it should be the contractor who needs to carry the warranty an extra three months in lieu of the owner having to accept a nine month warranty.

This methodology can effectively work with the warranty of the mechanical subcontractor, provided that it has been clearly stated as such in the specifications;

however, issues of fairness should also be incorporated. For example, if the building is accepted by the owner while systems were operating in the cooling mode, manufacture's warranties for equipment such as chillers and direct expansion rooftop equipment should be permitted to remain in effect from the date of acceptance regardless of the results of the subsequent seasonal commissioning. After all, failure of the heating system should not adversely affect the manufacturer's warranty of the chiller since the chiller had been proven to operate properly and was accepted by the owner. The review of warranty statements is part of the commissioning agent's responsibilities per paragraph K above.

3.04 CONTRACTOR COMMISSIONING PROCEDURES: The general contractor and all relevant subcontractors shall, in addition to being responsive to the procedures cited for execution by the commissioning agent in paragraph 3.03, perform as follows to achieve satisfactory HVAC systems commissioning. The contractor shall:

A. Demonstrate the performance of each piece of equipment to the commissioning agent and owner's representative after completion of construction. Schedule the TAB, HVAC controls, energy management, and other subtrade representatives as may apply to demonstrate the performance of the equipment and systems.

It is critical to the success of the commissioning process that all parties responsible for the installation of a specific system be present during the demonstration. Invariably, a failed performance test will be blamed on the individual not present at the demonstration. For example, in testing an air handling unit, at a minimum, the mechanical contractor and the controls contractor need to be present for the initial test as the system is operated through its various control modes. If satisfactory performance cannot be achieved, and cause of the failure not readily recognized and remedied, responsibility for the failure may be laid on the balancing contractor, the consulting engineer, the equipment supplier, etc. A subsequent visit will be required which should additionally involve all parties that had a hand in putting the system together. Because much time and effort can be wasted in repeated calls back to revisit a poorly operating system, it behooves all parties concerned to get it right the first time, and failing that, assuredly get it right the second time. This becomes a cumbersome but essential process of scheduling the right people at the right time. Similarly, one does not want to waste the time of a lot of people by calling forces en masse to the site to needlessly wait while systems for which they had no responsibility are checked; i.e., the boiler representative need not be on site while chiller operations are being inspected.

The commissioning agent can save all parties a lot of time by listing those parties that should be present during any particular test in the commissioning plan presented at the outset of construction.

B. At a minimum, the performance and operation demonstrations of the following equipment and/or systems will be required:

Here is where the scope of work is defined, specific to the project. All systems to be commissioned need to be mentioned. The nature of the tests and demonstrations to be performed would be further amplified by the line items shown on the Commissioning checklists. The systems, equipment, and percentages listed herein after are offered as examples.

    1. Incremental equipment (fan coil units, cabinet heaters, unit ventilators, etc.); approximately 20 percent of units installed.

2. Air balancing: major trunk duct flow and pressure checks, air terminals, variable volume boxes, outdoor air damper settings. The TAB trade representative shall identify all places where temperature, pressure and/or velocity readings were taken in major duct systems, and performance shall be demonstrated on up to 20 percent of the locations. Up to 5 percent of air terminals shall have performance demonstrated. Outdoor air settings shall be verified on all major air handling units and on up to 10 percent of incremental equipment and self-contained air conditioning units. Flows shall be verified on up to 20 percent of variable air volume terminals.

3. Water balancing: flow settings at pumps and in distribution piping, up to 20 percent of the locations of flow fittings.

4. Fire and smoke damper installation and operation, up to 30 percent.

5. AHU coil performance; all coils during both cooling and heating.

6. Fan and motor performance; all major air handling units and up to 30 percent of exhaust fans.

7. Pump performance.

8. Chiller system performance.

9. Boiler performance.

10. HVAC controls systems; complete control sequence of central plant equipment and major air handling units, and up to 20 percent of incremental and terminal equipment (unit ventilators, fan coils, cabinet heaters, variable air, volume boxes).

C. In addition to the foregoing, the contractor shall repeat any other measurement contained in the TAB report where required by the commissioning agent for verification or diagnostic purposes. Should any verification test reveal operation or performance not in accordance with contract documents, the Contractor shall rectify the deficiency, and reinspection shall be performed. Should operation or performance still not be as specified on the reinspection, the time and expenses of the commissioning agent to make further reinspections shall be considered as additional cost to the owner. The total sum of such costs shall be deducted from the final payment to the contractor.

This can get very dicey. There needs to be some mechanism to protect the commissioning agent from losing his shirt due to shoddy workmanship by the contractor. The commissioning agent will have entered into a contractual agreement with the owner to provide a service. The fee for that service is not open-ended, nor is the time that the commissioning agent gives in performing that service. If the commissioning agent is required to return to the project time and again to test systems that the contractor has not made to function properly, he cannot be expected to absorb the costs of these revisits under his base fee. To obtain recompense for exceptional expenditures of time, the failures of the systems have to be well documented to indicate that the commissioning agent was not frivolous in his rejection of systems performance in order to gain a higher fee. Also, the owner should not bear the additional expenses incurred by the commissioning agent due to failures of the contractor. The hourly rates of the commissioning agent should be a matter of record with the owner and the contractor so that the costs of additional services come as no surprise if invoked.

3.05 DEFICIENCY RESOLUTION

A. Deficiencies identified during the commissioning process shall be corrected in a timely fashion. The commissioning agent has no authority to dictate ways and means of deficiency resolution other than enforcing the dictates of contract draw-ings and specifications. Resolution of deficiencies that require interpretations or modifications to the contract documents shall be the responsibility of the Architect and Engineers. Project completion date shall not be delayed due to lack of timely resolution of deficiencies unless authorized contract extensions have been executed.

B. Written responses shall be made to deficiencies correspondence issued by the commissioning agent. The commissioning agent shall issue such correspondence as deemed appropriate during the commissioning process with original provided to owner and copies to the general contractor, architect, and consulting mechanical engineer. The general contractor, architect, and/or consulting mechanical engineer shall provide the Owner with a written response to each deficiency item cited by the commissioning agent as to corrective actions implemented. The writ-ten response shall be provided to the owner within 2 weeks of the date of the com-missioning agent's deficiency correspondence; copies shall be provided to the commissioning agent, general contractor, architect, and consulting mechanical engineer. Deficiencies which have not been fully resolved within the 2-week period shall be noted as such with explanation of intended resolution, and subse-quent status reports of the continued deficiency resolution shall be made in writ-ing at 2-week intervals until such time as the deficiency has been fully rectified. The owner reserves the right to withhold partial payment for construction contract or professional services until satisfactory resolution of mechanical deficiencies have been documented and verified.

The foregoing will go a long way in minimizing repeated observations of failures and deficiencies. By dictating that written acknowledgement be made, tracking of resolutions becomes a matter of record, and a lot of wasted time can be avoided. On large projects, this is essential as there will be many observations made, and it will be impossible to keep track of resolutions informally. For example, suppose at a point in the construction of the project, the commissioning agent observes that a balancing fitting has been omitted in a piping loop. He makes this obser-vation known to all concerned parties and awaits a response. Should there not be a requirement for written acknowledgement and response, the contractor could overlook or ignore the notification. Perhaps the balancing subcontractor makes the mechanical contractor again aware of the missing fitting, but by this time the mechanical contractor has had systems tested and filled and is driving toward project completion. With completion date approaching, the contractor asserts that he has completed his job and is ready for inspection, hoping that the missing balancing fitting was not really necessary anyway. The commissioning agent may have repeatedly written about the missing fitting as a continuing deficiency, but he has no enforcement capability until such time as he performs operation tests. The commissioning agent goes to the job site and finds that an air handling unit is not performing in accordance with design. The commissioning agent may or may not be able to assess the deficiency due to a lack of adequate water flow. In any event, subsequent visits will be required, and may eventually lead back to the absence of the balancing fitting.

By having written responses, it becomes more difficult for the contractor to ignore or overlook a deficiency. Without suitable written documentation from the contractor that the missing balancing fitting has been installed or acknowledg-ment from the design engineer that systems performance will not be adversely affected by the omission, the commissioning agent may refuse to test any systems affected by the missing fitting until appropriate notification from the contractor or the engineer is in hand.

3.06 SATISFACTORY COMPLETION: The contractor's personnel shall be made available to execute all aspects of the commissioning process until the owner accepts final results. Commissioning tasks and meetings may be repeated until the owner is satisfied and will not be fixed as one-time, one-chance events for the contractor.

This precludes the contractor from taking the position that he responded to the deficiency cited, remedied it, and sees no need in revisiting the issue. Actually, in some cases, this may be an entirely satisfactory response. For example, suppose an exhaust fan was found to be inoperative by the commissioning agent. The contractor found a broken fan belt, replaced it, and stated that the fan now runs fine. The commissioning agent need not convene a meeting to verify the fan's operation, but can do so himself incidental to one of his site visits.

On the other hand, there are many instances where fixing one problem may subsequently reveal another that could not previously be discerned as it was masked by the first. For example, suppose the fan with a broken belt was a return fan on a variable air volume system controlled by a variable frequency drive. Replacing the fan belt and making the fan operational in no way satisfies the need to establish that the fan runs properly under automatic controls and the variable frequency drive. Suppose that upon inspecting the operating return fan, it was found that the VFD is not properly tracking the input signal from the controls system. The controls or VFD problems need to be rectified and the system revisited.

The key to this paragraph is that the contractor is responsible for as many visits as may be necessary to get it right, and the final arbiter of when enough is enough is the owner, not the contractor.

3.07 PHASING OF CONSTRUCTION AND COMMISSIONING: Where project completion is performed in stages, the commissioning plan will take into account the staged start-up of each phase as shown on the drawings and/or specifed.

3.08 CLOSE-OUT SUBMITTALS: Close-out documents, consisting of but not limited to As-Built drawings, certificates of inspections, warranties, final testing and balancing reports, Operating and Maintenance Manuals as submitted to the A/E shall be copied to the commissioning agent for concurrent review.

3.09 DEMONSTRATIONS AND TRAINING: Commissioning agent shall be notified in advance of the dates and times of demonstrations for and training of the owner's operating and maintenance personnel, and the commissioning agent shall be present at all sessions.

3.10 COMMISSIONING REPORTS

A. The commissioning agent shall document commissioning milestones with reports. The documents shall acknowledge acceptance at the milestone or separately list deficiencies observed or discovered. The document shall be distributed to commissioning team members.

It is equally as important to acknowledge successful completion of a commissioning milestone as it is to cite deficiencies.

B. The commissioning agent shall prepare a final formal report to the owner that will include a narrative in the form of an executive summary of the results of commissioning process, impressions of the demonstration and training sessions, and a certification that the verification of each item is complete and all systems are operating as intended.

TABLE 6.1   Commissioning Checklist: Pump

Number: _____   Serving: _____

	GPM	Head (Ft.)	Hp	v/ph/fr	RPM	Type	Manufacturer	Model No.	Comments
Specified									
Submitted									
Installed									

Comments:
1.
2.
3.

Installation Check	Designed or Specified		Provided		Installation Check	Designed or Specified		Provided	
	Yes	No	Yes	No		Yes	No	Yes	No
Pressure Gauges					Suction Diffuser				
Suction					Triple Duty Valve				
Discharge					Strainer				
Isolation Valves					Drain				
Balancing Valve					Thermometers				
Check Valve					Inertial Base				
Flexible Connectors					Vibration Pads				
Unions									

Comments:
1.
2.
3.

Operational Check	Contractor		Commissioner		Comments
	Recorded	Date	Observed	Date	
Pump Discharge Pressure (PSIG)					
Pump Suction Pressure (PSIG)					
Pump Rotation					
Alignment					
Motor Amps					

Comments:
1.
2.
3.

COMMISSIONING CHECK LISTS: See Tables 6.1 to 6.5. *Two options are given for this paragraph; only one should be used in the project specifications. The first paragraph applies where all of the detailed check lists are included with the specifications. This is the most desirable approach as it most clearly defines scope and expectations. However, recognizing that there may be occurrences where, for one reason or another, completely detailed checklists cannot be provided at the time that the project specifications are printed and the project bid, the latter paragraph can be used to achieve a functional equivalent. Though the check lists should be as all encompassing as possible while being practical, there are always cases where something may have been missed and other cases*

TABLE 6.2 Commissioning Checklist: Boiler

Number: _____   Serving: _____

	Duty: Steam HW	Fuel				Output								
		Gas, Oil Elec, Dual	Gas Input MBH	Oil Input GPH	Elec Input KW	Net IBR	MBH	Steam Pres. PSIG	Water GPM	EWT °F	LWT °F	Mfr.	Model
Specified													
Submitted													
Installed													

Comments:
1.
2.
3.

Installation Check	Designed or Specified		Provided		Installation Check	Designed or Specified		Provided	
	Yes	No	Yes	No		Yes	No	Yes	No
Gas Train (UL, FM, etc.)					Isolation Valves				
Gas Train Vented					Flues/Vents				
Oil Transfer Pumps					Flues/Vents Dampers				
Oil Preheater					Drain Valves				
Relief Valve					ASME Stamp				
Relief Valve Discharge to Drain					Step Electric Control				
Relief Valve Discharge to Atmos					Step Burner Control				
Combustion Air Intake					Modulating Burner Control				

Comments:
1.
2.
3.

Operational Check	Contractor		Commissioner		Comments
	Recorded	Date	Observed	Date	
Start-Up					
Burner Set-Up/Test/Adjustment					
Flue Gas Analysis					
Combustion Air Intake Operation					
Boiler Sequencing					
Burner Sequencing					
Safeties/Alarms					
Emergency Shut Off					

Comments:
1.
2.
3.

where items that are not included in the project are specified. Therefore, an escape clause is provided to allow the commissioning agent and owner flexibility in the use and application of the checklists. Checklists should be developed specific to each project.

Option 1.   The following commissioning checklists are to be completed by the contractor and certified by the commissioning agent. Completion of the requirements of the checklists shall be viewed as a minimum requirement in satisfying the commissioning process. The checklists shall not be used to exclude any additional work required to

TABLE 6.3 Commissioning Checklist: Hydronic System Accessories

Number: _____   Serving: _____

	Air Separator	Expansion Tank	Strainer					
Specified								
Submitted								
Installed								

Comments:
1.
2.
3.

Installation Check	Designed or Specified		Provided		Installation Check	Designed or Specified		Provided	
	Yes	No	Yes	No		Yes	No	Yes	No
Air Separator					Make-Up Water				
With Strainer					Pressure Regulating Valve				
Without Strainer					Back-flow Preventer				
Pressure Gauges					Isolation Valves				
Isolation Valves					Strainer				
Drain					By-Pass				
Expansion Tank									
Tank Fitting					Pressure Relief Valve				
Site Glass									
Drain					Unions				

Comments:
1.
2.
3.

Operational Check	Contractor		Commissioner		Comments
	Recorded	Date	Observed	Date	
Air Separator Pressure Drop (PSIG)					
System Fill Pressure (PSIG)					
Relief Valve Setting (PSIG)					

Comments:
1.
2.
3.

commission systems and equipment as herein before specified or as deemed necessary by the commissioning agent and/or owner.

Option 2.   The following commissioning checklist applies to a pump. This is a sample sheet, and the contractor will be provided checklists by the commissioning agent specific to the project at the first project progress meeting. Sheets are to be completed by the contractor and certified by the commissioning agent. Completion of the requirements of the checklists shall be viewed as a minimum requirement in satisfying the commissioning process. The checklists shall not be used to exclude any additional work required to commission systems and equipment as herein before specified or as deemed necessary by the commissioning agent and/or owner.

TABLE 6.4 Commissioning Checklist: Air Handling Unit (Constant Volume, Heating/Cooling)

Number: _____   Serving: _____

Supply Fan

	Total CFM	Min. OA Cfm	E.S.P. In.WG	RPM	Drive Belt/ Direct	HP	Power V/Ph/Freq	Mfr.	Model	Style-Horiz,Vert, Draw/Blow-thru AF/FC etc.	Unit Mtg. Floor, Hung, Roof
Specified											
Submitted											
Installed											

Cooling Coil (Water or Glycol)

	E.A.T. °F db/wb	L.A.T. °F db/wb	Air P.D. In.WG	GPM	Water P.D. Ft.	Rows/Fins	FV FPM	Total Clg MBH	Sens Clg MBH	E.W.T. °F	L.W.T. °F
Specified											
Submitted											
Installed											

Heating Coil (Water, Steam, or Electric)

	E.A.T. °F	L.A.T. °F	Air P.D. In.WG	GPM, Lbs/Hr or KW	Water P.D. Ft.	Rows/Fins	FV FPM	Total MBH	E.W.T. °F	L.W.T. °F	Ent. Steam PSIG	Electric Steps & V/ph/Freq
Specified												
Submitted												
Installed												

Cooling Coil (Direct Expansion)

	E.A.T. °F db/wb	L.A.T. °F db/wb	Air P.D. In.WG	FV FPM	No. Ckts., Type	Rows/Fins	Total Clg MBH	Sens Clg MBH	Refrig Type	Sat Suction Temp°F	Refrig Temp °F
Specified											
Submitted											
Installed											

Prefilter / Final Filter / Coil Circulating Pump

	Type	% Eff	Mfr	Model	Type	% Eff	Mfr	Model	GPM	Head	HP	V/Ph/Fr	Model
Specified													
Submitted													
Installed													

Return/Relief Fan

	CFM	E.S.P. In.WG	RPM	Drive B/D	HP	Power V/Ph/Freq	Mfr.	Model	Style AF/FC etc.	Comments
Specified										
Submitted										
Installed										

Comments:

6.4 COMMISSIONING COMPUTER-BASED CONTROLS SYSTEMS

6.4.1 General

The controls systems is at the heart of the successful operation of all mechanical systems, yet it is perhaps the least understood. Controls design, installation, and operation are of such

TABLE 6.4 Commissioning Checklist: AHU (*Continued*)

Number: _____   Serving: _____

Installation Check	Designed or Specified		Provided		Installation Check	Designed or Specified		Provided	
	Yes	No	Yes	No		Yes	No	Yes	No
Supply Fan					Prefilter Section				
Isolation (External/Internal)					Type (Flat, Angle, Bag, etc.)				
Access Door					Access				
Discharge Flex Connection					Differential Pressure Gauge				
Return/Relief Fan					Frame				
Isolation (External/Internal)					Final Filter Section				
Access Door					Type (Flat, Angle, Bag, etc.)				
Inlet Flex Connection					Access				
Water/Glycol Cooling Coil					Differential Pressure Gauge				
Control Valve					Frame				
Face & By-Pass Dampers					Outdoor Air Intake				
Condensate Drain					Wall Louver, Size				
Coil Pull Access					Roof Intake, Size				
Reheat Coil (Water, Steam, Elec)					Ducted, Size				
Control Valve					Damper				
Face & By-Pass Dampers					Mixing Box				
Coil Pull Access					Factory Furnished				
Steam Trap					Field Constructed				
Electric Control (Step/SCR)					Dampers				
Preheat Coil (Water, Steam, Elec)					Return Air Size				
Control Valve					Relief/Exhaust Air Size				
Face & By-Pass Dampers					Outdoor Air Intake Size				
Coil Pull Access					Exhaust/Relief Air				
Steam Trap					Wall Louver, Size				
Electric Control (Step/SCR)					Roof Outlet, Size				
Coil Circulating Pump					Ducted, Size				
Access Section					Motor Operated Damper				
Freezestats					Barometric/Gravity Damper				
Smoke Detectors					Starters				
Supply Air					Vibration Isolation				
Return Air					Base (Spring, Pad, Etc.)				
Casing					Hanger (Spring, Rubber, Etc.)				
Double Wall					Curb Isolators				
Single Wall					Roof Curb				
Insulated					Spare Belts				
					Spare Filters				

paramount importance that assuring properly operating controls consumes more time than any other single activity in the commissioning of mechanical systems. With the advances of computer-based direct digital controls and energy management automation systems (EMS), controls systems have become more complex while simultaneously simplifying their commissioning. To commission a control system that is not equipped with central intelligence capable of storing data and printing trends of operations, much more legwork is required to exercise controls through their various sequences and modes of operations. The use of an energy management automation system greatly facilitates viewing mechanical systems operations and performance; however, to make full use of the automation system

TABLE 6.4 Commissioning Checklist: AHU (*Continued*)

Number: _____ Serving: _____

Operational Check	Contractor Recorded	Date	Commissioner Observed	Date	Comments
Dampers Operation					
Normally Open/Normally Closed					
Minimum Damper Positions					
Dampers Tracking					
Safeties (Freeze, Fire, Low Limit)					
Leakage					
Economizer Cycle					
Filters					
Clean					
Differential Pressure Clean					
Differential Pressure Loaded					
Control Valves					
Normally Open/Normally Closed					
Safeties					
Face & By-Pass Dampers					
Normally Open/Normally Closed					
Safeties					
Condensate Drain, Trapped/Operating					
Clean Coils					
Smoke Detectors Operation					
Freezestat Operation					
Low-Limit Operation					
Temperature Controls Operations					
Space Temperature					
Discharge Air					
Mixed Air					
Return Air					
Start/Stop Control					
Heating Sequence - Occupied					
Heating Sequence - Unoccupied					
Cooling Sequence - Occupied					
Cooling Sequence - Unoccupied					
Economizer Cycle					
Purge Cycle					

Comments:
1.
2.
3.

requires substantial work from the automation subcontractor and a deft understanding of the information that is presented in trended histories. The commissioning agent needs to assure that the project specifications are suitably written to provide for data acquisition appropriate to his needs in commissioning the mechanical systems. Further, the commissioning agent needs to be adept at verifying the automation system's energy and facility management operations in addition to the system's temperature control functions. The following sections of this chapter offer guidance for establishing that a project's automation system will be used to the fullest value in commissioning of the mechanical systems and also provides guidance for commissioning of the automation system itself.

TABLE 6.5   Commissioning Checklist: Master

Number: _____   Serving: _____

Specified								
Submitted								
Installed								

Comments:
1.
2.
3.

Installation Check	Designed or Specified		Provided		Installation Check	Designed or Specified		Provided	
	Yes	No	Yes	No		Yes	No	Yes	No

Comments:
1.
2.
3.

Operational Check	Contractor		Commissioner		Comments
	Recorded	Date	Observed	Date	

Comments:
1.
2.
3.

6.4.2   The Contract Documents

The successful operation of mechanical systems starts with good design drawings and specifications. The design drawings show how the mechanical systems component parts are to fit together and be installed. The specifications further amplify quality of materials, installation methods, and performance. The controls system specifications state how everything is supposed to work.

Review and understanding of the controls system specification by the commissioning agent is critical to the success of the project. Often, not all controls work is specified in the controls section of the specifications, but appears in a section for equipment where controls are to be furnished by the equipment manufacturer, i.e., the boiler controls specified with the boiler, chiller controller with the chiller, rooftop air conditioning units may be specified with factory installed controllers, etc. These manufacturer furnished controls become an integral part of the overall building controls system, and it is essential that the work specified in the controls section of the specifications is harmonious with the manufacturer furnished controls. For example, the controls section of the specification may call for the automation system to provide a chilled water reset signal to the chiller, but unless the chiller was specified to be capable of receiving and using the reset signal from the EMS, the manufacturer may ship a chiller incapable of operating with an externally generated reset signal. The interaction between the automation system and manufactured equipment via common protocols such as BACNet or LonWorks requires that intended control strategies, and where those control strategies reside, be clearly spelled out in the project specifications.

Review and understanding of the contract drawings with regard to the specified control sequences is critical, i.e., do the drawings provide the mechanical hardware and components to execute the operational sequences specified. A simple example of a conflict is where an air handling unit was specified to have economizer cycle control but the drawings only provide an outdoor air intake duct to the AHU of sufficient size to introduce minimum outdoor air.

To get the fullest diagnostic benefit from an energy management automation system on projects so equipped, provisions must be made in the contract specifications to provide for the submission of operating trends for review (trends are also called histories). The importance of trends cannot be overstated. A tremendous amount of information can be provided that enables the commissioning agent to verify systems operation and even perform diagnostics without having to visit the site, saving great amounts of time in the commissioning process. However, unless the contract specifications have provided very detailed requirements for the accumulating and formatting of trends, the trends submitted by the controls contractor may be of little benefit. A common statement found in specifications may read "Contractor shall submit trends for all control points identified in the Points List". This is totally inadequate as it does not tell the contractor how many trend submittals are required, nor does it provide for format of the trends. What most likely will result from this specification is a long list of trends showing temperatures and occurrences of changes of values. Trends of this nature are difficult to diagnose as the simultaneous interaction of various mechanical systems components is not clear, and cause and effect relationships are not readily evident.

The place in the contract specifications for detailing the requirements of trend submittals is in the controls section where the EMS is specified. Should the controls section not provide for trending to the satisfaction of the commissioning agent, the trending requirements could be included in the commissioning section of the specifications. The requirements included in the controls section of the specification should be similar to the following sample, modified as necessary to the specific requirements of the project. If wording similar to the following is not provided, the commissioning specification should augment the controls specification to obtain the trends needed. Parenthetical italic statements are explanations and not part of the specification.

SAMPLE SPECIFICATION

Energy management automation system test and guarantee:

A. The Energy Management Automation System (EMS) shall be guaranteed free from all mechanical, electrical, and software defects for a period of 2 years. During this 2-year period, the EMS subcontractor shall be responsible for the proper adjustments of all systems, equipment, and apparatus installed by him and do all work necessary to ensure efficient and proper functioning of the systems hardware and software.

This is a fairly common warranty statement.

The contractor (*general and/or mechanical*) shall arrange to meet with the EMS subcontractor, the engineer, the architect, the owner, and the commissioning agent (*if so contracted*) within 30 days prior to the specified end of the guarantee period for the purpose of compiling a list of items which require correction under specified guarantees. Should the contractor fail to schedule the final meeting, then the guarantee shall be automatically extended until such time as the meeting takes place, and the contractor shall be fully responsible for correcting such deficiencies as if they occurred under the original guarantee period.

This has tremendous value to the owner. Over the 2-year warranty period, the operation of the building should be fairly well known by the owner's operating and maintenance personnel. Habitual problems or idiosyncrasies of systems operations may have been discovered. This is the owner's last chance to get systems operating satisfactorily prior to relieving the contractors of their responsibilities. Some latent defects may require rectification under the terms of the contract, or some improvements to systems operations may be requested by change order.

B. Placing in service: Upon completion of the EMS installation, calibrate equipment and verify transmission media operation before the system is placed on line. All testing, adjusting, and calibrating shall be completed and systems shall be properly operating prior to acceptance by owner. Cross-check each control point within the EMS by comparing the control command and the field controlled device and/or equipment.

C. Prior to final acceptance and authorization for final payment by the owner, Energy Management Automation System "punch list" inspections shall be made by the engineer, the commissioning agent, and the owner's representatives.

The owner's representatives should include operating and maintenance personnel that will operate and service the mechanical equipment as well as those that are responsible for the operations of the automation system.

This is not to preclude that punch lists shall be made by the contractor to check the completion of his work prior to final "punch lists'" inspection.

This statement puts the contractors on notice that they need to perform their own inspection of systems installation and operation prior to amassing forces on site for a formal punch-out.

The punch lists' inspections shall be in three parts.

1. Installation Punch List: An inspection shall be performed and punch list prepared regarding the physical installation of the EMS equipment, wiring, etc.

This involves observation of installation techniques—Is wiring properly strung and supported? Are panels properly located and labeled? Is control tubing the proper type and properly supported? Has conduit been used as specified? Are the control devices submitted in shop drawings those that are installed on the project? etc.

2. Site Programming Punch List: A separate punch list shall be prepared regarding the software programming at the site. At a minimum, prior to requesting a Site Programming Punch List inspection, all specified software and control strategies shall have been loaded into the computers at the facility to be inspected.
3. Head-End Programming Punch List: A third punch list shall be prepared regarding the software programming at the head end. Prior to requesting a Head-End Programming Punch List inspection, all required "head end" software shall be installed.

This is primarily for campus type systems where a central head end monitors several remote buildings.

4. It is recognized that not all software programming may be fully "debugged" at the time of Punch List inspections; however, the EMS contractor shall be able to demonstrate that all required software strategies are installed and that equipment is being controlled by those strategies. To facilitate punch-outs of installations, balancing shall be complete and the contractor shall have their software completely loaded and functional at least 1 week prior to the submission of a complete set of histories to the engineer and the commissioning agent for review. The intent of this is to determine whether installed software is functioning as intended. As there may be more points required for trending than a single history program's capability, more than one run of histories may be necessary to provide all required data. Prior to submitting a trend, the Contractor shall perform a self-review to identify and correct problems. Histories shall be presented at hourly intervals for a 24-h period, unless directed otherwise.

There are several important issues in the foregoing paragraph.

The contractor is required to have at least a week of system operations under the control of the programmed EMS after systems have been balanced. Normally, balancing and controls are the last components of mechanical systems installation that are completed. Balancing should be complete prior to running trends so that balancing may be eliminated as a cause when operational or performance discrepancies are discerned in the trends analysis.

On large projects, the amount of trending required may exceed the computer's memory capability. This cannot be an excuse for the EMS vendor to submit less than the required trends. Either a computer with sufficient memory needs to be provided or the contractor needs to make multiple runs of trends.

The review of trends can be quite time consuming. The contractor needs to perform a self-review of trends for proper operation to correct obvious malfunctions prior to submitting trends to the engineer or commissioning agent. This usually entails an initial trend run and in-house diagnosis by the contractor. Failures revealed may require hardware or software corrections or both. Once the contractor has made corrections, he needs to run another set of trends for his own review. The contractor needs to go through this process as many times as necessary prior to submitting trends to engineer or commissioning agent to assure that his system is performing properly. The purpose of trend review by the engineer or commissioning agent should be that of verification, not debugging.

The requirement that trends be presented at hourly intervals for a 24-h period tells the contractor the format of the trends. This is far more valuable in diagnosing systems operations than a Change of Value format. Unless specified otherwise, the contractor will most likely submit trends in the Change of Value format as it is often the default within the automation system and requires less work than the 24-h format. The 24-h format is a minimum time interval; some manufacturer's systems can present trends in this format for more than 24 h. The hourly increment yields manageable data; trends run at smaller time intervals, say 15 min, can be unmanageable. Smaller time intervals are often used for more detailed analysis, diagnostics, and troubleshooting but are not advantageous for the initial commissioning process.

A further restriction can be placed on trends to assure that they are of optimal value. Dictating that trends cannot be submitted unless run for days when outdoor air temperatures have risen to at least 85°F (30°C) or dropped to at least 40°F (5°C) during the 24-h trended period assures that the trends will provide a good representation of cooling and heating operations. Trends run for a 24-h period when outdoor air temperatures only ranged from 65°F to 75°F (12–21°C) will be of limited value.

5. During punch-out and/or if malfunctions are discerned in guarantee period histories, systems being monitored shall be operated with an occupancy schedule, i.e., indications that a system was scheduled off for the 24 h of the history and remained off are of no value.

   For example, if a schedule places a facility in the unoccupied mode over a weekend, such as at a school, and trends are submitted for Saturday/Sunday operation, all equipment may be off and little knowledge of systems operations can be discerned. Yet the contractor could contend that he fulfilled his obligation of submitting trends for review.

6. All system setpoints (both calculated and manual inputs) shall be provided with the trends.

   It is essential to know what a setpoint is to determine whether the controls are functioning properly. Trends may indicate that a space is being maintained at 72°F (20°C), and this may appear proper unless it is known that setpoint is 76°F (21°C). Also, software generated setpoints need to be presented in trends, such as heating water reset values. Diagnosis can then be made as to whether the reset setpoint is being properly calculated and whether control devices are properly tracking the setpoint.

7. Punch-out of software shall not occur without the central plant operating in either the mechanical heating or cooling mode. If scheduled delivery of the project falls during a period in which the central plant is not operating, EMS punch-out will be delayed until such time as the central plant has been operating under EMS control for the afore specified 1 week.

   This places the onus on the contractor to deliver the project on time. Suppose that a project was scheduled for delivery in August. It would be expected that a suitable punch-out could be performed with the central cooling plant in operation. But due to late completion of the work by the contractors, the facility is not ready for punch-out until late September. It is possible that the weather has cooled to the point that the cooling plant is no longer in operation, but it is not cold enough to run the heating plant. The owner should not be expected to attempt to accept the control system without being able to witness the operations of the central plant.

8. Initial acceptance of the EMS to start the warranty in either the heating or cooling mode shall be conditional; final acceptance shall be predicated on acceptance of the EMS in the subsequent seasonal operation, i.e., to maintain the warranty, the EMS must function properly in both heating and cooling modes.

   This places the contractor on notice that the control system must operate properly in both the heating and cooling modes. Acceptance of operations in one mode does not automatically extend to acceptance of the other mode. The controls system is given conditional acceptance in anticipation that the systems will perform acceptably in the next seasonal changeover. Should the systems perform acceptably, the initial acceptance is honored as are the terms of the warranty period. Should the systems not perform acceptably, the initial acceptance is rescinded and a new date of acceptance established. The warranty commences on the new acceptance date.

9. The following constitute those items that, at a minimum, shall be included in a trend submittal:

The following is a sample. Listings should be specific to the project requirements. This type of list can also be used to cross-check the controls specifications and points list to assure that all trended points required have been specified and/or shown on the points list.

a. Outdoor air:

   1) Dry-bulb temperature
   2) Wet-bulb temperature and/or relative humidity
   3) Calculated enthalpy

b. Phase, Voltage, Frequency: Histories should indicate time and duration of any occurrence of power interruption and normal power resumption.

c. Control air compressor:

   1) Control air pressure (house side)
   2) Compressor run time

d. Demand: kW

e. Boilers:

   1) Scheduled run status
   2) Actual run status
   3) Burner status
   4) Boiler water temperature (as sensed in the boiler shell)

f. Heating water systems:

   1) Supply water temperature
   2) Return water temperature
   3) 3-way mixing valves position
   4) Calculated reset temperature

g. Heating water pumps:

   1) Command
   2) Actual status
   3) Log alarmed outages
   4) Provide history for main and standby pumps

h. Chilled water system:

   1) Chiller command
   2) Chiller status
   3) Amperage
   4) Log alarmed outages
   5) Calculated chilled water reset temperature
   6) Reset signal
   7) Chilled-water-supply temperature
   8) Chilled-water-return temperature

i. Condenser water pumps:

   1) Command
   2) Actual run status

    3) Log alarmed outages
    4) Provide history for main and stand-by pumps
    5) Flow status

 j. Condenser water systems:

    1) Cooling tower fans command
    2) Cooling tower fans status
    3) Amperage
    4) Log alarmed outages
    5) Condenser-water-supply temperature
    6) Condenser-water-return temperature
    7) Condenser-water-calculated reset temperature
    8) Mixing valve position

 k. System water pumps:

    1) Command
    2) Actual run status
    3) Log alarmed outages
    4) Provide history for main and stand-by pumps

 l. Water cooling and/or heating coils:

    1) Space or discharge (controlling) temperature
    2) Valve position

 m. Air handling units:

    1) Supply fan command
    2) Supply fan status
    3) Supply air temperature
    4) Outdoor, return, and relief air damper positions
    5) Space air temperature
    6) Return air temperature and enthalpy when enthalpy optimization is performed
    7) Mixed air temperature
    8) Duct static pressure (on VAV units)
    9) Vane position or variable frequency drive speed (on VAV units)

 n. Fan powered mixing boxes:

    1) Supply fan command
    2) Supply fan status
    3) Space air temperature
    4) Air damper position
    5) Heating coil valve position
    6) Air flow
    7) Supply air temperature

 o. Variable air volume terminals

    1) Space air temperature
    2) Air valve position
    3) Heating coil valve position
    4) Supply air temperature

 p. Incremental units (Unit ventilator, Fan coil units, Cabinet heaters, etc.):

    Space temperature

CHAPTER 7
SYSTEM CONTROL EQUIPMENT

Terry Hoffmann
Johnson Controls, Inc., Milwaukee, Wisconsin

7.1 INTRODUCTION

In order to discuss HVAC system control equipment, it is important to understand their purpose. Fundamentally, a control system does four things:

• Establishes a final condition

• Provides safe operation of the equipment

• Eliminates the need for ongoing human attention

• Ensures economical operation

Maintaining a set final condition is often considered the only purpose of the HVAC system because it has a direct impact on occupants of the facility. The comfort level of each individual is the end product of all control efforts, yet ASHRAE design standards are set to satisfy only 90 percent of all occupants. Well-designed systems that are operated correctly and maintained to their original delivered condition are capable of doing much better than this.

Safety pertains to both individuals and property. The most obvious safety-related HVAC controls are pressure and temperature cutout switches that shut off equipment to prevent damage to both the device and the operator. It is interesting to note that levels of safety consciousness vary throughout the world. While the general pattern would seem to indicate more concern for safety in economically developed countries, there are variances from country to country based on culture and operational practices. For example, it is quite normal for a motor control system installed in Germany to monitor each position of a hand-off-auto switch, as well as all electrical overload devices. Elsewhere, the same system would only monitor the auto and hand positions to be sure the operator is aware of override vs. automatic operation. As we develop into a completely global society based on a single set of standards, which are then tailored to the needs of a particular geographic area, it will be important to understand these differences and adapt our operational practices to meet them.

Elimination of the need for human attention is the cornerstone of the modern control system. In fact, we often call it an *automatic temperature control system* or a *building automation system* because the device or system is designed to operate without the aid of a person. The first thermostat, dating back to the 1880s, was not designed for the *comfort* of occupants, but for their *convenience* so they did not have to notify the boiler room staff whenever they wanted more heat or less heat. A good automatic control system reduces human attention to making sure that commissioning has been correctly accomplished, and occasionally making setpoint adjustments to better meet the current situation if it is outside of normal design conditions (see Chap. 6). These variances might include extremes of heat and cold, as well as building upgrades and retrofits that affect control variables either temporarily or long term.

Assurance of economical operation has become another important factor due to two important global trends. First, the need to conserve our global natural resources places a responsibility on each individual and organization to utilize them in the most efficient manner. If these resources are depleted without the development of suitable alternatives, it would have a tremendous impact on the world economy. Second, the advent of global competition has made *productivity* a central piece of the strategic operation of both manufacturing and service industries. Automatic control systems provide the vehicle for this conservation and productivity. It is notable that many specifications refer to the controls as an Energy Management and Control System.

7.2 FUNDAMENTAL CONTROL DEVICES

By looking at a functional block diagram of the fundamental control loop (Fig. 7.1), we can easily identify the individual pieces of equipment common to HVAC control systems.

Sensors at the beginning of the diagram sense the controlled variable (i.e., the final condition we are attempting to impact). They send a signal to the controller's input. For HVAC controls, these devices usually sense temperature, humidity, flow (velocity), speed (percent or RPM), position or state (on/off, high/low).

The *controller* analyzes the sensor's input and makes a comparison against the desired final condition—the setpoint. The controller then outputs a signal to the controlled device assembly to maintain the setpoint value. In networked systems, the controller is also responsible for reporting information to a higher level controller on the network, which then aggregates the information and interfaces to a higher level information system. In turn, this information can be accessed on a personal computer interface or a dedicated display device. (See the Control Networks section later in this chapter.)

The *actuator* often is part of a controlled device assembly, and is responsible for taking the output from the controller and transforming it—with the aid of some energy source—into a change of position in the controlled device. Actuators are usually powered electrically or pneumatically.

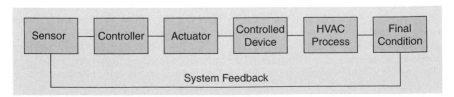

FIGURE 7.1 Fundamental control loop block diagram.

The *controlled device* regulates the flow of steam, water, or air according to the commands of the controller as applied by the actuator. The regulation of these fluids helps to balance the load, gains, or losses in the HVAC system. For example, the system may control temperature in a room affected by the number of occupants (load), sunlight (gain), or window open to a cooler outside temperature (loss). These devices include motors, valves, and dampers.

The *HVAC process* refers to the actual mechanical equipment such as ducts, coils, fans, and pumps that impact the *final condition* in accordance with mechanical design.

In summary, we are concerned with the operation and maintenance of sensors, controllers, actuators, valves, dampers, motors, the infrastructure connecting them to the mechanical equipment, and finally, when installed, all related network components that feed into the human operator of a computerized management and control system.

7.3 CONTROL SYSTEM TYPES

HVAC control systems are usually categorized by the energy source required to change the final condition:

- Self-contained
- Pneumatic
- Electromechanical
- Electronic
- Digital

Self-contained control systems combine the sensor, controller, and controlled device into one unit. In this system, a change in the controlled variable is used to actuate the controlled device. A self-contained pressure relief valve is an example of this type of control device. If the pressure in the line exceeds the maximum setpoint—which may be fixed or adjustable—it is vented to relieve pressure and protect the end devices, along with ensuring the safety of the operator and occupants.

Pneumatic control systems use compressed air to modulate the controlled device. In this system, the air is applied to the controller at a constant pressure, and the controller regulates the output pressure to the controlled device according to the desired rate of change. Typical air pressure for an HVAC control system is 20 psi (136 kPa).

Electromechanical control systems use electricity to power a mechanical device. These systems may have two-position action (on/off) in which the controller switches an electric motor, solenoid coil, or heating element; or, the system may be *proportional* so that the controlled device is modulated by a motor.

Electronic control systems use solid-state components in electronic circuits to create control signals in response to sensor information. Greater reliability of electronic components and reduced power requirements have allowed electronic systems to supplant electromechanical systems in all but the simplest applications and where digital control is not required.

Digital control systems (direct digital control, or DDC) utilize computer technology to detect, amplify, and evaluate sensor information. This evaluation can include sophisticated logical operations such as rule-based and fuzzy logic operations. The resulting digital output signal is then converted to an electrical or pneumatic signal, which is capable of operating the controlled device.

Table 7.1 compares the five different types of control systems.

TABLE 7.1   Control System Types

Classification	Source of power	Output signal
Self-contained	Vapor liquid filled	Expansion as a result of pressure
Pneumatic	15 psi (104 kPa)	0–15 psi (0–104 kPa)
	20 psi (136 kPa)	0–20 psi (0–136 kPa)
Electromechanical	24 VAC	0–24 VAC
	120 VAC	0–120 VAC
	220 VAC	0–220 VAC
Electronic	5, 12, or 15 VDC	0–5 VDC (typical)
	12 VAC	0–12 VAC
	24 VAC	0–24 VAC
Digital	24 VAC	0–20 milliamps
	120 VAC	0–10 VDC
	220 VAC	Direct digital

7.4   TYPICAL APPLICATIONS

Since this book is about HVAC control, we are interested primarily in the following applications, or subsets, of HVAC control:

- Temperature controls
- Flow controls
- Pressure controls

While the mechanical equipment may vary greatly in these applications, the control elements are largely the same. In the case of electromechanical or electronic controls, the only
difference between each application might be the sensing element. A bulb for temperature,
a bellows for pressure, or an impeller for flow are typical elements. Digital control systems
have great flexibility in handling any of these applications. In many cases, a single digital
controller assumes responsibility for temperature, flow, and pressure in order to control a
single piece of mechanical equipment. For a detailed description of each application and
system design fundamentals, refer to the latest *ASHRAE Applications Handbook*.

7.5   IMPACT OF SYSTEM FAILURE
ON SYSTEM PERFORMANCE

Failure of the control system has an obvious impact on the performance of the HVAC system. The most noticeable impact is on occupant comfort due to the loss of temperature,
humidity, or ventilation control. Older pneumatic and electronic systems are designed for
a single "best alternative" failure mode. In the case of temperature control, failure usually
defaults to a *Full Heat* setting to prevent freezeups in the northern climates, and to a *No
Flow* setting where freezeups are not a problem.

Both alternatives are concerned strictly with safety, not comfort. Modern digital control
systems are designed for fail-safe or fail-soft operation to lessen the impact of component
failures. For example, if a sensor is determined to be out of range or nonoperational, the digital controller can continue to operate using the last good reading that it obtained prior to the

failure, or it can be programmed to fail to a particular output. It also might be programmed to look at another input, such as outside air temperature, and make the best decision possible given that information. Any of the three alternatives are probably superior to the older "on/off" scenario.

Modern digital controllers, with their built-in failure modes, also have eliminated much of the domino effect that plagued less sophisticated systems. The chances of melting all of the chocolate bars in the cafeteria because temperature control failure defaulted to *Full Heat* over the weekend are greatly diminished. By the same token, it is less likely that any failure in a system will result in the failure or destruction of components in another, complementary system. That is because operators can preprogram and select a sophisticated array of checks and balances.

Another significant step toward eliminating catastrophic failures is the growing trend of system manufacturers to include not only an adjustable alarm limit for critical values, but also a prealarm limit which notifies the operator of an impending failure. This form of predictive feedback fits well with the concept of both predictive maintenance and reliability-centered maintenance. For older systems with less built-in intelligence, it is the operator's responsibility to spot these anomalies.

7.6 HISTORICAL EXPERIENCE CONCERNING FAILURE RATES

When comparing the mean-time-between-failure (MTBF) of a digital HVAC control system to that of a pneumatic or electromechanical system, it is appropriate to examine the system as having four levels of technology, each with its own characteristics (see Table 7.2).

Since the control and monitoring system requires all the layers to be operating correctly, system reliability is equal to that of the least reliable component, which often is the *personal computer*. Therefore, it is important to make sure that a networked system is configured with multiple PCs having duplicate functionality. A recommended plan is to have access to a spare monitor and to maintain a daily backup of the system on a second hard drive or a high-density removable backup device (tape or disk).

Actuators also present concerns because they are often the next least-reliable devices. Most buildings have older style actuators that are not very reliable, although many high-quality actuators being built today have 1-year reliability rates of over 99 percent. It is reasonable to ask the system manufacturer for reliability information as part of the original purchase decision. The information is also valuable when determining how many to have on hand as spares. There is a high probability that any manufacturer maintaining

TABLE 7.2 Mean-Time-Between-Failure (MTBF)

Layer	MTBF characteristic
Sensors/actuators	Very long MTBF for sensors (usually passive devices) Medium MTBF for actuators (mechanical components)
Digital controllers (microprocessor-based)	Long MTBF (failures usually related to power supply)
Wiring network	Very long MTBF (passive components)
Personal computers	Short MTBF (mechanical and high-voltage components in disk drives and monitors)

ISO 9000 certification of its design and manufacturing facilities will be able to provide these data.

The *wiring network* is not noted as having a high probability of failure. However, this is true only *if the network does not rely on a PC-based server and if the network is not shared with other information technology organizations within the facility.* Network maintenance is best handled by a computer network management specialist, so it lies beyond the scope of HVAC controls. The only way to make sure that network failures do not adversely impact system operations is to design and maintain a system with no single point of failure. In HVAC applications, the impact of a single failure can be minimized by using distributed control techniques and increasing the level of intelligence as far down the control ladder as possible. When adding devices to an existing system, always evaluate the impact of a controller failure on the operation of the system given its new tasks. *In many cases, it is better to add a new controller than to overburden an existing one.*

7.7   OPERATOR TRAINING REQUIREMENTS

Depending on the size of the facility and the complexity of the HVAC control system, it is possible to have vast differences in the educational requirements for those who operate and maintain the equipment. Anyone with responsibility for the system should have a working knowledge of HVAC basics, as well as familiarity with the system itself and the technology behind it. In addition, the operator should be able to accomplish the following tasks:

• Program/change setpoints and schedules as required to meet the operational needs of the organization
• Optimize energy usage and efficiency
• Ensure occupant comfort
• Understand the impact of system changes and maintenance on the core business functions of the organization
• Assist with building systems safety, health and environmental standards, requirements and codes
• Assist with planning for system upgrades and improvements designed to increase safety, improve reliability, and improve occupant comfort levels, thereby increasing productivity

In order to accomplish these tasks, the operator will require training above and beyond that which is available on the job. The following recommendations are to be viewed as consecutive and incremental. Each level builds on the next, leading to the ultimate goal of an informed, productive control system manager.

7.7.1   HVAC Basic Training

This course of study should include the following information:

• HVAC system types and piping systems
• Fan systems and fan characteristics
• Dampers and damper actuators

- Valves and valve actuators
- Boilers and related equipment
- Fundamentals of refrigeration
- Chiller types and applications
- Cooling towers
- Heat exchangers and miscellaneous equipment

Sources for this training include local technical colleges and universities, ASHRAE professional development seminars, mechanical systems and control systems manufacturers, and self/group teaching via books, videos, and study guides available from a variety of sources.

7.7.2   Control Systems Fundamentals

A typical course outline should include:

- Basic control theory
- Proportional control fundamentals
- Proportional plus integral control
- Proportional plus integral plus derivative control
- Economizer systems
- Mixed air control systems
- Variable air volume control systems
- Building pressurization control strategies
- Fan capacity control methods
- Control fundamentals for water systems

Sources for this training are similar to those for HVAC systems, with a heavier emphasis on materials obtained from control systems manufacturers.

7.7.3   System-Specific Operation and Maintenance (Pneumatic)

If pneumatic systems or devices are prevalent in the facility, it is important to understand:

- Pneumatic controls and air supply systems
- Air supply maintenance
- Thermostats: single and dual setpoint
- Controllers
- Valve and damper actuators
- Valve and damper actuator service
- Auxiliary pneumatic devices
- Temperature, pressure, and humidity transmitters
- Pneumatic receiver/controllers
- Calibration and adjustment of receiver/controllers
- Safety devices and procedures

Hands-on courses are best suited for this training and are available from local technical colleges and control system manufacturers.

7.7.4 System-Specific Operation and Maintenance (Digital)

The differences between the pneumatic and digital courses are technology- and terminology-based:

- Control systems evolution
- Computers, programming, and direct digital control (DDC)
- Systems terminology
- Basic hardware, software, and user input devices
- Analog and binary sensor types and applications
- Output types, actuators, and transducers
- Software data types
- Control loop types
- Programming direct digital controllers
- Communication to automation systems

Training of this nature should be provided directly from the manufacturer of the digital controls or from an alternate source familiar with the equipment.

7.7.5 Energy Management Concepts

This course of study ensures the understanding of basic energy management concepts and features relating to heating, cooling, air distribution, electrical, and lighting systems:

- Energy management fundamentals
- The building envelope
- Modeling the building environment
- Opportunities for conservation:
  Cooling systems
  Heating systems
  Air distribution systems
  Water distribution systems
  Electrical systems
  Lighting systems
- Performing an energy audit

The Association of Energy Engineers (AEE) has courses available which lead to examination and granting of a certified energy manager (CEM) certificate. This line of study is suitable for both operators and aspiring managers. ASHRAE also offers professional development seminars on this topic, as do local colleges and universities. Manufacturers of mechanical and control systems are likely to offer courses in applying their products to manage energy usage.

7.7.6 Building Automation and Control Systems Operation and Maintenance

With a firm foundation of mechanical, systems, and energy knowledge in place, it is possible to apply the features and functions of the building automation system to the needs of the enterprise:

- Basic system hardware configurations
- Communications network architecture
- Operator workstations and I/O devices
- System software

    On-line help, point types, menus

    Passwords, commands, summaries

    Reports and alarm management

    Scheduling points and summaries

    Point history, trending, operator transactions

    Energy management features

    Defining/creating systems and points

    Programming control loops and functions

- Hardware maintenance

It is strongly suggested that this course be taken on site using the actual installed equipment, or at a well-equipped lab operated by the control system manufacturer.

Given the breadth and depth of training required to properly maintain a complex HVAC control system, a decision must be made whether to provide training in-house or to augment the operation and maintenance team with an external control system specialist. There are three ways to obtain this assistance:

1. Include maintenance of the control system in a service contract with the mechanical contractor or controls company. This provides a limited amount of time each month to perform important tasks and observe system trends that may impact operation.

2. Outsource a mechanical and control systems specialist from an organization that has experienced staff people available. The individual is full-time and blends into the organization, while not burdening the organization with training and development requirements.

3. The operation and maintenance of the mechanical and control systems can be handled by an outside supplier as a task under a total facility management contract. This allows management to concentrate on planning and developing strategies for improvements in operation and productivity, while ensuring competent operation and maintenance at a price that is constant for a fixed period of time.

7.8 CONTROL PRODUCTS PREVENTIVE/ PREDICTIVE MAINTENANCE PROCEDURES

Numerous tasks are recommended for a comprehensive inspection and calibration of the HVAC control system. They may be accomplished on a rotating basis or several times throughout the year depending on seasonal conditions, winter shutdown, spring changeover, etc. Many of the tasks are system-specific and may not be required in a particular facility.

It must also be noted that the type of maintenance program adopted in the facility may have a significant impact on the frequency of many tasks. For example, a program that includes predictive maintenance information may reduce the frequency of tasks requiring hardware checks for wear or accuracy. Likewise, a good reliability-centered maintenance program will reduce the frequency of most tasks, while increasing the frequency of other tasks performed on mission critical equipment.

Finally, the word *clean* is used sparingly in all the procedures to describe situations where dirt adversely affects the operation of the device. *It is good general policy to maintain all control equipment to a high standard of cleanliness.* The interval between general cleaning can be evaluated on a yearly basis and extended if it is determined to be acceptable. For a detailed description of maintenance procedures refer to Part D: *Managing Maintenance.*

7.9 CONTROL NETWORKS

Modern controllers communicate over control networks designed to provide a fully distributed control environment, where every device has access to the information it requires to do its job efficiently and effectively. Likewise, each controller is responsible for providing information to a higher (supervisory) level that maintains order and provides widespread programming functions. In many cases, individual controllers can communicate on a peer-to-peer communications network that does not rely on a higher-level device. This protects the controllers from a failure at the supervisory level.

7.9.1 Network Devices

Workstations—Formerly designated as the head end of the system prior to distributed controls, the workstation usually contains the graphic user interface and programming tools required to configure both the network itself and the attached control devices. It may also contain the historical data collected from the networked controllers for long-term storage and display or printed historical reports.

Network Server—When networks grow too large to depend on a workstation to maintain data and deliver information to a large group of workstations, a network server may be necessary.

Supervisory Controllers—Often referred to as a building controller or network controller, these devices provide a grouping function and serve as a temporary repository for trend information, schedules and intersystem programming functions.

Programmable Field Controllers—These devices provide an open programming functionality for custom and complex applications such as chilled water pumping or pressure equalization.

Application Specific Controllers—These task-specific devices provide functionality unique to the application for which they were intended. Most systems provide a wide variety of application specific controllers for built-up air-handling units, rooftop units, packaged air conditioning units, terminal units, VAV boxes, fan coil units, and unit ventilators.

I/O Devices—While many controllers utilize directly connected devices for inputs such as temperature, pressure, and humidity sensing—and outputs like relays and actuators—others provide a network connection for these functions.

7.9.2 Network Layers

Control systems are usually configured into networks with several layers. The most common of these configurations uses three major layers.

The *enterprise layer* is the actual working network of the enterprise that is used for personal and server-based computing and applications such as finance, human resources, and quality control. It allows information from the HVAC control network to be distributed to individuals and software applications requiring it. In some cases, this layer shares the same physical connection as the supervisory layer. Workstations and servers may be connected at this level.

The *supervisory layer* (sometimes known as the network layer) provides peer-to-peer communications between the supervisory controllers and also may be the connecting point for programmable field controllers—with or without supervisory functions.

The *field layer*, usually called the field bus, is the connecting and communications vehicle from the application specific controllers (ASCs) to the supervisory controllers. ASCs may also use the field bus in a peer-to-peer fashion. Some systems connect programmable field controllers and intelligent I/O devices here.

A fourth layer may be provided exclusively for input and output devices (I/O). This *I/O layer* is usually limited in speed and capacity. Figure 7.2 shows a typical three-layer network that also provides for I/O on both the field bus and supervisory control levels. All system controllers and components are shown in their normal locations.

The supervisory and the enterprise network layers both use Ethernet as the physical layer. However, they often share network segments or are attached to each other with a network bridge.

Some control networks claim to be "flat," which permits any device to be attached anywhere on a single layer. The device can use several physical connections, yet not rely on

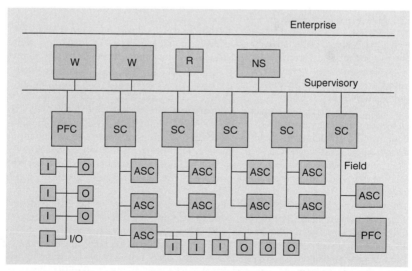

W = Workstation. R = Router. NS = Network Server. PFC = Programmable Field Controller.
SC = Supervisory Controller. I/O = Input or Output Device. ASC = Application Specific Controller.

FIGURE 7.2   Three-layer controls network.

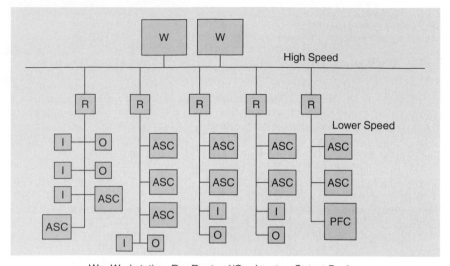

W = Workstation. R = Router. I/O = Input or Output Device.
ASC = Application Specific Controller. PFC = Programmable Field Controller.

FIGURE 7.3   "Flat" controls network.

supervisory controllers to link the supervisory and field networks (or the field network bus and I/O bus). Figure 7.3 is representative of a "flat" network configuration.

7.9.3   Network Protocols

The enterprise layer is almost always a derivative of the Internet Protocol (IP). Network protocols for the supervisory, field, and I/O layers are most often categorized as proprietary, differentiated by speed, object types, and other properties that are defined by the manufacturer. Many were developed prior to the widespread use of personal computers and ubiquitous networks.

Open networks can be proprietary or standard in nature, but have the distinction of being available for other manufacturers to develop connectable products. Many manufacturers of specialty products, such as variable frequency speed drives, are interested in having their products connect to as many systems as possible. They utilize open protocols to achieve this.

Standard protocols combine openness and availability. A standard protocol is certified by a national or international standards body, and is equally available to anyone who wishes to use it for its intended purpose.

One of the two most common standard protocols in the HVAC controls industry is the Building Automation and Controls Network (BACnet), which also is registered as a standard by the International Standards Organization (ISO) and the American National Standards Institute (ANSI). The other common protocol is LonTalk, developed by the Echelon Corporation and registered as an ANSI standard. LonTalk is a flat protocol, while BACnet is based upon a multilayer model that can be used at any or all of the supervisory, field, and I/O levels. Refer to *ASHRAE Guideline 13-2000, Specifying DDC Systems* for more information on BACnet. Refer to the *LonMark Interoperability Guidelines Rev. 3.3* for information on LonTalk (see Bibliography).

7.10 NETWORK SYSTEMS CURRENT AND FUTURE DIRECTIONS

7.10.1 Controllers

Controller functionality is always being enhanced, while the size and relative cost of controllers are gradually being reduced. Controller pricing reflects the increase in processor power for the same or less cost, as well as the decreasing cost of memory. As processor power and memory capacity increase, there is more opportunity to expand the applications and features that are included in the device itself or the programming library that it accesses. Three of the most common of these are self-tuning, self-configuration, and self-commissioning.

The *self-tuning* algorithm continuously monitors the setpoint, feedback, and commanded output value to optimize the proportional and integral terms of the PID loop equation. This is especially important for control loops involving significant changes in the load on a control process. For example, while a modern computer facility may require cooling all year, the differences in cooling load between summer and winter are great. If the control loop were not self-tuning, it might result in hunting and overshoot in one case, while not being sensitive enough to maintain accurate control in another. Without this self-tuning capability, the parameters would have to be manually adjusted two to four times a year.

Self-configuration is a controller feature that recognizes the I/O devices attached to it and automatically assumes an application profile that matches the configuration. In most cases, this is based on the use of proprietary input and output devices from a single vendor to simplify the decision-making process. However, it is still an effective feature. On large projects, this can result in reduced engineering and installation labor. On smaller projects, there may be less need for professional engineering resources to install the controllers and make them operational.

The *self-commissioning* feature performs all tests commonly required for commissioning without necessitating a controls technician to perform them manually. For example, a fan coil unit controller with variable speed control on the fan—and analog positioning of the control valve—might proceed through a sequence that starts the fan at minimum speed, proceeds to ramp it to maximum speed, strokes the valve from full closed to full open, then verifies the results of all commands. Adjustments to control loop tuning may be performed as part of the process. The result is an application that is fully tested and operational. All functions operate correctly and effectively. See Chap. 6.

Another significant application is the addition of system diagnostics to the controller, which helps optimize control and reduce energy costs. For example, a VAV controller might include diagnostic values such as the average actuator runtime, deviation from room temperature setpoint on a weighted moving average basis, average damper position, and amount of time the damper is at maximum (or full open) position. Each of these values can be reported on a real-time basis, or trended in database for evaluation and comparison purposes.

Undoubtedly there will be additional features and functions that take advantage of increased controller power and memory. Fuzzy logic and true artificial intelligence are two improvements at the top of the list for implementation in commercial controllers on a widespread basis. Hybrid control, state machine control, and near optimal control are other new developments of interest.

7.10.2 Network and Operating Systems

For more than a decade, the trend in control networks has been from proprietary to open, and from open to standard. Another trend has been from private networks to shared networks.

Likewise, it is obvious that the operating systems and networking hardware employed are also moving quickly from proprietary to standard.

As indicated previously, the two most common HVAC network protocols are BACnet and LonTalk. While there is considerable discussion as to which is superior or will become the single predominant standard in the industry, it is much more important to recognize the industry trend to reduce the emphasis on user workstations, and to rely on standard laptop or desktop computers accessing systems information and programming capabilities through the internet. While current systems rely on HTML pages and rudimentary connectivity to information, future systems will take advantage of newer standards for Web-based information transfer like XML (eXtensible Markup Language), SOAP (Simple Object Access Protocol), and the concept of Web services for access and delivery of data to enterprise systems outside of the mechanical and electrical system realm.

Perhaps a university professor wishes to perform a laboratory test that requires precise temperature and humidity control, as well as an audit trail of these values throughout the test as a proof source for publication of the results. By applying the published Web services programming model provided by the controls company, a graduate student could write the required program to access the data and format it to meet the needs of the academic community.

In order to take full advantage of these opportunities, equipment and networks would be best implemented with the standards of the IT community taking priority wherever practical.

7.11 PROGRAMMING DIGITAL CONTROLS

A modern digital controller requires complex programming because it combines multiple-loop controllers to determine application-dependent control solutions. For example, a VAV controller may have separate states of control for standby, unoccupied, occupied, cooling, heating, and warm-up. To program these functions, there must be a control sequence of operation to describe the states and actions of the loop controllers for all functions. A typical sequence of operations for a VAV controller follows:

7.11.1 Sequence of Operation

Occupied-Unoccupied Mode Control.   The occupied mode will be scheduled or manually commanded at the operator workstation. In the occupied mode, the air-handling unit will run continuously. In the unoccupied mode, the air-handling unit will be off, and the room temperature will be monitored and compared to the night low limit and night high limit setpoints. Upon a fall in room temperature below the night low limit setpoint, or a rise in room temperature above the night high limit setpoint, the air-handling unit will be started and remain on until the differential is satisfied.

Start-Stop Control.   The air-handling unit will be started and stopped as determined by the occupied-unoccupied mode control program. Upon receiving a start command, the supply fan will be started, a 12-sec (adjustable) delay will occur, and then the return fan will be started.

Mixed Air Damper Control.   When the outside air temperature is below the economizer switch-over setpoint (adjustable), the unit will be in economizer mode. The mixed air dampers will be modulated open from minimum position (adjustable) to 100 percent outside air on a call for cooling from the discharge temperature control signal. If the discharge temperature setpoint cannot be maintained by the mixing dampers, the chilled water valve will be modulated open in sequence. A mixed air low limit program will modulate the mixing dampers closed on a fall in mixed air temperature below setpoint (adjustable). When not in economizer mode, the mixing dampers will be at minimum position (adjustable) and the chilled water valve will be modulated to maintain discharge temperature setpoint.

Room Reset of Discharge Temperature Setpoint. The room temperature will inversely reset the discharge air temperature setpoint according to the following adjustable schedule:

Room temperature	Discharge air temperature
80°F (27°C)	55°F (13°C)
60°F (15°C)	90°F (33°C)

Discharge Temperature Control. The hot water valve and chilled water valve will be modulated in sequence as required to maintain the calculated discharge air temperature setpoint.

Fan Shutdown. The DDC controller will sense the status of the supply fan and the return fan via current sensing switches. Upon sensing that the supply fan is off, the DDC controller will close the outside air damper and exhaust air damper, open the return air damper, close the chilled water valve, and open the hot water valve.

Safeties. A fire alarm shutdown relay will stop the unit upon receiving a signal from the fire alarm system. A temperature low limit will stop the unit upon sensing a fall in temperature below setpoint.

Similar sequences are required for rooftop units, package air conditioning units, built-up air-handling units, and terminal controllers. Application specific controllers might be programmed using a programming language such as a control-adapted form of Basic or C+, through a graphical block programming language (compiled or interpreted), or via a question-and-answer programming tool that determines the correct programs to run based on questions tailored to the application.

Programmable field controllers usually require a graphical programming tool or a programming language because a question-and-answer session for every possible programming opportunity could never be developed. Supervisory controllers are more likely to use a proprietary programming interface, as well as the graphical tool or language, because they contain database information for all connected controllers, as well as information on network operation.

Regardless of how the devices are programmed, there must always be a repository for the completed programs so that, in case any single device fails, a replacement can be installed and downloaded quickly to resume system operations. Most networked systems use a workstation or server to perform this task, while many provide some form of redundancy so that even a loss of the server or workstation can be recovered from with minimal interruption of the performed functions.

7.12 SUMMARY

Controls for HVAC have made great strides in the past 30 years, from systems that relied predominantly on pneumatic controls to the digital control networks of today. The importance of any system, though, is to provide comfort, safety, and a productive environment with the greatest efficiency and the least cost. In order to accomplish this, our controls must be correctly chosen, engineered, installed, commissioned, and maintained. The greatest devices in the fastest and most connectable networks are worthless if they are not correctly programmed to perform the application for which they were intended. The future will bring additional technology breakthroughs, but it is the engineer, the installer, the operator and the maintainer who ensure the success of HVAC control systems.

INTERNET RESOURCES

The following links were active as of January 1, 2004 and may be of assistance to HVAC professionals with an emphasis on controls and automation.

XREF Publishing Company, HVAC/R Information Source, Unbiased cross-referenced database for the HVAC/R industry; www.xrefpub.com

Continental Automated Buildings Association, Advanced technologies for the automation of buildings, Articles, links, standards, training, publications; www.caba.org

LonMark Interoperability Association, Information on the LonTalk protocol and LonMark standards, Forums, links, downloads; www.lonmark.org

Naval Facilities Engineering Service Center, U.S. Navy's center for specialized facility engineering, Publications and links to many other government sites; www.nfesc.navy.mil

ASHRAE, American Society of Heating, Refrigerating, and Air-Conditioning Engineers, Learning institute, publications, standards; www.ashrae.org

AEE, Association of Energy Engineers, Books, seminars, links, news; www.aeecenter.org

Automated Buildings, Online magazine for automated buildings, Products, systems, articles; www.automatedbuildings.com

The HVAC Mall, HVAC/R directory, index, resource, Products, contracts; www.hvacmall.com

Johnson Controls, Controls manufacturer's home page, Products, training courses, articles; www.johnsoncontrols.com/cg

Facilities Net, Home page for *Building Operating Management* and *Maintenance Solutions* magazines, Products, articles, forums; www.facilitiesnet.com

As links change on a regular basis due to technology upgrades and corporate acquisitions/mergers, the best way to update these references is to use a search engine and search on an appropriate string such as "HVAC controls."

CHAPTER 8
WATER TREATMENT

Colin Frayne
The Metro Group, Inc., Long Island City, New York

Richard T. Blake
The Metro Group, Inc., Long Island City, New York

8.1 INTRODUCTION

This chapter discusses the technology of treating water for use in commercial and industrial heat-transfer equipment, with specific emphasis on heating, ventilating, and air-conditioning (HVAC) systems. Additionally, this chapter includes an overview of the environmental maintenance and surveillance of water systems, including cooling towers and evaporative condensers, in view of the now significantly increased awareness of the risks of infection that may result from systems that are poorly designed, dirty, and/or badly maintained.

It should be noted that the risks of infection (from bacterial organisms such as *Legionella sp.*) are not limited to cooling systems, but can emanate from almost any type of building facilities water systems, including domestic hot and cold water services, pools, spa baths, showers, and humidifier pans. (Similarly, it is recognized that risks of infection may also develop in all types of air handling and ventilation systems that are likewise poorly maintained or not cleaned on a regular basis. In air handling systems, infection and related problems typically involve mold organisms such as *Aspergillus sp.* or *Stachybotrus sp.*, the same organisms that can proliferate behind damp sheetrock walls and insulation.) *However, it is beyond the scope of this chapter to discuss in any great detail the environmental maintenance, surveillance, and infection control of facilities water and air systems. Additionally, it is also beyond the scope of this chapter to cover all aspects of industrial and process water treatment, since specific designs for each type of process are generally required. For fuller coverage of industrial water treatment, see the references and bibliography at the end of the chapter.*

8.2 WHY WATER TREATMENT?

Water treatment for corrosion and deposit control is a specialized technology. Essentially, it can be understood when one first recognizes why treatment is necessary to prevent serious failures and malfunction of equipment that uses water as a heat-transfer medium. This is seen more easily when one observes the problems water can cause, the mechanism by which water causes these problems—which leads to solutions—and the actual solutions or cures available.

8.2.1 Water Is a Universal Solvent

Whenever it comes into contact with a foreign substance, there is some dissolution of that substance. Some substances dissolve at faster rates than others, but in all cases a definite interaction occurs between water and whatever it contacts. It is because of this interaction that problems occur in equipment such as boilers, or cooling-water systems in which water is used as a heat-transfer medium. In systems open to the atmosphere, corrosion problems are made worse by additional impurities picked up by the water from the atmosphere.

Most people have seen the most obvious examples of corrosion of metals in contact with water and its devastating effect. Corrosion alone is the cause of failure and costly replacement of equipment and is itself a good reason why water treatment is necessary.

8.2.2 Cost of Corrosion

The direct losses due to corrosion of metals for replacement and protection are reported to be at least $20 to $25 billion annually in the U.S. alone; over $8 billion is spent for corrosion-resistant metallic and plastic equipment, almost $5 billion for protective coatings, and over $1 billion for corrosion inhibitors (year 2000 dollars). All this is just to minimize the losses due to corrosion! Typical examples of these losses resulting from failures of piping, boiler equipment, and heat-exchanger materials because of corrosion and deposits are depicted in this chapter. Only with good design and correct application of water treatment programs, including corrosion inhibitors, will HVAC equipment, such as heating boilers and air-conditioning chillers and condensers, provide maximum economical service life. However, even more costly than failures and replacement costs, and less obvious, is the more insidious loss in energy and operating efficiency due to corrosion, fouling, and deposits.

In heat-transfer equipment, corrosion and deposits will interfere with the normal efficient transfer of heat energy from one side to the other. The degree of interference with this transfer of heat in a heat exchanger is called the *fouling factor*. In the condenser of an air-conditioning machine, a high fouling factor causes an increase in condensing temperature of the refrigerant gas and thus an increase in energy requirements to compress the refrigerant at that higher temperature. The manufacturer's recommended design fouling factor for air-conditioning chillers and condensers is 0.0005. This means that the equipment cannot tolerate deposits with a fouling factor greater than 0.0005 without the efficiency of the machine being seriously reduced.

Figure 8.1 graphically illustrates the effect of scale on the condensing temperature of a typical water-cooled condenser. From this graph, we see that the condensing temperature increases in proportion to the fouling factor. An increase in condensing temperature requires a proportionate increase in energy or compressor horsepower to compress the refrigerant gas. Thus the fouling factor affects the compressor horsepower and energy consumption, as shown in Figure 8.2. Condenser tubes are quickly fouled by a hard water supply, which deposits calcium carbonate on the heat-transfer surface. The explanation of the mechanism of this type of fouling is given in a later section.

Table 8.1 lists the fouling factors of various thicknesses of a calcium carbonate type of scale deposit most frequently found on condenser watertube surfaces where no water treatment or incorrect treatment is applied. The additional energy consumption required to compensate for a calcium carbonate type of scale on condenser tube surfaces of a refrigeration machine is illustrated in Fig. 8.3. The graph shows that a scale thickness of only 0.025 in (0.635 mm) [fouling factor of 0.002] will result in a 22-percent increase in energy consumption, which is indeed wasteful.

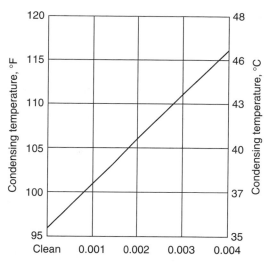

FIGURE 8.1 Effect of scale on condensing temperature. (*From Carrier System Design Manual, part 5, "Water Conditioning," Carrier Corporation, Syracuse, NY, 1972, p. 5-2. Used with permission*).

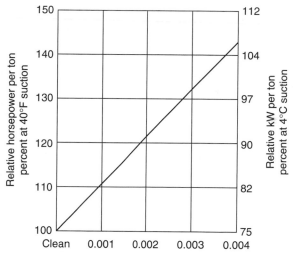

FIGURE 8.2 Effect of scale on compressor horsepower. (*From Carrier System Design Manual, part 5, "Water conditioning," Carrier Corporation, Syracuse, NY, 1972, p. 5-2. Used with permission*).

8.2.3 Cost of Scale and Deposits

The actual cost of scale is even more surprising. For example, a 500-ton air-conditioning plant operating with a scale deposit of 0.025 in (0.635 mm) of a calcium carbonate type will increase energy requirements by 22 percent if the same refrigeration load is maintained and cost approx. $4752 in additional energy consumption required for only 1 month (720 h)

TABLE 8.1   Fouling Factor of Calcium Carbonate Type of Scale

Approximate thickness of calcium carbonate type of scale, in (mm)	Fouling factor
0.000	Clean
0.006 (0.1524)	0.0005
0.012 (0.3048)	0.0010
0.024 (0.6096)	0.0020
0.036 (0.9144)	0.0030

Source: *Carrier System Design Manual,* part 5, "Water Conditioning," Carrier Corp., Syracuse, NY, 1972, p. 5-3. Used with permission.

of operation. This is based on an efficient electric-drive air-conditioning machine requiring 0.75 kW/(h ton) of refrigeration for compressor operation. Note that the average cost for this energy in early 2000 was 8.0 cents/kWh, thus $500 \times 720 \times 0.22 \times \$0.08 \times 0.75 = \$4752$.

With proper care and attention to water treatment, wasteful use of energy can be avoided. Likewise, in a boiler operation for heating or other purposes, an insulating scale deposit on the heat-transfer surfaces can substantially increase energy requirements.

Boiler scale or deposits can consist of various substances including iron, silica, calcium, magnesium, carbonates, sulfate, and phosphates. Each of these, when deposited on a boiler tube, contributes in some degree to the insulation of the tube. That is, the deposits reduce the rate of heat transfer from the hot gases or fire through the boiler metal to the boiling water. When this occurs, the temperature of the boiler tube metal increases. The scale coating offers

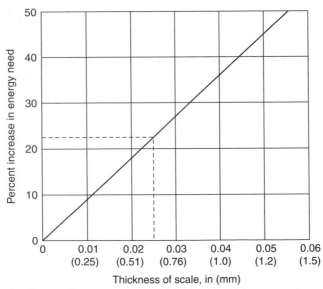

FIGURE 8.3   Effect of condenser tube scale on energy consumption $K = 1.0$ Btu/(h ft^3-°F). *Example.* Scale that is 0.025 in (0.6 mm) thick requires 22 percent increase in energy.

TABLE 8.2   Boiler Scaled Thickness vs. Energy Loss

Normal scale, calcium carbonate type, in (mm)	Dense scale (iron silica type)	Energy loss, %
$^1/_{32}$ (0.794)	$^1/_{64}$ (0.397)	2
$^1/_{16}$ (1.588)	$^1/_{32}$ (0.794)	4
$^3/_{32}$ (2.381)	$^3/_{64}$ (1.191)	6
$^1/_8$ (3.175)	$^1/_{16}$ (1.588)	8
$^3/_{16}$ (4.763)	$^3/_{32}$ (2.381)	12
$^1/_4$ (6.350)	$^1/_8$ (3.175)	16

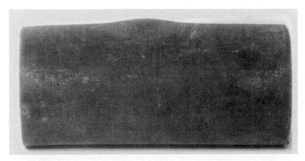

FIGURE 8.4   Boiler tube blister. (*Courtesy of The Metro Group, Inc.*)

a resistance to the rate of heat transfer from the furnace gases to the boiler water. This heat resistance results in a rapid rise in metal temperature to the point at which the metal bulges and eventual failure results. This is the most serious effect of boiler deposits, since failure of such tubes causes boiler explosions.

Figure 8.4 shows a boiler tube blister caused by a scale deposit. Table 8.2 shows the average loss of energy as a result of boiler scale. A normal scale of only 1/16-in (1.588-mm) thickness can cause an energy loss of 4 percent. For example, a loss of 4 percent in energy as a result of a scale deposit can mean that 864 gal (3270.6 L) more of No. 6 fuel oil than is normally used would be required for the operation of a steam boiler at 100 boiler hp (bhp) (1564.9 kg) for 1 month (720 h). The costs of additional fuel due to efficiency losses can easily run into many thousands of dollars each year.

8.3   WATER CHEMISTRY

Water and its impurities are responsible for the corrosion of metals and formation of deposits on heat-transfer surfaces, which in turn reduce efficiency and waste energy. Having seen the effects of corrosion and deposits, let us see how this can be prevented. The path to their prevention can best be approached through understanding their basic causes, why and how they occur.

Water, the common ingredient present in heat-transfer equipment such as boilers, cooling towers, and heat exchangers, contains many impurities. These impurities render the water supply more or less corrosive and/or scale forming.

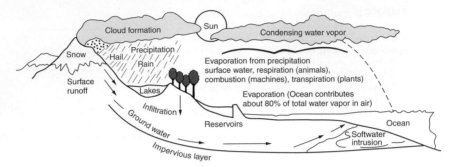

FIGURE 8.5 Hydrologic cycle.

8.3.1 Hydrologic Cycle

The hydrologic cycle (Fig. 8.5) consists of three stages: evaporation, condensation, and precipitation. This cycle begins when the rays of the sun heat surface waters on the earth, which vaporize and rise into the troposphere, a thin layer of air and moisture approximately 7 mi (11 km) thick surrounding the earth. Clouds of condensed moisture form in the troposphere and when carried over land by the wind, they contact cold-air currents. This causes precipitation or rain or snow. In this manner, water returns to the earth's surface, only to repeat the cycle. Throughout the hydrologic cycle, the water absorbs impurities. While falling through the atmosphere, water dissolves the gases, oxygen, nitrogen, carbon dioxide, nitrogen oxides, sulfur oxides, and many other oxides present in the atmosphere in trace amounts.

The quantity of these gases in the atmosphere depends on the location. For example, in large urban areas rainwater often contains high concentrations of carbon dioxide, sulfur oxides, and nitrogen oxides. In rural areas, water contains lesser amounts of these gases. Studies have noted that in the recent past the acidity of our rainfall has steadily increased. This is caused by the increased amounts of sulfur and nitrogen oxide gases that pollute the atmosphere (acid rain).

8.3.2 Water Impurities

In contact with the earth surface, rainwater will tend to dissolve and absorb many of the minerals of the earth. The more acidic the rainfall, the greater the reaction with the earth's minerals. This reaction includes hydrolysis and hydration. As water passes over and through gypsum, calcite, dolomite, and quartz rock, it will dissolve calcium, silica, and magnesium minerals from these rocks (Table 8.3). In similar manner, other minerals present in the earth's crust can be dissolved and taken up by the water. Table 8.4 shows some

TABLE 8.3 Reactions of Water with Minerals

Hydrolysis is the chemical reaction between water and minerals in which the mineral dissolves in the water:

| $NaCl$ | + | $H_2O \rightarrow Na^+$ | + | Cl^- | + | H_2O |
| Sodium chloride | + | Water = Sodium ion in solution | + | Chloride ion in solution | + | Water |

Hydration is the absorption of water by minerals, changing the nature of the mineral:

| $CaSO_4$ | + | $2H_2O \rightarrow CaSO_4 + 2H_2O$ |
| Calcium sulfate | + | Water = Calcium sulfate hydrate |

TABLE 8.4 Mineral Groups

Silicates	Quartz, augite, mica, chert, feldspar, hornblend
Carponates	Calcite, dolomite, limestone
Halides	Halite, fluorite
Oxides	Hematite, ice, magnetite, bauxite
Sulfates	Anhydrite, gypsum
Sulfides	Galena, pyrite
Natural elements	Copper, sulfur, gold, silver
Phosphates	Apatite

of the minerals present in the earth's surface, which by reaction with water dissolve and become impurities in water. Water accumulates on the earth's surface in lakes, rivers, streams, and ponds and can be collected in reservoirs. These surface water supplies usually contain fewer minerals but are more likely to contain dissolved gases.

Underground water supplies are a result of surface waters percolating through the soil and rock. The water supplies usually contain large quantities of minerals and not much dissolved gases, although there are numerous exceptions to this general rule. Table 8.5 lists the various sources of water. Figures 8.6 through 8.10 show typical analyses of surface waters and underground well waters. A brief observation of the analyses of these different water supplies shows that the natural impurities and mineral content do indeed vary with location. In fact, many well water supplies in a very proximate location exhibit vast differences in mineral content. Let us examine each of the basic impurities of water to see how they contribute to corrosion and deposits.

8.3.3 Dissolved Gases

Oxygen.   One of the gases in the atmosphere is oxygen, which makes up approximately 20 percent of air. Oxygen in water is essential for aquatic life; however, it is the basic factor in the corrosion process and is, in fact, one of the essential elements in the corrosion process of metals. Therefore, dissolved oxygen in water is important to us in the study of corrosion and deposits.

Carbon Dioxide.   Carbon dioxide is present in both surface and underground water supplies. These water supplies absorb small quantities of carbon dioxide from the atmosphere. Larger amounts of carbon dioxide are absorbed from the decay of organic matter in the water and its environs. Carbon dioxide contributes significantly to corrosion by making

TABLE 8.5 Sources of Water

Surface water	Lakes and reservoirs of fresh water
Groundwater	Water below the land surface caused by surface runoff drainage and seepage
Water table	Water found in rock saturated with water just above the impervious layer of the earth
Wells	Water-bearing strata of the earth—water secps and drains through the soil surface, dissolving and absorbing minerals of which the earth is composed (thus the higher mineral content of well water)

METROPOLITAN REFINING COMPANY INC.
50-23 23rd STREET • LONG ISLAND CITY • N.Y. 11101

CERTIFICATE OF ANALYSIS

NEW YORK CITY

DATE_____
SAMPLING DATE_____
REPRESENTATIVE_____

ANALYSIS NUMBER			339568			
SOURCE			CITY			
pH			6.9			
P ALKALINITY	$CaCO_3$,	mg/L				
FREE CARBON DIOXIDE	CO_2,	mg/L				
BICARBONATES	$CaCO_3$,	mg/L	12.			
CARBONATES	$CaCO_3$,	mg/L				
HYDROXIDES	$CaCO_3$,	mg/L				
M (Total) ALKALINITY	$CaCO_3$,	mg/L	12.			
TOTAL HARDNESS	$CaCO_3$,	mg/L	16.			
SULFATE	SO_4,	mg/L				
SILICA,	SiO_2,	mg/L	1.5			
IRON	Fe,	mg/L	Trace			
CHLORIDE	NaCl,	mg/L	13.			
ORGANIC INHIBITOR		mg/L				
PHOSPHATE	PO_4,	mg/L				
CHROMATE	Na_2CrO_4,	mg/L				
NITRITE	$NaNO_2$,	mg/L				
ZINC	Zn,	mg/L				
SPECIFIC CONDUCTANCE		mmhos/cm				
TOTAL DISSOLVED SOLIDS		mg/L	33.5			
SUSPENDED MATTER						
BIOLOGICAL GROWTHS						
SPECIFIC GRAVITY @ 15.5º/15.5ºC						
FREEZING POINT ºC/ºF						
	% BY WEIGHT					

NOTES:

1. ANALYTICAL RESULTS EXPRESSED IN MILLIGRAMS PER LITRE (mg/L) ARE EQUIVALENT TO PARTS PER MILLION (ppm). DIVIDE BY 17.1 TO OBTAIN GRAINS PER GALLON (gpg).

2. CYCLES OF CONCENTRATION = $\dfrac{\text{CHLORIDES IN SAMPLE}}{\text{CHLORIDES IN MAKEUP}}$

TREATMENT	TREATMENT CONTROL	FOUND	RECOMMENDED	FOUND	RECOMMENDED

REMARKS:

RTB:MW

SPEEDIPLY® PAT D MCP® PAT D MBF 28

FORM 1078-1

FIGURE 8.6 New York City (Croton Reservoir) water analysis. *(Courtesy of The Metro Group, Inc.*

water acidic. This increases its capability to dissolve metals. Carbon dioxide forms mild carbonic acid when dissolved in water, as follows:

$$CO_2 \quad + \quad H_2O \quad \rightarrow \quad H_2CO_3$$

carbon dioxide	water	carbonic acid

Sulfur Oxides.   Sulfur oxide gases are present in the atmosphere as a result of sulfur oxides absorbed from the atmosphere, in which they are present as pollutants from the combustion

METROPOLITAN REFINING COMPANY INC.
50-23 23rd STREET • LONG ISLAND CITY • N.Y. 11101

CERTIFICATE OF ANALYSIS

SYRACUSE, N.Y. (OTISCO LAKE)

DATE_____
SAMPLING DATE_____
REPRESENTATIVE_____

ANALYSIS NUMBER			57627			
SOURCE			CITY			
pH			7.4			
P ALKALINITY	CaCO₃,	mg/L	0.0			
FREE CARBON DIOXIDE	CO₂,	mg/L				
BICARBONATES	CaCO₃,	mg/L	85.			
CARBONATES	CaCO₃,	mg/L				
HYDROXIDES	CaCO₃,	mg/L				
M (Total) ALKALINITY	CaCO₃,	mg/L	85.			
TOTAL HARDNESS	CaCO₃,	mg/L	132.			
SULFATE	SO₄,	mg/L				
SILICA,	SiO₃,	mg/L	1.0			
IRON	Fe,	mg/L	0.0			
CHLORIDE	NaCl,	mg/L	21.			
ORGANIC INHIBITOR		mg/L				
PHOSPHATE	PO₄,	mg/L				
CHROMATE	Na₂CrO₄,	mg/L				
NITRITE	NaNO₂,	mg/L				
ZINC	Zn,	mg/L				
SPECIFIC CONDUCTANCE		mmhos/cm	245.			
TOTAL DISSOLVED SOLIDS		mg/L	148.			
SUSPENDED MATTER			Trace			
BIOLOGICAL GROWTHS			Trace			
SPECIFIC GRAVITY @ 15.5°/15.5°C						
FREEZING POINT °C/°F						
		% BY WEIGHT				

NOTES:

1. ANALYTICAL RESULTS EXPRESSED IN MILLIGRAMS PER LITRE (mg/L) ARE EQUIVALENT TO PARTS PER MILLION (ppm). DIVIDE BY 17.1 TO OBTAIN GRAINS PER GALLON (gpg).

2. CYCLES OF CONCENTRATION = $\dfrac{\text{CHLORIDES IN SAMPLE}}{\text{CHLORIDES IN MAKEUP}}$

TREATMENT	TREATMENT CONTROL	FOUND	RECOMMENDED	FOUND	RECOMMENDED

REMARKS:

RTB:MW

SPEEDIPLY® PAT'D MCP® PAT D MBF 28

FORM 1078-I

FIGURE 8.7 Water analysis of Syracuse. NY (Otisco Lake). (*Courtesy of The Metro Group, Inc.*)

of fuels containing sulfur, such as coal and fuel oil. In large urban areas, the quantity of sulfur oxides that are absorbed by surface water supplies and aerated waters used in cooling towers can be significant. Also, when dissolved in water, sulfur oxides form acids which create a corrosive atmosphere.

$$SO_3 \quad + \quad H_2O \quad \rightarrow \quad H_2SO_4$$

sulfur   water   sulfuric
trioxide   acid

METROPOLITAN REFINING COMPANY INC.
50-23 23rd STREET • LONG ISLAND CITY • N.Y. 11101

CERTIFICATE OF ANALYSIS

WASHINGTON, D.C. (POTOMAC RIVER)

DATE_____
SAMPLING DATE_____
REPRESENTATIVE_____

ANALYSIS NUMBER			20197			
SOURCE			CITY			
pH			7.7			
P ALKALINITY	CaCO₃,	mg/L				
FREE CARBON DIOXIDE	CO₂,	mg/L				
BICARBONATES	CaCO₃,	mg/L	90.			
CARBONATES	CaCO₃,	mg/L				
HYDROXIDES	CaCO₃,	mg/L				
M (Total) ALKALINITY	CaCO₃,	mg/L	90.			
TOTAL HARDNESS	CaCO₃,	mg/L	140.			
SULFATE	SO₄,	mg/L				
SILICA;	SiO₂,	mg/L	7.0			
IRON	Fe,	mg/L	0.0			
CHLORIDE	NaCl,	mg/L	41.			
ORGANIC INHIBITOR		mg/L				
PHOSPHATE	PO₄,	mg/L				
CHROMATE	Na₂CrO₄,	mg/L				
NITRITE	NaNO₂,	mg/L				
ZINC	Zn,	mg/L				
SPECIFIC CONDUCTANCE		mmhos/cm				
TOTAL DISSOLVED SOLIDS		mg/L	195.			
SUSPENDED MATTER						
BIOLOGICAL GROWTHS						
SPECIFIC GRAVITY @ 15.5°/15.5°C						
FREEZING POINT °C/°F						
	% BY WEIGHT					

NOTES:
1. ANALYTICAL RESULTS EXPRESSED IN MILLIGRAMS PER LITRE (mg/L) ARE EQUIVALENT TO PARTS PER MILLION (ppm). DIVIDE BY 17.1 TO OBTAIN GRAINS PER GALLON (gpg).
2. CYCLES OF CONCENTRATION = CHLORIDES IN SAMPLE / CHLORIDES IN MAKEUP

TREATMENT	TREATMENT CONTROL	FOUND	RECOMMENDED	FOUND	RECOMMENDED

REMARKS:

RTB:MW

SPEEDIPLY® PAT'D MCP® PAT'D MBF 28   FORM 1078-1

FIGURE 8.8   Potomac River (Washington, DC) water analysis. (*Courtesy of The Metro Group, Inc.*)

Nitrogen Oxides.   Nitrogen oxides are also present in the atmosphere both naturally and from pollutants created by the combustion process. These too form acids when absorbed by water, and contribute to the corrosion process.

$$3NO_2 \ + H_2O \ \rightarrow \ 2HNO_3 \ + \ NO$$

nitrogen   water   nitric   nitric
dioxide   acid   oxide

METROPOLITAN REFINING COMPANY INC.
50-23 23rd STREET • LONG ISLAND CITY • N.Y. 11101

CERTIFICATE OF ANALYSIS

EFFICIENCY THROUGH CHEMISTRY — METROPOLITAN EFFICIENCY PRODUCTS ®

JAMAICA, N. Y. (WELLS)

DATE
SAMPLING DATE
REPRESENTATIVE

ANALYSIS NUMBER		38140		
SOURCE		CITY WATER		
pH		7.0		
P ALKALINITY	$CaCO_3$, mg/L	0.0		
FREE CARBON DIOXIDE	CO_2, mg/L			
BICARBONATES	$CaCO_3$, mg/L	30.		
CARBONATES	$CaCO_3$, mg/L			
HYDROXIDES	$CaCO_3$, mg/L			
M (Total) ALKALINITY	$CaCO_3$, mg/L	30.		
TOTAL HARDNESS	$CaCO_3$, mg/L	60.		
SULFATE	SO_4, mg/L			
SILICA,	SiO_2, mg/L	14.3		
IRON	Fe, mg/L	0.07		
CHLORIDE	NaCl, mg/L	29.		
ORGANIC INHIBITOR	mg/L			
PHOSPHATE	PO_4, mg/L			
CHROMATE	Na_2CrO_4, mg/L			
NITRITE	$NaNO_2$, mg/L			
ZINC	Zn, mg/L			
SPECIFIC CONDUCTANCE	mmhos/cm	154.		
TOTAL DISSOLVED SOLIDS	mg/L	106.		
SUSPENDED MATTER				
BIOLOGICAL GROWTHS				
SPECIFIC GRAVITY @ 15.5°/15.5°C				
FREEZING POINT °C/°F				
	% BY WEIGHT			

NOTES: 1. ANALYTICAL RESULTS EXPRESSED IN MILLIGRAMS PER LITRE (mg/L) ARE EQUIVALENT TO PARTS PER MILLION (ppm). DIVIDE BY 17.1 TO OBTAIN GRAINS PER GALLON (gpg).
2. CYCLES OF CONCENTRATION = CHLORIDES IN SAMPLE / CHLORIDES IN MAKEUP

TREATMENT	TREATMENT CONTROL	FOUND	RECOMMENDED	FOUND	RECOMMENDED

REMARKS:

RTB:MW

SPEEDIPLY® PAT'D MCP® PAT'D MBF 28

FORM 1078-I

FIGURE 8.9 Water analysis of Jamaica, NY (wells). (*Courtesy of The Metro Group, Inc.*)

Hydrogen Sulfide. The odor typical of rotten eggs which is found in some water is due to the presence of hydrogen sulfide. This gas comes from decaying organic matter and from sulfur deposits. Hydrogen sulfide forms when acidic water reacts with sulfide minerals such as pyrite, an iron sulfide commonly called "fool's gold":

FeS	+	$2H^+$	+	→	Fe^{2+}	+ H_2S
ferric sulfide		acid in solution +			iron in solution	hydrogen sulfide

METROPOLITAN REFINING COMPANY INC.
50-23 23rd STREET • LONG ISLAND CITY • N.Y. 11101

CERTIFICATE OF ANALYSIS

YELLOW SPRINGS, OHIO (WELLS)

DATE_____
SAMPLING DATE_____
REPRESENTATIVE_____

ANALYSIS NUMBER		47588		
SOURCE		CITY WATER		
pH		7.6		
P ALKALINITY	$CaCO_3$, mg/L	0.0		
FREE CARBON DIOXIDE	CO_2, mg/L			
BICARBONATES	$CaCO_3$, mg/L	300.		
CARBONATES	$CaCO_3$, mg/L			
HYDROXIDES	$CaCO_3$, mg/L			
M (Total) ALKALINITY	$CaCO_3$, mg/L	300.		
TOTAL HARDNESS	$CaCO_3$, mg/L	454.		
SULFATE	SO_4, mg/L			
SILICA,	SiO_2, mg/L	9.5		
IRON	Fe, mg/L	0.0		
CHLORIDE	NaCl, mg/L	58.		
ORGANIC INHIBITOR	mg/L			
PHOSPHATE	PO_4, mg/L			
CHROMATE	Na_2CrO_4, mg/L			
NITRITE	$NaNO_2$, mg/L			
ZINC	Zn, mg/L			
SPECIFIC CONDUCTANCE	mmhos/cm	840.		
TOTAL DISSOLVED SOLIDS	mg/L	514.		
SUSPENDED MATTER		Abs.		
BIOLOGICAL GROWTHS		Abs.		
SPECIFIC GRAVITY @ 15.5°/15.5°C				
FREEZING POINT °C/°F				
	% BY WEIGHT			

NOTES:
1. ANALYTICAL RESULTS EXPRESSED IN MILLIGRAMS PER LITRE (mg/L) ARE EQUIVALENT TO PARTS PER MILLION (ppm). DIVIDE BY 17.1 TO OBTAIN GRAINS PER GALLON (gpg).

2. CYCLES OF CONCENTRATION = $\dfrac{\text{CHLORIDES IN SAMPLE}}{\text{CHLORIDES IN MAKEUP}}$

TREATMENT	TREATMENT CONTROL	FOUND	RECOMMENDED	FOUND	RECOMMENDED

REMARKS:

RTB:MW

SPEEDIPLY• PAT D MCP• PAT D MBF 28   FORM 1078-1

FIGURE 8.10   Water analysis of Yellow Springs, OH (wells). (*Courtesy of The Metropolitan Refining Co., Inc.*)

Hydrogen sulfide reacts with water to form hydrosulfuric acid, a slightly acidic solution. Its presence in water is also due to the decomposition of organic matter and protein, which contain sulfur. Hydrogen sulfide is also a constituent of sewer gas, marsh gas, and coal gas. It can be present in water and also comes from these sources. Because of its acidic reaction in water, hydrogen sulfide is very corrosive and must be removed or neutralized.

8.3.4 Dissolved Minerals

Alkalinity.   Alkalinity is the quantity of dissolved alkaline earth minerals expressed as calcium carbonate. It is the measured carbonate and bicarbonate minerals calculated as calcium carbonate since that is the primary alkaline earth mineral contributing to alkalinity. Alkalinity is also measured and calculated as the hydroxide when that is present. All natural waters contain some quantity of alkalinity. It contributes to scale formation because its presence encourages deposition of calcium carbonate, or lime scale.

pH Value.   The *quality* of alkalinity, or the measure of the relative strength of acidity or alkalinity of a water, is the pH value, a value calculated from the hydrogen-ion concentration in water. The pH scale ranges from 0 to 14. A pH of 7.0 is neutral. It indicates a balance between the acidity and alkalinity. As the pH decreases to zero, the alkalinity decreases and the acidity increases. As the pH increases to 14, the alkalinity increases and the acidity decreases.

The pH scale (Fig. 8.11) is used to express the strength or intensity of the acidity or alkalinity of a water solution. This scale is logarithmic so that a pH change of one unit represents a tenfold increase or decrease in the strength of acidity or alkalinity. Hence water with a pH value of 4.0 is 100 times more acid in strength than water with a pH value of 6.0. Water is corrosive if the pH value is on the acidic side. It will tend to be scale-forming if the pH value is alkaline.

Hardness.   Hardness is the total calcium, magnesium, iron, and trace amounts of other metallic elements in water, which contributes to the "hard" feel of water. Hardness is also calculated as calcium carbonate, because it is the primary component contributing to hardness. Hardness causes lime deposits or scale in equipment.

Silica.   Silica is dissolved sand or silica-bearing rock (such as quartz) through which the water flows. Silica is the cause of very hard and tenacious scales that can form in heat-transfer equipment. It is present dissolved in water as silicate or suspended in very fine, invisible form as colloidal silica.

Iron, Manganese, and Alumina.   These dissolved or suspended metallic elements are present in water supplies in varying quantities. They are objectionable because they contribute to a flat metallic taste and form deposits. These soluble metals, when they react with oxygen in water exposed to the atmosphere, form oxides that precipitate and cause cloudiness, or "red water." This red color, particularly from iron, causes staining of plumbing fixtures, sinks, and porcelain china and is a cause of common laundry discoloration.

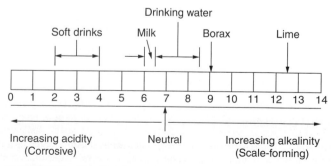

FIGURE 8.11   The pH scale.

Chlorides. Chlorides are the sum total of the dissolved chloride salts of sodium, potassium, calcium, and magnesium present in water. Sodium chloride, which is common salt, and calcium chloride are the most common of the chloride minerals found in water. Chlorides do not ordinarily contribute to scale since they are very soluble. Chlorides are corrosive, however, and cause excessive corrosion when present in large volume, as in seawater.

Sulfates. Sulfates are the dissolved sulfate salts of sodium, potassium, calcium, and magnesium in the water. They are present due to dissolution of sulfate-bearing rock such as gypsum. Calcium and magnesium sulfate scale is very hard, difficult to remove, and greatly interferes with heat transfer.

Total Dissolved Solids. The total dissolved solids (TDS) reported in water analyses are the sum of dissolved minerals including the carbonates, chlorides, sulfates, and all others that are present. The dissolved solids contribute to both scale formation and corrosion in heat-transfer equipment.

Suspended Matter. Suspended matter is finely divided organic and inorganic substances found in water. It is caused by clay silt and microscopic organisms, which are dispersed throughout the water, giving it a cloudy appearance. The measure of suspended matter is turbidity. Turbidity is determined by the intensity of light scattered by the suspended matter in the water.

8.4 CORROSION

Corrosion is the process whereby a metal, through reaction with its environment, undergoes a change from the pure metal to its corresponding oxide or other stable combination. Usually, through corrosion, the metal reverts to its naturally occurring state, the ore. For example, iron is gradually dissolved by water and oxidized by oxygen in the water, forming the oxidation product iron oxide, commonly called rust. This process occurs very rapidly in heat-transfer equipment because of the presence of heat, corrosive gases and dissolved minerals in the water, which stimulate the corrosion process.

The most common forms of corrosion found in heat-transfer equipment are as follows:

- General corrosion
- Oxygen pitting
- Galvanic corrosion
- Concentration cell corrosion
- Stress corrosion
- Erosion-corrosion
- Condensate grooving
- Microbiologically influenced corrosion (MIC)

8.4.1 General Corrosion

General corrosion is found in various forms in heat-transfer equipment. In a condenser-water or cooling-tower water circuit, it can be seen as an overall deterioration of the metal surface, with an accumulation of rust and corrosion products in the piping and water boxes.

On copper condenser tubes, it is observed most frequently as a surface gouging or a uniform thinning of the tube metal.

In boilers, general corrosion is observed in the total overall disintegration of the tube metal surface in contact with the boiler water. (See Figs. 8.12 and 8.13.)

General corrosion occurs when the process takes place over the entire surface of the metal, resulting in a uniform loss of metal rather than a localized type of attack. It is often, but not always, accompanied by an accumulation of corrosion products over the surface of the metal (Fig. 8.14).

FIGURE 8.12   General corrosion on condenser tube. (*Courtesy of The Metro Group, Inc.*)

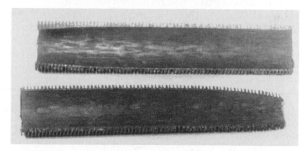

FIGURE 8.13   Pitting corrosion on condenser tubes. (*Courtesy of The Metro Group, Inc.*)

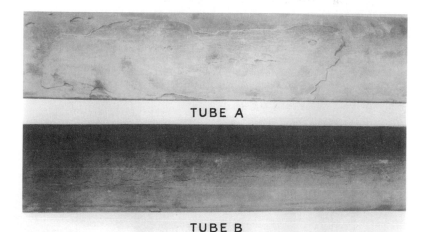

FIGURE 8.14   Boiler tube corrosion. (*Courtesy of Babcock & Wilcox Co.*)

Iron and other metals are corroded by electrochemical reaction resulting in the metal going into solution in the water. It is necessary, therefore, to limit corrosion of these metals by reducing the activity of both hydroxyl ions and hydrogen ions, i.e., by maintaining a neutral environment. Another important factor in the corrosion process is dissolved oxygen. The evolution of hydrogen gas in these reactions tends to slow the rate of the corrosion reaction and indeed, in many instances, to stop it altogether by forming an inhibiting film on the surface of the metal which physically protects the metal from the water.

Accumulation of rust and corrosion products is further promoted by the presence of dissolved oxygen. Oxygen reacts with the dissolved metal, eventually forming the oxide, which is insoluble and in the case of iron builds up a voluminous deposit of rust. Since the role of dissolved oxygen in the corrosion process is important, removal of dissolved oxygen is an effective procedure in preventing corrosion.

8.4.2 Oxygen Pitting

The second type of corrosion frequently encountered in heat-transfer equipment is pitting. Pitting is characterized by deep penetration of the metal at a small area on the surface with no apparent attack over the entire surface as in general corrosion. The corrosion takes place at a particular location on the surface, and corrosion products frequently accumulate over the pit. These appear as a blister, tubercle, or carbuncle, as in Fig. 8.15.

Oxygen pitting is caused by dissolved oxygen. It differs from localized pitting due to other causes, such as deposits of foreign matter. Following are examples of pitting caused by dissolved oxygen (Figs. 8.16 and 8.17). Oxygen pitting occurs in steam boiler systems where the feedwater contains dissolved oxygen. The pitting is found on boiler tubes adjacent

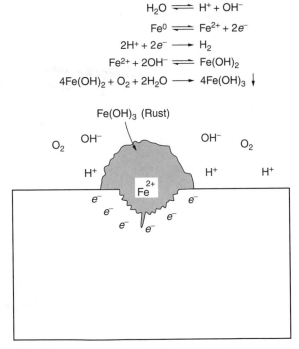

$$H_2O \rightleftharpoons H^+ + OH^-$$

$$Fe^0 \rightleftharpoons Fe^{2+} + 2e^-$$

$$2H^+ + 2e^- \longrightarrow H_2$$

$$Fe^{2+} + 2OH^- \rightleftharpoons Fe(OH)_2$$

$$4Fe(OH)_2 + O_2 + 2H_2O \longrightarrow 4Fe(OH)_3 \downarrow$$

FIGURE 8.15 Reactions forming blisters over pit.

FIGURE 8.16   Pitting on boiler tube. (*Courtesy of The Metro Group, Inc.*)

FIGURE 8.17   Blister over pits on boiler tubes. (*Courtesy of Babcock & Wilcox Co.*)

FIGURE 8.18 Pitting in boiler feedwater.
(*Courtesy of The Metro Group, Inc.*)

to the feedwater entrance, throughout the boiler, or in the boiler feedwater line itself. One of the most unexpected forms of oxygen pitting is commonly found in boiler feedwater lines following a deaerator. It is mistakenly believed that mechanically deaerated boiler feedwater will completely prevent oxygen pitting. However, quite to the contrary, water with a low concentration of dissolved oxygen frequently is more corrosive than that with a higher dissolved oxygen content. This is demonstrated by the occurrence of oxygen pitting in boiler feedwater lines carrying deaerated water.

Mechanical deaerators are not perfect, and none can produce a feedwater with zero oxygen. The lowest guaranteed dissolved oxygen content that deaerators produce is $0.0005 \text{ cm}^3/\text{L}$. This trace quantity of dissolved oxygen is sufficient to cause severe pitting in feedwater lines or in boiler tubes adjacent to the feedwater entrance. This form of pitting is characterized by deep holes scattered over the surface of the pipe interior with little or no accumulation of corrosion products or rust, since there is insufficient oxygen in the environment to form the ferric oxide rust. (See Fig. 8.18.)

8.4.3 Galvanic Corrosion

Corrosion can occur when different metals come in contact with one another in water. When this happens an electric current is generated similar to that of a storage battery. The more active metal will tend to dissolve in the water, thereby generating an electric current (an electron flow) from the less active metal. A coupling of iron and copper, as in Fig. 8.19, develops this current.

This tendency of a metal to give up electrons and go into solution is called the "electrode potential." This potential varies greatly among metals since the tendency of different metals to dissolve and react with the environment varies.

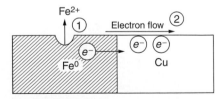

FIGURE 8.19 Galvanic corrosion caused by dis-similar-metal couple. (1) Iron going into solution loses two electronos: $Fe^0 \rightarrow Fe^{2+} + 2e^{-1}$. (2) electrons flow to copper, the less reactive metal.

In galvanic corrosion, commonly called *dissimilar-metal corrosion*, there are four essential elements:

1. A more reactive metal called the *anode*

2. A less reactive metal called the *cathode*

3. A water solution environment called the *electrolyte*

4. Contact between the two metals to facilitate electron flow

The rate of galvanic corrosion is strongly influenced by the electrode potential difference between the dissimilar metals. The galvanic series is a list of metals in order of their activity, the most active being at the top of the list and the least active at the bottom. *The further apart two metals are on this list, the greater will be the reactivity between them and, therefore, the faster the anodic end will corrode.* The galvanic series is shown in Fig. 8.20.

Corroded end (anodic, or least noble)

Magnesium, alloys (1)
Zinc (1)
Beryllium
Aluminum alloys (1)
Cadmium
Mild steel, wrought iron
Cast iron, flake or ductile
Low-alloy high-strength steel
Nickel-resist, types 1 & 2
Naval bronze (CA464), yellow bronze (CA268) aluminum bronze (CA687), Red bronze (CA230).
    Admiralty bronze (CA443) manganese bronze

Tin
Copper (CA102, 110), silicon bronze (CA655)
Lead-tin solder
Tin bronze (G & M)
Stainless steel, 12–14% chromium (AISI Types 410, 416)
Nickle silver (CA 732, 735, 745,752, 764, 770, 794)
90/10 Copper-nickel (CA 706)
80/20 Copper-nickel (CA 710)
Stainless steel, 16–18% chromium (AISI Type 430)

Lead
70/30 Copper-nickel (CA 715)
Nickel Aluminum bronze
Inconel* alloy 600
Silver braze alloys
Nickel 200

Silver
Stainless steel, 18 chromium, 8 nickle (AISI Types 302, 304, 321, 347)
Monel* Alloys 400, K-500
Stainless steel, 18 chromium, 12 nickel-molybdenum (AISI Types 316, 317)
Carpenter 20† stainless steel, incoloy* Alloy 825
Titanium, Hastelloy‡ alloys C & C276, Inconel* alloy 625
Graphite, graphitized cast iron

Protected end (cathodic, or most noble)

---

*International Nickle Trademark.
†Union Carbide Corp. Trademark.
‡The Carpenter Steel Co. Trademark.

FIGURE 8.20   Galvanic series.

If one or more of these four essential elements are eliminated, the corrosion reactions will be disrupted and the rate of corrosion slowed or halted altogether.

One method of preventing this type of corrosion is to eliminate contact of dissimilar metals in HVAC equipment by using insulating couplings or joints, such as a dielectric coupling which interferes with the electron flow from one metal to the other. Other forms of protection involve the removal of dissolved oxygen and use of protective coatings and inhibitors, which provide a barrier between the corroding metal and its environment.

8.4.4 Concentration Cell Corrosion

Concentration cell corrosion is a form of pitting corrosion that is a localized type of corrosion rather than a uniform attack. It is frequently called *deposit corrosion* or *crevice corrosion* since it occurs under deposits or at crevices of a metal joint.

Deposits of foreign matter, dirt, organic matter, corrosion products, scale, or any substance on a metal surface can initiate a corrosion reaction as a result of differences in the environment over the metal surface. Such differences may either be differences of solution ion concentration or dissolved oxygen concentration.

With concentration cell corrosion, the corrosion reaction proceeds as in galvanic corrosion since this differential also forms an electrode potential difference. Maintaining clean surfaces can best prevent this form of corrosion.

8.4.5 Stress Corrosion

Stress corrosion is a combination of exposure of a metal to a corrosive environment and application of stress on the metal. It is frequently seen on condenser tubes and boiler tubes in the area where the tubes are rolled into the tube sheets. In steam boilers, stress corrosion has been referred to as "necking and grooving." It is seen as a circumferential groove around the outside of a firetube where it enters the tube sheet. Figure 8.21 shows this type of corrosion.

FIGURE 8.21 Necking and grooving on boiler firetube. (*Courtesy of The Metro Group, Inc.*)

The corrosion failure is a result of a corrosive environment and stresses and strains at the point of failure. Usually it occurs at the hottest end of the tube at the beginning of the first pass against the firewall. It concentrates at the tube end because of strains from two sources. First, when tubes are rolled in, stresses are placed on the metal, expanding the metal to fit the tube sheet. Second, when a boiler is fired, the heat causes rapid expansion of the tube, and consequently strains are greatest at the tube ends, which are fixed in the tube sheets. The actually causes a flexing and bowing of the tube, and sometimes the expansion is so severe that the tubes loosen in the sheets. During this bending of the tube, the natural protective iron oxide film forming at the tube ends tends to tear or flake off, exposing fresh steel to further attack. Eventually, the tube fails due to both corrosion and stress.

Stress corrosion can also occur on condenser tubes and heat-exchanger tubes from heat expansion that causes stresses in the metal at tube supports or tube sheets. This problem is reduced by more gradual firing practices in boiler, which allow more gradual temperature changes, and by using proper inhibitors to correct the corrosive environment.

8.4.6 Erosion-Corrosion

Erosion-corrosion is the gradual wearing away of a metal surface by both corrosion and abrasion. It is also commonly called *impingement corrosion.*

Water moving rapidly through piping can contain entrained air bubbles and suspended matter, sand, or other hard particulates. This is not uncommon in cooling tower waters where such particles are washed from the atmosphere. These abrasive particles remove natural protective oxide films present on the surface of the metal and cause general corrosion of the exposed metal. The higher the velocity of the impinging stream, the greater the rate of erosion-corrosion.

8.4.7 Condensate Grooving

Condensate grooving is a particular phenomenon of steam condensate line corrosion in HVAC equipment. It is found in steam condensate piping on all types of equipment, heat exchangers, steam-turbine condensers, unit heaters, steam absorption condensers, radiators, or any type of unit utilizing steam as a heat-transfer medium.

Condensate grooving is a direct chemical attack by the steam condensate on the metal over which it flows and is identified by the typical grooves found at the bottom of the pipe carrying the condensate. This is shown in Fig. 8.22. *The primary cause of condensate grooving is carbon dioxide.* The dissolved carbon dioxide forms a mild carbonic acid. The methods available to prevent this type of corrosion include removal of bicarbonate and

FIGURE 8.22   Steam condensate line corrosion. (*Courtesy of The Metro Group, Inc.*)

carbonate alkalinity from the boiler makeup water (dealkalization) and use of carbonic acid neutralizers and filming inhibitors.

8.4.8 Microbiologically Influenced Corrosion

Since the early 1980s the phenomenon of microbiologically influenced corrosion (MIC) has become a very serious problem in building HVAC recirculating water systems. MIC is the term given to corrosion involving the reaction of microbiological species with metals. It is corrosion caused or influenced by microbiological organisms or organic growths on metals. There are many forms and mechanisms of MIC involving many types of microbiological organisms.

The basis causes of MIC found in recirculating water systems are as follows:

- Iron-related bacteria (IRB)
- Sulfate-reducing bacteria (SRB)
- Acid-producing bacteria (APB)
- Biological deposits (BD)

Iron-Related Bacteria.   A major group of organisms that are a direct cause of corrosion of iron and steel in recirculating water systems is the *iron-related bacteria*, such as *Gallionella ferrugine* and *Ferrobacillus sp*. This class of organisms is responsible for causing corrosion of iron and steel by direct metabolism of iron. Some of these organisms actually consume iron by using it in their metabolic process and then deposit it in the form of hydrated ferric hydroxide along with the mucous secretions. Iron bacteria are commonly found in all types of cooling system, especially in low flow areas.

Sulfate-Reducing Bacteria.   The best-known group of organisms involved in MIC is the sulfate-reducing bacteria. This group of organisms basically falls into three kinds—the *Desulfovibrio, Desulfotomaculum*, and *Desulfomonas* genera of organisms—all of which metabolize sulfur in one form or another. All are anaerobic (they live without directly metabolizing oxygen). The most widely known organism is the *Desulfovibrio* and is often found in cooling systems, especially where oil or sludge is present.

Acid-Producing Bacteria.   Another group of bacteria which cause MIC is the acid-producing bacteria. There are many types of APB, most of which are the slime forming bacteria such as *Pseudomonas, Aerobacter*, and *Bacillus* types, which exude various organic acids in their metabolic process. Organic acids, such as formic acid, acetic acid, and oxalic acid, have been identified in deposits of slime containing APB. These organic acids cause low pH conditions at local sites, resulting in corrosion at these sites.

One APB that is commonly responsible for MIC is the *Thiobacillus*. These organisms oxidize sulfur compounds forming sulfuric acid, which is extremely corrosive and leads to localized under-deposit and pitting corrosion, often resulting in pinholes in pipework.

Biological Deposits.   MIC can also be caused by other forms of organic growths such as algae, yeast, molds, and fungus, along with bacterial slimes. Even in the absence of specific corrosive organisms such as the IRD, SRB, or APB, biological deposits provide the environment for corrosion through establishment of concentration cells resulting in under-deposit corrosion. Biological deposits in general act as traps and food for other organisms resulting in rapid growth. This complex matrix sets up a corrosion potential between adjacent areas of a metal surface that may have a different type of deposit.

To control MIC it is important to understand the processes that cause it and thereby understand how to prevent it. It is clear that an essential control program will include control of all types of biological growths in recirculating water systems.

8.5 SCALE AND SLUDGE DEPOSITS

The most common and costly water-caused problem encountered in HVAC equipment is scale formation. The high cost of scale formation stems from the significant interference with heat transfer caused by water mineral scale deposits.

8.5.1 Mineral Scale and Pipe Scale

At this point, we should differentiate between mineral scale and pipe scale. *Mineral scale* is formed by supersaturation and deposition of the more insoluble minerals naturally present in water, the heat-transfer medium (Fig. 8.23). *Pipe scale* (Fig. 8.24), or *mill scale*, is the name given to the formation of natural iron oxide coating or corrosion products that form on the interior of piping (especially during construction projects where the pipe may be exposed to the elements), which can flake off and appear as a scale.

Mineral scale in steam boilers, heat exchangers, and condensers consists primarily of calcium carbonate, the least soluble of the minerals in water. Other scale components, in decreasing order of occurrence, are calcium sulfate, magnesium carbonate, iron, silica, and manganese. Present also in some scales are the hydroxides of calcium, magnesium, and iron, as well as the phosphates of these minerals, where phosphates and alkalinity are used as a corrosion or scale inhibitor. Sludge is a softer form of scale and results when hard-water minerals react with phosphate and alkaline treatments forming a soft, pastelike substance rather than a hard, dense material. In most cases, scales contain a complex mixture of mineral salts, because scale forms gradually and deposits the different minerals in a variety of forms.

The major cause of mineral scale is the inverse solubility of calcium and magnesium salts. Most salts or soluble substances, such as table salt or sugar, are more soluble in hot water than in cold. Calcium and magnesium salts, however, dissolve more readily and in

FIGURE 8.23 Pipe scale and iron corrosion products. (*Courtesy of The Metro Group, Inc.*)

FIGURE 8.24 Mineral scale deposits of water minerals. (*Courtesy of The Metro Group, Inc.*)

greater quantity in cold water than in hot, hence inverse solubility. This unique property is responsible for the entire problem of mineral scale on heat-transfer surfaces in HVAC equipment. From this property alone, we can readily understand why mineral scale forms on hot water generator tubes, condenser tubes, boiler tubes, etc. It is simply the fact that the hottest surface in contact with the water is the tube surface of this type of equipment.

In condenser water systems using recirculating cooling tower water or once-through cooling water, the water temperature is much lower than that in steam boiler or hot-water systems. At these lower temperatures most of the scale-forming minerals will remain in solution, but the tendency will be to deposit calcium carbonate on the heat-transfer surfaces where there is a slight rise in temperature. The primary factors that affect this tendency are:

• Alkalinity

• Hardness

• pH

• Total dissolved solids

• Temperature

The higher the alkalinity of a particular water, the higher the bicarbonate and/or carbonate content. As these minerals approach saturation, they tend to come out of solution.

Likewise, a higher concentration of hardness will increase the tendency of calcium and magnesium salts to come out of solution. The pH value reflects the ratio of carbonate to bicarbonate alkalinity. The higher the pH value, the greater the carbonate content of the water. Since calcium carbonate and magnesium carbonate are less soluble than the bicarbonate, they will tend to precipitate as the pH value and carbonate content increase.

Also affecting this tendency are the total dissolved solids and temperature. The higher the solids content, the greater the tendency to precipitate the least soluble of these solids. The higher the temperature, the greater the tendency to precipitate the calcium and magnesium salts because of their property of inverse solubility.

8.5.2 Langelier Saturation Index (LSI)

The Langelier index is a *calcium carbonate saturation index* that is very useful in determining the scaling or corrosive tendencies of a water. It is based on the assumption that any given water with a scaling tendency will tend to deposit a corrosion-inhibiting film of calcium carbonate and hence will be less corrosive. By inference, a water with a nonscaling tendency will tend to dissolve protective films and be more corrosive. This is not entirely accurate since other factors are involved in corrosion, as we have seen in Sec. 8.4 on corrosion, but, although a relatively simple tool, it provides a valuable index in determining the tendency of a water to directly influence scaling or corrosive actions.

In the 1950s, Eskell Nordell arranged five basic variables into an easy-to-use chart to quickly determine the pH of saturation of calcium carbonate and the Langelier index. This index is based on the pH of saturation of calcium carbonate.

The pH of saturation (or "saturation pH," pH_S) of calcium carbonate is the theoretical pH value of any particular water if that water is saturated with calcium carbonate. As, for example, the actual pH of a recirculating water approaches or even exceeds the pH of saturation of calcium carbonate under certain specific conditions, the tendency is to form a scale of calcium carbonate. If the actual pH is well below the pH of saturation of calcium carbonate, the tendency is to dissolve minerals and therefore to be corrosive. The Langelier index of a recirculating water, therefore, is determined by comparing the actual pH with the saturation pH of calcium carbonate under the same specified conditions.

To determine the Langelier index, the actual pH of the water must be measured, and the pH of saturation of calcium carbonate is calculated from a measure of the *total alkalinity, hardness, total dissolved solids*, and *temperature*.

A useful shortcut calculation of pH_S can be made for cold well or municipal water supplies that are used for once-through cooling or service water. The reason why this rapid calculation is valid is that these supplies are usually consistent in temperature [typically between 49 to 57°F (10 to 14°C)] and total dissolved solids (50 to 300 mg/L). If a water supply has these characteristics, the following formula can be used (see Fig. 8.25).

$$pH_S \text{ @ } 50°F \text{ } (10°C) = 11.7 - (C + D)$$

Likewise for hot-water supplies at 140°F (60°C), a short-form calculation of the pH of saturation of calcium carbonate can be done with the following formula:

$$pH_S \text{ @ } 140°F \text{ } (60°C) = 10.8 - (C + D)$$

Once the pH of saturation of calcium carbonate has been calculated, the LSI can be determined from the formula

$$LSI = pH - pH_S$$

where pH is the actual measured pH of the water and pH_s is the pH of saturation of calcium carbonate as calculated from Fig. 8.25. Figure 8.26 can also be used to determine the pH of saturation.

A positive index indicates scaling tendencies, while a negative one indicates corrosion tendencies. A very handy guide in predicting the tendencies of a water by using the LSI is shown in Table 8.6.

8.5.3 Ryznar Index (Ryznar Stability Index) (SI)

Another useful tool for determining the tendencies of water is the Ryznar Index. This index is also based on the pH of saturation of calcium carbonate and was intended to serve as a

A		C		D	
Total solids (mg/L)	A	Calcium hardness (mg/L of CaCo$_3$)	C	M Alkalinity (mg/L of CaCo$_3$)	D
50–300	0.1				
400–1000	0.2	10–11	0.6	10–11	1.0
		12–13	0.7	12–13	1.1
B		14–17	0.8	14–17	1.2
Temperature		18–22	0.9	18–22	1.3
		23–27	1.0	23–27	1.4
°F (°C)	B	28–34	1.1	28–35	1.5
32–34 (0–1.1)	2.6	35–43	1.2	36–44	1.6
36–42 (2.2–5.5)	2.5	44–55	1.3	45–55	1.7
44–48 (6.7–8.9)	2.4	56–69	1.4	56–69	1.8
50–56 (10.0–13.3)	2.3	70–87	1.5	70–88	1.9
58–62 (14.4–16.7)	2.2	88–110	1.6	89–110	2.0
64–70 (17.8–21.1)	2.1	111–138	1.7	111–139	2.1
72–80 (22.2–26.7)	2.0	139–174	1.8	140–176	2.2
82–88 (27.8–31.1)	1.9	175–220	1.9	177–220	2.3
90–98 (27.8–31.1)	1.8	230–270	2.0	230–270	2.4
100–110 (37.8–43.3)	1.7	280–340	2.1	280–350	2.5
112–122 (44.4–50.0)	1.6	350–430	2.2	360–440	2.6
124–132 (51.1–55.6)	1.5	440–550	2.3	450–550	2.7
134–142 (56.7–63.3)	1.4	560–690	2.4	560–690	2.8
148–160 (64.4–71.1)	1.3	700–870	2.5	700–880	2.9
162–178 (72.2–81.1)	1.2	800–1000	2.6	890–1000	3.0

$pH_S = (9.3 + A + B) - (C + D)$
$SI = pH - pH_S$
If index is 0, water is in chemical balance.
If index is positive, scale-forming tendencies are indicated.
If index is negative, corrosive tendencies are indicated.

FIGURE 8.25   Data for calculations of the pH of saturation of calcium carbonate. (*From Eskell Nordell, Water Treatment for Industrial and Other Uses, 2ed.,* © *1961 by Litton Educational Publishing, Inc. reprinted with permission of Nostrand Reinhold Co.*)

more accurate index of the extent of scaling or corrosion in addition to the tendency. This index was derived from the LSI by observation of actual cooling water conditions and is calculated as follows:

$$\text{Ryznar Index} = 2(pH_S) - pH$$

where pH_S is the pH of saturation of calcium carbonate, as calculated from Fig. 8.25, and pH is the actual measured pH of the water. Table 8.7 can be used to determine the tendency and extent of corrosion or scaling with the Ryznar Index.

Let us see how these indices can help us in analyzing a particular water supply. Figure 8.8 depicts an analysis report on the Washington, DC water supply. The Langelier saturation index at 50°F (10°C) is determined by using this analysis and the data shown in Fig. 8.25 as follows:

$$pH_S = 9.3 + A + B - (C + D)$$
$$= 9.3 + 0.1 + 2.3 - (1.8 + 2.0)$$
$$= 8.2$$

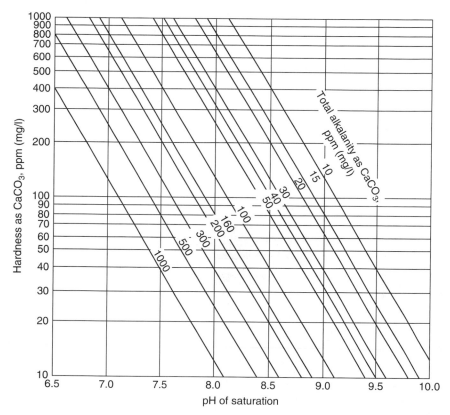

FIGURE 8.26 The pH of saturation for waters 49 to 57°F (10 to 14°C) and total dissolved solids of water 50 to 300 mg/L.

and

$$\text{LSI} = \text{pH} - \text{pH}_S = 7.7 - 8.2 = -0.5$$

From Table 8.6, according to the LSI this water supply is slightly corrosive and, therefore nonscale-forming. In fact, it is in an "uncertainty zone," where there is no clear indication of

TABLE 8.6 Prediction of Water Tendencies by the Langelier Index

Langelier saturation index	Tendency of water
2.0	Scale-forming and for practical purposes noncorrosive
0.5	Slightly corrosive and scale-forming
0.0	Balanced, but pitting corrosion possible
−0.5	Slightly corrosive and non-scale-forming
−2.0	Serious corrosion

Source: *Carrier System Design Manual*, part 5, "Water Conditioning," Carrier Corp., Syracuse, NY, 1972, p. 5–12.

scale or corrosion, and the likelihood is that both phenomena will occur, although not to any strong degree. Typically, the LSI uncertainty zone extends from +0.5 to –0.5. A very high tendency for scale is indicated by perhaps +3.0 and strong corrosion by perhaps –3.0.

To learn more about this water, the Ryznar index (SI) can be calculated in the same manner:

$$SI = 2(pH_S) - pH = 16.4 - 7.7 = -8.7$$

According to Table 8.6, this water supply tendency indicates "heavy corrosion." The Ryznar index, being more quantitative, indicates that the degree of corrosion would be greater than we would anticipate from the tendency shown by the qualitative LSI.

In practice, the SI often tends to give "worst-case" scenario, when compared to LSI (see Table 8.7) and may range from +3.0 (strongly scaling) to, perhaps, +11 or +12 (strongly corrosive). Thus, in an examination of a water supply, often both the Langelier and the Ryznar indices are used to determine the scale-forming or corrosion tendencies.

In open cooling tower condenser water systems and steam boilers, however, there is a constant accumulation of minerals as a result of evaporation of pure water, such as distilled water, and makeup water containing the various mineral impurities. Therefore, in these systems the pH, concentration of hardness, total dissolved solids, and alkalinity are constantly changing, making a study of the Langelier and Ryznar indices relatively complex and potentially subject to gross inaccuracies. These indices are useful indicator tools but other indicators should also be employed, such as mass balance checks on total minerals in (from makeup water) compared to total minerals out (with the bleed water).

8.5.4 Boiler Scale

Scale in boilers is a direct result of precipitation of the calcium, magnesium, iron, and silica minerals present in the boiler feedwater. Scale can be prevented by removing a portion of the scale-forming ingredients prior to the boiler with external water-softening equipment or within the boiler itself with internal boiler water treatment.

One of the most troublesome deposits frequently encountered in steam boilers is iron and combinations of iron with calcium and phosphate used in boiler water treatment. These sticky, adherent sludge deposits are caused by excessive amounts of iron entering the boiler with the feedwater. The iron is in the form of iron oxide or iron carbonate corrosion products. It is a result of corrosion products from the sections prior to the boiler, such as steam and condensate lines, condensate receivers, deaerators, and boiler feedwater lines.

TABLE 8.7 Prediction of Water Tendencies by the Ryznar Index

Ryznar stability index	Tendency of water
4.0–5.0	Heavy scale
5.0–6.0	Light scale
6.0–7.0	Little scale or corrosion
7.0–7.5	Significant corrosion
7.5–9.0	Heavy corrosion
9.0+	Intolerable corrosion

Source: *Carrier System Design Manual*, part 5, "Water Conditioning," Carrier Corp., Syracuse, NY, 1972, p. 5–14.

A program for preventing scale deposits must include treatment to prevent this troublesome type of sludge deposit.

8.5.5 Condensate Scale

In recirculating cooling tower condenser water systems for air-conditioning and refrigeration chillers, scale deposits are a direct result of precipitation of the carbonate, calcium sulfite, or silica minerals due to such an overconcentration of these minerals that their solubility or pH of saturation is exceeded and the minerals come out of solution. Scale in this equipment can include foreign substances such as corrosion products, organic matter, and mud or dirt. These are usually called *foulants* rather than *scale*. Treatment to prevent mineral scale should, therefore, include sufficient dilution of the recirculating water to prevent the concentration of minerals from approaching the saturation point, pH control to prevent the pH from reaching the pH of saturation of calcium carbonate, and chemical treatments to inhibit and control scale crystal formation.

8.6 FOULANTS

In addition to water mineral scale, other deposits of mud, dirt, debris, foreign matter, and organic growth are a recurrent problem in recirculating water systems. Deposits of foreign matter plug narrow passages, interfere with heat transfer and foul heat-transfer surfaces, causing inefficient performance of the equipment and higher-energy consumption than otherwise necessary.

8.6.1 Mud, Dirt, and Clay

Open recirculating cooling tower systems are most subject to deposits of mud, dirt, and debris. A cooling tower is a natural air washer with water spraying over slats and tower fill washing the air blown through either naturally or assisted by fans.

Depending on the location, all sorts of airborne dust and debris end up in cooling tower recirculating water systems. These vary from fine dust particles to pollen, weeds, plant life, leaves, tree branches, grass, soil, and stones. The fine particles of dust and dirt tend to collect and compact in the condenser water system, especially in areas of low circulation. At heat-transfer surfaces, the dust and dirt can deposit and compact into a sticky mud and seriously interfere with operating efficiency.

Muddy foulants are a common occurrence and form with the combination of airborne particles, corrosion products, scale, and organic matter. *Very rarely can one identify a foulant as a single compound because it is usually a complex combination of all these things.*

In closed recirculating water systems, foulants are not nearly as varied and complex as in open systems, but they are just as serious when they occur. Deposits in closed systems are usually caused by dirt or clay entering with the makeup water or residual construction debris. A break in an underground water line can result in dirt, sand, and organic matter being drawn into a system and is a common source of fouling.

Makeup water containing unusual turbidity or suspended matter is usually treated at the source by coagulation, clarification, and filtration so as to maintain its potability. Suspended matter and turbidity, therefore, are not common in makeup water in HVAC systems since the makeup water usually comes from a municipal or local source, over which there is a water authority responsible for delivery of clear, potable water.

Where a private well water, pond, or other nonpublic source of water is available for use as makeup water to recirculating water systems and boilers, it should be carefully examined for turbidity and suspended matter. The suspended matter measured as turbidity should be no more than the maximum of one turbidity unit for drinking water recommended by the Environmental Protection Agency. When the supply is excessively turbid, some form of clarification such as coagulation, settling, filtration, and/or fine strainers should be used to remove the suspended matter and reduce the turbidity to below one unit.

The more common problem with suspended matter and turbidity results from makeup water that is temporarily or occasionally dirty. This may occur when the local water authority is cleaning sections of a distribution main or installing new mains or when water mains are cut into during some nearby construction project. This kind of work creates a disturbance of the water mains, causing settled and lightly adherent pipeline deposits to break off and be flushed into the water supply. These deposits consist mostly of iron oxide corrosion products and dirt, clay, or silt.

8.6.2 Black Mud and Mill Scale

One of the most common and difficult foulants found in closed systems is a black mud made up of compacted, fine, black magnetic iron oxide particles. This black mud not only deposits at heat-transfer surfaces, but also clogs or blocks narrow passages in unit heaters, fan-coil units, and cooling, reheat, and heating coils in air-handling units.

The interior of black iron piping, commonly used for recirculating water, has a natural black iron oxide protective coating ordinarily held intact by oil-based inhibitors used to coat the pipe to prevent corrosion during storage and lay-up. This natural iron oxide protective coating is called *mill scale*, a very general term that can be applied to any form of pipe scale or filings washed off the interior of the pipe. This mill scale film becomes disturbed and disrupted during construction due to the constant rough handling, cutting, threading, and necessary bettering of the pipe. After construction, the recirculating water system is filled and flushed with water, which removes most of the loosened mill scale along with any other construction debris. However, very fine particles of magnetic iron oxide will continue to be washed off the metal surface during operation, and in many instances this washing persists for several years before it subsides. Mill scale plugging can be a serious problem. It is best alleviated in a new system by thorough cleaning and flushing with a strong, low-foaming detergent-dispersant cleaner. This, however, does not always solve the problem. Even after a good cleanout, gradual removal of mill scale during ensuing operation can continue.

8.6.3 Boiler Foulants

In steam boilers, foulants other than mineral scale usually consist of foreign contaminants present in the feedwater. These include oil, clay, contaminants from a process, iron corrosion products from the steam system, and construction debris in new boiler systems. Mud or sludge in a boiler is usually a result of scale-forming minerals combined with iron oxide corrosion products and treatment chemicals. Such foulants are commonly controlled by using modern polymeric dispersants, which prevent adherence on heat-transfer surfaces.

In heating boilers, the most frequent foulants other than sludge are oil and clay. Oil can enter a boiler system through leakage at oil lubricators, fuel oil preheaters, or steam heating coils in fuel oil storage tanks. When oil enters a boiler, it causes priming and foaming by emulsifying with the alkaline boiler water. Priming is the bouncing of the water level that eventually cuts the boiler off at low water due to the very wide fluctuation of this level. Oil can also carbonize at hot boiler tubes, causing not only serious corrosion from concentration corrosion cells but also tube ruptures, as a result of overheating due to insulating carbon deposits. Whenever oil enters a boiler system, it must be removed immediately to prevent these problems. This is usually done by blowing down the boiler heavily, and often, by additionally "boiling out" with an alkaline detergent cleaner.

Clay is a less frequent foulant in boilers, but it, too, can form insulating deposits on tube surfaces. Clay enters a boiler with the boiler makeup water that is either turbid or contaminated with excessive alum, used as a coagulant in the clarification process. Clay can be dispersed with the use of dispersants in the internal treatment of the boiler, but makeup water should be clear and free of any turbidity before it is used as boiler feedwater. Where turbidity and clay are a constant problem, filtration of the boiler feedwater is in order.

8.6.4 Construction Debris

All new systems become fouled and contaminated with various forms of foreign matter during construction. It is not uncommon to find these in the interior of HVAC piping and heat exchangers: welding rods, beads, paper bags, plastic wrappings, soft drink can rings, pieces of tape, insulation wrappings, glass, and any other construction debris imaginable.

It is necessary not only to clean out construction debris from the interior of HVAC systems prior to initial operation, but also to clean the metal surfaces of oil and mill scale naturally present on the pipe interior. This oil and mill scale, as has been shown, can seriously foul and plug closed systems and cause boiler tube failures, if the oil is carbonized during firing. Every new recirculating water system and boiler must be cleaned thoroughly with a detergent-dispersant type of cleaner or, as in steam boilers, with an alkaline boilout compound. This initial cleanout will remove most of the foulants and prevent serious operational difficulties.

8.6.5 Organic Growths

Organic growths in HVAC equipment are usually found in open recirculating water systems such as cooling towers, air washers, and spray coil units. Occasionally closed systems become fouled with organic slimes due to foreign contamination. Open systems are constantly exposed to the atmosphere and environs, which contain not only dust and dirt but also innumerable quantities of microscopic organisms and bacteria. Cooling tower waters, because they are exposed to sunlight, operate at ideal temperatures, contain mud as a medium and food in the form of inorganic and organic substances, and are a most favorable environment for the abundant growth of biological organisms. Likewise, air washers and spray coil units, as they wash dust and dirt from the atmosphere, collect microscopic organisms, which then tend to grow in the recirculating water due to the favorable environment. The organisms that grow in such systems consist primarily of *algae, fungi,* and *bacterial slimes.*

8.6.6 Algae

Algae are the most primitive form of plant life and together with fungus form the family of thallus plants. Algae are widely distributed throughout the world and consist of many different forms. The forms found in open recirculating water systems are the blue-green algae, green algae, and brown algae. The blue-green algae, the simplest form of green plants, consist of a single cell and hence are called unicellular. Green algae are the largest group of algae and are either unicellular or multicellular. Brown algae are also large, plantlike organisms that are multicellular.

Large masses of algae can cause serious problems by blocking the air in cooling towers, plugging water distribution piping and screens, and accelerating corrosion by concentration cell corrosion and pitting. Algae must be removed physically before a system can be cleaned since the mass will provide a continuous source of material for reproduction and biocides will be consumed only at the surface of the mass, leaving the interior alive for further growth.

8.6.7 Fungi

Fungi are also thallus plants similar to the unicellular and multicellular algae. They require air, water, and carbohydrates for growth. The source of carbohydrates can be any form of carbon. Fungi and algae can grow together; the algae living within the fungus mass are furnished with a moist, protected environment, while the fungus obtains carbohydrates from the algae.

8.6.8 Bacteria

Bacteria are microscopic unicellular living organisms that exhibit both plant and animal characteristic. They exist in rod-shaped, spiral, and spherical forms. There are many thousands of strains of bacteria, and all recirculating waters contain some bacteria. The troublesome ones, however, are bacterial slimes, iron bacteria, sulfate-reducing bacteria, and pathogenic bacteria. Control is accomplished by developing a program of addition of pesticides, called *microbiocides*, or simply *biocides*. There are many different types, both oxidizing and nonoxidizing, which are discussed later in this chapter.

8.6.9 Pathogenic Bacteria

Pathogenic bacteria are disease-bearing bacteria. Cooling-tower waters, having ideal conditions for the growth of bacteria and other organisms, can promote the growth of pathogenic bacteria, which may then give rise to infection and possibly death (such as in cases of growth of *Legionella* organisms and the development of Legionnaires disease). In very many instances around the world, where operating conditions are right (temperature, pH, etc.) and cleaning and maintenance is poor, pathogenic bacteria have been found growing in cooling-tower waters. *Therefore, it is as important to keep these systems clean and free of bacterial contamination, to inhibit growth of pathogenic bacteria, as it is to prevent growth of slime-forming and corrosion-promoting bacteria.* All cooling systems should be drained, cleaned, and disinfected at least twice per 12 month period, irrespective of whether the cooling systems operate on a summer cycle only or all year round. In addition to good maintenance, a surveillance program should also be instituted and all records kept in a logbook. Often, a risk assessment program is an additional and necessary precaution prior to starting up a new or winterized tower, or where design or location is suspect (e.g., poor drift elimination, or the tower is sited too close to a ventilation system fresh air intake).

8.6.10 Legionnaires' Disease and Pontiac Fever

These diseases are both forms of Legionellosis, are caused from the inhalation, by susceptible individuals, of certain serotypes (varieties) of the Gram-negative bacilli, *Legionella pneumophila*, and also by some other species/serotypes. The mode of transmission is usually through water droplets containing the organism.

Cooling systems with open recirculating cooling towers, or evaporative condensers, along with other types of equipment which can produce a water aerosol, that have not been properly designed, cleaned or maintained, may easily create the growth conditions required for transmission. Organisms can also be found in building potable water storage and distribution systems, such as hot water storage tanks (calorifiers), especially those designs that permit stratification of hot stored water and cold makeup water. Also, spa baths, hydrotherapy pools and others.

Legionellosis is a type of pneumonia. It was first identified after the 1976 Convention of the American Legion in Pennsylvania, when over 200 people were taken ill and 34 subsequently died. The species most commonly associated with disease outbreaks is *Legionella pneumophila serotype 1*, which appears to be the most pathogenic of the genus.

Legionnaires' disease is frequently characterized as an "opportunistic" disease, meaning that it most frequently attacks individuals who have an underlying illness or have weakened immune systems. Good control over cooling systems in hospital and other healthcare premises are therefore particularly important, in order to control infection risks. At its onset, Legionnaires' disease is characterized by high fever, chills, headaches, and muscular pain. A dry, nonproductive, cough develops and most patients suffer breathing difficulties. A third of the patients suffer from diarrhea and/or vomiting and half become confused or delirious. (In outbreaks, the fatality rate is typically around 20 percent). Although most people make a complete recovery, some can suffer long-term symptoms. If recognized in its early stages, the disease can be effectively treated with antibiotics. Legionnaires' disease is much more common than generally thought. Although significant efforts are now made to regularly clean and maintain water systems, the organism is ubiquitous and many cases of Legionnaires' disease still occur each year. Unfortunately most cases still go undiagnosed and, where identified, these may result in considerable adverse publicity.

Pontiac fever: Legionella bacteria may also cause other forms of Legionellosis such as Pontiac fever, which is a short, self-limiting, flulike illness that has no long-term effects. Pontiac fever develops rapidly, (5 h to 3 days) and lasts for 3 to 5 days. (The attack rate is high, typically some 95 percent.) It produces headaches, nausea, vomiting, aching muscles, and a cough, but no pneumonia develops.

Aerosol formation is the consequence of the normal operation of evaporative condensers, cooling towers and certain domestic water services equipment such as shower/spray systems. Bacteria from these water systems may be inhaled whenever an *aerosol* is produced, increasing the risk of exposure. Illness and fatalities are usually preventable, because aerosol formation is typically associated with poor water-system designs, installations, maintenance and management.

However, it has to be said that there is no evidence that well managed and maintained cooling systems provide any significant risk of exposure to Legionellosis. The consequence of regulations for the control of *Legionella* organisms in cooling systems and other water systems (and the penalties that could be imposed) has meant, and continues to mean, that the control of *Legionella* has become an even more important consideration for many engineers, water treatment managers, and the water management industry as a whole.

8.7 PRETREATMENT EQUIPMENT

Prior to internal treatment of HVAC equipment, it is frequently necessary to use mechanical equipment to remove from the feedwater supply damaging impurities such as dissolved oxygen, excess hardness, or suspended solids.

The choice of proper equipment and its need can be determined by studying the quality and quantity of makeup water used in a boiler, condenser water system, and an open or a closed recirculating water system.

8.7.1 Water Softeners

Hardness in the makeup water is the cause of scale formation. In equipment using large volumes of hard water, a substantial amount of scale can form on heat-transfer surfaces in a short time. In these circumstances, it may be economical to remove the hardness from the water supply before it is used in the equipment.

Determining the Need for Water Softeners. In open cooling tower condenser water, evaporative condensers, and surface spray units, removal of the hardness with pretreating equipment is usually not economical. These systems operate at pH values close to the neutral point, where the hardness can be kept in solution with the aid of antiscalants and pH control chemical treatments. This treatment will be required even if most of the hardness is removed. Therefore, it is not economically advantageous to install water softeners in this equipment unless the makeup water is so hard that it cannot be used at all in the particular system without some form of pretreatment to at least partially soften the water.

Waters with a hardness of more than 300 or 400 mg/L require some form of pretreatment to partially soften the water so that it can be used for cooling tower makeup water. Likewise, it is not economical to install water-softening equipment on closed chilled-water and low-temperature hot-water heating systems since such systems use very little makeup water and internal treatment with antiscalants can prevent scale on heat-transfer surfaces.

With steam and high-temperature hot-water boilers, however, removal of hardness from the makeup water is frequently required. The determining factor usually is the quality of available makeup water. With high-temperature hot-water boilers, if the hardness of the makeup water exceeds 100 mg/L, the initial fill and makeup water should be softened to remove the hardness and prevent excessive scale or sludge deposits on heat-transfer surfaces.

With steam boilers, the determining factors are both the hardness and the amount of makeup water used. In low-pressure heating applications where steam is used for heating only and possibly small amounts of humidification requiring less than 10 percent of the steam generated, the boiler feedwater will consist of 90 percent or more of return steam condensate. In instances such as this, the makeup water will not require external hardness removal since the small amount of hardness entering the system can be controlled with internal treatment.

In steam boiler systems that use more than 10 percent raw makeup water, it may prove economical and practical to install a water softener. Usually if the makeup water in these systems exceeds 100 mg/L hardness and the amount of makeup water is more than 1000 gal/day (3785 L/day), a water softener will be required. *This can be justified by comparing the cost of internally treatment the boiler water to control scale and sludge deposits with the operating cost of an external water softener.*

Another useful guideline is that if the hardness of water entering a boiler exceeds 1000 grains per hour, a water softener usually is required.

Note: Grains per hour are determined by multiplying the hardness in grains per gallon (17.1 mg/L per 1.0 gr/gal) by the makeup rate in gallons per hour.

$$\text{Grains per hour} = \text{hardness (gr/gal)} \times \text{makeup (gal/h)}$$

Ion-Exchange Water Softeners.   The water softener used for boiler makeup water is a synthetic zeolite softener containing an ion-exchange water softener resin. This ion-exchange resin adsorbs calcium and magnesium ions from the water passing over the resin bed. The resin at the same time releases sodium, hence the term "ion exchange." Figure 8.27 shows a typical ion-exchange water softener used for boilers.

The size of the softener required depends on the rate of makeup water and the amount of hardness to be removed. The softener should have a minimum delivery rate of 6.6 gal/min (25.2 L/min) per 100 bhp [3450 lb/h (1564.9 kg/h) steam rate].

8.7.2 Dealkalizers

Another ion-exchange water conditioners that may be required is the dealkalizer. A steam boiler that operates with makeup water containing excessive quantities of carbonate and bicarbonate alkalinity not only will develop excess alkalinity in the boiler (causing priming, foaming, and carryover), but also will generate large quantities of carbon dioxide as a result of decomposition of the carbonates and bicarbonates. This process results in an acid steam and condensate, which, as noted previously, cause severe corrosion of steam and condensate return lines.

The alkalinity in these cases can be reduced by 90 percent by passing the makeup water through a dealkalizer following the water softener. Usually a dealkalizer cannot be economically justified unless the total alkalinity as calcium carbonate exceeds 100 mg/L and the makeup rate exceeds 100 gal/h (378.5 L/h). With lesser quantities, the effects of the carbon dioxide generated can be controlled by the use of steam and return-line treatments. The economics of using a dealkalizer can be determined by comparing the costs of the steam and return-line treatments with the costs of installing and operating a dealkalizer.

FIGURE 8.27   Ion-exchange water softener. (*Courtesy of Ermco Inc.*)

In installations where the steam comes in direct contact with food products, the U.S. Food and Drug Administration permits use of certain treatments for control of carbonic acid–induced corrosion under limited conditions (21 CFR § 173.310) When it is not possible to control the treatment within these specified limitations, installation of a dealkalizer would be justified to remove 90 percent of the alkalinity, to reduce the excessive corrosion tendencies of the steam and return condensate.

The dealkalizer contains an ion-exchange resin similar to the water softener with the capability of exchanging carbonate, bicarbonate, sulfate, and other anions for chloride, hence the name chloride anion dealkalizer. Figure 8.28 shows a typical installation of a chloride anion exchange unit.

The dealkalizer installed must be sized to remove the alkalinity from the makeup water and to deliver dealkalized water at a rate of 6 percent gal/min (25.2 L/min) per 100 bhp [3450 lb/h (1565 kg/h) steam rate] so that it would be able to deliver the maximum amount of makeup water required at any given moment of operation. It usually accompanies the ion-exchange water softener and is installed as a complete packaged softener–dealkalizer. With this equipment, alkalinity is reduced and corrosion of steam and return condensate lines can be controlled.

8.7.3 Deaerators

To prevent serious corrosion and pitting of steam boiler feedwater lines and boiler tubes, it is necessary to remove the dissolved oxygen from the boiler feedwater, which may be done

FIGURE 8.28   Dealkalizer. (*Courtesy of Cochrane Division, Crane Co.*)

by the use of chemical treatment. In many installations, however, it is neither practical nor economical to use chemical treatment alone. In these circumstances, it is necessary to remove most of the dissolved oxygen mechanically by using feedwater heaters or deaerators followed by small quantities of treatment to remove the last traces of corrosive gases.

Whenever a steam boiler system is open to the atmosphere through vented condensate receivers, feedwater tanks, etc., the air absorbed will result in high quantities of dissolved oxygen. This increases in direct proportion to the amount of makeup water used because the cold raw makeup water, high in dissolved oxygen, not only will increase the dissolved oxygen content, but also will lower the temperature of the feedwater in the return condensate tank, enabling more oxygen from the atmosphere to be dissolved in the feedwater. This happens because the solubility of oxygen in water is inversely proportional to temperature. Figure 8.29 shows the solubility of oxygen with respect to temperature.

The feedwater temperature of low-pressure heating boilers operating with theoretically all return condensate will remain close to 180 to 200°F (82 to 93°C) since little or no cold-water makeup is used. In systems such as this, a deaerator or feedwater heater is not usually required. Hence the current design custom is to avoid the use of preheaters or deaerators with low-pressure heating boilers. However, if the low-pressure steam heating boiler were to operate with any steam loss such as at steam humidifiers in air-handling units or steam tables or pressure cookers in a cafeteria or restaurant, the cold-water makeup replacing that loss increases the dissolved oxygen content. *In these cases it is essential that a mechanical preheater or deaerator be used to remove the dissolved oxygen from the feedwater.* Without it, the cost for chemical oxygen scavengers alone is excessive.

The simplest mechanical deaerator is the open feed-water heater or preheater. This consists of open or closed steam coils placed in the vented condensate tank. The coils are thermostatically controlled to maintain the temperature of the feed water at 180 to 200°F (80 to 95°C) or the highest possible temperature without causing boiler feed-water pump difficulties such as cavitation or steam shock.

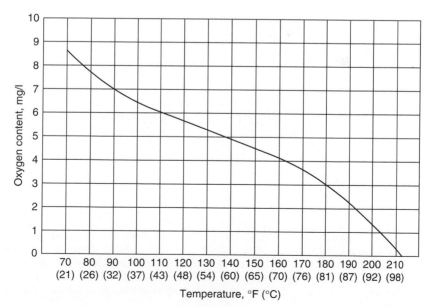

FIGURE 8.29 Solubility of oxygen from air at atmospheric pressure.

The deaerator is a more complex device that utilizes steam injection to scrub the incoming makeup water and condensate of the dissolved corrosive gases, oxygen and carbon dioxide. In the deaerator, water is sprayed over inert packing, such as glass beads or plastic fill, or is trickled through baffles or trays to break up the water and provide for intimate fixing of the feed water with the incoming steam. The steam is injected counter to the flow of the incoming water, which drives the gases, oxygen and carbon dioxide, upward out the vent. The vent releases only noncondensible gases and steam losses are at a minimum.

To completely protect the equipment, the last trace of oxygen must be removed by superimposing a chemical oxygen scavenger. Some deaerators can produce a feed water with dissolved oxygen content as low as 0.005 cm^3/L. Also, packaged deaerating heaters can provide a feed-water quality of 0.03 cm^3/L dissolved oxygen that is more than adequate for HVAC boiler systems. These packages include a factory-assembled unit complete with deaerating heater, controls, boiler feed-water pumps, and level controls. The deaereating heater consists of a storage or collection tank for condensate and raw makeup water and the deaerator section. The feed water is pumped from the storage section through a spray manifold into the top of the deaerating section. Steam is injected through a "sparge" pipe into the bottom of the deaerating section and is bubbled through the feed water. The steam drives out the oxygen and carbon dioxide gases through the vent.

Ideally, mechanical equipment should be employed to remove dissolved corrosive gases from boiler feed water for both low- and high-pressure boilers (in addition to chemical treatments). Similarly, to reduce corrosion risks, cold-water makeup should ideally be preheated.

8.7.4 Abrasives Separators

In closed hot water circuit, dual-temperature chilled/hot water circuit, or so-called secondary recirculating water systems, one of the major problems is the presence of suspended matter. Suspended matter not only causes deposit corrosion and fouling of heat-transfer surfaces, but also can seriously damage mechanical seals and shafts on pumps. Hard, abrasive particles will score shafts and mechanical seals, causing leaks and premature failures.

Mechanical seal failures are frequently attributed to the water treatment used in the recirculating water whereas they are actually caused by abrasives, finely divided iron particles, grit, sand, or other foreign particles. Water treatment and other minerals dissolved in the water will not cause seal failures unless excessive operating temperatures are encountered. This causes flashing of the water lubricant at the seal interface, resulting in precipitation of dissolved minerals which end up scoring the seal surface in the absence of the water lubricant. This occurs at temperatures in excess of 160°F (71°C).

To minimize these failures, very inexpensive abrasives separators are used. These not only remove the suspended abrasive particles from the water lubricant at the seal interface, but also force clean, clear water at the pump discharge pressure back into the seal cavity, preventing flashing due to excessive temperatures.

A typical abrasives separator is pictured in Fig. 8.30. The device shown removes suspended abrasive particles by centrifugal force as the water piped from the pump discharge enters the separator and is rotated in the cone-shaped bore at a high velocity developed by the pressure differential. Clear water is taken from the middle of the top outlet and piped to the pump housing for flushing the seal faces. This clear-flush water under pump discharge pressure prevents flashing at the seal surfaces. In larger installation, magnetic separators may additionally be installed.

FIGURE 8.30   Abrasives separator. (*Courtesy of Crane Packing Co.*)

8.7.5   Strainers and Filters

It is common practice to install strainers prior to pumps on open recirculating condenser water systems to protect the pump internals, vanes, shaft, and impellers from large damaging flakes of rust, suspended dirt, or other foreign particles that can enter an open system. Similarly, strainers should be installed on closed systems since they are frequently plagued with suspended black magnetic iron oxide mill scale as well as foreign particles.

The bucket strainer can be used as a coarse filter to remove larger particles before the pumps and heat exchangers. Iron and steel particles, however, can be so small that they could often pass through the finest mesh screen. Magnetic inserts attached to the strainer's bucket catch the fine iron particles, which can then be cleaned out by flushing with a high-pressure hose. Figure 8.31 shows a typical magnetic insert strainer bucket used for this purpose.

8.7.6   Free Cooling

One of the most interesting applications of strainers and/or filters on open cooling tower water systems is described in U.S. Patent 3,995,443 dated December 5, 1976. This patent describes a process commonly called *free cooling* or *crossover cooling*. During seasons of the year when the cooling tower water temperature can be held at 60°F (15.5°C) or below, the tower water is diverted to the chilled-water circuit bypassing the refrigeration machine. In this way, cooling is provided by naturally cool cooling tower water rather than by artificially chilled water. This procedure can provide substantial savings in energy since the refrigeration machine does not have to operate during the off-peak spring and fall months in certain areas of the country.

The dangers of diverting cooling tower water to the chilled-water circuit are obvious. Suspended matter, airborne particles, dirt foulants, mud, slime, etc., commonly present in

FIGURE 8.31   Strainer with magnetic inserts. (*Courtesy of Hayward Manufacturing Co.*)

open cooling tower waters can seriously foul chilled-water circuits. These systems have much narrower passageways in fan-coil units, terminal units, air-handling units, etc., with delicate controls, often needle valves, which are readily fouled by even the slightest amount of suspended matter. Therefore, it is necessary to maintain the cooling tower water free of suspended matter and foulants if such a procedure is to be successfully applied.

Methods of keeping the system clean include use of modern polymeric dispersants and deposit inhibitors along with very fine separators, strainers, bag, sand, or multimedia filters. The use of a highly efficient heat exchanger in place of strainers frequently solves the contamination problem by isolating the closed circuit from the open circuit in free-cooling systems.

8.7.7   Nonchemical Treatments (Magnetic, Electrostatic, and Similar Devices) and "Gadgets"

From time to time over the past 50 years or more, various forms of nonchemical devices and gadgets have appeared for which exaggerated claims have often been made. All are said to eliminate and prevent all forms of deposits in recirculating water systems and boilers without the use of chemical treatment. Some claim to also control corrosion, and others biological fouling. Some of these technologies are summarized below:

Magnetic devices. A magnet(s) is fixed onto, or plumbed into, a system, along the parallel axis of water flowing in a pipe. It is claimed that with careful sizing and fitting these devices inhibit the formation of scale.

Electrolytic devices. These devices consisting essentially of a *zinc anode* and a *copper cathode* are plumbed in a system and in contact with the water (electrolyte); they are claimed to prevent scale by acting in a similar way to a battery.

Library, Nova Scotia Community College

Electrostatic devices. These devices provide a static field within the water via the production of a very high, localized voltage (but low current). It is claimed that this field physically affects the dissolved mineral ions and inhibits the precipitation of scale.

Electronic devices. There are various designs, including *square-wave generators, sonar effect,* and *frequency modulation (FM) devices.* The signal generated is typically applied to a closed coil, formed by wrapping a signal cable around the outside of a water pipe. These devices are claimed to transfer energy to the ions of dissolved bicarbonate minerals, creating *nanocrystals* and preventing scaling of metal surfaces and holding the minerals in suspension until they can be bled from the system.

Catalytic devices. These nonmagnetic devices use a perforated nonferrous tube to encourage small calcite seed crystals to form and reduce the risk of bulk water scaling. They are promoted for use in hard waters under conditions where supersaturation can easily occur.

Pressure-changing devices. In these devices, it is claimed that when water flows through a perforated bar or tube, the pressure drops drastically, causing dissolved *carbon dioxide* to escape, raising the pH slightly and resulting in the water being supersaturated with millions of tiny calcites (crystalline *calcium carbonate*). These grow to subcolloidal size and are eventually dislodged into the bulk water by the shearing force of the water flow and remain suspended as discrete particles, rather than agglomerating as scale.

Many of these devices are rightly called "gadgets" and simply do not work. Others may have some limited effect on controlling scale, but even when seed crystals, nanocrystals, or sludge is produced, *unless this material is regularly removed from the water system,* deposition, fouling, and loss of heat-exchange will still occur. None of the devices have any practical control over corrosion or biological fouling. For many buyers or operators of building and similar HVAC type water systems, the allure of nonchemical devices is very appealing; however, such appeal may be significantly outweighed by the often very high equipment capital purchase or leasing costs.

8.8 TREATMENT OF BOILERS, OPEN COOLING, AND CLOSED LOOP SYSTEMS

8.8.1 General

The chemicals, equipment, and method of treatment required to optimize water treatment systems to reduce the corrosion, scaling, and fouling to accepted minimums must be tailored not only to each system but also to each geographic location. To obtain the optimum treatment of each system, it is strongly recommended that a reputable water treatment company and/or water treatment consultant be retained to select the chemicals, equipment, and method of treatment for each system required and to provide effective monitoring, control, and review services.

8.8.2 Treatment of Boiler Water Systems

General. Internal treatment of HVAC boilers is required to prevent the problems of corrosion, pitting, scale deposits, and erratic boiler operation due to priming, foaming, and carryover. To prevent these problems, correct blowdown and treatment must be applied.

Blowdown. Blowdown of a boiler is the spontaneous removal of some concentrated boiler water from the boiler under pressure. The recommended maximum concentrations of these impurities, which must be properly controlled, are outlines in Table 8.8. These limits are for boilers with an operating pressure up to 250 psig (1724 kPa) and are used as a guide only. Actual operating experience will determine the true limits with any specific boiler operation.

The number of times that the solids in the makeup water have been accumulated in the boiler water is called the *cycles of concentration*. To determine the required amount of blowdown, it is necessary to examine the makeup water analysis and compare it to the maximum allowable concentration of solids as outlines in Table 8.8. The maximum cycles of concentration permitted for each of the items listed in Table 8.8 is determined by dividing the maximum values in this table by the amount of each given in the makeup water analysis.

Let us examine the water analysis of Omaha, NE (Fig. 8.32), and compare it with the maximum concentration of solids allowed in a boiler (Table 8.8), to determine the maximum cycles of concentration allowed:

Analytical results	Maximum cycles of concentration
Silica, mg/L:	
$\dfrac{\text{Maximum}}{\text{Makeup water}} = \dfrac{150 \text{ mg/L}}{7.7 \text{ mg/L}}$	$= 19.5\times$
Suspended solids (hardness), mg/L:	
$\dfrac{\text{Maximum}}{\text{Makeup water}} = \dfrac{600 \text{ mg / L}}{159 \text{ mg / L}}$	$= 3.8\times$
Total dissolved solids, mg/L:	
$\dfrac{\text{Maximum}}{\text{Makeup water}} = \dfrac{3000 \text{ mg / L}}{414 \text{ mg / L}}$	$= 7.2\times$

From the above we see that the maximum allowable cycles of concentration are 3.8×. The lowest value obtained is used, for if this value were exceeded, difficulty with that particular impurity would result. If the makeup water were softened, hardness would be removed and no longer considered a limiting factor. In this case, the maximum allowable cycles of concentration would then be increased to 7.2×, the next lowest value. In practice, a water softener would be essential.

After the maximum allowable cycles of concentration in a boiler are ascertained, the blowdown rate required to maintain the solids accumulation below this maximum level can be calculated. Blowdown is used to remove accumulated boiler water solids.

The amount of solids present in the concentrated boiler blowdown water is equal to the amount of solids in the makeup water multiplied by the cycles of concentration. This can be expressed mathematically as *BCX*, where

B = blowdown, gal (L)

C = cycles of concentration

X = total solids concentration of makeup water, ppm (mg/L or g/gal)

The amount of solids entering the boiler with the makeup is expressed mathematically as *MX*, where *M* is makeup water, gal (L). Since blowdown is designed to maintain a specific

Library, Nova Scotia Community College

TABLE 8.8 Maximum Concentration of Boiler
Water Solids for Boilers up to 250 lb/in^2 (1124 kPa)*

Silica	150 mg/L
Suspended solids	600 mg/L
Total dissolved solids	3000 mg/L

*Adapted from American Boiler and Affiliated Industries
Manufacturers Association and American Society of Heating,
Refrigeration, and Air-Conditioning Engineers' guidelines.

METROPOLITAN REFINING COMPANY INC.
50-23 23rd STREET • LONG ISLAND CITY • N.Y. 11101

CERTIFICATE OF ANALYSIS

OMAHA, NE.

DATE_____
SAMPLING DATE_____
REPRESENTATIVE_____

ANALYSIS NUMBER				
SOURCE	AVERAGE ANALYSIS FLORENCE PLANT			
pH		9.2		
P ALKALINITY	CaCO₃, mg/L	10.		
FREE CARBON DIOXIDE	CO₂, mg/L			
BICARBONATES	CaCO₃, mg/L	52.		
CARBONATES	CaCO₃, mg/L	20.		
HYDROXIDES	CaCO₃, mg/L			
M (Total) ALKALINITY	CaCO₃, mg/L	72.		
TOTAL HARDNESS	CaCO₃, mg/L	159.		
SULFATE	SO₄, mg/L	181.		
SILICA,	SiO₂, mg/L	7.7		
IRON	Fe, mg/L			
CHLORIDE	NaCl, mg/L	23.		
ORGANIC INHIBITOR	mg/L			
PHOSPHATE	PO₄, mg/L			
CHROMATE	Na₂CrO₄, mg/L			
NITRITE	NaNO₂, mg/L			
ZINC	Zn, mg/L			
SPECIFIC CONDUCTANCE	mmhos/cm	678.		
TOTAL DISSOLVED SOLIDS	mg/L	414.7		
SUSPENDED MATTER				
BIOLOGICAL GROWTHS				
SPECIFIC GRAVITY @ 15.5°/15.5°C				
FREEZING POINT °C/°F				
	% BY WEIGHT			

NOTES:
1. ANALYTICAL RESULTS EXPRESSED IN MILLIGRAMS PER LITRE (mg/L) ARE EQUIVALENT TO PARTS PER MILLION (ppm). DIVIDE BY 17.1 TO OBTAIN GRAINS PER GALLON (gpg).
2. CYCLES OF CONCENTRATION = CHLORIDES IN SAMPLE / CHLORIDES IN MAKEUP

TREATMENT	TREATMENT CONTROL	FOUND	RECOMMENDED	FOUND	RECOMMENDED

REMARKS:

R. V. Blake

SPEEDIPLY• PAT D MCP• PAT D MBF 28

FORM 1078-1

FIGURE 8.32 Water analysis of Omaha, NE (*Courtesy of The Metro Group, Inc.*)

level of cycles of concentration, that level can be kept consistent only if the amount of solids leaving the boiler is precisely equal to the amount of solids entering the boiler. This is expressed mathematically as

$$BCX = MX$$
$$\text{(solids leaving)} \quad \text{(solids entering)}$$

Solving this mathematical equation for blowdown B, we obtain $B = M/C$. This formula is used to determine a blowdown rate with respect to the makeup rate. In percent, it can be expressed as

$$\% \text{ Blowdown} = \frac{100}{C}$$

In actual practice, however, it is not usually possible to measure the blowdown rate even though it is possible to calculate the amount required. Therefore, to determine if the blowdown rate is sufficient, the cycles of concentration are measured through the use of a simple chloride test.

Chlorides are the most soluble minerals and are always present in the makeup water in some degree. In addition, chlorides are only added to a boiler with the makeup water and not with treatment or from any other source. The cycles of concentration are found by comparing the chlorides of the makeup water with the chlorides of the boiler water:

$$C = \frac{\text{chlorides in boiler}}{\text{chlorides in makeup}}$$

This very simple and practical test is used by operating engineers to control the blowdown rate.

8.8.3 Internal Treatment of Boiler Water Systems

Scale and Sludge Control.   After the maximum allowable cycles of concentration are determined and a blowdown rate is established to prevent accumulation of minerals beyond the maximum allowable limit, treatment to prevent deposits and maintain precipitated solids in suspension must be considered. As outlined previously, the hardness minerals, calcium and magnesium, are precipitated in the boiler water and tend to build a scale on the heat-transfer surfaces unless some treatment is used. Without treatment, these minerals will eventually precipitate as the insoluble carbonate and sulfate salts.

Similarly, silica and complex silicates will form hard, dense scales if silica is present in excess of its solubility. Some treatment for preventing these hard, dense scales includes phosphate to preferentially precipitate the calcium as phosphate, and in the presence of excess alkalinity phosphate is precipitated as calcium hydroxy phosphate, also known as hydroxyapatite $[Ca_3(PO_4)_2 \, Ca(OH)_2]$. Magnesium with hydroxide alkalinity present in the boiler water will form the hydroxide precipitating as brucite $(MgOH_2)$. In these forms, the particles are more easily dispersed and held in suspension.

Nonphosphate treatments may consist of carbonate and silicate salts to preferentially precipitate calcium carbonate and magnesium silicate, which are more readily dispersed and held in suspension by polymers. The formulated boiler water treatment may include soluble organic polymers such as lignins, tannins, and polyelectrolytes that promote formation of the insoluble precipitate within the boiler water rather than on the heat-transfer surface. These materials act as nucleating sites, or places for the soluble ions to meet and form the insoluble particle dispersed throughout the boiler water. Some polymers are called *polyelectrolytes* because of their many positive and negative electrolyte sites on the polymer chain.

The polyelectrolytes, such as polyacrylamides, polyacrylates, polymethacrylates, polymerized phosphonates, polymaleates, and polystyrene sulfonates, also distort the crystal growth of the scale particle, rendering it less adhesive to heat-transfer surfaces and more readily dispersed with reduced tendency to compact into a dense scale. Some modern polymers also sequester hardness similar to the chelates, and additionally act to redissolve existing deposits. These polymers, therefore, act to prevent and remove scale by a threefold mechanism of dispersion, crystal distortion, and sequestration.

Another type of scale inhibitor is the chelating agent. Chelants are organic materials capable of solubilizing calcium, magnesium, and iron, preventing formation of the insoluble salts. Both EDTA (ethylenediaminetetraacetic acid) and NTA (nitrilotriacetic acid) are used in boiler water treatment as chelates to prevent scale formation and in some cases to remove existing scale deposits. The use of these materials in HVAC equipment is limited because they require close control and are particularly corrosive to iron when they are not controlled. Some proprietary formulations contain small amounts of chelants to provide very efficient scale control. Combinations of chelating agents and polymers have been widely applied with excellent results.

Boiler Section Corrosion Control by Alkaline Treatments. Corrosion in boilers may be controlled with oxygen scavengers, alkalinity boosters, and corrosion inhibitors. Corrosion of iron is greatly influenced by the pH value of the water in contact with the iron. The lower the pH value, the higher the corrosion rate; and the higher the pH value, the lower the corrosion rate. At a pH of 11.0, the corrosion rate of iron will be virtually nil, provided oxygen and other corrosive gases are removed. Treatment of boiler water to control general corrosion, therefore, will include an alkaline substance to raise the pH value to 10.5 to 11.5. This range is compatible with normal operation without foaming or carryover, and at the same time, it will promote good corrosion control. The alkaline materials include potassium and sodium hydroxide (commonly called *caustic potash* and *caustic soda*, respectively) and sodium carbonate (commonly called *soda ash*).

Boiler Section Corrosion Control by Oxygen Removal. As we have seen, dissolved oxygen in boiler water causes localized corrosion called pitting, and to prevent pitting, the oxygen must be removed. This is another way to alter or stabilize the environment. Deaerators are used for this purpose, but no deaerator is perfect, and the best ones produce a boiler feed water with a dissolved oxygen content of 0.0005 cm^3/L This very low oxygen content can cause serious pitting failures, especially in boiler feed-water lines and at boiler tubes adjacent to the feed-water entrance to the boiler. To prevent this, chemical oxygen scavengers are used to absorb the oxygen from the water and to ensure a completely oxygen-free environment. The most common oxygen scavengers are sodium sulfite and hydrazine. These materials have a strong affinity for oxygen and will absorb it from the water.

Sodium sulfite is widely employed, and approximately 8 ppm is needed for every 1 ppm of oxygen. As a fairly high sulfite reserve is required (typically 20 to 50 ppm or more), additional sulfite must be added and so the general rule is 10 ppm of sodium sulfite per 1 ppm oxygen. The maintenance of a slight excess of the oxygen scavenger in the water provides assurance that there is no dissolved oxygen present. Complete water treatment formulations containing sodium sulfite for this purpose will also include a catalyst, ensuring that the reaction between the dissolved oxygen and the oxygen scavenger is instantaneous, even in cold water.

Hydrazine is not widely used as an oxygen scavenger due to its toxicity. Other oxygen scavengers currently used as a replacement for hydrazine include diethylhydroxylamine

(DEHA), sodium erythorbate, carbohydrazide (CHZ), methylethylketoxime (MEKO), and hydroquinone (HQ). Typical reserves for these materials are very low, perhaps only 0.1 to 0.2 ppm.

Tannins are also used. They have been employed for over a century and have significant benefits over other oxygen scavengers in many HVAC situations, as, being organic, they do not contribute to TDS as when using sulfite, and they are much less expensive than DEHA or erythorbate. The hydrolysable tannin component of Quebracho and other tannins, under alkaline boiler water (BW) conditions produce a number of oxygen scavengers, primarily as the sodium salts. These include, tannic acid, ellagic acid, gallic acid, quinic acid, pyrogallol, hydroquinone, and catechol. In addition, they are very good passivators and reasonable sludge dispersants.

Tannins do impart a brown color to the boiler water, which can sometimes make testing for hardness, chlorides or alkalinity a little difficult when looking for subtler color changes, but the mere presence of a strong brown color confirms its presence and provides reassurance that oxygen corrosion is under control.

Boiler Section Corrosion Control by Passivating Agents.   Corrosion inhibitors are substances that do not necessarily alter the environment or conditions involving the corrosion process, but act as a barrier between the corrosive medium and corroding metal surface, Physical barriers such as protective coatings and galvanizing immediately come to mind as a common application of a corrosion inhibitor. These physical bathers actually separate the corrosive atmosphere containing water, oxygen, and acid gases from the base metal.

Corrosion inhibitors that form a protective film on the metal surface can be added to water, acting as a barrier to the corrosion process, i.e., inhibiting the corrosion reaction. Such barriers form by a chemical reaction between the metal surface and the inhibitor or by a physical attraction and adsorption on the metal surface. With this type of inhibitor, the film is not visible and nonaccumulative. This type of inhibitor has a thickness of only one molecule of the inhibiting film, hence, it is called a *monomolecular film*. With a film of this thickness, there is no interference with heat transfer, and therefore the inhibitors are found to be very effective in heat-transfer equipment. The films formed can be either adsorbed films, as in some organic inhibitors, or a chemically formed reaction product of the inhibitor and metal surface. The inhibiting film may also be a combination of both adsorption and reaction. The result is a reduction of the corrosion rate and passivity.

Passivity is described as a state of an active metal in which reactivity is substantially reduced, resisting corrosion, or when its electrochemical behavior becomes that of a less active metal. That is, the metal becomes passive or is passivated. Inhibitors in this sense are also called *passivators*. Inorganic inhibitors used in boilers for this purpose are molybdates, nitrites, borates, silicates, and phosphates. Some organic inhibitors used are phosphonates, polyacrylates, phosphinocarboxylic acids, and nitrogen-containing organics such as triazoles and amines.

Tannins are particularly useful as passivators and undergo a chemicophysical reaction at the boiler metal/water interface, forming an iron-tannate passivated (corrosion-inhibiting) film.

Post-Boiler Section Corrosion Control.   Soda ash, although still sometimes used for pH control in boilers, is not desirable for HVAC boiler equipment. The carbonates added to boiler water, although they increase the boiler water pH, will decompose, forming carbon dioxide gas, which is a cause of corrosion of steam and condensate lines and heaters. The carbon dioxide dissolves in the steam condensate forming carbonic acid. This is very corrosive to steel and copper and can be a cause of failure of unit heaters and condensate handling lines and equipment. Even where carbonates are not added, natural bicarbonates and carbonates enter the boiler system in the makeup water and the same problems of carbon dioxide corrosion can develop.

The carbon dioxide generated from the carbonates and bicarbonates naturally present in the boiler makeup water can be neutralized by using neutralizing amines, mild alkaline materials related to ammonia but much milder than it. Amines used for this purpose are morpholine, cyclohexylamine, and diethylaminoethanol. These materials, liquid at room temperature, boil and vaporize at approximately the same temperature as water, thereby going off with the steam, a mixture known as an *azeotrope*. As the steam condenses into water, the amines also condense, rendering that initial condensate alkaline and much less corrosive to the metal surface. The flowing condensate containing the neutralizing amine will then retain higher pH values, even with the adsorption of carbon dioxide gases. For the flowing condensate, a pH range of 7.0 to 9.0 is the most favorable for good corrosion control of all metals in heating systems including iron, steel, brass, and copper.

In addition to the use of neutralizers in steam and return condensate lines, film-forming inhibitors are also used. These are organic amines or amides such as octadecylamine and mixtures of octadecanol and stearamide. These inhibitors form insoluble films on the seam and condensate piping. They will steam distill from the boiler; i.e., they carry over with the steam and deposit in the steam and condensate system. The disadvantage of such inhibitors is that this film is not self-limiting, and heavy deposits plugging steam traps, strainers, etc., can result unless particular care is taken to avoid this problem. The filming-type steam and condensate line corrosion inhibitors are employed most successfully in systems where there is little or no return condensate. In heating boilers where most condensate is returned, the neutralizing amines are more suitable. Combinations of neutralizing and filming amines are also used.

8.8.4   Control Measures and Additional Boiler Water Treatment Measures

Priming, Foaming, and Carryover.   Priming of boiler water is the bumping and bouncing of the water level of the boiler during operation. Foaming, however, is a less violent activity. It consists of the formation of small bubbles in the surface of the boiling water like the soap foam in a washer.

Carryover essentially is the contamination of the steam with boiler water. It is a result of priming and foaming and can be a more subtle entrapment of boiler water with the steam, causing steam contamination without the evidence of priming and/or foaming.

The causes of priming, foaming, and carryover can be many and varied. Most frequently, they are caused by contamination of the boiler water with oil or other foreign substances. Other causes are excessive solids accumulation due to the lack of blowdown, high alkalinity, overtreatment, or mechanical malfunction.

Adequate blowdown and certain antifoam treatments can reduce the problems of priming, foaming, and carryover.

Treatment of Low-Pressure Steam Heating Boilers.   Most low-pressure steam heating boilers have a unique type of operation in that *all* the steam produced is for heating purposes *only*. Therefore, little makeup water is required as well as little accumulation of solids. The only makeup water is for slight loss at vents, leaks, or overflow at condensate receivers.

Without makeup water and accumulation of minerals, even with a very hard water supply, problems of scale formation are significantly reduced. However, corrosion under these conditions can become aggravated. The treatment program will usually consist of a corrosion inhibitor for the boiler and neutralizing or filming amine for the steam and condensate system.

This program, however, is limited to "closed" systems where all steam is returned to the boilers as condensate and where *no* steam is used for humidification or cooking, as in cafeterias or restaurants. Whenever steam is consumed, makeup water will tend to accumulate minerals in the boiler, and other treatment programs will have to be used, as described below.

The corrosion inhibitors most widely used in low-pressure steam boilers are sodium molybdate and sodium nitrite. To be certain that these inhibitors provide a continuous protective film and that no area of the metallic surfaces remains exposed, certain minimum levels must be maintained. Inhibitors are required to provide buffering, so that the pH can be maintained at 7.0 to 10.0 for maximum effectiveness.

For treating the steam and condensate of low-pressure heating systems, the neutralizing amines can be added gradually at regular intervals in sufficient quantity to maintain the pH value of the condensate at 7.5 to 8.5. The neutralizing amines should not be used in systems containing nitrites or where steam is used for humidification.

Treatment of High- and Low-Pressure Process-Steam Boilers. Because high-pressure steam systems operate at higher boiler water temperatures than low-pressure steam heating boilers, they cannot be treated with the corrosion inhibitors mentioned above. The method of treatment of high-pressure steam boilers involves heat-stable substances such as environmental stabilizers that alter the condition responsible for corrosion and deposits.

With low-pressure steam boiler operating with some makeup water due to steam losses at steam humidifiers, kitchen steam tables, dishwashers, or for any other purpose, the conditions favoring use of corrosion inhibitors are altered. With makeup water required because of steam losses, minerals present in the makeup will tend to accumulate. This requires increased blowdown to reduce the mineral buildup and additional treatment to replace that lost in blowdown. Likewise, a different type of treatment including scale inhibitors will be required to prevent hardness in the makeup water from forming scale. Therefore, such systems must be treated as high-pressure steam boilers with environmental stabilizers, rather than with corrosion inhibitors.

Treatment for corrosion control will consist of adjustment of the alkalinity and pH value with caustic soda to maintain the pH value at 10.5 to 11.5 consistent with efficient operating conditions. The boiler feed water should be deaerated to remove dissolved oxygen and to prevent oxygen pitting, followed by superimposing of an oxygen scavenger to remove the last traces of dissolved oxygen. Materials used for this purpose are sodium sulfite at levels of 20 to 50 mg/L or the organic materials diethylhydroxylamine and sodium erythorbate.

In most HVAC applications, sodium sulfite is preferred because of cost and ease of storage, handling, and control. Organic materials are preferred when dissolved solids are objectionable, such as in electrode boilers. These do not add to the mineral content of the boiler water. Diethylhydroxylamine, however, has the disadvantage of forming ammonia or amines, which are corrosive to copper, brass, and other copper alloys commonly used in heating systems. The choice of oxygen scavengers, therefore, must be carefully considered.

For control of scale and deposits, boiler water treatment includes a zeolite softener on makeup water when required, followed by internal treatment with phosphates, chelating agents, polymers, and/or dispersants. The basic program includes a phosphate polymer combination used to maintain a sodium phosphate concentration in the boiler water at 20 to 60 mg/L, ensuring that all the calcium has precipitated as phosphate.

The nonphosphate program includes polymers, phosphonates, or combinations of polymers, phosphonates, and chelants with the polymer in the range of 2 to 100 mg/L, depending on the type of polymer and formulation used. Chelates should not be used in excess of 2 to 3 mg/L free chelate because corrosion of iron can be accelerated in the boiler water.

Antifoam agents are used in the boiler water treatment formulations to prevent foaming and carryover. They enhance more efficient nucleate boiling and prevent foaming and carryover. Antifoam agents for boilers include polyalkylene glycols and polyamides used at 10 to 100 mg/L. Finally, the treatment program for high-pressure steam boilers and low-pressure process boilers should include a steam and return condensate line corrosion inhibitor. This should be either a neutralizing amine or a mixture of neutralizing amines to maintain the pH of the condensate at 7.0 to 9.0. Where the steam loss is high and the use of neutralizing amines is too costly, a film-forming inhibitor may be used to protect the condensate piping.

8.8.5 Feed and Control of Boiler Water Treatment Programs

Test Controls. The test control of water treatment is one of the most important aspects of the water treatment program. Without proper testing, more harm than good can result.

A typical analysis of a low-pressure steam boiler using a sodium molybdate inhibitor is shown in Fig. 8.33. With high-pressure steam boilers and low-pressure process boilers, the tests must be made more frequently. Testing three times a day may be required if load variations are such that conditions can change that rapidly. Daily testing with field test kits is more common and the preferred frequency for best results.

A typical test analysis of a high-pressure steam boiler water, with recommended controls, is shown in Fig. 8.34. Field test kits are available for easy testing of boiler water samples. Figure 8.35 depicts a typical drop test for chloride and sulfite used for boiler water. More elaborate test cabinets are used in the large boiler room where a watch engineer is on duty to make tests at frequent intervals. Figure 8.36 illustrates this type of equipment.

Feed Methods. Water treatment is applied to boilers by various means, ranging from shock dosage to highly sophisticated chemical proportioning pumps and controllers. These methods are briefly described.

Shock Feed Methods. Shock feeders are most convenient and effective with low-pressure heating boilers, which require only periodic addition of a corrosion inhibitor and perhaps weekly addition of a steam condensate neutralizing amine. One of the handier devices for this purpose is the force hand pump, which adds treatment directly to the boiler through a 3/4-in (20-mm) drain cock. Figure 8.37 illustrates this application.

Another shock device is the pot type or bypass feeder. This kind of feeder should be installed on the makeup water line to the boiler to inject treatment directly into the boiler with the pressure of the boiler feedwater pump. If it is installed on a raw makeup water line, an approved backflow preventer must be used to avoid the possibility of raw water contamination with boiler water treatment. A typical installation is shown in Fig. 8.38.

Proportioning Feed Methods. With high-pressure steam boilers and low-pressure process boilers, it is desirable to feed treatment to other boiler continuously in proportion to actual need. Proportioning feed systems will feed treatment based on feedwater flow.

Basic Proportioning Feed System. A simple proportioning feed system consists of a chemical feed pump interlocked with the boiler feedwater pump to transfer treatment directly from the drum into the feedwater tank prior to the boiler. This method is ideal where a single liquid boiler treatment formulation is used that contains all the required corrosion and deposit inhibitors *without* phosphates.

The feed system illustrated in Fig. 8.39 is applicable to any size boiler system provided the above conditions are met.

METROPOLITAN REFINING COMPANY INC.
50-23 23rd STREET • LONG ISLAND CITY • N.Y. 11101

CERTIFICATE OF ANALYSIS

DATE_____
SAMPLING DATE_____
REPRESENTATIVE_____

ANALYSIS NUMBER			341247			
SOURCE			LP STEAM BOILER			
pH			7.9			
P ALKALINITY	CaCO₃,	mg/L	0.0			
FREE CARBON DIOXIDE	CO₂,	mg/L				
BICARBONATES	CaCO₃,	mg/L	250.			
CARBONATES	CaCO₃,	mg/L				
HYDROXIDES	CaCO₃,	mg/L				
M (Total) ALKALINITY	CaCO₃,	mg/L	250.			
TOTAL HARDNESS	CaCO₃,	mg/L	86.			
SULFATE	SO₄,	mg/L				
SILICA,	SiO₂,	mg/L				
IRON	Fe,	mg/L				
CHLORIDE	NaCl,	mg/L	35.			
ORGANIC INHIBITOR		mg/L				
PHOSPHATE	PO₄,	mg/L				
CHROMATE	Na₂CrO₄,	mg/L	2800.			
NITRITE	NaNO₂,	mg/L				
ZINC	Zn,	mg/L				
SPECIFIC CONDUCTANCE		mmhos/cm	3831.			
TOTAL DISSOLVED SOLIDS		mg/L	3448.			
SUSPENDED MATTER						
BIOLOGICAL GROWTHS						
SPECIFIC GRAVITY @ 15.5º/15.5ºC						
FREEZING POINT ºC/ºF						
	% BY WEIGHT					

NOTES:
1. ANALYTICAL RESULTS EXPRESSED IN MILLIGRAMS PER LITRE (mg L) ARE EQUIVALENT TO PARTS PER MILLION (ppm). DIVIDE BY 17.1 TO OBTAIN GRAINS PER GALLON (gpg)

2. CYCLES OF CONCENTRATION = $\dfrac{\text{CHLORIDES IN SAMPLE}}{\text{CHLORIDES IN MAKEUP}}$

TREATMENT	TREATMENT CONTROL	FOUND	RECOMMENDED	FOUND	RECOMMENDED
--	CHROMATE	2800	2000 - 2500		
--	pH	7.9	7.0 - 10.0		

REMARKS:

R. V. Blake

SPEEDIPLY• PAT'D MCP• PAT'D MBF 28   FORM 1078-1

FIGURE 8.33   Analysis of low-pressure steam boiler water. (*Courtesy of The Metro Group, Inc.*)

Proportioning Water Treatment System for Steam Boilers without Deaerator.   This more complex program requires a more sophisticated feed system. The treatments with alkalinity and phosphates must be added directly to the boiler to prevent deposits in the feedwater line and damage to feedwater pumps. Figure 8.40 outlines this method of feeding treatment directly into each boiler in proportion to feedwater flow.

Proportional Water Treatment System for Steam Boilers with Deaerators.   When deaerators are installed, it is best to feed treatments for corrosion and pitting directly to the storage section of the deaerator. Some deaerator manufacturers warn against feed of treatment

METROPOLITAN REFINING COMPANY INC.

50-23 23rd STREET • LONG ISLAND CITY • N.Y. 11101

CERTIFICATE OF ANALYSIS

EFFICIENCY THROUGH CHEMISTRY — M METROPOLITAN EFFICIENCY PRODUCTS ®

DATE

SAMPLING DATE

REPRESENTATIVE

ANALYSIS NUMBER		374129		
SOURCE		HIGH PRESSURE BOILER		
pH		11.3		
P ALKALINITY	$CaCO_3$, mg/L	153.		
FREE CARBON DIOXIDE	CO_2, mg/L			
BICARBONATES	$CaCO_3$, mg/L			
CARBONATES	$CaCO_3$, mg/L	122.		
HYDROXIDES	$CaCO_3$, mg/L	92.		
M (Total) ALKALINITY	$CaCO_3$, mg/L	214.		
TOTAL HARDNESS	$CaCO_3$, mg/L	0.0		
SULFATE	SO_4, mg/L			
SILICA,	SiO_2, mg/L	23.		
IRON	Fe, mg/L			
CHLORIDE	NaCl, mg/L	841.		
ORGANIC INHIBITOR	mg/L			
PHOSPHATE	PO_4, mg/L	50.		
CHROMATE	Na_2CrO_4, mg/L			
NITRITE	$NaNO_2$, mg/L			
ZINC	Zn, mg/L			
SPECIFIC CONDUCTANCE	mmhos/cm	2244.		
TOTAL DISSOLVED SOLIDS	mg/L	1818.		
SUSPENDED MATTER				
BIOLOGICAL GROWTHS				
SPECIFIC GRAVITY @ 15.5°/15.5°C				
FREEZING POINT °C/°F				
	% BY WEIGHT			

NOTES:

1. ANALYTICAL RESULTS EXPRESSED IN MILLIGRAMS PER LITRE (mg/L) ARE EQUIVALENT TO PARTS PER MILLION (ppm). DIVIDE BY 17.1 TO OBTAIN GRAINS PER GALLON (gpg).
2. CYCLES OF CONCENTRATION = $\frac{\text{CHLORIDES IN SAMPLE}}{\text{CHLORIDES IN MAKEUP}}$

TREATMENT	TREATMENT CONTROL	FOUND	RECOMMENDED	FOUND	RECOMMENDED
	pH	11.3	10.5 - 11.5		
	P ALKALINITY	153.	200 - 400		
	PHOSPHATE	50.	30 - 50		
	SULFITE	35.	30 - 50		
	TOTAL DIS. SOLIDS	1818.	3000 MAX.		

REMARKS:

R. V. Blake

SPEEDIPLY® PAT'D MCP® PAT'D MBF 28

FORM 1078-I

FIGURE 8.34 Analysis of high-pressure steam boiler water. (*Courtesy of The Metro Group, Inc.*)

to the storage section of the deaerator if phosphates and other alkaline compounds are used, because they can cause precipitation of deposits prior to the boiler and can damage the deaerator tank, boiler feedwater pump, and lines. In this case, the phosphate and alkaline treatments should be fed directly to the boiler, and only the oxygen scavenger and amine, or other non-deposit-forming materials, should be fed into the preboiler section. When this is explained to the deaerator manufacturer, the objection is usually removed. Figure 8.41 describes this method.

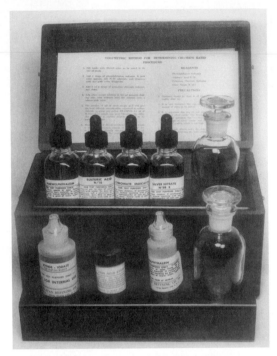

FIGURE 8.35 Field test kit for boilers. (*Courtesy of The Metro Group, Inc.*)

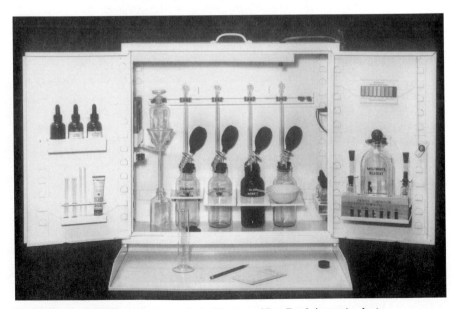

FIGURE 8.36 Test cabinet for water analysis. (*Courtesy of True Test Laboratories, Inc.*)

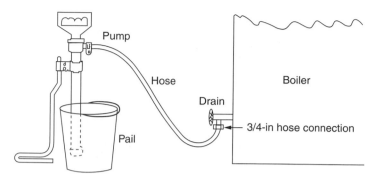

FIGURE 8.37   Hand pump for feeding inhibitors into boilers.

Automatic Blowdown and Feed controllers.   Since the ASME (American Society of Mechanical Engineers) code restrictions prohibit installing automatic valves on the bottom blowdown line, use of controllers and automatic valves is restricted to the side or top continuous blowdown line. The side or top blowdown is strictly used for controlling dissolved solids, because it will not remove suspended solids settled in the bottom of the boiler. Even when top blowdown controllers are installed, the bottom blowdown still must be administered to remove the settled and suspended solids. The top blowdown must never be used as a substitute for the bottom blowdown.

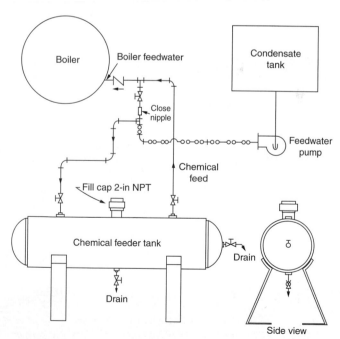

FIGURE 8.38   Bypass feeder for boilers. (*Courtesy of the Metro Group, Inc.*)

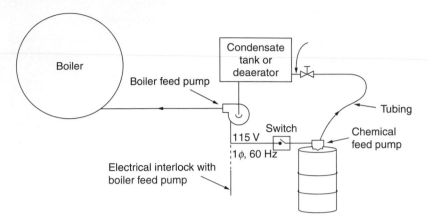

FIGURE 8.39   Simple proportioning water treatment feed. (*Courtesy of The Metro Group, Inc.*)

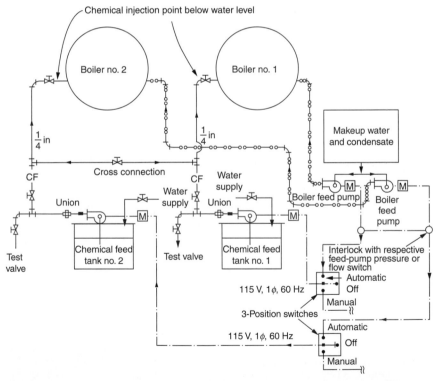

FIGURE 8.40   Proportioning water treatment feed system without deareator. (*Courtesy of The Metro Group, Inc.*)

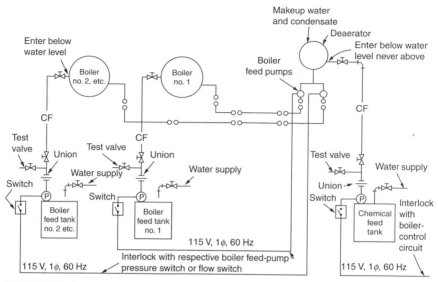

FIGURE 8.41 Proportioning water treatment feed system with deaerator. (*Courtesy of The Metro Group, Inc.*)

8.8.6 Treatment of Open Recirculating Water Systems

The major problems encountered in open cooling water systems are corrosion and deposits of scale, dirt, mud, and organic slime or algae growths. In open systems, there is a loss of water caused by both evaporation and windage drift. The water lost by drift will be of the same quality as the recirculating water; i.e., it will contain the same amount of dissolved minerals and impurities. The water lost by evaporation, however, will be of a different quality.

Evaporation from the recirculated water will be pure water vapor similar to water produced by distillation containing, theoretically, no minerals or dissolved solids unless droplets of water spray are carried over with the water vapor. This pure water vapor leaves the dissolved minerals behind in the recirculating water since they do not vaporize with the water.

The water lost by evaporation is replaced by makeup water, which contains dissolved minerals and impurities. As a result, there is an accumulation of minerals that constantly increases as the water in the system makeup water is introduced. Eventually the recirculating water will become so saturated with minerals that the most insoluble salts will come out of solution and form a scale on heat-transfer surfaces or within other parts of the system. To avoid this, the mineral content of the recirculating water must be controlled and limited. This is done by bleed-off.

Bleed-off. Bleed-off is the continuous removal of a small quantity of water concentrated with minerals from an open recirculating cooling water system. The water lost by windage drift also contains concentrated minerals and can be considered as bleed-off water. The term "zero bleed-off" refers to the most ideal situation where the windage drift is such that the minerals and suspended matter lost with the drift water droplets are sufficient to eliminate any need for an additional continuous bleed-off. In most cases, however, an additional bleed-off is required to limit the solids accumulation.

The most troublesome mineral scale is calcium carbonate, since it is the least soluble of the salts present in the recirculating water. Inhibitors are used to increase the solubility of calcium carbonate and some other minerals, but even when they are used, it is necessary to limit excessive concentration of minerals by bleed-off. Bleed-off also limits concentrations of alkalinity, total hardness, and silica.

Alkalinity should be limited to prevent precipitation of calcium carbonate scale. Alkalinity is present in the form of bicarbonates and carbonates, which combine with calcium and magnesium to form calcium carbonate and magnesium carbonate. (Calcium carbonate, being less soluble than magnesium carbonate, will form first). Alkalinity should be limited not only to prevent precipitation of calcium carbonates, but also to prevent high pH conditions, which may be damaging to some system components such as galvanized steel, brass, or cooling tower lumber.

With a total alkalinity of 500 mg/L, the pH value of a cooling tower water is expected to be approximately 8.7. This is based on the average pH of cooling tower water, as shown in Fig 8.42. This chart shows the expected average pH value, which will vary according to atmospheric conditions. For example, if the atmosphere has a considerable quantity of carbon dioxide and/or sulfur dioxide, these gases will tend to neutralize the alkalinity of the cooling tower water. For this reason, the alkalinity of the cooling tower water is not always a mathematical total of alkalinity from the makeup water; therefore, the condition may vary from location to location, and the maximum limitation of 500 to 800 mg/L of alkalinity as calcium carbonate is used only as a guide.

White Rust. One of the most frequently occurring problems caused by high alkalinity is that of "white rust." White rust is the deposit of zinc corrosion product, zinc carbonate ($ZnCO_3$) and basic zinc carbonate [$3Zn(OH_2)ZnCO_3H_2O$]. These products accumulate as a white waxy-like substance on the surface of galvanized metal of cooling towers. White rust is a complex phenomenon caused by high alkalinity o the cooling tower water and highly reactive poorly passivated galvanized sheet metal as a result of changes in the galvanizing process in the mid 1980s (Ref. 22).

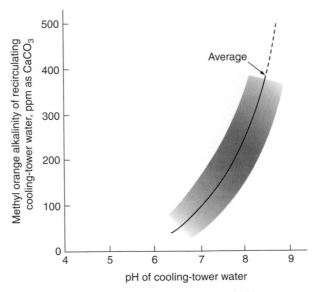

FIGURE 8.42 Average pH of cooling tower water. Normally the pH level for 90 percent of towers is within the shaded area. (*From Carrier System Design Manual, part 5., "Water Conditioning." Carrier Corporation, Syracuse, NY 1972, p5–15. Data from Betz Laboratories.*)

This problem can be controlled by cleaning and passivation of the zinc galvanized tower prior to starting up with a cleaner containing phosphate passivators. Treating with corrosion inhibitors containing phosphonate or phosphate and maintaining pH value of the tower water at 7.0 to 8.0 pH should follow this program.

Total hardness should be limited to prevent calcium sulfate scale. Sulfates are naturally present in the recirculating water or are added in the form of sulfuric acid for pH control.

When the concentration of calcium hardness exceeds the solubility of calcium sulfate, it will come out of solution. Inhibitors can be used to extend this solubility by forming a supersaturated solution. But bleed-off must still be used to maintain the maximum limit.

Where all the hardness salts are present as bicarbonate or carbonate, the limitation of total hardness as calcium carbonate is typically 400 to 600 mg/L. However, where sulfuric acid is employed or where sulfates (the next least soluble of the calcium salts) are naturally predominant (as in some well waters) the limit is somewhat higher.

If calcium sulfate exceeds 1200 to 1500 mg/L (calculated as calcium carbonate), it tends to precipitate, or, depending upon the water chemistry, a supersaturated solution may form. Scale inhibitors can extend this solubility beyond this natural maximum, perhaps up to 1.6× saturation, but the guideline of 1200 mg/L maximum is typically used to limit the cycles of concentration and control stress operating conditions.

Silica must be limited to prevent silica scale, and the limit is the solubility of silica at the temperatures encountered. The silica limitation is based on the solubility of silicon dioxide (SiO_2) at the temperatures encountered in HVAC cooling water systems. Silica is typically soluble up to approximately 150 mg/L. Beyond this concentration it will tend to come out of solution. Other factors, such as presence of scale inhibitors, the presence of magnesium and other mineral salts, and the alkalinity of the water will have an influence on this, but where silica is a factor, this limitation is typically used as a guide in determining the maximum cycles of concentration permissible.

The maximum levels of alkalinity hardness and silica that are suggested with scale inhibitors present are shown in Table 8.9.

Once the maximum limitations for certain dissolved minerals permitted in a recirculating cooling water system have been established, it is necessary to limit these minerals by bleed-off. See Table 8.9 for maximum concentrations. The maximum cycles of concentration recommended for each of the items listed in Table 8.9 are determined by dividing the maximum values in this table by the amount of each found in the makeup water analysis. To demonstrate how this is done, let us examine a city water analysis from Harrisburg, PA (Fig. 8.43), and determine the maximum cycles of concentration to be recommended in a cooling tower system water.

Shown below is a comparison of each of the limiting impurities, with that impurity in the makeup water, to determine the maximum cycles of concentration.

Analytical results	Maximum cycles of concentration
Alkalinity:	
$\dfrac{\text{Maximum}}{\text{Makeup water}} = \dfrac{500 \text{ mg/L}}{167 \text{ mg/L}}$	$= 3.0\times$
Hardness:	
$\dfrac{\text{Maximum}}{\text{Makeup water}} = \dfrac{1200 \text{ mg/L}}{149 \text{ mg/L}}$	$= 8.0\times$
Silica:	
$\dfrac{\text{Maximum}}{\text{Makeup water}} = \dfrac{150 \text{ mg/L}}{10 \text{ mg/L}}$	$= 15.0\times$

TABLE 8.9 Maximum Concentration of Mineral Solids for Cooling Towers, Evaporative Condensers, Air Washers, and Spray Coil Units

Total alkalinity, calcium carbonate	500 mg/L
Total hardness as calcium carbonate	1200 mg/L
Silica as silicon dioxide	150 mg/L

 ™ **THE METRO GROUP, INC.**

50-23 Twenty-Third Street
Long Island City, NY 11101
(718) 729-7200
FAX: (718) 729-8677

CERTIFICATE OF ANALYSIS
WATER ANALYSIS

Divisions:
Metropolitan Refining
Consolidated Water Conditioning
Cosmopolitan Chemical
Petro Con Chemical

CLIENT: _____ DATE: _____

ADDRESS: _____ REPRESENTATIVE: _____ SAMPLE DATE: _____

HARRISBURG, PA   ANALYSIS NO.: 359065   SOURCE: CITY

pH			7.5	PHOSPHATE	PO₄	mg/L	
P ALKALINITY	CaCO₃	mg/L		MOLYBDATE	Na₂MoO₄	mg/L	
FREE CARBON DIOXIDE	CO₂	mg/L		NITRITE	NaNO₂	mg/L	
BICARBONATES	CaCO₃	mg/L	167.	ZINC	Zn	mg/L	
CARBONATES	CaCO₃	mg/L		SPECIFIC CONDUCTANCE	msiemens/cm		818.
HYDROXIDES	CaCO₃	mg/L		TOTAL DISSOLVED SOLIDS		mg/L	501.
M (Total) ALKALINITY	CaCO₃	mg/L	167.	SUSPENDED MATTER			
TOTAL HARDNESS	CaCO₃	mg/L	149.	BIOLOGICAL GROWTHS TOTAL BACTERIA COLONIES/ML			
SULFATE	SO₄	mg/L		SPECIFIC GRAVITY @ 15.5°/15.5°C			
SILICA	SiO₂	mg/L	10.	FREEZING POINT			
IRON	Fe	mg/L		% BY WEIGHT			
CHLORIDE	NaCl	mg/L	187.				
ORGANIC INHIBITOR	PHOSPHONATE	mg/L					

RESULTS:

TREATMENT	TREATMENT CONTROL	FOUND	RECOMMENDED

REMARKS:

NOTES:
ANALYTICAL RESULTS EXPRESSED IN MILLIGRAMS PER LITRE (mg/L) ARE EQUIVALENT TO PARTS PER MILLION (ppm).
DIVIDE BY 17.1 TO OBTAIN GRAINS PER GALLON (gpg).
CYCLES OF CONCENTRATION = CHLORIDES IN SAMPLE/CHLORIDES IN MAKEUP

SW:dm

SAM WILDSTEIN, MANAGER LABORATORY SERVICES

Water Experts Since 1926 / Sales • Service • Solutions

FIGURE 8.43 City water analysis of Harrisburg, PA. (*Courtesy of The Metro Group, Inc.*)

From this table we see that the maximum cycles of concentration recommended are 3.0x. The lowest value obtained is used, for if it were exceeded, difficulty with that particular impurity would result. In this case, it would be the alkalinity, an excess causing precipitation of carbonate salts or excessive pH conditions. If a neutralizing acid were used in a system with this water, the alkalinity would be reduced and no longer considered a limiting factor. In that case, the maximum cycles of concentration would be increased to 8.0, the next lowest value.

After the maximum cycles of concentration recommended in an open recirculating cooling water system have been determined, the bleed-off rate required to maintain the solids accumulation below this maximum level can be calculated. The mathematical formula for the bleed-off rate with respect to the evaporation rate of an open cooling water system is

$$B = \frac{E}{C-1}$$

where B = bleed-off rate, gal/min (L/min)
E = evaporation rate, gal/min (L/min)
C = cycles of concentration

The purpose of bleed-off is to remove dissolved solids in order to maintain a maximum level, determined by the maximum recommended cycles of concentration. To maintain this constant maximum level of solids, the amount of solids entering the system must be equal to the amount of solids leaving the system. This can be expressed mathematically as

$$\begin{matrix} BCX & = & MX \\ \text{(solids leaving)} & & \text{(solids entering)} \end{matrix} \qquad (8.1)$$

where C = cycles of concentration
B = bleed-off rate, gal/min (L/min)
M = Makeup water, gal/min (L/min)
X = Solids concentration of makeup water (ppm, mg/L, or g/gal)

The makeup rate to an open cooling water system is proportional to the load on the system and is equivalent to the evaporation rate plus any losses of drift, overflow, or bleed-off. This is expressed as

$$M = E + B \qquad (8.2)$$

where M is the total makeup rate and E is the evaporation rate. Substituting in Eq. (8.1), we obtain

$$BCX = (E + B)X$$

Solving for B gives

$$B = \frac{E}{C-1}$$

The bleed-off is a total of all water losses from the recirculating systems such as leaks at pumps, overflow, windage drift, and actual bleed. All these together should be sufficient to keep the cycles of concentration below the recommended maximum. To determine if the total bleed-off from a system is adequate, the cycles of concentration are measured by a simple chloride test. Therefore, a measure of the chlorides in the recirculating water will tell very accurately how much the makeup water has concentrated in that system.

The cycles of concentration are found by comparing the chlorides of the makeup water with the chlorides of the recirculating water:

$$\text{Cycles of concentration} = \frac{\text{chlorides in recirculating water}}{\text{chlorides in makeup water}}$$

Scale Control.   Bleed-off is necessary to prevent deposits of mineral salts as a scale. Scale inhibitors must be used, however, to minimize bleed-off water losses and to permit operation at as high a solids concentration as possible, as outlined in Table 8.9. Without scale inhibitors, it would not be possible to accumulate bicarbonate alkalinity and hardness beyond limits, which are much lower than those outlined in Table 8.9.

Acid Feed.   One method of inhibiting scale is by using acid to control the Langelier saturation index. Acids neutralize the bicarbonate alkalinity. Acids can be used to maintain the pH value as close to 7.0 as possible with the alkalinity below the saturation level. This method, however, requires very careful pH and alkalinity control and, most important, careful and controlled feed of acid to avoid excessive corrosive conditions.

Water Softeners.   Another way to control scale in open recirculating water systems is by using ion-exchange water softeners. By reducing the hardness, higher cycles of concentration can be established without exceeding the pH of saturation. However, to be effective, this method must include simultaneous feed of acid to reduce the alkalinity. Therefore, it offers no further advantage over alkalinity reduction without softening. Ion-exchange softeners operate at less than 100 percent efficiency throughout the softening cycle. Therefore, trace quantities of hardness bypass the softeners before regeneration.

8.8.7   Chemical Inhibitor Treatment of Open Recirculating Water Systems

Scale Inhibitors, Dispersants, and Deposit Control Agents.   The most economical and effective method to control scale is to use scale inhibitors with or without alkalinity reduction. Scale inhibitors, when added to the recirculating water at very low levels, 1 to 20 mg/L, will reduce the scaling tendency. This is accomplished by holding the scale-forming minerals in solution beyond their saturation level, forming supersaturated solutions, by interfering with the growth of scale crystals and formation of scale, or by dispersing the particles of scale and preventing them from adhering to heat-transfer surfaces. In practice, scale inhibitors most likely act in all three ways, thereby preventing scale formation of carbonate, silicate, and sulfate salts of calcium and magnesium. Modern inhibitors also act to control general and specific deposits and risks of fouling, including biofouling.

The first scale inhibitor used was sodium hexametaphosphate, which when added to water at only 2 to 3 mg/L, was found to inhibit precipitation of calcium carbonate from supersaturated solutions. This phenomenon was called *threshold treatment* since the polyphosphate holds calcium carbonate in solution at the threshold of precipitation. This was later applied to cooling water treatment when polyphosphate became widely used as an inhibitor for scale as well as corrosion. Polyphosphates are still used today, in the range of 0.5 to 5 mg/L, to effectively inhibit scale in condenser water systems. There are, however, some disadvantages to polyphosphate scale inhibitors. These products tend to hydrolyze and revert to the orthophosphate form, which is ineffective as a threshold scale inhibitor. In fact, precipitates of calcium and magnesium orthophosphates can form, causing soft scales on heat-transfer surfaces called *sludge*. In addition, overfeed of polyphosphates in excess of

20 mg/L can result in the formation of insoluble calcium polyphosphate sludge. Finally, the ability of polyphosphates to hold calcium carbonate in solution diminishes significantly with high pH conditions (above 8.0) in recirculating cooling water systems.

Since the 1960s other (superior) scale inhibitors have been developed and new materials regularly enter the market. Polyacrylate (and their co- and ter-polymer derivatives), phosphonate, polymaleic acid and phosphinocarboxylic acid chemistry have proved to be significantly more effective than polyphosphates without the above disadvantages. Often, synthetic polymers, such as low-molecular-weight polyacrylates, polymethacrylates, polymaleates, polyhydroxy alkyl acrylates, and polyacrylamides are formulated with polyphosphate and other inorganics to provide multipurpose inhibitors.

Some of the most widely used organic substances for both scale and corrosion control are the various organic phosphonates. In particular, ATMP, HEDP, and PBTC phosphonates are used at very low levels (3 to 5 mg/L as "active" products), as threshold treatments and crystal modifiers to prevent precipitation of calcium carbonate, sulfate, and phosphate.

All these polymers, phosphonates, and organic phosphates have the distinct advantage of excellent scale-inhibiting properties over a wide range of pH and temperature conditions without the disadvantages of inorganic polyphosphates mentioned above. Many of the polymers, copolymers, and terpolymers used today are very effective zinc, iron, and phosphate sludge inhibitors, stabilizers, or deposit-control agents, as well as basic calcium carbonate scale inhibitors.

Corrosion Inhibitors.   Just as in boilers, corrosion in open cooling water systems is controlled with both environmental stabilizers and corrosion inhibitors. In open cooling water systems, however, oxygen cannot be economically removed with oxygen scavengers because these systems are constantly aerated. Therefore, corrosion inhibitors that are effective in oxygen-containing environments must be utilized.

Also unlike boilers, recirculating cooling water systems contain various types of metals and alloys, and inhibitors and environmental conditions maintained must be compatible with these multimetallic systems. A particular problem associated with these systems is galvanic corrosion caused by bimetallic couples.

Treatment of recirulating cooling water must first include control of the pH value for both scale and corrosion control. A pH value within the range of 6.5 to 9.0 is usually maintained depending on the type of corrosion and scale inhibitor used. Lower pH values tend to render the recirculating water excessively corrosive, while higher pH values will result in both amphoteric metal corrosion (as with zinc, brass, and aluminum) and scale-forming conditions.

In areas where the atmospheric environment is such that acid conditions develop in open cooling water systems, neutralizing substances, such as caustic soda and sodium carbonate (soda ash), can be used to maintain the desired pH value in the neutral range. These conditions can be found in large urban areas or at locations subject to acid fumes from an adjacent boiler stack or incinerator. Conversely, in areas where the makeup water is excessively alkaline, it may be necessary to add acid to maintain the desired pH range for control of scale, as previously outlined. Within the pH range of 6.5 to 9.0, corrosion is controlled by corrosion inhibitors added to the recirculating water. Inhibitors are substances that do not necessarily alter the environment, but do act as a barrier between the corrosive medium and the metal surface. These materials, when added to the recirculating water, form a protective barrier on the metal surface either by chemical reaction with the metal surface or by physical or chemical adsorption on the metal surface. An actively corroding metal can be rendered passive through the use of inhibitors that react in this manner. There are many types and combinations of corrosion inhibitors used for open cooling water systems. Basically they fall under four categories, *molybdate, zinc, phosphate,* and *organic inhibitors.*

Molybdate Inhibitors.   Molybdate inhibitors depend upon sodium molybdate as a weak oxidizer for ferrous metals promoting a complex passive oxide film. Molybdates are used in combination with other inhibitors such as zinc, phosphate, phosphonates, and organics for synergistic effect to produce highly effective corrosion inhibitor formulations. These formulations of inhibitors have a total effect that is greater than the sum total effectiveness of each individual component and are called synergistic blends. These blends also will include nonferrous metal inhibitors such as mercaptobenzothiazole, benzotriazole, and tolyltriazole. These are necessary to protect copper and copper alloys such as yellow brass, bed brass, and admiralty metal.

The molybdate inhibitor blends are usually used at low levels in open cooling water systems, typically maintaining 5 to 10 mg/L as sodium molybdate. However, there are many programs employing organic polymers (all organics) that contain extremely low levels of molybdenum. Here, the typical level in the cooling system is, perhaps, only 1.0 to 1.5 ppm as Mo. The molybdenum is acting primarily as a *tracer*, as it is capable of being easily and accurately tested. There is a secondary benefit, in that the Mo acts as a synergiser to enable other (organic) inhibitors, such as totyltriazole (TTA) for copper and hydroxyphosphinocarboxylic acid (HPCA), for iron/steel, to function better.

Zinc Inhibitors.   Some of the most effective inhibitors use zinc as a cathodic inhibitor in combination with other inhibitors that synergize the overall effectiveness of the inhibitor blend. Some blends are zinc-silicate, zinc-phosphonate, zinc-molybdate, zinc–polyphosphate-organic, and various zinc–organic-phosphate combinations.

The concentration of zinc is usually held to a maximum of 2 to 5 mg/L, depending upon the type of formulation. The United States Public Health Service Drinking Water Standards limit the zinc of drinking water to 5.0 mg/L, and where this criterion is observed, the zinc should be held to this maximum level. In some cases, zinc discharged even at this level can be toxic to aquatic life. Therefore, discharge to fishponds, streams, and other public waters should not include even this amount of zinc.

Some discharge criteria limit the permissible zinc concentrations to 1.0 mg/L or less. In this case, non-zinc inhibitors should be used.

Phosphate Inhibitors.   Phosphate inhibitor blends are used in alkaline corrosion inhibitor systems combined with polymers that inhibit formation of phosphate sludge. With the development of very effective phosphate scale inhibitors and dispersants such as the copolymers of acrylic acid, sulfonic acid polymer, sulfonated polystyrene, and maleic acid phosphate-based inhibitors have gained wide acceptance. The orthophosphate levels maintained vary from 5 to 20 mg/L. These inhibitors are most effective at high pH and alkalinity and are usually used at pH 8.0 to 9.0. The formulations require increased scale control from improved polymer and phosphate blends. The organic azole nonferrous metal inhibitors are also necessary for complete protection.

Organic Inhibitors.   These are basically fully organic inhibitors that are biodegradable and relatively nonpolluting. Of course, they will add some foreign organic substances to cooling water discharge, but the relative tolerance is better than molybdate, phosphate, or zinc compounds. The organic inhibitors include combinations of various organic and inorganic compounds that have been known to be effective corrosion inhibitors and that tend to be synergistic when used in combinations. Materials include the azoles and hydroxyphosphinocarboxylic acid.

Corrosion Testing.   The relative performance of the types of programs available is based on actual field experience and corrosion monitoring. This can be determined by inserting corrosion test coupons in the open recirculating cooling water system. This corrosion test

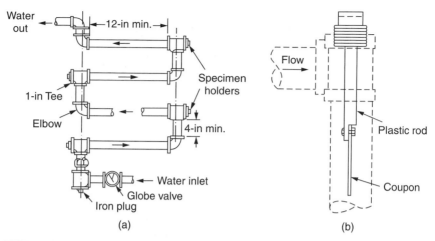

FIGURE 8.44 Corrosion test coupon assembly. (*Courtesy of The Metro Group, Inc.*)

method has been described by the National Association of Corrosion Engineers (NACE) and is consistent with the ASTM standard "Corrosivity Testing of Industrial Cooling Water (Coupon Test Method)."

The test coupon is placed in the recirculating water system in a test coupon "rack" described in Fig. 8.44. The corrosion test report will include the calculated corrosion rate in mils per year or micrometers per year and other pertinent data such as depth of pits, variations, and types of corrosion deposits. The corrosion rate of mild steel for a 30-day exposure time in an open recirculating cooling water system is generally rated in accordance with Table 8.10. The corrosion rate on mild steel can also be evaluated based on the lifespan of a standard 6-in (152.4-mm) Schedule 40 steel pipe assuming failure would occur first at a threaded joint (see Table 8.11).

Likewise the corrosion rate on standard 16-gauge copper condenser tubing, based on the expected lifespan of the tubing (assuming the corrosion is uniform), can be evaluated by using Table 8.12. The performance expected with modern (nonchromate) corrosion inhibitors today would be as follows:

• carbon steel: 0–2.0 mil/year
• copper: 0–0.2 mil/year

TABLE 8.10 Mild-Steel Corrosion Rates, 30-Day Exposure

Corrosion rate, mils/(MDD)*	Corrosion control
>5 (>27.3)	Poor
2–5 (10.9–27.3)	Good
0–2 (0–10.9)	Excellent

*MDD = milligrams per square deciliter per day.
Source: See Ref. 10.

TABLE 8.11 Evaluation of Corrosion Rates for Mild Steel

Corrosion rate, mils/yr (MDD)*	Half-life of standard 6-in (152, 4-mm) Schedule 40 pipe, years	Corrosion control
0–2 (0–10.9)	>70	Excellent
2–5 (10.9–27.3)	70–28	Good
5–8 (27.3–43.7)	28–17$\frac{1}{2}$	Fair
8–10 (43.7–54.6)	17$\frac{1}{2}$–14	Poor
>10 (>54.6)	<14	Intolerable

*MDD = milligrams per square deciliter per day.

TABLE 8.12 Evaluation of Corrosion Rates for Standard Copper Condenser Tubing

Corrosion rate, mils/yr (MDD)*	Expected life of standard 16-gauge copper condenser tubing, years	Corrosion control
0–1 (0–6.2)	>65	Excellent
1–2 (6.2–12.4)	65–32$\frac{1}{2}$	Good
2–3 (12.4–18.6)	32$\frac{1}{2}$–21$\frac{2}{3}$	Fair
3–4 (18.6–24.8)	21$\frac{2}{3}$–16$\frac{1}{4}$	Poor
>4 (>24.8)	<16$\frac{1}{4}$	Intolerable

*MDD = milligrams per square decimeter per day.

8.8.8 Control of Organic Growths in Open Recirculating Water Systems

Open recirculating water systems are continually exposed to the ambient air and its contaminants. Among those contaminants are microbiological organisms which, under favorable conditions, will grow and flourish in the warm cooling tower water. The major forms of biological organisms that thrive in cooling tower waters are the bacterial slimes, algae, and fungi. The spores and seeds of these organisms are present in the atmosphere attached to particles of dust, dirt, pollen, and other airborne particulates. When washed out of the air at the cooling tower they find a favorable environment to grow and multiply. These growths cause serious problems of corrosion, fouling, blockage, and accelerate buildup of deposits. Such growths not only interfere with water flow but also reduce heat transfer and increase energy losses.

Biological growths can be controlled by substances capable of limiting their growth or outright killing the organisms without harming the system or the environment. These substances are called microbicides, biocides, slimicides, algicides, and fungicides. They fall into the general classification called pesticides, which are controlled and registered by the Environmental Protection Agency. Pesticides used in cooling water treatment are more commonly referred to as microbicides.

A *disinfecting dosage* is one from which a complete kill can be expected within a few hours, provided the sterilizing solution comes into contact with the organism. An *inhibiting dosage* is the level at which the growth of the organism can be prevented or minimized once the dosage is removed from the system. A *biostatic dosage* is one that is applied (usually continuously) to maintain microorganisms at tolerably low levels. In general, only oxidizing biocides are applied as biostats, due to their low feed rate and generally lower cost compared to nonoxidizers.

Oxidizing biocides.   The most widely used microbiocide for recirculating water systems is chlorine, an oxidizing biocide. Chlorine used in excess will damage wood and organic fill in cooling towers. Usually it is only used in systems large enough to justify equipment for its controlled feed. Where chlorine is used on a continuous basis, the concentration of free chlorine should be maintained at 0.3 to 0.5 mg/L to minimize the attack on materials of construction. For cleaning purposes, shock feed of up to 50 mg/L can be used, provided this high chlorine content is held for no more than 8 h and the system is thoroughly flushed and drained, to remove dead organic matter and excess chlorine. The use of chlorine on an intermittent basis has been found to be effective in inhibiting the growth of a particular biological organism while at the same time minimizing the disadvantages of continuous chlorine feed.

Chlorine is a very strong oxidizing agent, and so it is excessively corrosive to metals and damaging to cooling tower lumber and organic fill. Furthermore, it is very difficult to control the feed of chlorine at levels that will be effective in limiting organic growth while not attacking the materials of construction. Bromine, less corrosive than chlorine, is more commonly used and is more effective than chlorine at the operating pH range of cooling towers today.

The microbicides most extensively used in cooling towers, air washers, and other open recirculating water systems for HVAC are bromine, quaternary ammonium compounds, organosulfur compounds and aldehydes. The most effective combinations include hydroxy phosphonoacetate with the organic polymers and triazoles. Other effective microbicides are organotin copper compounds used with the quaternary ammonium compounds, also dibromo-3-nitriloproprionamide mixed isothiazolins and 1-bromo-3-chloro dimethyl hydantoin.

Nonoxidizing microbicides are best applied to clean systems to inhibit, i.e., prevent, an organic growth rather than to kill or remove an existing growth. Sterilizing dosages of these biocides can be used to kill an organism, but usually penetrating agents are needed to assist the biocides in penetrating the mat of the organism, especially when the growth is of the algae type, which forms layers of thick mats on the surface of the equipment.

Cleaning and disinfecting a contaminated cooling water system requires a significant amount of physical flushing and removal of organic matter to prevent reinfestation. Care must be taken to circulate the solution through many nooks and crannies in these systems where growths can adhere and eventually grow to reinfest a system. Table 8.13 lists several of the more common microbicides and their effective sterilizing dosage and inhibiting dosage for most of the organisms found in HVAC open recirculating water systems.

The manner of feeding biocides to a system is very important. Often the continuous feed at low-inhibiting dosages is not particularly effective and can be very costly. Organisms tend to build up immunity to a single biocide, and low dosages may even encourage the growth of some organisms. Therefore, a shock dosage at a level high enough for a bacterial kill is more effective than a continuous dosage at a low level. In addition, occasional change of the type of biocide used will prevent development of microbial immunity.

Treatment of HVAC systems should include periodic shock dosages of a biocide only when experience indicates a need.

Bacteria Test.   Test methods are available for monitoring the recirculating water to obtain a total bacteria count and predetermine the need for biocides. Field methods include test strips, dipsticks, IME glass vials, and ATP meters. Laboratory methods include plate counts. Tests made periodically during operation of the system will provide a history that

TABLE 8.13 Common Microbicides Used in Treating Open Recirculating Water Systems

	Sterilizing dosage, ppm (mg/L)	Inhibiting dosage, ppm (mg/L)	Organisms controlled
Quaternary ammonium compounds			
n-Alkyl (60% C_{14}, 30% C_{16}, 5% C_{12}, 5% C_{18})-Dimethyl benzyl ammonium chloride	100–200	5–50	Algae, mold, and bacteria
n-Alkyl (98% C_{12}, 2% C_{14}) Dimethyl 1-naphthylmethyl ammonium chloride	100–200	2–20	Algae and bacterial slimes
Poly (oxyethylene) (dimethylimino) ethylene (dimethylimino) ethylene dichloride)	50–100	1–10	Algae, fungus, and bacterials slimes
Mixture of alkyl dimethyl (benzyl ammonium chlorides and ethylbenzyl ammonium chlorides)	50–100	1–10	Algae, fungus, and bacterial slimes
Organosulfur compounds			
Dimethyl dithiocarbamate salts of sodium or potassium	50–100	10–20	Bacterial and fungal slimes and molds
Methylene bis thiocyanate	50–100	2–10	Bacterial slime, fungi, and algae
N-methyl dithiocarbamate salts of sodium or potassium	50–100	10–20	Bacterial and fungal slimes, molds, and algae

will indicate if there is an increase in the bacteria count during any particular season, showing the possible need of a biocide. These tests are also used to determine the effectiveness of a biocide program over time and indicate when treatment programs should be changed or improved. The biocide program performance should be evaluated as follows:

Total bacteria count	Condition
0–10^2 colonies/mL	Excellent control
10^2–10^3 colonies/mL	Warning, but no serious fouling
10^4–10^5 colonies/mL	Maximum 10^4 for healthcare, 10^5 for commerce/industry
10^5–10^6 colonies/mL	Fouling anticipated and possible health risk
Above 10^6 colonies/mL	Serious risk of fouling and pathogens

8.8.9 Infection Control in Open Recirculating Water Systems

The control of infection from water systems requires environmental maintenance and surveillance. Awareness of the part played in the spread of infectious diseases by pathogenic organisms, by dirty and poorly designed or maintained cooling systems and other HVAC systems, has grown significantly over recent years. Publicity, litigation, and general environmental and health concerns have added to the awareness. Consequently, in the United States, many rules and codes now exist requiring owners and operators to improve operations by retro-design changes and to regularly clean and disinfect their cooling systems. Also required is the application of a comprehensive water treatment program and record keeping in a suitable logbook.

Such rules are, as yet, not as stringent as those in some parts of Western Europe and Australia and New Zealand, but are likely to become so in the not-so-distant future. As a result, regular cleaning and disinfection using the Wisconsin Protocol, or similar, is desirable.

Protocols for Cleaning and Disinfection Programs.   These may vary depending upon whether the disinfection procedure is part of a scheduled maintenance program, or an emergency decontamination response. Commonly, however, the disinfection procedures fall into three parts.

1. *Prechlorination stage* is designed to provide disinfected water, either remotely stored, or obtained by the chlorination of the water in the cooling tower, prior to cleaning work being carried out. Typically water is considered disinfected if a 20-ppm free reserve of chlorine is maintained for at least 1 h. Lower levels of free reserve may require a much longer period for initial disinfection.

2. *Draining and cleaning stage.* Here, the disinfected water is circulated for 1 to 4 h, then drained and the system physically cleaned. Biodispersants may be employed.

3. *Disinfection, recommissioning, and water treatment reinstatement stage.* The system is refilled and disinfected for the required period, then put back into service. Standards again vary. Typically for *routine maintenance* the water is considered disinfected if a 5-ppm free reserve of chlorine is maintained for at least 1 h. If the system is being *decontaminated* because of positive *Legionella* results, then the requirement is often 10 ppm free reserve for at least 1 h.

Note that during any cleaning and disinfection procedure, the application of disinfectant and dispersant may dislodge sufficient amounts of solids to clog screens and filters. All screen and filters should be inspected periodically during the emergency procedure, and during general use they should be cleaned as needed. Also, to assure adequate chlorine contact throughout the tower, be sure that spray nozzles and other mechanical components are operating properly. During the water treatment reinstatement stage, implement the chemical treatment program in accordance with instructions from your water treatment contractor. The deposit and corrosion inhibitor products used should not interfere with the action of the biocidal compounds in regards to their effectiveness against *Legionella*.

Legionella Testing.   Monitoring the total bacterial count in the cooling system water is a useful indicator of the general effectiveness of the biocide treatment program and to the possible need for modification; however, there is no definitive correlation between total plate count and the presence of *Legionella*.

With regard to *Legionella* testing, a positive result of below 1×10^2 in cooling waters tends to reflect only that the system is a potential amplifier for higher numbers and some inspection and future retesting may be necessary. Where the positive result is above 1×10^2, this usually indicates a cleaning and disinfection program is necessary. This is especially true if the positive result is for *Legionella pneumophila sero-group 1*. Actions based on absolute numbers are not necessarily useful; rather, if during a sampling and testing program, more than 30% of samples are proved positive, this is cause for concern.

Legionella species do not respond readily to most traditional biochemical tests used for identification of bacteria. Confirmation of isolates is achieved by serological methods, or the use of more sophisticated tests. There are several tests currently employed, each with their own benefits and limitations, including

- Standard culture method
- *Legionella* rapid assay method
- Direct fluorescent antibody (DFA) method

Many authorities still do not advocate regular testing and although unplanned and once-off tests are of minimal value it should be remembered that *Legionella organisms could kill you. Consequently, it is recommended that all new cooling systems and cooling systems that have shut down for an extended period to time be tested for the presence of* Legionella. *It is also recommended that all cooling systems be assessed to determine risks associated with* Legionella. *Those cooling systems found to be at high to moderate risk should be tested a minimum of four times a year.* All other cooling systems should have a test performed twice a year, as part of a planned program. This testing will allow for up-to-date record keeping and any trend to be identified. Also, it will help confirm the quality, or otherwise, of the overall infection control program.

Testing for *Legionella* requires special sampling techniques and usually requires a period of 14 working days from receipt of the sample to obtain a result. A positive result will include an identification of the *Legionella* species, and typically also of the sero-groups 1 to 6, plus a determination of the number of the bacterium per 1 mL of the original sample.

8.8.10 Control of Muds and Other Treatments

Mud and Dirt Control. Deposits of mud, dirt, and foreign suspended matter washed out of the atmosphere into open recirculating water systems can be as troublesome as scale and must be controlled for efficient operation of heat-transfer equipment. In the past, treatment for mud, dirt, and silt consisted only of physical removal with filters and separators or manual cleaning.

Now soluble polymers used in water clarification are applied to cooling waters with significant success. These polymers are high-molecular-weight polyelectrolytes that attract small dirt particles, forming a larger mass. This process is called coagulation. The larger mass, called an agglomerate, has a tendency to "float" or remain nonadherent to piping surfaces, and it does not compact into a mud. Coagulant-type polyelectrolytes can be used for mud and dirt control at dosages as low as 0.5 to 1 mg/L. Coagulants usually are best applied in a shock manner on a periodic basis rather than continuously, while dispersants are applied on a continuous basis for maximum results. These may be included in the proprietary corrosion and deposit inhibitor formulation.

Another type of mud and dirt control agent is the dispersant. Dispersants have the opposite effect and prevent formation of mud deposits because of their ability to "repel" particles. Dispersants prevent the formation of larger agglomerates and thus their compacting into a mud. This causes the particles to be held in suspension, so they are more readily removed with normal bleed-off.

Cooling Tower–Evaporative Condenser Treatment. A complete treatment program for condenser water systems for HVAC equipment should include all the materials required to prevent corrosion and deposits. In addition, parameters should be given for bleed-off control to avoid excessive accumulation of minerals in the cooling tower and evaporative condenser circuit.

Air Washer and Surface Spray Treatment. Treatment of air washers and surface spray units will be similar to that of open cooling tower and evaporative condenser systems. The essential difference is that the former systems are subject to a wide variation of conditions with respect to temperature and humidity.

8.8.11 Feed and Control of Cooling Water Treatment Programs

Test Controls. Simple, easy-to-use test kits (similar to that shown in Fig. 8.35) are available for regular testing of inhibitor levels and chlorides for cycles of concentration to control bleed-off. The tests should be made daily to ensure adequate control. Additional tests are made by the water service company to verify the accuracy of field test results.

Feed Methods. Treatments may be fed into systems simply by shock feed through any convenient opening, usually in the tower or condenser pan. This requires frequent additions at high levels to ensure maintenance of minimum levels of inhibitor in the recirculating water. For example, to maintain a minimum of 300 mg/L of a corrosion inhibitor in a cooling tower will require shock dosages at levels up to 664 mg/L every day. Figure 8.45 shows the treatment level after each initial shock dosage every 24 h in a 1000-gal (3800-L) system operating at 100-ton (352-kW) capacity, 24 h/day, seven cycles of concentration, 0.5 gal/min (1.9 L/min) bleed-off rate.

As shown in Fig. 8.45 this method can be very costly and wasteful. The shaded area indicates overtreatment required to maintain the minimum level of 300 mg/L. To conserve treatment, therefore, proportional feed methods should be used.

Shock methods of feed are required, however, for certain types of treatment such as cleaners and biocides. These are fed at high levels on an intermittent basis to effect a cleanup action or biological kill, rather than to continuously maintain minimum levels.

1. *Canister-type feeders:* To obtain a more continuous or gradual feed, the canister-type feeder is used. This feeder comes in various designs. All consist of a can, or reservoir, containing treatment that is gradually dissolved or diluted so that it is fed slowly over a long time.

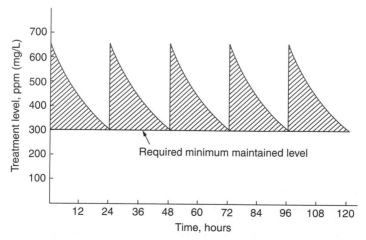

FIGURE 8.45 Variation of inhibitor level with shock treatment of cooling towers.

2. *Proportioning feed methods:* A more accurate method of feeding treatment is by means of a proportioning chemical feed pump. Proportioning feed pumps add treatment directly in proportion to the loss of treatment from the system, thereby maintaining a precise and consistent minimum treatment level. This avoids the excessive losses of shock feed or less accurate canister feeders.

A simple proportioning feed method utilizes a chemical feed pump interlocked with the recirculating water pump on the cooling tower, evaporative condenser, or spray cooler. The pump adds treatment directly from the drum or mixing tank. Figure 8.46 outlines this method of feed. This system automatically injects the necessary treatment into the recirculating condenser water line, based on an assumed continuous-load factor, Sufficient water is bled from the system on a continuous basis when the condenser water pump is on. This maintains the cycles of concentration at a safe operating level. Treatment is then fed continuously in proportion to the bleed rate so as to make up for loss of treatment in the bleed-off.

One of the most basic, fundamental, and universally adaptable fully automatic systems uses the water flow meter as a control. Since the rate of makeup water is proportional to evaporation and the actual load on a condenser water system, this is a fully automatic control system, independent of the quality of makeup water that requires no additional instrumentation for accuracy other than pH control, if needed. With this system, sufficient water can be bled automatically from the recirculating water to maintain the maximum cycles of concentration permitted. Treatment is then fed in direct proportion to bleed to replace any that is lost in the bleed-off. Figure 8.47 outlines this type of system.

Another widely used automatic proportioning feed-and-bleed system uses a total dissolved solids (TDS) analytical instrument as a control. This is commonly called the "conductivity controller" or "TDS controller." Since the dissolved mineral solids in water are directly proportional to the electrical conductivity, the conductivity measurement can be used to determine any variation in dissolved solids. Because the dissolved solids content

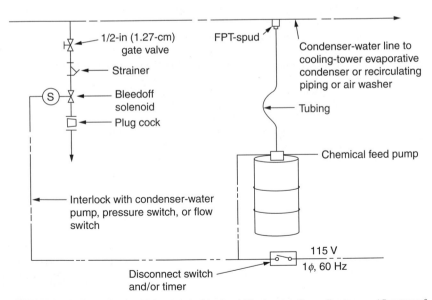

FIGURE 8.46 Simple proportioning chemical feed-and-bleed system for cooling towers. (*Courtesy of The Metro Group, Inc.*)

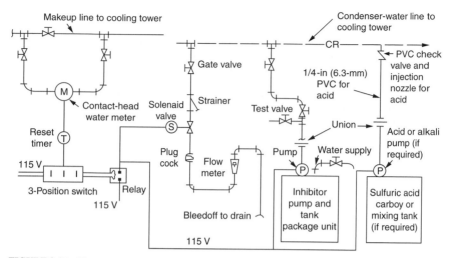

FIGURE 8.47 Water-meter-activated water treatment feed-and-bleed system for condenser water systems. (*Courtesy of The Metro Group, Inc.*)

increases in proportion to the evaporation and load on the cooling system, this system is fully automatic. In this system, bleed-off is activated when the TDS controller detects high TDS content so that the maximum permissible cycles of concentration are maintained. Treatment is then fed in direct proportion to bleed, to replace any that is lost in the bleed-off. Figure 8.48 outlines a system of this type.

Another controller used for HVAC systems, the pH controller, is needed where pH control is required because of atmospheric conditions and/or the quality of makeup water or the type of program used. The controller automatically maintains the pH value of the recirculating water at the desired level and, when necessary, activates a chemical feed pump to add acid or alkali (Figure 8.49). Figure 8.50 shows the type of TDS and pH controllers, chemical feed pumps, and package mixing tanks available.

Controllers are capable of being interconnected with building management systems (BMS) for rapid real-time remote monitoring and control through the computer. Chemical treatment can be controlled by constant monitoring of treatment levels in the cooling tower directly using auto analyzers, which then control chemical feed. Other systems are available which control chemical feed based on actual corrosion rate. The corrosion rate monitor is set to control and add treatment as needed to maintain a specific corrosion rate setting. Biocides are added using timer controls, which add biocides at specific times and days of the week as needed. All of this can be monitored and controlled through the computer.

8.8.12 Treatment of Closed Recirculating Water Systems

General. The need for treatment in closed recirculating water systems has frequently been questioned. It is often observed that because such systems are closed, they do not take on the impurities, excess minerals, and dissolved corrosive gases, which are the prime causes of scale and corrosion.

Operating experience, however, indicates that much more makeup is used in closed systems than one would anticipate. For example, a very slow drip past packing glands of

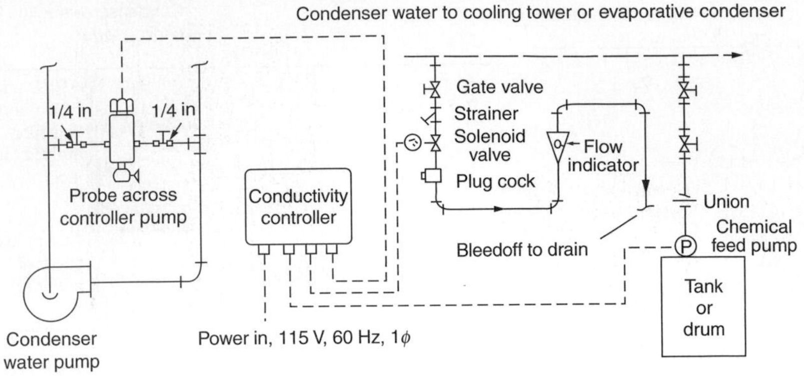

Condenser water to cooling tower or evaporative condenser

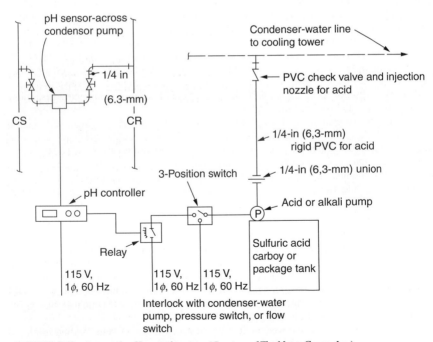

FIGURE 8.48 Conductivity-controlled automatic water treatment feed-and-bleed system. (*Courtesy of The Metro Group, Inc.*)

recirculating pumps to maintain lubrication at the rate of 1 oz/min (30 mL/min) will require a makeup rate of 337.5 gal/month (1260 L/month), or 4050 gal/year (15,000 L/year).

Experience has sometimes shown that the makeup rate to closed systems can average 100 percent of the volume per month. The makeup water is required because of water losses at packing glands and draining for maintenance and repairs of pumps, valves, or controls. With makeup water, dissolved minerals and oxygen are introduced. The calcium and magnesium

FIGURE 8.49 Automatic pH control system. (*Courtesy of The Metro Group, Inc.*)

FIGURE 8.50 Automatic water treatment system. (*Courtesy of Electrosystems, Inc.*)

hardness will deposit at heat-transfer surfaces and will accumulate as more makeup is introduced. All closed systems must "breathe," and expansion tanks are a source of atmospheric air and corrosive gases.

Galvanic couples of copper, iron, brass, and aluminum are commonly used in closed systems without dielectric fittings, which result in a higher rate of corrosion of the less noble metal at the couple. Therefore, it is generally accepted that treatment of these systems is required to prevent problems of corrosion and deposits as well as to maintain clean, efficient heat-transfer surfaces.

Before closed recirculating water systems are treated to prevent scale and corrosion, it is essential to clean them thoroughly. All types of debris may enter a system during construction, and these must be removed. In addition, protective oil-based coatings and mill scale must be removed from the interior of the piping to avoid corrosion and damage from these sources. Black magnetic iron mill scale is a particular problem in closed systems and must be removed not only to prevent corrosion and heat-insulating deposits but also to avoid damage to pumps, packing, and mechanical seals.

Materials most effectively used for such cleanouts are low-foaming detergents built with strong dispersants to prevent resettling. Dispersants are an important part of the closed system cleaner because they help to remove fine, black magnetic iron oxide particles by suspending them throughout the body of the recirculating cleaning solution and preventing them from settling and plugging. The detergent-dispersant cleaner should be added to the recirculating water at the dosage recommended by the manufacturer for 3 to 4 h. Then drain, refill, and recirculate as fast as possible, all at the same time, to obtain maximum flow through all portions of the system and to avoid pockets and deadend locations where sediment can be trapped. The system should be flushed out until all traces of residual detergent and suspended matter are gone.

After cleanout, closed systems must be treated for control of corrosion and deposits. First, the makeup water should be analyzed to determine if pretreatment, such as a water softener, is required. Then treatment must be added to prevent further corrosion and deposits. Closed systems are filled with fresh water and corrosion and deposit inhibitors or, under freezing conditions, with a solution of water and an inhibited glycol antifreeze compound. Antifreeze solutions are being used with increasing frequency to protect exposed piping of closed hydronic hot-water or chilled-water systems from freezing and rupturing. Where antifreeze solutions are required, an inhibited ethylene glycol or propylene glycol could be used at a minimum concentration of 30 percent. Specific HVAC-grade glycol antifreeze should be used, not automotive grade. Samples of solutions should be analyzed on a periodic basis to ensure that the inhibitor has not been depleted or decomposed. Closed systems filled with raw water must be treated further for corrosion and deposit control.

High-Temperature Hot-Water Heating Systems. High-temperature hot-water (HTHW) heating systems are defined as those operating above 350°F (178°C) and 450 lb/in² (3000 kPa). These systems are frequently used to provide central heating for large complexes or campus-like situations. Unless zeolite-softened water is used in these systems, the high temperature encountered, combined with the large volume of water in the system, will cause scale deposits to accumulate at the heat-transfer surfaces of the HTHW boiler. The ratio of water volume to heat-transfer surface area is such that even the initial scale deposits would be sufficient to cause significant interference with heat transfer.

Additional treatment should include an oxygen scavenger such as diethylhydroxylamine or sodium sulfite. The latter is preferable in systems containing nonferrous metals. Caustic soda should be used to maintain the pH value of the water at 8.0 to 10.0.

Dispersants and scale inhibitors are used to maintain clean heat-transfer surfaces. These consist of thermally stable inhibitors such as polyacrylates, polymethacrylates, and phosphonates.

Medium-Temperature Hot-Water Systems. Medium-temperature hot-water (MTHW) heating systems are defined as those operating at 250°F (120°C) up to 350°F (177°C) with pressures above 30 lb/in² (206 kPa). This kind of system is treated in the same way as HTHW systems. Oxygen scavengers, such as sodium sulfite or diethylhythoxyl amine, control corrosion and pitting, and the pH value is maintained at 8.0 to 10.0 with caustic soda or a similar alkali. The system should be filled with zeolite-softened water to avoid scale deposits on heat-transfer surfaces. Scale inhibitor and dispersant should be added in the form of thermally stable inhibitors, as described above.

Low-Temperature Hot-Water Systems. Low-temperature hot-water (LTHW) systems are defined as those operating below 250°F (120°C) with maximum water pressure of 30 lb/in² (206 kPa). In low-temperature systems, bimetallic couples of steel, copper, brass, and aluminum are frequently used. The recirculating water in an LTHW system is best treated with corrosion inhibitors such as a sodium nitrite–borax combination. This inhibitor and similar inhibitors are not used at temperatures above 250°F (120°C) because of their questionable stability at higher temperatures. Apparently the film-forming mechanism is severely restricted at higher temperatures so that there is spotty pitting and even decomposition of the inhibitor in the case of nitrite inhibitors.

Sodium nitrite–borax inhibitor systems and other nitrite-based compositions are used at 3000 to 4000 mg/L at temperatures above 180°F, while 1000 to 2000 mg/L is sufficient for systems operating below 180°F. With nitrite-based treatments, it is important to use specific copper and brass inhibitors such as mercaptobenzothiazole or phenyltriazoles to minimize copper attack and copper-induced pitting of steel.

Other corrosion inhibitors that are available in proprietary blends include phosphate organics, anides, silicates, benzoates, phosphonates, and molybdates, as well as various combinations of these with other organics. The concentrations recommended by the suppliers of these inhibitors should be followed to obtain good corrosion control.

Inhibitors for control of scale and deposit are used along with the corrosion inhibitors and are usually included in the proprietary blend. Scale inhibitors for LTHW heating systems include sodium polyacrylates, polymethacrylates, polymaleates, sulfonated polystyrene, carboxymethylcellulose, lignins, and phosphonates. Some of these materials also act as dispersants that prevent sedimentation of fine particulate matter and provide excellent scale and deposit control. When these inhibitors are used, it is typically not necessary to soften the makeup water required to compensate for "normal" losses such as drips, leaks, repairs, etc.

Chilled-Water Systems. Chilled-water systems operate below 140°F (60°C), usually in the range of 44 to 54°F (10 to 12°C) for comfort cooling. Sodium nitrite–borax inhibitors and other nitrite-based blends are used in the range of 500 to 1000 ppm with the pH maintained within the range of 7.0 to 10.0. It is important that nitrite-based inhibitors include nonferrous-metal inhibitors to reduce nitrite attack on solder and other nonferrous metals. Among these inhibitors are borax, sodium benzoate, phosphates, and copper inhibitors, mercaptobenzothiazole, or phenyltriazoles. Chilled-water systems invariably contain components of multimetallic construction, which include iron, steel, copper, brass, and aluminum, and which require balanced formulated inhibitors to ensure that all metals are individually protected, as well as protecting against the galvanic couples of these metals. Other inhibitors for chilled-water systems are organics blended with phosphates, silicates, or molybdate, sodium benzoate, organic phosphonates, and organic amines such as methylglycine and imidazoline. The corrosion inhibitor or inhibitor blend may be combined with dispersants and deposit inhibitors for a complete treatment. The dispersants are the same as those used in hot-water systems.

Closed-Circuit Coolers. Closed-circuit coolers are closed condenser water systems, which are combined with an evaporative cooler. The closed condenser water circuit is treated the same as the closed chilled-water circuit with formulated inhibitors for control of corrosion and deposits.

Solar Heat-Exchange Systems. Another type of closed system being used with growing frequency is the energy-saving solar heat-exchange circuit. These systems also require treatment for control of corrosion and deposits. The most effective treatments are those used for chilled- and hot-water heating circuits.

Test Controls. Quality tests on samples of closed system treated water should be made at a minimum of monthly intervals to record conditions and be assured correct treatment levels are being maintained. At a minimum tests should be made for inhibitor level and pH. The testing services offered by the water treatment supplier should be monthly, as a minimum, and include a complete water analysis, including inhibitor levels.

Method of Feed. Treatments are applied to closed systems by several convenient means. They may be added directly through any opening into the system, such as at the expansion tank or open reservoir.

For very small but "tight" systems, the use of a simple stirrup pump or force pump (similar to Fig. 8.37) may be employed to inject treatment directly through a drain valve or hose cock.

FIGURE 8.51   Chemical feed pumps and mixing tanks with control panel skid-mounted.
(*Courtesy of Neptune Chemical Pump Co.*)

In most instances, however, it is more convenient to install a bypass feeder across the
recirculating pump as shown in Fig. 8.51. With a differential in pressure between the discharge side and the suction side of the recirculating pump, water in the system is forced
through the bypass feeder, and the treatment is injected directly into the system. The maintenance of a treatment program for closed systems is relatively simple, and the benefits
obtained in corrosion protection and efficient operations are substantial.

REFERENCES

1. American Public Health Association (APHA): *Standard Methods for the Examination of Water
   and Waste Water*, APHA, Washington, DC, 1980.

2. American Water Works Association: *Water Quality and Treatment*, McGraw-Hill, New York, 1971.

3. *ASHRAE Handbook and Product Directory, Systems*, ASHRAE, Atlanta, GA, chap. 36, p. 36.16,
   1980.

4. Atkinson, I. J. N., and Van Droffelaar, H.: *Corrosion and Its Control*, National Association of
   Corrosion Engineers, Houston, 1982.

5. *Betz Handbook of Industrial Water Conditioning*. Beta Laboratories, Trevose, PA, 1980.

6. Blake, Richard T.: *Water treatment for HVAC and Potable Water Systems*, McGraw-Hill, New
   York, 8th ed., 1980.

7. Butler, G., and Ison, H. C. K.: *Corrosion and Its Prevention in Waters*, Litton Educational Publishing Co. (Reinhold), Chicago, 1966.

8. *Drew—Principles of Industrial Water Treatment*, Drew Chemical Corporation, Boonton, NJ, 11th ed., 1994.

9. Frayne, Colin: Cooling Water Treatment Principles and Practice, Chemical Publishing, New York, 1999.

10. Frayne, Colin: Boiler Water Treatment Principles and Practice, Chemical Publishing, New York, 2002.

11. Goddard, Hugh P. (ed.): "Materials Performance," vol. 13, no. 4, p. 9, April 1974 (Reprinted by permission).

12. Hausler, R. H.: "Economics of Corrosion Control," *Materials Performance*, National Association of Corrosion Engineers, Houston, June, p. 9, 1978.

13. Hinst, H. E.: "12 Ways to Avoid Boiler Tube Corrosion," Babcock and Wilcox (Reprint).

14. Kemmer, Frank N. (ed.): *The Nalco Water Handbook*, McGraw-Hill, New York, 2nd ed., 1988.

15. Kilbaugh, W. A., and Pocock, P. J.: "Pointers on the Care of Low Pressure Steam Steel Boilers," *Heating/Piping/Air Conditioning*, February 1962.

16. Likens, Gene E.: "Acid Precipitation," *Chemical and Engineering News*, Nov. 22, p. 29, 1976.

17. Mayer, W. F., and Larsen, Russell: "An Evaluation of White Rust and Cooling Tower Metallurgy," AWT Analyst, March 1992.

18. McCoy, J. W.: *Chemical Analysis of Industrial Water*; Chemical Publishing, New York, 1969.

19. McCoy, J. W.: *The Chemical Treatment of Cooling Water*; Chemical Publishing, New York, 1974.

20. Nathan, C. C. (ed.): *Corrosion Inhibitors*, National Association of Corrosion Engineers, Houstin, 1973.

21. Nordell, Eskel: *Water Treatment for Industrial and Other Uses*, Litton Educational Publishing (Reinhold), Chicago, 1961.

22. Pincus, L. I.: *Practical Boiler Water Treatment*, McGraw-Hill, New York, 1962.

23. Powell, Sheppard T.: *Water Conditioning for Industry*, McGraw-Hill, New York, 1954.

24. Speller, Frank N.: *Corrosion Causes and Prevention*, McGraw-Hill, New York, 1951.

25. Sussman, Sidney: "Hot Water Heating Needs Water Treatment," *Air Conditioning, Heating and Ventilating*, August 1965.

26. Sussman, Sidney: "Causes and Cures of Mechanical Shaft Seal Failures in Water Pumps," *Heating/Piping/Air Conditioning*, September 1963.

27. Uhlig, H. H. (ed): *The Corrosion Handbook*, Wiley, New York, 1948.

28. Uhlig, H. H.: *Corrosion and Corrosion Control*, Wiley, New York, 1971.

CHAPTER 9
ENERGY CONSERVATION

Peter M. Fairbanks, P.E.
Bluestone Energy Services, Inc., Braintree, MA

9.1 INTRODUCTION

There are opportunities in almost every facility to reduce energy consumption and energy cost, particularly in HVAC systems. Although facilities vary greatly in their use of energy and the potential for reducing energy consumption, one or more of the following conditions generally applies at any facility, creating opportunities to conserve energy:

- Systems are designed for worst case "design conditions" and operate most of the time at less than design conditions.
- System designers add margin on top of design conditions to be confident the system will perform. *Energy efficiency is not always the first priority in system design.*
- Space usage or system requirements have changed over time.
- Equipment is purchased on a lowest first cost basis rather than a life-cycle basis.
- Higher efficiency equipment has evolved since the existing equipment was installed.
- Equipment is worn and operating at low efficiency.
- Controls have failed and equipment is operating inefficiently.
- Personnel are unaware of the energy use incurred from unnecessary operation of equipment and systems.

Varying degrees of energy conservation can be achieved in facilities depending on the existing potential and the level of facility commitment to this goal. Some measures, such as shutting off lights when not needed, require little analysis or management support to implement. Other measures, such as chiller plant or refrigeration system upgrade, may require metering and significant analysis to demonstrate life-cycle cost effectiveness in order to obtain commitment of project funding.

There is insufficient space in this chapter to provide a comprehensive energy conservation treatment that would sufficiently address all of the variables regarding facilities and energy measures. Different facilities will require different levels of documentation of savings and costs. Implementation of one measure may affect the savings and viability of another measure. Various caveats apply regarding implementation of more sophisticated measures and there are issues that are site specific that will affect feasibility of measures.

Many measures will require careful integration into existing systems to avoid adverse effects on operation. In order for a comprehensive energy conservation program to be implemented effectively at most facilities, it should ideally be performed on a site specific basis under the direction of or with the assistance of engineers skilled in such projects. The benefits from such a program can be significant. For facilities where energy conservation has not been addressed, a comprehensive program will typically save from 10 to 30 percent of facility energy costs with projects that have simple paybacks of 5 years or less. If there are utility energy conservation programs in the area that provide incentives for energy conservation, qualifying projects will typically have paybacks of 3 years or less. A 3-year payback project has an internal rate of return of 20 percent. There are few opportunities where a facility can receive this level of return for such a relatively low-risk investment. In many instances energy conservation measures provide additional benefits such as reduced maintenance costs, improved working conditions, or increased production. Improvements to the environment also accrue, as less energy use means less pollution.

The focus of this chapter is to provide information that will be helpful in assessing the potential for energy conservation in your facility and organizing a comprehensive energy conservation program as well as to provide information on evaluation techniques for specific energy conservation opportunities (ECOs) and identification of some of the potential problems to avoid.

9.1.1 Energy Conservation Program Preparation

In order to develop a successful comprehensive energy conservation program for your facility the following preparation is recommended:

- Identify if there are applicable electric or gas utility energy conservation programs that would provide technical assistance to identify ECOs and incentives to install approved measures (contact your utility representatives). *Do this first.* If there are programs with significant incentive payments, you will need to follow the utility protocol to qualify for their assistance. Utility programs can be a great source of technical assistance and project funding; however, it should also be understood that the assistance provided is usually oriented toward that utility's energy and may not result in optimum energy conservation for *your* facility. For example, an electric utility may provide engineering analysis and incentives for optimizing an electric chiller upgrade, where a gas fired two-stage absorption chiller could be the most cost-effective option for the facility.

- Identify if there are sufficient in-house resources that can be committed to the program or if outside technical assistance is needed.

- Identify the threshold of return and sophistication of analysis necessary for your organization to obtain commitment of funds for project implementation. For many facilities, go/no-go decisions for less sophisticated measures are made on a simple payback basis (project invested money "paid back" in savings over 2 years, 3 years, etc.). For higher cost projects, projects with alternative installation options, or projects involving benefits occurring on different timelines, a more sophisticated life-cycle analysis may be appropriate. With this analysis, the value of avoided energy costs (and other avoided costs such as maintenance and labor) over the expected useful life of the installation are discounted to present worth by using an appropriate interest rate that represents rate of return for alternative investment of the project money or avoided borrowing cost of the project money. On an "apples to apples" present worth basis, the benefits of the project can be compared to the installed cost or to alternative projects.

- Identify the level of documentation and savings verification necessary. As a minimum, this typically includes the following:
  - Description of existing conditions and proposed change
  - Savings analysis
  - Cost estimate for project installation

For higher cost projects, more sophisticated projects, or for facilities that require a higher comfort level, preinstallation metering of operating parameters may be required. This could include spot measurement of amps, volts, kW, kWH, flow (Gpm), temperatures, Btus, units of oil or gas consumption as applicable with assumptions made to address parameter variation as a function of time of day, outside conditions, or production level. The highest level of metering and demonstration of savings potential would involve recorded measurement over several weeks or a month's time of applicable parameters as a function of an appropriate variable such as time of day or outside temperature. The operating performance can then be reasonably extrapolated out over a year using historical weather data for the variable parameter(s).

- Identify level of savings verification required. Utility programs or in-house management may require postinstallation savings verification in the form of either one-time measured parameters or recorded metered data collected over a period of time to verify savings calculations.

After completion of the preparation groundwork, an energy conservation program can be implemented that meets the requirements of your facility. A program structure that is recommended and that has been successful at many facilities consists of the following phases:

1. Determine the energy conservation potential (ECO) of the facility
2. Refine the program
3. Implement approved measures
4. Maintain measures

This format allows the program to start slow (minimizes time and cost) and allows management "buy-in" on an informed basis. For example, Phase 1 requires relatively minimal investment. At its completion, the overall potential of the program and potential ECOs will be defined. Phase 2, a more in-depth analysis of measures, will require a larger investment of time and resources; decisions regarding the scope and investment commitment for this phase can be made on an informed basis with the information developed in Phase 1. At the end of Phase 2, the feasibility of measures will be identified with associated annual energy savings, any associated positive (or negative) operating and maintenance impacts, potential utility incentive payments, and associated installed costs accurately defined. Decisions can now be made regarding measure implementation and priority of measure implementation in Phase 3. Phase 4 is protection of the investment through savings maintenance.

9.2 PHASE 1: DETERMINE ENERGY CONSERVATION POTENTIAL

A useful first step in a comprehensive energy conservation program is to determine the facility potential for becoming more energy efficient. This is a preliminary analysis that takes a macro-look at the facility energy usage profile supplemented with "walk-through" identification of likely ECOs. In order to understand how systems are operating, it may be necessary to perform some preliminary diagnostics on systems by spot measurement of amps, kW, flows and temperatures, and observing equipment operation in response to control manipulation. (More discussion of diagnostic technique is included later.) Depending on the size and complexity of a facility, a list of potential ECOs and the facility potential for cost effective energy conservation (5 to 10 percent of total energy cost, 20 to 30 percent of total energy cost, etc.) can be determined by a knowledgeable energy engineer in several days to several weeks. Ideally this analysis is performed with a "holistic" approach that, with input from operating and maintenance personnel, identifies operating and maintenance problems, indoor air quality issues, equipment at the end of useful life, etc. The preliminary analysis is described in Sections 9.2.1–9.2.3.

9.2.1 Facility Energy Consumption Data

Information is gathered on the energy consumption and cost for the major types of energy (electricity, gas, oil) used at the facility. Many facilities track these data on a monthly basis for budgeting and performance monitoring purposes. If the data are not being tracked to show demand, energy consumption, and cost, billing data should be obtained to create it. Ideally, several years of utility bills will be available. The desired result is to obtain sufficient data to provide an annual profile of energy use and cost for each type of energy. The billing structure for each type of energy should be identified (on peak energy cost, off peak energy cost, applicability of "demand ratchet" where the highest monthly demand for electricity is charged for the entire year, etc.) For purposes of the preliminary analysis of electricity conserving ECOs, unless there are significant pricing mechanisms such as demand ratchet that should be accounted for, using average cost of energy on a $/kWh basis is usually sufficiently accurate. The seasonal average cost for fossil fuels is also sufficiently accurate for preliminary analysis of fossil fuel conserving ECOs.

9.2.2 Energy Use Reconciliation

As a filter and a guide, it is sometimes useful to roughly reconcile end use of energy with annual facility energy billings. For some facilities, especially large complex facilities, the value of the reconciliation may not be worth the effort to accurately achieve it and the time would be better spent analyzing discrete systems for preliminary ECO potential. If it is decided that a reconciliation is desirable, and metered data, submetered data, operator log data, etc. are available, these can be used to provide a higher degree of accuracy; however, for purposes of this preliminary analysis, simple calculations based on operating system assumptions will generally suffice. Reconciliation of end energy use with total annual energy use based on billing should be achieved to a ±10 percent accuracy. If this level of accuracy doesn't occur, operating data should be refined and assumptions should be changed regarding loads, efficiencies, etc. until reconciliation is achieved. This exercise can result in a basic understanding of how energy is used in the facility and can help prioritize where the best opportunities for conservation exist.

End user energy consumption should be broken down into major categories by utilizing the following assumptions and calculations:

Electricity (Convert all usage to billing units—kWh)

- *Lighting*—Wattage of light fixtures multiplied by rough counts of lighting and hours of operation.
- *Electric Chiller Plant*—Average chiller load (tons) multiplied by efficiency at that load (0.5 to 0.7 kW/ton typical) multiplied by annual hours of operation. Associated chiller

pump and cooling tower fan energy determined by motor nameplate horsepower multiplied by assumed load factor (85 percent) multiplied by 0.746 kW/hp multiplied by annual hours of operation.

- *Air Handling Units*—Motor nameplate horsepower multiplied by assumed load factor (85 percent) multiplied by 0.746 kW/hp multiplied by annual hours of operation. Associated direct expansion cooling energy determined by average cooling load multiplied by efficiency (1 to 1.2 kW/ton typical) multiplied by annual hours of cooling load operation.

Fossil Fuel (Convert all usage to billing data units—therms gas, gallons oil, etc., or convert all energy into common units such as MMBtu—million Btu)

- *Building Skin Loss Heating Load*—Based on building construction and insulation characteristics, determine average heating load on an average heating season day and multiply by heating season hours. Publications from the American Society of Heating Refrigeration and Air Conditioning Engineers (ASHRAE), (www.ASHRAE.org) as well as many other heating, ventilating, and air conditioning (HVAC) manuals provide information on typical building construction heat transfer factors, average winter temperatures in various parts of the country and formulas for heat load calculation. Divide heat load by boiler system or if individual heaters, burner efficiency (typical 60 to 80 percent) to determine purchased heat energy quantity.

- *Outside Air Heating Load*—Identify amount of outside air brought into the building through analysis of drawing schedule data for air-handling units (AHUs), observation of AHU operation, assumptions based on ft^2 of AHU intakes and typical design air flow velocities (500 to 800 ft/min at intake, 500 ft/min at coil, etc). Using differential temperature based on average heating season outdoor temperature and average AHU discharge air temperature, and annual hours of heating season operation, calculate amount of heating Btus required by the following formula:

$$1.08 \times \text{CFM} \times \text{differential temperature} \times \text{annual hours}$$

If the building has more exhaust than makeup air and is negative, determine the quantity of infiltration air as a percent of the shortfall and calculate its heat load as well. Adjust for boiler/burner efficiency.

- *Steam Absorption Chiller*—Average chiller load (tons) multiplied by efficiency (20 lbs/ton –30 lbs/ton single stage absorber, 11 lbs/ton to 14 lbs/ton two stage absorber) multiplied by annual hours of operation. Adjust for boiler/burner efficiency.

9.2.3 Identify Potential ECOs

In the preliminary analysis, the focus is to identify all significant energy conservation opportunities. There may be additional smaller opportunities that can be explored in Phase 2 or 3, but the Phase 1 emphasis is typically to "get the most bang for the buck" and identify major opportunities. For each ECO, a cursory estimate of savings and installed cost should be developed with assumptions made as necessary regarding operating conditions, load, etc. A list of candidate ECOs that have typically been found to be cost effective based on "normal" operating conditions and New England energy pricing is listed below. The basis for normal operating conditions are office building systems operating 3000 plus h/yr, commercial facilities operating 4000 h/year, small manufacturing facilities with single shift operation, and large manufacturing facilities operating two or three shifts. The energy cost basis is electricity at an average $.09/kWh and fossil fuel cost ranging from $3.00 MMBtu for larger users with No. 6 oil and interruptible gas contracts to $8.00 MMBtu for facilities burning firm gas and No. 2 oil.

Lighting. *Retrofitting lighting is usually one of the first and best ways to conserve energy in a facility.* It is not uncommon to be able to reduce lighting energy by 40 percent or more while maintaining acceptable light levels. With reduction in lighting energy there is an associated reduction in air-conditioning load and a lesser negative impact on building-heating load. Because newer technology lighting, such as T-8 lamps with electronic ballasts, have higher light output and better color rendition than older T-12 lamps with magnetic ballasts, it is possible to significantly reduce lighting energy while improving light quality. Many areas, especially those with office/cubicle task lighting and computer use, are overlit. It is useful to spot meter with a foot candle meter to determine existing light levels and potential for delamping. Lens treatment as part of the retrofit, although adding expense, may be desirable to avoid computer screen glare or to eliminate aged prismatic lenses. It may be feasible to improve the efficiency of a fluorescent fixture by installing a reflector and reducing its energy level by delamping. Decisions regarding retrofitting existing fixtures vs. installing replacement fixtures (higher cost) will depend on issues such as lens treatment and fixture efficiency. Higher operating hours will make more expensive treatment more feasible. It is recommended that as part of Phase 2, sample lighting retrofits be installed and light levels measured before installation and after installation with sufficient time allowed for new lamp burn in. Lighting energy conservation is very site specific; however, the following are typical measures that have been found feasible and cost effective:

- Standard $2' \times 4'$ four lamp fixtures with T-12 (1.5″ diameter lamps) and magnetic ballasts—change to T-8 (1″ diameter lamps) and electronic ballasts. Delamp as appropriate to three lamps with or without reflector, two lamps with or without reflector, new three or two lamp fixtures.
- Standard $2' \times 2'$ T-12 "U" tube fixtures—change to T-8 U-tube/electronic ballast or three 2-ft T-8 lamps and electronic ballast, with or without reflector.
- High output fluorescent fixtures—replace with new fixtures containing T-8 lamps/electronic ballasts and reflectors.
- Two lamp strips containing 4′ or 8′ T-12 lamps—retrofit with reflector and single strip of 4′ T-8 lamps or replace with new fixtures containing single 4′ T-8 lamp strips with electronic ballasts and reflectors.
- Incandescent fixtures, recessed and surface, 150 W or less—retrofit with hardwired compact fluorescent 13 or 26 W retrofit kits (some allow dimming if required) or new compact fluorescent fixtures.
- Incandescent fixtures, 200 W or more—replace with T-5/electronic ballast fluorescent fixtures.
- Mercury vapor, high pressure sodium, or metal halide fixtures—replace with T-5/electronic ballast fluorescent fixtures.
- Office lighting redesign—Replace 3 to 4 lamp troffers with pendant mounted direct/indirect two lamp fixtures.

Lighting Controls.   Lighting control strategies that have been found effective include:

- Motion sensor control for spaces with sufficient lighting wattage controlled (200 W minimum) manual on, automatic off, ultrasonic sensor.
- Photocell control for outside lighting or lighting near window walls or under skylights.

- Energy Management System (EMS) control of building lighting on an area or floor basis allowing scheduling of lighting. Local override and emergency lighting circuits would be required.

HVAC. Heating, ventilating, and air-conditioning energy conservation strategies

Controls

- Install an EMS that provides the following control strategies for HVAC equipment: scheduling control, set back control, optimum start/stop, air-handling unit discharge temperature reset, air-side economizer control.
- For buildings with electric heat, install CO_2 sensor control of outside air dampers through the EMS.

Air-Handling Systems

- On variable air volume (VAV) systems install variable frequency drives (VFDs) to control supply fan and return fan speed as a function of system pressure set point in place of inlet vane control. This is applicable for air foil or backward curve fans, not forward curve fans, which have relatively efficiently unloading with inlet vane control.
- Convert dual duct systems to cooling-only VAV systems, if separate perimeter heating is installed.
- Convert primary electric heat in large air-handling units to hot-water coils served by new gas fired condensing boiler(s).
- Eliminate any simultaneous heating and cooling (lock out reheat in cooling season, isolate steam heat and reheat in cooling season, etc.)

Hydronic Systems

- Install VFDs on heating, reheating, or cooling pump sets (10 hp (7 W) and larger) that serve end users with two-way valves. The VFD would control pump speed as a function of system pressure in place of pressure bypass control.
- Install outside temperature based reset control, especially on systems without VFD control.

Chilled-Water Plant

- Replace inefficient chillers with efficient chillers. As a rule of thumb, cooling from a single stage steam absorption chiller costs two to three times as much per ton as cooling from a new efficient electric centrifugal chiller or a two stage steam absorption chiller. An electric centrifugal chiller installed before 1990 would have an efficiency on the order of 0.65 to 0.8 kW/ton, compared to 0.5 kW/ton or better (depending on size) for a new electric centrifugal chiller. If the chiller is inefficient and the boiler is inefficient and the facility is heated with hot water, a direct fired two stage steam absorption chiller could be cost-effective.
- Install a VFD on a chiller with inefficient unloading characteristics.
- Consolidate chilled-water systems and schedule chillers to use most efficient chiller first.
- Install plate and frame "free cooling" heat exchanger for facilities with cooling towers and cold weather chilled water loads (data centers, process, absence of air side economizer capability).

- Reduce chilled-water pumping energy by conversion to primary/secondary pumping or if compatible with chillers and chiller controls, variable pumping through chillers. This measure requires two-way valve control at the end users.
- Install a VFD on the condenser water pump(s) responding to chiller head pressure or differential condenser water temperature.
- Install a VFD(s) on the cooling tower fan(s) in place of single speed cycling fan control.
- Reset condenser water temperature lower as a function of wet bulb temperature down to chiller manufacturer's recommended minimum, typically 65° F (18°C).

Heating System

- Upgrade boiler/burner controls. On larger boilers, 100 hp plus, replace jackshaft "fixed" fuel/air relationship control with distributed digital/servo motor control and exhaust gas analysis system (O_2, CO, CO_2). With this type of control system, excess air can be precisely controlled.
- If boiler stack temperature exceeds 400°F (205°C) and boiler heat transfer surfaces are not fouled, install heat recovery to heat feed water or combustion air.
- If tubes are fouled, remove scale and improve water treatment to avoid scale deposits. If present, remove soot and deposits on fire side of tubes.
- Insulate exposed steam system piping and valves.
- Install a VFD on the boiler feed pump.
- Install VFDs on forced draft fans and induced draft fans to control fan speed in lieu of damper or vane control.
- Replace steam traps where applicable with steam ventures. These devices are engineered and must be accurately sized for the application. The benefit is that there are no moving parts to wear out, reducing maintenance and steam losses.
- Reduce boiler steam pressure if compatible with system requirements.
- Install gas fired radiant heating for large open space heating in place of heating and distributing air.

9.3 PHASE 2: REFINE THE PROGRAM

Based on the information gathered in Phase 1, decisions can be made regarding the scope of budgeting for energy conservation opportunities, the relative attractiveness of identified opportunities, and the subsequent extent of Phase 2 analysis. This decision-making process will result in a list of candidate ECOs for further analysis in Phase 2. The intent of Phase 2 is to prove or disprove assumptions made in Phase 1 regarding measures and to perform sufficient engineering analysis, metering and information-gathering so that measure-cost, savings, and benefits will be accurately defined to support implementation decisions in Phase 3. Issues regarding refrigerant selection, appropriate disposal of hazardous waste, safety issues, potential VFD harmonic distortion, VFD/motor compatibility, etc. should be addressed. Sufficient engineering design should be performed to obtain contractor pricing or highly accurate cost estimates. Analyses should be performed based on metered data, as necessary to verify savings. Historical weather data are often used for such analysis of system operation that are a function of outside conditions. These are engineering weather data that have been collected for as much as 30 years at representative locations across the country. One source for these weather data is the Department of Defense Engineering Weather Data,

USAF Handbook 32-1163, recently updated and available in draft form at the Air Force Civil Engineer Support Agency web site (www.afcesa.af.mil/publications/drafts/default.htm) and is available for purchase in CDROM format from the National Climate Data Center (www.ncdc.noaa.gov, click on Climate, Products & Publications, CDROMS, Engineering Data). Data are provided for 511 U.S. locations and 292 international locations and are in a bin-temperature format that summarizes on a monthly basis the average number of hours each location experiences a specific dry-bulb temperature range (i.e., 80–84°F (27–32°C) with mean coincident wet-bulb temperature for the range at specific time periods (i.e., 5 AM to 8 AM). Another source of engineering weather data is ASHRAE's updated Weather Year for Energy (WYEC-2) software and data that allow typical monthly weather profiles to be developed for 77 U.S. cities. This is available as a CDROM or as a download from ASHRAE (www.ashrae.org).

The following are some diagnostic procedures that have been found useful in verifying existing operation of equipment and systems:

9.3.1 Diagnostics

- An amp probe or preferably a kW-measuring device can be used to measure amps and volts or kW or existing motors verifying existing energy levels.
- An ultrasonic flow meter allows the flow measurement of water and other liquids from the outside of piping.
- A multichannel recorder capable of measuring and recording variable parameters over a period of time provides information on equipment operation as a function of a variable. Typical parameters measured include kW, kWh, temperature, and with the use of the ultrasonic flow meter, flow. With this equipment, for example, it is possible to record chiller load and energy use as a function of outdoor air temperature, by recording chilled water supply and return temperature, chilled water flow, chiller kW, and outdoor air temperature.
- A multichannel recorder is also useful to meter the kW load of air compressors over a period of time. Using compressor manufacturer's performance data, observed maximum and minimum load, unloading point, etc., the facility air load profile can be established. If determined to be necessary, a more accurate profile of energy and air use can be established by incorporating an air-flow meter, but this requires the time and expense of installation in the piping.
- A portable CO_2 meter is useful for determining potential for reducing outside air in a facility.
- A multimeter that measures dry-bulb temperature and wet-bulb temperature is useful in diagnosing air-handling system performance. With this device, percentage of outside air can be established; performance of heating and cooling coils, presence of simultaneous heating and cooling, etc. can be assessed.
- An infrared temperature recorder is used to obtain spot measurements of hot or cold surface temperatures to verify operation.

9.4 PHASE 3: IMPLEMENTATION

With the completed analyses of Phase 2, it should now be clear which ECOs meet the facility criteria for project implementation and what order of implementation is desirable. Many of the ECOs will require final engineering to address typical design issues. Depending on the scope of the ECO and the facility preference, decisions can be made regarding method of implementation—full engineered plans and specifications, design/build, third party performance contract, etc. Issues of phasing the work in and around facility operation to minimize disruption should be addressed. Any savings verification requirements should be clearly identified and incorporated into the project before the work is started. Operating requirements to maintain savings levels should be clearly defined and documented.

9.5 PHASE 4: SAVINGS MAINTENANCE

After the facility has incorporated energy conservation measures that meet its criteria of cost and value, the measures must be maintained. Periodic reviews of performance should be conducted to verify that set points have not changed, schedules are still appropriate, and equipment is maintained and not operating in "temporary off" design condition. Over time, if the energy conservation measure operating requirements are not vigilantly maintained, savings will typically erode. It is also usually possible after gaining experience with the new systems and equipment, to actually "tweak" performance and increase savings.

Going forward, energy efficiency should be a component of the decision-making process for replacing equipment that is at the end of its useful life or purchasing equipment for facility or process expansion. Motors that wear out should typically be replaced with premium efficiency motors. For more sophisticated equipment, such as chillers or boilers, it may be desirable to perform a life-cycle analysis and identify the present value of the energy savings for each increment of efficiency improvement over the life of equipment. As an example, consider a proposed chiller installation with an expected useful life of 20 years, an average load of 300 ton over a 2500-h cooling season, electricity cost of \$.07/kWh, escalating at 3 percent, and value of money at 8 percent per year. The life-cycle energy savings for each 0.01 kW/ton chiller efficiency improvement under this scenario would be \$6431. This value can be listed as an evaluation criterion in the purchase specification, and a fair evaluation can be made of a lower first cost chiller with a lower efficiency compared to a higher first cost chiller with a higher efficiency. With this type of purchasing analysis, a performance guarantee with penalties equal to the present worth value of life-cycle efficiency should be specified with efficiency demonstrated by a witnessed performance test.

CHAPTER 10
NOISE CONTROL

Martin Hirschorn
President, Industrial Acoustics Company,
Bronx, New York

10.1 INTRODUCTION

Is noise control engineering a science or an art? It is a bit of both.

Acoustic theory helps explain the acoustic world we live in and enables us to establish general design parameters for engineered noise control solutions and products, but it does not always do so very accurately. For instance, it is impossible to calculate the noise reduction of barriers, walls, enclosures, rooms, and silencers or the propagation of sound waves over open surfaces with the degree of accuracy needed for reliability. There are just too many variables. Consequently, we cannot rely on theory for more than directional indicators.

Optimum noise control solutions must therefore be based on engineered products with performance characteristics obtained from repeatable laboratory tests and/or extensive field data—because if we overdesign, it costs too much money, and if we do not adequately provide for noise control, we may have an unacceptable job. For critical jobs, where there are significant uncertainty factors, model testing is essential. This may include power plants, aviation terminals and test facilities, industrial factories, and air-handling units in high-rise buildings. Furthermore, apart from economics considerations, the structural, mechanical, aerodynamic, and thermodynamic engineering aspects of the solution to a noise control problem are often more complex than its acoustic components, so in many instances a multidisciplinary approach is essential.

This chapter is concerned primarily with basic acoustic engineering principles and how they can be applied to solve noise problems inherent in HVAC systems. The chapter first discusses the theory of sound, with emphasis on the acoustic engineering aspects, and then examines the nature of noise in HVAC systems and the means available for controlling noise.

10.2 THE NATURE OF SOUND

Sound is essentially the sensation produced through the ear by fluctuations of pressure in the adjacent air, and "noise" can be defined as sound that annoys, usually because the sound pressure level is too high. High noise levels not only interfere with direct voice

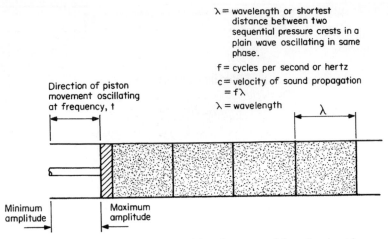

λ = wavelength or shortest
distance between two
sequential pressure crests in a
plain wave oscillating in same
phase.

f = cycles per second or hertz

c = velocity of sound propagation
= fλ

λ = wavelength

Direction of piston
movement oscillating
at frequency, t

Minimum
amplitude

Maximum
amplitude

FIGURE 10.1   Sound wave being propagated through a compressible medium in a tube.

communications and electronically transmitted speech; they are also considered a health hazard in both the working and living environments.

Sound waves are propagated in air as compressional waves. Although compressional waves are generally caused by vibrations of solid bodies, they can also be caused by pressure waves generated by the gas discharge of a jet engine or the subsonic velocities in an air-conditioning duct. When these waves strike solid bodies, they cause the bodies to vibrate, or oscillate.

To illustrate what happens, we can think of sound being generated by a piston oscillating back and forth in an air-filled tube (Fig. 10.1). This action of the compressor causes the air molecules adjacent to the piston to be alternately crowded together (or compressed) and then moved apart (or rarefied). The oscillation generated by the piston in this manner is referred to as "simple harmonic motion." And as shown in Fig. 10.2, a plot of the piston displacement can be presented as a sinusoidal function; that is, the sound wave generated in its purest form for a discrete sound is sinusoidal and has a frequency equal to the number of times per second that the piston moves back and forth.

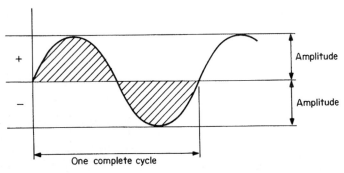

+

−

Amplitude

Amplitude

One complete cycle

FIGURE 10.2   Sine wave of the simple harmonic motion characterizing a pure tone.

10.2.1 Displacement Amplitude and Particle Velocity

Specifically, sound is transmitted through individual vibrating air particles. The vibration causes the particles to move, but they do not change their average positions if the transmitting medium itself is not in motion. The average maximum distance moved by individual particles is called the "displacement amplitude," and the speed at which they move is called the "particle velocity."

In air, the displacement amplitude may range from 4×10^{-9} in (10^{-7} mm) to a few millimeters per second. The smallest amplitude would be the lowest discernible sound, and the largest amplitude would be the loudest sound the human ear can perceive as a proper sound.

10.2.2 Frequency

The frequency of a sound wave is expressed in hertz (Hz). The range of human hearing extends from 20 to 20,000 Hz, but 12,000 to 13,000 Hz is the limit for many adults (and the exposure of teenagers to noisy rock music is likely to result in "old-age deafness" before they reach the age of 30).

10.2.3 Wavelength

The wavelength of sound is the distance between analogous points of two successive waves. It is denoted by the Greek letter λ and can be calculated from the relationship

$$\lambda = \frac{c}{f} \tag{10.1}$$

where c is the speed of sound in ft/s (m/s), and f is the frequency in Hz.

10.2.4 Sound Level

For convenience in measuring sound without having to take data at a large number of discrete frequencies, sound levels are often measured in one-third octave bands or in full octaves, as per Table 10.1. This table shows, for instance, that the one-third octave band

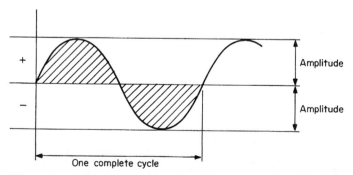

FIGURE 10.3 Threshold of hearing for young adults with normal hearing. [*C. M. Harris (ed.), Handbook of Noise Control, 2d ed., McGraw-Hill, New York, 1979, part 1, p. 8–4.*]

TABLE 10.1　Nominal One-Third Octave Band Center Frequencies and Ranges

Third octave band no.	Center frequency, Hz	Frequency range, Hz	Corresponding full octave band
14	25	22–28	Suboctave 22–45
15	—31.5—	28–36	
16	40	35–45	
17	50	45–56	1 45–89
18	—63—	56–71	
19	80	71–89	
20	100	89–112	2 89–178
21	—125—	112–141	
22	160	141–178	
23	200	178–224	3 178–355
24	—250—	224–282	
25	315	282–355	
26	400	355–447	4 354–709
27	—500—	447–563	
28	630	562–708	
29	800	708–892	5 707–1414
30	—1000—	891–1123	
31	1250	1122–1413	
32	1600	1412–1779	6 1411–2822
33	—2000—	1778–2240	
34	2500	2238–2819	
35	3150	2817–3549	7 2815–5630
36	—4000—	3547–4469	
37	5000	4465–5625	
38	6300	5621–7082	8 5617–11234
39	—8000—	7077–8916	
40	10000	8909–11225	

Note: Band numbers and center frequencies (nominal for ordinary use) are per ANSI/IEC standards. Frequency band limits are rounded to the nearest hertz.

with a center frequency of 63 Hz has a range from 56 to 71 Hz, while the corresponding full octave has a range from 45 to 89 Hz.

10.3　THE SPEED OF SOUND IN AIR

The speed of sound in air can be calculated from the expression

$$c = \begin{cases} 49.03\sqrt{460 + °F} & \text{in English units} \\ 20.05\sqrt{273 + °C} & \text{in metric units} \end{cases} \tag{10.2}$$

where c is the speed of sound in ft/s (m/s). Note that for all practical purposes, the speed of sound in air is independent of pressure.

For example, if the temperature is 70°F (21°C), the speed of sound is

$$c = 49.03\sqrt{530} = 1129 \text{ ft/s } (344 \text{ m/s})$$

TABLE 10.2 Wavelengths of Sound in Air at 70°F (21°C) $c = 1129$ ft/s

f, Hz	63	125	250	500	1000	2000	4000	8000
λ, ft	17.92	9.03	4.52	2.26	1.129	0.56	0.28	0.14
λ, m	5.46	2.75	1.38	0.69	0.34	0.17	0.085	0.043

We can then use this value of c to compute the wavelength λ at various frequencies f at 70°F (21°C). At a frequency of 1000 Hz, for instance, we find from Eq. (10.1) that $\lambda = c/f = 1129/1000 = 1.129$ ft (0.344 m); likewise, at 20 Hz the wavelength would measure about 56.5 ft (17.2 m), and at 20,000 Hz it would measure about $^2/_3$ in (17.2 mm). For 70°F (21°C), Table 10.2 gives the wavelengths of sound in air at several frequencies.

As a practical matter, because the thickness of walls and the absorptive sections of silencers are small in relation to the wavelengths of low frequencies, such structures generally attenuate sound much better in the middle and high frequencies than in the low ones. Larger and more complex structures are required for reducing low-frequency noise.

10.4 THE SPEED OF SOUND IN SOLIDS

The speed of sound in longitudinal waves in a solid bar can be shown to be

$$c_s = \sqrt{\frac{E}{\rho}} \qquad (10.3)$$

where c_s is the speed of sound in solids (m/s), E is the bar's modulus of elasticity (N/m²), and ρ is its density (kg/m³). This obviously means that sound travels faster through media of high modulus of elasticity and of low density. Accordingly, because rubber has a much higher elasticity and lower density than steel does (as one example), a rubber insert in a steel pipe will tend to slow down sound transmission along the pipe. Table 10.3 shows the speed of sound in various media.

One can speculate that since the elasticity and density in an absolute vacuum are zero, theoretically no sound waves should be able to travel through it. An absolute vacuum may thus be the ultimate noise barrier. However, no one is yet known to have been able to come up with a practical earthborn design.

TABLE 10.3 Speed of Sound in Various Media (Shown in Descending Order of Magnitude)

Medium	Speed ft/s	Speed m/s	Medium	Speed ft/s	Speed m/s
Steel	16,500	5029	Concrete	10,600	3231
Aluminum	16,000	4877	Water	4,700	1433
Brick	13,700	4176	Lead	3,800	1158
Wood (hard)	13,000	3962	Cork	1,200–1,700	366–518
Glass	13,000	3962	Air	1,129	344
Copper	12,800	3901	Rubber	130–490	40–149
Brass	11,400	3475			

10.5 THE DECIBEL

In using the term "decibel" it is important to understand the difference between sound power levels and sound pressure levels, since both are expressed in decibels.

Sound pressure levels, which can readily be measured, quantify in decibels the intensity of given sound sources. Sound pressure levels vary substantially with distance from the source, and they also diminish as a result of intervening obstacles and barriers, air absorption, wind, and other factors.

Sound power levels, on the other hand, are constants independent of distance. It is very difficult to establish the sound power level of any given source because this level cannot be measured directly, but must be calculated by means of elaborate procedures; thus, as a practical matter, sound power levels are converted to sound pressure levels, which form the basis of practically all noise control criteria.

(As one example, Sec. 10.24 illustrates how the sound power level of a fan in an HVAC system is a critical element in the silencer selection procedure to meet specified sound pressure level criteria in an office or space.)

10.5.1 Sound Power Level

The lowest sound level that people of excellent hearing can discern has an acoustic power, or sound power, of about 10^{-12} W. On the other hand, the loudest sound generally encountered is that of a jet aircraft, with a sound power of about 10^5 W. Thus the ratio of loudest to softest sounds generally encountered is 10^{17}:1.

A tenfold increase is called a "bel," so the intensity of the jet aircraft's noise can also be referred to as "17 bels," This cuts the expression of immense ranges of intensity down to manageable size. However, since the bel is still a rather large unit, it is divided into 10 sub-units called "decibels" (dB). Thus the jet noise is 170 dB, and to avoid confusion with any other reference intensity, we can say that it is 170 dB with reference to 10^{-12} W.

Sound power level L_W in decibels is therefore defined as

$$L_w = 10 \log \frac{W}{10^{-12}} \text{ dB re } 10^{-12} \text{ W} \tag{10.4}$$

where W is the sound power in watts. The sound power level in decibels can also be computed from

$$L_w = 10 \log W + 120 \tag{10.5}$$

Since 10^{-12} as a power ratio corresponds to -120 dB, we can see that by definition 1 W is equivalent to a 120-dB power level. Table 10.4 shows the sound power levels of typical sources.

Note that certain older literature may contain sound power level data referenced to 10^{-13} W, an obsolete standard. Where this is the case, deduct 10 dB to convert to the current standard of 10^{-12} W.

The question now is, How does one measure sound power W? This is where another way of looking at sound power helps. As shown in Fig. 10.4, consider a simple nondirectional source located at the center of a spherical surface (or at the center of a number of expanding spherical surfaces). Here the total sound power in watts is equal to sound intensity I (W/m^2) times the surface area S (m^2):

$$W = IS \tag{10.6}$$

TABLE 10.4 Sound Power Level L_W of Typical Sources

Source	Sound power W, W	L_W, dB re 10^{-12} W
Saturn rocket	100,000,000	200
Afterburning jet engine	100,000	170
Large centrifugal fan at 500,000 ft³/min (849,500 m³/h)	100	140
Seventy-five-piece orchestra/vaneaxial fan at 100,000 ft³/min (169,900 m³/h)	10	130
Large chipping hammer	1	120
Blaring radio	0.1	110
Centrifugal fan at 13,000 ft³/min (22,087 m³/h)	0.1	110
Automobile on highway	0.01	100
Food blenders—upper range	0.001	90
Dishwashers—upper range	0.0001	80
Voice—conversational level	0.00001	70
Quiet-Duct silencer, self-noise at + 1000 ft/min (5.1 m/s)	0.00000001	40
Voice—very soft whisper	0.000000001	30
Quietest audible sound for persons with excellent hearing	0.000000000001	0

where S for a spherical surface is $4\pi r^2$. Of course, as the sound waves move farther from their source, the surrounding spherical surface will become larger, and less power will pass through any unit element of the surface.

If the sound source is directional, the intensity will vary over the surface and the radiated power must be found by integration:

$$W = \int_s IS^{dS} \qquad (10.7)$$

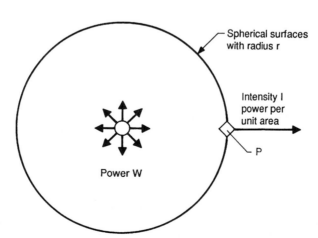

FIGURE 10.4 Ideal nondirectional sound source radiating W watts and producing a sound intensity I in watts per unit.

Since sound intensity I is rather difficult to measure, we measure sound pressure p instead. The relationship between sound pressure and intensity is

$$I = \frac{p_{rms}^2}{\rho c} \text{ W/m}^2 \tag{10.8}$$

where p = root-mean-square (rms) sound pressure, N/m^2
ρ = density of air, kg/m^3
c = speed of sound in air, m/s

The form of this equation will be familiar to many since it is analogous to the formula relating to electric power, voltage, and resistance:

$$P = \frac{E}{R}$$

where P = power, W
E = voltage, V
R = resistance, Ω

Sound intensity level L_i is defined as

$$L_i = 10 \ \log \frac{I}{I_{ref}} \text{ dB re } I_{ref} \tag{10.9}$$

where I_{ref} is 10^{-12} W/m^2.

10.5.2   Sound Pressure Level

Since sound-measuring instruments respond to sound pressure, the word "decibel" is generally associated with sound pressure level, but it is also a unit of sound power level. The square of sound pressure is proportional to, though not equal to, sound power.

Assuming a point source of sound radiating spherically in all directions, Eq. (10.6) tells us that $W = IS$. Accordingly, $10 \log W = 10 \log I + 10 \log S$, where S is the surface of the radiating sphere in ft^2 (m^2). Equation (10.5), however, tells us that $10 \log W = L_W - 120$. It can also be shown that $10 \log I = L_i - 120$. As a result, we get

$$L_w - 120 = L_i - 120 + 10 \log S$$

or

$$L_w = L_i + 10 \log S$$

Since $L_i = 10 \log(I/I_{ref})$ and $I = p^2/\rho c$, and since L_i can also be expressed as $10 \log(p^2/p_{ref}^2)$, which is also referred to as sound pressure level L_p, then

$$L_p = 10 \ \log \left(\frac{p}{p_{ref}} \right)^2 = 20 \ \log \frac{p}{p_{ref}} \tag{10.10}$$

Accordingly,

$$L_w = L_p + 10 \ \log \ S \tag{10.11}$$

10.6 DETERMINATION OF SOUND POWER LEVELS

The concept of the imaginary radiating sphere emanating from the sound source will be referred to again in Sec. 10.8, Propagation of Sound Outdoors. Here, on the other hand, without considering imaginary spheres, we are concerned with measuring the sound power of a source that is confined within a structurally rigid space; for very large pieces of equipment and operating machinery in a plant, this approach may be the most practical way to estimate sound power.

The best method for determining the sound power level of a source is to measure it inside a good reverberant room with a truly diffuse sound field. With the sound power thus contained within the room, and with its intensity evenly distributed throughout the room, often only one sound pressure level measurement has to be taken. Then the sound power level can be calculated from $L_W = L_p + K$, where K is a constant dependent on the room volume, on the reverberation time at a given frequency or frequency band, and on the humidity.

Another method consists of containing the sound within the rigid walls of a pipe or duct equipped with an anechoic termination to minimize end reflections. Here all the sound energy must travel through the duct, and its sound field can be measured at a suitable measuring plane by averaging the sound pressure level across it. Equation (10.11) can then be used for calculating the power level of the noise-maker. Figure 10.5 illustrates such an arrangement for a ducted fan, which is also the basis of U.S. and British standards. (The 1986 U.S. standard was published jointly by ASHRAE, ANSI, and AMCA: ANSI/ASHRAE 68–1986 and ANSI/AMCA 330-86.) Although the anechoic duct method must overcome some practical difficulties, such as allowing for aerodynamically induced noise at the microphone, it clearly illustrates the relationship between measurements of sound pressure level and sound power level.

A test code of the U.S. Air Movement and Control Association (AMCA) requires the use of a reverberant or semireverberant room for determination of fan sound power levels. In such an arrangement the microphone would not be affected by aerodynamic flow.

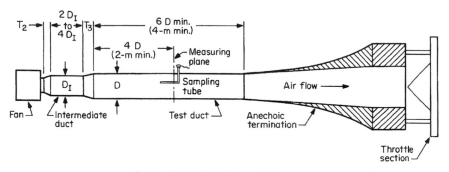

Test on fan outlet with open inlet

FIGURE 10.5 Anechoic duct method for fan sound power level determination. (*British Standard 848, Methods of Testing Fans, 1966, Part 2; ASHRAE/AMCA, Laboratory Method of Testing In-Duct Sound Power Measurement Procedure for Fans, 1986.*)

These two methods can yield comparable results, but relative fan sound power levels are likely to be most comparable if they have been determined under identical conditions.

In the British Standard 848, *Methods of Testing Fans*, 1966, part 2, the sound power level L_W in each frequency band would be calculated after averaging the sound pressure level L_p across the duct area according to $L_W = L_p + 10 \log A$, where A is the cross-sectional area in ft^2 (m^2) at the plane of measurement. The U.S. standard (as in Fig. 10.5) uses $L_W = L_p + 20 \log D - 1.1$, where D is the diameter in ft (m) of the test duct.

10.7   CALCULATING CHANGES IN SOUND POWER AND SOUND PRESSURE LEVELS

10.7.1   Sound Power Level

Let L_{w1} be the sound power level corresponding to sound power W, and let L_{w2} be the sound power level twice as great, or $2W$. Then from Eq. (10.4)

$$L_{w1} = 10 \log \frac{W}{W_{\text{ref}}}$$

and

$$L_{w2} = 10 \log \frac{2W}{W_{\text{ref}}} = 10 \log \frac{W}{W_{\text{ref}}} + 10 \log 2 = L_{w1} + 3 \text{ dB}$$

Note: In Eq. (10.4), $W_{\text{ref}} = 10^{-12}$ W = 1 pW (picowatt).

10.7.2   Sound Pressure Level

Assume L_{p1} to correspond to sound pressure p, and L_{p2} to sound pressure $2p$. Then from Eq. (10.10),

$$L_{p1} = 20 \log \frac{p}{p_{\text{ref}}}$$

and

$$L_{p2} = 20 \log \frac{2p}{p_{\text{ref}}} = 20 \log \frac{p}{p_{\text{ref}}} + 20 \log 2 = L_{p1} + 6 \text{ dB}$$

The addition of two equal sound pressures results in an increase of 6 dB, and the addition of two equal sound powers results in an increase of 3 dB. However, when two equal sound pressure levels are added, we are adding in effect two equal sound power levels, therefore:

$$L_{p1} + L_{p1} = 10 \log \left(\frac{p}{p_{\text{ref}}} \right)^2 + 10 \log \left(\frac{p}{p_{\text{ref}}} \right)^2$$

$$= 10 \log \left(\frac{p}{p_{\text{ref}}} \right)^2 \times 2$$

$$= 10 \log \left(\frac{p}{p_{\text{ref}}} \right)^2 + 3 \text{ dB}$$

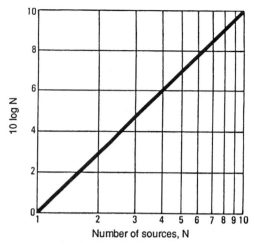

FIGURE 10.6 Predicting the combined noise level of identical sources.

Similarly, it can be said that when N identical sound sources are added,

$$L_p(\text{total}) = L_p(\text{single source}) + 10 \log N \qquad (10.12)$$

where N is the number of sources; $10 \log N$ is plotted as a function of N in Fig. 10.6.

Table 10.5 shows how to add two unequal decibel levels, and Fig. 10.7 presents Table 10.5 graphically. Examples:

1. Two fans produce an L_p of 95 dB each in the fourth octave band at a given location. The combined L_p in that band would then be 98 dB.
2. If one of these fans is slowed down to produce an L_p of 90 dB, the combination L_p would then be 96 dB.

TABLE 10.5 Addition of Sound Levels

Difference between the two levels, dB	Add to the higher level, dB
0	3
1	2.5
2	2
3	2
4	1.5
5	1
6	1
7	1
8	0.5
9	0.5
10 or more	0

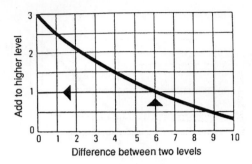

Example: 80 dB + 74 dB = 81 dB

FIGURE 10.7   Decibel addition.

10.8 *PROPAGATION OF SOUND OUTDOORS*

Section 10.5 (and Fig. 10.4) introduced the concept of sound propagating through a series of spheres increasing in size as the distance r from its source increases. We now need to differentiate between hemispherical and spherical sound sources. If the source is considered hemispherical, the surface area $S = 2\pi r^2$; if the source is spherical, $S = 4\pi r^2$.

A fully spherical source would not be encountered frequently in a practical situation (examples would be an aircraft flying overhead, a rocket in flight, or noise emanating from the top of a tall building or a vertical stack or from a bird flying through air). If the source radiates hemispherically, as most sources do when close to the ground or to other reflective surfaces, then for a uniformly directional source (such as a siren), the relationship between sound pressure and sound power would be

$$
\begin{aligned}
L_p &= L_w - 10 \ \log \ 2\pi r^2 \\
    &= L_w - 20 \ \log r - 10 \ \log \ 2\pi \\
    &= \begin{cases} L_w - 20 \ \log r + 2.3 & \text{if } r \text{ is in feet} \\ L_w - 20 \ \log r - 8 & \text{if } r \text{ is in meters} \end{cases}
\end{aligned}
\tag{10.13}
$$

For a spherical source, the relationship would be

$$
\begin{aligned}
L_p &= L_w - 10 \ \log \ 4\pi r^2 \\
    &= L_w - 20 \ \log r - 20 \ \log \ 4\pi \\
    &= \begin{cases} L_w - 20 \ \log r + 0.7 & \text{if } r \text{ is in feet} \\ L_w - 20 \ \log r - 11 & \text{if } r \text{ is in meters} \end{cases}
\end{aligned}
\tag{10.14}
$$

It will be noted that the sound pressure level for a hemispherical source is 3 dB higher than for a spherical source because the same sound intensity is considered to pass through an area half the size of a full sphere.

Not all sound sources radiate uniformly. If a sound source has a marked directional characteristic, this characteristic has to be taken into account; it is called the "directivity index" (DI). Figure 10.8 illustrates how noise emanating from an opening, stack, or pipe will vary with the directivity angle.

Other factors affecting the radiation of sound might be barriers and the attenuation of sound due to atmospheric conditions (such as molecular absorption in the air, wind, and rain)

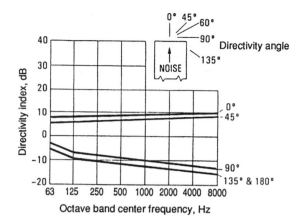

The noise emanating from an opening, stack or pipe, will vary with directivity angle between the point of measurement and the conduit centerline. Data shown for stack or pipe diameter of approximately 10-ft (3.05-m) equivalent diameter.

FIGURE 10.8 Directivity index from openings, stacks, or pipes. (*NEMA Standards Publication SM 33, Directivity in Openings, Stacks or Pipes, 1964.*)

and ground conditions (including grass, trees, shrubbery, snow, paving, and water). Attenuation due to such factors is generally significant in the high frequencies and over long distances and makes reliable and reputable outdoor measurements very difficult to obtain.

For a directional noise source, we can therefore estimate sound pressure levels by modifying Eq. (10.13) as follows:

$$L_p = \begin{cases} L_w - 20 \log r + \mathrm{DI} - A_a - A_b + 2.3 & \text{if } r \text{ is in feet} \\ L_w - 20 \log r + \mathrm{DI} - A_a - A_b - 8 & \text{if } r \text{ is in meters} \end{cases} \quad (10.15)$$

where DI = directivity index
A_a = attenuation due to atmospheric conditions
A_b = attenuation due to barriers
r = distance from source, ft (m)

For instance, if we know or estimate the L_w of a fan (now provided by many manufacturers), we can also estimate the L_p at a distance r by taking into account directivity and the other factors indicated in Eq. (10.15).

10.9 THE INVERSE-SQUARE LAW

From Eq. (10.15) we can see that if the sound pressure level of a source is measured at two different distances from the source, the difference in sound pressure levels at those locations is

$$L_{p2} - L_{p1} = 10 \log \left(\frac{r_2}{r_1} \right)^2 = 20 \log \frac{r_2}{r_1} \quad (10.16)$$

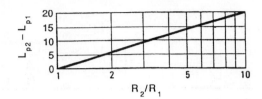

R₁ = distance from source to location 1

R₂ = distance from source to location 2

L_p1 = sound pressure level, location 1

L_p1 = sound pressure level, location 2

FIGURE 10.9 Inverse-square law.

where L_{p1} = sound pressure level at location 1, dB
L_{p2} = sound pressure level at location 2, dB
r_1 = distance from source to location 1, ft (m)
r_2 = distance from source to location 2, ft (m)

The relationship between $(L_{p2} - L_{p1})$ and r_2/r_1 is shown in Fig. 10.9.

Equation (10.16) shows that the sound pressure level varies inversely with the square of the distance from the source, with L_p decreasing by 6 dB for each doubling of distance from the source. This relationship is known as the "inverse-square law."

At locations very close to a sound source, a measurement point will be in what is known as the "near field" or the source. In the near field, neither Eq. (10.15) nor Eq. (10.16) applies, and L_p will vary substantially with small changes in position. As the distance increases, however, L_p will decrease according to the inverse-square law; Eqs. (10.15) and (10.16) will apply, and a measurement point can be said to be in the "far field" of the source.

For all practical purposes, the inverse-square law functions only in a "free field," which is defined as a space with no reflective boundaries or surfaces. Outdoors, such conditions are likely to exist only in an open field. In a reverberant field, such as might exist in the courtyard of a building or in a narrow street, the sound pressure level may decrease by a factor of less than 6 dB for each doubling of the distance. On the other hand, in a field of freshly fallen snow the decrease may be more than that predicted by the inverse-square law.

10.10 PARTIAL BARRIERS

Unobstructed sound propagates directly along a straight-line path from the source. If a barrier is interposed between that source and a receiver, some of the sound will be reflected back toward the source. These reflections can, of course, be attenuated by placing sound-absorptive surfaces on the barrier side facing the source.

Another portion of the sound emanating from the source is transmitted through the barrier (Fig. 10.10). To meet structural and wind loading criteria, however, most barrier designs significantly inhibit noise transmission to the extent that sound reaches the receiver

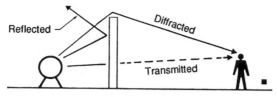

FIGURE 10.10   Barrier reflection, diffraction, and transmission.

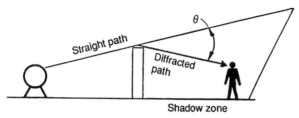

FIGURE 10.11   The shadow zone behind a barrier.

primarily by diffracting over and around the barrier. As shown in Fig. 10.11, the presence of the barrier creates a "shadow zone" in which diffraction attenuates the noise reaching the receiver; the extent of this attenuation is the angle Θ between the straight and diffracted sound paths. Angle Θ (and thereby barrier attenuation) increases if the receiver or source is placed closer to the barrier or (assuming that the barrier is long enough to prevent sound from diffracting around the ends) if the barrier height is increased.

The theoretical relationship between barrier height, source and receiver position, and barrier attenuation from diffraction can be mathematically expressed as a function of Fresnel number N, as shown in Fig. 10.12.

10.11   PROPAGATION OF SOUND INDOORS

Assume that a sound source is on the floor of an enclosed space and that there are no partitions or barriers between the source and the receiver, and assume further that none of the sound leaves the space and reaches the receiver by a flanking path. Under these conditions, the sound in the space will reach the receiver by two paths: a direct sound path and a reverberant sound path.

10.11.1   Direct Sound Path

In the far field of the source, sound from a source on or near the center of a wall or floor in a room will propagate to the receiver according to the inverse-square law:

$$L_{pd} = \begin{cases} L_w - 20 \ \log r + 2.3 & \text{if } r \text{ is in feet} \\ L_w - 20 \ \log r - 8 & \text{if } r \text{ is in meters} \end{cases} \quad (10.17)$$

where L_{pd} is the sound pressure level from direct sound.

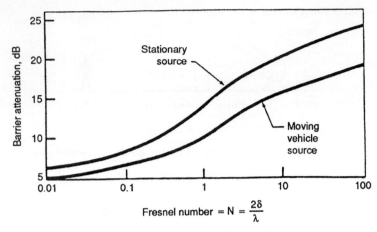

$$\text{Fresnel number} = N = \frac{2\delta}{\lambda}$$

where λ = wavelength of sound, ft or m

$\delta = A + B - d$, ft or m

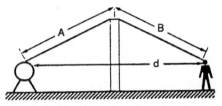

Note: N is a dimensionless number and can be used in English
or metric units on a consistent basis.

FIGURE 10.12 Barrier attenuation as a function of Fresnel number. (*Z. Maekawa,*
"Shielding Highway Noise," Noise Control Engineering, vol. 9, no. 1, July–Aug, 1977.)

10.11.2 Reverberant Sound Path

Reverberant sound will reach the receiver after reflecting off surfaces in the space. If the
sound in the space is diffuse (essentially equal at all locations), Eq. (10.18) applies:

$$L_{pr} = \begin{cases} L_w - 10 \ \log A + 16.3 \text{ if } A \text{ is in sabins} \\ L_w - 10 \ \log A + 6.0 \quad \text{if } A \text{ is in metric sabins} \end{cases} \tag{10.18}$$

where L_{pr} is the sound pressure level from reverberant sound, and A is the total absorption.
The "total absorption" of a surface is the product of the surface area S and the absorption
coefficient α of that surface:

$$A = S\alpha \tag{10.19}$$

where the units of A are sabins if S is in ft^2, and metric sabins if S is in m^2; α is the sound
absorption coefficient, the dimensionless ratio of sound energy absorbed by a given surface
to that incident upon the surface (see Sec. 10.22 and Table 10.43).

Total room absorption can be calculated as follows:

$$A = \Sigma S\overline{\alpha} = S_1\alpha_1 + S_2\alpha_2 + S_3\alpha_3 + \cdots + S_n\alpha_n \qquad (10.20)$$

where
A = total absorption in room, sabins (metric sabins)
S = total surface area in room, ft² (m²)
$\overline{\alpha}$ = average room absorption coefficient
$S_1, S_2, S_3, \ldots, S_n$ = surface area of different segments of wall, ceiling, and floor surfaces in room
$\alpha_1, \alpha_2, \alpha_3, \ldots, \alpha_n$ = corresponding sound absorption coefficients

Reverberant sound may be reduced by adding sound-absorptive materials to reflective room surfaces. The theoretical reduction in reverberant sound due to adding sound-absorptive treatment to the surfaces of a room containing a diffuse sound field is equal to

$$\text{Reduction in reverberant sound} = 10 \log\frac{A_2}{A_1} \qquad (10.21)$$

where A_1 is the total room absorption after adding sound-absorptive treatment, and A_2 is the total room absorption before adding treatment. This is illustrated in Fig. 10.13.

10.11.3 Effects of Direct and Reverberant Sound

The effects of direct and reverberant sound are shown in Fig. 10.14. Direct sound predominates close to the source, but direct sound diminishes with distance. Thus, farther from the source, reverberant sound predominates; under ideal conditions this occurs when the L_p in the room levels off with increasing distance from the source.

The quantitative relationship between L_w and L_p from both direct and reflected sound paths is shown in Fig. 10.15 as a function of distance from the source and total room absorption. Add 3 dB to L_p if the source is on the wall or floor of the room, add 6 dB if the source is at the intersection of two walls (or a wall and ceiling), and add 9 dB if the source is in a corner (Ref. 1).

Note in Fig. 10.15 [and also Eq. (10.17)] that increasing the total room absorption has no effect on direct sound; accordingly, adding sound-absorptive materials to room surfaces will show maximum reduction in L_p in areas where reverberant sound predominates. Also, note that small increases in total room absorption will not produce

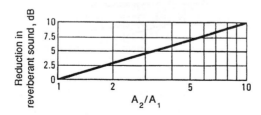

A_2 = Total room absorption after adding sound absorptive treatment

A_2 = Total room absorption before adding treatment

FIGURE 10.13 Effect of increasing room absorption.

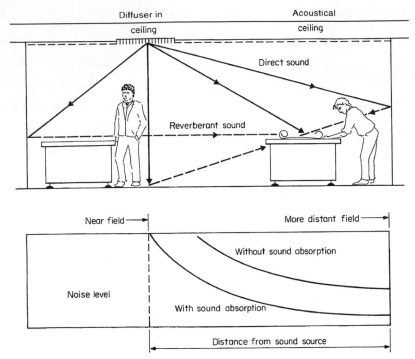

FIGURE 10.14 Effects of direct and reverberant sound on listeners in the source's near and far fields. Close to the source, direct sound predominates; at a distance, reverberant sound predominates.

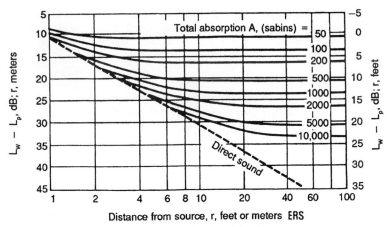

FIGURE 10.15 Effects of direct and reverberant sound in rooms. [*C. M. Harris (ed.), Handbook of Noise Control, 2d ed., McGraw-Hill, New York, 1979, part 1, p. 8–4.*]

significant decreases in sound pressure level; even in a location dominated by reverberant sound, doubling the room absorption will decrease the sound pressure level by only 3 dB.

10.12 SOUND TRANSMISSION LOSS

Figure 10.16 shows that when a sound path is broken by a partition, part of the sound is reflected, part is absorbed, and part is transmitted through the partition.

Ten times the logarithmic ratio of incident sound power to transmitted sound power is defined as "sound transmission loss" (TL). As shown in Fig. 10.17,

$$\text{TL} = 10 \log \frac{W_i}{W_t} = 10 \log \frac{1}{\tau} \tag{10.22}$$

where τ = sound transmission coefficient
W_i = incident sound power, W
W_t = transmitted sound power, W

10.12.1 The Mass Law

The mass law provides a theoretical relationship between the sound transmission loss of a single-wall (solid) partition, its weight, and the frequency of sound being transmitted through it. For normal incidence (NI), the relationship is

$$\text{TL} = \begin{cases} 10 \log w + 20 \log f - 33.5 & \text{if } w \text{ is in lb/ft}^2 \\ 10 \log w + 20 \log f - 47.5 & \text{if } w \text{ is in kg/m}^2 \end{cases} \tag{10.23}$$

where w is the weight (or mass density), and f is the frequency in hertz.

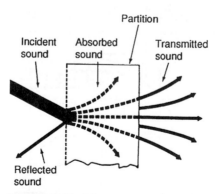

FIGURE 10.16 Effect of partitions on incident sound. (*Noise Control: A Guide for Workers and Employees, U.S. Department of Labor, Occupational Safety and Health Administration, 1980.*)

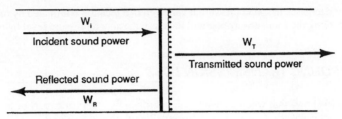

FIGURE 10.17 Incident sound power versus reflected and transmitted sound power.

Equation (10.23) tells us that for each doubling of the barrier's weight, the transmission loss increases by 6 dB. Equally, by doubling or halving the frequency, a 6-dB shift in TL occurs.

Equation (10.23) is commonly known as "the mass law," but more accurately it is an approximation. Actual data can deviate from mass law predictions by 10 dB or more, and the law generally does not apply to nonhomogeneous structures. As will be shown in Sec. 10.21 for example, multilayer walls or double walls separated by an air space generally provide greater TL than that predicted by the mass law.

10.12.2 The Effect of Openings on Partition TL

Windows, access ports, door seals, wall-to-ceiling joints, cutouts for wiring or plumbing, and other openings can significantly diminish the TL capabilities of a structure. As an example, if a 100-ft^2 (9.3-m^2) partition has a TL rating of 40 dB at a given frequency, a 1 percent [or 1-ft^2 (0.093-m^2)] opening in that partition will reduce the overall TL to 20 dB unless noise control measures are applied to the opening. The theoretical effect of openings in partitions or complete enclosures is shown in Fig. 10.18.

10.12.3 Single-Number TL Ratings: STC Ratings

For engineering rating purposes, the TL of partitions is frequently defined in terms of a single-number decibel rating known as "sound transmission class" (STC). STC ratings are

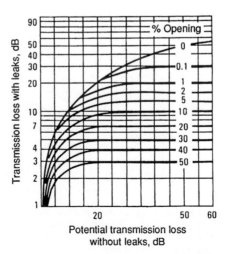

FIGURE 10.18 Effect of openings on partition TL.

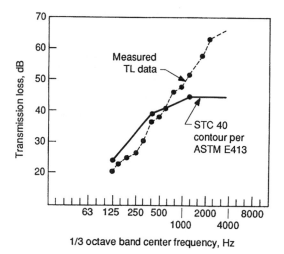

FIGURE 10.19 ASTM E413 contours for sound transmission class (STC) and noise isolation class (NIC). (*ASTM E413, Standard Classification for Determination of Sound Transmission Class, 1973.*)

determined by plotting contours of TL versus frequency in one-third octave bands from 125 to 4000 Hz and comparing the results with standard contours defined in ASTM E413 (Fig. 10.19). The total deficiencies must not be greater than 32 dB, but any single band's deficiency cannot be greater than 8 dB.

10.13 NOISE REDUCTION AND INSERTION LOSS

As shown in Fig. 10.20, "noise reduction" (NR) is simply the difference in sound pressure level between any two points along the sound path from a noise source:

$$NR = L_{p1} - L_{p2} \qquad (10.24)$$

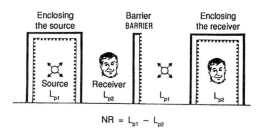

FIGURE 10.20 Illustration of noise reduction: $NR = L_{p1} - L_{p2}$. (*Lawrence G. Copley, "Control of Noise by Partitions and Enclosures," Tutorial Papers on Noise Control for Inter-Noise, Institute of Noise Control Engineers, 1972.*)

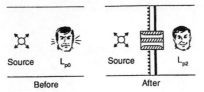

FIGURE 10.21   Illustration of insertion loss: $IL = L_{p0} - L_{p2}$. (*Lawrence G. Copley, "Control of Noise by Partitions and Enclosures," Tutorial Papers on Noise Control for Inter-Noise, 1972.*)

"Insertion loss" (IL), on the other hand is the before-versus-after difference at the same measurement point, brought about by interposing a means of noise control between the source and the receiver (Fig. 10.21):

$$IL = L_{p0} - L_{p2} \tag{10.25}$$

Like TL, NR and IL are typically rates as a function of full octave bands or one-third octave bands. The NR ratings of several types of soundproof room are listed in Table 10.42. A single-number NR rating system called "noise isolation class" (NIC) is often used for such rooms. Similar to the STC ratings described in Sec. 10.12.3, NIC ratings are established by plotting NR as a function of frequency and comparing the results against standard contours defined in ASTM E413.

10.14   THE EFFECTS OF SOUND ABSORPTION ON RECEIVING-ROOM NR CHARACTERISTICS

Figure 10.22 shows a receiver located within a room outside of which is a noise source. The relationship between the NR and TL characteristics of such a room can be shown to be represented by

$$NR = TL + 10 \log \frac{\overline{\alpha}_2 A_2}{S} \tag{10.26}$$

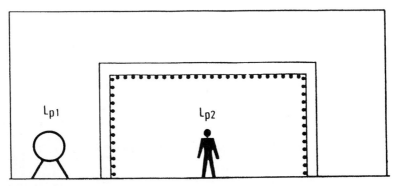

FIGURE 10.22   Noise source in outer room, and receiver in inner room.

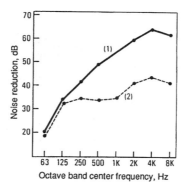

1. 2-in (51-mm) sound absorptive materials on inside wall surface
2. Reflective steel walls

FIGURE 10.23 Additional NR demonstrated by adding sound absorption to the inside surface of a reflective receiving room. (*Martin Hirschorn, Noise Control Reference Handbook, Industrial Acoustics Company, 1982.*)

where NR $= L_{p1} - L_{p2}$
  L_{p1} = sound pressure level in source room, dB
  L_{p2} = sound pressure level in receiving room, dB
  TL = transmission loss of receiving-room walls, dB
  $\overline{\alpha}_2$ = average sound absorption coefficient in receiving room
  A_2 = total wall area in receiving room, ft^2 (m^2)
  S = surface area separating the two rooms, ft^2 (m^2)

If the source room is highly reverberant and if the receiving room is highly absorptive such that $\overline{\alpha}_2$ is close to unity, then NR = TL. In the event that the receiving room is highly reflective, however, $\overline{\alpha}_2$ will be very low; for instance, if $S = A_2$ and if $\overline{\alpha}_2 = 0.01$, then NR = TL $- 20$ dB.

Accordingly, a highly absorptive receiving room can be seen to have a potential of 20-dB more noise reduction than a reflective receiving room with the same TL. This effect is illustrated in Fig. 10.23, which shows the NR of a 6-ft 4-in by 6-ft 0-in by 6-ft 6-in (1930- by 1829- by 1981-mm) room, which could be a fan plenum, tested with and without 2 in (51 mm) of sound-absorptive materials on the otherwise highly reflective steel inside walls. The sound absorption coefficient of a 2-in (51-mm) liner is relatively low at low frequencies, so the liner has little effect on NR. At the higher frequencies, however, NR is approximately 20 dB higher with the absorptive liner in place.

10.15 FAN NOISE

Fans are the primary source of noise generation in HVAC systems. It is always best to use fan L_W data provided by the fan manufacturer. However, if these data are not available, Eq. (10.27) can be used to predict the estimated fan L_W (dB re $1p_w$); see note in section 7.1 (Ref. 2):

$$L_w = K_w + 10 \log \frac{Q}{Q_1} + 20 \log \frac{P}{P_1} + C + \text{BFI} \qquad (10.27)$$

where K_W = specific sound power level, dB re $1p_w$ (from Table 10.6)
      Q = flow rate, ft³/min (L/s)
      Q_1 = 1 when Q is in ft³/min, 0.472 when Q is in L/s
      P = fan pressure head, in WG (Pa) [in WG is "inch water gauge"]
      P_1 = 1 when P is in inches WG, 249 when P is in Pa
      C = correction factor for point of operation, dB
    BFI = blade frequency increment to be added only to octave band containing blade
           pass frequency

The values of K_W and BFI are shown in Table 10.6, and Table 10.7 shows the octave band in which the BFI is likely to occur for different fan types. Values for C are given in Table 10.8.

Fans can generate high-intensity noise levels of a discrete tone at the blade pass frequency (BPF). The noise level's intensity will vary with the type of fan. The BPF can be established if the rpm and number of blades of the fan are known; the following equation can then be used:

$$\text{BPF} = \frac{\text{rpm} \times \text{number of blades}}{60} \text{ Hz} \qquad (10.28)$$

For example, if the rpm is 1200 and the number of blades is 8, BPF = 160 Hz.

These discrete tones at the BPF are usually the most predominant noises emanating from large fans, but such discrete frequencies may not show up in an octave band analysis. To find them, narrower frequency ranges may have to be measured, such as one-third octave bands or even one-tenth octave bands.

BFIs vary from 2 dB for centrifugal radial-blade fans to 8 dB for centrifugal pressure-blower fans (see Table 10.6). BFIs and second harmonic frequencies generally occur in the 63- to 500-Hz region (see Table 10.7).

Once a decision has been made as to the type of fan to be used, it is best to select one that operates close to the peak of its efficiency curve. Such a fan will typically generate the lowest noise level. The correction factor C for off-peak operation is shown in Table 10.8.

Example 10.1   A 35.5-in-diameter vaneaxial fan with eight blades has a 20,000-ft³/min flow rate, develops a 4-in WG head at a speed of 1765 r/min, and operates at 95 percent of peak efficiency. Determine the fan's L_W and BPF.

Solution   Calculate the fan's total L_W from Eq. (10.27):

- For K_W, Table 10.6 gives a range of octave band center frequencies for a vaneaxial fan with a diameter (or wheel size) under 40 in.
- Flow rate Q = 20,000 ft³/min, and Q_1 = 1. Thus 10 log (Q/Q_1) = 43.
- Fan pressure head P = 4 in WG, and P_1 = 1. Thus 20 log (P/P_1) = 12.
- Correction factor C comes from Table 10.8; at 95 percent peak efficiency, C = 0.
- From Table 10.6, for vaneaxial fans, BFI = 6. Furthermore, Table 10.7 and its note show that this BFI occurs in the 250-Hz octave band.

These data are tabulated in Table 10.9, which shows the total L_W.

To calculate the fan's BPF, use Eq. (10.28). Given 1765 r/min and eight blades, BPF = $(1765 \times 8)/60$ = 235 Hz, and Table 10.9 shows that the nearest octave band to 235 Hz in this example is 250 Hz.

TABLE 10.6 Specific Sound Power Levels K_W (dB re $1p_w$) and Blade Frequency Increment (BFI) for Various Types of Fans

Fan type	Wheel size	Octave band center frequency, Hz							BFI
		63	125	250	500	1000	2000	4000	
Centrifugal									
Airfoil, backward-curved, backward-inclined	Over 36 in (900 mm)	32	32	31	29	28	23	15	3
	Under 36 in (900 mm)	36	38	36	34	33	28	20	
Forward-curved	All	47	43	39	33	28	25	23	2
Radial blade and pressure blower	Over 40 in (1000 mm)	45	39	42	39	37	32	30	8
	From 40 in (1000 mm) to 20 in (500 mm)	55	48	48	45	45	40	38	
	Under 20 in (500 mm)	63	57	58	50	44	39	38	
Vaneaxial	Over 40 in (1000 mm)	39	36	38	39	37	34	32	
	Under 40 in (1000 mm)	37	39	43	43	43	41	28	6
Tubeaxial	Over 40 in (1000 mm)	41	39	43	41	39	37	34	
	Under 40 in (1000 mm)	40	41	47	46	44	43	37	5
Propeller									
Cooling tower	All	48	51	58	56	55	52	46	5

Note: These values are the specific sound power levels radiated from either the inlet or the outlet of the fan. If the total sound power level being radiated is desired, add 3 dB to the above values.

Source: 1987 *ASHRAE Handbook, Systems,* ASHRAE, Atlanta, 1987, chap. 52, "Sound and Vibration Control."

TABLE 10.7 Octave Band in Which Blade Frequency Increment (BFI) Is Likely to Occur

Fan type	Octave band in which BFI occurs*
Centrifugal	
Airfoil, backward-curved, backward-inclined	250 Hz
Forward-curved	500 Hz
Radial blade and pressure blower	125 Hz
Vaneaxial	125 Hz
Tubeaxial	63 Hz
Propeller	
Cooling Tower	63 Hz

*Use for estimating purposes. For speeds of 1750 r/min (29 r/s) or more, move the BFI to the next higher octave band. Where the actual fan is known, use the manufacturer's data.
Source: *1987 ASHRAE Handbook, Systems*, ASHRAE, Atlanta, 1987, chap. 52, "Sound and Vibration Control."

TABLE 10.8 Correction Factor C for Off-Peak Operation

Static efficiency, % of peak	Correction factor, dB
90 to 100	0
85 to 89	3
75 to 84	6
65 to 74	9
55 to 64	12
50 to 54	15

Source: *1984 ASHRAE Handbook, Systems*, ASHRAE, Atlanta, GA, 1984, chap. 32, "Sound and Vibration Control."

TABLE 10.9 Calculation of Total Fan L_W in Example 49.1

Calculation	Octave band center frequency, Hz						
	63	125	250	500	1000	2000	4000
Specific fan K_W	37	39	43	43	43	41	28
$10 \log (Q/Q_1) + 20 \log (P/P_1)$	55	55	55	55	55	55	55
C	0	0	0	0	0	0	0
BFI	—	—	6				
+3 dB to get total L_W (see note below Table 49.6)	3	3	3	3	3	3	3
Total L_W	95	97	107	101	101	99	86

10.16 COOLING TOWER NOISE

In the typical mechanically induced–draft cooling tower (Fig. 10.24), noise is generated by fan noise and water impact; at most locations of interest, however, fan noise predominates. For evaluation and control of cooling tower noise, see Refs. 3 to 6. A typical cooling tower noise control installation, consisting of air-intake and -discharge silencers, is shown in Fig. 10.25.

Cooling tower fan noise, if not available from the manufacturer, can be estimated from Eq. (10.27) and Tables 10.6 to 10.8. It should be noted, however, that the intake noise must propagate upstream against the air flow, make a 90° turn, divide as it disperses through the side of the tower, and pass through the louvers. This tortuous path results in the cooling tower fan's intake noise being less than its discharge noise. Typical fan attenuation at the air intake can

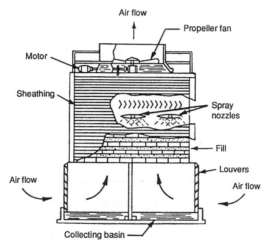

FIGURE 10.24 Mechanically induced–draft cooling tower.

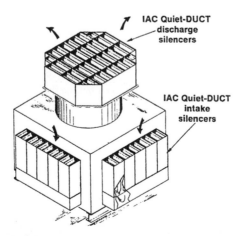

FIGURE 10.25 Silencers for cooling towers. (*Application Manual for Duct Silencers, Bulletin 1.0301.4, Industrial Acoustics Company, 1989.*)

amount to as much as 3, 7, 11, and 9 dB in the first four octave bands, respectively; however, in the last four bands water noise predominates. Clearly, wherever possible, data based on actual measurements and provided by the cooling tower manufacturer should be used.

10.17 DUCT SILENCERS—TERMINOLOGY AND TYPES

Duct silencers reduce the air-flow noise inside air-handling systems that is caused by the following:

- The fan—the air's prime mover
- The passage of air through straight ducts
- The impact of air flowing through duct components, such as elbows, branches, mixing boxes, rods, and orifices

We can generalize that any form of air movement will generate noise. If V is the velocity of air flow in a straight duct, the sound power level may be a function of V^5 to V^7, depending on the frequency and the duct component. This means that the noise generated by air flow inside a duct may increase or decrease by 15 to 21 dB every time the velocity is doubled or halved.

Six principal parameters are generally used to describe the aeroacoustic characteristics of silencers:

1. *Dynamic insertion loss (DIL):* The DIL is the difference between two sound power levels or intensity levels when measured at the same point in space before and after a silencer has been inserted between the measuring point and the noise source.

2. *Self-noise (SN):* The SN is the sound power level in decibels generated by a given volume of air flowing through a silencer of stated cross-sectional area.

3. *Air flow:* Accurate aerodynamic measurements are essential in describing any component of an air-handling system. DIL and SN data are always reported as a function of silencer face air-flow velocity.

4. *Static pressure drop:* This is generally related to silencer face velocity and volumetric air-flow capacity for a given silencer face area. For energy conservation considerations, it can also be related to the horsepower (kilowatts) required to overcome the pressure drop.

5. *Forward flow:* This applies to DIL and SN ratings with the air flow moving in the same direction as the noise propagation, such as in a fan discharge system.

6. *Reverse flow:* This applies to DIL and SN ratings with the air flow and noise propagation moving in opposite directions, such as in a fan inlet system.

There are many types of silencers, including the following:

Reactive Silencers.   These have tuned cavities and/or membranes and are designed mainly to attenuate low-frequency noise in diesel, gasoline, and similar engines. Such silencers, however, are rarely used in HVAC systems.

Diffuser-Type Silencers.   These are used primarily for jet engine test facilities and pneumatic cleaning nozzles in manufacturing operations. They often employ perforated "pepper pots" that slow down the flow velocities and/or prevent the generation of low-frequency noise.

Active Attenuators.   Much work has been done during the last 10 years on "active" silencers. These attenuate noise by means of electronic cancellation techniques involving microphones, speakers, synchronizing sensors, and microprocessors.

Such silencers are effective at low frequencies under 300 Hz but are not suitable for broadband noise reduction without the addition of a dissipative silencer.

Moreover, this cost and maintenance requirements do not make such silencers a practical proposition. However, they might constitute an answer in unusual situations where there is no room for conventional silencers and where very low frequency noise must be controlled.

Packless Silencers.   These can be used where the acoustic infill of conventional silencers could become a breeding ground for disease-carrying bacteria or where particulate matter from fiber erosion can contaminate streams of air or gas. This makes packless silencers particularly suitable for microchip manufacturers, food processing plants, hospitals, and pharmaceutical and other manufacturing plants requiring clean-room environments.

The absence of acoustic materials also reduces fire hazards where flammable materials could saturate the infill. Other applications therefore include engine test cell, kitchen exhausts, and facilities in general where fuels, grease, acids, and solvents might be carried in streams of air or gas.

Packless silencers could well become more important for general use if it becomes established that fiberglass causes lung illnesses.

Dissipative Silencers.   These are widely used in HVAC duct systems. Figures 10.26 and 10.27 show the general configuration of rectangular splitter silencers. The splitter, consisting of a strong, perforated-steel envelope containing sound-absorptive materials, divides the air or gas flow into smaller sound-attenuating passages. Rectangular silencers are used in rectangular ducts and are sometimes set up in very large tiers, or banks, on the intakes and exhausts of fans.

Figure 10.28 shows a tubular, or cylindrical, silencer. At first sight it looks similar in cross section to the rectangular silencer, but it consists of an outer cylindrical shell and an inner concentric bullet. Cylindrical silencers are often used in circular duct systems in conjunction with vaneaxial fans.

Dissipative silencers are available in a variety of executions, lengths, and cross sections to meet almost any noise-reduction and pressure-drop requirement of an HVAC system. The use of dissipative silencers is further discussed and illustrated in Secs. 10.23 to 10.29 in terms of applications. For discussions of the principles of silencer performance and duct break-out noise, respectively, see Sec. 10.18 and 10.20.

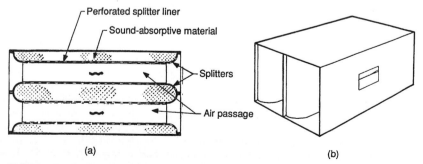

FIGURE 10.26 "Round-nosed" rectangular silencer. (*a*) Cross section; (*b*) external view. (*Application Manual for Duct Silencers, Bulletin 1.0301.4, Industrial Acoustics Company, 1989.*)

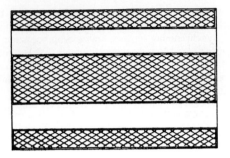

(Frequently used in Europe. Constitutes poor aerodynamic
and self-noise design. See Sect. 49.23.3 and Fig. 49.59.)

FIGURE 10.27 "Flat-nosed" rectangular silencer. (*Martin Hirschorn, "The Aero-Acoustic Rating of Silencers for 'Forward' and 'Reverse' Flow of Air and Sound," Noise Control Engineering, vol. 2, no. 1, Winter 1974.*)

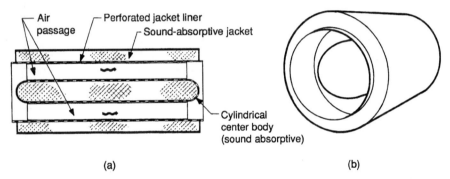

FIGURE 10.28   Cylindrical silencer. (*a*) Cross section; (*b*) external view. (*Application Manual for Duct Silencers, Bulletin 1.0301.3, Industrial Acoustics Company.*)

10.18   EFFECTS OF FORWARD AND REVERSE FLOW ON SILENCER SN AND DIL

The self-noise (SN) of a silencer varies by 7 to 26 dB for each doubling and halving of flow velocity, depending on the frequency, on the silencer's configuration, and on whether the noise and air flow are traveling in the same direction (i.e., forward or reverse flow).

As explained in Sec. 10.17, forward flow occurs if the air flow is traveling in the same direction as the sound propagation, as on the supply side of an HVAC system, and reverse flow occurs when air is traveling in a direction opposite to the direction of sound propagation, such as in a duct's return-air system. Both are illustrated in Fig. 10.29.

Figure 10.30 illustrates the effects of forward and reverse flow on silencer SN. Low-frequency SN is the greatest in the forward-flow mode, while high-frequency SN is the greatest in the reverse-flow mode.

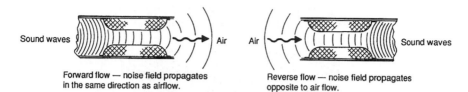

Forward flow — noise field propagates in the same direction as airflow.

Reverse flow — noise field propagates opposite to air flow.

Note: If velocity of air through silencer is 70 ft/s (21.3 m/s), the speed of sound in the forward-flow direction would be 1100 + 70 = 1170 ft/s (335.3 + 21.3 = 356.6 m/s). Similarly in the reverse-flow direction, the speed of sound through the silencer would be 1100 − 70 = 1030 ft/s (335.3 − 21/3 = 314 m/s). Approximate velocity of sound at sea level = 1100 ft/s (335.3 m/s).

FIGURE 10.29 Schematic of reverse flow versus forward flow. (*Application Manual for Duct Silencers, Bulletin 1.0301.4, Industrial Acoustics Company, 1989.*)

Because of the forward- and reverse-flow phenomena, silencer performance is best rated with air flow in terms of dynamic insertion loss (DIL) determined in accordance with ASTM E477 (Ref. 7) in a reverberant room in the reverse and forward modes. The test arrangement is shown in Figs. 10.31 and 10.32. See Ref. 8.

10.18.1 Brief Theory of the Effects of Air-Flow Direction on Silencer Performance

In examining the influence of air flow on the acoustic DIL, observers have found that air flow affects sound transmission in three major ways: (1) convection, (2) refraction, and (3) flow modification of the acoustic impedance of the duct walls. Since the third

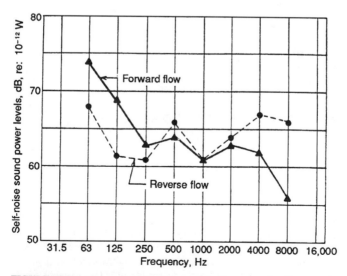

FIGURE 10.30 Characteristic self-noise spectra for rectangular silencers with 30 percent free area. (*M. Hirschorn, "Acoustic and Aerodynamic Characteristics of Duct Silencers for Airhandling Systems," ASHRAE Paper CH-81-6, 1981.*)

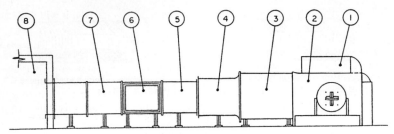

1. Air flow measurements station
2. System fan
3. System silencer
4. Signal source chamber
5. Upstream pressure test station
6. Silencer under test
7. Downstream pressure test station
8. Reverberation room

FIGURE 10.31 Typical facility for rating duct silencers with or without air flow. (*ASTM E477, Standard Method of Testing Duct Liner Materials and Prefabricated Silencers for Acoustical and Airflow Performance, American Society for Testing and Materials, 1973.*)

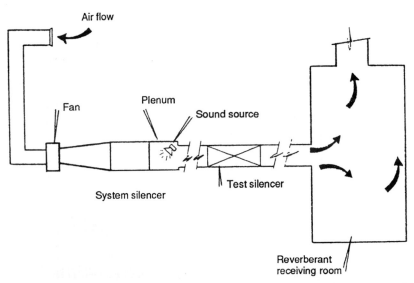

FIGURE 10.32 Schematic of the facility shown in Fig. 10.31; forward flow illustrated. (*M. Hirschorn, "Acoustic and Aerodynamic Characteristics of Duct Silencers for Airhandling Systems," ASHRAE Paper CH-81-6, 1981.*)

effect is rather insignificant for silencers using absorptive materials, it will not be discussed here.

Convection.   The term *convection* signifies that the speed of sound in the forward direction is greater than in the reverse direction. As a result, the sound waves (previously referred to as the "noise field") maintain longer contact with the absorptive boundary in the silencer in the reverse direction than in the forward-flow mode. This results in higher attenuation in the reverse direction than in the forward direction. Quantitatively, this difference between reverse- and forward-flow attenuation depends on the Mach number M in the duct, which is defined as

$$M = \frac{V}{c} \qquad (10.29)$$

where V is the velocity of air, and c is the velocity of sound. At sea level, c is approximately 1100 ft/s (335.3 m/s), and V in an air-conditioning silencer might typically be on the order of 70-ft/s (21.3-m/s) throat velocity, or a Mach number of about 0.064.

This dependence on the Mach number is modified by whether the air-flow pattern in the flow sublayer close to the boundary is streamlined or turbulent. If the pattern is streamlined, the ratio between reverse- and forward-flow attenuation can be shown to be $(1 + M)/(1 - M)^1$; if the pattern is turbulent, the ratio is expected to be $(1 + M^2)/(1 - M^2)^2$. If the Mach number is about 0.064 and if the turbulent sublayer is streamlined, this would correspond to a theoretical ratio between reverse- and forward-flow attenuation of about 14 percent; however, much wider fluctuations have been measured under actual test conditions.

Where turbulent flow conditions control, the ratio between reverse- and forward-flow attenuation might then be on the order of 30 percent of more; consequently, it follows that shape and construction can have a major effect on silencer attenuation values and that it cannot be concluded that all silencers will necessarily behave alike. There is only one way to be sure that silencers will provide the performance specified, and that is on the basis of actual test data.

(It is interesting to note that if the velocity of air through a duct equals Mach 1, then theoretically no noise at all should be transmitted in the reverse-flow direction. In fact, experimental jet engine intake silencers have been constructed on this principle.)

Refraction.   At higher frequencies, refraction begins to be significant, and it works in opposition to the effect of convection. That is, refraction tends to increase high-frequency attenuation in the forward-flow direction and decrease it in the reverse-flow direction. This situation is illustrated schematically in Fig. 10.33. As a sound wave travels in the forward-flow direction, there is a tendency for it to be refracted toward the boundary, which leads to smaller attenuation in the reverse-flow direction. This effect is significant only at higher frequencies when the wavelength is smaller than the cross-sectional dimensions of the duct.

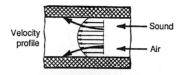

Under forward-flow conditions, high-frequency sound is refracted into the duct-silencer walls.

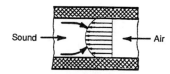

Under reverse-flow conditions, sound is refracted away from the walls and toward the center of the duct silencer.

FIGURE 10.33   The refraction of sound under forward- and reverse-flow conditions. (*Application Manual for Duct Silencers, Bulletin 1.0301.3, Industrial Acoustics Company.*)

It will be noted from the data in Fig. 10.57 that in the reverse-flow mode, silencer attenuation falls off markedly from the sixth octave band upward and increases for the forward-flow mode (Refs. 9 to 11).

10.19 COMBINING ACTIVE AND DISSIPATIVE SILENCERS

Active noise control presently is not a broadly used method for achieving HVAC noise control because of relatively higher costs compared with dissipative silencers. However, for selected applications, there may be significant benefits in combining the active technology with the broad band performance of dissipative sound absorptive silencers.

In active noise cancellation, sound is cancelled by destructive interference. The basis of all active attenuation systems is that the noise from a secondary source is generated with a mirror image wave form of the primary sound field to cancel unwanted sound downstream of the attenuator. The secondary noise source must also be controlled.

The secondary source must be of the same order of magnitude as the noise to be cancelled and must also be controlled. Figure 10.34 shows how noise and anti-noise sources cancel each other out.

A simple active noise control system, Fig. 10.35, utilizes two microphones; one for input and one for error corrections, a loudspeaker and a controller. Unlike dissipative silencers, active silencers will consume small amounts of electrical power and will require

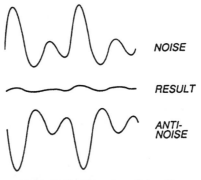

FIGURE 10.34   Noise Cancellation Theory. (Ref. 20)

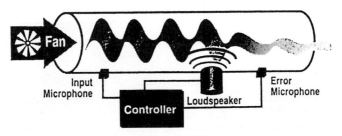

FIGURE 10.35   Active Duct Silencer. (Ref. 21)

TABLE 10.10   DIL of Dissipative Silencers and TCM (Tight Coupled Monopole) Active Attenuator

	63	125	205	500	1K	2K	4K	8K
3 m (10 ft)	13	26	42	52	55	53	51	42
900 mm (3 ft)	2	7	11	15	21	26	18	11
TCM	10	11	16	13	1	0	0	0

equipment maintenance from time to time. This may entail replacement of loudspeakers, which must often operate continuously in sometimes rugged environments.

Hybrid Active Dissipative Silencers.   The most effective application of the active silencer principle in HVAC Systems is a "hybrid" combination of active and dissipative silencers. Table 10.10 shows performance of one combination. Active duct silencers for frequencies in excess of 500 Hz are generally not considered practical due to increasingly complex "cross modes" at the higher frequencies.

The active silencer performance shown in Table 10.10 provides attenuation up to 500 Hz. The acoustical characteristics of dissipative silencers for 3 m and 900 mm long silencers, respectively, provide additional-low frequency attenuation as well as greater amounts of mid- and high-frequency attenuation. Other selections of dissipative silencers can be combined with the active silencer where the dissipative silencer provides a larger amount of low frequency but most of the attenuation above 500 Hz. Depending on space considerations, these can also be designed with very minimal pressure drop.

For most active silencer systems, performance can be limited by the presence of excessive turbulence in the airflow detected by the microphones. Manufacturers recommend using active silencers only where duct velocities are less than 1500 fpm and where the duct configuration is conducive to smooth, evenly distributed airflow. (Ref: 1995 ASHRAE Handbook, HVAC Applications).

10.20 SOUND TRANSMISSION THROUGH DUCT WALLS—DUCT BREAK-OUT AND BREAK-IN NOISE

The break-out phenomenon in particular illustrates the importance of reducing fan noise by means of silencers directly after the fan. Otherwise, duct runs that lack an adequate acoustic design may radiate unacceptably high noise levels into occupied spaces.

Air ducts are commonly manufactured from light-gauge sheet materials, which provide only partial containment of the sound field within the duct. Internal noise can be transmitted into the surrounding space (break-out), and in some cases external noise can pass into the duct (break-in), which then becomes a path for noise to travel into other occupied areas.

The phenomena of break-out and break-in sound transmission are illustrated in Figs. 10.36 and 10.37.

The magnitude of the sound transmission loss (TL) of a duct wall differs from that of a plenum wall panel due to the frequency-dependent nature of the sound propagation within the duct. If the cross-sectional dimensions of the duct are smaller than one-half of the wavelength, only plane waves can propagate within the duct. The vibration response of the duct walls and the pattern of radiation of sound from the duct are governed by the directional characteristics of the internal sound field. The forced response of the duct wall

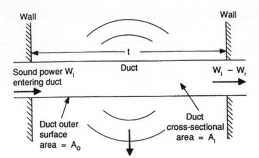

FIGURE 10.36 Break-out sound transmission through duct walls. (*ASHRAE Handbook 1987 Systems, chap. 52, "Sound and Vibration Control," American Society of Heating, Refrigerating and Air-Conditioning Engineers, Inc., Atlanta, 1987.*)

is proportional to the local sound pressure, which propagates in an axial direction at a speed that is equal to or greater than the speed of sound.

Practical TL curves have been developed that are divided into two regions (Ref. 2): one where plane-mode transmission within the duct predominates, and another where cross-modes prevail.

For break-out, the limiting frequency f_1 between these curves is given by

$$f_1 = \frac{24,134}{(ab)^{0.5}} \tag{10.30}$$

where a and b are the duct cross-sectional dimensions in inches; when working with metric units, convert millimeters to inches before using Eq. (10.30) (mm divided by 25.4).

For break-in, the lowest acoustic cross-mode frequency is used as the limiting frequency:

$$f_1 = \frac{6764}{a} \tag{10.31}$$

where a is the larger duct dimension in inches; when working with metric units, convert millimeters to inches before using Eq. (10.31) (mm divided by 25.4).

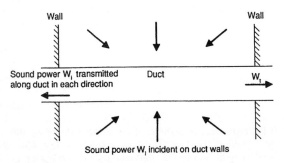

FIGURE 10.37 Break-in sound transmission through duct walls.

Below the limiting frequencies, break-out TL is given by

$$TL_{out} = 10 \log\left(\frac{fq^2}{a+b}\right) + 17 \qquad (10.32)$$

and break-in TL is given by the larger of

$$TL_{in} = TL_{out} - 4 - 10 \log\frac{a}{b} + 20 \log\frac{f}{f_1} \qquad (10.33)$$

or

$$TL_{in} = 10 \log\left[12l\left(\frac{1}{a} + \frac{1}{b}\right)\right] \qquad (10.34)$$

where f = frequency, Hz
$\quad q$ = mass per unit area of duct wall, lb/ft^2 (kg/m^2 × 0.2048)
$\quad l$ = duct length, ft (m)

Above the limiting frequencies, break-out TL is given by

$$TL_{out} = 20 \log qf - 31 \qquad (10.35)$$

and break-in TL is given by

$$TL_{in} = TL_{out} - 3 \qquad (10.36)$$

Air ducts are frequently installed above suspended ceilings or under access flooring. These confined spaces have the effect of modifying the radiating pattern of sound around the duct. Close proximity of the duct to a concrete slab will modify the response of the duct wall; the slab will also act as a reflecting plane to the overall sound radiation. Rigid partitions perpendicular to the axis of the duct may cause standing waves at frequencies where the wavelength is equal to the distance between the partitions. These standing waves can raise the local sound pressure levels in the occupied space by as much as 10 dB.

Circular ducts provide much higher TL than do rectangular ducts in the low frequencies, where most duct break-out problems occur. At higher frequencies, however, circular ducts can exhibit a resonance phenomenon at the duct's so-called "ring frequency," where the TL is sharply reduced.

Examples of break-out and break-in TL values are shown in Table 10.10 for an unlined rectangular duct made from 22-gauge [0.034-in (0.85-mm)] sheet steel and measuring 44 in (1118 mm) wide by 12 in (305 mm) deep; the break-out values for an equal-area circular-section duct 26 in (660 mm) in diameter made from spiral-wound 24-gauge [0.028-in (0.7-mm)] steel are shown for comparison. More comprehensive listings of break-out and break-in TL are shown in Tables 10.11 to 10.17.

Lagging on the outside of ductwork is often used to increase the TL values. The increase in performance due to the lagging will depend on the type and rigidity of the lagging material. A hard outer layer of sheet metal or gypsum board may not be a very effective means of reducing low-frequency noise caused by resonance effects in rectangular

TABLE 10.11   Examples of Duct Break-out and Break-in TL versus Frequency

Duct and TL type	Octave band center frequency, Hz							
	63	125	250	500	1000	2000	4000	8000
Rectangular,* TL_{out}	19	22	25	28	31	37	43	45
Rectangular,* TL_{in}	14	14	22	25	28	34	40	42
Circular,† TL_{out}	45	50	26	26	25	22	36	43

*Duct size: 44 by 12 in (1118 by 305 mm), 22 ga [0.034 in (0.85 mm)].
†Duct size: 26-in (660-mm) diameter, 24 ga [0.028 in (0.7 mm)].

ducts. Limp covering materials that effectively add mass to the duct wall may improve the TL values by reducing wall response without adding stiffness. In critical situations, it may be necessary to apply panels with air spaces to the duct surfaces for maximum noise reduction.

Undoubtedly, more correlation between field and empirical data is required on break-out and break-in noise. Some acoustic consultants and engineers consider that the TL data presented here (from Ref. 2) may be overstated when translated to field installations; for instance, the break-out noise sound levels are likely to be higher than would be arrived at by using the TL figures in Tables 10.10 to 10.16. However, in the meantime, the above procedures (including Tables 10.10 to 10.16) can be used, bearing in mind that the introduction of safety factors might be in order.

TABLE 10.12 TL_{out} versus Frequency for Various Rectangular Ducts

Duct size*		Octave band center frequency, Hz							
in (mm)	Gauge in (mm)	63	125	250	500	1000	2000	4000	8000
12 × 12 (300 × 300)	24 ga 0.028 (0.7)	21	24	27	30	33	36	41	45
12 × 24 (300 × 600)	24 ga 0.028 (0.7)	19	22	25	28	31	35	41	45
12 × 48 (300 × 1200)	22 ga 0.034 (0.85)	19	22	25	28	31	37	43	45
24 × 24 (600 × 600)	22 ga 0.034 (0.85)	20	23	26	29	32	37	43	45
24 × 48 (600 × 1200)	20 ga 0.04 (1.0)	20	23	26	29	31	39	45	45
48 × 48 (1200 × 1200)	18 ga 0.052 (1.3)	21	24	27	30	35	41	45	45
48 × 96 (1200 × 2400)	18 ga 0.052 (1.3)	19	22	25	29	35	41	45	45

*All duct lengths are 20 ft (6 m).

TABLE 10.13 TL_{in} versus Frequency for Various Rectangular Ducts

Duct size*		Octave band center frequency, Hz							
in (mm)	Gauge in (mm)	63	125	250	500	1000	2000	4000	8000
12 × 12 (300 × 300)	24 ga 0.028 (0.7)	16	16	16	25	30	33	38	42
12 × 24 (300 × 600)	24 ga 0.028 (0.7)	15	15	17	25	28	32	38	42
12 × 48 (300 × 1200)	22 ga 0.034 (0.85)	14	14	22	25	28	34	40	42
24 × 24 (600 × 600)	22 ga 0.034 (0.85)	13	13	21	26	29	34	40	42
24 × 48 (600 × 1200)	20 ga 0.04 (1.0)	12	15	23	26	28	36	42	42
48 × 48 (1200 × 1200)	18 ga 0.052 (1.3)	10	19	24	27	32	38	42	42
48 × 96 (1200 × 2400)	18 ga 0.052 (1.3)	11	19	22	26	32	38	42	42

*All duct lengths are 20 ft (6 m).

10.21 NOISE CRITERIA

Noise is unwanted or objectionable sound, and numerous standards define its limits for specific types of noisemakers, specify how the sound is to be measured, and in certain instances specify when. These standards are published by local and national government agencies, national and international standards organizations, the military, professional societies, and others. A few of these standards (or criteria) are given below.

10.21.1 dBA Criteria

One way of rating sounds is by means of the A scale, a sound-level meter weighing network that approximates the response of the human ear to sound. Both the human ear and the A-weighing network are more sensitive to high-frequency than low-frequency sound. In decibels, A-scale levels are expressed as dBA. Typical noise source dBA levels are shown in Table 10.18.

A dBA sound level can be determined in two ways: (1) by using an instrument that reads directly in dBA or (2) by applying weighing factors to measured octave band sound pressure levels (SPLs). Table 10.19 shows the A-scale weighing factors as well as an example of their use.

The upper part of Table 10.20 (from James Botsford, Ref. 12) permits estimates of sound pressure levels when the dBC and dBA, or C- and A-scale sound-level meter, values are known (see also Ref. 13). By taking corresponding (C–A) values from −1 to +20, Botsford developed curves that show their average octave band relationships based on

TABLE 10.14 TL$_{out}$ versus Frequency for Various Circular Ducts*

Duct size and type	Octave band center frequency, Hz							
	63	125	250	500	1000	2000	4000	8000
8-in (200-mm) diam, 26 ga [0.022 in (0.55 mm)], long seam, length = 15 ft (4.5 m)	>45	(53)	55	52	44	35	34	26
14-in (350-mm) diam, 24 ga [0.028 in (0.7 mm)], long seam, length = 15 ft (4.5 m)	>50	60	54	36	34	31	25	38
22-in (550-mm) diam, 22 ga [0.034 in (0.85 mm)], long seam, length = 15 ft (4.5 m)	>47	53	37	33	33	27	25	43
32-in (800-mm) diam, 22 ga [0.034 in (0.85 mm)], long seam, length = 15 ft (4.5 m)	(51)	46	26	26	24	22	38	43
8-in (200-mm) diam, 26 ga [0.022 in (0.55 mm)], spiral-wound, length = 10 ft (3 m)	>48	>64	>75	>72	56	56	46	29
14-in (350-mm) diam, 26 ga [0.022 in (0.55 mm)], spiral-wound, length = 10 ft (3 m)	>43	>53	55	33	34	35	25	40
26-in (650-mm) diam, 24 ga [0.028 in (0.7 mm)], spiral-wound, length = 10 ft (3 m)	>45	50	26	26	25	22	36	43
26-in (650-mm) diam, 16 ga [0.064 in (1.6 mm)], spiral-wound, length = 10 ft (3 m)	>48	>53	36	32	32	28	41	36
32-in (800-mm) diam, 22 ga [0.034 in (0.85 mm)], spiral-wound, length = 10 ft (3 m)	>43	42	28	25	26	24	40	45
14-in (350-mm) diam, 24 ga [0.028 in (0.7 mm)], long seam with two 90° elbows, length = 15 ft (4.5 m) plus elbows	>50	54	52	34	33	28	22	34

*In cases where background noise swamped the noise radiated from the duct walls, a lower limit in the TL is indicated by a > sign. Parentheses indicate measurements in which background noise has produced a greater uncertainty than usual in the data.

TABLE 10.15 TL_{in} versus Frequency for Various Circular Ducts*

Duct size and type	Octave band center frequency, Hz							
	63	125	250	500	1000	2000	4000	8000
8-in (200-mm) diam, 26 ga [0.022 in (0.55 mm)], long seam, length = 15 ft (4.5 m)	>17	(31)	39	42	41	32	31	23
14-in (350-mm) diam, 24 ga [0.028 in (0.7 mm)], long seam, length = 15 ft (4.5 m)	>27	43	43	31	31	28	22	35
22-in (550-mm) diam, 22 ga [0.034 in (0.85 mm)], long seam, length = 15 ft (4.5 m)	>28	40	30	30	30	22	22	40
32-in (800-mm) diam, 22 ga [0.034 in (0.85 mm)], long seam, length = 15 ft (4.5 m)	(35)	36	23	23	21	19	35	40
8-in (200-mm) diam, 26 ga [0.022 in (0.55 mm)], spiral-wound, length = 10 ft (3 m)	>20	>42	>59	>62	53	53	43	26
14-in (350-mm) diam, 26 ga [0.022 in (0.55 mm)], spiral-wound, length = 10 ft (3 m)	>20	>36	44	28	31	32	22	37
26-in (650-mm) diam, 24 ga [0.028 in (0.7 mm)], spiral-wound, length = 10 ft (3 m)	>27	38	20	23	22	19	33	40
26-in (650-mm) diam, 16 ga [0.064 in (1.6 mm)], spiral-wound, length = 10 ft (3 m)	>30	>41	30	29	29	25	38	33
32-in (800-mm) diam, 22 ga [0.034 in (0.85 mm)], spiral-wound, length = 10 ft (3 m)	>27	32	25	22	23	21	37	42
14-in (350-mm) diam, 24 ga [0.028 in (0.7 mm)], long seam with two 90° elbows, length = 15 ft (4.5 m) plus elbows	>27	37	41	29	30	25	19	31

*In cases where background noise swamped the noise radiated from the duct walls, a lower limit in the TL is indicated by a > sign. Parentheses indicate measurements in which background noise has produced a greater uncertainty than usual in the data.

TABLE 10.16 TL$_{out}$ versus Frequency for Various Flat-Oval Ducts

Duct size*		Octave band center frequency, Hz							
$a \times b$	Gauge								
in (mm)	in (mm)	63	125	250	500	1000	2000	4000	8000
12 × 6 (300 × 150)	24 ga 0.028 (0.7)	31	34	37	40	43			
24 × 6 (600 × 150)	24 ga 0.028 (0.7)	24	27	30	33	36			
24 × 12 (600 × 300)	24 ga 0.028 (0.7)	28	31	34	37				
48 × 12 (1200 × 300)	22 ga 0.034 (0.85)	23	26	29	32				
48 × 24 (1200 × 600)	22 ga 0.034 (1.85)	27	30	33					
96 × 24 (2400 × 600)	20 ga 0.04 (1.0)	22	25	28					
96 × 48 (2400 × 1200)	18 ga 0.052 (1.3)	28	31						

*All duct lengths are 20 ft (6 m).

TABLE 10.17 TL$_{in}$ versus Frequency for Various Flat-Oval Ducts

Duct size*		Octave band center frequency, Hz							
$a \times b$	Gauge								
in (mm)	in (mm)	63	125	250	500	1000	2000	4000	8000
12 × 6 (300 × 150)	24 ga 0.028 (0.7)	18	18	22	31	40			
24 × 6 (600 × 150)	24 ga 0.028 (0.7)	17	17	18	30	33			
24 × 12 (600 × 300)	24 ga 0.028 (0.7)	15	16	25	34				
48 × 12 (1200 × 300)	22 ga 0.034 (0.85)	14	14	26	29				
48 × 24 (1200 × 600)	22 ga 0.034 (1.85)	12	21	30					
96 × 24 (2400 × 600)	20 ga 0.04 (1.0)	11	22	25					
96 × 48 (2400 × 1200)	18 ga 0.052 (1.3)	19	28						

*All duct lengths are 20 ft (6 m).

TABLE 10.18 Typical Noise Source dBA Levels

Noise source	dBA
Noise at ear level from rustling leaves	20
Room in a quiet dwelling at midnight	32
Soft whisper at 5 ft (1.52 m)	34
Large department store	50–65
Room with window air conditioner	55
Conversational speech	60–75
Self-service grocery store	60
Busy restaurant or canteen	65
Within typing pool (nine typewriters in use)	65
Passenger car at 50 ft (15.2 m)	69
Vacuum cleaner in private home at 10 ft (3.05 m)	69
Ringing alarm clock at 2 ft (0.61 m)	80
Loudly reproduced orchestral music in large room	82
Buses, trucks, motorcycles at 50 ft (15.2 m)	82–85
Pneumatic tools at 50 ft (15.2 m)	85
Eight-hour OSHA criteria—hearing conservation programs	85
Medium-size automatic printing-press plant	86
Bulldozer at 50 ft (15.2 m)	87
Jackhammer at 50 ft (15.2 m)	88
Eight-hour OSHA criteria—engineering or administrative noise controls	90
Heavy city traffic	90
Heavy diesel-propelled vehicle at 25 ft (7.6 m)	92
Grinders	93–95
Small air compressor	94
Hammermill	96
Plastic chipper	96
Cutoff saw	97
Home lawn mower	98
Multiple spot welder	98
Turbine condenser	98
Drive gear	103
Banging of steel plate	104
High-pressure gas leak	106
Magnetic drill press	106
Air chisel	106
Positive-displacement blower	107
Large air compressor	108
Jet aircraft at 500 ft (152 m) overhead	115
Human pain threshold	120
Inside jet engine test cell	150

about 1000 noises. He stated that these curves can be used to predict the levels of five out of eight octave bands within 3 dB for two-thirds of all noises. The lower part of Table 10.20 evaluates his octave band relationships in terms of actual measurements.

With the A-scale weighing factors shown in Table 10.19, dBA design-guide sound pressure levels can be established on the basis of equivalent octave band levels, as shown in

TABLE 10.19   A-Scale Weighting Factors

Octave band center frequency, Hz	31.5	63	125	250	500	1000	2000	4000	8000
Weighting factor	−39	−26	−16	−9	−3	0	+1	+1	−1

Example of dBA calculation from octave band levels								
Octave band center frequency, Hz	63	125	250	500	1000	2000	4000	8000
SPL spectrum, dB	83	85	82	81	76	60	50	44
A-scale weighting factor	−26	−16	−9	−3	0	+1	+1	−1
Spectrum adjusted to A-scale	57	69	73	78	76	61	51	43

Logarithmic Decibel Addition

     60   79   76   51.5

         79   76

                81 dBA

Table 10.21, where the speech interference levels (SILs) are the arithmetic average of the sound pressure levels at the 500-, 1000-, and 2000-Hz center frequencies (see Sec. 10.21.4).

Figure 10.38 shows the statistical expectations of community response to noise. The parameter L_{dn} is a day-night equivalent A-weighted sound level with an additional 10-dB penalty imposed on noise exposure between 10 p.m. and 7 a.m. In addition to L_{dn}, several other terms involve the use of dBA ratings:

L_{eq} Equivalent sound level, the dBA of a steady-state sound that has the same dBA-weighted sound energy as that contained in the actual time-varying sound being measured over a specific period.

$L_{eq(x)}$ L_{eq} over a period of x hours. That is, if $x = 24$ h, we have $L_{eq(24)}$.

L_d The equivalent A-weighted sound level between 7 a.m. and 10 p.m. Also known as "daytime equivalent sound level."

L_n The equivalent A-weighted sound level between 10 p.m. and 7 a.m. Also known as "nighttime equivalent sound level."

L_x The time-varying dBA level that will be or is exceeded x percent of the time.

The relationships between L_d, L_n, L_{dn}, and $L_{eq(24)}$ are defined by the following equations and are summarized in Table 10.22. Typical L_{dn} levels at various locations are shown in Table 10.23.

$$L_{dn} = 10 \log \frac{1}{24} \left(15 \times 10^{Ld/10} + 9 \times 10^{(L_n + 10)/10} \right) \tag{10.37}$$

$$L_{eq(24)} = 10 \log \frac{1}{24} (15 \times 10^{Ld/10} + 9 \times 10^{L_n/10}) \tag{10.38}$$

TABLE 10.20 Estimating Octave Band Values if the C- and A-Scale Sound-Level Meter Readings Are Known

Determine the C-scale minus A-scale weighting factors, and deduct from the C scale the corresponding values shown below to obtain the approximate octave band sound levels:

	Octave band center frequency, Hz								
C–A	31.5	63	125	250	500	1000	2000	4000	8000
−1	−26	−24	−23	−20	−17	−10	−6	4	−8
0	−20	−19	−17	−15	−13	−7	−6	−7	−9
2	−13	−12	−11	−10	−8	−6	−8	−11	−14
5	−9	−8	−7	−7	−8	−10	−13	−17	−22
10	−6	−5	−6	−8	−13	−17	−20	−26	−32
15	−5	−4	−6	−14	−19	−23	−28	−33	−41
20	−5	−4	−6	−19	−26	−31	−38	−44	−50

The above relationships were checked against actual measurements, and here are the results:

	Octave band center frequency, Hz							
Comparison	63	125	250	500	1000	2000	4000	8000
Generator, 40 hp (30 kW)								
Actual measurement	77	83	73	62	60	57	49	43
Botsford prediction	80	78	70	65	61	56	51	43
Difference	−3	+5	+3	−3	−1	+1	−2	0
Generator, 20 hp (15 kW)								
Actual measurement	74	77	74	69	64	59	51	41
Botsford prediction	77	75	67	62	58	53	48	40
Difference	−3	+2	+7	+7	+6	+6	+3	+1
Vaneaxial fan, 25,000 ft³/min								
(11.8 m³/s), no load*								
Actual measurement	73	83	84	89	86	82	69	63
Botsford prediction	81	82	83	85	87	85	82	79
Difference	−8	+1	+1	+4	−1	−3	−13	−16

*If this comparison is typical, the fan prediction seems close in five octave bands but is on the high side, particularly in the 4000- and 8000-Hz bands.

10.21.2 Community and Workplace Noise Regulations

Municipalities, states, and agencies of the U.S. government have established noise criteria for a broad range of conditions.

Many HVAC noise considerations relate to indoor space, but the compressors, chillers, fans, and cooling towers associated with air-conditioning systems can have a significant impact on the acoustic environment of a building. Accordingly, some of the criteria are for outdoor areas or are referenced to property boundary lines. Several are shown in the tables noted below.

Local, State, Highway, and Navy Regulations.   The New York City Noise Control Code establishes three ambient noise quality zones, as shown in Table 10.24. The Chicago noise control ordinance is shown in Table 10.25, Minnesota noise control regulations are shown

TABLE 10.21   dBA Octave Band Design-Guide Table

		Octave band center frequency, Hz								
dBA	SIL	31.5	63	125	250	500	1000	2000	4000	8000	16,000
115	—	142	131	122	115	109	106	105	104	104	112
110	—	137	126	117	110	104	101	100	99	99	107
105	—	132	121	112	105	99	96	95	94	94	102
100	92	127	116	107	100	94	91	90	89	89	97
95	87	122	111	102	95	89	86	85	84	84	92
90	82	117	106	97	90	84	81	80	79	79	87
85	77	112	101	92	85	79	76	75	74	74	82
80	72	107	96	87	80	74	71	70	69	69	77
75	67	102	91	82	75	69	66	65	64	64	72
70	62	97	86	77	70	64	61	60	59	59	67

Source: M. Hirschorn, "Noise Level Criteria and Methods of Engineering Noise Control," *National Safety News*, September 1972.

TABLE 10.22   Relationships between L_d, L_n, L_{dn}, and $L_{eq(24)}$ Sound Levels

$L_d - L_n$	Add to L_d for L_{dn}	Add to L_d for $L_{eq(24)}$
−4	10	+2
−2	8	+1
0	6.5	0
2	5	−0.7
4	3.5	−1
6	2	−1.5
8	1	−1.7
10	0	−1.8

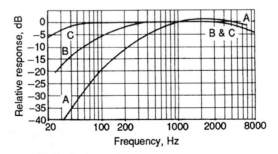

FIGURE 10.38   Frequency response of A, B, and C scales.

TABLE 10.23 Typical L_{dn} Sound Levels at Various Locations

$L_d - L_n$	Add to L_d for L_{dn}	Add to L_d for $L_{eq(24)}$
−4	10	+2
−2	8	+1
0	6.5	0
2	5	−0.7
4	3.5	−1
6	2	−1.5
8	1	−1.7
10	0	−1.8

TABLE 10.24 New York City Noise Control Code *Measured for any one hour*

	Standards in $L_{eq(1)}$	
Ambient noise quality zone	7 a.m.–10 p.m.	10 p.m.–7 a.m.
Low-density residential	60 dBA	50 dBA
High-density residential	65 dBA	55 dBA
Commercial and manufacturing	70 dBA	70 dBA

Note: All noise measurements shall be made at the property line of the impacted site. When instrumentation cannot be placed at the property line, the measurement shall be made as close thereto as is reasonable. However, noise measurements shall not be made at a distance less than 25 ft (7.6 m) from a noise source.

TABLE 10.25 Chicago Environmental Control Ordinance—Maximum Sound Pressure Levels at Residential and Business-Commercial Boundaries

Manufacturing zoning districts	31.5	63	125	250	500	1000	2000	4000	8000	dBA
				Octave band frequency, Hz						
				Residential boundaries						
Restricted	72	71	65	57	51	45	34	34	32	55
General	72	71	66	60	54	49	44	40	37	58
Heavy	75	74	69	64	58	52	47	43	40	61
				Business-commercial boundaries						
Restricted	79	78	72	64	58	52	46	41	39	62
General	79	78	73	67	61	55	50	46	43	64
Heavy	80	79	74	69	63	57	52	48	45	66

A. Districts as defined in City of Chicago Zoning Ordinance.

B. Maximum levels in "Restricted" Manufacturing Zoning Districts apply at the lot boundaries.

C. Maximum levels in "General" or "Heavy" Manufacturing Zoning Districts apply at the boundary of the Residence, Business or Commercial District, or at 125 ft (38.1 m) from the nearest property line of a plant or operation, whichever distance is greater from that plant or operation.

Note: The New York City Noise Control Code and the Chicago Environmental Control Ordinance also contain allowable noise limits for various noise sources. Reference each document for complete details.

TABLE 10.26   State of Minnesota Noise Control Regulations

	Day (7 a.m.–10 p.m.)		Night (10 p.m.–7 a.m.)	
NAC*	L_{50}	L_{10}	L_{50}	L_{10}
1	60	65	50	55
2	65	70	65	70
3	75	80	75	80

*NAC stands for Noise Area Classification system according to land activity at receiver.

Acceptable sound levels for the receiver are a function of the intended activity in that land area. The Noise Area Classifications are grouped and defined as follows:

*NAC-1: Residential areas, hotels, hospitals, schools, resorts, etc.

*NAC-2: Urban shopping areas, rapid transit terminals, finance, insurance, legal trade areas.

*NAC-3: Manufacturing areas.

in Table 10.26, the Federal Highway Administration design noise levels are shown in Table 10.27, and general specifications for ships of the U.S. Navy in regard to noise levels are shown in Table 10.28.

OSHA Regulations.   People working in noisy environments, such as mechanical-equipment rooms, may also be concerned about the standards set by the U.S. Occupational Safety and Health Administration (OSHA). OSHA criteria for noise levels exceeding 85 dBA for an 8-h day are listed in Table 10.29; in essence, these criteria mandate hearing-conservation measures (including annual audio-metric testing and provision of hearing protectors). OSHA criteria for workplace exposures exceeding 90 dBA for an 8-h day (Table 10.30) require engineered noise control measures or administrative procedures that would limit the time employees are exposed to excessive noise levels.

HUD Site Acceptability Standards.   The standards of the U.S. Department of Housing and Urban Development (HUD) are shown in Table 10.31.

TABLE 10.27   Federal Highway Administration (FHWA) Design Noise Levels, dBA

$L_{eq(1)}$	L_{10}	Area
57 (exterior)	60 (exterior)	Within parks, open spaces, and other tracts of land requiring special qualities of serenity and quiet
67 (exterior)	70 (exterior)	Playgrounds, recreation and picnic areas, and outside of residences, motels, public meeting rooms, schools, churches, libraries, and hospitals
72 (exterior)	75 (exterior)	Other developed lands
52 (exterior)	55 (exterior)	Within residences, motels, hotels, public meeting rooms, schools, houses of worship, libraries, hospitals, and auditoriums

TABLE 10.28   General Specifications for Ships of the United States Navy, Section 073 *Airborne noise levels in decibels*

Ship Space Category	Octave band center frequencies, Hz									SIL value
	32	63	125	250	500	1000	2000	4000	8000	
A	115	110	105	100		SIL value requirement		85	85	64
B	90	84	79	76	73	71	70	69	68	
C	85	78	72	68	65	62	60	58	57	
D	115	110	105	100	90	85	85	85	95	
E	115	110	105	100		SIL value requirement		85	85	72
F	115	110	105	100		SIL value requirement		85	85	65

Category A: Spaces, other than category E spaces, where intelligible speech communication is necessary.

Category B: Spaces where comfort of personnel in their quarters is normally considered to be an important factor.

Category C: Spaces where it is essential to maintain especially quiet conditions.

Category D: Spaces or areas where a higher noise level is expected and where deafness avoidance is a greater consideration than intelligible speech communication.

Category E: High noise level areas where intelligible speech communication is necessary.

Category F: Topside operating stations on weather decks where intelligible speech communication is necessary.

10.21.3   Noise Criteria (NC) Curves

One of the most commonly used ways to rate the noisiness of an indoor space is of established octave band spectra known as "NC curves" (Ref. 2). These curves are plotted in Fig. 10.40 and tabulated in Table 10.32.

The projected or measured NC level within an occupied space is determined by the highest NC level corresponding to the sound pressure level in any octave band. A sound pressure level of 57 dB in the 63-Hz band, for example, corresponds to NC 30, whereas 57 dB in the 125-Hz band corresponds to NC 40.

If a sound pressure level spectrum in the eight octave bands were 57, 60, 62, 54, 51, 44, 38, and 32 dB, the corresponding NC levels for each octave band would therefore be 30, 45, 55, 50, 50, 45, 40, and 35. However, since the highest NC level is 55, it is also the controlling one for this single-number rating. Not necessarily an intentional justification is the fact that spectra with strong peaks can be more objectionable than NC comparisons would indicate.

Typical NC design levels for a variety of indoor space usages are shown in Table 10.32.

NC curves have been in widespread use since the late 1950s. Since then, similar systems have been proposed: preferred noise criteria (PNC) curves, room criteria (RC) curves, and noise rating (NR) curves. These have not enjoyed the popularity of NC curves, but they have been used occasionally and should be understood.

Preferred Noise Criteria (PNC) Curves.   When researchers created a spectrum corresponding to an NC curve in all octave bands, they found that the resultant sound was objectionable in terms of low-frequency rumble as well as high-frequency hissing.

The preferred noise criteria (PNC) curves developed in 1971 are generally 4 to 5 dB lower than the NC curves in the 63-Hz band; 1 dB lower in the 125-, 250-, 500-, and 1000-Hz bands; and 4 to 5 dB lower in the 2000-, 4000-, and 8000-Hz bands. PNC curves are shown in Figs. 10.39–10.41 and tabulated in Table 10.33. They are used the same way as NC curves.

TABLE 10.29   OSHA Criteria for Hearing Conservation Programs

Employers shall administer continuing, effective hearing conservation programs wherever employee noise exposures equal or exceed an 8-hour time-weighted average of 85 dBA or, equivalently, a dose of 50% measured according to the following equation:

$$D = 100\left(\frac{C_1}{T_1} + \frac{C_2}{T_2} + \cdots + \frac{C_N}{T_N}\right)$$

where D = workday dose, %
1, 2, 3 = periods of exposure to different levels
    C = actual exposure time at different levels
    T = permissible exposure time at a given level in accordance with
        the following table

A-weighted sound level L, dB	Reference duration T, h	A-weighted sound level L, dB	Reference duration T, h
80	32.0	92	6.2
81	27.9	93	5.3
82	24.3	94	4.6
83	21.1	95	4.0
84	18.4	96	3.5
85	16.0	97	3.0
86	13.9	98	2.6
87	12.1	99	2.3
88	10.6	100	2.0
89	9.2	101	1.7
90	8.0	102	1.5
91	7.0	103	1.4

Examples:

1. Assume exposure of:
        85 dBA for 5 h   87 dBA for 2 h   80 dBA for $\frac{1}{2}$ h

$$D = 100\left(\frac{5}{16} + \frac{2}{12.1} + \frac{0.5}{32}\right) = 49.34\%$$

(*acceptable, since D is less than 50%*)

2. Assume exposure of:
        100 dBA for 1 h   90 dBA for 4 h   85 dBA for 3 h

$$D = 100\left(\frac{1}{2} + \frac{4}{8} + \frac{3}{16}\right) = 118.75\%$$

(*unacceptable, since D exceeds 50%*)

Note: The exposure in example 2, when evaluated in reference to OSHA criteria for engineering or administrative controls, can be shown to be acceptable, since levels below 90 dBA do not enter into those criteria. However, exposures exceeding a 50% dose still require implementation of hearing conservation programs.

TABLE 10.30 OSHA Criteria for Engineering or Administrative
Controls

Feasible administrative or engineering controls shall be utilized if
noise dose D is greater than 1.0 in accordance with the following
equation:

$$D = \frac{C_1}{T_1} + \frac{C_2}{T_2} + \frac{C_3}{T_3} + \cdots + \frac{C_N}{T_N}$$

where D = daily noise dose (must not exceed unity)
$\quad\quad C$ = actual exposure time at a given noise level
$\quad\quad T$ = permissible exposure time at that level in accordance
$\quad\quad\quad$ with the following table

Duration per day, h	Permissible exposure "slow" response, dBA
8	90
6	92
4	95
3	97
2	100
1.5	102
1	105
0.5	110
0.25 or less	115

*Exposure to impulsive or impact noise should not exceed 140 dB peak
sound pressure level.*

Examples:

1. For an 8-h day at constant noise levels, 90 dBA is the maximum
   allowable level.
2. Assume exposure of:
   $\quad\quad$ 100 dBA for 2 h $\quad\quad\quad\quad$ 90 dBA for 6 h

   $$D = \frac{2}{2} + \frac{6}{8} = 1.75$$

   Engineering or administrative controls are necessary to reduce
   noise dose to unity.
3. Assume exposure of:
   $\quad\quad$ 100 dBA for 1 h $\quad\quad$ 90 dBA for 4 h $\quad\quad$ 85 dBA for 3 h

   Exposure below 90 dBA does not contribute to OSHA noise "dose"
   for administrative or engineering controls to be employed. Therefore

   $$D = \frac{1}{2} + \frac{4}{8} = 1.00$$

   (*acceptable*)

TABLE 10.31 U.S. Department of Housing and Urban Development (HUD) Site Acceptability Standards

L_{dn} at 6.5 ft (2 m) from building setback line nearest noise source	Acceptability
Not exceeding 65 dBA	Normally acceptable
Above 65 dBA but not exceeding 75 dBA	Normally unacceptable
Above 75 dBA	Unacceptable

Room Criteria (RC) Curves. The use of NC curves can result in undesirable rumble if the NC level is determined primarily by sound pressure levels at the lower frequencies or in undesirable hissing if the NC level is determined at higher frequencies.

To establish a more balanced sound, RC curves have been established for which the objective is to design sound spectra to within ±2 dB of an RC curve at all frequencies. In this regard, a spectrum exceeding an RC curve by more than 5 dB below 250 Hz may result in too much rumble, whereas a spectrum more than 5 dB higher at 2000 Hz would probably have an unacceptable hissing quality. RC curves are shown in Fig. 10.42. Recommended RC curve design goals are the same as those shown in Table 10.33.

Noise Rating (NR) Curves. NR curves were developed by the International Organization for Standardization (ISO) in 1971 to rate noisiness with the 1000-Hz octave band as the reference point. At NR 70, for instance, the curve has a level of 70 dB at 1000 Hz.

Compared to NC curves (Table 10.34), the NR curve generally permit higher levels at lower frequencies and lower levels at higher frequencies, although there are considerable variations in the NC/NR 20 to NC/NR 70 range. An NR level is determined by the highest level corresponding to any of the octave bands. NR curves are shown in Fig. 10.43 and tabulated in Table 10.35.

10.21.4 Speech Interference Levels

Speech interference levels (SILs) are the arithmetic average of the sound pressure levels at the 500-, 1000-, and 2000-Hz center frequencies. Table 10.36 shows the SILs and distances

TABLE 10.32 NC Curve Tabulations

Noise criteria	Octave band number							
	1	2	3	4	5	6	7	8
NC 20	51	41	33	26	22	19	17	16
NC 25	54	45	38	31	27	24	22	21
NC 30	57	48	41	35	31	29	28	27
NC 35	60	53	46	40	36	34	33	32
NC 40	64	57	51	45	41	39	38	37
NC 45	67	60	54	49	46	44	43	42
NC 50	71	64	59	54	51	49	48	47
NC 55	74	67	62	58	56	54	53	52
NC 60	77	71	67	63	61	59	58	57
NC 65	80	75	71	68	66	64	63	62
NC 70	83	79	75	72	71	70	69	68

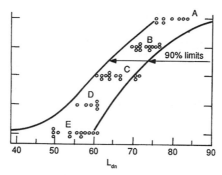

Normalized outdoor day/night sound level
of intruding noise, dB

Community reaction

A. Vigorous action
B. Several threats of legal action or strong appeals to local
   officials to stop noise.
C. Widespread complaints or single threat of legal action.
D. Sporadic complaints.
E. No reaction although noise is generally noticeable.

Corrections for background sound — apply for the season of operation and the
ambient sound characteristics of the nearby neighborhood (see refs. 1 and 2).

Type of correction	Description	Amount correction
Seasonal	Summer (or year-round operation) Winter only (or windows always closed)	0 +5
Background sound	Quiet suburban or rural community (remote from large cities and from industrial activity and trucking)	-10
	Normal suburban community (not located near industrial activity)	-5
	Urban residential community (not immediately adjacent to heavily traveled roads and industrial areas)	0
	Noisy urban residential community (near relatively busy roads or industrial areas)	+5
	Very noisy urban residential community	+10

These corrections are based on reported typical residual noise levels as shown
above. If measured data at the site under investigation differs significantly from the
table, different corrections may be warranted. The residual sound level is that sound
level exceeded 90% of the time.

Source: "Community Noise," U.S. Environmental Protection Agency Report NTID
300.3, December 1971, Washington, DC.

FIGURE 10.39 Statistical expectations of community response to noise.
(*Gas Turbine Installation Sound Emissions, ANSI B-133.8, American Society
of Mechanical Engineers, 1977.*)

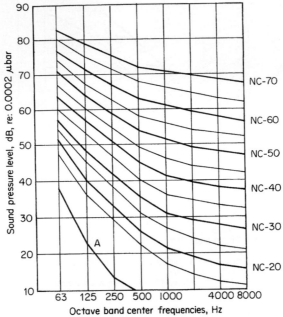

A: approximate threshold of hearing for continuous noise.
(Refer to Table 49.31 for numerical values.)

FIGURE 10.40   Indoor noise criteria (NC) curves. [*Leo Beranek (ed.), Noise and Vibration Control, McGraw-Hill, New York, 1971, part 2: "NC Curves," fig. 18.2, p. 565.*]

TABLE 10.33   Preferred Noise Criteria (PNC) Curve Tabulations

PNC curve	Octave band center frequency, Hz								
	31.5	63	125	250	500	1000	2000	4000	8000
PNC 15	58	43	35	28	21	15	10	8	8
PNC 20	59	46	39	32	26	20	15	13	13
PNC 25	60	49	43	37	31	25	20	18	18
PNC 30	61	52	46	41	35	30	25	23	23
PNC 35	62	55	50	45	40	35	30	28	28
PNC 40	64	59	54	50	45	40	36	33	33
PNC 45	67	63	58	54	50	45	41	38	38
PNC 50	70	66	62	58	54	50	46	43	43
PNC 55	73	70	66	62	59	55	51	48	48
PNC 60	76	73	69	66	63	59	56	53	53
PNC 65	79	76	73	70	67	64	61	58	58

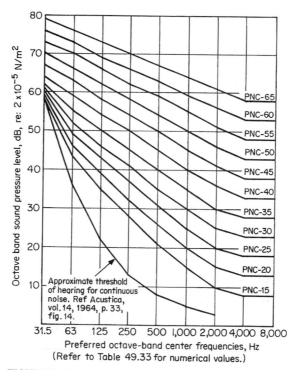

FIGURE 10.41   Preferred noise criteria (PNC) curves. [*Leo Beranek (ed.), Noise and Vibration Control, McGraw-Hill, New York, 1971, part 2: "NC Curves," fig. 18.3, p. 567.*]

at which the average person would have to talk in a normal, raised, and very loud (or shouting) voice to be understood. Table 10.37 shows how the relative difficulty of telephone usage in noisy areas varies with SIL and dBA levels.

Assuming a spectral content of the NC curves in Fig. 10.38, Table 10.38 shows the approximate relationships between NC level, SIL, and dBA. These relationships will vary with other spectral shapes, however, and should be checked accordingly.

10.21.5   Ambient Noise Levels as Criteria

In the absence of standards, the simplest approach to establishing criteria is to stipulate that the noise from equipment such as fans or other HVAC machinery shall not cause the ambient, or background, sound levels to increase in any octave band. This assumes, of course, that these levels are satisfactory and are not affected by other existing noisemakers.

The combined effect of ambient sound and equipment sound can be additive. Suppose the existing sound pressure level is 60 dB in a given octave band; if other equipment contributes a sound pressure level of 60 dB in the same octave band, the resultant theoretical sound pressure level would be approximately 63 dB (see Fig. 10.6). Ideally, equipment sound levels should be 10 dB below the ambient sound level so as not to raise the combined sound level.

Table 10.39 shows the octave band ambient sound levels typically found in a variety of outdoor environments. Indoor ambient levels can be estimated from the data in Table 10.34.

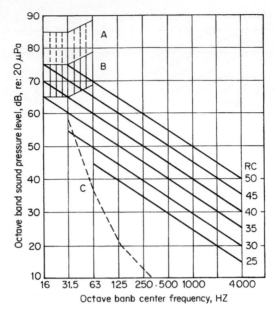

FIGURE 10.42 Room criteria (RC) curves. (*ASHRAE Handbook 1987 Systems, Chap. 52, "Sound and Vibration Control," American Society of Heating, Refrigerating and Air-Conditioning Engineers, Inc., Atlanta, 1987.*)

10.22   ENCLOSURE AND NOISE PARTITION DESIGN CONSIDERATIONS

10.22.1   Actual versus Predicted Sound Transmission Losses

One rule of noise control engineering is that theoretical prediction schemes for acoustic structures or silencers are useful only as guidelines. Substantial and costly errors can develop if not checked out in the laboratory and/or field. For instance, air-handling units (Sec. 10.28) for acoustically critical situations are usually checked out in specially constructed full-scale test facilities, which try to encompass all variables before assembling 91 identical units, as in one application.

The sound transmission loss (TL) of partitions is generalized by the mass law; Eq. (10.23). This equation is valid only when incident sound is normal to the partition's surface and is within a frequency range where the TL is not affected by the partition's stiffness or internal damping. For multilayer and double-wall structures separated by air spaces, the mass law is not applicable.

This disparity between theory and practice is illustrated in Fig. 10.44, which compares the measured and predicted performance levels of commercially available double- and single-wall nonabsorptive noise control partitions 4 in (102 mm) thick.

The measured TL of the 10-lb/ft^2 (48.8-kg/m^2) single-wall partition is in some instances more than 10 dB greater than predicted by the mass law. To achieve a TL of 40 dB in the 125-Hz octave band, for example, the mass law indicates a surface density of 38 lb/ft^2 (185.5 kg/m^2).

TABLE 10.34 Typical Room Design NC Criteria, dB

Type of area	Low	Average	High
Residences			
Private homes (rural and suburban)	20	25	30
Private homes (urban)	25	30	35
Apartment houses, two- and three-family units	30	35	40
Hotels			
Room, suites, banquet rooms, ballrooms	30	35	40
Halls, corridors, lobbies	35	40	45
Kitchens, laundries, garages	40	45	50
Hospitals and clinics			
Private rooms	25	30	35
Operating rooms, wards	30	35	40
Laboratories, halls, lobbies, waiting rooms	35	40	45
Washrooms and toilets	40	45	50
Offices			
Boardroom	20	25	30
Conference rooms	25	30	35
Private offices, reception rooms	30	35	40
General open offices, drafting rooms	35	40	50
Halls and corridors	35	45	55
Tabulation and computation	40	50	60
Auditoriums and music halls			
Concert and opera halls, sound studios	20	22	25
Legitimate theaters, multipurpose halls	25	27	30
Movie theaters, TV audience studios, semi-outdoor amphitheaters, lecture halls	30	32	35
Lobbies	35	40	45
Houses of worship and schools			
Sanctuaries	20	25	30
Libraries, classrooms	30	35	40
Laboratories, recreation halls	35	40	45
Corridors, halls, kitchens	35	45	50
Public buildings			
Public libraries, museums, courtrooms	30	35	40
Post offices, banking areas, lobbies	35	40	45
Washrooms and toilets	40	45	50
Restaurants, cafeterias, lounges			
Restaurants, nightclubs	35	40	45
Cocktail lounges	35	45	50
Cafeterias	40	45	50
Stores, retail			
Clothing and department stores	35	40	45
Department stores (main floor), small stores, supermarkets	40	45	50
Sports activities indoors			
Coliseums	30	35	40
Bowling alleys, gymnasiums	35	40	45
Swimming pools	40	50	55

(Continued)

TABLE 10.34 Typical Room Design NC Criteria, dB (*Continued*)

Type of area	Low	Average	High
Transportation (rail, bus, plane)			
Ticket sales offices	30	35	40
Lounges and waiting rooms	35	45	50
Manufacturing areas			
Supervisor's office	40	45	50
Assembly lines, light machinery	45	60	70
Foundries, heavy machinery	55	65	75

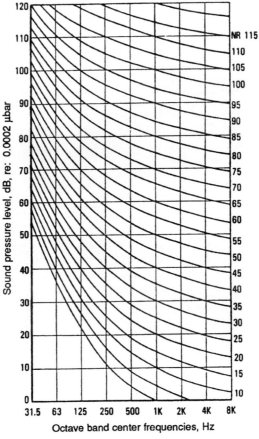

FIGURE 10.43 Noise rating (NR) curves. (*"Noise Assessment with Respect to Community Response," ISO Draft Standard 1996, November 1969.*)

TABLE 10.35 Noise Rating (NR) Curve Tabulations

NR curve	Octave band center frequency, Hz								
	31.5	63	125	250	500	1000	2000	4000	8000
0	55.4	35.5	22.0	12.0	4.8	0	−3.5	−6.1	−8.0
5	58.8	39.4	26.1	16.6	9.7	5	+1.6	−1.0	−2.8
10	62.2	43.4	30.7	21.3	14.5	10	6.6	+4.2	+2.3
15	65.6	47.3	35.0	25.9	19.4	15	11.7	9.3	7.4
20	69.0	51.3	39.4	30.6	24.3	20	16.8	14.4	12.6
25	72.4	55.2	43.7	35.2	29.2	25	21.9	19.5	17.7
30	75.8	59.2	48.1	39.9	34.0	30	26.9	24.7	22.9
35	79.2	63.1	52.4	44.5	38.9	35	32.0	29.8	28.0
40	82.6	67.1	56.8	49.2	43.8	40	37.1	34.9	33.2
45	86.0	71.0	61.1	53.6	48.0	45	42.2	40.0	38.3
50	89.4	75.0	65.5	58.5	53.5	50	47.2	45.2	43.5
55	92.9	78.9	69.8	63.1	58.4	55	52.3	50.3	48.6
60	96.3	82.9	74.2	67.8	63.2	60	57.4	55.4	53.8
65	99.7	86.8	78.5	72.4	68.1	65	62.5	60.5	58.9
70	103.1	90.8	82.9	77.1	73.0	70	67.5	65.7	64.1
75	106.5	94.7	87.2	81.7	77.9	75	72.6	70.8	69.2
80	109.9	98.7	91.6	86.4	82.7	80	77.7	75.9	74.4
85	113.3	102.6	95.9	91.0	87.6	85	82.8	81.0	79.5
90	116.7	106.6	100.3	95.7	92.5	90	87.8	86.2	84.7
95	120.1	110.5	104.6	100.3	97.3	95	92.9	91.3	89.8
100	123.5	114.5	109.0	105.0	102.2	100	98.0	96.4	95.0
105	126.9	118.4	113.3	109.6	107.1	105	103.1	101.5	100.1
110	130.3	122.4	117.7	114.3	111.9	110	108.1	106.7	105.3
115	133.7	126.3	122.0	118.9	116.8	115	113.2	111.8	110.4
120	137.1	130.3	126.4	123.6	121.7	120	118.3	116.9	115.6
125	140.5	134.2	130.7	128.2	126.6	125	123.4	122.0	120.7
130	143.9	138.2	135.1	132.9	131.4	130	128.4	127.2	125.9

TABLE 10.36 Approximate Speech Interference Levels (SILs)

Distance		SIL, dBA			
ft	mm	Normal	Raised	Very loud	Shouting
1	305	77	83	89	95
3	914	67	73	79	85
6	1829	61	67	73	79
12	3658	55	61	67	73

TABLE 10.37 Telephone Usage in Noisy Areas

SIL, dB	dBA	Telephone use
Less than 65	72	Satisfactory
65–80	72–87	Difficult
Above 80	87	Impossible

TABLE 10.38 Approximate Relationships between NC Level, SIL, and dBA

NC	20	30	40	50	60	70
SIL	22	32	42	51	61	71
dBA	31	40	49	58	68	77

Source: M. Hirschorn, *Noise Control Reference Handbook,* Industrial Acoustics Company, Bronx, NY, 1989.

Figure 10.45 compares the TL of 20-lb/ft^2 (97.6-kg/m^2) modular steel partitions 4 in (102 mm) thick separated by a 4-in (102-mm) air space with that of a 150-lb/ft^2 (732.3-kg/m^2) concrete wall. The concrete-wall data are presented both on a calculated and measured basis. However, modular partitions, regardless of their materials, will be only as good as the construction of their joints.

Table 10.40 shows the measured TL of commercially available modular steel noise control partitions and other building materials (see Sec. 10.12).

10.22.2 Joints

To facilitate handing, acoustic panel components for the construction of modular soundproof rooms, machinery enclosures, and fan plenums, for instance, usually measure no more than 48 in (1219 mm) wide by 144 in (3658 mm) high. Their weight ranges from 380 to 480 lb (172 to 218 kg), depending on their acoustic and structural characteristics. When employed as wall sections, they must be joined together to protect the acoustic integrity of the panels. Figure 10.18 shows that an opening as small as 0.1 percent can reduce the TL of a partition from 40 to 30 dB, as one example.

TABLE 10.39 Estimated Outdoor Ambient Sound Levels, dB

	Octave band center frequency, Hz							
	63	125	250	500	1000	2000	4000	8000
	Octave band number							
Condition	1	2	3	4	5	6	7	8
Nighttime								
Rural*	42	37	32	27	22	18	14	12
Suburban*	47	42	37	32	27	23	19	17
Urban*	52	47	42	37	32	28	24	22
Business or commercial area	57	52	47	42	37	33	29	27
Daytime								
Business or commercial area	62	57	52	47	42	38	34	32
Industrial or manufacturing area	67	62	57	52	47	43	39	37
Within 300 ft (91 m) of continuous heavy traffic	72	67	62	57	52	48	44	42

*No nearby traffic of concern.

Note: The sound levels listed here are generally applicable for various outdoor locations and thus can be used as design criteria.

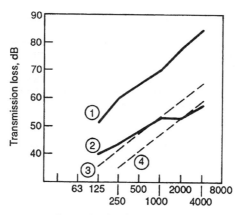

Octave band center frequency, Hz

1. Measured data for two 4-in (102-mm) acoustic panels separated by 4-in (102-mm) air space. Total weight: 20 lb/ft² (98 kg/m²).
2. Measured data for single panel, 4-in (102-mm) thick. Weight: 10 lb/ft² (49 kg/m²).
3. Mass law prediction for 20 lb/ft² (98 kg/m²) panel.
4. Mass law prediction for dense concrete: 10-lb/ft² (49-kg/m²) panel.

FIGURE 10.44 Sound transmission loss of single- and double-wall acoustic panels compared to mass law predictions. (*Martin Hirschorn, Noise Control Reverence Handbook, Industrial Acoustics Company, 1982.*)

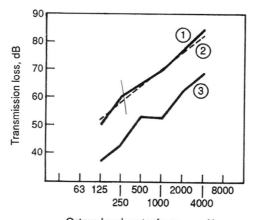

Octave band center frequency, Hz

1. Measured data for two 4-in (102-mm) acoustic panels separated by 4-in (102-mm) air space. Total weight: 20 lb/ft² (98 kg/m²).
2. Mass law prediction: 150-lb/ft² (723-kg/m²) panel.
3. Measured data for dense concrete: 150-lb/ft² (723-kg/m²) panel.

FIGURE 10.45 Sound transmission loss of double-wall acoustic panel compared to single concrete wall and mass law prediction. (*Martin Hirschorn, Noise Control Reference Handbook, Industrial Acoustics Company, 1982.*)

TABLE 10.40 Transmission Loss (TL) Data and STC Ratings for Commercially Available Noise Control Partitions and Commonly Used Building Materials

Product	Octave band center frequency, Hz								STC	Weight	
	63	125	250	500	1000	2000	4000	8000		lb/ft²	kg/m²
Acoustic partition											
Noishield regular panel	20	21	27	38	48	58	67	66	40	8.0	39.1
Noise-Lock I panel	25	27	31	41	51	60	65	66	44	9.5	46.4
Noise-Lock II, Fire–Noise-Lock panel	27	30	32	41	50	59	67	71	45	10.5	51.3
Super-Noise-Lock panel	31	34	35	44	54	63	62	68	48	15.0	73.2
Noishield hard panel	22	33	45	52	58	68	75	65	56	9.5	46.4
Noishield septum	21	19	23	35	50	60	68	72	37	9.0	43.9
Trackwall (industrial regular)	18	25	35	45	52	51	56	58	46	10.0	48.8
APR single-leaf personnel door	22	22	28	39	33	31	35	37	33	7.0	34.2
Standard Noise-Lock door	—	26	42	43	47	52	56	—	47	7.0	34.2
Single Trackwall (architectural)	—	40	44	48	53	53	58	—	51	10.0	48.8
Double Trackwall (architectural)	—	51	60	65	70	78	85	—	70	20.0	98.0
Single Trackwall (architectural, absorptive on one side)	—	28	40	50	53	53	58	—	50	14.0	68.3
Quadraseal Noise-Lock door	—	47	52	58	66	70	67	—	63	20.0	98.0
Building material											
Concrete, 12 in (305 mm) thick	30	37	43	53	53	63	69	—	53	150.0	732.0
Plasterboard, $^3/_8$ in (9.5 mm) thick	—	12	18	22	28	32	25	—	26	1.6	7.8
Plaster, $^1/_2$ in (12.7 mm) thick, over $^3/_8$-in (9.5-mm) gypsum lath, both sides 2- by 4-in (51- by 102-mm) studs on 16-in (406-mm) centers	—	30	37	42	44	39	51	—	39	13.4	65.4
Galvanized steel, 22 ga (0.7 mm)	—	13	17	22	28	34	38	—	27	1.4	6.8
Solid-core wood door, normally closed	—	23	27	29	27	26	29	—	27	3.9	19.1

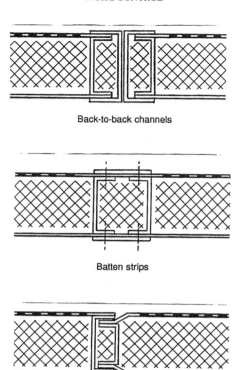

Back-to-back channels

Batten strips

Tongue and groove joint

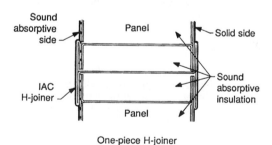

Sound
absorptive
side
Panel
Solid side
IAC
H-joiner
Sound
absorptive
insulation
Panel

One-piece H-joiner

FIGURE 10.46   Acoustic panel joints. (*Courtesy of Industrial Acoustics Company.*)

Also, HVAC plenums must be airtight because they can be subjected to air-pressure differentials on the order of 10 in WG (2491 Pa). In addition, they must be designed to withstand seismic upsets, and outdoor installations must be able to withstand snow covers and wind-velocity forces in excess of 100 mi/h (161 km/h).

Figure 10.46 shows several acoustic panel joint designs. Note the following about these designs:

- Batten strips can work acoustically but have limited structural strength.
- Back-to-back channel joiners have greater strength but are difficult to seal against airflow or noise leakage.

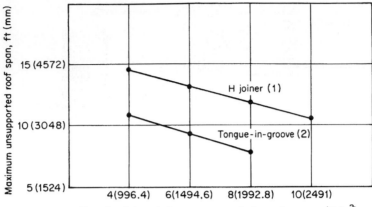

FIGURE 10.47   Strength of H joiner versus tongue-in-groove panel joints.

- Tongue-in-groove joints can provide acceptable acoustic seals without using separate joiners and can provide reasonable structural strength for relatively lower pressure applications.
- H joiners, roll-formed from single lengths of steel, form center-supported box beams and constitute excellent panel joints for high-pressure applications. They also provide excellent acoustic joints.

Figure 10.47 shows the load-bearing strength of panels 4 in (102 mm) thick and 48 in (1219 mm) wide connected with H sections. A thinner tongue-in-groove panel 3 in (76 mm) thick and 40 in (1016 mm) wide is also shown, though 2-in (51-mm) tongue-in-groove panels are also used in low-pressure fan plenum installations. All of these are of a welded-steel acoustic plenum panel design, and all are commercially available.

The data in Fig. 10.47 indicate that the heavier and wider panels connected by H joiners can be installed in longer unsupported spans than can the narrower and thinner tongue-in-groove structures.

Adjacent wall and roof sections are not the only locations where close attention must be paid to airtight, acoustically tight joiners. Wall-to-ceiling, wall-to-floor, and corner joints are equally important, and examples of such good joint designs are shown in Fig. 10.48.

10.22.3   Windows and Seals

Here again, joints and seals play an important role. Window panes must be securely sealed within their frame perimeters, and the frame itself must be sealed in place within the wall. Desiccant should be provided within double-glazed windows to dry up moisture between the panes.

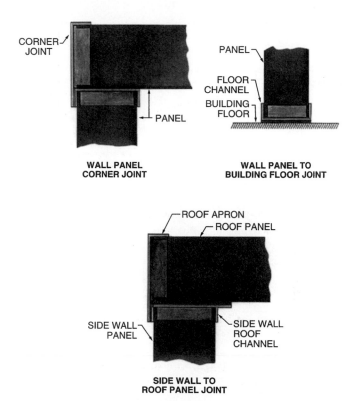

FIGURE 10.48 Wall-to-ceiling, wall-to-floor, and corner joints. (Moduline for New Construction and Renovation, *Bulletin 6.0513.0, Industrial Acoustics Company.*)

Figure 10.49 shows the sound transmission loss (TL) characteristics of 0.236- and 0.472-in- (6- and 12-mm-) thick single-glazed windows and of one double-glazed window with a 7.87-in (200-mm) air space and sound-absorbent material inside the air space.

For single-glazed windows, the data in Ref. 14 show that attenuation up to about 500 Hz increases by 3 to 5 dB when doubling the glass thickness from 6 mm to 12 mm; however, resonance and coincidence effects cause significant TL dips at specific frequencies. Coincidence effects occur because of the excitation of a panel by an airborne sound wave. A coincidence effect's frequency is reduced by increasing the thickness of the glass. Laminated windows of the same thickness will significantly reduce the reduction in TL caused by coincidence effects.

For double-glazed windows, air spaces between two glass panes will not produce significant TL improvements at spacings less than approximately 4 in (101.6 mm). The use of sound-absorptive materials inside the air space is definitely helpful, as is the use of glass panes of dissimilar thicknesses

Figure 10.50 shows a modern acoustic window design that uses high-strength aluminum-alloy framing, tempered or laminated glass, extruded aluminum mullions, and weather-and-acoustic seals. Multiple modules as large as 48 by 48 in (1219 by 1219 mm) can be installed in lengths practically without limit to form a "window wall" (Fig. 10.51), such as in a sewage treatment plant.

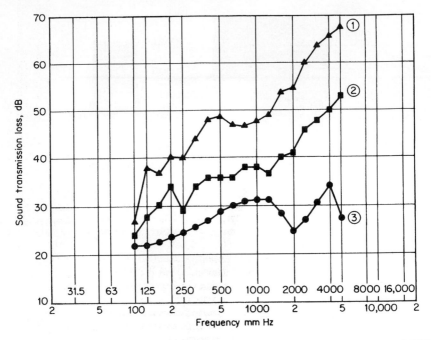

	*Thickness, mm	Airspace, mm	Thickness, mm	STC	Rw
1. Double-glazed windows with 200-mm airspace	12	200	6	50	49
2. Single-glazed windows	12	...	...	31	31
3. Single-glazed windows	6	...	...	29	29

*Multiply dimension in millimeters by 0.03937 to get dimensions in inches.

FIGURE 10.49   Sound transmission loss for double-glazed windows with air space and for single-glazed windows. (*Curve 1: Industrial Acoustics Company. Curves 2 and 3: Technical Advisory Service, Pilkington Flat Glass, Ltd., England.*)

Acoustically, the design shown in Fig. 10.50 exhibits a sound transmission class (STC) performance of STC 39 when single-glazed with $1/2$-in (12.7-mm) glass, of STC 49 when double-glazed with $1/4$-in (6.35-mm) glass and a 7.9-in (200-mm) air space, and of STC 50 when the double glazing uses glass thicknesses of $1/2$ in (12.7 mm) and $1/4$ in (6.35 mm). Capable of withstanding hurricane-force winds of up to 182 mi/h (293 km/h), the structure has a thermal U-value of 0.40 Btu · h/(ft^2 · °F) [2.27 W/(m^2 · °C)]. And it allows no water penetration when tested in accordance with ASTM Specification 331-70 under conditions equivalent to a 68-mi/h (109-km/h) wind load and a heavy rain.

10.22.4   Doors and Seals

Relying on silenced duct systems for ventilation, many acoustically designed structures use fixed (nonopening) windows. The acoustic doors are operable, of course, but their

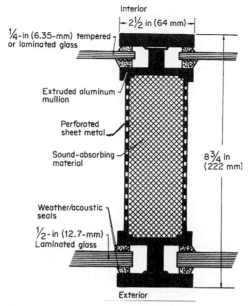

Interior

$2\frac{1}{2}$ in (64 mm)

$\frac{1}{4}$-in (6.35-mm) tempered
or laminated glass

Extruded aluminum
mullion

Perforated
sheet metal

Sound-absorbing
material

$8\frac{3}{4}$ in
(222 mm)

Weather/acoustic
seals

$\frac{1}{2}$-in (12.7-mm)
Laminated glass

Exterior

FIGURE 10.50 Acoustic window design. (IAC Noise-
Lock Vision Wall Window Wall System, *Bulletin 3.0301.1,
Industrial Acoustics Company.*)

seals must be designed to provide reliable perimeter tightness while allowing easy
access.

Still another requirement in many applications is a flush sill. Sills raised to create a seal
can obstruct access and may not be suitable when "barrier-free" doorways are required.
"Sweep" seals, which have continuous contact with the floor, can acoustically eliminate the
need for sills but may be subject to high wear and may make a door more difficult to operate.
Crank-down seals can also be used, but they are inconvenient for doors requiring quick and
easy access.

The state-of-the-art swinging door (Fig. 10.52) uses cam-action hinges that gradually raise and lower the leaf approximately $\frac{1}{2}$ in (12.7 mm) from fully closed to fully
open and back again. A simple seal at the bottom of the leaf compresses against the floor
as the door closes, and it lifts up as the door opens. At the jambs and head of the door,
magnetic seals are used. Self-adjusting against the steel frame, these seals compensate
for slight misalignments due to plumb tolerances or lack of absolute flatness. Note that
these are double seals, providing closure at both the inside and outside surfaces of the
opening.

In sliding doors, bottom closure can be accomplished by using a crank-down or
pneumatically operated bottom seal or by aligning the door on an incline so that a
compression seal lifts on opening and seals on closing. The best head and jamb seals for
sliding doors are pneumatic devices that are actuated only when the door is in the closed
position (Fig. 10.53).

The performance of commercially available acoustic doors is shown in Table 10.41.
Sound transmission loss versus frequency for a range of STC contours is shown in
Table 10.42.

FIGURE 10.51 Acoustic "window wall." (*IAC Vision Wall, Bulletin 3.1232.1, Industrial Acoustics Company.*)

10.22.5 Transmission Loss of Composite Structures

The transmission loss of an acoustic structure (partition or enclosure) can be no greater than the TL of its elements. If wall panels have a TL of 50 dB in a given octave band, for example, an overall TL of 50 dB could not be achieved unless all doors, windows, and joiners exhibit corresponding TL characteristics. The TL of a composite partition consisting of acoustically dissimilar elements (such as walls and doors) can be approximated by the following theoretical relationships:

$$\text{TL}_c = 10 \log \frac{1}{\tau_c} \qquad (10.39a)$$

$$\tau_c = \frac{\tau_1 S_1}{S_T} + \frac{\tau_2 S_2}{S_T} + \frac{\tau_3 S_3}{S_T} + \cdots + \frac{\tau_N S_N}{S_T} \qquad (10.39b)$$

where
TL_c = sound transmission loss of composite partition, dB
τ_c = sound transmission coefficient of composite partition
$S_1, S_2, \ldots, S_N$ = area of individual elements in partition, ft^2 (m^2)
$\tau_1, \tau_2, \tau_3, \ldots, \tau_N$ = sound transmission coefficients of individual elements in partition
S_T = total partition area = sum of $S_1, S_2, S_3, \ldots, S_N$

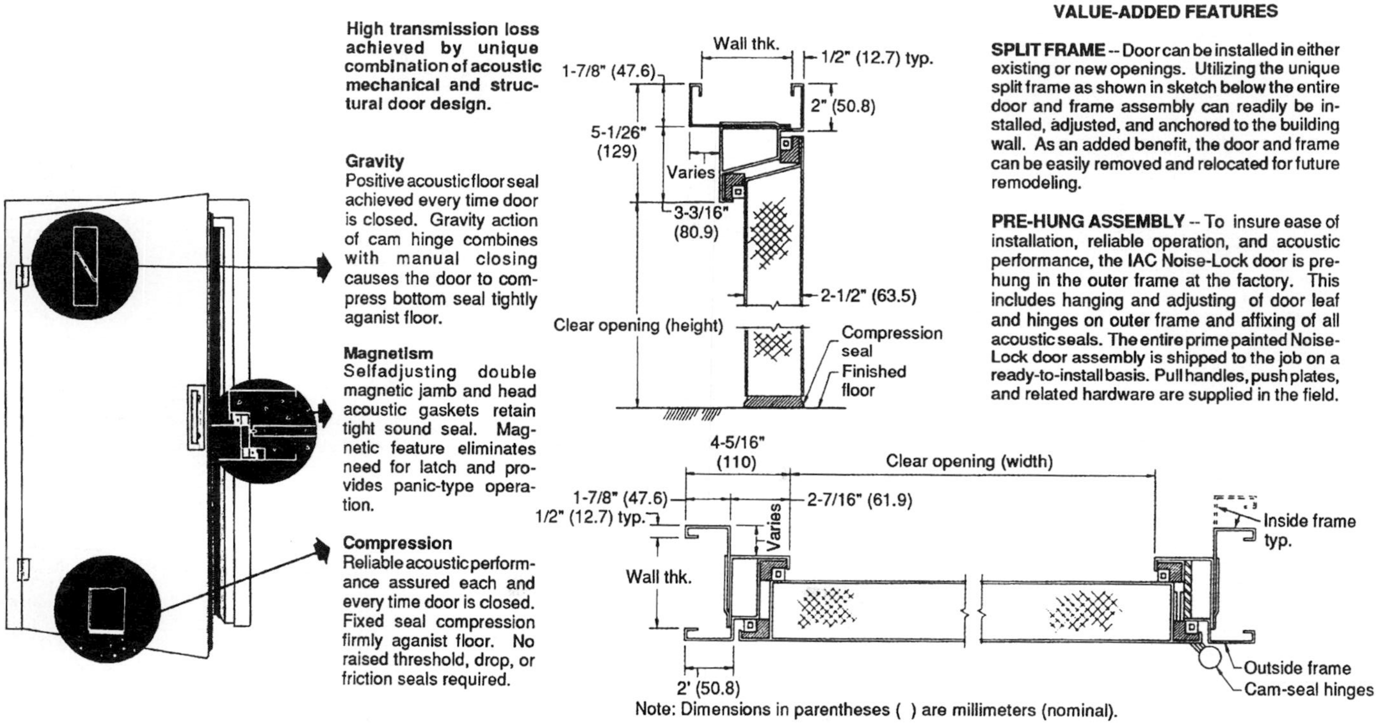
High transmission loss achieved by unique combination of acoustic mechanical and structural door design.
Gravity
Positive acoustic floor seal achieved every time door is closed. Gravity action of cam hinge combines with manual closing causes the door to compress bottom seal tightly aganist floor.
Magnetism
Selfadjusting double magnetic jamb and head acoustic gaskets retain tight sound seal. Magnetic feature eliminates need for latch and provides panic-type operation.
Compression
Reliable acoustic performance assured each and every time door is closed. Fixed seal compression firmly aganist floor. No raised threshold, drop, or friction seals required.
Wall thk.
1-7/8" (47.6)
1/2" (12.7) typ.
2" (50.8)
5-1/26" (129)
Varies
3-3/16" (80.9)
2-1/2" (63.5)
Clear opening (height)
Compression seal
Finished floor
VALUE-ADDED FEATURES
SPLIT FRAME -- Door can be installed in either existing or new openings. Utilizing the unique split frame as shown in sketch below the entire door and frame assembly can readily be installed, adjusted, and anchored to the building wall. As an added benefit, the door and frame can be easily removed and relocated for future remodeling.
PRE-HUNG ASSEMBLY -- To insure ease of installation, reliable operation, and acoustic performance, the IAC Noise-Lock door is prehung in the outer frame at the factory. This includes hanging and adjusting of door leaf and hinges on outer frame and affixing of all acoustic seals. The entire prime painted Noise-Lock door assembly is shipped to the job on a ready-to-install basis. Pull handles, push plates, and related hardware are supplied in the field.
4-5/16" (110)
Clear opening (width)
1-7/8" (47.6)
1/2" (12.7) typ.
2-7/16" (61.9)
Varies
Wall thk.
Inside frame typ.
2' (50.8)
Outside frame
Cam-seal hinges
Note: Dimensions in parentheses ( ) are millimeters (nominal).

FIGURE 10.52 Design features of swinging acoustic door with cam-action hinges. (Noise-Lock Doors, *Bulletin 3.0501.3, Industrial Acoustics Company.*)

TABLE 10.41 Sound Transmission Loss of Commercially Available Acoustic Doors

Swinging-door type	Door leaf thickness, in (mm)	Transmission loss, dB Center frequency, Hz						STC	UL label[a]
		125	250	500	1000	2000	4000		
Magnetic D seal[b]	2.5 (63.5)								1 h and 1.5 h
Laboratory test		26	42	43	47	52	56	47	
Field performance		26	40	38	39	48	54	43	
Quadraseal, laboratory test[c]	2.5, 2.5 (63.5, 63.5)	47	52	58	66	70	67	63	
		Sound absorption coefficient						NRC	
Absorptive one side, magnetic seal, one surface perforated, 23% open area	2.5 (63.5)	0.50	0.68	1.03	1.05	1.00	0.99	0.95	

Sliding-door type	Door panel thickness, in (mm)	Transmission loss, dB Center frequency, Hz						STC	UL label
		125	250	500	1000	2000	4000		
Standard design I, 14 ga, solid both sides[d]	4 (102)	36	41	47	53	51	58	51	1 h
Standard design II, 14 ga, solid both sides	4 (102)	36	41	47	53	51	58	51	1.5 h
Double-wall standard design, 4-in air space, solid all sides	12 (305)	53	61	65	74	78	88	70	
Absorptive one side, 14 ga solid one side, 18 ga perforated one side	4 (102)	25	36	45	53	50	56	46	

		Sound absorption coefficient						NRC
Absorptive one side with septum, 14 ga solid one side, 18 ga perforated one side[e]	4 (102)	26	43	49	54	52	58	50
Absorptive septum design, perforated one side, 23% open area[f]	4 (102)	0.45	0.95	0.85	0.85	0.85	0.85	0.90

Application information[g]

1. All transmission-loss tests conducted in accordance with ASTM E90-61, -66, or -70 and E413-73 by independent acoustic consultant or laboratory. Copies available on request.

2. Sound absorption test conducted in accordance with ASTM C423. Copies available on request.

3. Sound absorption is an *optional* feature with the system. It is obtained by perforating one side of the door leaf with $3/32$-in (2.38-mm) holes on $3/16$-in (4.76-mm) staggered centers. This feature may lower transmission loss by approximately 6 dB. Sound absorption will help reduce reverberation time and noise-level buildup on the perforated side of the door.

4. Underwriters' Laboratory fire resistance and water hose tests conducted in accordance with UL 10B procedure. Certification available on request.

5. Maximum glass area for fire-rated doors is 100 in^2 (645 cm^2), with no dimension exceeding 12 in (305 mm); $1/4$-in (6-mm) wire-reinforced glass must be specified.

6. UL labels available for single-leaf doors up to 3- by 7-ft (914- by 2134-mm) clear opening and for 7-ft by 7-ft 6-in (2134- by 2286-mm) double-leaf designs for flush $2\frac{1}{2}$-in (63.5-mm) doors.

7. $1\frac{1}{2}$-h UL label for single-leaf doors up to 3 ft 6 in by 7 ft 6 in (1067 by 2286 mm) and for double-leaf openings up to 7 ft 0 in by 7 ft 6 in (2134 by 2286 mm). Certifications are available for sizes greater than listed above.

8. UL label available for opening up to 13 ft 6 in by 12 ft 7 in (4115 by 3835 mm). Certifications also available for larger openings.

9. IAC heavy-duty industrial and commercial doors are available with sound transmission class (STC) ratings of 50 and noise isolation class (NIC) ratings of up to 75.

Note: Arithmetic average of sound absorption coefficients in $1/3$ octave bands centered at 250, 500, 1000, and 2000 Hz. By convention, maximum NRC used is 0.95.

[a]Fire-rated doors use D-type compression seals and can be supplied with blast pressure ratings. Extensive laboratory testing of doors with magnetic seals and compression D-type seals, respectively, has established that the noise-reduction and sound-transmission-loss characteristics of these seals are closely comparable.

[b]See Fig. 49.50 for a description of the magnetic seal.

[c]A Quadraseal door is two parallel doors acting in tandem.

[d]14 ga ≈ 1.9 mm; 18 ga ≈ 1.2 mm.

[e]A septum door leaf has a solid metal sheet across the entire face as part of the acoustic infill.

[f]Higher sound-absorption data are available without septum.

[g]See ASTM E90, *Standard Method for Laboratory Measurement of Airborne Sound Transmission Loss of Building Partitions,* 1975; and ASTM E413, *Standard Classification for Determination of Sound Transmission Class,* 1973.

Source: Industrial Acoustics Company.

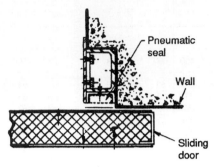

FIGURE 10.53 Pneumatic seal for sliding acoustic door. (*Courtesy of Industrial Acoustics Company.*)

Example. A 22.3-m² partition consists of IAC Noishield paneling and a Noise-Lock door. For the paneling, $\tau = 1.58 \times 10^{-6}$ and $S = 20.3$ m²; for the door, $\tau = 6.31 \times 10^{-6}$ and $S = 2$ m². Calculate TL_c at 2000 Hz.

Solution

$$\tau_c = \frac{1.58 \times 10^{-6} \times 20.3}{22.3} + \frac{6.31 \times 10^{-6} \times 2}{22.3} = 2.0 \times 10^{-6}$$

$$\mathrm{TL}_c = 10 \log \frac{1}{2 \times 10^{-6}} = 57 \text{ dB}$$

10.22.6 Flanking Paths

An acoustic "flanking path" circumvents noise control silencers or partitions. Typical examples are floors and ceilings that have a lower TL than the partitions between rooms (Fig. 10.54). Another example is break-out noise transmitted through a duct wall into the surrounding space (see Sec. 10.20). Flanking paths can dramatically diminish the effectiveness of the best silencers, acoustic enclosures, and partitions.

10.22.7 Room Performance

Given the variables of panel and door design, joints and door seals, flanking paths, and ventilation systems, the noise reduction provided by an enclosure or room is most reliably determined by building and testing it as an entity. This has proven to be feasible for modular audiometric rooms, music practice rooms, quiet havens for noisy areas, and many others. Some are often shipped completely preassembled (Table 10.43).

Table 10.44 shows the performance of several commercially available modular soundproof rooms. Comparing the acoustic performance of different walls, windows, and doors is meaningful only if all data are taken on the same basis.

The U.S. standard for such building components is ASTM E90, which defines specific test and calculation procedures to be followed in the determination of TL and NR ratings (Ref. 15). For complete rooms, ASTM E596 applies (Ref. 16).

TABLE 10.42 Sound Transmission Loss (TL) versus Frequency for a Range of STC Contours*

						Frequency, Hz									
125	160	200	250	315	400	**500**	630	800	1000	1250	1600	2000	2500	3150	4000
44	47	50	53	56	59	**60**	61	62	63	64	64	64	64	64	64
43	46	49	52	55	58	**59**	60	61	62	63	63	63	63	63	63
42	45	48	51	54	57	**58**	59	60	61	62	62	62	62	62	62
41	44	47	50	53	56	**57**	58	59	60	61	61	61	61	61	61
40	43	46	49	52	55	**56**	57	58	59	60	60	60	60	60	60
39	42	45	48	51	54	**55**	56	57	58	59	59	59	59	59	59
38	41	44	47	50	53	**54**	55	56	57	58	58	58	58	58	58
37	40	43	46	49	52	**53**	54	55	56	57	57	57	57	57	57
36	39	42	45	48	51	**52**	53	54	55	56	56	56	56	56	56
35	38	41	44	47	50	**51**	52	53	54	55	55	55	55	55	55
34	37	40	43	46	49	**50**	51	52	53	54	54	54	54	54	54
33	36	39	42	45	48	**49**	50	51	52	53	53	53	53	53	53
32	35	38	41	44	47	**48**	49	50	51	52	52	52	52	52	52
31	34	37	40	43	46	**47**	48	49	50	51	51	51	51	51	51
30	33	36	39	42	45	**46**	47	48	49	50	50	50	50	50	50
29	32	35	38	41	44	**45**	46	47	48	49	49	49	49	49	49
28	31	34	37	40	43	**44**	45	46	47	48	48	48	48	48	48
27	30	33	36	39	42	**43**	44	45	46	47	47	47	47	47	47
26	29	32	35	38	41	**42**	43	44	45	46	46	46	46	46	46
25	28	31	34	37	40	**41**	42	43	44	45	45	45	45	45	45
24	27	30	33	36	39	**40**	41	42	43	44	44	44	44	44	44
23	26	29	32	35	38	**39**	40	41	42	43	43	43	43	43	43
22	25	28	31	34	37	**38**	39	40	41	42	42	42	42	42	42
21	24	27	30	33	36	**37**	38	39	40	41	41	41	41	41	41
20	23	26	29	32	35	**36**	37	38	39	40	40	40	40	40	40
19	22	25	28	31	34	**35**	36	37	38	39	39	39	39	39	39
18	21	24	27	30	33	**34**	35	36	37	38	38	38	38	38	38
17	20	23	26	29	32	**33**	34	35	36	37	37	37	37	37	37
16	19	22	25	28	31	**32**	33	34	35	36	36	36	36	36	36
15	18	21	24	27	30	**31**	32	33	34	35	35	35	35	35	35
14	17	20	23	26	29	**30**	31	32	33	34	34	34	34	34	34

*A particular STC contour is identified by its TL value at 500 Hz.

10.23 SOUND ABSORPTION IN ROOMS

The theory of acoustically absorptive surfaces on indoor sound has been summarized in Sec. 10.11 and 10.14. Table 10.44 gives sound absorption coefficients of commercially available acoustic panel products and commonly used building materials.

In many instances, however, actual sound absorption coefficient data may not be readily available for the specific wall, floor, and ceiling surfaces being used. When the sound sources are HVAC terminals, Tables 10.44, 10. 45, and 10.46 can be used in a general way to predict how absorption can affect sound pressure levels at a given location.

For rooms of various sizes and absorption categories ("hard," "average," and "soft"), Table 10.45 estimates the expected attenuation when a listener within the room is within 5 ft (1.5 m) of a terminal within the room. At such a close distance, the direct sound field from

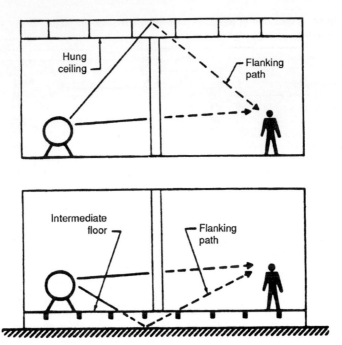

FIGURE 10.54   Flanking paths. (*Martin Hirschorn, Noise Control Reference Handbook, Industrial Acoustics Company, 1982.*)

TABLE 10.43   Noise Reduction (NR)* and Noise Isolation Class (NIC) Characteristics of Commercially Available "Quiet Rooms"

Product	Octave band center frequency, Hz								NIC
	63	125	250	500	1000	2000	4000	8000	
All-purpose room									
Single-glazed	17	20	28	31	36	35	37	45	34
Double-glazed	17	23	30	38	44	45	47	49	41
Music practice room or educational study room									
Outside to inside	—	28	37	38	48	54	58	—	45
Inside to outside	—	23	44	42	50	55	62	—	47
Room to room, 4-in (102-mm) space	—	37	72	74	90	>92	>91	—	61
Medical room									
Single-walled	33	31	39	50	57	61	68	62	53
Double-walled	37	52	64	80	93	>93	>93	>93	75
"Mini" Series 250 audiometric booth	—	18	32	38	44	51	52	50	39

*NR ratings are outside to inside unless otherwise stated. Where background noise swamped the noise exceeded from the walls, a lower limit on the NR is indicated by a > sign.

Source: ASTM E90, *Standard Method for Laboratory Measurements of the Noise Reduction of Complete Sound Isolating Enclosures,* 1978.

TABLE 10.44 Typical Sound Absorption Coefficients for Acoustic Products and Commonly Used Building Materials

| Product | Octave band center frequency | | | | | | | NRC |
	125	250	500	1000	2000	4000	8000	
Acoustic product*								
Noishield	0.89	1.20	1.16	1.09	1.01	1.03	0.93	0.95
Noise-Lock II, Super-Noise-Lock, and Fire-Noise Lock	0.94	1.19	1.11	1.06	1.03	1.03	1.04	0.95
Noishield Septum	0.50	0.68	1.03	1.05	1.00	0.99	—	0.95
Noise-Foil/Varitone (flush-mounted)	0.94	1.19	1.11	1.06	1.03	1.03	1.04	0.95
Trackwall	0.45	0.95	0.85	0.85	0.85	0.85	0.80	0.90
Noishield Hard	0.045	0.03	0.02	0.02	0.015	0.02	—	0.02
Noise-Foil/Varitone, 2-in (5-mm) thick (flush-mounted)	0.57	0.86	1.15	1.07	0.94	0.92	—	0.95
Building material								
Concrete block, coarse	0.36	0.44	0.31	0.29	0.39	0.25	—	0.35
Concrete block, painted	0.10	0.05	0.06	0.07	0.09	0.08	—	0.05
Brick	0.03	0.03	0.03	0.04	0.05	0.07	—	0.05
Plaster	0.14	0.10	0.06	0.05	0.04	0.03	—	0.05
Glass, heavy plate	0.18	0.06	0.04	0.03	0.02	0.02	—	0.05
Glass, window grade	0.35	0.25	0.18	0.12	0.07	0.04	—	0.15
Wood	0.15	0.11	0.10	0.07	0.06	0.07	—	0.10

*All products are 4 in (102 mm) thick unless otherwise noted. Numbers greater than 1.0 are caused by edge diffraction effects dependent on panel size and geometry.

the terminal predominates over the reverberant sound field, which, however, decreases with increasing room size; therefore, sound-absorptive treatment provides less attenuation in smaller than in large rooms. Table 10.46 gives examples of rooms that are typically "hard," "average," or "soft" from a sound absorption standpoint. Table 10.47 is the same as Table 10.45, except that it also estimates attenuation characteristics in the far field up to 20 ft (6.1 m) from an HVAC terminal.

Figure 10.55 also illustrates how sound is attenuated in accordance with room size, distance from sound source, and degree of sound-absorptive treatment.

TABLE 10.45 Room Attenuation in Direct Sound Field as a Function of Room Size and Absorption Characteristics
For listener within 5 ft (1.5 m) of terminal

| Room surface | | Room absorption characteristic, dB | | |
ft²	m²	Hard	Average	Soft
500	46.5	0	3	5
1,000	92.9	3	5	6
2,000	185.8	5	6	7
5,000	464.5	7	7	8
8,000	743.2	7	8	8
10,000+	929.0+	8	8	8

TABLE 10.46   Room Absorption Characteristics as a Function of Room Type

Type of room	Absorption rating
Radio and TV studios, theaters, lecture halls	Soft
Concert halls, stores, restaurants, offices, conference rooms, hotel rooms, school rooms, hospitals, private homes, libraries, businesses, machine rooms, churches, reception rooms	Average
Large churches, gymnasiums, factories	Hard

TABLE 10.47   Attenuation Estimates at Various Distances from Terminal Noise in Accordance with Room Sound Absorption Characteristics

Room surface		Distance*		Room absorption characteristic, dB		
ft²	m²	ft	m	Hard	Average	Soft
500	46.5	5	1.5	0	3	5
1,000	92.9	7	2.1	3	6	8
2,000	185.8	9	2.7	6	9	11
5,000	464.5	15	4.6	10	13	15
8,000	743.2	18	5.5	12	15	17
10,000	929.0	20	6.1	13	16	18

*Approximate distance r of listener.

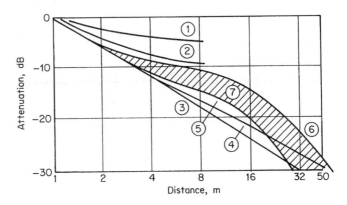

1. Small room, minimal absorption
2. Small room, strong absorption
3. Sound propagation in free field - inverse square law
4. Large office with heavily absorptive ceiling
5. Same as 4, but with reflective diffusers
6. Same as 5, but with weak absorptive ceiling
7. Range of most applications

FIGURE 10.55   The sound pressure level decreases with a doubling of the distance inside different rooms. (*Courtesy of Dr. Ing. E. Schaffert, BeSC, GmbH, Berlin.*)

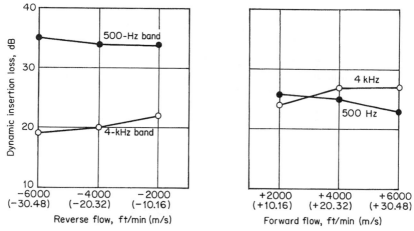

FIGURE 10.56 DIL variations as a function of face velocity for 24-in (610-mm) cylindrical silencers.

10.24 SILENCER APPLICATION

Sections 10.17 and 10.18 discussed silencer terminology, types of silencers, and the effects of forward and reverse flow on silencer self-noise (SN) and dynamic insertion loss (DIL). This section considers more specifically the effects of flow velocity on silencer attenuation; it also discusses the practical significance of interaction between SN and DIL, pressure drop and energy consumption, and the importance of silencer location for maximum noise reduction and minimum pressure drop. Section 10.25 will then present a procedure for selecting silencers for HVAC duct systems.

10.24.1 Specific Effects of Flow Velocity on Silencer Attenuation

Figure 10.56 shows dynamic insertion loss (DIL) variation in the 500- and 4000- Hz octave bands as a function of air flow for a 24-in (610-mm) cylindrical silencer. As the face velocity (volumetric air flow divided by silencer cross section) varies from 2000-ft/min (10.2-m/s) reverse flow to 2000-ft/min (10.2-m/s) forward flow in the 500-Hz octave band, the insertion loss varies from 34 to 26 dB, or by 8 dB. In the 4000-Hz band the corresponding change is from 22 to 24 dB.

Figure 10.57 shows the insertion loss for a rectangular silencer at 2000-ft/min (10.2-m/s) face velocity in the forward-flow and reverse-flow modes. Up to 1000 Hz the DIL is higher in the reverse-flow mode, with a difference of 6 dB in the 250-Hz band.

Figure 10.58 shows the same effects in cylindrical silencers as a function of diameter at 6000-ft/min (20.5-m/s) face velocity. (Cylindrical silencers can be designed for higher flow velocities because their pressure is significantly lower than that of rectangular silencers.)

For a cylindrical silencer with a 24-in (610-mm) diameter, Fig. 10.59 shows the change in DIL that results from using different acoustic designs while essentially maintaining the same pressure-drop characteristics (Ref. 17). (The model without a sound-absorptive jacket has a slightly lower pressure drop.)

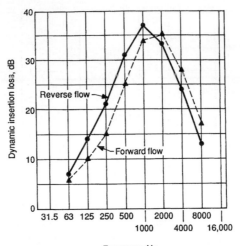

FIGURE 10.57   Characteristic DIL patterns for a dissipative rectangular silencer at 2000-ft/min (10.2-m/s) face
velocity. (*Martin Hirschorn, "The Aero-Acoustic Rating
of Silencers for 'Forward' and 'Reverse' Flow of Air and
Sound," Noise Control Engineering, vol. 2, no. 1, Winter
1974.*)

10.24.2 Interaction of DIL with Self-Noise

The self-noise (SN) of a silencer can be thought of as a noise "floor" that may limit the ability of the silencer to achieve desired sound levels.

Figure 10.60 shows how silencer SN can preclude meeting a noise control criterion (in
this case the Chicago Environmental Control Ordinance) unless the face velocity and self-
noise are decreased. The condition here is particularly sensitive to flow velocity noise
because of the low high-frequency sound pressure level specification.

Figure 10.61 shows the SN for two types of rectangular silencers, and Table 10.48 shows the
DIL and SN data at three octave bands for several commercially available rectangular silencers.

10.24.3 Pressure Drop

To push or pull air past any ductwork component requires fan energy, which is usually
described in terms of pressure head in inches water gauge (inWG) or pascals. Any duct
component offers a pressure ΔP that must be overcome by the fan.

Obviously, the objective in any system is to keep the fan pressure head to a bare minimum. As the pressure head increases, more power is required, which increases both the initial cost and the operating costs. Most HVAC fans have a pressure head ranging from 4 to
6 inWG (996.4 to 1494.6 Pa).

Silencers are generally allowed pressure drops up to a maximum of about 0.5 in WG
(124.6 Pa), though often they will be less. A silencer has three primary areas of pressure drop:

1. At the silencer inlet, the flow passage contracts to the space between the splitters
   (Fig. 10.28). Good practice here is to design the splitters with rounded noses to
   minimize the abruptness of the contraction.

2. Friction losses in the splitter passages are directly proportional to silencer length.

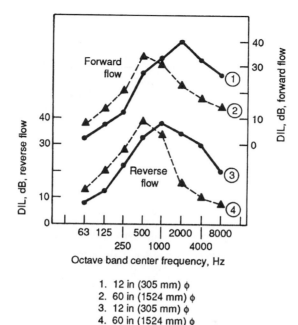

1. 12 in (305 mm) φ
2. 60 in (1524 mm) φ
3. 12 in (305 mm) φ
4. 60 in (1524 mm) φ

Change in DIL with diameter for cylindrical silencers at face velocity. Note that high frequency DIL decreases with diameter whereas low-frequency DIL increases. Reverse flow DIL generally is higher than forward flow in first five octaves.

FIGURE 10.58 Change in DIL with diameter for cylindrical silencers at 600-ft/min (3.05-m/s) face velocity. (*Martin Hirschorn, "The Aero-Acoustic Rating of Silencers for 'Forward' and 'Reverse' Flow of Air and Sound." Noise Control Engineering, vol. 2, no. 1, winter 1974.*)

3. An exit loss occurs as the flow expands. To minimize these losses, the splitters' trailing edges can be rounded or fitted with aerodynamic tail sections.

Accordingly, the pressure drop of a silencer at a given air flow is a function of length, the free-area ratio, and the aerodynamic design of splitter leading and trailing edges.

Figure 10.62 shows a cylindrical silencer with an attached tail cone, and Fig. 10.63 illustrates the effect of the cone, which can reduce silencer pressure drop by as much as 33 percent.

At a given temperature, the pressure drop of a silencer varies as the square of the velocity through the silencer. Mathematically, this means that

$$\Delta P_2 = \Delta P_1 \left(\frac{Vp_2}{V_1} \right)^2$$

(10.40)

where subscript 1 represents conditions at which the pressure drop is known, and subscript 2 represents conditions at which it is to be determined. For example, if a silencer has a ΔP of 0.1 inWG (24.9 Pa) at a face velocity of 1000 ft/min (5.1 m/s), the ΔP at 2000 ft/min (10.2 m/s) will be $0.1(2000/1000)^2 = 0.1 \times 4$, or 0.4 inWG (99.64 Pa).

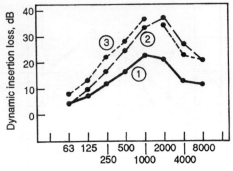

1. Type Ns — No absorptive jacket
2. Type Cs — 4-m (102-mm) absorptive jacket
3. Type FCs — 8-in (203-mm) absorptive jacket

DIL of three 24-in (610-mm) diameter cylindrical silencers at 4000-ft/min (20.3-m/s) face velocity.

FIGURE 10.59 Effect of absorptive jacket on cylindrical silencer performance. (*M. Hirschorn, "Acoustic and Aerodynamic Characteristics of Duct Silencers for Airhandling Systems," ASHRAE Paper CH-81-6, 1981.*)

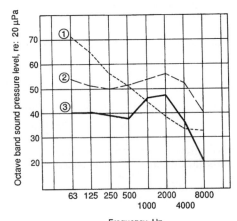

1. Chicago 55-dBA criterion
2. L_p to self-noise of 24 ft² (2.23 m²) of 30 percent free area intake silencer at 1000-ft/min (5.1-m/s) flow velocity
3. L_p due to self-noise of 24 ft² (2.23 m²) of 60 percent free area intake silencer at 1000-ft/min (5.1-m/s) flow velocity

FIGURE 10.60 Effect of self-noise on silencer performance with silencer in zoning boundary. (*M. Hirschorn, "Acoustic and Aerodynamic Characteristics of Duct Silencers for Airhandling Systems" ASHRAE Paper CH-81-6, 1981.*)

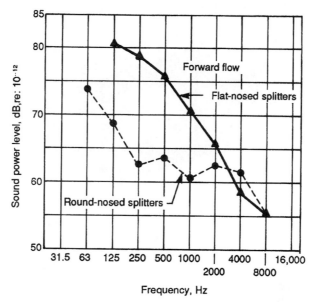

FIGURE 10.61 Self-noise for flat- and round-nosed rectangular silencers. (*Martin Hirschorn, "The Aero-Acoustic Rating of Silencers for 'Forward' and 'Reverse' Flow of Air and Sound," Noise Control Engineering, vol. 2, no. 1, Winter 1974.*)

TABLE 10.48 Dynamic Insertion Loss (DIL) and Self-Noise (SN) Data for Rectangular Silencers 3 ft (914 mm) Long

1000 ft/min (5.1 m/s) face velocity forward flow at 60°F (15.6°C)

Type	DIL, dB			SN, dB		
	250 Hz	500 Hz	2000 Hz	250 Hz	500 Hz	2000 Hz
Steel and acoustic fill standard splitter silencers						
S	16	28	35	49	47	49
Es	15	25	34	33	32	33
Ms	13	20	22	36	34	32
M_L	10	17	13	30	27	28
L	9	14	23	37	32	36
Clean flow*						
Hs	14	19	28	49	47	49
Packless splitters†						
XL	17	21	11	44	46	57
XM	10	17	12	44	46	57
KL	13	12	7	36	43	46
KM	6	13	7	38	43	46

*Absorptive material protected to minimize adsorption of foreign matter.
†Reactive design using no absorptive material.

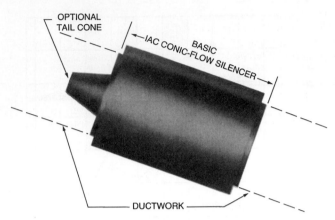

FIGURE 10.62 Silencer tail cone. (*Application Manual for Duct Silencers, Bulletin 1.0301.3, Industrial Acoustics Company.*)

The pressure drop at a given temperature can be calculated if the pressure drop is known at another temperature. If the pressure drop is known at subscript 1 conditions, the ΔP at subscript 2 conditions can be determined by

$$\Delta P_2 = \Delta P_1 \frac{T_1}{T_2} \qquad (10.41)$$

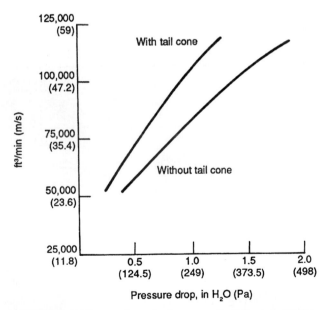

FIGURE 10.63 Pressure-drop reduction as a result of tail-cone installation. (*Application Manual for Duct Silencers, Bulletin 1.0301.3, Industrial Acoustics Company.*)

where T_1 and T_2 are in consistent units of absolute temperature in degrees Rankine (°R = °F + 460) or kelvin (K = °C + 273). For example, if a silencer has a ΔP_1 of 0.3 inWG (74.7 Pa) at 70°F (530°R), the ΔP_2 at 300°F (760°R) is 0.3 × (530/760) = 0.21 inWG (52.3 Pa).

Most companies in Europe design silencers with flat-nosed splitters (Fig. 10.29), as distinct from round-nosed ones (Fig. 10.28). The flat-nosed configuration constitutes a poor aerodynamic and self-noise design (Fig. 10.61); its pressure drop for the same air flow can be three times higher than for a round-nosed silencer with an equal free area.

The significantly higher self-noise generation of flat-nosed splitters results from a higher exit velocity. Noise due to air flow is a function of its velocity V and varies from V^5 to V^7, depending on frequency and duct components.

10.24.4 Energy Consumption

The cost element in operating air-flow system components such as filters, coils, terminal devices, grilles, registers, and silencers can be determined from

$$\Delta\$ = \begin{cases} \dfrac{1.03 \times \Delta P \times Q \times U \times C}{E} & \text{in English units} \\[3mm] \dfrac{0.002434 \times \Delta P \times Q \times U \times C}{E} & \text{in metric units} \end{cases} \qquad (10.42)$$

where $\Delta\$$ = annual energy cost, dollars
  ΔP = pressure drop, inWG (Pa)
  Q = air flow, ft³/min (m³/h)
  U = utilization, percentage of total hours per year (i.e., 50 for 50 percent)
  C = cost, \$/kWh
  E = combined system fan and motor operating efficiency, percent (i.e., 75 for 75 percent)

For example, say that silencer A handles an air flow of 24,000 ft³/min at a pressure drop of 0.55 inWG and that silencer B handles the same air flow at a pressure drop of 0.09 inWG. If $U = 50$, $C = \$0.05$/kWh, and $E = 75$, the difference between the annual energy costs of silencers A and B is calculated as follows:

$$\Delta\$_A = \frac{1.03 \times 0.55 \times 24{,}000 \times 50 \times 0.05}{75} = \$453.20$$

$$\Delta\$_B = \frac{1.03 \times 0.09 \times 24{,}000 \times 50 \times 0.05}{75} = \$74.16$$

By using silencer B instead of silencer A, the saving in annual energy cost is therefore 453.20 − 74.16 = \$379.04. Installing silencer B on each floor of a 50-story building would thus save 50 × 379.04 = \$18,952 per year, and over 30 years the energy cost savings would be \$568,560 without taking inflation into account. If C were \$0.15/kWh instead of \$0.05/kWh, the energy cost savings in 30 years would be \$1,705,680, and this amount would be significantly greater if an annual inflation factor were applied.

10.24.5 Effects of Silencer Length and Cross Section

Silencer length and cross section are important because space is usually at a premium in HVAC systems. However, splitter silencers (for instance) of a given design and cross

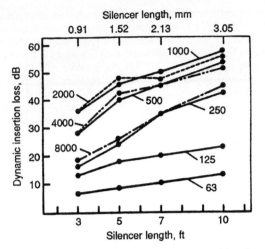

FIGURE 10.64   Silencer DIL as a function of length.
(M. Hirschorn, "Acoustic and Aerodynamic Characteristics of Duct Silencers for Airhandling Systems," ASHRAE Paper CH-81-6, 1981.)

section are usually available with different free areas, with the lower pressure drop and lower attenuation usually going with the larger free area. Therefore, if cross section can be traded off for more length, the required attenuation can be obtained by selecting a longer silencer. Figure 10.64 shows the DIL as a function of length for a rectangular silencer with 30 percent free area.

The reverse trade-off is not usually practical when the length is not available, because a short silencer with high attenuation characteristics and large cross section would require long transition sections to keep the pressure drop down.

Figure 10.65 shows that when two silencers with 60 percent free area are directly installed in series, the measured DIL approximates the linearly added ratings of two silencers tested individually.

10.24.6   Impact on Silencer ΔP of Proximity to Other Elements in an HVAC Duct System

The pressure-drop and DIL characteristics of silencers in the United States are generally determined by tests conducted in accordance with ASTM E477, which mandates the use of straight duct lengths "no less than five equivalent diameters from the entrance," with "the downstream duct no less than ten equivalent diameters from the exit" of the silencer under test (Ref. 7).

Table 10.49 shows the ΔP factors (multipliers relative to ASTM E477 test data) that can be applied to silencer pressure drop when a silencer is located less than 2.5D ($\sqrt{4a/\pi}$) for a rectangular duct of dimensions a and b) upstream or downstream of other duct elements.

10.24.7   Duct Rumble and Silencer Location

In duct systems, duct rumble is generally caused by poor aerodynamic design and/or by excessive low-frequency fan noise within the duct that breaks out of the duct construction when there are no silencers close to the fan. That is, duct rumble can occur when duct

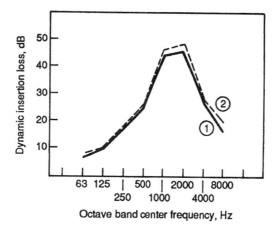

1. Measured DIL (not sensitive to spacing ranging from
   24 in (610 mm) to 192 in (4880 mm) between silencers).
2. Linearly added DIL rating of two silencers tested individually.

FIGURE 10.65 Comparison of measured results and linear addition of DIL ratings for two silencers with 60 percent free area at 2000-ft/min (10.2-m/s) forward flow. (*M. Hirschorn, "Acoustic and Aerodynamic Characteristics of Duct Silencers for Airhandling Systems," ASHRAE Paper CH-81-6, 1981.*)

sections get excited by vibrational impulses developed as a result of aerodynamic turbulence and/or by pulses generated by the prime mover (usually a fan). Even with good airflow design, duct rumble can arise when the fan blade pass frequency or one of its harmonics coincides with the natural resonance of a given section of a duct system; therefore, rumble is most likely to occur in duct systems without silencers.

If a silencer with adequate attenuation characteristics in the critical frequencies is installed right after the fan, duct rumble should not develop or will be reduced. Preferably, the silencer installation should be made with aerodynamically well-designed transition sections; there should be no duct runs before the silencer in which resonances could develop. Because the flow velocities after a silencer are generally low, duct rumble is not likely to develop farther downstream. Nevertheless, a smooth aerodynamic design—with no abrupt changes in cross sections and flow directions—should be maintained throughout the duct system.

Where an adequate duct silencer cannot be installed immediately after the fan inside the equipment room, some of the low-frequency noise will radiate out of the duct wall into the equipment room. In this case, the duct and the room may effectively act as a kind of low-frequency silencer with unknown acoustic performance characteristics. The equipment room itself must therefore have adequate sound transmission loss characteristics to prevent fan noise from entering critical office space. Also, a silencer must be installed in the duct immediately outside the equipment room to prevent fan noise from exiting the mechanical space into critical office-space ceiling plenums.

10.24.8 Effect of Silencer Location on Residual Noise Levels

Figure 10.66 shows a laboratory test arrangement in which a fan discharges air into a room through a duct containing a silencer together with an abrupt contraction-and-expansion section. To determine the effects on residual room noise, separate tests were conducted

TABLE 10.49 Pressure-Drop Factors When Silencer Is Located Less than 2.5D Upstream or Downstream from Other Duct Elements

The air-flow and pressure-drop data for all IAC duct silencers is based on tests run in accordance with ASTM E477 and other codes that specify minimum lengths of straight duct connections upstream and downstream.

However, in practice, these corrections may vary significantly from the test procedure. The modifiers in the following are based on arrangements frequently encountered in air-handling systems and are based on conservative practice.

The factors shown apply only to silencer pressure drop and do not include the pressure drop for the other system components (elbows, transitions, etc.) adjacent to which the silencer is located.

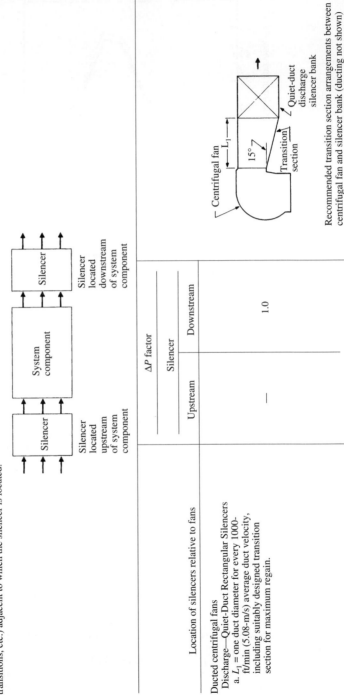

	Silencer located upstream of system component	System component	Silencer located downstream of system component

	ΔP factor	
	Silencer	
Location of silencers relative to fans	Upstream	Downstream
Ducted centrifugal fans		
Discharge—Quiet-Duct Rectangular Silencers	—	1.0
a. L_1 = one duct diameter for every 1000-ft/min (5.08-m/s) average duct velocity, including suitably designed transition section for maximum regain.		

Recommended transition section arrangements between centrifugal fan and silencer bank (ducting not shown)

Centrifugal fan
L_1
15°
Transition section
Quiet-duct discharge silencer bank

Elbows (without turning vanes)		
Distance of silencer from elbow:		
$D \times 3$	1.0	1.0
$D \times 2$	1.5	1.5
$D \times 1$	2.0	2.0
Elbows (with turning vanes)		
Distance of silencer from elbow:		
$D \times 3$	1.0	1.0
$D \times 2$	1.2	1.2
$D \times 1$	1.75	1.75
$D \times 0.5$	3.0	3.0
Directly connected	4.0	Not advised
Terminal Devices		
Mixing boxes, pressure reducing valves, terminal reheats, etc.	—	1.0
Grilles, Registers, and Diffusers		
Deflecting Type		
Allow 24 in (610 mm) upstream and $2\frac{1}{2}D$ downstream of silencer.	1.0	1.0
Nondeflecting Type		
Allow at least 12 in (305 mm).	1.0	1.0
Transitions		
With 15° included angle (7.5° slope)	1.0	1.0
With 30° included angle (15° slope)	1.25	1.0
With 60° included angle (30° slope)	1.5	1.0
Coils and Filters		
Silencer downstream—12 in (305 mm) from face	—	1.0
Silencer upstream—24 in (610 mm) from face	1.0	

Quiet-duct silencers

Downstream

Upstream

Silencers before and after elbows

Note: Silencer baffles should be parallel to the plane of elbow turn.

Grille

Quiet-duct silencer

Mixing box

Discharge silencer downstream of mixing box and upstream of grille

15° Transition

30° Transition

Quiet-duct silencer

Silencer between upstream and downstream transitions

Quiet-duct silencer

Quiet-duct silencer downstream from coil

Quiet-duct silencer upstream from filter

(Continued)

TABLE 10.49 Pressure-Drop Factors When Silencer Is Located Less than 2.5D Upstream or Downstream from Other Duct Elements (*Continued*)

Quiet-duct discharge silencers

Quiet-duct intake silencers

	ΔP factor	
	Silencer	
	Upstream	Downstream
Cooling Towers and Condensers Type L or Type ML Silencers	2.0	2.0
	This multiplier includes typical allowance for intake and discharge dump losses.	

The pressure-drop increase due to the addition of silencers to a cooling tower is partially offset by the resulting decrease in the entrance and discharge losses of the system.

Immediately at System Entrance or Exit	Silencer at intake	Silencer at discharge
Silencer type or model:		
CL, FCL	2.0	5.0
NL	4.0	4.0
ML	1.5	3.5
Cs, FCs, Ns, L	1.5	3.0
Ms	2.5	2.0
S, Es	1.5	1.5

The relatively higher multipliers for the lower-pressure-drop silencers, such as the CL and L types, for instance, are due to the dump losses to the atmosphere being significantly higher relative to their rated values.

Quiet-duct intake silencers

Quiet-duct discharge silencer
Silencers immediately at intake and discharge of equipment room

0.2 D min

D

Quiet-duct intake silencer

Pressure-drop factors for silencers at the entrance to the system can be materially reduced by the use of a smoothly converging bellmouth with sides having a radius equal to at least 20% of its outlet dimension.

Notes applying to Table 49.48:
1. For maximum structural integrity, Quiet-Duct Silencer splitters should be installed vertically. Where vertical installation is not feasible, structural reinforcement is required for silencers wider than 24 in (610 mm).
2. Unless otherwise indicated, connecting ductwork is assumed to have the same dimensions as fan intake or discharge openings.
3. When elbows precede silencers, splitters should be parallel to the plane of elbow turn.
4. L_1 = distance from fan exhaust to entrance of discharge silencer; L_2 = distance from fan inlet to exit of intake silencer.
5. ΔP factor = multiplier relative to silencer laboratory-rated pressure-drop data.
6. D = diameter of round duct or equivalent diameter of rectangular duct.
7. Unless otherwise noted, the multipliers shown do not include the pressure drop of other components (elbows, transitions, dump losses, etc.), which must be calculated separately.

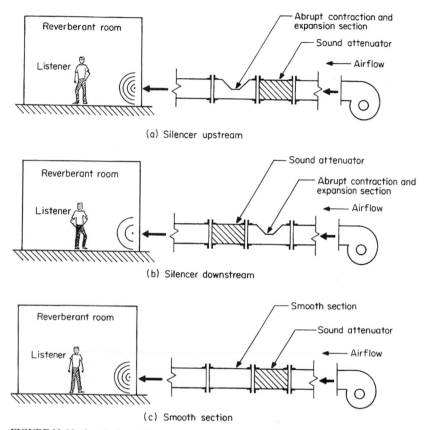

FIGURE 10.66   Installation of silencer upstream and downstream of a duct element generating high noise levels. (*Courtesy of Industrial Acoustics Company.*)

with the silencer on each side of the contraction-expansion section. A third test was conducted with the contraction-expansion section replaced by a straight run of duct.

The results from these tests (Fig. 10.67) show how noise can be regenerated after a silencer in an aerodynamically poorly designed duct system. Clearly the silencer gives the best results in an aerodynamically smoothly designed duct system when care is taken not to include sudden contractions, expansions, or turns.

The duct system must also be designed for minimum self-noise, and silencers must be positioned between the listener and the noise source. Otherwise, the benefits of a duct silencer may be canceled out.

10.25   SYSTEMIC NOISE ANALYSIS PROCEDURE FOR DUCTED SYSTEMS

To avoid HVAC noise problems in an enclosed or open space, the noise reduction (NR) requirements (if any) to meet the criteria for that space must be determined. If silencers are required, care must be taken to ensure that silencer pressure drop is not excessive and that silencer self-noise is not so high as to prevent meeting the criteria.

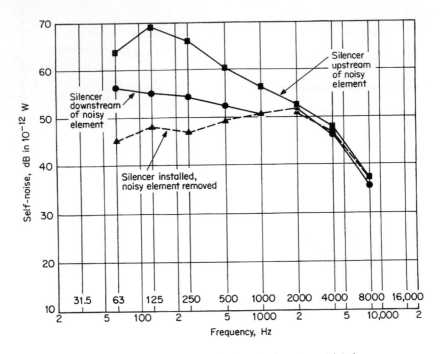

Note: Noisy element in these tests was a restricted duct section having abrupt contraction followed by sudden expansion (see Fig. 10.64).

FIGURE 10.67   Effect of silencer location on duct element noise. (*Courtesy of Industrial Acoustics Company.*)

Some noise reduction is inherently provided by the acoustic characteristics of ductwork. And in a room where criteria must be met, absorption and distance from terminals also influence the sound pressure level at any given location. If the ductwork (lined and unlined) and room effects are not sufficient to meet the criteria, any remaining attenuation must be provided by silencers.

10.25.1 Procedure

Silencer NR requirements can be determined by the following procedure, which consists of calculating the items in lines 1 to 11 of Table 10.59. The data used to illustrate this procedure come from Fig. 10.68, these data are summarized in Table 10.58, and the results of the calculations based on these data are given in Table 10.59.

Line 1: Noise Criteria.   Using Sec. 10.21, establish the sound pressure level criteria in each octave band by selecting the noise criteria (NC) values from Table 10.32. From Table 10.33 an "average" criterion for a reception room is NC 35; accordingly, use the NC 35 values from Table 10.31. Enter these values on line 1 of Table 10.59.

Line 2: Room and Terminal Effect.   This is a factor for converting the L_p (sound pressure level) criteria of line 1 to the L_W (sound power level) at each terminal. This factor accounts for the size and acoustic characteristics of the space as well as for the number and location

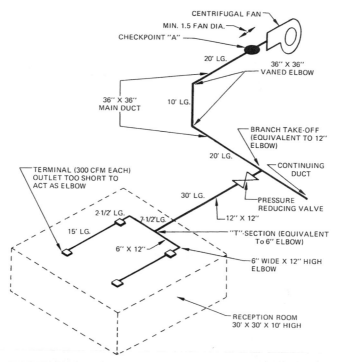

FIGURE 10.68 Air distribution diagram: Noise analysis example for ducted systems. (*Application Manual for Duct Silencers, Bulletin 1.01103, Industrial Acoustics Company.*)

of the terminals. Consider only those terminals connected directly to fans by ductwork; include the supply and return fans.

To obtain this factor, follow these steps:

1. Refer to Tables 10.33 and 10.46 to determine the characteristics of the room. For the reception room in the example, select "average" sound absorption ratings (in accordance with the criterion used in line 1).

2. Determine whether the listener is likely to be in the direct or reverberant field. This determination is made by assessing the individual's probable location in the space. For example, in a theater the key is the seat, not the lobby; in an office the key is the seated position, not the standing position.

3. Find the room absorption characteristic in either Table 10.45 or 10.47. For this example, use Table 10.47 and interpolate to find the absorption characteristic for a room surface area of 3000 ft^2; assume that the absorption characteristic is 10 dB.

4. From the value obtained in step 3, subtract the terminal effect given in Table 10.50. In the example the listener is likely to be within 11 ft of two terminals but probably not within 11 ft of more than two terminals. Therefore, subtract 3 dB from the 10 dB obtained in step 3, and enter 7 dB on line 2 of Table 10.59.

Line 3: Allowance for End Reflection. End reflection accounts for the fact that some low-frequency noise is reflected back into the duct. As shown in Table 10.50, end

TABLE 10.50 Correction Factor for Number of Terminals

Number of terminals providing primary influence	1	2	3	4	6	8
dB to be subtracted from room absorption effect	0	3	5	6	8	9

reflection can add significantly to system noise attenuation. Use the dimensions of the duct leading to the terminal, and interpolate as necessary. For the 6- by 12-in rectangular duct in the example, the square root of the duct area is $\sqrt{6 \times 12} = 8.5$ in; using this in Table 10.51, interpolate attenuation values of 13, 9, 5, 2, and 0 dB. Enter these values on line 3 of Table 10.59.

Line 4: Ductwork Attenuation.   Determine for each octave band the attenuation provided by unlined ducts (no inner sound-absorptive lining) and, if applicable, by ducts internally lined with sound-absorptive materials. Then enter the attenuation values on line 4 of Table 10.59.

In the example there are 40 ft of 6- by 12-in duct, 40 ft of 12- by 12-in duct, and 50 ft of 36- by 36-in duct between the nearest terminal and checkpoint A. These are all unlined sheet-metal ducts, so interpolate the attenuation between the sizes listed in Table 10.52; note that this table lists the attenuation values in dB/ft. Add these calculated values for the ducts to find each band's total value, and enter the bands' totals on line 4 of Table 10.59.

For acoustically lined duct runs, the attenuations per foot for common duct sizes are shown in Tables 10.53 and 10.54 for linings 1 and 2 in thick, respectively. Multiply the unit attenuation for each octave band by the length of the lining to get the total attenuation; however, as per the *1987 ASHRAE Handbook* (HVAC Systems and Applications, chapter 52), usable high frequency attenuation values (above 500 Hz) are limited to 10-ft (3048-mm) lengths.

To illustrate the use of Table 10.53, 10 ft of 18- by 54-in duct lined with 1.5-lb/ft³-density fiberglass 1 in thick is estimated to provide the following attenuation:

Hertz	63	125	250	500	1000	2000	4000
Attenuation	0.4	1.1	3.1	8.4	16.5	13.2	11.6

Except for small cross sections, the use of duct lining alone cannot sufficiently attenuate the low-frequency noise from air-handling equipment. Moreover, effective noise reduction

TABLE 10.51 Attenuation Due to End Reflection at Ductwork Openings

Values in decibels

Duct dimension*		Octave band number					
in	mm	1	2	3	4	5	6 and higher
5	127	17	12	8	4	1	0
10	254	12	8	4	1	0	0
20	508	8	4	1	0	0	0
40	1016	4	1	0	0	0	0
80	2032	1	0	0	0	0	0

*Diameter of round duct or square root of area of rectangular duct.

TABLE 10.52 Attenuation of Unlined Sheet-Metal Ducts
Values in dB/ft (dB/0.3 m)

Type	Duct size in	cm	Octave band number 1	2	3	4 and higher
Small	6 × 6	15 × 15	0.2	0.2	0.15	0.1
Medium	24 × 24	61 × 61	0.2	0.2	0.1	0.05
Large	72 × 72	1.8 × 1.8	0.1	0.1	0.05	0.01
Round	4 to 12	10 to 30.5	0.03	0.03	0.03	0
Round	Over 12	Over 30.5	0	0	0	0

close to the principal noise maker, the fan, is essential to prevent unacceptable duct break-out noise into offices and other rooms. The insertion loss of a 5-ft-long commercially available duct silencer is as follows:

Hertz	63	125	250	500	1000	2000	4000	
DIL	8	18	24	40	45	46	41	26

To provide equivalent attenuation in the 250-Hz octave band, a 1-in-thick lined duct 80 ft long would be required.

Line 5: Elbow Attenuation. Determine from Table 10.55 the allowances for elbows and branch takeoffs between the terminal and noise source. In the example we have one 12- by 12-in (305- × 305-mm) branch takeoff, two elbows without vanes 6 in (152 mm) wide by 12 in (305 mm) high, and two 36- by 36-in (0.91- × 0.91-m) vaned elbows between the nearest terminal and checkpoint. A. Branch takeoffs may be evaluated similarly to elbows, using the width of the branch. Total the attenuation values and enter them on line 5 of Table 10.59.

Line 6: L_W Split—Branch to Terminals. Determine the allowance from Table 10.56. In the example there are four terminals; therefore, enter a value of 6 dB on line 6 of Table 10.59.

Line 7: L_W Split—Main Duct to Branch Ducts. Determine from Table 10.57 the allowance for the division of sound power to the duct branches located between the terminals and checkpoint A, excluding the terminal supply branch (which was accounted for in line 6). In the example the allowance is 10 dB since the area ratio between the 12- by 12-in branch and the 36- by 36-in main duct is approximately 11 percent. Enter 10 dB on line 7 of Table 10 58.

Line 8: Safety Factors. Enter a safety factor of −3 dB.

Line 9: permissible L_W. Arithmetically add the line 1 criteria to the sum of the phenomena considered in lines 2 through 8. Make sure to subtract the 3-dB safety factor. The result is the permissible sound power level spectrum that cannot be exceeded at checkpoint A (Fig. 10.68) without exceeding the line 1 criteria

Line 10: L_W of Fan. Enter the manufacturer's or the calculated fan sound power level in each octave band.

TABLE 10.53 Sound Attenuation in Straight, Lined, Sheet-Metal Ducts of Rectangular Cross Section in dB/ft (dB/0.3 m); 1-in (25-mm) Lining Thickness*; No Air Flow

Internal cross-sectional dimensions		Octave band center frequency, Hz						
in	mm	63	125	250	500	1000	2000	4000
4 × 4	100 × 100	0.16	0.44	1.21	3.32	9.10	10.08	3.50
4 × 6	100 × 150	0.13	0.37	1.01	2.77	7.58	8.26	3.13
4 × 8	100 × 200	0.12	0.33	0.91	2.49	6.82	6.45	2.57
4 × 10	100 × 250	0.11	0.31	0.85	2.32	6.37	5.02	2.07
6 × 6	150 × 150	0.12	0.34	0.93	2.56	7.01	7.50	3.17
6 × 10	150 × 250	0.10	0.27	0.75	2.04	5.61	5.67	2.67
6 × 12	150 × 300	0.09	0.26	0.70	1.92	5.26	4.80	2.33
6 × 18	150 × 460	0.08	0.23	0.62	1.70	4.67	2.95	1.51
8 × 8	200 × 200	0.10	0.28	0.77	2.12	5.82	6.08	2.95
8 × 12	200 × 300	0.09	0.24	0.65	1.77	4.85	4.98	2.64
8 × 16	200 × 410	0.08	0.21	0.58	1.59	4.37	3.89	2.17
8 × 24	200 × 610	0.07	0.19	0.52	1.42	3.88	2.39	1.41
10 × 10	250 × 250	0.09	0.24	0.67	1.84	5.04	5.17	2.79
10 × 16	250 × 410	0.07	0.20	0.55	1.49	4.10	4.04	2.41
10 × 20	250 × 510	0.07	0.18	0.50	1.38	3.78	3.30	2.05
10 × 30	250 × 760	0.06	0.16	0.45	1.23	3.36	2.03	1.34
12 × 12	300 × 300	0.08	0.22	0.60	1.64	4.48	4.52	2.67
12 × 18	300 × 460	0.07	0.18	0.50	1.36	3.74	3.71	2.39
12 × 24	300 × 610	0.06	0.16	0.45	1.23	3.36	2.89	1.97
12 × 36	300 × 910	0.05	0.15	0.40	1.09	2.99	1.78	1.28
15 × 15	380 × 380	0.07	0.19	0.52	1.42	3.88	3.84	2.53
15 × 22	380 × 560	0.06	0.16	0.43	1.19	3.27	3.20	2.29
15 × 30	380 × 760	0.05	0.14	0.39	1.06	2.91	2.46	1.86
15 × 45	380 × 1140	0.05	0.13	0.34	0.94	2.17	1.51	1.21
18 × 18	460 × 460	0.06	0.17	0.46	1.26	3.45	3.37	2.42
18 × 28	460 × 710	0.05	0.14	0.38	1.03	2.84	2.69	2.13
18 × 36	460 × 910	0.05	0.13	0.34	0.94	2.59	2.15	1.78
18 × 54	460 × 1370	0.04	0.11	0.31	0.84	1.65	1.32	1.16
24 × 24	610 × 610	0.05	0.14	0.38	1.05	2.87	2.73	2.26
24 × 36	610 × 910	0.04	0.12	0.32	0.87	2.39	2.24	2.02
24 × 48	610 × 1220	0.04	0.10	0.29	0.78	1.90	1.75	1.66
24 × 72	610 × 1830	0.03	0.09	0.25	0.70	1.06	1.07	1.08
30 × 30	760 × 760	0.04	0.12	0.33	0.91	2.49	2.32	2.14
30 × 45	760 × 1140	0.04	0.10	0.28	0.76	1.88	1.90	1.91
30 × 60	760 × 1520	0.03	0.09	0.25	0.68	1.35	1.48	1.57
30 × 90	760 × 2290	0.03	0.08	0.22	0.60	0.76	0.91	1.02
36 × 36	910 × 910	0.04	0.11	0.29	0.81	2.01	2.03	2.04
36 × 54	910 × 1370	0.03	0.09	0.25	0.67	1.42	1.66	1.83
36 × 72	910 × 1830	0.03	0.08	0.22	0.60	1.02	1.30	1.50
36 × 108	910 × 2740	0.03	0.07	0.20	0.54	0.57	0.80	0.98
42 × 42	1070 × 1070	0.04	0.10	0.27	0.73	1.59	1.81	1.97
42 × 64	1070 × 1630	0.03	0.08	0.22	0.60	1.11	1.47	1.75
42 × 84	1070 × 2130	0.03	0.07	0.20	0.55	0.81	1.16	1.45
42 × 126	1070 × 3200	0.02	0.06	0.18	0.49	0.45	0.71	0.94
48 × 48	1220 × 1220	0.03	0.09	0.24	0.67	1.30	1.65	1.90
48 × 72	1220 × 1830	0.03	0.07	0.20	0.56	0.92	1.35	1.70
48 × 96	1220 × 2440	0.02	0.07	0.18	0.50	0.66	1.05	1.40
48 × 144	1220 × 3660	0.02	0.06	0.16	0.45	0.37	0.65	0.91

*Based on measurements of surface-coated duct liners of 1.5-lb/ft³ (24-kg/m³) density. For the specific materials tested, liner density had a minor effect over the nominal range of 1.5 to lb/ft³ (24 to 48 kg/m³).
Source: 1987 ASHRAE Handbook, HVAC Systems and Applications, Chap. 52.

TABLE 10.54 Sound Attenuation in Straight, Lined, Sheet-Metal Ducts of Rectangular Cross Section in dB/ft (dB/0.3 m); 2-in (50-mm) Lining Thickness*; No Air Flow

Internal cross-sectional dimensions		Octave band center frequency, Hz						
in	mm	63	125	250	500	1000	2000	4000
4 × 4	100 × 100	0.34	0.93	2.56	7.02	19.23	10.08	3.50
4 × 6	100 × 150	0.28	0.78	2.13	5.85	16.03	8.26	3.13
4 × 8	100 × 200	0.26	0.70	1.92	5.26	14.42	6.45	2.57
4 × 10	100 × 250	0.24	0.65	1.79	4.91	13.46	5.02	2.07
6 × 6	150 × 150	0.26	0.72	1.97	5.40	14.81	7.50	3.17
6 × 10	150 × 250	0.21	0.58	1.58	4.32	11.85	5.67	2.67
6 × 12	150 × 300	0.20	0.54	1.48	4.05	11.11	4.80	2.33
6 × 18	150 × 460	0.17	0.48	1.31	3.60	8.80	2.95	1.51
8 × 8	200 × 200	0.22	0.60	1.64	4.49	12.31	6.08	2.95
8 × 12	200 × 300	0.18	0.50	1.36	3.74	10.26	4.98	2.64
8 × 16	200 × 410	0.16	0.45	1.23	3.37	9.23	3.89	2.17
8 × 24	200 × 610	0.15	0.40	1.09	2.99	5.68	2.39	1.41
10 × 10	250 × 250	0.19	0.52	1.42	3.89	10.66	5.17	2.79
10 × 16	250 × 410	0.15	0.42	1.15	3.16	8.66	4.04	2.41
10 × 20	250 × 510	0.14	0.39	1.06	2.92	7.22	3.30	2.05
10 × 30	250 × 760	0.13	0.34	0.95	2.59	4.04	2.03	1.34
12 × 12	300 × 300	0.17	0.46	1.26	3.46	9.48	4.52	2.67
12 × 18	300 × 460	0.14	0.38	1.05	2.88	7.62	3.71	2.39
12 × 24	300 × 610	0.13	0.35	0.95	2.59	5.47	2.89	1.97
12 × 36	300 × 910	0.11	0.31	0.84	2.31	3.06	1.78	1.28
15 × 15	380 × 380	0.15	0.40	1.09	2.99	7.64	3.84	2.53
15 × 22	380 × 560	0.12	0.34	0.92	2.52	5.55	3.20	2.29
15 × 30	380 × 760	0.11	0.30	0.82	2.25	3.89	2.46	1.86
15 × 45	380 × 1140	0.10	0.27	0.73	2.00	2.17	1.51	1.21
18 × 18	460 × 460	0.13	0.35	0.97	2.66	5.79	3.37	2.42
18 × 28	460 × 710	0.11	0.29	0.80	2.19	3.95	2.69	2.13
18 × 36	460 × 910	0.10	0.27	0.73	2.00	2.94	2.15	1.78
18 × 54	460 × 1370	0.09	0.24	0.65	1.78	1.65	1.32	1.16
24 × 24	610 × 610	0.11	0.29	0.81	2.21	3.73	2.73	2.26
24 × 36	610 × 910	0.09	0.25	0.67	1.84	2.65	2.24	2.02
24 × 48	610 × 1220	0.08	0.22	0.61	1.66	1.90	1.75	1.66
24 × 72	610 × 1830	0.07	0.20	0.54	1.48	1.06	1.07	1.08
30 × 30	760 × 760	0.09	0.25	0.70	1.92	2.65	2.32	2.14
30 × 45	760 × 1140	0.08	0.21	0.58	1.60	1.88	1.90	1.91
30 × 60	760 × 1520	0.07	0.19	0.52	1.44	1.35	1.48	1.57
30 × 90	760 × 2290	0.06	0.17	0.47	1.28	0.76	0.91	1.02
36 × 36	910 × 910	0.08	0.23	0.62	1.70	2.01	2.03	2.04
36 × 54	910 × 1370	0.07	0.19	0.52	1.42	1.42	1.66	1.83
36 × 72	910 × 1830	0.06	0.17	0.47	1.28	1.02	1.30	1.50
36 × 108	910 × 2740	0.06	0.15	0.41	1.14	0.57	0.80	0.98
42 × 42	1070 × 1070	0.07	0.21	0.56	1.54	1.59	1.81	1.97
42 × 64	1070 × 1630	0.06	0.17	0.47	1.28	1.11	1.47	1.75
42 × 84	1070 × 2130	0.06	0.15	0.42	1.16	0.81	1.16	1.45
42 × 126	1070 × 3200	0.05	0.14	0.38	1.03	0.45	0.71	0.94
48 × 48	1220 × 1220	0.07	0.19	0.52	1.42	1.30	1.65	1.90
48 × 72	1220 × 1830	0.06	0.16	0.43	1.18	0.92	1.35	1.70
48 × 96	1220 × 2440	0.05	0.14	0.39	1.06	0.66	1.05	1.40
48 × 144	1220 × 3660	0.05	0.13	0.34	0.94	0.37	0.65	0.91

*Based on measurements of surface-coated duct liners of 1.5-lb/ft³ (24-kg/m³) density. For the specific materials tested, liner density had a minor effect.
Source: 1987 ASHRAE Handbook, HVAC Systems and Applications, Chap. 52.

TABLE 10.55   Attenuation of Unlined Sheet-Metal Elbows

Values in decibels per elbow[*]

Duct diameter or width†		Octave band number							
in	cm	1	2	3	4	5	6	7	8
5–10	12.7–25.4	0	0	0	0, 0, 1	1, 3, 5	2, 4, 7	3, 4, 5	3
11–20	27.9–50.8	0	0	0, 0, 1	1, 3, 5	2, 4, 7	3, 4, 5	3	3
21–40	53.3–101.6	0	0, 0, 1	1, 3, 5	2, 4, 7	3, 4, 5	3	3	3
41–80	104.1–203.2	0, 0, 1	1, 3, 5	2, 4, 7	3, 4, 5	3	3	3	3

*Where three values are shown, use: *First*, round elbows; *second*, square elbows with vanes; *third*, square elbows without vanes, or branch takeoffs with flow diverter. Where one value is given, apply it to any of these three conditions.

†Width is dimension in plane of turn. If area transition occurs between inlet and outlet of elbow, use width of smaller cross section.

Line 11: DIL Required.   Compare the maximum permissible sound power levels (line 9) with the predicted fan sound power levels (line 10). If the latter is greater than the former in any octave band, the difference is the dynamic insertion loss (DIL) required to meet the criteria in that band.

The data in Table 10.59 are graphically illustrated in Fig. 10.69, where the shaded area shows the required silencer DIL.

10.25.2   Silencer Selection

A silencer must be selected to provide both the necessary attenuation and DIL, and the self-noise level should be no greater than the permissible levels in line 9 of Table 10.59.

A large range of silencer types and sizes is available so that selections can be made to meet specified pressure drops.

Of course, the lower the pressure drop, the lower the operating costs, although the initial cost may be higher. (See Secs. 10.24.3, Pressure Drop, and 10.24.4, Energy Consumption.)

10.25.3   Calculating the Attenuation Effects of Lined Ducts

The sound attenuation of lined ducts may be estimated by the use of empirically derived equations that express the insertion loss of a given length of duct in terms of the duct cross section and the sound absorption properties of the lining material (Ref. 18). Absorption depends on the type of lining material, but in general it is related to density and thickness at each frequency of interest.

TABLE 10.56   Allowance for L_W Split—Branch to Terminals

Number of terminals	1	2	3	4	8	10	20	40	100
Allowance factor, dB	0	3	5	6	9	10	13	16	20

TABLE 10.57 Division of Sound Power—Main Duct to Branches

Area of branch in % of area of main duct	$\frac{1}{5}\%$	$\frac{1}{2}\%$	1%	2%	5%	10%	20%	50%
dB to be added to branch L_W to obtain main duct L_W	27	23	20	17	13	10	7	3

The equations take the form of two intersecting curves defining the low-frequency attenuation A_1 and the high-frequency attenuation A_2:

$$A_1 = \frac{t^{1.08} h^{0.356} (P/A) L \cdot f^{1.17 - K_2 d}}{K_3 d^{2.3}} \ \text{dB} \tag{10.43}$$

$$A_2 = \frac{K_4 (P/A) L \cdot f^{K_5 - 1.61 \ \log(P/A)}}{w^{2.5} h^{2.7}} \ \text{dI} \tag{10.44}$$

TABLE 10.58 Summary of Acoustic Input Data for "SNAP" Calculations in Table 10.59

Line	Data
1	Sound pressure level criterion: NC 35
2	Room and terminal effect: General case = Individual terminals Room size = 30 ft long, 30 ft wide, 10 ft high Number of terminals = 4 Distance from reference point to terminal 1 = 11 ft Distance from reference point to terminal 2 = 11 ft Distance from reference point to terminal 3 = 15 ft Distance from reference point to terminal 4 = 15 ft
3	Allowance for end reflection: Duct cross-sectional dimensions = 6 by 12 in
4	Ductwork attenuation—terminal to checkpoint: Duct size = 12 by 12 in; duct length = 30 ft Duct size = 36 by 36 in; duct length = 50 ft
5	Elbow attenuation—terminal to checkpoint: Branch take-off with flow diverter; width = 12 in; quantity = 1 Elbow type = square, without turning vanes; width = 6 in; quantity = 1 Elbow type = square, without turning vanes; width = 36 in; quantity = 2
6	Power-level split—branch to terminals: An allowance of 6 dB is being made to account for the sound power level division to each outlet based on four terminals (from line 2).
7	Power-level split—main duct to branch duct: Main duct = 36 by 36 in Branch duct = 12 by 12 in
8	Safety factors: A 3-dB safety factor was selected.
10	Sound power level of fan: The fan power level was input by the user.

Source: *Systemic Noise Analysis Procedure,* Bulletin 1.0110.3, Industrial Acoustics Company, Bronx, NY, 1976.

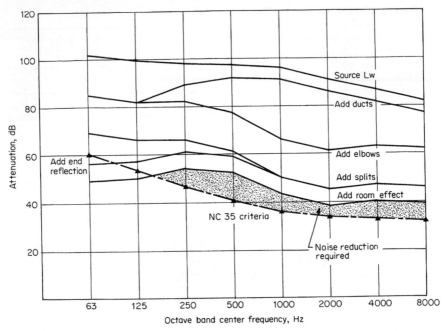

FIGURE 10.69   Graphic representation of attenuation of systemic noise: Noise analysis example for ducted system. (*Courtesy of Industrial Acoustics Company.*)

TABLE 10.59   Summary of Acoustic "SNAP" Calculations

		Octave band center frequency, HZ							
		63	125	250	500	1000	2000	4000	8000
		Octave band number							
Line	Calculation	1	2	3	4	5	6	7	8
1	Noise criteria at NC 35	60	53	46	40	36	34	33	32
2	Room and terminal effect	7	7	7	7	7	7	7	7
3	Allowance for end reflection	13	9	5	2	0	0	0	0
4	Ductwork attenuation	17	17	9	5	5	5	5	5
5	Elbow attenuation	0	0	7	15	25	25	19	15
6	L_W split—branch to terminals	6	6	6	6	6	6	6	6
7	L_W split—main duct to branch	10	10	10	10	10	10	10	10
	Totals	113	102	90	85	89	87	80	75
8	Safety factors	−3	−3	−3	−3	−3	−3	−3	−3
9	Permissible L_W	110	99	87	82	86	84	77	72
10	L_W of fan	102	99	98	97	96	91	87	82
11	DIL required	0	0	11	15	10	7	10	10

Source: Systemic Noise Analysis Procedure for Air Handling Systems, Industrial Acoustics Company, 1979.

where $K_2 = 0.19$ (0.119) [metric shown in parentheses]
$\quad K_3 = 1190$ (5.46×10^{-3})
$\quad K_4 = 2.11 \times 10^9$ (3.32×10^{18})
$\quad K_5 = -1.53$ (−3.79)
$\quad d$ = liner density, lb/ft³ (kg/m³)
$\quad t$ = liner thickness, in (mm)
$\quad h$ = the smaller inside duct dimension, in (mm)
$\quad w$ = the larger inside duct dimension, in (mm)
$\quad P$ = inside duct perimeter, in (mm)
$\quad A$ = inside area of duct, in² (mm²)
$\quad L$ = duct length, ft (m)
$\quad f$ = one-third octave band center frequency, Hz

The equations may be used directly to calculate attenuation in any one-third octave band, using the lower of A_1 or A_2. Due to the effect of break-out, the maximum attenuation should be limited to 40 dB. Also, for frequencies above 1000 Hz, the lined length L should be empirically limited to 10 ft (3.05 m).

To express octave band attenuation, use the lowest of the one-third octave frequencies in the octave band of interest to calculate A_1. Use the highest of the one-third octave frequencies to calculate A_2.

10.26 ACOUSTIC LOUVERS

In an HVAC fan room, acoustic louvers can be used to allow fresh air in while reducing the fan noise outside the building. Acoustic louvers are designed to keep face velocities and pressure drops low and can occupy relatively large building wall sections. Louvers are therefore treated as wall components with openings and are therefore also acoustically rated on the basis of sound transmission loss (ASTM E90) rather than rated as silencers (ASTM E477).

A typical louver installation is shown in Fig. 10.70, and the transmission loss characteristics of commercially available acoustic louvers are listed in Table 10.60.

Systemic analysis of acoustic louvers is accomplished in seven steps (as illustrated in Table 10.63):

1. Establish the octave band sound pressure level criterion from Sec. 10.21 and Table 10.39.
2. Calculate the divergence, taking into account the face area of the louver as well as the distance from the louver to the location (external to the fan room) where the criterion

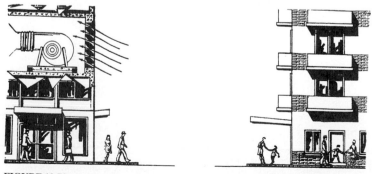

FIGURE 10.70 Acoustic louver installation, SNAP Form II, Bulletin 1.0303.2. (*Courtesy of Industrial Acoustics Company.*)

TABLE 10.60 Sound Transmission Loss of Acoustic Louvers

Louver	Octave band center frequency, Hz							
	63	125	250	500	1000	2000	4000	8000
Standard model	5	7	11	12	13	14	12	9
Low-pressure-drop model	4	5	8	9	12	9	7	6

must be met. Table 10.61 shows the divergence for a range of face areas and distances. Enter the table at the left in the "Distance" column and read across to the right to establish the divergence for the appropriate area; interpolate as necessary. For example, at a distance of 50 ft (15.24 m) the divergence for a 36-ft^2 (3.34-m^2) louver would be interpolated as 26 dB. (For more on divergence, see Sec. 10.8, Propagation of Sound Outdoors, and Sec. 10.9. The Inverse-Square Law.)

3. Determine the fan noise attenuation due to the sound-absorptive characteristics of the fan room. Table 10.62 summarizes fan-room effects. Enter this table under the "Distance" column and read across to the right, interpolating as necessary to establish the room absorption for the size and type of room. For example, at a distance of 10 ft (3.05 m) a "soft" room of 2000 ft^2 (185.8 m^2) shows a 15-dB factor.

4. Add the values obtained in steps 1 to 3, and subtract from their total a 3-dB safety factor. The result is the permissible noise-source sound power level that will meet the criterion.

5. Enter the manufacturer's fan sound power level. If it is not available, calculate it by using the procedure described in Sec. 10.15.

6. Compare the values obtained in step 5 with the values obtained in step 4. If the fan sound power level is greater than the permissible sound power level in any octave band, the difference is the sound transmission loss (TL) that the louver must provide.

7. Compare the required TL to the TL of the louver to be used. Select the lowest-pressure-drop lever that will meet the requirements.

Table 10.63 shows these steps for an installation where a 24,000-ft^3/min (670 m^3/min) fan is located 10 ft (3.05 m) from a 36-ft^2 (3.34-m^2) air-intake louver in a 2000-ft^2 (185.8-m^2)

TABLE 10.61 Divergence Factor for Acoustic Louvers, dB

Distance from louver to point of interest, ft	Face area of louver, ft^2							
	12 $^1/_2$	25	50	100	200	400	800	1600
12$^1/_2$	19	16	13	0	0	0	0	0
25	25	22	19	16	13	0	0	0
50	31	28	25	22	19	15	13	0
100	37	34	31	28	25	22	19	16
250	45	42	39	36	33	30	27	24
500	51	48	45	42	39	36	33	30
1000	57	54	51	48	45	42	39	36
2000	63	60	57	54	51	48	45	42

Note: When working in metric units, convert to English units before entering this table: (1) Multiply the distance in meters by 3.281 to get the distance in feet; (2) multiply the area in square meters by 10.764 to get the area in square feet.

TABLE 10.62 Fan-Room Absorption Factors

Distance from louver to fan, ft	Room surface, ft²																	
	500			1000			2000			4000			8000			16,000		
	Acoustic characteristic of room, dB*																	
	S	A	H	S	A	H	S	A	H	S	A	H	S	A	H	S	A	H
5	9	5	2	10	7	4	11	8	5	11	9	7	12	10	8	12	10	9
10	12	6	2	14	8	4	15	9	6	15	10	8	16	12	10	17	13	10
15	12	6	2	15	8	4	17	10	6	18	11	8	19	13	10	20	15	12
20	12	6	2	15	8	4	18	10	6	19	12	8	20	13	10	22	15	12

*S = soft (walls and ceiling covered with 4-in-thick absorptive panels); A = average (30 percent of walls and ceiling covered with 4-in-thick panels); H = hard (walls and ceiling made of concrete, brick, metal, or similar material).

Notes:

1. Bottom factors in each column are maximum for any greater distance.
2. Where direct line of sight from the noise generator to the opening exists, use one-half of the values shown in this table.
3. When working in metric units, convert to English units before entering this table: (a) Multiply the distance in meters by 3.281 to get the distance in feet; (b) multiply the area in square meters by 10.764 to get the area in square feet.

Source: *Systemic Noise Analysis Procedure,* SNAP II, Bulletin, 1.0503.2, Industrial Acoustics Company, 1975.

TABLE 10.63 Systemic Noise Analysis Procedure for Louver Application

Step	Calculation	Data source	Octave band center frequency, Hz							
			63	125	250	500	1000	2000	4000	8000
			Octave band number							
			1	2	3	4	5	6	7	8
1	Ambient, dB	Table 10.39	72	67	62	57	52	48	44	42
2	Divergence, dB	Table 10.61	26	26	26	26	26	26	26	26
3	Room absorption, dB	Table 10.62	15	15	15	15	15	15	15	15
4	Total, steps 1 to 3	Steps 1 to 3	113	108	103	98	93	89	85	83
	Minus 3-dB safety factor		−3	−3	−3	−3	−3	−3	−3	−3
	Result—noise source permissible L_W, dB		110	105	100	95	90	86	82	80
5	Noise source L_W re 10^{-12} W, dB	Manufacturer's data	101	101	100	98	97	92	84	76
6	Transmission loss required, dB	Step 5 minus step 4	—	—	—	3	7	6	2	
7	Transmission loss of louver, dB	Table 49.60	4	5	8	9	12	9	7	6

fan plenum, which is lined with absorptive materials and can be classified as "soft." The louver is 50 ft (15.24 m) from an apartment house. The installation is close to continuous heavy traffic and the criterion is taken from Table 10.39. The fan sound power level was supplied by the manufacturer. The intake louver's face area was selected on the basis of a maximum permissible pressure drop of 0.5 inWG (124.6 Pa).

10.27 HVAC SILENCING APPLICATIONS

An HVAC system may have several locations where noise control is required. Figure 10.71 shows some of the different types of noise control devices used in HVAC systems:

1. *Cylindrical silencers:* Shown also in Fig. 10.28, these control the noise of axial-flow fans.

1. Cylindrical silencers	4. "Through-the-wall" vent silencers	7. Cooling-tower silencers
2. Rectangular silencers		8. Noise control enclosures
3. Fan plenums and air-handling systems	5. "Over-the-wall" vent silencers	9. Acoustic louvers
	6. Exhaust vent silencers	10. "Clean-flow" and packless silencers
		11. Slim louvers

FIGURE 10.71 Typical HVAC silencing applications. (*Application Manual for Duct Silencers, Bulletin 1.0301.3, Industrial Acoustics Company.*)

2. *Rectangular silencers:* Shown also in Fig. 10.26, these are modular units often built up in banks immediately before fan intake and discharge sections with aerodynamic transition sections. Table 10.48 illustrates some ways in which rectangular silencers are used in HVAC systems. Additional applications are shown in Fig. 10.72.

3. *Acoustic plenums and quiet air-handling systems:* Acoustic plenums control fan noise in the air stream and in adjacent areas and generally provide excellent thermal

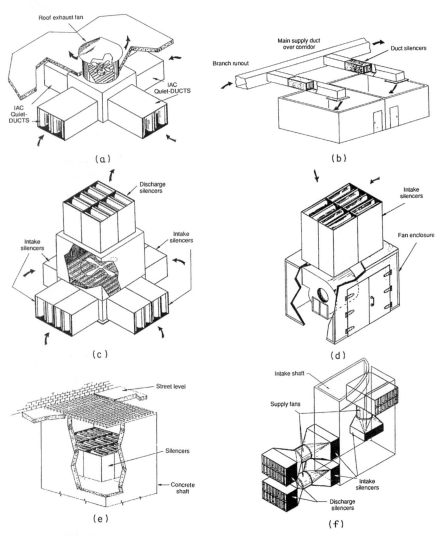

FIGURE 10.72 Rectangular silencers in ventilation HVAC systems. (*a*) Roof exhaust silencer arrangement for propeller fans; (*b*) duct silencers prevent sound transmission from room to room through connecting duct system; (*c*) silencers for air-cooled condenser; (*d*) centrifugal-fan enclosure with intake silencers; (*e*) duct silencer bank with intake or discharge shaft; (*f*) underground garage intake and discharge silencers. (Application Manual for Duct Silencers, *Bulletin 1.0301.4, Industrial Acoustics Company, 1989.*)

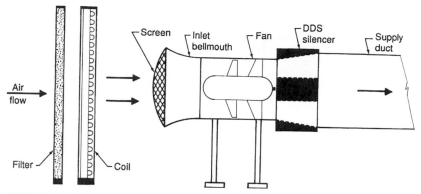

FIGURE 10.73 Diffuser-silencer for axial-flow fans. (Application Manual for Duct Silencers, *Bulletin 1.0301.4, Industrial Acoustics Company.*)

insulation. Fan plenums may also include silencers; for instance, Fig. 10.73 shows a diffuser-silencer that attenuates vaneaxial fan exhaust noise while facilitating aerodynamic pressure regain. Other acoustically designed plenum housings to control fan noise in the air stream and in adjacent areas are illustrated in Figs. 10.83 and 10.84.

4. (and 5 and 6) *Vent silencers:* These provide conversational privacy between rooms while allowing air circulation. They are commercially available in exhaust-vent configurations as well as in "through the wall" and "over the wall" designs behind acoustic ceilings (Figs. 10.74 and 10.75).

7. *Silencers for cooling towers and roof exhaust fans:* These silencers prevent acoustic annoyance to neighboring buildings. Figure 10.25 illustrates the use of silencers with cooling towers, and Fig. 10.72 shows other fan intake and discharge silencer arrangements.

8. *Modular air-handling units and built-up fan plenums:* Discussed in Secs. 10.29 and 10.30, these shield office occupants against machinery noise and help create quiet office spaces. Figures 10.81, 10.83, and 10.84 show several types of air-handling units used in buildings.

9. *Acoustic louvers:* These control noise, permit air flow, and provide decorative protection against weather and forced entry. They are also used as noise barriers for cooling towers and other equipment. Acoustic louvers are illustrated in Fig. 10.69 and 10.71.

FIGURE 10.74 "Through the wall" vent silencer. (Quiet Vent Silencers for Air Transfer Systems, *Bulletin 1.0601.1, Industrial Acoustics Company.*)

FIGURE 10.75 "Over the wall" vent silencer. (Quiet Vent Silencers for Air Transfer Systems, *Bulletin 1.0601.1, Industrial Acoustics Company.*)

FIGURE 10.76a Anechoic chamber for testing and calibrating electronic equipment: a modular soundproof room with silenced HVAC system. (*Courtesy Industrial Acoustics Company.*)

10. *Clean-flow silencers:* These offer protected acoustic infill for silencers in hospitals, clean rooms, operating suites, and pollution control research facilities.

11. *Packless silencers:* Having no acoustic fill, they can be cleaned with steam, chemicals, hot water, or vacuuming. Packless silencers are ideal for silencing microchip plants, corrosive environments, food and dairy operations, clean rooms, hospital operating rooms, and research facilities.

12. *Slim louvers:* Only 4 in (102 mm) deep, these allow for ventilation and provide noise control.

Figure 10.76a–f shows some types of rooms in which soundproofing and the silencing of HVAC equipment are important.

10.28 SELF-NOISE OF ROOM TERMINAL UNITS

The best noise control design for an HVAC system can be undone if the self-noise from the terminal units selected exceeds the sound pressure criteria selected. (The silencer selection procedure described in Sec. 10.25 takes into account correction factors for terminal units, room absorption characteristics, duct end reflection, and other variables.) In general, terminal devices do not generate significant low-frequency noise; rather, they are more likely to produce a discrete high-frequency sound.

We know that sound levels can increase with the doubling of velocities from 15 to 21 dB, depending on the frequency. Therefore, to ensure desired sound levels, velocities must be

FIGURE 10.76b   One of nine audiometric testing rooms in London, England.

FIGURE 10.76c   One of 121 music practice rooms at the University of Western Michigan.

FIGURE 10.76d   One of three studios at KFOG Radio, San Francisco.

appropriately selected to ensure that the desired criterion is not exceeded by the flow noise from the terminal unit.

Manufacturers of ceiling diffusers, grilles, registers, and mixing boxes now publish sound level data, usually presented in the form of an NC rating (such as NC 30 at given conditions of air-flow volume, pressure drop, and size). Such ratings usually assume a room round absorption of 8 or 10 dB and a certain minimum distance from the sound source.

For more detailed sound pressure level L_p calculations, the sound power level L_W of a terminal device must be known; the sound pressure level at a given distance from the

FIGURE 10.76e   Reverberant room used for the development of air conditioning systems.

FIGURE 10.76f   Shelters for torch cutting controls in a continuous casting mill.

source can then be estimated by referring to Fig. 10.15. However, we must also know the room constant, or total room absorption A (see Sec. 10.11) and the directivity factor, which in this case assumes a diffuser located essentially in the middle of the ceiling. If S is the area of all the surfaces in a room and if $\overline{\alpha}$ is the average absorption coefficient of the room (taking into account also objects and people), then $A = S\ \overline{\alpha}$ (where the units of A are sabins if S is in ft^2, and metric sabins if S is in m^2) [Eq. (10.19)]. Using (10.15), if the room constant is 1000 sabins (92.91 metric sabins) and if the distance from the source is 20 ft (6.1 m), then $L_W - L_p = 14$ dB. So if a source's L_W at a given frequency is 50 dB, the L_p would be 36 dB.

Chapter 52 of the *1987 ASHRAE Handbook, Systems,* provides the following empirical equation for "normal rooms" when there is a single source in the room:

$$L_p = \begin{cases} L_w - 5\ \log V - 3\ \log f - 10\ \log r + 25\ \text{dB} & \text{in English units} \\ L_w - 5\ \log V - 3\ \log f - 10\ \log r + 12\ \text{dB} & \text{in metric units} \end{cases} \qquad (10.45)$$

room sound pressure level of chosen reference point, dB re 0.08 inWG (20 Pa)

L_W = sound power level of source, dB re 10^{-12} W
V = room volume, ft^3 (m^3)
f = octave band center frequency, Hz
r = distance from source to reference point, ft (m)

A general equation for an overall sound power level prediction from air-conditioning diffusers is (see Ref. 19)

$$L_w = 10 + 10S + 30\ \log \xi + 60\ \log u \qquad (10.46)$$

where S = area of duct cross section prior to diffuser, m^2
  ξ = normalized pressure drop, derived from $\Delta P/0.5\rho u^2$
  ρ = density of air, kg/m^3
  u = mean-flow velocity in duct prior to the grid, m/s
  ΔP = pressure drop across diffuser, PA

Use Eq. (10.46) with metric units only; convert English to metric units as follows:

$$\text{ft} \times 0.3048 = \text{m} \qquad \text{ft}^2 \times 0.0929 = \text{m}^2$$
$$\text{in}^2 \times 0.0006452 = \text{m}^2 \quad \text{inWG} \times 249.1 = \text{Pa}$$

Equation (10.46) tells us in effect that when the velocity is doubled, the sound power level L_W increases by 18 dB. A doubling of the diffuser area will increase the L_W by 3 dB, while a doubling of the pressure drop across the diffuser (assuming that the duct velocity prior to the grid can be kept constant) will increase the diffuser's self-noise by 9 dB.

The above theory and Eq. (10.46) are intended only as guidelines to explain how the variables of pressure drop, flow velocity, and diffuser area affect the acoustic performance characteristics of diffusers; they are not intended for calculation purposes. For instance, a duct's end reflection and number of terminals would be part of an HVAC system calculation for silencer requirements, as in Sec. 10. 25; they are not directly related to the diffuser area as given in Eq. (10.46).

Essential guidelines to terminal unit selection therefore involve attention to flow velocity and pressure drop. An increase in area is more than offset by reductions in sound power levels due to lower flow velocities and pressure drops.

Flow-noise sound levels from terminal units also depend on correct air-approach configurations (Ref. 2). Manufacturers' ratings apply only to outlets with a uniform air-velocity distribution throughout the neck of the unit. If these conditions cannot be duplicated, then increases in sound levels up to 12 dB can occur, as shown in Fig. 10.77. Also, the misalignment of flexible duct connections can increase sound levels significantly, as in Fig. 10.78.

Pressure drop is also a factor in flow-noise generation from volume control dampers. Figure 10.79 shows the decibels to be added to outlet sound for throttled damper pressure ratios close to the outlet. In acoustically critical spaces, balancing dampers, equalizers, and similar devices should never be installed directly behind terminal devices or open end outlets. They should be located 5 to 10 diameters from the opening, followed by silencers or lined ducts to the terminal or open duct end. Figure 10.80 shows the decibels to be added to sound ratings for volume dampers located at different distances from diffusers and throttled at various pressure ratios.

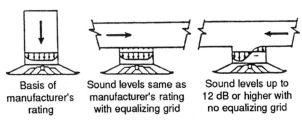

| Basis of manufacturer's rating | Sound levels same as manufacturer's rating with equalizing grid | Sound levels up to 12 dB or higher with no equalizing grid |

FIGURE 10.77 Proper and improper air-flow conditions to an outlet. (ASHRAE Handbook Applications *chap. 52, "Sound and Vibration Control," American Society of Heating, Refrigerating and Air-Conditioning Engineers, Inc., Atlanta, 2003.*)

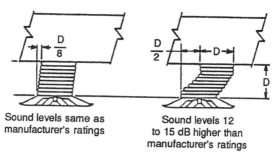

Sound levels same as manufacturer's ratings

Sound levels 12 to 15 dB higher than manufacturer's ratings

FIGURE 10.78 Effect of proper and improper alignment of flexible duct connector. (ASHRAE Handbook Applications, Chap. 47 *"Sound and Vibration Control,"* American Society of Heating, Refrigerating and Air-Conditioning Engineers, Inc., Atlanta, 2003.)

The noise generated by multiple air-distribution devices of equal sound intensity follows the 10 log N rule (where N is the number of diffusers), which means that 3 dB must be added for two terminal units, and 10 dB for 10 units. See Fig. 10.6 for other additions.

10.29 THE USE OF INDIVIDUAL AIR-HANDLING UNITS IN HIGH-RISE BUILDINGS

During the last decade, factory-assembled and -tested individual floor air-handling units (AHUs) have been used in many high-rise office buildings. An AHU with a 20,000-ft³/min (570 m³/min) air-flow capacity, for example, can be placed on each building floor, which may have an area of 20,000 ft² (1858 m²).

AHUs are often located in individual mechanical-equipment rooms in tightly configured floor layouts and may be adjacent to office spaces that must be kept quiet. NC 30 or NC 35 criteria are not uncommon.

Such individual AHUs are taking the place of large, central air-handling systems, where one system might serve a 2,000,000-ft² (185,800-m²) high-rise office building from one

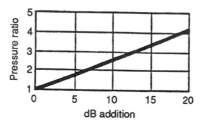

Pressure rating (PR) = $\dfrac{\text{Throttled pressure (drop)}}{\text{Catalog pressure (drop)}}$

FIGURE 10.79 Decibels to be added to outlet sound for throttled damper close to the outlet. (ASHRAE Handbook Applications, Chap. 47 *"Sound and Vibration Control,"* American Society of Heating, Refrigerating and Air-Conditioning Engineers, Inc., Atlanta, 2003.)

Pressure ratio (PR) = $\dfrac{\text{Throttled pressure}}{\text{Minimum pressure}}$						
Location of volume damper	1.5	2	2.5	3	4	6
(A) OB damper in neck of lenear diffuser	5	9	12	15	18	24
(B) OB damper in plenum side inlet	2	3	4	5	6	9
(C) Damper in supply duct at least 5 ft (1.5 m) from plenum	0	0	0	2	3	5

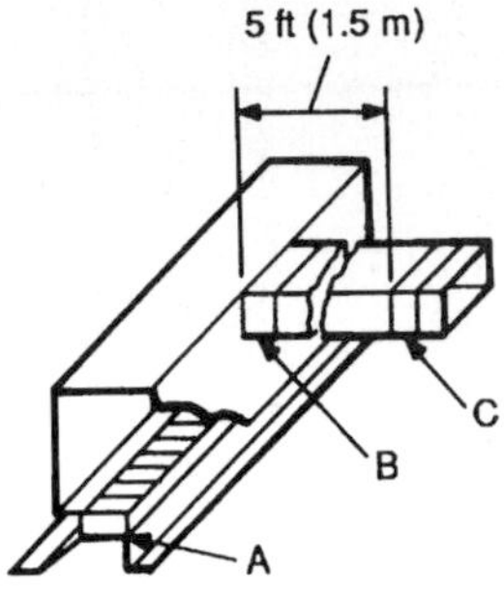

5 ft (1.5 m)

FIGURE 10.80 Decibels to be added to diffuser sound rating to allow for throttling of volume damper. (*ASHRAE Handbook Applications, Chap. 47 "Sound and Vibration Control," American Society of Heating, Refrigerating and Air-Conditioning Engineers, Inc., Atlanta, 2003.*)

mechanical-equipment-room floor located in the middle of the building. The core of the central air-handling system would consist of four 450,000-ft³/min (212.25-m³/s) supply fans, each with an 800-hp (597-kW) drive. It is clear that the energy demand and operating costs would be very high if such a system had to be turned on to accommodate overtime work on Saturday and Sunday work in just a few offices.

Individual floor AHUs can therefore provide a high degree of flexibility and lower operating costs. A quiet AHU is illustrated in Fig. 10.81. The noise levels for such units inside a mechanical-equipment room and just outside adjacent offices are shown in Fig. 10.82.

10.30 BUILT-UP ACOUSTIC PLENUMS

The following applies to field-assembled air-handling units, as distinct from the factory-assembled and -tested AHUs discussed in Sec. 10.28. In an acoustic sound-absorbent plenum, noise from the fan inlet, fan housing, and drive motor is radiated within the plenum, resulting in a reduction of the system sound pressure levels, which at present cannot be predicted accurately.

In a typical "blow through" plenum arrangement there will be no return-air system fan attenuation. Similarly, in a "draw through" plenum there will be no supply duct system attenuation. On the other hand, in a typical "combination" plenum there will be attenuation in both systems. All three plenum types are shown in Fig. 10.83, and a cutaway view of a combination plenum is shown in Fig. 10.84.

The main parameters affecting fan plenum attenuation are the connecting duct area, the surface area of the plenum, the average absorption coefficients of the plenum walls and floors, the plenum dimensions, the angle between the fan inlet and the plenum opening, and the distance between the fan inlet and the plenum discharge. These variables are largely reflected in a formula in Ref. 2. This formula is based on dimensionally small-scale experimental models not directly related to air-conditioning fan plenums; obviously, the

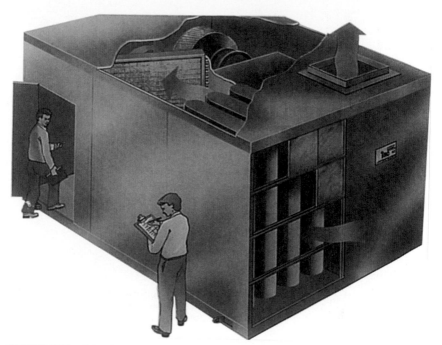

FIGURE 10.81   Quiet air-handling unit. (*Courtesy of Industrial Acoustics Company.*)

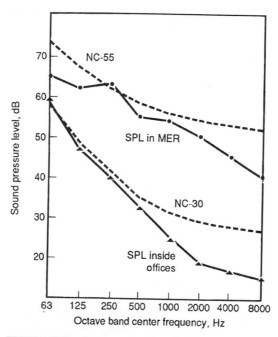

FIGURE 10.82   Sound pressure levels of quiet airhandling unit in mechanical-equipment room and outside of adjacent offices. (*Courtesy of Industrial Acoustics Company.*)

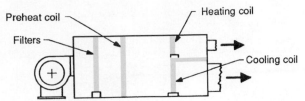

In a typical 'blow-thru' plenum arrangement, noise from the fan outlet may propagate through the plenum and then through the supply air duct system.

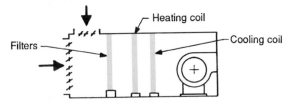

In a 'draw-thru' system, noise in the supply duct may be due to fan inlet and fan housing noise propagating through the plenum and into the return air system.

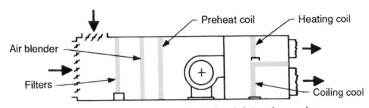

In 'combination' plenums, both fan inlet noise and exhaust noise propagate through the plenum.

FIGURE 10.83   Plenum types.

FIGURE 10.84   "Combination" plenum, cutaway view. (*Courtesy of Industrial Acoustics Company.*)

configuration and internal elements will vary in a full-size plenum. No full-scale data are presently available to verify this formula, which usually results in relatively high numbers.

Plenum silencer characteristics should therefore be regarded merely as *supplemental,* thus providing a safety factor for the supply-air and/or return-air duct silencers.

10.31 FIBERGLASS AND NOISE CONTROL—IS IT SAFE?

On June 24, 1994, the U.S. Department of Health and Human Services (HHS) made the bombshell announcement that fiberglass will be listed as a material "reasonably anticipated to be a carcinogen." This has naturally created considerable concern among the many industries which use fiberglass extensively; however, we are here only interested in the use of fiberglass for noise control purposes (Ref 23).

The announcement has contributed to concerns that there may be a new health crisis on hand similar to the one based on asbestosis, which resulted in a large number of lawsuits for several hundred million dollars in compensation claims. These caused the bankruptcies of several large companies, with Johns Manville being the most prominent.

Of course, there is considerable evidence that fiberglass is not similar to asbestos. A press release from NAIMA (North American Insulation Manufacturers Association), states: "The Canadian Government, after reviewing all the available scientific data, determined in 1993 that glass wool is 'unlikely to be carcinogenic to humans.'" They concluded that glass wool "is not entering the environment in quantities or under conditions that may constitute a danger in Canada to human life or health." Kenneth D. Mentzer, executive vice president of NAIMA, adds, **"It is critical to keep in mind HHS's statement that the listing . . . of a substance, such as glass wool, does not mean that it is a risk to a person in their daily life"** (Ref 24).

Also, Owens Corning Fiberglass, a multibillion dollar international manufacturer of fiberglass, has possibly staked its future on offering to indemnify customers against judgments, settlements and reasonable legal fees, allegedly due to bodily injury resulting from their products. The company has also published an extensive report on fiberglass and health (Ref 25).

Nevertheless, disquiet about Fiberglass fibers is not limited to the United States; there is also considerable medical and regulatory discussion on this issue in Europe particularly Germany and Scandinavia. At the May 31, 1995 session of the Acoustical Society of America on "Health and Legal Aspects regarding the Use of Porous Materials," a well-known European acoustician expressed the opinion that exposed Fiberglass today would not be used in Germany for school or hospital projects.

10.31.1 The Concern with Indoor Air Quality (IAQ)

As a consequence, there has been much interest not only in Fiberfree silencers but also in Fiberfree sound absorbers which may be used in duct lining, rooms or wherever noise control or acoustical conditioning is required.

Of course, there have always been reasons unrelated to the carcinogen issue why fibrous infill in HVAC silencers, duct lining, fan plenums, sound proof enclosures, quiet rooms and other noise control products may not be desirable. Fiberfree silencers were developed more than 40 years ago for jet engine noise suppressors because the fibrous materials could not withstand the high temperatures, vibrations, heat and velocities they were subjected to (Ref 26).

Any kind of particulate matter, whether from plants, animals, human activity, building products, paints, carpets, consumer products or fibrous materials can affect the IAQ, or indoor air quality, and the sterility of hospital operating rooms, patient care areas, microchip manufacturing, food preparation, cosmetic, pharmaceutical manufacturing plants and wherever clean-rooms are required.

In fuel laboratories, kitchens, gasoline handling facilities and other chemical plants there is the risk that the fiberglass infill in HVAC silencers could become saturated with fuel and pose a fire hazard.

10.31.2 Legionnaire's Disease

In the 1960s and 1970s concern developed about the relationship between HVAC Systems and IAQ (Indoor Air Quality). Anxiety about IAQ was fueled by the deadly outbreak of Legionnaire's disease in 1976. Concerns, among many other contributors, included the possible introduction of particulate fibrous matter into air/gas streams from acoustical duct lining and silencers.

10.31.3 Encapsulating the Fibrous Acoustical Infill

For many applications in hospitals and clean-rooms, "Clean-Flow" silencers and sound absorber panels can be used where the acoustical infill is encapsulated inside polymer sheets. Encapsulations prevent the erosion of inorganic particulate matter from the silencer as long as the protective polymer sheeting is intact. There have been no reports of failures but they are of course possible and may even have to be expected with time.

10.31.4 Packless Fiberfree HVAC Silencers

To eliminate any risk whatsoever of particulate fibers from silencers entering the air/gas stream, IAC developed a line of all-metal completely Fiberfree Packless Silencers. They use no infill or fibrous material of any kind and work by virtue of acoustically resistive patented perforations in the splitters.

We developed several standard silencer types with test data for dynamic insertion loss, self-noise and pressure drop with lengths from 3 ft to 9 ft (914 mm to 2743 mm). Figure 10.85 shows a Fiberfree Cylindrical All-Metal Silencer being tested at IAC's Aero-Acoustic Laboratory. Figures 10.86*a* and 10.86*b* show Packless Fiberfree Rectangular Duct and Elbow Silencers while Figures 10.87, 10.88, and 10.89 show installations of Fiberfree silencers in a pharmaceutical plant, nuclear fuels plant and a firing range.

In Figure 10.90, we compare the acoustical dynamic insertion loss (DIL) of an all-metal Fiberfree silencer (Model 9XL), a fiberglass encapsulated silencer (Model 7HLFM) and a "standard" 7MS silencer, where the fiberglass is protected with 22 percent free area perforated metal. All comparisons are at the same silencer face velocity of 750 fpm (3.8 m/s).

Acoustically, the packless and encapsulated fiberglass silencers work well up to about 500 Hz but then lose more than 20 dB in the 1000 and 2000 Hz octave bands compared with silencers where the fiberglass is protected by perforated metal only. However, high-frequency attenuation often is not needed in air-conditioning systems because the high frequencies are frequently attenuated significantly by the duct system.

FIGURE 10.85 Testing of Fiberfree cylindrical all-metal silencer at IAC's Aero-Acoustic Laboratory.

FIGURE 10.86a Packless Fiberfree rectangular duct silencer.

FIGURE 10.86b Tubular Fiberfree elbow silencer.

FIGURE 10.87 Packless Fiberfree silencers in laboratory of pharmaceutical company.

FIGURE 10.88 Packless Fiberfree silencers are being used extensively at British Nuclear Fuels Laboratories, Warrington, England.

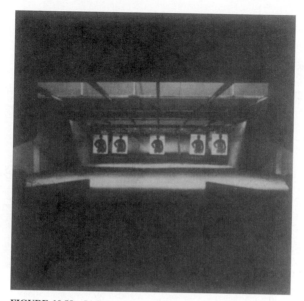

FIGURE 10.89 IAC Packless Fiberfree all metal silencers in firing range reduce airborne noise passing through suspended exhaust system thus enabling the Irving TX Police Department to meet municipal codes.

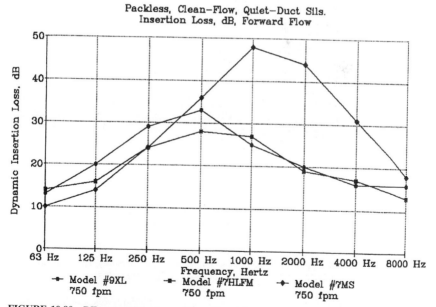

FIGURE 10.90 DIL comparison for packless, Clean Flow Fiberglass encapsulated and "Standard" silencers.

FIGURE 10.91a One type of Fiberfree sound absorbers looks like this conventional noise-lock ceiling.

FIGURE 10.91b Embossed Fiberfree sound absorber for walls and ceilings.

10.31.5 Fiberfree Sound Absorbers

The technology for engineering Fiberfree sound absorptive panels has been available for some time and a number of designs using sintered aluminum and acrylic plastics with microperforations are being offered. However, most are very expensive. Figure 10.91 shows an economical Fiberfree all-metal design with noise reduction coefficients, NRCs, of 0.8 and Figure 10.92 gives sound absorption coefficients vs. frequencies of 2.5 in (64 mm) fiberfree sound absorber.

IAC's impervious Clean-Flow Fiberfree Panel Modules can be steam-cleaned or hosed down and kept sterile through chemical treatments. Panel surfaces have no perforations. (Figs. 10.93*a* and 10.93*b*). They are available with and without fiberglass with NRCs of 0.80 and 0.60, respectively.

Conclusions. Fiberglass free and fiberglass encapsulated silencers and acoustical panel products are available; also new products can be developed for specific applications.

To avoid the risk of fiber contamination entirely, it is best to use all-metal Fiberfree silencers, duct liners and sound absorbers. If there is concern that internal cavities will trap dirt and water, drainage openings and accessibility for periodic cleaning can be provided. Self-cleaning devices can also be engineered.

Acoustic performance tests for fiberglass free silencers, panels and enclosures indicate they are as reliable as conventional designs when subjected to the same rigorous laboratory procedures. Moreover, several thousand Fiberfree silencers are presently demonstrating their reliability in field installations.

Finally, the evidence classifying fiberglass as a possible carcinogen is highly controversial. Most importantly, a number of studies conducted by responsible researchers, such

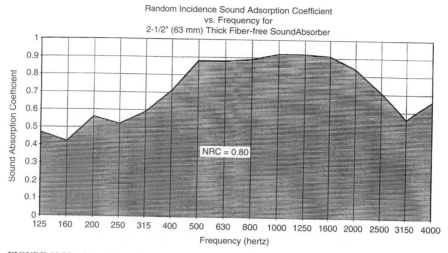

FIGURE 10.92   Sound absorption coefficients of 2.5 in (635 mm) deep all-metal sound absorber.

as the World Health Organization, a United Nations agency, the Canadian Government, and the Owens-Corning Fiberglass Company conclude that fiberglass is unlikely to be a carcinogen.

Also IAC with more than six hundred thousand HVAC silencers in the field, has never encountered an HVAC installation where fiberglass has visibly eroded through

FIGURE 10.93   Clean Flow Fiberfree Modules with impervious facings can be steam cleaned.

the perforated sheetmetal facings. However, with so many different non-inspected installations in the field, applied in varying ways, it is impossible to be categorical on this subject.

Nevertheless, based on the above scientific findings, we believe the use of fiberglass in well-engineered noise control products remains a safe combination.

REFERENCES

1. C. M. Harris (ed.), *Handbook of Noise Control*, 2d ed., McGraw-Hill, New York, NY, 1979.
2. *1987 ASHRAE Handbook, Systems*, ASHRAE, Atlanta, GA, 1987, chap. 52, "Sound and Vibration Control."
3. L. Miller and F. M. Long, *"A Practical Approach to Cooling Tower Noise Evaluation," Heating, Piping and Air Conditioning*, vol. 34, no. 6, June 1962.
4. H. Seelbach, Jr., and F. M. Oran, "Control of Cooling Tower Noise," *Heating, Piping and Air Conditioning*, vol. 35, no. 16, June 1963.
5. M. Hirschorn, "Silencing Cooling Towers," *Heating, Piping and Air Conditioning* 24, issue 8, August 1952.
6. M. Hirschorn, "Some Considerations Involved in the Silencing of Air Conditioning and Ventilating Air Intakes, Roof Exhausters and Cooling Towers," *Noise Control*, January 1957.
7. ASTM E477, *Standard Method of Testing Duct Liner Materials and Prefabricated Silencers for Acoustical and Airflow Performance*, American Society for Testing and Materials, Philadelphia, 1973.
8. M. Hirschorn, "Acoustic and Aerodynamic Characteristics of Duct Silencers for Airhandling systems," ASHRAE Paper CH-81-6, 1981.
9. *Application Manual for Duct Silencers*, Bulletin 1.0301.4, Industrial Acoustics Company, Bronx, NY, 1989.
10. U. Ingard, "Influence of Fluid Motion Past a Plan Boundary on Sound Reflection, Absorption and Transmission," *Journal of the Acoustical Society of America*, vol. 31, no. 7, July 1959.
11. F. Mechel, "Schalldämpfung in absorbierend ausgekleideten Kanalen mit überlagerter Luftstromung," *Proceedings of Third International Congress on Acoustics*, Stuttgart, 1959.
12. James H. Botsford, "dBA Reduction of Barriers," *Proceedings of Noise Expo*, Chicago, Illinois, 1973.
13. Laymon N. Miller, "Noise Control for Buildings and Manufacturing Plants," Lecture Notes, Bolt Beranek & Newman, Inc., Cambridge, MA, 1981.
14. *Technical Advisory Service* (publication), Pilkington Flat Glass, Ltd., St. Helens, Merseyside, England, 1974.
15. ASTM E90, *Standard Method for Laboratory Measurement of Airborne Sound Transmission Loss of Building Partitions*, 1975.
16. ASTM E596, *Standard Method for Laboratory Measurements of the Noise Reduction of Complete Sound Isolating Enclosures*, 1978.
17. Martin Hirschorn, "The Aero-Acoustic Rating of Silencers for 'Forward' and 'Reverse' Flow of Air and Sound," *Noise Control Engineering*, vol. 2, no. 1, Winter 1974.
18. H. L. Kuntz and Robert M. Hoover, "The Interrelationships Between the Physical Properties of Fibrous Duct Lining Materials and Lined Duct Sound Attenuation," preliminary ASHRAE paper, 1987. See *1987 ASHRAE Handbook, Systems*, ASHRAE, Atlanta, GA, Chapter 52.
19. Leo L. Beranek (ed.), *Noise and Vibration Control*, McGraw-Hill, New York, NY, 1971.
20. "Technology for Noise Cancellation Expands," *Diesel Progress Engines & Drive*, Feb 1993.

21. ASHRAE Handbook—HVAC Applications, Chap. 47 Sound & Vibration Control ASHRAE, Atlanta, GA, 2003.

22. "Energy Conservation by Active Noise Attenuation in Ducts by K. H. Eghtesadi, WKW Hong, HG Leventhall," *Noise Control Engineering Journal,* Nov–Dec 1986.

23. *Ref. 1* Statement by Donna Shalala. Secretary of U.S. Department of Health and Human Services, June 24, 1994 as communicated in letter by Schuller International to its Employees and Business Partners, June 27, 1994.

24. *Ref. 2* NAIMA (North American Insulation Manufacturers Association) press release June 27, 1994.

25. *Ref. 3* "The Facts about Fiberglass and Health" as published by Owens-Corning Fiberglas Company.

26. *Ref. 4* "Fiberglass and Noise Control: Is It a Safe Combination?" by Martin Hirschorn, Sound & Vibration/Oct 1994.

MANAGING MAINTENANCE FOR HVAC

CHAPTER 11
STRATEGIC PLANNING

Ron Moore

The RM Group, Knoxville, Tennessee

11.1 INTRODUCTION

Competitive pressures—regional, domestic, and, increasingly, global—are driving companies to change their historical ways of doing business, to become more competitive, to reduce costs, and to continuously seek ways to increase revenues and profits. As you might reasonably expect, HVAC systems and their operating efficiency, costs, etc. are coming under increasing scrutiny as to performance. After all, if the HVAC system fails, not much else happens in today's business environment. Because of this, many companies have taken a very strong risk-averse position in their businesses, making certain that the HVAC system works better than 99.9 percent of the time (less than 8 h per year of lost performance). This often requires things like redundant capacity, an extraordinary level of spare parts, mechanics on call, rapid, responsive, and often reactive maintenance. Given the risk of loss to the business and the probability of failure, this may be an appropriate strategy under historical paradigms. However, it may *not* be appropriate given the technology and methods that are now proven and available. This chapter outlines the basis for developing a maintenance strategy within your organization which has a strong reliability focus for assuring optimal performance. As you will see, the best companies have come to recognize, among other things, that maintenance is a *reliability* function, not a repair function.

11.2 BENCHMARKING—DEFINING AND MEASURING THE OBJECTIVES OF YOUR STRATEGY

One of the first steps necessary in developing your strategy is to define key objectives and develop the basis for measuring achievement of those objectives. One method used to do so is to seek out other organizations believed to be successful, and emulate them. This is often referred to as *benchmarking*. It is an increasingly common practice used to help companies improve their performance, and it is described below.

According to Dr. Jack Grayson of the National Center for Manufacturing Sciences,[1] benchmarking involves "seeking out another organization that does a process better than yours and learning from them, adapting and improving your own process . . ." Benchmarks, on the other hand, have come to be recognized as those specific performance measures which reflect a 'best in class' standard. *Best practices*, as the name implies, are those practices which are determined to be best for a given process, environment, etc., and allow a company to achieve a benchmark level of performance in a given category. Benchmarking, then, is a process for identifying benchmarks, or measures of best in class. Best practices are those practices which lead to benchmark performance. A good maintenance organization will be driven by a reliability strategy, and will define the key performance measures by which they should be judged, and will work diligently to assure that best practices are put in place which will lead to superior, or benchmark, performance. Caution: simply finding benchmark data and making arbitrary decisions without a clear understanding of the underlying practices, will often yield disastrous results—*don't do this!* Below we will explore the benchmarking and best practices relationship, and how to effectively use these principles to improve productivity, reduce costs, and support improved profitability.

11.3 FINDING BENCHMARKS OR BEST PERFORMERS

The first step in benchmarking is to define those processes for which benchmark metrics are desired, and which are believed to reflect the performance objectives of the company. While this may seem simple, it can often be quite complicated. You must first answer the question "What processes are of concern?" and "What measures best reflect my company's performance for those processes?" Then you must answer the question "What measures best reflect my department's performance which, in turn, supports organizational objectives?" and so on. These decisions will vary from industry to industry, and even from department to department. Let's suppose we have an overall objective to reduce the cost for maintaining HVAC systems, while concurrently assuring high reliability and availability. Performance measures might include cost per unit of "facility," e.g., *square feet, horsepower*, (m^2, w^2) etc., and dollars lost from HVAC equipment downtime. Additional suggestions for the measures of success are provided in Table 11.1. *The key is to select those measures which reflect the performance objectives of the organization for a given process*, and subsequently to determine how the organization compares to others, generally within your industry. It is generally best when no more than five key metrics are used in each category of interest, and to make sure they can relate one to the other in a supportive, integrated relationship.

Comparing your company to other companies is the next step. Having selected the key metrics, you must now make sure that you are properly measuring these metrics within your company, that you understand the basis for the numbers being generated, and that you can equitably apply the same rationale to other information received. Developing comparable data from other companies may be somewhat difficult. You have a number of choices:

- Seek publicly available information through research.
- Set up an internal benchmarking group which will survey multiple facilities within a corporation, assuring fair and equitable treatment of the data. Benchmarks will then be defined in the context of the corporation.
- Seek the assistance of an outside company to survey multiple facilities, usually including your corporation, but also expanded to other companies within the company's industry.
- Seek the assistance of an outside company to survey multiple facilities, many of which may be outside your industry, in related, or even unrelated fields.
- Some combination of the above, or other alternative.

TABLE 11.1 Sample Performance Measures

General Measures
Unit cost of supply ($\$/ft^2$, $\$/hp$, $\$/ft^3$, $\$/m^2$, $\$/w$, $\$/m^3$, etc.)
Productivity (*$ per person per asset, per sq ft, etc.*)
Equipment availability by *area, unit, component*
Equipment reliability—*average life or mean time between repair (MTBR)*
Energy consumed (*$ or units, e.g., KWH, BTU, etc.*)
Utilities consumed (*e.g., water, wastewater, nitrogen gas, distilled water, etc.*)
Spare Parts/MRO (maintenance repair order) inventory turns

Losses and "Bad Practice" Measures
Production or business losses from breakdown (*$, quantity, no. of events*)
Overtime rate (*%$ or %h*)
Rework rate (*$, units, number*)
Unplanned downtime
Reactive work order rate—*emergency, run-to-fail, breakdown, etc. (%, %h, %$)*

Personnel Measures
OSHA injury rates (*Recordables, Lost time per 200K h*)
Training (*time, $, certifications, etc. per craft*)
Personnel attrition rate (*staff turnover in %/year*)

PM (preventive maintenance) Measures
Planned and scheduled work/total work (%)
PM/work order schedule compliance (% on schedule)
Hours covered by work orders (%)
PM work by operators (%)
PM's per month
Cost of PM's per month
"Wrench" time (%)
"Backlog" (weeks of work available)
Mean time to repair, including commissioning

Condition Monitoring Measures
Average vibration levels (*overall, balance, align, etc.*)
Average lube contamination levels
Schedule compliance for condition monitoring
PDM (predictive maintenance) effectiveness (*accuracy of predictive maintenance by technology*)
Process measures trended for *pressure, temperature, flow, current, etc.*

Census Measures
Mechanics, electricians, etc. per:
   Support person
   First line supervisor
   Planner
   Maintenance engineer
   Total site staff
Total number of crafts

Normalized Cost Measures
Maintenance cost/facility replacement value (%)
Facility replacement value per mechanic (*$/mechanic*)
Stores value/facility replacement value (%)
Stores service level—stock out (%)
   Critical spares
   Normal spares
Contractor expenditure/total maintenance expenditure (%)

The Building Operation and Management Association (BOMA), provides an extensive database for certain performance measures, e.g., cost per square foot for various types of facilities, and may make a useful beginning. If the decision is made to actually do a benchmarking study, it is recommended that the Benchmarking Code of Conduct outlined in *The Benchmarking Management Guide*, published by Productivity Press, or similar standard, be followed. This assures that issues related to confidentiality, fairness, etc. are followed. For benchmarking outside the company, particularly within your industry, it is strongly recommended that you use an outside firm. This will help avoid any problems with statutes related to unfair trade practices, price fixing, etc. Benchmarking is an excellent tool to use to improve performance, but should not be used as a means to garner information otherwise not available on specific competitors, or to place your company at risk of violating the law.

Benchmarking has become associated with "marks," as opposed to finding someone who does something better than you do and doing what they do, which may more accurately be called best practices. Notwithstanding the semantics, benchmarking and application of best practices are powerful tools to help improve operational and financial performance. The ultimate benchmark, of course, is consistently providing the lowest cost per unit of supply, over the long term. Applying benchmarking and best practices can help assure that position.

Finally, you don't need to do benchmarking to make improvements, so don't let the lack of specific benchmarks stop the improvement process. Select those measures which you believe to be critical to your success, measure them, and then put in place practices which you believe will improve them, now.

11.4 RELIABILITY AND BEST PRACTICES—HOW TO ACHIEVE A BENCHMARK LEVEL OF PERFORMANCE

We've discussed the process for benchmarking of key performance indicators. Now we're going to explore those practices which will support superior performance. Maintenance practices, when combined with good operations and design practices, will determine the reliability of HVAC equipment for a given facility. Therefore, it is incumbent upon us to assure good design, operation, and maintenance practices, and to use the experience of the maintenance function to improve operations and design, and vice versa. We must use an integrated approach built on teamwork to assure maximum reliability of our facility. Another way of thinking about reactive maintenance is to ask "What work did I do last week that I didn't plan to do last Monday morning?" All that is reactive.

From a 1992 study,[2] the typical U.S. manufacturer was shown to have the following approximate level of maintenance (reliability) practices in their manufacturing facilities:

- Reactive maintenance   50%
- Preventive maintenance   25%
- Predictive maintenance   15%
- Proactive maintenance   10%

Some definitions would be useful:

1. *Reactive maintenance* was defined by practices such as run-to-fail, breakdown, and emergency maintenance. Its common characteristics were that it was unanticipated and typically had some level of urgency.

2. *Preventive maintenance* were those practices that were time based, that is, on a periodic basis prescribed maintenance would be performed. Examples include: annual overhauls, quarterly calibrations, monthly lubrication, etc.

3. *Predictive maintenance* practices were those that were equipment condition based. Examples include changing a bearing long before it fails based on vibration analysis, changing lubricant based on oil analysis showing excess wear particles, replacing steam traps based upon ultrasonic analysis, etc. In the best facilities, it also included review of process parameters as part of the condition-based approach.

4. *Proactive maintenance* practices were those that were root cause based, or that sought to extend machinery life. Examples were varied, and include modification of operation, design, or maintenance practices, but all sought to eliminate the root cause of particular problems. Specific examples might include root cause failure analysis, precision alignment and balancing of machinery, precision equipment installation commissioning, and improved design and vendor specifications. The overall goal was to ensure that the equipment was "fixed forever."

These typical facility maintenance practices were then compared to so-called benchmark facilities of the study. These benchmark facilities were characterized as those who had achieved extraordinary levels of improvement and/or performance in their operation.

The striking characteristics of these benchmark facilities, as compared to the typical facility, were twofold:

1. Reactive maintenance levels differed dramatically. The typical facility incurred some 50 percent reactive maintenance, while the benchmark facilities typically incurred less than 10 percent, principally because they were reliability focused in their maintenance practices.

2. The benchmark facilities were driven by an integrated combination of preventive, predictive, and proactive methods, with a strong emphasis on predictive, or condition-based, maintenance practices. Further, preventive and predictive methods were used as tools to be proactive and eliminate the cause of failures.

These conclusions are reinforced by R. Ricketts in *Organizational Strategies for Managing Reliability*,[3] wherein he states that the best refineries were characterized by, among other things, the "religious pursuit of equipment condition assessment"; and that the worst were characterized by, among other things, "staffing . . . designed to accommodate rapid repair," and failures being "expected because they are the norm." In the same study, he also provides substantial data to support the fact that as reliability increases, maintenance costs decrease, and vice versa. This has also been the experience of the author.

11.5 WORST PRACTICES

Reactive maintenance, at levels typically beyond 10–20 percent, should be considered a practice to avoid, a "worst practice." Reactive maintenance tends to cost more and lead to longer periods of downtime. In general, this is due to the ancillary damage which often results when machinery runs to failure; the frequent need for overtime; the application of extraordinary resources to "get the equipment back on line, NOW"; the frequent need to search for spares; the need for expedited (air freight) delivery of spares, etc. Further, in a reactive mode, the downtime period is often extended for these very same reasons. Moreover, in the rush to return the equipment to operation, many times no substantive effort will be made to verify the quality of the repair and equipment condition at start up. Hudachek and Dodd[4] report that reactive maintenance practices for general rotating machinery costs some 30 percent more than preventive maintenance practices, and about twice as much as predictive maintenance practices.

As noted above, some small level of reactive maintenance is acceptable, and is generally that which has minimal consequence as to cost, uptime, safety, and environmental impact.

11.6 BEST PRACTICES

In the author's observation, organizations that employ best practices—that is those which facilitate operating at benchmark levels of performance (world class, best in class, etc.) are those which have a *reliability culture*, whereas those that are mediocre and worse, have a *repair culture*.[5]

In a repair culture, the maintenance department is viewed as someone to call when things break. They can even become very good at crisis management and emergency repairs; they often have the better craft labor—after all, the crafts are called upon to perform miracles; but they will never rise to the level achieved in a reliability culture. They often even view themselves as second class employees, since they are viewed and treated as "grease monkeys," repair mechanics, and so on. They often complain of not being able to maintain the equipment properly, only to be admonished when it does break, and placed under extraordinary pressure to "get it back on line." They are rarely allowed to investigate the root cause of a particular problem and eliminate it, or to seek new technologies and methods for improving reliability—"it's not in the budget" is a familiar refrain. They are eager, sometimes desperate, to contribute more than a repair job, but are placed in an environment where it's just not possible. They are doomed to repeat the bad practices and history of the past, until the company can no longer afford them, or to stay in business.

In a reliability culture, reliability is the watchword. No failures is the mantra. Machinery failures are viewed as failures not of the machine, but of the system which allowed the failure to occur. In a reliability culture, preventive, predictive, and proactive maintenance practices are blended into an integral strategy. To the maximum extent possible, maintenance is performed based on condition of equipment and of process. Condition diagnostics are used to analyze the root cause of any failures, and methods are sought to avoid the failure in the future. The methods for achieving and supporting this culture are discussed in Chap. 12.

REFERENCES

1. National Center for Manufacturing Sciences (NCMS) Newsletter, November 1991.

2. R. Moore, F. Pardue, A. Pride, and J. Wilson, *The Reliability-Based Maintenance Strategy: A Vision for Improving Industrial Productivity*, Computational Systems, Industry Report, Knoxville, TN, September, 1993.

3. R. Ricketts, Organization Strategies for Managing Reliability, National Petroleum Refiners Association, Washington, DC: New Orleans, LA, Annual Conference, May 1994.

4. R. Hudachek and V. Dodd, Progress and Payout of a Machinery Surveillance and Diagnostic Program, American Society of Mechanical Engineers, New York, NY, 1985.

5. R. Moore, Reliability, Benchmarks, and Best Practices, *Reliability Magazine*, Knoxville, TN, December 1994.

BIBLIOGRAPHY

J. Ashton, "Kev Maintenance Performance Measures," *Dofasco Steel-Industry Survey*, Hamilton, Ontario, Canada, 1993.

E. Jones, "The Japanese Approach to Facilities Maintenance," *Maintenance Technology*, Barrington, IL, August 1991.

K. Kelly and P. Burrows, "Motorola: Training for the Millennium," *Business Week*, March 1994.

Information from DuPont and Other Large Manufacturing Companies.

CHAPTER 12
TYPES OF MAINTENANCE

Ron Moore
The RM Group, Knoxville, Tennessee

12.1 PREVENTIVE MAINTENANCE

Preventive maintenance takes on a new meaning when using a reliability-based strategy. It encourages time-based maintenance, but focuses on doing those things which help us to: (a) avoid failures, e.g., regular tightening, lubricating, cleaning, (TLC), filter changes, instrument calibrations, and other basic care efforts; (b) detect onset of failure in sufficient time to manage the development problem, e.g., routine tours to listen, smell, feel, monitor condition, do conformance inspections, etc., or finally, (c) do those which are based on known, highly consistent, wear- or age-related failure modes. Preventive maintenance is used principally as a tool for basic care, analysis and planning, and almost always includes comprehensive use of a computerized maintenance management system (CMMS). The maintenance function establishes and analyzes machinery histories, performs cost analyses of historical costs and impacts, and uses these cost and machine histories to perform Pareto analyses to determine where resources should be allocated for maximum reliability. Preventive maintenance performs maintenance planning, assuring coordination with other functions, assuring that spare parts are available, that tools are available, that stores inventories and use histories are routinely reviewed to minimize stores, yet maximize the probability of assuring equipment uptime. Preventive maintenance staff have recognized that doing all maintenance on a strict interval basis is not optimal. They recognized that PM's are often based on arbitrary data, from the manufacturer or otherwise, and that most machines will exhibit wide variation in their mean time to failure. *Few machines are truly average*, a basic assumptions of many PM schedules. Any group of equipment may follow a wide variation of failure modes, frequencies, and effects. Good preventive maintenance practices, then, include:

- Understanding equipment failure modes, frequencies, and effects.
- Strong statistical base for those PM's done on a time interval.
- Exceptional planning and scheduling capability.
- Solid machinery history analysis capability.
- Strong and flexible cost analysis capability.
- Comprehensive link to stores and parts use histories.
- Routine analysis of stores and its support of cost effective reliability.

- Comprehensive training in the methods and technologies required for success.
- Comprehensive link to the predictive or condition-based program.

12.2 PREDICTIVE MAINTENANCE

Predictive or condition-based maintenance is typically at the heart of a good reliability program. Knowing machinery condition, through the application of vibration, oil, infrared, ultrasonic, motor current, and particularly to process parameters (pressure, temperature, flow, etc.), drives world class reliability practices, and assures maximum reliability. It is generally best if all technologies are under single leadership, allowing for synergism and routine, informal communication teamwork, that focuses on maximizing reliability. For example, using modern vibration analysis tools we could identify a developing problem with a bearing long before it actually fails. This would allow the maintenance on the bearing (perhaps used in a fan, pump, etc.) to be planned and orderly. That is, we could check to assure we have the proper parts, tools, and resources, and we could work with operations to perform the maintenance so as to minimize the impact of the repair on the facility's requirements. Further, the vibration analysis could also preclude what might have historically resulted in a catastrophic failure. Finally, in a proactive use of vibration analysis, it could also facilitate a diagnosis of the root cause of the bearing failure, and then be used in the commissioning of the bearing to specific standards at startup to assure excellence in the repair. Other applications include using condition monitoring to plan overhauls more effectively, since it is known what repairs will be needed. Overhauls can be done for what is necessary, but only what is necessary. Therefore, they take less time—much of the work is done before shutdown, and only what is necessary is done. Condition monitoring allows for validation of the precision and quality of the repair at start up to assure the equipment is in like new condition. Condition monitoring allows better planning for spare parts needs, thus minimizing the need for excess inventory in stores. Good predictive maintenance practices include:

1. Routine use of equipment condition assessment.
2. Applying all appropriate and cost effective technologies.
3. Avoiding catastrophic failures and unplanned downtime by knowing of problems in machinery long before it becomes an emergency.
4. Diagnosing the root cause of problems and seeking to eliminate the cause.
5. Defining what maintenance jobs need to be done, when they need to be done—no more, no less.
6. Planning overhaul work more effectively, and doing as much of the work as possible before shutdown.
7. Setting commissioning standards and practices to verify equipment is in like new condition during start up.
8. Minimizing inventory, through knowledge of machine condition and subsequent planning of spare parts needs.
9. Comprehensive communications link to maintenance planning, equipment histories, stores, etc. for more effective teamwork.
10. Comprehensive training in the methods and technologies required for success.
11. An attitude that failures are failures of the design, operation, and maintenance processes, not the individual.
12. Continuously seeking ways to improve reliability and improve equipment performance.

12.3 PROACTIVE MAINTENANCE

Proactive maintenance at the best facilities is the next step in reliability. At facilities which have a strong proactive program, they have gone beyond routine preventive maintenance, and beyond predicting when failures will occur. They aggressively seek the root cause of problems, actively communicate with other departments to understand and eliminate failures, and employ methods for extending machinery life. Predictive maintenance, or more appropriately, condition monitoring, is an integral part of their function, because these methods and technologies provide the diagnostic capability to understand machinery behavior and condition. They have come to understand, for example, that alignment and balancing of rotating machinery can dramatically extend machinery life and reduce failure rates. They have also learned that doing the job perfectly in the facility is not sufficient, and that improved reliability also comes from their suppliers. Therefore, they have a set of supplier standards which require reliability tests and validation. In the net, they constantly seek to *design* equipment for reliability, to specify and *purchase* reliable equipment, to *store* it reliably, to *install* it reliably and verify the quality of the installation effort, to *operate* the equipment to maximize reliability, and finally to *maintain* the equipment for reliability. These issues are discussed further below.

Good *design* practice requires consideration of life cycle costs, including maintenance and operating costs, not just initial cost. It also requires maintainability, e.g., ease of access, isolation valves, lifting lugs, jacking bolts, inclusion of tools, etc.

Good *purchasing* practice requires good specifications from, and good communications with, maintenance and engineering in order to assure comprehensive and high quality supply. For example, a good motor specification might require that all motors (1) be balanced to within 0.10 in (2.5 mm) per second vibration at one time turning speed, (2) have no more than 5 percent difference in cross phase impedance at load, (3) have co-planar feet not to exceed 0.003 ($\times 760$ µm) in.

Good design and procurement practices also require the application of reliability-centered maintenance principles. That is, when we design and procure, we must consider the most common and consequential failure modes, and work with our suppliers to eliminate those in the design phase. If that's not possible, then we must work to assure that we can detect those failure modes in sufficient time to take action to mitigate or manage them more effectively.

Good *stores* practices requires a number of support functions which assure retaining the reliability of the equipment, and providing it when and where it is needed. Additional detail regarding the purchasing and stores function is provided in the chapter on stores and parts management.

Good *installation* requires precision in installation practices, and a process for validating the quality of the effort using specific measurement techniques to verify process quality (proper temperatures, flows, pressures, etc.), as well as equipment quality (proper vibration levels, oil quality, amperage levels, etc.). This should also include a process for resolving non-conformances.

Good *operation* practices assures that operators work with precision, and within the limits of the capability of the equipment. For example, not running pumps dead headed or in a cavitation mode, not trying to re-start a motor 10 times in a row without success (and burning out the motor), and so on. In the experience of the author, some 60–70 percent of equipment failures are a direct result of poor practices in areas other than maintenance which resulted in a maintenance or repair requirement. Good maintenance practices go well beyond simple maintenance issues.

Proactive *maintenance* practices which have yielded extraordinary gain include:

- Root cause failure analysis.
- Precision alignment and balancing.

- Supplier standards for reliability.
- Training of purchasing staff in reliability standards.
- Installation commissioning standards to verify proper installation.
- Precision operation—process control within all design limits.
- Comprehensive training in the methods and technologies required for success.
- Continuous communication with production, engineering, and purchasing to maximize reliability.

The key is to assure that the knowledge base within maintenance and operations which represent problems (otherwise called opportunities for improvement) are in fact used effectively for the continuous improvement process. The best facilities have developed and implemented these practices to assure success of their operation.

12.4 KEY PERFORMANCE INDICATORS

In order to apply the best practices described above, and to become a benchmark facility, you must have measures of your progress and success, i.e., you must have benchmarks by which you are judged. As Dr. Joseph Juran, a leading quality expert, said "If you don't measure it, you don't manage it." Put in a more positive sense, if you measure it, you will manage it, and it will improve. These measures, or benchmarks, by which you will judge your performance then, are critical for your success—you can hardly afford to be managing the wrong measures. The measures then should reflect operational and financial performance, and a listing of suggestions for consideration is provided in Table 11.1.

Table 12.1 provides a sampling of prospective benchmarks data. These are offered hot as benchmarks which resulted from a comprehensive benchmarking study, but rather as based on a review of the literature, and the experience of the author. These measures, or benchmarks, will facilitate a measure of the effectiveness of the best practices espoused above.

Finally, overall success of any good HVAC reliability strategy requires:

Knowledge of equipment performance and condition

Training to assure understanding

Focus on the right goals

TABLE 12.1   Nominal Benchmarks for Various Measures

Indicator	Definition	World class level
Planned maintenance	Planned maintenance/total maintenance	>90%
Reactive maintenance	Run to fail, emergency, etc. (1, 3)	<10%
Maintenance overtime	Maint OT/total maintenance time (1, 3)	<5%
Maintenance rework	Work orders reworked/total WO's (3)	~0%
Inventory turns	Turns ratio of spare parts (3, 4)	>2–3
Skills Training	Workers receiving >40 hr/yr (3)	>90%
	Spending on worker training (% of payroll) (5)	~4%
Safety performance	OSHA injuries per 200,000 labor hour (6)	<0.5
Maintenance cost/yr	Percent of replacement value (4)	<1–3%

Source: R. Moore, Reliability, "Benchmarks, and Best Practices," *Reliability Magazine*, Knoxville, TN, December 1994.

Teamwork for synergy

Communication to assure understanding and cooperation

Benchmarks to provide metrics of performance and feedback on effectiveness

Leadership to create a continuous improvement environment and assure knowledge, training, focus, teamwork, communication, and benchmarks.

Reliability, benchmarking, and applying best practices are keys to your success. By learning what the best facilities are achieving and how you compare (benchmarking), by emulating how they are achieving superior results (best practices), and by judiciously applying the practices described above, adding a measure of your own common sense and management skill, your facility can exceed the performance levels of the current benchmarks. This will raise its overall effectiveness, lower operating and maintenance costs, improve reliability, and achieve world class performance.

CHAPTER 13
MANAGING MAINTENANCE: IN-HOUSE VERSUS OUTSOURCING

William N. Berryman
Berryman Consulting, Gilbert, Arizona

13.1 INTRODUCTION

13.1.1 In-House versus Outsourcing Strategies

The decision of whether to outsource maintenance is based on many factors. Does the company provide added value in the following categories over your present operating model: cost, financial stability, technical capability, core competency, past performance as a supplier to other companies (ask for references), resource availability, equipment availability, regional, national, and global capabilities, quality programs, reporting capability, contract terms and conditions (must be in alignment with your legal dept. requirements), management capability and approach, and overall value additions?

Of these factors, you must consider certain elements when deciding what to outsource. Some of the decisions that need to be made are: What is the right size of your company? What is the required environment for your company, i.e., is this environment critical? For example, environment is critical in data centers, in electronic manufacturing, or in a manufacturing line where any downtime is costly. Another area to be considered before deciding what to outsource: Is there specialized equipment that is critical to the operation that requires a specialist to maintain?

One of those factors is the accessibility of the outsourcing company. This applies only when you are outsourcing certain elements of your equipment, i.e., HVAC overhauls.

The next strategic decision to be made: what are the goals of outsourcing? One of the common goals is to reduce costs by outsourcing and negotiating the contractual elements so that the cost reduction goals can be obtained. Some of the strategy requirements are: what equipment maintenance will be outsourced and at what frequency level do you wish to have this maintenance accomplished? For example, many companies will outsource major equipment overhauls due to the specialized equipment, knowledge, and expertise required to accomplish this task, rather than employ this level of expertise on a full-time basis. The cost of this type of specialized activity becomes one of your financial considerations.

13.1.2   Outsourcing Goals

Outsourcing goals need to be established up front. Cost reduction is not the only element to consider. Other elements are available technical expertise, consistency, and geographic location.

13.2   RECOMMENDED METHODOLOGIES FOR OUTSOURCE DECISION

13.2.1   Methodology 1

One of the methodologies for outsourcing is to develop a list of all the items you maintain within *your* area of responsibility. Develop this list into a matrix and within that matrix develop three columns: the first column is to be used for outsourcing to a service provider, the second column is for your company provision and administration (supplier and in-house), and the third is for your company provision and service provider administration activities. By developing this matrix and putting these three columns in place, you have the ability to select the requirements for each activity by placing an X in the appropriate column. The items in the table should be grouped by category, i.e., grounds, custodial, fire protection, electrical maintenance & operations, general maintenance, etc. Develop a matrix for each category as seen in Table 13.1.

This allows for selection consistency in large companies where multiple sites are outsourcing their maintenance activities to the same outsourcing agent. This also allows for the consistency of the measurements/metrics used to drive the business and assures that the service provider is complying with contractual requirements.

TABLE 13.1   Outsourcing Matrix

	Service provider	Your company provides + administration	Your company provides + service provider administration
1. The facilities service provider shall be responsible for providing all maintenance of electrical distribution systems and equipment. Included in the maintenance and operation contract shall be:			
• Annual preventive maintenance program			
• High voltage components, main switchgear, transformers, etc.	————	————	————
• Low voltage components, main switchgear, transformers, etc.	————	————	————
• Monthly emergency lighting maintenance	————	————	————
• Response to service calls, i.e., lights, no power, etc.	————	————	————

Note: Place an X on the appropriate line.

Small companies can also use this process because it simplifies the identification of the items required for outsourcing.

13.2.2   Methodology 2

Critical Environment.   Utilize an existing equipment list to establish criticality, and outsource only the maintenance of equipment that would not directly impact the operations of the critical environment. The equipment that is usually outsourced in this methodology is what is known as *creature comfort* equipment, i.e., office area HVAC, or other office environment equipment. This methodology usually applies to small companies with limited assets to be maintained.

13.3   UTILIZATION OF SUBCONTRACTORS BY THE OUTSOURCING CONTRACTOR

In many cases, subcontractors are used by service providers, but the owner company may want to manage some subcontractors for a variety of reasons. For example, the subcontractor has serviced the critical equipment for the owner company and has the experience and has provided excellent service for some time.

A good method for determining management of the subcontractors is as follows:

- Define the subcontractors
- Define their core competence
- Then categorize the subcontractors as A, B, or C
  - A = managed by owner company
  - B = managed by outsourcing contractor, but must provide service due to special expertise required to support
  - C = outsourcing contractor will manage these subcontractors, but can replace them if more competent subcontractor can be found, or corporate agreements allow for cost leveraging.

13.3.1   Contract Maintenance

There are multiple methods of contract maintenance. Several are listed below.

Individual contracts for Maintenance of Systems or Structures (i.e., HVAC Systems, Painting/Roofing Management).   This method allows the company to manage the scheduling and cost of individual companies' activities.

This requires extensive management time and inspection of contractor quality, and could become difficult from a scheduling perspective if the specific contractor is not available when the work is scheduled. Then rescheduling would be required and this could add cost to the maintenance of facilities. Another additional cost is the contract management. If the facility is large and requires multiple contractors, then each contractor will require a contract that defines the activities, cost, and timelines. This usually requires a contract administrator and multi-invoice management. It also requires additional financial administrative support.

This methodology would be effective if planned properly and contractors selected carefully. Some of the keys to sources using multiple contractors are: defining the processes and procedures for scheduling, quality control, billing, and service requirements for facilities customers, and communicating specific requirements in the contract.

Note: We will use HVAC systems and assume we know the number of equipment sets.

1. Contact local suppliers of HVAC maintenance (4 to 5).
2. Request a list of their existing customers.
3. Request financial information (assure they have a stable business for continued support).
4. Provide HVAC equipment information to them for bids (types and numbers).
5. Provide your process and procedure requirements (i.e. you may provide the PM procedures to them from your CMMS, they may need to provide spares, their invoices must come to you for approval, etc.).
6. Develop contract language that is specific.
7. Set expectations for PM schedule adherence, emergency calls response, documentation, billing, escalation processes, etc.
8. Select the contractor.

Outsourcing.   This is another method of contract maintenance. In this model, the work to be outsourced, usually all or most of facility operations, is defined. A single contractor is selected for maintenance of systems and structures. This contractor will usually provide the management, support, maintenance management software, and technical support personnel (craft personnel); these are all employees of the outsource contractor.

This methodology provides on-site support personnel; scheduling is defined by the contractor as agreed upon by customer requirements, and is usually a long-term contract (3 to 5 years). It also has the advantage of having a single point of contact and the support to provide administration of financial, scheduling, documentation, and personnel management. This model endows the customer company management and the contractor management with the ability to strategically plan the maintenance of the facility and equipment, but can reduce the customer involvement in day-to-day operations.

The key to success in outsourcing is to plan well by understanding the scope of work with attention to detail. Define a selection process, understand the scope, budget, and communicate this clearly in a request for proposal. Understanding the outsource contractor's ability to fully support all facets of the contract is very important, so be sure to understand what you need accomplished and then define their core competencies. Also, ask for specifics and completed successful projects. In this model, the outsource contractor may manage subcontractors for special maintenance items in which they may not have core competencies, i.e., elevator maintenance, roofing repairs, crane, lifting device load testing, etc. These must be specific to the project and support capabilities.

13.4 SETTING EXPECTATIONS AND DELIVERABLES WITH THE OUTSOURCING CONTRACTOR

This is usually accomplished by defining the requirements of the project. The following are some of the elements:

• Date of transition start
• The personnel hiring practices (transition process, if required)

- The development of the PM program elements
  - Data collection
  - Criticality
  - Safety requirements (see transition schedule)
  - Organization development (org chart)

Next is setting the metrics, or measurements. Measurements take on many faces. Some companies measure many different elements of the project. In many cases, efficiency of the outsource contractor is compromised due to the customer requesting numerous measurements too frequently.

A good rule of thumb for metrics/measurement development is: *Measure what will improve your business or what you need to provide your customers.*

Do not request measurements that are not important, as you are defeating the purpose of having an outsourcing company perform your maintenance. It will cost more money to measure these items that may not provide value.

13.5 OUTSOURCING CONTRACTOR SELECTION

A very important element of a successful outsourcing plan is the selection of the contractor who will manage and perform the maintenance activities.

This relationship is very important, so set criteria for selection other than the usual purchasing process requirements, i.e., insurance, bonding, etc. The criteria should be aligned with your company's business philosophies of safety, quality, and process execution.

One other important element is the *working relationship.* If there is an incompatibility in this area due to lack of expectation setting and understanding, then the path to a successful program may contain additional challenges. These challenges may be costly and could permanently damage the relationship or at the very least cause delay in project execution.

Many contractors have convincing sales personnel that may tell you that they can provide all the elements of a perfect outsourcing program, but do your own research and use a scorecard or some process in the selection of the contractor. Make sure that the outsourcing contractor has qualified, certified, or licensed (as required) personnel. If the contractor is going to hire personnel for your project, ensure the selection process conforms with your company requirements for hiring practices.

One of the cautions in this area is the risk of co-employment. So, as you develop the relationship, keep the co-employment issue in mind.

13.6 ELEMENTS OF AN OUTSOURCING PROPOSAL

The first element is a prequalification questionnaire: Request for Information (RFI). This questionnaire is submitted when looking for qualified sources to provide a required service. In addition to the questionnaire, there is usually a requirement for a company profile. This document does not guarantee that any work will be provided, but is used for information associated with the proposals. The key requirements for this document are:

1. *General information.* For example, business name, address, telephone numbers, web page, and key contact for bidding.
2. *Type of business.* Is it a corporation, partnership, full proprietorship, etc., location of corporate office, when business was founded, and name of the parent company, if different.

3. Requirement for names of the officers and their titles, their Dunn & Bradstreet rating and their federal tax ID.

4. Financial information going back 5 years, including any federal or state licensing, and any liens against the company.

5. A list of qualifications or experience on projects of the type you are bidding and any references from other companies where this type of work has been performed. Also include labor relations information, trade agreements.

6. Safety information is one of the most important elements of this document. This is generally based on number of OSHA recordable incidents and is rated against every 200,000 labor-hours along with the number of recordable incidents. Other elements may include lost work-day cases and restricted work days. Generally, your company should know if this company has a written safety program and safety training. Do they have drug policies? Request to see if they have any federal or state OSHA citations.

7. Insurance requirements that your company may need.

8. Our company's minimum typical insurance is listed in detail in the contract. If you are awarded any work, you will be required to provide official certificates of insurance meeting or exceeding the stated limits PRIOR to commencement of any work on site.

Create a form to gather this information. For an example, see Fig. 13.1.

13.7 REQUEST FOR PROPOSAL

The second element of an outsourcing proposal is the actual Request for Proposal (RFP) document. The RFP should include the following:

1. *An invitation to submit the proposal.* The invitation to submit the proposal is only sent to requested, prequalified companies.

2. *Instructions to the bidders.* Where the RFP is to be picked up by the contractor and when. Where the prebid meeting will be held (if applicable). A project walk through, if required. A deadline date for questions and clarification. A deadline for response to those questions and clarifications. Any response due dates and times. Pre-award clarifications and final award date.

   Note: these dates and times once defined must be held in order to provide for a fair bid process.

3. *Definition of the scope of work.* This generally includes specifications and/or drawings that define all the details of the proposal.

4. *A schedule of the work to be performed.*

5. *Bid forms.*

6. *Agreement and general conditions.* Defining compensation strategy. There are various types of compensation, i.e., lump sum, time and materials, guaranteed maximum price, cost plus fee, unit pricing, and others.

7. *Any supplementary conditions.*

   When submitting the RFP you must ensure that the bidder has reviewed and understands all the information provided. This document is binding by the representations made in the proposal and if they are awarded the contract it will become part of the contract documents.

PRE-QUALIFICATION QUESTIONNAIRE SUBCONTRACTOR DATE:____

The following questions relate to (your company name) key requirements. Please answer each question in writing, as your company believes it would pertain to these requirements. If your program manual addresses any of the following questions, you may provide the manual and location of applicable information. Limit your response to 25 pages, excluding corporate brochures.

1. General Information

Name of Business:	
Street Address:	
Post Office Box:	
City, State, ZIP:	
Telephone Number:	Fax Number:
Web Page URL:	
Key Contact for Bidding:	
Email Address:	

2. Organization of Business

This firm is a: [ ] Corporation [ ] Partnership [ ] Sole Proprietorship [ ] Other
For Corporation, State of Corporation: _____
Date Founded: _____ Under Present Management Since: _____
Name of Parent Company (if applicable): _____

Names of officers or principals	Title

Dunn & Bradstreet rating: _____ Effective Through: _____
Bonding Rate: _____ Insurance Multiplier: _____
Federal Tax Identification Number: _____
Has this organization, or other organization with which the officers or partners were involved during the past 5 years, ever failed to complete any work awarded to them?
[ ] Yes [ ] No
If yes, attach a separate sheet to explain.
Are there any outstanding claims against this organization, or other organization with which the officers or partners were involved during the past 5 years?
[ ] Yes [ ] No
Has this organization, or other organization with which the officers or partners were involved during the past 5 years, ever filed for bankruptcy protection?
[ ] Yes [ ] No
If yes, attach a separate sheet to explain.
Has this organization ever operated under a different name?
[ ] Yes [ ] No
If yes, list the names.

FIGURE 13.1   Request for information.

3. Financial Information

Provide your annual revenue for the last 3 years and a forecast for the current year.

Year	Corporation/Parent	Subsidiary/Division
200_	$	
200_	$	
200_	$	
200_(forecast)	$	

Financial Statements: Provide copies of your firm's audited financial statements (income statement, balance sheet, cash flow statements) for the last 3 years. Be sure to include all supplementary financial notes provided by your internal or external accounting representative. If you are a subsidiary of a parent company, we will require that the parent company of any subsidiary execute a performance guarantee as a condition of awarding a contract. Therefore, the financially liable entity must provide the financial statement.

4. License Information

Attach a copy of State License Qualification Form, and other evidence of licensing in locations where you are able to perform work (please include international locations).

License number	Location	Type of license of work licensed for

5. Qualification & Experience

Projects: Describe projects or contracts of comparable size, scope, schedule, and complexity your firm has completed within the past 3 years in the _____ industry. At least one project or contract must be for a company other than this one. Describe the level of involvement of any personnel you may propose for a project of this nature.

Project name	Cleanroom Type/Class	General scope	# of Tools/ Contract amount	Schedule	Project environment (existing/new)

References: Provide a minimum of three project management (clean room, tool install, and sustaining, if available) references for your top three projects. At least one reference must be for a company other than this one.

Client name/ Project name	Contract amount	Contract strategy	Current reference name	Telephone number

FIGURE 13.1   *(Continued)*

Capabilities: Describe your capabilities (continuous air particle monitoring systems, bulk chemical delivery systems, wall, ceiling, floor systems, local electrical distribution systems, life safety systems, laterals, etc.) and list any exclusion(s) to standard capabilities.

Fabrication: List the locations of your fabrication facilities (if applicable).

6. Labor and Labor Relations

If applicable, list the union labor organizations with which you have signatory agreements.

Trade agreement with	Expires	Trade agreement with	Expires

Current Number of Employees:

Type	Total	Union (yes/no)
Field Employees		
Shop Employees		
Office Employees		N/A

7. Bidding/Contracting

Preferred Type of Work:

List trades normally performed by your own forces or items normally furnished by your firm. Please indicate any contracting strategies that you are not interested in bidding.

8. Safety Data and Record

List your Worker's Compensation Interstate Experience Modification Rate for the 3 most recent years.

Year	Corporation/Parent	Subsidiary/Division

Corporation/Parent Safety Info

Year	# Recordable incidents	# Lost workday cases	# Restricted workday cases	# Labor hours worked

FIGURE 13.1   *(Continued)*

Subsidiary/Division Safety Info

Year	# Recordable incidents	# Lost workday cases	# Restricted workday cases	# Labor hours worked

Does your company have a written safety program? [ ] Yes [ ] No If yes, please attach.
Does your company have a safety-training program? [ ] Yes [ ] No If yes, please attach.
Does your company have a written drug policy? [ ] Yes [ ] No If yes, please attach.
Please list AND provide a brief summary of any Federal and/or State OSHA citations your company has received in the last 3 years.

Company Safety Manager:

Name:	Length of service with company:
Address:	
City, State, ZIP:	
Telephone Number:	Fax Number:
Pager Number:	Email address:

9. Additional Information

Please list any additional information that you feel will help us determine the qualifications and expertise of your firm as they would apply to this project(s).

_____

_____

_____

_____

This prequalification questionnaire was completed by:

Date:	
Name:	
Signature:	
Title:	
Address:	
City, State, ZIP:	
Telephone Number:	Fax Number:
Email address:	

FIGURE 13.1 *(Continued)*

13.7.1 Selecting the Bidders

The main consideration is usually cost, but include all the elements, as defined earlier, as it will provide for a greater success opportunity.

13.8 TRANSITION

The transition of the outsourcing contractor is one of the most important elements of a successful program, if you have decided what to outsource. The transition plan listed here uses a phased approach that will allow for success.

13.8.1 Phase 1

Develop a detailed schedule with ownership for each element and deliverables dates. Some of the major elements of a good plan are listed below.

Communications. Conduct the initial meeting with the new contractor, present the draft of the transition plan and introduce the transition team, develop a communications strategy for the on-site personnel, establish a meeting schedule for progress reports, announce the new partnership to your company, develop a plan to notify vendors/subcontractors of changes.

Legal. Draft the contract, negotiate the contract, get the contract signed, review any insurance requirements, and then obtain the required insurance coverage.

Human Resources. Create a retention plan for incoming employees, identify key management positions, create a contingency plan, complete all job descriptions, conduct employee "get acquainted" sessions, and conduct new employee orientations.

Administrative. Establish an address for your contractor, order signage, establish basic office forms/files, provide for any special licensing requirement (if needed), develop key customer contact list, create personnel files, establish training records, establish uniform standards and place order, obtain corporate manuals.

Financial Plan. Meet with the financial team and develop contract requirements, acquire the operations budget for review, acquire capital budget for review, acquire chart of accounts, implement plan for meeting reporting requirements, develop approval matrix, process transition invoices, document financial control process, provide on-site financial support, analyze site financial support requirements and make recommendations for staffing, provide sample invoice format, develop site specific and deliver training for time cards, purchase orders, and approval authority.

Safety Plan. Have safety manager on-site for start-up, meet with safety representative, develop safety metrics, supply project workers compensation strategy, provide unemployment and workers compensation contract, train project personnel on safety and worker compensation, designate and train site safety coordinator, post all safety required information, conduct site audit, complete project initial safety overview.

13.8.2 Phase 2

Regulatory Requirements. Reduce operating costs, improve regulatory compliance, and review justification of SOP (Standard Operating Procedure) coordinator.

Internal Customer Satisfaction.   Develop employee survey, survey employees, develop action plan on basis of employee survey, design "service call" leave-behind cards, implement customer call back satisfaction program, and reduce utility costs.

Work Management.   Develop operational plan for approval, review current organization and recommend new organization staffing, develop PM teams, develop cross-functional teams, merge all PM and call-in-maintenance into one system.

CMMS Development/Turnover.   Review work flow and major processes, establish priority and response time, develop call escalation procedure, provide emergency telephone numbers, verify equipment and systems listing, group equipment into critical/noncritical categories, assess equipment condition, establish deficiency reporting, identify critical systems and equipment. Perform skills analysis by functional area, i.e., carpentry, electronics, electrical, HVAC, etc. Perform equipment data analysis.

Benchmarking.   Conduct customer/site presentation, supply data collection requirements, and establish baseline performance metrics.

Subcontractor Management.   Develop subcontract or self-performance decision tree, review subcontract contracts for cost opportunities, obtain/create approved supplier list, design emergency response support plan, develop and institute subcontract partnership program, negotiate/renegotiate contracts.

Materials Management.   Define materials, establish management process, order materials, utilize management software, load data, communicate and train personnel.

Metrics/Measurements Development.   Preventive maintenance procedures scheduled versus complete, corrective activities on equipment, and measure cost of maintaining equipment.

Note: This is not a comprehensive list of measurements.

13.9   CONCLUSION

This chapter has defined some of the decisions you need to make when considering outsourcing maintenance along with some tools to incorporate if you embark on the outsourcing trail.

Once the outsourcing contractor selection has been made, the next step is *developing an execution plan*. This is very important to the successful start-up and relationship. Make sure the plan and schedule is very detailed as this transition is usually more difficult than anticipated.

Define each action, assign an owner, and define timelines for the deliverables. Review this schedule on a frequent basis with the transition team. During this phase, it is also very important to keep accurate lessons-learned documentation and any scope changes that may occur.

CHAPTER 14
COMPUTERIZED MAINTENANCE MANAGEMENT

Howard J. McKew, P.E., CPE
RDK Engineers, Andover, ME

George Player, CPE
RDK Engineers, Andover, ME

14.1 INTRODUCTION

Preventive maintenance, planned maintenance, asset management, or any term used to define the process of maintaining equipment and facility systems has become a high tech business endeavor. These systems have evolved and are now utilized for maintenance, project tracking, work order repairs, labor, and material tracking. In addition, computerized maintenance management software (CMMS) systems can be linked to the accounting department software and electronically bill a department and recover the costs for services provided. Web-based CMMS systems are now a reality and not a product for future consideration. Throw in the final arrival of bar-code technology via handheld computer technology and collectively, these systems, their features, and options are out of the prototype phase and into the day-to-day application phase.

Analogous to facility managers who attend project meetings and ask for a design intent document to clearly understand how a proposed building program space will be utilized, what systems will be involved, and what parameters we need to operate under, the same process should be employed when a new CMMS system is to be purchased. With so much potential and so many features, options, and benefits, strategic planning is the cornerstone to implementation success. These business management tools have finally arrived and proper implementation is essential to the success of CMMS application.

Preventive maintenance (PM) is planned maintenance while *predictive maintenance (PDM)* is real-time maintenance. For the educated building owner, these two proactive initiatives contribute significantly to a well-balanced facility management/operating budget. It is strategic investing today for tomorrow. an intangible insurance policy that promises equipment reliability, extended equipment life, and efficient use of energy.

Managing a facility should be considered as maintaining an asset versus putting capital into a cost center. Much has been written about the subject of facility management and the quest to do a better job at maintaining this asset has come under intense corporate scrutiny in recent years. An example of proactively maintaining a building's assets has been through the introduction of computer technology as the new cornerstone of the facility management process. Applying today's CMMS technology can transform this cornerstone into a foundation of facility management success. Complementing this technology with bar-code tagging simply enhances this business tool.

For a health care facility, the patient care environment and system reliability are two strategic issues that should be an integral part of the facility management mission. For an educational institution, customer comfort and system reliability become part of the facility strategic plan. Industrial facilities count on equipment reliability and are the most likely to go beyond PM to PDM to contribute to product delivery. Whether it is health care, academia, and/or industry, all facility management entities share a third common agenda and that is, to maintain cost-effective operations. The CMMS system can be the tool to organizing this process and achieving these strategic requirements.

Those who invest in planned maintenance are people who recognize and appreciate the need to routinely care for the building systems. You don't get something for nothing! Analogous to the upkeep of your automobile, PM/PDM are something that you would rather not do but know is essential if your car is going to provide you with a reliable means of travel. If the car does not start; if it runs but doesn't get really good gas mileage; or the engine overheats, then investing in a PM tune-up is something the owners can relate to. They know they will receive an immediate return on their investment.

Routinely changing the oil, oil filter, and air filter have less obvious returns on their investment. The automobile owners do not experience the same immediate feeling of satisfaction from this investment as they do with the tune-up of the engine. They quite often miss that "money well spent" experience.

It is this empty feeling that raises the question "When do you change the oil in your refrigerant system?" The car manufacturer may recommend the oil be changed every 3000 mi and the refrigeration equipment manufacturer may recommend changing oil every year. The end user is not provided statistical data to support these directives so most people will not strictly adhere to the manufacturer's recommendations. Instead they will select PM "tasking" based on their experience and/or what they can afford within their operating budget. *Planned maintenance is not an exact science but rather an educational experience.*

Once building owners recognize that they must invest in a PM program, the next issue will be to recognize the scope of work, the cost to start, and the cost to continually manage this program. Today's technology is inherently driving the planned maintenance process via the use of CMMS systems. Manual PM systems are antiquated and CMMS systems are the processes for today that comes in good, better, and best software. *What may be good for a small simplistic application may not be good for a large multibuilding complex.* A more sophisticated CMMS system may be the best solution for the large campus settings and yet, as a first step the process of choice frequently decrees "learn to walk before you can run."

14.2 PLANNING THE NEW CMMS SYSTEM

Returning to the analogy of the automobile, a person planning a long trip would certainly refer to a roadmap as an integral tool for the trip. Similarly, a CMMS plan will become an integral tool to reaching the facility management goal and doing so on time and in budget. Under the umbrella of "commissioning of facility management," the roadmap for CMMS system success begins with the writing of the design intent document (DID) for this new system (Table 14.1).

So how do you start a DID-CMMS? Data collection and not just data collection of equipment. Data collection begins with determining the day-to-day and annual *deliverables*.

Step 1: Identify what the deliverables will be at the end of the CMMS system implementation program. To do so, facility management should ask their boss (i.e., Vice President of Support Services) what is needed from facility operations to meet the corporate operating goals. Goals such as minimum operating cost when benchmarked to the competition, client satisfaction, and safe, comfortable space environment are three very reasonable goals that CMMS system performance can contribute to goal success.

To begin the Step 1 process, meet with the appropriate staff to determine what data need to be collected. Equipment and resources will be some of the data required to implement and maintain the CMMS system. Benchmark sources will be needed, as well as lessons learned from past manual and/or computerized systems. Capital project policy and procedures, scheduled and unscheduled work order historical data, and current response process will need to be collected. Computer software, hardware, and computer-literate personnel resources will also need to be determined. And finally, available funding and potential for additional financial support will need to be identified.

Step 2: Ask questions with the first question being "will the CMMS system be utilized for capital projects and the project managers only?" In most cases, and in an ideal world, a CMMS system is best utilized to manage all the work for the facility support department. This would include planned/preventive maintenance work, unscheduled work, and emergency work. There are, however, some considerations. First, under the umbrella

TABLE 14.1 Design Intent Documents & CMMS Needs Matrix

Customer issues	Priority	Comments
CMMS has several options	PM work orders	Consider other department(s)
Compatible with . . .	Microsoft/Apple	Consider AutoCAD interface
"Windows" driven	Yes	Internet interface preferred
Purchase or lease	Lease	Budget for $
On-site training	2 days	Unlimited telephone support
Automatic upgraded	Preferred	Maintenance agreement needed
On-line support	Required	Third party support option
Customized reports	Required	"Microsoft" preferred
Graphics software	Required	"PowerPoint" preferred
CMMS administrator	Required	$___, salary
Handheld computer	Required	Budget 2 @ $ ___ each
Bar code adaptable	Required	Budget for $ ___
Automatic scheduling	Required	Automatic load-leveling
Data retrieval survey hours	$ ___ budget	Based on ___ hours, third party
Database software	$ ___ budget	
Data input hours	$ ___ budget	Based on ___ hours
Scheduling hours	$ ___ budget	Based on ___ hours
Customized reports	$ ___ budget	Based on ___ hours
Capital Project Module	Preferred	Budget for $ ___
Regulatory Compliance Module	Preferred	Budget for $ ___
Asset Management Module	Preferred	Budget for $ ___

A Powerful Tool: The customer needs matrix is a tool that can be modified by the user to outline costs and hours associated with the implementation of a CMMS. It should also be used to establish the timeline and milestones for implementation.

of planned work, should project work be included? Planned work is not always preventative maintenance. It may be an overhaul of equipment, or could involve requested work on items not even in an equipment database. It may also be a capital project such as a build-out of existing space. It is important to determine whether the existing or proposed new CMMS system will be able to handle these types of planned work request. It is also imperative to ensure that this system can not only assign requests to the project but can also document and manage all the tasks associated with the project. It will be necessary to not forget to determine other applications in the institution/facility already set up to handle these types of requests, such as project management software. Based on data collection, determine which application would be best suited for managing and balancing the workload. If it is the CMMS system, continue data collection.

Unscheduled work orders differ from emergency work orders in the fact that the request is not life threatening and/or need not to be done immediately. Still, the unscheduled work order does require action at some planned moment in time. Here again, capital projects can fall into this category and the same analysis needs to be followed to determine whether the CMMS system is the optimum business tool for managing these work requests.

In the case of emergency requests, besides the fact that the request is *urgent*, many of the activities are still the same. Resources, material and tool inventory, available funds, who will be billed for the work, and project timeline are all integral to the work order.

Examples of how CMMS is used are in the manufacturing environment, where most requests would apply to equipment and, in the case of maintenance, it would make sense to use the CMMS program to manage the work and maintain an accurate history of work performed on this equipment. In an institutional environment, this may not always be the case. This is particularly so when the CMMS does not have a good equipment database. Here much of the work may be on equipment or items not maintained in this database. In this case, it will have to be determined how much information is too much. On the other hand, if a facility manager tracks absolutely everything in one system, it may end up that the CMMS system accumulates so much data that he/she begins to miss the wood for the trees. The solution is to try and determine up front a clear line on what data will be maintained by the CMMS and what other applications will be either replaced or supplemented by the CMMS.

Step 3:   In determining the fields of application it is not unusual to run multiple applications within the CMMS system. To do so, it is important to evaluate common fields in the applications. The most common examples would be labor names (both shop and employee names), location codes and descriptions, and equipment designations. In matching up these fields, or even better, extracting the information in existing programs as source data for the CMMS system, a facility manager will not only save time, but ensure that the programs will be able to communicate through joint application reports.

Step 4:   Integrating other CMMS modules of the system will require the user to have the choice of subprograms/modules that will allow facility managers to better manage the facility. The individual responsible for setting up the CMMS system will assign the task or job to a shop or person. The shop or person will request parts or services from purchasing. The shop or person will also take stock out of inventory. In determining the hierarchy of the facility organization, one must consider establishing relative shops, crews (made up of workers from multiple shops), supervisors, shifts, and work location. Through the CMMS system, these areas can be categorized relative to resources in multiple ways to ensure that the facility and employees in the methods can be managed.

Step 5:   Joining together the CMMS with another facility or computerized accounting system is not an unusual request. Most large institutions are departmentalized. Facilities management may not always possess the authority to administer purchasing and receiving. In these cases, the facility manager should follow the same practices and methods regarding purchasing and receiving. First, determine if you have a choice in whether you use the CMMS application or the existing applications. In many cases, departments are reluctant to give up control over areas they consider their jurisdiction. In these cases, measure carefully whether this is a battle worth fighting.

When taking into account the benefits of sharing data between software systems, a strong reason for following this approach may be in-house budget constraints. The three most common modules for merging software data are labor (noted above), inventory, and purchasing. Based on this commonality, quite often, the decision to limit the CMMS is made because other shared data will be needed due to internal billing procedures. In most cases, labor is a required module that is integrated in all aspects of a CMMS and is also billable to other departments for services rendered.

On the other hand, this CMMS system and its capabilities may be taking on more work than resources are prepared to handle. Sit down and evaluate honestly both the existing applications, the resources/ability to manage the new module(s), and the current working relationship with other involved departments to determine the best course of action.

Whatever the outcome, an exchange of common field data is important. An example of this common field data is when setting up the facility group inventory module. In this case, the purchasing department software may be able to provide upfront historical data of past department purchases. In sharing existing purchasing information through the purchasing department instead of the CMMS, this other group may be able to supply more accurate information or descriptions of materials and services required by the facility management department. At the same time, the facility manager may be able to supply the more appropriate labor file or authorized requestors so that in the end, integrated reports bridge both systems while meeting the needs of the CMMS application.

Another consideration for sharing data between software may be redundant running of applications. In many instances, today's CMMS modules are designed to work with industry standard applications. These perform at various levels of integrations. In some instances, it is just mirrored data that require a program being run to sync data between applications. In more advanced software, the programs are actually linked and perform seamlessly. It is important to note that this blending of data requires care when setting up data integrity (and edit validation) as well as security privileges.

14.3 SYSTEM REQUIREMENTS

In purchasing a CMMS system and routinely referring to the DID matrix, it is important to know what hardware requirements are needed to operate the new system. Will it be standalone or be tied into a network at the facility? Will the system consist of workstations, servers, or both? Who will have access to enter and edit data? Will facility departments have access to enter work requests? Will the maintenance staff utilize hand held devices to document completed preventive maintenance, unscheduled and emergency work requests?

A comprehensive review of the maintenance agreement from the CMMS vendor should include:

- *Upgrade path:* Does having a maintenance agreement allow a discount on future upgrades to the application?

- *Patches and fixes:* These should be free from the vendor for a predetermined time after purchase regardless of whether there is a maintenance agreement. For maintenance agreements, patches and fixes should be included for the life of the agreement.

- *Are there third party developers/supporters of the system* that may provide less expensive support agreements?

What level of support does the maintenance agreement provide? Does it offer support only for their software, or will *they* help resolve conflicts between their application and hardware or other software?

The involvement of a representative of the information system (IS) management department is essential. It is imperative to always get a full hardware/software specification from the CMMS vendor prior to signing any agreement. Workstation operating systems specifications as well as network requirements should be included. The question needs to be asked up front if there are any known conflicts with other common applications. Also, it is imperative to be aware of topology issues such as line requirements (i.e., will the application bog down because the existing network runs only on a 4meg token ring?), network protocols, and security features. Assemble this information and meet with the in-house IS person and/or the institution's IS consultant to review these issues. At the same time, procure a similar list from the IS group detailing what their requirements are for applications sitting on their network. Have them list all common software loaded onto local workstations and network servers to determine if there is any possible point of conflict. Include in this list routers, firewalls, and antivirus software. Submit this list to the CMMS vendor to determine if there are any known conflicts.

In requesting information from the CMMS vendor there is a frequently documented list of commonly asked questions, concerns, and answers from the vendor pertaining to the CMMS software. Often, this list can raise a red flag to IS personnel about possible concerns. For some software providers, web site chat sessions are available where the web site bulletin board provides discussion of the software. This is a way to obtain and read unsolicited opinions of the program, and again, anticipate any possible problems. When selecting a new software program, it is also a good idea to "speak with your peers" concerning how well their system operates. What are the strengths and weaknesses of the software and support from the vendor? Do the reporting features meet their requirements? Can reports be customized or are they standard? What type of maintenance agreement, if any, is required and at what cost? Also, a facility manager should ask his/her colleagues for a laundry list of things they consider essential to a program to perform their duties, and a separate wish list of options they would like in a system.

Determining support assignments for the CMMS system can often multilayered. For the facility group, an office person may provide level 1 support for the application only. This person is the one to go to if there is a need to ask how to enter a work request or print a report. If there is a more significant error, this person will hopefully also be able to determine who should be alerted. This is level 2 support and will apply to both hardware and software issues. In many institutions, the IS department will handle these problems and so it is imperative that IS be an integral partner in the CMMS investment to ensure that this is the case. The downside of this need to partner is that the IS group may be reluctant to support another department specific program. Even if they do support the facility group at this level, it is unlikely they will support CMMS software problems. This level 3 supports what may be a contractual agreement with either the CMMS supplier or a third-party contractor.

In establishing as much expertise among the facility management staff, training level 1 support administration can be very cost effective. If this person or persons can be trained to level 2 support (to some extent) and basic level 3, then it will potentially save money in both service charges and down time of the CMMS application. It is important to keep in mind that in many cases CMMS programs are just a loss leader for a company. They may just break even on the purchase price of the product, but they make their profits through support agreements and customization of programs or reports. The more the facility manager can control the cost—first cost, operating cost, and maintenance agreement cost—the better the facility manager will be over the life of the software.

Important to note is that work with the IS department, and all other departments with which facility management has a high degree of interaction can help determine what level of interaction these other groups may have with the CMMS program. In some instances, facility managers may grant some direct access to the software (i.e., entering work requests). In these cases users will need to consider both software licensing and security, as well as network security. The facility manager should also plan on working with the CMMS supplier and IS to remain abreast and take advantage of any new technology currently (or soon to be) available. If this agreement is set up for this routine awareness in advance, it may save additional upgrade costs in the future.

To effectively plan CMMS system requirements it will be important to determine the capabilities of the department's software report writer. In most applications, the CMMS will come with some basic canned reports. It should also come with at least a basic report writer to either modify these canned reports or create simple custom reports. The buyer needs to review the offered reports and see if they meet needs. If they fulfill basic requirements, a facility manager may still want to have the CMMS provider modify them (i.e., to reflect the institution identity including company name in the header). It is also important to assess whether this report writer will be capable of supporting future custom report requirements.

In assessing the CMMS potential, ease of use and output capabilities should be thoroughly reviewed. The reports should not just be outputted to a printer or screen, they should be able to export to common file formats such as MSWord, Excel, or in today's Internet environment, HTML. If the report writer does not support these features, or seems awkward to use, inquiry should be made as to potential for upgrades that will improve these situations. While inquiring, the question needs to be asked about charges for manufacturing custom reports for the end user's needs. In some instances, the facility manager may know the custom reports that will be needed up front and can possibly negotiate that these reports be part of the purchase.

Another option to customizing reports and/or providing the report writer support can be the separate purchase of a third-party report writer. If this is the direction to be taken, the CMMS provider will need to confirm that they can/will assist during installation to set up table and field mapping. An in-house person should also be trained in the use of program(s) that are settled upon.

14.4   APPLYING CMMS

In addition to the traditional applications of CMMS systems, if the facility operations group does in-house projects and bills other departments for services provided, the finance department will most likely assemble tables of cost centers, project numbers, or other billing information to ensure that the costs are recovered for all labor, material, and equipment due to the department. To effectively implement this business management tool, beyond the routine preventive maintenance module, the following questions should be asked:

- What output is required from the CMMS system?
- Can the CMMS system produce accurate labor report to be utilized during budget preparation?
- Can repeat work orders be flagged to indicate a potential system problem?
- Can these reports be utilized when dealing with an oversight authority such as the Joint Commission on Accreditation of Healthcare Organizations in hospital?

An important aspect of CMMS systems for administrators is not the ability to issue a job request to a tradesman to perform a task. A dispatcher with a pencil can perform that chore. Administrators have to demonstrate to their directors, vice presidents, various state and federal agencies, and in some instances the legal community that there is control within the facility management operation. They must show that they have maintained the facility in the past, are aware of the current state of the institution, and understand what will be required of them in the future. The only way a facility manager can demonstrate this is through the ready availability of data. To do so and to implement a CMMS system, it must be determined what information must be available to the department in the future, while setting up and managing this system today.

Implementation also includes reviewing all labor and budget reports currently required by the department and determining which suits the needs currently and what needs to be tightened up as a means of demonstrating the department's performance or requirements.

For many institutions, implementation can also mean managing the paper trail of regulatory requirements of the department. It has been determined and proven that validation of fulfillment of these requirements can be performed through CMMS system reporting.

Another implementation process of CMMS is the trending, monitoring, and analysis that can be administered through CMMS. This opportunity can perform this task in two ways:

1. In some applications, a common task performed/requested on a common location or piece of equipment will activate a system alarm in the CMMS application. In these cases, it is important to set up three aspects of the data entry. The type of calls that you wish to monitor (a problem list), a list of locations or equipment, and a strict training procedure for data entry clerks to ensure that the flags set up are not circumvented by inaccurate data entry.

2. In most CMMS systems, the automatic flag is not available at the time of data entry. Instead, trending and monitoring is performed through output reports. The setup process is the same as far as problem codes, location/equipment specification for tracking, and data entry training, but reporting is often more flexible. It is often detrimental to an application's performance to perform complicated validation checks every time a record is saved, and that in essence is what's being done when the system has automatic flagging during the data entry process. The more complicated the flagging requirements are, the more it will slow down the data entry. Reports, however, can be run at one's convenience with a higher degree of detail. It can present the facility manager with a report in an understandable format informing of a violation of preset conditions. Since the report is being run as its own process, and is set up in advance, in most cases it will have no effect on the systems, and the conditions can be set as convoluted or as simple as the administrator determines necessary.

14.5 MANAGING THE IMPLEMENTATION PROCESS

Almost all facilities utilize some type of preventive maintenance system, whether a current computerized system, manual/paper file system, outsourced service work, and/or a combination of these processes. The process to transfer the current data into a new CMMS needs to be carefully evaluated.

Will this process be completed in-house? By utilizing in-house staff the implementation of the new CMMS can be less expensive but staff will be taken away from their daily responsibilities. Will this process be shared by a number of staff or will one individual manage it? Will the person responsible for the conversion of the CMMS later be responsible to maintain and update the database? The alternative to this approach is to out-task the collection of equipment data. It has been proven that an outside firm can facilitate this process while using engineering students as part of the data collection team.

The benefit of this approach is that the skilled in-house facility management labor is better used while unskilled college students with technical training grasp the opportunity to contribute to the process while learning more about the business they will be entering. Get from the CMMS provider an estimate to perform the service, along with a work plan and a timeline.

In implementation, time can be the most important aspect. Probably most purchasers of CMMS systems don't totally grasp the magnitude of the data collection aspect of the work. The purchase of the software is only the tip of the iceberg; the real cost to implement a CMMS system is surveying the facility to identify and document all the equipment and/or assets. This level of effort can take weeks and even months depending on the size of the facility.

In researching the optimum solution to equipment data collection, this migration of data process can be a daunting task. On the other hand, the cost of contracting with the CMMS provider will often be much higher than anticipated. The facility manager must determine what resources are available within the department, within the institution, and outside the facility apart from the CMMS provider. Inquiring with other similar companies using the same CMMS to evaluate their procedures during the startup phase can shed some light on the issues, concerns, and cost—What would they have done differently, and what worked well for them?

In many instances, the CMMS provider will develop a cost, timeline, and other "deliverables," but the facility manager may have to pay for this service. In these cases, it may be possible to have a portion of the evaluation charge credited to the cost of the implementation should the facility manager decide to use the provider. Another option would be to see if there is a third-party developer of options for the CMMS application. This is not uncommon for high-end industry applications and by contacting these firms, one can find that migration and support are often also part of their business.

If the software provider is contracted to implement the new system, this cost can be included in the purchase of the new CMMS. For some, there can be a concern that the software provider may not share the same sense of ownership to complete the task as the in-house staff. For many CMMS software firms, they are not in the business of going out and collecting the equipment data. They are software providers and so the option to contract with an independent contractor with dedicated resources to implement the new system may be the best overall option. No matter which option is selected for the migration of data, a central contact will need to be appointed within the department to monitor the progress and relay the requests for information. Meetings should be held regularly to keep the process on schedule.

In addition to going out and collecting data, consideration must be given to any existing CMMS or manual database. The challenges associated with this transfer of information can also be a daunting task. Quite often these existing systems may not be as complete as one would expect; so check and balance of equipment database and survey equipment data must be scrutinized. Consideration has to be given to how the data gathering process will take place. Will future capital projects with general contractors, or a construction management firm delivering the information electronically, be able to transfer their data into the CMMS? Will staff be assigned to gather the technical information via a data retrieval form (Fig 14.1) in order to include the new equipment into the system? Will a third-party survey team, gathering the technical information via a handheld computerized version of this data-retrieval form, be used to include the new equipment into the system?

Using today's technology, and the comfort level people now have with the PDA type of handheld computer, data collection has taken a new twist. Whether the in-house staff performs the survey work or a third party does the legwork, handheld computer technology can be the most efficient means to build the database.

MECHANICAL – ELECTRICAL DATA RETRIEVAL				
FACILITY:		COST CENTER:		Bar Code:
EQUIPMENT DESCRIPTION:				I.D. #:
SERVICE:				PRIORITY:
LOCATION	BUILDING / FLOOR:			ROOM:
	DESCRIPTION:			
EQUIPMENT	MAKE:			
	MODEL NUMBER:			
	SERIAL NUMBER:			
	CAPACITY:			
DATE INSTALLED: / /		VENDOR / MFG:		WARRANTY:
COMMENTS:				

COMPONENTS (Fuses, Filters, Belts, Etc.)			
	INV. CODE #	DESCRIPTION	QUANTITY
1.			
2.			
3.			
4.			
Special Instructions: MSDS Information:			

FIGURE 14.1   Data retrieval forms (front).

Integral with this technology comes the time tested capability to apply bar coding to the equipment data process (Fig 14.2). This management and inventory tool has been around for decades and only now does it seem that bar coding is ready to be embraced by facility management. Simultaneous with this application is a need to recognize the limits, restrictions, and cautions that go with bar coding. The process goes beyond simply sticking a label to a piece of equipment. First, can the CMMS system being considered by the facility manager accept bar coding? Can the handheld bar coder recognize the specific code selected? What about the material being used? Will it discolor from the heat, sunlight, water, and/or the surrounding environment?

In looking beyond the survey and bar coding phase, who will have access to a handheld computer to efficiently use the technology? This is a study within itself to determine the effective means by which bar coding will be used. At the same time, for the facility management industry, this application is long overdue and only now with the handheld computers of today with their attachment option for a bar coder has CMMS reached a plateau

Motor			
Vendor / Mfg.			
Make:		Model:	
Serial #:		Frame:	
HP	RPM	Type	
Cycle	Phase	Design	Serv. Fac.
Voltage	Amperage	Code	NEMA Desg.
Dr. End Bear.		Opp End Bear.	
Ambient Temp:		Rating:	
Pressure Vessel Specification			
NB#	Certified By:		MAWPs
Year Built	CRA#		Receivers
Head	Shell		
Pressure Specifications			
PSI	MAX WP	FT To Head	
Max Work Press Shell		Max Working Pressure Tubes	
Compressor Air Max WP			
HVAC			
Refrigerant		LBS	OZS
PUMP			
GPM	RPM	Pump Head	
Max WP			
Miscellaneous			
Belt	Motor Sheave	Fan Sheave	Controller
Relief Valve	Regulator		

Figure (2) Back

FIGURE 14.1 (*Continued*). Data retrieval form (back).

in managing data that will offer several other opportunities to provide premium facility management.

14.6 CONCLUSION

There have been numerous books written about computerized maintenance management software systems and even more individual articles. Much can be learned from those who have already been up and running with CMMS systems. At the same time, much can be learned from others in technology and businesses associated with facility management. IS departments, senior management, third-party facility assessment companies and third-party CMMS support firms can provide a wealth of information about investing in a CMMS application. Because computer technology is so intertwined, and inherently more powerful than a single computer user, application begins with one step at a time. Learning from a small investment and limited application is a good way to start. Investing in the experience from a third party is also a viable way to begin the process as well as maintain ongoing support for many a CMMS system owner.

(a)

(b)

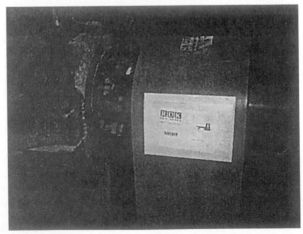

(c)

FIGURE 14.2   Bar coding applications.

Facility management has changed dramatically over the past 20 years, as computer technology has gone from cutting-edge to everyday application. Prior to the 1980s and prior to the energy embargo of 1973–74 operation and maintenance of mechanical and electrical systems were simpler. Back then, selection of equipment was based on the belief that bigger was better along with ample redundancy. A facilities engineer would have 50 percent (if not 100 percent), boiler and chiller backup capacity sitting there ready to come online. When it was time to pull maintenance on one of these big units, there was another unit ready to go.

The operating strategy back then was also quite simple: on-off control. Energy and labor were relatively inexpensive, water was perceived as clean and plentiful, and environmental guidelines were not very stringent. The *Facility Engineer* title was probably *Chief Engineer* or *Plant Engineer*. Certainly, the individual was not considered as "Manager." If anything, the entire operation and maintenance staff was considered a necessary evil, an overhead cost of doing business.

Maintenance procedures, like system operations, were also relatively simple then. Preventive maintenance consisted of a file cabinet full of neatly typed, alphabetized equipment record cards. Each card noted what maintenance needed to be done and when this work should be done. No practice measures, predictive maintenance, or run-time action agenda. Simple and to the point, like the systems themselves.

Computers and more specifically, computer software has changed all that. At the same time, the cost of energy and the cost of doing business have had a dramatic impact on facility management. Electronics slowly replaced pneumatic automatic controls, and now there are direct digital controls and computers that you can hold in your hand. In addition, digital camera application, bar coding, and laser technology software offers the facility manager numerous opportunities to automate planned maintenance and move from the dark ages to the twenty-first century.

Although it can be a panacea for building management and the "cure-all" to operation and maintenance, the CMMS system is only the tip of the iceberg when it comes to facility management. This exciting tool has dramatically changed the way everyone does business but is only the cornerstone to technology. For the facility manager, however, it has become a double-edged sword. It has placed maintenance on the "cutting edge" of its industry and the "bleeding edge" of performance. Because the computer could inherently offer so much from within such a small box, facility managers were expected to use its software options as soon as possible. Maintenance was now going high tech, and the people doing the work were expected to act accordingly.

It has been documented often than planned maintenance traditionally takes a back seat to "fire fighting" unscheduled work orders. Purchasing a CMMS system will not put out these fires unless there is a mission, a plan, and a commitment to methodically change how maintenance is done. We have seen how operation and maintenance have changed significantly over the past 20 years and we know how that has adversely affected the skill levels of these crews. Planned maintenance can't be achieved unless facility management staff "make the time" and commit themselves to a quality process that will keep them current with computer-age technology. In addition, they need to be patient and plan their venture into CMMS, one step at a time.

When discussing operation and maintenance performance with vice presidents of support services who have succeeded today, they will point out three priorities: first, train the staff to be the best they can be; second, reverse the trend of unscheduled work over planned maintenance; and third, reduce operating cost. Nowhere do they say CMMS is essential to successful facility management, but it can be used as a tool to improved performance.

Getting started with CMMS parallels these same three priorities. Training begins with a reintroduction to one's job description. Responsibilities have changed over the last 20 years, and personnel must keep abreast of computer technology. Those who don't embrace this mandate will find someone else doing their job. With more qualified people capable of doing a wider range of services proficiently, a proactive approach to maintenance can be the next step. Enter the CMMS system and a focus on planned maintenance only! A commitment to applying the preventive maintenance module of a CMMS software and follow-through on completing the work orders will be a positive first step to accomplishing the next priority, reversing the trend of unscheduled works.

It has been found that implementing a strong, planned maintenance program will initially increase the number of unscheduled work orders. This is because the workforce, now performing proactive maintenance, will find other tasks that need to be done. In time, the unscheduled work decreases, because systems are being maintained and equipment isn't breaking down. Reliable system operation equates to reduce operating cost, energy costs, and no product disruptions, and third priority.

With that much work ahead, how can facility managers consider applying all those other CMMS computer software modules too? Keeping up with computer-age technology and getting started with CMMS requires careful planning and a commitment to take one step at a time. After all, it took us 20 years to get to this version of facility management and now management tools are literally in our hands.

CHAPTER 15
THE HANDHELD COMPUTER

Howard J. McKew, P.E., CPE
RDK Engineers

15.1 INTRODUCTION

Handheld computers with software are here; they are proven and ready to be counted on to assist in building management. Unfortunately, the building industry, made up of the design community, construction community, facility management, and property management, has been slow to embrace this technology. Also long overdue is the day-to-day use of the bar coder attachment and portable document format (pdf) files, as well as the digital camera attachment for these handheld computers. This chapter is dedicated to raising awareness of these business tools and how they can overcome issues and concerns and provide a better way to succeed on a regular basis in the building industry.

During the past 25 years computer application has infiltrated the building industry but progress can be considered slow relative to certain areas of this technology. This may be attributed to:

- A generation of professionals in responsible positions who did not grow up with computers

- A computer industry that was continually changing faster than the building industry could keep pace

- Computer technology that was not user friendly and/or compatible with other computer software

- The first cost of purchasing computer hardware, software, and ongoing annual computer costs

- The cost of becoming computer-literate and keeping up with computer technology

Integral to the success of handheld computers will be to adjust and/or overcome these obstacles. One key to this feat will be access to simple but effective software programs so that the learning curve is kept to a minimum. At the same time, the ability of this new software to interact with other programs will be critical. The days of sole-source software are gone as users expect software to interact with other software. In addition, and to the credit of many, knowledge and familiarity with hardware and software is growing; this will help embrace these mini-computers in the near future. As in the past, users expect computer technology to continue to change at a quick pace with current know-how being replaced with new know-how.

Simultaneous with this continued need to computerize, contract documents are in dire need to be changed to accommodate these twenty-first century applications. Building programs require their projects incur costs that historically result in insufficient equipment, panel, valve, and device labeling; incomplete computer-aided drawings (CAD); and inadequate operation and maintenance (O&M) data within O&M manuals that are delivered late. Today's technology can hold the key to these issues but the building industry must change its way of delivering these tools to the building owner and do so with handheld computer application.

Complete, useful facility documentation delivered in a timely and cost-effective manner is essential. Just like the building industry cannot afford to overlook the benefits of handheld computers, building owners cannot afford to begin the operation, maintenance, and facility management of a new or renovated building without having the tools in place on Day 1 to succeed. Handheld computers and construction document compliance go hand in hand. An analogy of this is how construction "first costs" today can begin at $100/ft^2 and dramatically jump up from there to $200, $300, and $400/ft^2. The burden to efficiently manage the annual utilities costs and labor costs is simply heightened when construction is completed and occupancy occurs. A rule-of-thumb for operating cost of owning a building over the useful service life of this facility breaks out to "20% in first cost + 80% in O&M cost." With antiquated contract requirements and designers and builders who are not knowledgeable of their facility management counterpart, the operating and maintenance costs can exceed the industry benchmarks for efficient and effective building management. While focused on the first-cost tip of the iceberg, the design and construction community has failed to look below the waterline to see what their counterpart, the building support services, needs from construction documents. The initial $100/ft^2 first cost, as a rule, doesn't address or include any operating cost into the construction budget.

Construction specifications typically draw upon past requirements with little visionary perspective toward a more modern approach to contract document deliverables. Tuning up the specification writing process will require communication between the project designers, mentored by the facility manager. To date, these two professions have had limited collaboration and, as a result, have mutually undermined the progress of operation and maintenance. It is not appropriate to just provide documents on a computer disk any more. More intelligent data are needed as well as a more efficient way of bringing this information to the technician or worker outside the main office. Handheld computers and simple, cost-effective ways of accessing these data are available today. What is needed now is a paradigm change that embraces this technology by the entire building industry.

15.2 HANDHELD COMPUTER APPLICATION ONE: RECORD DRAWINGS

Since the early 1900s the methods for operating and maintaining a facility have remained the same. Only recently has computer technology influenced facility management, policies, and procedures in this twenty-first century. For the most part, facility management, and how the building technician goes about his or her job, has not changed very much in 50 years. At the same time, little has changed for the design community or builders who have stayed closed to their own, time-tested procedures for project closeout, delivering record drawings and O&M manuals.

During the period between the 1980s and 2000 one significant change has been the *introduction of PDA technology*. In the late 1990s the use of this limited and simple handheld computer technology was embraced by many an individual. These handy devices were used by children and adults alike to play computer games, maintain contact and things-to-do lists, to mention three features. Soon they became the computerized version of a daily planner.

The adoption of handheld technology is only now being recognized as a powerful business tool that can enhance facility management. Here the tip of the computer technology iceberg is demonstrating that their application capacity and available options increase annually. By the time the twentieth century closed out, handheld computers were on the rise as manufacturers such as Hewlett Packard recognized this next generation of business technology. Soon handheld computers were offering a variety of attachment features including a bar code scanner, a digital camera, and wireless capabilities. Collectively they can now provide a powerful tool for the building industry as well as the facility management industry. In parallel with this surge in hardware came a need to also offer a wide range of software applications. Software firms such as BuildingSmartSoftware, Inc. (www.BuildingSmartSoftware.com) and InterPro (www. InterProsoft.com) are just two firms striving to help make the change to handheld application by developing software programs to meet today's needs in the building industry. The mission of firms such as these is to make available services to program the design, construction, operation, maintenance, and facility management experience and to make obsolete current contract specifications.

Beginning with the record drawings specification, when a building is completed, the building operators receive these documents but with questionable value for addressing the day-to-day operating needs of the building. In the closing years of the twentieth century these record drawings did come to the building owner as CAD, but the facility department was ill equipped to take advantage of the benefits of CAD documents. Like the paper version record drawings of the past, these computerized versions would be stored away for future reference when a building alteration or building expansion was being planned. Operating engineers and service technicians did not have convenient access to these drawings as they toured their building.

Contract document specifications regarding record drawings reflect the design community working in a vacuum when it comes to specifying the contract criteria for as-built conditions at the end of a construction project. In surveying building owners, the majority of them have no training in the use of CAD documents. As a result, CAD record drawings serve little purpose for most building operators. This problem is compounded by CAD software and associated license requirements should a facility manager choose to make use of these record drawings on a day-to-day basis. Add the dilemma of maintaining as-built drawings as various alterations and renovations are made and the facility manager tries to cost-effectively manage these documents and the value of CAD documents becomes a question. The result is often an ill-prepared facility staff for efficiently using CAD record drawings.

What handheld computers bring to CAD documents is a means to place these drawings in the hands of the technician touring the facility. At the same time, it can do this in a very simple manner by loading preselected drawings that are pertinent to day-to-day operation and maintenance needs and procedures. For example, a technician needs to know the location of a valve and/or its application. Pulling out the handheld computer, this O&M person can quickly locate the valve by clicking on the appropriate pdf-CAD file (Fig 15.1) and respond accordingly. Once the valve is located, the handheld computer's bar code scanner can quickly pull up the pertinent valve information for the database such as operating instructions (Fig. 15.2).

Another feature of handheld computer technology when it comes to CAD documentation is the global positioning system (GPS) where a device can be quickly located in the same

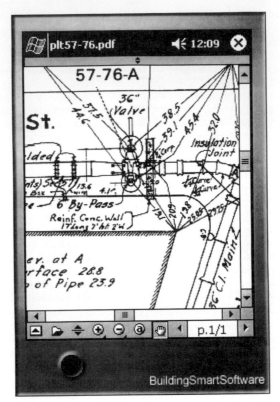

FIGURE 15.1   CAD/pdf—Computer-aided design/portable document files.

way fishermen have been using GPS in their own business. An important link to GPS can be the bar code reader on a handheld computer that targets the device and then a program can pinpoint the location on a record document in the computer. This tool can accelerate the accessing of a record drawing.

Interestingly, while handheld computers loaded with CAD pdf files are available now, the design community has not addressed what CAD layers a facility manager will need, once the record drawings have been transferred to the building owner. "Layers" are just that: documents stacked one above the other as a means to organize the information being placed on the CAD drawing. An example of this document, all the emergency shutoff valves are on a layer by itself. When used by a facility manager, the handheld computer can pull up the CAD pdf file that shows the architectural background and the isolation valves. Other mechanical data are available but located on other layers as needed.

To date, the building design community has been focused on layering CAD drawings appropriate for building construction documents. This concentration on design and install has caused them to frequently overlook "operate and maintain." Only now, approximately 25 years after CAD application began to be embraced by the building industry have CAD layering standards for facility management begun to be recognized as needed.

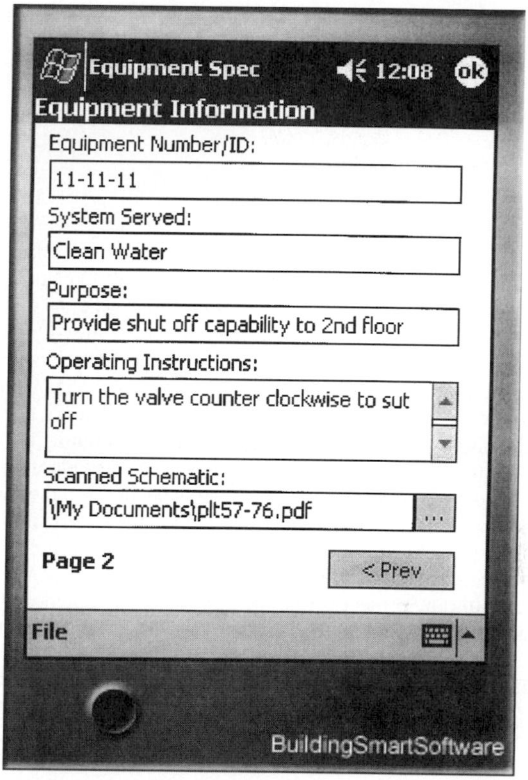

FIGURE 15.2   Operating instructions.

15.3   HANDHELD COMPUTER APPLICATION TWO: O&M MANUALS

Another "deliverable" of the construction process has been to provide O&M manuals specified by design engineers with little to no experience in the day-to-day requirements of operating a building. A construction company would compile the O&M manual per the contract specification, also with little experience in the management of a facility. As a result, the deliverables were provided late in the building process and flawed by insufficient data. Again, the operating engineers and service technicians did not have convenient access to these documents as they toured their building.

Speaking for the industry spanning the past 38 years, it is safe to say that many of the design specification standards in 2003 have not changed much since 1965. Engineers today continue to specify O&M manuals with little insight into what is really needed. Seldom does a building system designer ask the question. "What type of maintenance management system do you have? Is this system a "manual" process requiring several steps in documentation or is it a computerized maintenance management software (CMMS) system?" Instead, it has been safer to imply operation and maintenance criteria rather than specify exactly what is needed. An example of this is the requirement that the contractor furnish O&M manuals with the following:

- Parts list
- Lubricants
- Maintenance requirements
- Troubleshooting checklist
- Warranty

If a facility manager questioned the designer responsible for the O&M manual specification, the design engineer would be hard pressed to explain in great detail what all these criteria meant. Instead of researching what a building operator does, how a work order is issued, tracked, and closed out, how inventory is managed, and/or what special instructions are required when performing maintenance or repair, the engineer relies on antiquated specifications that offer a safe haven for spec writing. Lots of words but not a lot of substance!

The problem of lack of accuracy stems from the design community failing to partner with the operation and maintenance community and, as a result, standard specifications today continue to reflect a wealth of misinformation. What is needed is a thorough understanding of the operation and maintenance process and then putting the criteria into the contract specifications. With this approach, equipment data would be collected in an orderly manner so that the information could be downloaded into the building owner's CMMS system. All this can be done before a building project is completed if correctly planned and there is follow-through on the process. Once completed, handheld computers are ready to serve.

The solution is based on time management: eliminate those antiquated contract specifications that required the O&M data be provided near the end of the construction project. This current criteria has existed for at least 50 years if not longer. Why the designer's specification should call for O&M data to be compiled at the end of a construction project instead of immediately following acceptance of the associated equipment shop drawing is a mystery. Long before the construction job is finished, the equipment manufacturers have been paid and have gone on to procure more work from other project bids. To go back 12–18 months later for O&M data associated with a piece of equipment delivered earlier routinely creates a problem between the contractor and the equipment provider. As a result, it is not unusual for O&M manuals to be completed and delivered months after a job has been done and the occupants have moved into the space. The timeline immediately placed the facility staff in a "fire-fighting" mode of operation with little information and training on how the equipment and systems are to be operated and maintained.

The new contract specification needs to require that all operation and maintenance data be provided immediately after the associated equipment shop drawing has been accepted by the design team. Refer to Chap. 14, "Computerized Maintenance Management," for more information. Combine this new approach with handheld computers and the facility group can now download all that O&M data into a software program so that the technician will have in hand:

- Log-out/Tag-out instructions
- Special instructions
- Equipment specification
- Preventive maintenance work order
- Commissioning (functional performance test) procedure

With O&M being procured earlier in the construction process, the next step is to get this information delivered on computer disk so that the information can be downloaded into the handheld computer for "information at your fingertips." Like accessing CAD record drawings, equipment data noted above can streamline as well as accelerate the operation and maintenance process. The technology of specifications documented on Microsoft has been around for more than 25 years. Now handheld computers provide the tool to make the equipment specification, as well as system data and sequence of operation readily available to a technician when needed rather than filed away in a cabinet in the engineering department.

15.4 HANDHELD COMPUTER APPLICATION THREE: LABELING

Progress in the building industry continues to be hampered simply because "old habits are hard to change." In assessing the building industry's reluctance to change, one need only go back to the construction of the Henry Ford Fairlane mansion built at the start of 1900. The remote boiler plant was efficiently designed and included valve tags along with a valve tag chart that documented the purpose of the valve and the system served. Today, over a hundred years later, design engineers are still specifying valve tags!

The simple task of specifying the correct method for labeling of equipment, panels, valves, and devices is archaic, with the design engineer again working in a vacuum, not communicating with their facility counterpart. The building industry, and more specifically the heating, ventilating, and air-conditioning engineer, routinely identifies the project's central air-handling systems starting with AC-1 or AHU-1 and pumps with P-1. The next project, at the same facility will repeat this labeling with another AC-1, AHU-1, and P-1. The identification criteria for other components, such as valve tags, have been noted to be over 100 years old with the requirement that a valve is tagged and a valve chart mounted on a wall. The design community needs to learn how to correctly and comprehensively label equipment. Currently, the industry standard for labeling equipment is of no value when used as part of a CMMS database. This database identification number process needs to be far more specific than simply stating AC-1. Based on the facility manager's labeling procedures, AC-1 may need to be 01-L2-QA-AC-1-E (building 1, level 2, quadrant A, air-conditioning unit 1, on emergency power).

Integral to proper labeling is the *addition of a bar code to the label process*. Handheld computers with a bar code scanner can be the link between the label and all the features of software programming including:

- Access to the CAD drawing as highlighted previously
- Access to the equipment operation data as highlighted previously
- Access to the equipment maintenance data as highlighted previously

With labeling that can span numerous applications as well as bar coding, handheld application can become limitless.

15.5 PROBLEMS FACING HANDHELD COMPUTERS

Application of handheld computers begins with leader of the group "champion" and using this technology first before it is handed off to the workforce. There has to be first-hand experience and "buy-in" at the top before it can be delegated to others. The problems that will face this leader once he or she begins to introduce this technology will start at the work-desk and extend out to the individual technicians. Initial use can appear to be complicated and someone familiar with the use of this tool can provide reinforcement so that in a short time this device, like its "cousin" the cell phone, becomes part of the culture and day-to-day way of doing business.

A common problem in the past with computers and their use has been the office support. It was not unusual for administrative personnel to lack adequate training on basic computer skills before they were thrown into the day-to-day application of computer software programs. Users who included the administrative assistants, technicians, and other staff were prepared to use neither software nor hardware nor did they possess the basic understanding of this technology. Fortunately today these individuals are far more computer-literate than in the early days of this equipment. At the same time, training and availability of on-line support is essential if the transition from twentieth-century computer technology to today's technology is to be successful.

Unlike the initial software programs of the past, handheld programs should, for the most part be simple, smart checklists requiring little operating instructions for the user. Pdf files should provide only that information needed for a person to do their job outside of the main office. This handheld tool should be primarily for on-tour, on-site, quick response. The software should not try to displace the large and often complex business programs that require someone sitting down and spending time in front of a computer screen. Handheld computers are mobile business tools. This new software, in many cases, will simply be a means for quick access to information.

Through a collaboration of designer and builder, with operation and maintenance, handheld computer programs would be better prepared to embrace the products of construction documents. A good example of this quick access is the bar code scanner. For more than 25 years now, scanners have been used for materials management. It was a key component to inventory control. With new handheld programs that utilize bar code scanning, this business tool will become the link between a valve tag, nametag, and/or piece of equipment. "Turn the key," which in this case is "scan the bar code," and needed information can be brought to the computer screen. Consider the following computer screen options when a bar code is scanned:

1. Equipment specification
2. Lock-out/Tag-out instructions
3. Preventive maintenance work order
4. Daily/hourly log of operating parameters
5. Special seasonal instructions for startup or shutdown
6. Trouble-shooting checklist
7. Emergency action plan instructions and contact list
8. CAD drawing showing location of the piece of equipment
9. Functional performance test checklist for recommissioning
10. Global positioning system to pinpoint location

These 10 functions downloaded into the handheld computer for quick reference offer the users a menu of software checklist programs during a typical workday. In addition, the work performed when associated data is brought up onto the computer screen can be easily downloaded at the end of the day as the information flows seamlessly in the main computer database. What is done with this information becomes the by-product of handheld software programming. Action response being as simple as touching the handheld computer screen with a pointer allows the information to be documented as part of operating, maintaining, and managing the facility.

The same scenario can be applied to a handheld construction inspection program that can be used for efficiently documenting punch-listing a construction job. Another handheld software application available is the equipment startup checklist (i.e., Commissioning 1-2-3) that can be used by a trade contractor or equipment manufacturer. In the near future it will not be unreasonable to have equipment specifications downloaded onto a handheld computer for a service technician to quickly reference when repairing this equipment. Transfer of documentation will also be done via a wireless connection between two points.

The combination of programming knowledge paired up with building technology will accelerate the application of handheld computer programs. Additional programs to efficiently allow individuals to perform their work and do so cost-effectively while making the documentation process a more recognizable seamless process will allow computer use to continue to expand its boundaries. What is needed is a visionary perspective of where technology can take this business and the expertise of software programmers followed by a response by hardware capabilities that can accommodate the capacity and speed required of these programs.

Reporting as a business tool through computer technology just gets easier with a building organization that is proactive in monitoring, measuring, benchmarking, and reporting on building management. Because handheld computers and their associated software are tools of the trade, it will be imperative that the partnership of programmer and technician be focused on the "big picture," which is to use the information stored and gathered. With no management plan to capitalize on this twenty-first century technology, handheld programs serve little use. Analogous to the CAD drawings that have been turned over to building owners for the last 20 years only to be stored and not used, asset database offers little return on investment unless all those involved step back and look at the "big picture." What good is handheld technology if there is not a long-term, grand scheme for these invaluable tools?

15.6 THE LONG-TERM SUCCESS OF HANDHELD TECHNOLOGY

This chapter has so far stated what has been standard practice in the building industry dating back more than 100 years. The building industry is in dire need of changing the methods of the design and construction phases of building programs. Professionals in this business need to rewrite the standards and redesign their process of delivering the product. The concept coined as "commissioning of facility management" is directed to focus on facilitating this change on a job-by-job basis.

In the past 10 years building system commissioning (documentation-verification-training) has grown significantly as owners demand means to assure a building will operate as designed. What has been overlooked during this period has been the commissioning of the building management process (documentation-verification-implementation) of O&M tools. Without this new quality-assured application of computerized facility management, the building industry will continue to languish in the twentieth-century policies and procedures. In addition, building operation will continue to be compromised on Day 1 of occupancy.

Commissioning of facility management is analogous to a ship that is built and then commissioned before it is sent to sea. It is imperative that the facility management organization and its management tools be commissioned along with building system commissioning if it is to operate in an efficient, trouble-free, and self-sufficient manner over the life of the facility.

The process is intended to be a quality-control initiative to link the designer, the builder, and the facility operator together within the building program process. An extension to traditional building system commissioning, commissioning of facility management focuses on the operation and maintenance process necessary to "steer the ship" through troubled waters. Its intent is to focus on making the startup of operation an extension of the design and construction team effort to complete a building based on design intent and programming needs.

This continuation of the quality process focuses on implementation of a computerized building management business plan. It also facilitates data collection sufficient to establish organization structure, job descriptions, shift cover, policy and procedure, and computer software necessary to maximize this building support service organization using:

• Industry-standard computer software (i.e., Microsoft's suite of software programs)
• Twenty-first century specification requirements based on computerized facility management
• CMMS systems in lieu of manual maintenance management systems
• Bar code technology to link and merge equipment, panels, valves, devices, etc., to facility database to handheld software programs
• Computer software based on development of programs that use handheld computer technology
• Programming the experience of various professional engineers, certified plant engineers, and builders
• Use of handheld computers with the power to have equipment database, CAD drawings, operating instructions, preventive maintenance, etc., in the palm of the technician's hand
• Reporting as a business tool through computer technology

The very foundation of computerized technology tools, the platform on which facility management software programs need to operate will be with programs that interact with other programs. For operation and maintenance groups, their business tool platform is the CMMS system that holds the asset database of the facility, the preventive maintenance work order system, unscheduled work order system, manpower requirements, etc. (refer to "Computerized Maintenance Management," Chap. 14, for more in-depth discussion on CMMS). It is this system that is the cornerstone of the building management process because it contains the majority of databases including equipment assets, work order and work history, parts and material inventory, employee records, and reports.

If CMMS systems are the vehicle to twenty-first century facility management, then handheld computers are the driver of this vehicle. With the power of data access in a technician or manager's hand, response to routing or emergency situations are the same; quick, efficient, and based on data, not assumptions. Handheld computers are simply the next generation of computer technology that has taken all that information stored in file cabinets and placed it at one's disposal as quickly as one can scan a bar code.

The long-range success of this new technology is dependent on two issues: First, there is an abundance of programmers to work in concert with engineers and facility managers. In a reverse mentoring process, programmers possess the skills to computerize another's technical knowledge into software programs. While the engineers or builders with 25 or more years of proven experience can do their job as they did in the twentieth century, it is this new generation of software specialists who hold the expertise to shape the engineer's or builder's process into handheld technology. The programmer may not understand the building industry process, but this individual can still computerize it so that the process becomes the next generation of application through software. Here is where a visionary point of view will be needed to shape this next generation of computers, software, and how the building industries do their jobs. Today there is an abundance of programmers available to upgrade that in-house IT specialist position many companies filled several years ago. Computer junkies just can't provide the next generation of support services.

The second issue will be easy access to software programs for use in these handheld computers. Collectively, this new generation of software technology will be built to flow seamlessly from office computer to handheld device and back. Here again, building experience merged with program experience will overcome this hurdle. As both entities work for a common cause, speed and capacity of computers will push this technology to further enhance our high-tech way of running our business. The "anchor" that can hold back this fast-moving business will be those who are reluctant to embrace handheld technology.

To overcome obstacles such as buy-in by end users of handheld technology application, one must begin with developers of computer software. Currently there are few handheld programs being created for facility management. Those that are being developed do so lacking coherence and consistency as these entrepreneurial startup firms work independently, taking the first steps to establishing the next generation of computerized facility management. Lacking at this point is a distribution center where awareness and access to purchasing these software programs can be made. Searching the Internet for handheld programming firms such as BuildingSmartSoftware, Inc. is one way to locate these new software businesses.

Assuming commissioning of facility management can overcome the antiquated means by which O&M tools (i.e., record drawings, etc.) are delivered to the building management group, the responsibility to embrace handheld technology will lie with the managers of facilities. How handheld software programs transcend their counterpart, the laptop computer, will require the facility management organization to have a strong foundation to work from and the confidence to implement these new programs. This will mean the organization has invested the time and money in computer backbone technology from the point of view of hardware, software, skill levels, and proficiency. Here stands the Achilles' Heel of handheld technology. Currently, there are very few standardized programs available on the market. Unknown to many, there are numerous handheld software programs available on the market. As noted, the effort is currently fragmented, needing an organization to consolidate the products and solidify the building industry programs.

As handheld technology begins to be embraced, support will be needed by software programmers, who are proficient in trouble-shooting software problems and assisting in making sure these programs can interact with other programs. Searching the Internet, a viewer will find many of these new generation programmers such as www.InterProsoft.com who can provide this new service. Other startup programming firms are coming online offering handheld programs and the expertise to assist the building industries with making the transition to this next generation of computer technology and business process. As people become more comfortable and proficient with on-line technology, facility management via web sites, and virtual reality, handheld computers will be convenient tools to access information as easy as one uses cell phones today. Soon these startup firms will challenge the more established CMMS system firms for a piece of the documentation business.

15.7   THE COMPETITIVE EDGE

The phrase "knowledge is power" is inaccurate unless rephrased to "knowledge, if used, is power." Today, we recognize valve tags and valve charts serve little use when compared to their application back in the nineteenth century. Equipment labeling is inadequate for immediate input into a CMMS system. CAD record drawings also are of little use and their application inadequate for today's "bottom-line" management mindset. We also should admit that current contract specifications are antiquated in many areas as they relate to operation and maintenance of building systems after the project is built.

A study published in *Association of Facility Engineers Journal* some years ago indicated that only 15 percent of the CMMS systems purchased actually get implemented. Today that percentage has increased, but not significantly. There are few resources committed to populating a CMMS database with little value perceived from this initiative. Instead, outsourcing the operation often appears to be a better return on investment than investing within. At the same time most, if not all, outsource firms will quickly highlight that they have a CMMS system and they use it as an integral part of their business.

Firms competing for the business of running other people's buildings recognized years earlier that they could market their services based on "knowledge is power." Data collection and use of that information can be what differentiates one operation and maintenance group from several other operation and maintenance groups (their competition). One financial goal of end users who apply checklists to their everyday jobs is the ability, through handheld technology, to create "smart checklists." The difference between a technician using a paper version log versus a computerized check log can make the difference between cost-effectively reporting and, in some cases, not getting around to reports. It is very possible that handheld computer technology can be the competitive edge between how one group operates and maintains a facility and how another doesn't do so as efficiently.

The handheld computer user tours a facility, inspects a system, and documents the observations. The handheld computer is not intended, at least at this time, to displace a person's desktop computer. It is more appropriate to consider these programs to enhance any standardized checklist and not to displace other computer programs that require the convenience of a larger screen and a larger keyboard while sitting in a comfort chair, doing the work. These handheld programs are intended to be mobile and quick-reference listings that can be filled out with the touch of a pointer while touring an installation.

Computer software users, in the facility management industry, recognize that advancements in computer technology offer a multitude of opportunities. These smart programs and a bar code scanner transform standard checklists into "smart lists," making knowledge more powerful because they provide a wide variety of business activities in a "pick & click" movement such as:

- Time management through action-reaction checklists
- Daily and/or weekly safety tour checklist for insurance compliance
- Wireless communication and email messaging
- Less chance of error in transferring the documents from the field to the office
- Seamless transfer of documentation, downloaded from the handheld computer into the server database
- Seamless transfer of the documentation out of the server database and into another software application
- Out of the database can come value-added, dollars-saved, and/or cost-avoidance financial incentives
- Trending and "lessons learned" when checklist data are downloaded into a database
- Quality initiatives through similar database knowledge
- Historical trends, benchmarking, and business improvements

Since the inception of computers, changing the way we do business using today's technology has become an annual event. What will help the building industries continue to change and grow is contracting out computer system engineers to sit next to a design engineer, walk a job site with a job superintendent, or provide support services to a property manager. Envision the facility management team made up of the standard job roles and now add in a systems engineer consultant as an integral part of the project team. With more knowledge and information shaping how we do business, the sky is the limit as "experience-meets-technology" through software.

15.8 SUMMARY

The application of handheld computers is as broad as the application of standardized checklists. The applications are not limited to just the building industries. Regulatory inspectors, insurance inspectors, anyone who uses a checklist as part of their business process will have a need for a handheld, computerized version of their checklist. Equipment manufacturers should be interested in downloading their standard information (warranty, startup instructions, trouble-shooting instructions, preventive maintenance, etc.) over the Internet and onto a handheld computer. Once this manufacturer's information is made available to the builder or trade contractor, equipment startup, operating instructions and maintenance can be documented and emailed back to the vendor for receipt of warranty paperwork.

Collecting this information from the equipment manufacturer over the Internet and downloading on to a handheld computer will allow the end user to load up their handheld computer with portions of these data for day-to-day needs, seasonal needs, and/or other information mission from past data collection efforts. The advertising and procurement of available programs will most likely be from these equipment vendors with more generic programs being purchased through national organizations (i.e., engineers, builders, owners, operators, agencies, etc). In addition, with an ability to buy software programs online, prepackaged with a handheld computer will make purchasing easier and reduce the troubles associated with loading a program on to a computer at a later date. Instructions could also be downloaded with the purchase.

Publishing firms will most likely align themselves with software programming firms who will develop simple, brief software programs for handheld computers. A central clearing house could be set up to respond to specific questions relative to the program purchased. Another alignment will be between software program firms and/or publishing firms with handheld computer manufacturers to offer preloaded programs for ease of purchase and use of the programs. This partner-purchasing approach will be of interest in the beginning where many first-time buyers will necessarily have a handheld computer. It will also eliminate potential nuisance problems associated with downloading the information by an inexperienced handheld computer user. The market edge for publishing firms, whose business to date has been books, will be to position the publishing firm as the first "on the street" with a family of smart checklists for the handheld computer user. Because the program categories are limitless, the potential buyers have no boundaries. From coast to coast, the market will mirror the buyers of the computers and books.

HVAC COMPONENTS— SELECTION AND MAINTENANCE

CHAPTER 16
HEAT GENERATION EQUIPMENT

16.1 BOILERS

E.D. Cancilla

Chairman, Ajax Boiler Inc., Santa Ana, California

16.1.1 INTRODUCTION

Boilers, water heaters, and hot air furnaces convert energy derived from a fossil fuel source into fluid thermal energy. Boiler output is in the form of heated fluids such as water, steam, or thermal oil. Unfired boilers use a tube bundle or electric coil to convert thermal energy to a fluid. Fossil fuel energy is converted into mechanical energy by internal combustion engines and gas turbines.

The first known development of a crude steam boiler occurred in approximately 150 B.C. and consisted of a boiler drum with steam pipes connected to the axis of a hollow sphere supported over a fire. Two open steam pipes mounted at right angles to the axis produced a reactive force, rotating the sphere.

Early boiler development was focused on steam boilers to drive steam engines. The first steam boiler patent, issued in 1766, was for a water tube boiler with alternately inclined water tubes at opposite angles that were arranged in a furnace and connected to a steam drum. The use of boilers expanded to providing hot water and low-pressure steam for heating buildings and high-pressure steam for use in industry and for the generation of electricity.

The development of the copper fin boiler started when Rudd produced a multicoil smooth copper water tube boiler in the 1920s. Burkay developed a boiler with a smooth copper coil tube combined with a copper finned radiator. A copper fin cone coil water tube boiler design was developed by Ajax in the early 1940s. In the early 1960s an inexpensive horizontal multitube copper fin pool boiler was introduced.

Boilers are an assembly of a combustion system, comprising a combustion chamber furnace and a pressure vessel/heat exchanger that transfers heat from the products of combustion into the fluid contained in the pressure vessel. Operating control and safety devices complete the boiler.

Combustion systems have evolved over the years from coal, wood, or combustible waste materials to primarily gas, oil, and coal fuels. In the year 2000 approximately 52 percent of the electricity generated in the United States employed coal-fired high-pressure steam boilers. The trend is toward gas fired boilers because of air pollution conditions and the increased availability of natural gas. Most commercial and industrial boilers today are gas or oil fired. The state of California no longer allows the use of coal or oil fired boilers, except, when specifically authorized, limited oil firing of boilers for emergency facilities is allowed.

Combustion chambers in boilers generally are (1) a combination of water tubes and refractory surfaces or (2) a combination of water-backed surfaces, such as a large diameter furnace and fire tubes in a boiler drum with some refractory surfaces. Packaged boilers are equipped with (1) operating controls, (2) flame safeguard controls, and (3) pressure and temperature safety devices. Field-erected boilers will have these devices and controls mounted during the installation of the boiler. Stack emission devices are used to reduce the emission of pollutants. Stack heat recovery systems can be used to conserve energy and reduce operating costs.

Boilers, in most cases, are required by state and national law to meet the boiler and pressure vessel code requirements of the American Society of Mechanical Engineers (ASME), National Board of Boiler and Pressure Vessel Inspectors (NB), Underwriters Laboratories (UL), and the Canadian Standards Association (CSA) plus state and local codes as well as many energy and environmental requirements. The codes require independent inspection of boiler materials, manufacturing processes, and testing of boilers. This also includes inspection of installations, operation, and repairs of boilers. Testing, listing, and labeling of boilers by independent third-party National Recognized Testing Laboratories (NRTL) provide assurance that the boiler's design and performance is in compliance with the ASME, UL, and/or CSA code requirements.

Boilers used for heating and air conditioning are primarily water boilers or 15 lb per square inch (psi) low-pressure steam boilers. Municipal and central plant steam distribution systems usually provide from 75 to 120 psi (51500 to 83000 N/m^2) steam to reduce the piping cost.

Boilers are operated as closed systems with treated and/or deaerated water. In nonheating applications steam boilers may consume steam and require treated make-up water. Boilers are generally made of ferrous metals since the materials are subject to minimum corrosion conditions.

16.1.2 BOILER TYPES

Boiler designs can be broadly separated into four classifications, *water-tube, fire-tube, drum*, and *cast-iron sectional*.

16.1.2.1 Water Tube

In 1856 Stephen Wilcox of Babcock & Wilcox developed the first inclined water tube design that connected the boiler water spaces in the front and rear with a steam space above. The design was patented in 1867. The water tube boiler design has the advantage of lower pressure and temperature differential stresses. This is because small diameter drums can be used with the water tubes. The design successfully handles expansion and contraction over the wide range of temperatures of the combustion flame, the flue gases, and the water and steam. The water tube boiler is more suitable for larger capacities and higher pressures while still maintaining a high margin of safety.

Straight and bent water tube boilers use tubes with exterior surfaces that are *plain* (smooth). Copper fin boilers have exterior *finned* surfaces. The copper fins greatly improve the tubes' convection *heat transfer* capabilities. Multiple rows of inclined straight steel tubes are arranged in a triangular pitch layout so that the hot flue gases pass around the water tubes at turbulent flow velocities providing excellent heat transfer. Bent water tubes depend primarily on convection heat transfer. However, tube bending, offsetting, welding, and flexing of the tubes do not provide the uniform tube spacing needed to provide flue gas velocities to obtain the best convection heat transfer. Higher boiler water tube flow is needed in the first row of bent tubes to handle the intense radiant heat exposure. Most **condensing boilers** are designed so that the condensing occurs outside of the combustion chamber. By using a water flue gas "counter flow" design the existing flue gases pass over the cold entering water. Higher boiler efficiencies are obtained with low entering boiler water temperatures below 50°F to 60°F (11°C to 18°C) and low boiler exit temperature 120°F to 130°F (49°C to 55°C). Special stainless steel is used in condensing boilers to handle the condensate. When stack flue temperatures are above condensation temperature, that is, above approximately 135°F (56°C), stainless steel vent material must be used. At lower vent stack temperatures plastic vent pipe may be used.

The condensing boiler opens up a whole new range of applications such as low-temperature hydronic, low-temperature water source heat pump systems, snow melting, heating large commercial pools, etc.

Water-tube boiler products of combustion, flue gases, pass over the exterior of the water-filled boiler tubes. The water tubes are connected to boiler water or steam drums (headers).

1. **Straight inclined tube** boilers have triangular pitch water tubes that are rolled into a tube sheet and header box at each end. (see Figs. 16.1.1 and 16.1.2.)
2. **Bent-tube** boiler designs have several tube shapes.
   - Bent tubes are connected top and bottom to one or more drums (see Fig. 16.1.3). In some designs, the tubes are connected to the drums using mechanical taper joints.

FIGURE 16.1.1 Atmospheric water tube boilers. (*Courtesy of Ajax Boiler Inc.*)

FIGURE 16.1.2   Straight tube water-tube boiler.

- D-style: This unit consists of an upper drum and a lower drum connected by tubes (see Fig. 16.1.4).
- A-style: Fig. 16.1.5 shows a typical A-style boiler comprising a single upper drum and the lower drums in symmetrical pattern.
- O-style: Similar to A-style, but with one lower drum (see Fig. 16.1.6).
- Figure 16.1.7 shows an upper internal steam drum with steam separator.

Copper fin tube cone coil water tube boiler design for commercial applications and an inexpensive horizontal straight multitube copper fin pool boiler (see Fig. 16.1.8).

Fire-tube boiler designs developed with the addition of fire tubes into the shell or drum boiler. Initially it was developed as a Firebox boiler with a large combustion chamber, for coal firing, located under the boiler drum. One end of the drum and the fire tubes were open to the combustion chamber with the hot combustion gases traveling through the fire tubes and exiting into a flue box connected to the vent stack (see Fig. 16.1.9). In principal, the fire-tube boiler is the reverse of the water-tube boiler. The hot combustion gases are in contact with the inside surfaces of the tubes and the water and steam are in contact with the outside surfaces of the tubes. Fire-tube boiler designs usually require forced draft burners. The Scotch marine forced draft boiler incorporates a cylindrical 'Morrison' tube, combustion chamber into the boiler drum providing a self-contained boiler (See Fig. 16.1.10). Scotch boiler combustion chamber loadings are generally very high with furnace loading up to 200,000 btu/ft^3 not uncommon. Scotch boiler tube sheets

FIGURE 16.1.3   Bent-tube watertube boiler. (*Courtesy of Bryan Steam Corp.*)

must be designed to handle high thermal stresses due to the combustion cylinder gases exiting the furnace at approximately 2000°F (approx 1100°C) and the fire-tube "pass" gases progressively decreasing to exit at temperature of 450°F to 500°F (240°C to 333°C). The combustion tube and the fire tubes are welded to large tube sheets at each end of the boiler. Fire-tube boilers are limited in size and pressure rating due to the combined effects of pressure and thermal stresses and the need for thicker shell and tube sheet plate as the size increased. Some fire-tube designs, when continuously operated outside a limited temperature differential band (outlet temperature minus inlet temperature) will experience fatigue stress failures. In most fire-tube designs, thermal stresses cause large differential expansion forces which often loosen tube joints and, in extreme cases, result in rupture of the boiler vessel.

Compared to the Firebox design, the Scotch design provides increased output capacity and efficiencies due to the water-backed combustion chamber absorbing the radiant heat from the flame. Water treatment is important since scale from hard water will develop. The Scotch boiler has been used for many years as the mainstay of marine propulsion boilers; however, its application to water heating is limited.

Fire-tube boilers are built to channel hot combustion gases through the inside tube passages.

- Scotch, in which the horizontal fire tubes are housed within a horizontal cylindrical pressure vessel or "shell."

- Firebox, where the horizontal tubes inside a round or box-shaped shell are mounted above a refractory-lined "firebox" or combustion chamber.

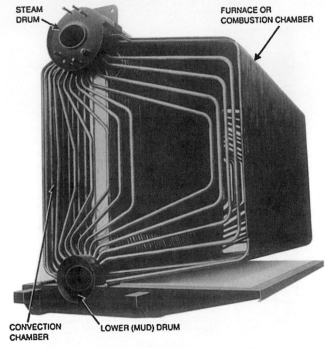

STEAM
DRUM

FURNACE OR
COMBUSTION CHAMBER

CONVECTION
CHAMBER

LOWER (MUD) DRUM

FIGURE 16.1.4   D-style water-tube boiler. (*Courtesy of Cleaver-Brooks.*)

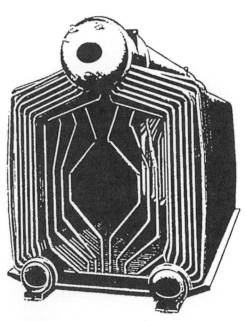

FIGURE 16.1.5   A-style boiler. (*Courtesy of Cleaver-Brooks.*)

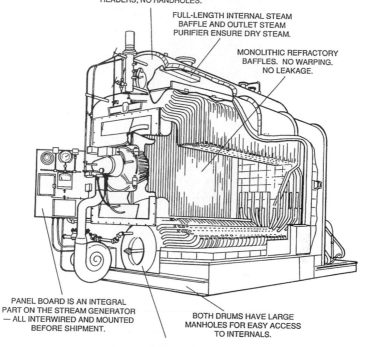

TWO-DRUM BOILER — ALL TUBES
TERMINATE IN DRUMS. NO
HEADERS, NO HANDHOLES.

FULL-LENGTH INTERNAL STEAM
BAFFLE AND OUTLET STEAM
PURIFIER ENSURE DRY STEAM.

MONOLITHIC REFRACTORY
BAFFLES. NO WARPING.
NO LEAKAGE.

PANEL BOARD IS AN INTEGRAL
PART ON THE STREAM GENERATOR
— ALL INTERWIRED AND MOUNTED
BEFORE SHIPMENT.

BOTH DRUMS HAVE LARGE
MANHOLES FOR EASY ACCESS
TO INTERNALS.

ENTIRE GENERATOR IS MOUNTED
ON A RIGID STRUCTURAL BASE
EXTENDED TO FORM THE REAR
FAN PLATFORM.

FIGURE 16.1.6   O-style boiler. (*Courtesy of Cleaver-Brooks.*)

Heater control over varying loads. This big 42" O. D. steam drum comes with a full complement of steam dryers, plus Cleaver-Brooks' patented water level control baffles. This combination results in a dry steam product even when load swings far beyond the ordinary. The baffles prevent diluting of the entering steam/water mixture through reservoir water. This results in more effective steam separation and greatly improves water level control in the drum.

Cleaver-Brooks' exclusive patented steam purifiers are also available to meet the solids concentration requirements of central station installations. Extra storage capacity, easier access. Two 24", I. D. lower drums mean that CA steam generators keep more water on reserve to meet sudden load demands. The steam drum and the lower water drums have 12" × 16" manways at each end — providing access for servicing and eliminating troublesome leaking handhole plates normally required with header-type drums.

FIGURE 16.1.7   Steam separator-drum internals. (*Courtesy of Cleaver-Brooks.*)

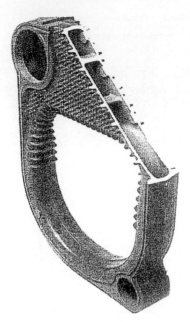

FIGURE   16.1.8   Typical   cast-iron
boiler section.

• Vertical fire-tube boilers, generally are smaller in size where the fire-tubes are
mounted in a vertical cylindrical shell. Residential water heaters are vertical fire-tube
designs.

16.1.2.2 Cast-Iron Sectional

The cast-iron boiler evolved from a vertical water-tube design. The vertical water tubes
were replaced with cast-iron sections to take advantage of the corrosion resistant

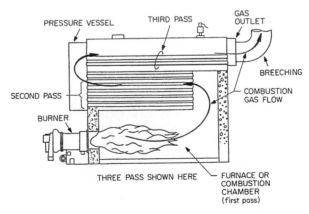

FIGURE 16.1.9   Firebox boiler. (*Courtesy of Cleaver-Brooks.*)

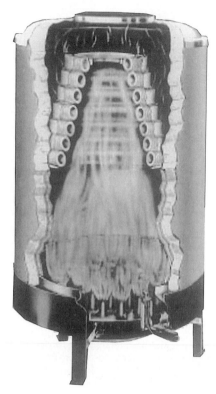

FIGURE 16.1.10   Fire tube boiler-scotch marine.

properties of cast iron. Sectional castings have water-filled vertical risers with horizontal headers at the top and bottom or top and middle. Cast-iron sections have studs that separate the sections and provide a path for the flow of flue gases that transmits the heat to the water-filled sections. The sections are connected with tapered nipples held together with tie rods or bolts and sealed with gasket materials. See Fig. 16.1.8. Cast-iron boilers, due to their relatively heavy (thick) sections, are intrinsically slow in heating up. Because of the mass of iron, these boilers will retain heat more than water-tube boilers. Water-filled casting sections must withstand temperature differentials ranging from 2400°F (approx 1300°C) combustion temperatures at the bottom to 500°F (approx 250°C) stack vent gas temperatures at the top. In addition cast-iron sections must withstand the thermal expansion and contraction stresses of boiler temperature cycling and the mechanical stresses from the bolting required to hold the sections together without leaks. Fatigue fractures of cast-iron sections may occur due to the thermal stress conditions. Thermal shock will cause cast-iron sections to crack whenever cold water is fed into the hot cast-iron boiler.

Wet-base cast-iron boilers use forced draft burners with the flame enclosed on the top, bottom, and sides by water-backed castings and by refractory on each end. Dry-base cast-iron sectional boilers use under-fired atmospheric burners. Cast-iron boilers are popular in residential and small commercial applications.

Boiler Construction by Output Classifications

Water boilers

Low-pressure steam boilers

High-pressure steam boilers

Thermal fluid boilers

The boiler industry manufactures a broad range of types and sizes of boilers ranging from small packaged residential hot water boilers through huge field-erected utility power generating boilers in excess of 200 ft (60 m) tall.

Water boilers are generally constructed to ASME Section IV with maximum allowable working pressure to 160 lb/in^2 (11 bar) and maximum temperature 250°F (121°C). Water boilers exceeding these Section IV limits are classified as medium or high-temperature hot water (MTHW or HTHW) boilers, are designed per Section I ASME code.

Steam boilers are classified as (1) low-pressure boilers with maximum allowable working pressure (MAWP) of 15 lb/in^2 (1.03 bar), constructed to ASME Section IV, or (2) high-pressure boilers, generally 150 lb/in^2 (10.3 bar) MAWP, constructed to ASME Section I.

For HVAC purposes, most boilers are constructed as "packaged boilers." They are completely shop assembled with fuel burner, draft system, insulation and jacket, and all controls. The advantages of the packaged boilers are as follows:

1. Boiler is mounted on an integral base ready to be moved into place on the foundation pad. Connections required are (1) sources of water, fuel, and electricity; (2) steam and condensate, blow down, relief valve, and drain piping (or hot water supply and return); (3) a stack for vent gases; and (4) foundation anchor bolts.

2. Responsibility for design and performance is from a single source, the manufacturer. The boiler is usually test-fired prior to shipping. Third-party NRTL, ASME, UL, and/or CSA label is further evidence of design approval and proper quality control.

3. Standard boiler sizes are constructed in the manufacturer's plant with lower costs, fast lead times, and optimum quality.

4. Typical package boiler input-to-output efficiency varies from 77 to 84 percent over a firing range of 40 to 100 percent. The ratio between maximum and minimum firing rates is known as *turndown ratio*. A boiler with 50 percent minimum firing rate is said to have a 2:1 turndown ratio.

5. Packaged boilers save space and are adaptable to a wide variety of locations from subbasements to penthouses.

6. Packaged boiler designs are available for outdoor rooftop operation.

16.1.3   OPERATING PRESSURE

Low-pressure heating boilers in the United States are fabricated in accordance with Section IV of the ASME Code, which limits the maximum allowable working pressure of low-pressure steam boilers to 15 psig (10.3 bar) and low-pressure hot water boilers to 160 psig (11 bar) at temperatures not exceeding 250°F (121°C).

FIGURE 16.1.11   Cross-section through typical packaged firebox boiler. (*Courtesy of Ajax Boiler Inc.*)

The practical boiler pressure and temperatures *operating* limits are lower to reduce nuisance shut downs by the high limit pressure or temperature controls or the opening of the relief valves. Realistic maximum operating values are:

- Low-pressure steam boilers 13.5 psig (0.93 bar)
- Low-pressure hot water boilers 140 psig (9.6 bar) at 230°F (110°C)
- For operating pressures or temperatures above these values, the boiler should be constructed to ASME Code Section 1.

16.1.4   SELECTING A PACKAGED BOILER

There are several criteria involved in selecting a packaged boiler. These include:

1. Type of fluid to be heated (low-pressure steam, high-pressure steam, hot water, and high-temperature hot water)
2. Boiler size (the rate of heat transfer)
3. End use of boiler heat, i.e., space heating, humidification air-reheat, laundry, kitchen, or domestic water system use.
4. Importance of boiler output reliability required and the need for redundant capacity. Generally, it is preferable to provide redundancy by having multiple boilers with a total capacity exceeding design load. For example, two boilers each capable of providing 75 percent of the required energy output would provide complete redundancy (100 percent backup) for a large part of the heating season.
5. Type of fuel, primarily natural gas or No. 2 fuel oil. The use of heavy fuel oil, grades 4 through 6, has declined. In remote locations, propane is used. Other types of fuel are available, e.g., coal, wood, biomass, but these are seldom used in conventional applications.
6. Atmospheric combustion burners for gas firing only. Atmospheric boilers are simpler and less expensive to purchase and maintain. Smaller gas-fired boilers sold in the United States are atmospheric units. Larger gas-fired atmospheric boilers are becoming popular because of reduced maintenance, operating costs, and first costs.

7. Forced draft burners are required for oil firing and are available for gas and combination gas/oil fuels. These burners provide better fuel-air ratio control during modulating service, but the linkage mechanism requires periodic tightening and adjustments to retain this control.

8. Induced draft systems on atmospheric boilers have a blower mounted in the flue outlet to draw the flue gas through the boiler. The blower is handling hot flue gases and must be constructed for high-temperature operation and corrosion resistance. The required volumetric flow from a draft inducer is approximately double that of the equivalent forced draft blower.

   Additional design criteria include:

1. Controls system complexity

2. Emissions control requirements

3. Location, available space, and access limitations

4. Noise levels

5. Life cycle costing, including warranty coverage

16.1.5   GENERAL DESIGN CONSIDERATIONS

There are several design considerations which are applicable to all boilers.

16.1.5.1   Combustion

Proper combustion is influenced by the size and type of the combustion chamber, the mixing and control of the fuel and air, the heat transfer passages, and the venting.

Combustion chambers can be refractory or water-backed surfaces or a combination of the two. Refractory provides hot surfaces that assist the combustion. Water-backed cold surfaces can quench the flame edges deterring complete combustion. The combustion chamber must be large enough to obtain the complete mixing and burning of the fuel.

The combustion process begins with the mixing of the fuel with the primary combustion air. Secondary combustion air is introduced into the flame to provide additional oxygen to the fuel to complete the combustion.

Atmospheric burner primary combustion air is aspirated by the gas stream flowing from the gas orifice into a venturi mixing tube. Secondary combustion air is added to the flame at the burner ports by the draft action of the vent stack. The boiler vent is an important and critical component of the combustion process. Primary combustion air openings in the boiler room are necessary for good combustion.

Forced draft burners use an air blower to provide both primary air and secondary combustion air. It is mounted on the inlet to the combustion chamber. They use sizeable electric motors and are more costly to operate than atmospheric burners.

Combustion air supply must be adequate for the complete combustion of the fuel. Theoretical or stoichiometric perfect combustion does not include excess air. Therefore, excess air is required since perfect and complete mixing of the fuel with the oxygen is not possible. "Excess air" is the air entering the combustion process whose oxygen content is not consumed in burning the fuel. A small amount of excess air is needed since (1) none of the available combustion processes provides completely homogeneous fuel-air mixing and (2) allowance must be made for the effects of wear on the forced draft burner air-fuel ratio linkage controls. This excess air is measured as the percent oxygen in the boiler stack gases and is expressed as a percentage of the air required for stoichiometric combustion. The amount of excess air is a measure of wasted energy. The wasted energy includes the extra fuel that is required to heat the excess air. In the case of forced draft units, additional electrical power is required to blow the excess air through the boiler. In some recent low-emission designs, a high level of excess air have been used to lower combustion chamber temperatures, reducing formation of nitrogen oxides, but with reduced combustion efficiency.

Boiler codes have a maximum limitations for carbon monoxide (CO) of 400 ppm (parts per million) and nitrogen oxides (NO_x) of 250 ppm. Refer to Section 9 of this chapter for more information on local air quality emissions requirements.

Furnace heat release rate per unit of furnace volume has, for many years, been a governing factor in the selection of boilers. Current Scotch packaged fire-tube boiler designs utilize furnace heat release rates as high as 200,000 Btu/ft^3 (2066 kW/m^3).

Water-tube boiler designs use heat release volumes from 60,000 to 100,000 Btu/h/ft^3 (620 to 1033 kW/m^3).

Refractory combustion chambers, with their hot surface temperatures, contribute to the complete combustion of the fuels.

Water-backed combustion chamber surfaces may quench the outer flame envelope before complete combustion, especially in designs with high furnace loadings. The boiler surfaces exposed to the flame are usually a combination of high-temperature refractory and water-backed surfaces.

16.1.5.2 Heat Transfer

The transfer of heat to boiler fluids occurs as radiant, convection, and conduction heat transfer.

Radiant is a highly intense transfer of heat energy to water-backed surfaces. The surfaces absorbing radiant heat must have a low heat loading and sufficient fluid circulation to continuously conduct the heat energy from the metal surfaces to keep the **radiant heat transfer density**; within the temperature limits of the metal.

Hot gases leaving the combustion chamber pass across the water-backed **convective heating surfaces**; they must have sufficient velocity and turbulence to effectively transfer heat through the hot surface wall(s) to the fluid being heated. Convection heat transfer effectiveness increases as the flue gas turbulence and velocity increases. Since flue gas volume decreases as it cools, the boiler design must use baffling and reduced pass area to maintain heat transfer efficiency.

Closely spaced triangular pitch water tubes maximize the use of the flue gas velocity and turbulence. In addition, the use of flue gas baffles obtain excellent heat transfer efficiencies. Fire-tube boilers may use spiral inserts (turbulators) in the tubes to maintain flue gas velocity through the length of the fire tube as the flue gas cools. Progressively reducing the number of fire tubes in the passes compensates for the reduced flue gas volume and velocities as the temperatures decrease.

Conduction heat transfer occurs as the heat flows through the water-backed surfaces (water of fire tubes) into the boiler fluid. The highest conduction heat transfer occurs with larger temperature differential between the fluid being heated and the flue gasses. Forced circulation of the boiler fluids is needed to move the heated fluid away from the water-backed surfaces and allow the cooler boiler fluid to contact the hot surfaces. Fluid circulation will improve the heat transfer and ensure that the water-backed metal surfaces do not exceed their temperature limit.

Heating surface as defined in ASME Code Sections I and IV. The code does not fully recognize the design improvement factors that utilize flue gas velocities and flue gas turbulence to increase the convective heat transfer process. The code does now recognize fin tube heating surfaces.

Water circulation within boilers must be adequate to carry heat away from localized high-temperature areas (hot spots) and thus prevent damage from overheating. Hot spots may result in the localized generation of steam bubbles which, on moving to lower temperature areas, collapse, resulting in noise and vibration. In steam boilers, circulation is complicated by the need to provide proper "disengaging" surface for the steam bubbles to rise to the surface of the water and break free. Since the heat transfer rate from flue gases to steam is less than to water, boiler design must provide correct heating surface loading to prevent hot spots. *Inclined water-tube* steam boilers with large headers have good thermal water circulation. In *fire-tube boilers* the heat in the fire tubes provides heat for thermal water circulation; however, the fire tubes cause restriction to the water circulation. In bent-tube steam boilers thermal circulation of the water is generated through heated 'riser' bent tubes and unheated "down-corner" pipes.

Condensation.   The temperature at which water vapor in combustion products gases will condense is approximately 135°F (57°C). However, depending upon atmospheric pressure, humidity, and temperature, condensation can occur at temperatures as low as 125°F to 130°F (53°C to 55°C). Condensation will occur anytime combustion products come into contact with boiler metal surfaces at or below this temperature. Condensation will corrode the boiler metals and damage boiler refractory. In addition, condensation droplets will quench the flame preventing complete combustion, reducing the combustion efficiency. Carbon soot from the incomplete combustion will congest the flue gas passages, further hampering combustion and reducing the convection heat transfer.

Thermal shock will damage a boiler when cold or cool inlet water temperature causes temperature-induced stresses in boiler pressure vessel components. Conflicting expansion-contraction loads can result in failure of the pressure vessel requiring substantial repairs or even complete vessel replacement.

• Water-tube boilers are resistant to thermal shock.

• Cast-iron boilers, and to some extent fire-tube boilers, must be protected from low inlet water temperature.

• Boilers should not be connected to two- or three-pipe heating and chilled water cooling systems where a common return pipe will allow chilled water to return to the boiler each time the building switches from air conditioning to heating. This hookup can cause problems when used with cast-iron or fire-tube boilers. Boiler condensing and fatigue stress can also occur with common return piping systems.

Fatigue fractures.   Fatigue fractures occur in boilers due to the thermal stress cycling conditions. Boiler components must be able to handle temperature differences and the stresses they place on heavy wall tube sheets and castings. Fire-tube boiler tube sheets, with temperature differentials of from 2000°F furnace exit temperatures to 500°F (approx 430°C) flue gas exit temperatures and are welded to the boiler drum with welded fire tubes across its face, are subject thermal stresses. Boiler cycling contributes to the metal thermal fatigue stresses. The same metal fatigue condition occurs from bottom to top on cast iron sections that are bolted tightly together. Water-tube boilers are not subject to these thermal fatigue stresses.

- Water in steam boilers must be heated from entering temperature to boiling temperatures.
- Water level controls must be properly applied, installed, and maintained. The failure to maintain a high enough boiler water line probably will result in damage to the pressure vessel with possible failure. Too high a water level in steam boilers may result in abnormally wet steam and carryover of water into the steam-piping system, degrading the heat transfer system and overworking condensate traps.
- The boiler must, when completely **assembled,** be capable of being installed in the space available, including allowance for access areas and periodic maintenance functions such as inspection and tube replacement.
- The following **installation features** must be properly designed: foundations, electrical supply, water supply, relief valve drains, venting, combustion air supply, noise parameters, and alarm systems.

16.1.6   WATER-TUBE BOILERS

16.1.6.1   Operating Pressure

Water boiler standard design pressures are 30, 60, 125, and 160 lb/in^2 (2.1, 4.1, 8.6, 11.0 bar).

High-temperature water boilers standard design pressures are 300, 400, and 500 lb/in^2 (20.7, 27.6, 34.5).

Steam boiler standard design pressures are 15, 150, 200, 250, and 300 lb/in^2 (1.03, 10.3, 13.8, 17.2, 20.7 bar); water-tube boilers are available for all operating pressures from 15 psi (103 kPa) through the ultra-high pressures used in utility boilers, which often exceed 3500 psi (241 bar).

16.1.6.2   Water Tube Boilers

Water-tube boilers are available in all sizes from residential through large utility power generation boilers. Above 800 bhp (7849 kW), water-tube boilers are used almost exclusively since the large rolled shell of the scotch fire-tube boiler becomes prohibitively expensive, both the manufacture and to transport.

In recent years, small packaged water-tube boilers, ranging to 800 bhp (7849 kW) have become the preferred design for hot water space heating applications. This preference has developed because, unlike fire-tube boilers, water-tube boilers are largely impervious to *thermal shock* and *fatigue stress failures.*

16.1.6.3   Water Tube Boiler Design

1. **Pressure Vessel: Straight inclined water tube** boilers have large fabricated steel headers with removable head plates that allow full inspection, cleaning, or replacement of the tubes. The tubes are rolled into the header tube sheets. Two headers and the tube bundle make up the boiler pressure vessel. Large 2 in (50.8 mm) diameter straight steel tubes (SA-178) or copper tubes (SB-75) provide excellent unobstructed low pressure drop boiler water circulation.

Bent water tube boilers have upper and lower drums that are connected by two or more rows of bent tubes. Down comers pipes should be located outside of the combustion chamber of the boiler to assist boiler circulation. In larger boilers, the upper drum is generally connected to the lower drum only by the boiler tubes. In steam boilers, the upper drum will contain baffling to direct and dry the steam before it exits the boiler. Bent water tube boilers use drums fabricated from steel pipe or rolled steel plate. Small drums are equipped with inspection openings at each end. Large drums requiring entry for internal inspection and maintenance are equipped with manways.

2. **Tubes and tube attachments:** The most commonly used bent water tube material is SA-178 steel and tube sizes vary between 1in (25.4 mm) and $1^1/_2$ in (38.1 mm) outside diameter. For bent tubes, good design practice requires long radius bends without thinning of wall tube thickness and without flow restrictions. Smaller bent tube boilers use mechanical tube fittings and replacement of the tubes requires special tools. Straight tubes are generally expanded into the drum(s) and/or tube sheets. Scotch fire tubes may be welded in the hot passes and rolled in the cooler passes.

3. **Bent tubes must** be sloped for proper flow: The exact amount of slope depends on the location of the tubes in the boiler. Boilers with bent tubes, which are 1 in (25.4 mm) diameter, or less, should be pitched with minimum slope from horizontal as follows: All furnace floor tubes must have a minimum slope of 6.5° to the horizon to achieve good circulation and drainage; and all furnace roof tubes must have a minimum slope of 7.5° to the horizon to permit good circulation and maximum steam-relieving capacity.

4. **Furnace and heat exchanger designs:** Furnace and heat exchanger design is important for good combustion. Radiant heat transfer occurs within the furnace. Hot flue gas convection heat transfer occurs in the heat exchanger. Several surfaces are used to contain the heat of the combustion process and channel it to the heat-absorbing surfaces (see Fig. 16.1.12).

   a. *Furnace design:* Refractory walls and/or floors combined with various water tube arrangements comprise the combustion chamber, also called the furnace. Refractory provides hot surfaces that contribute to the combustion process. Water tube surfaces can quench the flame edges if they contact them. Refractory furnace materials are usually preformed panels backed with high-temperature insulation. The hot surface material may also be formed refractory clay material or ceramic fiber product.

   b. *Tangent tube* walls provide a single row of tubes placed adjacent to one another. This design needs boiler water circulation to absorb the intense radiant heat load.

   c. *Multirow bent tube* design should provide high boiler water flow in the first row to handle the intense radiant heat load and proper flue gas flow velocities for convection heat transfer in the second and subsequent tube row(s).

   d. *Multirow inclined straight tubes:* Large diameter straight steel tube arranged in triangular close pitch efficiently extracts the radiant and convection flue gas heat. The triangular pitch tube bundle provides flue gas flow around each tube with flue gas velocities and turbulence to achieve outstanding convection heat transfer. Large headers and large inclined straight boiler tubes ensure outstanding boiler water circulation.

   e. *Copper fin tube:* The extended fin surface provides excellent flue gas convection heat transfer performance. Pumped boiler water circulation ensures efficient transfer from both the radiant and convection heat sources.

   f. *Steel finned tube:* Fins are welded to the tubes to extend external heating surface. The tube wall temperature is higher with this type of wall because less cooling water is available per unit of heat-absorbing surface, unless pumped boiler water circulation is provided.

   g. *Membrane tube walls:* Solid steel strips are welded between tubes in this construction. The tube wall temperature is higher; therefore, boiler water circulation is required to absorb the intense radiant heat load.

A)   Tangent Tube Walls

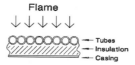

B)   Multiple-row Tube Walls

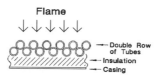

C)   Finned Tube Walls

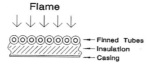

D)   Membrane Tube Walls

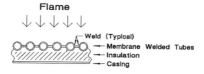

E)   Refractory Walls

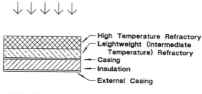

FIGURE 16.1.12   Furnace wall construction. (*Courtesy of Ajax Boiler Inc.*)

Convection heating surface: Convection heating surface is designed to incorporate the maximum number of tubes in the smallest possible space consistent with flue gas pressure drop limitations. Convection heat transfer effectiveness increases as the flue gas turbulence and velocity increases. Since flue gas volume decreases as it cools, the boiler design must use baffling and reduced pass area to maintain heat transfer efficiency.

Adequate accessibility should be provided to clean and, if necessary, replace tubes. Soot blowers are sometimes needed in convection sections when sooting occurs from firing heavy oil or solid fuels.

5. Boiler casing and **insulation:** Water tube boilers with forced draft combustion systems may use pressurized furnaces to maximize flue gas pressure drop across the convection tube banks. Two types of casing are used; membrane and double-wall.

   a. Membrane construction: Membranes between the tubes in the outermost tube rows, or a continuous membrane casing outside the tubes provide a means of containing the hot combustion gases. The membrane is backed by insulation or an insulation/air gap combination (see Fig. 16.1.12d)

   b. Double-wall construction (Fig. 16.1.13). Double-wall constructions consist of an inner and outer casing with either insulation or circulated combustion air between the casings. The inner casing is welded or otherwise sealed to provide a leak proof containment for the combustion gases.

   Insulation is laid over the inner casing to reduce heat losses, or in some cases, the gap between the inner and outer casings is arranged to form a channel for combustion air flow. By this means, the heat energy which would have been lost to the boiler room is captured by the combustion air and returned to the furnace.

   The outer casing provides additional strength, a cover for the insulation, and an aesthetic appearance.

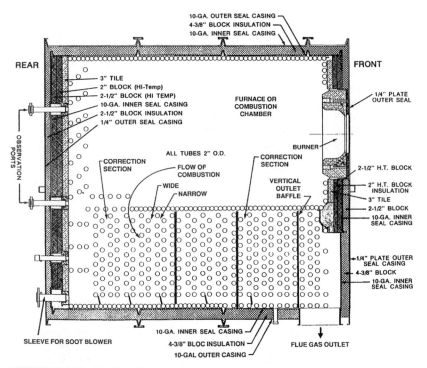

FIGURE 16.1.13   Double-wall construction.

Note: This is the plan of a D-type boiler. (*Courtesy of Cleaver-Brooks.*)

16.1.7 FIRE-TUBE BOILERS

Fire-tube boiler designs route the flue gases through the inside of the boiler tubes that are surrounded by boiler water in the pressure vessel/drum. Fire-tube boilers are limited in size and pressure due to the high thermal stresses and the requirement for thicker pressure vessel steel.

16.1.7.1 Operating Pressure

Fire-tube boilers are commonly available for maximum allowable working pressures up to 150 psi (10.3 bar). Some manufacturers build custom scotch units to 300 psi (20.6 bar); however, these are generally limited in size to 250 bhp (2453 kW) because of the high cost of producing the rolled cylindrical outer heavy wall shell.

16.1.7.2 Size Ranges

Fire-tube boilers are generally available in the range 20 to 800 bhp (196 to 7848 kW) and in pressure up to 150 psi (10.3 bar). The larger units, 150 hp (1471 kW) and above, tend to use the scotch design.

16.1.7.3 Types of Fire-Tube Boilers

1. In the *scotch boiler* the burner fires into a cylindrical steel combustion chamber after which the hot gases pass through one, two, or three tube passes before leaving the boiler. Two, three, and four pass boiler gas flows are identified in Fig. 16.1.14. The combustion chamber and all of the tubes are welded to large tube sheets at each end of the boiler and are immersed in boiler water inside a larger cylindrical pressure vessel, or shell.

   Scotch boilers are further classified into "dry back" and "wetback" types. In the dry back boiler, the combustion gases are directed from combustion chamber to tube-pass and from tube-pass to tube-pass by a steel refractory insulated "turn around chamber." In the wetback design, the turn around chamber is water cooled.

2. The *firebox* or *horizontal return tube* (HRT) boilers (see Fig. 16.1.9) comprises a bank of fire tubes immersed in boiler water mounted above a high-temperature refractory lined combustion chamber. Large firebox boilers were often factory-built pressure vessels mounted on field erected fireboxes, but smaller current sizes are factory packaged with combustion chamber. Packaged firebox boilers are produced in widely varying formats and in sizes up to around 100 bhp (981 kW). In the typical current design, the pressure vessel is rectangular in shape (see Fig. 16.1.11) with water legs framing the sides and rear of the combustion chamber.

16.1.7.4 Fire-Tube Boiler Design

1. *Pressure vessel:* Scotch fire-tube boilers use round drums rolled from steel plate with full diameter heavy tube sheets welded on each end. Scotch boilers are fabricated in sizes up to 1000 bhp (9810 kW). Firebox boilers have rectangular drums and headers fabricated and welded from steel plate. In all cases, the relatively large expanses of flat pressure vessel surfaces often require stay rods for support against internal pressure.

2. *Tubes and tube attachments:* The most commonly used tube material is SA178 steel and tube sizes vary between $1\frac{1}{2}$ in (38 mm) and 3 in (76 mm) outside diameter. Straight tubes

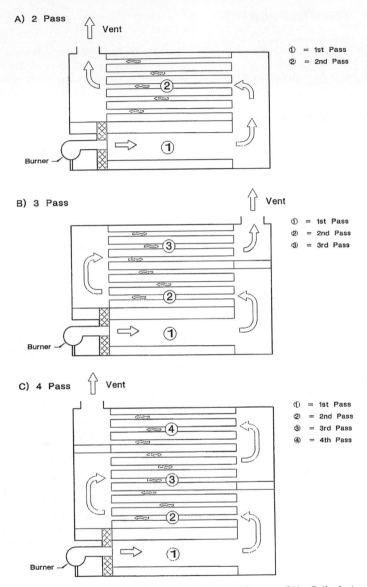

FIGURE 16.1.14   Firetube boiler pass arrangements. (*Courtesy of Ajax Boiler Inc.*)

facilitate inspection and mechanical cleaning. Tube joints exposed to combustion gases leaving the combustion chamber are usually welded and in some cases, on cooler passes, the tubes may be expanded into the tube sheets.

3. *Furnace designs:* Fire-tube boilers utilize widely varying furnace designs. In the scotch boiler, the cylindrical combustion chamber is completely water-backed. Very high furnace loading of up to 200,000 but/ft³ is common.

Firebox boilers have a rectangular five-sided refractory-lined box (four walls and a floor). When this box is constructed from bricks, the boiler is said to be *brick set*. In an adaptation of the design for packaging purposes, the firebox is mounted inside a steel casing to which the boiler pressure vessel is bolted or welded. Adaptations for low pressure boilers use water legs descending from the boiler pressure vessel to the floor of the firebox to form the water-backed walls for the combustion chamber.

16.1.8   CAST-IRON BOILERS

16.1.8.1   Operating Pressure

Cast-iron boilers are limited to low-pressure steam up to 15 lb/in^2 (1.03 bar) and hot water up to 30 lb/in^2 (2.06 bar) and 250°F (121°C).

16.1.8.2   Size Ranges

The primary use for cast-iron boilers lies in the residential market; however, small commercial sizes are available up through 200 bhp (150 kW).

16.1.8.3   Types of Cast-Iron Boilers

Cast-iron boilers are cast-iron pressure vessel sections with internal push-nipples or sectional with external cast-iron headers (drums). Cast sections with *internal push-nipple* boilers are joined with tapered nipples and held tightly together with tie rods or bolts. The gas joints between sections are sealed with gasket material. The nipples are sized to provide the required free circulation of water and/or steam from section to section.

Cast sections with external header boiler are connected to the headers (drums) with screwed nipples. There are three headers, a supply manifold centered at the top and two at the bottom, one at each side. With this type of boiler, it is possible to temporarily isolate a damaged section and put the boiler back into service when a section becomes cracked. Cast-iron boilers that are built with water-filled spaces across the bottom of the combustion chamber are called wet-base boilers. This design permits circulation of water within the base and sides of each section and reduces boiler heat loss to the mounting pad. The boilers have refractory front and rear walls.

Sectional boilers are made up of vertical sections and look much like a loaf of sliced bread. The number of sections determines boiler capacity. Each of the inside sections of a particular model are identical. The end sections are, of course, different from the inside sections. Because of the sectional construction mode, cast-iron boilers' individual sections can be moved into the boiler room and then assembled. Even so, individual sections are heavy and may weigh as much as 2000 lb (910 kg). Cast-iron boilers can be field assembled or factory assembled, complete with burner, fuel train, operating controls, safety devices, and prewired panel-mounted controls. During assembly care must be taken to properly tighten push-nipples, tapered nipples, threaded nipples, and the bolting together of the assembly to prevent leaky joints and to minimize expansion and contraction stresses.

16.1.9   ADDITIONAL BOILER DESIGN CRITERIA

Several additional boiler design criteria should be examined during the selection process. These are efficiency, emissions control equipments, control system complexity, and life cycle costing.

16.1.9.1 Efficiency

In HVAC terms, efficiency is generally defined as the ratio between heat output in the fluid being heated and heat input in the form of fuel being consumed. Typical efficiencies for contemporary boiler packages range from 77 to 80 percent and some in the 82 to 84 percent range. There are several caveats related to efficiency-based selection of boiler products.

1. Condensing boilers with efficiencies over 84 percent will have partial or complete condensation of the water vapor in the flue gas. Condensing boilers can accommodate inlet water temperatures below 135°F (57°C), as found in low-temperature heating systems, heat pumps, snow melting, and pool heating applications.

2. Boilers which will operate in a partial or fully-condensing mode require corrosion-resistant stack or chimney and provisions for drainage of the slightly acidic condensate.

3. In general, the higher the claimed operating efficiency, the higher the first cost of the boiler. In some cases, first cost increases by a factor of 2 to 4 times. As fuel cost and utility rebate incentives increase, the fuel savings payback will become more attractive. High-efficiency boilers are more complex than standard efficiency units and maintenance costs will be slightly higher.

16.1.9.2 Emissions Control Requirements

Specific emissions control requirements are mandated nationally and locally, or may be specified by the owner or design engineer. Emissions control falls into the following categories: carbon monoxide, nitrogen oxides, smoke, and sulfur oxides.

1. Carbon monoxide (CO) is a poisonous pollutant emitted from all carbon-based combustion processes. Stationary boilers are required to emit no more than 400 ppm by volume of carbon monoxide in the flue gas, but most properly designed boilers provide CO levels of 50 ppm or less.

2. Nitrogen oxides (NO_x) are a constituent of photochemical smog and acid rain. In southern California, NO_x levels are limited to 30 ppm for small boilers and 12 ppm for larger boilers measured by volume, corrected to 3 percent oxygen. The correction factor is

$$\text{Corrected } NO_x \text{ at 3\% } O_2 = \frac{\text{Measured } NO_x \text{ (in ppm)} \times 18}{21 - (\text{measured } O_2)} \qquad (16.1.1)$$

NO_x generation is dependent on both fuel and combustion system type. Natural gas premix-type combustion systems can attain the lowest NO_x levels (10 to 20 ppm) followed by forced draft natural gas boilers with flue gas recirculation (FGR) (25 to 30 ppm). Forced draft No. 2 oil systems (no longer permitted in California) typically run 60 to 80 ppm with blue gas recirculation and 80 to 120 ppm without FGR. Low nitrogen No. 2 fuel oils are available, which can give NO_x levels in the 35 to 50 ppm range with FGR and are permitted for short term emergency (medical) service.

3. Smoke is primarily a fuel-based issue. Natural gas and propane are inherently smoke free. Fuel oil is inherently smokier as fuel grade number increases. The normal benchmark for smoke measurement is the Bacharach smoke scale where "0" represents a clean stack and "9" represents a smoke level approximately equivalent to 10 percent obscuration through the stack plume. Note that the Bacharach scale covers the lower 10 percent of the older Ringelmann scale and thus provides a far more sensitive means of measurement. The pump/filter apparatus used in the Bacharach test also provides an easy means of measurement at the boiler outlet as opposed to visual observations at the stack.

4. Sulfur oxides (SO_x) are emitted when fuels containing sulfur are burned. Fuel oils and coal contain sulfur in varying degrees and, in HVAC systems, the primary form of SO_x control is by fuel selection. Clean fuels, such as natural gas, contain no sulfur and therefore produce no SO_x emissions. No. 2 fuel oil contains only limited amounts of sulfur and is generally considered *clean* in terms of SO_x emissions. Coal and heavy oil burning systems generally require "back and cleanup," usually in the form of wet scrubbers to meet SO_x requirements.

16.1.9.3 Control System Complexity

The level of controls complexity is primarily mandated by insurance code and/or by local, state, or national codes. The owner's insurer will require that safety controls on the boiler (fuel shut-off valves, water level controls, limit controls, etc.) conform to one of the available safety codes (ASME, UL, CSA, FM, IRI, etc.). Many states have adopted the ASME Controls and Safety Devices Code CAS-1 and some states and local jurisdictions have their own specific safety requirements. Once safety requirements have been satisfied, controls complexity may be limited to simple on-off automatic controls or may expand to include staged or fully-modulated firing, outdoor reset temperature control of water boilers, and or lead-lag control of multiple boilers. Microprocessor burner management/flame safe-guard systems will provide diagnostics and status and fault communication and communication with building energy management systems.

16.1.10 SYSTEMS AND SELECTIONS

16.1.10.1 Load Analysis

Once the heating and/or process loads have been calculated, a decision must be made to use a steam, water, or high-temperature hot water system. A comparative study of capital, operational, and maintenance costs should be made, including electrical requirements.

The frequently cited advantages attributed to hot-water systems are:

1. Lower operational temperatures reduce system heat losses and improve boiler efficiency.
2. Easier to maintain.
3. No condensate return systems required.
4. Need for steam traps (and associated maintenance costs) eliminated.
5. Smaller pipe size for a given load.
6. Quieter operation than steam.
7. More uniform space heating environment.
8. Minimizes water treatment and make up water requirements.
9. Lower levels of piping expansion and contraction.

Advantages cited for steam systems are:

1. Higher temperature available for user.
2. Constant temperature to user at condensing condition.
3. Improved heat transfer performance allows smaller user equipment.
4. Fast response to load change.
5. Reduced pumping effort required (less electrical power).

16.1.10.2 Number of Boilers Required

In most uses, a complete loss of heat cannot be tolerated and multiple boilers with redundant capacity must be used. Use of a single boiler to provide 100 percent of the required system heat is not good practice unless the loss of the boiler for a time will not seriously curtail operation of the facility. Sizing options include selecting two boilers each sized for two-thirds or three-quarters of the required capacity. For most of the plant operating load range, one boiler will satisfy critical load requirements. In the event of a breakdown, or for servicing, one boiler will carry the main load. Boiler sizing must include normal and peak load conditions as well as seasonal load variations.

Heating Boilers: The total or "gross" load, in Btu/h or W, for the building comprises:

1. Conventional heating load including building heat loss and process requirements.

2. Domestic water heating requirements, if boiler water is to be used to generate hot potable water. Adjust for storage versus semi-instantaneous water heaters that will add to the boiler peak load requirement. Design guidelines for estimating hot water usage are given in Table 16.1.1.

3. Air-conditioning load, which may, in some locations, give a summer boiler water demand that exceeds the winter heating demand.

4. Pickup allowance, which covers the need to heat the building from cold within a limited period of time. This is typically a factor in mild climates where the boilers are often shut off for 12 or more hours each day. In the absence of unusual factors, a maximum of 10 percent of known loads may be added.

5. *Piping Losses:* Allowance should be made to cover unusable heat loss from piping. When this amount is not specifically calculated, 15 percent of known loads may be added.

6. Maximum instantaneous demand includes sudden short duration peak load requirements, other than pickup allowances. Steam boilers, in particular, should be protected to keep steam flow within the boiler's rated capacity by over sizing or by control orifices, valves, or other means. Water boilers must be protected from low-temperature operation

TABLE 16.1.1   Hot-Water Demands and Use for Various Types of Buildings

	Nominal capacity range in $I = B = R$ net rating			
	Push-nipple boiler		Header (drum) boiler	
Fuel	Steam, lb/h (kg/h)	Hot water, MBtu/h (kW)	Steam, lb/h (kg/h)	Hot water, MBtu/h (kW)
Gas	25.2–4,035 (11.4–1,836.3)	26–4,160 (7.6–1,220)	1,008.8–11,349 (457.6–5,147.9)	1,040–11,700 (304.8–3,429.3)
Oil	76.5–407 (34.7–184.6)	79–420 (23–123)	1,008.8–11,349 (457.6–5,147.9)	1,040–11,700 (304.8–3,429.3)
Dual fuel, gas or oil	412–3,900 (186.9–1,769)	425–4,020 (125–1,180)	1,008.8–11,349 (457.6–5,147.9)	1,040–11,700 (304.8–3,429.3)
Anthracite coal	97–254 (44–115.2)	100–262 (29.3–76.8)	—	—
Bituminous coal	142.6–292 (64.7–132.5)	147–301 (43–88.2)	—	—

Note: 1 gal = 3.7 L.
*Per day of operation.

during peak demand and pickup periods to prevent condensation damage and, in some cases, thermal shock.

16.1.10.3 Operating Pressure

Boilers cannot be operated at their design pressure (MAWP) because the safety valves and controls must be set to operate at or below the design pressure by code and law. Boiler design pressures must therefore be selected in excess of normal system operating pressure.

For 15 lb/in^2 (1.03 bar) steam boilers, use 12 lb/in^2 (0.83 bar) maximum system pressure. For hot water boilers, the total operating pressure includes building or site static pressure (due to height of system), thermal expansion pressure, and pressure due to pump head working against pipe and system resistance. Boiler operating pressure must not exceed the boiler design pressure minus an allowance for operation of safety valves. This is generally 5 to 8 lb/in^2 (0.35 to 0.55 bar) for 30 lb/in^2 (2.06 bar) design boilers; and 10 to 15 percent for boilers with design pressures of 45 lb/in^2 (3.09 bar) and over. Allowance for pump head is not usually made when circulator pumps are mounted between the boiler and the system resistance.

In steam systems, the boiler generates the pressure to distribute steam through the piping system to the user equipment. The boiler selected must be capable of operating at the pressure required by the system heat user.

If the system is designed for hot water, sufficient pressure must be applied throughout the system to prevent the hot water from flashing to steam. Steam flashing results in noisy operation and mechanical damage. The recommended safety margin in system pressure is to use the saturation pressure for 40°F (22°C) above the maximum expected system water temperature.

System operating pressure is created by cold fill pressure plus the thermal expansion resistance provided by the expansion tank. The expansion tank uses captive air or another gas to provide a cushion of gas to absorb the increased volume of the water in the system after heating from cold. Expansion tanks are available, which contain an elastomeric diaphragm that separates the air cushion from the system water, thus reducing the opportunity for corrosion due to air dissolving in the water. On high-temperature hot water systems nitrogen is generally used as the cushion gas as a means of preventing corrosion in the boiler.

16.1.10.4 Operating Temperatures

Operating temperatures for water heating systems fall into three major categories.

1. Normal cold climate heating systems in which the generally accepted inlet and outlet temperatures are 160°F (71°C) and 180°F (82°C), respectively. In cold U.S. climates, the boiler generally runs continuously and condensation problems do not occur as long as the boiler inlet water temperature is always above the condensing range. Water circulation rates through hot water generators are given in Table 16.1.2.

2. Warm climate heating systems where afternoon sun requires that the boiler be shut down for a few hours a day. In such cases, system operating temperatures should not be allowed to drop to the extent that boilers would run in a condensing mode, causing corrosion damage and shortening the life of the boiler. To protect the boiler, provisions can be made to allow it to run at noncondensing temperatures with a blend water system loop to produce the desired system operating temperature.

3. **Heat pump loops** run at very low temperatures, generally around 65°F (18.3°C). The condensing boiler is an excellent application for this low-temperature loop system. The return loop blend system for noncondensing boilers returns boiler output water to blend

TABLE 16.1.2 Water Circulation Rates Through Hot-Water Generators

bhp	Gross boiler output at nozzle, MBtu/h	System temperature drop, °F (°C)					
		10 (5.5)	20 (11.1)	30 (16.7)	40 (22.2)	50 (27.7)	60 (33.3)
		Circulation rate gal/min					
20	670	134	67	45	33	27	22
30	1,005	200	100	67	50	40	33
40	1,340	268	134	89	67	54	45
50	1,675	335	168	112	84	67	56
60	2,010	400	200	134	100	80	67
70	2,345	470	235	157	118	94	78
80	2,680	536	268	179	134	107	90
100	3,350	670	335	223	168	134	112
125	4,185	836	418	279	209	168	140
150	5,025	1005	503	335	251	201	168
200	6,695	1340	670	447	335	268	224
250	8,370	1675	838	558	419	335	280
300	10,045	2010	1005	670	503	402	335
350	11,720	2350	1175	784	587	470	392
400	13,400	2680	1340	895	670	535	447
500	16,740	3350	1675	1120	838	670	558
600	20,080	4020	2010	1340	1005	805	670
750	25,100	5025	2512	1675	1255	1006	838
800	26,800	5360	2680	1790	1340	1070	900
1000	33,500	6700	3350	2230	1680	1340	1120

Note: Conversion factors: 1 hp = 746 W and 1 gal = 3.8 L.

with the (below 135°F) low-temperature loop water, to obtain boiler return water temperatures of 140°F. A sufficient (reduced) amount of 180°F water is returned to the loop to maintain its desired design low loop temperature. The loop pump will normally provide sufficient circulation and a standard gate valve can be used to adjust the bypass flow. The return loop blend system eliminates, for low loop temperature application, the need for positive recirculation systems with blending control valves or the need to use an indirect water heat exchanger or coil.

16.1.10.5 Boiler Capacity

For water heating boilers, the boiler size may be selected from the manufacturer's gross output ratings. For steam boilers, the calculated system load must be adjusted to the "from and at 212°F (100°C)" basis prior to selection from the manufacturer's gross output settings using the "factor of evaporation" information given in Table 16.1.3. Use the following equation:

$$\text{Required boiler capacity, from and at 212°F (100°C), bhp} = \frac{\text{design load at design temperature, lb/h}}{\text{factor of evaporation from Table 16.1.3, lb/bhp}} \qquad (16.1.2)$$

Select factor of evaporation from Table 16.1.3 using design feed water temperature and design *operating* pressure.

TABLE 16.1.3 Factor of Evaporation, lb/bhp Dry Saturated Steam

Feed-water temp., °F	Gauge pressure, psig																	
	0	2	10	15	20	40	50	60	80	100	120	140	150	160	180	200	220	240
30	29.0	29.0	28.8	28.7	28.6	28.4	28.3	28.2	28.2	28.1	28.0	28.0	27.9	27.9	27.9	27.9	27.9	27.8
40	29.3	29.2	29.1	29.0	28.9	28.7	28.6	28.5	28.4	28.3	28.2	28.2	28.2	28.2	28.2	28.1	28.1	28.1
50	29.6	29.5	29.3	29.2	29.1	28.9	28.8	28.8	28.7	28.6	28.5	28.5	28.4	28.4	28.4	28.3	28.3	28.3
60	29.8	29.8	29.6	29.5	29.4	29.2	29.1	29.0	28.9	28.8	28.8	28.7	28.7	28.6	28.6	28.6	28.6	28.5
70	30.1	30.0	29.9	29.8	29.7	29.5	29.4	29.3	29.2	29.1	29.0	29.0	28.9	28.9	28.9	28.8	28.8	28.8
80	30.4	30.3	30.1	30.0	30.0	29.8	29.6	29.6	29.5	29.3	29.2	29.2	29.2	29.2	29.1	29.1	29.1	29.0
90	30.6	30.6	30.4	30.3	30.2	30.0	29.9	29.8	29.7	29.6	29.5	29.5	29.4	29.4	29.4	29.3	29.3	29.3
100	30.9	30.8	30.6	30.6	30.5	30.3	30.2	30.1	30.0	29.8	29.8	29.8	29.7	29.7	29.7	29.6	29.6	29.6
110	31.2	31.2	30.9	30.8	30.8	30.6	30.4	30.3	30.2	30.0	30.0	30.0	30.0	30.0	29.9	29.9	29.8	29.8
120	31.5	31.4	31.2	31.2	31.1	30.8	30.7	30.6	30.5	30.4	30.3	30.3	30.2	30.2	30.2	30.1	30.1	30.1
130	31.8	31.7	31.5	31.4	31.4	31.1	31.0	30.9	30.8	30.7	30.6	30.6	30.5	30.5	30.4	30.4	30.4	30.4
140	32.1	32.0	31.8	31.7	31.6	31.4	31.3	31.2	31.1	31.0	30.9	30.8	30.8	30.8	30.8	30.7	30.7	30.6
150	32.4	32.4	32.1	32.0	31.9	31.7	31.6	31.5	31.4	31.2	31.2	31.2	31.1	31.1	31.0	31.0	30.9	30.9
160	32.7	32.7	32.4	32.4	32.3	32.0	31.9	31.8	31.7	31.5	31.4	31.4	31.4	31.4	31.3	31.3	31.2	31.2
170	33.0	33.0	32.7	32.6	32.6	32.3	32.2	32.1	32.0	31.8	31.7	31.7	31.7	31.6	31.6	31.6	31.5	31.5
180	33.4	33.3	33.0	33.0	32.9	32.6	32.5	32.4	32.3	32.2	32.1	32.0	32.0	32.0	31.9	31.9	31.8	31.8
190	33.8	33.7	33.4	33.3	33.2	32.9	32.8	32.7	32.6	32.5	32.4	32.4	32.3	32.3	32.2	32.2	32.1	32.1
200	34.1	34.0	33.7	33.6	33.5	33.2	33.1	33.0	32.9	32.8	32.7	32.6	32.6	32.6	32.6	32.5	32.4	32.4
212	34.5	34.4	34.2	34.1	33.9	33.6	33.5	33.4	33.3	33.2	33.1	33.0	33.0	33.0	32.9	32.9	32.8	32.8

Note: These metric conversion factors can be used: 1 psig = 0.069 bar, 1 lb = 0.45 kg, and °C = $\frac{5}{9}$ (°F − 132), 1 bhp = 9.81 kW.

16.1.11   HIGH-TEMPERATURE WATER SYSTEMS

Medium temperature water (MTW) systems operating between 250 and 325°F (121 to 163°C) with corresponding operating pressures between 50 and 180 psi can use ASME Section I boilers. High-temperature water (HTW) systems are generally operated between 350 and 430°F (177 to 221°C) with corresponding generator operating pressures between 200 and 525 psi (13.8 and 36.2 bar).

Boilers operating as hot water generators must be pressurized to prevent the formation of steam. Means must also be provided to allow for expansion of the water as it is heated to operating temperature. The appropriate operating pressure levels required for various operating temperatures are shown in Fig. 16.1.15. System pressurization, which is crucial to proper system operation, is accomplished through the use of a properly designed expansion tank. The design of this tank depends on system size and operating temperatures and pressures.

16.1.11.1   Circulation in HTW Boilers

Because of the elevated temperatures in HTW boilers, circulation of water in HTW boilers is crucial to boiler longevity. Circulation types are:

- Pumped jet-induced circulation (see Fig. 16.1.16)
- Forced circulation with or without separation drum
- Once through orifice controlled (see Fig. 16.1.17)

Jet-Induced Circulation.   The D-type two drum water tube HTW generators have proved their dependability by operating trouble-free for more than 30 years. They have low first, operating, and maintenance costs. Output capacities range to 65 million Btu/h (19 MW) for pressures to suit water temperatures up to 430°F (221°C) with supply-to-return temperature differences up to 200°F (93°C). Figure 16.1.17 shows the water circulation through the HTW unit. The return water flows from the circulation pump through the internal distribution pipe into the plenum chamber and then rises through all furnace tubes to the top drum. Here water mixes with the drum water. From there, a part is taken out as the supply flow requires and the rest circulates down to the lower drum through the down-comer tubes in the gas outlet pass. The drum cross-section provides for free longitudinal water flow.

Large, Jet-Induced Circulation, Multidrum Units.   Shop-assembled, large-capacity HTW generator units are an advanced multidrum type of design. Figures 16.1.18 and 16.1.19 show units for capacities of 70 to 150 Btu/h (20.5 to 44 MW) and 430°F (221°C) supply water temperature. These large units have two lower drums and one upper drum, whereas the medium-size units have two drums. The return water enters the lower drums through a jet pipe that discharges into the mixing chamber of both lower drums. The "return" water is forced by the system or the boiler circulation pumps directly into the bottom drum (Fig. 16.1.18). An internal distributing pipe with multiple jet outlets discharges into a plenum chamber around the furnace tube inlets. The multiple-jet action of the return water induces the boiler water already in the drum to flow into the plenum chamber. There, both waters mix and enter the furnace tubes, flow up through the tubes, and discharge into the upper drum, completing the first stage of circulation.

The second stage of circulation results from controlling the gas flow by gas pass baffling. The baffles define a hot-gas pass in which the water tubes are risers and the cold-gas pass in which the water tubes are down comers. The combined action of return jets, plenum chamber, water movement in the lower drum, and hydraulic heat self-balancing effect increases the down flow of water in the down comers located in the cold-gas pass.

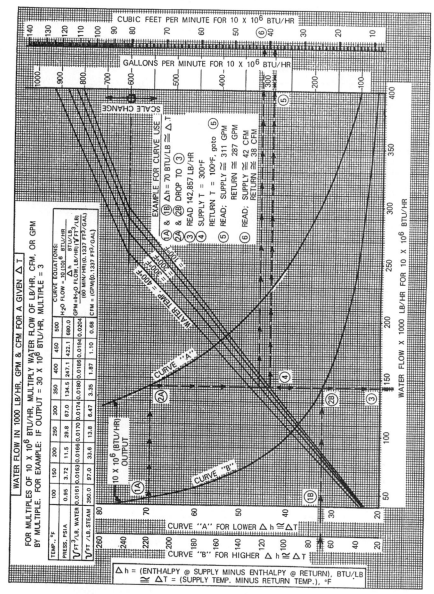

FIGURE 16.1.15 Water flow for a given temperature change. (*Courtesy of Cleaver-Brooks.*)

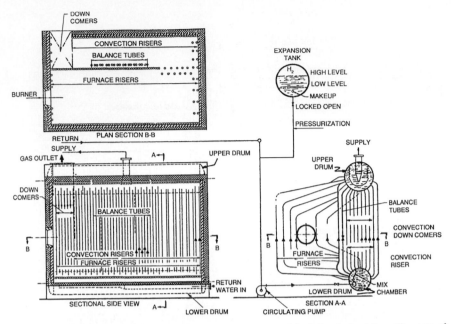

FIGURE 16.1.16   Industrial watertube HTW generator—pumped jet-induced circulation. (*Courtesy of Cleaver-Brooks.*)

Unheated down comers at the front and rear of the boiler for large units normally use additional circulation safeguards.

Breakup screens and separation baffles mix the discharge water from the furnace tubes with the water present in the upper drum. The water leaving the upper drum of the generator does not contain steam because the operating drum pressure is higher than the operating

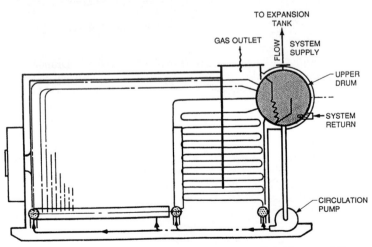

FIGURE 16.1.17   Industrial watertube boiler—orifice controlled. (*Courtesy of Cleaver-Brooks.*)

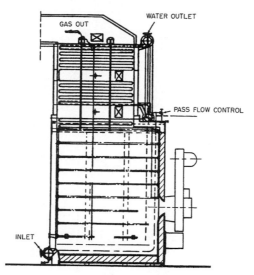

FIGURE 16.1.18 Forced-circulation boiler—orifice-controlled. (*Courtesy of Cleaver-Brooks.*)

water temperature saturation pressure. The difference is usually provided by nitrogen pressurization of the expansion tank. If the boiler or system pumps fail, the plenum chamber baffle and nozzle arrangement in the lower drums will allow natural water circulation by difference of the density of the water in riser tubes and down-comer tubes. Furnace tubes, therefore, are never without a flow of water. Separation baffles in the upper drum provide a channel for the longitudinal return flow toward the down flow end of the upper drum.

Forced-Circulation Boilers with Release Drum. Figure 16.1.17 illustrates a unit where the water is continuously circulated through the HTW generator by pumps, which take the water from the upper drum and force it through single-pass or multipass, continuous-tube heat-absorption surfaces. These surfaces can form the furnace and the convection section, either in parallel, or in series. Pump capacity must be sufficient to provide the required water velocity and mass flow in the tubes.

Once-Through Orifice-Controlled Forced-Circulation Boilers. Drumless: Orifice-controlled once-through forced-circulation boilers are available in many designs. Typical designs are La-Mont, corner-tube, multipass, and combinations thereof. The boiler convection section can be in back of, or above, the furnace (Fig. 16.1.17). Some units employ multipass tube groups. Each group is equipped for automatic flow control by use of aquastat(s) located in manifold blocks, which actuate the control valves.

Water is forced by pump pressure through the tubes forming the heat-absorbing surfaces in the furnace and in the convection section. These pumps can be either the boiler circulation pumps or the system pumps. The water flow quantity through these types of generators cannot be varied by more than 10 percent from design value.

16.1.11.2 Tube Sizes and Orifices

The jet-induced first-stage forced-circulation water tube HTW generators use tubes 2 in (51 mm) in diameter and larger. The tubes have unrestricted internal flow area. Orifices in tube inlets or tube outlets are not used as in some generators.

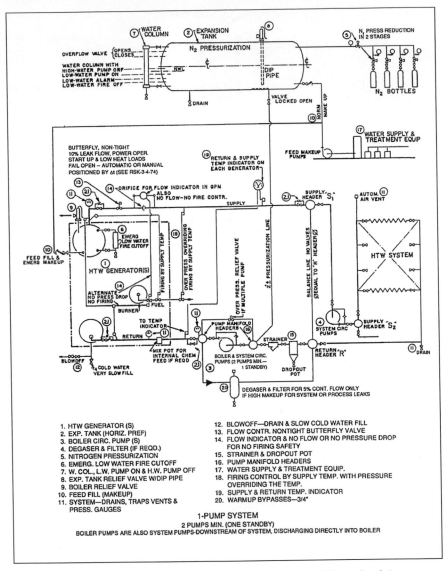

FIGURE 16.1.19   HTW system elements—one-pump design. (*Courtesy of Cleaver-Brooks.*)

Once-through orifice-controlled forced-circulation HTW generators may use tubes 2 in (51 mm) in diameter, but generally they are smaller. To ensure equal flow through each tube of a multitube group, inlet (or outlet) orifices are used. Such orifices may be in each tube, or in the inlet, or outlet manifold of a group of tubes. An orifice anywhere in fluid-carrying elements—whether individual tubes, groups of tubes, or manifold headers—can be obstructed by foreign material. Dirt or scale in a system can build up at an orifice and restrict or stop flow, causing overheating of tube metals and eventual tube failure.

16.1.11.3 HTW Design Considerations

Casing Corrosion. Casing corrosion which occurs when high-ash and sulfur fuels are fired can be eliminated by seal-welded inner casings.

Orificed Generator. In an orificed generator, the tube distribution headers provide a dropout point for solids which, after some accumulation, flow into the orifices, where they can cause plugging. Unless firing is stopped at once (within seconds), plugged tubes will be damaged or will even rupture. A water-flow interruption in an orifice-controlled once-through flow generator would, of course, lead to furnace tube rupture if the fire is not immediately put out.

Natural-circulation provisions should be provided in the generator so as to give the flow control ample time to shut off the burner(s) if a no-flow incident by pump or generator should occur.

An HTW generator designed to accommodate natural circulation is less likely to be damaged by plugging, can be cleaned independent of the external system, can be easily inspected internally, can be easily isolated for short- or long-term storage, and is much less susceptible to damage caused by failure of external system components. One further advantage of such generator design is that the system circulation flow rate can be adjusted to meet varying system requirements without causing any operational difficulty in the boiler.

Makeup Water. Equipment to provide properly treated and deaerated makeup water should include pumps, level regulator, storage, deaerator, chemical treatment, etc. This equipment should be located close to the HTW generator. (For a detailed discussion of water treatment see Chap. 8). In a plant where steam of 5 to 10 psig (0.34 to 0.68 bar) pressure is available, a small deaerator for makeup water is advisable.

Although an HTW system is a closed system and theoretically would require no makeup water, in practice, provisions must be made for the supply and treatment of 1 to 3 percent of the system's water content as the hourly makeup water capacity.

When no plant steam is available for deaeration, chemical means must be used to rid the makeup water of oxygen. For a large system, a small, low-pressure steam boiler for steam to a small deaerator may be necessary.

Generator Arrangements within the System. The system can be a one-pump or two-pump type. Figure 16.1.19 shows the one-pump design where the water is pumped directly into the bottom drum of the generator.

The pressure of the pump(s) moves the water through the generator, into the system's supply line, through the system load, and back through the return line to the inlet of the pump(s). One pump (group of pumps) provides the system flow. We thus have a one-pump system design.

Figure 16.1.20 shows a two-pump system. The generator circulating pumps move the water from the return line through the generator and into the supply line. A second pump or group of pumps takes the water from the supply line leaving the generator and pushes the water through the system into the return line. This type of system divorces the system flow from the boiler flow requirement. Either system is satisfactory for a jet-induced generator, but is likely to be required for orifice flow control generators.

For either the one-pump or the two-pump system, it is advisable to use pump manifold headers which permit operation with any one pump to any one generator.

System Expansion. Expansion space maybe provided as follows:

- Expansion space in the generator drum (this is not recommended and is not used except in very small systems), pressurized by steam or by inert gas, usually nitrogen (air, conducive to corrosion) is not recommended

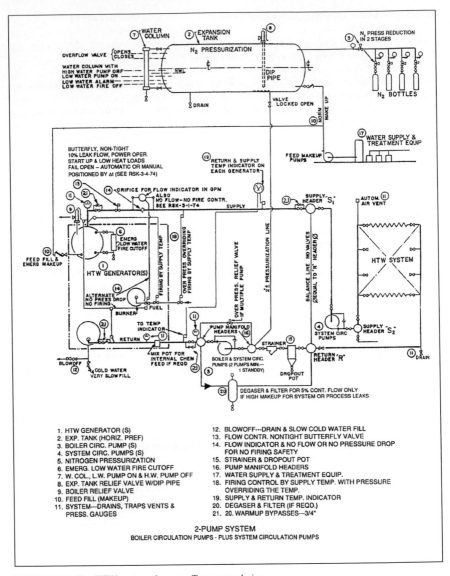

FIGURE 16.1.20   HTW system elements: Two-pump design.

- System expansion tank, with or without overflow tank, pressurized by inert gas or by steam (Fig. 16.1.20), N_2 always preferred
- Overflow tank, atmospheric or pressurized
- Any combination of the above

The total water content of the system including generators, heat users, and all connecting piping will expand in volume as the water is heated from the system fill temperature to system operating conditions. This expansion must be accommodated. Large systems need large expansion tanks external to the generators. A small system, where the system water content is less then two times the water content of the boilers, may not need a separate expansion tank, provided the generator is a drum-type unit and that the upper drum has sufficient space for system and generator water expansion at operating-temperature variations. The tabulation in Fig. 16.1.21 shows the available expansion space in the top drums of standard HTW jet-induced circulation, D-style, and multidrum units.

Water expansion beyond the volume that the specific generator drum can accommodate will require a separate expansion tank. Operational difficulties of a few early systems generally were caused by insufficient expansion space, lack of expansion tanks, or a restricted permissible water-level variation within the generator drum or expansion tank.

Expansion Vessel. Expansion space for larger system volume expansions should be provided in a separate vessel and should be not less than two times the maximum volume of expansion of the water in the system, including the generator. Consideration must be given in this calculation to the probability that the generator may be operating with a greater temperature rise from inlet to outlet than the temperature change which occurs in the external system. The total water expansion volume must be taken by the expansion space provided in the expansion drum above normal drum water level.

The system water expansion under consideration would be only for normal operating-temperature variations and need not take into account the expansion from a cold-start temperature.

EXPANSION SPACE IN								"DELTA" and "DL" JET-INDUCED CIRCULATION HTW GENERATORS												
SIZE DL TYPE	SIZE DELTA TYPE	UPPER DRUM DIA. in ID	WATER CONTENT FILLED TO LWL			DRUM LENGTH (ft)	EXPANSION SPACE OF UNIT		VOLUME REQUIRED FOR WATER EXPANSION FROM 80°F to...... (ft³.)				RATIO OF EXPANSION SPACE TO VOLUME REQUIRED FOR EXPANSION OF BOILER WATER WHEN WATER IS HEATED FROM 80°F to......				PERMISSIBLE MAXIMUM WATER CONTENT OF SYSTEM WHEN WATER IS RAISED FROM 80°F to......				RATIO OF DRUM EXPANSION SPACE TO WATER CONTENT FILLED TO LWL @80°F
			lb.@ 80°F	ft³.	gal		ft³.	gal	°F				°F				°F				
									300	350	400	450	300	350	400	450	300	350	400	450	
	26	36	6,480	104	780	8.2	36.2	270	8.85	12.4	16.55	21.4	4.10	2.9	2.20	1.70	2,380	1,500	920	530	0.35
	34	36	8,380	135	1,000	10.6	45.6	340	11.5	16.0	21.5	27.8	4.00	2.85	2.15	1.65	3,000	1,850	1,130	650	0.34
	42	36	10,270	165	1,230	13.0	55.0	412	14.0	19.7	26.2	34.0	3.90	2.80	2.10	1.60	3,610	2,240	1,365	770	0.33
	52	36	12,630	203	1,517	16.0	68.0	510	17.5	24.2	32.3	42.0	3.88	2.82	2.10	1.62	4,440	2,770	1,680	955	0.33
60	60	36	14,540	233	1,746	18.5	78.0	585	20.0	27.8	37.0	48.0	3.90	2.81	2.10	1.62	5,075	3,180	1,930	1,100	0.33
68	68	36	16,435	264	1,974	21.0	86.5	648	22.5	31.5	42.0	54.5	3.85	2.75	2.06	1.58	5,660	3,470	2,100	1,175	0.33
76	76	36	18,360	295	2,200	23.3	95.5	715	25.0	35.2	47.0	61.0	3.82	2.72	2.04	1.57	6,175	3,800	2,300	1,270	0.32
86	86	36	20,715	333	2,490	26.3	107	800	28.5	40.0	53.0	68.5	3.76	2.70	2.02	1.56	6,950	4,240	2,540	1,400	0.32
94	94	36	22,610	364	2,715	28.7	116	870	31.0	43.5	58.0	75.0	3.74	2.66	2.00	1.55	7,475	4,600	2,760	1,510	0.32

- FL Water Content = D Water Content × 1.17

- Table shows required and available expansion volumes for water at 80° F temperature heated to operating temperature. This is seldom so provided and to take care of this large temperature difference, overflow tanks should be provided.

- Expansion space generally is provided only for water temperature variations under operating conditions applied to boiler water, water in expansion drum, water in supply line, water in heat-absorbing apparatus, water in return line, and marginal water in the expansion drum.

- A supply water temperature of 400° F might vary between 350° F and 400° F, or about by 50°. The return line water temperature might vary for a return of 200° F from 150° F to 250° F, or by about 100°F. This ratio holds good for most of the HTW systems in that the water in the supply line and expansion drum may vary by some 50° F and the water in the return line may vary by some 100° F to 200° F.

- Note: When using metric, convert to USCS units before applying this chart. Use the following values: in = mm/25.4; °F = 1.8 °C + 32; ft³ = m³ × 35.311; ft = m × 3.28.

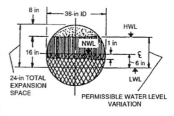

FIGURE 16.1.21   Expansion space in jet-induced HTW generator. (*Courtesy of Cleaver-Brooks.*)

For cold-start expansion, an overflow storage tank is recommended to save the treated water.

An inert-gas or steam-pressurized expansion drum should pressurize the system on the generator circulation pump inlet side, and the bottom of such an expansion tank should be above the generator top-drum elevation. This is recommended so that a minimum static head continues to be available if a failure of the pressuring action occurs. The expansion tank should be arranged horizontally for minimum variation of water level and maximum hydraulic pump suction head. The pressure in the drum maintains a positive suction head at the inlet to the circulation pumps and should be sufficient to provide the required pump suction head for the pumps selected; if it is less, cavitation will occur.

The system is pressurized by loading the expansion space of the expansion drum with steam or an inert gas such as N_2. The pressure is then transmitted by a small pipeline to the inlet side of the generator circulation pump. See Figs. 16.1.19 and 16.1.20. This pressurization line to the pump inlet connects to the bottom of the expansion drum with a locked-open shutoff valve near the drum. An inert gas such as nitrogen is used to eliminate the possibility of saturating the cool water contained in the drum with oxygen, thus minimizing corrosion of the generator and system components.

For good operational system and generator control, water-flow indicators should be used with a flow orifice located in the supply line from each unit where the water temperature is constant. A nontight-closing motor, or a manually actuated butterfly valve in the supply line leaving the generator, is a desirable flow-changing means. For generator startup a $^1/_2$- or $^3/_4$-in (12.7- or 19.1-mm) valve bypass line should be installed around generator shutoff valves to permit slow warm-up of a cold generator.

Water-Flow Quantity. Figure 16.1.15 shows a curve from which we can read water weights and volumes for different water-flow rates through the system or the HTW generator as they result from varying temperatures and enthalpies. The curve is based on 10 TBtu/h (2.9-MW) heat flow. Thus for 30 TBtu/h (8/8 MW), the flow values for water weight and volume must be multiplied by 3.

16.1.12   HEAT-RECOVERY BOILERS

16.1.12.1   Introduction

Heat-recovery boilers are used to reclaim heat from high-temperature exhaust gases resulting from thermal and/or chemical processes. Such regained heat is used to generate steam, heated water, or heated organic fluids. Table 16.1.4 lists a few processes and typical exhaust temperatures. Heat-recovery boilers are normally considered economical for exhaust gas temperatures above 600°F (315°C).

Heat-recovery boilers may be either fire-tube or water-tube design. Selection depends on the type of gas, gas temperature, operating and design pressures, plant requirements, etc. Water-tube boilers are suggested for higher steam or hot-water output and design pressures and will cause lower gas pressure losses. Fire-tube design can be furnished to operate with higher waste gas pressures and will normally cause a greater pressure loss in the gas-flow stream.

16.1.12.2   Design Considerations

Before any heat-recovery boiler is designed, an analysis of the economics involved should be conducted. Physical and chemical behavior of the hot gas stream and gas velocity should

TABLE 16.1.4 Typical Process Exhaust Temperatures

Chemical and thermal process	Temperature, °F (°C)	
Chemical oxidation	1350–1475	(730–800)
Annealing furnace	1100–1200	(590–650)
Diesel-engine exhaust	1000–1200	(540–650)
Forge and billet heating furnace	1700–2200	(920–1200)
Refuse incinerator	1550–2000	(840–1100)
Open-hearth steel furnace, air-blown	1000–1300	(540–700)
Open-hearth steel furnace, oxygen-blown	1300–2100	(700–1150)
Basic oxygen furnace	3000–3500	(1650–1900)
Glass-melting furnace	1200–1600	(650–870)
Gas-turbine exhaust	700–1100	(370–590)
Fluidized-bed combustor	1600–1800	(870–980)

be studied to avoid erosion and corrosion of tubes. The pressure and temperatures of the waste gas stream are important, so as to avoid the need for filters, inducers, blending systems, etc. Possible effects on the basic process must be considered, and a value must be placed on the energy absorbed.

Tables 16.1.5 and 16.1.6 provide information related to types of heat-recovery boilers which are normally recommended for a variety of waste gas streams as well as a definition of waste materials and some problem constituents which result from their incineration.

Physical and Chemical Behavior of Gases. Incinerator flue gas may contain a significant amount of abrasive solid particulates, which in time may damage the tube's surface and weaken the mechanical strength. Chemical behavior concerns involve the presence of corrosive or fouling gases. Table 16.1.7 indicates the flue gas composition for a municipal waste incinerator. The presence of SO_2, SO_3, or chlorine compounds is a serious concern unless metal surfaces are maintained above the dew point temperatures. The best operating range of boiler metal where such gases are present is from 350 to 450°F (175 to 230°C). Corrosion can occur above and below these temperatures.

Flue Gas Fluid Dynamic Behavior. Velocity distribution and the magnitude of the gas stream play a vital role in achieving high thermal performance and prolonged boiler life. Uniform velocity distribution through or around the tubes results in a uniform temperature gradient, proper internal circulation, and minimal stresses. Velocity control is important to avoid tube erosion and to maintain limited pressure drop. Table 16.1.8 lists recommended velocities for both fire-tube and water-tube boilers. The investigators suggest that a gas velocity less than 35 ft/s (11 m/s) allows particulates to deposit inside the tubes of fire-tube boilers, which compromises the effectiveness of the heat-transfer area. Similarly, insufficient gas velocities in a water tube design allow particulate matter to collect in the gas passageways and may affect the soot blowers' effectiveness.

Fire-tube heat-recovery boilers are designed in various configurations:

- Single-, double-, and four-pass boilers for straight heat recovery from hot flue gases
- Double- or three-pass boilers with supplementary firing

Water-tube heat-recovery boilers are tailor-made to effectively absorb the heat available from the waste gas stream while maintaining appropriate gas velocities and minimizing erosion.

TABLE 16.1.5   Heat Sources and Application of Heat-Recovery Boilers

Process	Type of boiler	Comments
Incinerator	Firetube and watertube	Care must be taken to prevent corrosion. Hoppers, soot blowers, and superheaters may be used. Fins are never used. Induced-draft fan is needed.
Gas turbine	Finned watertubes	Superheater, economizer, pressure drop critical. Must be within 10 to 12 inWG (2.5 to 3 kPa) exhaust gas pressure range.
Petroleum refinery (catalytic regeneration process)	Firetube or watertube	High-pressure gas requires special shell design. Produced steam used for the process itself. Installed space limited.
Heat-treat/metal refining processes (aluminum ore and copper ore refining)	Firetube and watertube	For small flow rates, firetube is preferred. Space limitation. Watertube boilers also used with soot blowers, greater tube thickness. Induced-draft fan needed.
Chemical processes (sulfur combustion)	Firetube and watertube	Corrosion prevention, metal temperatures above dew point. Induced-draft fan needed. Fan blades must be refractorized or of high metal alloy.
Rice hull combustion	Firetube boilers	With furnace to cool down high-temperature combustion gas. Automatic to soot blowers required to clean tube. Gas velocity 45 to 50 ft/s (13 to 15 m/s) to avoid erosion.
Diesel engine	Firetube or watertube	For small engines, firetube boilers; for bigger engines, watertube with finned tubes or bare tubes. Superheater can be used in watertube design. Pressure drop critical.
Fluidized-bed combustor	Firetube and watertube	Hoppers, soot blowers, superheater, and economizer. Average gas velocity is 50 ft/s (15 m/s) for wood or coal fuels.
Glass melting	Watertube and firetube	Slag buildup and fouling can be severe because of fluxing material (alkaline or acidic).
Miscellaneous (gasification, papermill)	Watertube and firetube	Gasified products cooled to lower temperature before being fired in any thermal process.

TABLE 16.1.6 Classification of Wastes to Be Incinerated

Waste type, description*	Principal components	Approximate composition, % by weight	Moisture content, %	Incombustible solids, %	Value of refuse as fired, Btu/lb (kJ/kg)
0. Trash	Highly combustible waste, paper, wood, cardboard cartons. Including up to 10% treated papers, plastic or rubber scraps; commercial and industrial sources	Trash, 100	10	5	8500 (19.8)
1. Rubbish	Combustible waste, paper, cartons, rags, wood scraps, combustible floor sweepings; domestic, commercial, and industrial sources	Rubbish, 80 Garbage, 20	25	10	6500 (15.1)
2. Refuse	Rubbish and garbage; residential sources	Rubbish, 50 Garbage, 50	50	7	4300 (10)
3. Garbage	Animal and vegetable wastes, restaurants, hotels, markets; institutional, commercial, and club sources	Garbage, 65	70	5	2500 (5.8)
4. Animal solids and organic	Carcasses, organs, solid organic wastes, hospital laboratory, abattoirs, animal pounds, and similar sources	Animal and human tissue, 100	85	5	1000 (2.3)
5. Gaseous	Industrial process wastes	Variable	Depends on predominant components	Variable according to wastes survey	Variable according to wastes survey
6. Semisolid and solid wastes	Combustibles requiring hearth, retort, or grate	Variable	Depends on predominant components	Variable according to wastes survey	Variable according to wastes survey

*The figures on moisture content, ash, and Btu values as fired have been determined by analysis of many samples. They are recommended for use in computing heat release, burning rate, velocity, and other details of incinerator designs. Any design based on these calculations can accommodate minor variations.

TABLE 16.1.7   Composition of Flue Gas from Combustion of Solid Waste and Hazardous Materials

Chemical composition	Municipal solid	Hazardous
N_2	78–88%	78–88%
O_2	6–10%	6–12%
CO_2	6–12%	6–12%
SO_2	2–200 ppm	1200–1400 ppm
HCl	5–100 ppm	(Based on waste)
HF	0–1 ppm	(Based on waste)
CO	5–50 ppm	0–50 ppm
NO_x	50–100 ppm	100–200 ppm

Note: ppm = parts per million.

Design calculations for heat transfer from waste gases follow the routine procedures well documented in other publications. For gases burdened with high levels of particulate matter, the gas-side coefficients may be reduced by as much as 10 percent in consideration of the surface fouling that is likely to occur. Overall coefficients between 5 and 16 Btu/(h · ft^2 · °F) [28.4 and 90.9 W/(m^2 · K)] are realistic. For gases containing erosive matter, the gas velocity must necessarily be low, therefore dictating a low coefficient. For very clean waste gas,

TABLE 16.1.8   Heat-Recovery Design Guidelines

	Maximum gas velocity, ft/s (m/s)			Fouling factor	Maximum fins per in (mm)	Fin thickness in (mm)	Soot blowers
	Firetube	Inline	Water-tube staggered				
Clean: Combustion products of natural gas	200 (61)	120 (36.6)	100 (30.5)	0.001	6 (0.24)	0.05 (1.27)	No
Average: combustion products of No. 2 oil and wood, fume incinerators	170 (51.8)	100 (30.5)	80 (24.4)	0.002	5 (0.2)	0.05 (1.27)	Provision
Dirty: Combustion products of Nos. 5 and 6 oil, liquid incinerators, municipal sludge, fluid cat, cracker	130 (39.6)	80 (24.4)	65 (19.8)	0.004	4 (0.16)	0.06 (1.52)	Yes
Dirty and abrasive: Combustion products of coal	100 (30.5)	60 (18.3)	40 (12.2)	0.004	3 (0.12)	0.075 (1.91)	Yes

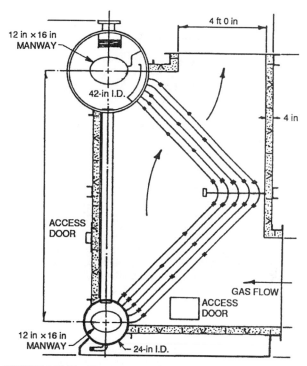

FIGURE 16.1.22   Heat-recovery boilers for gas-turbine applications. (*Courtesy of Cleaver-Brooks.*)

designs can be based on heat-transfer coefficients which may approach 20 Btu/(h · ft² · °F) [113.6 W/(m² · K)].

For those instances where the gas particulate loading is very heavy, consideration should be given to the provision of a dust drop area. This may be installed in the ductwork carrying the gas to the boiler or within the boiler itself. The boiler or system designer should be consulted for these details before the total system design is begun.

For installations where large quantities of gas are to be cooled [25,000 lb/h (11,300 kg/h) of gas or greater], a water tube design is likely to be specified unless the waste gas is available at the source of pressure of 1 psig (0.15 kPa) or more. Figures 16.1.22, 16.1.23, and 16.1.24 all depict water tube-type equipment. Note that super heaters (available only with this type of equipment) provide steam temperatures higher than saturation temperatures. Economizers, although more regularly furnished with water tube-style boilers, may be used in conjunction with fire tube design, to enhance the total recovery of heat.

16.1.13 SOLID-FUEL BOILERS

Solid fuel such as coal, wastes, etc., when readily available, can substantially reduce the cost of producing steam and hot water, now and in the future (Fig. 16.1.25). Such comparisons may vary over time and should be rechecked periodically.

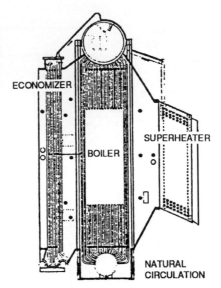

FIGURE 16.1.23   Heat-recovery boilers for incinerator application. (*Courtesy of Cleaver-Brooks.*)

Because solid fuels burn much more slowly, a large combustion zone is required for more difficult combustion conditions. To ensure complete combustion of the fuel, larger quantities of excess air are supplied, creating the need for larger flue gas flow areas. Therefore, the boilers are nearly always substantially larger for equal output. Solid fuels normally require a higher-temperature environment in the combustion zone to facilitate complete fuel combustion.

To accomplish this, often the chamber is completely lined with refractory, surrounded by insulation and seal casing. The convection zone or heat exchanger is normally mounted above this chamber. Solid-fuel boilers are thus supplied to the site in a number of components, and final assembly is completed, at the installation.

A boiler with an external furnace is shown in Fig. 16.1.25. This boiler is also known as a horizontal return tubular (HRT) boiler. The gases going through the horizontal tubes could be two or three passes. This will cool the flue gases, decreasing the gas exit temperature and thus increasing boiler efficiency.

The fuel-burning mechanism is located within the refractory combustion chamber. Access doors provide a means for ash removal, boiler cleaning, and inspection. Fuel-burning equipment as well as ash removal can be automated.

The fuel must be physically hand-delivered to the burner equipment, and because ash results from the combustion of these fuels which must be frequently removed from the combustion zone, operators must be in frequent attendance.

For these reasons, solid-fuel burning installations are normally found only at intermediate to large facilities. A boiler appropriate for an intermediate-size solid-fuel installation might be as shown in Fig. 16.1.26. For anthracite fuels which are very low in volatility, extensive refractory arches are required to ensure continued ignition of the fuel. Relatively low levels of excess air are required, and the boiler assembly might be quite compact. See Fig. 16.1.27. However, high-quality bituminous coals can be pulverized and injected into a furnace through properly designed burner hardware (Fig. 16.1.28). The boiler design can

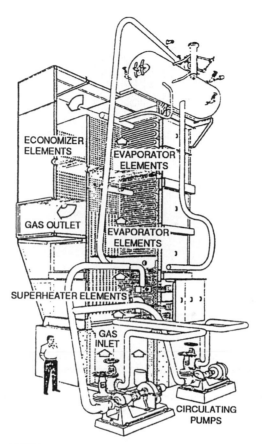

FIGURE 16.1.24 Heat-recovery boilers for positive waste heat application. (*Courtesy of Cleaver-Brooks.*)

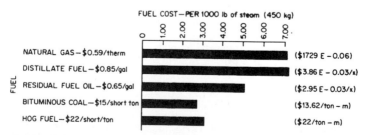

FIGURE 16.1.25 Cost of steam generator with various fuels. (1990 dollars). (*Courtesy of Cleaver-Brooks.*)

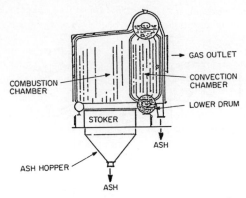

FIGURE 16.1.26   Water tube D-shaped two-drum boiler. (*Courtesy of Cleaver-Brooks.*)

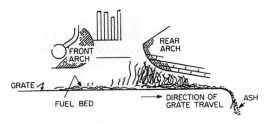

FIGURE 16.1.27   Arches for low-volatility fuels. (*Courtesy of Cleaver-Brooks.*)

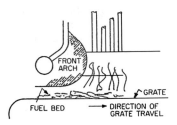

FIGURE 16.1.28   Arches for high-volatility fuels. (*Courtesy of Cleaver-Brooks.*)

be quite similar to that intended for gas or oil, except that modifications are needed for the removal of ash. A detailed discussion of boilers and burner equipment for use with solid fuels is beyond the scope of this chapter.

16.1.14 UNFIRED BOILERS

Unfired boilers are those which use a higher temperature fluid to generate steam at a lower temperature. Typical high-temperature fluids are steam, high-temperature hot water, and high-temperature thermal fluid. Unfired steam boilers fall into two categories:

1. Steam boilers used to generate steam for heating or process requirements where the high-temperature fluid is readily available and/or is not cost-effective to use combustion-based equipment. See Fig. 16.1.29.

2. Stainless steel boilers used to produce "clean" or "ultrapure" steam for humidification purposes and critical environments such as "clean rooms" and semiconductor manufacturing facilities. An all-stainless steel unfired steam boiler is shown in Fig. 16.1.30.

FIGURE 16.1.29   Vertical unfired steam boiler. (*Courtesy of Ace Boiler Inc.*)

FIGURE 16.1.30   Stainless steel unfired boiler for ultra-pure steam. (*Courtesy of Ace Boiler Inc.*)

16.1.15   OPERATION AND MAINTENANCE

Boilers are primarily constructed to be unmanned, fully automatic systems. It is important to review operation and maintenance features of the types of boiler being considered during the selection process. Several pertinent issues are:

1. Most states and jurisdictions have regulations governing the level of operator qualifications in and attendance for steam boilers and, in some cases, water boilers.

2. For water boiler systems, operators must be properly trained in the use of equipment provided to maintain inlet water temperatures at above-condensing levels.

3. Operation and maintenance costs should be carefully examined for major maintenance expense items. Cost of proprietary (most bent tubes) tube replacement or cast-iron section replacement should be compared for the types of boiler being considered. Most auxiliary systems that offer improved efficiency come with a cost of maintenance which must be factored into the selection process.

4. Boiler water treatment is of paramount importance to the continued operation of the boiler. Water treatment to prevent oxidation of the boiler steel and sealing of the internal surfaces is a necessary part of ongoing maintenance operations. Water treatment is the province of local water treatment specialists who generally contract with the boiler owner to provide water treatment services.

16.1.16   ELECTRIC BOILERS

16.1.16.1   Applications

Electric boilers are compact and clean, but are very costly sources for the production of hot water or steam and are available in ratings from 5 to over 50,000 kW. Vessel design, controls safety valves, and selection criteria are all similar to fired water and steam boilers.

16.1.16.2   Classifications

Electric boilers are usually classified as either resistance or electrode type with further distinctions as high or low voltage and steam or hot water.

Resistance Boilers.   Resistance-type boilers use met-sheathed resistance elements which are submerged in the boiler water. The water is heated by convection from the surface of the sheaths of the heating elements. The electric current flows through a resistance wire centered inside, but insulated from, the element sheath. The element sheath has no electric potential.

Electrode Boilers.   Electrode boilers use the water as a resistor, and the electrodes only provide an electric contact with the water.

Voltage Classifications.   Voltages on which boilers are designed to operate are classified as low, 600 V or less; or high, greater than 600 V. Resistance-type boilers are always low-voltage. Electrode boilers are available in both low- and high-voltage designs. Both resistance and electrode boilers are available in designs to generate either steam or hot water.

16.1.16.3   Resistance Boilers

Resistance boilers consist of a pressure vessel and resistance-type heating elements of sufficient number and rating to obtain the desired total boiler kilowatt rating. The boilers have controls to regulate the power needed by either the steam or hot-water demand, a low-water cutoff, and other items as appropriate to the boiler's design (steam or hot water), as required by ASME and local codes.

Resistance Steam Boilers.   Resistance steam boilers (see Fig. 16.1.31) have vessels sized to provide a steam-to-water ratio of about 40:60 by volume.
  Ratings Available.   Resistance-type steam boilers are available in ratings from 5 to over 4000 kW at voltages from 208 to 600 V. Pressure ratings typically offered by manufacturers are 15 lb/in^2 (103 kPa) for low-pressure heating applications, 100 lb/in^2 (690 kPa) for "miniature" boilers (usually less than 100 kW), and 125 lb/in^2 (862 kPa) for boilers over 100 kW. Pressure rains up to and over 2000 lb/in^2 (13,793 kPa) are available.

Resistance Hot-Water Boilers.   Boilers for hot water are similar to but smaller than steam boilers (for the same kilowatts). Controls consist of at least a low-water cutoff, temperature control, and at least one high-temperature limit. Figure 16.1.32 shows a typical resistance-type hot-water boiler.
  Residential Boilers.   Hot-water boilers used for residential heating usually use a single-phase 12/240-V supply and provide for energizing the heating elements in increments of 3 to 6 kW with a short (5- or 10-s) delay between increments. Such boilers are usually not over 30 kW.

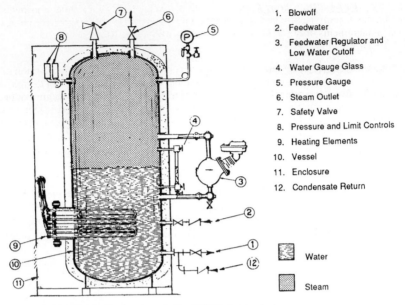

1. Blowoff
2. Feedwater
3. Feedwater Regulator and Low Water Cutoff
4. Water Gauge Glass
5. Pressure Gauge
6. Steam Outlet
7. Safety Valve
8. Pressure and Limit Controls
9. Heating Elements
10. Vessel
11. Enclosure
12. Condensate Return

Water

Steam

FIGURE 16.1.31   Resistance steam boiler. (*CAM Industries, Inc.*)

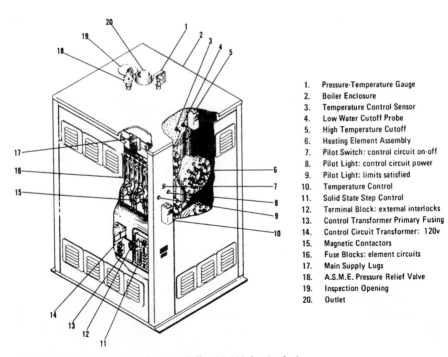

1. Pressure-Temperature Gauge
2. Boiler Enclosure
3. Temperature Control Sensor
4. Low Water Cutoff Probe
5. High Temperature Cutoff
6. Heating Element Assembly
7. Pilot Switch: control circuit on-off
8. Pilot Light: control circuit power
9. Pilot Light: limits satisfied
10. Temperature Control
11. Solid State Step Control
12. Terminal Block: external interlocks
13. Control Transformer Primary Fusing
14. Control Circuit Transformer: 120v
15. Magnetic Contactors
16. Fuse Blocks: element circuits
17. Main Supply Lugs
18. A.S.M.E. Pressure Relief Valve
19. Inspection Opening
20. Outlet

FIGURE 16.1.32   Resistance hot-water boiler. (*CAM Industries, Inc.*)

Commercial and Industrial Boilers. Boilers designed for commercial and industrial installations usually subdivide the elements into groups of 15 to 36 kW and control the boiler in increments only when the total power exceeds 50 to 60 kW.

16.1.16.4 Electrode Boilers

Electrode boilers use water as a resistor. Protection against low-water burnout is inherent. They are generally much more compact than resistance-type boilers of equivalent kilowatt and voltage ratings. Only electrode boilers can be designed to operate on high voltages (above 600 V). This makes it possible and feasible to attain boiler ratings of 50,000 kW or more and to use voltages up to 25,000 V. Electrode boilers require properly conditioned water for successful operation. Water-quality requirements are critical. The treatment that may be needed to make the available water suitable for the boiler should be determined as part of the boiler selection process. Table 16.1.9 lists the water-quality limits typical for electrode boilers.

Controls for Electrode Boilers. Steam pressure and current flow (or kilowatts) are the variables used to control these boilers. Pressure is the primary variable with current flow being used as a feedback signal and as an override control variable, to limit the boiler power consumption to its rating or to some lower value.

Back-Pressure Valves (Steam Boilers). Because these boilers require steam pressure to force water out of the electrode chamber, a back-pressure valve is usually needed on the boiler steam outlet to ensure that the steam pressure will not drop below that required to operate the boiler. On high-voltage boilers, the back-pressure valve is sometimes used to control the boiler pressure and to isolate the boiler from steam line pressure variations caused by rapid fluctuations in the rate of steam consumption.

Electrode Boiler Types. There are three basic designs according to the control method used:

1. Totally immersed electrode with insulating-sleeve control

2. Variable-immersion electrode, water-level control

3. Jet or spray type with variable flow to electrodes

Totally Immersed Electrode Boilers. Totally immersed electrode boilers are available for both hot-water and steam applications and for both low and high voltages. Kilowatt ratings are determined by the voltage, number, and size of the electrodes; the conductivity of the boiler water at the design operating temperature; and the water circulation rate past the electrodes. Figure 16.1.33 shows a low-voltage hot-water boiler, Fig. 16.1.34 shows a high-voltage hot-water boiler, and Fig. 16.1.35 shows a high-voltage steam boiler.

Controls. Control systems for totally immersed electrode boilers employ a movable, insulating sleeve or shield which is interposed between the energized cylindrical electrode and a concentric metallic neutral assembly to which the electric current flows.

Turndown Ratio. These boilers can be turned down to 5 to 10 percent of their rated maximum output. Further output reductions require that power to the boiler electrodes be interrupted by a contactor or circuit breaker rated for the anticipated frequency of operation.

Pressures and Ratings. Operating pressures are usually between 75 lb/in^2 (517 kPa) and 160 lb/in^2 (1103 kPa). Output ratings from 1 to 45 MW are available.

Selection Considerations. High-voltage boilers of this design are well suited for use on voltages at the lower distribution ratings, that is, 2300 to 7200 V, and for operation where low conductivity boiler water is both available and desired.

Variable-Immersion Electrode Boilers. Electrode boilers using a variable water level to change the depth of electrode immersion are used only to generate steam. These boilers

TABLE 16.1.9 Typical Recommended Water-Quality Limits for Electric Boilers

Boiler type and water classification	Hardness, ppm maximum	pH range	p* alkalinity, ppm maximum	Iron, ppm maximum	Conductivity,* MMho	TDS,† ppm maximum	Oxygen, ppm maximum	Notes
I. Resistance								
A. Steam								
1. Feedwater	5	≥7.5	50	1	450	300	0.03	
2. Boiler water	0	7.5–10.5	500	3	4500	3000	0	
B. Hot water								
1. Makeup water	5	≥7.5	400	5	500	300	4	
2. Boiler water	5	8.5–10.5	500	5	6000	3000	0	
II. Electrode								
A. Low-voltage								
1. Steam								
a. Feedwater	2	≥7.5	50	1	200	100	.03	
b. Boiler water	0	7.5–10.5	400	3	2000	2000	0.005	‡
2. Hot water								
a. Makeup water	5	≥7.5	350	2	200	100	4	
b. Boiler water	5	8.5–10.5	400	2.5	1000	500	0	‡

TABLE 16.1.9 Typical Recommended Water-Quality Limits for Electric Boilers (*Continued*)

Boiler type and water classification	Hardness, ppm maximum	pH range	p* alkalinity, ppm maximum	Iron, ppm maximum	Conductivity,* MMho	TDS,† ppm maximum	Oxygen, ppm maximum	Notes
B. High-voltage								
1. Totally immersed electrode								
a. Steam								
(1) Feedwater	1.0	≥7.5	25	0.5	50	25	0.005	‡ ‡
(2) Boiler water	0	7.5–10	125	1	<500	<250	0	‡ ‡
b. Hot water								
(1) Makeup	1.0	≥7.5	100	0.5	250	125	4	
(2) Boiler water	1.0	8.5–10	125	0.5	<500	<250	0	‡
2. Variable-immersion electrode								
a. Steam								
(1) Feedwater	0.1	≥7.5	25	0.5	20	10	0.005	
(2) Boiler water	0	7.5–10	125	1	<200	<100	0	‡
3. Jet type								
a. Steam								
(1) Feedwater	1	≥7.5	50	1	300	150	.005	
(2) Boiler water	0	7.5–10	400	2	<3000	<1500	0	§

*Limit of this variable in feedwater is determined by boiler water limit and acceptable blowdown rate.
†TDS is total dissolved solids.
‡Conductivity limits vary with voltage and water temperature; consult the boiler manufacturer.
§Water must be nonfoaming.
Source: CAM Industries, Inc./Precision Parts Corp.

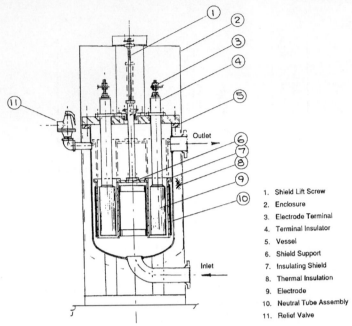

1. Shield Lift Screw
2. Enclosure
3. Electrode Terminal
4. Terminal Insulator
5. Vessel
6. Shield Support
7. Insulating Shield
8. Thermal Insulation
9. Electrode
10. Neutral Tube Assembly
11. Relief Valve

FIGURE 16.1.33 Electrode hot-water boiler, low-voltage type. Explanation: When more heat is required, temperature control system raises insulating shield (7) to increase exposure of electrode (9) to neutral (10). (*CAM Industries, Inc.*)

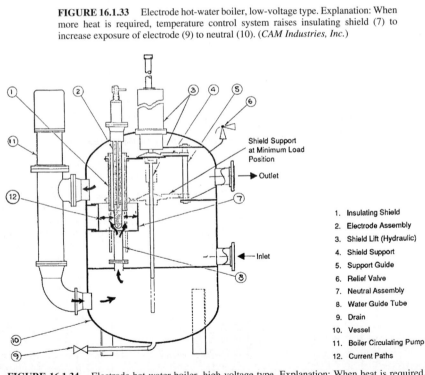

1. Insulating Shield
2. Electrode Assembly
3. Shield Lift (Hydraulic)
4. Shield Support
5. Support Guide
6. Relief Valve
7. Neutral Assembly
8. Water Guide Tube
9. Drain
10. Vessel
11. Boiler Circulating Pump
12. Current Paths

FIGURE 16.1.34 Electrode hot-water boiler, high voltage type. Explanation: When heat is required, temperature control system causes insulating shields (1) to be lifted to expose electrodes (2) to neutral (7). The boiler circulating pump (11) operates continuously. System water is circulated by system pumps. (*CAM Industries, Inc.*)

16.54

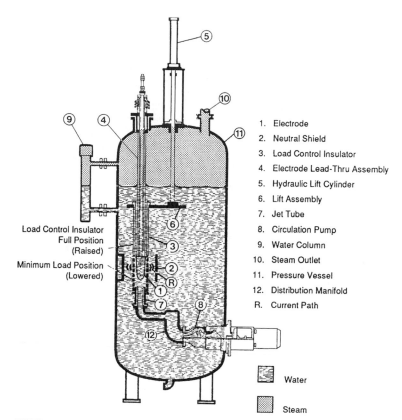

1. Electrode
2. Neutral Shield
3. Load Control Insulator
4. Electrode Lead-Thru Assembly
5. Hydraulic Lift Cylinder
6. Lift Assembly
7. Jet Tube
8. Circulation Pump
9. Water Column
10. Steam Outlet
11. Pressure Vessel
12. Distribution Manifold
R. Current Path

Load Control Insulator
Full Position
(Raised)

Minimum Load Position
(Lowered)

Water

Steam

FIGURE 16.1.35 High voltage steam boiler with totally immersed electrode. Explanation: The rate of steam generation is controlled by raising or lowering the load control insulator (3) to change the exposure of the electrode (1) to the neutral shield (2). The boiler circulation pump (8) forces water past electrodes to carry steam bubbles away from electrodes. (*CAM Industries, Inc.*)

are available in both low- and high-voltage designs. The control principle involved is that the flow of current through the water will be proportional to the wetted area of the electrodes, which is in turn controlled by the depth of electrode immersion in the water.

Control Systems. In all designs the electrode is fixed, and the water level is made to rise or fall as needed to develop the required steaming rate. Current flow in the low-voltage boilers of this type is from electrode to electrode. Figures 16.1.36 and 16.1.37 show the basic elements of the two systems as applied to low-voltage electrode steam boilers.

High-voltage electrode boilers using the variable-immersion principle are usually of single-vessel, two-chamber construction (see Fig. 16.1.38). Current flow is from electrode to neutral. Most high-voltage boilers of this type incorporate a circulation pump to increase the rate of water flow past the electrodes. In some cases the pump is also used as part of the water-level regulation system.

Pressure and Kilowatt Ratings. Low-voltage boilers of this type are available for operation over a range of pressures from as low as 5 lb/in^2 (34 kPa) to as high as 1200 lb/in^2 (8275 kPa). Ratings cover the range of 15 to 2800 kW. Operating pressures of high-voltage boilers of this type are typically limited to the range of 75 lb/in^2 (517 kPa) to 400 lb/in^2 (2757 kPa). Ratings cover the range of 2000 to 50,000 kW at voltages from 4.16 to 13.8 kV.

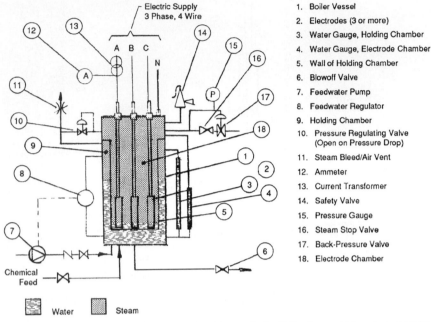

1. Boiler Vessel
2. Electrodes (3 or more)
3. Water Gauge, Holding Chamber
4. Water Gauge, Electrode Chamber
5. Wall of Holding Chamber
6. Blowoff Valve
7. Feedwater Pump
8. Feedwater Regulator
9. Holding Chamber
10. Pressure Regulating Valve (Open on Pressure Drop)
11. Steam Bleed/Air Vent
12. Ammeter
13. Current Transformer
14. Safety Valve
15. Pressure Gauge
16. Steam Stop Valve
17. Back-Pressure Valve
18. Electrode Chamber

FIGURE 16.1.36   Single-vessel, low-voltage, electrode steam boiler. Boiler shown at high-pressure condition. Explanation: Water-level regulator (8) controls feed pump (7) to maintain a minimum water level in holding chamber (9). Pressure in top of holding chamber is controlled by pressure regulating valve (10) to be either equal to or less than the pressure in the electrode chamber (18). When electrode chamber pressure exceeds setting of valve (10), the valve throttles and pressure in holding chamber is reduced by bleeding through vent (11). Water from electrode chamber is displaced into holding chamber, reducing depth of electrode immersion and steam generated until pressure returns to set point. On pressure drop in electrode chamber, valve (10) opens to equalize pressure in both chambers and to increase depth of water on electrodes and steam output. (*CAM Industries, Inc.*)

Water Requirements.   Low-voltage boilers of this type require water which has been softened. The conductivity of the boiler water must be limited to the manufacturer's recommendation. See Table 16.1.9. High-voltage boilers of this type require very high-quality, low-conductivity feed water. Demineralization is usually required. Feed water flow should be proportional to steam flow rather than on/off.

Jet or Spray Electrode Steam Boilers.   Jet or spray electrode boilers have electrodes located in the steam space above the boiler waterline. The boiler is fitted with one or more circulating pumps which draw water from the bottom of the boiler, forcing it up a central column and through nozzles directed at the electrodes. Each of the streams of water becomes a current path between the energized electrode and the neutral-potential nozzle header. The rate of circulation is approximately 30 times the steaming rate. The excess water passes through the bottom of the electrodes and falls onto a perforated plate located a fixed distance below the electrode. This plate is also at neutral potential. A second current path is thus established.

Jet-type boilers are offered for high-voltage use only.

Control Systems.   Control of the boiler output is accomplished by varying the number of streams of water allowed to strike the electrode. The control occurs in one of two ways:

1. *Constant flow:* A regulating sleeve which is concentric with the nozzle header is raised or lowered. The sleeve shunts unwanted water streams back to the bottom of the boiler.

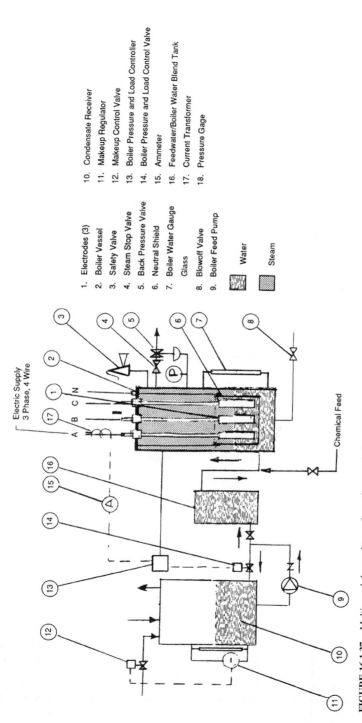

1. Electrodes (3)
2. Boiler Vessel
3. Safety Valve
4. Steam Stop Valve
5. Back Pressure Valve
6. Neutral Shield
7. Boiler Water Gauge Glass
8. Blowoff Valve
9. Boiler Feed Pump

10. Condensate Receiver
11. Makeup Regulator
12. Makeup Control Valve
13. Boiler Pressure and Load Controller
14. Boiler Pressure and Load Control Valve
15. Ammeter
16. Feedwater/Boiler Water Blend Tank
17. Current Transformer
18. Pressure Gage

Water
Steam

Electric Supply
3 Phase, 4 Wire

Chemical Feed

FIGURE 16.1.37 Multivessel, low-voltage, electrode steam boiler. Boiler shown at normal operating condition. Explanation: Pressure and load controller (13) controls boiler feed pump (9) to run continuously and causes valve (14) to bypass all feedwater in excess of steam generation rate back to condensate receiver (10). Water level in boiler is regulated to generate steam needed to maintain steam pressure or maximum kilowatts desired, as applicable. If steam demand decreases, water from boiler is displaced back to blend tank (16) and cold feedwater from blend tank is returned to condensate receiver (10). If no steam is needed, feed pump (9) stops. (*CAM Industries, Inc.*)

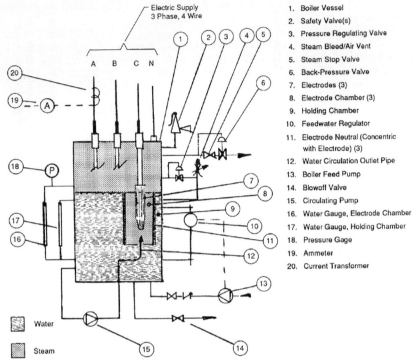

1. Boiler Vessel
2. Safety Valve(s)
3. Pressure Regulating Valve
4. Steam Bleed/Air Vent
5. Steam Stop Valve
6. Back-Pressure Valve
7. Electrodes (3)
8. Electrode Chamber (3)
9. Holding Chamber
10. Feedwater Regulator
11. Electrode Neutral (Concentric with Electrode) (3)
12. Water Circulation Outlet Pipe
13. Boiler Feed Pump
14. Blowoff Valve
15. Circulating Pump
16. Water Gauge, Electrode Chamber
17. Water Gauge, Holding Chamber
18. Pressure Gage
19. Ammeter
20. Current Transformer

FIGURE 16.1.38   Variable-immersion, single-vessel, high-voltage electrode steam boiler. Boiler shown in high-pressure operating condition. Explanation: Water-level regulator (10) controls feed pump (13) to maintain a minimum water level in holding chamber (9). Pressure in top of holding chamber is controlled by pressure-regulating valve (3) to be either equal to or less than pressure in electrode chamber (8). When electrode chamber pressure exceeds setting of valve (3), valve throttles and pressure in holding chamber is reduced by bleeding through vent (4). Water from electrode chamber is displaced into holding chamber (9), reducing depth of electrode immersion and rate of steam generation until pressure returns to set point. On pressure drop in electrode chamber, valve (3) opens to equalize pressure in both chambers and to increase depth of water on electrodes and steam output. Circulating pump (15) operates full-time to wash tips of electrodes via water circulation outlet pipe (12). (*CAM Industries, Inc.*)

2. *Variable flow:* A valve is installed in the circulating pump discharge to regulate the water flow to the nozzle header.

The position of the regulating sleeve or flow control valve is controlled either to maintain the desired steam pressure or to generate steam at a preselected rate at constant kilowatts.

Jet-type boilers operate with a constant water level, and feed water flow should be modulated at the same rate as the steam flow. Conductivity controllers are required on all high-voltage electrode steam boilers and should be arranged to initiate surface blow-down when the conductivity is too high and to activate a chemical feed system when the conductivity is too low.

Water Requirements.   The jet-type boiler was developed to allow the use of higher-conductivity water than other high-voltage designs. This reduces the blow-down requirements by a factor of 10 or more and permits the use of water which has been merely softened rather than demineralized. See Table 16.1.9.

Pressures and Ratings.　Jet-type boilers operate best at pressures of 75 lb/in^2 (517 kPa) or higher. Operation at lower pressures requires special designs and significant derating. Jet-type boilers are offered for high voltage only. Ratings up to 78 MW at 25 kV are available. Figures 16.1.39 and 16.1.40 show a jet-type boiler.

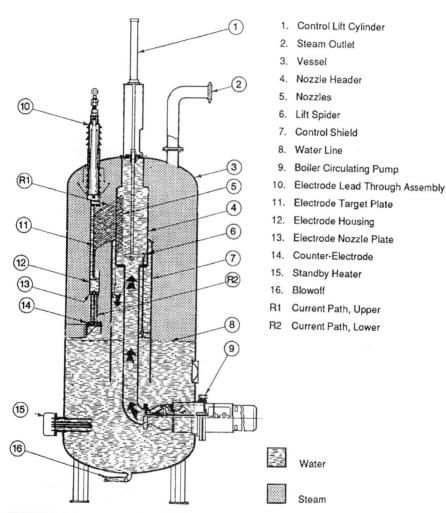

1. Control Lift Cylinder
2. Steam Outlet
3. Vessel
4. Nozzle Header
5. Nozzles
6. Lift Spider
7. Control Shield
8. Water Line
9. Boiler Circulating Pump
10. Electrode Lead Through Assembly
11. Electrode Target Plate
12. Electrode Housing
13. Electrode Nozzle Plate
14. Counter-Electrode
15. Standby Heater
16. Blowoff
R1　Current Path, Upper
R2　Current Path, Lower

Water

Steam

FIGURE 16.1.39　Jet-type high-voltage boiler. Explanation: Pump (9) fills nozzle header (4), water flows through nozzles (5) to strike electrode target plate (11), water falls to bottom of electrode housing (12) and through plate (13) to counterelectrode (14). R_1 and R_2 are current paths from electrode to nozzles (5) and counterelectrode (14), both of which are neutral. To control steam output, control shield (7) is lifted to divert nozzle discharge to inside of control shield. Steam output is proportional to water flow to electrode. (*CAM Industries, Inc.*)

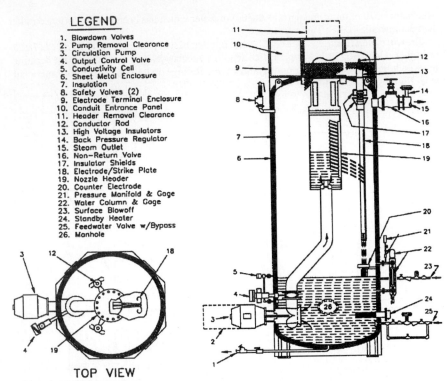

LEGEND

1. Blowdown Valves
2. Pump Removal Clearance
3. Circulation Pump
4. Output Control Valve
5. Conductivity Cell
6. Sheet Metal Enclosure
7. Insulation
8. Safety Valves (2)
9. Electrode Terminal Enclosure
10. Conduit Entrance Panel
11. Header Removal Clearance
12. Conductor Rod
13. High Voltage Insulators
14. Back Pressure Regulator
15. Steam Outlet
16. Non-Return Valve
17. Insulator Shields
18. Electrode/Strike Plate
19. Nozzle Header
20. Counter Electrode
21. Pressure Manifold & Gage
22. Water Column & Gage
23. Surface Blowoff
24. Standby Heater
25. Feedwater Valve w/Bypass
26. Manhole

TOP VIEW

FIGURE 16.1.40   Jet-Type 'Variable Flow' High Voltage Boiler. Explanation: Water from the lower part of the boiler shell is pumped by either an internal or external circulation pump (3) to the nozzle header (19) and flows by gravity through the jets to strike the electrode (18), thus creating a path for the electrical current. As the unevaporated portion of the water flows from electrode (18) to the counter electrode (20), a second path for current is created. Primary voltage connections are made directly to the electrode terminals (12), often eliminating the need for a step-down transformer. At maximum rated conductivity, approximately 3% of the flowing water is evaporated. Regulation of the boiler output is accomplished by a butterfly valve (4), which regulates the amount of water reaching the nozzle header. This control valve is positioned by the boiler pressure and load control system either to hold the steam pressure constant or to stay within an adjustable KW limit. (*Precision Parts Corp.*)

16.1.16.5 Installation

Planning the installation of electric boilers required attention to both mechanical and electrical details, as well as to the water to be used in the boiler.

Mechanical Requirements.   Same requirements as a fossil-fuel-fired boilers. Figure 16.1.41 shows pressure drops typical for hot-water boilers.

Clearances.   Boilers must be installed with adequate clearances at the sides and top to permit access to all valves and piping and to meet the requirements of the local boiler or building code. Clearances in front of the electric cabinets are also prescribed by electrical and/or building codes.

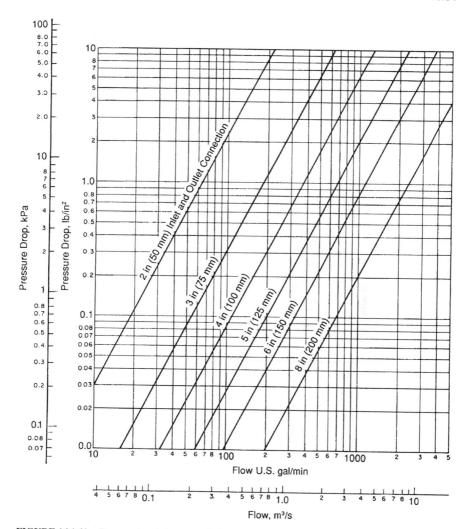

FIGURE 16.1.41   Pressure drop in hot-water boilers.

Electric Wiring.   Supply circuits for electric boilers *must meet the requirements of the local electrical code.* In the absence of a local code, the National Electrical Code (NEC) (ANSI/NFPA 70) should be used as a guide; it is available from the National Fire Protection Association (NFPA). Conductors of the circuit supplying the boiler should be sizes for 125 percent of the load served, as should be circuit protective devices.

Figure 16.1.42 shows the relation of amperes to kilowatts for the voltages commonly encountered in the United States. In all cases, the boiler manufacturer's recommendations should be followed.

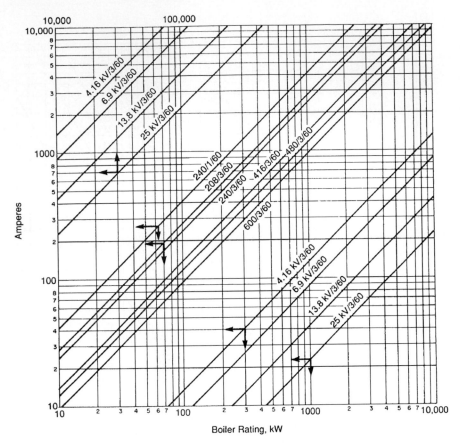

FIGURE 16.1.42   Boiler amperage draw.

Water Quality.   Table 16.1.9 shows the recommended water-quality limits for boiler feed water and makeup water and for water in the boiler or heating system. Water-quality requirements are more critical for electrode boilers than for the resistance type. In addition, the minimum water-quality requirements for electrode boilers vary for the different designs and are usually more stringent as the voltage increases.

Recovery of condensate for return to the boiler is recommended both as an energy conservation measure (Fig. 16.1.43 shows the increase in pounds of steam per kilowatt with boiler feed water temperature) and as a source of high-quality feed water. Leakage from hot-water heating systems not only wastes heat but also increases corrosion in the system and can lead to scale deposits in the boiler.

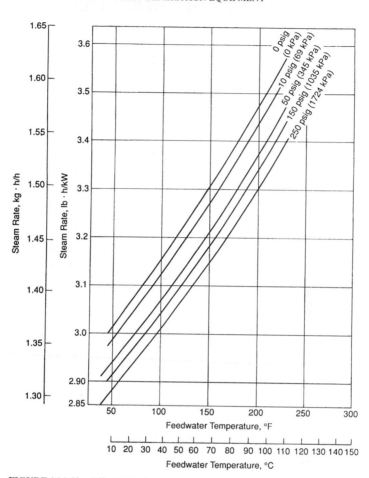

FIGURE 16.1.43   Effect of feedwater temperature on steam rated per kilowatt.

16.2 BURNERS AND FUELS

E.D. Cancilla

Chairman, Ajax Boiler Inc., Santa Ana, California

16.2.1 INTRODUCTION

Burner design for commercial boilers is divided into four types: atmospheric (or natural draft) burners, forced draft burners, premix low No_x burners, and coal stoker burners. Table 16.2.1 shows fuel capability with these burner types.

16.2.1.1 Atmospheric Burner

The atmospheric burner is the oldest and simplest type of burner that requires only adequate draft and gas pressure to function. Atmospheric designs exclusively use gaseous fuels such as natural gas, propane, digester gas, or manufactured gas. The atmospheric burner is shown schematically in Fig. 16.2.1 and a complete burner assembly is shown in Fig. 16.2.2. Atmospheric burners are generally constructed from cast-iron or stainless or galvanized steel sheet.

Atmospheric burners provide good, dependable combustion performance and efficiency. Since they do not have linkages and louvers, they do not need frequent service and adjustments to provide consistent performance. They are quiet operating, highly reliable with low maintenance and low operating costs.

Atmospheric burners are ignited by electric or electronic ignition systems and their operation is monitored by electronic or microprocessor-based "flame safeguard" systems. Atmospheric burners can be equipped as on-off, staged firing, or modulating burners. There are no moving parts in an atmospheric burner and electrical demand is low. Control systems can be provided to meet all insurance and local safety codes.

Adequate combustion air supply, venting, and draft control are very important. Atmospheric boilers are normally supplied with either draft diverter's or barometric dampers (see Fig. 16.2.3). The draft diverter may be designed into the boiler or may be a separate assembly which is attached to the boiler casing at the installation. The breeching that conducts the flue gas to the stack or chimney is attached to the draft diverter or barometric tee.

A comparison between the draft diverter and the barometric damper yields the following: Draft diverters and barometric dampers both reduce high draft levels by allowing air from the boiler room to enter the chimney. In the case of the draft diverter, the amount of

TABLE 16.2.1 Burners and Fuels

Burner type	Natural gas or propane	Light oil	Heavy oil	Combination gas/oil	Solid fuel coal/wood
Atmospheric	Yes	No	No	No	No
Forced draft	Yes	Yes	Yes	Yes	Yes*
Stoker	No	No	No	No	Yes

*Fuel must be in pulverized form.

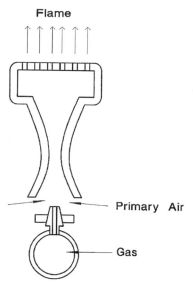

FIGURE   16.2.1 Atmospheric   burner. (*Courtesy Ajax Boiler Inc.*)

air taken from the boiler room is a large, relatively fixed amount. The barometric damper will adjust to compensate for varying draft conditions. From this perspective, the barometric damper is a more efficient draft control since heated boiler room air loss is limited to the amount required to control the draft.

The draft diverter, however, as its name implies, diverts downdrafts from the boiler and thus makes boiler operation, especially ignition, more reliable under gusting wind conditions.

FIGURE 16.2.2 Atmospheric burner assembly. (*Courtesy Ajax Boiler Inc.*)

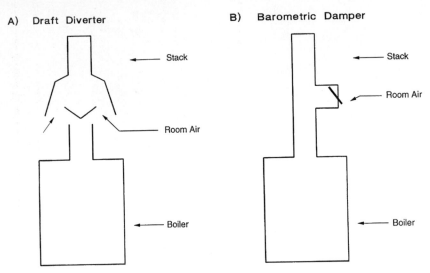

FIGURE 16.2.3   Draft Controls. (*Courtesy Ajax Boiler Inc.*)

16.2.1.2 Forced Draft Burner

The forced draft burner utilizes an electric fan to deliver combustion air and the fuel/air mixture into the boiler. A typical packaged forced draft gas/oil burner is shown in Fig. 16.2.4. These burners provide slightly better combustion efficiencies but the linkages and louvers must be kept in adjustment for proper combustion performance. First cost, operating, and maintenance costs are substantially higher. Noise levels and electric energy consumption are also much higher than with atmospheric burners. More expensive pressure vent stacks are usually required. Forced draft burners can commonly fire gaseous or liquid fuels.

Forced draft burners can be equipped for on-off, staged, and modulating firing with electronic or microprocessor-based flame safeguard control systems. In the modulating burner, fuel flow and air flow are controlled together to maintain the proper air-fuel ratio throughout the firing range; however, the control linkage must be frequently adjusted to compensate for linkage wear and slippage.

Packaged, force draft burners are available in sizes up to 60 million Btu/h (18 mm) input. Forced draft boilers may require draft controls if a chimney height exceeds 20 ft (6 m) for a single boiler. The simplest form of draft control is the barometric damper, but more sophisticated "sequenced draft controls" are available which, in addition to controlling drat, close off the boiler flue gas outlet when the boiler shuts down to minimize stack energy losses.

16.2.1.3 Low NO$_x$ Burners

A new advanced "premix" burner mixes gaseous fuel homogeneously with the correct amount of air to support combustion. The fuel air mixture passes through a screen at the burner surface. The screen is designed to provide the continuous flow of fuel air mixture and also keeps the burner surface cool. This low temperature, along with the very even air-fuel mixing, results in NO$_x$ levels in the 10 to 20 ppm range (corrected to 3 percent O$_2$). Figure 16.2.5 shows a typical premix burner arrangement.

FIGURE 16.2.4   Forced draft gas/oil burner. (*Courtesy of Ajax Boiler Inc and Power Flame Inc.*)

FIGURE 16.2.5   Low No$_x$ Induced fuel gas recirculation burner. (*Courtesy Ajax Boiler Inc. and Power Flame Inc.*)

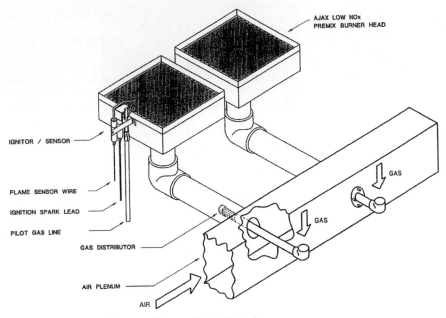

FIGURE 16.2.6   Premix burner. (*Courtesy Ajax Boiler Inc.*)

Flue gas recirculation forced draft burners are available as low emissions burners. Inert boiler exhaust (flue) gas is drawn into the blower inlet to vitiate (dilute) the combustion air. NO_x levels in the order of 25 ppm (corrected to 3 percent O_2) are available by this method. Figure 16.2.6 shows this type of burner.

16.2.1.4 Stoker

Solid fuel stoker-type combustion systems are seldom used in HVAC projects unless very special circumstances exist. Stokers are primarily natural draft combustion systems where fuel is delivered to a fixed or moving grate and ash is removed from beneath the grate. In some stoker systems, fuel is delivered to the grate by use of a pneumatic, fan-driven air fuel flow. Solid fuel stoker systems burning wood or coal have become virtually nonexistent in recent years because of the increased need for smoke and particulate control equipment and because of the increased availability of low-cost natural gas.

16.2.2 *FUELS*

The amount of heating given off during complete combustion of a known amount of fuel is the *heat of combustion*. For liquids and solids, the heat of combustion is usually given in units of Btu/lb (mJ/kg), while gaseous fuels are given in units of Btu/(s·ft³) [kcal/(s·m³)]. Heating value, calorific value, and fuel Btu (kcal) value are other terms frequently used.

Typical heating values and specific gravities for various gaseous fuels are given in Table 16.2.2, while Table 16.2.3 gives their chemical composition.

Alternate fuels are very seldom used. However, with the ever-changing cost and supply of standard boiler fuels, one needs to know what fuels are available and some of their basic properties. New boilers are being designed and existing ones retrofitted in order to burn fuels that in the past were not normally burned in a boiler. Some of the more popular new fuels used are solvents, alcohols, kerosene, used oils, digester gas, landfill gas, and wood. In the following pages various liquid fuels, gaseous fuels, and solid fuels are compared and discussed.

As shown in Table 16.2.4, several grades (1, 2, 4, 5, and 6) of fuel oil are available commercially, and each has its own specific properties. These oils must pass particular tests specified by Commerical Standard CS12 8 or ASTM D-396. The remaining fuels shown in Table 16.2.4 are not normally considered boiler fuels. Because of this, standards such as those mentioned above are not available.

The higher heating value (gross heating value) of a fuel is the heat obtained from combustion where the combustion products are cooled to a temperature at which all combustion-generated water is condensed. The lower heating value (net heating value) of a fuel is the heat obtained from combustion where the combustion products *are not cooled* to the point of condensation. In the United States, the higher heating value is the one normally reported and used in calculations.

The "flash point" of a liquid fuel may be defined as the temperature to which it must be heated to give off sufficient vapor to form a flammable mixture with air. This temperature varies with the apparatus and procedure employed, and consequently both must be specified when the flash point of an oil is stated. To determine the flash point, either the closed-cup (Pensky-Martens) test or the open-cup (Cleveland) test is used.

The sulfur present in any boiler fuel is a concern because of its corrosive and polluting effects. In boilers the sulfur dioxide and water vapor in the combustion products may unite to form acids that can be highly corrosive to the breeching. The presence of some gaseous sulfur compounds may lower the dew point of water vapor in the flue gases, further aggravating corrosion problems.

The viscosity of an oil is the measure of its resistance to flow. Commercial oils have maximum limits placed on this property because of its effect on both the rate at which oil will flow through pipelines and the degree of atomization that may be secured in any given equipment. Viscosity versus temperature curves for commercial fuel oils are shown in Fig. 16.2.7.

TABLE 16.2.2 Gaseous Fuels—Heating Values and Specific Gravity

Fuel	Heating value		Specific gravity
	Btu/(s·ft³)	kJ/(s·m³)	
Carbon monoxide	321	11,960	0.967
Coal gas	149	5,552	0.84
Coke oven gas	569	21,200	0.4
Digester gas	655	24,405	0.86
Hydrogen (H₂)	325	12,109	0.0696
Landfill gas	476	17,735	1.04
Natural gas	974–1129	36,290–42,065	0.60–0.635
Propane	2504–2558	93,296–95,308	1.55–1.77

TABLE 16.2.3 Chemical Composition of Gaseous Fuels

	Methane (CH$_4$)	Ethane (C$_2$H$_6$)	Propane (C$_3$H$_8$)	Butane (C$_4$H$_{10}$)	Carbon monoxide (CO)	Hydrogen (H$_2$)	Hydrogen sulfide (H$_2$S)	Oxygen (O$_2$)	Carbon dioxide (CO$_2$)	Nitrogen (N$_2$)
Carbon monoxide	—	—	—	—	100	—	—	—	—	—
Coal gas*	0.2	—	—	—	28.4	17.0	—	0.3	3.8	50.6
Coke oven gas	32.3	—	—	—	5.5	51.9	0.8	—	2.0	4.8
Digester gas†	64.0	—	—	—	—	0.7	—	—	30.2	2.0
Hydrogen H$_2$	—	—	—	—	—	100	—	—	—	—
Landfill gas	47.5	—	—	—	0.1	0.1	0.01	0.8	47.0	3.7
Natural gas	82–93	0–15.8	—	—	0–0.45	0–1.82	0–0.18	0–0.35	0–0.8	0.5–8.4
Propane gas‡	—	2.0–2.2	73–97	0.5–0.8	—	—	—	—	—	—

*Contains 0.02 percent sulfur.
†Contains 2.0 percent water.
‡Contains up to 24.3 percent C$_3$H$_6$.

TABLE 16.2.4 Grades of Fuel Oil

Fuel	Heating value, Btu/lb (MJ/kg)	Specific gravity	Flash point, °F (°C)	Sulfur, wt %	Viscosity, SSU at 100°F (38°C)	Kinematic viscosity at 38°C, cSt	Ash, wt %
No. 1 fuel oil	19,670–19,860 (45.8–46.2)	0.805–0.845	100 (37.8)	0.5		1.4–2.2	—
No. 2 fuel oil	19,170–19,750 (44.6–45.9)	0.855–0.876	100 (37.8)	0.5	32–38	1.8–3.6	—
No. 4 fuel oil	18,280–19,400 (42.5–45.1)	0.887–0.910	130 (54.4)	0.3	60–300	10.3–64.6	0.1
No. 5 fuel oil	18,100–19,020 (42.1–44.3)	0.922–0.946	130 (54.4)	0.3	20–40 @ 122°F (50°C)	39.5–81.3 @ 50°C	0.1
No. 6 fuel oil	17,410–18,990 (40.5–44.2)	0.959–0.986	150 (65.6)	0.3	45–300 @ 122°F (50°C)	92–647 @ 50°C	0–0.3
No. 1 diesel*	18,540 (43.1)	0.81–0.85					
No. 2 diesel*	19,440 (45.2)	0.841					
JP-4*	18,540 (43.1)	0.87–0.90		0.4			
JP-5*	18,540 (43.1)	0.82	140 (60.0)	0.4			
Kerosene*	18,540 (43.1)	0.80					
Alcohols*	8,419–18,810 (196–438)	0.79–0.85					
Ketones*	10,400–16,100 (24.2–37.4)	0.80–0.81					
Solvents*	12,250–17,558 (28.5–40.8)	0.79–0.88					
Motor oil (used)*	18,460–19,270 (42.9–44.8)	0.84–0.96					

*Nonstandard liquid boiler fuel.

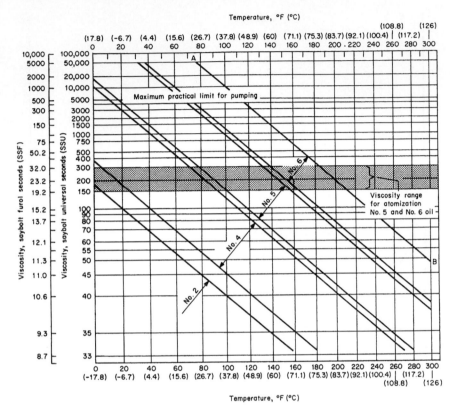

FIGURE 16.2.7   Viscosity vs. temperature curves for commercial fuel oils.

The ash test is used to determine the amount of noncombustible impurities in the fuel. These impurities come principally from the natural salts present in the crude oil or from chemicals that may be used in refinery operations. Some ash-production impurities in fuels cause rapid deterioration of refractory materials in the combustion chamber, particularly at high temperatures; some impurities are abrasive and destructive to pumps, valves, control equipment, and other burner parts.

Coal is still the most widely used solid fuel for boilers. Properties of three classes of coal are given in Table 16.2.5. The composition of coal is reported by either the proximate analysis (Table 16.2.5) or the ultimate analysis, both expressed in weight percent.

The ash fusion temperature is a measurement of when the fuel's ash becomes molten. If the ash in the fuel has a low melting or fusion temperature, the ash will coalesce into masses and deposit on the firing surface or on boiler surfaces. Such deposits are difficult to remove and will interfere with boiler and burner operation.

Burning of wood in boilers is, in some instances, a feasible boiler fuel. Normally the wood used for fuel is a waste or by-product of some process of manufactured products. Table 16.2.6 gives pertinent information for typical hardwood and softwood.

TABLE 16.2.5 Composition of Coal (Proximate Analysis)

	Heating value, Btu/lb (MJ/kg)	Bulk density, lb/ft³ (kg/m³)	Fixed carbon, wt %	Moisture, wt %	Volatile matter, wt %	Sulfur, wt percent	Ash, wt %	Ash fusion temp., °F (°C)
Anthracite	—	50–58 (800–929)	80.5–85.7	2.8–16.3	3.2–11.5	0.6–0.77	9.7–20.2	
Bituminous	—	42–57 (673–913)	44.9–78.2	2.2–15.9	18.7–40.5	0.7–4.0	3.3–11.7	2030/2900+ (1110/1543+)
Lignite	7440 (17.3)	40–54 (641–865)	31.4	39.1	29.5	0.4	4.2	1975/2070 (1079/1132)

TABLE 16.2.6 Composition of Wood

| | Heating value, Btu/lb (dry) (kJ/kg) | Weight percent on a day basis | | | | | Water, wt % | Fusion temp., °F (°C) |
		Carbon	Hydrogen	Oxygen	Nitrogen	Ash		
Hardwoods	8732 (20,300)	50.9	6.3	41.9		0.84	32.7	2200–2750 (1204–1510)
Softwoods	8903 (20,700)	52.1	6.0	40.8	0.1	1.07	54.4	2200–2750 (1204–1510)

16.3 BURNER SYSTEMS*

Webster Engineering & Manufacturing Co, Inc.
Winfield, Kansas

16.3.1 INTRODUCTION

The purpose of a burner is to supply energy, in the form of heat extracted from fuel, to a boiler or similar vessel in order to generate steam, hot water, hot air or some other heated medium for space comfort or a production process. Regardless of fuel, three required ingredients for combustion are *fuel, heat*, and *oxygen*. For proper combustion, the well designed burner will provide necessary turbulence and feed velocities.

16.3.2 GAS BURNERS

The optimum air-fuel ratio for natural gas, at standard conditions, is approximately 9.5:1 (9.5 m³ air/1 m³ fuel) for perfect combustion. At 10 percent excess air, the air-fuel ratio is about 10.5:1. Furthermore, the fuel must be put into the furnace in such a way that all the gas comes in contact with enough air to burn it completely.

Also the air and gas must come together at the right time so that the mixture is ready for spontaneous ignition and a stable flame pattern is established.

The gas pressure requirements should be at a minimum, and the power requirements for the fan should be reasonable. The pressure developed by the fan should be high enough so that the amount of air delivered for combustion is not appreciably affected by changes in stack conditions. Thus, a change in pressure at the boiler stack outlet of as much as ±0.5 in water column (in WG) (24.5 Pa) should not appreciably affect the air-fuel ratio. Significant changes in combustion settings can result from draft variations in boilers designed for low draft loss.

There are, of course, many controls, both operating and safety, in connection with the burner, and these are certainly important to good burner design. This section, however, deals primarily with the mechanical functioning and design of the burner itself.

The color of the flame in good gas burner design should be mentioned even though it is a controversial issue. Colors range from blue transparent to yellow luminous. Whatever the flame color, one must rely on instruments which analyze the stack gases. If the instruments show no unburned fuel in the stack and there is no deposition on the boiler tube walls from minimal amounts of excess air, the flame is good—regardless of color.

Available Gases and Characteristics.   The most common fuel gas is natural gas. Some other less common gases are listed in Table 16.3.1.

Flame Speeds and Temperatures.   The most common gases are natural, propane, and butane. Fortunately the flame speed (ignition velocity) is about the same [2.3 ft/s (0.7 m/s)] for each of these gases for a mixture of the theoretical amount of air for complete combustion. The flame temperatures for these gases at around 3700°F (2030°C) are also similar. But these gases differ greatly in density and heating value. Thus the piping and porting must be different for each gas, to keep the same combustion characteristics. For example, the values in Table 16.3.1 show the variation in gas flow velocity and pressure, assuming that 1 MBtu/h (143 kW) is to be released through a $^1/_2$-in (12.7-mm) orifice.

---

*Edited by Cleaver Brookes, Division of Aqua Chem, Inc, Milwaukee, WI.

TABLE 16.3.1   Contents of Gases

Gas	Btu/ft³ (kJ/m³) gross	ft³ air/ft³ gas (m³ air/m³ gas) (no excess)	Ultimate CO₂, %	Specific gravity
Natural	1020 (38,000)	10.0 (2.83)	10.0	0.65
Blast furnace	92 (3,428)	0.68 (0.02)	25.5	1.02
Produces bituminous	163 (6,073)	1.24 (0.04)	18.6	0.86
Coke oven	569 (21,200)	5.45 (0.15)	10.8	0.40
Propane (refinery)	2504 (93,296)	23.20 (0.66)	14.0	1.77
Butane (refinery)	3184 (118,632)	30.00 (0.85)	14.3	2.00

TABLE 16.3.2   Flow Characteristics of Gases for 1 MBtu/h (14.3 kW)

Gas	Flow, ft³/h (m³/h)	Orifice size, in (mm)	Velocity, ft/s (MPa)	Pressure, in H₂O (kPa)
Natural	980 (27.7)	0.5 (12.7)	200 (1.02)	11.9 (2.96)
Propane	400 (11.3)	0.5 (12.7)	81 (0.41)	5.3 (1.32)
Butane	314 (8.9)	0.5 (12.7)	64 (0.33)	3.7 (0.92)

Although the values in Table 16.3.2 are approximate, they clearly show that the designer must use care in selecting burners which are to burn more than one gas.

16.3.3 OIL BURNERS

A well-designed oil burner will have the following functions and/or features:

- The air-fuel ratio must be held constant.
- All the oil must come in contact with air at the right time.
- There must be a proper flame pattern.
- Oil pressure and power requirements should be minimal.
- The fan pressure should be adequate.
- The oil burner must atomize the oil sufficiently well.

16.3.3.1 Methods of Atomization

In oil burner design, probably no other subject is as controversial as the method of atomizing the fuel. The process of atomization is to break up the oil into fine droplets, not so fine as to be a vapor, but small enough to obtain intimate air and fuel mixture. Every method has its staunch supporters and there is no one best method.

The size of burner, type of oil, and some other factors such as "turndown" ratios enter into the decision making process for the best method to use.

16.3.3.2 Types and Characteristics of Oils

Fuel oils range from No. 6, a thick and black oil, to No. 1, a very thin colorless oil similar to kerosene. The characteristics of oils can be found in Table 16.3.3.

The heaviest and thickest oils must be heated to be pumped, and heated still more to reduce the viscosity for adequate atomization. The lighter, less viscous oils have their own handling and atomizing problems. For example, No. 1 oil will vaporize quicker in suction lines to pumps with less vacuum (less negative) than more viscous oils resulting in fuel feed problems. Also, steam atomization of light oil might not be the best *if the oil is stored outside in cold weather*. Steam may condense in the oil gun in advance of reaching the nozzle, thus loss of atomization. The solution is to heat the oil, atomize with air, or atomize mechanically with pressure.

16.3.3.3 Basic Combustion Principles

For any fuel to burn, it must combine with a specific amount of air (actually oxygen) and the temperature of the mix must be sufficiently high to start the reaction, which then becomes self-sustaining. If there is insufficient air, the fuel-air mix will not burn, regardless of temperature. Similarly, if there is too much air, the fuel will not burn, regardless of temperature. However, between these limits the fuel will burn completely or partially, depending on how well the fuel and air are mixed and the amount of air in the mix.

The amount of oxygen left over after combustion is the "excess O_2" or "excess air." One objective in burner design is to have a minimum of excess air in the combustion products.

To find the theoretically correct quantity of air required to burn a fuel, the following equation may be used:

$$W = 11.52C + 34.56\left(H - \frac{O}{8}\right) + 4.32S \qquad (16.3.1)$$

where W = lb (kg) air/lb (kg) of fuel
$\quad C$ = decimal percentage of carbon in fuel
$\quad H$ = decimal percentage of hydrogen in fuel
$\quad O$ = decimal percentage of oxygen in fuel
$\quad S$ = decimal percentage of sulfur in fuel

The air requirements are often expressed in $(ft^3/min\ air)/(MBtu/h)$ or $(m^3/s)/kW$.

The calculated value for air required is in pounds of air per pound (kilogram) of fuel. On average, because of varying environmental conditions and fuel quality differences, burners are nominally adjusted for fuel oil to operate in the 20 percent excess-air range. For air flow required for burning most oil fuels, the rule of thumb is 200 $(ft^3/min)/(MBtu/h)$ [3.22E − 0.04 $(m^3/s)/kW$].

For most typical automatic fuels, Eq. (16.3.1) yields a value of W of 12 to 15. This value is also referred to as the "fuel-air ratio." For rough approximations, a fuel-air ratio of 15 is adequate.

The air required per gallon (liter) of fuel burned in a boiler can be readily worked out on an approximate basis. The turndown ratio is related to the air-fuel ratio. Suppose that a boiler is rated for 50 TBtu/h (14.655-kW) input. The air flow is about 50 × 200 $(ft^3/min)/(TBtu/h)$ (at 20 percent excess air) or 10,000 ft^3/min (4.72 m^3/s). If now, with all air dampers closed, the fan delivers 2000 ft^3/min (0.94 m^3/s), the air turndown ratio is 10,000/2000 (4.72/0.94), or 5:1. If the air-fuel ratio stays constant, the low-fire boiler input is about 10 TBtu/h (2931 kW) for a fuel turndown of 5:1 also. Now if it is desired to have

TABLE 16.3.3a Commercial Standard CS12–48: Detailed Requirements for Fuel Oils[a] (Part 1)

No.	Grade of fuel oil[b]	Sayb. Univ. at 100°F (39°C) Max.	Min.	Sayb. Furol at 122°F (50°C) Max.	Min.	Kin. C.S. at 100°F (39°C) Max.	Min.	Kin. C.S. at 122°F (50°C) Max.	Min.	Gravity degrees API Min.
1.	A distillate oil intended for vaporizing pot-type burners and other burners requiring this grade.[d]					2.2	1.4			35
2.	A distillate oil for general purpose domestic heating for use in burners not requiring No. 1.	40				(4.3)				26
4.	An oil for burner installations not equipped with preheating facilities	125	45			(26.4)	(5.8)			
5.	A residual type oil for burner installations equipped with preheating facilities		150	40			(32.1)	(81)		
6.	An oil for use in burners equipped with preheaters, permitting a high-viscosity fuel			300	45			(638)	(92)	

[a]Recognizing the necessity for low sulfur fuel oils used in connection with heat-treatment, non-ferrous metal, glass and ceramic furnaces and other special uses, a sulfur requirement may be specified in accordance with the following table.

Grade of Fuel Oil	Sulfur, Max, Per Cent
No. 1	0.5
No. 2	1.0
Nos. 4, 5, and 6	no limit

Other sulfur limits may be specified only by mutual agreement between buyer and seller.

[b]It is the intent of these classifications that failure to meet any requirement of a given grade does not automatically place an oil in the next lower grade unless in fact it meets all requirements of the lower grade.

[c]Lower or higher pour points may be specified whenever required by conditions of storage or use. However, these specifications shall not require a pour point lower than 0°F under any conditions.

[d]No. 1 oil shall be tested for corrosion in accordance with par. 15 for 3 h at 122°F. The exposed copper strip shall show no gray or black deposit.

[e]The 10 percent point may be specified at 440°F (230°C), maximum for use in other than atomizing burners.

[f]The amount of water by distillation plus the sediment by extraction shall not exceed 2.00 percent. The amount of sediment by extraction shall not exceed 0.50 percent. A deduction in quantity shall be made for all water and sediment in excess of 1.0 percent.

16.78

TABLE 16.3.3b Commercial Standard CS12-48: Detailed Requirements for Fuel Oils[b] (Part 2)

No.	Grade of fuel oil[b]	Flash point °F Min.	Pour point °F Max.	Water and sediment percent Max.	Carbon residue on 10 percent residuum, percent Max.	Ash percent Max.	Distillation temperatures °F		
							10 percent point Max.	90 percent point Max.	End point Max.
1.	Distillate oil intended for vaporizing pot-type burners and other burners requiring this grade.[d]	100 (30 °C) or legal	0	trace	0.15		420 (240°C)		625 (380°C)
2.	A distillate oil for general purpose domestic heating for use in burners not requiring No. 1	100 (39°C) or legal	20 (−16°C)	0.10	0.35		[e]	675	(380°C)
4.	An oil for burner installations not equipped with preheating facilities	130 (55°C) or legal	20 (−16°C)	0.50		0.10			
5.	A residual type oil for burner installations equipped with preheating facilities	130 (55°C) or legal		1.00		0.10			
6.	An oil for use in burners equipped with preheaters permitting a high-viscosity fuel	150 (65°C) or legal		2.00[f]					

a lower input than 10 TBtu/h (2931 kW), this can be accomplished by reducing the low-fire fuel input. However, since the air flow cannot be reduced (the dampers are already closed), the air-fuel ratio will rise as low fire is approached. Therefore, air turndown for this boiler-burner package is always 5:1 maximum. However, fuel turndown may be much greater with only flame stability or perhaps flame chilling and subsequent unburned fuel generation as the lower limit.

16.3.3.4 Flame Stability

Every burner generates a flame front in the boiler furnace, and it is important for this flame front to remain stable for a wide range of firing rates and air-fuel ratios. A good burner should maintain a stable flame front with maximum air and minimum fuel without blowoff. A practical limit, however, may be the electronic flame scanner. The scanner may lose sight of a flame which is shrunk by excessive air.

16.3.3.5 Products of Combustion and Control or Pollutants

The basic products of combustion are carbon dioxide (CO_2) and water (H_2O). However, there are generally small amounts of carbon monoxide (CO), nitric and nitrous oxides (NO_x), unburned hydrocarbons, carbon particulates, sulfur, and ash. It is always good to keep these products to an absolute minimum, and this is a big part of the burner designer's task. These limits are generally spelled out by local, state, or federal authorities.

16.3.3.6 Control of Fuels

The basic devices used for fuel control are

- Pumps
- Relief valve
- Pressure-reducing valve
- Metering valve
- Fuel temperature controls

The final design of a fuel-burning system includes a blower with sufficient capacity to provide the air required for the design capacity at the pressure needed to overcome the pressure losses in the burner, boiler, and auxiliaries.

The designer must select or design fuel-flow controls which very closely meter fuel quantity flowing to the burner. Since the fuel input flow is only about one-fifteenth of the air flow, small variations in fuel flow will have a substantial effect on the fuel-air ratio. Therefore, these components which are selected to control such flow must singly or in combination provide a fairly precise and dependable function.

The pump selection is important in that the pump must deliver dependably a flow of fuel oil to the burner system in sufficient quantity to meet the burner maximum capacity, plus a small excess amount which returns to storage.

The fuel oil from the pump must be routed through the burner piping at a consistent pressure. This task is normally handled by carefully selected relief valves, back-pressure valves, or pressure-reducing valves. Quite often this is accomplished through the use of two or more such devices. The fuel, at constant pressure, is then allowed to flow through a metering valve to the burner nozzles. The metering valve is usually designed so that it is little affected by minor changes in oil viscosity. The valve incorporates a variable-size port which is actuated to provide varying fuel throughout to the burner in accordance with heat demand imposed on the boiler.

Figure 16.3.1(a) indicates the piping schematic for firing oil with air atomization. The schematic also shows the oil preheat housing for assurance of warm oil on start cycle with heated oil.

Figure 16.3.1(b) indicates the piping schematic for pressure atomization. The metering valve is in the return, by-pass line to meter the flow of oil through the nozzle by returning excess oil back to the oil storage tank.

16.3.4 SOLID-FUEL BURNERS

During the era of bountiful supply of natural gas and oil fuels, it was not economically prudent to use solid fuels. Although solid fuels have always been available at a lower cost per unit of energy, the equipment required to deliver the fuel to the burner, as well as to control the combustion process, is more costly. Solid fuels normally contain significantly more noncombustible material (ash) which must be extracted from the combustion zone and disposed of. Larger boiler equipment is required to provide significantly greater combustion space for the slower burning process when solid fuels are used.

However, in recent years, a substantial effort has been made to improve solid-fuel burner systems, in an effort to economize on the cost of energy. There are vast improvements not only in burner systems but also in fuel-handling hardware. Also considerable research is being done on the concept of converting solid fuels to gaseous and liquid forms. In this section we discuss only presently available equipment of proven design.

Solid fuels, both coal and wood, are being burned in boilers in a number of system designs. The most popular and best known is the *stoker*. This combustion process is frequently referred to as "mass burning." The fuel is fed onto a grate from a fuel hopper from above or pumped upward onto a grate from below. The air required for the combustion of the fuel is normally delivered from below, being forced through the bed of fuel on the grate surface. The fuel bed thickness varies and depends on the specific type of stoker system employed, type of fuel being used, and load on the boiler.

The most commonly used stoker system in small to intermediate industrial boilers is a chain grate or traveling grate. See Fig. 16.3.2. These stokers are formed of a series of endless chains of special configuration that provide a flat surface onto which the fuel is fed. The grate moves from the front boiler face toward the rear of the combustion zone. The fuel is fed onto the top surface of the grate in a layer of uniform thickness. The bed of fuel traverses the length of the furnace during which time it burns, releasing its energy to the heat-absorbing surfaces.

The speed at which the fuel bed travels is automatically adjusted under the control of the boiler pressure. The combustion air fed to a plenum chamber beneath the stoker flows upward through the fuel bed and is simultaneously controlled by fan dampers in such a manner as to provide a relatively constant air-to-fuel ratio. The depth of fuel carried on the stoker is normally controlled manually by adjusting the height of a gate at the fuel feed. The design of the system is such that the combustion process is completed before the fuel on the bed reaches the end of the stoker. The stoker links traverse a drum where the ash is dropped off into a hopper. The grates then move to the front or fuel feed end where a new bed of fuel is laid down. The grate heat-release rates for stokers of this type are moderately high.

There are many variations on this principle that employ mechanisms other than chain grates, where the fuel bed is moved through the combustion zone by reciprocating grates,

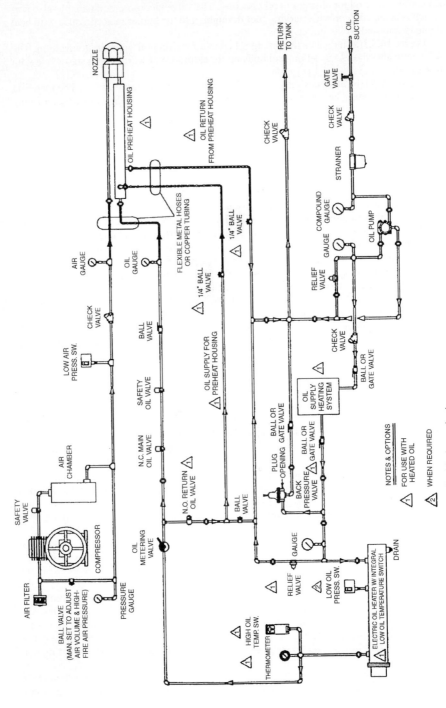

FIGURE 16.3.1a Piping schematic for firing oil with air atomization.

16.82

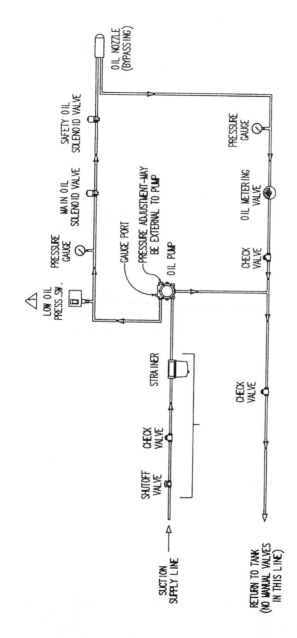

NOTES & OPTIONS

⚠ WHEN REQUIRED

FIGURE 16.3.1b Piping schematic for pressure atomization.

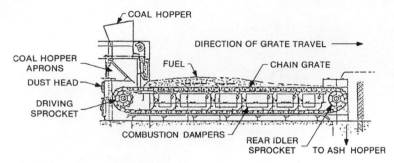

FIGURE 16.3.2   Typical chain grate stoker. (*Courtesy of Cleaver-Brooks.*)

vibrating grates, and sliding systems. Grate systems may be horizontally oriented within the combustion zone or inclined to allow the fuel to move downward toward the ash discharge zone. (See Fig. 16.3.3).

Chain grates or traveling grates may have the fuel fed onto them by a spreader mechanism that throws the fuel onto the grate system. Where spreader systems are used, the fuel is injected from a point some distance above the grate bed. The traveling grate in this design usually travels from the rear of the combustion chamber toward the front, with the ash discharging into a hopper at this point. The solid fuels used for these systems may contain a higher level of fine particles, which will burn in suspension. The larger fuel pieces, being more dense, will fall to the grate, where the combustion process is completed. (See Fig. 16.3.4.)

A smaller grate surface is required for this system design, permitting the use of a combustion chamber of smaller floor dimensions. However, the height dimension is usually greater, to provide sufficient volume in the combustion zone so that ample time is available to complete the combustion of all fuel delivered to the furnace. Spreader stoker systems are frequently used in boilers designed to generate steam in quantities of 75,000 lb/h (34,000 kg/h) and larger. Coal-burning systems of this design can be operated at relatively high heat-release rates per square foot (meter) of grate area.

Smaller boilers, having capacities of 10,000 lb/h (4536 kg/h) of steam or less, are frequently equipped with underfeed stokers, where the fuel is forced into the combustion zone

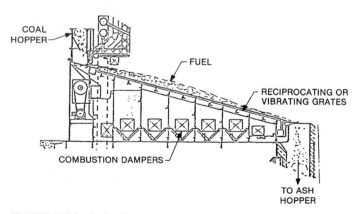

FIGURE 16.3.3   Inclined vibrating grate stoker. (*Courtesy of Cleaver-Brooks.*)

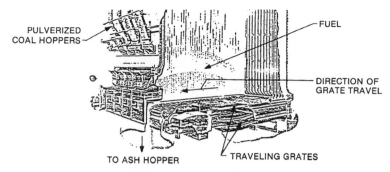

PULVERIZED
COAL HOPPERS

FUEL

DIRECTION OF
GRATE TRAVEL

TO ASH HOPPER　　　TRAVELING GRATES

FIGURE 16.3.4　Full spreader or traveling grate. (*Courtesy of Cleaver-Brooks.*)

by a ram or screw. (See Fig. 16.3.5.) When the fuel reaches the end of the feed conduit, the fuel moves upward over casting surfaces containing holes or slots through which the combustion air is delivered.

The combustion process is completed as the fuel moves in a sideward direction over the stoker surfaces, called "tuyeres." The ash from the spent fuel falls into channels or hoppers at each side of the stoker, from which it is easily removed.

The procedure for extraction of the ash varies and depends greatly on the quantity of ash generated, facilities available at the jobsite, quality of fuel to be burned, and means for disposal. Small boilers obviously will generate small quantities of ash. It is not unusual to expect the boiler operator to manually remove the accumulated ash as required, from the points of collection, using simple tools. However, on large boilers, significant quantities of ash will collect in hoppers on the boiler. The number of hoppers and their location will vary according to boiler design. Ash can be extracted from these hoppers manually through seal valves, or automatic extraction systems can be provided. A variety of systems are available including pneumatic, water, and mechanical converters. These systems all include closed conduits which carry the material to a safe point of discharge. Such systems are normally installed in the field after the boiler has been installed on its foundation.

Many other solid-fuel burner systems are in use. For very large boilers typical of electric utilities, pulverized-fuel burners are common. For these systems, the fuel is prepared by grinding to powder form. The fuel is then blown into the combustion zone, along with the combustion air through air registers, not unlike large oil and gas burner systems. Efficient combustion with low levels of excess air is routinely achievable. These systems

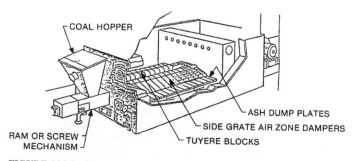

COAL HOPPER

ASH DUMP PLATES
SIDE GRATE AIR ZONE DAMPERS
RAM OR SCREW
MECHANISM
TUYERE BLOCKS

FIGURE 16.3.5　Underfeed stoker. (*Courtesy of Cleaver-Brooks.*)

are normally restricted to very large boilers, since the cost of fuel preparation and equipment maintenance is high and can be justified only when large amounts of fuel are used. Systems of this type are complex, and they include fuel distribution and air-handling systems and elaborate control systems which require skilled boiler operating technicians. But in large plants the benefits are significant in terms of energy recovered per dollar of fuel, and the improved efficiency justifies the added equipment, maintenance, and labor costs.

Later technology and research have produced fluidized-bed systems and fuel gasification and liquefaction. The benefit of these systems is that they provide the potential for the burning of marginal-type fuels without polluting the atmosphere. Coals with high ash and/or sulfur content can be burned directly in a fluidized bed. Use of an appropriate bed material provides for the capture of sulfur compounds within the bed, minimizing the escape of acid-forming gases to the environment. The collection of particulate matter from the flue gases is a proven technology, making the fluidized-bed combustion systems a viable concept. However, control systems and fuel-handling and combustion air supply equipment are more complex than other available systems, making fluidized-bed equipment more expensive to install and operate.

Gasification and liquefaction designs holding promise for the future are emerging from the laboratories, but they will be broadly adapted only after more extensive commercialization.

16.3.5   CONTROLS

16.3.5.1   Controls Systems—Solid-Fuel-Fired Boilers

The use of solid fuels can provide significant savings in purchased-fuel costs, but there are some other costs that must be considered when these fuels are burned. Fuel handling is an important factor in the design of a system if a dependable, efficient installation is to result. Even under the best conditions, some monitoring is required to ensure that fuel is being delivered dependably and at the appropriate rate to the boiler. Therefore, although control systems are provided to control the boiler automatically with optimum fuel-to-air ratios on smaller industrial installations, some degree of control sophistication may be omitted because an operator in periodic attendance is common.

The boiler water level control systems used are of the same type and quality as those on boilers intended for use with other fuels. Burner operating controls receive a signal from the boiler pressure for steam boilers or from the temperature control for hot-water boilers. The fuel feed rate is controlled to satisfy the boiler pressure sensor for steam or the temperature sensor for hot water. Depending on the firing system, the operating control(s) will increase the speed of the gate and the spreader feeders, if used, or will change the frequency of operating cycles of the fuel feed mechanism if an underfeed system is used. Simultaneously the dampers controlling the flow of combustion air on both overfire and underfire and adjusted to provide the appropriate air flow in a fixed relationship to the rate of fuel delivered. Solid-fuel-fired boilers are normally operated with a furnace pressure of zero to slightly negative. This condition is created by draft inducers and damper systems under the control of a furnace pressure sensor.

On grate burning systems, air delivered beneath the grate is ducted from the fan through an automated damper into a plenum chamber, which can be divided into a number of zones. Each zone is equipped with a manually adjusted damper to provide the optimum air to the underside of the grate, in accordance with the needs of the fuel bed. The need for frequent manual adjustment of these individual dampers is rare and will depend more on the specific fuel quality than on the steam load. (See Figs. 16.3.2 and 16.3.3.)

Stoker-fired boiler systems are well adapted to relatively stable, slowly changing steam loads. Greater time is required to bring a bed of solid fuel up to the temperature needed to increase the incidence of volatiles. The larger furnace envelope and boiler water content also have an impact on this process. Rapid startup and rapid shutoff are not easily achievable.

For more responsive steam supply using solid fuels, pulverized fuel may be employed. Burner systems designed for these fuels can provide a rapid increase or decrease of heat release in the boiler furnace, since this prepared fuel will volatilize very quickly in the combustion chamber environment. There is some lag in the fuel supply between pulverizer and burner, but this will cause few problems since only large boilers are used with this quality fuel and since the stored energy available in the boiler is usually sufficient to cope with minor load swings.

Pulverized-coal-burning systems are usually controlled by fairly sophisticated control networks for several reasons. The fuel is very combustible, and fast response flame sensors are needed to continuously monitor the flame activity, just as is required for oil or gas flames. Since the furnaces for coal-burning systems are usually large, multiple monitors are common. Also, since the boilers using this fuel are large and the continuous energy release in the boiler is great, any sudden changes in the flows to the boiler, either fuel or air, are accompanied by a change in chamber pressure. The sidewalls of the combustion zone are usually formed of tubes carrying boiler water. Large volumes containing this combustion environment cannot be subjected to rapidly fluctuating internal pressures without running the danger of causing some failure. Therefore, the internal environment needs to be monitored with sensitive instrumentation to provide a signal to the boiler operator when upsets in the combustion process begin.

Clearly burners and control systems can vary from a very simple ram pushing fuel into a small chamber on cycles dictated by a steam-pressure-monitoring device, to a very complex computer control which monitors everything from the speed of conveyers delivering fuel to pulverizers to the temperature and pressure of the gas exiting the flue gas cleanup device.

Recommendations and requirements for control systems, particularly for larger boilers, are provided in a number of codes which are recognized by the American National Standards Institute (ANSI), insurance carriers, and equipment suppliers.

16.3.5.2 Combustion Fuel-Air-Ratio Controls

To achieve good combustion, the burner must provide a sufficient amount of air to burn the fuel. To improve combustion efficiency, the amount of air should be regulated, so that only enough air to completely burn the fuel is supplied. For safety reasons, the system should be designed and set up to provide more air than is actually required. If there is insufficient air, then the unburned fuel could be combined with additional air later and could be an explosive mixture.

To provide the proper amount of air for combustion, various fuel-air-ratio controls are used. The simplest control provides a fixed flow of air and fuel, and the burner simply turns on and off. Other types of burners are capable of varying the rate of fuel input and have a means of providing variable air and fuel flow rates based on heat requirements.

In addition to the basic types of controls, different systems are available which provide greater or lower accuracy in flow control. Feedback systems are sometimes added to more precisely control excess air and to prevent fuel-rich or fuel-smoking conditions. The more accurate controls and feedback systems allow the burner to operate closer to the optimum excess air level, which increases the efficiency of the unit.

16.3.5.3 Flow Control Equipment

Fuel-air-ratio control equipment can be broken down into three general types: atmospheric, fixed-flow, and variable-flow equipment.

Atmospheric Equipment. An atmospheric burner relies on the natural draft through a chimney to provide the required air flow. The proper design of the chimney is critical in obtaining the required air flow and preventing downdrafts or other conditions that may affect air flow. With this equipment, the air flow occurs at all times, whether the burner is on or off. The actual amount of air flow varies according to temperatures which determine the draft.

The fuel flow, except for solid fuels, is controlled by a valve, which is generally operated electrically to provide automatic operation. The fuel rate is regulated by a combination of a pressure control valve and an orifice.

The atmospheric burner is used in most homes for heating and hot water. Because few components are needed for this type of burner, it is both inexpensive and reliable. The accuracy of air-flow control is very poor, however, and because of this it is relatively inefficient.

One device that is sometimes added to increase efficiency is a draft control damper. This prevents air flow up the chimney when the burner is not operating and prevents the warm air in the room from escaping.

Fixed-Flow Equipment. Fixed-flow controls provide a specific flow of both air and fuel. A fan is used to supply combustion air, and the quantity can be regulated by adjusting a register or damper. The fuel flow is obtained in the same manner as for the atmospheric burner.

With this type of control, the quantity of air flow can be regulated with a fair degree of accuracy. The draft effect and changes in the draft are not important factors in determining air flow. (It is important, though, to provide proper designs to prevent large changes in stack pressure, which can prevent proper air flow.) The combustion air fan is usually cycled with the fuel valve, so that air flow only occurs when the burner is operating.

The fixed-flow control can be staged so that it can operate at more than one flow rate. For an oil burner, this can be done by using multiple valves and nozzles. Opening more valves increases the total fuel input. This can be done with single burner (one air fan) or multiple burners. Another method is to use a single nozzle and to vary the pressure, where the fuel flow increases with pressure. Again, this can be done with valves where the different valves are connected to different fuel pressure regulators. There are limits on the turndown available with pressure, since the square law requires large pressure variations to obtain small flow change. The multiple-nozzle approach can easily provide large turndowns.

The air-flow control for staged fixed input is generally achieved by positioning a damper. Typically an electric actuator is used to pen or close the damper, to provide more or less air. In some cases, the air flow is fixed at the maximum input requirements, and only the fuel flow is adjusted. This approach is also used on some atmospheric burners.

Variable-Flow Equipment. Variable-flow or modulating controls provide the ability to infinitely vary both fuel and air flow between some minimum and maximum values. In most cases, variable-flow valves are used to regulate the flow of fuel and air. This valve is generally called an "air damper" when it is used to control air flow. The air is supplied by a fan which operates at a constant speed and is considered a constant-volume device.

The fuel is supplied to the valve at constant pressure, so that the area change in the valve causes a specific known change in fuel flow. The fuel system (both gas and oil) is designed so that there is a relatively large pressure drop across the fuel valve and a small drop in the nozzle. This allows the fuel valve to be the primary regulator of flow, with the nozzle sized to accommodate large changes in flow.

With natural gas, a butterfly valve is generally used to control flow. The valve is opened or closed to change flow rates. On oil, a number of different variable-orifice valves are used. Since the pressure drop at the nozzle must be low, air or steam is generally used to atomize the fuel.

16.3.5.4   Fuel-Air Ratio Control Systems

The control system used to operate this equipment also varies by type of burner and accuracy requirements. Simple on/off burners (atmospheric or fixed-position) use switches that will cycle the burner on and off. These switches can be driven by any number of different parameters that reflect the need for heat. In a home furnace, the room temperature is used. For a boiler, the outlet water temperature or steam pressure is generally used. Because these burners are designed for simplicity and low cost, the operating controls tend to be very limited.

For staged input, a series of switches are used. The settings (temperature or pressure) are staged so that the input is increased as the need for heat increases.

Modulating systems come in three basic forms: single-point, parallel, and fully metered. Each system has advantages and disadvantages. All three systems, however, are limited in that they are used only for controlling between the minimum and maximum burner operating valves. Below this point, the burner operates as an on/off unit (described above).

The most common control is the single-point positioning system. A measurement of the error (actual minus desired) in temperature or pressure is used to operate a positioning actuator. The greater the error, the more the actuator moves. This actuator drives the fuel and air control valve via mechanical linkage. The linkage is adjusted so that the appropriate amounts of air and fuel are delivered to the burner at all firing rates. A jackshaft control system employs a shaft as the primary mechanical link between the fuel and air valves.

A parallel control system uses two actuators, one for air and one for fuel. These actuators operate from a single error measurement, and they independently drive the air and fuel valves. Operation of the system is adjusted so that the proper amounts of air and fuel are delivered to the burner at all firing rates. The primary difference between this system and the single-point positioning system is that the mechanical linkage between the air and fuel valve is replaced by an electric or pneumatic signal used to operate the individual actuators. Parallel positioning systems offer the advantage of allowing large differences between the air and fuel valve positions when mechanical linkage would be impractical.

The metering system is similar to the parallel positioning system, except that it actually measures air and fuel flow rates. With both parallel and single-point positioning, the flow rate is assumed, based on the position of the valve. The metering system positions the valve to obtain the desired flow rate. The flow measurement accuracy, temperature, pressure, viscosity, and other values are added to increase the accuracy.

16.3.5.5   Feedback Systems

In the past few years, a number of improved stack measurement systems have been used, either to provide fine tuning of the fuel-air ratio or to prevent the burner from operating in an improper manner. By directly measuring the products of combustion, all the possible variables that can affect the fuel-air ratio are taken into account.

The typical measurements, what the value represents, and how these values are used, are listed below. These systems are offered on new equipment and often are packaged for use on existing control systems. The two most commonly measured combustion products are oxygen and carbon monoxide (CO).

Oxygen.   The oxygen measurement is the most common and is representative of the excess air in combustion. Products are offered for test instruments which are used to periodically monitor and set up the fuel-air ratio and for continuous control. As a controller, the oxygen trial system will provide a small change in either air flow or fuel flow to maintain the desired excess-air (or oxygen) level. *The primary flow rate control is obtained by one of the standard systems.*

Carbon Monoxide. The presence of this gas indicates incomplete combustion. A small amount of CO, measured in ppm (mg/L), is common. If this value increases, it represents a decrease in efficiency, since not all the fuel is converted to heat. Although CO has been used to control excess air in a similar way to oxygen, it is more frequently used as a monitor to detect problems. New low-excess-air burners do not experience a gradual change in CO with changes in excess air, which makes it difficult to use them for controlling. This excess-air value is also limited to gas fuel, with opacity being the measurement of incomplete combustion for liquid and solid fuels.

A modulating control generates the input necessary to match the load and is quick to respond to load changes. Its cost is obviously higher, but the major cost factor is the type of modulating control. The efficiency varies with the accuracy of the modulating control system.

A single-point positioning system is the lowest-cost modulating control. Generally the accuracy is the same as for good parallel positioning control and lower than that of a typical metering system. Parallel positioning costs more than the single-point method, but there is a dramatic increase in cost with the metering system.

A metering system that uses only a pressure differential does not offer any major increase in accuracy over the single-point control. Without correction for changes in air density and fuel density or makeup water, the same variable will limit the accuracy of all systems.

Oxygen trim systems have generally been recognized as the best method of increasing the accuracy of all systems, and they are used with all the different control systems. With oxygen trim, all variables that could change air or fuel flow are taken into account, and so the accuracy is limited only by the accuracy of the oxygen reading.

There has been a general shift away from metering systems to single-point positioning with oxygen trim. The cost of this package is well below that of a metering system, and the accuracy is the same. One advantage offered by some metering systems is the ability to correct for lead-lag that occurs when the boiler modulates. During modulation, the dynamics of the rate of change in air and fuel flow plus some mechanical factors result in a slightly different excess-air level. By having the air change lag the fuel change, the potential of going too low in excess air is eliminated.

In the final analysis, a number of factors determine performance, cost, and efficiency, so that this chapter is only a guide in understanding the various controls and systems. Often the designer of the process of heating system can overcome shortfalls, or they place tighter restrictions on the boiler. As with each component, it must be viewed with the total system and not by itself.

CHAPTER 17
HEAT DISTRIBUTION EQUIPMENT

17.1   STEAM

Lehr Associates
New York, New York

17.1.1   INTRODUCTION TO STEAM

Nearly any material, at a given temperature and pressure, has a set amount of energy within it. When materials change their physical state, i.e., go from a liquid to a gas, that energy content changes. Such a change occurs when water is heated to a gaseous state—steam. When steam is used for heating, a cycle of different energy states occurs. First, water is heated in a boiler to its vaporization point, when it boils off as steam. The vapor is carried to the desired estimation where it is allowed to cool, giving off heat. Usually, the water, now cooled back to a liquid, is returned to the boiler to be revaporized.

The heat content of water is usually measured in British thermal units (Btu) or calories. Knowing the temperature is not sufficient to determine the energy content of steam—the pressure must also be known, as well as the amount of actual vapor or condensate (moisture). "Steam" can exist as saturated (containing all the vapor it can), dry (at the saturation point or above), wet (below the saturation point), and superheated (capable of holding even more vapor). Wet steam—containing condensate—has less energy than dry steam.

These conditions are specified for water in a chart called the Mollier diagram (see Fig. 17.1.1). The Mollier diagram specifies the energy content for steam at various vaporization levels. On the two axes of the diagram are enthalpy (a measure of the heat content of a volume of steam) and entropy (a measure of the energy available for work). Rigorous analysis of the thermodynamics of a heating system involves measurements of the specific volume of steam available; its pressure, temperature, and moisture values; and the efficiencies of heat transfer of the elements of the heating system. Usually vendors of steam equipment provide details of their systems based on saturated-steam conditions, which simplifies their sizing and use. Saturated-steam tables (see Table 17.1.1) give the values that are necessary to determine the amount of energy the steam has available for heating.

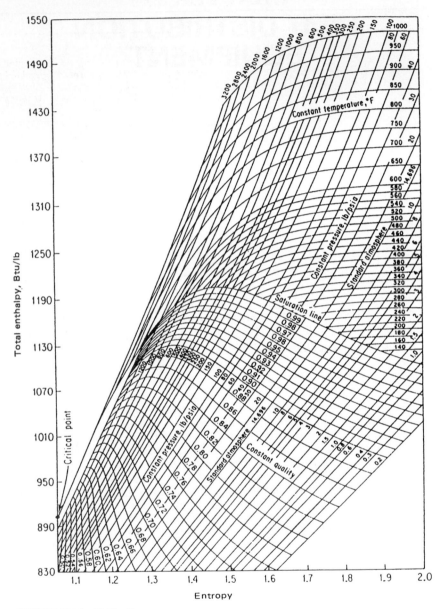

FIGURE 17.1.1 Mollier diagram.

TABLE 17.1.1   Saturated-Steam Tables

Gauge pressure		Absolute pressure, psig	Temperature, °F	Heat content			Specific volume of steam V_g, ft³/lb
in Hg vacuum	psig			Sensible (h_f), Btu/lb	Latent (h_{fg}), Btu/lb	Total (h_g), Btu/lb	
27.9		1	101.7	69.5	1032.9	1102.4	330.0
25.9		2	126.1	93.9	1019.7	1113.6	173.5
23.9		3	141.5	109.3	1011.3	1120.6	118.6
21.8		4	153.0	120.8	1004.9	1125.7	90.5
19.8		5	162.3	130.1	999.7	1129.8	73.4
17.8		6	170.1	137.8	995.4	1133.2	61.9
15.7		7	176.9	144.6	991.5	1136.1	53.6
13.7		8	182.9	150.7	987.9	1138.6	47.3
11.6		9	188.3	156.2	984.7	1140.9	42.3
9.6		10	193.2	161.1	981.9	1143.0	38.4
7.5		11	197.8	165.7	979.2	1144.9	35.1
5.5		12	202.0	169.9	976.7	1146.6	32.4
3.5		13	205.9	173.9	974.3	1148.2	30.0
1.4		14	209.6	177.6	972.2	1149.8	28.0
	0	14.7	212.0	180.2	970.6	1150.8	26.8
	1	15.7	215.4	183.6	968.4	1152.0	25.2
	2	16.7	218.5	186.8	966.4	1153.2	23.8
	5	19.7	227.4	195.5	960.8	1156.3	20.4
	10	24.7	239.4	207.9	952.9	1160.8	16.5
	15	29.7	249.8	218.4	946.0	1164.4	13.9
	20	34.7	258.8	227.5	940.1	1167.6	12.0
	25	39.7	266.8	235.8	934.6	1170.4	10.6
	30	44.7	274.0	243.0	929.7	1172.7	9.5
	40	54.7	286.7	256.1	920.4	1176.5	7.8
	50	64.7	297.7	267.4	912.2	1179.6	6.7
	60	74.7	307.4	277.1	905.3	1182.4	5.8
	70	84.7	316.0	286.2	898.8	1185.0	5.2
	80	94.7	323.9	294.5	892.7	1187.2	4.7
	90	104.7	331.2	302.1	887.0	1189.1	4.3
	100	114.7	337.9	309.0	881.6	1190.6	3.9
	125	139.7	352.8	324.7	869.3	1194.0	3.2
	150	164.7	365.9	338.6	858.0	1196.6	2.8
	175	189.7	377.5	350.9	847.9	1198.8	2.4
	200	214.7	387.7	362.0	838.4	1200.4	2.1

Note: Metric conversion factors are: 1 in Hg = 25.4 mm Hg; 1 lb/in² = 0.07 bar; °F = 1.8 × °C + 32; 1 Btu/lb = 554 cal/kg; 1 ft³/lb = 0.06 m³/kg.

To calculate the steam consumption of a heating device, the following equation should be employed:

$$Q = \frac{H}{SP_{wv}(T_e - T_v) + h_{fg} + SP_w(T_v - T_c)} \qquad (17.1.1)$$

where H = heating load, Btu/h (W)
h_{fg} = latent heat of vaporization, Btu/lb (kJ/kg)
T_e = entering steam temperature, °F (°C)
T_v = steam temperature at vaporization, °F (°C)
SP_{wv} = specific heat of water vapor, Btu/(lb·°F) [cal/(g·°C)]
SP_w = specific heat of water, Btu/(lb·°F) [cal/(g·°C)]
T_c = leaving temperature of condensate, °F (°C)
Q = steam rate, lb/h (kg/h)

$$Q = \frac{H}{0.45(T_e - T_v) + h_{fg} + T_v - T_c} \qquad (17.1.2)$$

or in International System (SI) units,

$$Q = \frac{H}{0.52(T_e - T_v) + 0.28h_{fg} + 1.16(T_v - T_c)} \qquad (17.1.3)$$

When saturated steam is supplied to the heating unit, $T_e = T_v$, so $T_e - T_v = 0$. Normally T_c is maintained at or near T_v so that the factor $T_v - T_c$ can be omitted from the calculation without significantly affecting the outcome.

For a system supplying saturated steam we can simplify the calculation to

$$Q = \frac{H}{h_{fg}} \quad \text{or} \quad Q = \frac{H}{0.28h_{fg}} \text{ (SI units)} \qquad (17.1.4)$$

The following formula converts the steam rate Q into gallons per minute (liters per second) so that the condensate will be in units normally associated with the flow of liquids:

$$\frac{Q}{500} = \text{gal/min} \quad \text{or} \quad \frac{Q}{3600} = \text{L/s (SI units)} \qquad (17.1.5)$$

17.1.2 INTRODUCTION TO STEAM HEATING SYSTEMS

Steam systems are used to heat industrial, commercial, and residential buildings. These systems are categorized according to the piping layout and the operating steam pressure. This section discusses steam systems which operate at or below 200 psig (14 bar).

17.1.3 GENERAL SYSTEM DESIGN

The mass flow rate of steam through the piping system is a function of the initial steam pressure, pressure drop through the pipe, equivalent length of piping, and size of piping. The roughness of the inner pipe wall is a variable in determining the steam's pressure drop. All the charts and tables in this section that outline the performance of the steam transmitted through the piping assume that the roughness of the piping is equal to that of new, commercial-grade steel pipe.

17.1.4 PRESSURE CONDITIONS

Steam piping systems are usually categorized by the working pressure of the steam they supply. The five classes of steam systems are high-pressure, medium-pressure, low-pressure, vapor, and vacuum systems. A high-pressure system has an initial pressure in excess of 100 psig (6.9 bar). The medium-pressure system operates with pressures between 100 psig (6.9 bar) and 15 psig (1 bar). Systems that operate from 15 psig (1 bar) to 0 psig (0 bar) are classified as low-pressure. Vapor and vacuum systems operate from 15 psig (1 bar) to vacuum. Vapor systems attain subatmospheric pressures through the condensing process, while vacuum systems require a mechanically operated vacuum pump to attain subatmospheric pressures.

17.1.5 PIPING ARRANGEMENTS

The general piping scheme of a steam system can be distinguished by three different characteristics. First, the number of connections required at the heating device describes the system. A one-pipe system has only one piping connection which supplies steam and allows condensate to return to the boiler by flowing counter to the steam in the same pipe. The more common design is to have two piping connections, one for the supply steam and one for the condensate. This arrangement is known as a two-pipe system.

Second, the direction of the supply steam in the risers characterizes the piping design. An up-feed system has the steam flowing up the riser; conversely, a down-feed system supplies steam down the riser.

Third, the final characteristic of the piping design is the location of the condensate return to the boiler. A dry return has its condensate connection above the boiler's waterline, while a wet-return connection is below the waterline.

17.1.6 CONDENSATE RETURN

By analyzing how the condensate formed in the heating system is returned to the boiler, an understanding of how the system should operate is achieved. There are two commonly used return categories: mechanical and gravity.

If devices such as pumps are used to aid in the return of condensate, the system is known as a *mechanical return*. When no mechanical device is used to return the condensate, the system is classified as a *gravity return*. The only forces pushing the condensate back to the boiler or condensate receiver are gravity and the pressure of the steam itself. This type of system usually requires that all steam-consuming components be located at a higher elevation than the boiler or the condensate receiver.

With either mechanical or gravity return systems, the mains are normally pitched $\frac{1}{4}$ in (6.3 mm) for every 10 ft (3 m) of length, to ensure the proper flow of condensate. The supply mains are sloped up away from the boiler, and the return mains are pitched down toward the boiler. This allows condensate to flow back to the boiler. See Sec. 18.8 for more on condensate control.

17.1.7 PIPE-SIZING CRITERIA

Once the heating loads are known, the steam flow rates can be determined; then the required size of the steam piping can be specified for proper operation. The following factors must be analyzed in sizing the steam piping:

TABLE 17.1.2   Pressure Drops for Steam Pipe Sizing

Initial steam pressure		Total pressure drop in supply piping		Pressure drop for mains and risers		Total pressure drop in return piping	
psig	bar	lb/in²	bar	(lb/in²)/100 ft	bar/100 m	lb/in²	bar
Vacuum		1–2	0.069–0.138	$^1/_8$–$^1/_4$	0.028–0.057	1	0.069
0	0	$^1/_6$–$^1/_4$	0.004–0.017	$^1/_{32}$	0.007	$^1/_{16}$	0.004
2	0.138	$^1/_4$–$^3/_4$	0.017–0.052	$^1/_8$	0.028	$^1/_4$	0.017
5	0.345	1–2	0.069–1.38	$^1/_4$	0.057	1	0.069
15	1.03	4–6	0.276–0.414	1	0.228	4	0.276
30	2.07	5–10	0.345–0.069	2	0.455	5	0.345
50	3.45	10–15	0.069–1.03	2–5	0.455–1.14	10	0.69
100	6.90	15–25	1.03–1.72	2–5	0.455–1.14	15	1.03
150	10.3	25–30	1.72–2.07	2–10	0.455–2.28	20	1.37

- Initial steam pressure
- Total allowable pressure drop
- Maximum steam velocity
- Direction of condensate flow
- Equivalent length of system

For different initial pressures, the allowable pressure drop in the piping varies. Table 17.1.2 gives typical values in selecting pressure-drop limits. To ensure that the parameters from the table are suitable for an application, check that the total system pressure drop does not exceed 50 percent of the initial pressure, that the condensate has enough steam pressure to return to the boiler, and that the steam velocity is within specified limits to ensure quiet and long-lasting operation.

When steam piping is sized, there is a trade-off between quiet, efficient operation and first-cost considerations. A good compromise point exists when the steam supply pipe is sized for velocities between 6000 and 12,000 ft/min (30.5 and 61 m/s). This allows quiet operation while offering a reasonable installed cost. If the piping is downsized so that the velocity exceeds 20,000 ft/min (101 m/s), the system may produce objectional hammering noise or restrict the flow of condensate when it is counter to the steam's direction. It is recommended that the piping be sized so that the velocity will never approach 20,000 ft/min (101 m/s) in any leg.

As condensate flows into the return line, a portion of it will flash into steam. The volume of the steam-condensate mixture is much greater than the volume of pure condensate. To avoid undersizing the return lines, the return piping should be sized at some reasonable proportion of dry steam. A maximum size would be to assume that the return is 100 percent saturated steam. An acceptable velocity for the design of the return lines is 5000 ft/min (25.4 m/s).

17.1.8   DETERMINING EQUIVALENT LENGTH

The "equivalent length" of pipe is equal to the actual length of pipe plus the friction losses associated with fittings and values. For simplicity's sake, the fitting and valve losses are stated as the equivalent length of straight pipe needed to produce the same friction loss. Values for common fittings and valves are stated in Table 17.1.3.

The equivalent length—*not the actual length*—is the value used in all the figures and charts for pipe sizing. Common practice is to assume that the equivalent length is 1.5 times

TABLE 17.1.3 Length of Pipe to Be Added to Actual Length of Run—Owing to Fittings—to Obtain Equivalent Length

Size of pipe, in	Length to be added to run, ft*				
	Standard elbow	Side outlet tee†	Gate valve‡	Globe valve‡	Angle valve‡
$1/2$	1.3	3	0.3	14	7
$3/4$	1.8	4	0.4	18	10
1	2.2	5	0.5	23	12
$1 1/4$	3.0	6	0.6	29	15
$1 1/2$	3.5	7	0.8	34	18
2	4.3	8	1.0	46	22
$2 1/2$	5.0	11	1.1	54	27
3	6.5	13	1.4	66	34
$3 1/2$	8	15	1.6	80	40
4	9	18	1.9	92	45
5	11	22	2.2	112	56
6	13	27	2.8	136	67
8	17	35	3.7	180	92
10	21	45	4.6	230	112
12	27	53	5.5	270	132
14	30	63	6.4	310	152

*Metric conversion: 1 in = 2.54 cm and 1 ft = 0.31 m.
†Values given apply only to a tee used to divert the flow in the main to the last riser.
‡Valve in full-open position.
Example: Determine the length in feet of pipe to be added to actual length of run illustrated.

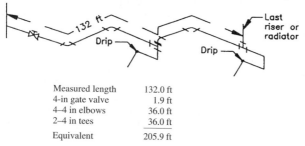

Measured length	132.0 ft
4-in gate valve	1.9 ft
4–4 in elbows	36.0 ft
2–4 in tees	36.0 ft
Equivalent	205.9 ft

Source: *Reprinted by permission from ASHRAE Handbook—1989 Fundamentals.*

the actual length when a design is first being sized. After the initial sizing and layout are completed, the exact equivalent length should be calculated and all the pipe sizes checked.

17.1.9 BASIC TABLES FOR STEAM PIPE SIZING

Figure 17.1.2 is used to determine the flow and velocity of steam in Schedule 40 pipe at various values of pressure drop per 100 ft (30.5 m), based on 0 psig (1-bar) saturated steam. By using the multiplier curves at the bottom of the figure it may also be used at all saturated pressures between 0 and 200 psig (1 and 14 bar). Figure 17.1.2 is valid only when steam and condensate flow in the same direction.

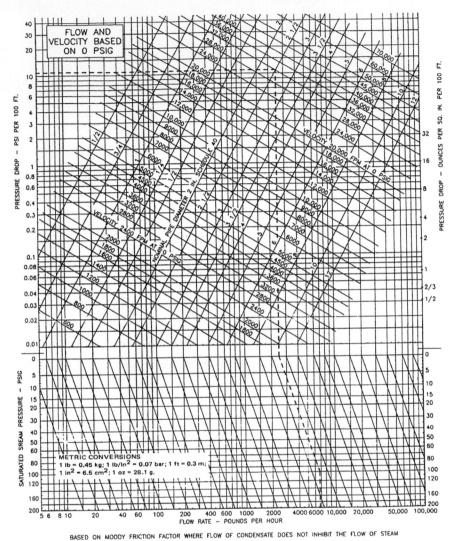

FIGURE 17.1.2   Basic chart for flow rate and velocity of steam in Schedule 40 pipe, based on saturation pressure of 0 psig (0 bar). (*Reprinted by permission from ASHRAE Handbook—1989 Fundamentals.*)

17.1.10   TABLES FOR LOW-PRESSURE STEAM PIPE SIZING

Table 17.1.4, derived from Fig. 17.1.2, gives the values needed to select pipe sizes at various pressure drops for systems operating at 3.5 and 12 psig (0.24 and 0.84 bar). The flow rates given for 3.5 psig (0.24 bar) can be used for saturated-steam pressures from 1 to 5 psig (0.07 to 0.34 bar), and those for 12 psig (0.84 bar) can be used for saturated pressures from 8 to 16 psig (0.55 to 1.1 bar) with an error not to exceed 8 percent.

TABLE 17.1.4 Flow Rate of Steam, lb/h, in Schedule 40 Pipe* at Initial Saturation Pressure of 3.5 and 12 psig[†]

Nom. pipe size, in	$1/16$ lb/in² (1 oz) Sat. press., psig		$1/8$ lb/in² (2 oz) Sat. press., psig		$1/4$ lb/in² (4 oz) Sat. press., psig		$1/2$ lb/in² (8 oz) Sat. press., psig		$3/4$ lb/in² (12 oz) Sat. press., psig		1 lb/in² Sat. press., psig		2 lb/in² Sat. press., psig	
	3.5	12	3.5	12	3.5	12	3.5	12	3.5	12	3.5	12	3.5	12
$1/4$	9	11	14	16	20	24	29	35	36	43	42	50	60	73
1	17	21	26	31	37	46	54	66	68	82	81	95	114	137
$1\,1/4$	36	45	53	66	78	96	111	138	140	170	162	200	232	280
$1\,1/2$	56	70	84	100	120	147	174	210	218	260	246	304	360	430
2	108	134	162	194	234	285	336	410	420	510	480	590	710	850
$2\,1/2$	174	215	258	310	378	460	540	660	680	820	780	950	1,150	1,370
3	318	380	465	550	660	810	960	1,160	1,190	1,430	1,380	1,670	1,950	2,400
$3\,1/2$	462	550	670	800	990	1,218	1,410	1,700	1,740	2,100	2,000	2,420	2,950	3,450
4	640	800	950	1,160	1,410	1,690	1,980	2,400	2,450	3,000	2,880	3,460	4,200	4,900
5	1,200	1,430	1,680	2,100	2,440	3,000	3,570	4,250	4,380	5,250	5,100	6,100	7,500	8,600
6	1,920	2,300	2,820	3,350	3,960	4,850	5,700	7,000	7,200	8,600	8,400	10,000	11,900	14,200
8	3,900	4,800	5,570	7,000	8,100	10,000	11,400	14,300	14,500	17,700	16,500	20,500	24,000	29,500
10	7,200	8,800	10,200	12,600	15,000	18,200	21,000	26,000	26,200	32,000	30,000	37,000	42,700	52,000
12	11,400	13,700	16,500	19,500	23,400	28,400	33,000	40,000	41,000	49,500	48,000	57,500	67,800	81,000

*Based on Moody friction factor, where flow of condensate does not inhibit the flow of steam.

[†]The flow rates of 3.5 psig can be used to cover saturated pressure from 1 to 6 psig, and the rates at 12 psig can be used to cover saturated pressure from 8 to 16 psig with an error not exceeding 8 percent. The steam velocities corresponding to the flow rates given in this table can be found from the basic chart and velocity multiplier chart, Fig. 17.1.2.

[‡]Metric conversions: 1 in = 2.54 cm, 1 lb/in² = 0.07 bar, and 1 lb = 0.46 kg.

Source: *Reprinted by permission from ASHRAE Handbook—1989 Fundamentals.*

TABLE 17.1.5　Steam Pipe Capacities for Low-Pressure Systems, lb/h
For use on one-pipe systems or two-pipe systems in which condensate flows against the steam flow

Nominal pipe size, in	Two-pipe systems		One-pipe systems		
	Condensate flowing against steam		Supply risers up-feed	Radiator valves and vertical connections	Radiator and riser runouts
	Vertical	Horizontal			
A	B*	C†	D‡	E	F†
$3/4$	8	7	6	—	7
1	14	14	11	7	7
$1^1/_4$	31	27	20	16	16
$1^1/_2$	48	42	38	23	16
2	97	93	72	42	23
$2^1/_2$	159	132	116	—	42
3	282	200	200	—	65
$3^1/_2$	387	288	286	—	119
4	511	425	380	—	186
5	1,050	788	—	—	278
6	1,800	1,400	—	—	545
8	3,750	3,000			
10	7,000	5,700			
12	11,500	9,500			
16	22,000	19,000			

*Do not use column B for pressure drops of less than $1/_{16}$ lb/in^2 per 100 ft of equivalent run. Use Fig. 17.1.2 or Table 17.1.4 instead.

†Pitch of horizontal runouts to risers and radiators should be not less than $1/_2$ in/ft. Where this pitch cannot be obtained, runouts over 8 ft in length should be one pipe size larger than called for in this table.

‡Do not use column D for pressure drops of less than $1/_{24}$ lb/in^2 per 100 ft of equivalent run except on sizes 3 in and over. Use Fig. 17.1.2 or Table 17.1.4 instead.

Note: Steam at an average pressure of 1 psig is used as a basis of calculating capacities. Metric conversion factors of 1 in = 2.54 cm and 1 lb = 0.46 kg can be used.

Source:　*Reprinted from ASHRAE Handbook—1989 Fundamentals.*

Table 17.1.5 is used for systems where the condensate flows counter to the supply steam.

To size return piping, Table 17.1.6 is used. This table gives guidelines for return piping for wet, dry, and vacuum return systems.

17.1.11　TABLES FOR SIZING MEDIUM- AND HIGH-PRESSURE PIPE SYSTEMS

Larger, industrial-type space-heating systems are designed to use either medium- or high-pressure steam at 15 to 200 psig (1.03 to 14 bar). These systems often involve unit heaters and/or air-handling units. Figures 17.1.3 to 17.1.6 provide tables for sizing steam piping for systems of 30, 50, 100, and 150 psig (2, 3.5, 6.9, and 10.5 bar).

TABLE 17.1.6 Return Main and Riser Capacities for Low-Pressure Systems, lb/h

Pipe size, in	$\frac{1}{32}$ lb/in² or $\frac{1}{3}$-oz drop per 100 ft			$\frac{1}{24}$ lb/in² or $\frac{2}{3}$-oz drop per 100 ft			$\frac{1}{16}$ lb/in² or 1-oz drop per 100 ft			$\frac{1}{8}$ lb/in² or 2-oz drop per 100 ft			$\frac{1}{4}$ lb/in² or 4-oz drop per 100 ft			$\frac{1}{2}$ lb/in² or 8-oz drop per 100 ft		
	Wet	Dry	Vac.	Wet	Dry	Vac.	Wet	Dry	Vac.	Wet	Dry	Vac.	Wet	Dry	Vac.	Wet	Dry	Vac.
G	H	I	J	K	L	M	N	O	P	Q	R	S	T	U	V	W	X	Y
Return main																		
¾	—	—	—	—	—	42	—	—	100	—	—	142	—	—	200	—	—	283
1	125	62	—	145	71	143	175	80	175	250	103	249	350	115	350	—	—	494
1¼	213	130	—	248	149	244	300	168	300	425	217	426	600	241	600	—	—	848
1½	338	206	—	393	236	388	475	265	475	675	340	674	950	378	950	—	—	1,340
2	700	470	—	810	535	815	1000	575	1,000	1400	740	1,420	2,000	825	2,000	—	—	2,830
2½	1180	760	—	1580	868	1,360	1680	950	1,680	2350	1230	2,380	3,350	1360	3,350	—	—	4,730
3	1880	1460	—	2130	1560	2,180	2680	1750	2,680	3750	2250	3,800	5,350	2500	5,350	—	—	7,560
3½	2750	1970	—	3300	2200	3,250	4000	2500	4,000	5500	3230	5,680	8,000	3580	8,000	—	—	11,300
4	3880	2930	—	4580	3350	4,500	5500	3750	5,500	7750	4830	7,810	11,000	5380	11,000	—	—	15,500
5	—	—	—	—	—	7,880	—	—	9,680	—	—	13,700	—	—	19,400	—	—	27,300
6	—	—	—	—	—	12,600	—	—	15,500	—	—	22,000	—	—	31,000	—	—	43,800
Riser																		
¾	—	—	—	—	—	—	—	—	—	—	—	—	—	—	—	—	—	—
1	—	48	—	—	48	143	—	48	175	—	48	249	—	48	350	—	—	494
1¼	—	113	—	—	113	244	—	113	300	—	113	426	—	113	600	—	—	848
1½	—	248	—	—	248	388	—	248	475	—	248	674	—	248	950	—	—	1,340
2	—	375	—	—	375	815	—	375	1,000	—	375	1,420	—	375	2,000	—	—	2,830
2½	—	750	—	—	750	1,360	—	750	1,680	—	750	2,380	—	750	3,350	—	—	4,730
3	—	—	—	—	—	2,180	—	—	2,680	—	—	3,800	—	—	5,350	—	—	7,560
3½	—	—	—	—	—	3,250	—	—	4,000	—	—	5,680	—	—	8,000	—	—	11,300
4	—	—	—	—	—	4,480	—	—	5,500	—	—	7,810	—	—	11,000	—	—	15,500
5	—	—	—	—	—	7,880	—	—	9,680	—	—	13,700	—	—	19,400	—	—	27,300

Note: This table is based on pipe size data developed through the research investigations of The American Society of Heating, Refrigerating and Air-Conditioning Engineers. Metric conversion factors of 1 in = 2.54 cm, 1 lb in² = 0.07 bar, and 1 ft = 0.31 m can be used.

Source: Reprinted by permission from ASHRAE Handbook—1989 Fundamentals.

17.11

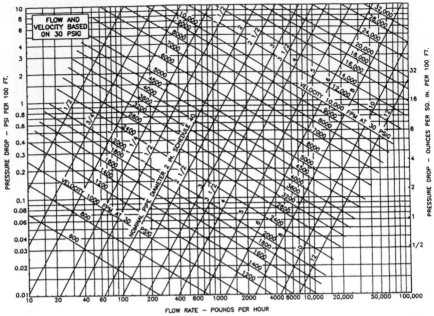

BASED ON MOODY FRICTION FACTOR WHERE FLOW OF CONDENSATE DOES NOT INHIBIT THE FLOW OF STEAM
(MAY BE USED FOR STEAM PRESSURE FROM 23 TO 37 PSIG WITH AN ERROR NOT EXCEEDING 9%)

METRIC CONVERSIONS:
1 lb = 0.45 kg; 1 lb/in^2 = 0.07 bar; 1 ft = 0.3 m;
1 in^2 = 6.5 cm^2; 1 oz = 28.1 g.

FIGURE 17.1.3   Chart for flow rate and velocity of steam in Schedule 40 pipe, based on saturation pressure of 30 psig (2.1 bar). (*Reprinted by permission from ASHRAE Handbook—1989 Fundamentals.*)

17.1.12   AIR VENTS

The presence of air in the steam supply line impedes the heat-transfer ability of the system due to the high insulating value of air. Air also interferes with the flow of steam by forming pockets at the ends of runs that prevent the steam from reaching the system's extremities.

A valve that releases air from the system while restricting the flow of all other fluids is known as an "air vent." Air vents should be located at all system high points and where air pockets are likely to form. Venting should be done continually to prevent the buildup of air in the system.

Air enters the system by two means. First, when cold makeup feed water is supplied to the boiler, air is present in the water. As the water is heated, the air tends to separate from the water. Second, when the system is turned off, steam is trapped in the pipes. Eventually the steam cools and condenses. Since the volume of the condensate is negligible compared to the initial volume of the steam, a vacuum is formed in the piping. Air leaks into the system through openings in the joints until the internal pressure equalizes. Upon restarting the system, the air is swept along with the steam and becomes entrained in the system.

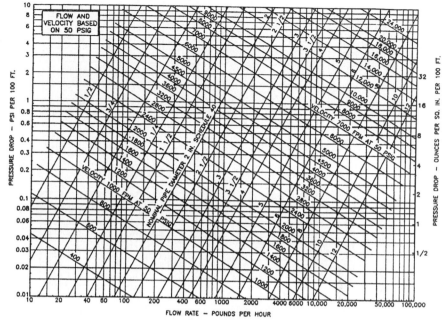

BASED ON MOODY FRICTION FACTOR WHERE FLOW OF CONDENSATE DOES NOT INHIBIT THE FLOW OF STEAM
(MAY BE USED FOR STEAM PRESSURE FROM 40 TO 60 PSIG WITH AN ERROR NOT EXCEEDING 8%)

METRIC CONVERSIONS:
1 lb ≈ 0.45 kg; 1 lb/in² ≈ 0.07 bar; 1 ft ≈ 0.3 m;
1 in² = 6.5 cm²; 1 oz ≈ 28.1 g.

FIGURE 17.1.4 Chart for flow rate and velocity of steam in Schedule 40 pipe, based on saturation pressure of 50 psig (3.5 bar). (*Reprinted by permission from ASHRAE Handbook—1989 Fundamentals.*)

17.1.13 STEAM TRAPS

When steam is transmitted through the piping or the end-user equipment, it loses part of its heat energy. As heat is removed from saturated steam, a vapor-liquid mixture forms in the pipe. The presence of liquid condensate in the steam lines interferes with the proper operation of the system. See Sec. 18.8 "Condensate Control." Liquid condensate derates the system's heating capacity because water has a much smaller amount of available energy than steam does. Furthermore, the accumulation of water in the supply steam piping can obstruct the flow of the steam through the system.

A valve that permits condensate to flow from the supply line without allowing steam to escape is known as a "steam trap." All steam traps should be located such that condensate can flow via gravity through them. *Through mechanical means*, the steam trap recognizes when steam is present by sensing the density, kinetic energy, or temperature of the fluid at the trap. When conditions indicate that steam is absent, the trap opens and allows the condensate to drop to the return line. As soon as the trap senses the presence of steam, it slams shut.

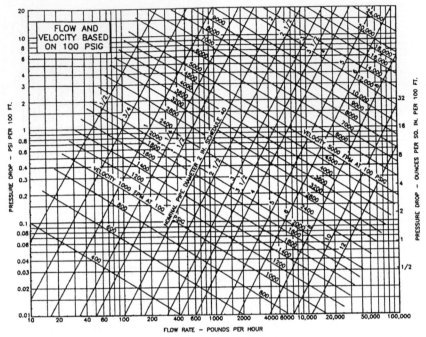

BASED ON MOODY FRICTION FACTOR WHERE FLOW OF CONDENSATE DOES NOT INHIBIT THE FLOW OF STEAM
(MAY BE USED FOR STEAM PRESSURE FROM 85 TO 120 PSIG WITH AN ERROR NOT EXCEEDING 8%)

METRIC CONVERSIONS:
1 lb = 0.45 kg; 1 lb/In^2 = 0.07 bar; 1 ft = 0.3 m;
1 in^2 = 6.5 cm^2; 1 oz = 28.1 g.

FIGURE 17.1.5   Chart for flow rate and velocity of steam in Schedule 40 pipe, based on saturation pressure of 100 psig (7 bar). (*Reprinted by permission from ASHRAE Handbook—1989 Fundamentals.*)

17.1.14   STEAM TRAP TYPES

There are six types of steam traps normally employed in the heating, ventilating, and air-conditioning (HVAC) industry. Since traps differ in their operational characteristics, selection of the proper trap is critical to efficient operation of the system. Different applications require specific types of traps, and no one type of trap will perform satisfactorily in all situations.

Three of the six basic types of traps operate thermostatically by sensing a temperature difference between subcooled condensate and steam: liquid-expansion, balanced-pressure thermostatic, and bimetallic thermostatic traps. Two other types—the bucket trap and the float-and-thermostatic trap—are activated by differences in density between steam and condensate. These are also known as blast type traps. Finally, the thermodynamic steam trap operates on the differences in the velocity at which steam passes through the trap. This velocity difference can also be considered as a change in kinetic energy.

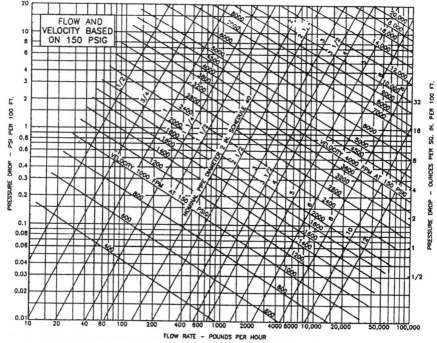

BASED ON MOODY FRICTION FACTOR WHERE FLOW OF CONDENSATE DOES NOT INHIBIT THE FLOW OF STEAM
(MAY BE USED FOR STEAM PRESSURE FROM 127 TO 180 PSIG WITH AN ERROR NOT EXCEEDING 8%)

METRIC CONVERSIONS:
1 lb = 0.45 kg; 1 lb/in² = 0.07 bar; 1 ft = 0.3 m;
1 in² = 6.5 cm²; 1 oz = 28.1 g.

FIGURE 17.1.6   Chart for flow rate and velocity of steam in Schedule 40 pipe, based on saturation pressure of 150 psig (10.5 bar). (*Reprinted by permission from ASHRAE Handbook—1989 Fundamentals.*)

17.1.15 BALANCED-PRESSURE STEAM TRAPS

The balanced-pressure steam trap (Fig. 17.1.7) employs a bellows filled with a fluid mixture that boils below the steam temperature. When steam is present at the trap inlet, the liquid in the bellows is vaporized and expands to seal the trap. Condensate accumulates at the trap and starts to subcool. When the condensate cools enough to condense the fluid in the bellows, the trap opens and the condensate flows through the trap.

This type of trap has two possible drawbacks. First, it must allow condensate to subcool 5 to 30°F (2.8 to 16.7°C) below the steam temperature to operate. Second, it discharges condensate intermittently.

Advantages of the balanced-pressure trap are that it is freeze-proof, can handle a large condensate load, does a good job of air venting, and is self-adjusting throughout its operating range. These traps are typically used in conjunction with steam radiators and sterilizers.

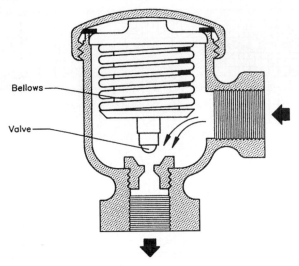

FIGURE 17.1.7   Balanced-pressure steam trap.

17.1.16 BIMETALLIC THERMOSTATIC STEAM TRAPS

These traps operate on the same principle as the balanced-pressure steam trap. The bellows mechanism is replaced by a bimetallic strip formed from two dissimilar metals that have very different coefficients of expansion. As the bimetallic strip is heated, the difference in the expansion rate of the metals causes the strip to bend. The trap is fabricated so that when the strip is heated to the steam's temperature, there is enough movement to close off the valve. The bimetallic thermostatic trap (Fig. 17.1.8) has a slow response to load conditions, requiring as much as 100°F (55.5°C) of subcooling, and is not self-adjusting to changes in inlet pressure.

These traps are suited for superheated steam applications and situations where a great deal of condensate subcooling is required to prevent flashing in the return line. Normally these traps are applied to steam-tracing lines that can tolerate partial flooding.

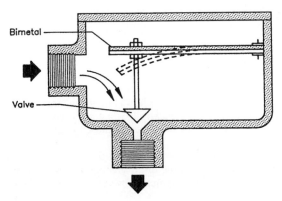

FIGURE 17.1.8   Bimetallic steam trap.

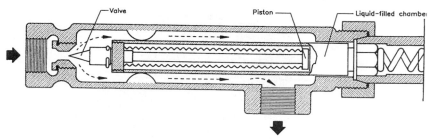

FIGURE 17.1.9   Liquid-expansion steam trap.

17.1.17   LIQUID-EXPANSION STEAM TRAPS

The liquid-expansion steam trap (Fig 17.1.9) is designed with an oil-filled cylinder which drives a piston. When steam is present, the oil expands, thrusting the piston out. The end of the piston acts as the valve and seals the port to the return line. As condensate collects in the trap and cools, the oil starts to contract. The contraction of the oil causes the piston to move away from the port and permits the flow of condensate from the trap.

These traps are freeze-proof and are used for freeze protection of system low points and heating coils. Their limitations are that they are not self-adjusting to changes of inlet pressure and that they require condensate subcooling by 2 to 30°F (1.1 to 16.7°C).

17.1.18   BUCKET STEAM TRAPS

Bucket traps operate by gravity, utilizing the density difference between liquid and vapor. When the body of the trap is filled with liquid and a vapor enters the bucket, the bucket will float. As the bucket fills with liquid, the bucket sinks. The bucket's movement activates a value. If the bucket rises due to the vapor pressure, the valve closes; and when the bucket sinks, the valve opens, permitting condensate to flow from the trap. The most common type of bucket trap is the inverted bucket (Fig. 17.1.10), so named because the bucket has its open side facing down.

Bucket traps are capable of working at very high pressures, can discharge condensate at the saturated-steam temperature, and are resistant to water hammer. Unfortunately, if the water seal is lost, the bucket trap will continuously allow steam to pass through. Other disadvantages of these traps are their susceptibility to freeze-up, their lack of good air-venting capability, and their intermittent discharge.

Inverted-bucket traps are usually installed on high-pressure indoor steam main drips.

17.1.19   FLOAT-AND-THERMOSTATIC STEAM TRAPS

A float-and-thermostatic steam trap (Fig. 17.1.11) is actually two distinct traps in one unit. The balanced-pressure steam trap, outlined previously, is located at the top of the trap body and acts as an air vent. The rest of the unit consists of a float that rises and falls based on the level of condensate in the trap. The trap inlet is located about the outlet. The float position

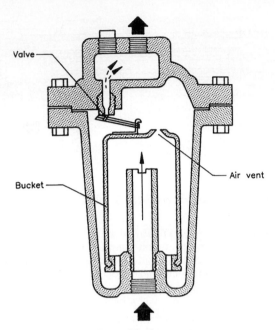

FIGURE 17.1.10   Inverted-bucket steam trap.

operates a valve that controls flow to the return line. As the condensate level rises above the outlet, the float causes the valve to open. If the condensate level drops enough, the float causes the valve to close. Since the float allows the valve to open only when the condensate level is above the outlet, a water seal is maintained to prevent steam from passing through the outlet when the valve is open.

The float-and-thermostatic steam traps cannot be used on a superheated-steam system unless they are modified and are usually not installed outdoors because they are subject to

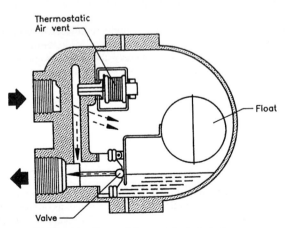

FIGURE 17.1.11   Float-and-thermostatic steam trap.

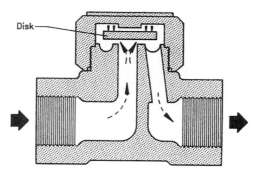

Disk

FIGURE 17.1.12 Thermodynamic steam trap.

freeze-up. These types of traps will continuously vent air. They do not require subcooling of condensate and are unaffected by changes in system pressure. Typically float-and-thermostatic traps are used in conjunction with heating devices, such as unit heaters, water heaters, and converters.

17.1.20 THERMODYNAMIC STEAM TRAPS

The design of the thermodynamic steam trap (Fig. 17.1.12) is based on the theory that the total pressure of fluid passing through the trap will remain constant. Since the total pressure equals the sum of the static and dynamic pressures, any increase in dynamic pressure will cause a decrease in the static pressure, and vice versa. These traps have only one moving part, a disk that can seal off both the inlet and the outlet of the trap. Steam entering the trap accelerates radially over the disk, causing a reduction in static pressure under the disk. As the steam dead-ends above the disk, the static pressure above the disk increases. This difference in pressure induces the disk to seal off the trap's openings. The trap will remain closed until the steam in the trap condenses sufficiently to reduce the pressure above the disk to an amount less than the inlet steam pressure. At that point, the disk moves away from the inlet port.

Thermodynamic steam traps should not be used on systems operating below 5 psig (0.34 bar) or on those that have back pressures equal to or greater than 80 percent of their supply pressure.

These traps are compact and have a long life due to the simplicity of their design. They can operate under high pressures, responding quickly to load and pressure variations while discharging condensate without requiring subcooling. Thermodynamic traps are usually installed in main drips and steam tracer lines.

17.1.21 STEAM TRAP LOCATION

Steam traps are located either in the return line or in drip legs. A "drip leg" (shown in Fig. 17.1.13) is a piping assembly that hangs below the supply main; its purpose is to remove condensate and sediment from the main. Gravity allows condensate and sediment to leave the main and accumulate in the drip leg. When the condensate in the leg

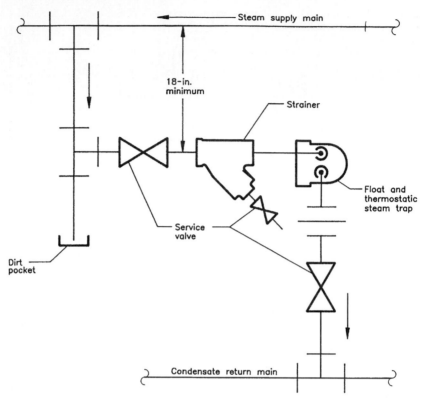

FIGURE 17.1.13   Typical drip-leg piping assembly.

rises to the level of the trap intake, the trap fills and then discharges the condensate to the return line. The drip leg pipe should be of sufficient size to permit condensate to drain freely from the main. For mains of 4 in (102 mm) or less in diameter, the drip leg should be the same size as the main pipe. For mains larger than 4 in (102 mm), the pipe diameter of the drip leg should be half of the main's size, but not less than 4 in (102 mm). Where possible, all drip legs should be at least 18 in (45.7 cm) long. A trap should be installed in the return line after every steam-consuming device. Each device should have its own trap to prevent possible "short-circuiting" that could occur if multiple devices share a common trap. A drip leg should be located before risers, expansion joints, bends, valves, and regulators. System low points, end of mains, and untrapped supply runs of over 300 ft (100 m) are additional locations where drip legs should be installed.

17.1.22   STEAM TRAP SIZING

A steam trap must be properly sized to handle the full load of condensate. (See "determining condensate load for a system," Section 17.1.24) Mains have their largest condensate loads during startup. Table 17.1.7 gives values for the condensate load of mains at startup.

TABLE 17.1.7 Startup Condensate Loads in Steam Mains, lb/h per 100-ft Length

Pipe size, in	Steam pressure, psig*†							
	0	5	15	30	50	100	150	200
2	6	7	8	9	10	13	15	16
$2^1/_2$	10	11	12	14	16	20	23	25
3	13	14	17	19	22	27	30	33
4	18	20	23	27	31	38	43	47
5	25	28	32	37	42	51	58	64
6	32	36	41	48	55	67	75	83
8	48	54	62	72	82	100	113	125
10	68	77	88	102	116	142	160	177
12	90	101	116	134	153	188	212	234

*Based on 70°F (21°C) ambient air. Schedule 40 pipe uninsulated.
†For metric equivalents, use the following conversion factors: 1 in = 2.54 cm = 25.4 mm; 1 lb/in^2 = 0.07 bar.

TABLE 17.1.8 Steam Trap Selection: Safety Factor

Trap type	Safety factor multiplier
Balanced-pressure thermostatic trap	3
Bimetallic trap	2.5
Liquid-expansion trap	3
Inverted-bucket trap	2.5
Float-and-thermostatic trap	2
Thermodynamic trap	1.5

The performance of a steam trap is affected by the inlet pressure and back pressure of the system. Therefore, when a trap is chosen, it is prudent to oversize the trap by a reasonable amount. Table 17.1.8 gives a guideline on how large to size traps. Grossly oversizing a trap will cause the system to operate improperly.

17.1.23 STEAM TRAP SELECTION

Once the size of the steam trap is known, the type of trap which will provide the best performance must be selected. When a trap is chosen, care must be taken to select a type that will operate over the full range of pressures that the system will exert.

The best operating economy based on trap life and minimization of waste steam must be considered. If the trap will be subjected to low ambient temperatures, it should be of a freeze-proof design. For traps serving heating devices, continuous gas-venting capability is desirable. When the application is examined, the need for steam trap construction which is resistant to corrosion and water hammering should be considered.

17.1.24 DETERMINING CONDENSATE LOAD FOR A SYSTEM

The steam consumption of a system over time is equal to the amount of condensate formed during that period. Unfortunately, only when traps of the modulating type (such as float-and-thermostatic traps) are employed does the condensate return simultaneously equal the steam consumption.

If a blast type, say a bucket trap, is installed, the flow of condensate will be intermittent and equal to the trap's discharge rate, not the steam consumption rate. Since blast-type traps discharge intermittently, you can safely assume that not all the traps will discharge at once. For sizing purposes, the rule of thumb is that no more than two-thirds of the blast-type traps will discharge at any given time. This condensate load and the design steam consumption for the equipment utilizing modulating-type traps should be combined to determine the peak condensate load of the entire system. When the piping is sized, consider oversizing the condensate return main by one pipe size. This can be beneficial when future increases in the system's steam consumption are anticipated.

17.1.25 WATER HAMMER

Water hammering is a phenomenon that occurs when condensate remaining in a pipe flashes into steam. The sudden expansion of the condensate causes a vibration in the pipe which can lead to premature failure of joints and can cause an objectional noise throughout the structure the pipe is serving. A more dangerous situation can develop if enough condensate accumulates in the pipe to block the passage of steam. The steam pressure behind the blockage will build up. Eventually the blockage may be transmitted through the pipe at a speed approaching the design velocity of the steam. When water travels at such a high velocity, it can damage the first obstruction it comes to, such as a valve or elbow. Both water hammering and damage from blockages can be prevented by proper trapping and pitching of the steam lines.

When certain gases, such as carbon dioxide (CO_2), are trapped in steam lines, the gases tend to mix with the condensate and form unwanted by-products, such as mild acids. These by-products will accelerate the rate of erosion in the system and cause premature failure in the system's components. Proper air venting will reduce the amount of gas in the system and increases its operating life.

17.1.26 WATER CONDITIONING

The formation of scale and sludge deposits on boiler heating surfaces creates a problem in generating steam. Water conditioning in a steam generating system should be under the supervision of a specialist. Refer to Chap. 8 of this handbook for a discussion of water treatment.

17.1.27 FREEZE PROTECTION

Whenever a steam system is servicing an area whose outdoor temperature will drop below 35°F (1.7°C), the designer must make provisions to prevent freezing. An alarm should be installed to alert the building operator of a loss of steam pressure or exceptionally low condensate temperatures. If air-handling units are used, the alarm should also terminate the supply fan's operation. The following recommendations will help to minimize freezing problems in steam systems:

1. Select traps of nonfreezing design if they are located in potentially cold areas.
2. Install a strainer before all heating units.
3. Do not oversize traps.
4. Make sure that condensate lines are properly pitched.
5. Keep condensate lines as short as possible.
6. Where possible, do not use overhead return.
7. If heating coils are used, allow only the interdistributing tube type.
8. Limit the maximum tube length of heating coils to 10 ft (3 m).
9. All coils and lines should be vented and drainable.

17.1.28 PIPING SUPPORTS

All steam piping is pitched to facilitate the flow of condensate. Table 17.1.9 contains the recommended support spacing for piping. The data are based on Schedule 40 pipe filled with water and an average amount of valves and fittings.

TABLE 17.1.9   Recommended Hanger Spacing

Pipe size, in	Distance between supports, ft
	Length
$3/4$	4
1	7
$1^{1}/_{4}$	7
$1^{1}/_{2}$	9
2	10
$2^{1}/_{2}$	11
3	12
$3^{1}/_{2}$	12
4	14
5	15
6	17
8	19
10	20
12	23
14	25

Note: Figures are based on Schedule 40 steel pipe filled with water including a normal amount of valving and fittings. These conversion factors can be used: 1 in = 2.54 cm and 1 ft = 0.3 m.

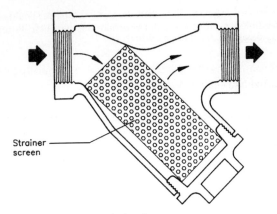

FIGURE 17.1.14   Typical strainer.

17.1.29   STRAINERS

Strainers (Fig. 17.1.14) should be located in the supply main before all steam-consuming devices and as part of the drip-leg assembly to collect particles and sediment carried in the system. Strainers located in areas not susceptible to freeze-up should extend down directly under the steam lines to allow sediment and particles to collect at the bottom of the strainer. In areas where freezing is possible, strainers should be installed at about a 20° angle below the horizontal plane. This will form an air pocket which will allow for expansion if the water in the strainer freezes.

The strainers should be cleared regularly as part of a routine maintenance schedule.

17.1.30   PRESSURE-REDUCING VALVES

As steam pressure increases, both the specific volume of the steam and the heat of vaporization increase.

Many times the boiler is designed to operate at a higher steam pressure than the heating components. The higher boiler pressure allows the supply-main size to be reduced because of the smaller specific volume of the steam. At a convenient point in the main near the heating devices, a "pressure-reducing valve" is installed. This valve reduces the pressure and allows the steam to expand. As the steam expands, its heat of vaporization increases, allowing for greater system efficiency. The pipe size directly downstream of the pressure-reducing valve should be increased to accommodate the steam's expansion. This should be done even if the reducing-valve connections for the inlet and outlet are the same size.

17.1.31   FLASH TANKS

A reservoir where condensate accumulates at low pressure before it returns to the boiler is normally provided. Another name for this reservoir is the *flash tank*. As the hot condensate reaches a low-pressure area, some of the liquid will flash into steam.

At the top of the flash tank, a steam line routes the steam that has just formed back into the system to be utilized. The flash tank improves the efficiency of the system and guarantees that only liquid condensate is returned to the boiler.

17.1.32   STEAM SEPARATORS

The need for pure steam without the presence of water droplets is imperative to permit control devices to operate properly. A device that allows vapor to pass while knocking water droplets from the stream is known as a *steam separator*.

Steam separators should be installed before all control devices and anywhere else in the system where small water droplets cannot be tolerated. Obviously, steam separators are not required on superheated-steam installations.

17.2 HOT-WATER SYSTEMS

Lehr Associates
New York, New York

17.2.1 INTRODUCTION

The predominant method of heating today's buildings, whether single-family dwellings or large structures, uses hot water to convey heat from a central generating source throughout the building. In nearly all new construction, the water is circulated through a piping distribution network by an electrically driven pump; this type of system is classified as a forced-circulation system. Heat from the circulating water is transferred to radiators, finned tubes, cabinet heaters, or other types of terminal units distributed strategically throughout the structure.

Older systems used gravity to circulate the hot water, by utilizing the difference in density between supply and return columns of the piping network. Since this type of system is rarely installed today, this chapter confines itself to forced-circulation systems. As a matter of fact, the latest American Society of Heating, Refrigeration, and Air-Conditioning Engineers (ASHRAE) guide refers readers to editions published before 1957 for details on designing gravity hot-water systems.

All hot-water heating systems rely on some form of central generating facility as the source of heat. This facility can be in the form of a boiler that consumes oil, gas, or electricity as the prime energy source or steam-to-water and water-to-water heat exchangers that derive heat from a utility or district-heating network.

This chapter gives details on the basic types of hot-water systems, as characterized by their temperature rating, general principles of system design, and special considerations of the equipment that comprises hot-water systems.

17.2.2 CLASSES OF HOT-WATER SYSTEMS

Hot-water systems are classified by operating temperature into three groups: low, medium, and high temperature. The *1987 ASHRAE Handbook* provides the following distinctions among these systems:

1. *Low-temperature water (LTW) system:* A low-temperature hot-water system operates within the pressure and temperature limits of the American Society of Mechanical Engineers' (ASME) *Boiler Construction Code* for low-pressure heating boilers. The maximum allowable working pressure for such boilers is 160 lb/in² (11 bar) with a maximum temperature of 250°F (121°C). The usual maximum working pressure for LTW systems is 30 lb/in² (2 bar), although boilers specifically designed, tested, and stamped for higher pressures frequently may be used with working pressures to 160 lb/in² (11 bar). Steam-to-water or water-to-water heat exchangers are often used, too.

2. *Medium-temperature water (MTW) system:* MTW hot-water systems operate at temperatures of 350°F (177°C) or less, with pressures not exceeding 150 psia (10.5 bar). The usual design supply temperature is approximately 250 to 325°F (121 to 163°C), with a usual pressure rating for boilers and equipment of 150 lb/in² (10.5 bar).

3. *High-temperature water (HTW) system:* When operating temperatures exceed 350°F (177°C) and the operating pressure is in the range of 300 lb/in² (20.7 bar), the system is

an HTW type. The maximum design supply water temperature is 400 to 450°F (205 to 232°C). Boilers and related equipment are rated for 300 lb/in² service (21 bar). The pressure and temperature rating of each component must be checked against the system's design characteristics.

LTW systems are generally used for space heating in single homes, residential buildings, and most commercial- and institutional-type buildings such as office structures, hotels, hospitals, and the like. With a heat-transfer coil or similar device inside or near the boiler, LTW systems can supply hot water for domestic water supplies. Terminal units vary widely and include radiators, finned-tube fan-coil units, unit heaters, and others. Typically overall heat loads do not exceed 5000 to 10,000 MBtu/h (1.5 to 3 MW).

MTW systems show up in many industrial applications for space heating and process-water requirements. Overall loads range up to 20,000 MBtu/h (6 MW). Generally HTW systems are limited to campus-type district heating installations or to applications requiring process heat in the HTW range. System loads are generally greater than 20,000 MBtu/h (6 MW).

The designs of MTW and HTW systems resemble each other closely. The systems are completely closed, with no losses from flashing. Piping can run in practically any direction, since supply and return mains are kept at substantial pressures. Higher temperature drops occur in MTW and HTW systems, relative to LTW systems, while a lesser volume of water is circulated (depending on the heat load of the system). LTW systems lend themselves better to combined hot-water/chilled-water heating/cooling systems. Extra care and expense must be devoted to fittings, terminal equipment, and mechanical components, especially for HTW systems.

Finally, often a combined system is desirable: an MTW or HTW circuit for process heat and an LTW circuit for space heating. The hot water for the LTW system can be obtained via a heat exchanger with the main heating system.

17.2.3 DESIGN OF HOT-WATER SYSTEMS

Design of hot-water systems involves a complex interplay of heat loads and the type of generating system. A traditional starting point, primarily for residential LTW systems, was the assumption of a 20°F (11°C) temperature drop through the circuit, from which the overall flow rate could be determined. A more recent practice is to perform a rigorous analysis, because the 20°F (11°C) assumption can lead to oversized pipes and flow rates.

System design can be broken down into five elements:

1. Determining the heat load.
2. Selecting terminal units or convectors based on the average water temperature and temperature drop and locating them on the architectural plan.
3. Developing a piping layout, including the choice of return system.
4. Locating mains, side branches, and other piping elements.
5. Specifying mechanical components, the expansion tank, and the boiler.

A good initial point is to run the flow main from the boiler to the terminal unit or units with the largest heat load and then to select branch runs to connect other terminal units. Two common space-heating terminal elements are convectors and wall fins, both of which contain a length of finned tube over which air can be fanned if desired. The air entering temperature is usually assumed to be 65°F (18°C). Most manufacturers supply tables showing heat ratings of the convectors, based on the assumed temperature drop, and the average entering water temperature (AWT). See Table 17.2.1 for an example for finned-tube convectors.

TABLE 17.2.1   Typical Ratings of Wall Fin Elements

Element type	Rows	Hot-water capacity, Btu/(h·ft), * at 65°F (17.4°C), entering air with average water temperature of:					
		220°F 104.4°C	210°F 98.9°C	200°F 93.3°C	190°F 87.8°C	180°F 82.2°C	170°F 76.7°C
Steel, 1¹/₄ in (32 mm)}	1	1260	1140	1030	940	830	730
	2†	2050	1850	1680	1520	1350	1190
Copper-aluminum, 1 in (25.4 mm)	1	1000	900	820	740	660	580
	2†	1480	1340	1210	1100	970	860
Steel, grilled enclosure, 1 in (25.4 mm)	1	1310	1190	1080	980	860	760
	2†	2080	1880	1700	1540	1370	1210

*1 Btu/(h·ft) = 0.0768 kcal/(h·m).
†4-in (10.2-mm) center-to-center gap.

An alternative approach is to assume a constant-temperature water flow (based on the leaving temperature of each class of terminal equipment) and to compute the required flow rate.

Both daily and annual variations in heat loads should be evaluated in order to arrive at a suitable design. This is especially true when LTW systems combining hot-water heating and cool-water cooling are envisioned. Figure 17.2.1 shows the seasonal effects of outside temperature on one type of piping design, the two-pipe system.

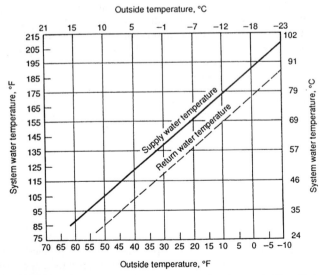

FIGURE 17.2.1   Seasonal operating characteristics of a two-pipe forced hot-water system. (*Courtesy of The Industrial Press.*)

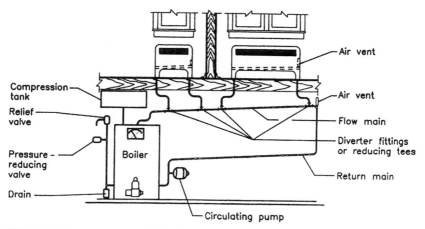

FIGURE 17.2.2 Arrangement of piping for a one-pipe forced hot-water system with closed expansion tank. (*Courtesy of The Industrial Press.*)

17.2.4 PIPING LAYOUT

Once a preliminary evaluation of heat load and terminal units has been performed, a piping layout can be undertaken. The usual starting-point options—running the flow main by the shortest and most accessible route to the larger heat loads—can be explored for the type of overall piping arrangement desired.

Pipe circuits generally are organized into one- or two-pipe arrangements. One-pipe systems with radiators or similar terminal units often have a feed and return pipe that diverts water from the flow main to the radiator and back to the flow main; even though two pipes are present, the system is still considered a one-pipe arrangement (see Fig. 17.2.2). Finned-tube heating elements running along the outer walls of small residences—a common arrangement—are true one-pipe systems, as shown in Fig. 17.2.3. Each terminal unit in the circuit receives progressively lower water temperature; thus the units are sized larger as they are located farther from the heat source.

Two-pipe systems allow for parallel heating arrangements, whereby terminal units can receive hot water at roughly similar inlet temperatures. The cooled water returns via a second pipe. The flow of this pipe can be specified to run in direct or reverse fashion back to the heat generator. Choosing between these options allows for better balancing of heat supplies among various terminal units and for some variation in overall system capital cost. Reverse-return systems specify that the distance that the water travels to a particular unit is the same as the return distance from that unit (Fig. 17.2.4)

17.2.5 PRESSURE DROP AND PUMPING REQUIREMENTS

All hot-water systems require some type of pumping to overcome friction losses of the flowing water, because whatever head is developed by the height of the water system (static pressure) is offset by the return pressure. Some more complex systems are better served economically by two or more pumps strategically located, rather than one large pump.

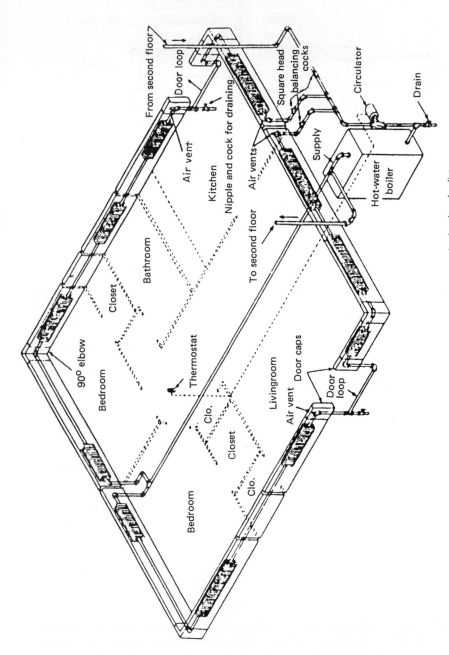

From second floor

Door loop

Air vent

Kitchen

Nipple and cock for draining

Air vents

Square head
balancing
cocks

Circulator

Drain

Supply

Hot-water
boiler

To second floor

Bathroom

Closet

90° elbow

Bedroom

Thermostat

Clo.

Closet

Clo.

Bedroom

Livingroom

Air vent   Door caps

Door
loop

Air vent

FIGURE 17.2.3 Typical installation of one-pipe forced-circulation "loop" hot-water system using baseboard radiators. (*Courtesy of The Industrial Press.*)

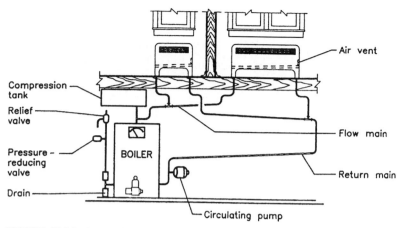

FIGURE 17.2.4 A two-pipe forced hot-water heating system with reverse-return piping. (*Courtesy of The Industrial Press.*)

Standard charts provide data on friction loss for runs of common types of piping (Fig. 17.2.5). To this should be added pressure losses from elbows, fittings, and other elements (Table 17.2.2). Similarly, manufacturers of radiators and other terminal units provide data on friction losses through their equipment.

Pump specifications are arrived at by first computing the overall pressure drop and the amount of desired water flow. "Pump curves"—charts which show the pressure developed by pumps as a function of the flow rate—can be used to arrive at the correct sizing. Many designers prefer to work with mass flow rate [pounds (kilograms) per hour] rather than gallons per minute (liters per second), units common to pump curves. The conversion between the two is temperature-dependent; two quick conversions commonly used are

Water at 40°F (4.4°C): 1 lb/h = 0.002 gal/min   (1 kg/h = 1.26 E – 4 L/s)

Water at 400°F (204.5°C): 1 lb/h = 0.0023 gal/min   (1 kg/h = 1.45 E – 4 L/s)

The next step is to determine the system curve for the hot-water circuit. The following formula is employed:

$$\frac{H_1^{0.5}}{W_1} = \frac{H_2^{0.5}}{W_2} \qquad (17.2.1)$$

where H_1 = known or calculated head, ft (m)
W_1 = design flow rate, gal/min (L/s)
H_2 = system curve head point, ft (m)
W_2 = system curve flow-rate point, gal/min (L/s)

With this equation, various system curve points can be plotted on the pump curve. The point where the system curve and the pump curve intersect is the operating point of the pump. Pump manufacturers specify optimum operating conditions (in terms of energy consumption, efficiency, and capacity of the pump) for their equipment.

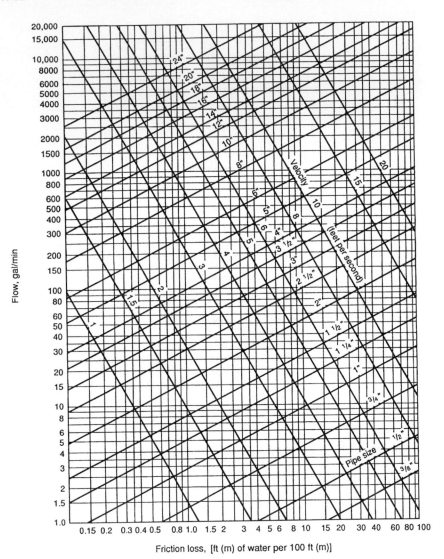

FIGURE 17.2.5 Friction loss for open-system piping. (*From Carrier Air Conditioning Company, Handbook of Air Conditioning System Design, McGraw-Hill, New York, © 1995. Used with permission.*)

17.2.6 PIPE SIZING

Hot-water system piping must be sized to carry the maximum desired amount of heating water throughout the system, while accounting for the static head of the elevation of the system and friction losses from pipe and fittings. Pipe sizes generally step down as water flows from the main(s) to branch circuits or individual heating units.

TABLE 17.2.2 Fitting Losses in Equivalent Feet* of Pipe
Screwed, welded, flanged, flared, and brazed connections

Nominal pipe or tube size, in*	Smooth bend elbows						Smooth bend tees						Mitre elbows		
								Straight-through flow							
	90° Std.†	90° Long Rad.‡	90° Street†	45° Std.†	45°† Street†	180° Std.†	Flow through branch	No reduction	Reduced 1/4	Reduced 1/2	90° Ell	60° Ell	45° Ell	30° Ell
3/8	1.4	0.9	2.3	0.7	1.1	2.3	2.7	0.9	1.2	1.4	2.7	1.1	0.6	0.3
1/2	1.6	1.0	2.5	0.8	1.3	2.5	3.0	1.0	1.4	1.6	3.0	1.3	0.7	0.4
3/4	2.0	1.4	3.2	0.9	1.6	3.2	4.0	1.4	1.9	2.0	4.0	1.6	0.9	0.5
1	2.6	1.7	4.1	1.3	2.1	4.1	5.0	1.7	2.3	2.6	5.0	2.1	1.0	0.7
1 1/4	3.3	2.3	5.6	1.7	3.0	5.6	7.0	2.3	3.1	3.3	7.0	3.0	1.5	0.9
1 1/2	4.0	2.6	6.3	2.1	3.4	6.3	8.0	2.6	3.7	4.0	8.0	3.4	1.8	1.1
2	5.0	3.3	8.2	2.6	4.5	8.2	10	3.3	4.7	5.0	10	4.5	2.3	1.3
2 1/2	6.0	4.1	10	3.2	5.2	10	12	4.1	5.6	6.0	12	5.2	2.8	1.7
3	7.5	5.0	12	4.0	6.4	12	15	5.0	7.0	7.5	15	6.4	3.2	2.0
3 1/2	9.0	5.9	15	4.7	7.3	15	18	5.9	8.0	9.0	18	7.3	4.0	2.4
4	10	6.7	17	5.2	8.5	17	21	6.7	9.0	10	21	8.5	4.5	2.7
5	13	8.2	21	6.5	11	21	25	8.2	12	13	25	11	6.0	3.2
6	16	10	25	7.9	13	25	30	10	14	16	30	13	7.0	4.0

(*Continued*)

TABLE 17.2.2 Fitting Losses in Equivalent Feet* of Pipe
Screwed, welded, flanged, flared, and brazed connections (Continued)

Nominal pipe or tube size, in*	Smooth bend elbows						Smooth bend tees				Mitre elbows			
	90° Std.†	90° Long Rad.‡	90° Street†	45° Std.†	45°† Street†	180° Std.†	Flow through branch	Straight-through flow			90° Ell	60° Ell	45° Ell	30° Ell
								No reduction	Reduced 1/4	Reduced 1/2				
8	20	13	—	10	—	33	40	13	18	20	40	17	9.0	5.1
10	25	16	—	13	—	42	50	16	23	25	50	21	12	7.2
12	30	19	—	16	—	50	60	19	26	30	60	25	13	8.0
14	34	23	—	18	—	55	68	23	30	34	68	29	15	9.0
16	38	26	—	20	—	62	78	26	35	38	78	31	17	10
18	42	29	—	23	—	70	85	29	40	42	85	37	19	11
20	50	33	—	26	—	81	100	33	44	50	100	41	22	13
24	60	40	—	30	—	94	115	40	50	60	115	49	25	16

*Conversion factors: 1 ft = 0.31 in; 1 in = 25.4 mm.
†R/D approximately equal to 1.
‡R/D approximately equal to 1.5.
Source: Carrier Air Conditioning Company, *Handbook of Air Conditioning System Design,* McGraw-Hill, New York, © 1965.
Used with permission.

Once the overall heating demand and the operating temperature of the heating system are known, calculations can be made for pipe sizes. The relationship between Btu demand and water flow rate is

$$\text{Btu/h} = \text{gal/min} \ (500 \ \Delta T°F) \tag{17.2.2}$$

A rough calculation of the overall friction head for the main can be done by measuring the longest main circuit and adding an equivalent length of 50 percent of the main to account for fittings. More accurate determinations are made by adding the equivalent pipe lengths of the fittings on the main to the length of the longest main. The manufacturer's literature usually includes charts similar to Table 17.2.2 showing equivalent lengths of common fittings.

Various methods have been worked out to determine the suitable pipe diameters to provide a sufficient flow rate. Usually the procedure must be iterated several times to select the best combination of flow rate, fluid velocity, and pressure drop. Table 17.2.3 shows these relationships for various pipe sizes if one assumes a maximum pressure drop of 4 ft per 100 ft (1.2 m per 30.5 m) and a maximum velocity of 10 ft/s (3 m/s). Once the pipe sizes have been determined, the system's pressure head should be compared to the head developed by the circulation pump. The pump may have to be resized, necessitating another iteration of the pipe sizing.

TABLE 17.2.3   Allowable Flow Rates for Closed System Piping, Standard-Weight Steel Pipe

Pipe size, in	Flow range, gal/min	Pressure drop range, ft per 100 ft
$\frac{1}{2}$	0–2	0–4
$\frac{3}{4}$	3–4	2.5–4
1	5–7.5	2.0–4
$1\frac{1}{4}$	8–16	1.25–4
$1\frac{1}{2}$	17–24	2–4
2	25–48	1.25–4
$2\frac{1}{2}$	49–77	2–4
3	78–140	1.5–4
4	141–280	1.25–4
5	281–500	1.5–4
6	501–800	1.75–4
8	801–1700	1.0–4
10	1701–2500	1.25–2.75
12	2501–3600	1.25–2.25
14	3601–4200	1.25–2.0
16	4201–5500	1.0–1.75
18	5501–7000	0.9–1.50
20	7001–9000	0.8–1.25
24	9001–13,000	0.6–1.00

Note: The above capacities are based on a maximum pressure drop of 4 ft per 100 ft and a maximum velocity of 10 ft/s. Conversions: 1 in = 25.4 mm, 1 gal = 3.8 L, and 1 ft = 0.31 m.

17.2.7 VENTING AND EXPANSION TANKS

Hot-water systems require pressures greater than atmospheric at all times to prevent air infiltration. Flashing or boiling of water is also minimized by maintaining the system above the water vapor pressure—preventing this also minimizes water hammer.

Maintaining this pressure, as well as allowing for the expansion and contraction of water as it is heated and cools, is most frequently carried out by means of an expansion tank. The expansion of medium- or high-temperature water systems can be calculated by consulting steam tables. The specific volume of water at its initial conditions is subtracted from its volume at the highest temperature, to calculate the volume change. To a certain limited extent, the water's expansion and contraction are offset by the similar changes that system piping and heating units undergo. These changes can be calculated from coefficients of expansion of the materials of the piping.

The simplest type of expansion tank is open to atmosphere at an elevation that provides the pressurization (head) the system requires. Open tanks have the disadvantage of allowing air to enter the system via absorption in the water. Closed tanks are more common now, especially with larger systems. Three common types of expansion/pressurization tanks are in use today:

1. *Adjustable expansion tank.* This tank employs an automatic valve along with a closed tank that has water and air feeds. As the temperature in the system rises, the pressure rises. A control valve releases air in the tank to the atmosphere. When the pressure and the water level drop, high-pressure air is injected into the tank. High-temperature systems should use nitrogen rather than plain air to reduce corrosive effects.

2. *Pump-pressurized cushion tank.* This design involves a makeup tank which is fed by a pump and a back-pressure control valve. For small systems (depending on local codes and on the water pressure available) the pump is skipped and city water pressure is used to feed a makeup tank that pressurizes the heating circuit. In principle, either type of pressurized tank can be roughly sized by assuming the expansion and contraction rates of the water to be equal.

3. *Compression tank.* A compression tank employs a specified volume of gas within an enclosure. As the water temperature and volume increase, the pressure on the gas volume rises, causing that gas volume to decrease. In this manner, the tank accommodates changing water volumes while keeping the system within a specified range of upper and lower pressures.

In low-temperature systems, the compression tank is usually connected to the system through an air separator situated between the boiler exit and the suction inlet of the circulating pump. Air separated from the water will rise into the compression tank. When the compression tank is located at a system's high point, it can be smaller in volume since the pressure is at its lowest. Tank sizing is also dependent on the location of the circulation pump relative to the tank.

One commonly used formula for sizing the compression tank, when operating temperatures are below 160°F (71.1°C), is

$$V_t = \frac{EV_s}{P_0/P_1 - P_0/P_2}$$
(17.2.3)

where V_t = compression tank volume
V_s = volume of circulating system, exclusive of compression tank
E = coefficient of expansion from initial to operating temperature
P_0 = absolute pressure in compression tank prior to filling
P_1 = absolute static pressure after filling
P_2 = absolute pressure at system operating temperature

For operating temperatures between 160 and 280°F (71.1 to 137.8°C), this formula is used:

$$V_t = \frac{(0.00041t - 0.0466)V_s}{P_0/P_1 - P_0/P_2} \quad \text{(USCS units)} \tag{17.2.4}$$

$$V_t = \frac{(0.000738t - 0.03348)V_s}{P_0/P_1 - P_0/P_2} \quad \text{(metric units)} \tag{17.2.5}$$

where t is the maximum operating temperature.

Compression tanks can be supplied with an impermeable membrane (diaphragm) to prevent air from being drawn into the circulating water when the system temperature drops. The diaphragm also allows the compression tank to be smaller in volume.

Diaphragm compression tanks are equipped with sight glasses or similar devices to monitor the water level. Too low a water level prevents the air behind the diaphragm from affecting the system's pressure.

17.2.8 MECHANICAL AND CONTROL EQUIPMENT

Mechanical components for low-temperature systems are under less severe service than those for medium- and high-temperature units; correspondingly, the care with which components are specified should increase with the higher-temperature systems.

ASME and ASHRAE rules should be observed for dealing with pressure vessels. Specifically, the chemical condition of the circulating water in high-temperature systems should be checked periodically by an expert. Pressure gauges should be located at both ends of the circulation pump. Modulating combustion controls, rather than straight on/off controls, are necessary to minimize pressure swings that lead to flashing. Where compression-type expansion tanks are used, an interlock with the system's heat generator should be installed to prevent operation when the water level in the tank is too low or insufficient air is present to maintain the tank compression. Valves and fittings for high-temperature systems should be specified with materials that resist corrosion and erosion, such as stainless steel.

The primary control factor for a hot-water system is the operating temperature range, which in turn is based on outside air temperatures. Electronic thermosensors and thermostats function to keep the room air temperature within the desired range. The system should also be equipped with a manual on/off control.

The electronic control for moderating the room air temperature can be of several types. Most are based on a solenoid device, which sends a signal current on the basis of a temperature reading. The control can be a simple on/off device or can have various modulating schemes to minimize large temperature swings. Temperature controls can also be set for zone heating of certain rooms, or areas within a large room, depending on the piping layout. In this case, the electronic control is connected with various flow control valves that will reduce or expand water flow to the heating units.

17.3 INFRARED HEATING

Lehr Associates
New York, New York

17.3.1 INTRODUCTION

Infrared heating takes advantage of the fact that light can be used to convey heat, just as sunlight can warm a cold surface. Infrared rays heat the objects they strike rather than the air surrounding the objects. When infrared is used for total heating, the surrounding air conducts heat away from the objects and the space eventually reaches an equilibrium temperature. When it is desirable to heat only a small area within a larger one or in an open space (such as an entrance canopy), infrared heating is unsurpassed because the energy required is only that used to warm objects or people, and not all the surrounding space.

17.3.2 TYPES OF HEATERS AND APPLICATIONS

A large variety of uses are suited to infrared heating, and many emphasize heating selected areas of large spaces. Examples include the following:

Bus shelters	Milking parlors
Canopies	Parking lots
Garages	Skating rinks
Gymnasiums	Snow melting
Indoor tennis courts	Stadiums
Loading docks	Work areas
Manufacturing areas	

Infrared heaters offer the user total flexibility in heating a building. Infrared systems can be used to maintain desired temperatures inside an entire building. They can also be used for supplementary areas or spot comfort heating within a building and are unmatched for outside area heating.

Infrared heaters are commonly designed to radiate energy by either burning a fuel or radiating heat from electrical-resistance wires. The details of common types are as follows:

17.3.2.1 Fueled Heaters

Gas-Fired.   These units use either natural gas or propane, delivered under pressure. The gas flows along a tube or within a mesh-metal or ceramic "emitter." The emitter elements are mounted in a panel in parallel rows, each up to several feet in length. Tubular heaters are larger, with several tubes mounted in parallel and running up to 20 ft (6 m) in length. The heat output of a tubular heater can range upward of 100,000 Btu (29.3 kW), while panel-type heaters range from 20,000 to 200,000 Btu (5.9 to 58.6 kW). Both types are normally mounted high on walls or suspended by wires from ceilings. Small portable units, fueled from a pressurized liquefied petroleum gas (LPG) tank, can be floor-mounted.

17.3.2.2 Electric Heaters

Single- or Double-Element Tubular Units.   These units come with reflectors that limit the radiation pattern to a 30°, 60°, or 90° angle. Double-element units can be symmetric or asymmetric. These heaters could be conduit- or chain-mounted, adjustable 45° in either direction. Metal-sheath, quartz tube, or quartz lamp elements may be used interchangeably.

Portable Units.   These provide warm, sunlike heat for hard-to-heat spots or for construction or maintenance work outdoors. This type of heater is usually made of extruded aluminum, with gold-anodized reflectors, heavy-duty metal-sheath heating elements, and safety screens. The unit provides rapid heatup and is operated from power distribution centers, a disconnect switch, etc. Usually units up to 5 kW are provided with a stand; 6-kW units and up are mounted on two-wheeled carts.

Heavy-Duty Overhead Units Designed for Spot Heating Applications.   These are operated manually from a disconnect switch or automatically controlled.

Glass-Panel Units.   The heating element is fused onto specially tempered glass which will withstand a pressure of 27,000 lb/in^2 (1901.4 kg/cm^2). The fused-on heating element is covered with enamel for corrosion protection. The whole assembly is mounted in silicone rubber to completely eliminate hum and expansion noise.

17.3.3   PHYSIOLOGY OF INFRARED HEATING

Because infrared heating can be used so selectively—heating only a small space—its physiological effects are worth considering. A person under an infrared lamp is warmed by the absorption of the light; a portion of the light is reflected. Light reflected from surfaces around a person—the floor, furniture, and the like—is also absorbed by one's skin. Absorptance and reflectance are set by the color temperature (corresponding to the light's frequency) of the light and the properties of the surface on which the light falls. A person's skin has an absorptance of around 0.9 at 1200 K (color temperature), falling to around 0.8 at 2500 K.

A person feels comfortable or uncomfortable based on the surrounding temperature, humidity, air flow, and degree of activity. The "mean radiant temperature" (MRT) has been defined as the temperature at which an imaginary isothermal black enclosure would exchange the same amount of heat by radiation as in an actual environment. As room temperature falls below normal ranges, or as air flow increases heat flow away from a person, the MRT rises. The designer using infrared heating for comfort purposes can use a set of calculations by the American Society of Heating, Refrigeration, and Air-Conditioning Engineers (ASHRAE) to determine the MRT needed to provide desired comfort levels. ASHRAE has also established comfort charts that determine comfort levels on the basis of room temperature, relative humidity, and radiated heat.

Infrared heating can also be used, of course, for *total* space heating. Then comfort-level calculations become secondary to normal calculations of heat loss from a structure and the amount of heat provided by heating units.

17.3.4 SPACING AND ARRANGEMENT OF ELECTRIC HEATERS

17.3.4.1 Total Heating

To achieve total heating with infrared heat, calculate the normal heat loss of the building or room to be heated and supply an equivalent number of watts of infrared heat to replace what is being lost. Heaters should be arranged in a grid pattern to achieve the most even coverage of the floor area.

Infrared heaters heat the exposed floor and the objects in the room. The objects in turn radiate heat to the cooler surrounding air until an equilibrium is reached. Thermostats may be used to maintain the desired comfort level in the room.

Two parallel heaters aimed at an angle toward the section of a room are more efficient than a single heater positioned directly overhead. Two heaters can cover a greater area, with a more even coverage, than one heater.

17.3.4.2 Determining the Number and Size of Heaters

Determine the required watt density (WD) from Table 17.3.1.

> **Example**   A tight, uninsulated building, such as a pole barn, steel shed, or concrete-block warehouse, located in an area with an ambient temperature of 45°F (7.22°C) would require 30 W/ft² (323 W/m²) of occupied floor space to obtain a

TABLE 17.3.1   Required Watt Density*

	Local spot and small-area heating				
Desired† temperature rise, °F	Tight, uninsulated building	Drafty‡ indoors or large glass area	Loading area, one end open	Outdoor, shielded; less than 5 mi/h wind	Outdoor,¶ unshielded; less than 10 mi/h wind
20	30	40	50	55	60
25	37	50	62	70	75
30	45	60	75	85	90
35	52	70	87	100	105
40	60	80	100	115	120
45	67	90	112	130	135
50	75	100	125	145	150
60	90	120	150	175	N.R.
70	105	140	175	205	N.R.
80	120	160	200	235	N.R.
90	135	180	N.R.	N.R.	N.R.
100	150	200	N.R.	N.R.	N.R.

*Watts of input to heaters per square foot of occupied floor space.
†Temperature rise is the increase above the existing temperature in the room. For outdoor use, subtract temperature rise from 70° to determine lowest outside temperature at which comfort can be achieved with the watt density listed under the appropriate column.
‡Doors frequently open.
¶Normal outdoor clothing; people stationary. N.R. = not recommended.
Note: Use the following conversion factors: °C = (°F − 32)/1.8; 1 mi/h = 1.609 km/h.

comfortable working temperature of 65°F (18.3°C) [assuming a 20°F (11.1°C) temperature rise].

From job conditions, determine the minimum mounting height available. The lower the mounting height, the greater the watt density attainable with the fewest heaters. However, the lowest recommended mounting height is 10 ft (3.04 m), to prevent too high a watt density on the upper part of one's body. Higher mounting will provide coverage of a greater area at a lower watt density.

Using the selected mounting height, refer to A or B in Table 17.3.2 to obtain the watt density and area of coverage at three-fourths overlap or full overlap, respectively, at the indicated distance S between parallel heaters.

Example At a mounting height of 14 ft (4.27 m), two heaters mounted with centerlines 8 ft (2.44 m) apart for full overlap would cover an area 14 ft (4.27 m) wide and 16 ft (4.88 m) long at a watt density of 22 W/ft² (237 W/m²).

Table 17.3.2 is based on the use of 60° asymmetric heaters placed a distance S apart for three-fourths or full overlap.

17.3.4.3 Increasing the Watt Density

To increase the watt density, use additional heaters between the two parallel heaters. One additional heater equals 100 percent greater watt density; three additional heaters equal 150 percent greater watt density. Using Table 17.3.3, determine the watt density and

TABLE 17.3.2 Spacing and Coverage

	A			B		
Mounting height, ft	WD	$W \times L$, ft	3/4 Overlap S, ft	WD	$W \times L$, ft	Full overlap S, ft
10	33	13×12	$8^{1}/_{2}$	38	11×12	6
11	27	14×13	9	33	12×13	$6^{1}/_{2}$
12	24	15×14	10	29	12×14	7
13	21	16×15	$10^{1}/_{2}$	25	13×15	$7^{1}/_{2}$
14	18	17×16	11	22	14×16	8
15	16	18×17	12	20	15×17	$8^{1}/_{2}$
16	14	20×18	$12^{1}/_{2}$	18	16×18	9
17	13	21×19	13	16	17×19	$9^{1}/_{2}$
18	11	22×20	14	14	18×20	10
19	10	23×21	15	13	19×21	$10^{1}/_{2}$
20	9.6	24×22	$15^{1}/_{2}$	11	20×22	11
21	8.8	25×23	16	10	20×23	$11^{1}/_{2}$
22	8.0	26×24	17	9.6	22×24	12
23	7.2	27×25	$17^{1}/_{2}$	8.8	22×25	$12^{1}/_{2}$
24	6.8	28×26	18	8.4	23×26	13
25	6.4	30×27	19	8	24×27	$13^{1}/_{2}$

Note: 1 ft = 0.304 m. WD = watt density; W = width; L = length; S = spacing.

TABLE 17.3.3   Single-Heater Watt Density and Coverage

Mounting Height, ft	90°		60°		60° Asymmetric	
	WD	$W \times L$, ft	WD	$W \times L$, ft	WD	$W \times L$, ft
10	18	12×12	25	$8\,^1/_2 \times 12$	19	11×12
11	15	13×13	22	9×13	16	12×13
12	13	14×14	18	10×14	14	12×14
13	11	15×15	16	10×15	13	13×15
14	10	16×16	14	11×16	11	14×16
15	9	17×17	13	12×17	9.6	15×17
16	8	18×18	11	12×18	8.8	16×18
17	7	19×19	10	13×19	8.0	17×19
18	6.4	20×20	8.8	14×20	7.2	18×20
19	5.6	21×21	8.0	15×21	6.4	19×21
20	5.2	22×22	7.6	15×22	5.8	20×22
21	4.8	23×23	7.0	16×23	5.4	20×23
22	4.4	24×24	6.2	17×24	4.8	22×24
23	4.0	25×25	5.8	17×25	4.5	22×25
24	3.7	26×26	5.4	18×26	4.1	23×26
25	3.5	27×27	5.0	19×27	3.9	24×27

Note: 1 ft = 0.304 m. WD = watt density; W = width; L = length.

coverage of the 60° and 60° asymmetric heater at the selected height. Add a sufficient number of heaters to attain the required watt density.

Example   At a 14-ft (4.27-m) mounting height, a watt density of 33 W/ft² (355 W/m²) is required. With full-overlap spacing (*B* in Table 17.3.2), two asymmetric heaters provide only 22 W/ft² (237 W/m²) on 10-ft (3.04-m) centers, covering an area of 14 ft (4.27 m) by 16 ft (4.88 m). From Table 17.3.3, at a 14-ft (4.27-m) mounting height, a 60° heater will provide the necessary addition (14 W) over an area of 11 ft × 16 ft (3.35 × 4.88 m) when mounted between the two 60° asymmetric heaters. [*Note:* A 90° heater will add 10 W and cover an area of 16 × 16 ft (4.88 × 4.88 m). Select the heater that will best suit the needs of the application.]

17.3.4.4   Greater Length Coverage

To obtain coverage for lengths greater than those listed in Table 17.3.2, use additional parallel groups. Use a length centerline the same as the width centerline (*S*) listed, thereby increasing the length coverage by the same footage while retaining the same watt density.

Example   For the example above with a 14-ft (4.27-m) mounting height, a watt density of 33 W/ft² (355 W/m²), but an area 14 ft × 24 ft (4.27 m × 7.32 m) to be covered, use an additional group of two parallel asymmetric heaters, plus the one "fill-in" heater, spaced on 8-ft (2.4-m) centers. This will provide an additional 8 ft (2.44 m) of coverage and maintain the watt density at 33 W/ft² (355 W/m²).

17.3.4.5 Additional Width Coverage

Use side-by-side sets of parallel heaters. Very little overlap is required. At a 10-ft (3.04-m) mounting height, centerlines of back-to-back heaters should be 2 ft (0.61 m) apart and an additional 1 ft (0.3 m) apart for every 5 ft (1.52 m) of mounting height above 10 ft (3.04 m).

17.3.5 GAS INFRARED RADIANT HEATING

Gas-fired infrared units have many of the same advantages as electrically powered units, with the bonus of cost savings where the price of gas is less than the Btu equivalent amount of electricity. Gas-fired units can be used for both local (zone) heating of a workspace and total space heating.

Gas-fired units require that several precautions be taken to protect against fire hazards or damage from intense radiation. Generally these standards are detailed in local building codes. Among them are the following:

1. *Adequate combustion air:* Sufficient air must be available for complete combustion. In addition, the presence of flammable dusts or vapors must be kept low. Venting to prevent buildup of combustion gases must be allowed for; generally a positive air displacement of 4 ft^3/min per 1000 Btu/h (387 L/min per kW) of gas input is desired.

2. *Humidity control:* Among the combustion by-products is water vapor; the venting must be planned to allow for this.

3. *Clearance to combustibles:* Both the roof and articles or materials below and beside the heating unit must be kept a sufficient distance from the unit. This distance depends on the heat output of the unit and the flammability of the materials near it. Equipment manufacturers provide a clearance-to-combustibles chart that specifies minimum distances. For example, for a 50,000-Btu/h (14.65-kW) units, the top clearance is around 9 in (22.9 cm), the side clearance is around 10 in (25.4 cm), and the bottom clearance is 36 to 42 in (91.4 to 106.7 cm).

4. *Controls and shutoff:* Gas units are generally started by electric glow plugs or similar devices. Differential-pressure gauges are usually included to cut off gas flow in the event of a combustion air on flue blockage.

17.3.5.1 Design of Total-Heating Gas-Fired Systems

The selection of the proper heater size is determined by the available mounting height. There must be sufficient clearance from floor to roof to allow the heater to be installed at the recommended mounting height with enough clearance above the unit to prevent overheating of the roof.

From an economical standpoint, the largest heater is used for the available mounting height as long as the recommended maximum distance between heaters is not exceeded. Normally heaters are mounted 2 to 3 ft (0.6 to 0.9 m) below overhead obstructions and the clearance to combustibles is maintained from the roof regardless of the combustibility of the roof.

17.3.5.2 Louvers

Inlet louvers may be necessary if the building is extremely tight and well insulated. If inlet louvers are used, they should be small and well distributed on at least two outside walls and on all outside walls in larger installations.

17.3.5.3 Annual Estimated Fuel Cost

Fuel consumption will be influenced by a number of variables over which the designer has little control. Consumption calculations should be treated as estimates.

A simplified formula to approximate the annual gas consumption in therms is:

$$\frac{HL \times DD \times 24}{TD \times 100,000}$$

where HL = heat loss, Btu/h (kW)
     DD = degree days
     TD = temperature differential, °F (°C)
  100,000 = heating value of 1 therm of gas

17.3.5.4 Heater and Thermostat Placement

The placement and the spacing of the heaters are made easy by referring to Fig. 17.3.1 and Table 17.3.4. It is common practice to set in from the corner of a building 15 to 20 ft (4.57 to 6.10 m) with the first heater along an outside wall.

If clearances below the units necessitate mounting closer to the roof than suggested, a piece of noncombustible millboard is frequently utilized above the unit.

In most buildings a single perimeter loop system as shown in Figure 17.3.1 will satisfy the heat loss. If the maximum distance between heater rows is exceeded, an interior row or partial row may be needed.

In buildings where an interior row is indicated, the only heat losses are the roof and air losses. Center-row spacing is frequently double that used along the perimeter.

Control systems are available in three voltages: 120 V, 25 V, and millivolt. The selection of the voltage may be a matter of personal preference, or in some cases the code may dictate the system used. Millivolt control is used primarily for small systems since each heater has its own thermostat. The 25-V system is normally the most economical system

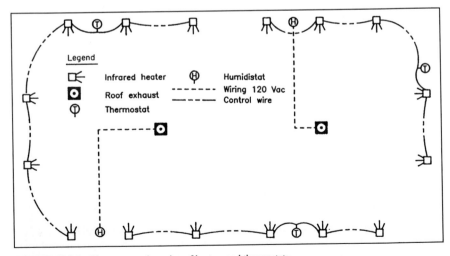

FIGURE 17.3.1 Placement and spacing of heaters and thermostats.

TABLE 17.3.4 Heater Placement Guide for Full Building Heat

	Input, Btu/h*							
	25,000	30,000	50,000	60,000	75,000	90,000	110,000	120,000
Mounting height, ft*								
10°	9–13	10–14	14–18	15–20	17–22	20–25	22–28	24–28
30°	7–10	7–12	11–15	12–16	14–19	16–22	18–25	18–26
Range between units, ft	7–18	8–20	12–25	15–30	18–34	20–42	25–46	28–50
Between rows ft* (max.)	60	60	80	80	100	100	110	110
Heater to wall, ft* (max.)	5	6	8	10	12	12	14	14
Required exhaust,† ft³/min	100	120	200	240	300	360	440	480
Clearance to combustibles, in								
Top	30	30	36	36	48	48	54	54
Below	72	72	88	88	104	104	120	120
Sides	30	30	36	36	42	42	48	48
Back	24	24	33	33	39	39	48	48

*Conversion factors: Btu/h × 0.293 = W; ft × 0.304 = m; ft³ × 0.0283 = m³.
†Air exhaust based on 1000 Btu/h of gas and 4 ft³/min per 1000 Btu input.

to install. On most applications using the 25-V system, the 18/2 control wire is taped to the gas piping with a drop at each heater. If the code requires that 25-V control wiring be installed in conduit, then no savings advantage is experienced. Three thermostats for zone control are common practice with infrared systems and allow for flexibility of control. Thermostats should not be located directly in the rays of the infrared generators because they will not read the ambient temperature of the building correctly.

On 25-V systems, take special note of the volt-ampere ratings of the various control systems and the size of the transformers utilized. Some systems will allow the use of eight units on one 75-VA transformer while another system will set a limit of four units on the same transformer.

17.4 ELECTRIC HEATING

Lehr Associates
New York, New York

17.4.1 INTRODUCTION

Electric energy is ideally suited for space heating. It is relatively simple to control and distribute. In many applications, the cleanliness and compactness of electric heaters are a very attractive design alternative. Such heaters require no fuel storage, produce no fumes or exhaust emissions, and, depending on the specific setup, present a safer alternative to combustion heaters.

Cost and energy conservation are the dominating design factors in electric heating. Usually, compared with other heating methods, electric heating has a lower installation cost, requires less maintenance, has lower insurance rates, and is easier to zone. Electric space heating is often used when minimal initial cost is a dominant factor. However, electricity is a relatively expensive form of energy. The operating costs of electric systems are normally higher than those of heating with gas or fossil fuels.

17.4.2 SYSTEM SELECTION

A careful analysis is necessary to select a building heating system that is both effective and cost-efficient. The effectiveness is determined by the ability of the system to meet the heating needs of the building. This is done by determining the power requirements to maintain an indoor design temperature over a range of out-door temperatures. This power requirement is called the *load* on the system and is useful when presented in a load profile or load duration curve. A load profile curve shows the power required to maintain an indoor temperature plotted against the time of day. The load duration curve shows the same information in a different fashion. It plots the number of hours for which the system is at each load level. The total energy requirements can be determined by finding the areas under these curves.

Costs associated with the system must be estimated over the life cycle of the equipment. The initial expense of the equipment and its installation cost are obtainable, but the daily costs, long-term maintenance, and replacement value must be estimated. All these costs must be included in an efficiency analysis, if it is to be complete.

Operating costs are particularly difficult to determine. The estimated annual heating requirement is used along with estimates of electricity prices to determine yearly operating costs. The pricing structure of electricity must be accounted for, especially if demand changes or time-of-day rates apply. Local utility companies may provide estimates of future energy prices. The prices of other energy sources, such as oil, and gas, should also be considered. The following cost comparison can be used to make an initial appraisal of the relative costs of electric versus gas or oil heating. (See Figs. 17.4.1 and 17.4.2.)

Electric heating systems are either centralized or decentralized. Centralized systems serve the heating needs of an entire structure. Hot-water, steam, and warm-air systems are examples of centralized systems. Decentralized systems come in various designs: natural-convection units, forced-air units, radiant units, or radiant panel-type systems.

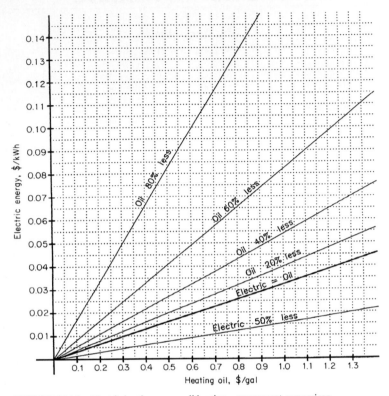

FIGURE 17.4.1 Electric heating versus oil heating—energy cost comparison.

These designs are geared to the control of individual heating sections of a structure. The following systems are all presently in use:

I. Decentralized systems

    A. Natural-convection units

        1. Floor drop-in units
        2. Wall insert and surface-mounted units
        3. Baseboard convectors
        4. Hydronic baseboard convectors with immersion elements

    B. Forced-air units

        1. Unit ventilators
        2. Unit heaters
        3. Wall insert heaters
        4. Baseboard heaters
        5. Floor drop-in heaters

    C. Radiant units (high-intensity)

        1. Radiant wall, insert, and surface-mounted
        2. Metal-sheathed element with reflector

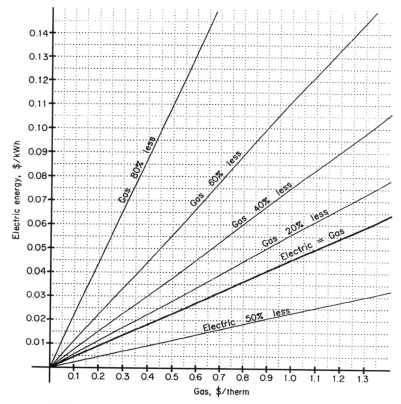

FIGURE 17.4.2 Electric heating versus gas heating—energy cost comparison.

3. Quartz-tube element with reflectors
4. Quartz lamp with reflector
5. Heat lamps
6. Valance (cove) unit

D. Radiant panel-type systems (low-intensity)

1. Radiant ceiling-mounted with embedded conductors
2. Prefabricated panels
3. Radiant floor-mounted with embedded conductors
4. Radiant convector panels

II. Centralized systems

A. Heated-water systems

1. Electric boiler
2. Electric boiler with off-peak storage
3. Heat pumps
4. Integrated heat-recovery systems

B. Steam systems: electric boiler, immersion, or electrode
C. Heated-air systems

1. Duct heater
2. Electric furnace
3. Heat pump
4. Integrated heat-recovery system
5. Unit ventilator
6. Self-contained heating and cooling unit
7. Storage unit (ceramic, water)

17.4.3 CENTRAL HOT-WATER SYSTEMS

A central hot-water system comprises an electric boiler and a series of radiators, convectors, unit heaters, cabinet heaters, or heating and ventilating units. Water is heated in the boiler by elements or electrodes and is distributed throughout the structure. Multiple elements are combined to achieve the total capacity required. Multiple-element systems are arranged to energize or de-energize each unit in sequence to avoid large fluctuations in voltage. It is also important that the elements not be energized when the circulating pump is not operating.

Some boilers are designed for space efficiency and can be wall-mounted. A 20-kW boiler is approximately 1.5 ft^3 (0.04 m^3). Larger applications with capacities up to 5000 kW can be achieved with electrode boilers rated at 5 kV and higher. Electric boilers are space-efficient and clean. They have no need for flues, smoke-stacks, or fuel storage. See Chap. 16 for a detailed discussion of electric boilers.

Hot-water systems can be tailored to specific applications. Some boilers are particularly adaptable to multiple zoning. A control system energizes the number of elements needed to match the heat requirements of each zone. Hydronic water storage is also possible with hot-water systems. During normal system operation, water is stored at high temperature. This water can be used during off-peak operation. A large storage tank can hold water with temperatures of 200 to 275°F (93.3 to 135.0°C) at pressures up to 75 psig (5.2 bar). An automatic valve can supply water at a lower temperature by mixing it with cooler water from a main supply line. Steam can be obtained by withdrawing hot water into a low-pressure separating chamber.

17.4.4 WARM-AIR SYSTEMS

A central blower and system of ducts comprise a simple warm-air system. Electric heating units are installed in ducts near the blower to regulate the temperature of the entire structure. Individual room control may be achieved by heating air at the outlets. Warm-air systems provide a convenient means for fresh air intake, which ensures good ventilation. Other important functions of this type of system include circulation and filtering.

Heating elements for ductwork are fabricated in either slip-in or flanged designs. Electrical codes usually specify that at least 48 in (122 cm) of straight duct be installed between the element and fan outlets, duct elbows, baffles, or similar obstructions. This measure minimizes overheating of sections of the heating element that get uneven air flow.

The heating elements can be sized by looking at heat-transfer calculations. For easier design, nomograms have been developed that relate air flow, temperature rise, and kilowatt rating of the heating element (Fig. 17.4.3).

Electric furnaces consist of heating coils and a blower in an insulated housing. Air is drawn in through the bottom of the furnace and is filtered. The blower then forces the air over the heating coils, transferring heat from the coils to the air. Electric furnaces are compact, requiring no flue or fuel storage.

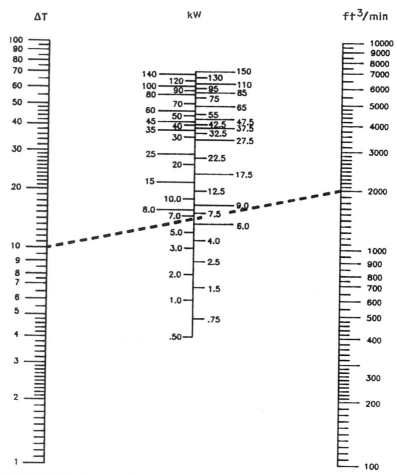

FIGURE 17.4.3   Nomogram for determining the kilowatts required as related to air flow and temperature rise. (*Courtesy of Industrial Engineering & Equipment Co., St. Louis, MO.*)

17.4.5 CONVECTOR WITH METALLIC HEATING ELEMENT

Certain convective heating units use metallic heating elements. Heating is provided by bare-wire, low-temperature bare-wire, or sheathed elements. Air is heated by heat transfer from the elements. A reflector panel radiates heat away from the unit, minimizing the casing temperature rise and maximizing heat transfer to the room. The space between the elements and the reflector serves as an air passage that promotes convection. It is important that the location of electric convectors be such that air movement across the elements is not impeded.

Metallic heating element convectors can be wall-mounted, recessed, or surface-mounted. Wall- or surface-mounted types can be equipped with a circulating fan to force convection. Recessed versions do not use fans.

17.4.6   UNIT VENTILATORS

The function of a unit ventilator is to provide heat, ventilation, and cooling to an area. The ventilator comprises an inlet grille, air filter, fan, heating elements, damper, and diffuser. The total capacity of the unit is the sum effect of many small electric heating elements. The temperature is controlled by activating and deactivating individual elements. Unit ventilators use a certain percentage of outdoor air. This type of unit has many applications: classrooms, motels, offices, and nursing homes.

17.4.7   UNIT HEATERS

Unit heaters are available in three types: cabinet heaters, horizontal projection heaters, and vertical projection heaters. The cabinet type can be floor-, wall-, or ceiling-mounted. Recessed unit heaters are also available. Unit heaters are designed with a fan that circulates room air over the heating elements. Unit heaters require no venting or piping, which makes them a convenient source of supplemental heating. Unit heaters can be used to heat occupied rooms in unheated buildings or unattended equipment enclosures in which a certain temperature must be maintained.

17.4.8   BASEBOARD HEATERS

Electric baseboard heaters contain one or more horizontal heating elements in a metal casing. Elements are made of various materials and are available in many sizes. Some of these elements include finned, sheathed, cast grid, ceramic, extended-surface, and coated glass.

Baseboard heaters are placed on the floor along the base of a wall. Proper positioning is important to achieve uniform heating.

Electric hydronic baseboard heaters contain immersion heating elements. Antifreeze models are also available.

17.4.9   INFRARED HEATERS

Infrared heaters use high-temperature elements to provide heating. Other designs use quartz tubes or lamps instead. Heaters are either suspended from the ceiling or mounted on the wall. Elements are placed in a trough-type casing surrounded by reflectors.

Infrared heaters have many applications. When convection heating is made impractical by high ventilation, infrared heaters can be used. They can be focused to heat specific sections of a room, and they provide comfort despite low ambient temperature. Outdoor spot heating is possible with infrared heaters. Industry uses modified designs for heating and industrial drying. See Sec. 17.3 for a discussion of infrared heating.

17.4.10   VALANCE, CORNICE, OR COVE HEATERS

These heaters are similar in shape to baseboard heaters. They are usually installed several inches below the ceiling on an upper outside wall. Heating elements come in various forms: metal-sheathed, coated-glass, or metal panels. Units are available up to 6 in (152 mm) high and project from the wall less than 3 in (76 mm). With these units the main source of heating is the room ceiling, which is heated by the convection flow of air over the heater. Also the heated panel provides some direct radiation into the room.

Hydronic valance units can provide either heating or cooling by circulation of either hot or cold water. Hydronic valence heating/cooling systems have a drain through that collects and disposes of condensate during the cooling mode of operation. The fins are designed to allow the flow of condensate into a drain.

17.4.11 RADIANT CONVECTOR WALL PANELS

Radiant convector wall panels have glass electric heating panels. The panels are supported on insulators within a metal frame, with a reflector arranged to provide space for air circulation. Protective guards are provided for safety.

By passing a current through a thin coat of conductive material on glass, heat is produced. The conductive material is usually sprayed-on aluminum or printed metallic oxide grids. Some radiant panels use tubular elements welded to extended aluminum panels. These units have emissivity characteristics similar to those of glass panels. See Sec. 17.10 for a further discussion.

17.4.12 INTEGRATED HEAT RECOVERY

An integrated heating and cooling system takes account of all the heat sources and sinks within a structure. The system is modeled around these sources and sinks for maximum effectiveness and efficiency. It is becoming increasingly common for commercial and industrial settings to have a custom-designed system to meet their energy needs. A system considers all heat gains, including those from such sources as lights, people, machinery, and the sun. By transferring heat from hot areas to areas that require heating, the dual purposes of heating and cooling are served. Generally refrigeration equipment is incorporated into such systems to produce a cooling effect.

Integrated system design requires a careful accounting of the building's energy needs. Many sources of heat are available only during normal working hours, and supplemental heating is needed during off hours. Electric heaters are ideal for this purpose and can be used to provide heat for both personnel and machinery. In other applications heat is accumulated during periods of excess and stored for later use.

17.4.13 HEAT PUMPS (See also Sec. 18.6)

A heat pump is a device that operates in a cycle, requires work, and accomplishes the task of transferring heat from a low-temperature area to a high-temperature area. A heat pump uses the mechanical refrigeration cycle to cool and reverses the roles of the evaporator and the condenser to provide heating.

Heat pumps use electricity more efficiently than resistance heaters. With resistance elements, 1 kW of heat is produced for each kilowatt of electricity. For each kilowatt of electricity used by a heat pump, 1 kW of compression heat is produced plus a refrigeration effect. The refrigeration effect of the heat pump can vary from 10 to as high as 50 percent of the electric energy input. This effect is dependent on the temperatures involved.

The "efficiency" of a heat pump is expressed by the coefficient of performance (COP). COP is used as an indicator of how a heat pump operates under specified temperature conditions. The most significant variable in the equation for the efficiency of the heat pump is the temperature of the heat sink. An equation for the COP is

$$\text{COP} = \frac{\text{heat of compression} + \text{refrigeration effect}}{\text{heat of compression}}$$

To ensure proper equipment selection, some quantitative relationships must be determined and analyzed. Heating capacity versus outdoor air temperature (OAT) is one such relationship. Figure 17.4.4a shows the output of a system operating at a series of outdoor air temperatures; the capacity drops off as the outdoor air temperature falls.

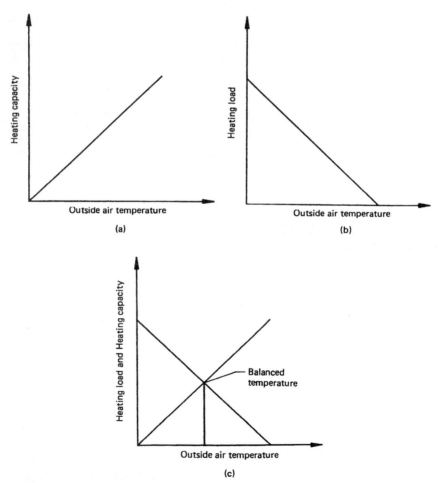

FIGURE 17.4.4 Relationships between heating capacity and heating load at various outside air temperatures.

Another important relationship is heating load versus outside air temperature. Figure 17.4.4b indicates the load required to maintain an internal temperature at various outdoor air temperatures. With a reduction in outside air temperature, the heating load increases.

Applying this kind of analysis to a heat pump shows that under certain conditions a heat pump's capacity equals the heating load of a space. The temperature at this point is called the *balanced point,* or *balanced temperature,* determined graphically by the intersection of the capacity versus outside air temperature curve and the load versus outside air temperature curve (Fig. 17.4.4c).

When the outdoor air temperature is below the balanced temperature, the heating needs of a space will not be satisfied by the heat pump alone. Supplemental heating must be used, and its minimum capacity must make up the heating load of the space. The heat pump is most efficient when supplemental heat is not required. It is more important for a heat pump to satisfy an area's cooling requirement, because supplemental heating can be easily provided.

Heat pumps come in four types. They are designated by what they use as a heat source and sink. Air-to-air systems use outdoor air as heat sink. The heat pump treats the air, and, when needed, a supplemental electric heater is activated to heat the supply air.

The air-to-water system uses a condenser water loop as the heat sink. Auxiliary devices such as a cooling tower and boiler are used to keep the temperature within limits, so that the heat pump will operate efficiently. When cooling is necessary, the tower is operated; and the boiler is used when heating is required. Heat is sometimes rejected to ground wells instead of cooling towers. In other instances, many heat pumps are attached to the same condenser water circuit; so when heating and cooling are required simultaneously in different areas, the heat added or rejected to the loop by the auxiliary devices is minimized.

A water-to-water heat pump is called a "heat reclaim chiller." It can provide both heating and cooling. It provides hot water at the condenser and cool water at the evaporator. The water-to-air heat pump is not very common. It is an air-cooled chiller that can reverse its cycle to produce warm water.

17.4.14 SPECIFYING ELECTRIC HEATING SYSTEMS

Generally, once the heat load of a building or room is known, the appropriate electric heating system or unit can be sized, just as for hydronic or other heating systems. The most efficient systems usually are decentralized ones, since the temperature conditions of each room can be adjusted individually to meet variations in air infiltration, heat generation by lights or appliances, or sunlight.

Heaters with direct electrical-resistance elements transform power to heat with 100 percent efficiency. Disregarding for the moment the subsequent inefficiencies of conveying that heat to air, water, or some other heating medium, the overall electricity consumption of an electric heating system can be estimated according to the following formula:

$$E = \frac{\text{HL(DD) (24)}}{\text{TD}(kV)} C_D$$

where E = energy required for auxiliaries, Btu (kWh)
   HL = design heat loss including infiltration and ventilation, Btu/h (W)
   DD = number of 65°F (18.3°C) degree-days for estimate period
   TD = design temperature difference (indoor—outdoor), °F (°C)

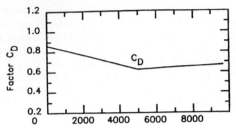

FIGURE 17.4.5   Correction factor C_D versus degree-days.

k = heating value of fuel, units consistent with HL and E; for electric heat k can be approximated at 1

v = heating value of fuel, units consistent with HL and E; for electric heat, 3413 Btu/kWh (100100 W/kW)

C_D = empirical correction factor effect versus 65°F (18.3°C) degree-days; determined by the graph in Fig. 17.4.5.

24 = constant used in unit conversion, that is, 24 h/day

17.4.15   ELECTRIC CIRCUIT DESIGN

The National Electrical Code®, supplemented by local code restriction, provides very strict and specific standards for connecting electric heating systems with a building's power supply.

Depending on the type of heating unit—particularly in the case of forced-air heaters—electrical codes usually require two types of overtemperature control. Power cutouts are the bimetallic type or liquid-filled bulbs. Both are designed to sense air blockage by means of a temperature rise, to interrupt the power circuit, and to reset automatically.

Many types of small-capacity heating units have integral power and thermostat controls. Thermostats carry line voltage to the unit and are designed to cycle the unit on a predetermined basis. Wall- or remote-mounted thermostats are supplied for larger heating units or for temperature control of large interior spaces. These thermostats are selected with a view toward how the room will perform during heatup and cooldown and whether radiant, convective, or forced-air heating is being employed.

When the outdoor air temperature is below the balanced temperature, the heating needs of a space will not be satisfied. Supplemental heating must be used, and its minimum capacity must make up the heating load of the space. The heat pump is most efficient when supplemental heat is not required. It is more important for a heat pump to satisfy an area's cooling requirement because supplemental heating can be easily provided.

17.4.16   HEAT PUMP TYPES*

Heat pumps come in four types. They are designated by what they use as a heat source and sink. Air-to-air systems use outdoor air as a heat sink. The heat pump treats the air, and, when needed, a supplemental electric heater is activated to heat the supply air.

The air-to-water system uses a condenser water loop as the heat sink. Auxiliary devices such as a cooling tower and boiler are used to keep the temperature within limits, so that

---

*See also Sec. 18.6

the heat pump will operate efficiently. When cooling is necessary, the tower is operated; and the boiler is used when heating is required. Heat is sometimes rejected to ground wells instead of cooling towers. In other instances, many heat pumps are attached to the same condenser water circuit; so when heating and cooling are required simultaneously in different areas, the heat added or rejected to the loop by the auxiliary devices is minimized.

A water-to-water heat pump is called a "heat reclaim chiller." It can provide both heating and cooling. It provides hot water at the condenser and cool water at the evaportor. The water-to-air heat pump is not very common. It is an air-cooled chiller that can reverse its cycle to produce warm water.

17.5 SOLAR SPACE HEATING

Lehr Associates
New York, New York

17.5.1 INTRODUCTION

Solar energy systems for heating, ventilating, and air-conditioning (HVAC) applications can make use of incident energy from the sun to heat a transfer fluid or to energize photovoltaic materials. Photovoltaic materials are considered beyond the scope of this chapter; electricity generated by such solar panels can be used in many of the same ways as electricity from the local power grid, for electric heating, infrared heating, etc. Solar energy systems can be further separated into *passive designs*, which entail constructing and locating walls, roofs, windows, etc., in such a fashion to maximize the effects of incipient solar energy; or *active designs*, which involve transferring energy derived from sunlight to other parts of a structure or from one medium to another. Active systems are the focus of this chapter.

Solar energy is received by collection systems in several forms: direct radiation, reflected radiation, and diffuse radiation. The sum of these is termed "insolation." Direct radiation usually accounts for around 90 percent of the radiant energy. Tables of insolation totals for various latitudes, times of day and year, and collector tilts are available from standard handbooks. These tables allow estimates of how much energy is available to a solar collector, which in turn can be used to size the unit to meet the expected energy demand. Further corrections are made for weather patterns, collector efficiencies, and other factors.

Solar energy collected by active systems can be used for several purposes: space heating, space heating plus hot-water heating, space heating plus hot-water storage, and other combinations. The more common type of system is liquid-circulating flat-plate collector; this is the focus of the remainder of this chapter. Flat-plate collectors (henceforth called "collectors") employ a black absorber plate backed with insulation, tubes, or other passageways for heat-transfer fluid, transparent covers, and a frame. Plumbing connections at the back or side of the frame allow the unit to be hooked up to a circulation network running inside a house or other structure (see Fig. 17.5.1).

The collector operates by absorbing radiant energy from sunlight on the black absorber plate. Tubes carrying fluid pass along the surface of the plate, heating the fluid. Temperatures within the collector can reach a maximum of 400°F (204.4°C).

17.5.2 TYPES OF DISTRIBUTION SYSTEMS

Solar collectors can be arranged in a variety of configurations to suit various applications. Generally, the following types have undergone extensive development and commercialization:

1. *Thermosiphon systems:* This design is one of the simplest (Fig. 17.5.2) A storage tank is placed a certain height above the collector plate. As fluid in the collector warms, it becomes buoyant, causing a convective movement from the collector to the top of the storage tank (a pipe directs the fluid to that location). In time, the tank's reservoir becomes completely warmed, and the fluid can be released from the tank to space heating tubes below the tank or to an auxiliary storage tank in the basement of the building.

Sunlight

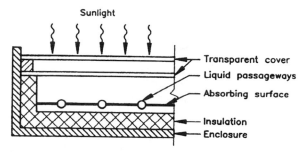

FIGURE 17.5.1   Flat-plate collector section.

Electrically driven pumps or subsequent heating with fuel causes the fluid to be distributed through the building's heating system.

2. *Direct forced-circulation systems:* Forced-circulation systems (Fig. 17.5.3). depend on a pump to move the fluid. A direct system uses solar energy in direct contact with the fluid of interest (usually water being heated). Such direct systems are often employed in climates where freezing is not a problem. However, various provisions for automatic drain-down or other measures can minimize freezing risks.

3. *Indirect forced-circulation systems:* These units are identical to direct systems except that a working fluid (i.e., an antifreeze solution) is used to collect solar energy (Fig. 17.5.4).

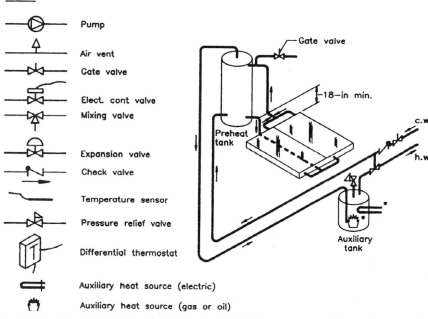

FIGURE 17.5.2   Thermosiphon system schematic diagram. Asterisk means either auxiliary energy source is acceptable. [*From Jan F. Kreider and Frank Kreith (eds.), Solar Energy Handbook McGraw-Hill, New York, © 1981. Used with permission.*]

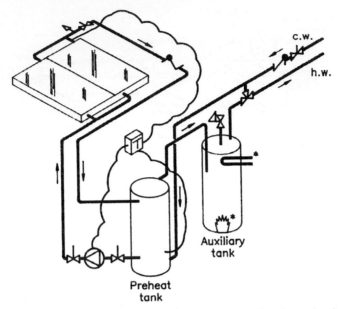

FIGURE 17.5.3   Direct forced-circulation water heating with preheat tank and separate auxiliary tank. Asterisk means either auxiliary energy source is acceptable. For explanation of other symbols, see Fig. 17.5.2. [*From Jan F. Kreider and Frank Kreith (eds.), Solar Energy Handbook McGraw-Hill, New York, © 1981. Used with permission.*]

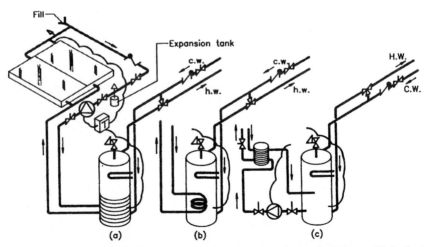

FIGURE 17.5.4   Indirect forced-circulation water heating with a single tank. (*a*) External jacket heat exchanger, (*b*) internal coil heat exchanger, and (*c*) external shell-and-tube heat exchanger with pump. For explanation of symbols, see Fig. 17.5.2. [*From Jan F. Kreider and Frank Kreith (eds.), Solar Energy Handbook, McGraw-Hill, New York, © 1981. Used with permission.*]

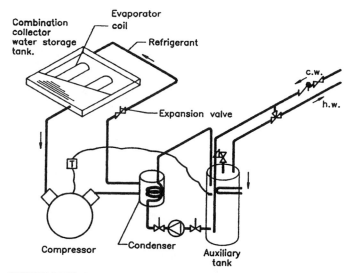

FIGURE 17.5.5   Indirect forced-circulation solar water heating using electric heat pump and combined collection/storage unit that is permitted to freeze. For explanation of symbols, see Fig. 17.5.2. [*From Jan F. Kreider and Frank Kreith (eds.), Solar Energy Handbook, McGraw-Hill, New York, © 1981. Used with permission.*]

A heat exchanger then transfers the energy to water or some other circulation fluid. Thus indirect systems are somewhat more complex and usually require a larger collector surface area to make up for inefficiencies in heat transfer.

4. *Combined collector-storage systems:* These units are indirect, forced-circulation heating systems with a heat pump or similar device (Fig. 17.5.5). The energy efficiency is improved by employing the heat pump to draw energy from the surroundings even when solar energy is limited. Another common feature is some means of heat storage, usually a supplementary hot-water tank or rock beds or similar media.

The design of solar heating systems is composed of three interconnected design tasks: Solar panel selection, energy storage and heat transfer to the distribution system, and design of the primary heat distribution system. This section covers the distribution system design only. References to solar collectors and storage devices are made to clarify the design methodology and requirements.

17.5.3   GENERAL DESIGN

In general, the design of the solar distribution system is similar to that of the standard heating system. The solar heating system conforms with accepted plumbing, heating, ventilating, and air-conditioning engineering practices.

17.5.3.1   Pipe Sizing

The hot-water pipe should be big enough to provide heat to the building in the quantities specified by the load calculation. The maximum velocity in the system should not exceed

10 ft/s (3 m/s) at any time. Refer to Sec. 17.2 for a discussion of hot-water distribution pipe sizing.

17.5.3.2 Solar Hot Water Operating Temperatures

A fundamental constraint of solar systems for space heating is that the maximum temperature of the solar unit is often below the normally used hydronic space-heating range. Obviously, if the solar system is to provide heating, the temperature it develops must be sufficient to deliver the necessary heat. A good rule of thumb is that in winter the maximum temperature of the solar unit should be 100°F (55.6°C) more than the actual outdoor ambient temperature. This rule assumes a 30 to 40 percent collector efficiency. Thus, when it is, say, 15°F (−9.4°C) outside, the solar collector can produce 115°F (46.1°C) water.

Such temperatures limit the operating range of space heating systems, most of which can operate well above 200°F (93.3°C). The system designer has several alternatives to consider:

1. The space heating system may be oversized to produce the same Btu/h (watt) output as a smaller, but hotter, conventional system.

2. The hot water can be used in conjunction with an air-heating fan-coil system. This coil can be sized appropriately for a required design supply water temperature in the order of 120 or 140°F (60 or 48.9°C). Alternatively, conventional floor panel systems (tubing embedded in the floor), which are usually designed to operate at a supply temperature of 100°F (37.8°C), can use hot water from the solar collector at the required outdoor design temperature.

3. Slab-on-grade houses can utilize a modified hydronic floor panel perimeter loop in combination with the preheat air duct (supplementary standby) system. The perimeter loop (one or two copper tubes embedded in concrete close to the outside wall) prevents cold floors and provides for a relatively high percentage of required house heat at a low water temperature [on the order of 100°F (37.8°C)].

A hydronic system with a low water temperature has several advantages when viewed from the perspective of good solar collector design:

1. A low required water temperature in the hydronic system will increase the range of outdoor temperatures at which the solar system will be operable.

2. A low temperature in the hydronic system will allow operation of the solar collector at a lower mean liquid temperature and will increase the collection efficiency.

17.5.4 HEAT-TRANSFER MEDIA

Liquid solar heating systems are generally more efficient than air-based ones, simply because of the greater heat capacity of liquids. Among liquid systems, there are a variety of constraints, such as the presence of freezing conditions or the need for certain temperature ranges, that dictate what type of heat-transfer medium to use.

Generally, water-based systems are most common, because of both the simplicity of the overall design and the design similarities to conventional hot-water heating systems. But water has the great disadvantage of freezing and harming the plumbing components of the solar system. Therefore, antifreeze-type solutions, such as ethylene glycol, are often utilized in solar space heating. A 50:50 water-ethylene glycol mixture is commonly specified. A heat-exchange coil is used to transfer heat from this solution to a hydronic space heating system.

In terms of the collector operating efficiency, the water system is best based on the following:

1. The glycol solution generally must run at a higher operating temperature to overcome the inefficiencies of the heat exchanger (while accomplishing the same ultimate heat output). Solar collectors lose overall efficiency as their operating temperature rises.

2. All-water systems usually have a drainback feature to prevent freezing damage when the system is not running or it is night. The drainback allows all the energy collected by the system to be used. Systems without drainback have a certain amount of inherent reverse heat loss to atmosphere when they are not running.

Ethylene glycol is a toxic material. When glycol systems are used for domestic hot water, local building codes usually specify a double-walled containment between the glycol and water.

Other heat-transfer fluids that have been used in solar heating systems include propylene glycol, which is less toxic than ethylene glycol. Hydrocarbon oils, commonly used in industrial heat-transfer applications, have been tried but run the risk of igniting. Silicone fluids show low toxicity and reactivity but tend to be viscous and to have low specific heat.

17.5.5 WATER DRAINBACK SYSTEMS

All-water solar collector systems incorporate a drainback feature to prevent freezing damage. The overall system is usually pressurized. Collectors are pitched at angles that permit unimpeded downward flow when the system is shut off. Some means of air venting must be included to allow the system to refill upon being restarted. A compression/expansion tank (essentially, a vent tank) is used to store the drawn-down water. Overall, the system is equipped with differential thermostats that start a pump when a preset ΔT is reached between the exit and return points on the storage tank. When the return temperature reaches a low (i.e., when the onset of darkness prevents solar heating), the thermostat will automatically shut off the pump, allowing the drainback step to begin.

There are two types of drainback designs: open-drop and siphon return. Salient features of the two are as follows:

1. Open-drop designs use a pump to force water up to the solar collectors and then to an oversized vertical down-comer that returns the water to the storage tank. The down-comer is oversized by about two pipe sizes, and it has an air vent, to prevent any back-pressure filling of the down-comer. When the pump stops, the system automatically drains. Air-balancing valves are required at various points throughout the system to prevent bypasses or back pressures. The pump is sized and selected on the basis of overcoming the elevation difference between the collector and the storage tank.

2. Siphon returns are designed such that the down-comer fills rapidly. This setup balances the fluid head with the fluid in the down-comer, so that pumping requirements are essentially independent of system height. Balancing valves permit the eventual drainback when low temperatures are experienced. In addition, added insulation on plumbing parts near the solar collector minimizes ice formation.

17.5.5.1 Heat Exchangers and Multiple Heat-Transfer Fluids

Another alternative to freeze prevention is the use of a subsystem containing an antifreeze-based fluid (usually a glycol). Some designs call for subsystem to be used only to keep the water flowing in the collectors warm during low-light, cold periods. Alternatively, the antifreeze subsystem can be used exclusively in the collector and related outdoor components, with the collected heat transferred to water indoors through a heat exchanger. In either case, the heat exchanger must be designed to work efficiently and must be properly sized.

In general, the heat exchanger should be sized generously to extract the maximum amount of heat from the fluid in the collector. An undersized exchanger will waste heat that has already been collected, decreasing the overall system efficiency. Glycol-to-water exchangers are commonly designed with straight-tube or U-tube configurations. The exchanger can also be placed in the storage tank at the base of the solar heating system, in which case glycol is flowing throughout the system's lines.

17.5.6 PUMPING CONSIDERATIONS

Glycol has an important effect on system flow rates.

A 50 percent glycol mixture has a slightly higher density—but lower specific heat—than water. The net relationship is that a greater fluid flow rate is required to satisfy the system's design. The increased flow rate for a 50 percent glycol mixture as compared with water and for the same heat conveyance is shown in Table 17.5.1. As an example, suppose that a particular glycol collector subsystem is to be designed on the basis of 140°F (60°C) mean fluid temperature. The initial design base (stated to water) requires a flow of 10 gal/min (0.63 L/s). The new flow-rate requirement will then be 11.5 gal/min (0.73 L/s).

Above roughly 30°F (-1.1°C), the viscosity of a 50 percent glycol solution becomes higher than that of pure water, by a factor of around 8 to 16 cP (0.008 to 0.016 (N · s/m^{20}). This fact necessitates correction of the pressure drop at various operating temperatures. Table 17.5.2 shows the pressure drop iterated through two corrections; the first is the correction at equal water or glycol flow rates, and the second shows the correction at a glycol flow rate already corrected upward to account for the lower heat capacity of glycol solutions. In practice, the designer of a glycol system would then take a pure-water design and raise the flow rate by the factor indicated in Table 17.5.1 and raise the pressure drop by the "Combined factor" column in Table 17.5.2. The designer could then consult pump curves to specify the correct pump. Generally, the pump curve can be used as is (i.e., for pure water), as long as the flow-rate and pressure-drop corrections for glycol solutions have already been calculated.

TABLE 17.5.1 Increased Flow Requirement for Same Heat Conveyance for 50 percent Glycol Compared with Water

Fluid temperature, °F (°C)	Flow increase needed for 50 percent glycol compared with water
40 (4.4)	1.22
100 (37.8)	1.16
140 (60.0)	1.15
180 (82.2)	1.14
220 (104.4)	1.14

TABLE 17.5.2 Pressure-Drop Correction Factors for 50 Percent Glycol Solution Compared with Water

Fluid temperature, °F (°C)	Pressure-drop correction with flow rates equal	Combined pressure-drop correction; 50 percent glycol flow increased per Table 9.1
40 (4.4)	1.45	2.14
100 (37.8)	1.10	1.49
140 (60.0)	1.00	1.32
180 (82.2)	0.94	1.23
220 (104.4)	0.90	1.18

17.5.7 ADDITIONAL FLUID SYSTEM CONSIDERATIONS

As with conventional hot-water heating systems, the performance of a solar heating system is improved by the addition of a compression tank. In particular, the compression tank helps minimize corrosion damage caused by dissolved oxygen, which is soon neutralized during operation. For an all-water system, this tank can be sized according to conventional procedures. A compression tank for an all-water system will experience greater vapor pressure changes than that for glycol systems, since the glycol system has a higher boiling point. Thus correct sizing is important. Conversely, a glycol mixture has a larger expansion rate than water alone. The general guideline is to size the tank 20 percent larger than if a water system were used.

17.5.7.1 Leak Protection

Leaks are a serious threat to the long-term operation of any solar heating system. To minimize their occurrence, welded or sweated joints should be used wherever possible, and threaded joints should be tightly wound with the appropriate sealing tape. Glycol systems should not contain an automatic makeup valve, as this will lead to dilution of the mixture over time. Also pumps with mechanical seals, rather than gland seals, should be used.

17.5.7.2 Glycol Sludge

Sludge often forms where glycol is present. This is due to a number of factors, including reactions between the glycol and oils, fluxes, or other hydrocarbon contaminants; reactions with chromate-based water treatments (which should never be used with a glycol system) or with galvanized piping; and gradual degradation of the glycol solution. To minimize sludge generation, galvanized pipe should not be used and the system should be cleaned and flushed prior to startup.

17.5.7.3 Storage Subsystem

As has been mentioned, many solar heating systems employ storage tanks that contain heated water. The tank serves to moderate swings in system demands, to store heat for periods when few solar rays are available, and to provide a convenient means of heat exchanging with a glycol or other fluid subsystem. Storage subsystem sizing plays a significant role in the overall performance of a solar heating system. A common rule of thumb is to size the storage tank in the range 1.5 to 2.0 gal of water for each 1 ft^2 of solar collector area (61 to 81 L/m^2). Larger systems are rarely cost-effective.

Flows from storage tank may be arranged to feed directly into space heating systems (a supplementary setup) or may be used as a standby system, operating through heat exchangers, air coils, or similar devices. The size of the storage system can be calculated more exactly by treating it as a means of storing surplus heat after space heating duties have been fulfilled. Thus, a large heat drawdown would allow for a larger storage unit.

17.5.8 MATERIALS AND EQUIPMENT

Equipment intended for heating the water or storing the hot water should be protected against excessive temperatures and pressures by approved safety devices and in accordance with one of the following methods:

1. A pressure relief valve and a temperature relief valve.
2. A combination temperature/pressure relief valve.
3. A pressure relief valve, a temperature relief valve, and an energy cutoff device.
4. A combination temperature/pressure relief valve and an energy cutoff device.

All safety devices specified by the plumbing designer should meet the criteria set forth by the American Society of Mechanical Engineers or Underwriters' Laboratories.

17.6 UNIT HEATERS

Ronald A. Kondrat
Robert L. Linstroth
Hal M. Sindelar
Product Managers, Heating Division,
Modine Manufacturing Co., Racine, Wisconsin

17.6.1 INTRODUCTION

Although there may be various definitions for the term "unit heater," it is described here as a self-contained heating package usually suspended overhead or mounted from walls or columns, generally intended for installation in the space to be heated. Elements include a fan, motor, heating element, and an enclosure. Depending on the type, additional items may include filters, power venters, directional air outlets, and duct collars.

Unit heaters have three principal characteristics: (1) relatively large heating capacities in compact casings, (2) the ability to project heated air in a controlled manner over a considerable distance, and (3) a relatively low installed cost per Btu (watt) of output. They are, therefore, usually employed in applications where the heating capacity requirements, the physical volume of the heated space, or both are too large to be handled adequately or economically by other means. By eliminating extensive duct installations, the space is freed for other use.[1]

Frequently, however, the unit heater's role is relegated to providing supplementary heating. Properly applied, the unit heater can be the *primary* source of heat in many types of buildings. The ideal applications for unit heaters exist in buildings having large open areas such as factories, warehouses, and other industrial and commercial buildings; showrooms, stores, and laboratories; and auxiliary spaces, where they may be used singly or in groups in corridors, vestibules, and lobbies, or to blanket areas around frequently opened doorways, for spot heating, and for maintaining air temperatures year-round in greenhouse applications.

Unit heaters are also employed where filtration of the heated air is required. They may also be modified to provide ventilation by using outside air.

Unit heaters may be applied to a number of industrial processes, such as drying and curing, in which the use of heated air in rapid circulation with uniform distribution is of particular advantage. Unit heaters may be used for moisture absorption applications to reduce or eliminate condensation on ceilings or other cold surfaces of buildings in which process moisture is released. When such conditions are severe, exhaust fans and makeup air units may be required.[2] Today hydronic unit heaters are also being used with increasing frequency to utilize waste heat in order to conserve energy.

Consulting engineers, plant engineers, architects, contractors, and building owners often judge unit heaters on the basis of observed jobs. Sometimes these jobs are poorly engineered, and as a result the unit heater is unfairly assumed to be the problem contributing to an unsatisfactory heating system. Here are a few examples of common misapplications:

Unit heater motors are generally designed to operate at a maximum ambient temperature of 104°F (40°C). If the unit heater is so oriented that it does not discharge its heat into the space to be heated (away from the unit heater), heat builds up around the unit and its motor may fail prematurely.

If a unit heater's mounting height is specified for high-speed motor operation and the unit is operated at low speed, then the heated air may not reach floor level, thus sacrificing comfort and wasting heat.

Selection of the wrong unit heater type, over- or undersizing a unit, mounting the unit too high or too low, and other factors can contribute to poor comfort conditions and wasted heat.

The information contained in this chapter can be helpful to heating system design engineers, specifiers, and users in selecting and applying unit heaters for the overall upgrading of unit heater installations and in obtaining the most comfortable conditions.

17.6.2   UNIT HEATING SYSTEM DIFFERENCES

Steam, hot-water, gas-fired, oil-fired, and electric heating systems all have unique characteristics, and they can be compared only if the advantages and disadvantages of each are known.

If we compare initial unit heater costs by using comparable Btu (watt) capacities, oil-fired unit heaters are the most expensive and are followed in order by gas, electric, and steam/hot-water unit heaters. If we rank by installation costs, the order again begins with oil-fired units because of added piping, tanks, pumps, filters, valves, and venting. Similar installation costs accrue with steam/hot-water units requiring supply and return piping, traps, strainers, and valves. Gas-fired unit heaters, with one-way piping and venting, can be installed for about one-half the cost of hydronic units. Last, electric unit heaters with no piping or venting have lower installation costs.

No economic comparison would be complete without a review of operating costs projected to at least a year of future operation. Because operating or fuel costs vary locally, they must be interpreted individually. All local energy rates can be translated to cost per Btu (kWh). Natural or propane gas is normally sold by therms (100,000 Btu or 29.3 kWh) or by cubic feet (cubic meters). Your local utility or dealer will know your calculated rate of Btu/ft^3 of gas (usually 800 to 1150 for natural gas or 2500 for propane). Electric unit heaters deliver 3413 Btu/kW (3,600,000 J/kW). For oil heaters, as a reference, No. 2 fuel oil is about 140,000 Btu/gal (40×10^6 J/L). Overshadowing all these costs, however, is the fuel availability in each locality. Where fuel or power supplies are interruptible, the value of a dependable fuel supply could offset operating-cost advantages.

Negative pressures within a building will not affect the operation of hydronic or electric unit heaters because they are not connected to flue systems that lead to the outside of the building. But when vented gas or oil units are used in a building with negative pressure, there could be a problem. If the gas-fired unit heaters are gravity vented, a negative pressure within the building will adversely affect the draft required for efficient operation of the unit. Negative building pressures can cause flue gases, which are normally vented to the outside, to spill into the building causing the blocked vent safety switch to trip. Pilot flame outage is another result of negative pressure within a building. Where negative pressures exist, installation of mechanical draft systems may be required for applications using gas- or oil-fired unit heaters. For gas-fired "separated combustion" units can also be used.

Hydronic unit heaters usually require some type of water treatment system. In existing steam or hot-water heating systems, water treatment has usually been provided. If the chemical condition of the condensate and/or supply water to be used in a heating system is unknown, it is recommended that the condensate and/or supply water be tested by a reputable testing firm and evaluated to determine if treatment is required.

The amount of maintenance a unit requires during its lifetime depends on the conditions under which the unit heater operates. Steam units, e.g., installed on high-pressure steam systems which supply steam for processing operations, are often subject to corrosion from acids and oxygen present in makeup water introduced into the system. This is because processing operations require greater quantities of boiler makeup water. Proper installation and maintenance of deaerating equipment and adequate boiler-water treatment can do much to reduce corrosion in steam unit heater condensers as well as piping, valves, and traps. For similar reasons, adequate boiler-water treatment must be provided and maintained in hot-water boilers.

Certain airborne contaminants can cause metal parts to corrode. The corrosion attack can be rapid or slow, depending on the concentration of the contaminant. Gas- or oil-fired heat exchangers, flue or vent pipes, and metal chimneys are most vulnerable. When contaminating vapors exist, steps to correct or relieve the condition must be taken. For example, proper venting of degreasing or cleaning vats will reduce the distribution of the contaminants throughout the heated areas of the building.

Corrosive agents in the atmosphere can shorten the life of unit heaters (particularly gas- and oil-fired units). Minute particles of trichlorethylene and carbon tetrachloride, e.g., when heated and combined with water vapor in the products of combustion, produce an acid solution harmful to heat exchangers and other metal parts. Reduce corrosive atmospheric contaminates by installing exhaust hoods above tanks containing solvents and acids.

Oil, gas, and electric unit heaters are self-contained heating units; failure of one or more unit heaters will not affect the entire heating system. Failure of the entire system (except for an electric power failure or a cutoff of the main gas supply to the building) is highly improbable. Should a steam or hot-water boiler system fail, all the equipment supplied by that system is useless until the boiler is repaired and placed in operation. If there is only one boiler supplying the facility, then the entire facility will be without heat.

Properly installed, a single unit heater (oil, gas, hydronic, or electric) can be removed from the system for repair or replacement without shutting down the entire system.

Since a majority of gas and oil unit heaters draw combustion air from the building they heat, the air they consume must be replenished. This is not the case with hydronic, electric or gas-fired separated combustion unit heaters.

In the event of a power failure, not only will electric unit heaters fail to operate, but so will gas, oil, and hydronic unit heaters, since these units are powered by electric motors.

17.6.3   CLASSIFICATION OF UNIT HEATERS

Unit heaters can be classified as follows:

- By type of air mover (either propeller or centrifugal blower)
- By airflow configuration (vertical or horizontal air delivery)
- By heating medium (steam, hot water, gas, oil, or electricity)
- By unit profile (standard or low profile)

17.6.3.1   By Type of Air Mover

In making your selection, be sure to consider both the heating capacity and the air delivery capacity of the unit, to be sure that the air is delivered precisely where you want it to go.

Propeller Fan Unit Heaters.   Propeller fan units are classified as zero-static-pressure type and cannot be used with air discharge nozzles or in connection with ductwork. This unit heater has a fixed air delivery with a temperature rise (difference between inlet and discharge air temperatures) usually in the range of 55 to 65°F (30.5 to 36.0°C). It is the most widely used of all unit heaters primarily because of its broad application, versatility, lower cost, and compact size.

It is used in free-air delivery applications where the heating capacity and distribution requirements can best be met by units of moderate output, used singly or in multiples, and where filtration of the heated air is not required. Horizontal blow units are usually associated with low to moderate ceiling heights. Downblow units are used in high ceiling spaces and where floor and wall space limitations dictate an out-of-the-way location for the heating equipment. Downblow units may be supplied with adjustable diffusers.[2]

Centrifugal Blower Fan Unit Heaters.   Designed for use with ductwork or discharge air nozzles, the blower unit is classified as a high-static unit [up to 0.7 in (1.8 cm) water column external static pressure]. Since its operation is generally quieter than a propeller-type unit heater having the same air delivery, it is frequently installed in areas where lower sound levels are desirable. Also this unit has a variable-pitch motor sheave which permits the air volume to be adjusted to accommodate a range in air temperature rise. Equipped with air discharge nozzles, the unit is ideal for use in spot-heating applications. In general, a blower unit heater is classified as a heavy-duty unit.

17.6.3.2   By Airflow Configuration

Unit heaters are produced in two basic types; horizontal air delivery and vertical air delivery. Each has its own distinctly different heat-throw and heat-spread characteristics. For any comfort application, selection of the proper type is very important.

Horizontal Delivery Unit Heaters.   Characterized by its horizontal air discharge, this type is widely used for general industrial and commercial heating applications. Horizontally positioned louvers attached to the air discharge opening are usually standard and can be adjusted to lengthen or shorten heat throw. Adjustable vertical louvers (when available, they are optional) used in combination with horizontal louvers permit complete directional control of heated air output.

Vertical Delivery Unit Heaters.   Due to their directly downward air discharge, vertical units are particularly desirable for heating areas with high ceilings and where craneways and other obstructions dictate high mounting of heating equipment. Air distribution devices (optional) that attach to the air discharge openings provide distinctly different heat-throw patterns to meet specific heat-throw and heat-spread requirements.

17.6.3.3   By Heating Medium (Steam, Water, Gas, Oil, or Electric)

Selection is usually dictated by the type of energy which is most prevalent, more reliable, or least costly. For example, in an industrial plant where steam is generated by in-plant processing operations, steam unit heaters may be selected if the existing boiler capacity is adequate to handle the heating load. In most areas electricity is more costly than gas. In certain areas gas may be in critical supply, and electricity, though more expensive, may constitute a more dependable energy source. Selection is also determined by economics which involve an examination of initial cost, operating cost, and conditions of use, including the size, location, and environment of the proposed installation.

It is standard practice among trade associations to rate all types of unit heaters on the basis of the amount of heat delivered in Btu/h (watts) at an entering air temperature of 60°F

(15.6°C).[2] If the units are operated under other than rated conditions, the manufacturers' literature should be reviewed or the manufacturer contacted directly. Most manufacturers' literature also provides data on other than standard rated performances.

Steam or Hot Water. There is generally no major difference in the construction of a steam or hot-water unit heater, as shown in Fig. 17.6.1. Both use a condenser which consists of fins attached to tubes through which steam or hot water produced by a remotely located boiler is circulated. Air moved across the condenser by a fan removes heat from the fins and discharges it to the space being heated. Although there is usually no difference between a steam unit and hot-water unit, the dissimilarity in piping arrangements of the two systems is discussed elsewhere in this chapter.

Rating of steam unit heaters has been standardized according to the following basis: dry saturated steam at 2-psig (13.78-kPa) pressure at the heater coil, air at 60°F (15.6°C) and 29.92 inHg (101.04 kPa) barometric pressure entering the heater, and the heater operating free of external resistance to air flow.[2]

The capacity of a heater increases as the steam pressure or hot-water temperature increases, assuming the entering air temperature is constant. The capacity decreases as the entering air temperature increases. The heating capacity for any condition of steam pressure and entering air temperature, other than standard, may also be determined by means of a standard procedure.[2]

A standard for the rating of hot-water unit heaters has been established according to the following basis: entering water at 200°F (93.3°C), water temperature drop of 20°F (11.1°C), entering air at 60°F (15.6°C) and 29.92 inHg (101.04-kPa) barometric pressure, and the heater operating free of external resistance to air flow. Variations in entering water temperature, entering air temperature, and water flow rate will affect capacity. A standard

FIGURE 17.6.1 Low profile gas-fired unit heater. A propeller fan is the air mover pushing air through the heat exchanger. Typically residentially certified allowing for use in residential garages. (*Courtesy of Modine Manufacturing Co.*)

method is used to translate the heating capacity, as obtained under standard rating conditions, to other conditions of air and water temperature.[2]

Gas (Conventional Type).   Heat exchanger tubes (aluminized or stainless steel) are fired by a natural or propane gas burner (Fig. 17.6.2). Flames produced by the burner burn within the tubes of the heat exchanger. Heat transferred from the flames and conducted to exterior surfaces of the tubes is removed by the air stream produced by a fan or blower, passing over the heat exchanger tubes.

This type of unit heater is similar to the conventional two different features: a closed collector box for flue gas (open collector box) and a power exhauster (rather than a gravity vent). The power exhauster functions like a damper to prevent heated room air from escaping through the chimney during the "off" cycle. Thanks to these two features, power-exhausted gas-fired unit heaters are roughly 13-percent higher in actual efficiency than conventional gas-fired unit heaters. They also consume less fuel than conventional types and are more flexible in venting arrangements. Unlike gravity-vented unit heaters, which must be vented vertically, power-exhausted gas-fired unit heaters can be vented through the side wall. They can also use smaller and less expensive flue gas vents. Power-vented gas-fired unit heaters can also be used to overcome negative air pressure in a room or space.

FIGURE 17.6.2   Gas-fired propeller or blower fan unit heater. A propeller or blower fan is the air mover pushing room air through the heat exchanger. Horizontally positioned adjustable air deflectors permit heated air to be directed up, down, or straight out. Vertically positioned deflector blades (optional) may be added for complete directional control of heated air. (*Courtesy of Modine Manufacturing Co.*)

Gas (Separated Combustion Type).   Gas-fired separated combustion units have the same basic design characteristics as conventional gas unit heaters except for the addition of (1) an enclosed burner compartment which is isolated from the heated space, and (2) a power exhaust system. These units are designed to take all of their combustion air directly from outside of the building. Because no indoor air is used for combustion, this type of unit is ideally suited for hard to heat applications, which may include buildings with mildly dusty or dirty atmospheres, high humidity, and negative pressure problems, public automobile repair facilities and parking garages, institutional buildings, and buildings where indoor air quality is of the utmost importance. Because separated combustion unit heaters are power vented, they can be vented through a side wall or vertically, usually using much smaller flue gas vent pipes than conventional unit heaters. High-efficiency models are available.

Gas-fired unit heaters are rated in terms of both input and output, in accordance with the approval requirements of the Canadian Standards Association.

Fuel Oil.   In operation, No. 1 or No. 2 fuel oil is atomized under pressure by a gun-type burner and is mixed with high-velocity combustion air to burn in a solid-cone flame in an aluminized-steel heat exchanger (Fig. 17.6.3). Resultant radiated heat is transferred to the exterior surfaces of the heat exchanger and is removed by the fan-produced air stream passing over the surfaces of the exchanger.

Ratings of oil-fired unit heaters are based on heat delivered at the heater outlet in Btu/h (watts).[2]

FIGURE 17.6.3   Oil-fired unit heater. A motor-driven propeller fan directs room air over the exterior surfaces of the heat exchanger. When the burner is ignited, a combination fan and limit control prevents fan operation until the heat exchanger has warmed up, and after the burner is shut down, it allows the fan to run until the heat exchanger has cooled. Horizontal and optional vertical louvers provide complete directional control of heated air. (*Courtesy of Modine Manufacturing Co.*)

Electricity.   Heating elements consist of spirally wound fins on steel tubes which encase electrical-resistance wire (Fig. 17.6.4). Heat generated by the electric current is transferred through the tubes to the fin surfaces, where it is removed by the air movement produced by a fan.

Electric unit heaters are rated on the energy input to the heating element, expressed in terms of kilowatts, or Btu/h.[2]

17.6.3.4 By Unit Profile

Gas unit heaters are available in two configurations, standard and low profile. When selecting units for a particular application, be sure to take the physical characteristics into consideration.

Standard Unit Heaters.   Standard unit heaters are configured in a vertical fashion. This makes the units narrower than a low profile unit and also increases the height. As a result, these units are used where headroom is not an issue. Since the units are in a vertical configuration, they can either use atmospheric or in-shot burners depending on the units' design. Typically standard unit heaters are used in commercial and industrial applications, since most are not residentially certified.

Low Profile Unit Heaters.   Low profile unit heaters are relatively new to the market compared to standard unit heaters. Low profile units are ideal for low headroom applications due to the horizontal design, as shown in Fig. 17.6.1. To achieve the horizontal configuration most low profile unit heaters use an in-shot burner, which requires the use of a power

FIGURE 17.6.4 Electric unit heater. Vertical and horizontal delivery units are available Finned-tube heating elements located in the air stream heat the room air blown through the unit by the propeller fans. Horizontal delivery units are equipped with adjustable louver-type air deflectors. Vertical delivery models may be used with or without air deflector devices (optional). Three air deflector assemblies are offered, each with its own distinctive air distribution pattern. Heavy-duty, long-heat-throw, horizontal delivery models are also available. (*Courtesy of Modine Manufacturing Co.*)

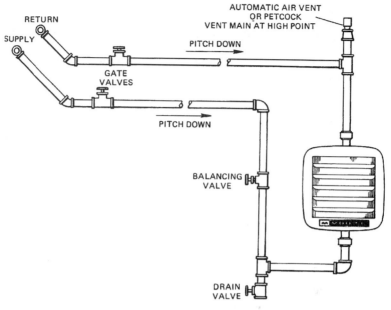

AUTOMATIC AIR VENT
OR PETCOCK
VENT MAIN AT HIGH POINT

RETURN

SUPPLY

PITCH DOWN

GATE
VALVES

PITCH DOWN

BALANCING
VALVE

DRAIN
VALVE

HORIZONTAL UNIT HEATER CONNECTED TO
OVERHEAD HOT WATER MAINS

FIGURE 17.6.5 Hot-water heating. Piping should be sized to adequately accommodate the required water flow. Valves and unions on supply and return piping facilitate the takedown of the unit when necessary without shutting down the system. Return lines should be pitched for drainage. (*Courtesy of Modine Manufacturing Co.*)

exhauster to evacuate the flue products. In addition to the commercial certification low profile unit heaters typically have residential certification for the smaller size units allowing them to be used in residential garages.

17.6.4 TYPICAL UNIT HEATER CONNECTIONS

These connections are illustrated in Figs. 17.6.5 to 17.6.9.

17.6.5 CALCULATING HEAT LOSS FOR A BUILDING

Prior to selecting a unit heater, first determine the heat loss for the area to be heated. For a more exact method of calculating the heating load, see Chap. 2.

1. Determine both the inside temperature to be maintained and the design winter outside temperature for the locality. The difference between these two temperatures is called the "design temperature difference."

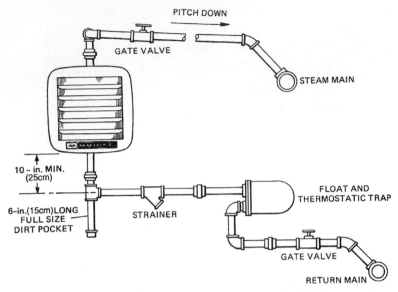

FIGURE 17.6.6   Steam heating. Pipe sizes should be adequate to handle both steam and condensate under maximum load conditions. Traps should be selected to handle the condensate when the unit heater is operating at maximum steam pressure and minimum entering air temperature. Provision should be made for elimination of entrapped air in system. (*Courtesy of Modine Manufacturing Co.*)

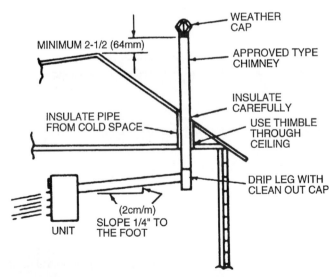

FIGURE 17.6.7   Gas heating. Piping should conform with local codes and requirements for type and volume of gas burned and pressure drop allowed in the line. Consult local codes for venting regulations. Vent pipe should be at least 6 in (15 cm) from combustible material and insulated at the junction with the roof or where the vent pipe passes through insulated spaces.

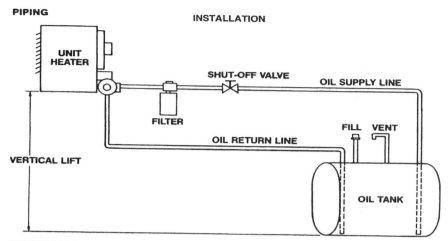

FIGURE 17.6.8   Oil heating. Single oil-fired unit heaters with two-stage oil pumps have a suction lift of 20 ft (6.1 m) on a two-pipe system and can be easily supplied from a separate tank. For multiple units, install a centralized oil distribution system with a separate booster pump connected to a common tank to feed fuel oil in either a looped or pressurized oil supply system. (*Note:* These circuits do not satisfy any particular local code requirements. Compliance with local codes is the responsibility of the installer.)

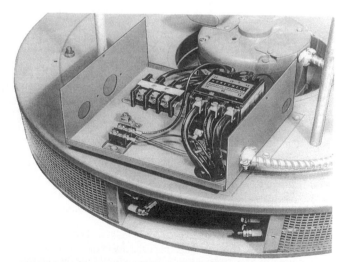

FIGURE 17.6.9   Electric heating. Wiring must be in accordance with the National Electric Code (ANSI CI) and applicable local codes. Power supply to unit heaters must be protected by a fused disconnect switch. Unit heater should not be located closer than 12 in (0.3 m) to combustible material. (*Courtesy of Modine Manufacturing Co.*)

2. Calculate net areas in square feet (meters) of glass, floor, walls, and roof exposed to outside temperature or unheated spaces. Include all door areas as glass.

3. Select overall heat-transfer coefficients from Table 17.6.1 (or appropriate sections of Chap. 2), and compute the heat-transfer loss for each type of area in Btu/h (watts) by multiplying each area by its heat-transfer coefficient and by the temperature differential.

4. Add 10 percent to the heat-loss figures for areas exposed to prevailing winds.

5. Calculate the volume of the room or area in cubic feet (cubic meters), and multiply by the estimated number of air changes per hour due to infiltration (usually one or two). Determine the number of cubic feet per hour (cubic meters per second) of air exhausted by ventilating fans or industrial processes. Substitute the larger of these two figures into the formula to determine the heat required to raise the air from outside to room temperature.

Infiltration

$$\text{Btu/h} = [\text{volume (ft}^3 \times \text{number of air changes per hour} \times \Delta\,\text{TF}°)] / 55$$
$$W = [\text{volume m}^3 \times \text{number of air changes per hour} \times \Delta\,\text{TC}°] / 3$$

For exhaust air

$$\text{Btu/h} = \text{cfm} \times 1.08 \times \Delta\,\text{TF}°$$
$$W = J/s \times 1.2 \times \Delta\,\text{TC}°$$

6. The totals of Btu/h (watt) losses from steps 3, 4, and 5 will give the total Btu/h (watts) to be supplied by unit heaters. [*Note:* If processes performed in the room release considerable amounts of heat, this amount should be determined as accurately as possible and may be subtracted from the total Btu/h (watt) loss if they are operated 24 h/day during the peak heating season.]

17.6.6 SELECTING UNIT HEATERS

Once the building heat loss and the heating medium to be used have been determined, other factors must be considered during the unit heater selection process. First, establish the final air temperature and air delivery rate desired in cubic feet per minute (cubic meters per second).

The final air temperature, as stated earlier, should be between 55 and 65°F (30.5 and 36.0°C) higher than the desired room temperature that will be maintained. By locating final air temperatures of unit heaters operating at standard conditions in the manufacturers' literature, computation of these values is unnecessary, unless the condition varies from the standard. Should this be the case, consult the manufacturer or the manufacturer's representatives.

The ratings in ft³/m (m³/s), which are also based on standard conditions, are included in the manufacturer's literature. Again, if the conditions under consideration are other than standard, the manufacturer should be consulted because the volume of air that the unit heaters must provide should be a known factor.

An important point to consider with every heating medium is the change in entering air temperature. This change affects the total heating capacity in most unit heaters and the final air temperature in all units. The differences in temperature of the air entering the unit and of the air being maintained in the heated area are important, especially regarding downblow units.[2] As entering air temperatures increase, the outlet air temperature also increases, which makes the exiting air more buoyant and thus more difficult for it to penetrate the occupied zone near floor level.

TABLE 17.6.1 Common Heat-Transfer Coefficients*†

Building material	U factor‡	Building material	U factor‡
Walls		Wood, 1-in	
Poured concrete, 80 lb/ft³§		(uninsulated) w/³/₈-in built-up roofing	0.48
8 in	0.25	w/1-in blanket insulation	0.17
12 in	0.18	Wood, 2-in	
Concrete block, hollow cinder aggregate		(uninsulated) w/³/₈-in built-up roofing	0.32
8 in	0.39	w/1-in blanket insulation	0.15
12 in	0.36	Concrete slab, 2-in	
Gravel aggregate		(uninsulated) w/³/₈-in built-up roofing	0.30
8 in	0.52	w/1-in insulation board	0.16
12 in	0.47	Concrete slab, 3-in	
Concrete block, w/4-in face brick		(uninsulated) w/³/₈-in built-up roofing	0.23
Gravel, 8 in	0.41	w/1-in insulation board	0.14
Cinder, 8 in	0.33	Gypsum slab, 2-in	
Metal (uninsulated)	1.17	(uninsulated) w/¹/₂-in gypsum board	0.36
w/1-in blanket insulation	0.22	w/1-in insulation board	0.20
w/3-in blanket insulation	0.08	Gypsum slab, 3-in	
Roofing		(uninsulated) w/¹/₂-in gypsum board	0.30
Corrugated metal (uninsulated)	1.50	w/1-in insulation board	0.18
w/1-in bolt or blanket	0.23	Windows	
w/1 ¹/₂-in bolt or blanket	0.16	Vertical, single-glass	1.13
w/3-in bolt or blanket	0.08	Vertical, double-glass, ³/₁₆-in air space	0.69
Flat metal		Horizontal, single-glass (sky light)	1.40
w/³/₈-in built-up roofing	0.90	Doors	
w/1-in blanket insulation under deck	0.21	Metal—single sheet	1.20
w/2-in blanket insulation under deck	0.12	Wood, 1-in	0.64
		2-in	0.43

*Heat loss through concrete floors on ground in large buildings may be ignored.
†Refer to *1985 ASHRAE Handbook, Fundamentals*, ASHRAE, Atlanta, GA, 1985, for expansion of this table.
‡Equivalent values for U factor: —Btu/(h)(ft²)(°F) = 5.675 W/(m²)(K) = U factor   1 K = 1°C   1 in = 2.54 cm = 0.254 m
§For metric equivalents, see appendix B.
Source: Modine Manufacturing Co. Used with permission.

Higher-velocity units and units with lower discharge air temperatures maintain lower temperature gradients than units with higher discharge temperatures. Valve-controlled or bypass-controlled units employing continuous fan operation maintain lower temperature gradients than units that employ intermittent fan operation. Lower temperature gradients lead to a general reduction in overall heat loss and improved comfort conditions. Directional control of the discharged air from a unit heater can also be an important factor in producing satisfactory distribution of heat and in reducing floor-to-ceiling temperature gradients.[2]

The use and character of the space to be heated influence the selection of the unit heater type(s) best suited to fulfill the heating requirement. Most industrial and commercial heating applications will benefit from the use of a combination of horizontal and vertical air delivery unit heaters. Each type differs principally in the method by which air and/or its direction (horizontal or vertical) is delivered.

The cost of equipment and its installation is less when fewer unit heaters are used, but comfort can be compromised. However, the ultimate system could utilize a greater number of units than might be economically practical. Generally, in large areas where maximum comfort is not necessarily the most important factor, it may be desirable to select the largest unit or units, such as in a warehouse-type building. In smaller areas or areas where the comfort of occupants is a prime consideration, specify more heaters of lower capacity.

A heated space occupied by a large number of people requires a greater number of air changes per hour than spaces occupied by relatively few people. As mentioned earlier, this should be considered when the heat loss is calculated for the space to be heated.

17.6.6.1 Mounting Height Is Critical

Improper mounting height is probably the cause of more unsatisfactory unit heater installations than any other single factor. When units are mounted higher than the maximum height recommended by the manufacturer, heated air may never reach the floor and thus drafts are created and fuel dollars are wasted. If units are mounted too low, hot blasts and excessive air velocities may cause discomfort to room occupants.

Horizontal delivery unit heaters usually are the best choice when ceilings are low and there are few obstructions in the path of air streams from the units. Standard louver-type air deflectors horizontally positioned in the air discharge openings of the unit heaters can be adjusted to direct the air up, down, or straight out (Fig. 17.6.2). If complete directional control of heated air is desired, add optional louvers vertically positioned in the air discharge openings to achieve the desired effect.

Vertical delivery units are often used without optional air deflector assemblies (Fig. 17.6.10). However, several types of air deflectors are generally offered to accommodate the need for air distribution patterns not attainable when vertical units are installed without them. These air deflectors are illustrated together with the heat-throw and heat-spread patterns they produce in Figs. 17.6.11 to 17.6.14.

The use of air deflector assemblies influences heat throw and the recommended mounting heights of units. Therefore, when vertical delivery unit heaters are equipped with air deflectors, follow the manufacturer's mounting height recommendations for the unit heater with the specific deflector.

When it is necessary to install units at low mounting heights, select models with lower ft³/min (m³/min) ratings, because the greater volume of air handled by larger units can create excessive air movement.

Better air distribution and economy of heating system operation is realized when a greater number of smaller unit heaters are used (instead of fewer but larger units).

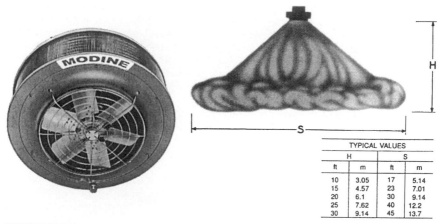

TYPICAL VALUES			
H		S	
ft	m	ft	m
10	3.05	17	5.14
15	4.57	23	7.01
20	6.1	30	9.14
25	7.62	40	12.2
30	9.14	45	13.7

FIGURE 17.6.10 Without deflector. Installed without an air deflector, a vertical delivery unit heater (*left*) has a heat spread similar to that shown at right. Where human comfort is not a major consideration and air control is not critical, vertical units are frequently installed without air deflectors. (*Courtesy of Modine Manufacturing Co.*)

When standard steam unit heaters are installed on high-pressure systems, the temperature of the delivered air may be too high for comfort. Furthermore, high-temperature air is very buoyant, making proper heat distribution difficult. Some manufacturers offer a line of low-outlet-temperature unit heaters designed primarily for use on steam systems operating at steam pressures of about 30 lb/in^2 (2 bar) or more. These units supply heated air at more normal temperatures and are recommended for use on systems operated at high steam pressures, particularly in warehouses and storage areas, where adequate air volume and air velocity may be more important than maintaining temperatures high enough for maximum comfort. Low-pressure steam and conventional hot-water units are usually selected for smaller installations and for those concerned primarily with comfort heating.

For extremely hard-to-heat areas in a building, such as around frequently opened large doors and in other locations having extreme exposure, use a heavy-duty unit heater with a high-velocity horizontal air stream. This type of unit is referred to as a "door heater." One such unit heater will often replace two or more smaller units. When a unit of this capability is installed, caution should be used in directing the high-velocity air stream away from room occupants. In addition, use blower-type gas unit heaters with nozzles equipped with adjustable air deflector blades to direct heat downward or obliquely into hard-to-heat areas.

17.6.7 NOISE LEVELS*

Noise levels in workplaces may be a concern in some applications. Wherever fans and motors move air, sound is produced. Of course, such is the case with unit heaters. Sound emissions of certain large unit heater models may limit their use in applications where sound-level requirements may be critical. In such instances, smaller models should be selected, which in total meet the heat load criteria of a larger single unit heater. The air velocity ratings are generally indicative of sound levels, i.e., the higher the ft^3/min (m^3/s) and ft/min (m/min) ratings of a unit heater, the greater the sound emission.

---

*See also Chap. 10

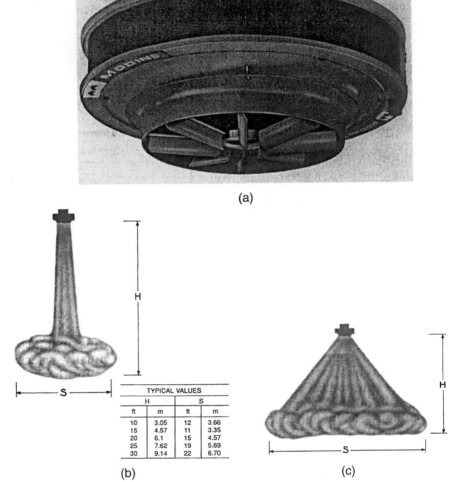

(a)

TYPICAL VALUES			
H		S	
ft	m	ft	m
10	3.05	12	3.66
15	4.57	11	3.35
20	6.1	15	4.57
25	7.62	19	5.69
30	9.14	22	6.70

(b)   **(c)**

FIGURE 17.6.11   With cone-jet deflector. (*a*) Adjustable louver blades permit a variety of air distribution patterns, from (*b*) a high-velocity beam of heated air into a confined space on the floor to (*c*) a broad gentle cone of air to cover a much broader area. This permits heater to be mounted higher when louver blades are positioned vertically. (*Courtesy of Modine Manufacturing Co.*)

The absence of a standard sound testing and rating code in the unit heater industry makes it impossible to compare manufacturers' catalogued sound ratings. Unit heater manufacturers either develop their own procedures or adopt one of several sound testing standards developed by independent testing laboratories or other industry groups such as American National Standards Institute (ANSI S1.21-1972).

Publication 303, Application of Sound Power Level Ratings, by the Air Movement and Control Association (AMCA) states:

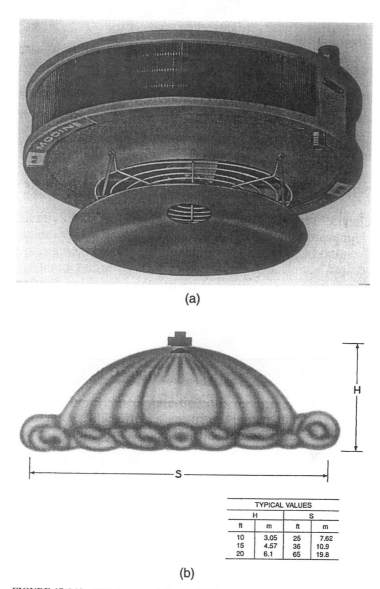

(a)

(b)

TYPICAL VALUES			
H		S	
ft	m	ft	m
10	3.05	25	7.62
15	4.57	36	10.9
20	6.1	65	19.8

FIGURE 17.6.12   With truncone deflector. (*a*) This cone-shaped air deflector may be raised or lowered to increase or decrease the heat spread of the unit heater. Generally it is used to provide (*b*) gentle air-flow patterns when unit heater is mounted low. (*Courtesy of Modine Manufacturing Co.*)

Sound measurements cannot be made as precisely as those used to establish air movement or heat-transfer ratings. Within the present state of the art, differences in sound power levels ±2 dB or less are not considered significant. In comparing products of different manufacturers, it is good practice to disregard differences of less than 4 dB. This is particularly true in the first octave band where the difference of ±6 dB or less should be disregarded.[3]

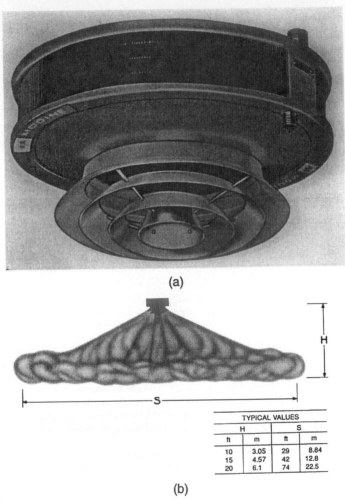

(a)

(b)

TYPICAL VALUES			
H		S	
ft	m	ft	m
10	3.05	29	8.84
15	4.57	42	12.8
20	6.1	74	22.5

FIGURE 17.6.13   With anemostat, (*a*) Comprising several concentric cones in fixed position, this assembly produces (*b*) an exceptionally wide heat spread at very low air velocities. For example, a unit heater equipped with a four-cone anemostat may be mounted as low as 7 ft (2.1 m). (*Courtesy of Modine Manufacturing Co.*)

Even though sound levels cannot be measured as accurately as air movement or as precisely as heat loss, the procedure for measuring sound is comparable to that for calculating the heat loss of a given space.

Wall, ceiling, floor surface, the kind of activity performed, and the type of equipment occupying the space all affect the heat loss. Similarly, these same factors affect the direct radiation of sound from its source to the human ear. All factors influence the sound absorption value of the space in which the unit heaters operate. Sound emissions from unit heaters are also affected by the type of mounting used to suspend them. Certain mounting

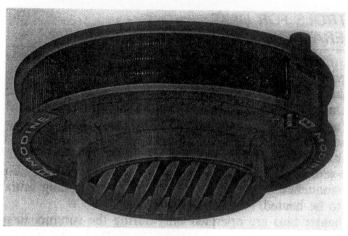

(a)

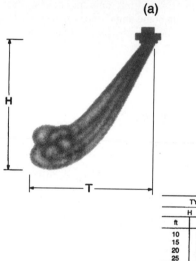

(b)

TYPICAL VALUES			
H		S	
ft	m	ft	m
10	3.05	28	8.58
15	4.57	44	13.4
20	6.1	58	17.7
25	7.62	72	21.9

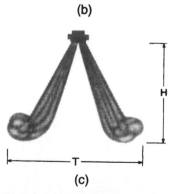

(c)

FIGURE 17.6.14 With louvers (*a*) Louver blades may be adjusted to discharge air directly downward, all air to (*b*) left or right of the unit heater, or by splitting the air stream, air may be discharged to (*c*) both left and right of the unit heater simultaneously. (*Courtesy of Modine Manufacturing Co.*)

arrangements will promote the transmission and amplification of operation noise and vibration through building structural members.

If sound criteria constitute a critical aspect in unit heater selection, complete details of the application requirement should be referred to the manufacturer of the unit heaters for evaluation and recommendation.

17.6.8 CONTROLS FOR UNIT HEATER OPERATION

The control device common to all unit heaters is the thermostat. It cycles the flow of water or steam to unit heaters on some hydronic systems on which fans operate continuously. On other hydronic systems where steam or water flow is continuous, the thermostat activates unit heater fans intermittently when heating is required.

Oil, gas, and electric unit heating systems rely on a thermostat to activate the burners or the electric heating elements. On some units, once the heat exchangers are hot, a heat-sensing or time-delayed switch activates the fan. This may be the most desirable sequence since it prevents cold air from being initially discharged into the space to be heated.

When unit heater fans are operated only during the summer to provide air circulation, a manual summer/winter switch is required to prevent the heater from delivering unwanted heat.

Unit heaters lend themselves ideally to zone heating. Equipment and installation costs can be reduced by using a single thermostat to control the operation of several unit heaters.

The controls for a steam or hot-water unit heater can provide either *on/off operation* of the unit fan or *continuous operation* of the fan. Continuous fan operation interrupts and lessens the vertical floor-to-ceiling air temperature stratification patterns which may occur.

One type of thermostat control used with unit heaters is designed to automatically return the warm air, which would normally rise to the ceiling, down to the occupied level. Two thermostats are required for this type of control arrangement.

The lower thermostat is placed in the zone of occupancy and is used to control the heating medium; it allows the unit heater to operate whenever the thermostat set point is not satisfied. The higher thermostat is placed near the unit heater at the ceiling or roof level, where the warm air tends to stratify. This thermostat allows the unit heater fan only to operate and deliver warm stratified air from the ceiling or roof level to the zone of occupancy, thus reducing heat stratification. The fan will continue to run until the set-point temperature at the higher level is reached.

Numerous control options can be used to improve unit heater performance. Adding controls to a heating system will increase equipment and installation costs. However, the convenience, refinements in comfort, and energy savings often outweigh the extra costs. For example, when a pilot flame outage occurs, it is very inconvenient to manually relight the pilot flame on a gas unit heater mounted 20 ft (6 m) above the floor. In this instance an intermittent pilot ignition system is well worth the extra cost and can save the gas otherwise consumed by conventional standing pilot systems.

Another example: Low-voltage wiring need not normally be enclosed in conduit. This can be a real savings. But when line voltage is used, circuit wiring *must* be in conduit.

Because of the wide range of controls available, control selection can be confusing. Therefore, it is recommended that a unit heater or control manufacturer be consulted on this specialized subject. In addition, refer to Chap. 7.

17.6.9   LOCATING UNIT HEATERS

The following observations may be helpful:

- Use as few unit heaters as possible to give proper heat coverage of the area. The number of units selected will depend on the heat throw or heat spread of the individual heaters.

- More than any other single factor, improper mounting height is responsible for most unsatisfactory unit heater installations. When unit heaters are installed at heights higher than recommended improper heat distribution results and comfort conditions are either difficult or impossible to maintain. But excessive air movement may cause discomfort when units are installed too low.

- Horizontal delivery unit heaters should be located so that the air streams of the individual units wipe the exposed walls of the building with either parallel or angular flow without blowing directly against the walls. Heaters should be spaced so that each supports the air stream from another heater. This sets up a circulatory air movement around the area to produce a blanket of warm air along the cold walls.

- It is advisable to locate unit heaters so that their air streams are subjected to a minimum of interference from columns, machinery, partitions, and other obstacles.

- Unit heaters installed in a building exposed to a prevailing wind should be located so as to direct a large portion of the heated air along the windward walls.

- Large expanses of glass, or large doors that are frequently opened, should be covered by long-throw unit heaters such as large horizontal delivery unit heaters or door heaters.

- In buildings having high ceilings, vertical delivery unit heaters equipped with the correct air distribution devices are recommended to produce comfort in central areas of the space to be heated. Horizontal delivery units are generally used to heat the peripheral areas of the same building.

- Horizontal delivery units should be arranged so that they do not blow directly at occupants. Their air streams should be directed down aisles, into open spaces on the floor, or along exterior walls of the building.

- When only vertical delivery units are used, they should be located so that exposed walls are blanketed by their warm air streams.

- To obtain the desired air distribution and heat diffusion, unit heaters are commonly equipped with directional outlets, adjustable louvers, or fixed or adjustable diffusers. For a given unit with a given discharge temperature and outlet velocity, the mounting height and heat coverage can vary widely with the type of directional outlet, adjustable louver, or diffuser.

- Several unit heaters may be operated by a single thermostat. In large, open spaces where similar activities are carried on, zonal heating will improve comfort and generally reduce fuel costs. Unit heaters may also be controlled individually, either manually or by a thermostat.

- For spot heating of individual spaces in larger unheated areas, single unit heaters may be used, but allowance must be made for unheated air from other spaces. The use of gas-fired infrared heaters is preferable for this situation. Other advantages of infrared heaters are that installation is easy, connection to the fuel supply is simple, installation costs are low, maintenance is inexpensive, and operation is noiseless. Infrared heating is fast and economical. Refer to Sec. 17.3 for additional information on infrared heaters.

Various spacing and arrangements of unit heaters for providing adequate heating coverage are demonstrated and described in Figs. 17.6.15 to 17.6.20. Installation costs will be simplified and reduced if unit heaters are properly placed in the space to be heated.

The following precautions should be taken:

- *Do not install gas- or oil-fired units in potentially explosive or flammable atmospheres laden with grain dust, sawdust, or similar airborne materials.* In such applications a blower-type heater is recommended in a separate room with ducting to the potentially explosive area. Room air should be separated from the unit by the use of a back draft damper.

- Consult piping, electrical, and venting instructions in the unit heater installation manual before installation.

- *Do not locate any gas- or oil-fired unit heater in areas where chlorinated, halogenated, or acid vapors are present in the atmosphere.*

- Unit heaters in an occupied zone [less than 7 ft (2 m) above the floor level] *must* have fingerproof guards for all moving parts (fans, belts, sheaves, etc.). High-temperature surfaces, such as flue pipes, must be insulated or protected by safeguards to prevent body contact.

- Installation of gas-fired unit heaters in high-humidity or saltwater atmospheres may cause accelerated corrosion, reducing the normal lifespan of the units.

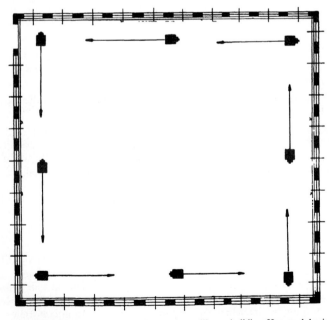

FIGURE 17.6.15 Locating unit heaters in a mill-type building. Here each horizontal delivery unit supports the air stream from another to produce circulatory air movement around the perimeter of the building where heat loss is greatest.

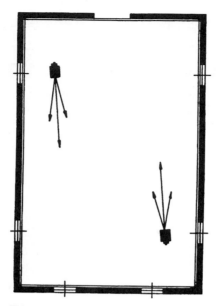

FIGURE 17.6.16 Locating unit heaters in a warehouse. Propeller unit heaters can provide maximum heat coverage of an area with a minimum number of units.

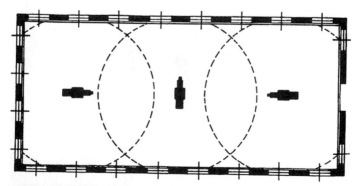

FIGURE 17.6.17 Locating unit heaters in a narrow building. An arrangement of vertical delivery unit heaters illustrates one of the advantages of a vertical (or down-blow) air stream. Notice how heat spread from one unit overlaps the spread of another and how together they blanket cold outside walls with warm air to counter heat loss.

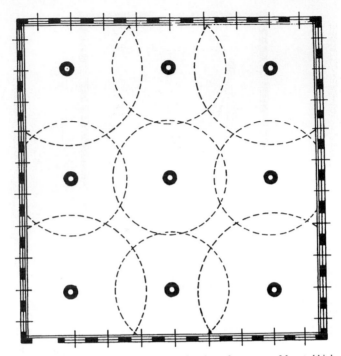

FIGURE 17.6.18   Locating unit heaters in a large factory area. Mounted high above machinery, assembly lines, and other obstructions on the floor, vertical delivery unit heaters beam heat directly downward to cover the entire floor area.

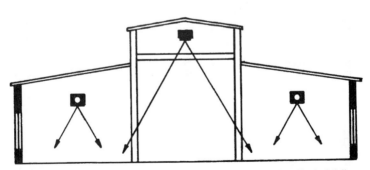

FIGURE 17.6.19   Locating unit heaters in a monitor-type building. Vertical delivery unit heaters installed in the high-ceiling central section of the building clear the craneway below them.

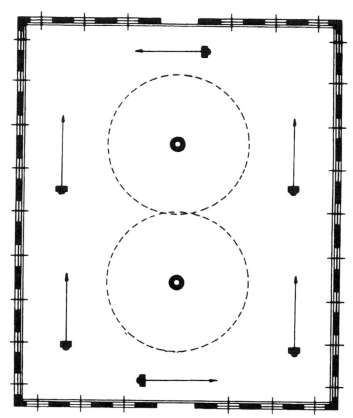

FIGURE 17.6.20 Locating unit heaters in a monitor-type building. In the same building, notice how horizontal delivery units in the two areas having low ceilings wipe outside walls with warm air.

17.6.10 REPAIRING VS. REPLACING UNIT HEATERS

1. Repairing a worn-out unit heater is costly. Labor and parts are not inexpensive. Repair the same unit twice, and chances are the price of a brand new unit heater will be exceeded, not including the added cost for takedown and reinstallation of the old unit heater each time it is repaired.

2. Repairs are unreliable. Repair a heat exchanger today, and perhaps next month, or on a cold day next winter, a motor or an overheat control on the same unit will need replacing.

3. It costs as much to reinstall a repaired unit heater as it does to install a new one. Labor costs can be a significant part of the bill. For a steam or hot-water unit, e.g., depending on the size and mounting height, it will take from 3 to 35 labor-hours to reinstall the unit. Add repair costs to installation costs, and the bill is substantial.

4. Replace an old, worn-out heater with a new unit heater, and many years of trouble-free service can be expected.

5. Replace an old "repairable" unit heater with a new unit, and get more usable heat from fuel dollars. New units conserve fuel and cut heating bills because they are more efficient and throw heat farther—new motors are more efficient and use electricity frugally.

6. Replace old depreciated units with new, more efficient types to conserve energy.

7. Replace old units with new units of an alternate fuel type when production losses occur due to interruptible fuel supply.

REFERENCES

1. 2002 *ASHRAE Handbook, HV Systems and Equipment*, American Society of Heating, Refrigerating and Air-Conditioning Engineers, Atlanta, GA.

2. Ibid., p. 28.7.

3. AMCA Bulletin 303, "Application of sound power level ratings for fans," Air Movement and Control Association, Arlington Heights, IL, June 1979, p. 1.

17.7 HYDRONIC CABINET HEATERS

Ronald A. Kondrat

Product Manager, Modine Manufacturing Co., Racine, Wisconsin

Hal M. Sindelar

Product Manager, Modine Manufacturing Co., Racine, Wisconsin

A cabinet heater is a specific type of unit heater which employs a centrifugal blower, a finned-coil heat exchanger, and an air filter. This package is housed in a "cabinet" which may be mounted on floors, walls, or ceilings. (See Fig. 17.7.1 which shows several types.) One of its chief benefits is that the cabinet heater can be used in building modernizations, where existing piping or wiring may be reused, to reduce construction costs.

The basic cabinet heater is primarily a heating-only appliance using 100 percent recirculated air. Modifications include cooling with chilled water (fan-coil units) and heating/cooling with 0 to 100 percent outside air (unit ventilators). Cabinet heaters may be exposed or recessed and can have a variety of air-flow arrangements through louvers, grilles, or ductwork.

17.7.1 CABINET UNIT HEATERS—HEATING ONLY

Figure 17.7.2 illustrates a floor-mounted cabinet heater with a pedestal mount. The steam/hot-water heating coil (1) consists of aluminum fins mechanically bonded to copper tubing. Air movement is via a centrifugal blower/fractional-horsepower motor package (2) mounted on a blower platform (5) which serves as the barrier between the positive and negative pressure sections of the core. The filter (3) can be disposable or a permanent, washable type. Air discharge is through the top louvers (4) and return air is through the pedestal (6). End compartments (7) provide space for piping connections and controls (8). Top-mounted doors (9) provide access to these compartments without removing the front panel.

17.7.1.1 Coil Types

Steam Coils. Steam coils are rated under standard conditions of 2 psig (13.8 kPa) steam pressure and 60°F (15.6°C) entering air temperature. Other steam pressure and entering air conditions may be extended from standard conditions as directed by the unit's manufacturer or by using classical thermodynamic principles. One method for testing air-heating coils is given in American Society of Heating, Refrigeration, and Air-Conditioning Engineers (ASHRAE) Standard 33–78.[1] Typical applications utilizing copper tubing are from 0 to 15 psig (0 to 103 kPa).

Hot-Water Coils. Standard ratings of hot-water cabinet heater coils are at the following conditions: 200°F (93.3°C) entering water temperature (EWT), 20°F (11.1°C) water temperature drop (WTD), and 60°F (15.6°C) entering air temperature (EAT). Capacities at various water flow rates and at entering air and water temperatures are extended from these conditions. Typical hot-water applications range from 160 to 24°F (71.1 to 115.6°C).

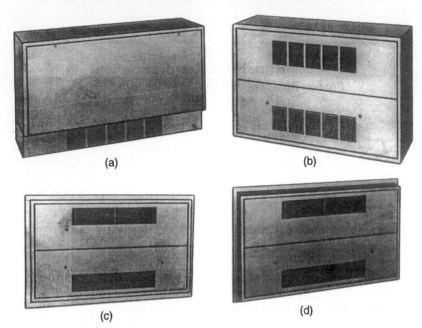

(a)   (b)

(c)   (d)

FIGURE 17.7.1 Cabinet unit heaters, (*a*) exposed floor unit, model C; (*b*) exposed wall or ceiling unit, model CW; (*c*) fully recessed wall or ceiling unit, with aluminum bar grille, model CW; (*d*) partially recessed wall or ceiling unit, with aluminum bar grille, model CW. (*Courtesy of Modine Manufacturing Co.*)

A common practice is to utilize a half-serpentine, single-row coil for both steam and hot-water use. This configuration has the benefit of minimal pressure drop across the coil in hot-water applications. Higher capacities require multiple-row, single-tube serpentine coils. This circuitry not only extracts the maximum amount of heat from the water, but also provides increased water velocity to ensure turbulent flow.

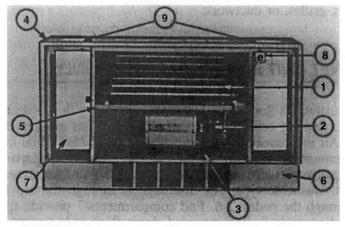

FIGURE 17.7.2 Model C cabinet unit heater. See main text for explanation of part numbers. (*Courtesy of Modine Manufacturing Co.*)

Finned tubular heater with standard terminals.

Finned tubular heater with standard terminals and bushings for duct mounting.

FIGURE 17.7.3   Finned tubular heaters. (*Courtesy of Wellman Thermal Systems Corporation.*)

Electric Coils.   Electric heating coil ratings are based on the energy input to the coil in kilowatts (1 W = 3.413 Btu/h). Capacities range from fractions of kilowatts to tens of kilowatts depending on the size and number of elements used. In the United States, single-phase voltages of 120, 208, 240, and 277 V are common for residential use, while commercial applications may require three-phase 208, 240, or 480 V.

As the primary heating mode, sheathed tubular elements may be inserted into finned copper tubes, much like a hydronic heating coil. This configuration provides greater heat-transfer surface. As an intermediate system, either straight or U-bent finned elements (Fig. 17.7.3) can be used in addition to hydronic coils.

17.7.2   FAN-COIL UNITS—HEATING AND COOLING

By varying the selection of coils, piping, and controls and with the addition of a condensate removal system, the cabinet heater may serve the dual purpose of heating and cooling. The control plant includes a water chiller, boiler, or other heat source, circulating pumps, and temperature controls.

The cooling capacities of fan-coil units are rated at 80°F (26.7°C) dry-bulb (DB) temperature, 67°F (19.4°C) wet-bulb (WB) temperature, 45°F (7.2°C) EWT, and a 10°F (5.6°C) water temperature rise. One method for testing and rating is given in American Refrigeration Institute's (ARI) Standard 440.[2]

Fan-coil units may use single or multiple coils and are adaptable to two- or four-pipe systems. Each combination has varying effects on initial costs, operational costs, and the amount of control provided.

17.7.2.1   Two-Pipe Single-Coil System

This system requires the selection of hot water or chilled water at the control plant. Advantages are lower installation costs and the ability to use lower hot-water temperatures. This is due to the oversized water coils that are required for the smaller temperature difference between chilled water and room air temperatures. Disadvantages include frequent changeovers during intermediate seasons, loss of control during mild weather when cooling is required during the day and heat in the morning and evening, and additional controls required to segregate chilled water from the boiler and hot water from the chiller.

17.7.2.2 Two-Pipe Double-Coil System

A common method of handling the intermediate seasons is to have an auxiliary electric heating coil which will provide for a small heating load while chilled water is available for cooling demand. Another variation is a dedicated chilled-water coil and a total electric heating coil.

17.7.2.3 Four-Pipe, Single- or Multiple-Coil System

This system eliminates the changeover difficulties of the two-pipe system by providing a continuous supply of both chilled and hot water. The coil configuration can be either two separate, dedicated coils or a split coil where two or three rows are used for cooling and a single row is used for heating.

Variations on this theme allow for cooling via chilled water or direct expansion and heating via hot-water, steam, or electric heating.

A typical control system for a chilled-water/hot-water four-pipe system would consist of a three-way motorized valve on the supply side of the coil and separate two-way motorized valves on the return side. A low-voltage heating/cooling thermostat with a fan control subbase would be used. On a call for heat, the three-way motorized valve opens to the hot-water supply, the hot-water return valve opens, the chilled-water return valve closes, and the blower is energized through a relay. As the room temperature increases to satisfy the thermostat, the hot-water return valve closes. With increasing room temperature, the three-way motorized supply valve moves from its normal position (open for hot water) to the cooling position (closed for hot water), the hot-water return valve is closed, and the chilled-water return valve opens. Blower operation can be either automatic or manual.

17.7.3 UNIT VENTILATORS—HEATING, COOLING, AND VENTILATING

The introduction of outdoor air through outside air dampers enables a fan-coil unit to heat, ventilate, or cool a room. Unit ventilators, sometimes called *classroom air conditioners*, are used in heavily occupied areas where a large amount of ventilation air is required. Damper operation can be either manual or motorized, with 0 to 25 or 0 to 100 percent of the unit's rated capacity introduced as outside air. Coils are sized to heat or cool the room loss plus ventilation air requirement.

The standard air rating of a unit ventilator is based on the delivery in cubic feet per minute (liters per second), converted to standard air at 70°F (21.1°C).

The three basic control cycles commonly used are cycle I, cycle II, and cycle III, which are briefly defined as follows:

Cycle I. Except during warmup, 100 percent outdoor is admitted at all times.

Cycle II. A minimum amount of outdoor air (normally 25 to 50 percent) is admitted during the heating and ventilating stage. The percentage is gradually increased to 100 percent, if needed, during the ventilation cooling stage.

Cycle III. Except during the warmup stage, a variable amount of outdoor air is admitted as needed to maintain a fixed temperature of the air entering the heating element. This is controlled by the air-stream thermostat, which is set low enough, often 55°F (13°C), to provide ventilation cooling when needed.

17.7.4 SELECTION

Selection of a cabinet heater to match the calculated room design loads requires the following information:

Heating: room heat load and heating medium to be used

In addition to the above input information, a manufacturer's ratings and specifications sheet is used (Table 17.7.1). The rating sheets provide unit capacities under the standard conditions listed previously in this chapter. Coil descriptions, blower configurations, motor information, and physical data are also provided. Note that sizes are based on nominal value in ft³/min (L/s), say 200, 400, etc.

17.7.4.1 Selection

Steam capacities for cabinet unit heaters are based on standard conditions of 2 psig (13.8-kPa) steam and a 60°F (15.6°C) EAT. Correction factors for other than standard conditions are listed in Table 17.7.2. Steam capacities are derived as follows:

$$\text{Unit capacity} = \text{capacity [at 2 psig (13.8 kPa) and 60°F (15.6°C) EAT]} \times \text{correction factor}$$

To simplify selection at nonstandard conditions, actual required capacities should be converted to the equivalent capacity at 2 psig (13.8 kPa) and 60°F (15.6°C) EAT. Selection is made from Table 17.7.3.

Determine the equivalent capacity at 2 psig (13.8 kPa) and 60°F (15.6°C) EAT as follows:

$$\text{Equivalent capacity} = \frac{\text{actual required capacity}}{\text{correction factor}}$$

The final air temperature should be greater than 100°F (37.8°C) to maintain a feeling of warmth. The final air temperature (FAT) at other than standard conditions is derived as follows:

$$\text{Final air temperature} = \frac{\text{actual capacity}}{1.085 \, (\text{air flow})} + \text{EAT}$$

where the air flow is in cubic feet per minute (liters per second).

TABLE 17.7.1 Ratings and Specifications

Unit size	2	3	4	6	8	10	12	14
(1) Heating capacity—hot water								
Total MBH								
Standard coil	12.4	22.1	28.1	44.3	48.8	55.7	72.3	79.2
High capacity coil	22.4	30.4	44.2	59.7	71.9	84.6	105.6	118.3
(2) Heating capacity—steam								
Total MBH—standard coil	19.8	28.3	42.5	52.1	67.7	76.0	95.4	102.6
Cond. Lb/Hr.	21.0	29.0	44.0	54.0	70.1	78.7	98.8	106.2
(3) Heating range—hot water								
MBH—standard coil	8.13	12.23	16.31	28.46	36.48	43.55	54.72	61.79
MBH—high capacity coil	16.24	21.31	28.44	36.60	54.72	65.84	81.105	91.118
(4) Heating range—steam								
MBH—standard coil @2 PSI, 60° EAT	14.20	18.29	29.43	42.52	52.67	60.75	75.95	81.102
Coil								
Rows/FPI								
Standard coil	$1/2$	$1/11$	$1/11$	$1/11$	$1/11$	$1/11$	$1/11$	$1/11$
High capacity coil	$2/14$	$2/14$	$2/14$	$2/14$	$2/9$	$2/9$	$2/9$	$2/9$
Face Area—Ft^2								
Standard coil	1.0	1.3	1.6	2.3	3.4	3.4	4.6	4.6
High capacity coil	1.1	1.5	1.8	2.7	3.6	3.6	4.8	4.8
Coil Conns.								
Standard coil	3/4" MPT	3/4" MPT	3/4" MPT	3/4" MPT	1" MPT	1" MPT	1" MPT	1" MPT
High capacity coil	5/8" IDS	5/8" IDS	5/8" IDS	5/8" IDS				
Blowers								
No./Dia./width in.	1/5 1/4/7	1/5 1/2/7	2/5 1/4/7	2/5 1/2/7	3/5 1/2/7	3/5 1/2/7	4/5 1/2/7	4/5 1/2/7
Drive	Direct	Direct	Direct	Direct	Direct	Direct	Direct	Direct
Speed RPM								
High	1090	1050	1090	1050	1090	1050	1090	1050
Medium	850	820	850	820	850	820	850	820
Low	650	625	650	625	650	625	650	625

CFM								
High	250	330	450	620	840	1050	1240	1430
Medium	195	265	350	485	655	820	970	1120
Low	150	195	270	370	545	685	805	930
(5) Motor HP	1/30	1/30	1/15	1/15	1/10	1/10	1/8	1/8
Volts/Phase/Hertz	115/1/60	115/1/60	115/1/60	115/1/60	115/1/60	115/1/60	115/1/60	115/1/60
FL Amps. Standard Shaded Pole	1.4	1.4	1.9	2.1	3.3	3.5	3.8	4.2
Motors								
FL Amps. option #140	.5	.6	1.3	1.3	1.8	1.9	2.6	2.6
FL Amps. option #141	3.2	3.2	3.2	3.2	6.4	6.4	6.4	6.4
FL Amps. option #141A	4.4	4.4	4.4	4.4	8.8	8.8	8.8	8.8
Filters								
Type Supplied	Permanent	Permanent	Permanent	Permanent	Permanent	Permanent	Permanent	Permanent
Width/Length/Thickness	8.5/21/.5	8.5/26/.5	8.5/31/.5	8.5/44/.5	11/50/.5	11/50/.5	11/62/.5	11/62/.5
Length/Height/Depth Model C, CW, CR	39/25/9.75	44/25/9.75	49/25/9.75	62/25/9.75	72/28/12	72/28/12	84/28/12	84/28/12
Weights								
Shipping WT.–Lbs.								
Model C	80	90	110	120	160	165	185	190
Model CW	90	100	120	130	170	175	195	200
Model CR	92	103	124	135	175	180	202	207

(1) Rating with 200° Entering Water Temperature, 60° EAT, and 20° TD at high fan speed.

(2) Rating with 2 lbs. steam and 60° EAT at high fan speed.

(3) Range with 200° EWT at low to high fan speed.

(4) Range at low to high fan speed.

(5) Sizes 2–6 have 1 motor. Sizes 8–14 have 2 motors.
Horse power and amperage ratings represent both motors for sizes 8–14.

Notes: MBH = 100 Btu/h.

See Appendix B for conversion factors.

Source: Modine Manufacturing Co. Used with permission.

TABLE 17.7.2 Conversion Factors for Steam

Steam pressure, lb/in²	Steam temperature, °F	Latent heat of Steam, Btu/lb	Entering air temperature, °F										
			0	10	20	30	40	50	60	70	80	90	100
0	212.0	970	1.450	1.368	1.274	1.190	1.110	1.030	0.970	0.880	0.808	0.734	0.659
2	218.5	966	1.494	1.408	1.318	1.232	1.153	1.074	1.000	0.924	0.852	0.780	0.708
5	227.2	961	1.550	1.464	1.376	1.288	1.211	1.130	1.057	0.980	0.907	0.838	0.767
10	239.4	953	1.627	1.542	1.456	1.370	1.289	1.210	1.136	1.057	0.986	0.917	0.848
15	249.7	946	1.690	1.608	1.523	1.437	1.356	1.278	1.200	1.122	1.050	0.979	0.908

Note: See Appendix B for metric equivalents.
Source: Modine Manufacturing Co. Used with permission.

TABLE 17.7.3   Steam Selection, Standard Coil, 2 psig, 60°F EAT

Unit size	ft³/min	r/min	MBtu/h	EDR*	FAT, °F	Condensate, lb/h
2	250	1050	19.8	82	133	21
	150	625	14.6	61	156	15
3	330	1050	28.3	118	139	29
	195	625	18.0	75	145	19
4	450	1050	42.5	177	147	44
	270	625	29.3	122	160	30
6	620	1050	52.1	217	137	54
	370	625	42.7	178	166	44
8	840	1050	67.7	282	134	70
	545	625	52.9	220	149	55
10	1050	1050	76.0	317	126	79
	685	625	60.6	253	141	63
12	1240	1050	95.4	398	130	99
	805	625	75.1	313	146	78
14	1430	1050	102.6	428	126	106
	930	625	81.7	340	141	85

*Equivalent direct radiation.
Note: For each size, the top row indicates high fan speed and the bottom row indicates low fan speed.
See Appendix B for metric equivalents.
Source: Modine Manufacturing Co. Used with permission.

EXAMPLE

Heating load, 29,000 Btu/h (8494 kW)
Entering air temperature, 60°F (15.6°C)
Entering steam pressure, 5 psig (34.5 kPa)

Solution   From Table 17.7.2 the correction factor is 1.057.

$$\text{Equivalent capacity} = \frac{29{,}000}{1.057} = 27{,}500 \text{ Btu/h} \ (8.054.8 \text{ kW})$$

Table 17.7.3 shows that selection of model 3 with 28,300 Btu/h is sufficient for application requirements.

$$\text{Actual capacity} = 28{,}300 \times 1.057 = 29{,}900 \text{ Btu/h} \ (8757.7 \text{ kW})$$

$$\begin{aligned}
\text{Condensate rate} &= \frac{\text{actual capacity}}{\text{latent heat of steam}} \\
&= \frac{29{,}900 \text{ Btu/h}}{961 \text{ lb/h}} = 31 \text{ Btu/lb}
\end{aligned}$$

$$\text{Final air temperature} = \frac{29{,}900 \text{ Btu/h}}{1.085 \ (330 \text{ ft}^3/\text{min})} + 60°F = 144°F$$

17.7.4.2 Hot-Water Selection

Hot-water heating capacities for cabinet unit heaters are based on standard conditions of 200°F (93.3°C) EWT, 60°F (15.6°C) EAT, and a 20°F (11.1°C) WTD. Correction factors for other than standard conditions are listed in Table 17.7.4

Hot-water capacities are derived as follows:

$$\text{Unit capacity} = \text{capacity at 200°F (93.3°C) EWT and 60°F (15.6°C) EAT}$$
$$\times \text{correction factor}$$

To simplify selection at nonstandard conditions, actual required capacities should be converted to the equivalent capacity at standard conditions. Then the model required can be quickly approximated from the capacity ranges in the rating and specification table (Table 17.7.1). Final selection is made from capacity tables, Tables 17.7.5 and 17.7.6. Capacities at 180°F (82.2°C) EWT and 60°F (15.6°C) EAT may be read directly from Tables 17.7.7 and 17.7.8.

EXAMPLE

Heating load, 27,000 Btu/h (7908.3 kW)

EAT, 60°F (15.6°C)

EWT, 210°F (98.9°C)

Water flow, 2 gal/min (0.126 L/s)

Solution   From Table 17.7.4 the correction factor is 1.071. Thus

$$\text{Equivalent capacity} = \frac{27,000}{1.071} = 25,200 \text{ Btu/h (7381.1 kW)}$$

Table 17.7.5 shows selection of model 4 with 25,800 Btu/h (7556.8 kW) is sufficient for application requirements.

$$\text{Actual capacity} = 25,800 \times 1.071 = 27,600 \text{ Btu/h (8084 kW)}$$

$$\text{FAT} = \frac{27,600}{450 \times 1.08} + 60°F = 117°F$$

17.7.4.3 Electric Heating Coils

Electric heating coils are chosen by matching the room heat loss with the equivalent Btu/h (watt) rating of the coil. Table 17.7.9 provides capacities for auxiliary electric heating coils.

TABLE 17.7.4 Hot-Water Correction Factors for Conditions Other than 200°F EWT and 60°F EAT

Entering water temperatures, °F	Entering air temperatures, °F										
	0	10	20	30	40	50	60	70	80	90	100
160	1.229	1.139	1.049	0.962	0.880	0.795	0.715	0.634	0.568	0.484	0.410
170	1.307	1.212	1.124	1.036	0.954	0.869	0.785	0.704	0.628	0.552	0.478
180	1.383	1.290	1.199	1.110	1.024	0.940	0.859	0.774	0.698	0.622	0.546
190	1.460	1.362	1.272	1.182	1.100	1.011	0.929	0.845	0.768	0.690	0.615
200	1.537	1.440	1.349	1.259	1.171	1.085	1.000	0.917	0.838	0.760	0.684
210	1.613	1.515	1.424	1.331	1.249	1.158	1.071	0.988	0.908	0.829	0.753
220	1.690	1.591	1.500	1.408	1.318	1.230	1.141	1.058	0.978	0.898	0.820
230	1.767	1.669	1.572	1.482	1.391	1.301	1.215	1.129	1.048	0.967	0.889
240	1.844	1.745	1.648	1.554	1.468	1.374	1.285	1.200	1.118	1.036	0.957

Note: See Appendix B for metric conversions.
Source: Modine Manufacturing Co. Used with permission.

TABLE 17.7.5 Heating Capacities for Standard Coil, 20°F EWT, 60°F EAT

Unit size	gal/min	Water pressure drop, ft	High fan speed				Low fan speed			
			ft³/min	MBtu/h	WTD, °F	FAT, °F	ft³/min	MBtu/h	WTD,* °F	FAT,† °F
2	0.50	0.1	250	10.7	44.0	99	150	7.8	31.9	111
	1.00	0.1		12.0	24.7	104		9.1	18.6	119
	1.28	0.1		12.4	20.0	105		9.5	15.3	122
	1.50	0.2		12.6	17.3	106		9.7	13.3	123
	2.00	0.2		12.9	13.3	107		10.2	10.5	127
3	1.00	0.1	330	19.9	36.8	110	195	12.2	25.1	117
	1.50	0.2		20.1	27.6	116		13.6	18.7	124
	2.23	0.3		22.1	20.0	121		14.9	13.8	130
	2.50	0.4		22.8	18.8	123		15.2	12.5	131
	3.00	0.5		23.0	15.7	124		15.6	10.7	133
4	1.00	0.1	450	20.1	41.3	101	270	16.1	33.1	115
	2.00	0.3		25.8	26.6	113		19.2	19.7	125
	2.87	0.5		28.1	20.0	117		20.5	14.7	130
	4.00	0.7		29.9	15.4	121		21.5	11.1	133
	5.00	0.9		31.0	12.8	123		22.1	9.1	135
6	2.00	0.3	620	36.9	36.9	115	370	28.2	28.2	130
	3.00	0.6		41.0	27.3	121		30.4	20.3	136
	4.43	0.9		44.3	20.0	126		31.8	15.9	139
	5.00	1.1		45.1	18.0	127		32.6	13.0	141
	6.00	1.4		46.3	15.4	129		33.2	11.0	143

	2.00	0.1		41.7	42.6	105		32.5	33.2	115
	4.00	0.2		47.5	24.2	112		35.5	18.1	120
8	5.00	0.4	840	48.8	20.0	113	545	36.2	14.8	121
	6.00	0.5		49.7	16.9	114		36.7	12.5	122
	8.00	0.9		51.3	13.1	116		37.2	9.5	122
	3.00	0.1		50.8	34.6	104		39.8	2.71	113
	5.00	0.4		54.9	22.5	108		42.7	17.5	117
10	5.10	0.5	1050	55.7	20.0	108	685	43.3	15.5	118
	7.00	0.7		57.1	16.7	110		44.5	13.0	119
	9.00	1.1		58.2	13.2	111		45.3	10.3	121
	4.00	0.2		68.1	34.8	110		51.4	26.3	118
	6.00	0.6		71.0	24.2	112		53.7	18.3	121
12	7.40	0.8	1240	72.3	20.0	113	805	54.7	15.1	122
	10.00	1.4		74.0	15.1	115		55.4	11.3	123
	12.00	1.8		74.9	12.8	115		55.9	9.5	124
	4.00	0.2		73.7	37.7	107		57.0	29.1	116
	6.00	0.6		77.2	26.3	109		59.4	20.2	118
14	8.10	1.0	1430	79.2	20.0	111	930	61.2	15.5	120
	10.00	1.4		80.5	16.5	111		62.0	12.7	121
	12.00	1.8		81.3	13.9	112		62.4	10.6	121

*Water temperature drop.
†Final air temperature.
Note: See Appendix B for metric conversions.
Source: Modine Manufacturing Co. Used with permission.

TABLE 17.7.6 Heating Capacities for High-Capacity Coil, 200°F EWT, 60°F EAT

Unit size	gal/min	Water pressure drop, ft	High fan speed				Low fan speed			
			ft³/min	MBtu/h	WTD, °F	FAT, °F	ft³/min	MBtu/h	WTD,* °F	FAT,† °F
2	1.00	0.7	250	20.1	41.3	134	140	13.5	27.8	148
	1.50	1.1		21.8	29.9	140		14.2	19.5	153
	2.00	1.7		22.7	23.4	144		14.5	15.0	155
	2.40	2.8		23.2	20.0	146		14.7	12.7	156
	3.00	4.3		23.8	16.4	148		14.9	10.3	158
3	2.00	1.4	325	31.4	32.1	149	195	21.2	21.7	160
	2.50	2.1		32.4	26.5	151		21.9	17.9	163
	3.00	2.8		33.0	22.5	154		22.3	15.2	165
	3.60	3.7		33.4	20.0	155		22.6	13.4	167
	4.00	4.6		33.8	17.3	156		22.9	11.7	168
4	2.00	1.5	440	41.9	42.9	148	270	28.7	29.4	158
	2.50	2.3		43.2	35.4	151		29.6	24.3	161
	3.00	3.0		44.1	30.1	152		30.2	20.6	163
	4.00	5.0		45.2	23.1	155		31.0	15.9	166
	4.70	8.8		45.8	20.0	156		31.4	12.7	168
6	2.00	1.4	615	49.8	49.8	135	370	35.9	35.9	149
	3.00	2.8		54.3	36.2	141		38.1	25.4	155
	4.00	4.8		56.8	28.4	145		39.3	19.6	158
	5.00	7.2		58.5	23.4	148		40.1	16.0	160
	6.00	10.1		59.7	20.0	150		40.6	13.5	161

8	5.00		1.0	67.0	27.4	134		50.9	20.8	147
	7.00		1.7	71.4	20.9	138		53.9	15.7	152
	7.40	835	1.9	71.9	20.0	139	540	54.4	15.0	153
	10.00		3.2	74.9	15.3	143		56.5	11.6	157
	12.00		4.3	76.5	13.0	144		57.6	9.8	158
10	6.00		1.4	79.8	27.2	131		62.2	21.2	144
	8.00		2.2	83.6	21.4	134		65.0	16.6	148
	8.70	1040	2.5	84.6	20.0	135	680	65.8	15.5	149
	12.00		4.3	88.0	15.0	138		68.1	11.6	152
	14.00		5.8	89.1	13.0	139		68.9	10.1	153
12	7.00		1.7	99.8	29.2	135		77.2	22.6	150
	9.00		2.6	103.5	23.5	138		79.9	18.2	153
	10.80	1220	3.6	105.6	20.0	140	790	81.5	15.4	155
	13.00		5.0	107.6	16.9	141		82.9	13.0	157
	15.00		6.5	109.1	14.9	142		83.9	11.4	158
14	9.00		2.6	114.2	25.9	135		87.8	20.0	148
	11.00		3.7	117.1	21.8	137		89.9	16.7	151
	12.10	1410	4.4	118.3	20.0	137	915	91.1	15.4	152
	15.00		6.5	120.2	16.4	139		92.4	12.6	153
	17.00		8.2	121.8	14.7	140		93.3	11.2	154

*Water temperature drop.

†Final air temperature.

Note: See Appendix B for metric conversions.

Source: Modine Manufacturing Co. Used with permission.

TABLE 17.7.7 Heating Capacities for Standard Coil, 180°F EWT, 60°F EAT

Unit size	gal/min	Water pressure drop, ft	High fan speed				Low fan speed			
			ft³/min	MBtu/h	WTD, °F	FAT, °F	ft³/min	MBtu/h	WTD,* °F	FAT,† °F
2	0.50	0.1	250	7.9	31.7	89	145	6.3	25.5	102
	1.00	0.1		10.4	20.8	98		7.8	15.7	112
	1.50	0.2		11.7	15.6	103		8.5	11.4	116
	2.00	0.2		12.5	12.5	106		9.0	9.5	119
3	1.00	0.1	330	14.4	28.8	100	195	12.3	24.6	118
	1.50	0.2		16.5	22.0	106		13.8	18.4	125
	2.00	0.3		17.9	17.9	110		14.7	14.7	129
	2.50	0.4		18.8	15.0	112		15.3	12.2	132
	3.00	0.5		19.5	13.0	114		15.7	10.5	134
4	1.00	0.1	450	17.9	35.9	97	270	14.5	28.9	109
	2.00	0.3		22.7	22.7	106		17.4	17.4	119
	3.00	0.6		25.1	16.7	111		18.7	12.4	124
	4.00	0.7		26.5	13.2	114		19.5	9.7	126
	5.00	0.9		27.5	11.0	116		20.0	8.0	128
6	2.00	0.3	620	31.2	31.2	106	370	23.9	23.9	119
	3.00	0.6		34.7	23.2	111		25.9	17.2	124
	4.00	0.8		36.9	18.4	115		27.0	13.5	127
	5.00	1.1		38.3	15.0	117		27.8	11.1	129
	6.00	1.4		39.4	13.1	118		28.3	9.4	131

8	2.00	0.1		35.0	42.6	98		27.3	27.9	106
	4.00	0.2	840	39.9	20.4	103	545	29.8	15.3	110
	5.00	0.4		41.0	16.8	105		30.4	12.4	111
	6.00	0.5		41.8	10.2	105		30.8	10.5	112
	8.00	0.9		43.1	11.0	107		31.3	8.0	112
10	0.50	0.1		7.9	31.7	89		6.3	25.5	102
	3.00	0.1		42.7	29.1	97		33.5	22.8	105
	5.00	0.4	1050	46.1	18.9	100	685	35.9	14.7	108
	5.70	0.5		46.8	16.8	101		36.4	13.1	109
	7.00	0.7		48.0	14.0	102		37.4	10.9	110
	9.00	1.1		48.9	11.1	102		38.1	8.7	111
12	4.00	0.2		57.2	29.3	102		43.2	22.1	109
	6.00	0.6	1240	59.7	20.3	104	805	45.1	15.4	111
	7.40	0.8		60.8	16.9	105		46.0	12.7	112
	10.00	1.4		62.2	12.7	106		46.6	9.5	113
	12.00	1.8		62.9	10.7	106		47.0	8.0	113
14	4.00	0.2		61.9	31.7	99		47.5	24.5	107
	6.00	0.6	1430	64.9	22.1	101	930	49.5	17.0	109
	8.10	1.0		66.6	16.8	102		51.0	13.1	111
	10.00	1.4		67.7	13.8	103		51.6	10.7	111
	12.00	1.8		68.3	11.6	104		52.0	8.9	112

*Water temperature drop.

†Final air temperature.

Note: See Appendix B for metric conversions.

Source: Modine Manufacturing Co. Used with permission.

TABLE 17.7.8 Heating Capacities for High-Capacity Coil, 180°F EWT, 60°F EAT

Unit size	gal/min	Water pressure drop, ft	High fan speed				Low fan speed			
			ft³/min	MBtu/h	WTD, °F	FAT, °F	ft³/min	MBtu/h	WTD,* °F	FAT,† °F
2	1.00	0.7	250	17.1	35.0	123	140	11.5	23.6	135
	1.50	1.1		18.6	25.4	128		12.1	16.6	139
	2.00	1.7		19.4	20.0	132		12.4	12.8	141
	2.50	2.9		20.0	16.4	133		12.6	10.4	143
	3.00	4.3		20.3	13.9	135		12.8	8.8	144
3	2.00	1.4	325	26.0	26.6	134	195	17.6	18.4	143
	2.50	2.1		26.8	22.0	134		18.1	15.1	145
	3.00	2.8		27.4	18.7	137		18.5	13.1	147
	3.50	3.7		27.8	16.2	138		18.8	11.5	148
	4.00	4.6		28.1	14.4	139		19.0	10.0	150
4	2.00	1.5	440	34.6	36.6	132	270	24.1	25.0	142
	2.50	2.3		35.7	29.4	134		24.6	20.7	145
	3.00	3.0		36.5	24.9	136		25.4	17.5	147
	4.00	5.0		37.4	19.2	138		25.7	13.4	149
	4.70	8.8		37.9	16.6	139		26.0	9.9	151
6	2.00	1.4	615	42.4	42.4	124	370	30.7	30.7	136
	3.00	2.8		46.4	30.9	129		32.6	21.7	141
	4.00	4.8		48.6	24.3	133		33.6	16.8	144
	5.00	7.2		50.0	20.0	135		34.3	13.7	145
	6.00	10.1		51.0	17.0	136		34.7	11.5	147

			835 / 1040 / 1220 / 1410				540 / 680 / 790 / 915			
5.00		1.0		56.3	23.0	122		42.8	17.5	133
7.00		1.7		60.0	17.5	126		45.3	13.2	137
7.40	8	1.9	835	60.9	16.7	127	540	45.7	12.6	138
10.00		3.2		62.9	12.9	129		47.5	9.7	141
12.00		4.3		64.3	11.0	131		48.4	8.2	142
6.00		1.4		67.1	22.9	119		52.3	17.8	130
8.00		2.2		70.3	18.0	122		54.6	14.0	134
8.70	10	2.5	1040	71.1	16.7	123	680	55.3	13.0	135
12.00		4.3		74.0	12.6	125		57.2	9.8	138
14.00		5.8		74.9	10.9	126		57.9	8.5	138
7.00		1.7		83.9	24.5	123		60.7	17.7	131
9.00		2.6		87.0	19.8	126		67.1	15.2	138
10.80	12	3.6	1220	88.7	16.8	127	790	68.5	13.0	140
13.00		5.0		90.4	14.2	128		69.7	11.0	141
15.00		6.5		91.7	12.5	129		70.5	9.6	142
9.00		2.6		96.0	21.8	123		73.8	16.8	134
11.00		3.7		98.4	18.3	124		75.6	14.0	136
12.10	14	4.4	1410	99.4	16.8	125	915	75.9	12.9	136
15.00		6.5		101.0	13.8	126		76.9	10.6	137
17.00		8.2		102.4	12.3	127		77.7	9.4	138

*Water temperature drop.
†Final air temperature.
Note: See Appendix B for metric conversions.
Source: Modine Manufacturing Co. Used with permission.

17.7.4.4   Cooling Coil Selection

EXAMPLE   Here 100 percent recirculated air is used. The application requirements for cooling are:

Total cooling load, 6700 Btu/h (1962.4 kW)

Sensible cooling load, 5400 Btu/h (1581.7 kW)

Indoor design conditions, 80°F (26.7°C) dry bulb and 67°F (19.4°C) wet bulb

Entering water temperature, 45°F (7.2°C)

Ventilation provided by infiltration

17.7.4.5   Cooling Selection

With 100 percent recirculated air, coil entering-air conditions will be essentially 80°F (26.7°C) DB and 67°F (19.4°C) WB. Referring to the cooling capacity in Table 17.7.10, choose size 3 and by interpolation find that 1.05 gal/min will provide the required 5400 Btu/h (1581.7 kW) sensible capacity at 80°F (26.7°C) DB and 67°F (19.4°C) WB conditions. Total capacity is 5500 Btu/h (1610.9 kW), which is inadequate. To provide 6700 Btu/h (1962.4 kW) total capacity requires 1.25 gal/min (0.079 L/s), which provides 6100 Btu/h (1786.7 kW) sensible capacity.

17.7.4.6   Heating, Cooling, and Ventilation

EXAMPLE   Here we use 25 percent outside air and 75 percent return air. The application requirements for cooling are:

Total cooling load, 23,750 Btu/h (6956.4 kW) (room only)

Sensible cooling load, 16,620 Btu/h (4868 kW) (room only)

Indoor design conditions, 80°F (26.7°C) DB, 67°F (19.4°C) WB

Outdoor design conditions, 95°F (35°C) DB, 75°F (23.9°C) WB

Entering water temperature, 40°F (1.7°C)

Ventilation air required, 200 ft³/min (94.4 L/s)

Outside air damper set at 25 percent outside air (oa) maximum, ft³/min (L/s)

TABLE 17.7.9   Heating Capacities of Electrical-Resistance Coil

Unit size	Rating, kW	Equivalent rating, MBtu/h	Max. unit amperage @115/60/1
2	1.0	3.4	10
3	1.5	5.1	15
4	2.0	6.8	20
6	2.5	8.5	25

Note: See Appendix B for metric conversions.
Source: Modine Manufacturing Co. Used with permission.

TABLE 17.7.10 Cooling Capacities for Two-Row Coil, 45°F EWT, High Fan Speed, Btu/h

Unit size	gal/min	Pressure drop, std. coil, ft	Entering air temperature, °F											
			75 DB, 63 WB			78 DB, 65 WB			80 DB, 67 WB			84 DB, 69 WB		
			Sens. heat	Total heat	H₂O temp. rise	Sens. heat	Total heat	H₂O temp. rise	Sens. heat	Total heat	H₂O temp. rise	Sens. heat	Total heat	H₂O temp. rise
2	0.5	0.80	2.62	2.62	10.5	2.89	2.89	11.5	3.07	3.07	12.3	3.43	3.43	13.7
	1.0	1.55	3.17	3.17	6.3	3.49	3.49	7.0	3.71	3.71	7.4	4.15	4.15	8.3
	1.5	3.20	3.52	3.52	4.7	3.88	3.88	5.2	5.07	5.89	7.8	5.66	6.58	8.8
	2.0	5.50	4.40	5.01	5.0	4.84	5.69	5.7	5.12	6.39	6.4	5.71	7.15	7.1
	2.5	8.30	4.42	5.35	4.3	4.87	6.08	4.9	5.15	6.84	5.4	5.75	7.64	6.1
3	1.0	1.70	4.46	4.46	8.9	4.92	4.92	9.8	5.23	5.23	10.5	5.84	5.84	11.7
	1.5	3.60	4.98	4.98	6.6	6.53	7.27	9.6	6.91	8.17	10.9	7.72	9.14	12.2
	2.0	6.02	5.98	6.98	7.0	6.58	7.93	7.9	6.97	8.91	8.9	7.78	9.96	10.0
	2.5	9.30	6.02	7.47	6.0	6.62	8.49	6.8	7.01	9.55	7.6	7.83	10.67	8.5
	3.0	13.00	6.04	7.91	5.3	6.65	8.99	6.0	7.04	10.11	6.7	7.86	11.29	7.5
4	1.5	4.00	7.25	8.04	10.7	7.87	9.07	12.1	8.12	10.11	13.5	9.08	11.20	14.9
	2.0	6.80	7.64	9.02	9.0	8.31	10.19	10.2	8.61	11.39	11.4	9.61	12.63	12.6
	2.5	10.00	7.93	9.73	7.8	8.64	11.00	8.8	8.97	12.30	9.8	10.01	13.66	10.9
	3.0	14.40	8.16	10.26	6.8	8.89	11.61	7.7	9.25	13.00	8.7	10.32	14.44	9.6
	4.0	24.50	8.47	11.01	5.5	9.25	12.46	6.2	9.65	13.97	7.0	10.76	15.54	7.8
6	2.0	1.84	9.50	10.99	11.0	10.43	12.32	12.3	10.75	13.68	13.7	12.05	15.10	15.1
	3.0	3.82	10.38	12.98	8.7	11.31	14.58	9.7	11.72	16.24	10.8	13.11	17.96	12.0
	3.5	5.10	10.67	13.68	7.8	11.63	15.38	8.8	12.08	17.14	9.8	13.50	18.97	10.8
	4.0	6.52	10.91	14.26	7.1	11.90	16.04	8.0	12.37	17.89	8.9	13.82	19.81	9.9
	5.0	9.85	11.28	15.15	6.1	12.31	17.06	6.8	12.83	19.04	7.6	14.33	21.10	8.4
8	4.0	1.00	16.03	18.36	9.2	16.80	20.80	10.4	17.30	23.25	11.6	18.60	25.50	12.7
	5.0	1.50	16.69	19.98	8.0	17.60	23.00	9.2	18.10	25.60	10.2	19.50	28.00	11.2
	6.0	2.20	17.21	21.22	7.1	18.20	24.60	8.2	18.80	27.40	9.1	20.50	30.00	10.0
	8.0	4.00	17.96	23.02	5.8	19.10	26.60	6.7	19.50	29.50	7.4	21.50	32.80	8.2

Note: See Appendix B for metric conversions.
Source: Modine Manufacturing Co. Used with permission.

TABLE 17.7.11 Cooling Capacities for Two-Row Coil, 40°F EWT, High Fan Speed, MBtu/h

Unit size	gal/min	Pressure drop, std. coil, ft	75 DB, 63 WB			78 DB, 65 WB			80 DB, 67 WB			84 DB, 69 WB		
			Sens. heat	Total heat	H$_2$O temp. rise	Sens. heat	Total heat	H$_2$O temp. rise	Sens. heat	Total heat	H$_2$O temp. rise	Sens. heat	Total heat	H$_2$O temp. rise
	0.5	0.80	3.02	3.02	12.1	3.28	3.28	13.1	3.46	3.46	13.8	3.82	3.82	15.3
	1.0	1.55	3.65	3.65	7.3	3.97	3.97	7.9	4.19	4.19	8.4	4.62	4.62	9.2
2	1.5	3.20	4.05	4.05	5.4	5.43	6.16	8.2	5.70	6.81	9.0	6.29	7.49	9.9
	2.0	5.50	5.06	6.02	6.0	5.49	6.69	6.7	5.77	7.39	7.4	6.36	8.13	8.1
	2.5	8.30	5.10	6.43	5.1	5.53	7.14	5.7	5.81	7.89	6.3	6.41	8.68	6.9
	1.0	1.70	5.14	5.14	10.3	5.60	5.60	11.2	5.90	5.90	11.8	6.51	6.51	13.0
	1.5	3.60	6.82	7.70	10.3	7.41	8.56	11.4	7.79	9.45	12.6	8.59	10.41	13.9
3	2.0	6.02	6.89	8.38	8.4	7.49	9.32	9.3	7.87	10.29	10.3	8.68	11.33	11.3
	2.5	9.30	6.94	8.97	7.2	7.53	9.98	8.0	7.92	11.02	8.8	8.73	12.13	9.7
	3.0	13.00	6.97	9.50	6.3	7.57	10.56	7.0	7.96	11.66	7.8	8.78	12.83	8.5
	1.5	4.00	8.08	10.09	13.5	8.69	11.13	14.8	8.93	12.19	16.3	9.87	13.30	17.7
	2.0	6.80	8.60	11.30	11.3	9.26	12.49	12.5	9.54	13.71	13.7	10.53	14.97	15.0
4	2.5	10.00	8.97	12.17	9.7	9.67	13.46	10.8	9.99	14.79	11.8	11.02	16.17	12.9
	3.0	14.40	9.26	12.82	8.5	9.98	14.19	9.5	10.33	15.61	10.4	11.39	17.08	11.4
	4.0	24.50	9.67	13.73	6.9	10.44	15.21	7.6	10.83	16.74	8.4	11.93	18.34	9.2
	2.0	1.84	10.67	13.68	13.7	11.49	15.03	15.0	11.79	16.41	16.4	13.07	17.86	17.9
	3.0	3.82	11.69	16.12	10.7	12.60	17.76	11.8	13.00	19.44	13.0	14.37	21.19	14.1
6	3.5	5.10	12.06	16.98	9.7	13.01	18.72	10.7	13.44	20.51	11.7	14.84	22.37	12.8
	4.0	6.52	12.37	17.69	8.8	13.34	19.50	9.8	13.80	21.38	10.7	15.23	23.33	11.7
	5.0	9.85	12.85	18.77	7.5	13.87	20.71	8.3	14.37	22.73	9.1	15.86	24.82	9.9
	4.0	1.00	17.94	22.98	11.5	18.30	25.40	12.7	19.00	27.70	13.8	20.50	30.00	15.0
8	5.0	1.50	18.80	24.98	10.0	19.60	27.70	11.1	19.80	30.05	12.0	21.70	32.80	13.1
	6.0	2.20	19.47	26.52	8.8	20.40	29.30	9.8	21.00	32.10	10.6	22.70	35.05	11.7
	8.0	4.00	20.44	28.70	7.2	21.50	31.75	7.9	22.30	35.00	8.7	24.00	37.80	9.5

Note: See Appendix B for metric conversions.
Source: Modine Manufacturing Co. Used with permission.

17.7.4.7 Cooling Selection

Since 25 percent outside air (oa) is specified, a size 8 fan coil [nominal 800 ft^3/min (377.5 L/s)] will be required to meet the 200 ft^3/min (94.4 L/s) ventilation requirement. Sensible heat (SH) gain from the outside air [abbreviated in what follows as SH (oa)] must be added to the room sensible heat load as follows:

$$SH \ (oa) = ft^3/min \ (oa) \times 1.08 \times (DB°F \ outside - DB°F \ inside)$$

$$SH \ (oa) = 200 \ ft^3/min \ (oa) \times 1.08 \times (95°F \ DB \ outside - 80°F \ DB \ room)$$

$$SH \ (oa) = 3240 \ Btu/h \ outside \ air$$

The total heat (TH) of the outside air is derived by

$$TH \ (oa) = ft^3/min \times 4.5 \ (Btu/lb \ outside - Btu/lb \ inside) \ ^*$$
$$= 200 \ ft^3/min \times 4.5 \times 7.0 \ Btu/lb \ oa$$
$$(total \ heat \ of \ outside \ air - total \ heat \ of \ room \ air)$$
$$= 6300 \ Btu/h$$

$$Unit \ sensible \ cooling \ load = 16,620 + 3240 = 19,860 \ Btu/h$$
$$Unit \ total \ cooling \ load = 23,750 + 6300 = 30,050 \ Btu/h \ ^\dagger$$

The conditions of the air mixture entering the coil are determined from a psychrometric chart to be 84°F (28.8°C) DB and 69°F (20.6°C) WB.

Referring to the cooling capacity table, Table 17.7.11, shows that the required sensible capacity at 84°F (28.8°C) DB and 69°F (20.6°C) WB will be obtained with 4.0 gal/min (0.25 L/s) [20,500 Btu/h (6004.4 kW)]. The total capacity is normally sufficient [30,000 Btu/h (8787.0 kW)].

17.7.5 APPLICATIONS

Due to the many configurations available, selecting a cabinet heater involves more than matching load requirements with capacities. Location is of prime importance in planning a project. Figure 17.7.4 shows some of the standard air-flow arrangements available. Some of these installation requirements also affect performance, e.g., mounting height and throw versus output.

Figure 17.7.5 indicates the performance of a single cabinet unit heater operating at high fan speed in a uniformly heated space at the single mounting height shown. Strong opposing drafts, large obstructions in the air stream of the unit, and higher than normal discharge air temperatures (resulting from high steam pressures) can prevent the heated air discharged by the cabinet unit from reaching the floor.

Under unfavorable conditions such as these, allowances must be made to ensure maintenance of desired comfort.

---

*The Btu/lb value is determined from a psychometric chart.
†For metric equivalents, use conversion factors given in Appendix B.

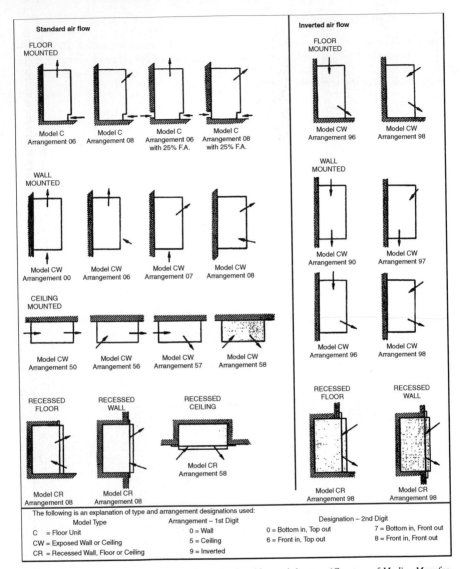

FIGURE 17.7.4 Air-flow arrangements for hydronic cabinet unit heaters. (*Courtesy of Modine Manufacturing Co.*)

Correction Factor R for Steam Units. These correction factors are to be used as multipliers to correct the recommended mounting heights and throw of cabinet units installed under conditions other than 2 psi (0.14 bar) steam pressure and 60°F (15.6°C) entering air.

Duct Applications. Recessed ceiling and wall units may be installed in a duct system. The external static pressure effect must be considered when the cabinet heater is sized. Table 17.7.12 provides conversion factors for hot-water applications. Information on high static motors is found on the ratings and specifications sheet, shown in Table 17.7.1.

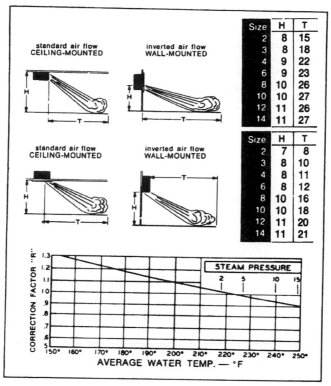

FIGURE 17.7.5   Mounting height and heat throw. (See Appendix B for metric conversions.) (*Courtesy of Modine Manufacturing Co.*)

TABLE 17.7.12   External Static Pressure,* Standard Coil

Unit size	0.0 ESP*		0.05 ESP*		0.1 ESP*		0.125 ESP*	
	ft³/min	Btu factor	ft³/min	Btu factor	ft³/min	Btu factor	ft³/min	Btu factor
2	250	1.00	210	0.90	170	0.80	150	0.77
3	330	1.00	280	0.90	230	0.80	200	0.77
4	450	1.00	400	0.90	325	0.80	285	0.77
6	620	1.00	540	0.90	430	0.80	380	0.77
8	840	1.00	720	0.90	590	0.80	510	0.77
10	1050	1.00	905	0.90	735	0.80	640	0.77
12	1240	1.00	1065	0.90	870	0.80	755	0.77
14	1430	1.00	1230	0.90	1000	0.80	870	0.77

*For 0.1 ESP and for 0.125 ESP, specify high static motor.
Note: See Appendix B for metric conversions.
Source: Modine Manufacturing Co. Used with permission.

REFERENCES

Note: Standards are continually updated. Refer to latest edition.

1. *Method of Testing for Rated Forced Circulation Air Cooling and Heating Coils under Defrosting Conditions*, ASHRAE Standard 33-78, American Society of Heating, Refrigeration, and Air-Conditioning Engineers, Atlanta, GA, 1984.

2. *Room Fan Coil Air Conditions*, ARI Standard 440, Air Conditioning and Refrigeration Institute, Arlington, VA.

17.8   HEAT EXCHANGERS

Susan Perdue Hammock
Alfa Laval Inc., Richmond, Virginia

17.8.1   INTRODUCTION

Heat exchangers have long been recognized as an effective means of transferring heat from one fluid to another while protecting delicate and costly HVAC system components from damaging contact with dirty and sometimes corrosive transfer fluids. Maintaining and operating the heat exchanger within its original design parameters and implementing an effective preventative O&M program will result in lower heat exchanger maintenance cost, better heat exchanger performance, reduced system down time, and lower overall HVAC system operational cost.

The guidelines that follow are intended to help maximize those results.

17.8.2   HEAT EXCHANGER OVERVIEW

Since its introduction in 1930, the plate heat exchanger (PHE) has gained wide acceptance as a viable means of transferring heat between liquids such as water to water and steam to water. In recent years its acceptance has spread to numerous fluid applications such as water to coolants in applications such as refrigerant and ammonium chiller systems. Typical applications include:

- Cooling Tower and Natural Cooling Water Isolation Applications where the PHE is used to transfer heat and isolate the HVAC system from the dirty, potentially fouling cooling liquids

- Free Cycle Cooling—Chiller Bypass Applications where the PHE is used to bypass the chiller, transferring heat from the condenser cooling media to a natural, free cooling source such as a local lake or river

- High Rise Pressure Interceptor Applications where the PHE is used in high rise building designs to create multiple pressure loops that allow the entire system to be designed for normal design pressures

- Thermal Storage Applications where the PHE is used in both the thermal charging and discharge modes

- Water Heating Applications where the PHE is used to transfer heat from an impure heat source to the pure water source

- District Heating and Cooling Applications where the PHE is used to reliably divide up the heating and cooling network with minimal energy loss

- Coolant Evaporation and Condensing Applications where the PHE is used as a compact means of enhancing refrigerant coolant evaporation and condensation

In all applications, the primary function of the plate heat exchanger (Fig. 17.8.1) is to create the optimum conditions for the transfer of heat between two fluids. This is accomplished by passing the fluids through a series of thin corrugated plates which are pressed

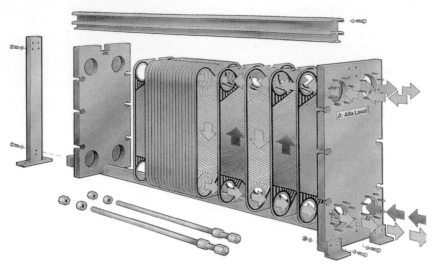

FIGURE 17.8.1   Typical plate heat exchanger. (*Courtesy of Alfa Laval, Inc.*)

together and sealed with gaskets in the traditional gasketed PHE, by welds and gaskets in a semiwelded PHE, or by brazing in a brazed heat exchanger (BHE). The fluids are pumped counter currently between the thin channels that are formed between the plates. Hot fluid flows on one side and the cold fluid on the other. The results are effective transfer of heat between the fluids through the plates without intermixing of the fluids.

The advantages of plate heat exchangers over other types of heat exchangers, such as shell-and-tube exchangers, include:

- *Compact heat exchange:* One-sixth the weight and one-third the foot print for the same heat load capacity.
- *Greater transfer efficiency:* High heat transfer coefficients can provide transfer efficiencies in excess of 95 percent and can achieve 1°F (0.6°C) degree approaches between the process fluids.
- *Flexible design:* Plates can be easily added to accommodate future expansions and changing load demands.
- *Low maintenance cost:* Fast and easy cleaning and change out of parts, if needed.
- *Lower initial cost:* Low capital equipment cost and two-thirds less valuable building space required.

These advantages account for the plate heat exchangers ever growing popularity in HVAC and refrigeration-type applications.

If properly operated and maintained, plate heat exchangers can provide many years of trouble-free operation. The keys to this success lie in properly selecting and designing the exchanger for the application, operating it within its design limits, and properly maintaining it. *A deficiency in any one of these areas will result in reduced thermal performance, reduced operating life, and needless repair, maintenance, and operational cost.* The impact can range from an increase in operating and maintenance cost to full system failure, loss of building comfort, and sometimes loss of valuable refrigerated products.

17.8.3 PHE AND SYSTEM OPERATION

All equipment should be operated and maintained in accordance with the instructions provided by the original equipment supplier. A preventative maintenance program that incorporates those recommendations is the most cost effective long-term means of realizing low maintenance and operating cost. It also minimizes overall system on or operational cost by maintaining the exchanger and associated system components operating at peak performance levels.

17.8.3.1 Start

Before initial start-up of the HVAC system, it (including the interconnecting piping) must be flushed of all foreign matter prior to final tie in of the exchanger and operation of the system. The same holds true if the system has been worked on or modified after initial start-up.

The high turbulence design of the plate heat exchanger resists fouling between the plates. However, a plate heat exchanger will filter out any solids that are larger than the spacing between the plates. Thus, any large materials left in the system will likely end up being filtered out by the heat exchanger.

Depending on the type and quantity of material, the consequences of using the exchanger as a filter can vary from no operational impact to total system performance loss. The loss of performance is typically due to the loss of heat transfer area resulting from the plugging of the inlet ports and/or the heat exchanger. Tell tale signs of pluggage include a lack of or reduction in thermal performance, high or increased pressure loss across the exchanger, no or reduced flow through the exchanger, or a combination of all three. The system should be designed with an in-line basket filter or inlet port filter to prevent any large debris from entering the exchanger.

All systems should be equipped with pressure surge suppression devices and slow operating valves to protect the heat exchanger and other system components from the potentially devastating effects of pressure shocks. During start-up, fluid systems need to be charged slowly to prevent pressure shocks from occurring. Water hammer must be completely avoided. External leaks, reduction of thermal performance, internal leaks, and inter-mixing of fluids are indicators of pressure surge damage.

Prior to start-up, the isolation valve on the inlets to the heat exchanger must be closed completely, and those on the outlets and the exchanger vents, completely opened. Also, inside distance between the exchanger pressure plate and the frame plate, the "A" dimension, should also be checked and verified against the name plate data or with the manufacturer (Fig. 17.8.2).

Unless specified otherwise by the manufacturer or the system operating procedures, the low pressure side of the exchanger should be charged first, followed by charging of the high pressure side. Pumps should be started slowly and the exchanger inlet valves opened slowly after pump start-up to prevent pressure shocks.

The outlet vents should be left opened until all the air in the side of the exchanger which is being charged is vented. Afterwards, the vent is closed and charging of the other side of the exchanger is completed in the same manner.

17.8.3.2 Shut Down

Prior to shut down, review the system shut down procedures and those provided by the exchanger manufacturer. Be sure to establish whether one side should be shut down before

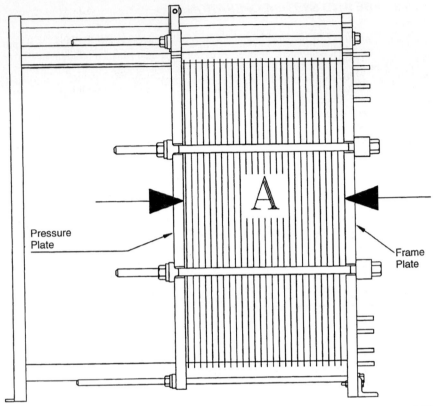

FIGURE 17.8.2   The "A" Dimension, the distance between the pressure and frame plate. (*Courtesy of Alfa Laval Inc.*)

the other. Slowly close the exchanger inlet valve or the pump discharge valve for the side of the exchanger which will be shut down first. After the valve is completely closed, shut the pump down. Do the same for the other side.

If the exchanger is to be shut down for more than a few days, it should be drained. Depending upon the process fluids, it may be advisable to rinse and dry the exchanger as well.

Before draining, close the exchanger outlet valves to prevent the loss of fluid downstream of the valve. Once the outlet valve is closed, open the exchanger vent followed by slowly opening the PHE drain valve. If the unit is not equipped with drain valves, draining can be accomplished by disconnecting the lower PHE flange connection. The PHE vent connection can be used to introduce rinse water and dry hot air sequentially, when rinsing and drying is recommended.

17.8.3.3   Normal Operation

During normal operation, the heat exchanger must also be protected against large solids and pressure surges. Solids build-up results in a loss of transfer surface and an associated loss of thermal performance. Pressure surges will result in damaged plates, gaskets, weld seams, and/or brazed seams which in turn result in loss of thermal performance, leaks, and potential intermixing of fluids.

In open systems, such as cooling tower and cooling water isolation applications, the cooling liquid is typically dirty and may contain high, unpredictable solids levels and debris. To prevent loss of exchanger performance, the heat exchanger should be protected against large solids through the use of an in-line strainer or similar device. The strainer should be installed upstream of the heat exchanger to capture any large solids prior to introduction to the exchanger.

Any strainers or filters need to be inspected and cleaned on a regular basis (Fig. 17.8.3 a, b, c). Some strainers are supplied with automatic flushing mechanisms which are either electronically or pneumatically operated. The mechanism periodically activates to flush the strainer with a small amount of process water to remove accumulated solids, to limit the pressure drop build-up in the strainer, and to maintain its overall effectiveness.

In high-solids applications, heat exchanger back flushing valves are also used to back flush the exchanger. The back flushing valve is hard-piped into the exchanger's inlet and outlet piping. It is used to reverse the flow of liquid through the exchanger, on an "as needed" basis, to flush out any materials that may have accumulated in the unit.

Pressure surges are one of most devastating operation upsets. As with any piece of equipment, the heat exchanger must be protected from these high-energy surges. *Unless properly protected, the plates and gaskets in an exchanger can be irreparably damaged.* Internal and external leakage of fluids may occur, as well as cross contamination of fluids due to interleakage. Both conditions may require that gaskets, welded cassettes, or whole-brazed units be replaced. Plates can also be deformed and require replacement. In extremely severe situations, the complete plate pack may need replacement.

Operational changes must be made slowly and smoothly. Sudden starting and stopping of pumps and the quick opening or closing of valves can send instantaneous high-energy shocks through the system. Quickly closing valves downstream of the exchanger can also send devastating shock waves back through the system. These shock waves can damage the heat exchanger and other system components.

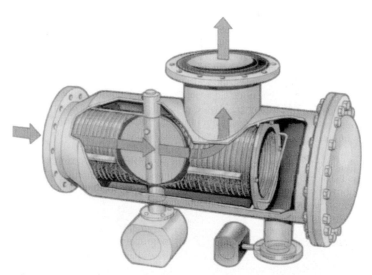

FIGURE 17.8.3a   In-line strainer with automatic flushing and backflushing system. Normal operation. (*Alfa Strainer, courtesy of Alfa Laval, Inc.*)

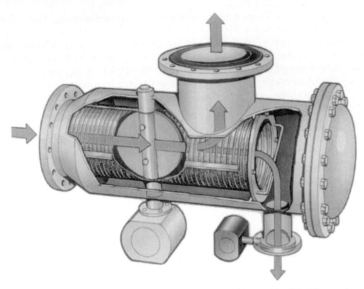

FIGURE 17.8.3b   In-line strainer with automatic flushing and backflushing system. Flushing. (Alfa Strainer, courtesy of Alfa Laval, Inc.)

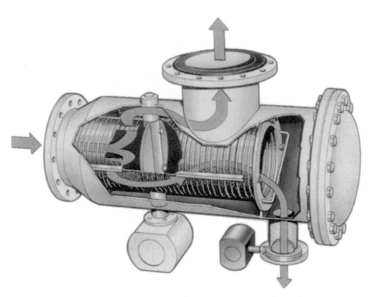

FIGURE 17.8.3c   In-line strainer with automatic flushing and backflushing system. Back-flushing. (Alfa Strainer, courtesy of Alfa Laval, Inc.)

Pressure surges can also occur when using steam, as in water heating applications. If the steam outlet of the heat exchanger is located on the top of the heat exchanger or if a steam trap has been installed above the elevation of the exchanger steam outlet, condensate will accumulate in the exchanger. When it does, severe and repeated hammering of the system and the exchanger occurs. The results are the same as with any other form of high pressure surge; plate deformation, gasket blowout, and leakage.

For steam applications, the steam outlet of the heat exchanger should always be located on the bottom of the exchanger and the outlet piping sloped away from the outlet to facilitate condensate drainage. A steam trap should also be located below the elevation of the exchanger outlet, to trap the condensate and avoid its backup into the exchanger.

17.8.4 PHE MAINTENANCE

A heat exchanger is a static device which *cannot* create pressure, pressure surges, energy, temperature, plugging solids, or scale. If properly designed for the service and operated within its original design parameters, it will provide many years of trouble free, high efficiency service. When it is abused and operated outside of the conditions it was designed for, problems may occur. When problems occur, it is important that the source of the problem be identified and eliminated as quickly as possible. Doing so will minimize the impact of the problem and prevent it from growing into a bigger, more costly problem.

In addition to the operating procedures outlined above, simple preventative techniques such as periodic visual inspections, performance monitoring, and an appropriate cleaning regimen can further help extend the operating life of the exchanger.

17.8.4.1 Inspection and Monitoring

The exchanger and the area immediately adjacent to it should be inspected periodically for external signs of problems. Leaks, drips, and/or dry solids accumulated on the outside of the exchanger or the surrounding area are indications that there may be problems with the sealing system, the gaskets, the welded seams or the brazed seams.

The performance of the unit should also be monitored. Flow, pressure, and temperature gauges or monitors should be installed in the inlet and outlet piping of the exchanger. These gauges should be checked on a regular basis and any changes noted. Changes in exchanger flow rates, temperature differences, and changes in pressure are primary indicators that a problem may be developing. An operational log should be maintained, so trends and operational changes in pressure, temperature, and flow rate can be readily spotted.

Any problems or suspected problems should be reviewed with the exchanger manufacture or a qualified PHE service organization. This review should include the operating history of the exchanger, the overall system and the specifics of the performance variance. All are necessary to help identify the cause of the change and to develop recommendations to prevent the cause from happening again. Once the cause is isolated, an action plan needs to be developed and implemented to eliminate the problem and to repair any associated damage.

17.8.4.2 Cleaning

Depending upon the nature of the transfer fluids and the application, performance of the heat exchanger may degrade over time. This decline in performance is typically due to the build-up of scale, sediment, and/or biological mass on the plates. Fouling of the exchanger manifests itself as a decrease in thermal performance, an increase in pressure drop across the exchanger, and/or a reduction in the flow through the exchanger.

Two methods are currently available for cleaning the exchangers: the plates are either removed from the exchanger and mechanically cleaned or the plates are chemically cleaned while still installed in the exchanger. The first method can be done either at the site by or under the guidance of qualified field service personnel or off-site in a qualified PHE service center. Either way, the plates are removed from the exchanger and are cleaned external to the system.

An effective alternative to external cleaning is chemically cleaning the plates while installed in the exchanger. Cleaning in Place (CIP) is an economical method of maintaining the exchanger at peak performance and extending its operating life. It is recommended that regular CIP cleaning be included in a preventative maintenance program to maximize exchanger performance and minimize system operational cost and overall maintenance cost.

Under no circumstances should hydrochloric acid be used to clean stainless steel plates, nor should hydrofluoric acid be used to clean titanium plates. If these acids are used on these plates, the plates will be corroded and need to be replaced. Cleaning agents containing ammonia and organic acids such as nitric acid must not be used to clean copper-brazed plate exchangers. Doing so will result in corrosion and pitting of the brazed joints and ultimate failure of the exchanger.

Quality water of known makeup should be used in preparing all cleaning agents. Water with a chlorine content of 300 PPM or higher must not be used.

In all cases, care must be taken to properly dispose of all materials used in the cleaning process, which can sometimes complicate on-site cleaning.

17.8.4.3   On-Site Cleaning

On-site cleaning can also be cumbersome, messy, and not necessarily the safest or most reliable alternative. The cleaning as well as the disassembly and reassembly of the exchanger should be done under the direction of a qualified field service engineer. If not, it should be done by a service organization which is specifically qualified in plate exchanger cleaning. The leading exchanger manufacturers provide such services and should be consulted. They can tailor a service program to best suit your needs.

Nonmetallic brushes, high pressure washing, and various cleaning agents can be used to clean the plates on-site. The combination used will depend on the nature and degree of fouling. Common cleaning agents for encrusted scales and sedimentation include:

- Hot water
- Nitric, sulfuric, citric, or phosphoric acid
- Complexing agents such as EDTA or NTA
- Sodium polyphosphates

It is recommended that the concentration of these agents not exceed 4 percent and that a maximum temperature of 140°F (60°C) be used.

For biological growth and slime, alkaline cleaning agents such as sodium hydroxide and sodium carbonate are usually effective. Recommended maximum concentrations and temperatures for these agents are, respectively, 4 percent and 176°F (80°C). Cleaning can sometimes be enhanced by the addition of small quantities of complex foaming agents or surfactants.

Care must be taken during the cleaning process *not to damage the gaskets*. All gaskets should be thoroughly inspected after cleaning and any damaged gaskets replaced. If more than a couple of gaskets need to be replaced, *all* gaskets should be replaced to assure uniform gasket hardness, sealing force, and extended operation. If the exchanger has been in operation for a number of years and/or the unit is opened frequently, all the gaskets should be replaced at the same time.

For glue-free, clip-on, or snap-on gaskets (Fig. 17.8.4) make sure that the gasket groove under the gasket is free of any foreign matter before installing the plates back in the unit. Regardless of gasket type, after hanging the plates in the exchanger, the gasket (and the groove it will seat in) should be wiped down prior to tightening of the exchanger to remove any foreign matter that may have gotten on the gaskets and/or grooves during hanging.

17.8.4.4 Off-Site Cleaning

A recommended cleaning alternative to on-site cleaning is to have the plates cleaned off-site in a qualified PHE service center. The well-qualified service centers provided by leading PHE manufacturers are suitably equipped and staffed to cost-effectively clean, inspect, and replace, as necessary, plates and gaskets. In those situations where the plates and gaskets need to be replaced, the manufacture's service centers can also provide plate and gasket failure analysis and subsequent recommendations on how to improve long-term operation of the exchanger.

Qualified service centers should be used whenever gasketed plates are regasketed. Nondestructive gasket removal techniques, such as the use of low-temperature nitrogen, are used to remove the gasket without damaging the valuable plates. Once the gaskets are removed, a combination of high-pressure washing and chemical baths are used to clean the plates.

Using dye penetration techniques, the centers then inspect the plates for pin holes and cracks that would not otherwise be detected. Plates passing the dye penetrant inspection are fitted with gaskets using oven-cured glues and two-part epoxy to assure a high quality, high

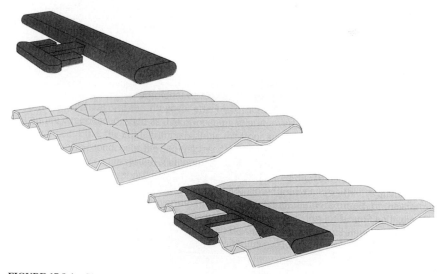

FIGURE 17.8.4   Glue free, clip-on gasket. (*Courtesy of Alfa Laval, Inc.*)

strength fit. The reconditioned plates are shipped from the centers, in plate pack order to facilitate reinstallation in the frame and minimize the time required to do so at the site.

Where access has been provided for removal of the heat exchanger, the need to remove and reinstall the plates in the frame can be eliminated by sending the entire exchanger to the service centers for reconditioning and repair. The exchanger is inspected, disassembled, plates reconditioned, and the exchanger reassembled. The exchanger can also be hydrotested at the center prior to it being shipped back to the site.

17.8.4.5   Cleaning in Place (CIP)

An alternative to external plate cleaning is on-site cleaning of the plates while still installed in the exchanger. CIP is a procedure used to clean the heat exchanger while it is installed in the system without having to open the exchanger. CIP procedures are particularly effective in maintaining the performance of brazed exchangers and need to be performed before the exchanger is highly fouled. During CIP cleaning, specific chemicals at prescribed temperatures, typically 140 to 170°F (60 to 77°C), are circulated at low velocities through the fouled side of the exchanger to chemically clean the unit of fouling deposits. Sometimes CIP methods are used to clean entire systems.

CIP cleaning is achieved through a number of mechanisms, namely:

- Chemical or partial dissolving of the deposit through chelation or chemical dissolution
- Decomposition of the deposit containing sand, fiber, or other foreign particles
- Reducing the adhesive force between the deposit and the plate
- Killing the biological mass present in the mass

CIP cleaning is accomplished at low velocities, typically 10 to 20 percent of the exchanger's normal operating velocity. Thus it is important to place the exchanger back in service shortly after CIP cleaning to completely flush the unit of loosened deposits. If the exchanger is placed in stand-by service, the deposits loosened during CIP may harden and reattach to the plates. In that case, the full benefit of the cleaning will not be achieved.

When there is only an occasional need to clean the exchanger a portable CIP unit can be rented from some of the leading PHE manufacturers (Fig. 17.8.5). For installations where

FIGURE 17.8.5   Mobile CIP unit. (*Courtesy of Alfa Laval, Inc.*)

there are a large number of installed exchangers or where regular cleaning is required, purchase of a suitably sized portable unit should be considered.

The CIP unit is connected to the inlet and outlet piping of the exchanger after shut down, isolation, and drainage of the exchanger. The cleaning solution is circulated through the exchanger in reverse flow to the operating flow direction, at a prescribed temperature, until the scale has been loosened and removed. The type, concentration, and temperature of chemicals to be used are dependent upon the type of scale to be removed.

To achieve the maximum benefit from CIP, the fouling deposit and circulating fluid should be analyzed to determine the best chemistry to be used. This analysis and CIP technical support is provided by some of the leading exchanger manufacturers. These manufacturers also provide CIP chemicals, CIP units, and field service specialists to assure that the methods and chemistry to be used are right for the application and exchanger.

17.8.4.6 Exchange Plates and Start Packs

A time-saving external cleaning alternative that is offered by some of the leading exchanger manufacturers is an exchange plate program. In the exchange plate program, reconditioned plates of the same model and materials as those in the exchanger are sent to the site and swapped out for the existing plates. The existing plates are then sent to the local center for cleaning and replenishing of the exchange plate inventory. In this approach, downtime is limited to the time it takes to change out the plate pack.

In critical applications where the time to receive the exchange plates cannot be tolerated, it may be advisable to have a "Strat Pack," plate pack onsite. A Strat Pack is made up of enough plates to change out the complete plate pack for one of the critical heat exchangers at the site. By having the Strat Pack on site, the pack can be changed out as soon as the need arises and the existing plates cleaned later at a more convenient time.

17.8.5 PHE GASKET AND PLATE PROTECTION

If operated within design conditions and not abused by improper operation and maintenance, an exchanger should provide many years of trouble-free, high-efficiency heat transfer. A heat exchanger should *never be opened except for repair or cleaning*. The fewer times it is opened the longer the gaskets will last before needing to be replaced and the lower the potential the plates will be damaged.

17.8.5.1 Gaskets

In most cases, external leaks are a sure sign of gasket failure. Intermixing of transfer fluids is a possible sign of gasket failure, but because of the design of the gasket system, it more commonly results from a plate problem. The most common causes for gasket failure include:

• Gasket blow-out
• Crushing
• Permanent set in old gaskets

In all three failure modes, the affected gasket(s) need to be replaced. In the case of gasket blow-out and crushing, the cause of the failure needs to be established and eliminated. Gasket blow-out is commonly caused by the exchanger not being tightened enough or by pressure surges. Both causes must be eliminated to protect the exchanger and other system components.

Gasket crushing can be caused by swelling of the gasket or plate misalignment. If swelling is suspected, compatibility of the gasket material with the transfer fluid should be checked to make sure that the right gasket material is being used. The composition of the transfer fluid should be checked against the design specs to establish whether there have been any changes. Small changes in some components, even trace amounts, can result in swelling and failure of the gasket.

Crushing of gaskets due to plate misalignment is typically caused by failure to install the plates correctly after cleaning, damaged hangers positioning the plate incorrectly during tightening and/or installation of deformed plates. The first can be avoided by taking care during the installation process, using qualified personnel to reassemble the exchanger.

During regasketing and prior to reinstalling any plates, the hangers should be inspected. Any damaged hangers need to be repaired or replaced. Care should also be taken not to damage the hangers during the hanging of the plates and exchanger tightening process.

Plate deformation, particularly in or around the gasket grove, can cause crushing of adjacent gaskets, plate misalignment, and leakage. In most cases plate deformation cannot be repaired and the plate and affected gaskets have to be replaced.

With time, all gaskets take on a permanent set, which causes a reduction in sealing force. The decline in sealing force usually takes place slowly, over many years. This decline in sealing force is a result of natural oxidation and/or chemical attack. It is also dependent upon the system operating conditions—pressure, temperature, and thermal cycling. The greater the extreme, the quicker the decline.

Gasket service life varies with each application. Replacement of the gaskets should be considered every time an exchanger is opened, particularly in high temperature, high pressure applications, if the exchanger has been in operation for a number of years, or the exchanger is frequently opened.

Because of the difference in gasket sealing force between an old and a new gasket, the mixing of old and new gaskets should be avoided.

17.8.5.2 Plates

Intermixing of transfer fluids is often a sign of plate failure. The most common forms of plate problems include:

- Corrosion
- Erosion
- Fatigue
- Deformation

Corrosion is caused by the presence or concentration of corrosive components in the transfer fluids. Examples are: chlorine attack of stainless steel, fluoride attack of titanium, and ammonia attack of copper in a copper-brazed heat exchanger. The most common forms of corrosion are pitting, crevice, and stress.

The rate of corrosion is primarily dependent upon the concentration of the corrosive component, the fluid pH, and the fluid temperature. Plate corrosion is generally more predominant in the high-temperature areas of the plate such as in the inlet and flow distribution area.

Chloride corrosion is directly proportional to the chloride concentration, the fluid temperature, and inversely proportional to the fluid pH. Chloride corrosion of stainless steel becomes more aggressive with the accumulation of materials on the plate such as calcium carbonate and silica scale. As the oxygen levels under fouling deposits decrease, the corrosion rate accelerates.

Erosion of plates can take place if high concentrations of an abrasive solid, such as sand, is present in the transfer fluids. Tell-tale signs of erosion are a smooth, shiny finish on the plates, typically in the inlet neck and distribution area. In severe cases the inlet port of the plate can be eroded away, which can lead to leakage.

The best way to prevent erosion is to eliminate the abrasive solids from the transfer fluid using a strainer or similar device. If erosion is encountered, the fluid velocity through the inlet port and neck area of the plate should also be reviewed by a qualified service engineer. Adjustments in the fluid flow rate may be necessary to reduce the velocity in these high-velocity areas by either reducing the fluid flow and/or by increasing the number of plates in the exchanger.

Repeated flexing of the plate will result in fatigue cracks and inter-leakage of the transfer fluids. The most common cause fatigue cracking is due to subjecting the plates to frequent and regular pulsations. Fatigue-causing pulsations are typically associated with piston and chemical dosage pumps or reoccurring water hammer. The best means of preventing this form of fatigue is by protecting the exchanger and other system components from the pulsations.

Another potential cause of fatigue in plates is tightening of the exchanger to an improper "A" dimension, the inside distance between the frame plate and pressure plate of the exchanger (refer back to Fig. 17.8.2). Plates are designed to have metal-to-metal contact once tightened to the proper "A" dimension. This metal-to-metal contact strengthens and helps support the plate to minimize flexing and vibration during operation and assure proper sealing by partially encapsulating the gaskets on gasketed exchangers. Whenever the exchanger is opened, it is important that the plate pack be retightened to the proper "A" dimension designated by the manufacturer. If the plates are too loose, premature failure of the plates and blow-out of gaskets resulting in leaks are likely results.

It is just as important not to over tighten the exchanger. Tightening the plate pack down to a dimension which is shorter than 1 percent below the specified "A" dimension may result in gasket crushing and plate deformation. Leaks may result and plates and gaskets will likely need to be replaced.

Care must also be taken when reinstalling plates to make sure that the plate hanger has not been damaged and the plates hang properly in the frame. The hanger serves to position the plate so that it seats properly against the next plate's sealing surface. In doing so, the plates and gasket are suitably positioned and compressed preventing plate deformation and leaks.

Another cause of plate deformation is high-pressure surges such as water hammer. A sudden high-pressure surge can render a plate useless. The best prevention is to protect the exchanger and other system components from pressure surges and pulsations, as noted earlier.

Once deformed, a plate is rarely salvageable and needs to be replaced. Exceptions are plates manufactured from high cost, exotic materials such as titanium. Depending upon the extent of the damage, it may be cost effective to have the plates repressed. Due to the high cost of repressing, it is typically less expensive to replace standard type 304 and 316 stainless steels with new plates than to have them repressed. This option and the cause of the deformation should be discussed with the exchanger manufacturer and the cause of the deformation eliminated.

There is no reliable means of repairing corroded, eroded, or fatigued plates. The only way to handle this kind of plate problem is to eliminate the cause of the problem and replace the damaged plate(s). If plates are replaced without identifying the cause and correcting it, money should be budgeted for the future replacement of the plates.

17.8.6 SHELL AND TUBE HEAT EXCHANGER OPERATION AND MAINTENANCE

In many ways the operation and maintenance procedures of a shell and tube (S&T) heat exchanger are similar to those of a PHE. Both are started-up, operated, and shut-down in the same manner and the same operating precautions such as protecting the exchanger from pressure shocks apply. As with a PHE, a S&T exchanger should not be operated outside the operating conditions originally specified without prior consultation with the exchanger manufacturer. In all cases, the S&T exchanger should be operated and maintained in accordance with the instructions provided by the original equipment supplier.

17.8.6.1 Operation

During start-up, the flow of the coldest fluid should be established first, after which the hot fluid should be gradually introduced to prevent thermal shock. All vents should remain open until all air has been purged from both sides.

In shutting the exchanger down, the flow of the hot fluid should first be reduced gradually, followed by the systematic shutdown of the cold side. The speed with which each side is started-up or shutdown depends upon the nature of the fluids involved, operating conditions, and the differential temperatures and pressures.

If the fluids are corrosive or susceptible to freezing, the exchanger should be drained completely after shutdown. All fluids drained from the exchanger should be recovered, stored, or disposed of properly.

Prior to start-up and during operation, the external bolts should be checked to assure they are tightened to the proper specification. Bolted joints should be tightened uniformly in a diametrically staggered pattern to the torque values specified by the original equipment manufacturer. Overtightening may damage the gaskets and result in leaks.

17.8.6.2 Maintenance

The interior and exterior of the exchanger should be inspected on a regular basis. The exterior shell, flanges, bonnets, and hub should be inspected for signs of damage, such as dents, bulges, stress marks, and corrosion. Gaskets should be checked for leaks and displacement. Instrumentation should be checked and operating conditions noted for comparison with the previous reading. All variations and problems should be noted and corrected.

Interior inspections should be attempted only after the unit has been fully drained and allowed to cool. A visual inspection of the tube ends for thinning, nozzle threads for damage, and erosion of the tubes should be done during the interior inspection. One or all of these conditions could be a indication of high fluid velocities, excessive particles entrained in the fluid and/or corrosion. Depending upon the damage found, the unit may need to be replaced.

Where removable bundles are used, removal and replacement of the damaged bundle should be done by a qualified service organization. Similarly, the detection and repair of tube joint leaks and tube splits as well as the internal cleaning of the exchanger should be done by a qualified service organization.

Mechanical cleaning of the tubes may be required to restore the operating efficiency of the exchanger. Rotary, nonmetallic, electric, or pneumatically driven brushes should be used to mechanically loosen tube deposits and scale. Use of metallic brushes may damage the interior surface of the tubes, which could accelerate tube corrosion and failure. Once brushed, a tube should be flushed with clean water to remove lose deposits and scale.

During the reassembly of the exchanger, all gaskets should be replaced and all bolts tightened per the manufacturer's recommendations. Care should be taken to ensure that all gaskets and seals are properly positioned and that the bolts are not overtightened, to prevent gasket crushing and leaks.

In some applications, the exchanger can be chemically cleaned without having to disassemble the exchanger. The CIP techniques described above for a PHE can also be used on S&T exchangers. CIP cleaning should be performed on a regular basis prior to the tubes becoming heavily fouled and in accordance with the recommendations provided by the CIP specialist.

The use of in-line strainers and back-flushing valves as described above for PHEs may also have a beneficial effect on S&T operation and maintenance. Each should be evaluated on a case-by-case basis and incorporated into the overall system layout and design.

17.9 RADIATORS FOR STEAM AND HOT WATER HEATING

Mike O' Rourke

Business Unit Manager, Sterling Hydronics, Westfield, Massachusetts

17.9.1 INTRODUCTION

Comfort heating with steam or hot water from boilers is known as "hydronic" heating. The most common form of hydronic heating uses heat distributing units in each room to warm the occupants. Free-standing cast iron radiators (Fig. 17.9.1), once the main type of radiation, can still be found but have been supplanted by finned tube radiation. Typically shorter in length, installed either free standing or recessed in the walls, radiators distributed heat to the room by both radiation and convection. Made of thick metal castings, and containing relatively large quantities of water, they continued to supply heat to the space between pump on times.

Finned tubes and convectors, typically mounted inside an enclosure, provide heat by convection currents passing over the fins. (Fig. 17.9.2) These units contain less water and mass than radiators and so respond to heat requirements more quickly. Since they generally occupy more wall area than radiators, the heat output is spread more evenly throughout the room. Using modern control designs, such as continuous circulation and outdoor reset controllers, continues to offer the comfort and economy of the original systems.

Additional updating to the original cast iron radiator also occurred in Europe in the form of steel panel radiators, which are now also manufactured in North America (Fig. 17.9.3). They are similar to the old style of radiators, have a large surface area, and provide much of their heat through radiation as well as convection. These radiators come in two different forms. The pressed steel type radiators are formed from sheet metal, which is stamped to provide water channels, and two halves are pressed and spot-welded together. Typically these radiators are much lighter in mass and react more quickly to changes in temperature than cast iron radiators. The second form of steel radiator is the steel flat tube panel radiator using tubes that have been shaped into a flat surface. These tubes contain much less water than a typical cast iron tubular radiator and are slim in profile, usually from 2 to 4 inches in depth. Much like fin-tube radiators, this second form of radiator can be spread over much more of the wall area allowing more heating throughout the room.

17.9.2 HEATING ELEMENTS

Common to modern hydronic heat distributing equipment is the element. Elements consist of tubing or pipe with attached fins which serve to increase the heat transfer area. The size and construction of the element is matched to the housing and application.

For pressed steel or flat tube panel radiators there is usually no additional enclosure required, except in cases where surface temperatures maybe a concern. The element is in the enclosure, as was often the case for cast iron radiators.

Residential baseboard elements generally consist of $^1/_2$ in (1.25 cm) to 1 in (2.5 cm) tubing with aluminum fins. These units are rated for hot water systems and generally

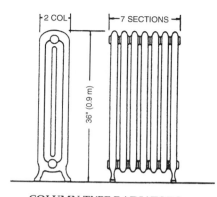

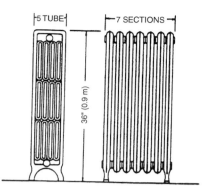

COLUMN TYPE RADIATORS   **TUBE TYPE RADIATORS**

FIGURE 17.9.1   Typical cast iron radiators.

FIGURE 17.9.2   Finned-tube element in slope-top cover.
(*Courtesy of Sterling Hydronics.*)

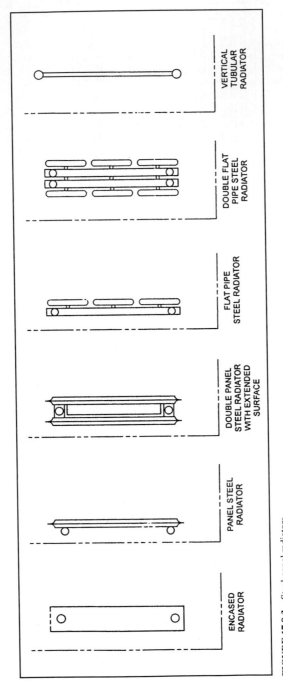

FIGURE 17.9.3   Steel panel radiators.

contain only one element. Low in height, the housings covering these elements are surface mounted next to the floor.

Cabinet convectors (Fig. 17.9.4) use convector elements (Fig. 17.9.5) which consist of multiple, small diameter $3/8$ (1 cm) to $1/2$ in (1.25 cm) tubes and multiple aluminum fins arranged vertically. Convector elements use either cast bronze or cast iron headers to connect to the tubing of the coil. Used due to space considerations, the housings may be surface mounted or recessed.

Finned tube convector elements (Fig. 17.9.6) are characterized by larger diameter tubing or pipe, usually $3/4$, 1, or $1^1/4$ in (2, 2.5, 3 cm) for copper tubing and $1^1/4$ or 2 in (2.5 or 5 cm) for iron pipe with standard lengths available up to 8 or 12 ft (2.4 or 3.7 m). Other distinguishing characteristics are a higher mounting height and higher output ratings. These units are rated for use with steam with conversion tables available to determine heating capacities with water (Table 17.9.1).

Fin dimensions vary in height and depth; the larger ones are used for greater heat output, and the narrower are useful where a slim enclosure is desired.

Fin spacing on the tube varies from 24 to about 60 fins per foot (79 to 197 fins per meter), with the larger quantity usually providing more heat transfer at greater material cost. In some cases, such as a low enclosure, the larger number of fins may produce a lower output because of resistance to air flow.

17.9.3 ENCLOSURES

Residential enclosures, referred to as baseboard, are generally available in standard lengths between 2 and 10 ft (0.6 to 2.5 m) and range in height from $6^7/8$ in to 11 in (7.5 to 28 cm). The output from these units ranges from a low of 400 to a high of 900 Btu/h ft.

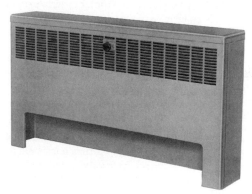

FIGURE 17.9.4   Cabinet convector. (*Courtesy of The Hydronics Institute.*)

FIGURE 17.9.5   Convector elements. (*Courtesy of Sterling Hydronics.*)

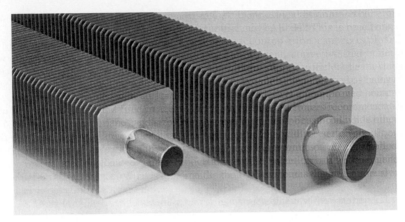

FIGURE 17.9.6   Finned-tube elements.

(35 to 80 W/h m). With a preferred mounting location under windows or on an outside wall, the units are unobtrusive and some are equipped with dampers to regulate the output (Fig. 17.9.7).

Cabinet convectors as the name implies function within metal enclosures of varying dimensions and outlet configuration. Common to these units is the mounting location of the convector low and horizontal in the cabinet a fact that the heat output benefits greatly from the "chimney effect" of the cabinet. Since the height depth, length, and outlet configuration determine the output, specific design considerations need to be looked at to select one of these units.

Commercial finned tube enclosures. Rated output for these units ranges from a low of 990 Btu/h ft (88 W/m) to a high of 4110 Btu/h ft (366 W/m). For surface-mounted units, the slope-top enclosure (Fig. 17.9.8) is popular because it discourages sitting or blocking with packages. Other familiar production models are the flat-top front outlet (Fig. 17.9.9), flat-top top outlet (Fig. 17.9.14), and expanded metal cover for industrial applications (Fig. 17.9.10). Some models are available with dampers for temperature control by the room's occupants.

Steel enclosures, either in enamel or prime-coat finish, are available in 16 or 18 gauge (1.52 or 1.2 mm). They may be supported by the same bracket as the heating element or by separate back plates or hanger strips.

Standard enclosures are generally open at the bottom and have an outlet at or near the top. In some cases a grille is available for the bottom inlet, and it is particularly useful where fin-tube radiation is installed above eye level. Convector enclosures are in the shape of self-contained boxes while fin-tube enclosures form a horizontal pattern along the wall, with sheet-metal accessories to provide access to valves and to adjust the length to fit specific spaces. Other accessories adjust the appearance to fit irregularities of the wall.

The covers are available in various heights, either to enhance the output of a heating element or to enclose two or three tiers of element for high output (Fig. 17.9.11). Covers are manufactured to fit most commercial and industrial dimension requirements, either as a self-contained unit with end caps (Fig. 17.9.12) or extending wall to wall, appearing visually as a continuous unit abutting the partitions (Fig. 17.9.13).

17.9.4 ARCHITECTURAL ENCLOSURES

Where desired by the architect or engineer, fin-tube manufacturers can design and produce enclosures to match adjacent design details at the windows or to conform to other aesthetic or practical needs of the installation. Extruded shapes are often integrated into the enclosure

TABLE 17.9.1 Correction Factors for Steam Pressures, Water Temperatures, and Air Temperatures Other than Standard

Water temperatures and air temperatures

Entering air temperature, °F (°C)

Average water temperature, °F (°C)	45 (7.2)	55 (12.7)	Std. 65 (18)	70 (20.9)	75 (23.7)	80 (26.4)	85 (29.1)	90 (32.2)	95 (35.0)	100 (37.7)	110 (43.3)	120 (48.4)	130 (53.9)	140 (59.4)	150 (64.9)
90 (32.2)	0.19	0.13	0.08	0.06											
100 (39.7)	0.25	0.19	0.13	0.11	0.08	0.06									
110 (43.3)	0.31	0.25	0.19	0.16	0.13	0.11	0.08	0.06							
120 (48.4)	0.38	0.31	0.25	0.22	0.19	0.16	0.13	0.11	0.08	0.06					
130 (53.9)	0.45	0.38	0.31	0.28	0.25	0.22	0.19	0.16	0.13	0.11	0.06				
140 (59.4)	0.53	0.45	0.38	0.34	0.31	0.28	0.25	0.22	0.19	0.16	0.11	0.06			
150 (64.9)	0.61	0.53	0.45	0.42	0.38	0.34	0.31	0.28	0.25	0.22	0.16	0.11	0.06		
160 (70.4)	0.69	0.61	0.53	0.49	0.45	0.42	0.38	0.34	0.31	0.28	0.22	0.16	0.11	0.06	
170 (75.9)	0.78	0.69	0.61	0.57	0.53	0.49	0.45	0.42	0.38	0.34	0.28	0.22	0.16	0.11	0.06
180 (81.4)	0.86	0.78	0.69	0.65	0.61	0.57	0.53	0.49	0.45	0.42	0.34	0.28	0.22	0.16	0.11
190 (86.9)	0.95	0.86	0.78	0.73	0.69	0.65	0.61	0.57	0.53	0.49	0.42	0.34	0.28	0.22	0.16
200 (92.4)	1.05	0.95	0.86	0.82	0.78	0.73	0.69	0.65	0.61	0.57	0.49	0.42	0.34	0.28	0.22
210 (97.9)	1.14	1.05	0.95	0.91	0.86	0.82	0.78	0.73	0.69	0.65	0.57	0.49	0.42	0.34	0.28
215 (Std.) (100.7)	1.20	1.09	1.00	0.95	0.91	0.86	0.82	0.78	0.73	0.69	0.61	0.53	0.45	0.38	0.31
220 (103.4)	1.25	1.14	1.05	1.00	0.95	0.91	0.86	0.82	0.78	0.73	0.65	0.57	0.49	0.42	0.34
230 (108.9)	1.39	1.25	1.14	1.09	1.05	1.00	0.95	0.91	0.86	0.82	0.73	0.65	0.57	0.49	0.42
240 (114.4)	1.44	1.39	1.25	1.20	1.14	1.09	1.05	1.00	0.95	0.91	0.82	0.73	0.65	0.57	0.49
250 (119.9)	1.54	1.44	1.39	1.32	1.25	1.20	1.14	1.09	1.05	1.00	0.91	0.82	0.73	0.65	0.57
260 (125.4)	1.64	1.54	1.44	1.41	1.39	1.32	1.25	1.20	1.14	1.09	1.00	0.91	0.82	0.73	0.65
300 (130.9)	2.09	2.00	1.87	1.81	1.76	1.70	1.64	1.59	1.54	1.50	1.41	1.32	1.20	1.09	1.00

FIGURE 17.9.7 High-trim baseboard. (*Courtesy of Weil-McLain Crop.*)

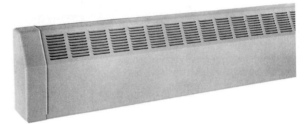

FIGURE 17.9.8 Slope-top enclosure. (*Courtesy of Sterling Hydronics.*)

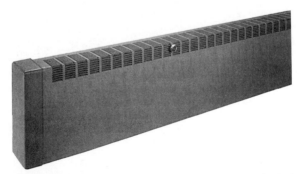

FIGURE 17.9.9 Flat-top front outlet. (*Courtesy of Sterling Hydronics.*)

FIGURE 17.4.10   Expanded-metal industrial cover. (*Courtesy of Sterling Hydronics.*)

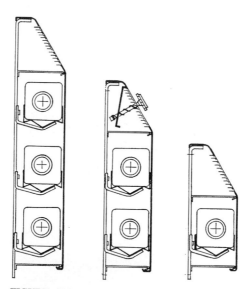

FIGURE 17.9.11   Single-tier and multitier elements. (*Courtesy of Slant/Fin Corp*).

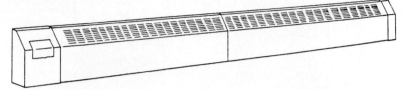

FIGURE 17.9.12   Unit with end caps. (*Courtesy of Slant/Fin Corp.*)

FIGURE 17.9.13   Wall-to-wall finned tube in front outlet enclosure. (*Courtesy of Sterling Hydronics Equipment.*)

design. Shorter, wider, or narrower configurations to meet almost any job requirement, instead of standard production enclosures, can be obtained given suitable lead time (Figs. 17.9.14 and 17.9.15).

Where low-profile high output is desired two elements can be mounted side by side. Tamper-resistant covers, beyond simple looking devices, can be developed in consultation with the design engineer. Typical enclosures are made of 18 gauge (1.2 mm) steel strongly braced to withstand physical damage, but they can also be obtained in 16 and 14 gauge (1.52 and 1.9 mm) where required for special protection.

17.9.5   RATINGS

The output of radiation in Btu/h (watt) varies with the type of element and the type of enclosure, and it should be obtained from manufacturers' catalogs. The $I = B = R$ ratings of baseboard and fin-tube radiation are based on actual tests of the equipment in the laboratory of

FIGURE   17.9.14   Architectural   finned-tube   enclosure—high top. (*Courtesy of Sterling Hydronics Equipment.*)

FIGURE 17.9.15 Architectural finned-tube enclosure—low top. (*Courtesy of Sterling Hydronics Equipment.*)

The Hydronics Institute Div of GAMA, and all such products currently certified by that organization, are listed in their published ratings book.[1] In addition to the initial tests to establish the Btu/h (watt) output, certified products are retested at specified intervals to ensure that the product continues to merit the published rating and is manufactured in accordance with catalog specifications.

Any manufacturer can submit a product for testing regardless of whether the company is a member of The Hydronics Institute Div of GAMA. Certification tests are also performed at the laboratory on finned tube units of special design when required by the engineer or architect for specific installations, to ensure the product will produce the output required.

Other determinants of Btu/h (watt) output, in addition to the design and dimensions of the product, include circulating water or steam temperature, rate of flow, and incoming air temperature. The standards for these variables are specified in the "I = B = R Testing and Rating code for Finned Tube Radiation." Standard tests are based on 1-lb (0.45-kg) steam at 215°F (102°C) with 65°F (18°C) entering air. Correction factors are listed for other steam and water temperatures in Table 17.9.1.

17.9.6 SELECTION

Two of the selection criteria are the heat output of the unit and the physical space requirements. For example, a single-tier copper-tube element in a low enclosure can have the same output per linear foot (meter) as a two-tier iron-pipe element in a higher cover. Considerations of the length of available wall and wall height beneath the windows, in relation to the required

Btu/h (watt) rating, are some of the factors taken into account. The pipe diameter chosen will be mainly determined by permissible pressure drops.

Another major consideration in determining the Btu (watt) output per foot (meter) is the temperature drop of the circulating water. It has been the pattern to design the system on the basis of a 20°F (11°C) drop from the boiler supply to the boiler return, but greater drops of 30, 40, or 50°F (16, 22, or 28°C) may be used, with possible reductions in pipe diameters for the mains. A design guide containing tables and worksheets is available from The Hydronics Institute.[2] The savings in piping costs must be balanced against the possible need for a higher-head circulating pump.

For example, with 180°F (82°C) boiler water supply and a 40°F (22°C) design drop, the average water temperature in the first third of the piping would be approximately 170°F (77°C) with 160 and 150°F (71 and 65°C) in the latter two-thirds, respectively. The active lengths of elements would be selected on the basis of those water temperatures. This applies to larger installations and to temperature drops of 30°F (17°C) or more on series-loop arrangements. The temperature drop has no effect on the selection of boiler size.

A 1°F (0.55°C) temperature drop per 1 gal/min (3.7 L/min) of water flowing emits 8.34 Btu (8797 per min.). In 1 h the output is approximately 500 Btu (528,000 J), and for a 20°F (11°C) drop each 1 gal/min (3.7 L/min) will provide 10,000 Btu/h (10.6 MJ/h or 2830 W). If the system is designed for a 40°F (22°C) drop, the heat output per 1 gal/min (3.7 L/min) is twice as much, so for a given design output, only half the volume of water will have to be circulated. This permits a reduction in the pipe diameter of the main.

To ensure proper heat transfer through the elements, an adequate minimum flow rate must be maintained: for $^3/_4$-in (1.9-cm) diameter, 0.5 gal/min (1.9 L./min); 1-in (2.5-cm) diameter, 0.9 gal/min (3.4 L/min); and $1^1/_4$-in (3.2-cm) diameter, 1.6 gal/min (6.1 L/min).

17.9.7 APPLICATION

The choice of element and cover is related to the type of building and the purpose of the hydronic equipment. Finned tube radiation is widely used for total comfort heating, particularly in buildings of moderate area for each floor, where the perimeter heat can properly reach interior areas. This also applies where floor space is divided into small areas.

For heating of industrial buildings and warehouses, it is good practice to use fin-tube elements in expanded-metal covers around the perimeter, supplemented at major loss areas (such as loading docks) with unit heaters. In cold climates where radiant floor panels are used in one-story buildings, finned tube elements can provide quick pickup or supplement the radiant heat to supply the adequate amount of Btus not provided by the floor panel.

Where a central ventilation system is required, tempered air at moderate temperature is introduced, often at the indoor design temperature of 70°F (21°C), and the perimeter finned tube element supplies the necessary Btus to satisfy the transmission heat loss.

Often in modern office buildings where air cooling is required in summer, the air system is used for core heating in winter, in conjunction with fin-tube radiation to satisfy the envelope losses. This is particularly useful for after-hours occupancy when the air system is shut down, because control of individual perimeter offices can be maintained via room thermostats and zone valves. This is especially advantageous because it eliminates the noise of blowers and ducts when all else in the building is quiet. In a hot-water system which has been purged of air, and with the usual provision for expansion and contraction, there is little expectation of sound coming from a proper finned tube installation.

An economical operation for combination cooling and heating is the use of cabinet fan-coil units using chilled water in summer and the use of perimeter finned tube elements, in conjunction with the cabinet units, using warm water in winter. The cabinet units can provide quick heat during morning pickup and individual office control for limited areas, while the finned tube radiation is used to overcome downdrafts from the cold windows and walls all along the exterior.

Another combination widely used where quiet operation is required, such as in libraries and churches, is valence cooling in summer, using chilled water in finned elements installed behind enclosures along the upper wall near the ceiling, and the finned tube in its normal position near the floor for winter heating. The valance unit can also be used for heating, with low-output commercial baseboard installed in high-heat-loss areas at the floor line. This provides natural convection of air in both summer and winter, thereby eliminating the noises of forced air and the discomforts of unbalanced systems.

17.9.8 PIPING ARRANGEMENTS

Various convenient and economical piping arrangements depend on the configuration of the building, the degree of local temperature control desired, and many other considerations. Finned tube is eminently suitable for series-loop systems where each segment is piped directly to the next segment, room to room, or covering a wide space of wall in an industrial building.

In small buildings a single loop may suffice, running from the boiler and back to it directly. See Fig. 17.9.16. A larger building can have two or more individual loops connected

SINGLE ONE PIPE LOOP BASEBOARD SYSTEM

One pipe systems offer the benefit of zoning each room at will. All of the water flowing through a one pipe circuit flows from the circulator, through the main and each one pipe-connected baseboard, through the boiler, and back to the circulator.   A one pipe loop layout is shown in the following sketch.

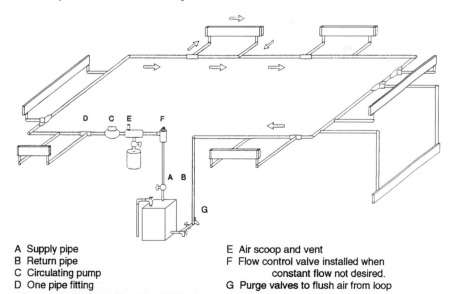

A Supply pipe
B Return pipe
C Circulating pump
D One pipe fitting

E Air scoop and vent
F Flow control valve installed when
constant flow not desired.
G Purge valves to flush air from loop

FIGURE 17.9.16 Single one pipe loup baseboard system.

MULTIPLE SERIES CIRCUIT SYSTEM

A multiple circuit series loop system offers flow control and/or temperature control in several living areas within a residence. All of the water flowing through a multiple series circuit flows through a trunk to every series-connected baseboard loop, into a return trunk, through the boiler, and back to the circulator.

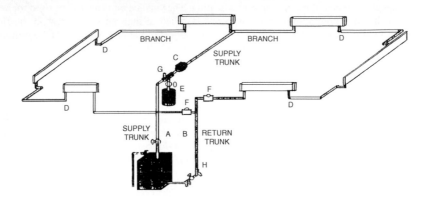

A Supply pipe

B Return pipe

C Circulating pump

D Nipple and cap installed in tee to provide for drainage

E Expansion Tank

F Balancing valves to regulate flows in loops.

G Air vents

H Purge valves to flush air from loop

FIGURE 17.9.17   Multiple series circuit system.

to supply and return headers at the boiler or joining the return main at a convenient tee. Such loops can be balanced with a manual valve to adjust the water flow for proper heat emission, or they can be automatically controlled by zone valves or individual circulators to respond to temperature needs at various sides of the building. See Figs. 17.9.17 and 17.9.18. A typical piping layout for a small manufacturing plant is shown in Fig. 17.9.19.

Larger still are the buildings which use two-pipe reverse-return mains, or primary and secondary hookups. With two-pipe mains, each heating unit can be individually valved to throttle or turn off the flow. The reverse-return arrangement of the mains permits a balanced pressure drop through various parts of the system. Primary-secondary pumping includes a master pump to keep water circulating through the main, while the secondary circulators supply specific zones of the building, or individual risers and returns, in response to their respective thermostarts. (See Fig. 17.9.20.)

17.9.9   AUTOMATIC CONTROL

Many control arrangements are possible with hydronic fin-tube installations, to provide comfort to the occupants and economy to the owner. These vary from simple on/off operation, to more efficient two-stage thermostat control, to modulating water temperature in

Installing a Pump in Each Loop

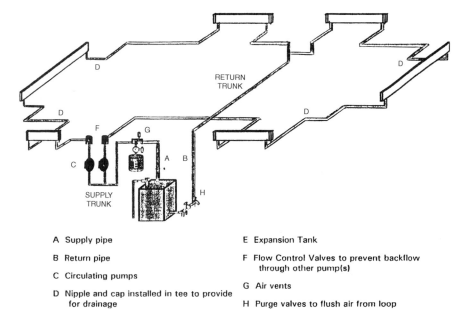

A Supply pipe

B Return pipe

C Circulating pumps

D Nipple and cap installed in tee to provide
   for drainage

E Expansion Tank

F Flow Control Valves to prevent backflow
   through other pump(s)

G Air vents

H Purge valves to flush air from loop

FIGURE 17.9.18 Installing a pump in each loop.

response to outdoor conditions. The more sophisticated type senses indoor versus outdoor temperatures, maintains a steady condition of comfort at all occupied hours, and schedules appropriate setback periods for energy conservation.

The accurate sensing of modern thermostats is best served by the accurate response of circulating water. This is particularly effective where continuous circulating mains, with modulating water temperatures, automatically maintain a proper comfort balance between the heat output of the system, the heat loss of the space, and the fuel input to the boiler. (See Chap. 7, System Control Equipment)

For all types of equipment discussed in this chapter, minimal maintenance and upkeep are required. Since most systems (hot water) are closed systems, no special treatments for corrosion are required. Any system water treatment chemistry as recommended by the boiler manufacturer should suffice for maintenance of these units. If scratches or abrasions reveal any of the underlying bare metal, exterior surfaces of these units may be touched up with most metal-finishing paints. The only caution is that, generally, it is not advisable to use metallic paints, as these will tend to hold the heat inside the radiator and reduce its output. For any units with fin surfaces, it is advisable to vacuum out the fins at periodic intervals to remove accumulated dust and debris. Most manufacturers recommend this be done prior to each heating season. If a higher degree of sanitation is required, most units can be flushed with a mild detergent and/or disinfected and then wiped or blown dry. In addition, radiation should be periodically inspected, looking for loose assemblies, fasteners, or covers, and accessories that should be adjusted or tightened as necessary. Also, check the brackets and fasteners to ensure that they are maintaining their attachment to the mounting surface and make any corrections necessary.

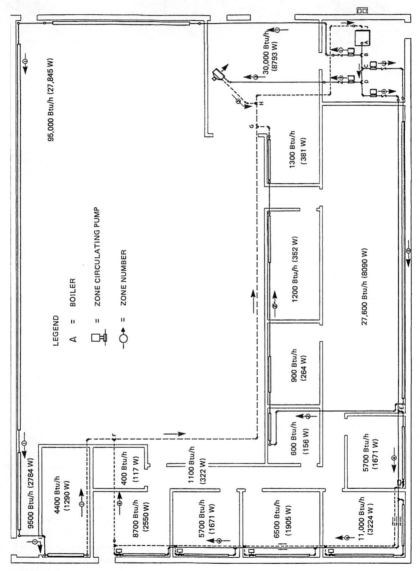

FIGURE 17.9.19 Piping layout—small manufacturer plant. (*Courtesy of Hydronics Institute.*)

17.148

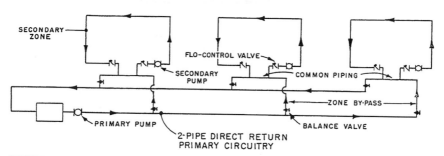

FIGURE 17.9.20 Primary-secondary piping.

REFERENCES

1. *I = B = R Ratings for Boilers, Baseboard Radiation, and Finned Tube Commercial Radiation.* The Hydronics Institute Div. of GAMA, Berkeley Heights, 2003.

2. *Advanced Installation Guide for Hydronic Heating Systems* (pub no. 200). The Hydronics Institute Div. of GAMA, Berkeley Heights, NJ, 1991.

3. *ASHRAE Handbook HVAC Systems and Equipment,* (Chap. 32, Hydronic Heat-Distribution Units and Radiators, Atlanta, GA, 2000.

17.10 RADIANT PANEL HEATING*

Radiant Panel Association
Loveland, Colorado

17.10.1 INTRODUCTION

The purpose of this chapter is to provide a basic understanding of the design and installation of hydronic radiant floor heating systems in residential and commercial applications. While it is recognized that there are other forms of radiant panels, such as ceiling, wall, and plate panels, details of these are not covered here. It should be noted that while many types of residential and commercial buildings are covered, it is not intended to encompass the larger type of buildings which are normally within the province of the architect or heating engineer. It is acknowledged that there are various heat sources which can be used with hydronic radiant floor heating; this chapter will limit its scope to the application of boilers designed for use in low temperature hydronic heating systems.

17.10.2 DEFINITIONS AND TERMS

17.10.2.1 Radiant Panel Heating

By definition, radiant panel space heating and/or cooling systems use controlled temperature less than 300°F (150°C) surfaces mounted on, or embedded in floors, walls, or ceilings, from which 50 percent or more of the heat transfer is by radiation to other surfaces seen by the panel.

It is important to understand the principles of heat transfer under which a radiant panel operates. Unlike convection heaters which heat air that in turn fills the room, radiant panels heat objects by radiation. Heat will leave a warmer object and travel to a cooler object, if one is present. The greater the difference in temperature between the two objects, the faster the heat will transfer. The sun is warmer than the earth, therefore it radiates its heat to the earth in the form of short wave energy. Empty space between the sun and the earth is not heated. It is only when the short wave energy strikes the earth that it converts into heat and long wave radiant heat energy.

Another example of radiant energy is that produced by a wood stove. If a person stands in front of a hot wood stove, the part of the body facing the stove will be warm while the part facing away will be cool. This is because the long wave radiation given off by the stove only strikes the front of the person where it converts to heat. It does not heat the air therefore the back of the person senses the cooler air temperature. The only air that is heated by the stove is that air which comes in direct contact with the stoves surface or surfaces which have been heated by the stove's radiant energy.

Surface area is an important factor in radiant heat transfer. A large surface area at a warm temperature can radiate as much heat as a small surface area at a high temperature. These three factors determine radiant energy transfer: (a) the size of the radiating surface area, (b) size of the receiving surface area, and (c) surface temperature difference between

---

*Adapted by permission from IBR Guide #400: "Radiant Panel Heating" published by The Hydronics Institute, a Div. of GAMA, Berkeley Heights, N.J.

the radiating surface and the receiving surface. Heat does not rise, hot air does. Radiant heat travels in all directions. Although convection does occur when cool air comes in contact with a warm floor, more than half of the heat generated by a typical floor heating system is radiant.

Wall Panels. Radiant panels designed for use on walls are generally manufactured units which are surface mounted to the wall or stand upright on the floor. They are constructed either of steel, cast iron, or aluminum (with copper or plastic pipe embedded in channels). Radiant wall panels (or "radiators" as they are commonly referred to) differ from baseboard convectors in that their primary heat transfer is by radiation. While common baseboard convectors do provide some radiant heat, their finned tube configuration is designed to heat air and circulate it through the room by natural convection.

Wall radiators come in all sizes and shapes. They can take the form of a rectangular, flat box mounted on the wall or a long slim panel running around the perimeter of the room. As with all radiant panels, the design has its foundations in surface area and surface temperature regardless of the shape.

Because wall radiators are most often manufactured units and their surfaces are unobstructed, calculating output is straightforward. The manufacturer can publish output based on square feet, lineal feet, or per manufactured panel at various water temperatures and flow. The designer simply selects the amount of radiant panel required to meet the heating load. Calculating pipe sizing and pressure drop is done in the same manner as used with conventional hot water heating systems.

In order to obtain peak performance from wall radiators, they should be positioned to allow a maximum viewing field of the space they are heating. Radiant heat is transferred in straight lines similar to light. Objects placed between the radiator and the space will "shade" the radiant output of the radiator and reduce its ability to provide maximum comfort.

Radiators are most often plumbed in parallel or a single pipe system using diverter tees, but can be plumbed in series provided the designer takes into consideration the drop in water temperature and increase in pressure as it passes from one radiator to the next. Connection can be made with copper pipe or approved plastic pipe such as crosslinked polyethylene (PEX), Composite Pipes of PEX Aluminum PEX or Rubber Aluminum Rubber.

Ceiling Panels. The ceiling has some definite advantages when it comes to radiant heating and/or cooling. It has an unobstructed view of the entire space (except under desks, tables, chairs, etc.), the surface is unimpaired by furniture or covering, and it is a large area. Tubing can be placed above sheetrock, sandwiched between layers or embedded in ceiling plaster. Ceiling panels can also be premanufactured to any size and surface mounted or in T-channel as a part of a suspended ceiling.

Since radiant energy does not convert to heat until it strikes a surface, panels mounted on the ceiling will radiate down and warm the floor or any other surface it "sees." In this way the floor warmed by the ceiling radiant actually becomes a secondary radiant panel.

Plumbing and controlling radiant ceiling panels is very similar to floors. Water temperature and flow control are dealt with in almost the same manner. Water temperature and panel surface temperatures can be greater than that in floor panels because there is no actual human contact. Supply and return piping can be copper composite pipes, or PEX (crosslinked polyethylene). Because higher surface temperatures also produce more output per square foot, ceiling panels are often positioned at the perimeter of the building rather than covering the entire ceiling. They can also be zoned to heat specific areas.

Ceiling panel output is generally specified by the manufacturer on the basis of water temperature and flow. Sizing becomes a matter of determining the square feet required based on the manufacturer's specifications and the heat load of the space.

Floor Panels.   The floor radiant panel is almost always constructed on the job site. As this manual will show, radiant floors can consist of tube embedded in concrete slabs, attached to the underside of subfloors, or sandwiched between sub-floor layers. Because of the variety of factors such as floor construction, floor covering, and area use, output is not as easily determined as wall or ceiling panels.

Floor coverings can dramatically alter the water temperature and output characteristics of a floor panel. The manner chosen to install the tubing in the floor also affects the floor output. For this reason each installation should be examined carefully and customized for the application.

As this manual will demonstrate, a properly designed radiant floor system can deliver an extremely even and efficient heat to the space.

17.10.2.2   Additional Definitions

British Thermal Unit (Btu): A unit for measuring quantity of heat. It is approximately the heat required to raise the temperature of a pound of water 1°F.

Btu per Hour (Btu/h): A unit for measuring the rate at which energy is transferred.

Convection: Heat transfer by movement of fluid (i.e., water, air). Natural convection is due to differences in density from temperature differences; warm air rises, cool air falls, causing a circular flow. Forced convection is produced by mechanical means.

Degree Day: A unit, based on temperature difference and time, used in estimating heating system energy consumption. For any one day whose mean temperature is below 65°F (19°C), the degree days for that day is the difference between 65°F (19°C) and the mean for that day. Degree days for any period is the sum of the degree days for each day in that period.

Design Temperature: The temperature an apparatus of a system is designed to maintain (inside design temperature) or operate against (outside design temperature) under the most extreme conditions to be satisfied.

Infiltration: Air flowing inward as through a crack between window and frame, or door and frame, or frame and wall, etc.

Oxygen Permeation: The ability of oxygen to pass through a material due to the materials molecular structure and a difference in oxygen pressure on each side.

Perm: The unit of measurement of permeance equal to 1 grain divided by ($ft^2 \times h \times in$ mercury vapor pressure difference).

Radiation: Energy radiated in the form of waves or particles.

Radiant Barrier: A membrane with a polished aluminum surface which reflects long wave radiant energy. It is typically a composite of aluminum laminated to a plastic film, which is available in rolls or as the face of fiberglass batt insulation.

Reactive Control: A mechanical or electromechanical device which controls flow in response to a temperature input resulting in a desired output water temperature.

Thermal Conductance (C): The number of heat units (Btu) that will pass through 1 ft^2 of nonuniform material in 1 h for each °F difference in temperature between the two bounding surfaces of the material.

Thermal Conductivity (k): The number of heat units (Btu) that will pass through 1 ft^2 of uniform material 1 in thick in 1 h for each °F difference in temperature between the two surfaces of the material.

Thermal Mass: A dense material used to store and transfer heat. Generally in the form of a concrete-like material poured as a finished floor or subfloor in which hot water tubes are embedded.

Thermal Resistance (R): The ability of a material or combination of materials to retard or resist the flow of heat. It is the reciprocal of U.

Thermal Resistivity (r): The ability of unit thickness of a uniform material to retard or resist the flow of heat. It is the reciprocal of thermal conductivity (1/k).

Transmission: In thermal load calculations, a general term for heat travel (by conduction, convection, or radiation, or any combination thereof).

U (Overall Coefficient of Heat Transfer): The amount of heat flow, expressed in Btu per hour per ft^2 per °F temperature difference, between the air on the inside and the air on the outside of a building section (wall, floor, roof, or ceiling). This term is frequently called the U value.

Vapor Barrier: A material which retards the transmission of water vapor. (Permeance not more than 1 perm.)

17.10.3   HISTORY AND APPLICATIONS

17.10.3.1   History

Radiant floor heating has been utilized for centuries by many cultures. The Romans used elaborate fire trenches and under floor ducting systems to heat the stone floors of bath houses. The Koreans have used a similar system of fire pits filled with stones under their homes for hundreds of years. In 1909, an Englishman filed a patent on a radiant heating system using tubing embedded in concrete or plaster.

While European and Far Eastern cultures adopted radiant floor heating as a viable heating system, it was not until the end of World War II that much interest was shown in the United States. During the 1940s and 50s a floor heating industry began to grow. There was a considerable amount of interest generated primarily by those soldiers that had been in other countries and had experienced the heating system first hand. The famous architect Frank Loyd Wright used radiant floor heating extensively. Unfortunately, the high cost of materials and labor to install copper or black iron pipe, coupled with some performance problems, caused interest to wane in the 60s. Lack of insulation in buildings often required floor surface temperatures above comfortable levels and also produced wide temperature swings within the space.

The 1970s brought higher standards of construction and insulation as well as the acceptance of new materials for piping such as plastic and rubber compounds. Germany, Switzerland, and other European countries reported an astounding increase in radiant floor activity. In the United States floor heating was rediscovered by the solar industry as a perfect match for solar collection systems. Hydronic radiant heating required water temperatures far less than that of more conventional fin-tube convectors or fan coils. This, combined with the outstanding comfort floor heating provides, has made radiant floor heating grow into an important part of the hydronics industry.

17.10.3.2   Applications

Hydronic radiant flooring (HRF) has many applications. This chapter is primarily concerned with residential and light commercial construction although the fundamentals can be applied to a variety of situations. It is not within the scope of this chapter to encompass snow melt, green houses, playing fields, barns, etc.

Typical applications which are within the scope of this chapter are single and multifamily residences, shops, warehouses, garages, small commercial buildings, and manufacturing facilities. Buildings built on a concrete slab are generally ideally suited to HRF. New construction techniques have made suspended floors and wood floors candidates as well.

17.10.4 DESIGN CONSIDERATIONS

Insulation. A well insulated building is the best assurance to designing a successful HRF system (Fig. 17.10.1). The poorer the insulation, the higher the heat loss. High heat loss requires warmer floor surface temperatures. Poorly insulated buildings may produce floor temperatures beyond human comfort levels and cause large temperature swings because of excessive heat build up in the floor.

A concrete HRF slab constructed on or below grade will have increased heat loss to the ground due to its higher than normal temperature. Insulation placed around the perimeter and/or under the slab will reduce that heat loss (Table 17.10.1). Perimeter losses are more significant than downward losses, therefore it is recommended that insulation be applied to or below the frost line. Buildings built in moderate climates on dry soil generally do not require insulation under the slab. The earth itself acts as an insulative barrier. Structures built in cold climates, on damp soil, or areas with high water tables should have under-slab insulation to prevent excess downward heat loss. Response time can also be increased in most applications by installing under-slab insulation.

Insulation under a suspended floor which is over a heated space is seldom required, although it can be used to improve thermal control. A highly insulative carpet and pad may create an upward thermal resistance greater than the downward resistance, causing an excess downward heat flow. Under-floor insulation can help counteract this situation. A radiant barrier may be installed under the floor as an inexpensive but effective means to reduce downward heat travel.

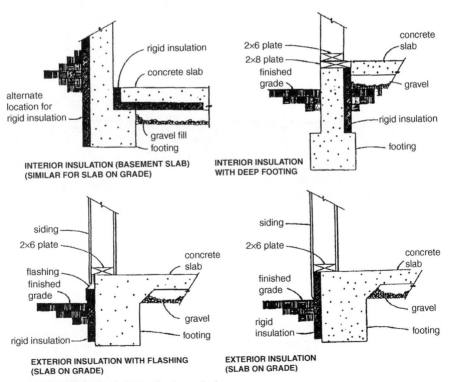

FIGURE 17.10.1 Insulation application methods.

TABLE 17.10.1   Slab Insulation

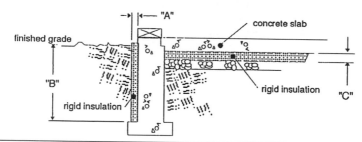

DEGREE DAYS			HARD FLOOR			CARPETED FLOOR		
			"A"	"B"	"C"	"A"	"B"	"C"
less	than	2000	1.0 in	1 ft		1.5 in	1 ft	
2000	to	3000	1.5 in	1 ft		2.0 in	1 ft	
3000	to	4500	2.0 in	1 ft		1.5 in	2 ft	
4500	to	6500	1.5 in	2 ft		2.0 in	3 ft	
6500	to	9000	2.0 in	3 ft		2.0 in	3 ft	1.0 in
more	than	9000	2.0 in	3 ft	1.0 in	2.0 in	3 ft	1.5 in

It is highly recommended that insulation be applied to suspended floors over unheated spaces such as vented crawl spaces, cantilevered floor, and unheated garages/basements. A radiant barrier in conjunction with stem wall foundation wall insulation may be used for a suspended floor over a controlled ventilation crawl space.

Floor Covering.   The temperature of the floor surface is all important. Hard surface floors such as concrete, tile, or linoleum are ideally suited to HRF. As thicker floor coverings are added, the thermal mass under the floor covering must increase in temperature to maintain the proper floor surface temperature. Although wood flooring and carpeting can be used quite successfully with HRF, calculations must be made to ensure that the proper surface temperatures are reached without exceeding acceptable temperatures in the thermal mass.

Hard surface floors are more desirable in front of large picture windows or often-opened outside doors. A hard surface has more heat output capacity and can react quicker to counteract cold drafts or brief influxes of cold air. Approximate *R* values of floor coverings are shown in Table 17.10.2.

Room Temperature.   Since HRF systems heat by radiation and not by heating air, the room air temperature is not a direct result of floor heating. The surfaces in the space are warmed by radiation from the floor. Air is warmed by coming in contact with room surfaces. Air temperatures remain fairly constant throughout the space. In a properly designed HRF system the floor surface temperature will be only a few degrees warmer than the air temperature.

Most conventional heating systems use convective means to transfer heat into the space. Forced air systems use a combination of forced convection and natural convection to achieve heat distribution. Hot air is forced into the space by means of a blower. This hot air rises to the ceiling, cools, and then falls, where it is returned to the furnace for reheating.

The effect of an HRF system on the air temperature is one of minimal drafts and temperature stratification. Fig. 17.10.2 illustrates the comparison of heating systems. Keep in mind

TABLE 17.10.2   *R*-Values of Floor Covering

Approx. R-values of various floor covering		
Floor covering	Thickness	R-Value
Carpet*	1/8" (3 mm)	0.6
	1/4" (6 mm)	1.0
	1/2" (13 mm)	1.4
	3/4" (19 mm)	1.8
	1" (25 mm)	2.2
Rubber pad	1/4" (6 mm)	0.3
	1/2" (13 mm)	0.6
Urethane pad	1/4" (6 mm)	1.0
	1/2" (13 mm)	2.0
Vinyl, tile		0.2
Hardwood	3/8" (9.5 mm)	0.5
	3/4" (19 mm)	1.0

*For wool carpets multiply *R*-value by 1.5.

that proper building and system design can minimize the differences. Structures with high ceilings can benefit greatly from a HRF system. Air temperatures at the ceiling will remain cool, while the air nearer the floor at the occupied level will maintain the desired temperature. The need for ceiling fans to force down wasted hot air from the ceilings is eliminated. The actual temperature stratification will vary but typically a structure with a 20 ft (6 m) ceiling would have ceiling air temperatures 2°F (1°C)/(1.5 m) cooler than at the 5-ft level.

Comfort.   Indoor climate comfort is difficult to describe. What is considered comfortable to one person can be uncomfortable to another. The same person can experience various degrees of comfort or discomfort at different times in the same environment, depending

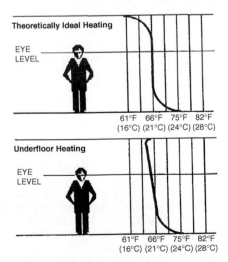
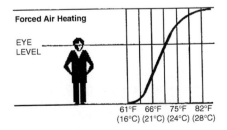

FIGURE 17.10.2   Heating stratification comparison.

on levels of activity or biological states. A person at rest after eating or exercising feels warmer than at other times. It is not possible to maintain a constant comfort level for everyone at all times, therefore compromises must be made.

The comfort of the human body is affected externally by three things—contact with the ambient air, evaporation of skin moisture, and radiation. Dependent on the ambient air temperature, the body will either absorb heat from the air or give up heat to the air by conduction. Air movement across the skin (drafts) causes increased evaporation of skin moisture, reducing skin temperature. This is desirable if a person is overheated, but detrimental to comfort if one is trying to keep warm.

The human body, as a surface, will either radiate its heat to cooler surrounding surfaces or absorb heat from warmer surfaces. A person standing near a cold picture window on a winter day can feel heat leaving his or her body. The perception is that the window is cold or that there is cold air coming from the window. In reality, it is the warm surface of the body radiating its heat to the cold surface of the window. The larger the temperature difference between the body and surrounding surfaces, the faster heat is lost or gained. If heat gain is rapid, the perception is one of a hot surface, such as is felt by a hot wood stove. However, if body heat loss is rapid, the perception is of a cold surface such as the illustration of the cold picture window.

Mean radiant temperature (MRT) is a term used to describe the collective effect on an occupant of all of the surrounding surfaces is an indoor environment. It is subject to surface area, surface temperature, and proximity. Simply put, the cooler the MRT of the surrounding environment, the faster a person's body heat will be lost. HRF systems help in counteracting a low MRT produced by cold outside walls, ceilings, and windows by raising the MRT, and thereby lowering the body heat loss.

The HRF system provides a high level of comfort by reducing drafts, creating even air temperature distribution, and raising the MRT. HRF is particularly useful in counteracting large windows and high ceilings where falling drafts can be problems. See Fig. 17.10.3.

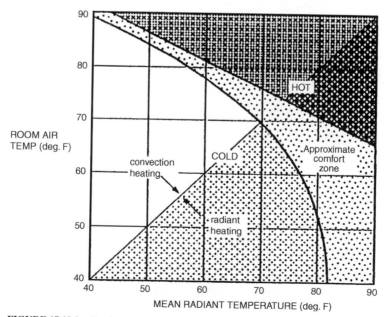

FIGURE 17.10.3 Comfort zone graph.

A factor which has a great deal to do with MRT, and therefore comfort, is the average unheated surface temperature (AUST). This is found by averaging together the surface temperatures of all the unheated surfaces in the space. The lower the AUST, the more the heated surface must compensate in order to raise the MRT. Consider standing in a green house on a cold day versus inside a well-insulated structure. At the same air temperature the green house environment will seem much colder because the average unheated surface temperature is quite low, causing the body radiant heat loss to increase. A warm radiant panel, such as a floor, can compensate by raising the MRT (the effect of the combination of the AUST and the heated surface).

Windows have the largest impact on AUST in modern, properly insulated buildings. Most rooms with typical windows have an AUST of only 1°F (0.56°C) or less below the room air temperature. As window area is added the AUST drops. A room with glass making up 40 percent of two outside walls would have an AUST approximately 4°F (2.2°C) below the room temperature at a 0°F (–18°C) outdoor temperature. A four-season porch with windows on three outside walls may have an AUST as low as 10°F (5.6°C) below the room temperature. If air temperature is the only source of heat, it would have to be raised significantly to compensate but, if a properly sized radiant panel is used, it will raise the MRT of the space instead of air temperature.

The output of a radiant panel heating systems is primarily governed by the surface temperatures surrounding the heated panel. Larger temperature differences between the heated panel and surrounding surfaces (such as walls, windows, and ceiling) create greater heat transfer to the space. The increased temperature difference between the panel and the AUST causes more heat to be transferred by both radiation and convection, particularly in a radiant floor or wall panel.

To get a close approximation of the effect these surfaces have, the unheated surface temperature must be averaged by area. Table 17.10.3 below shows a 12 ft × 12 ft (3.7 m × 3.7 m) room with R-19 walls, R-30 ceiling, and double glazing on the windows. The window area and outside temperature are varied. The result shows the AUST in °F below room temperature. This chart can be used to assist designers in approximating AUST for use in determining system output as shown later in this manual.

To use Table 17.10.3:

TABLE 17.10.3   AUST in °F Below Room Temperature

Outside walls	% Glass	Outside temperature					
		(–2°C) 30°F	(–7°C) 20°F	(–13°C) 10°F	(–17°C) 0°F	(–25°C) –10°F	(–29°C) –20°F
1	20	0.8	1.1	1.3	1.5	1.7	1.9
	40	1.3	1.6	1.9	2.2	2.6	2.9
	60	1.7	2.1	2.5	3.0	3.4	3.8
	80	2.1	2.6	3.2	3.7	4.2	4.8
2	20	1.5	1.9	2.3	2.6	3.0	3.4
	40	2.3	2.9	3.5	4.1	4.7	5.3
	60	3.2	4.0	4.8	5.6	6.3	7.1
	80	4.0	5.0	6.0	7.0	8.0	9.0
3	20	2.2	2.7	3.2	3.8	4.3	4.8
	40	3.4	4.3	5.1	6.0	6.8	7.7
	60	4.6	5.8	7.0	8.1	9.3	10.5
	80	6.9	7.4	8.8	10.3	11.8	13.3

- Find the number of outside walls in the left-hand column, then move to the next column and select the approximate percent of the total outside wall area which is glass.
- Continue horizontally to the right and read the number beneath the corresponding outside temperature.
- Subtract this number from the design room temperature to determine AUST.
- Use this AUST instead of room air temperature in Table D "Output Chart" when figuring panel output.

Floor Space. Of particular interest to the architect and interior designer is the increased use of floor space provided by the HRF system. Allowances do not have to be made for baseboard units or floor registers; the entire floor space is usable. Furniture and appliances can be set up to the wall. Floor registers are not required under windows and in front of sliding glass doors. Furniture or objects set on the floor will absorb heat from the floor and, in turn, their surface temperatures will be raised, adding to the raising of the MRT.

The use of HRF under cabinets is not recommended due to the excess heat buildup which can cause food to deteriorate at an accelerated rate. Closet installations are optional. It is seldom detrimental to apply HRF in closets except when food or garbage is present. Applying HRF under a bathtub or shower, or even set in surrounding tile can be very pleasing.

Response Time. Once the HRF control has called for heat in a space, time is required for the system to respond. Typically, HRF systems react slowly due to the fact that they incorporate a large mass at low temperatures. There is an inertia that must be considered both in heating up and cooling down. HRF systems operate best when designed to maintain a constant temperature under continual operation.

Modern construction techniques and improved insulation have greatly reduced the adverse effects of a slow response time.

Although it is true that a HRF system will take considerably longer than a forced air system to bring the air temperature of a space up to temperature, it will remain much more constant once the desired air temperature is achieved. As discussed above, comfort does not necessarily relate to air temperature. A HRF system will provide radiant comfort well before the desired air temperature is reached. The result is that a person may feel comfortable before the air temperature reaches the desired setting.

Because of the flywheel effect and slow response time of the thermal mass, HRF systems do not perform well with night setback thermostats. Rooms that are set back do not cool quickly. Rooms that are recovered must begin reheating early to be back at set point at the desired time. Little savings result and temperature overshoot can create a comfort problem. A steady state with gradual change is the most desirable mode for HRF systems.

A major advantage of the flywheel effect is experienced when sudden brief inrushes of cold air are experienced, such as that of opening a large overhead door. Two very important things happen. First, the cold air travels along the floor and is rapidly heated by contact with the warm floor. Second, once the door is closed the floor is instantly radiating heat to the occupants unlike a forced air system which must reheat all of the air within the space to achieve comfort. In this case, response time is dramatically improved due to the thermal mass.

Insulative floor coverings such as carpet and pad will adversely affect response time. The more insulative the floor covering, the higher the temperature required for the thermal mass. The R-value of a floor covering is its resistance to the flow of heat. The higher the R-value, the more the thermal resistance. High R-value floor coverings slow the heat transfer from the thermal mass to the floor surface and thus reduce recovery time as well as limit the output of the floor.

Zoning. A distinct advantage of HRF systems is their zoning flexibility. Each tube loop can be designed as an individual heating zone or an entire system can be operated from a single thermostat. Although a thermostat in every room may be the ultimate in control, it is seldom necessary. It is more common to group spaces with similar use, size, or location on one control.

The HRF system is unique in its ability to naturally self-regulate heat output. The rate of heat output is determined by the difference in the floor surface temperature and the MRT of the surrounding surfaces. As a result the output of the floor near a sliding glass window will be greater than that of a floor in an interior room, even though the floor surface temperatures are the same. The design challenge is to ensure that proper water temperatures and flow are allowed for each location.

As a general rule, it is best to group spaces on the same side of the building with similar floor coverings and heat losses. The HRF system's self-regulating ability, along with water temperature and flow design: lend flexibility to zone planning. Zone size is only limited by manufacturers' rated manifold and pump capacity. Some things to consider in zoning are: (a) client wishes, (b) physical separation of spaces, (c) floor levels, (d) frequency of use, (e) type of use, (f) heat loss characteristics, (g) solar gain, (h) internal gain, (i) types of floor coverings, and (j) types of floor construction. Consider that excessive zoning increases system cost, complexity, owner interaction, and may be of little benefit.

17.10.5 SYSTEM COMPONENTS

17.10.5.1 Tubing

Tubing can be divided into three categories: (a) metallic, (b) synthetic, and (c) composite. Each has advantages and disadvantages but all can be suitable for radiant floor heating when used with the proper components. As a general rule, all pipe, used for floor heating should be flexible enough to be laid out in continuous loops with no joints in the floor.

When selecting tubing, compare pressure and temperature ratings as well as the wall thickness for heat transfer and durability. Although most systems operate between psi 15 to 20/(103–137 kPa/s), some codes require a hydrostatic test of 100 psi (690 kPa/s) for 30 min. Others require that the tube withstand 180°F (83°C) water at 100 psi (690 kPa/s). Keep in mind weight and flexibility for installation. Check compatibility with embedding material. (See Table 17.10.4.)

Oxygen Permeation (Diffusion).   All hydronic heating systems are susceptible to oxygen entering the system through numerous sources such as threaded fittings, air vents, and gas permeable materials. Excess amounts of oxygen in a system can lead to premature failure of ferrous metal components due to corrosion. While copper tube is, for all practical purposes, impervious to oxygen migration through its walls, all synthetic tubes display a degree of permeation. Whether this characteristic can lead to problems for ferrous metals in radiant floor heating systems is dependent on a variety of factors. The temperature must be given the strongest consideration as a determining factor. The internal system pressure and the speed of flow are less important. Water quality can also play an important role.

Although oxygen permeation is measurable in the laboratory, there has been much debate as to the long-term effects in actual installations. With radiant floor heating systems numbering in the hundreds of thousands in the United States, there has not been strong evidence to indicate that oxygen permeation in synthetic tubing has contributed widely to system failures. On the other hand, many manufacturers offer synthetic tubing which incorporates an oxygen diffusion barrier, which dramatically reduces the measurable amount of oxygen permeation to the point where it is no longer in question. The Hydronics Institute recommends use of tubing with a rate of oxygen diffusion through the wall of the

TABLE 17.10.4 Tube Comparison Chart

Features	Metal	Plastic			Rubber		Other
	Copper	PE	PEX	PB	EPDM	Reinforced	Composite
Safe working temp.	180°F (82°C)		180°F (82°C)	180°F (82°C)	285°F	180°F (82°C)	180°F (82°C)
Safe working press.	100 psi (690 kPa)		100 psi (690 kPa)	100 psi (690 kPa)	30 psi (207 kPa)	100 psi (690 kPa)	100 psi (690 kPa)
Inside dia. restriction	1	5	4	2	9	8	3
ASTM Standard	B88	none	F876	D3309	none	none	none
Thermal conductivity	10	6	6	7	5	4	6
Suitable—tight bends	2	6	5	7	10	9	7
Flexibility @ 70°F (2°C)	1	5	4	6	10	9	4
Flexibility @ 32°F (0°C)	1	4	3	6	10	9	3
Kink resistance	2	5	6	5	8	9	4
Recoil memory	5	4	6	4	1	2	3
Corrosion inert	1	10	10	9	9	9	10
Scale resistant	1	10	10	10	9	9	10
Oil resistant	10	9	9	4	3	6	9
Field repair	solder	heat/mech	heat/mech	weld/mech	mech	mech	mech

Note: Values 1 to 10: 1 = least, 10 = most

The values represented below are typical and do not represent individual products.

tube which does not exceed 0.1 grams per cubic meter per day when ferrous metals are present in the system.

Where oxygen permeation is in question there are several alternatives. One alternative would be to use a tubing with an oxygen barrier as mentioned above. Because only ferrous metals are at risk, a heater exchanger can be used as a second alternative to separate the heat transfer fluid in the tubing from that in ferrous components such as cast iron boilers. Third, the system could be designed and installed without any corrodible ferrous components. Finally, corrosion inhibitors (water treatment) could be added to the heating water to control corrosion.

Metallic Tubing. The most common metal pipe used today is soft copper. Copper is highly conductive, meets all plumbing standards, stays put when bent, and is readily available. Although it has been used successfully, some of the disadvantages are: it is heavy and clumsy to work with, kinks easily, is susceptible to internal corrosion, and can be attacked externally by acids in cementitious thermal mass.

Synthetic Tubing. The vast majority of radiant floor heating systems installed in the last 20 years have used synthetic tubing. It is light weight, flexible, inert to corrosion, resistant to scaling, and not easily crushed, crimped, or kinked. The internal smoothness of synthetic tubing creates less friction, so there is less resistance to flow than in metallic tubing. Synthetic tubing is available in coils of relatively long lengths which allow for a variety of tubing layouts without joints.

Synthetic pipe can be divided into two distinct categories: (a) plastics and (b) rubber compounds. Plastics are more rigid and generally have thinner tube walls and better heat transfer characteristics than rubber tubes. On the other hand, rubber tubes can be extremely flexible and durable.

Rubber Compound Tubing. The base material for most rubber tubing is ethylene propylene diene monomer (EPDM). This material has a long and proven history of use in industry for such things as automotive hose, O-rings, window channeling, and other products requiring resistance to weathering. It can be extruded as a simple tube or can be reinforced with fiber webbing and jacketed with other materials. Temperature requirements for radiant floors are well within the range of EPDM. Unreinforced tubing has a pressure limit of approximately 40 psi (276 kPa), while reinforced tubing can withstand extremely high pressures dependent upon the degree of reinforcement. EPDM has a broad resistance to chemicals but not to oils and other hydrocarbon fluids. An oil-resistant rubber jacket may be extruded over the base EPDM to make it more compatible for installations where oil or asphalt is present.

Polyethylene (PE and PEX) Tubing. Polyethylene tubing for floor heating is extruded from high density polyethylene (HDPE) or cross-linked polyethylene (PEX). HDPE is more flexible than PEX, but is not as tolerant to temperature and pressure. The cross linking process of PEX links molecular chains into a three-dimensional network which makes the tube durable within a wide range of temperatures and pressures. Polyethylene tubing has been the most widely used tubing for floor heating in Europe. Although PEX is the most rigid of the synthetic tubing, it will meet or exceed any of the most stringent pressure and temperature tests required by the codes. It also has the property of returning to its original shape when the proper amount of heat is applied. This is particularly useful if the tube is accidentally kinked or crushed during installation. Polyethylene is highly resistant to chemicals, including oils.

Polybutylene (PB) Tubing. Polybutylene is a plastic developed in the United States and used almost exclusively in hot and cold water piping since the late 60s. It has had widespread use in radiant floor heating for over 20 years. PB tubing has a high strength to wall thickness ratio which results in the thinnest walls and greatest flexibility within the plastic tube group. The temperature and pressure properties of PB tubing fall in the range of code requirements and surpass 100 psi (690 kPa), 180°F (83°C) water test. It also retains its flexibility over the widest range of temperatures which makes it easy to work within extreme cold conditions. If kinked or damaged, PB may be fusion welded in the field using a special fitting and heat tool. PB has excellent chemical resistance but is susceptible to oils and other hydrocarbon fluids.

Composite Tubing. To address the issues of oxygen permeation, recoil memory, and bending, a new generation of tubing is entering the market place. This is a tube which uses the benefits of both metal and synthetic tubing. It is a tube which incorporates a thin metal layer, usually aluminum, sandwiched between layers of plastic. This gives the tube many good properties. It has the advantage of being inert to corrosion and impervious to oxygen permeation, yet semirigid. It bends more easily than soft copper, retains the bent shape, and stays put. It can meet or exceed all radiant floor temperature and pressure requirements. Damage in the field by kinking or crushing would require tube replacement or a mechanical joint.

17.10.5.2 Manifolds

The length of any particular tube circuit in the floor is limited by flow and pressure drop; therefore, many circuits are generally used to cover the floor area. The device used to distribute water to each individual tube from a single supply and return is a manifold (sometimes referred to as headers). The manifold can be as simple as a series of Ts or it can incorporate sophisticated controls.

Manifold Construction. The most common materials from which manifolds are constructed are brass, copper, and plastic. The manifolds may come in a set of one supply and one return or the supply and return may be molded together in a single piece. Some manifolds are available in fixed lengths serving a given number of circuits. For example, you may be able to order a manifold in 4, 6, or 8 ports (branches). Many times there are provisions which allow lengths to be joined together to provide longer lengths with more ports. Typical manifolds are shown in Fig. 17.10.4.

Some manifolds are available in building blocks which allow a single circuit at a time to be added. A mounting bracket for attaching it to the wall is often supplied with the manifold. There are also enclosed mounting cabinets available for a finished look.

Manifold Fittings. Copper pipe is generally used as the supply and return piping although plastic can be used where water temperatures do not exceed the rating of the tube. The two most common methods of joining supply and return piping to the manifold are pipe thread fittings and solder joints. In either case, installation of unions and isolation valves is recommended. Many manifolds come already equipped with unions and/or isolation valves.

The manifold fittings which attach the floor tubing are commonly of the compression variety. They can be as simple as slipping the tube over a nipple and applying a crimp ring with a special tool or they can be a combination of nipples, nuts, inserts, and O-rings. A gripper fitting is sometimes used which incorporates a one-way retainer ring with gripping fingers and an O-ring for sealing. Each manufacturer has developed and tested fittings which best suit their particular manifold and application. All fitting instructions should be followed carefully to ensure proper assembly.

Flow Control. Manifolds may be simple pipes with numerous ports or they may contain balancing valves on each of the floor circuits with temperature and flow meters. Balancing valves allow the installer to fine-tune the system to each application. They also allow adjustment after the system has been installed. A circuit which has an excessive temperature drop can be adjusted by closing down the balancing valves in adjoining circuits slowing flow and forcing more water through the affected circuit. In many cases, balancing the system is done in the design stage by determining proper circuit lengths, thus making balancing valves unnecessary. Balancing can also be accomplished manifold to manifold by placing a balancing valve on the return port of the manifold.

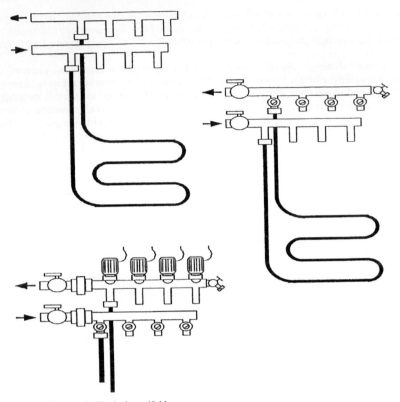

FIGURE 17.10.4   Typical manifolds.

Some manifolds have a provision for motorized flow control valves which can be installed on any or all of the individual circuits. This valve is used in conjunction with individual heating zone controls and is generally a flow-no-flow valve. It controls the flow in a particular circuit and allows one or more circuits to be closed while the others are open, providing flow to only those calling for heat.

17.10.5.3   Heat Transfer Media (Thermal Mass)

Unlike radiators or fin-tube convectors which transfer heat directly into the space, radiant floor tubing must have an additional medium to do that job. This medium is field-installed. Heat in the tubes must be transferred to the broad surface area of the floor to effectively heat the space. This means that the heat must travel out horizontally between the tubes and rise to the surface of the floor. The more evenly this heat can be distributed between the tubes, the more consistent the floor surface temperature will be across the floor. System design calculations are based on the assumption that the floor surface temperature is consistent. A poor heat transfer medium will not properly conduct the heat laterally, and therefore will cause hot spots directly above the tube and cold spots between the tubes.

Type of building construction has a bearing on what medium is used. The most common types of construction where floor heating is being used are slab-on-grade, slab-below-grade, wood suspended floors, and concrete suspended floors. The system designer must select the heat transfer medium that best suits the building construction and the particular application. Keep in mind that the greater the thermal resistance between the water in the tube and the surface of the floor, the higher the required water temperature and the longer the response time.

Air and Wood. A method of installing tubing in a suspended wood floor application is that of attaching the tube directly to the underside of the subfloor between the floor joists. A bright aluminum barrier is then installed a few inches below the tube, sometimes followed by fiberglass insulation. The barrier must be pure polished aluminum in order to be effective. The purpose of the aluminum is to redirect the downward heat loss by reflecting it back to the under surface of the subfloor. The barrier will do this effectively as long as it remains clean. In time, the surface of the barrier may become dusty and its effectiveness dramatically reduced (Fig. 17.10.5).

The tubing can also be laid between furring strips on top of the subfloor with a finished wood floor nailed above over it. Again, an aluminum barrier may be laid beneath the tube for its heat reflective qualities.

The primary heat transfer medium in both these installations is air. The airspace becomes heated by contact with the tube, and the hot air heats the underside of the floor. The floor is generally wood and wood has a high resistance to thermal conductivity. Couple this with the fact that air is also thermally insulative, add a carpet for floor covering, and the required water temperatures can become quite high.

This method has been used successfully and is particularly applicable in situations of low heat requirements and retrofits where other methods are not practical.

Metal Pans (fins). Metal pans can be used to transfer heat laterally and are primarily used on wood suspended floors. These pans are most often aluminum with a channel down the middle which accepts the hot water tube. The pans can be mounted flush against the underside of the subfloor or between furring strips on top of the subfloor with a finished floor nailed over them. Although the pans assist in spreading the heat laterally, insulation is generally placed directly in contact with the metal plates, thereby eliminating the air space. The conductivity of the metal plates provides the lateral transfer of heat. The heat must also travel through insulative wood to reach the surface. Carpet and pad increase the thermal resistance and can result in a high water temperature requirement. As with all floor heating design, the desired result is a floor surface temperature which is evenly distributed. Care must be taken in calculating water temperatures and floor outputs (Fig. 17.10.6).

Gypsum Cement. Tubing can be attached to the subfloor, wood, or concrete, and a thin gypsum cement poured over it (Fig. 17.10.7) as an underlayment to the finished floor. Gypsum cement is lighter than concrete and is extremely resistant to shrink cracking. It can be poured thinner than concrete and is pumped through a hose making it easy to apply in hard to reach areas. Gypsum cement provides a thin ($1/4$ in to $1 1/2$ in) [3.2 to 3.8 cm] thermal mass

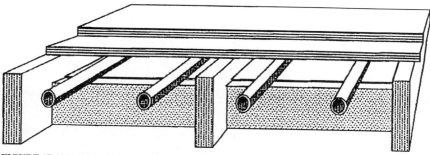

FIGURE 17.10.5   Tube under wood floor.

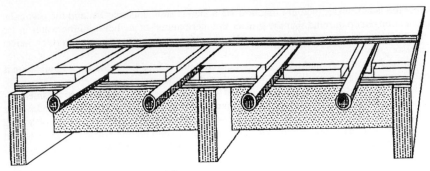

FIGURE 17.10.6   Metal pan on wood floors.

which does an excellent job of distributing the heat evenly between tubes. Although it can take up to two weeks for the moisture to completely leave the floor before floor coverings can be applied, gypsum cement can be walked on within an hour of the pour. Gypsum cement is quite hard, but requires some type of floor covering. It will accept vinyl, tiles, carpet, and glue-down wood floors. It is not suitable for use in areas of heavy wheeled traffic. Gypsum cement products used with radiant floor heating should be specially formulated to make them more heat resistant.

Lightweight Concrete.   Lightweight concrete is concrete which has had a foaming agent added to create tiny air bubbles and thereby make the product lighter. It is generally pumped onto the job through a hose and is finished like conventional concrete. Lightweight concrete is not nearly as strong as regular concrete and is prone to shrinkage cracks. A fiber filler is sometimes added to reduce the cracking. This product is used on suspended wood floors as a thermal mass poured over tubes attached to the subfloor. The heat transfer properties are not as good as conventional concrete due to the air bubbles, but will do an acceptable job.

Tile Grout.   For small areas where tile will be installed, the tube may be directly embedded in the grout for the tile. Grout has excellent heat transfer qualities. A crack isolation membrane is recommended between the grout and the tile because the tube can act similar to a control joint in concrete and cause cracking. Wire mesh placed over the tube in the grout will also help reduce crack potential.

Concrete.   Perhaps the most common heat transfer medium is conventional concrete. This is because many floor heating jobs are slab on grade or slab below grade. Tubing may

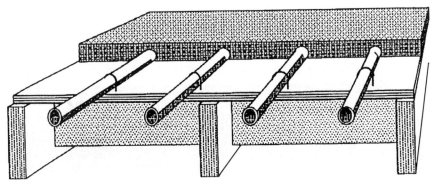

FIGURE 17.10.7   Gypsum underlayment on wood subfloor.

be embedded in a single pour, sandwiched between two pours, or installed in the sand beneath the concrete. While concrete is an excellent thermal conductor it also presents the greatest thermal mass. In some designs such as off peak utility applications, this is highly desirable. It is also a benefit to garages and warehouses where large overhead doors are opened and closed. With other applications the slow response time and residual heat buildup can cause undershoot and overshoot problems.

Single Pour. The tubing is usually tied to wire mesh (6 × 6 welded wire mesh is recommended) laid on top of the sand or gravel sub-base (Fig. 17.10.8). When under-slab insulation is used, the insulation can either be buried beneath the gravel or laid on top of the gravel sub-base with the wire mesh laid directly on it. The tube is tied with metal or plastic wire ties. Standoffs can be attached to the wire mesh to lift it up into the concrete slab or the wire mesh can simply be lifted into place as the concrete is being poured. While the generally recommended procedure is to keep the tube approximately 2 in (5 cm) below the surface of the slab, many contractors prefer to leave the tube at the bottom of the slab. In practical application it makes little difference.

To take advantage of a large heat storage mass, the tube may also be buried in sand beneath the slab. This method produces the longest response time but, on the other hand, produces a high flywheel which can be useful when supplemented with solar.

Double Pour. While more expensive, pouring a topping slab over a structural slab with insulation and tube sandwiched in between (Fig. 17.10.9) will result in easier tube installation and much quicker response times. The board insulation must be secured to the structural slab. The tube is then tied to wire mesh and laid down over the insulation. Some manufacturers have special anchoring devices to attach tube to the insulation to hold it in place. This eliminates the need for wire mesh. A concrete pour of $1\frac{1}{2}$ to 2 in (3.8 to 5 cm) tops off the finished slab.

17.10.5.4 Circulating Pumps

Movement of water through the HRF system is accomplished with one or more circulating pumps. Although there are several types of pumps, centrifugal pumps are the most common in HRF systems. These pumps use an impeller with curved vanes turning at high rpm to create a low pressure, which draws the water through the pump. If a large quantity of air enters the pump along with the water, the pump will cavitate and not circulate the water. Centrifugal pumps require a certain amount of water pressure with a minimum amount of air trapped in the water in order to function properly. A common problem with HRF pumps

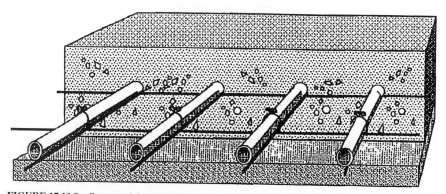

FIGURE 17.10.8   Concrete slab, single-pour.

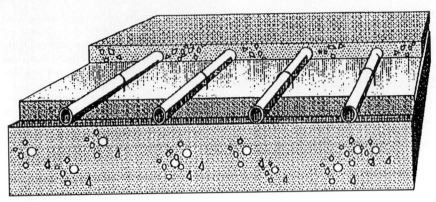

FIGURE 17.10.9   Concrete slab, double-pour.

is trapped air. The system should be designed to vent any air in the system before it reaches the pump.

Head Pressure.   The pump must be able to develop sufficient pressure to overcome the resistance, or friction, encountered as the water passes through the tubing, manifold, boiler, and other components of the system. This is known as pressure drop. The more the water that is forced through the system, the higher the pressure drop. The pressure developed by the pump can be expressed in two ways, pounds per square inch (psi) [kPa] or (m)(m) or feet of head (ft). The most common is feet of head.

Feet of head refers to the height, in feet, that the pump is capable of lifting a column of water in a vertical standpipe at a given flow (gpm) (1/m). A pump which can lift 28 gpm (1061/m) up a 6 ft (1.8 m) stand pipe and out the top is said to be capable of pumping 28 gpm (1061 m) at 6 (1.8 m) feet of head. That same pump may only be able to pump 12 gpm (45.1/m) out the top of a 13 ft (4 m) stand pipe (12 gpm at 13 ft head) [45.1/m at 4 m]. (See Fig. 17.10.10a next page). The friction in a system created by components, fittings, and lengths of pipe can be related to a comparable height of standpipe. It is important to understand that the physical height of the heating system above the boiler has no relation to the head pressure of the pump as expressed in feet of head. Since the heating system is a closed piping arrangement, the pressure due to height of the system is exerted on both sides of the pump and these counterbalance each other.

Pump manufacturers provide output charts with pump curves showing feet of head on the vertical axis and gpm on the horizontal axis (Fig. 17.10.10b next page). In order to select the proper pump, find the gpm required and the friction in head pressure produced by the HRH system. These two should intersect at a point below the pump curve. If they intersect above the pump curve, a larger pump is required. The method for determining gpm and feet of head is covered later in this chapter.

17.10.5.5   Boilers

Although there are many sources of hot water for HRF systems, by far the vast majority of systems are heated with a conventional boiler. Boilers are designed specifically for hydronic heating and are therefore readily adaptable to HRF. All boilers are tested by the manufacturers according to the Department of Energy (DOE) test procedures. Verification tests are conducted by the Hydronics Institute to confirm the manufacturers' testing.

a.

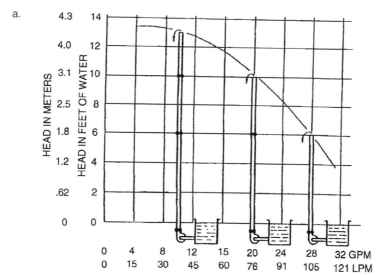

Example: Pump with 4 gpm (15 1/m) capacity. Enter at 4 gpm (15 1/m) capacity on lower horizontal line and then read upward to intersection of curves representing pumps A, B, C, and D. At each of these intersections, read to the extreme lefthand vertical line to obtain head pressure. For pump A the head pressure is 7 ft., (2.1 m pump B 8 ft., 2.5 m pump C 11 ft., 3.4 m, and pump D 13 ft., 4 m)

b.

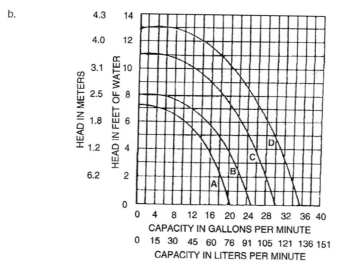

FIGURE 17.10.10   Head pressure and pump curves.

The tests measure the Btu input and steady state efficiency. In addition, each boiler model is tested to determine its annual fuel utilization efficiency (AFUE). This is based on cyclic tests which take into consideration off-cycle losses.

An essential component of a HRF system is an efficient, correctly-sized boiler. If a boiler is undersized, it is not quite obvious that it will not supply sufficient heat. Although it is not quite as obvious, an oversized boiler may prove inefficient, as well as being more costly to install. A boiler with an I = B = R rating will deliver the approved output efficiently if it is installed and fired in accordance with the manufacturers recommendations.

Boilers come in all shapes and sizes. Some are constructed of cast iron while others are copper or steel. There are models that can be hung on the wall or set on the floor in a mechanical room. Conventional through-the-roof flues are required on many, where others are force ventilated and can be vented through a side wall. Selection of a boiler is generally one of personal preference. What is important to the HRF system is that it be properly sized and that proper controls are used.

High Mass Boilers.   Boilers with wet walls constructed of cast iron are generally considered to be high mass boilers. They are termed high mass because the cast iron walls of the water jacket are thick and heavy, which allows them to retain heat. High mass boilers also hold a good deal of water within the water jacket, adding to the overall mass. This is a proven technology which has stood the test of time.

The mass within the boiler acts as a flywheel as it absorbs and dissipates heat. This allows the boiler to compensate for changes in flow or water temperature slowly. An HRH system containing several zones could have changing flow rates, depending upon the number of zones calling for heat. When the flow rate is slowed, the mass of the boiler absorbs the additional heat from the burner and prevents the boiler from short cycling or flashing to steam. Because HRF systems are comparatively slow to react to control inputs, high mass boilers are a good match.

Water temperatures used by HRF systems are usually on the low end of the operating range for high mass boilers. Operating the boiler at too low a water temperature can cause loss of efficiency, thermal shock to the boiler, and condensation of flue gases which produce corrosion. Operation of a higher mass boiler at water delivery temperature below 120°F. (50°C) is not recommended without the manufacturer's authorization. Mixing valves or other control apparatus can easily be added to the HRF system to allow the high mass boiler to operate in its most efficient range while producing the low water temperature required by the radiant floor.

Low Mass Boilers.   Boilers which contain relatively small quantities of water are considered low mass boilers. These units are quick to heat up and cool down. They are quite often force ventilated and make provisions for flue gas condensation. Unlike the high mass boilers, they cannot absorb dramatic changes in water flow and therefore require special attention to consistent flow rates.

Low mass condensing boilers can operate safely at the lower water temperatures required by HRH systems, but controls must be applied to prevent flashing (rapid buildup of steam pressure) due to changes in flow from zone operation.

Water Heaters.   Domestic water heaters are becoming more popular as a heat source for small HRF jobs. While not technically a boiler, they exhibit many of the characteristics of high mass boilers while having the ability to supply relatively low water temperatures. Water heaters are limited by output and generally do not have the longevity of boilers. Their efficiency ratings are also typically lower than boilers. Simplicity and low upfront costs have made the water heater attractive to some.

A separate dedicated water heater should be used for the HRF system. Mixing domestic water with heating water is not recommended. There are water heaters now available which have internal heat exchangers which keep the heating system and domestic water separate. This allows the same unit to do double duty.

Ground Source Heat Pumps. The ground source heat pump can be coupled with HRF resulting in improved performance of the GSHP due to the low water temperatures required, multiple zoning capabilities, and the heating efficiencies of HRF. A buffer tank should be used between the GSHP and the HRF system. This tank should be sized to approximately 1 gall (3.58 L) of water for each 1,000 Btu (kjoules) output of the GSHP. The GSHP should be designed to maintain the water temperature within the buffer tank while the HRF system draws off the buffer tank to supply the various zones of heating.

17.10.5.6 Controls

It is impossible, within the scope of this chapter, to completely cover the entire subject of controls. Full information regarding installation, wiring, and servicing should be obtained from the control manufacturer. A basic understanding of boiler controls is assumed. The purpose of this section is to indicate the function of controls required by HRF systems and the recommended method of arranging controls for proper operation.

Controls for HRF systems are made up of mechanical and electrical components. Mechanical components can either be set manually or driven by an electrical device, depending on the design of the system. The electrical portion is made up of switches, thermal sensitive elements, relays, and drive motors. Controls designed for use on circuits of approximately 25 to 30 V are known as low voltage controls. Controls designed for use on circuits of 110 to 115 V are designated as high voltage controls. High voltage controls are available for higher than 110 V but are not often required for residential purposes. Low voltage controls cannot be used on high voltage circuits.

Fluid Temperature Control. The temperature of the floor thermal mass is extremely crucial to the proper performance of an HRF system. Controlling the temperature of the water circulating through the floor is a primary factor in achieving proper floor temperatures. Water, which is hotter than the HRF system design calls for, can be pulsed through the floor at intervals, allowing the heat to dissipate laterally through the thermal mass. This technique can produce an average floor temperature equivalent to the required design temperature but may also cause hot spots over the tubes and cold spots between the tubes. Water cooler than the design calls for will not provide enough heat to maintain the proper temperature of the floor thermal mass. In some cases this can actually cause a heat drain while the energy pours into the earth below, never raising the floor temperature high enough to satisfy the room thermostat.

The ideal control would be one which would exactly match the water temperature to that required by the thermal mass of the floor, which maintains the floor surface temperature needed to radiate the appropriate amount of heat at any particular time. This, of course, is not possible with today's technology and the constantly changing output requirements of the floor. Contemporary controls range from simple fixed temperature controls to relational reset controls which adjust water temperature based on continuous outdoor and indoor temperature monitoring. Selection of any particular control is determined by the requirements of the boiler and the level of technical sophistication desired by the designer.

When using any type of mixing device, it is important that proper flow be maintained through the boiler at all times. Too much restriction of flow through the boiler can cause excessive cycling or flashing to steam. "Shunt" valves or pumps across the inlet and outlet of the boiler can be incorporated to help control boiler flow when necessary.

Fixed Boiler Temperature. The most basic control system is one that uses the temperature limiting device in the boiler to set the desired boiler supply water temperature. When the system calls for heat, the boiler fires and maintains a fixed water temperature within the minimum/maximum range of the internal boiler controls. (Fig. 17.10.11a).

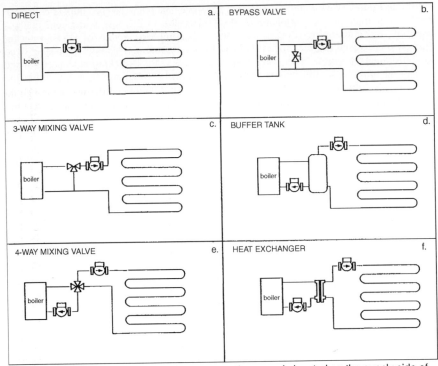

Unlike conventional hydronic heating systems the pump is located on the supply side of the floor heating loops. This forces water through the tubes for better purging of air on start-up and reduces the possibility of pump cavitation.

FIGURE 17.10.11   Boiler piping schematics.

Boiler Reset Control.   A boiler reset control sets the maximum water temperature of the boiler in proportion to the outside temperature. It consists of two temperature sensing elements and an adjustable control. One sensor is mounted outside to sense outdoor temperature. The other sensor is mounted on the supply side of the boiler (often clamped to the supply side pipe). The internal boiler controls are then wired to the reset control box to control the *on* and *off* switch to the boilers burners. When the reset control needs hotter water, it allows the burners to fire longer. The reverse is true when cooler water is required.

The reset control compares the outdoor temperature to the supply water temperature and adjusts the water supply temperature accordingly. The installer sets the desired ratio on the control. As the outdoor temperature drops the reset control increases the boiler supply water temperature. If the reset control increased the supply water temperature 1°F (1°C) for every 2°F (2°C) drop in outdoor temperature, the reset ratio would be 1:2. The correct way for determining the reset ratio is shown in the following formula:

$$\frac{\text{design water temp. } (-) \text{ indoor design temp.}}{\text{indoor design temp. } (-) \text{ outdoor design temp.}} = \text{reset ratio}$$

The minimum boiler supply temperature called for by the reset control must be adequate for the type of boiler used. Boiler manufacturers' instructions will specify the lowest water temperatures at which boilers can safely operate.

Boiler Bypass Valve. When a boiler is being used which must be operated at a higher temperature than the water temperature required by the floor, a simple solution may be a bypass valve. This method is primarily applicable when a fixed temperature is desired and there is no variation in the flow rate. A balancing valve is placed between the supply and return side of the boiler (Fig. 17.10.11b). A second valve may be required on the return side of the boiler to help restrict flow through the boiler.

When the system calls for heat, the pump will pull a portion of the water through the boiler and a portion from the water returning from the floor. The returning water will mix with the boiler supply water and temper the water. The resultant water temperature will depend upon the mix ratio. Opening the bypass valve will result in cooler water to the floor, while closing the bypass will direct more water through the boiler, resulting in hotter water to the floor. Care must be taken to provide adequate water flow through the boiler for proper boiler operation. Consult boiler manufacturers' instructions for minimum temperatures.

3-Way Mixing Valve. The basic 3-way mixing valve does essentially the same job as a boiler bypass valve. The valve is mounted in line on the supply side of the boiler with a connecting line to the return side of the boiler (Fig. 17.10.11c). By manually adjusting the valve, a mixture of boiler water and returning water can be balanced to the desired floor delivery temperature.

Another form of 3-way mixing valve is a tempering valve. This device is plumbed in the same manner as the 3-way manual valve but is designed to produce a constant temperature. The tempering valve has an internal thermal element which throttles the amount of return water that is allowed to mix with the boiler water. The thermal element is manually adjusted to a set temperature and will maintain a water mixture suitable to attain the set temperature at all times.

Motorized 3-way mixing valves are used with outdoor reset controls similar to the boiler reset control. A synchronous low voltage motor constantly modulates the 3-way valve to provide the proper mix of water to achieve floor delivery water temperatures called for by the reset control.

4-Way Mixing Valve. Maintaining a constant flow in the boiler is very desirable. The 4-way mixing valve makes this possible (Fig. 17.10.11e). It is positioned between a primary circulation loop (boiler and pump) and a secondary circulation loop (floor and pump). Boiler water and water returning from the floor are proportionally mixed to the desired floor delivery temperature and circulated back to the floor. Some of the returning floor water is also mixed with recirculating boiler water and returned to the boiler. Flow in both the primary loop and the secondary loop remains constant. The high water temperatures returned to the boiler reduce the chance of cold shocking the boiler and condensation.

4-Way mixing valves can be set to a fixed position manually or operated by an electric motor. This motor may be a wire wound synchronous motor or a thermal motor, either of which modulates the mixing valve to obtain the water temperature required by the control.

Buffer Tank. A buffer tank is another solution to the problem of maintaining a uniform flow, high temperature, and fixed water volume in the primary boiler loop while accommodating a low water temperature and variable flow rate in the secondary floor loop. Primary loop water is circulated from the supply side of the boiler to a pressure vessel (buffer tank), which typically holds 20 to 50 gal (75 to 190 L) of water, and returned to the boiler (Fig. 17.10.11d). An aquastat control senses the water temperature at the top of the buffer tank and turns the boiler and primary circulator on accordingly to maintain the set temperature. The secondary loop circulator draws water from the top of the buffer tank as called for by the room temperature controls and the returning water is fed into the bottom of the buffer tank.

The buffer tank reduces excessive cycling of the boiler by allowing it to react only to the mass of the buffer tank. The floor controls can call for heat at any time and any flow rate without interfering with the normal operation of the boiler.

Heat Exchanger.   The use of a heat exchanger (Fig. 17.10.11f) allows the primary heating boiler loop to be completely separated from the secondary floor loop. This is particularly useful when it is necessary to isolate the heat transfer fluids of the floor from those of the boiler. It is also an effective way to step down the temperature of the boiler to a temperature useful for HRF systems.

Two common heat exchanger construction types are tube-in-shell and plate. The tube-in-sell exchanger incorporates a primary boiler pipe inside a secondary floor pipe. The hot water flowing through the inner pipe conducts its heat through the walls of the pipe to the floor water flowing through the outer pipe. A plate exchanger is made up of a series of parallel chambers manifolded together and separated by metal plates. Hot boiler water circulates through every other chamber and transfers heat through the plate walls to the floor water circulating through the adjoining chambers.

Proper sizing of a heat exchanger is critical. Flow, exchange rate, and Btu output are important criteria that must be calculated before a selection can be made. Check the manufacturers' sizing procedures before choosing the heat exchanger.

Heat exchangers are generally very compact and can be mounted in close proximity to the boiler. Special attention must be paid when installing an exchanger to ensure that the inlet and outlet ports are plumbed properly.

Room Temperature Control.   Room temperature is the single most important control criteria in any heating control strategy. Although there are many other factors to be considered in designing and controlling an HRF system, all strategies hinge on controlling the room temperature. Some control strategies work directly by sensing room temperature while others achieve the desired room temperature by monitoring related factors which ultimately affect room temperature. Almost all control systems use some sort of room temperature-sensing device as the final control.

Room temperature controls regulate two things, system water flow and system water temperature. One strategy is to circulate water continuously through the floor while modulating the water temperature to accommodate room heating needs (temperature modulating system). A second strategy is to maintain a fixed water temperature but pulse it through the floor as required to keep the floor mass at the appropriate temperature (flow modulating system). A third strategy is to combine both flow modulation and temperature modulation into one system of control.

Flow Modulating Systems.   The simplest and most direct control is one that uses a room temperature-sensing thermostat to circulate water at a fixed temperature whenever heat is called for. This system could consist of a thermostat and relay combination which turns on a boiler and circulator when the thermostat senses a temperature drop in the room (Fig. 17.10.12a). A slightly more elaborate system incorporates zone valves which will open to allow flow through the floor when the thermostat calls for heat.

Zone valves may or may not have end switches. An end switch is a small switch located on the zone valve which can be used as a secondary control device. As an example, the thermostat may be connected to the zone valve. When the thermostat calls for heat, the zone valve opens. Upon opening, the end switch makes contact and turns on the boiler and circulator pump (Fig. 17.10.12b).

Systems which use zone valves without end switches generally have continuously circulating water while the boiler controls keep the water at the set level. The boiler and circulator can be controlled by an outdoor sensor, which will turn them off when the temperature outside rises above 65°F (18°C). In lieu of continuous operation, some manufacturers offer a special relay board to turn off the boiler/pump if all zones are satisfied.

Temperature Modulating Systems.   Temperature modulating systems operate on the premise that as the outdoor temperature goes down, room temperature can be maintained by raising the floor water temperature proportionately. These systems generally have continuously circulating water where the water temperature is modulated by some type of outdoor

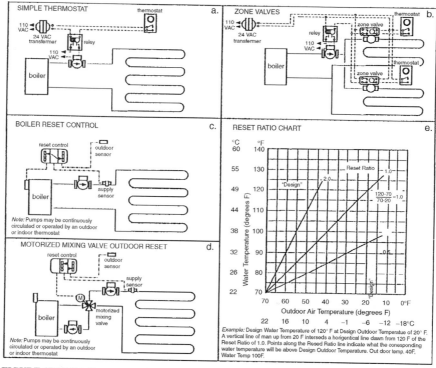

FIGURE 17.10.12   Room temperature controls.

reset control (Fig. 17.10.12c). Outdoor reset controls sense the outdoor temperature and the temperature of the water going to the floor or leaving the boiler. They then regulate the supply water temperature by operating a boiler control (see boiler reset control above) or a mixing device such as a 3-way or 4-way mixing valve (Fig. 17.10.12d). Each manufacturer of reset controls provides a procedure for calculating the proper reset ratio (Fig. 17.10.12e).

17.10.6   SYSTEM DESIGN

The key to the proper operation of any HRF system lies in good planning. Although the initial design steps seem cumbersome and time consuming, the time and money they save in the long run is well worth the effort. It is not necessary to go through a complete system sizing process for every job prospect. An estimate will help qualify the customer, but once the job becomes a fairly certain possibility, a thorough design process should be followed.

17.10.6.1   Floor Plan

The very first step in designing a successful HRF system is to make a study of the floor plans of the building (Fig. 17.10.13). Take notice of areas with unusual amounts of glass or

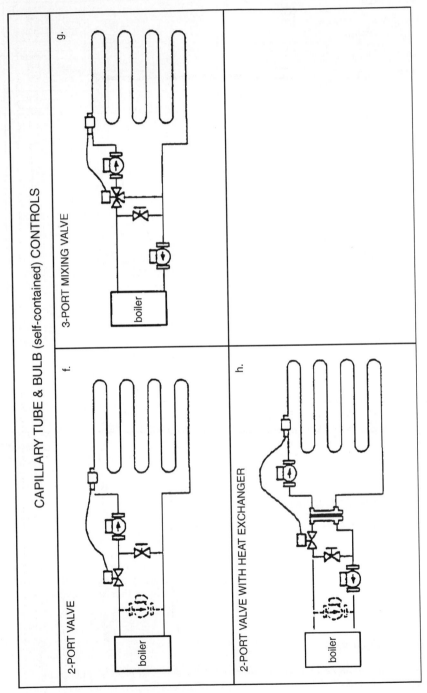

CAPILLARY TUBE & BULB (self-contained) CONTROLS

g. 3-PORT MIXING VALVE

f. 2-PORT VALVE

h. 2-PORT VALVE WITH HEAT EXCHANGER

boiler

FIGURE 17.10.12 (Continued).

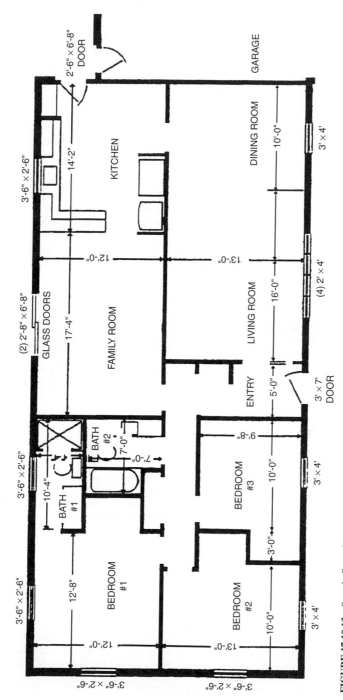

FIGURE 17.10.13 Sample floor plan.

very high ceilings. Also make note of areas with relatively small amounts of floor space and two or more outside walls. Consider special use areas such as shops or exercise rooms that may need to be treated differently. Identify the rooms and their respective floor coverings.

Begin the system sizing worksheet (Table 17.10.5) by identifying the job, location, and type of building (ie., single story, slab on grade residence). Also enter the inside temperature the HRF system will be required to maintain. Date the work-sheet for future reference. Enter the name of each room in the appropriate space on line 1 on the worksheet and indicate the type of floor covering on line 2. (A blank worksheet is provided, for convenience, at the end of this chapter (Table 17.10.10.)

Heat Loss Calculations.   A room by room heat loss calculation should be conducted from the information provided by the contractor and the floor plans. It is not within the scope of this chapter to instruct on the technique of performing such calculations. It is essential that the potential user of this section have a working knowledge of $I = B = R$ Heat Loss Calculation Guides No. H-21 or H-22, Ref. (1). Table 17.10.6 provides a sample heat loss calculation, which will be used in the examples in this section. A room by room calculation is necessary because tubing must be sized and spaced accordingly. Enter the heat loss figures for each room on line 3 of the sizing worksheet, Table 17.10.5. A blank form is available in Table 17.10.10.

Zoning.   Once the heat loss of each room is known, floor coverings are identified, and the floor plan studied, zoning can begin. Zones of heating are separate thermostatically-controlled spaces. Although prudent use of zoning can be very beneficial, excessive zoning increases system cost, complexity, and the owner interaction with little increase in benefit. To determine whether an area should be a separate zone, many things should be considered. The first consideration is always the client's wishes. Most HRF systems will perform adequately with little zoning. The client may be content to sacrifice some comfort and efficiency for simplicity and low cost. On the other hand, an owner may wish to control temperatures from room to room.

It several zones are being considered, look for logical separation of spaces by physical barriers such as walls, partitions, or varying floor levels. A two-story home may be easily zoned into two zones, one on the first level and one on the second. Also respect the type and frequency of use. Living areas may be grouped on one zone with sleeping areas on another. Heat loss and heat gain should also be regarded when deciding on zones. Large windows can cause a room to both lose heat and gain heat in greater proportions than the rest of the structure and therefore be a good candidate for a separate zone. Finally, a large difference in the R-value of floor coverings can be cause for zoning. Hot-water temperatures required for a heavy carpet and pad can cause hot spots and overheating in areas on the same zone which have hard surface floor coverings. Some of this disparity can be handled with tube spacing and flow, while in other areas this can actually be desirable, for example, bathrooms. The size of the zone may be limited by the manufacturer's rating on the manifold or pump capacity.

Outline the zones on the floor plan with a highlighting pen and number them. Make sure that all parties are in agreement with the zone selection.

Manifold Placement.   Each manufacturer has a recommended mounting procedure for manifolds, but most are in agreement that out of the way places are best. The back of a closet or behind a door mounted in the wall cavity are the most common locations. Always consider ease of access when placing a manifold. More than one installer has come back to do service work and found the manifold inaccessible.

Manifolds should be placed as close to the zone they are servicing as possible. The ideal location is one that allows the tube to fan out in all directions with little congestion near the manifold. Remember that the tube can usually be routed out the back of the wall cavity, as well as the front. Try to avoid locations that will require a concentration of tubes down a hallway or through a room. This is not always possible.

TABLE 17.10.5 System Sizing Worksheet

I=B=R HRM WORKSHEET

OWNER _____ ADDRESS _____ DATE _____

TYPE OF BUILDING _Slab on grade_   INSIDE AIR TEMP. _70°_ F   OUTSIDE AIR TEMP. _0°_ F

		ZONE #1	#1	#1	#2	#2	#3	#3	#3	#3	#3
1	ROOM	Living	Entry	Dining	Kitchen	Family	Bdrm #1	Bdrm #2	Bdrm #3	Bath #1	Bath #2
2	FLOOR COVERING	3/8" wood	tile	3/8" wood	tile	1/2" crpt	1/2" crpt	1/2" crpt	1/2" crpt	vinyl	vinyl
3	HEAT LOSS (BTU/HR)	3799	1650	2886	3887	4026	3335	3254	2212	1280	333
4	TOTAL AREA (SQ FT)	208	65	130	171	207	153	130	127	44	49
5	AREA WITHOUT TUBING	0	13	0	49	0	10	10	10	9	28
6	ACTUAL HEATED FLOOR AREA	208	52	130	122	207	143	120	117	35	21
7	FLOOR OUTPUT REQUIRED	18	32	22	32	19	23	27	19	37	16
8	FLOOR SURFACE TEMP REQ. (D)	80	86	82	86	81	82	84	81	88	79
9	R-VALUE OF FLOOR COVERING (B)	0.5	0.2	0.5	0.2	1.7	1.7	1.7	3.4	0.2	0.2
10	DESIGN WATER TEMP (D)	87	92	92	92	115	116	125	146	94	91
11	TUBE SIZE (NOMINAL I.D.) (E)	1/2"PB	1/2"PB	1/2"PB	1/2"PB	1/2"PB	1/2"PB	1/2"PB	1/2"PB	1/2"PB	1/2"PB
12	TUBE SPACING (E)	12"	6"	9"	6"	12"	9"	9"	9"	6"	(12") 9"
13	TUBE SPACING FACTOR (E)	1	2	1.3	2	1	1.3	1.3	1.3	2	2
14	ENTER VALUE FROM LINE 6	208	52	130	122	207	143	120	117	35	21
15	TOTAL TUBE REQUIRED (F)	208	104	169	244	207	185.9	156	152.1	70	42

		1	2	3	4
16	ZONE SUMMARY				
17	TOTAL TUBE REQUIRED	481	451	606	
18	MFR'S MAX LOOP LENGTH	300	300	300	
19	NUMBER OF TUBE LOOPS	2	2	3	
20	MANIFOLD SIZE				
21	LONGEST LOOP LENGTH	241	226	202	
22	TOTAL HEAT LOSS	8335	7913	10414	
23	GPM REQUIRED	1.1	1.1	1.4	
24	SUPPLY WATER TEMP	100	123	133	
25	PRESSURE DROP (F)	2.7 ft hd	2.5 ft hd	1.6 ft hd	

	SYSTEM TOTALS		
26	TOTAL TUBE REQUIRED	1538	
27	TOTAL NO. OF TUBE LOOPS	7	
28	LOOPS PER ROLL OF TUBE	2	600 ft roll
29	NO. OF ROLLS	3.5	
30	TOTAL HEAT LOSS	26662	btu/hr
31	SYSTEM GPM REQUIRED	3.6	gpm
32	SYSTEM PRESSURE DROP	2.7 + trunk	ft of head
33	SYSTEM VOLUME	15 + trunk	gallons
34	SYSTEM SUPPLY TEMP	133	deg F

TABLE 17.10.6 Sample Heat Loss Calculations

I·B·R CALCULATION FORM 1304

HEAT LOSS

Copyright 1968. The Hydronics Institue, Inc.
Berkeley Heights, NJ 07922

Customer _____

Name Joe Cool

Address _____

Telephone _____ Date _____

Heat Loss Calculated By _____

Consider the piping required for supply and return lines to the manifold from the boiler. The shorter the runs, the less head pressure the pump must overcome. It is better to have the manifold close to the zone with long supply and return lines, than to have the manifold far from the zone. Indicate the manifold locations on the floor plan with a marker for future reference (Fig. 17.10.14).

Boiler Location. Use the guidelines provided in I = B = R Installation Guide No. 200 (Ref. 1) for locating the boiler.

Area Calculations. Measure the total area of each room and enter this number on line 4 of the worksheet. The information may already be available from the heat loss calculations. Tubing should be installed only in areas where the floor surface will be exposed to the room. Review the floor plans and indicate all areas where the floor will be covered by built-in cabinets, book cases, entertainment centers, window seats, or large appliances such as a stove, refrigerator, washer, or dryer. Tubing is generally not installed on stairs or under prefab bathtubs, so these areas should also be identified. Calculate the area in each room where tubing will not be installed and enter this information on line 5 of the worksheet. Subtract the area without tubing (line 5) from the total area (line 4) to find the actual heated floor area (line 6). This is the actual floor area that will be required to support the heat loss of each room.

17.10.6.2 Floor Output Calculations

The ultimate output of the HRF system is determined by the difference between the surface temperature of the exposed floor and the surfaces of the walls, ceilings, and furniture (AUST) surrounding it. The greater this difference, the more output is required from the floor. Water temperatures and flows must be figured based on floor coverings and required outputs in order to assure that the surface temperature of the floor is adequate to maintain the proper room temperature. For calculation purposes in this guide the indoor room temperature will be used in place of the AUST. The result is quite adequate for most system-sizing applications. For rooms with unusual amounts of glass, AUST should be used in place of room air temperature to obtain a more accurate prediction of floor output and required water temperature. Consult the *ASHRAE 1992 HVAC Systems and Equipment Handbook*, Chapter 6 (Ref. 3) if a greater knowledge of AUST is desired.

Required Floor Output. Only that floor area which is exposed to the interior of the room and which contains tubes will contribute to heating the space. Divide the heat loss of the room (line 3) by the actual heated floor area (line 6) to find the floor output required to heat the space at the outdoor design temperature. This figure is expressed in Btu/h/ft². Enter the result of this calculation on line 7 of the worksheet. This is the maximum output the floor should be required to produce. During a normal heating season the floor will seldom be called on to deliver this amount of heat.

It is generally accepted that the maximum allowable output in a normal traffic area is 30 Btu/h/ft². (340 ks/h/m²). Higher outputs require higher surface temperatures, which may cause discomfort to individuals standing on the floor for extended periods of time (for example, a kitchen or recreation room). Higher outputs can be obtained around perimeters and in hard surface areas bordered by excessive glass. Rooms with required floor outputs greater than 30 Btu/h/ft² (340 kJ/h/m²) may require special attention in the following calculations, or even supplemental heat. If possible, have the building changed to reduce the heat loss. This can be accomplished by using more efficient windows, smaller windows, or increasing wall and/or ceiling insulation.

Floor Surface Temperature. As mentioned earlier, the output from HRF is primarily a function of the surface temperature of the floor. The greater the floor surface area is in comparison to the rest of the rooms' unheated surfaces, the lower the floor surface temperature required.

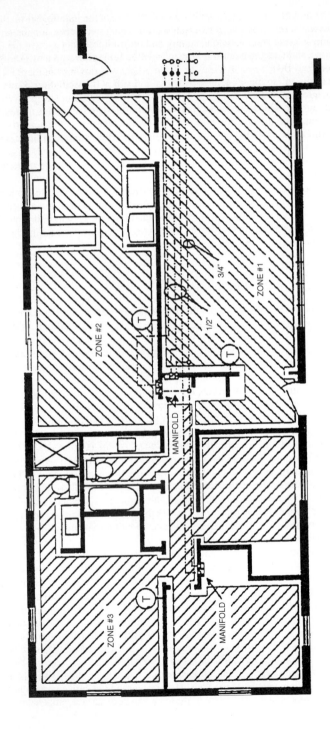

FIGURE 17.10.14 Zone identification and piping runs.

A small room with two outside walls, such as a bathroom, will require higher floor surface temperatures than a large room with two outside walls. The more the output required in Btu/h/ft^2 (kJ/h/m^2), the higher the surface temperature.

Using Fig. 17.10.15 locate the floor output required for each room and read the floor temperature required by moving to the right of the output and down from the floor surface line. Enter the required floor surface temperatures on line 8 of the worksheet. Keep in mind that these temperatures are only required when the outdoor temperature drops to the design level. The HRF system will operate well below the required floor surface temperatures the majority of the time.

HRF floor surfaces will only feel warm to the touch as they approach the 85°F range because that is the approximate surface temperature of the human body. When a person touches a surface which is colder than his body, heat flows from the body to the surface and the sensation is that of cold. When a bare foot comes in contact with room temperature ceramic tile, the warm foot immediately looses heat to the mass of the tile, which makes the tile feel cold. The closer the tile temperature comes to the surface temperature of the foot, the more neutral is the feeling. In most cases, the HRF floor will feel neutral to the touch. Surface temperatures warmer than 85°F (25°C) are generally not desirable except in areas such as bathrooms, entries, or around pools.

Floor surface temperature is regulated by the temperature and flow rate of the water through the embedded tubing. These factors are determined by tube size, pump size, and tube spacing. Close spacing of the tubes and high flow rates produce the most even floor surface temperature. Wide spacing and large tubing can result in warm and cool spots across the floor. In this case, the required floor surface temperature becomes an average of the warm areas over the tubes and the cool areas between the tubes. Insulative floor coverings, such as carpet and pad, will cause the heat to spread laterally between the tubes and result in a more even surface temperature. In this case, wider spacing will not affect surface temperatures.

When a higher than acceptable floor surface temperature is required, several solutions may be applied. First, look for ways of reducing the heat loss of the room by selecting better windows or insulation. Installing tubes closer together around the perimeter to raise the surface temperature in areas of little or no traffic can bring the average floor temperature up to meet the requirement. If the required floor temperature cannot be reduced sufficiently, supplemental heating may be required to reduce the required Btu output of the floor.

Floor Covering. Of all the variables that contribute to the successful operation of an HRF system, floor coverings can make the most difference. While HRF systems can operate well with most floor coverings, it is imperative that care be taken in selecting and designing around the floor coverings. Any covering which is placed over the subfloor will restrict the heat transfer from the tubing. Ceramic tile and vinyl have only a small effect on the overall operation of the HRF system, but carpet and pad can have a significant impact. Because surface temperature is all-important, anything placed between the floor surface and the tubes will require the water temperature to be raised. The more insulative the floor covering, the hotter the water temperature required.

Insulation qualities of floor covering are measured in R-value. Table B shows the average R-value for various floor coverings. To find the R-value of a particular floor covering, simply locate the floor covering and thickness on the left of the chart and read the R-value directly to the right. When using carpet and pad, be sure to add the R-values together. Note that using the conventional urethane pad can more than double the R-value of the carpet and pad combination. A $^1/_2$-in (13-mm) plush carpet with a $^1/_4$-in (6-mm) rubber pad can be a good compromise for both feel of the carpet and operation of the HRF system. This is an area where it is important for the owner to be involved and made to understand the reasoning behind floor covering choices. Once the floor coverings have been determined, enter R-values on line 9 of the worksheet.

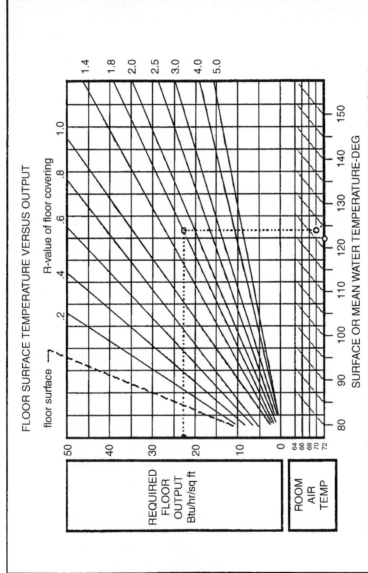

MEAN WATER TEMPERATURE: Find the required floor output on the left of the chart and move right to intersect the diagonal line representing the R-value of the floor covering. Extend a line down vertically to intersect the room air temperature. Follow the diagonal line to the left down the bottom of the chart and read mean water temperature.

FLOOR SURFACE TEMPERATURE: Follow the instructions above using the floor surface dingonal line rather than the R-value line and read the surface temperature at the bottom of the chart.

FIGURE 17.10.15   Floor output.

Because of the insulative properties of some carpet and pad combinations, consideration should be given to the downward flow of heat below the floor. As the temperature of the thermal mass is increased to penetrate the floor covering, a greater temperature difference is created between the thermal mass and the area below the floor. If the space below is unheated or a slab on grade, additional under floor and slab insulation should be considered. Should the space below be heated, such as a basement, downward heat loss should be taken into account. In some cases the downward flow of heat can be sufficient to heat the lower space.

Water Temperature.   As warm water travels through the tube embedded in the floor, it releases heat to the thermal mass. The larger the difference between the temperature of the water and the thermal mass, the more heat is transferred. Because of this transfer, the water entering the floor is at a higher temperature than that returning from the floor. This temperature drop can range from 5 to 40°F (2.7 to 22°C) during various phases of operation. For design purposes, the temperature drop from supply to return across an HRF floor is considered to be 15°F (8.3°C). To determine the average or mean design water temperature, locate the required floor output in the left-hand column of Fig 17.10.15. Also locate the R-value of the floor covering along the top or the right side of the chart. Extend a line horizontally from the floor output to intersect the R-value line. From that intersection drop vertically down to the appropriate room temperature line and then diagonally to the bottom of the chart and read the mean water temperature. Enter this temperature on line 10 of the worksheet.

The design water temperature is not required at all times. It is only required when the outdoor temperature is at or below the design temperature for which the heat losses were calculated or when the room is recovering from a set back of the thermostat. Use the lowest water temperature that will provide the required floor surface temperature. This will result in more even floor surface temperatures, minimize thermal stresses on the slab and floor coverings, and reduce overshooting of the desired room temperature.

17.10.6.3   Tubing Section

As demonstrated in the section on components, tubing comes in a variety of sizes and materials. The primary purpose of the tube is to deliver water to the thermal mass of the floor in sufficient quantities and at a predetermined temperature to provide the heat required to maintain the design floor surface temperature. While most tubes are capable of doing this job with proper design, this section will assist in determining the tube capable of performing the task most efficiently.

Tube Size.   Once the design water temperature has been established, the required output of the floor in Btu/h/ft^2 (ks/h/m^2) will determine the rate of flow of water. A small heat output requirement may be handled with a single small tube per square foot, whereas a large heat output may be handled with a single large tube or several small tubes per square foot. The selection of tube size is a function of both flow and heat distribution as illustrated by Table 17.10.7. Tube size is stated as the nominal inside diameter (ID) of the tubing. This can be a bit a confusing because, although the actual outside diameter (OD) of a $^1/_3$-in (13-mm) tube, for example, will remain constant with different tubing materials, the ID can vary significantly. This will affect the actual flow rates of tubes having the same nominal ID but made from different materials. Table 17.10.7 is based on averages and can be applied to most applications. For flow rates of a specific tube contact the manufacturer.

Follow the manufacturer's recommendations for tube size. When more than one size of tube is available, consider the following criteria for making a selection.

Zone Size.   Larger ID tubing can accommodate longer tube loops. In applications with large zones or long distances from the manifold, larger ID tubing may allow the use of less tubing and fewer manifold connections.

TABLE 17.10.7 Tube Spacing Selection

	TUBE SPACING (o.c.)							
Required Floor Output (Btu/sq ft/hr)	Tube Diameter (nominal I. D.)							
	3/8"		1/2"		5/8"		3/4"	
10-20	9"	1.3	12"	1.0	12"	1.0	15"	0.8
20-30	6"	2.0	9"	1.3	9"	1.3	12"	1.0
30-40	4"	3.0	6"	2.0	6"	2.0	9"	1.3

Spacing Factor - to find total tube required, multiply the spacing factor by the tube coverage area.

Slab Thickness.   Heat from tube embedded in a thick concrete slab will spread out laterally as it rises to the surface and will produce a fairly uniform surface temperature; therefore most tube sizes will work. On the other hand, tubes are much closer to the surface in a thin thermal mass. This can create warm and cold spots if tubing is spaced too far apart. Smaller ID tubing closer together will provide a more even heat distribution and reduce the required thickness of the thermal mass.

Tube Quantity.   When spacing is not the prime consideration, using larger tubing reduces the overall amount of tube and labor needed and thereby reduces cost. Use this cost-cutting measure only after it has been determined that it will not affect performance of the system.

Tube Material.   Consult the tubing information provided in this guide under the section entitled System Components.

After considering all the above factors, select the tube size and enter it on line 11 of the worksheet. Tube size may differ within the same job. For example, the basement slab may require $^1/_2$ ID (13 mm) tubing while $^3/_8$ in (9.6 mm) ID tubing would be a better choice for the suspended floor above. If possible, working with the same size tubing throughout the job is desirable from a design, materials, and labor standpoint.

Tube Spacing.   As the heating load per square foot is increased, more warm water must be delivered to the floor. This can be accomplished by (a) increasing the size of the tube, (b) increasing the pump pressure, or (c) increasing the amount of tubing per square foot. Table 17.10.7 takes all these factors into consideration. The term *on center* (o.c.) refers to the distance from the center of a tube to the center of the next tube lying parallel to it.

Some general guidelines pertaining to spacing selection are as follows:

Slab Thickness.   In thin topping slabs, 9 in (23 cm) o.c. or wider spacing used under hard surface floors, such as vinyl or tile, will result in noticeable floor surface temperature variations. Use 9 in (23 cm) o.c. or greater spacing only under wood floor or carpet. The high R-values of wood and carpet result in a greater lateral heat movement between the tubes, providing a more even floor temperature.

Heat Output.   Tube spacing can be an effective tool to use in areas of high heat loss. Use closer o.c. spacing in close proximity to areas of expected high heat loss. 4 or 6 in (10 or 15 cm) spacing can be used to cover the area 3 ft (0.9 m) in front of a glass patio door, picture windows, or entry way. These cold surfaces will draw more heat from the floor than normal because of the extreme differences in surface temperatures (lower AUST). Because of the low surface temperature of the glass, the floor will put out many more Btu/h/ft^2 than shown in Fig 17.10.15 and Table 17.10.7, therefore the water flow and temperature must be able to provide that amount of heat. Putting tubes closer together in front of the window or high heat loss area, in effect, increases the flow.

Warm Floors. At times it is desirable to have a floor warmer than normal. Tile bathroom floors or areas around hot tubs are two examples. Here closer spacing can increase floor temperature without increasing water temperature, which would affect the rest of the zone.

As shown in Table 17.10.7, for any given required floor output there are a variety of choices of tube spacing. Find the floor output required which was listed on line 7 of the worksheet and locate it on the tube spacing chart. Move to the right and find the tube spacing number under the appropriate tube diameter size. This is the maximum tube spacing allowable for the given criteria. Enter this number or a number less than this number (based on the above variables) on line 12 of the worksheet.

Total Tube Required. To calculate the actual amount of tubing which will be needed, enter the actual heated floor area from line 6 of the worksheet. Enter this figure on line 14 of the worksheet.

Table 17.10.7 contains a spacing factor next to each o.c. number. This spacing factor is derived by dividing 12 in (30 cm) by the o.c. number. For example, 12 in (30 cm) divided by a spacing of 6 in (15 cm) o.c. results in a spacing factor of 2.0. This means that there are 2 ft (2 m) of tubing per every square foot (qm^2) of actual tube coverage area. Enter the appropriate spacing factor on line 13 of the worksheet and multiply it by line 14, the actual heated floor area. The result of this calculation will produce the total tube required which, can be entered on line 15.

17.10.6.4 Zone Summary

The above procedure gives a room-by-room account of equipment and sizing requirements. This information must now be summarized and applied to each zone. Identify the areas contained in each zone and complete the summary section on the worksheet as follows:

Tubing Requirements. The information found in lines 11 through 15 of the worksheet is used in determining the material needs for each individual zone.

Total Tube Required. The total amount of tube required per zone can be found by simply adding together all the total tube required figures on line 15 of the worksheet which pertain to one zone.

Loop Lengths. As the length of a tube is increased, friction is also increased. More friction requires more pump pressure to overcome it. The pump size is one factor that limits than length of tube which can be used in an HRF loop. Another is the velocity of the water. Forcing water through tubing at too high a rate can cause erosion of the inner surface and can lead to tube wall failure. Most manufacturers publish limits as to tube flows and lengths. Enter the manufacturers maximum tube length on line 18 of the worksheet. If this is not available, a good rule of thumb is as follows: $^3/_8$ in (9.6 mm) ID – 200 ft (6 m), $^1/_2$ in ID (13 mm) – 300 ft (9 mm), $^3/_4$ (19 mm) ID – 500 ft (15 m).

Loop lengths may vary as long as they remain under the manufacturer's recommended maximum. The easiest system of balancing flow is to have all loops on a given manifold the same length. The flow in each tube will then be the same. Water will take the path of least resistance. When a manifold has a mixture of short and long tube loops, the longer loops will be deprived of adequate flow and the shorter loops will have excess flow. Manifolds equipped with balancing valves assist in dealing with challenge.

Some designers prefer dividing total tube required in the zone by one loop length equal to or less than the maximum allowable. For example, a zone which required a total of 1200 ft (370 m) of tube could be divided by a 300-ft (90-m) tube length [maximum length for $^1/_2$ m (12 mm) ID tube] resulting in four loops of equal length. These could then be distributed throughout the zone and may, in some cases, require a single loop to service one full room and a fraction of another.

A second technique for determining loop length is to limit loops to a single room utilizing only that length of tubing which is required for each specific room. Rather than using a 300-ft (90-m) loop to do all of one room and part of the next, as illustrated above, a 200-ft (60-m) loop would be used for that room only. This method requires more calculation and attention to detail but gives clear control of individual rooms at the manifold and makes tubing layout easier. Depending on the size and spacing of the tubing, a room may use two loops instead of one-and-a-half. Each room must be looked at separately. Take the total tube required by the room and divide it by the number of loops that will result in loop lengths less than the maximum allowable. Remember to consider the amount of tube required to get to and from the manifold location. For example, consider a room with a total tube requirement of 530 ft (163 m) of $^3/_8$ in (9.5 mm) ID tube, dividing the total tube by 2 would result in two loop lengths of 265 ft (82 mm). This in excess of the maximum allowable tube length for $^3/_8$ in (9.5 m) ID tubing (200/ft) (61 m). Dividing the total tube by 3 produces three loops 177 ft (55 m) long, which is well under the maximum loop length of 200 ft (61 m). There is also an additional 23 ft (7.3 m) of tube remaining (200 ft − 177 ft = 23 ft) to get to and from a manifold which is not located in the room.

Whichever method is chosen, enter the resultant number of tube loops on line 19 of the worksheet. Also, add all of the loop lengths together, divide by the total number of loops and enter the average loop length on line 21 of the worksheet.

Manifold Size. Most manufacturers provide manifolds for use with their tubing. The number of loops in a zone will determine the manifold to be used. A single manifold may have enough ports to accommodate all of the loops in a single zone or two or more manifolds may be required. This will depend on the physical capability and manufacturer's recommendations for each application. After checking the manufacturer's specifications, enter the ID size of the manifold and the number of loops it can service on line 20 of the worksheet.

Fluid Requirements. The most common heat transfer fluid in use is water. The calculations in this guide use water as the transfer medium. Additives such as rust inhibitors or oxygen scavengers have very little affect on the heat transfer properties of water. Glycol and other antifreeze solutions will reduce the heat transfer properties as well as require higher pump pressures, because of the increased viscosity of the fluid. If a 50/50 glycol/water mixture is used, add 50 percent to the system pressure drop at the same flow rate.

Flow Per Zone. The charts used to determine tube size and spacing are based on the heating requirement of the floor and flow rates. When 1 lb of water drops 1°F in 1 h, it gives off 1 Btu of heat. A gallon of water weighs approximately 8.3 lb, so it will give off 8.3 Btu if it drops 1°F, as it is circulated through the system. When the Btu/h requirement of a zone is known, the amount of water needed at a given temperature drop can be calculated in gallons per minute.

Design Water Temperature Drop (ΔT). The design water temperature drop is the difference between the temperature of the water entering in and departing from the floor as determined by the designer. The ΔT chosen for the system design provides the basis for determining zone the system GPM requirements and affects the pipe and pump size requirements. The design water ΔT is a criteria that is based on the outdoor design temperature. In actual practice, the ΔT across the supply and return will vary. As a normal practice, 15°F ΔT is used in calculations. This translates to 7500 Btu per GPM of flow. To find the flow needed per zone, add the heat losses of all the rooms in the zone and enter the number on line 22 of the worksheet. Divide the total zone loss on line 22 by 7500. The result is the number of GPM required for the zone and can be entered on line 23. Do this same calculation for each zone.

To calculate flow for a ΔT other than 15°F, multiply the desired ΔT times 500 (for ease of calculation). Five hundred is approximately the number of Btu delivered by the system when one GPM is circulated through the system. (10°F ΔT = 5000 Btu/h, 20°F ΔT = 10,000 Btu/h).

The higher the design water temperature drop, the more Btu/h delivered per GPM which, in turn, requires less water to be circulated to deliver a given amount of Btu. The lower the flow rate, the smaller the pipe size required. So, a larger ΔT means smaller pipe and pump sizes, and lower installation and operating costs. On the other hand, too large a ΔT can cause noticeable surface temperature variations, and thermal stresses on slabs, floor coverings, and boilers.

Zone Supply Water Temperature. A look at line 10 of the worksheet will reveal that the design water temperature varies from room to room. Without the ability to supply a separate water temperature to each room, the room with the highest design water temperature must be chosen to represent the zone. Because this temperature is a mean or an average between the supply and the return, the actual supply water temperature must be calculated using the design water temperature and the design water temperature drop. Simply divide the design water temperature drop by 2 and add the result to the design water temperature. For example, a design ΔT of 15°F divided by 2 = 7.5°F. Add this to a design water temperature of 110°F and the result is a supply water temperature of 117.5°F. This is the temperature of the water which must be provided to the supply side of the manifold. The water will return from the floor to the manifold at 102.5°F (117.5°F − 15°F) under design conditions.

Zone Pressure Drop. The length of a tube and the amount of water which must be forced through the tube determine the amount of pump pressure required. The longest tube loop length is used to establish the pressure drop for the entire zone because the pump must produce enough pressure to maintain proper flow through that loop. If this is accomplished, the other loops will also be satisfied. Manifolds that provide balancing valves for each loop can be adjusted to correct for differing loop lengths and flow requirements. A balancing manifold should always be used if loop lengths within a zone vary more than 15 percent. A longer loop generally covers more area than a shorter loop and needs to provide more heat. Therefore, the longer loop would require more flow than the shorter loop. Because water takes the path of least resistance and the shorter loop has less resistance due to its length, the flow will be greater in the short loop than the long loop where it is needed. A balancing manifold can be adjusted to restrict the flow to the short loop and force the water through the longer loop to provide the proper GPM. See the manifold manufacturer's instructions for details on balancing.

Table 17.10.8 provides the pressure drop in tubes of various sizes and materials. Select the longest tube loop within the zone and calculate the GPM required for that tube. This is done by first dividing the longest loop length by the total tube required for the zone from line 17 of the worksheet. Multiply the result of this calculation by the total GPM required for the zone listed on line 23 of the worksheet. The result is the GPM required for the longest loop. For example, if (a) the longest $^1/_2$ in ID PEX tube loop in a zone is 325 ft, (b) the total tube required for the zone is 1000 ft and (c) the zone GPM required 2.1 GPM, the GPM required for the longest loop would be 0.7 GPM (325 ft/1000 ft × 2.1 GPM = 0.7 GPM).

Locate this GPM on the left-hand side of Table 17.10.8, than move to the right and find the pressure drop in feet of head (Ft Hd) for $^1/_2$ in PEX tubing (approx. 1.9 Ft Hd). This is the pressure drop in 100 ft of tubing.

Because the loop in the example is 325 ft in length, multiply the pressure drop from the chart by 3.25 to get the total pressure drop for the loop (3.25 × 1.9 = 6.18 Ft Hd). This is also the head pressure for the zone. Enter this number on line 25 of the worksheet.

17.10.6.5 Total System Summary

The final step in the system design process consists of collecting the data from each zone and applying it to the overall system requirements. Although this section will cover some of the information needed to properly size boiler, pump, and pipe sizing for the primary

TABLE 17.10.8 Tube Pressure Drop

Pressure drop in various tubes per 100 ft of tube
Pressure drop shown in feet of head (psl = Ft Hd/2.3066)

Gpm/loop	203" ID	1/4" ID	3/8" nominal ID			1/2" nominal ID			5/8" nominal ID	
			EPDM	PB	PEX	EPDM	PB	PEX	EPDM	PEX
0.1	5.9	1.18		0.16	0.28			0.06		
0.2	17.1	4.26		0.58	0.80			0.20		
0.3	33.9	8.01		1.24	1.81		0.50	0.42		
0.4		18.80		2.11	3.09		0.80	0.69		
0.5		23.19	3.80	3.20	4.57			1.03		
0.6				4.46	6.24		1.10	1.43		0.54
0.7		31.53		5.93	8.32		1.50	1.89		0.71
0.8			6.20	7.60	10.53		1.90	2.39		0.90
0.9				9.44	12.89		2.40	2.99		1.11
1.0		17.30	12.70	11.62	15.70	3.50	2.90	3.60		1.35
1.2				16.08	21.58	4.65	4.00	4.99		1.86
1.4				21.39	28.60	6.60	5.40	6.57		2.46
1.6				27.38	36.35	8.65	6.90	8.33	2.90	3.12
1.8						10.63	8.50	10.34	3.15	3.85
2.0						13.30	10.40	12.48	4.10	4.62
2.2							12.40	14.50	5.40	5.25
2.4							14.50	17.10	6.45	6.18
2.6							16.80	19.80	7.50	7.15
2.8								22.45	8.80	8.20
3.0								24.70	9.70	9.20

Note: Pressure drops are approximate, consult manufacturer's specifications.

loop, a working knowledge of the I = B = R 200 Installation Guide (Ref. 2) or equal experience is required.

System Tube Requirements.   If the tube specified for the system is all one size, simply adding together the tube from each zone will result in the total tube required for the job. If two or more tube sizes are used, find the total of each size of tube used. Enter this information on line 26 of the worksheet. Also add the total number of loops for each size tubing from all of the zones and enter this number on line 27.

Determining the number of loops per roll of tubing will depend on the length of tubing within a roll as provided by the manufacturer. It is wise to know the roll sizes at the beginning of the sizing process, so that you can adjust the loop lengths accordingly to maximize the use of a roll. Splicing tubing in the floor is generally not an acceptable practice.

If loops are of varying lengths throughout the system, it may be necessary to look at loops individually and determine how they may be combined to get the most use from a single roll of tubing. Always allow at least a 10 percent overage for actual field applications. Coming up 15 ft (4.6 m) short after laying out a loop is not a pleasant experience. The number of loops calculated per roll of tubing can be entered on line 28 of the worksheet and the number of rolls required per tube size entered on line 29.

System Heat Loss.   Adding the heat loss totals from each zone will provide the total heat loss for the entire system. Enter this total on line 30 of the worksheet. The total system heat loss is used to select the boiler with an I = B = R rating which indicates that it will deliver sufficient output to match the heat loss of the system. Slightly oversizing the boiler by 10 percent to 15 percent is recommended to assist in cold start-up and improve recovery times. Any other equipment added to the system which will require additional heat from the boiler, such as an indirect water heater, should also be added to the system heat loss when choosing a boiler.

Total System Flow.   In most cases, adding the GPM required from each zone will result in the total GPM required for the system and can be entered on line 31 of the worksheet. If other devices such as baseboard convectors, fan coils, or indirect water heater are also included in the system, their flow requirements should be added as well.

System Pressure Drop.   Pressure drop in the primary loop, that loop containing the primary circulator pump, boiler, expansion tank, miscellaneous fittings, and the supply and return plumbing, must be calculated. As in the secondary loop (floor), friction in pipe and fittings causes restriction to water flow. If there are several zones with a supply and return to each zone, use the supply and return which has the greatest pressure drop as representative of the others. If the pump can overcome the resistance in that path, it can handle the others as well. Pressure drops across pipes, fittings, and the boiler must be added together with the longest supply and return path to provide the total pressure drop for the primary loop. The system pressure drop entered on line 32 of the worksheet is the total of (a) the pressure drop in the primary loop, (b) the pressure drop of the longest supply and return line, and (c) the zone with the highest pressure drop. Pressure drop calculation instructions are found in I = B = R Guide 200 (Ref. 2).

System Volume.   It is important to know the volume of heat transfer fluid in the system in order to properly size the expansion tank and plan for additives such as glycol or water treatment chemicals. Table 17.10.9 presents the approximate water volume of pipe and tubing per lineal foot. Find the tube size and multiply the volume represented by the total tube required on line 26 of the worksheet to calculate the volume of the water the tubing in the floor will hold. Do the same calculation for the various pipe sizes and lengths in the supply and return lines, and the primary loop. Finally add the volume of the boiler and any additional equipment from the information provided by the manufacturer. This is the total volume for the system and can be entered on line 33 of the worksheet. Use the expansion tank manufacturer's tables or charts to select an expansion tank that fits the system's volume, operating pressure, and temperature.

TABLE 17.10.9   Pipe and Tubing Volume

Pipe/Tubing Size (ID)	Volume (gal/linear ft)	Pipe/Tubing Size (ID)	Volume (gal/linear ft)
0.203"	0.0017	$3/4$"	0.023
$1/4$"	0.0025	1"	0.041
$3/8$"	0.0058	$1 1/4$"	0.064
$1/2$"	0.0100	$1 1/2$"	0.092
$5/8$"	0.0160	2"	0.164

17.10.6.6   Supplementary Heat

There are times when heating apparatus other than HRF is either required or requested as part of the system. It may be that the owner or designer is utilizing the HRF to do only a portion of the building heating, such as a basement or main floor, and using conventional devices, such as convectors or fan coils, to provide the balance of the heating requirements. There may also be cases in which HRF is not capable of delivering all the heat required in a space without exceeding generally accepted maximum surface temperatures. As a result, additional heating may be provided by convectors or wall radiators. In either case, special allowances must be made for this equipment when doing the system design calculations.

   Selection.   The type of equipment used for supplementary heating is largely a choice of personnel preference. The four obvious choices are fan coils, convectors, wall radiators, and radiant ceiling panels. Fan coils provide the most heat output in the smallest space, for example, the kick space under a cabinet in a bathroom, but they can also be noisy and cause drafts. Convectors are easily installed when wall space is available and can be effective under large windows. Wall-mounted radiators create a feel most like an HRF floor, and work well in bathrooms as well as other rooms which have available wall space. The ceiling can be utilized as a radiant panel by installing tubes in a plaster bed or using one of the manufactured electric radiant ceiling panels. The choice is up to the designer and the owner.

   Calculations.   Supplemental or primary hot-water heating devices other than HRF floors usually require much higher water temperatures than radiant floor heating. A mixing valve is usually installed in the system to allow the boiler to deliver the higher temperature to the conventional heating units while providing the appropriate lower temperature to the floor. Look at the floor surface temperatures (line 8 of the worksheet) for each room. If a room's floor surface temperature exceeds 85°F and is going to receive a lot of foot traffic, it may be a candidate for supplemental heat. First look at the possibilities of selective tube spacing and layout to redistribute the heat as illustrated previously in this chapter. If this is not possible, use Fig. 17.10.15 to determine the maximum output which the floor can contribute to the heating requirement of the room. This is done by finding the highest acceptable floor temperature on the chart and locating the corresponding floor output. Multiply this output by the actual heated floor area (line 6 of the worksheet) of the room. The result in Btu/h (J/h) is the floor contribution to the heating requirement. Finally, subtract this floor output from the heat loss for that room (line 3 of the worksheet).

   The remainder is the amount of supplemental heat required. Using manufacturer's specifications and the I = B = R Guide 200 (Ref 2), select an appropriate heating device to provide this additional heat.

   If supplemental heat is to be added, always remember to reduce the floor output required (line 7 of the worksheet) accordingly. The supplemental heat should be added to the total

appropriate range of temperature settings. The required temperature is listed on line 24 of the worksheet. As a general rule, flow setters and balancing valves should be specified as full-line size to reduce restriction. Ball valves or valves designed specifically for balancing are best for balancing purposes. Do not use gate or globe valves for this purpose.

Determine whether the system will be controlled by outdoor reset controls or a simple thermostat. Use the indoor air temperature and the outdoor design temperature located at the top of the worksheet along with the water temperatures on line 24 or 34 to compare manufacturers specifications for outdoor reset controls.

17.10.7   INSTALLATION METHODS

Methods of installation vary from manufacturer to manufacturer and installation to installation. This section presents the most commonly used methods and applications. *Always consult the manufacturers' instructions before beginning an installation.*

A good installation begins with good communication. *Everyone connected with the installation must know what is expected and when.* Provide the General Contractor with a schedule/check-list indicating each phase of the installation and what must be done in preparation for each phase. Supply each subcontractor who is affected by the installation with guidelines as to their responsibilities and things which they should be aware of. Always do a preinstallation inspection shortly before installation begins to avoid wasted time due to inadequate or incomplete preparation by the other trades.

17.10.7.1   Manifold Mounting

In many installations, the manifold is the first item to be installed, even before the supply and return piping is roughed in. This is particularly true in concrete slab applications. The manifold location is defined during the system sizing process and noted on the floor plan. If the builder must provide a special enclosure or wall cavity, shop drawings and specifications should be submitted to the builder well in advance of installation. If possible, assemble the manifolds prior to arriving at the job site. This will save time and reduce errors. At the job site, confirm the direction manifolds are to face for future access.

Concrete Slab.   When installing an HRF system in a concrete slab, the walls are generally not in place and therefore not available for mounting the manifold. Find the location of the future wall as accurately as possible and drive two pieces of rebar into the ground on the centerline of the wall (wood stakes are not recommended). Tape or wire the manifold to the rebar (Fig. 17.10.16) at a sufficient height to allow the tubing to be easily installed. Remember to take the finished floor height into consideration when mounting the manifold. Once the floor has been poured and the walls built, the rebar can be cut off flush with the slab.

An alternative to a wall-mounted manifold is to mount the manifolds in a plumbing pit. This is done by using concrete forms to create a box in the concrete floor which will contain the manifolds as shown in Fig. 17.10.17. Once the concrete is poured, the forms are removed. In this application the manifolds are usually mounted horizontal to the ground. Use rebar driven into the ground to support the manifolds. A step is provided at the lip of the pit to accept a lid or grate once the floor is poured.

When supply and return lines must be installed under the slab, have trenches dug by the general contractor and use type K soft copper pipe with as few joints as possible. Install Schedule 40 plastic pipe several sizes larger than the copper in the trench. Insulate the copper pipe and feed it through the plastic pipe. This will protect the copper pipe and allow the pipe

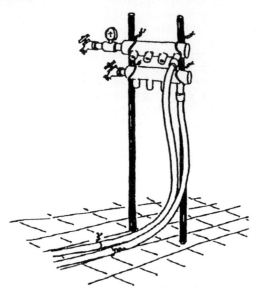

FIGURE 17.10.16   Temporary manifold support in concrete
slab applications.

to be removed and replaced in the future should a problem ever occur. Another option is to
use PEX or composite pipes with no joints, lay out the supply return lines, then insulate them
with standard foam-type pipe insulation. Fill the trench, covering the pipe with sand.

   Wood Framed Construction.   The manifold is generally installed on a wall or in a wall
cavity as shown in Fig. 17.10.18. The location and facing direction is established in the

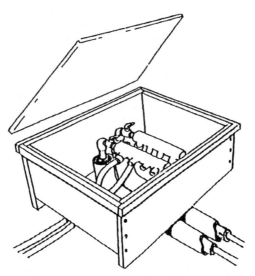

FIGURE 17.10.17   Pit-mounted manifolds, etc.

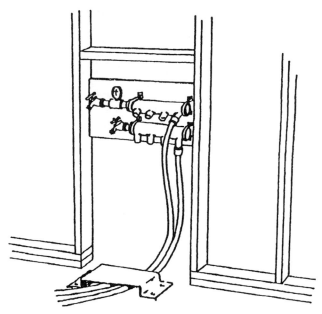

FIGURE 17.10.18   Mounting manifolds in a wall cavity.

system sizing procedure. Dimensions and a shop drawing of the wall cavity should be presented to the general contractor so that it will be provided as the wall is built. A board is installed at the back of the cavity to receive the mounting bracket screws. Some manufacturers provide a ready-made cabinet which fits into the wall cavity and has built-in manifold mounting brackets.

Supply and return lines can be roughed in before or after the manifolds are installed. It is a good practice to install isolation valves and unions for the supply and return lines at the manifold location. This will allow easy access to the manifold for future service work without disrupting the operation of the rest of the heating system. Also making provisions for control wiring to the wall cavity at this time is required.

Mount the manifold at a sufficient height to allow easy installation of the tube. Mounting a manifold too high is seldom a problem, but a manifold mounted too low can cause real problems with tubing installation. There must be enough room between the manifold and the finished floor to allow the tube to make a bend before exiting the wall cavity below finished floor height. Height allowances should also be made to permit the tube to come to a completely vertical position without bending stresses at the manifold fittings.

17.10.7.2   Tubing Layout

Tube selection is made during the system sizing procedure. The tube layout may be drawn on the floor plan in detail or simply marked as number of loops, lengths, and spacing for each room. Detailed drawings (Fig. 17.10.19) are sometimes required by building officials.

These drawings can also reduce wasted tube, installation time, and mistakes. Draw on a clean set of floor plans using different colors for alternating loops. Note flow direction, tube spacing, loop length, manifold locations, and nail guards.

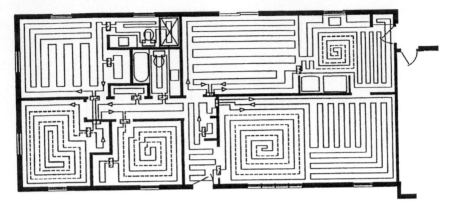

FIGURE 17.10.19   Sample detailed tube layout.

Layout Patterns.   The first consideration in planning a tubing layout is that of heat flow. As the water travels through the tube, it will lose heat to the floor, therefore the hottest water should be delivered to the area with the greatest heat loss first. Also, areas with high heat loss should be serviced by separate loops, where possible, for better control. Several layout (Fig. 17.10.20) patterns are recommended and each offers features to address specific applications. The two basic patterns are serpentine and counter-flow spiral. Each can be modified to adapt to individual installations.

Serpentine.   Figure 17.10.20a illustrates a simple serpentine pattern which consists of a single tube, which is laid out back and forth on parallel spacing. This pattern will produce a temperature gradient across the floor as the circulating water cools from one side of the pattern to the other. Using the serpentine pattern, the highest temperature water can be directed to the area of highest heat loss. The warmest water can be circulated to the outside wall first with the temperature diminishing as it moves inward to the center of the room. A serpentine pattern can be modified in many ways to accommodate odd size rooms or several outside walls (Fig. 17.10.20b).

Counter-Flow Spiral.   The counter-flow spiral layout pattern features an even distribution of heat and floor temperature over an entire zone. It is achieved by paralleling the supply and return (Fig. 17.10.20c), thereby balancing a higher temperature supply with its corresponding lower temperature return. The selection of a counter-flow spiral tubing layout assumes an even heat loss situation exists within the zone or there is no major heat loss to contend with. The counter-flow spiral can be weighted to side for dealing with areas with higher heat loss by having smaller spacing between tubes (Fig. 17.10.20d).

Tube Attachment.   When the tube is embedded in concrete or a cementitious topping, it must be secured so that it will not float to the top and become exposed. When tubing is installed in wood frame construction using metal heat transfer plates, the channels provided in the plates will secure the tube. There are many approaches to tube attachments this chapter will present a few of the most common (Fig. 17.10.21).

Wire Ties.   Small lengths of wire or nylon electrical ties are used to secure the tubing to wire mesh embedded in concrete. The metal wire ties often have a small loop on each end which accommodate the use of a twisting tool. This tool consists of a hook which swivels on a handle. The two loop ends of the wire tie are slipped over the hook and the handle is moved in a circular motion causing the wire ends to twist together. The grid of the wire mesh can be used as reference when laying out the tube. The wire mesh is then either supported by stand-offs or lifted to within 2 in (5 cm) of the surface as the concrete is poured.

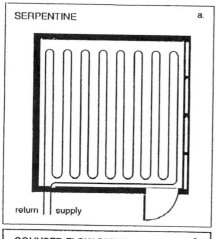

SERPENTINE a.

return | supply

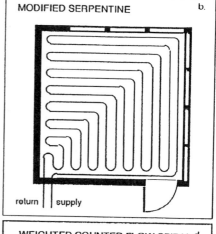

MODIFIED SERPENTINE b.

return | supply

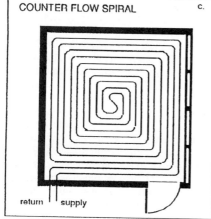

COUNTER FLOW SPIRAL c.

return | supply

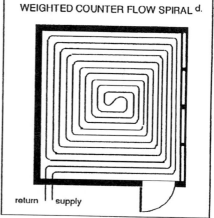

WEIGHTED COUNTER FLOW SPIRAL d.

return | supply

FIGURE 17.10.20 Tube layout patterns.

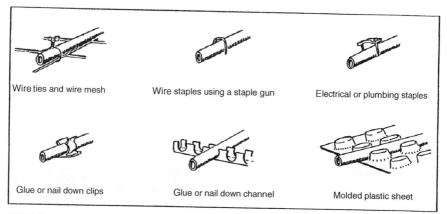

Wire ties and wire mesh

Wire staples using a staple gun

Electrical or plumbing staples

Glue or nail down clips

Glue or nail down channel

Molded plastic sheet

FIGURE 17.10.21 Typical tube attachment.

Staples.   Attaching tube to a wood subfloor is often done with staples. These can either be hammer driven, such as electrical cable or plumbing staples, or they can be installed with a staple gun modified for this application. Care should be taken not to damage the tube during installation. A staple driven into a tube may self seal and not show up as a leak until the staple has corroded away. Always use caution when installing staples.

Clips.   Various types of clips are available. Some are in the form of long strips or channels which are notched to receive and hold tubing. Other versions are single clips that can be laid out in a pattern on the floor; the tube is then snapped into place. Large plastic sheets, which cover the entire floor, with clipping devices molded into are also available; some are laminated to board insulation. Clip devices are nailed, stapled, or glued to the subfloor or insulation board. Once they are installed, the tubing is then laid out in the appropriate pattern and snapped into the clips. The choice of clipping device may determine the layout pattern. Always consult the manufacturer's instructions before proceeding.

Plates.   Metal heat diffusion plates are used in place of a cementitious thermal mass. They are either installed between wood stringers on top of the subfloor or attached to the underside of the floor between floor joists. In either case, the plates are generally constructed with a channel which will accept and hold the tubing in place.

Preparation.   It is a good idea to check the job site prior to installation to make sure that the other trades have done their part and the area is ready for tube installation. All debris and equipment should be removed from the installation site. Allow adequate time for installation without interference of traffic from other trades, which may hinder installation or damage tubing. Ensure that there is water on site for pressure testing. Take photographs for future reference as the installation progresses.

Concrete.   The subsoil or fill should be flat and smooth with adequate drainage. Insulation (perimeter and/or under slab) should be in place with the flat wire mesh on top. Locate and mark all interior walls and wall openings (unless nonbearing walls will be glued down) and all areas where no tubing is to be installed. Provide chases/passageways through stem walls or grade beams where required to permit installation of plumbing or tubing. Identify manifold locations and trenches for supply and return lines if required. Provide planking or plywood sheets to protect the tubing if a wheelbarrow is to be used. Instruct the concrete workers in the care of using tools such as rakes and shovels. These can damage tubing if used carelessly.

If tubing is to pass through a control joint or expansion joint in the concrete slab, either provide an indentation below the slab so the tube can dip below the joint, or sleeve the tube with insulation for 6 in (15 cm) either side of the joint (Fig. 17.10.22). It is best to plan the tube layout pattern to avoid joints in the slab whenever possible.

Cementitious Underlayment.   Cementitious underlayments are poured in place floor toppings made up of gypsum cement or light weight concrete. They can be poured over an

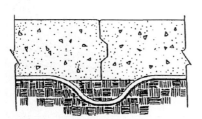 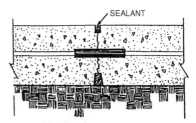

FIGURE 17.10.22   Tube through an expansion or control joint.

existing concrete slab or a wood subfloor. The underlayment will affect the finished height of the floor. Make sure everything that will be affected by the increase in finished floor height is adjusted. Some items that should be inspected are double wall plates, stair risers, door and window sill heights, plumbing fixtures, and floor vents.

Any damaged or delaminating subfloor material should be repaired or replaced. All cracks and voids should be filled with caulk, fiberglass insulation, or quick-setting drywall patching material to prevent the poured underlayment from leaking through to the space below. Gypsum floor underlayments require special floor sealers while concrete may require a slip sheet. Check with the underlayment contractor. Ensure that the manifold wall cavities are properly framed, end that the base plates are cut out in front of the manifolds and other areas where the tube must pass. Provide dams on stairways, entrances, and any other areas where necessary. Use a marker to indicate areas where no tubing will be installed, such as cabinet locations. Install any supply and return lines which must penetrate through the poured floor if not already done.

Tube Installation. In most applications it is recommended that the manifold be in place before tubing installation begins. This reduces errors and confusion as installation progresses. Schedule the installation so that all other trades will be out of the way. Inspect each roll of tubing for gouges, deep scratches, or kinks, and reject tubing found with these defects. Confirm the tube coverage area as identified during the system design procedure.

A tubing uncoiler is a worthwhile tool which will reduce installation time and manpower. Start at one end of the manifold, pull the tube from the uncoiler and attach it to the first supply port in the manifold. Start at one side of the tube coverage area and fan to the other side as loops are added and tubing layout patterns are installed. Follow the preplanned layout closely. It may be helpful to mark the layout pattern spacing and corners on the floor with a marking pen. Keep tubing a minimum of 6 in (15 cm) away from the edge of the slab, walls, and other locations where plates, fixtures, or cabinets might be fastened to the floor. Do not exceed the manufacturer's minimum bend radius and take care not to kink any tubing. Some tubing has more coil memory than others. Try to lay the tubing with the coil memory in the same direction as the bends in the layout pattern. This is not difficult and can easily be accomplished with a little care.

Attach the tube to the floor every 2 to 3 ft (0.6 to 0.9 m), at bends, and as required to prevent it from floating or moving. Work through the layout pattern and return to the manifold location without any tubes crossing over one another. Complete the first loop, returning it to the manifold and connecting it to the corresponding return port, before proceeding with the next loop. This prevents accidental mixing of return and supply ends of tubing which would result in no flow.

Do not attach the tubing to the floor within 3 or 4 ft (0.9 or 1.2 m) of the manifold until all of the loops have been installed. The tubes can then be shifted and arranged in a neat fashion to provide the proper curve up to the manifold, then attached all at the same time. Place nail guards over the tubing in front of the manifold, where tubing travels through a doorway, and any other location where a nail may be driven when installing floor coverings.

Pressure Testing. A pressure test is required on the tubing before it is embedded in a thermal mass or becomes inaccessible. This pressure test usually runs between 30 and 100 psi (206 and 690 kPa), depending on local code requirements and manufacturers' recommendations. It is also highly recommended that pressure be maintained on the floor tubing while the thermal mass is being poured. This pressure test can either be hydrostatic (water pressure) or air. The hydrostatic test has the advantage of easily locating leaks before the pour, as evidenced by water puddles. On the other hand, an air test will produce noticeable bubbles in the wet thermal mass when a leak is present during the pour and finishing process. The hydrostatic test is most commonly used, but water should not be left in the tubing after the pour is complete if freezing temperatures are anticipated. A glycol solution should be used in these conditions. Be careful when using an air-pressure test: Rapid decomposition can be explosive in nature, and dangerous.

If a leak in the tube is discovered before the pour, the tube should be removed and replaced. Splices in the floor should be avoided if possible. In many areas, mechanical codes do not allow splices in the floor. If a leak occurs during the pour, it may be impossible to replace the tubing. In this case, two holes may be drilled in the floor, the tube cut at the point of the leak, and the ends fed down through the hole to be spliced below the floor. At least this splice would be more accessible in the future should it become necessary. If a splice is the only alternative, use only manufacturer approved splices and follow the instructions carefully. Record the exact locations of the splice for future reference. If a problem should occur at some later date, the floor can be chipped away at that location and the splice replaced.

The pressure gauge for the test may be mounted at the manifold or in the mechanical room if the supply and return piping is plumbed in. Place the gauge in a location where it can be read from outside the pour. It is difficult to determine if the floor is holding pressure when the gauge is not viewable once the floor is poured.

17.10.7.3 Piping Arrangements

There are many ways that the primary supply and return piping can be plumbed. This guide presents some of the most common arrangements. Actual boiler plumbing is not illustrated here but can be found in the I = B = R Guide 200, "Installation Guide for Residential Hydronic Heating" (Ref 2). Consult the instructions provided with all equipment before the final installation. Each of the piping illustrations shown here can incorporate mixing valves, reset devices, or zone controls.

Supply and Return.   When a single manifold is used, the water is carried to the inlet of the manifold by a single supply pipe and return water is carried back to the boiler by a single return line. In this configuration (Fig. 17.10.23a), the water supplied to the first loop in the manifold travels a shorter distance than the water supplied to the last loop. In a single manifold the additional distance traveled is only the length of the manifold and generally has an insignificant affect on flow. The pressure drop is a combination of the main supply and return pipe added to the pressure drop of the manifold/tube.

Separate Supply Lines.   Water can be delivered to each manifold or zone by a separate pump (Fig. 17.10.23b). This ensures that each manifold/zone is receiving the proper pump pressure regardless of the operation of other zones. The pumps can also be used as zone flow control devices by turning them on and off with thermostat controls. Pressure drops for each pump supplied circuit are figured separately.

Reverse Return.   This piping arrangement (Fig. 17.10.23c) directs the water entering the first loop in the supply manifold to the last fitting on the return manifold. In this way, if each loop is the same length, the water path will be the same length for each loop. This arrangement is useful when there are no loop balancing valves available or when manifolds are very long, such as those used for tubing installed between joists under a suspended floor. Pressure drop is determined by combining the pressure drop of the main pipe and the manifold/tube set with the highest pressure drop.

Direct Return.   The most common plumbing arrangement used with HRF is direct return (Fig. 17.10.23d). It is the same configuration as the simple supply and return schematic with addition of multiple manifolds plumbed in parallel. In this configuration, the loop furthest from the boiler may receive the least amount of flow. If the pump and

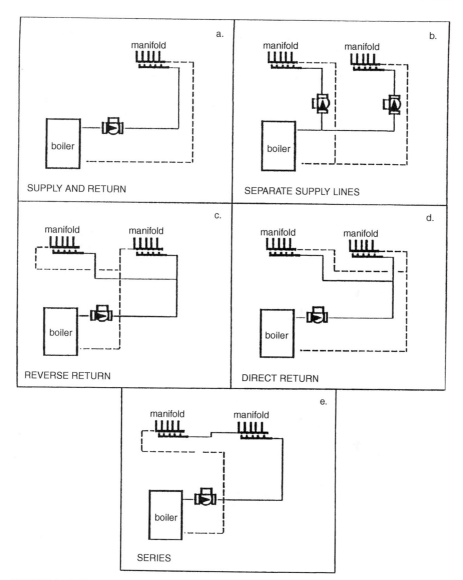

FIGURE 17.10.23 Piping arrangements.

piping are sized correctly, this will not present a problem. Should the system contain a large number of manifolds or zones, reverse return or separate supply lines should be considered to maintain a proper balance of flow.

Series. In this configuration a single pipe or main is used (Fig. 17.10.23e). This main enters directly into the supply side of the manifold, then continues from the return side of the manifold into the supply side of the next manifold. In this manner all of the water flowing through a circuit flows through each series-connected manifold. As the water travels through each manifold and related floor tubing, it drops in water temperature. Water reaching the last manifold in the series will be cooler than the first. This system also requires the highest pump pressure because the pressure drop of the system is the combined pressure drop of each manifold/tube set as well as the main pipe.

Combination Systems (Supplemental Heating).   Often a system is called upon to provide HRF heating as well as more conventional types of hot-water heating. An example would be an installation which incorporates radiant floor heating in the basement concrete slab while utilizing fin-tube baseboard units on the main floor. Another example would be the use of a radiator as supplemental heat in a small space which does not have sufficient floor area to handle the entire heating load. These situations require different water temperatures, and possibly flows, to be supplied by the same system.

Combination systems are not difficult to design. Each subsystem is plumbed in parallel with the others and utilizes whatever devices are required to achieve proper temperature and flow. Generally, the boiler will provide the proper temperature fin-tubes, radiators, or fan coils, and a mixing device is used to reduce the temperature for the HRF system. The simplest way to design and plumb a combination system is to layout the subsystem which will do the majority of the heating first, then branch off the supply and return just before the boiler to service the secondary subsystem (Fig. 17.10.24a).

When practical, use a single pump to service all the subsystems (Fig. 17.10.24b). Be sure to size the pump to handle the full load by adding all the subsystem flow requirements together and using the highest subsystem pressure drop. Most HRF manufacturers have instructions on how to incorporate a combination system into your design.

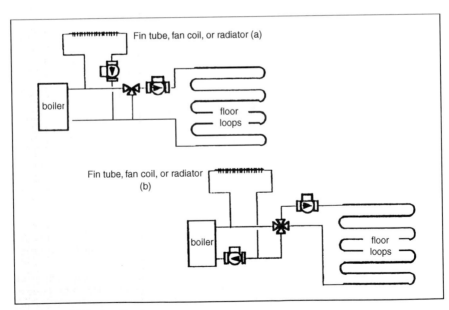

FIGURE 17.10.24   Combination systems.

17.10.8 SUMMARY

The combinations of tubing, manifolds, controls, boilers, and conventional hot water heating practices are endless. It is not possible to investigate every application or configuration in this chapter. A basic understanding of the theories and methods involved in the design and application of HRF will provide a foundation on which to make judgments and evaluations of manufacturer's equipment and system design.

REFERENCES

1. IBR Heat Loss Calculation guides H21 and H22, Hydronics Institute, Berkeley Heights, NJ.
2. IBR Insulation Guide No. 200.
3. *HVAC Systems and Equipment*, ASHRAE, Atlanta, GA, 2002.

17.11   SNOW-MELTING SYSTEMS*

Lehr Associates
New York, New York

17.11.1   INTRODUCTION

It has become common in recent years to use snow-melting systems in the vicinity of commercial installations, especially for such access areas as walkways, automobile ramps, and delivery areas. While obviating the need for snow shoveling, plowing, or blowing, snow-melting systems provide dry surfaces. These advantages also apply to residential applications. Usually no more than a standard hot-water boiler combined with the melting system is necessary. Wider use of residential snow-melting systems has been hampered by a lack of information and by a belief they are too expensive—but the costs need to be evaluated objectively before such a system is ruled out.

System costs are often inflated when the design is too conservative. Designers readily increase the capacity of the heating boiler or accessory pumping and piping equipment to cover poorly understood application criteria. Of course, this practice increases the cost, often unnecessarily. A better understanding of the application's parameters, combined with clearer knowledge of how snow-melting systems function, can be used to bring installation costs to a reasonable level.

Most snow-melting systems are one of three types: an embedded piping system carrying hot water or heating fluid (usually a glycol solution), embedded electrical resistance heating elements, or radiative systems using infrared lamps. The last system can provide illumination as well as space heating, if desired.

Of the three types, the heated-fluid system provides the best application of all the essential design elements for snow removal. We now examine this system in more detail. The basic design process comprises seven steps:

1. Determine the snow-melting load.
2. Decide on a piping layout.
3. Determine the gallon/minute (liter/second) flow rate, and specify the heat exchange unit.
4. Size and select the piping.
5. Size and select a pump.
6. Determine the appropriate specialties.
7. Select the heat source.

17.11.2   DETERMINATION OF THE SNOW-MELTING LOAD

The performance of a snow-melting system depends on many factors, ranging from the type of installation to the predicted volume of snow. Other factors include the outside air

---

*Information contained in Tables 17.11.1, 17.11.2, and 17.11.3 and some of the text in this chapter are used courtesy of ITT Fluid Handling Division.

temperature, snowfall rate and density, wind speed and exposure, humidity, and physical properties of the slab being heated. While it is possible to compute the effects of each of these factors, the calculations are laborious and time-consuming. A generalized method, presented here, offers adequate accuracy.

Basic physics tells us that the amount of heat needed to melt a load of snow is, first, the heat necessary to warm the slab to the snow-melting temperature. This value is called the *pickup load*. The second amount is the heat needed to melt the snow itself. This value depends on the rate of snowfall, the snow's density, and, indirectly, the outside air temperature. When the temperature is very low, snow is usually less dense; when the temperature is higher, especially near the freezing point, the snow is heavier and denser. The proper heating capacity for the snow-melting system will be the larger of the pickup load or the melting load. Both must be calculated to make the appropriate determination.

To calculate the pickup load, two things must be known: the initial slab temperature and the amount of heat necessary to bring the slab up to 32°F (0°C), the snow-melting temperature. The initial slab temperature can often be considerably below the outside air temperature; a good initial temperature to use is 10°F (−12.2°C). The amount of heat necessary to bring the slab up to the freezing point will depend on the depth of bury of the piping that conveys the heating fluid.

A deeper bury for the piping embedment will require proportionately more heat because a greater mass of concrete is present. Thus it is desirable to bury the pipes only as deep as is necessary for structural reasons. Usually this can be interpreted as a minimum depth equal to the diameter of the pipes themselves. A customary depth of bury is 2 in (50.4 mm) below the surface of the slab. When the concrete that will become the slab is poured, take the precaution of securing the pipes so that they do not rise or move. Table 17.11.1 provides values for pickup loads on the basis of varying depths of bury and for a range of initial slab temperatures.

Good engineering judgment is the most direct way of calculating the snowfall rate. In the case of residential systems, a good nominal value is 1 to 1.5 in/h (25.4 to 38.1 mm/h) snowfall rate. Industrial systems can be sized with a design rate of 2 to 4 in/h (50.8 to 101.6 mm/h). This figure is usually looked on as the maximum rate; higher rates would be onerous to the system's cost. However, it may be advisable to check on local snowfall rate patterns.

Snow density depends on the air temperature during the snowfall; as previously mentioned, a higher temperature implies a denser snowfall. The average temperature during a snowfall is around 26°F (−3.3°C), which yields a density of about 6 lb/ft³ (96.1 kg/m³). Table 17.11.2 lists the Btu (watt) snow-melting heating quantities for a number of snowfall rates. This table is based on an average snow density of 10 lb/ft³ (160.2 kg/m³).

A final step to sizing the snow-melting system is to obtain the area, in ft² (m²), of the slab to be heated. Multiply this value by the appropriate value from Table 17.11.1 for the pickup load; perform the same multiplication of the area by the appropriate value from Table 17.11.2 to obtain the snow-melting load. The greater value (in Btus or watts) is the appropriate one to use in sizing the overall system.

For our example, let us assume that the initial slab temperature is 10°F (−12.2°C) and that the tubes are buried such that the distance from the centerline of the tubes to the surface of the slab is 3 in (76.2 mm). From Table 17.11.1 the pickup load is 304 Btu/h·ft² (959.1 W/m²).

From Table 17.11.2, based on an average 1.5 in/h (38.1 mm/h) snowfall rate, 180 Btu/h (567.9 W) is the required melting load. Since the pickup load is greater than the snow-melting load, the basis for our design will be the pickup load or 304 Btu/ft² (959.1 W/m²). The area to be melted is 1000 ft² (92.9 m²):

$$1000\,\text{ft}^2 \times 304\,\text{Btu/ft}^2 = 304{,}000\,\text{Btu}$$

$$(92.9\,\text{m}^2 \times 959.1\,\text{W/m}^2 = 89{,}100\,\text{W})$$

Our snow-melting requirement becomes 304,000 Btu/h (89,100 W).

TABLE 17.11.1 Pickup Load, Btu/(h·ft²)*

Slab temperature at start,		Distance from centerline of tube to surface of slab, in (mm)											
°F	(°C)	¾ (19)	1 (25.4)	1½ (38.1)	2 (50.8)	2½ (63.5)	3 (76.2)	3½ (88.9)	4 (101.6)	4½ (114.3)	5 (127)	5½ (139.7)	6 (152.4)
30	(−1.1)	7*	9	14	18	23	28	33	37	41	46	51	56
20	(−6.7)	41	55	83	110	138	166	194	221	249	276	304	332
10	(−12.2)	76	101	152	203	250	304	355	406	456	507	558	609
0	(−17.8)	110	147	220	294	368	442	575	588	662	735	808	881
−10	(−23.3)	145	193	289	396	482	578	675	771	867	964	1060	1156
−20	(−28.9)	179	239	359	478	598	718	838	958	1079	1195	1215	1335

*Values given in table are in Btu/(h·ft²). The metric conversion factor is Btu/(h·ft²) × 3.155 = W/m².
Note: Bold figures indicate normal design areas.

TABLE 17.11.2   Snow Melting

in/h	(mm/h)	Btu/(h·ft²)	(W/m²)	
$\frac{1}{2}$	(12.7)	60	(189.3)	
1	(25.4)	120	(378.6)	
$1\frac{1}{2}$	(38.1)	180	(567.9)	Normal
2	(50.8)	240	(757.2)	used
3	(76.2)	360	(1135.8)	design
4	(101.6)	480	(1514.4)	range
5	(127.0)	600	(1893.0)	
6	(152.4)	720	(2271.6)	

17.11.3   PIPING LAYOUT

The same principles that are used in radiant panel systems (see Sec. 17.10) apply to snow-melting systems. Grid or coil tubing patterns are suitable. A good rule of thumb is to use a tube spacing of 12 in (304.8 mm), center to center. When pipes are closer, installation costs go up; when they are spaced more widely, there may be incomplete melting.

17.11.4   DETERMINE THE GALLONS/MINUTE (LITERS/SECOND) REQUIREMENT AND SPECIFY A HEAT EXCHANGER

If a heat exchanger is used—as would be the case if a heating fluid other than pure water were considered—the lesser heat-carrying capacity must be factored in. For antifreeze glycol solutions, an increased amount of solution must be conveyed to deliver the same amount of heat. The relationship between the two is expressed by the following equation:

$$\frac{\text{Btu/h}}{8600} = \text{gal/min}$$

Note: 8600 is an English-unit constant based on a 50/50 glycol solution with a 20°F (–6.7°C) temperature difference. To use this equation in metric designs, first carry out the calculation in English units, and then convert the result to metric.

$$\frac{304,000\,\text{Btu/h}}{8600} = 35.3\,\text{gal/min}$$

Note: Expansion joints in concrete walks or driveways cause certain piping problems. Avoid concentrated stresses on the coils or grid panels and on piping mains. Piping mains should be covered with an insulation blanket below the level of the slab.

17.11.5   SELECT SPECIALTIES

Generally, there is no need to add the snow-melting load requirements to the boiler load, since U.S. Weather Bureau records indicate that 26°F (–3.3°C) is the average snowfall temperature.

TABLE 17.11.3 Percentage of Boiler Capacity Required for Space Heating When the Outside Temperature is 25°F (–4°C)

Design temperature of heating system				
20°F (–6.7°C)	10°F (–12.2°C)	0°F (–17.8°C)	–10°F (–23.3°C)	–20°F (–28.9°C)
90%	75%	65%	55%	50%

At this temperature most heating systems are operating at less than maximum capacity. The design condition is a much lower temperature, in most cases. Table 17.11.3 shows the proportion of boiler capacity necessary for space heating at 25°F (–3.8°C) outside temperature.

In our example, the snow-melting system requires 304,000 Btu/h (89,100 W) for snow melting alone. If a 500,000 Btu/h (146,541 W) boiler is required for the comfort heating system at a design condition of –10°F (–23.3°C), Table 17.11.3 shows that only 55 percent, or 275,000 Btu/h (80,598 W), is required by the comfort heating system at 25°F (–3.8°C). Therefore, a balance of 225,000 Btu/h (65,944 W) is available for snow melting. Subtracting this 225,000 Btu/h (65,944 W) from the total snow-melting load of 304,000 Btu/h (89,100 W) shows that only an additional 79,000 Btu/h (23,153.6 W) must be added to the boiler size.

Compression Tank.   When a compression tank is sized, the snow-melting system should be treated in the same manner as a radiant panel system. The heat exchanger, when used, will take the place of the boiler and will be designed by following the guidelines for a flash-type boiler.

Refer to the manufacturer's literature for sizing criteria for the compression tank and fittings, boiler fittings, and relief valve.

17.11.6   ELECTRICAL SNOW MELTING

As in electrical space heating, resistance wires can be used to warm outdoor surfaces. The materials that deliver the heat are usually either mineral-insulated cable or resistance wires.

Heat load calculations for electric systems are the same as for hot-fluid designs. Electric heating inherently has a set heat output, so over- or underheating is minimized by the design. The amount of heat per unit area can be varied by the spacing of the heating elements. It is important that concrete slabs be properly designed when electric heating is used. When the elements are embedded in the slabs, expansion-contraction joints or other features create shear forces that could snap the elements. Slabs are sometimes poured in layers; in such cases, the uppermost layer of an element-containing slab should have a high degree of weatherability, including limited air content and aggregate size.

17.11.7   ELECTRIC HEAT OUTPUT

Designs for electric snow-melting systems differ slightly depending on whether mineral-insulated cable or embedded wires are used. Cable is laid out much as piping would be; embedded wires can be laid out in single strands or as prefabricated mats.

American Society of Heating, Refrigeration, and Air-Conditioning Engineers (ASHRAE) standards suggest cable layouts of 2 lin ft of cable per square foot of slab area (6.6 lin m/m²) for concrete slabs and 3 lin ft/ft² (9.8 lin m/m²) for asphalt surfaces. The difference is due to the heat conductance of the slab material.

The following procedure can be used to size the heating system with mineral-insulated cable:

1. Find the total power requirement W_t:

$$W_t = A \cdot w$$

2. Find the total resistance R:

$$R = \frac{E^2}{W_t}$$

3. Find the cable resistance per foot (meter) r_1:

$$r_1 = \frac{R}{L_1}$$

4. Calculate the required cable spacing S:

$$S = \frac{12A}{L}$$

5. Find the current I required for the system:

$$I = \frac{V}{R}$$

In these equations

W_t = total power needed, W
A = heated area of each heated slab, ft² (m²)
w = desired watt density input, W/ft² (W/m²)
V = voltage available, V
r_1 = calculated cable resistance, Ω/ft (Ω/m) of cable
L_1 = estimated cable length, ft (m)
R = total resistance of cable, Ω
L = actual cable length, ft (m)
r = actual cable resistance, Ω/ft (Ω/m) of cable
S = cable center-to-center spacing, in (mm)
I = total current per cable amperes

The calculated cable resistance is compared to the resistances available from manufacturers. If there is a significant variation from what is calculated, the system is recalculated with the closest available cable (it is possible for large installations to custom-manufacture the cable). In addition, a cold lead—a length of cable that does not generate heat—must be allowed for, to reach an aboveground junction box. The junction box should be located so that it can be kept dry.

Embedded wires or mats follow similar design procedures. The manufacturer's specifications for the wire determine the necessary ohm/foot (ohm/meter) resistance and wire diameter. Prefabricated mats must be selected with an eye toward how the heated surface is being constructed; if it is made of hot-poured asphalt, beneath which the mat will be placed, a mat that can withstand that heat should be used.

Electric snow-melting cable or wire can also be used to prevent snow or ice buildup on gutters and downspouts. The cabling can be mounted in a zigzag fashion along the edge of a roof and along the length of a gutter, and it can be looped for some distance down the downspout. The *ASHRAE Handbook* recommends cabling rated at 6 to 16 W/lin ft (19.7 to 52.5 W/lin m).

17.11.8   INFRARED (RADIANT) SNOW MELTING

Infrared systems tend to be somewhat more expensive than embedded-cable or wire designs, in terms of both installed cost and operating cost. This additional expense is offset by the advantages of providing comfort heating, in areas such as entrances or walkways, as well as additional illumination. The lamps can be installed under entrance awnings, along building faces, or on poles.

The primary design considerations for these systems are heating capacity and fixture arrangement. If snow melting is the paramount criterion, fixtures with narrow beam patterns can be specified; if comfort heating is also important, wider patterns may be in order. Manufacturer's literature customarily provides ratings in watts per square foot (meter) for the fixtures. A common practice is to take the table value of average watt density for snow-melting requirements in various regions of the country, multiply it by the target area in square feet (meters), and multiply by a correction factor of 1.6. For more precise designs, the arrangement of infrared fixtures is first determined, and then a table is compiled by which the intensity of light (heat) incident on each section of the area being heated is determined. In this manner, stray energy losses at the edges of the area to be heated are kept to a minimum. See Sec. 17.3 for a discussion on infrared heating.

17.11.9   SYSTEM CONTROLS

Either manual or automatic controls can be installed with snow-melting systems. Manual systems require the operator to start the system some amount of time before precipitation begins to fall. This preheating depends on the type of installation that was specified originally (a low-duty residential system, where some snow buildup may be permissible, versus a high-duty commercial system, where no buildup is desirable), weather conditions, and the materials of construction of the slab or surface being heated. Automatic control systems are generally available that either sense the presence of snow (which would allow preheating) or that start when the surface temperature falls below a preset limit.

Since both electric and infrared snow-melting systems are powered by electricity, the control setups are similar. Usually, it is a simple on/off setup. High-capacity snow-melting systems, which use substantial amounts of power, are often equipped with temperature sensors in the slab, permitting the system to cycle on and off during relatively warm, light-snow conditions. Finally, a number of fully automatic temperature and precipitation detection systems are becoming available.

17.12 HEAT TRACING

Lehr Associates
New York, New York

17.12.1 INTRODUCTION

A common problem of many commercial and industrial fluid systems is keeping the fluid at a constant temperature while it is circulating through the system. Hot water from a boiler, for example, is often needed at elevated temperatures, a considerable distance from the boiler. In many industrial plants, piping is exposed to the elements, and there is danger of freeze damage or difficulties with increased viscosity or other properties. The solution to many of these problems is either to continually recirculate fluid within the system (while replenishing heat losses) or to provide *heat tracing*.

Heat tracing may be defined as *the supplemental heating of fluid system piping through extraneous means*. Generally, the heat is provided by a system completely separate from whatever device is providing overall heat for the fluid in the pipes. In the past, heat tracing was commonly implemented with supplementary steam lines in close proximity to the main process-fluid lines. Today, heat tracing is usually accomplished by electrical tracings—resistance wires wrapped along and around pipes and pipe fittings. A relatively recent innovation in the field is the use of variable-wattage (self-regulating) resistance wires. These wires are designed with the inherent capability of drawing current until a desired temperature at any point along the length of the tracing is reached. Wires can even cross over each other, a situation that would normally present a fire hazard.

In principle, electric heat tracing is very similar to electric heating with mineral-insulated (MI) cable. Overall calculations are made of the heat loss and the area to be warmed; then the desired power input is arrived at, and the general layout of the heating cable is designed. Electric heat tracing can offer substantial energy and capital-cost savings over continuously recirculating fluid systems, since the energy consumption need be only what is needed at specific times; additional heating units, balancing valves, and similar equipment are avoided. For some piping networks, electric heat tracing is the only practical method. The remainder of this chapter deals with electric heat tracing only.

17.12.2 BASIC DESIGN CONSIDERATIONS

A variety of applications are tailor-made for heat tracing:

1. Exterior pipes carrying fuel oil or water
2. Gutters, downspouts, and other rain-handling piping
3. Circulating water pipes on external cooling towers
4. Underground pipe carrying hot fluid for relatively long distances
5. Hot-water supplies for commercial dishwashing
6. On-demand hot water for bathing or other purposes
7. Prevention of condensation buildup in air lines
8. Freeze protection for instruments or other sensitive equipment

Heat-tracing duty can be continuous or intermittent. Commonly available tracings are rated for temperatures from ambient to around 300°F (148.9°C) for continuous use and 400°F (204.4°C) for intermittent use. General-purpose MI cable can also be used for some tracing applications; this material is rated for continuous use even above 1500°F (815.6°C). The power output of the tracing cable is usually in the range of 10 W/ft (32.8 W/m); MI cable is available at up to 200 W/ft (656.2 W/m). Tracing cable is also defined by the maximum length of one circuit. For 10-W/ft (32.8-W/m) cables, this length ranges from 100 to 1000 ft (30.5 to 305 m). (Longer system lengths can be handled by multiple circuits.)

Physically, the cable is a flat double run of copper wire encased in an insulating barrier material (Fig. 17.12.1a, b). The wires and insulation are then wrapped in metal braid and/or a plastic jacket for mechanical and chemical protection. MI cable (Fig. 17.12.1c) is usually circular in cross section, with a magnesium oxide insulation core and plastic or metal jacketing. Self-regulating or variable-wattage tracing cable (Fig. 17.12.1a) operates through the use of a special carbon-based conducting polymer. As the temperature drops, the polymer contracts, allowing for a higher flow of current between the two wires (in essence, the cable is an infinitely parallel circuit). As the temperature rises, the polymer expands, with reduced current flow and higher resistance. This design permits high current flow at one point along the length of the cable and low flow at another.

In low-duty applications, heat tracing is simply run along the process pipe or tube and attached with a cable tie or tape around the circumference of the pipe. Figure 17.12.2 shows some common tracing arrangements. At pipe elbows, it is generally desirable to run the tracing along the outer radius of the bend. At pipe tees or other unions, the tracing can be

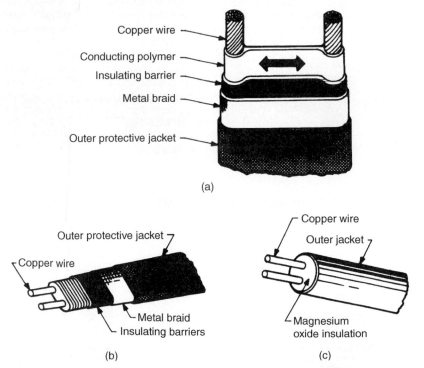

FIGURE 17.12.1   Heat cables. (*a*) Self-regulating, (*b*) constant-wattage, and (*c*) mineral-insulated. (*Courtesy of Chromalux, E. L. Wiegand Division, Emersion Electric Co. used with permission.*)

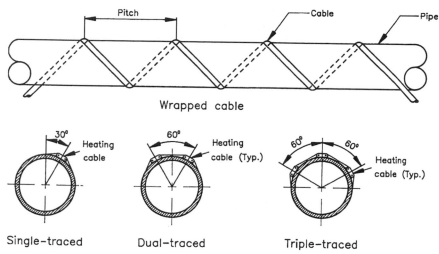

FIGURE 17.12.2   Common pipe-tracing arrangements.

run optionally in several loops around the entire joint. A similar configuration is possible for valve bodies. (Self-regulating tracing permits overlaps around these fittings.)

For higher-duty applications, the tracing is wrapped in a helical fashion along the length of the pipe. Details specifying the pitch for such a helix are given in Sec. 18.7.

It is customary to apply the tracing next to the pipe or tube to be heated, beneath any insulation or wrapping on the pipe. The type of insulation used on the pipe then becomes a design factor in specifying the heat tracing. In underground applications, the tracing must also be sealed against water seepage, Usually this can be accomplished through the use of pipe insulation with waterproof jacketing.

Special heat-conducting aluminum tapes are available where heat is to be conducted from the tracing to the rest of the circumference of the pipe. Conducting tapes may also be applied to wrapping storage tanks or other large structures. Some manufacturers also provide special tee joints or end seals for field installation of the heat tracing; others pre-engineer the tracing to specified lengths and joints.

17.12.3 *ELECTRIC HEAT-TRACING DESIGN*

Specifying a heat-tracing installation follows the same general outline as for any type of heating system. The heat losses are calculated. The desired warming is analyzed, taking into account the type of piping and the fluid it is carrying. Finally, the physical dimensions of the system are detailed, and the ancillary electrical equipment is specified.

17.12.3.1 Calculating Heat Demand

A typical heat-tracing application requires that the pipe be kept at some temperature above ambient conditions. In outdoor environments, wind conditions should also be factored in.

As an example, a pipeline carrying fuel oil, outdoors, is to be traced. The desired maintenance temperature is 100°F (37.8°C). The minimum temperature prevailing at the pipe

TABLE 17.12.1 Heat Losses from Insulated Metal Pipes, W/(ft·°F)*† (*For up to 20 mi/h wind speed*)

Pipe size, in/s	Insulation ID, in	Insulation thickness, in							
		½	¾	1	1½	2	2½	3	4
½	0.840	0.054	0.041	0.035	0.028	0.024	0.022	0.020	0.018
¾	1.050	0.063	0.048	0.040	0.031	0.027	0.024	0.022	0.020
1	1.315	0.075	0.055	0.046	0.036	0.030	0.027	0.025	0.022
1¼	1.660	0.090	0.066	0.053	0.041	0.034	0.030	0.028	0.024
1½	1.990	0.104	0.075	0.061	0.046	0.038	0.034	0.030	0.026
2	2.375	0.120	0.086	0.069	0.052	0.043	0.037	0.033	0.029
2½	2.875	0.141	0.101	0.080	0.059	0.048	0.042	0.037	0.032
3	3.500	0.168	0.118	0.093	0.068	0.055	0.048	0.042	0.035
3½	4.000	0.189	0.133	0.104	0.075	0.061	0.052	0.046	0.038
4	4.500	0.210	0.147	0.115	0.083	0.066	0.056	0.050	0.041
4½	5.000	0.231	0.161	0.125	0.090	0.072	0.061	0.054	0.044
5	5.563	0.255	0.177	0.137	0.098	0.078	0.066	0.058	0.047
6	6.625	0.300	0.207	0.160	0.113	0.089	0.075	0.065	0.053
7	7.625	0.342	0.235	0.181	0.127	0.100	0.084	0.073	0.059
8	8.625	0.385	0.263	0.202	0.141	0.111	0.092	0.080	0.064
9	9.625	0.427	0.291	0.224	0.156	0.121	0.101	0.087	0.070
10	10.750	0.474	0.323	0.247	0.171	0.133	0.110	0.095	0.076
12	12.750	0.559	0.379	0.290	0.200	0.155	0.128	0.109	0.087
14	14.000	0.612	0.415	0.316	0.217	0.168	0.138	0.118	0.093
16	16.000	0.696	0.471	0.358	0.246	0.189	0.155	0.133	0.104
18	18.000	0.781	0.527	0.401	0.274	0.210	0.172	0.147	0.115
20	20.000	0.865	0.584	0.443	0.302	0.231	0.189	0.161	0.125
24	24.000	1.034	0.696	0.527	0.358	0.274	0.223	0.189	0.147

Note: Values in table are based on the formula below, plus a 10 percent safety margin. The k factor of 0.25 for fiberglass at 50°F is assumed.

$$\frac{W}{\text{ft of pipe}} = \frac{2\pi k\, \Delta T}{Z \ln (D_o/D_i)}$$

where k = thermal conductivity, Btu·in/(h·ft²·°F)
$\quad D_o$ = outside diameter of insulation, in
$\quad D_i$ = inside diameter of insulation, in
$\quad T$ = temperature differential, °F
$\quad Z$ = 40.944 Btu·N/(W·h·ft)
†For any desired metric equivalents, refer to metric conversion factors in Appendix B.

Source: Chromalox, E. L. Wiegand Division, Emerson Electric Co. Used with permission.

is 0°F (−17.8°C), the maximum wind velocity is 20 mi/h (32.2 km/h). For various pipe diameters and insulation thicknesses, check Table 17.12.1 for the heat loss in watts per foot per degree Fahrenheit of temperature differential. Table 17.12.1 is based on the following formula:

$$\frac{W}{\text{ft (of pipe)}} = \frac{2\pi k\, \Delta T}{Z \ln (D_o/D_i)} \qquad (17.12.1)$$

where k = thermal conductivity of pipe insulation, $\text{Btu} \cdot \text{in}/(\text{h} \cdot \text{ft}^2 \cdot °\text{F})$
$\quad D_o$ = outside diameter of insulation, in
$\quad D_i$ = inside diameter of insulation, in
$\quad \Delta T$ = temperature differential °F
$\quad Z$ = 40.944 $\text{Btu} \cdot \text{in}/(\text{W} \cdot \text{h} \cdot \text{ft})$
$\quad W$ = watts

(See Appendix B for metric conversions.)

The table assumes a k factor of 0.25 for fiberglass at 50°F (10°C) and adds in a 10 percent safety margin.

Multiply the value from Table 17.12.1 by the temperature differential [in the example, 100°F (37.8°C)] to determine the heat loss per foot (meter) of pipe. For a 1.5-in (3.8-cm) metal pipe with 2-in (5.1-cm) insulation, the Table 17.12.1 figure is 0.038 W/(ft·°F). Multiplied by ΔT [100°F (37.8°C)], this leads to a heat loss of 3.8 W/ft (12.4 W/m) of pipe.

More rigorous calculation would determine the exact k factor of the insulation and add corrections based on higher or lower wind speeds. Underground piping would call for an additional safety factor of 25 percent, even with piping below the frost line. If plastic pipe is being used (rather than metal), higher heat capacity is needed to overcome the poorer heat-transfer characteristics of plastic pipe.

17.12.3.2 Determining Heat-Tracing Specifications

Consult the manufacturers' literature to find a cable product with the proper temperature range of 100°F (37.8°C) service. The cable chosen must have a heat output equal to or in excess of the heat demand of the pipe system. In the example, since a 3.8 W/ft (12.5 W/m) heat demand is well within the range of most heat-tracing cables, a selection can be made from a variety of constant-wattage or self-regulating tracings.

If the heat demand were higher, one option would be to spiral-wrap the tracing around the pipe. Table 17.12.2 provides a determination of the pitch (spacing between spirals) that the tracing would require to equal a given heat demand. Read the table by dividing the system's heat demand by the heat output of the tracing. Locate that number or the next higher one in the row specifying the pipe diameter of the system. Then read the pitch in inches at the top of that column.

Where straight lengths of tracing are used, the overall length of the tracing will equal the pipe system's length, plus allowances for fittings. Roughly 1 ft (0.304 m) extra should be allowed for valves below 2-in (5.08-cm) diameter; above that, consult the manufacturer's literature. For flanges, allow for twice the flange diameter; for pipe hangers, allow three times the pipe diameter for each hanger.

17.12.4 ACCESSORY AND CONTROL EQUIPMENT

Like most electric HVAC systems, electric heat tracing offers relatively straight-forward installation and control procedures. Care must be taken to prevent corrosion damage and to minimize fire or shock hazards. This section outlines the common industry practices.

The first control factor to consider with heat tracing is overcurrent protection. For constant-wattage units, the total current load can be calculated according to the following equation:

$$\text{Total current load (A)} = \frac{\text{cable length} \times \text{W/length at operating conditions}}{\text{operating voltage}} \quad (17.12.2)$$

TABLE 17.12.2 Wrapping Factor,* ft cable/ft pipe†

	Pitch, in																	
Pipe size, in	2	3	4	5	6	7	8	9	10	11	12	14	16	18	24	30	36	42
½	1.90	1.47	1.29	1.19	1.14	1.10	1.08	1.06										
¾	2.19	1.64	1.40	1.27	1.19	1.14	1.11	1.09	1.07	1.06								
1	2.57	1.87	1.55	1.38	1.27	1.21	1.16	1.13	1.11	1.09	1.07							
1¼	3.07	2.18	1.76	1.53	1.39	1.30	1.24	1.19	1.16	1.13	1.11	1.08	1.06					
1½	3.43	2.41	1.92	1.65	1.48	1.37	1.29	1.24	1.20	1.16	1.14	1.10	1.08	1.06				
2	4.15	2.86	2.25	1.90	1.67	1.52	1.42	1.34	1.28	1.24	1.20	1.15	1.12	1.10	1.05			
2½	4.91	3.36	2.61	2.17	1.89	1.70	1.56	1.46	1.39	1.33	1.28	1.21	1.17	1.13	1.08	1.05		
3	5.88	3.99	3.06	2.52	2.17	1.93	1.76	1.63	1.53	1.45	1.39	1.30	1.23	1.19	1.11	1.07	1.05	
4	7.43	5.01	3.82	3.11	2.65	2.33	2.09	1.92	1.78	1.67	1.58	1.45	1.36	1.29	1.17	1.11	1.08	1.06
5	9.09	6.10	4.63	3.75	3.17	2.77	2.47	2.24	2.06	1.92	1.81	1.63	1.51	1.42	1.25	1.17	1.12	1.09
6	10.75	7.20	5.44	4.40	3.70	3.22	2.86	2.58	2.36	2.19	2.04	1.83	1.67	1.55	1.34	1.23	1.16	1.12
8	13.88	9.28	6.99	5.63	4.72	4.08	3.60	3.23	2.94	2.71	2.51	2.22	2.00	1.83	1.53	1.36	1.26	1.20
10	17.20	11.49	8.65	6.94	5.81	5.01	4.41	3.95	3.58	3.28	3.03	2.65	2.37	2.15	1.75	1.52	1.38	1.29
12	20.34	13.58	10.21	8.19	6.85	5.89	5.18	4.62	4.18	3.83	3.53	3.07	2.73	2.40	1.97	1.68	1.51	1.39
14	22.30	14.89	11.18	8.97	7.49	6.44	5.66	5.05	4.57	4.17	3.85	3.34	2.96	2.67	2.11	1.79	1.59	1.46
16	25.44	16.98	12.75	10.22	8.53	7.33	6.43	5.74	5.18	4.73	4.35	3.77	3.33	3.00	2.34	1.97	1.73	1.57
18	28.58	19.07	14.31	11.47	9.57	8.22	7.21	6.42	5.80	5.29	4.86	4.20	3.71	3.33	2.58	2.15	1.88	1.69
20	31.71	21.16	15.88	12.72	10.61	9.11	7.99	7.11	6.42	5.85	5.38	4.64	4.09	3.66	2.82	2.34	2.03	1.81
24	37.99	25.34	19.02	15.22	12.70	10.90	9.55	8.50	7.66	6.98	6.41	5.52	4.85	4.34	3.32	2.72	2.33	2.07

*To determine the wrapping factor, divide the calculated heat loss by the heat output of the cable. Locate the value that is equal to or the next highest in the row for the pipe size in your application. The value at the top of the column is the pitch or spacing from center to center of the cable along the pipe.

†For any desired metric equivalents, refer to metric conversion factors listed in Appendix B.

Source: Chromalox, E. L. Wiegand Division, Emerson Electric Co. Used with permission.

The current load so determined should be multiplied by 125 percent to arrive at the correct rating for fuses, circuit breakers, or other overcurrent protection devices.

For self-regulating tracing, the calculation is different because the tracing experiences a large current inrush upon cold startup, followed by steadier current demand based on the system's operation. Inrush current is usually specified by the tracing supplier. The current load is then calculated according to the equation

$$\text{Total current load (A)} = \text{inrush current (A/length)} \times \text{cable length} \qquad (17.12.3)$$

The temperature of the tracing line may be controlled by several means. Either the ambient temperature or the temperature of the tracing line can be monitored. Ambient sensing is normally recommended for freeze protection. The sensing device(s) should be placed where temperatures are expected to be lowest; additional sensors may be placed at points considered most representative of the system.

CHAPTER 18
REFRIGERATION EQUIPMENT

18.1 REFRIGERANTS

H. Michael Hughes
Engineering Consultant, Lavonia, Georgia

Mark W. Spatz
Honeywell International, Inc., Buffalo, New York

Donald W. Batz
Senior Vice-President, Comm Air Mechanical Service, Oakland, California

PART A: REFRIGERANT SELECTION

Refrigerants are the working fluids for refrigeration cycles. They absorb heat from the medium to be cooled—air in the case of a direct expansion air conditioner or water for a chiller. The absorbed heat is then carried by the refrigerant to a heat rejector, e.g., condenser, where the heat can be given up. The refrigerant is then recycled in the system to absorb more heat. In most refrigeration systems, this is a continuous process, so heat is continually being absorbed and rejected as the refrigerant is moved around the cycle.

The most common type of refrigeration cycle is the vapor compression cycle. This is the type of refrigeration cycle used in household refrigerator/freezers, automobile air conditioning, most residential, commercial, and institutional air conditioning, and commercial (supermarket) refrigeration. Other types of refrigeration cycles include absorption, used in some large water chillers and a very small percentage of residential systems. Commercial aircraft use the Brayton cycle, an all-gas cycle using air as the refrigerant.

18.1.1 SELECTION CRITERIA

Almost any fluid can be made to function as a refrigerant in a variety of cycles. Many fluids, however, exhibit undesirable properties that limit their utility in refrigeration cycles. Traditionally, refrigeration system designers have based the selection of the refrigerant on three major criteria—safety, reliability, and performance. More recently, a fourth criterion has emerged—environmental acceptability.

18.1.1.1 Safety

Safety generally is broken down into two areas, flammability and toxicity. Both are complex issues, the details of which are beyond the scope of this text.

In general, toxicity addresses acute, subchronic, and chronic effects. Within these broad categories, the effects on future generations (mutations, birth defects, etc.), as well as exposed individuals, are evaluated before products are introduced into commerce. ASHRAE Standard 34-2001[1] broadly classifies refrigerants on the basis of chronic exposure limits as defined by TLV-TWA (Threshold Limit Value–Time Weighted Average) or equivalent indices. Class A refrigerants are those deemed to be of low toxicity with allowable exposure limits of 400 ppm or greater for an 8-h day and 40-h work week. Class B refrigerants are those which have a greater toxicity with exposure limits of less than 400 ppm. This does not mean that refrigerants with a B classification cannot be used safely. In fact, several refrigerants that have been assigned a B classification have been used successfully for many years. ASHRAE Standard 15-2001[2] addresses acute toxicity by limiting the quantity of refrigerant permitted in occupied space. For more detailed information review the Material Safety Data Sheet (MSDS) for the particular refrigerant in question. These are available from the refrigerant manufacturer as well as other sources.

Flammability is also classified by ASHRAE Standard 34-2001. This standard utilizes three classifications—1, 2, or 3—based on flammability. Class 1 refrigerants are nonflammable, Class 2 are moderately flammable, and Class 3 are highly flammable. Class 3 would include chemicals that are used as fuels, such as hydrocarbons. Most refrigerants in use for air conditioning applications are Class 1 fluids. There are other organizations that classify refrigerants based on flammability. Underwriters Laboratories (UL) is one of the most widely recognized in the U.S. because of its safety listing service for air conditioning and refrigeration equipment. For transportation purposes in the U.S., there is a third basis of classification promulgated by the Department of Transportation (DOT). Each of these classification schemes can yield differing results based on the criteria used, e.g., ammonia, which is classified as moderately flammable by ASHRAE (Class 2), is classified as flammable by UL and nonflammable by the DOT. In the past, even the test methods used to evaluate flammability differed between different organizations. Recently, there has been an effort by ASHRAE and UL to harmonize test methods and conditions.

In the past, most refrigerants were single component fluids plus a limited number of azeotropes. Now refrigerant blends that can fractionate substantially are being used commercially. These blends can shift composition under various leakage scenarios and have inherently differing compositions in the liquid and vapor phases that can alter the flammability characteristics. Refrigerant blends that are classified under the ASHRAE Standard are based on the worst case of fractionation. This classification is determined according to the same criteria as a single-compound refrigerant. "Worst case of fractionation" is defined as the composition during fractionation that results in the highest concentration of the flammable component(s) in the vapor or liquid phase.

18.1.1.2 Reliability

Reliability of a refrigeration system is largely dependent on the hardware design, installation, and application. The refrigerant can, however, affect the reliability of the system, and its properties are a part of the selection process.

Chemical stability is a very important property for a refrigerant. If the refrigerant decomposes due to the temperatures or pressures that it is exposed to in the manufacture of the refrigeration system, shipping, storage, or operation, it is unlikely that continued operation of the system will be satisfactory. The decomposed products will have property differences that can severely impact capacity, efficiency, or other operating characteristics.

Material compatibility is equally vital. If the refrigerant is corrosive to metals in the system, or if it dissolves or embrittles plastics and elastomers, unsatisfactory performance and/or life can be expected. In many cases, materials can be selected in the design process for which compatibility has been determined.

Lubricant miscibility/solubility is generally considered desirable because it is the primary mechanism for oil return to the compressor. It is possible to design systems that operate with lubricants, which are immiscible and insoluble with the refrigerant, but the complexity of the refrigerant piping is increased.

There are other properties that are important, including dielectric strength for systems using hermetic compressors and a freezing point well below the expected operating range (and also below the unit storage range).

18.1.1.3 Performance

The performance of a refrigeration system is characterized by its capacity and efficiency. The choice of the refrigerant can dictate the type of system as well as the size and configuration of most components including the compressor, condenser, evaporator, expansion device, and connecting lines. There are two types of properties which dictate performance—thermodynamic and transport. Together these properties are considered to be thermophysical properties.

If one were to select a single property with which to characterize the performance of a refrigerant, it would be the boiling point (understood to be at atmospheric pressure). There is a very strong relationship between boiling point and the theoretical capacity of a refrigeration system as illustrated by Fig. 18.1.1. This correlation is useful in selecting a replacement refrigerant for the same or similar equipment. There is an inverse relationship

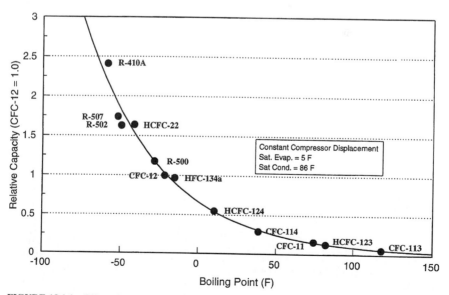

FIGURE 18.1.1 Effect of boiling point, on capacity.

between boiling point and vapor pressure. A refrigerant with a low boiling point will have a high vapor pressure and vice versa.

In general, refrigerants with a high boiling point have more favorable thermodynamic properties. If one analyzes two different refrigerants on Mollier (pressure-enthalpy) diagrams, the lower pressure (higher boiling point) refrigerant will generally exhibit a higher efficiency. However, real systems have losses associated with pressure drops and heat transfer resistances. These losses are associated with transport properties such as viscosity and thermal conductivity. Higher pressure refrigerants tend to have more favorable transport properties and therefore the losses from the ideal cycle are less. This means that there is a tradeoff between the thermodynamic and transport properties of alternative refrigerants. As a result, there is seldom an obvious choice of refrigerant and the selection of the best refrigerant often comes down to economics. Table 18.1.1 show selected refrigerants arranged by boiling point.

Additional properties that can affect performance include latent heat and vapor heat capacity. Latent heat capacity is simply a measure of the amount of heat per lb, absorbed or rejected during phase change. Obviously a high latent heat capacity will reduce the required mass flow rate, that will tend to reduce pressure drop losses. Either too high or too low a vapor heat capacity can have a negative impact on both efficiency and reliability of the compressor. If the vapor heat capacity is too high, so-called wet compression can occur. This means that liquid is formed in the compression process that can result in physical damage to the compressor. If too low, excessive superheating will occur during compression, with high discharge temperatures. Hermetic compressors usually rely on moderate vapor heat capacity to cool the motors and overheating can result if the vapor heat capacity is too low.

TABLE 18.1.1   Refrigerants in order of boiling point

ASHRAE number	Type of refrigerant	Class of refrigerant	Boiling point °F/(°C)
123	Single component	HCFC	82.2/ (27.9)
11	Single component	CFC	74.9 (23.8)
245fa	Single component	HFC	59.5 (15.3)
236fa	Single component	HFC	29.5 (−1.4)
134a	Single component	HFC	−15.1 (−26.2)
12	Single component	CFC	−21.6(−29.8)
401A	Zeotrope	HCFC	−27.7 (−33.2)
500	Azeotrope	CFC	−28.3 (−33.5)
409A	Zeotrope	HCFC	−29.6 (−34.2)
22	Single component	HCFC	−41.5 (−40.8)
407C	Zeotrope	HFC	−46.4 (−43.6)
502	Azeotrope	CFC	−49.8 (−45.4)
408A	Zeotrope	HCFC	−49.8 (−45.4)
404A	Zeotrope	HFC	−51.0 (−46.1)
507	Azeotrope	HFC	−52.1 (−46.7)
402A	Zeotrope	HCFC	−54.8 (−48.2)
410A	Zeotrope	HFC	−62.9 (−52.7)
13	Single component	CFC	−114.6 (−81.4)
23	Single component	HFC	−115.7 (−82.1)
508B	Azeotrope	HFC	−125.3 (−87.4)
503	Azeotrope	CFC	−126.1 (−87.8)

18.1.1.4 Environmental Acceptability

The generally accepted theory that the chlorine in fluorocarbon refrigerants can be a major contributor to stratospheric ozone depletion has resulted in international regulations to phase out the chlorine-containing types. It has also resulted in a heightened awareness of other potential environmental concerns. As a result, future fluorocarbon-based refrigerants will be restricted to hydrofluorocarbons (HFCs) with no chlorine.

Another environmental issue which has been raised is the effect of refrigerants on global warming; the so-called "greenhouse effect." This effect is related to the infrared absorption characteristics of various gases. On a global basis, both carbon dioxide and water vapor are by far the major contributors to this effect, but other gases, including fluorocarbons, can have an impact. On a per molecule basis, fluorocarbons tend to have a relatively high potential for contribution to atmospheric warming. The term *global warming potential* (GWP) has been introduced as a convenient way of indicating the direct contribution of various gases if emitted to the atmosphere, e.g., due to leakage or during servicing. For most air conditioning equipment, this direct effect is overshadowed by the indirect effect created by the introduction of carbon dioxide into the atmosphere at the electric power plant. The CO_2 is generated in the production of electric power to operate the air conditioner. A method to incorporate both of these effects into a single index number has been developed. This single number has been designated as LCCP[3] (Life Cycle Climate Performance). The LCCP value captures the total warming contribution over the life of a piece of refrigeration equipment.

18.1.2 REFRIGERANT TYPES

There are several ways one can classify refrigerants. One method is by chemical/molecular composition. Another useful way of distinguishing refrigerants is by physical composition, i.e., whether they are single component fluids or mixtures. They can also be classified according to the type of refrigeration system that they are used in.

18.1.2.1 Chemical Composition

Chemicals suitable for use as refrigerants can be broadly broken down into two major classifications—organic and inorganic. The organic group can be subdivided into hydrocarbons and halocarbons. Halocarbons can be further subdivided into chlorofluorocarbons (CFCs), hydrochlorofluorocarbons (HCFCs), and hydrofluorocarbons (HFCs).

Inorganic refrigerants include ammonia, water, and air, as well as other more obscure refrigerants. ASHRAE also classifies carbon dioxide as inorganic, although most chemists would disagree.

Hydrocarbons include propane, ethane, and isobutane. There are other hydrocarbons, but their use as refrigerants has been rare. In general, hydrocarbons have not found favor for use in air-conditioning systems because of their extremely flammable characteristics. They have been successfully applied as refrigerants in industrial systems and as a minor component in some refrigerant blends.

The halocarbons have found widespread use as refrigerants, because many members of this family of compounds have exhibited the desirable characteristics of stability, low toxicity, and nonflammability. They have also demonstrated excellent thermodynamic and transport properties, yielding high efficiencies and reliable systems in practice. The halocarbons are hydrocarbon-based molecules in which some or all of the hydrogens have been replaced by halogens. In general, the halogens are restricted to chlorine and fluorine, although a few bromine-containing compounds have limited usge as refrigerants, but not for air-conditioning applications. Bromine is considered to be a greater depleter of stratospheric ozone than chlorine, therefore the few bromine containing refrigerants were phased out even before the CFCs.

The CFCs are fully halogenated, i.e., all hydrogens on the base hydrocarbon have been replaced by halogens. The HCFCs contain chlorine, but still retain one or more hydrogen atoms on the molecule. The effect of the hydrogen is to reduce the atmospheric stability that shortens the lifetime of the molecule, resulting in substantial reductions in the potential to deplete stratospheric ozone. As an example, HCFC-22 is almost identical in structure to CFC-12. Both molecules have a single carbon atom surrounded by four atoms of which two are fluorine. The difference is that for CFC-12, the two remaining atoms are both chlorine, and for HCFC-22, only one is chlorine and the other is hydrogen. The result is that HCFC-22 has an ozone depletion potential (ODP) that is only 5 percent of that of CFC-12. The HFCs contain only hydrogen and fluorine atoms attached to the carbon backbone. Figure 18.1.2 illustrates typical molecular structures for the three major classes of halocarbons. A useful way to look at the various molecules that can be created from the substitution of chlorine and fluorine atoms on basic simple hydrocarbons is to arrange the compounds in a triangular pattern according to their molecular structure. Figures 18.1.3 and 18.1.4 show the methane- and ethane-based refrigerants, respectively. Each dot represents a possible combination of atoms, in which the top molecule is the base hydrocarbon and each successive row down represents an additional substitution of either chlorine or fluorine atoms. This particular arrangement provides one of the easiest ways of understanding the refrigerant numbering system that ASHRAE has adopted.

The triangular representation also allows some generalizations about properties. Those molecules above the midpoint are flammable at atmospheric pressure, i.e., where half or fewer of the hydrogens have been replaced, the molecule will be flammable. The bottom row of molecules is fully halogenated, and these molecules tend to have long atmospheric lifetimes that raise concerns from an environmental aspect. In fact, all the CFCs come from the bottom row. An observable impact of chlorine is that it raises the boiling point of the molecule. As one moves toward the lower left point of the triangles, the boiling point increases. Another property of interest is the toxicity of the chemicals. There is a less well-defined, but observable, trend toward greater toxicity as one moves to the left (higher in chlorine).

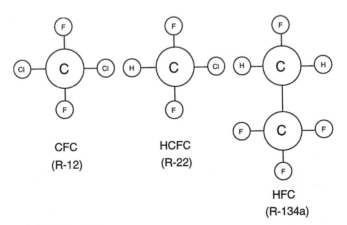

FIGURE 18.1.2   Molecular structures for hydrocarbons.

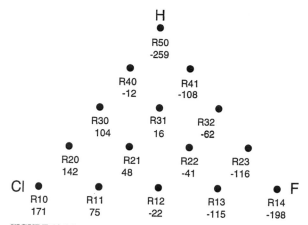

FIGURE 18.1.3 Methane series refrigerants.

In the ethane series, for some combinations, it is possible to arrange the same atoms in different ways. These variations are isomers and are designated by a lower case suffix of a, b, or c. R-134a is an isomer of R-134 and exhibits different properties. It is, therefore, important to include the full designation when referring to the refrigerant by its number. For further reading on this area, the reader is directed to an excellent article "Quest for Alternatives."[4]

18.1.2.2 Physical Composition

Most refrigerants are single component fluids, i.e., they consist of a single molecular species. It is possible, however, to mix two or more refrigerants. There are several reasons

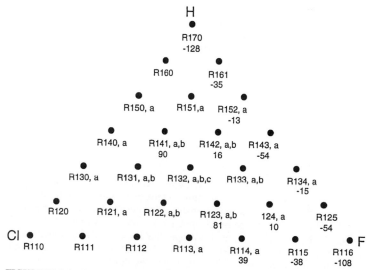

FIGURE 18.1.4 Ethane series refrigerants.

to consider blends of refrigerants. It is possible to modify the thermodynamic properties, e.g., change the boiling point of the refrigerant that affects capacity or change the discharge temperature to enhance reliability. In some cases, the flammability of an otherwise desirable refrigerant can be suppressed by the addition of a nonflammable component. Unfortunately it isn't possible to render a highly flammable refrigerant like propane nonflammable by the use of a minor additive. It is in fact difficult to maintain nonflammability of any mixture that contains more than a minor amount of a hydrocarbon such as propane. Other reasons to consider mixtures would include improvement of lubricant miscibility and solubility or to improve leak detectability.

Two types of mixtures can be formed with refrigerants. The most likely type of mixture, if one were to arbitrarily select two or more refrigerants, would be a simple mixture or zeotrope. The characteristics of a zeotrope are as expected, i.e., the blend would tend to average the properties of the components. This characteristic allows great flexibility in tailoring the properties to achieve a desired result. The composition of the liquid and vapor phases would be different because the more volatile component(s) would more readily evaporate, enriching the vapor phase in the lower boiling (higher pressure) constituents. Another characteristic of zeotropic mixtures is commonly described as temperature glide. Glide occurs as a result of the segregation or fractionation which zeotropes exhibit.

To better understand temperature glide, consider a binary mixture of two arbitrary refrigerants, A and B, in which A is the more volatile (lower boiling point). We will also assume that equal amounts of A and B are used. As evaporation commences, the initial vapor that boils off will have a higher percentage of A and less of B. It would not be unusual if the ratio were 70/30 rather than the 50/50, as originally formulated. The liquid that remained would then become enriched in B as the evaporation process continued. This shifting of the composition would also change the boiling point as the process continues. In evaporation, the temperature increases as the mixture evaporates. In condensation, the effect is reversed, i.e., the temperature decreases continually as condensation progresses. The generally accepted definition of temperature glide is the difference between the dew and the bubble points of the refrigerant at any given pressure. The dew point of a refrigerant is saturated vapor where the first drop of liquid forms in condensation (or conversely where the last drop of liquid evaporates). The bubble point is saturated liquid where the first bubble of vapor forms in boiling (evaporation). For single component refrigerants, these points are at the same temperature at any given pressure.

There has been considerable interest in attempting to take advantage of the temperature glide by matching the refrigerant glide to the temperature differential of the source and sink fluids. This should, in theory, reduce the log mean temperature difference in a heat exchanger; however, in practice it is difficult to match these glides. In addition, there is a mass transfer resistance with blends that tend to offset the benefit, due to the temperature glide match. In general, heat exchanger performance is degraded with refrigerant blends, and, as a result, energy efficiency is lower than anticipated.

As mentioned above, there is another type of mixture for refrigerants. These mixtures are azeotropes. Azeotropes are unique and do not behave as mixtures, but rather function almost like a single component refrigerant. The mixtures do not segregate and, in fact, are very difficult to separate by distillation. Another characteristic is their boiling point, always either higher or lower than either of the constituents. It is possible to form azeotropes with more than two components, but these are rare. Azeotropes do not exhibit the temperature glides of zeotropic mixtures and the heat transfer characteristics are not degraded due to the mass transfer resistance. As a result azeotropes are favored by equipment system designers over zeotropes. In the past, CFC-based azeotropes that have seen widespread acceptance in commercial use include R-500, R-502, and R-503. New HFC azeotropes are now starting to be accepted in commercial use. R-507, which was developed as a replacement for R-502 for frozen food applications, is widely accepted in supermarket applications. An azeotropic refrigerant is only a true azeotrope at one temperature for a given composition. At other temperatures, the composition of the azeotrope will differ. From an engineering standpoint, these minor deviations from true azeotropy can be ignored, since the deviations are so small that they can't be measured in systems. There are some zeotropes that are extremely close to azeotropes and, in fact, can be treated as azeotropes for engineering purposes. An example of such a zeotrope is R-410A, a 50/50 mixture of HFC-32 and HFC-125. These two compounds form an azeotrope at an 80/20 weight percent ratio. They also form azeotropes at other ratios slightly different from the 80/20. Although not a true azeotrope, this mixture is applied and handled like an azeotrope.

One way to illustrate the differences between azeotropes and zeotropic mixtures is by a phase diagram that plots either temperature or pressure as a function of the composition of a mixture. Figures 18.1.5 and 18.1.6 illustrate typical phase diagrams.

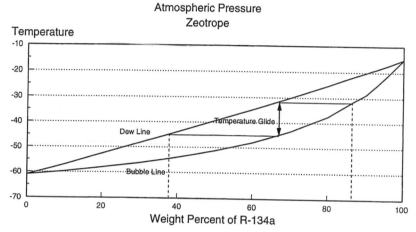

FIGURE 18.1.5   Phase diagram for R-32/134a.

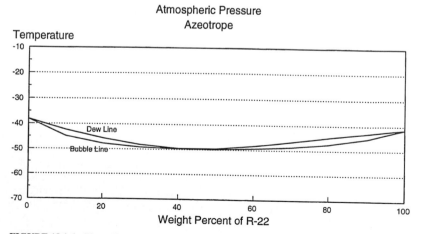

FIGURE 18.1.6   Phase diagram for R-22/115 (R 502).

18.1.3 REFRIGERATION SYSTEMS

For air conditioning applications in buildings, there are two types of refrigeration systems employed—vapor compression systems and absorption systems. By far, the greater majority are vapor compression, but for large water chillers, absorption represents a significant portion. This is especially true in Japan, where absorption systems dominate the chiller market; however, smaller unitary systems employ vapor compression systems almost exclusively. Almost all absorption chiller systems utilize water as the refrigerant, with lithium bromide used as the absorbent. Smaller systems have used ammonia as the refrigerant and water as the absorbent. Both water and ammonia share some common advantages as refrigerants. These include availability, low cost, and excellent heat transfer characteristics. Water has the disadvantage of high freezing and boiling points. The high boiling point results in a very low vapor pressure and, therefore, systems that employ water as the refrigerant have to operate in a vacuum. Ammonia has the disadvantages of moderate flammability and a relatively high level of toxicity. It is also incompatible with copper, which tends to offset the outstanding heat transfer properties of the fluid; it cannot take advantage of the high-performance surfaces available with copper tubing.

Vapor compression systems can employ a wide variety of refrigerants, although air-conditioning systems have concentrated around a very limited number of refrigerants. Building systems have utilized both low- and high-pressure refrigerants. Low-pressure refrigerants have been the exclusive province of centrifugal chillers. Until recently CFC-11 was the low-pressure refrigerant of choice, although a very limited quantity of systems used CFC-113 and CFC-114. The current low-pressure refrigerant is HCFC-123. It offers performance comparable to CFC-11, but with greatly improved environmental properties. HCFC-123 has a very short atmospheric lifetime and, therefore, has minimal environmental impact. It does, however, contain chlorine and is scheduled for phaseout eventually. There are few alternatives suitable for low-pressure applications, although R-245fa is a potential low-pressure refrigerant for centrifugal chillers.

Higher-pressure refrigerants can also be effectively utilized in centrifugal chillers. CFC-12 was widely used in the past but has been more recently supplanted by HFC-134a. Positive displacement compressor systems have primarily used HCFC-22. These types of systems encompass a wide variety of equipment ranging from PTACs, using small rotary compressors with capacities down to 6000 Btu/h (1758 W), to large screw chillers with capacities exceeding 6,000,000 Btu/h (1,758,000 W). In between is a large group of unitary equipment and chillers using both reciprocating and scroll type compressors. The majority of this equipment continues to use HCFC-22, but it is recognized that ultimately HFC replacements will be required. Three major HFC alternatives have emerged to cover the range of applications. HFC-134a appears to have utility for some large screw chiller applications. In the U.S., R-407C is expected to see duty primarily as a service fluid to replace HCFC-22 in existing equipment and for a very limited number of new systems, such as large rooftop units. In Europe, this refrigerant is being used as a replacement for R-22 in new air-conditioning and heat pump systems. This refrigerant was adopted because of the rapid phaseout of R-22 in this region. It doesn't require significant design changes, so equipment designed for R-22 can essentially be used for R-407C. In the United States and Japan, the majority of unitary systems, and some chillers that utilize hermetic compressors, will utilize R-410A. Even though R-410A is a higher-pressure refrigerant than R-22, it has been selected for the majority of applications because of its superior heat transfer characteristics. It is also more tolerant of pressure drop that minimizes these losses, especially in split systems. Extensive testing has demonstrated higher energy efficiencies than R-22 with similarly-sized heat exchangers.

18.1.4 MATERIALS COMPATIBILITY

A large body of data has been developed over the years on CFCs and R-22. With the advent of many new refrigerants, much of the conventional wisdom is no longer relevant or correct. There has been a concentrated effort by many researchers to develop equivalent data for the newer alternatives, such as interim HCFC-based and long term HFC-based refrigerants. The most comprehensive source of data is the ARTI Refrigerant Database.[5]

18.1.4.1 Lubricants

Virtually all vapor compression refrigeration systems require lubricants to permit reliable compressor operation. Historically, fluorocarbon-based refrigerants, such as CFCs, used refined mineral oils that exhibited a high degree of miscibility and solubility with the refrigerant. Ammonia has historically also used mineral oils, but was almost totally immiscible with the oils. Systems using CFCs took advantage of the miscibility/solubility characteristics to return the oil that was pumped into the system by the compressor. Most HCFC systems could also use mineral oil, although miscibility was incomplete. Synthetic oils (alkylbenzenes) were developed for use where greater miscibility and solubility was required.

The advent of HFC refrigerants required a new class of lubricants, since these refrigerants are almost totally immiscible with both mineral and alkylbenzene lubricants. Automobile air-conditioning systems have gravitated to polyalkyline glycol lubricants (PAGs) while stationary systems generally use polyol ester (POEs) or polyvinyl ethers (PVEs), because of compatibility problems between hermetic motor materials and PAGs.

Lubricants for refrigeration systems are offered in several viscosity grades, ranging from about 20 to 100 centistokes (cS). In general, the grade is specified by the compressor manufacturer. With mineral oils, various brands were usually considered to be interchangeable (but not different viscosities). With the new synthetic PAG, POE, and PVE lubricants, the viscosity grades may be different than in the past and different brands can no longer be considered interchangeable. Compressor manufacturers have expended considerable effort to evaluate the new lubricants and their recommendations should always be followed.

18.1.4.2 Plastics and Elastomers

The inherent chemical stability of fluorocarbons means that most plastics and elastomers can be used witth them. One notable exception is the fluoropolymer class of elastomers. Many of these were compatible with CFC-12, but are not compatible with HFCs, and also are not compatible with HCFCs such as R-22. It is important to note that within a particular class of materials, e.g., neoprene, there is a wide variation in formulations and it should not be assumed that all members of that class are suitable simply because most are suitable.

HCFC-123 exhibits more aggressive solvency than most other refrigerants. Therefore caution must be used before retrofitting an existing CFC-11 chiller. New materials have been developed which permit satisfactory use with HCFC-123.

The new lubricants, and particularly the POEs, are much stronger solvents than the mineral oils they replace. They are also much stronger solvents than the refrigerants, in most cases. In systems using HFC and HCFC refrigerants that develop compatibility problems, it is usually the lubricant rather than the refrigerant that is the culprit for the new alternate refrigerants.

DISCLAIMER

Although all statements and information contained herein are believed to be accurate and reliable, they are presented without guarantee or warranty of any kind, expressed or implied. Information provided herein does not relieve the user from the responsibility of carrying out its own tests and experiments, and the user assumes all risks and liability for use of the information and results obtained. Statements or suggestions concerning the use of materials and processes are made without representation or warranty that any such use is free of patent infringement and are not recommendations to infringe on any patents. The user should not assume that all toxicity data and safety measures are indicated herein or that other measures may not be required.

REFERENCES

1. ASHRAE 2001. Number Designation and Safety Classification of Refrigerants. *Standard* 34-2001. American Society of Heating, Refrigerating, and Air-Conditioning Engineers, Inc., Atlanta, GA.

2. ASHRAE 2001. Safety Code for Mechanical Refrigeration. *Standard* 15-2001. American Society of Heating, Refrigerating and Air-Conditioning Engineers, Inc., Atlanta, GA.

3. Arthur D. Little (ADL), Global Comparative Analysis of HFC and Alternative Technology for Refrigeration, Air Conditioning, Foam, Solvent, Aerosol Propellant, and Fire Protection Applications, 2002.

4. McLinden, M., Didion, D., Quest for Alternatives. *ASHRAE Journal*, December 1987, American Society of Heating, Refrigerating, and Air Conditioning Engineers, Inc., Atlanta, GA.

5. Calm, J., Refrigerant Database, ARTI 21-CR Research Project 605-00010 & 610-00010, 2001, Air-Conditioning and Refrigeration Technology Institute (ARTI), Arlington, VA.

PART B: REFRIGERANT MAINTENANCE

18.1.5 INTRODUCTION

One of the most pressing issues facing the refrigerant user today is the maintenance and management of the refrigerant itself. This issue reaches not only the contractors, but also the end users, whether as the actual equipment owner or a property management provider. The economics of today's refrigerant issues have forced the user to consider its maintenance and management long before the equipment is turned on. The choices of manufacturers, equipment, refrigerant, and installation practices all must be considered in the development of a refrigerant management and maintenance program. The development of a refrigerant maintenance program will involve several steps that all become critical in implementing an effective program. The major steps in this development process are (1) evaluating the goals of the refrigerant maintenance program; (2) selecting the equipment, type of refrigerant, installation, and maintenance procedures consistent with the goals of the program; (3) writing out and communicating the plan to all involved; (4) enforcing the maintenance program; and finally (5) reevaluating the program regularly.

18.1.6   EVALUATING GOALS

Evaluating the goals of the refrigerant management and maintenance program is the first and most critical step in establishing a successful program. The demands of not only government and environmental regulation, but also determining the long-term goals of the program help set the agenda for success. A number of factors that will influence the process from a regulation standpoint are the effects of future regulation, refrigerant supplies, and cost of replacement. It is well advised that each end user utilizes all sources available for information on government trends and industry lobby efforts. The uncertain nature of future additional regulation and/or taxes requires diligence in monitoring this arena. The factoring of long-term goals in relation to cost and capitalization of replacement equipment, expected life of equipment, and cost of refrigerant maintenance procedures versus replacement of all must be evaluated according to each individual user requirements. The choice of service contractors, and consultants in the case of a property manager/building owner, or the levels of training and equipping of a contractor, must be considered to successfully develop a workable program.

18.1.7   SELECTION OF EQUIPMENT AND PROCEDURES

Selecting the equipment, type of refrigerant, installation, and setting maintenance procedures consistent with the goals, is the next step in a successful program. Each individual owner, manager, or contractor should develop a written procedure for the equipment and refrigerants being dealt with. This should include the procedures an owner/manager will expect, not only from their own staff, but also from outside contractors and consultants. This must include a method of notification of work to be done, report of any refrigerant system problems and detailed record or refrigerant additions or losses, with causes. The days of sending a contractor to the roof and signing the tag on the way out are long gone. A higher level of expectation and communication must be demanded and received to protect the refrigerant resources. This communication also will build a greater understanding of the factors that influence the refrigeration equipment operation and maintenance. Each contractor must develop a training program of procedures to be followed with work on the refrigerant system. The potential liability and losses when an accidental release occurs have risen to the point that a contractor cannot ignore this critical area. Together with the proper record keeping of maintenance practices and constant training, a contractor can become an ally to the building owner/manager in the refrigerant maintenance program. In considering the installation of new equipment, consideration should be made of the level of refrigerant containment desired. The cost is typically much less if the additional maintenance tools are built in during new construction, rather than retrofitted later. These tools include isolation valving, transfer vessels, relief containment measures, and high-efficiency purges on larger equipment. If these are ordered and built in when the equipment is ordered new, vs. retrofitted in the field, a substantial savings can be realized. This planning of procedures, when weighed and balanced by the long term goals of the program, sets the stage for success.

18.1.8   COMMUNICATING THE PLAN

Writing out and communicating the plan to all involved is the third step in developing the process. Establishing the processes and reporting procedures is good, but useless if not communicated to all that come into contact with the equipment. Whether these are the owner/management staff or the contractors, letting all know that refrigerant containment and conservation is the heart of the maintenance program is critical. The plan should be simple, but comprehensive. Feedback on program improvement should be encouraged. It is a wise idea to post these refrigerant maintenance procedures and goals in the equipment areas or roof access and it is a good way of communicating the seriousness of the program to the service personnel as they work. Be specific in the procedures; simple, but direct.

18.1.9   IMPLEMENTING THE MAINTENANCE PROGRAM

Enforcing the maintenance program is where the rubber meets the road. A user must be proactive in monitoring compliance with the goals of the program. If the procedures require acid testing, leak detection, purge maintenance, and use of clean recovery equipment/ drums, for example, then follow up during a service visit and spot check the methods of work followed. It is, after all, the asset of the end user that is at potential risk for loss or contamination. When you consider the cost of replacement of 100 percent of the refrigerant charge due to mixing of refrigerant from the use of dirty recovery drums, then 15 to 20 min of time to visit with and observe the contractor or maintenance staff becomes a good investment. Time spent observing and evaluating the programs' effectiveness allows constant monitoring of the success of the methods chosen.

18.1.10   PROGRAM REEVALUATION

Reevaluating the program is the step that separates programs with one time success from a program that continues to improve over time and remain consistent in the face of changing situations. The gathered input of service personnel, owners, consultants, and others involved in the refrigerant maintenance program allows constant tuning to everchanging situations. For example, an owner may decide that 5 years of use of the existing equipment will be required before replacement can be considered. This may be due to the lack of capital funds, future plans for remodeling the building, or expectations that the loads in the building may change significantly in the future. The existing equipment is mechanically sound, so evaluation will center on refrigerant maintenance. The program will include installing custom leak detection systems, regasketing, adding purge containment, increasing moisture and frequency acid testing, and requiring all service personnel to follow the guidelines chosen. Recording of any venting or releases and work requiring opening of the refrigerant system is required. Contingency plans need to be made regarding the options available in case of major equipment failure, or the loss or contamination of the refrigerant charge. It may make economic sense to opt for replacement rather than to rebuild, due to the loss of the refrigerant charge and the mechanical damage.

18.1.11   SETTING UP THE PROGRAM

As you set up a refrigerant management and maintenance program, there are many items that must be taken into consideration. Start with the main information needed to give you the basis for a program that can work for you.

1. Equipment Type

   • New high efficient packaged equipment with environmentally friendly refrigerant
   • New built-up system with environmentally friendly refrigerant
   • Existing equipment with chlorofluorocarbon (CFC) refrigerant
   • Existing equipment converted to one of the new alternate refrigerants

2. Type of System

   • Package units   • Centrifugal
   • Built-up system   • Screw
   • Chilled water   • Absorption
   • Reciprocating   • Thermal storage

3. Type and Amount of Refrigerant

	High Pressure	Low Pressure
a. CFC—lb (kg)	_____	_____
b. HCFC—lb (kg)	_____	_____
c. HFC—lb (kg)	_____	_____
d. Alternates—lb (kg)	_____	_____

4. Type and Amount of Oil

   • Mineral oil—gal (L)   _____
   • Alkylbenzene oil—gal (L)_____
   • Polyester oil—gal (L)   _____

5. Piping and Accessories

   • Driers #_____   Type_____
   • Sight glass
   • Standard
   • Moisture indicating
   • Relief valve
   • Standard relief
   • Manifolded double relief
   • Rupture disk
   • Nonfragmenting double-resetting valve
   • Purge unit
   • Standard purge
   • High efficiency purge

- Hot water pressure pack
- Electric pressure blanket
- Quick disconnect service valves
- External oil filters with isolation valves
- Oil reclaim educator
- Oil quality monitor
- Receivers
- Pump out systems
- Recovery and recycle units
- Refrigerant monitoring system
- Record keeping system
- Certified technicians
- In-house
- Outside contractor

After completing this inventory of your equipment, systems, and other resources, put together your plan to manage and maintain your current supply of refrigerant, and budget ahead for replacement of equipment or conversion of units using CFCs or HCFCs as the need arises. (Most people have not realized that with a large unit containing 800 (360 kg) to 1000 lb (450 kg) of CFC 11 or 12 it could cost over $25,000 just to recharge a unit if the charge was lost or badly contaminated.) This is if the refrigerant type is available. Set up a plan and then work with the plan to save a lot of headaches.

Basic refrigerant management and maintenance just requires using common sense.

18.1.12 STARTING YOUR PROGRAM

1. Start a history card for each system and piece of equipment. This should include the following:
   - Make and model of unit
   - Capacity of unit
   - Start-up date of unit
   - Warranty status/date
   - Type of refrigerant
   - Amount of refrigerant in system
   - Location of equipment
   - Complete service records as performed including data and technician

2. All systems should be clearly marked as to type of refrigerant and amount of charge. These markings should be visible at any place that someone can access the system.

3. Refrigerant in larger systems should be checked for moisture or acid regularly.

4. Oil should be checked for contaminants and wear metals at least annually.

5. Refrigerant pressures and temperatures should be checked on regular basis.

6. Systems should be checked for visible signs of leaks on a regular basis. Repairs should be made immediately when a leak is discovered.

7. Complete leak check of system with a compatible leak detector should be accomplished annually.

8. Driers should be replaced any time a system is opened for service or repair. (Make sure to use driers approved for the type of refrigerant in the system.)

9. Consider installing refrigerant monitoring equipment in equipment rooms. This will give you 24-h leak checking.

10. Make certain that any technician working on your equipment is trained, certified to work on your units, and has the right tools to work with the refrigerant in your systems (gages, hoses, leak detectors, P/T charts, protective clothing, etc.).

11. Make every effort possible to be certain that you don't get cross contamination of refrigerants in your units. Be careful, especially when using recovery equipment.

12. Plan any conversions to alternate refrigerants to coincide with annual teardown or when a major failure occurs. Do your homework ahead of time.

13. If you do convert a system to one of the new refrigerants, plan to have the old refrigerant reclaimed and then save for use on other equipment that may use that type of refrigerant, or sell it back the recyclers.

14. Make sure that you keep records of refrigerant purchases and usages in case you are ever audited by the EPA.

15. Whenever you add or replace equipment, make sure that you look at system efficiency, type of refrigerant, and serviceability of the unit before you purchase a new piece of equipment.

16. Install refrigerant containment accessories on existing equipment whenever possible.

17. Learn what alternate refrigerants are available. The list continues to grow. Match your application to the refrigerant.

INFORMATION SOURCES

Refrigerant technology continues to change. Contact the following organizations and publications to obtain important up-to-date information concerning refrigerants:

1. Refrigerant manufacturer's catalogs, application manuals, and web sites

2. Equipment manufacturer's catalogs, application manuals, service manuals, and web sites

3. American Society of Heating, Ventilating, and Refrigerating Engineers (ASHRAE), 1791 Tullie Circle, N.E., Atlanta, GA 320329, www.ASHRAE.org

4. Mechanical Service Contractors of America (MSCA), 1385 Piccard Drive, Rockville, MD 20850, www.MCAA.org

5. Building Owners & Managers Association (BOMA), 1201 New York Avenue, N.W., Ste 300, Washington, DC 20005, www.BOMA.org

6. Environmental Protection Agency, www.EPA.gov

18.2 POSITIVE DISPLACEMENT COMPRESSORS/CHILLERS AND CONDENSERS

Chan Madan

Continental Products, Inc. Indianapolis, Indiana

18.2.1 INTRODUCTION

Positive displacement compressors include *reciprocating, screw, and scroll compressors.*

18.2.2 RECIPROCATING COMPRESSORS

A *reciprocating compressor* is a single-acting piston machine driven directly by a pin and connecting rod from its crankshaft. It is a positive-displacement compressor in which an increase in the pressure of the refrigerant gas is achieved by reducing the volume of the compression chamber through work applied to the mechanism.

Various combinations of piston size (bore), length of piston travel (stroke), number of cylinders, and shaft speed result in various compressor sizes, ranging from $^1/_{12}$ to 200 hp (0.06 to 149 kW).

The most commonly used reciprocating compressor is for refrigerants R-22 and R-134a. For heating, ventilating, and air conditioning (HVAC), and process cooling, the most practical refrigerant is R-134a; R-22 is the most practical refrigerant today. However, R-134a is gaining acceptance, in view of the CFC regulations worldwide. As a matter of fact, the same countries only accept R-134a today. Other environmentally acceptable refrigerants are R-404A and R-507 for low and medium temperature applications; R-407C for medium temperatures and air-conditioning applications. Recently, R-410A has gained acceptance as an environmentally acceptable substitute for R-22, but only for residential and small equipment. R-410A is not a drop-in refrigerant for R-22. See Sec. 18.1 for a thorough discussion of refrigerants.

Reciprocating compressors are of three types: *open drive, hermetic, semihermetic.*

18.2.2.1 Open-Type Compressor Units

Open-type compressors are designed to have the drive shaft extend outside the crankcase through a mechanical seal. This seal prevents outward leakage of refrigerant and oil and inward leakage of air and moisture. Figure 18.2.1 shows a cutaway of a typical open-drive compressor.

The drive shaft is adaptable to an electric-motor or gas-engine drive. Electric-motor drives are either *belt-driven* or *directly coupled* to the compressor by means of a flexible coupling. Gas-engine drives are usually directly coupled.

Figure 18.2.2 shows a typical belt-driven unit. The drive consists of (1) a flywheel mounted on the compressor shaft and (2) a small pulley mounted on the motor shaft. These are interconnected by one or more V-belts. The size of the flywheel is usually fixed by the manufacturer, and the size of the motor pulley can be varied to achieve the desired speed. Speed variations are also obtained by ranging the size of the compressor flywheel and motor pulley.

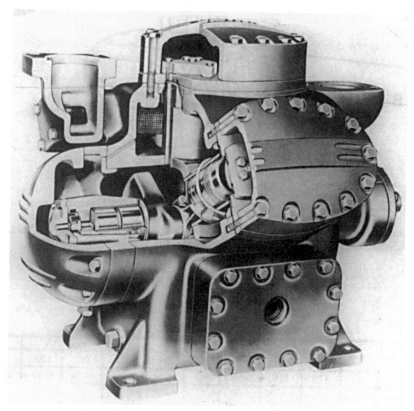

FIGURE 18.2.1   Open drive compressor. (*Courtesy of Carrier Corporation.*)

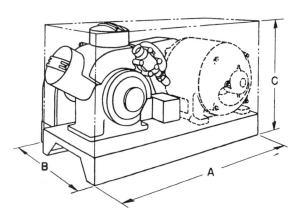

FIGURE 18.2.2   Belt-driven compressor. (*Courtesy of Carrier Corporation.*)

The compressor is rigidly mounted on a steel base. The motor is mounted on an adjustable rail. This allows alignment of the motor pulley and tightening of the belts. Proper belt alignment and belt tension are most important for efficient compressor operation.

Belt Alignment.   Belt alignment (Fig. 18.2.3) can be checked as follows:

1. Line up the compressor flywheel and motor pulley with a straightedge. Slide the motor pulley on the shaft to correct any parallel misalignment. For correct angular alignment, loosen the motor hold-down bolts and turn the motor frame as required.
2. When the alignment is completed, move the motor away from the compressor with the adjusting screws to tighten the belts. Tighten the belts just enough to prevent slippage.

Belt Tension.   Belt tension can be checked by checking the amount of deflection as the belt is depressed at the center of the span. The rule of thumb is that belts deflect approximately 1 in for every 24-in span (1 cm for every 24-cm span). A longer span will deflect proportionately more.

Figure 18.2.4 shows a typical *direct-drive* unit. The compressor shaft is connected directly to an electric motor through a flexible coupling and is designed to run at motor speed; this speed is 1750 rpm for a 60-Hz power supply and 1450 rpm for a 50-Hz power supply. Two-speed or variable-speed motors are sometimes used for closer capacity control, but for HVAC applications this is cost-prohibitive and not commonly used.

The compressor and motor are rigidly mounted on a steel base. Proper coupling alignment is essential for trouble-free operation. The maximum permissible angular or parallel misalignment for all couplings is 0.010 in (0.25 mm). The manufacturer's recommendations are necessary for alignment. Basically, there are two alignment methods employed:

• The dial-indicator method (Fig. 18.2.5)
• The straightedge-and-caliper method (Fig. 18.2.6)

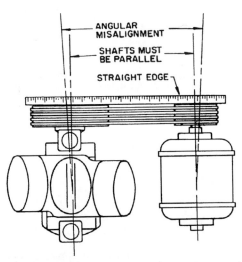

FIGURE 18.2.3   Correct belt alignment. (*Courtesy of Carrier Corporation.*)

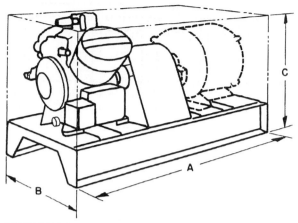

FIGURE 18.2.4   Direct-drive unit. (*Courtesy of Carrier Corporation.*)

18.2.2.2 Hermetic Compressors

Hermetic compressors are also known as "sealed" or "welded" compressors or as "cans," since the motor and compressor are mounted within a common pressure vessel, sealed by welding. Figure 18.2.7 shows a cutaway of a typical hermetic compressor.

The compressor consists of pistons, connecting rods, bearings, and a lubrication system, and is driven by a crankshaft that is common to both the compressor and the motor.

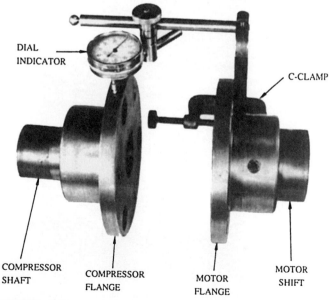

FIGURE   18.2.5   Alignment with a dial indicator. (*Courtesy of Carrier Corporation.*)

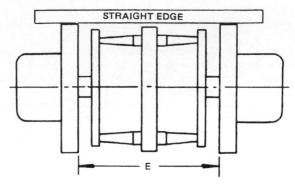

FIGURE 18.2.6 Alignment with a straight edge and calipers (*Courtesy of Carrier Corporation.*)

The motor-starter windings are of refrigerant gas-cooled design. Most hermetic compressors are internally spring-mounted. Some have built-in suction accumulators for protection against liquid floodback. Lubrication of the compressor is usually achieved by a careful design that allows lubrication of the bearing and motor surfaces.

Hermetic compressors are commercially available in sizes from $\frac{1}{12}$ to 24 hp (0.06 to 18 kw). Larger sizes are proprietary and are made by some manufacturers for use on their own packages.

FIGURE 18.2.7 Hermetic compressor. (*Courtesy of Copeland Corporation.*)

Hermetic Motors and Motor Protection.   Hermetic motors are specially designed by various manufacturers to keep the compressor losses to a minimum. This allows the compressor to operate effectively at maximum compression ratios.

Furthermore, these motors must have high dielectric strength, be resistant to abrasives, and be compatible with the refrigeration lubrication oil and refrigerant. Such other factors as insulation, efficiency, performance starting current, starting and breakdown torques, cost, and availability are also important.

Although the suction-gas-cooled design feature allows hermetic motors to be of considerably small sizes, it also poses a problem in protecting the motors from quick overheating or from drawing excessive current (amps). The most common method of motor protection is to have thermal overload devices embedded in the windings. These mechanisms trip when overloading, overheating, or any other abnormal condition occurs.

18.2.2.3 Semihermetic Compressors

Semihermetic compressors are also known as "accessible" hermetic compressors. The motor and compressor are mounted within a common pressure vessel and sealed by bolted plates so that the motor and compressor parts are accessible. In case of a compressor failure, these parts can easily be repaired or replaced, and the rebuilt compressor can be bolted back together. Except for this accessibility, all other features in the semihermetic design remain similar to the hermetic design: the compressor consists of pistons, connecting rods, and lubrication-system bearings, driven by a crankshaft common to both compressor and motor.

Semihermetic compressors are available in the following designs for compressor-motor cooling:

- *Air-cooled design.* This design uses air circulation for proper cooling of the compressor motor. A constant air flow is required across the compressor housing, and this is accomplished by direct impingement of air from fan discharge. Typically, air-cooled compressors are limited in size to $^1/_4$ to 3 hp (0.19 to 2.24 kW).

- *Water-cooled design.* In this design, water coils are wrapped around the compressor housing; compressor-motor cooling is provided by circulating water. Typically, this design is limited to water-cooled condensing units and is not practical in today's HVAC market.

- *Refrigerant-cooled design.* This design is the one most commonly used in HVAC applications. As in hermetic compressors, the motor is cooled by return suction gas. Refrigerant-cooled compressors are available in sizes from 2 to 10 hp (1.5 to 75 kW).

18.2.3   SCREW COMPRESSORS

A complete discussion of screw compressors is included in Sec. 18.5.

18.2.3.1 Screw Vs. Reciprocal and Centrifugal Compressors

Relative advantages of each:

- *Screw vs. Reciprocal*

  1. Screw compressors operate more or less like pumps, and have continuous flow of refrigerant compared to reciprocals, which have pulsations. This results in smooth compression with little vibration. Reciprocals, on the other hand, make pulsating sounds and vibrate.

2. Screw compressors have almost linear capacity-control mechanisms resulting in excellent part-load performance.
3. Due to its smooth operation, low vibration screw compressors tend to have a longer life than reciprocals.

- *Screw vs. Centrifugal*
  1. Centrifugals are constant-speed machines. These machines "surge" at certain operating conditions, resulting in poor performance and high power consumption at part-load.
  2. Screw compressors have proven themselves in tough refrigeration applications, including on-board ships. Today, screw compressors practically dominate refrigerated ships, transporting fruits, vegetables, meats, and frozen foods across the ocean with good reliability. These compressors have replaced the traditional shipboard centrifugals.

18.2.4   SCROLL COMPRESSORS

In 1886, an Italian patent was issued for the basic scroll concept. The first American patent was issued in 1905 and very little was done with the idea until the 1960s and the 1970s, when scroll development work was undertaken in Germany and France. The scroll idea was tried for various applications, such as vacuum pumps and expansion engines.

Although the scroll concept is rather old, it has only been in the last few years that machine tools have been developed to the point where components could be machined to the minimum clearances that are noted to produce high-efficiency scrolls.

Scroll technology is based on two scrolls; the first scroll is *fixed*, while the second scroll *orbits* around the fixed scroll. These scrolls are intermeshed and form a crescent-shaped space between them. When the second scroll orbits around the fixed spiral, the suction gas in the space is compressed until the gas reaches the maximum pressure in the center of the scrolls. This compressed gas is then discharged through a port in the center of the compression chamber.

Due to the intricate design of the scroll compressor, the gas is discharged smoothly, almost like a pump. This smooth compression reduces vibration (compared to reciprocating action) which is of a pulsating nature. Other features are as follows:

- Fewer moving parts than reciprocating compressors
- Less rotating mass and less internal friction
- Smooth compression cycle with low torque
- Low noise levels
- Low vibrations

Motors for scroll compressors are suction gas-cooled where the suction gas cools the motor, achieving high efficiency and long life. The motor is protected by an external protector which senses excessive current, disconnecting it before over-heating.

18.2.4.1   Lubrication

Lubrication to the scroll journal and shaft bearings is achieved by a centrifugal pump submerged in the oil sump. Oil is moved upwards through the passage to lubricate the upper and lower shaft bearing through parts in the shaft wall. It then leaves the upper end of the passage to lubricate the orbiting scroll journal bearing. Lubrication for the scroll contacting surfaces is provided by a small amount of oil entrained within the suction gas stream.

18.2.4.2 Design Types—Compliant or Noncompliant

In the *compliant* design, the surfaces of the scroll plates are allowed to touch lightly, thereby providing high efficiency and reliability. In the *noncompliant* design, the scrolls do not touch, maintaining a slight clearance.

18.2.5 POSITIVE DISPLACEMENT LIQUID CHILLER SYSTEMS

A liquid chiller system cools water, glycol, brine, alcohol, acids, chemicals, or other fluids. The most common use of a chiller system is as a water chiller for human-comfort cooling application. The chilled water generated by the chiller system is circulated through the cooling coil of a fan coil (or air-handling unit), as shown in Fig. 18.2.8.

The fan coil circulates air within the conditioned space. Air from the room moves over the chilled-water cooling coil of the fan coil, is cooled and dehumidified, and returns to the room. In this process the chilled water in the cooling coil picks up the heat and is returned to the chiller system for cooling. As the cycle is repeated, the chiller system maintains the conditioned space at comfort level.

Field-Assembled Liquid Chillers. Originally, liquid chillers were "field-assembled," with the components "field-matched" to develop a field-erected system. These systems were custom-built to perform a specific application. As a result the design depended on the

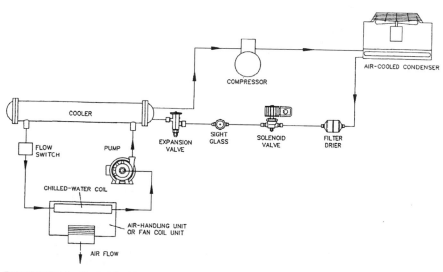

FIGURE 18.2.8 Typical liquid chiller system. (*Courtesy of Continental Products. Inc.*)

application, the availability of parts, the labor, and the field engineer. Some systems were well thought out, were carried out with detail and care, and performed very well for the particular application. Others were ill-conceived, had a poor choice of components, and resulted in a bad experience for the owners.

Field-assembled systems could not be pretested to check if they would perform properly. They depended entirely on the design concepts, the availability of matched parts, and the field experience of the labor force.

As the cost of field labor became prohibitive, and as owners had poor experiences from field-erected systems, the concept of *factory packaging* became popular.

Factory-Packaged Liquid Chillers.   The idea of a completely pre-engineered system is to assemble all the components on a common steel skid and to pipe, wire, pressure-test, evacuate, and charge the system with refrigerant (usually R-22). In this manner, all the system's components are preselected, heat-transfer balanced with each other, prepped, prewired, and factory run-tested before actually being installed on the job. Factory-packaged systems, if manufactured to good engineering standards and correctly capacity-rated, are very cost-effective, resulting in years of good service to the owners.

There have been many improvements from early factory packages to today's systems. Today it is reasonable to expect a reliable, fully factory-tested packaged liquid chiller system that has the following: compressor(s); condenser(s); a liquid chiller; refrigeration specialties, such as expansion valve(s), filter dryer(s), sight glass(es), and solenoid valve(s); and electrical components with power and safety controls. Chiller and condenser pumps (for water-cooled systems) and factory-mounted and -wired flow switches are also available. Most manufacturers test-run the system before its final, preshipment paint finish or cleanup. Field piping of water and additional electrical components are all that are needed before the system is ready for startup.

18.2.5.1 Packaged Liquid Chiller Systems

Packaged chillers are available with the following choices:

1. Scroll compressor chillers from 3 to 240 hp (2.2 to 180 kW) with multiple scrolls
2. Screw compressor chillers from 30 through 440 hp (22 to 328 kW)
3. Hermetic/semihermetic reciprocal chillers from 3 through 440 hp (2.2 to 328 kW)
4. Open-drive reciprocal or screw compressors for industrial applications

Major Components.   A typical liquid chiller system essentially consists of four components:

- Compressor(s)
- Condenser(s)—air cooled, water cooled, or evaporative cooled
- Expansion valve(s)
- Evaporator cooler(s)—direct expansion or flooded type

Other essential parts of the total system are the refrigerant charging valve, filter dryer, liquid solenoid valve, sight glass and moisture indicator, expansion valve, and electrical control center.

Electric Control Center.   The control center is an essential part of the total system, it includes power, operating, and safety controls, usually mounted in a common control panel.

The power controls are separated from the operating and safety controls by a divider plate or other means. The power controls include a starting contractor (in the case of hermetic and semihermetic compressors, which have internally protected compressor motors) or NEMA-rated starters with overload protection (in the case of open-drive compressors, which use NEMA-rated electric motors).

The operating control includes a chiller thermostat, which senses the incoming water temperature to the cooler.

The safety controls consist of the following:

1. *High-condenser-pressure switch.* This opens if the compressor discharge pressure reaches a preset value. It is usually of a manual-reset type; i.e., the operator will have to reset the control manually to restart the system. Manual reset control is used to give the operator a chance to discover the cause of high-condenser pressure.

2. *Low-refrigerant-pressure switch.* This opens when the evaporator cooler's pressure reaches a minimum safe limit preset by the manufacturer. This switch may be of a manual- or automatic-reset type; since it is also used for the pumpdown cycle, it is usually of an automatic-reset type.

3. *Oil-pressure control.* This switch, usually a manual-reset type, is provided to shut down the compressor when the compressor's lubrication oil pressure drops below a minimum safe value, as determined by the compressor manufacturer, or when insufficient oil pressure is developed after compressor startup. This switch is used only on semihermetic and open-drive compressors, not on hermetic types. (Hermetic compressors usually have no way to sense the oil pressure, but rely on internal means of lubrication.)

4. *Freeze-protection switch.* This switch (the so-called "low chilled-liquid temperature control") operates similarly to a thermostat, sensing the temperature of the chilled liquid leaving the cooler. In case of a freeze-up condition and of liquid leaving the cooler, this controller opens the circuit to stop the total system. The minimum value is preset at the factory to prevent cooler freeze-up in case other safety controls malfunction.

5. *Low-pressure freezestat.* This control is usually of a manual-reset type and has a 60 s built-in time delay. It senses the evaporator cooler's pressure, and if this pressure continues to drop below the preset value for a period of 60 s, the switch opens to shut down the total system. This acts as a protection against freeze-up, as well as against a loss of refrigerant.

6. *Flow switch.* This control, which can be either factory-mounted or field-furnished, is needed in the chilled-water piping to protect against a cooler freeze-up in the event of no liquid flowing through the system. This device is an interlock type and provides essential protection against pump failure or other malfunctioning of the total system.

7. *Motor overload protection.* This is provided with hermetic and semihermetic compressors as a built-in feature, whereas for open-drive compressors the over-load heaters are sized to protect the motor.

8. *Power-factor corrector capacitors.* These are for improved motor performance, reduced line losses, and lower utility costs.

9. *Indicator lights.* Various indicator lights show the system's operation and, in case of a system failure, provide diagnostics for ease of service.

10. *Pressure gauges.* High- and low-pressure and oil-pressure gauges are factory-installed and piped to the compressor.

11. *Compressor cycle meter, ammeter, and unit disconnect switches.*

Several other safety controls are available, as follows:

1. *Lock-out timer.* This control prevents the compressor from short-cycling on power interruptions to safety controls.

2. *Phase failure.* This control relay monitors the sequential loss and reverse of a three-phase power supply.

3. *Alarm-bell contacts.* These allow alarm-bell connections to the high-pressure control or to other safety controls for signaling if the unit fails on manual-reset safety controls.

4. *Low ambient controls.* These controls are used specially for air-cooled condenser chillers or evaporative-cooled chillers. The low ambient controls may consist of fan-cycling pressure controls or fan speed controls. The fan speed control usually has a solid-state controller and a single-phase condenser fan motor (which modulates the fan speed in response to condenser pressure). The low ambient controls allow operation of the chiller system on days of low ambient temperature.

5. *Relief valve.* The pressure-relief valve is set at a pressure above the high-pressure cutout, to relieve the system before the system reaches its maximum design working pressure. These valves should be piped and vented outside. Fusible plugs or rupture disks can be used in some instances.

6. *High-motor-temperature protection.* This control consists of a high-temperature thermostat or thermistor. It is located in the discharge gas steam of the compressor.

7. *High oil temperature.* This controller protects the compressor when there is a loss of oil cooling or when a bearing failure results in excessive heat generation.

Other Components.   Factory-mounted, -piped, and -wired pumps for chilled water and condenser water are becoming available as a part of the packaged chiller system. This has eliminated the need for field labor for plumbing, wiring, and inter-locking the pumps with the chiller control panels. Factory-mounted pumps are checked for pump rotation, which is phased in with an air-cooled condenser motor to ensure that the condenser fans operate vertically *up* and that the pumps operate in the correct direction.

Other accessories, such as filters, air eliminators, and storage tanks, can also be factory-mounted and -piped, eliminating the need for separate mechanical areas and the chance of incompatible field components.

18.2.5.2   Typical Chiller Refrigeration Cycle

In a typical chiller system (Fig. 18.2.8), as the water or other liquid flows through the system, the flow switch contact is made, and if the thermostat calls for cooling and all safety devices are closed, the compressor will start. The hot gas from the compressor is discharged into the condenser (air-cooled, water-cooled, or evaporative-cooled). As it travels through the condenser, this high-pressure refrigerant cools and changes its phase to high-pressure liquid. In the case of an air-cooled condenser, the condenser rejects the heat to the air; in the case of a water-cooled or evaporative-cooled condenser, the condenser rejects the heat to the water. The high-pressure liquid refrigerant now goes through a filter dryer. Then it goes through the liquid solenoid valve (which should be open now), sight glass, and moisture indicator, and into the expansion valve. The expansion valve meters the liquid refrigerant through the evaporator cooler. The cooler allows the water (or other liquid) to be cooled by the action of the evaporating liquid refrigerant. The refrigerant picks up the heat from the flowing liquid, is returned back to the suction side of the compressor as a low-pressure gas, and is then ready to be recycled again through the compressor.

18.2.5.3 Chiller-System Freeze Protection

If there is any danger of freezing in a closed-loop chilled-water system, it is recommended that the system be charged with a premixed industrial-grade heat-transfer fluid. *Automotive antifreeze or other commercial glycols are not recommended.* These may include silicates that can coat the cooler tubes, fouling the system prematurely and shortening the life of the pump seals.

For more details about heat-transfer fluids, consult your local industrial chemical supplier.

Fouling Factor.   Fouling results from scaling, corrosion, sediment, and biological growth (slime, algae, etc.); most water supplies contain dissolved or suspended materials that cause these problems. Such fouling causes thermal heat transfer to the water side of chiller systems; the measure for resistance to this heat transfer is called the *fouling factor*.

A general practice is to allow a fouling factor of 0.00025 (h · ft^2 · °F)/Btu [(m^2 · °C)/W] for coolers and water-cooled condensers. For seawater, or where the cooling water is untreated, a fouling factor of 0.001 (h · ft^2 · °F)/Btu [(m^2 · °C)/W] is recommended. In this case, the use of 304 or 316 stainless steel, 90/10 cupronickel, 70/30 cupronickel, or admiralty brass tubes may be considered; the condenser heads can be made of brass or can be treated with epoxy or other protective coating.

18.2.5.4 Types of Refrigerant

R-22 and R-134a are the most popular refrigerants for reciprocating liquid chillers. See Sec. 18.1.

18.2.5.5 Chiller Ratings

Capacity Rating Standards.   Most manufacturers rate their packaged chillers according to Air Conditioning and Refrigeration Institute (ARI) Standard 590 (see Table 18.1.1). ARI Standard 590 is based on the following:

- Air-cooled package at an ambient-air temperature of 95°F (34°C)
- Water-cooled package at a condenser-entering water temperature of 85°F (30°C) and at a condenser-leaving water temperature of 95°F (35°C)
- Evaporative-cooled package at a dry-bulb temperature of 95°F (35°C) or at a wet-bulb temperature of 75°F (24°C)
- Cooler water for all types at an entering temperature of 54°F (12°C) and at a leaving temperature of 44°F (7°C)
- Fouling factor for both the cooler and the condenser = 0.00025 (h · ft^2 · °F)/Btu [(m^2 · °C)/W]

ASHRAE Standard 30-77 is used for testing reciprocating liquid chillers for rating verification.

Energy Efficiency Ratio (EER)

$$EER = \frac{Btu/h \ output}{watts \ input}$$

TABLE 18.2.1   Typical Air-Cooled Chiller Rating

Models MBA use a single compressor, models DBA are dual-circuit with a dual compressor, and models FBA have four compressors with four independent circuits.

Model	Tons	kW	EER	Model	Tons	kW	EER
MBA 3	2.6	3.6	8.8	DBA 52	42.4	43.8	11.6
4	3.9	5.5	9.3	62	52.4	55.0	11.4
5	4.6	6.0	9.3	70	62.3	66.2	11.3
7	6.6	8.0	9.8	75	67.7	73.0	11.1
9	8.6	9.2	11.2	80	73.2	79.8	11.0
10	10.1	10.8	11.3	90	79.7	85.4	11.2
15	13.6	14.9	11.0	100	86.3	91.0	11.4
20	16.5	16.6	11.9	110	92.8	102.5	10.9
25	21.2	21.9	11.6	120	101.0	115.8	10.5
30	24.2	26.1	11.1	FBA 130	114.1	121.9	11.2
35	31.2	33.1	11.3	140	128.3	136.4	11.3
40	36.6	39.9	11.0	160	146.3	159.6	11.0
50	43.1	45.5	11.4	180	159.4	170.8	11.2
60	49.7	57.0	10.5	200	177.8	187.5	11.4
				240	198.8	228.0	10.5

Typically:

- The EER for air-cooled packages ranges from 8 to 12
- The EER for water-cooled packages ranges from 9 to 13
- The EER for evaporative-cooled packages ranges from 10 to 16

18.2.5.6   Chiller Selection Guidelines

To select a packaged chiller from the manufacturer's rating table, it is necessary to know at least four of the following five items:

1. Capacity in tons of Btu/h (kW)
2. Fluid flow rate in gal/min (L/min)
3. Entering fluid temperature in °F (°C)
4. Leaving fluid temperature in °F (°C)
5. Type of fluid (water or other)

Use the following formula to calculate the fifth variable if only four are known:

$$\text{Tons} = \frac{\text{gal/min} \times \Delta \times \text{cp} \times \text{SG}}{24}$$

where Δ = (entering fluid temperature, °F) − (leaving fluid temperature, °F)
    cp = specific heat of fluid, Btu/lb
    SG = specific gravity of fluid

18.2.5.7 Types of Chillers

Heat-Recovery Chillers.   Any HVAC or process application that has a simultaneous use for chilled and hot water is a potential heat-recovery installation. Typical applications are: buildings that need cooling on one side and heating on another; computer room cooling and reheating; restaurants; hotels; and hospitals.

Heat-recovery chillers extract heat from superheated gas vapor before it condenses in the condenser. Thus heat recovery offers "free heat" and eliminates, in certain instances, the need for separate heating equipment.

It is important that a heat-recovery heat exchanger not be oversized; otherwise, the advantage of high-temperature heat recovery is lost. A pressure-enthalpy diagram, as shown in Fig. 18.2.9, demonstrates the potential heat recovery for a typical chiller system.

Figure 18.2.10 shows a schematic of a typical heat-recovery chiller. For air-cooled chillers, a heat-recovery heat exchanger can be piped in series, as shown. For water-cooled systems, heat recovery can be in series or in parallel.

Factory-packaged heat-recovery chillers are available in sizes from 3 to 200 hp (2.24 to 150 kW).

Heat-Pump Chillers.   Heat-pump chillers are becoming more and more popular because of the following advantages:

1. They eliminate the use of a separate boiler or heating system.

2. They eliminate redundant piping for heating and cooling.

3. They use a two-pipe system but provide the comfort of a three-pipe system.

4. They use the same air handlers (or fan coils) for cooling and heating.

5. They use the same chiller-water pump for cooling and heating.

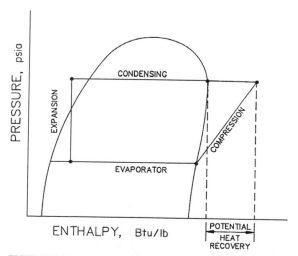

FIGURE 18.2.9   Pressure/enthalpy diagram. (*Courtesy of Continental Products. Inc.*)

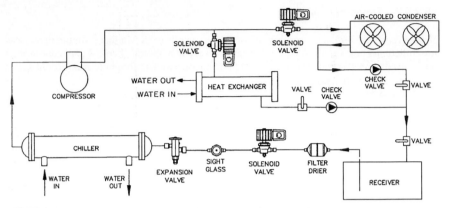

FIGURE 18.2.10   Typical heat recovery chiller. (*Courtesy of Continental Products, Inc.*)

Heat-pump chillers utilize the same heat exchanger for cooling water as they do for heating water. The principle of operation is that, during heating, a reversing valve directs the flow of the hot-gas refrigerant from the compressor to the water heat exchanger, instead of to the condenser. The heat exchanger is now being used as a condenser, and the condenser is being used as an evaporator. The gas is returned back to the compressor, through the reversing valve, with a common suction connection.

During heating, the same pump that circulates the water during cooling is used. This eliminates the need for separate hot-water and chilled-water pumps and piping. For the summer season, the same valve is reversed back to normal cooling.

Figure 18.2.11 shows a heat-pump chiller schematic. *Air-cooled* heat pumps use outside ambient air as the medium; therefore, they need wider fin spacing, as well as hot-gas defrosting. *Water-cooled* heat pumps can use groundwater, river water, or wastewater as the medium.

Available in sizes of 5 to 200 hp (3.7 to 150 kW), heat-pump chillers are suitable for most locations with a winter-design dry-bulb temperature of 20°F (−7°C). They are also available with auxiliary electric heaters, which are useful for unexpected cold spells or as a backup.

Low-Temperature Glycol, Brine, Alcohol, and Gas Chillers.   For low-temperature cooling with glycol, brine, alcohol, gases, or other fluids, several special features are necessary. Factory-packaged chillers for these applications are available. Field modifications of an HVAC chiller do not always produce the desired results. Some of the considerations for process chillers are:

1. Type of refrigerant (R-22 is recommended.)
2. Correct sizing of expansion valve
3. Temperature controller for low temperature
4. Low-pressure switch
5. Low-temperature cutout
6. Oil separator(s)
7. Suction accumulators

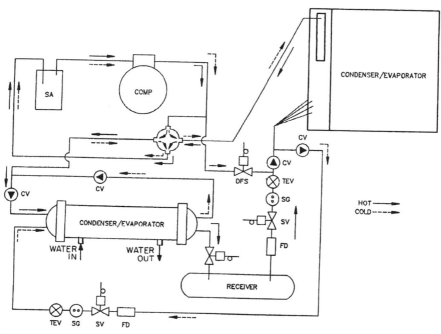

FIGURE 18.2.11 Heat pump chiller-schematic. (*Courtesy of Copeland Corporation.*)

  8. Dual-compressor system with common dual-circuit chiller for 50 percent redundancy

  9. Dual compressor, dual condenser, dual cooler, and dual electrical components for 100 percent redundancy

 10. Primary chiller, with secondary heat exchanger, for corrosives, chemicals, food products (e.g., wines and fruit juices), gas cooling and condensation, incineration, and environmental protection

Packaged Process Chillers. Packaged chillers used for HVAC are designed primarily for human-comfort conditions. A typical system is designed for chilled-water flow to produce a temperature difference of 10°F (5°C), cooling from 54 to 44°F (12 to 7°C) or from 52 to 42°F (11 to 6°C).

For process cooling, it is not always possible

1. to maintain a 10°F (5°C) temperature difference.

2. to work between a 40 and 50°F (4 and 10°C) temperature range.

3. to keep a steady load.

4. to use ambient-related controls the load may be constant year-round.

5. to use single-compressor systems.

6. to have a high return-water temperature.

7. to use standard electrical components, such as in explosion-proof atmospheres.

8. to use standard construction (steel or copper coolers or condenser) or copper or aluminum air-cooled condensers.

Typical applications for process chillers include the following:

Acid cooling	Machine-oil cooling
Bakeries	Marine systems
Breweries	Milk cooling
Candy and fruit glazing	Mushroom cooling
Chemicals and petrochemicals	Pharmaceuticals
Chicken and fish hatcheries	Photo labs
Computer and clean-room cooling	Plastics, injection and blow molding
Dough mixers	Plating and meal finishing
Electronic-cabinet cooling	Printing plants
Environmental test chambers	Pulp and paper
Explosion-proof chillers	Shrimp freezing
Flight simulators	Soil freezing
Foundries	Solvent recovery
Fruit-juice cooling	Steel mills
Ice rinks	Textile plants
Laser cooling	Welding
Lobster tanks	Wineries

18.2.6 CONDENSERS

Condensers are heat exchangers designed to condense the high-pressure, high-temperature refrigerant discharged by the compressor. In this process, the condensers reject the heat that was picked up by the evaporator cooler or chiller. At the same time, condensers convert the high-pressure, high-temperature gas into high-pressure, high-temperature liquid refrigerant, ready to be recycled through the expansion valve, evaporator, and back to the compressor.

There are three types of condensers—*air-cooled, water-cooled, and evaporative-cooled.* They are discussed in Secs. 18.2.6.1, 18.2.6.2, and 18.2.6.3, respectively. Some details are later in this chapter.

Condenser-Coil Circuiting. For multiple compressors, multiple-circuit condensers are used. For example, a two-, three-, or four-compressor unit has a two-, three-, or four-circuit condenser, respectively. In this manner, each circuit is independent of the other, thus providing redundancy in case of failure. Independent refrigerant circuits also ensure that refrigerant and lubrication oil for each section of the compressor system are not mixed. Typical two- and four-circuit condenser headers are shown in Fig. 18.2.12.

Parallel piping of two, three, or four compressors into a common hot-gas inlet connection, with a common liquid outlet connection, is also being utilized.

Compressors are cycled on and off for capacity reduction, but the condenser-coil surface remains the same. For example, if the condenser coil is sized for three compressors and one out of three compressors is shut down, the refrigerant gas from the other two compressors will continue to pump into the same coil. Now this coil will be oversized for the amount of hot gas being pumped into it. Although the condenser fans will also be cycled to

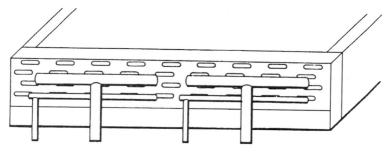

FIGURE 18.2.12   Two-and four-circuit condenser headers (typical).

reduce the air flow, the net effect is still a larger condenser surface, resulting in subcooled liquid. A more efficient system is to feed the subcooled liquid through the expansion valve.
  For parallel piping, the following need and drawback should be considered:

1. It is essential that there be some means to interconnect multiple compressors in order to maintain lubrication oil in each compressor; otherwise, one compressor may be starved of lubrication and thus become damaged.

2. Since multiple compressors have a common refrigerant circuit, even if one (hermetic or semihermetic) compressor fails or "burns out," it contaminates the complete system. To replace one compressor, the complete system has to be "cleaned," evacuated before refrigerant is charged, and put back into service. Condensers with independent refrigerant circuits do not have this problem.

Condenser Components

  Fans.   Condenser fans are of a propeller type and are statically and dynamically balanced for low-vibration operation. Propeller fans are made of aluminum, galvanized steel, stainless steel, or plastic materials and range in diameter from 18 to 30 in (46 to 76 cm). Direct-driven fans are mounted on the fan motor shaft, and belt-driven fans have belt-and-pulley combinations.
  For belt-driven condensers, larger-diameter fans are utilized and have the advantage of fewer fans, compared to several direct-driven smaller fans. Lower speeds (400 to 700 rpm) are achieved by belt drives. The motors are typically standard open drip-proof or totally enclosed NEMA-rated, four-pole, 1800 rpm. These are readily available.
  Belt-driven chiller packages are more suitable for high to medium ambient conditions where an on-off cycling of fans is not needed. Other means of low ambient control are utilized for medium to low ambient control conditions.
  If belt drives are used and the fans cycle often, the belt tension needs to be checked more often than usual. Typically, an access door is provided to access the motor bearings and belts. For additional convenience, extended lubrication lines are installed with external grease fittings.
  Motors.   Typical direct-driven fan motors are six-pole and operate at 1100-rpm speeds. These motors range in size from $1/3$ to 2 hp (0.25 to 1.5 kW) and have been specially designed for air-cooled condenser applications by various motor manufacturers. A typical condenser fan motor is of a 56-frame "totally enclosed air over" design with a built-in overload protector.
  Motor Speed Control for Low Ambient Operation.   For multiple fans on a medium- to large-size chiller package, the fan motor can be cycled on and off by sensing condenser

pressure or ambient temperature. This is adequate for medium ambient temperature operation. For lower ambients or a single-fan chiller package [3 to 9 hp (2.24 to 6.7 kW)], another choice is to modulate the fan motor speed. Typical fan-speed controllers sense condenser pressure or liquid temperature and modulate the motor in response to a rise or fall of pressure or temperature. At higher pressure or temperature, fans operate at higher speed; at lower pressure or temperature, fans modulate at lower speeds. All condenser fan motors are not suitable for fan speed control. Typically, a ball-bearing-type motor is needed to allow operation at lower speeds. Some single-phase motors, specially designed for speed control, have proved successful. Three-phase motors and controllers are being developed.

For a three-phase packaged chiller with multiple fans requiring fan speed control, a combination of three-phase and single-phase motors is used. For 208–230/3/60 power, a single-phase motor presents no problem, since the motor can be wired to two of the three power legs. For 460/3/60 power, single-phase motors require a step-down transformer or a separate single-phase power source.

Fan Venturi.   Its design is critical for optimum air flow with minimum air losses as well as for low outlet noise. See Fig. 18.2.13.

Fan Guards.   Fan guards are mounted around a fan venturi and are designed to meet the standards of the Occupational Safety and Health Administration (OSHA) so as to protect against accidents as well as to allow free air circulation.

18.2.6.1 Air-Cooled Condensers

Air is used as the medium to cool and condense the hot refrigerant vapor. Generally, an air-cooled condenser consists of copper tubes and of copper or aluminum fins, which are expanded on the tubes for maximum heat transfer. The tubes are arranged in parallel or staggered rows and circuited for low refrigerant and air-pressure drop. The complete tube-and-fin condenser coil has a hot-gas inlet and refrigerant-liquid outlet connections. Condenser fan(s) are either directly driven or belt-driven to allow ambient air to circulate over the condenser coil.

Air is drawn from the bottom, goes over the condenser coil and condenser fan motor, and discharges upward. High-pressure, high-temperature refrigerant is being pumped through the hot-gas inlet of the condenser coil and is distributed according to the coil circuit, moving in the opposite direction of the air movement. In this process, heat transfer takes place. The cooler ambient air is circulated over the hot refrigerant. The ambient air picks up the heat, gets warm, and is discharged to the atmosphere. The hot refrigerant gas gets cooled and condenses into liquid.

Medium to Low Ambient Controls.   The capacity of an air-cooled condenser is based on the temperature difference between the summer ambient-air temperature and the condensing temperature. When the packaged chiller is operated at temperature conditions lower than the design ambient-air temperature, the temperature difference between the condensing temperature and the ambient-air temperature is reduced, resulting in increased condenser capacity and lower condensing pressure.

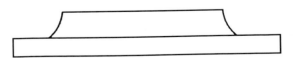

FIGURE 18.2.13 Typical fan venturi. (*Courtesy Continental Products Co.*)

If the ambient-air temperature falls below 60°F (15°C), the condensing pressure falls below a point where the expansion valve can no longer feed the cooler (evaporator) properly. Therefore, for 60°F (15°C) and below, it is necessary to use one or more of the following means to control the condensing pressure:

Fan Cycling. By cycling one fan of a two-fan system, two fans of a three-fan system, and so on, a reasonable condensing pressure can be maintained. The cycling of fans can be in response to ambient-actuated thermostats, sensing the ambient-air temperature entering the condenser coil, or in response to the actual condensing pressure. Fan cycling is reasonable and simple and is recommended for medium-temperature applications.

Fan Speed Control. By fan-speed modulation, single-fan chiller packages or multiple-fan systems can operate at medium to low ambient conditions.

Flooded-Head Pressure Control. Flooded-head pressure control holds back enough refrigerant in the condenser coils to render some of the coil surface inactive. This reduction of the effective condensing surface results in a higher condensing pressure, thus allowing enough liquid-line pressure for normal expansion-valve operation.

Typical head pressure-valve piping is shown in Figs. 18.2.14 (a nonadjustable-valve combination) and 18.2.15 (an adjustable-valve combination). Both valves require the use of a liquid-refrigerant receiver, as shown. The capacity of the receiver is critical in that it must be large enough to hold all the refrigerant during high ambient conditions. If the receiver is too small, liquid refrigerant will be held back in the condenser during high ambients, resulting in high discharge pressures. During low ambients the receiver pressure falls until it approaches the setting of the control valve orifice. This orifice throttles, restricting the flow of liquid from the condenser. Thus the liquid refrigerant is backed up in the condensing coil, reducing the surface area.

Flooded controls can maintain operation down to –40°F (–40°C) ambient or below. Under normal summer conditions the liquid side of the valve remains open, and the hot-gas side is fully closed. Under low ambient conditions, the liquid side remains closed on startup, causing the condenser to "flood." This flooding continues until the condensing pressure reaches the valve setting [typically 180 psig (1241 kPa) for R-22, or 100 psig (689 kPa) for R-12]. Meanwhile the gas-side valve is open, allowing a portion of the hot gas to flow directly to the receiver, maintaining high pressure of the liquid for proper expansion-valve operation. Once the preset pressure is achieved, the valve modulates to maintain high head pressure, regardless of the ambients.

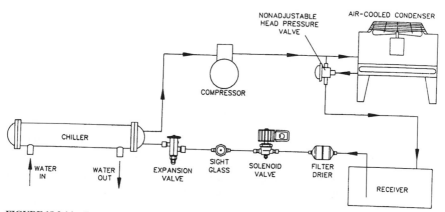

FIGURE 18.2.14 Typical head pressure-valve piping—nonadjustable. (*Courtesy of Continental Products Co.*)

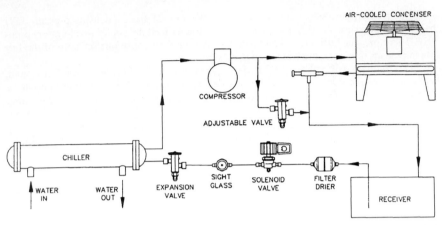

FIGURE 18.2.15   Typical head pressure-valve piping—adjustable. (*Courtesy of Continental Products Co.*)

Inlet-Fan Damper Control.   This control modulates the air flow through the coil by the movement of dampers, in response to the condensing pressure. Usually a combination of fan cycling and damper control is utilized. Inlet dampers are mounted on the inlet of the fan and are actuated by a damper motor. Outlet dampers have also been used, but not very successfully. Experience has shown that damper control is generally not as effective as other means to achieve same results, so its usage is limited.

18.2.6.2   Water-Cooled Condensers

Water-cooled condensers are of the following three types, all of which use water as the cooling medium:

1. *Shell-and-tube condensers.* These condensers, for chillers, are generally built with a steel shell and finned copper tubes. The cooling water circulates through the tubes, and the hot-gas refrigerant is on the outside of the tubes on the shell side. The condensation of refrigerant vapor takes place as the high-temperature hot gas comes in contact with the cool tube surfaces. The condenser water thus picks up the heat rejected by the compressor. Water circuiting is baffled so as to have two, three, four, or six passes.

2. *Shell-and-coil condensers.* In this arrangement, a copper or cupronickel coil is contained within a shell. This type of condenser is limited to smaller sizes and is not generally used for chiller packages.

3. *Tube-and-tube condensers.* This type (which cannot be cleaned easily) consists of two tubes, one contained within the other. The annular space is used for water flow, and the inner tube is used for refrigerant condensing. Because the refrigerant undergoes a considerable pressure drop in the single tube, the use of tube-and-tube condensers is limited to smaller chiller packages up to 10 tons.

Condenser tubes can be cleaned mechanically or chemically. In any case, it is important to have cooling-water treatment for an efficient overall chiller system. Cooling water for condensers can be obtained from a cooling tower or from city water, but because it is uneconomical to use city water, a cooling tower is commonly used. A cooling tower's

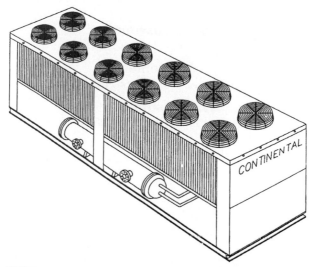

FIGURE 18.2.16 Typical evaporative-cooled condenser. (*Courtesy of Continental Products, Inc.*)

water temperature can vary according to the wet-bulb temperature of the ambient air. To keep the chiller system operating at a low water temperature, a water-regulating valve is installed on the condenser's water inlet. This valve modulates the flow of water in response to the condensing temperature or pressure.

18.2.6.3 Evaporative-Cooled Condensers

Evaporative-cooled condensers employ a copper, stainless-steel, or steel tube condensing coil that is kept continuously wet on the outside by a water-recirculating system. Simultaneously, centrifugal or propeller fan(s) move atmospheric air over the coil. A portion of the recirculated water evaporates, reusing heat from the condensing coil and thus cooling the refrigerant to its liquid state.

A typical evaporative-cooled condenser is shown in Fig. 18.2.16. The complete evaporative-cooled condenser consists of the following:

- Condensing coil (usually prime surface without fins)
- Centrifugal or propeller fan(s)
- Water distribution system
- Drift eliminator
- Water makeup and drains

Since it combines principles of both air-cooled and water-cooled systems, an evaporative-cooled condenser can be considered a combination of these. The driving force is the ambient wet-bulb temperature, which is usually 15 to 25°F (8 to 14°C) lower than the ambient dry-bulb temperature. The overall effect is that an evaporative-cooled condenser operates at a much lower condensing temperature than an air- or water-cooled system. This results in the lowest compressor energy input and hence in the most efficient packaged chiller system.

18.3 ABSORPTION CHILLERS

Nick S. Cassimatis
Gas Energy, Inc., Brooklyn, New York

Jay Kohler
York International Corp., York, Pennsylvania

Willis Schroader
Robur Corp., Evansville, Indiana

PART A: GENERAL

18.3.1 INTRODUCTION

The purpose of this section is to provide the engineer or designer with practical information that can be used in the application of absorption chillers to air-conditioning systems.

Absorption chillers are machines that utilize heat energy directly to chill the circulating medium, usually water. The absorption cycle utilizes an absorbent (usually a salt solution) and a refrigerant (water).

Absorption equipment using ammonia (as the refrigerant) and water (as the absorbent) has been used for cooling applications, including refrigeration applications. This type of equipment is not typically used for commercial chilling applications and is not covered in this chapter (see Part 9, chapter 3 of Ref. 3 for details).

Absorption chillers are usually classified according to the type of heat energy used as the input and whether it has a single-stage or two-stage generator. Absorption chillers using steam, hot water, or a hot gas as the heat energy source are referred to as "indirect-fired" machines, while those which have their own flame source are called "direct-fired" machines. Machines having one generator are called "single-stage absorption chillers," and those with two generators are called "two-stage absorption chillers."

Although small-tonnage absorption machines are produced for residential and small commercial applications in the 5- to 10-ton (17- to 144-kW) sizes, this chapter deals with large-tonnage equipment—machines ranging in capacity from 100 to 1500 tons (352 to 5280 kW).

The reader should keep in mind that the operating cycle of the single-stage absorption chiller is similar to that of the two-stage chiller except that there is not second stage generator. The following discussion applies equally to both single-stage and two-stage chillers.

18.3.2 DESCRIPTION OF CYCLE

The absorption cycle is similar to the more familiar vapor compression cycle.

18.3.2.1 Mechanical Refrigeration Cycle

In the mechanical refrigeration cycle (Fig. 18.3.1), refrigerant vapor is drawn in by the compressor (1), compressed to high temperature and high pressure, and discharged into the

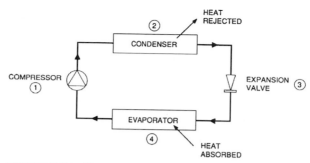

FIGURE 18.3.1 Mechanical refrigeration cycle. See text for explanation of circled numbers. (*Courtesy of Gas Energy, Inc.*)

condenser (2). In the condenser, the vapor is cooled and condensed to a high-pressure, high-temperature liquid by the relatively cooler water flowing through the condenser tubes.

The heat removed from the refrigerant is absorbed by the condenser water and is rejected to the atmosphere by the cooling tower.

The hot refrigerant liquid is metered through an expansion valve (3) into the low-pressure evaporator (4). The lower pressure causes some of the refrigerant to evaporate (flash), chilling the remaining liquid to a still lower temperature.

Heat is transferred from the warm system water (which is flowing through the evaporator tubes) to the cooler refrigerant. This exchange of heat causes the refrigerant to evaporate and the system water to cool.

18.3.2.2 Absorption Cycle

A parallel can be drawn between the mechanical refrigeration cycle and the absorption cycle. In the absorption cycle (Fig. 18.3.2), energy in the form of heat is added to the first-stage generator, and the solution pump (1) pumps the lithium bromide solution from the low-pressure absorber to the relatively higher-pressure first- and second-stage generators. In both these sections, heat is used to produce refrigerant vapor.

The refrigerant vapor is cooled and condensed into liquid by the cooling water flowing through the condenser tubes (2).

Liquid refrigerant from the condenser is metered through a metering orifice (3) (similar to the expansion valve of the mechanical system) and is pumped by the refrigerant pump to the evaporator (4), where it is sprayed over the evaporator tubes. The extremely low vacuum in the evaporator causes some of the refrigerant to evaporate. During this process, heat is transferred from the relatively warm system water (which is flowing through the evaporator tubes) to the cooler refrigerant.

In the absorber, the spray solution absorbs the refrigerant vapor and the resultant heat of absorption is removed by the condenser water (from the cooling water tower) flowing through the tubes. In this system, note that heat is removed in both the condenser and the absorber sections by the cooling water and that it is finally rejected to the atmosphere by the cooling tower in the same manner as in the mechanical system.

Figure 18.3.3 shows schematically the arrangement of the different components of a typical two-stage absorption chiller. To make the absorption cycle more understandable, Fig. 18.3.4 breaks it down into six steps.

Reference 1 contains a description of the operating principles and thermodynamics of the absorption cycle.

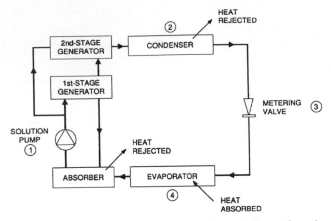

FIGURE 18.3.2   Two-stage absorption refrigeration cycle. See text for explanation of circled numbers. (*Courtesy of Gas Energy, Inc.*)

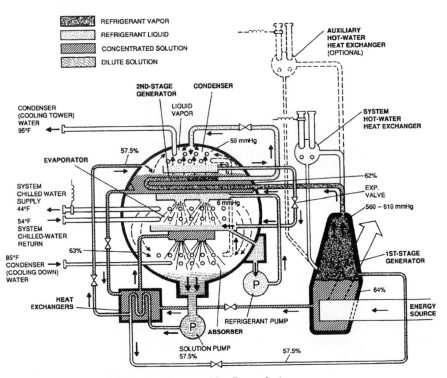

FIGURE 18.3.3   The chilling cycle. (*Courtesy of Gas Energy, Inc.*)

The two-stage absorption chilling cycle is continuous; however, for the sake of clarity and simplicity, it is divided into six steps.

1. Solution Pump/Heat Exchangers

A dilute solution (57.5%) of lithium bromide and water descends from the Absorber to the Solution Pump. This flow of dilute solution is split into two streams and pumped through heat exchangers to the First-Stage Generator and to the Second-Stage Generator.

Parallel flow type chillers with split of solution flow virtually eliminate the possibility of crystallization (solidification) by allowing the unit to operate at much lower solution concentration and temperatures than in series flow systems.

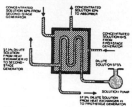

Note: There are two heat exchangers, but only one is shown for illustrative purposes.

2. First-Stage Generator

An energy source heats dilute lithium bromide solution (57.5%) coming from the Solution Pump/Heat Exchangers. This produces hot refrigerant vapor which is sent to the Second-Stage Generator, leaving a concentrated solution (64%) that is returned to the Heat Exchangers.

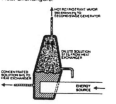

3. Second-Stage Generator

The energy source for the production of refrigerant vapor in the Second-Stage

---

Generator is the hot refrigerant vapor produced by the First-Stage Generator.

This is the heart of the efficient two-stage absorption effect. The refrigerant vapor produced in the First-Generator is increased by 40% — at no additional expense of fuel. The result is much higher efficiency than in conventional systems

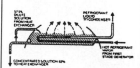

This additional refrigerant vapor (dotted arrow) is produced when dilute solution from the Heat Exchanger is heated by refrigerant vapor from the First-Generator. The additional concentrated solution that results is returned to the Heat Exchanger. The refrigerant vapor from the First-Stage Generator condenses into liquid giving up its heat, and continues to the Condenser.

4. Condenser

Refrigerant from two sources — (1) liquid resulting from the condensing of vapor produced in the First-Stage Generator, and (2) vapor (dotted arrows) produced by the Second-Stage Generator — enters the Condenser. The refrigerant vapor is condensed into liquid and the refrigerant liquid is cooled. The refrigerant liquids are combined and cooled by condenser water. The liquid then flows down to the Evaporator.

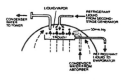

5. Evaporator

Refrigerant liquid from the Condenser passes through an Expansion Valve and flows down to the Refrigerant Pump, where it is pumped up to the top of the Evaporator. Here the liquid is sprayed out as a fine mist over the Evaporator tubes. Due to the extreme vacuum

---

(6 mmHg) in the Evaporator, some of the refrigerant liquid vaporizes, creating the refrigerant effect. (This vacuum is created by hygroscopic action — the strong affinity lithium bromide has for water — in the Absorber directly below.)

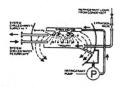

The refrigerant effect cools returning system chilled water in the Evaporator tubes. The refrigerant liquid/vapor picks up the heat of the returning chilled water, cooling if from 54 to 44°F (12 to 7°C). The chilled water is then supplied back to the system.

6. Absorber

As refrigerant liquid/vapor descends to the Absorber from the Evaporator, concentrated solution (63%) coming from the Heat Exchanger is sprayed out into the flow of descending refrigerant. The hygroscopic action between lithium bromide and water — and the related changes in concentration and temperature — result in the creation of an extreme vacuum in the Evaporator directly above. The dissolving of the lithium bromide in water gives off heat, which is removed by condenser water entering from the Cooling Tower at 85°F (30°C) and leaving for the Condenser at 92°F (33°C) (dotted lines). The resultant dilute lithium bromide solution collects in the bottom of the Absorber, where it flows down to the Solution Pump.

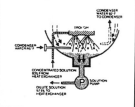

The chilling cycle is now completed and begins again at Step 1.

FIGURE 18.3.4 The six steps in the two-stage absorption chilling cycle. (*Courtesy of Gas Energy, Inc.*) Reference 1 contains a description of the operating principles and thermodynamics of the absorption cycle.

18.3.3 EQUIPMENT

The direct-fired two-stage absorption chiller (Fig. 18.3.5) can double as a heater. In the heating mode the direct-fired type operates in much the same manner as a boiler except that it operates under a vacuum. In this version, simultaneous heating and cooling is possible. In another version, the hot vapor is piped to the evaporator section, where it condenses.

FIGURE 18.3.5   Direct-fired chiller-heater. (*Courtesy of York International Corp.*)

The heated water is circulated through the evaporator tubes. This version cannot produce simultaneous heating and cooling.

The greatest advantage of chiller-heater types of machines is that one piece of equipment can meet both the cooling and heating needs of the facility and can offer substantial space savings, especially in buildings where space is at a premium.

Two-stage steam chillers (Fig. 18.3.6) do not typically have heating capability. Table 18.3.1 gives general performance information for single- and two-stage absorption units.

FIGURE 18.3.6   Steam chiller. (*Courtesy of York International Corp.*)

TABLE 18.3.1   Nominal Rating Conditions for Absorption Chillers

	Single-stage	Two-stage
Chilled water		
Leaving water temperature, °F (°C)	44 (6.7)	44 (6.7)
Flow rate, gal-min/ton (L/kJ)	2.4 (0.43)	2.4 (0.43)
Temperature range, °F (°C)	10 (5.5)	10 (5.5)
Condenser water		
Enter temperature, °F (°C)	85 (29.4)	85 (29.4)
Flow rate, gal-min/ton (L/kJ)	3.6 (0.065)	4.0–4.5 (0.72–0.081)
Temperature range, °F (°C)	17 (9.4)	10–12 (5.5–8.3)
Steam pressure range (dry and		
saturated), psig (kPa G)	5–14 (34–96)	43–130 (296–896)
Fouling factor		
Evaporator, h-ft^2-°F/Btu (m^2 K/W)	0.00010 (0.000018)	0.00010 (0.000018)
Absorber, h-ft^2-°F/Btu (m^2 K/W)	0.00025 (0.000044)	0.00025 (0.000044)
Typical full load energy consumption:		
Steam, lb/h-ton (kg/h-kW)	18.7 (2.6)	9.8 (1.3)
Natural gas, hhv, MBH/ton (kW/kW)	n/a	12.0 (1.0)

Most absorption equipment currently manufactured for use in air conditioning is water-cooled single- or two-stage equipment using the lithium bromide-water cycle. References 2, 3, and 4 provide more information on the different types of absorption equipment available.

18.3.4   APPLICATIONS

Since the energy input of these machines is in the form of heat, they are best-suited to applications where heat is available in a usable form and at a relatively low cost. The available heat source also dictates which type of machine is best-suited to a particular application. Table 18.3.2 shows different machines and their applications.

Absorption chillers are particularly suitable for cogeneration applications. Chapter 5 of this handbook discusses the application of absorption chillers to cogeneration systems.

18.3.5   ENERGY ANALYSIS

18.3.5.1   Energy Cost Calculations

There are many detailed computer programs that can help the engineer choose the best type of chiller for his or her particular application. However, a quick way to estimate the operating costs of different types of chillers is as follows:

$$C = H \times T \times E \times R$$

where C = annual energy costs, dollars
H = equivalent full load hours of operation
T = chiller rated capacity, tons
E = chiller energy usage at full load
R = local utility rate

TABLE 18.3.2 Applications of Absorption Chillers

Application	Direct-fired	Machine type steam or hot-water	Heat recovery
Areas where electricity costs are high	√	√	√
Areas where the primary fuel cost (gas, oil) is relatively low	√	√	
Areas where the cost of utility steam is low or where steam is available as a byproduct		√	
Areas where heat in the form of hot gases is available as a byproduct		√	√

Example Which type of chiller will be best for a building that requires 500 tons of chilling and which operates an average of 850 full load hours per year? The various energy costs and chiller energy usage are:

Electricity (including demand charges) = $0.14/kWh

Natural gas = $4.15/1000 ft^3

Utility steam = $8.75/1000 lb

Electric centrifugal chiller: energy usage = 0.65 kW/ton

Two-stage steam chiller: steam rate = 9.7 lb/(ton h)

Direct-fired chiller: gas usage = 11.73 ft^3/ton h) (LHV)

Solution

1. Electric centrifugal chiller:

$$C = H \times T \times E \times R = 850 \times 500 \times 0.65 \times 0.14 = \$38,675$$

2. Two-stage chiller:

$$C = H \times T \times E \times R = 850 \times 500 \times 9.7 \times 8.75 / 1000 = \$36,072$$

3. Direct-fired chiller:

$$C = H \times T \times E \times R = 850 \times 500 \times 11.73 / 0.90 \times 4.15 / 1000 = \$22,988$$

Where the 0.90 converts the 11.73 ft^3/(ton h) to HHV.

   This simple calculation indicates quickly that a direct-fired chiller will be the best choice if the only consideration is the operating cost. An economic analysis is needed to compare operating costs against initial costs. While the chiller cost is typically the largest initial cost, other costs need to be considered. Factors which may favor absorption chillers are (1) gas utility rebates, which may offset the initial chiller cost, and (2) the ability of direct-fired absorbers to double as a heater, which may eliminate the need for a boiler.

18.3.5.2 Hybrid Chiller Plants

Frequently, a simple comparison of absorption versus electric-driven chillers results in a nonfavorable evaluation for absorption chillers, due largely to the high initial capital cost.

In such cases, an alternative approach may be the use of a hybrid chiller plant. In hybrid chiller plants, both absorption and electric-driven chillers may be used. The capital cost of hybrid chiller plants is typically lower than that of an all-absorption plant. Hybrid chiller plants are attractive when electric rates vary. In many cases, electric rates can vary from low, during periods of low usage, to very high, during peak usage periods. In hybrid chiller plants, an operating schedule can be selected that utilizes the lowest operating cost. If large variations exist in utility pricing, the operating costs of a hybrid facility can be lower than either all-electric or all-absorption facilities.

An analysis of hybrid chiller plants was performed for a variety of utility rates and different hybrid and all-electric operating schedules (Ref. 5). The analysis was performed using a net present value (NPV) calculation, which considered capital and operating costs along with the time cost of money. The analysis concluded that hybrid chiller plants have the highest (best) NPV when considering peak-usage electric costs that are typical of many utilities.

An additional factor that favors hybrid chiller facilities is the uncertainty of future utility rates. In an environment where future rates are uncertain, hybrid facilities offer the flexibility to adjust operating schedules to accommodate rate changes.

18.3.6 UNIT SELECTION

For a given application, the absorption unit selected must provide the required chilling capacity with the smallest possible machine size. Machine size is usually based on specified chilled-water flow rates and temperatures, but it can also be influenced by the flow rate and temperature rise of the condenser water.

To arrive at the best system, both the initial system cost and the annual operating cost must be carefully examined. When selecting an absorption unit, the manufacturer's procedures should be followed.

Since chiller performance improves with increased leaving chilled-water temperature, the unit should be chosen carefully. Consideration should also be given to resetting this temperature during milder weather.

Figures 18.3.7 and 18.3.8 show the effect of water temperatures on energy input as the chilled-water and condenser-water temperatures vary.

The cooling tower used with absorption chillers is usually larger than that used with mechanical systems. The cooling tower cost, however, can be reduced if the chiller can operate with a greater condenser water temperature rise. Also, since chiller performance improves with lower entering condenser water temperature, the small additional cost of a slightly larger cooling tower can be offset by the energy-cost savings. If a source of good quality water is available, such as a river, lake, or well, it can be used in lieu of a cooling tower.

Chillers generally spend most of their operating time at part load conditions. Figure 18.3.9 shows the relationship between load and energy input for conditions of constant tower water temperature and for conditions where the tower water is reduced at part load conditions. Chiller efficiency increases as tower water temperature is reduced, up to a limit. Absorption chillers are restricted in their ability to operate with low tower water temperatures. Most installations require a cooling tower bypass valve, as shown in Fig. 18.3.11, to control the tower water temperature. Manufacturers should be consulted for their specific requirements.

Steam absorption chillers have the additional advantage of being able to operate over a wide range of conditions (see Fig. 18.3.10).

Hot-water operated single effect chillers are effective for heat recovery applications. Hot water can be provided as a byproduct of an application or it can be used as an intermediate

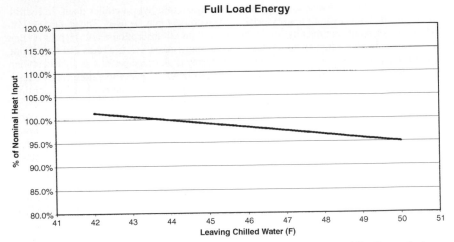

FIGURE 18.3.7   Energy output vs. leaving chilled-water temperature. (*Courtesy of Gas Energy, Inc.*)

fluid for recovering heat from a hot gas. The available cooling capacity of hot water operated chillers depends on the hot water temperature. The capacity is typically a maximum at inlet temperatures of around 250°F (121°C) and is significantly diminished at inlet temperatures of 190°F (88°C) or below.

The reader should consult manufacturer's engineering manuals and catalogs, which usually contain more detailed data on these machines. The final selection should be based on the best combination of system components and lowest operating costs. The designer

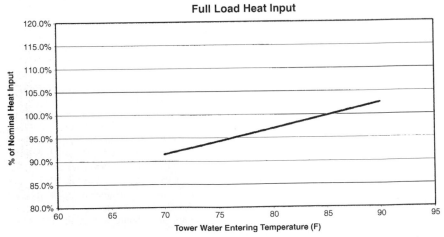

FIGURE 18.3.8   Energy input vs. condenser water inlet temperature. (*Courtesy of Gas Energy, Inc.*)

Absorption Chiller Part Load Heat Input

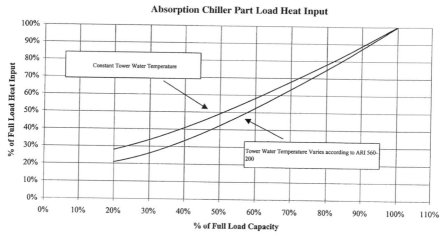

FIGURE 18.3.9   Energy input vs. output. (*Courtesy of Gas Energy, Inc.*)

should therefore consider carefully the chilled-water temperature, condenser water temperature, water quality, fuel supply, etc., before making the final selection.

Figure 18.3.11 shows a typical cooling tower piping schematic with a cooling tower bypass valve. Figure 18.3.12 shows a typical steam piping schematic for a steam absorption chiller. Figure 18.3.13 shows the basic elements of an absorption chiller used to recover the exhaust gas from a heat source such as an engine, a gas turbine, or some process.

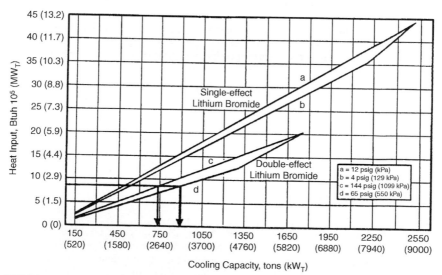

FIGURE 18.3.10   LiBr input vs. output capacities.

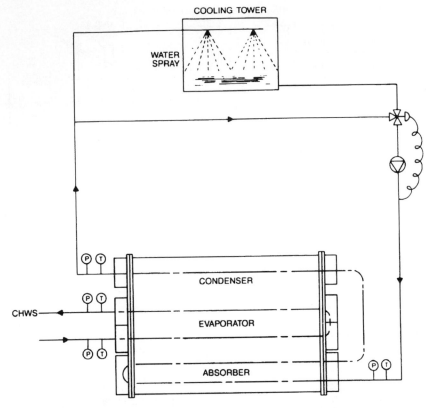

Note:
To ensure uniform and correct flow from the chiller, balancing valves should be installed downstream of the chilled water, and condenser water.

Locate pressure and temperature gauges between the chiller water outlets and the balancing valves.

FIGURE 18.3.11 Typical absorption chiller piping. (*Courtesy of Gas Energy, Inc.*)

Table 18.3.3 shows typical full load cycle conditions for single-stage and two-stage chillers. The two-stage cycle data are based on a parallel flow solution flow.

18.3.7 LOCATION

Because of their rather compact size, light weight, and absence of vibration, absorption chillers can be installed almost anywhere in the building: basement, mid-floor, or roof. Generally speaking, absorption chillers are not recommended for outdoor installation.

When selecting a site, consider structural support, access for service, and tube pull area. The machine-room floor should be able to support the full operating weight of the chiller. The chiller should be located so that all piping presents a neat appearance, with a minimum

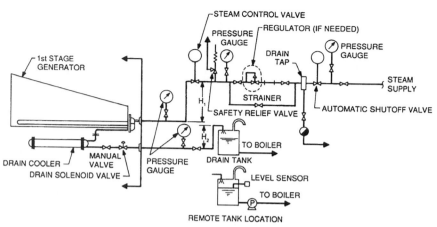

NOTES:

H_1 = 40-in (102 cm) MINIMUM TO PREVENT CONDENSATE BACKFLOW.

H_2 = 32.8-ft (10-m) MAXIMUM TO PREVENT EXCESSIVE BACK PRESSURE.

CONDENSATE LEAVES DRAIN COOLER AT APPROXIMATELY 14.3 psig (98.6 kPa),195°F (90.6°C).

MAXIMUM INLET STEAM PRESSURE 128 psig (882 kPa).

AUTOMATIC SHUTOFF VALVE TO BE FAIL SAFE TYPE (CLOSES UPON POWER FAILURE).

BOTH THE STEAM SUPPLY AND THE CONDENSATE DRAIN LINES MUST BE PROPERLY PITCHED TO PREVENT HAMMERING.

FIGURE 18.3.12   Typical steam chiller piping. (*Courtesy of Gas Energy, Inc.*)

number of fittings and turns and with full and easy access to the chiller for service and maintenance. The tube pull area is usually on either end of the chiller, and the tube pull space is approximately equal to the length of the chiller.

The location of the chiller will influence the overall installation cost of the system. The chiller should therefore be located so that the piping to the tower, flue stack (direct-fired chillers), building services, and fuel supply, and the electrical wiring, are kept to a minimum.

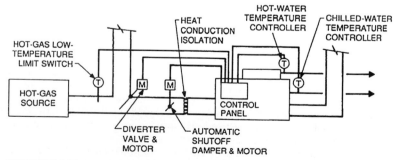

FIGURE 18.3.13   Typical heat-recovery absorption chiller ductwork. (*Courtesy of Gas Energy, Inc.*)

TABLE 18.3.3   Typical Full Load Cycle Conditions

	Single-stage	Two-stage
Water temperatures		
Leaving chilled water, °F (°C)	44.0 (6.7)	44.0 (6.7)
Entering condenser water, °F (°C)	85.0 (29.4)	85.0 (29.4)
Evaporator		
Fluid	Water	Water
Temperature, °F (°C)	41.3 (5.2)	40.7 (4.8)
Pressure, mmHg (kPa)	6.6 (0.88)	6.5 (0.86)
Condenser		
Fluid	Water	Water
Temperature, °F (°C)	111.2 (44.0)	103.0 (39.4)
Pressure, mmHg (kPa)	68.3 (9.1)	53.7 (7.2)
Absorber		
Fluid	Lithium bromide/water	Lithium bromide/water
Leaving temperature, °F (°C)	99.8 (37.7)	98.2 (36.8)
Pressure, mmHg (kPa)	6.1 (0.81)	5.5 (0.74)
Solution concentration, %	58.9	57.9
Low-temperature generator		
Fluid	Lithium bromide/water	Lithium bromide/water
Leaving temperature, °F (°C)	209 (98)	189 (87)
Pressure, mmHg (kPa)	68.4 (9.1)	53.7 (7.2)
Solution concentration, %	63.9	61.6
High-temperature generator		
Fluid	n/a	Lithium bromide/water
Leaving temperature, °F (°C)	n/a	307 (153)
Pressure, mmHg (kPa)	n/a	542 (72)
Solution concentration, %	n/a	63.6

Consider the type of fuel to be used when selecting the installation site. For instance, a gas chiller allows more flexibility, in terms of its placement, than oil or steam chillers, because no storage tank is required.

18.3.8   INSTALLATION

Since the absorption chiller generates very little vibration, vibration eliminators are generally not required. However, vibration-eliminating mounts or pads should be considered when the machine is installed in an area where even mild noise or vibration would be a problem, such as on a floor near a conference room or sleeping area, or on the roof of such an area.

Absorption chillers will operate properly and produce maximum capacity only if they are installed level. It is therefore important that the machine be leveled when it is installed in place and be checked again (and adjusted if necessary) after the piping, refrigerant, solution, and system water have been installed. Follow the manufacturer's instructions and observe the allowable tolerances.

Leave sufficient space, usually 40 to 60 in (1.0 to 1.5 m), around the chiller for service and maintenance work. The machine room must be well lighted and properly ventilated to keep its ambient temperature within manufacturer's specifications. Humidity in the machine room should also be held to a minimum to prevent corrosion and damage to electrical controls and components. Since the chiller contains large amounts of water (refrigerant), it should be protected from freezing; machine-room temperatures should be kept above 35°F (2°C). If the chiller is shipped with a factory charge, it is important to consider if storage in an unheated environment is required. If so, contact the manufacturer for storage information.

Follow standard engineering practices in designing the piping system and other services. Adequate support must be provided for system piping so that no weight is placed on the chiller's water boxes and connecting nozzles.

Direct-fired chillers require air for the combustion process. The amount of air depends on the amount of fuel consumed, the type of fuel, and the amount of excess air used in the combustion process. As a general rule, about 12 ft^3 of air is required for every 1000 Btu of fuel consumed. Combustion products are discharged from the equipment room through a chimney. Equipment manufacturers have specific requirements regarding the chimney breeching pressure which need to be considered as part of the chimney design.

Electrical power is needed for the chiller pumps and controls. Electrical power usage is typically in the range of 0.02 to 0.05 kW/ton. Reference 6 provides useful information regarding installation of absorption chillers. ANSI/ASHRAE Standard 15-2001 (Ref. 7), which has been incorporated into many building codes, provides safety requirements for installation of absorption chillers.

18.3.9   INSULATION

Absorption chillers are normally shipped to the job site uninsulated. After the machine has been installed, piped, operated, and tested, it must be insulated.

Absorption chillers must be insulated on the cold surfaces of the evaporator, on the refrigerant lines and pump, on the chilled-water boxes, and on the refrigerant tank. This prevents them from sweating, which causes corrosion.

The first stage generator, heat exchangers, and piping carrying hot solution must also be insulated to prevent heat loss, minimize machine-room temperature, protect personnel against injury, and protect the chiller against crystallization. The heat loss from an uninsulated two-stage chiller is approximately 5 percent of the total chiller heat input. Follow the manufacturer's recommendations and local codes on the insulation material to use and on methods of application.

18.3.10   OPERATION AND CONTROLS

Absorption chillers meet their load variations and maintain their leaving chilled-water temperature by varying the heat input to the chiller, which in turn varies the reconcentration rate of the absorbent solution. Chiller capacity is directly proportional to solution concentration in the absorber.

The nominal operating range of most absorption chillers can be adjusted from a range of 40 to 42°F (4.4 to 5.6°C) to a range of 48 to 50°F (8.9 to 10°C). However, machine efficiency will decrease as the leaving chilled-water temperature is reset below the design temperature. The manufacturer's recommendation should be followed.

The chiller will maintain the selected leaving chilled-water temperature even though the building's load changes during the course of operation. This is usually accomplished through a temperature controller, whose sensor is located in the leaving chilled-water outlet of the chiller. The controller, in turn, regulates the heat input to the chiller by modulating the energy input of the heat source. Energy-management-type control systems can accomplish chiller control by monitoring many other system temperatures, in addition to that of the leaving chilled water and by then resetting the leaving chilled-water temperature to maximize chiller efficiency.

Most chiller controls and components are factory-mounted, wired, and set, except for water flow switches and steam control valves (when required).

Operating characteristics of absorption chillers differ from other chillers. Because of the volume of solution which must be heated for cooling to occur, absorption chillers take longer to respond to changes in operating conditions. Absorption chillers typically take from 15 min to 1 h to reach full load from a cold start, depending on the chiller size. Also, absorption chillers use a dilution cycle after shutdown, which continues to chill water after the heat source is shut off. The manufacturer's recommendations should be followed for operation of chilled and cooling water during dilution operation.

18.3.10.1 Safety Controls

In addition to the operating controls, absorption chillers are typically equipped with factory-installed and -wired automatic controls. Typical safety devices are:

1. *Low-temperature cutout switch.* This is a thermostat that will stop the chiller if the evaporator temperature falls too low.
2. *Chilled-water flow switch.* This can be either of the pressure- or flow-sensitive type. It will stop the chiller if the chilled-water flow is reduced below the design level or is stopped.
3. *Cooling-water flow switch.* Similar to the chilled water flow switch, this will stop the chiller if the flow of cooling water is interrupted or reduced below the design limits.

The above switches will usually reset themselves when proper water-flow levels have been reestablished.

4. *Concentration limiters.* Found on some chillers, these are designed to prevent the solution from reaching a high absorbent concentration, which can lead to crystallization. This high concentration occurs when the cooling-water temperature becomes too high and/or when the solution flow within the machine is reduced.
5. *First-stage generator high-pressure and high-temperature sensors.* These sensors are designed to interrupt operation if either the temperature or pressure within the machine exceeds the preset limits. These switches may require manual reset by the operator after the cause has been determined and corrected.
6. *Alarms.* Some or all of the above devices may be wired so that, when activated, they will sound an alarm to alert the operator.
7. *Rupture disk.* Some absorption chillers are fitted with a rupture disk or some other type of pressure-relief device on the chiller's generator section to protect against overpressure.
8. *Other controls.* Special machines, such as the direct-fired or heat-recovery types, may employ additional controls to provide safe operation.

Direct-fired machines will normally be provided with all necessary controls, safeties, indicators, and monitoring devices needed for the proper and safe operation of the burner.

Heat-recovery machines will normally be provided with controls to interlock the chiller to the heat source and with an exhaust-gas low-temperature-limit switch, which will prevent the chiller from operating until the gas temperature reaches the design temperature.

18.3.10.2 System Interlock Controls

For proper operation, the absorption chiller must be electrically interlocked with the chilled-water and cooling-water (condenser) pumps and the cooling tower; special machines, such as the heat-recovery type, must also be interlocked to the heat source. Most chiller manufacturers provide a signal source at the chiller control panel to start the pumps when the chiller starts.

In addition to the control interlocks, most manufacturers include terminals in their control panels to permit the monitoring of machine operation from a remote location. Provision can also be made to sound an alarm or activate other warning devices when the chiller operation is abnormal or interrupted.

The installing contractor will usually be required to install flow switches of the chilled-water and condenser water piping to prove water flow to the chiller control circuitry. If water flows are not established within a short time, the chiller will cease operation and an alarm will sound to alert the operator.

The chiller manufacturer will normally provide sufficient information about where the interlocks should be located and how they should be wired. As a rule, external controls are not provided by the chiller manufacturer.

18.3.10.3 Lead-Lag Controls

In some installations, even though a single chiller can meet the building's needs, two or more units may be installed to provided flexibility, economy of operation, and at least partial redundancy. In cases where two machines are installed to provide either half of the building's total load, lead-lag controls can be supplied to allow the operator to choose the "lead" or "lag" machine to meet the load requirements.

Lead-lag control can be achieved whether the chillers are installed in parallel or in series. In either case, a return chilled-water temperature sensor may be used to cycle either the lead or the lag unit when one unit can carry the load. Figures 18.3.14 and 18.3.15 show typical lead-lag connections.

When the chillers are installed in parallel with one chiller water pump and when one chiller is shut down, the chilled water flowing through it will be higher than the set point.

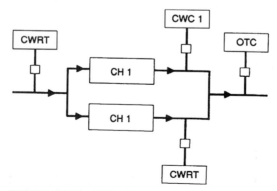

FIGURE 18.3.14 Chillers installed in parallel. CH = chiller; CWRT = chilled-water return temperature sensor; CWC = chilled-water temperature controller; OTC = override temperature controller. (*Courtesy of Gas Energy, Inc.*)

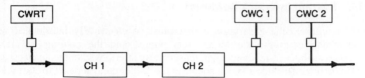

FIGURE 18.3.15   Chillers installed in series. See Fig. 18.3.14 for definitions of the abbreviations. (*Courtesy of Gas Energy, Inc.*)

To prevent this from happening, an override temperature controller can be used to reset the operating unit to a lower setting so that the mixture temperature is at or close to the original setting. A more energy-efficient way to handle parallel chilled water flow is to shut off the water flow through the inoperative chiller.

18.3.11   OPERATION AND MAINTENANCE

Absorption chillers presently manufactured in North America and in Japan are built to rigid standards of internal cleanliness and vacuum integrity. They are designed for maximum efficiency and long life. However, as with any piece of equipment of comparable size and complexity, it can only deliver its design output and efficiency if it is properly operated and maintained. Manufacturers can provide a schedule of recommended maintenance.

18.3.11.1   Corrosion Control and Leak Tightness

The two most important maintenance concerns for the operator of the absorption chiller are the need to maintain corrosion control and leak tightness. The useful life of the chiller is directly proportional to how tightly the chiller is maintained and how good the corrosion control is.

The lithium bromide solution used as the absorbent in absorption chillers is extremely corrosive and will attack steel, copper, and copper alloys in the presence of air and at temperatures above 300°F (150°C).

All absorption chillers use some type of corrosion inhibitor to protect the internal parts of the chiller against corrosive attacks. The lithium bromide solution that is provided with the chiller will typically include the corrosion inhibitor. Different inhibitors, each of which have their advantages and disadvantages, are used by different manufacturers. It is very important that any inhibitor use be maintained within the limits specified by the manufacturer. A solution sample must be obtained at least twice per season and analyzed to determine the presence of adequate amounts of inhibitors. It should be tested for alkalinity and the presence of ammonia, copper, iron, etc. Ideally these components should be absent or present only in very small amounts. The chiller manufacturer normally provides guidelines as to the acceptable ranges within which these components can be present; usually these are given as parts per million (ppm) or milligrams per liter (mg/L).

More significantly, however, than the numerical value of these components is the change that occurs from one sampling to the next. This change is a good indicator of the inhibitors' effectiveness. The laboratory test results will determine how much of which types of inhibitor should be added to maintain the solution within the manufacturer's specifications.

Whenever an absorption chiller is to be opened to the atmosphere for repair and maintenance, it is standard practice to break the vacuum with nitrogen and to bleed a small amount of nitrogen into the unit while the work is performed. Nitrogen is an inert gas, and its use will prevent corrosive attack while the chiller is exposed to the atmosphere.

As explained earlier, absorption chillers operate at a relatively high vacuum; therefore, unless the chiller is properly maintained, air can leak into the machine.

18.3.11.2 Purging

In addition to air leakage, there is the constant production of noncondensable gases that are created during the operation of the chiller. To prevent the accumulation of air and noncondensable gases, absorption chillers are equipped with either automatic or manual purging systems, which must be operated regularly. If purging is not carried out regularly, such accumulation can reduce the chilling capacity, permit deterioration of the chiller's internal parts through corrosion, and even cause crystallization within the chiller. If palladium elements are used, they require replacement every 2 to 5 years.

Crystallization is the precipitation of the lithium bromide salt crystals from the solution. The precipitate forms a slushlike mixture that plugs pipelines and other fluid passages within the chiller, rendering the machine inoperable. The most common causes of crystallization are:

- Insufficient purging of the chiller
- Condenser water temperature outside the specified range
- High solution concentration
- Power failure while the chiller is in operation

Present-day absorption chillers are not likely to experience crystallization if properly operated and maintained. Most manufacturers provide adequate controls to safeguard against crystallization. In addition to chiller controls, manufacturers often provide recommendations to prevent crystallization due to power failure, including insulation of chiller hot components and use of backup power.

On steam-type absorption chillers it is important that the steam's quality and pressure are as per the manufacturer's specifications. A condensate sample should be obtained and analyzed at least once per season.

The typical component content allowance is as follows:

pH	7 to 9 [at 75°F (24°C)]
Ni	Less than 0.17 ppm (mg/L)
Cu	Less than 0.40 ppm (mg/L)
Carbonic acid	Less than 3 ppm (mg/L)

18.3.11.3 Burner and Pumps

The burner of direct-fired absorption chillers should be checked and serviced at least once per season. This service should include a complete analysis of combustion products.

The solution and refrigerant pumps of the absorption chillers should periodically be disassembled, inspected, and rebuilt if necessary. Follow the manufacturer's schedule for the recommended intervals.

18.3.11.4   Water Treatment and Tube Cleaning

One of the most important elements of proper maintenance is the cleanliness of the tubes.

To prevent fouling and scaling, it is very important that the chiller's owner-operator engage the services of a reputable water-treatment specialist for both the initial charging of the system and its continuous monitoring and treatment. Improperly treated or maintained water can result in decreased efficiency, high operating costs, and premature failure of the chiller. See further discussion in Chapter 8 of this book.

It is equally important to clean the tubes of the absorber, condenser, and evaporator at the frequencies recommended by the manufacturer. In addition to periodic cleaning, the tubes must be inspected. Tube failures usually occur as a result of corrosion, erosion, and the stress corrosion and fatigue caused by thermal stress.

Eddy-current inspection of all heat-exchanger tubes is an invaluable preventive-maintenance method. It provides a quick method of determining tube condition at a reasonable cost.

As with all maintenance programs, follow the manufacturer's recommendations on inspection intervals and employ a reputable organization to perform the eddy-current testing.

REFERENCES

1. *2001 ASHRAE Handbook*, Fundamentals, ASHRAE, Atlanta, chap. 1.
2. *2002 ASHRAE Handbook*, Refrigeration, ASHRAE, Atlanta, chap. 41.
3. Handbook of Applied Thermal Design, Part 9, chapter 3 (Heat Driven Refrigeration).
4. Application Guide for Absorption Cooling/Refrigeration using Recovered Heat ASHRAE, Atlanta GA 1995.
5. Smith, B., 2002, Economic Analysis of Hybrid Chiller Plants, *ASHRAE Journal* (July) pp. 42–46.
6. Crowther, H., 2000, Installing Absorption Equipment, *ASHRAE Journal* (July) pp. 41–42.
7. ASHRAE. 2001. Safety Standard for Refrigeration Systems. ANSI/ASHRAE Standard 15-2001.

BIBLIOGRAPHY

Alefeld, G. and R. Radermacher, 1994, *Heat Conversion Systems*, CRC, Boca Raton, FL.

An Introduction to Absorption Cooling, 1999, UK Energy Efficiency Enquires Bureau, Harwell, Oxfordshire, UK.

Herold, K.E., R. Radermacher, and S.A. Klein, 1995, *Absorption Chillers and Heat Pump*, CRC, Boca Raton, FL.

PART B: ABSORPTION CHILLERS

18.3.12   INTRODUCTION

Absorption chillers are machines that utilize heat energy directly to chill the circulating medium, usually water. The absorption cycle utilizes an absorbent and a refrigerant.

18.3.13 *DESCRIPTION OF CYCLE*

The absorption cycle is not much different from the more familiar mechanical refrigeration cycle.

18.3.13.1 Mechanical Refrigeration Cycle

In the mechanical refrigeration cycle (Fig. 18.3.16), refrigerant vapor is drawn in by the compressor (1), compressed to high temperature and high pressure, and discharged into the condenser (2). In the condenser, the vapor is cooled and condensed to a high-pressure, high-temperature liquid by the relatively cooler water flowing through the condenser tubes.

The heat removed from the refrigerant is absorbed by the condenser water and is rejected to the atmosphere by the cooling tower.

The hot refrigerant liquid is metered through an expansion valve (3) into the low-pressure evaporator (4). The lower pressure causes some of the refrigerant to evaporate (flash), chilling the remaining liquid to a still lower temperature.

Heat is transferred from the warm system water (which is flowing through the evaporator tubes) to the cooler refrigerant. This exchange of heat causes the refrigerant to evaporate and the system water to cool.

18.3.13.2 Absorption Cycle

The gas fired absorption chiller operates on an ammonia-water (ammonium hydroxide—NH_4OH) absorption refrigeration cycle (see Fig. 18.3.17). Ammonia is the refrigerant and distilled water is the absorbent. The solution charge in the unit is approximately two parts water to one part refrigerant. *It is a two-pressure system.* Exact pressures are primarily controlled by the temperature of the ambient air drawn across the air cooled condenser-absorber section. Pressure separations are maintained during operation by restrictors in both the refrigerant and solution lines and a check valve on either side of the solution pump. Keep in mind that water has a very high affinity for the *refrigerant vapor. Strong solution* is solution which is strong in its refrigerant content. *Weak solution* is solution which is weak in its refrigerant content.

On cooling demand from an external control, heat from the *gas burners* is applied to the *generator,* causing the solution inside to boil. Since refrigerant vaporizes at a much lower

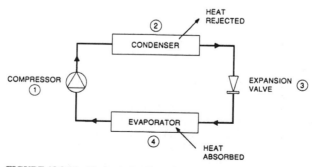

FIGURE 18.3.16 Mechanical refrigeration cycle.

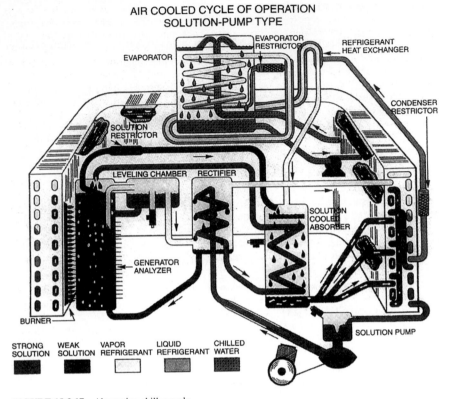

AIR COOLED CYCLE OF OPERATION
SOLUTION-PUMP TYPE

EVAPORATOR
EVAPORATOR RESTRICTOR
REFRIGERANT HEAT EXCHANGER
CONDENSER RESTRICTOR
SOLUTION RESTRICTOR
LEVELING CHAMBER   RECTIFIER
SOLUTION COOLED ABSORBER
GENERATOR ANALYZER
BURNER
SOLUTION PUMP

STRONG SOLUTION   WEAK SOLUTION   VAPOR REFRIGERANT   LIQUID REFRIGERANT   CHILLED WATER

FIGURE 18.3.17   Absorption chiller cycle.

temperature than water, a high percentage of the *refrigerant vapor* and a small amount of water vapor rises to the top of the generator. The remaining solution (*weak solution*) tends to gravitate to the bottom of the generator. As this boiling action takes place, high side pressure is increased. This pressure forces weak solution through a strainer and *restrictor* where it is reduced to *low side pressure*, and the weak solution is then metered into the solution-cooled *absorber*.

Now, lets go back and follow the path of the refrigerant vapor as it rises to the top of the generator. This vapor leaves the *generator* and enters the *leveling chamber*. The leveling chamber is designed to slow down the flow of vapor, causing it to drop some of the water which is carried from the generator. Leaving the leveling chamber, the vapor now passes into the *rectifier*. The rectifier is simply a heat exchanger. Inside the rectifier is a coil through which strong solution is flowing to the generator. This solution is cooler than the vapor passing across it. When the hot vapor makes contact with the cooler coil, it will condense any water vapor which it has carried from the generator and the leveling chamber. This weak solution drains to the bottom of the rectifier and returns through the condensate line to the generator. This is accomplished by thermal-siphoning action. Refrigerant vapor leaves the rectifier at *high pressure* and temperature. This vapor enters the U-shaped *condenser* coil. The condenser fan is moving ambient air across the outside of this coil, removing heat from the refrigerant vapor, causing a change of state from vapor to liquid.

The liquid refrigerant is subcooled as it passes from the condenser through the first of two refrigerant strainer restrictor assemblies into the refrigerant heat exchanger. The liquid refrigerant then passes through a second refrigerant strainer restrictor assembly, where it is reduced to low side pressure as it enters the *evaporator*. Flowing over the outside of the evaporator coil is water containing heat removed from the conditioned space. Heat from this water causes the refrigerant to boil at a low temperature. As the water gives up heat to the refrigerant, the refrigerant vaporizes and the water is chilled as it drains into the bottom of the chiller tank to be recirculated to the conditioned space. The refrigerant vapor leaves the evaporator and enters the inner section of the refrigerant heat exchanger. Here it picks up heat from the liquid refrigerant moving counterflow in the outer section of the *refrigerant heat exchanger*. The vapor then enters the solution cooled absorber where it is reunited with the weak solution. In the solution cooled absorber, the hot weak solution is distributed over a large coil area through which flows cooler strong solution. This temperature difference aids the absorption process, allowing refrigerant vapor to be entrained in the weak solution. This mixture now leaves the solution cooled absorber through multiple feeder tubes into the *air-cooled absorber*. This hot solution now passes through the air-cooled absorber, giving heat to the cooler ambient air being drawn across the coils by the condenser fan. To ensure complete absorption, the absorber coils cross over into the last passes of the outside coil exposing this circuit to the coolest ambient air. By the time the mixture passes through the absorber, the absorption process is completed, and we again have strong solution.

This strong solution enters the inlet tank of the *solution pump*. The solution pump is a volume displacement pump utilizing an inlet check valve, a discharge check valve, and a Teflon diaphragm operated in conjunction with a hydraulic pump designed to deliver a pulsating pressure from 0 to 400 psig (0 to 2757.9 kPa). At 0 psig, from the hydraulic pump, the inlet check valve is opened by unit pressure, allowing solution to fill the upper cavity of the solution pump flexing the diaphragm downward. When the pressure above and below it are equalized, the spring loaded valve will close. The hydraulic pump now delivers a positive pressure to the bottom side of the diaphragm forcing it upward. As it does so, it will force the solution out through the discharge check valve and into the discharge line. As soon as the pressure is equalized above and below it, the spring loaded valve will close.

As the strong solution leaves the solution pump, it enters the inside coil of the rectifier where it removes heat from the hot vapor being passed across the coil. The strong solution enters the inside coil of the solution-cooled absorber, where additional heat is gained from weak solution being distributed over the outside of this coil. The strong solution then returns to the generator, completing the solution and vapor flow cycle at design conditions.

18.3.14 MAINTENANCE

18.3.14.1 Care of Service Tools and Equipment

For maximum safety, it is imperative that all service tools and equipment be kept in a safe and in good operating condition.

Periodic inspections should be made on all service tools and equipment which handle refrigerant 717 (ammonia). All gauges, gauge manifolds, and hose fittings must be steel. Refrigerant 717 is corrosive to brass and copper.

After each use, all hoses, gauges, manifolds, and buckets should be flushed and cleaned with water. They should be stored in a manner to prevent damage when carried on the service truck.

Hoses should be checked at the beginning of each cooling season and periodically during use. They should be checked for the following:

- Possible weak spots
- Cuts and cracked areas
- Condition of the threads and fittings
- Condition of the gasket

Gauge manifold should be checked at the beginning of each cooling season. Check for leaks around the valve stems, gauge connections, and fittings.

Gauges should be checked prior to each cooling season for leaks and calibration.

Solution tank should be tested yearly by a company with the necessary test facilities. It should withstand a test pressure of 500 psig (3447.4 kPa). If there is a DOT number stamped on the tank, it must be tested in accordance with federal regulations.

Refrigerant 717 tank *must* have a DOT number stamped on the tank. It must be suitable for this type of refrigerant and tested in accordance with federal regulations.

Purging bucket should be no less than a 2-gal (7.57-L) bucket, with a handle, preferably strong plastic, and in a color other than yellow. Some method should be provided to hold the end of the purge line under water while purging, other than by hand.

It is recommended that a gas mask, approved for refrigerant 717, be available for use if needed, particularly at service shops.

18.3.14.2 Safety Information

Proper service and repair procedures are vital to the safety to those performing repairs. This manual outlines procedures for servicing and repairing chillers and chiller-heaters using safe, effective methods. These procedures contain many notes, cautions, and warnings which should be followed, along with standard safety procedures to eliminate the possibility of personal injury.

It is important to note that repair procedures and techniques, tools, and parts for servicing chillers and chiller-heaters, as well as the skill and experience of the individual performing the work, vary widely. Standard and accepted safety precautions should be used when handling refrigerant 717 (ammonia) fluids and safety goggles or other protection should be used during cutting, grinding, chiseling, prying, or any other process that can cause material removal or projectiles.

Some procedures require special tools. Before substituting another tool or procedure, you must be completely satisfied that personal safety will not be endangered.

While working on chillers or chiller-heaters, or any other equipment, *safety should be foremost in our minds and actions*. Using good common sense can prevent many unnecessary accidents. Certain safety precautions should be taken when working on the sealed refrigeration unit. To be prepared for any unexpected incidents, the following precautions should be taken:

- Any time the sealed refrigeration unit is entered, safety goggles must be worn to protect the eyes.
- All service tools and equipment should be in good condition.
- Any time the sealed refrigeration unit is entered, consult the chemical safety information supplied in this book.
- Any time the sealed refrigeration unit is entered, a garden hose attached to a nearby faucet or a bucket of clean water should be handy.
- All personnel not involved in the actual work on the equipment should be kept at a safe distance.

Like all refrigerants, refrigerant 717 (ammonia) should be treated with respect. If refrigerant or refrigerant solution gets into eyes, on the skin, is inhaled, or ingested, follow emergency and first-aid procedures in this section.

The excess refrigerant solution and/or water used for purging must be disposed of in accordance with local, state, and federal regulations.

If working on the chilled or chilled-hot water system and antifreeze is spilled, wash the area thoroughly with water, as it can be toxic.

18.3.14.3   Data on Aqua-Ammonia-Chromate Charging Solution

A. *Composition*   CAS No.

Ammonia	1336-21-6
Sodium chromate tetrahydrate	10034-82-9
Sodium hydroxide	1310-73-2
Water	N/A

B. *Physical Characteristics*
Aqua-ammonia-chromate is yellow in color and has a strong ammonia odor. Solution is 100 percent soluble in water.

C. *Fire and Explosion Information*
No special fire extinguisher required. Use any fire extinguisher required for surrounding fire. Use pressure demand self-contained breathing apparatus. Flash point, auto-ignition, and flammable (explosive) limits are not applicable to solution.

Unusual Fire and Explosion Hazards
Vapors in the range of 15 to 28 percent ammonia in the air can explode on contact with a source of ignition. Use of welding or flame-cutting equipment on a container *is not recommended* unless ammonia has been purged and container rinsed with water.

D. *Emergency and First Aid Procedures*
  a. *Eyes:* Immediately flush eyes with plenty of water for at least 15 min holding eyelids apart to assure complete irrigation. Washing eyes within 1 min is essential to achieve maximum effectiveness. Seek medical attention immediately.
  b. *Skin:* Flush area with soap and water for at least 15 min. Remove contaminated clothing and footwear and wash before reuse. Contact a physician for ammonia burns, if any, or any skin irritation.
  c. *Inhalation:* Get to fresh air immediately. If breathing has stopped, give artificial resuscitation. Oxygen may be administered by qualified personnel. If conscious, irrigate nasal and mouth with water. Seek medical attention immediately.
  d. *Ingestion:* Do not induce vomiting. If conscious, give the person large quantities of water or milk to drink immediately. If vomiting occurs spontaneously, keep airway clear. Seek medical attention immediately.

E. *Chronic Overexposure Effects*
Sodium chromate is listed as a carcinogen as per NTP (1981) and IARC (1982) and is on the TSCA inventory.

F. *Reactivity*
Aqua-ammonia-chromate is considered stable.

G. *Acute*

   a. *Sodium hydroxide:* Corrosive to all body tissues with which it comes in contact. The effect of local dermal exposure may consist of multiple areas of superficial destruction of the skin or of primary irritant dermatitis. Similarly, inhalation of dust, spray, or mist may result in varying degrees of irritation or damage to the respiratory tract tissues and an increased susceptibility to respiratory illness. These effects only occur when the TLV is exceeded.

   b. *Ammonia:* Ammonia is a severe irritant of the eyes, respiratory tract, and skin. It may cause burning and tearing of the eyes, runny nose, coughing, chest pain, cessation of respiration, and death. It may cause severe breathing difficulties which may be delayed in onset. Exposure of the eyes to high concentrations may produce temporary blindness (cataracts and glaucoma) and severe eye damage. Exposure of the skin to high concentrations of the gas may cause burning and blistering of the skin. Contact with liquid NH_3 may produce blindness and severe eye and skin burns.

   c. *Sodium chromate tetrahydrate:* Causes severe irritation to eyes and may cause blindness. May cause deep, penetrating ulcers on skin. May cause severe irritation of the respiratory tract and nasal septum, and possible perforation.

H. *Chronic Effects*

   a. *Sodium hydroxide:* No known chronic effects.

   b. *Ammonia:* No known chronic effects.

   c. *Sodium chromate tetrahydrate:* Prolonged or repeated eye contact may cause conjunctivitis. Prolonged or repeated skin contact, especially with broken skin, may cause "chrome sores."

I. *Cancer Hazard (Sodium Chromate Tetrahydrate)*

Hexavalent chromium compounds, as a group of chemicals, have shown sufficient evidence of carcinogenicity to humans and animals.

   a. *Reference:* IARC (International Agency for Research on Cancer) Monographs on the Evaluation of Carcinogenic Risks to Humans, Supplement 7, 1987. NTP (National Toxicology Program) Annual Report on Carcinogens, 1983.

   There is laboratory evidence that aqueous sodium dichromate administered directly into the lung, at the highest tolerated dose, over the lifetime of rats, causes a significantly increased incidence of lung cancer. It is suspected that if sodium chromate were tested in the same manner as aqueous sodium dichromate, it would give a similar response. Other laboratory animal tests indicate that this product is carcinogenic to laboratory test animals.

   b. *Reactivity:* Hazardous polymerization will not occur. Avoid contact with organic materials, oils, greases, and any oxidizing materials. The solution is incompatible with strong acids. Ammonia reacts with acetaldehyde, acreoin, chlorine, fluorine, bromine mercury, gold, silver, silver solder, and hypochlorite to form explosive compounds. Avoid use of nonferrous metals. When ammonia is heated above 850°F (454°C), hydrogen is released. The decomposition temperature may be lowered to 575°F (300°C) by contact with certain metals, such as nickel.

J. *Spill or Leak Procedures*

Contain spill, do not allow it to enter sewers or waterways. Ventilate area and allow ammonia to dissipate, if possible. Soak up with inert absorbent (sand) and place in labeled container(s) for proper disposal.

K. *Waste Disposal*

The solution must not be discharged into sewers or navigable waters or allowed to contaminate underground water sources. Waste solutions should be reclaimed, if possible. If reclamation is not possible, contact local waste disposal contractor for proper disposal.

L. *Safety Equipment and Health Tips*

a. Use chemical goggles, a face shield, and wear protective clothing: solution is very irritating to the eyes. Overexposure to chromate in the solution has the potential to permanently damage the eyes.

b. Wear rubber gloves: aqua-ammonia can damage skin, allowing exposure to chromium.

c. Work in a well-ventilated area, or if in a confined space, use a NIOSH/MSHA approved respirator. Ammonia is a strong irritant and may damage mucous membranes.

d. Do not swallow solution. This material is alkali and may damage tissue.

e. Good hygiene dictates washing hands after handling this material. Wash protective clothing before reuse.

f. If an emergency shower and eyewash are not available, keep a bucket of fresh water or a garden hose readily available.

18.3.15 UNIT ANALYSIS

18.3.15.1 Checking Externals

THE FOLLOWING EXTERNAL CHECKS SHOULD BE MADE IN THE EXACT ORDER THAT THEY APPEAR BELOW, PRIOR TO CHECKING THE INTERNAL OPERATION OF THE UNIT. ANY ATTEMPTS AT SHORTCUTS OR FAILING TO MAKE THESE CHECKS IN THE ORDER THAT THEY APPEAR BELOW COULD RESULT IN AN INCORRECT ANALYSIS OF UNIT OPERATION.

General Appearance.   Check the general appearance of the equipment for such things as proper support, clearance, etc. *A yellow residue or an ammonia odor, if detected, could indicate a sealed system leak.*

Air Flow through the Condenser.   Since the unit capacity is reduced by failure to properly remove heat from the unit by air passing over the condenser-absorber coils, it is recommended that these coils be inspected and cleaned as required.

When inspecting for debris on the condenser-absorber coils, *check the inner coil as well as the outer coil.*

Unit Level.   The unit should be level both front to back and side to side. The level can be checked by placing a level on the top of the unit. If the unit is not level, metal shims are recommended for use under proper corners to level. If more than $1/_2$-in (12.7-mm) thickness of shims is needed under one corner or end, support shims should be inserted at the center of the long frame rail.

The unit must be level to provide even water flow over the surface of the evaporator.

Air Flow—Air Handler or Surface.   Rated air flow across the coil or coils is 1200 CFM (34 m³/min) for 3-ton (11 kW) units, 1600 CFM (45.3 m³/min) for 4-ton (14 kW) units, and

2000 CFM (56.6 m³/min) for 5-ton (17.5 kW) units. For all sizes of air handlers, the air flow rate is 400 CFM (11.3 m³/min) per ton.

For the correct adjustment, set blower to obtain an 18 to 20°F (−7.78 to −6.67°C) drop in air temperature across the coil, with an 8 to 10°F (−13.33 to −12.22°C) drop in water temperature across the chiller.

Water Level and Flow Rate

a. Turn off power and gas to unit.

b. With electrical supply to the unit turned off, disconnect condenser fan motor wires at control panel and tape exposed ends.

c. Remove top panel. Remove four bolts attaching fan support to side panels and set condenser fan assembly on generator/leveling chamber assembly.

d. Remove the styrofoam chiller tank and the distribution pan cover.

NOTE: The chiller tank top is held in place with metal tabs pushed under a plastic band. To remove these tabs, place a small piece of metal, such as a 6-in (152-mm) ruler between the plastic band and metal hold-down tab. This releases the band from a notch on the metal tab. The tab can now be raised straight up and away from the top.

e. Start the chiller or chiller-heater electrically.

f. In order to establish correct chilled water level, with the unit operating electrically, drain excess water from chiller tank. This is done by extending fill test hose to horizontal position at the "0" notch (see Fig. 18.3.18).

g. After correct water level is established, flow rate must be adjusted.

h. Place a ruler on the chilled water return tube and adjust height of water in accordance with Table 18.3.4 below.

i. After setting correct water flow, turn power *off.* Check wiring diagram and replace condenser fan wires on correct terminals.

18.3.15.2   Purging Unit of Noncondensibles

Caution.   Always wear proper safety equipment and know safety data on the refrigerant solution charge.

Note 1.   The unit should be purged only after all externals have been checked and corrected and the unit is still not cooling properly, or anytime the unit has been off, on the generator high temperature limit switch. Purging should be done with the unit in operation (burner on) and with at least a purge valve gauge pressure of 5 psig (34.5 kPa). Proceed as follows:

1. Remove cap from purge valve on top of purge chamber (See Fig. 18.3.19). Check to be sure valve is completely closed. Remove pipe plug from valve.

2. Connect ¹/₄-in (6.35 mm) steel adapted fitting (¹/₄-in [6.35 mm] pipe thread to ¹/₄-in [6.35 mm] flare thread) to purge valve on solution pump.

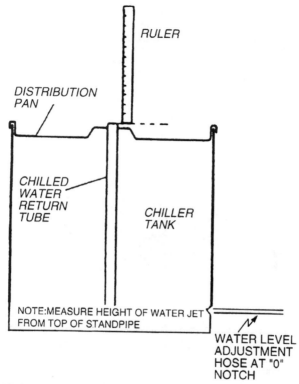

FIGURE 18.3.18 Water level adjustment.

3. Connect purge hose to adapter and immerse loose end at least 6-in (152.4 mm) below surface of water in bucket. If purge hose is not available, a $1/4$-in (6.35 mm) steel flare nut and a length of $1/4$-in (6.35 mm) aluminum tubing can be used.

4. Open valve just enough for bubbles to come out the end of the hose into the water. If noncondensible gases are present, bubbles will rise to surface of water. Ammonia vapor will be absorbed by the water and this is indicated by a cracking sound. Close purge valve immediately if yellow solution appears in the purge bucket.

5. Close purge valve when bubbles do not rise to surface of water.

6. Remove purge hose and adapter. Replace cap and plug.

TABLE 18.3.4 Water Weight Adjustment

Unit size	Flow rate	Water column
3 ton (11 kW)	7.2 GPM (27.4 L/min)	4" (101.6 mm)
4 ton (14 kW)	9.6 GPM (36.5 L/min)	7" (177.8 mm)
5 ton (17.5 kW)	12 GPM (45.6 L/min)	6 $1/4$" (158.8 mm)

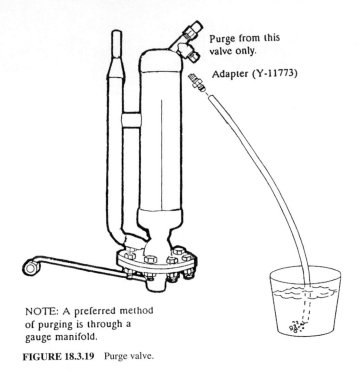

Purge from this valve only.

Adapter (Y-11773)

NOTE: A preferred method of purging is through a gauge manifold.

FIGURE 18.3.19   Purge valve.

Note 2.   Dispose of removed solution in accordance with local, state, and federal regulations.

18.3.15.3 Solution Pump Operation

Turn the unit off at the thermostat, or disconnect the "Y" wire at the low voltage terminal strip. Allow the unit to go through a normal time delay shutdown and leave out of operation for a minimum of 10 min.

During the 10 min "off" period, connect gauges:

- Use gauge manifolds with charging hoses.
- Connect the low pressure gauge to the low side valve located on the side of the SCA.
- Connect the high pressure gauge to the high side valve located on the leveling chamber.
- Connect the low pressure gauge to the purge valve located on the solution pump.

Close valves on gauge manifolds. Open the high side, low side, and purge valves on the unit $1/4$ turn.

This test simply determines the ability of the solution pump to move solution. If the purge gauge on the unit has continued deflection for more than 5 min, replace the solution pump before continuing to further unit analysis as stated below. If the purge gauge on the unit shows either of the other patterns shown below in Fig. 18.3.20, continue the diagnostic procedures.

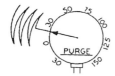

Deflection Stops within 5 minutes
Diagnosis:
Solution pump capable of emptying, continue check out procedure.

Wide Deflection - more than 5 minutes.
Diagnosis:
Solution pump malfunction, replace solution pump.

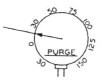

No Deflection within 5 minutes.
Diagnosis:
Possible store out, continue check out procedure.

FIGURE 18.3.20 Purge gauge patterns.

1. On pilot burner models, turn the gas valve to the "pilot" position. On DSI models, disconnect the RED wire from the chilled water limit switch.
2. Start the unit electrically.
3. Observe gauge patterns.

18.3.15.4 Solution Restrictor Flow Check and Gauge Readings

TURN THE GAS VALVE TO "ON" POSITION.

Having checked the operation of the solution pump, turn the gas valve on and reconnect the red (#33) wire to the chilled water limit switch.

Immediately grasp the solution strainer-restrictor line, as illustrated below (see Fig. 18.3.21). If there is weak solution flow through the line, it will get hot as heat travels from the generator toward the solution cooled absorber. This is a preliminary step to be used in

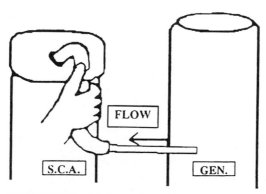

FIGURE 18.3.21 Solution restrictor check.

further analysis. The indication of solution flowing through the line does not eliminate the possibility of a partially plugged solution restrictor.

After determining that solution is flowing through the solution restrictor, observe the pressure patterns on the gauges that are connected to the unit. Once the pressure patterns are established, record both those pressures and the ambient temperature at the time the pressures were taken. The next section (Trouble-Shooting with Gauges) includes a temperature/pressure chart (see Tables 18.3.5 and 18.3.6) and gauge illustrations (Fig. 18.3.22) which, with this information, will enable you to determine the problem.

18.3.15.5 Trouble-Shooting with Gauges

Use Table 18.3.5 to determine what the pressures should be at the temperature you recorded in the Solution Restrictor Flow Check and Gauge Readings section. Also, at the time the ambient temperature was noted, you recorded the pressures shown on the gauges. Comparing those readings with pressures found in the table, determine whether the unit is operating at normal pressures, or above normal pressures.

If the pressures are above or below normal, the next step is to compare them with the gauge readings on the next two pages, eliminating all that do not match as possible problems. Once you narrow it down, observe the solution pump gauge to see how it is operating: normal with 2 to 4 lb (0.9 to 1.8 kg) of deflection, normal to above normal, normal to erratic to normal, or erratic. That should pinpoint the problem. After determining the problem, you will be interested in how high or how low the gauges are reading and how they are acting. Make notes on your worksheet as to how the pressures are acting.

Caution   Should the high-side gauge ever reach 350 lb (157.5 kg), shut the gas off immediately.

TABLE 18.3.5   Pressure/Temperature

Ambient air °F	Low side ±5 psig		Pump pressure normal deflection 2 to 4 lb			High side ±10 psig			Ambient air °F
	3 Ton 5 Ton	4 Ton	3 Ton	4 Ton	5 Ton	3 Ton	4 Ton	5 Ton	
60	20	17				195	175	180	60
65	25	22				210	190	195	65
70	29	26				225	205	210	70
75	33	30	8–12 PSIG	7–10 PSIG	10–15 PSIG	240	220	225	75
80	38	35	below Low side	below Low side	below Low side	225	235	240	80
85	43	40				270	250	255	85
90	48	45				285	265	270	90
95	53	50				300	280	285	95
100	57	54				320	300	305	100
105	61	58				335	315	320	105
110	65	62				350	335	340	110
115	68	65				365	350	355	115

TABLE 18.3.6 Pressure/Temperature (Metric)

Ambient air °C	Low side ±34.5 kPa		Pump pressure normal deflection .9 to 1.8 kg			High side ± 10 psig			Ambient air °C
	3 Ton 5 Ton	4 Ton	3 Ton	4 Ton	5 Ton	3 Ton	4 Ton	5 Ton	
15.6	137.9	117.2				1344.5	1206.6	1241.0	15.6
18.3	172.4	151.7				1447.9	1310.0	1344.5	18.3
21.1	200.0	179.3				1551.3	1413.4	1447.9	21.1
23.9	227.5	206.8	55.2–82.7 kPa below Low side	48.3–69.0 kPa below Low side	69.0–103.4 kPa below Low side	1654.7	1516.8	1551.3	23.9
26.7	262.0	241.3				1758.2	1620.3	1654.7	26.7
29.4	296.5	275.8				1861.6	1723.7	1758.2	29.4
32.2	331.0	310.3				1965.0	1827.1	1861.6	32.2
35.0	365.4	344.7				2068.4	1930.5	1965.0	35.0
37.8	393.0	372.3				2206.3	2068.4	2102.9	37.8
40.6	420.6	399.9				2309.7	2171.8	2206.3	40.6
43.3	448.2	427.5				2413.1	2309.7	2344.2	43.3
46.1	468.8	488.2				2516.5	2413.1	2447.6	46.1

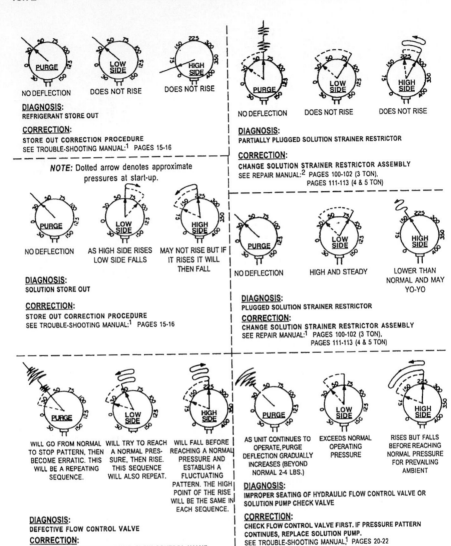

FIGURE 18.3.22a   Trouble shooting with gauges.

In Table 18.3.6 the pressures and temperatures of Table 18.3.5 have been converted to metric measures.

BIBLIOGRAPHY

1. Trouble-Shooting Manual, form AS42502, published by Robur Corp., Evansville, IN.
2. Repair Manual, form AS42451, published by Robur Corp., Evansville, IN.

NORMAL DEFLECTION   LOW AND STEADY   LOW AND STEADY

DIAGNOSIS:
LOW ON REFRIGERANT

CORRECTION:
ADD REFRIGERANT
SEE TROUBLE-SHOOTING MANUAL.[1] PAGE 27

STEADY TO ERRATIC   EXTREMELY LOW   STEADY CLIMB TO EXTREMELY HIGH. MAY UNSEAT RELIEF VALVE

DIAGNOSIS:
PLUGGED REFRIGERANT RESTRICTOR

OTHER SYMPTOMS:
FROST OR CONDENSATION AT POINT OF RESTRICTION

NOTE: IF NO VISUAL INDICATION OF RESTRICTION, ALWAYS REPLACE CONDENSER RESTRICTOR FIRST

CORRECTION:
REPLACE INDICATED RESTRICTOR
SEE REPAIR MANUAL.[2] PAGES 98-100 (3 TON) & PAGES 108-110 (4/5 TON)

MORE THAN 2-4 LBS. DEFLECTION   HIGHER THAN NORMAL. (MAY PULSATE IN UNISON WITH PUMP)   LOWER THAN NORMAL

DIAGNOSIS:
CROSS LEAK IN S.C.A.

OTHER SYMPTOMS:
MAY BE UNUSUAL NOISE IN SOLUTION COOLED ABSORBER

CORRECTION:
CHANGE S.C.A.
SEE TROUBLE-SHOOTING MANUAL.[1] PAGES 24-25 & REPAIR MANUAL

NORMAL DEFLECTION   HIGH AND STEADY   HIGH AND STEADY

DIAGNOSIS:
EXCESS REFRIGERANT

CORRECTION:
REMOVE EXCESS REFRIGERANT
SEE TROUBLE-SHOOTING MANUAL.[1] PAGE 30

ERRATIC DEFLECTION   BELOW NORMAL RANGE   BELOW NORMAL RANGE

DIAGNOSIS:
LOW ON SOLUTION

CORRECTION:
ADD SOLUTION
SEE TROUBLE-SHOOTING MANUAL.[1] PAGES 32-35

NOTE:
EXTREME UNDERCHARGE WILL CAUSE BOTH HIGH AND LOW SIDE PRESSURES TO SWING THROUGH A RANGE FROM BELOW NORMAL TO ABOVE NORMAL.

MORE THAN 2-4 LBS. DEFLECTION   HIGHER THAN NORMAL RANGE   HIGHER THAN NORMAL RANGE

DIAGNOSIS:
EXCESS SOLUTION

CORRECTION:
REMOVE EXCESS SOLUTION
SEE TROUBLE-SHOOTING MANUAL.[1] PAGE 38

NOTE:
EXTREME OVERCHARGE WILL CAUSE BOTH HIGH AND LOW SIDE PRESSURES TO SWING THROUGH A RANGE FROM BELOW NORMAL TO ABOVE NORMAL.

FIGURE 18.3.22b *(Continued)*

18.4 CENTRIFUGAL CHILLERS

John M. Schultz, P.E.
Retired Chief Engineer, Centrifugal Systems,
York International Corporation, York, Pennsylvania

George White, P.E.
Aggreko, Inc., Benicia, California

PART A: GENERAL

18.4.1 INTRODUCTION

A refrigeration system that uses a centrifugal compressor to cool water for air-conditioning purposes is called a "centrifugal chiller." The capacities of these machines range from 100 to 10,000 tons (350 kW to 35 MW) of refrigeration. Their reliability is high, and their maintenance requirements low, because centrifugal compression involves the purely rotary motion of only a few mechanical parts.

Most centrifugal chillers have water-cooled condensers. The source of the condenser water is usually a cooling tower, but lake or river water can also be used. Air-cooled condensers are employed in locations where cooling water is not available.

In sizes above 2000 tons (7000 kW), the major components of a water-cooled chiller (heat exchangers, compressor, etc.) must be shipped individually for mounting and connecting at the jobsite; these chillers are called "field-erected" units. Smaller sizes can be completely assembled, piped, and wired before they leave the manufacturer's plant; these units are called "factory packages."

Figure 18.4.1 is a photograph of a factory package which even includes a reduced-voltage motor starter. Only the water piping and power supply need to be connected to this unit in the field before startup. Figure 18.4.2 is a photograph of a large field-erected unit.

18.4.2 REFRIGERATION CYCLES

Many water-cooled chillers employ a single stage of compression in the basic vapor-compression refrigeration cycle shown in Fig. 18.4.3. Sometimes a liquid subcooler is added between the condenser and the expansion device to improve the coefficient of performance (reduce the power requirement) of the cycle. When a subcooler is used, it is usually built into the bottom of the condenser. Two or three stages of compression are employed in some water-cooled designs and in all air-cooled units. In these cases, some improvement in the coefficient of performance can be obtained by using the intercooled refrigeration cycle of Fig. 18.4.4. This figure shows two stages of compression and one intercooler. When three compression stages are used, two intercoolers are possible.

The degree to which subcooling and intercooling reduce the power requirement of the basic refrigeration cycle depends on which refrigerant is being used and how great the temperature difference (temperature lift) is between the evaporator and the condenser. The choice of refrigerant and cycle for a particular application involves many considerations. Various manufacturers choose different combinations to achieve the same overall results.

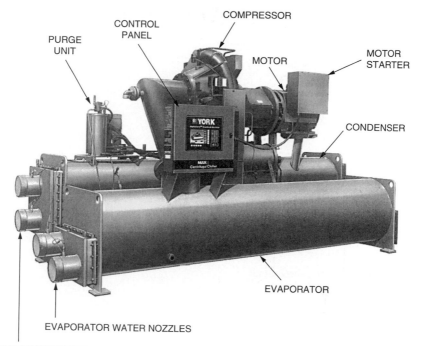

PURGE
UNIT

CONTROL
PANEL

COMPRESSOR

MOTOR

MOTOR
STARTER

CONDENSER

YORK

MAX
Centrifugal Chiller

EVAPORATOR

EVAPORATOR WATER NOZZLES

CONDENSER WATER NOZZLES

FIGURE 18.4.1 Factory-packaged unit, including motor starter.

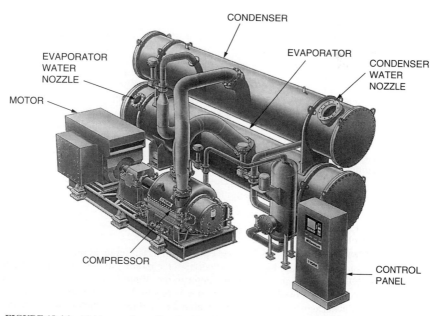

CONDENSER

EVAPORATOR
WATER
NOZZLE

EVAPORATOR

CONDENSER
WATER
NOZZLE

MOTOR

COMPRESSOR

CONTROL
PANEL

FIGURE 18.4.2 Field-erected centrifugal chiller. Capacity: 3000 tons (11 MW).

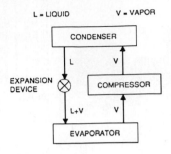

FIGURE 18.4.3 Single-stage vapor-compression refrigeration cycle.

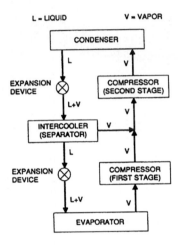

FIGURE 18.4.4 Intercooled vapor-compression refrigeration cycle.

18.4.3 COMPONENTS

In most centrifugal chillers, the compressor is driven by a squirrel-cage induction motor, either directly or through speed-increasing gears. In factor packages, the gears are built into the compressor; in field-erected units, a separate gearbox is usually employed. Because the optimum performance of a centrifugal compressor is strongly influenced by its rotating speed, different gear ratios are used with the same compressor for different operating conditions.

Steam turbines, gas turbines, and internal-combustion engines are also used to drive centrifugal compressors, with or without gears, generally in a field-erected configuration. Figure 18.4.5 is a photograph of a gas-engine-driven unit.

18.4.3.1 Motors

A motor that is built into the compressor and operates in a refrigerant atmosphere is called a "hermetic" motor. Refrigerant liquid and vapor cool the motor. No mechanical shaft seal is needed because the shaft that connects the motor to the compressor never penetrates the refrigerant envelope. Additional compressor power is required to provide the refrigeration effect that cools the hermetic motor and to overcome the windage loss that results from running the motor in a dense refrigerant atmosphere.

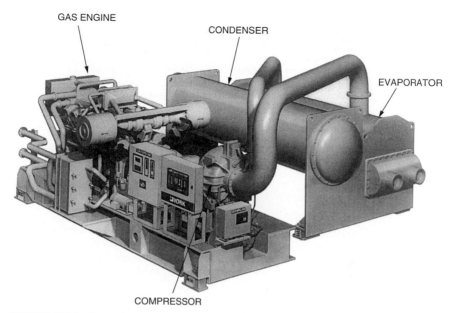

FIGURE 18.4.5   Gas-engine-driven centrifugal chiller. Capacity: 1000 tons (3500 kW).

An "open" motor is one that operates in atmospheric air and is air-cooled. It is often attached directly to the compressor, in which case it is called a "close-coupled" motor. Open motors tend to be more efficient than hermetic motors and are easier to maintain. If the winding of a hermetic motor fails, the entire refrigerant charge may be contaminated by the burning insulation.

18.4.3.2 Evaporators

Horizontal shell-and-tube evaporators are used in centrifugal chillers. These are of the flooded type with liquid refrigerant surrounding the tubes. The chilled water flows inside the tubes. The tube bundle occupies only the lower part of the shell; the upper part is a space in which liquid refrigerant drops out of the vapor that is flowing to the compressor. Mist eliminators or baffles are installed above the tube bundle to help remove the entrained liquid from the vapor.

The outside surface (refrigerant side) of the evaporator tubes is finned to increase the heat transfer area. The fins are usually modified to increase the boiling heat-transfer coefficient. The inside (water side) of the tubes usually has ribs to increase the heat-transfer coefficient of the water film.

18.4.3.3 Condensers

Horizontal shell-and-tube condensers are used in water-cooled units, with the cooling water inside the tubes. The tube bundle fills most of the shell. The outside surface of the tubes is finned, and the fins usually modified, to increase the refrigerant condensing heat-transfer rate. The inside of the tubes usually has ribs to enhance the water-side performance.

Air-cooled condensers contain coils of externally-finned tubes. The refrigerant condenses inside the tubes and drains into a receiver under the coils. Fans are used to force the flow of cooling air over the outside of the tubes. Some liquid subcooling almost always occurs in air-cooled condensers.

Liquid refrigerant flows from the condenser to the evaporator through an expansion device. The device may be a liquid-level control or float valve in the receiver of an air-cooled condenser or in a liquid well at the bottom of a water-cooled condenser. Alternatively, it may be a thermostatic expansion valve that controls the superheat of the vapor leaving the evaporator. Often, the expansion device is merely one or more orifices that have been sized and arranged so as to pass a negligible amount of vapor with the flowing liquid. In some units the liquid flows through a turbine that recovers a portion of the expansion energy to help drive the compressor.

18.4.3.4   Purge Units

Many packaged centrifugal chillers use refrigerant 123. The evaporators of these chillers operate in a vacuum. Some air inevitably leaks into the vacuum, bringing with it a small amount of water vapor. The leakage air and water vapor tend to collect in the condenser, from which they must be purged in order to maintain condenser heat-transfer performance and prevent acid formation in the refrigerant.

To remove these noncondensables, a purge unit is provided. This device draws refrigerant from the top of the condenser along with any air or water vapor that may be present. The purge unit compresses and condenses most of the refrigerant and returns the condensate to the system. The air does not condense and is vented, along with a small amount of uncondensed refrigerant, to the atmosphere. Some water vapor is expelled with the air, and some condenses with the condensed refrigerant. The condensed water is removed from the condensed refrigerant either by a filter-drier cartridge or by a gravity (density) separator. Water from the latter must be drained manually.

Purge units vary in their ability to separate refrigerant from the air that is expelled to the atmosphere. Efficiency in this respect is important because of the cost of replacing the lost refrigerant, and because releasing refrigerant to the atmosphere harms the earth's ozone layer and has the potential to increase global warming. In some units the purged air/refrigerant mixture flows through a canister of activated carbon before it reaches the atmosphere. The carbon absorbs most of the refrigerant. When the carbon becomes saturated with refrigerant, the carbon must be regenerated or the canister must be replaced. The absorbed refrigerant is not returned to the chiller when the canister is replaced.

Purge units operate automatically, and even include an alarm system to indicate when an air leak is causing an excessive amount of purging.

18.4.3.5   Refrigerant Transfer, Storage, Recycling

The U.S. Environmental Protection Agency (EPA) requires that before a centrifugal chiller is opened for repair, the refrigerant be evacuated from the chiller to minimize loss of refrigerant to the atmosphere. A vacuum pump or a positive displacement compressor is needed to achieve the specified vacuum levels.

The pump or compressor discharges to a water-cooled condenser that drains to a storage receiver where the refrigerant is held until the chiller has been repaired. In some circumstances the refrigerant must be filtered, dried, and/or distilled to remove oil, moisture, and other impurities before it is returned to the chiller. A separate, self-contained refrigerant recover/recycle unit is needed to accomplish these functions.

EPA requires that the performance of refrigerant recovery/recycling units be certified. The Air Conditioning and Refrigeration Institute (ARI) provides this certification. The ARI certification program involves the periodic random testing of standard production units relative to an ARI performance standard (Ref. 1). The tests are conducted by an independent testing laboratory that is under contract to ARI (Ref. 2). The model numbers of certified units are published in a directory (Ref. 3) that is updated semiannually to reflect the results of continual random testing.

18.4.3.6   Additional Information

More information about centrifugal compressors, heat exchangers, motors, engines, and turbines can be found in the *ASHRAE Handbook* (Ref. 4).

18.4.4   CAPACITY CONTROL

The volumetric flow capacity of a centrifugal compressor must continuously be adjusted for variations in evaporator loading and for changes in the temperature lift (compressor head) between the evaporator and the condenser. A set of movable guide vanes at the inlet to the first impeller provides this modulating control. The vanes are called "prerotation vanes" (PRVs) because their function is to create a swirling motion (rotation) in the refrigerant's flow stream just ahead of the impeller. Figure 18.4.6 is a photograph of a typical PRV design. A compressor that has more than one impeller may have more than one set of PRVs.

Figure 18.4.7 represents the performance of a centrifugal compressor operating at constant speed and using PRV control. Each vane position generates a line that terminates at a "surge" point above which the refrigerant alternately flows backward and forward through the compressor. This unstable operating condition, called "surging," can be avoided by recirculating some of the compressor discharge vapor back to the compressor suction. The recirculated vapor, called "hot-gas bypass" (HGBP), adds to the evaporator flow so that the total flow is in the region of stable operation to the right of the surge points.

HGBP is inefficient because it imposes additional load on the compressor. The need for HGBP is reduced in some compressors by using diffuser controls that move the surge points to the left. Methods that can be used to reduce the size of the surge area are (1) varying the width of the radial diffusion space that surrounds the impeller and (2) throttling the flow between the impeller and the diffuser.

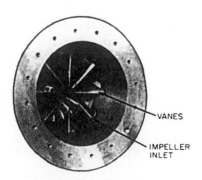

VANES

IMPELLER
INLET

FIGURE 18.4.6   Prerotation vanes (PRVs).

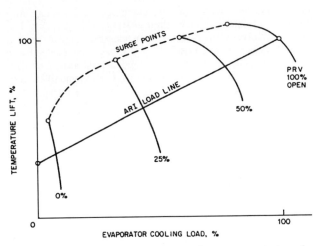

FIGURE 18.4.7   Centrifugal compressor performance at constant speed.

The most efficient way to control the capacity of a centrifugal compressor is to vary its speed. Turbine and engine drives lend themselves to this type of control rather easily. Wound-rotor motors can also be used. The most common method of varying speed, however, is to vary the frequency of the alternating current that is supplied to an induction motor that drives the compressor. Variable-frequency inverters are used to modulate the motor speeds of many centrifugal chillers with capacities up to 1400 tons (4900 kW). The capacity control system of most inverter drives combines both speed control and PRV control to optimize the compressor efficiency at every possible operating condition. Some variable-speed engine drives are also optimized by including PRV control. Figure 18.4.8 shows a variable-frequency inverter connected to a packaged centrifugal chiller.

18.4.5   POWER CONSUMPTION

Of primary importance in the design of an air-conditioning system is the optimization of initial cost and operating cost. The principal operating cost of a centrifugal chiller is the cost of the power that it consumes. At full load, the power requirement of most water-cooled, motor-driven chillers ranges from 0.5 to 0.7 kW/ton (0.14 to 0.20 kW/kW). The variation in electric power consumption is due to a number of factors that can be summarized by the equation

$$kW/ton \sim \frac{(lb\,/\,min\,/\,ton)(temperature\ lift,\,°F)}{\left(\begin{array}{c}compression\\efficiency\end{array}\right)\left(\begin{array}{c}mechanical\\efficiency\end{array}\right)\left(\begin{array}{c}electrical\\efficiency\end{array}\right)}$$

$$kW/kW \sim \frac{[kg\,/\,(s\,\cdot\,kW)](temperature\ lift,\,°C)}{\left(\begin{array}{c}compression\\efficiency\end{array}\right)\left(\begin{array}{c}mechanical\\efficiency\end{array}\right)\left(\begin{array}{c}electrical\\efficiency\end{array}\right)}$$

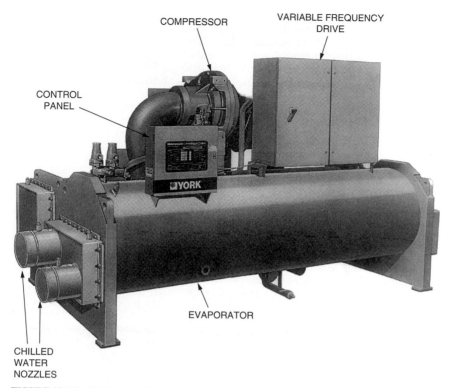

FIGURE 18.4.8 Variable-speed inverter drive for a centrifugal package.

The refrigerant mass flow rate {lb/min/ton [kg/(s · kW)]} depends mainly on which refrigerant and which refrigeration cycle are being used. The evaporating and condensing temperatures are of secondary importance. To the extent that these temperatures matter, reducing the difference between them (temperature lift) reduces the refrigerant flow rate.

Temperature lift is the chief factor in determining the compression work input (head) per unit of flow. It also influences the compressor's aerodynamic efficiency (compression efficiency). Lowering the temperature lift increases the compression efficiency at full load; it also increases the efficiency at part load when speed control is used, but not when PRV control is used. Temperature lift equals the difference between the water temperatures leaving the heat exchangers added to the temperature differences within the heat exchangers between the refrigerant and the leaving water:

$$\text{Temperature lift} = (\text{LCWT} - \text{LEWT}) + (\text{LEWT} - \text{ET}_p) + (\text{CT}_p - \text{LCWT})$$

where LCWT = leaving condenser water temperature, °F (°C)
　　 LEWT = leaving evaporator water temperature, °F (°C)
　　　 ET_p = refrigerant evaporating temperature, °F (°C)
　　　 CT_p = refrigerant condensing temperature, °F (°C)

Much can be done to lower the power consumption of a centrifugal chiller by increasing the chilled-water temperature, decreasing the condenser-water temperature, and using effective heat exchangers that have small leaving-temperature differences.

Compression efficiency depends primarily on the aerodynamic design of the centrifugal impeller(s) and diffuser(s). Additional considerations, when more than one stage of compression is used, are the design of the flow passages between the stages and the mixing of intercooler vapor flow(s) with interstage flow(s).

The mechanical efficiency of a centrifugal chiller is reflected by its oil-cooling requirement. The friction losses that are carried off by the lubricating oil are generated in the compressor and hermetic-motor bearings, the gearbox, and the shaft seal.

The electrical efficiency is the product of the motor's efficiency and (when used) the variable-frequency inverter's efficiency.

18.4.5.1 Part Load

When the evaporator's cooling load is reduced, the temperature lift is also reduced. This is because the heat-transfer rates are reduced in both heat exchangers, which lowers the leaving-temperature differences, and because (usually) the entering condenser water temperature (ECWT) or air temperature is lower. With constant-speed PRV control, the compression efficiency is reduced at part load, as are the other efficiencies as well. The net effect on electric power consumption (kW) when the leaving evaporator water temperature (LEWT) is held constant is illustrated in Fig. 18.4.9 for a water-cooled unit. The strong influence of temperature lift on power consumption is evident.

Except in the immediate vicinity of the full-load design point (100 percent load and power), the ECWT lines will always be lower with variable-speed capacity control than they will be with constant-speed PRV control. The difference becomes more pronounced as the ECWT values become less than the full-load value. The reason is that compression efficiency is adversely affected by low temperature lift when PRV control is used.

Because centrifugal chillers rarely operate at full load and design lift, part-load power is considerably more significant than full-load power in determining the total annual operating cost.

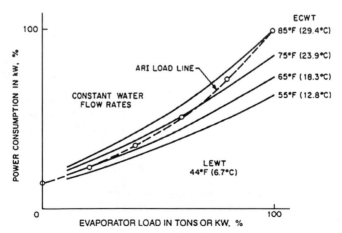

FIGURE 18.4.9 Centrifugal chiller performance with varying condenser water temperature.

18.4.5.2 Multiple Units

Two or more centrifugal chillers are often operated together to cool a large building. This not only provides some cooling capacity in the event of a chiller failure, it also saves power. When two machines are used, one machine alone can handle a cooling load that is less than half of the building's maximum load. The load will be high for the one operating machine, and so will its efficiency. The power consumption will be less than it would be if the load were being handled by one machine of twice the capacity.

Two machines can be operated with their evaporators and their condensers in series or in parallel. Water pressure drop prohibits operating more than two heat exchangers in series.

The choice between series and parallel evaporators depends on several factors. One is the part-load water flow and dehumidification requirements of the cooling coils in the air-handling units. At half-load or less, when only one machine is running, a series evaporator can always deliver the design chilled-water flow at the design LEWT to the air handlers. With evaporators in parallel, one machine running alone must overcool its chilled water if that water is to be combined with uncooled water to provide design flow at design temperature.

Another factor is chiller power consumption. Overcooling at part load increases the temperature lift and thus increases the combined annual power consumption of parallel units. On the other hand, series evaporators usually have only one water pass to avoid excessive pressure drop. Parallel evaporators generally have two passes. This confers a heat-transfer advantage on the parallel units that lowers their temperature lifts and power consumption. When only one machine is running, the heat transfer advantage of the parallel arrangement will partially offset the disadvantage of overcooling.

When series condensers are used with series evaporators, the condenser water should be in counterflow with the chilled water. Series condensers are rarely used, however, because parallel condensers can have more water passes and thus better heat transfer performance than series condensers.

Pumping power must also be considered. Parallel heat exchangers can each have their own water pumps. When only one machine is running, the condenser pump of the idle machine can be shut down. The chilled water pump can also be shut down if the air-handling units can operate with reduced flow at reduced load.

Two identical machines are often chosen, whether in series or in parallel, so that either unit can be operated without the other for equal periods of time over the year. But machines of different capacities, operating for different periods of time, will usually consume less total annual power.

18.4.5.3 Heat Recovery

In cool weather, part of a building may need heating at the same time as another part needs cooling. In that event, the condenser heat can be applied to the heating load instead of being dissipated to the atmosphere in a cooling tower. This mode of operation is called "heat recovery." The heating water is circulated in a closed loop that is separated from the cooling tower water in order to avoid contamination of the former by the latter. The two water circuits may pass through the same "double bundle" condenser, or the heating water may have a separate "auxiliary" condenser. Factory packages with heat-recovery condensers are available with cooling capacities up to 1300 tons (4600 kW).

For the heating to be effective, the temperature of the water leaving the condenser must be over $100°F$ ($38°C$). This makes the ECWT $95°F$ ($35°C$) or more. At the same time, the evaporator load may be only half the full-load value. Figure 18.4.9 shows that these conditions require considerably more power than would be the case with a normal ECWT of

72°F (22°C) at half load. The cost of this additional power is included in the economic evaluation of a heat-recovery application.

The temperature lift during heat-recovery operation is greater than it is during full-load cooling operation. Figure 18.4.7 indicates that at loads less than 50 percent of full load, the compressor will generally be operating in the surge area when heat-recovery temperature lift is needed. An automatic HGBP valve is required in heat-recovery units to prevent surging.

Heat-recovery operation occurs during cool weather when the cooling load can be met with an LEWT that is higher than the full-load value. By increasing the LEWT during heat-recovery operation, the temperature lift will be reduced. This will reduce the need for HGBP and its associated power penalty. Even without heat recovery, it always saves power to operate with the highest possible LEWT.

Because the compressor must be designed for the heat-recovery lift, its efficiency will be less during the summer-cooling season than it would be if it did not have this extra capability. The cost of the extra summertime power consumption that results from the lower compression efficiency is also part of the economic analysis. With speed control, the summer power penalty is significantly less than it is with PRV control.

18.4.6 RATINGS

A centrifugal chiller's rating is a statement of its refrigeration capacity, power consumption, and heat-exchanger water-pressure drops at a specific set of operating conditions. "Refrigeration capacity" is the net usable cooling capacity that remains after the chiller's own internal needs (hermetic-motor cooling, oil cooling, purge-unit cooling, etc.) have been satisfied. In a motor-driven unit, the "power consumption" is the total electric power that is required to operate the unit (motor, inverter, control panel, oil heater, oil pump, purge unit, etc.). Turbine and engine drives involve additional forms of energy consumption (steam rate, fuel flow). In these cases, only the input power to the compressor shaft may be known by the chiller manufacturer.

ARI has established a standard set of operating conditions at which centrifugal chillers are to be rated (Ref. 5) and a program to certify that a manufacturer's ratings have been verified by an independent testing agency (Ref. 6).

18.4.6.1 Full Load

The standard ARI conditions for full-load ratings are:

Evaporator. LEWT = 44°F (6.7°C). Water flow rate = 2.4 gal/min/ton [0.043 L/(s · kW)]. It follows from this that the temperature of the chilled water returning to the evaporator is 54°F (12.2°C).

Water-cooled condenser. ECWT = 85°F (29.4°C). Water flow rate = 3.0 gal/min/ton [0.054 L/(s · kW)]. The temperature of the cooling water leaving the condenser is about 95°F (35.0°C). The exact value depends on how much power is put into the refrigerant by the driver of the compressor and by the electrical auxiliaries of the unit.

Air-cooled condenser. Entering air temperature = 95°F (35.0°C). Barometric pressure = 14.7 psig (101 kPa).

Heat-exchanger fouling. Water-side fouling allowance = 0.0001 h · ft^2 · °F/Btu (m^2 · °C/kW) in the evaporator and 0.00025 h · ft^2 · °F/Btu (0.044 m^2 · °C/kW) in the condenser. Air and refrigerant-side fouling = zero.

18.4.6.2 Part Load

The standard ARI conditions for part-load ratings are the same as for full-load ratings except that the ECWT of water-cooled condensers varies linearly with load from 85°F (29.4°C) at full load to 65°F (18.3°C) at 50 percent wad and then remains constant at 65°F(18.3°C) to zero load. For air-cooled condensers, the entering air temperature varies linearly with load from 95°F (35.0°C) at full load to 55°F (12.8°C) at zero load. The standard part-load conditions for water-cooled condensers are indicated by the ARI load lines in Figs. 18.4.7 and 18.4.9.

The ARI standard includes a method for calculating an *integrated part-load value* (IPLV) for power consumption, which reflects the typical loads and conditions that are encountered over a full operating year. The IPLV is an index of total annual power consumption.

Temperatures, flow rates, and fouling allowances other than those of the ARI standard may be needed for some applications, but the standard values are typical of most applications. Standardized rating conditions are essential if valid comparisons are to be made among the offerings of different chiller manufacturers.

18.4.6.3 Testing

The ARI standard includes a method for verifying the rating of a centrifugal chiller by test. The standard specifies the allowable tolerances for tested values of full-load capacity, full-load power consumption, part-load power consumption at selected capacities, and the overall IPLV. The tolerance values are a function of the chilled-water temperature range at full load and the percent of full-load capacity at which the test is conducted. Typical tolerance values are 5 percent for full-load capacity and power consumption, 10 percent for half-load power, and 10 percent for the IPLV.

The ARI standard specifies the test procedures, instruments, measurements, and calculations that are needed to verify a chiller's rating within the specified tolerances. To comply at the jobsite with all the test requirements is usually difficult, if not impossible. To avoid field-test problems, specific chiller models are chosen at random by ARI and tested in factory test facilities that have been qualified by ARI. The tests are witnessed, and the results verified, by an independent testing agency that is under contract to ARI. The centrifugal chiller models whose ratings have been confirmed by these ARI tests are listed in an ARI directory of certified products (Ref. 7). The directory is revised and reissued semiannually to reflect any irregularities that may be disclosed by continual random testing.

When a new unit is tested at the factory, its heat-exchanger tubes are clean. This makes the water-side fouling resistance zero. To simulate the effects of rated fouling during the test, the temperature lift must be increased above its clean-tube value. To accomplish this, the chilled-water temperature is lowered, and the condenser water temperature raised, by amounts that are calculated by an equation in the ARI standard. In a field test, it is usually assumed that the water-side fouling resistances coincide with the rated fouling allowances, in which case no adjustments of the water temperatures are made. If the tubes are cleaned just before the test, however, the fouling resistances are zero and the fouling allowances must be simulated by adjusting the water temperatures.

18.4.7 CONTROLS

Centrifugal chillers have microcomputer control systems. These controls can be programmed to start and stop a chiller at different times of the day on different days of the week. The chiller controls can also operate the cooling-tower fans and the chiller's water pumps.

At startup, the chilled-water temperature pulldown rate, and the pulldown motor current, can be programmed to limit electrical power demand.

Microcomputer control systems can be interfaced with building automation systems to start and stop the chiller, and to vary the LEWT and the motor-current limit, to satisfy overall building energy needs and costs.

The status of any of the system-monitoring parameters (pressures, temperatures, etc.) can be obtained from an alphanumeric or graphical display on the microprocessor control panel. The set points of the controls can also be read and compared with the measured operating values. If one of the controls stops the compressor, the display indicates the cause of the shutdown and provides diagnostic information.

Operating and setpoint data, along with control shutdown information, can be logged and stored in the microcomputer control panel for periodic transmittal to a remote printer. The data and information can also be transmitted to an external service organization that is under contract to maintain the chiller.

18.4.7.1 Chilled-Water Temperature

The primary function of the control system is to maintain the LEWT at a given set point by modulating the compressor's speed and/or PRV position. At loads that are less than the compressor's minimum capacity (10 to 15 percent of full load), the modulating controls become ineffective. Under these conditions, the LEWT falls below the control set point and the compressor is automatically stopped. After the building's load has restored the chilled-water temperature to its normal value, the compressor is automatically started again. To prevent this off-on cycling from occurring so often that it harms the motor, a time delay of about 30 min is interposed between starts. When speed control of the electric motor is used, the time delay may be reduced to about 3 minutes.

If the evaporator pressure becomes too low, a low-evaporator-pressure control (or low-refrigerant-temperature control) will stop the compressor before the evaporating refrigerant can freeze the chilled water in the evaporator tubes. Because this is an abnormal condition, the compressor will not restart automatically after a low-evaporator-pressure or -temperature shutdown.

One of the causes of low evaporator pressure (temperature) is low chilled-water flow. A flow switch is interlocked with the chiller controls to prevent the compressor from starting or operating without sufficient chilled-water flowing. An interlock with the chilled-water pump may also be used to prevent the compressor from operating unless the chilled-water pump is operating. A strainer at the chilled-water inlet to the evaporator prevents individual tubes from experiencing low flow (and possible freeze-up) because of tube blockage.

18.4.7.2 Condenser

A high-discharge-pressure cutout will stop the compressor if the condenser pressure becomes too high. A condenser water flow switch and/or a condenser pump interlock is used to prevent the compressor from starting or operating without sufficient condenser water flow. A strainer at the water inlet to the condenser prevents individual tubes from being blocked.

If the condenser pressure becomes too low, the flow of liquid refrigerant to the evaporator may become inadequate. This can usually be prevented by cycling the cooling-tower fans, but in some cases a cooling tower bypass valve in the condenser water circuit must be used to control the minimum condenser pressure.

A cooling tower bypass is required by a heat-recovery unit. When both condenser water circuits are used simultaneously, the temperature of the tower water must be forced up to the temperature of the heating water by a controlled bypass of the cooling tower.

18.4.7.3 Motor

The maximum motor current can be set at a value that is less than the full-load motor current in order to reduce power costs and/or demand charges. In this control mode, the compressor capacity control is overridden by the motor-current control.

Other controls will stop the unit in the event of motor overload (excessive current), high voltage, low voltage, a momentary power fault, phase imbalance, or excessive winding temperature in a hermetic motor.

18.4.7.4 Compressor

Low-oil-pressure and high-oil-temperature cutouts are provided to stop the compressor if its proper lubrication is threatened. The oil-pump motor is protected by cutouts that respond to overload and/or a high winding temperature. When water-cooled oil coolers are employed, the flow of water may be controlled in order to prevent the oil temperature from becoming too low. A thermostatically controlled heater keeps the oil hot during standby (idle) periods in order to minimize the amount of refrigerant vapor that is absorbed by the oil.

A high-temperature cutout in the compressor discharge pipe stops the compressor if over-heating occurs. Some chillers also have a device that stops the compressor if surging occurs.

Excessive compressor speed must be prevented if a turbine or engine drive is used. The overspeed trip device that accomplishes this is mounted on the driver, not on the compressor.

18.4.8 INSTALLATION*

When a centrifugal chiller is installed in a building, sufficient space must be left around it and above it to allow for future maintenance and service needs. At one end of the heat exchangers, the space must be long enough to allow for cleaning and possible replacement of the heat-transfer tubes. The water piping should be arranged with access to the tubes in mind. For maximum accessibility, "marine" water boxes can be employed. These provide an access port through which the heat-transfer tubes can be serviced without disconnecting any of the water piping.

18.4.8.1 Noise and Vibration

The chiller is mounted on vibration isolator pads or springs to prevent the transmission of structure-borne noise and vibration to the building. Spring isolators are generally used only in upper-floor installations and other vibration-sensitive locations. When spring mountings are needed, the water piping is isolated from the chiller and the pipe hangers are isolated from the building.

Some applications are sensitive to airborne noise. In these cases the acoustic treatment of the equipment room is based on the noise characteristics of the particular unit to be installed.

---

*See also Part B.

The chiller manufacturer will supply the necessary information in the manner prescribed by an ARI standard for measuring and reporting machinery noise (Ref. 8).

18.4.8.2 Ventilation

The safety code for mechanical refrigeration (Ref. 9) requires that equipment rooms be ventilated to the outdoors by power-driven fans. The required capacity of the fans depends on the amount of refrigerant contained in the largest chiller in the room. The fans need only provide this much ventilating capacity when a large amount of refrigerant has leaked into the room. The normal (usually lower) operating capacity of the fans depends on the size of the room, the number of personnel in the room, and the amount of heat-generating equipment in the room.

The equipment room must have a refrigerant detection device that increases the ventilating rate of the fans to their full capacity, and sounds an alarm, if the concentration of refrigerant in the air in the room reaches the maximum regular daily exposure level that is safe for personnel in the room. The maximum safe concentration level is reported by the refrigerant manufacturer in his Material Safety Data Sheet. The safe concentration level is established by the American Conference of Governmental and Industrial Hygienists (ACGIH) and/or the Occupational Safety and Health Administration (OSHA).

Another requirement of the code is that the outlet of each chiller's pressure-relief device (bursting disc or relief valve) and purge unit (when applicable) be piped to the outdoors. The vent pipe should contain a cleanable dirt and moisture trap. Because R-123 condenses at temperatures below 83°F, the vent pipe of an R-123 chiller may need to be heat traced to prevent liquid refrigerant from collecting in the trap. Additional information about heat tracing can be found in Sec. 17.12 of this volume.

18.4.8.3 Thermal Insulation

The temperature of the evaporator, the low-side piping on the chiller, and the chilled-water piping will generally be lower than the dew point of the air in the equipment room. These components will "sweat" unless they are thermally insulated. Factory packages can be obtained with the necessary thermal insulation already in place.

18.4.8.4 Instrumentation

Pressure gauges and thermometers are often installed in the water piping to aid in operating and servicing the unit. To obtain the water pressure drop of a heat exchanger, one gauge that has valved connections to both the inlet and outlet water pipes of the exchanger is more accurate than two separate gauges.

Heat-exchanger pressure drop provides only a fair indication of the water flow rate. If a chiller's performance is to be measured in the equipment room with good accuracy, flow meters must be installed in the water piping.

18.4.8.5 Power Supply

The motor starter may be a part of a factory-packaged chiller, or it may be mounted separately from the unit. A fused disconnect switch is needed in either case. Capacitors for power-factor correction can be used with any starter except the star-delta open-transition type.

18.4.8.6 Safety Codes

The design and manufacture of centrifugal chillers are governed by an ANSI/ASHRAE safety code (Ref. 9). This code invokes other codes and standards, such as the National Electrical Code and the ASME Boiler and Pressure Vessel Code.

State and local codes govern the chiller's installation. These codes may differ from the ANSI/ASHRAE code, and always include additional requirements.

18.4.9 OPERATION*

For the most part, centrifugal chillers operate automatically. The only manual functions are starting the unit, setting the LEWT and the motor-current limit, and stopping the unit. In many cases, some or all of these functions are also automatic.

A human operator performs other functions, such as maintaining a log of operating information that will aid in the servicing of the unit, and even indicate the need for service in the first place. Such indications as excessive air leakage, loss of refrigerant, oil loss, insufficient oil temperature, plugging of the oil filter, fouling of the heat-exchanger tubes, internal water leakage, unusual vibration, and excessive surging can all be deduced from operator logs and observations. Of course, these indications are often automated too, by microcomputer data logging and programmed analysis.

18.4.10 MAINTENANCE*

Specific maintenance requirements vary from one chiller to another. Routine procedures for most chillers include annual changes of the oil-filter element and the refrigerant filter-drier cartridge(s). The filter-drier cartridge of a purge unit should be replaced several times per year. A chemical analysis of the lubricating oil should be made each year, but the oil need not be changed unless the analysis so indicates.

The oil-filter element should be checked for bearing wear particles and for an excessive amount of iron rust particles. The compressor bearings are not usually inspected unless there is an indication of need, such as wear articles in the oil. Most rust particles originate in the heat exchangers, not in the compressor. They are an indication of air and moisture in the system. A chemical analysis of the refrigerant will determine if any contaminants are present.

A check for refrigerant leaks is usually made twice per year unless excessive purging, reduced refrigerant liquid level, or sounding of the equipment room refrigerant detection alarm indicates an earlier need. The EPA requires that leaks be repaired if the rate of leakage equates to an annual loss of 15 percent of a chiller's initial refrigerant charge. High-pressure chillers are more likely to leak large amounts of refrigerant than are low-pressure chillers. That makes it desirable, for R-134a and R-22 units, to set the equipment room refrigerant detection alarm at a level that is lower than the maximum personnel exposure setting.

Control settings and instrument calibrations should be checked annually, as should the electrical resistances of the compressor's motor windings.

Heat-Exchanger Tubes.   Proper maintenance of the heat-transfer tubes begins with maintaining the quality of the water that flows through the tubes. The chilled water is almost always contained in a closed circuit, so one initial chemical treatment may last for several years. The condenser water circuit is almost always open to the atmosphere and is therefore

---

*See also Part B.

in continuous need of chemical treatment in order to prevent fouling (sedimentation, corrosion, scaling, etc.) of the inside surface of the condenser tubes. Samples from the condenser water circuit should be analyzed monthly to check the chemical quality of the condenser water. The chilled water should be analyzed every six months. Additional information about water chemistry and treatment can be found in Chap. 8 of this volume.

The water strainers are cleaned annually and the inside of the heat-transfer tubes inspected. The tubes are cleaned if the inspection indicates a need. Automatic tube-cleaning systems can be installed that will operate continuously, whenever the unit is running. Keeping the tubes clean reduces power consumption because water-side fouling increases the temperature lift.

The tunes can also be inspected for wear on the outside surface (at the tube supports) and for fatigue cracks in the tube wall. This is accomplished by passing an electromagnetic probe through each tube. The moving probe generates eddy currents in the tube wall and senses the impedance of the tube wall. The output of the sensing coil is displayed on a cathode-ray tube (CRT) or recorded on a strip chart. Fluctuations of the output signal indicate variations in the thickness or composition of the tube wall.

REFERENCES

1. "Standard 740-95 for Refrigerant Recovery/Recycling Equipment." ARI, Arlington, VA, 1995.
2. "Refrigerant Recovery/Recycling Equipment Certification Program, Operational Manual OM-740," ARI, Arlington, VA, 1991.
3. "Directory of Certified Refrigerant Recovery/Recycling Equipment," ARI, Arlington, VA.
4. "2000 ASHRAE Handbook, HVAC Systems and Equipment," ASHRAE, Atlanta, GA, 2000, Chapters 7, 34, 35, 37, 39.
5. "Standard 550/590-98 for Water Chilling Packages Using Water Vapor Compression Cycle," ARI, Arlington, VA, 1998.
6. "Certification Program for Centrifugal and Rotary Screw Water-Chilling Packages, Operational Manual OM-550," ARI, Arlington, VA, 2000.
7. "Directory of Certified Applied Air-Conditioning Products," ARI, Arlington, VA.
8. "Standard 575-94 Method of Measuring Machinery Sound within Equipment Space," ARI, Arlington, VA, 1994.
9. ANSI/ASHRAE Standard 15-2001, "Safety Code for Mechanical Refrigeration," ASHRAE, Atlanta, GA, 2001.

PART B: CENTRIFUGAL CHILLERS MAINTENANCE OF OPERATION

18.4.11 INTRODUCTION

In today's industrialized world, air conditioning is no longer a luxury; it is a necessity. Businesses are dependent on both comfort cooling and process cooling. Most of the large air-conditioning applications (high-rise office buildings, hospitals, universities, industrial facilities, manufacturing plants, chemical plants) involve centrifugal chillers. Centrifugal chillers have been used for heat loads of over 100 tons because of their reliability and performance. With the use of computerized design and manufacturing techniques among other tools, centrifugal chillers have become even more reliable and efficient. This trend of increased reliability should continue.

Most of these facilities are constructed by a general contractor, and the mechanical equipment is purchased through the mechanical contractor. A great deal of attention is focused on the price of the equipment. The more enlightened client will take this process one step further and consider the *total life cycle cost*. The life cycle analysis is important because the operating expenses (maintenance, service, repair, utilities) over the life of the equipment can be as much as 10 to 100 times the initial installed cost of the equipment. If risk and/or down time is considered, the figure could even be orders of magnitude larger.

In this chapter we will address philosophies, strategies, and tasks to manage operating costs, specifically related to centrifugal chillers and large chilled water plants. The proven results of these strategies are reduced downtime, increased reliability, full design capacity, and maximum efficiency.

In Sec. 18.4.12 we will discuss the various types of centrifugal chillers. We will also compare and contrast the components and materials that comprise the different styles and models of these chillers. At the end of the section we will describe a process that we have been using over the past few years in office buildings, hotels, sports arenas, manufacturing plants, hospitals, and universities to reduce operating and maintenance costs.

In Sec. 18.4.13 we will describe the various factors that affect reliability in centrifugal chillers. Reliability involves keeping the machine operating to meet the desired load. The reliability of a chiller is improved through maintenance and a good maintenance program can pay for itself through cost avoidance and downtime prevention. Energy savings provide additional cost reduction as a collateral benefit of a good preventive maintenance program.

In Sec. 18.4.14, we will discuss machine performance. Once a machine is running reliably, we can then start to look at its performance. The two measures of performance are capacity and efficiency. Capacity is the total refrigeration capacity of the machine. It is not unreasonable to expect a chiller that is 10, 20, or even 30 years old to operate at or near its full load design capacity, *if it has been properly maintained*. Once the machine is operating at its maximum capacity, an equipment owner can check for the second measure of performance: efficiency. Efficiency is often expressed as the electrical cost (kilowatts) per ton of air conditioning delivered. Even though a chiller could be meeting the load requirements, it could also be consuming 20 percent or 30 percent more energy to deliver the design tonnage—therefore wasting money. We will document specific methods and processes to identify and improve machine inefficiencies.

In Sec. 18.4.15 we will document specific upgrades that allow the equipment owner to squeeze additional reliability, capacity, or efficiency from a chiller plant. Equipment manufacturers are continually creating new products for use on existing systems. We would recommend these upgrades and improvements for chillers and central plants that are running reliably while efficiently meeting the capacity requirements.

In Sec. 18.4.16, we will list criteria useful in repair versus replace decisions. These decisions usually surface before or after expensive repairs. There are *external* factors to consider: government regulations, industry codes, utility costs, and interest rates. There are also *internal* factors to consider: building use, future needs, the availability of funds, internal rate of return, and other financial criteria.

In this era of "Do more with less," it is a challenge to balance cutting costs while maintaining performance. While the cost and risk associated with operating centrifugal chillers have increased, the quantity and quality of resources available to equipment operators has also increased. While we may not be able to change environmental regulations or macro-economic conditions, there are a multitude of tools, tasks, and resources readily available that we can use to keep mechanical equipment functioning reliably while reducing overall maintenance expenses. Implementing the procedures in this chapter is a good start toward improving your central plant performance.

18.4.12   COMPONENTS AND MATERIALS

In this section, we will describe the common components and materials that comprise centrifugal chillers. We will describe what each part or component is, what it does, and how it can fail or cause other parts and components to fail. Prediction and prevention of these failures will be addressed in a later section on reliability.

18.4.12.1   Refrigerant

First, a word concerning the refrigerant issue. Since December 31, 1995, CFC-based refrigerants are no longer manufactured in the United States, with some exceptions. Even with the phaseout, some equipment owners have chosen the strategy of containment to delay making a choice between converting or replacing their chillers. Regardless of the refrigerant class, for maintenance purposes there are two types of refrigerants: *high pressure* and *low pressure*.

For low-pressure refrigerants (HCFC-123, CFC-11, CFC-113), moisture infiltration is a key concern. In most equipment rooms there will be a vacuum inside the chiller while the chiller is not operating, because these refrigerants have boiling points above room temperature. When the chiller is operating, the evaporator is in a vacuum. When the chiller is under a vacuum, air can infiltrate the chiller if there is a leak.

To remove air, moisture, and other noncondensibles, equipment manufacturers add *purge units*. Problems associated with purge units can vary from incomplete moisture or noncondensible removal, purge compressor failure, or in a few extreme cases, older style purge units (pre-1990) could even pump out an entire refrigerant charge. The new style purge units, called high-efficiency purge units, have safety features to prevent losing an entire refrigerant charge. If a purge unit underperforms or fails, then, at the very least, moisture removal will be reduced and, at the very worst, a chiller will trip off on high head pressure.

For the high-pressure refrigerants (HFC-134a, HCFC-500, HCFC-502, HCFC-22, CFC-12), air infiltration is not as serious an issue. Since these refrigerants are under positive pressure, a leak will cause refrigerant to be expelled and the chiller will slowly lose its charge. To remove moisture in chillers that use high-pressure refrigerants, some equipment manufacturers apply a *dehydrator*, which is similar to a purge unit on low-pressure machines. If a dehydrator fails, moisture removal will be reduced, and contaminants will enter the system. Since there are fittings and screens on the dehydrator, it is another leak source.

High-pressure chillers are often equiped with a *pump out unit* (condensing unit) and a *utility vessel* for storing refrigerant. The pump out unit is often comprised of a reciprocating compressor and a condensing unit. The utility vessel is essentially a large pressure vessel with a relief valve. The utility vessel is usually sized large enough to hold an entire refrigerant charge and is used to store the refrigerant charge while the chiller is being serviced.

There is also an issue of safety with both low- and high-pressure refrigerants. Although refrigerants have been thoroughly tested and used safely for years, *they must be handled properly to avoid injury.* The most immediate *safety* concern with any refrigerant is the displacement of oxygen. The risk is that if refrigerant leaks into an equipment room that isn't properly ventilated, it can fill up a room and displace enough oxygen to suffocate someone who enters the room. This can be solved by complying with the ASHRAE standard for equipment room ventilation and sensors. There are other safety related issues such as cardiac sensitivity pertaining to short-term exposure and long-term exposure. Table 18.4.1 compares the short-term exposure risk for various refrigerants:

To appreciate this table, it must be understood that the higher the number, the safer the product is, i.e., it takes a higher level of refrigerant to produce the effect. With proper training and equipment, *refrigerants can be used safely.* For more information, consult the refrigerant manufacturers or your local refrigerant distributors.

Regardless of what refrigerant is used, *leak prevention is key to reliable performance.* The main sources of leaks in chillers are on *fittings,* (external refrigerant and oil lines, filter housings, valves) gaskets, and o-rings. The chillers should be checked frequently for refrigerant leaks. When minor leaks are found, they can be temporarily repaired by either reflaring a copper line or applying some type of high vacuum sealant, such as glyptol. For significant leaks, the machine must be regasketed. When the machine is reassembled, an additional sealant (like Loctite R #515) can be applied on the gaskets and o-rings to provide additional leak protection. Refrigerant leaks can cause several problems if left untreated. Leaks can cause low-pressure chillers to trip out on high condenser pressure and high-pressure chillers to trip out on low evaporator pressure.

With respect to leak management, your equipment must comply with the Clean Air Act of 1990, which documents acceptable leak rates. It specifies that chillers used for comfort cooling can lose 15 percent of their refrigerant charge per year, and chillers used for process cooling can lose 35 percent of their refrigerant charge per year. Any leak rate over these thresholds must be repaired within 30 days, or the machine must be shut down.

Leaks in centrifugal chillers are one of the most common problems and the root cause of secondary and tertiary damage. When air infiltrates a refrigerant system, it contains a certain amount of moisture. Some of the water (from the moisture in the air) mixes with refrigerant to form acids. Some of the water oxidizes the iron in the shells and castings to form rust. This rust and acid can attack surfaces inside the chiller and cause premature wear of bearings, seals, gears, and any other moving or sensitive parts. This can cause premature failure of other components or, in an extreme case, shorten equipment life. Moisture and contaminant removal are equally important in the lubrication circuit for the same reasons as in the refrigerant circuit.

TABLE 18.4.1 Comparison of Various Refrigerants

Acute	Low pressure		High pressure	
	CFC-11	HCFC-123	HCFC-22	HFC-134a
LC_{50} (40 hr)	26,000 ppm	32,000 ppm	>300,000 ppm	>500,000 ppm
Cardiac sensitization	5,000 ppm	20,000 ppm	50,000 ppm	75,000 ppm
Anesthetic effect	35,000 ppm (10 min)	40,000 ppm (10 min)	140,000 ppm (10 min)	205,000 ppm (4 h)
Pressure at 72°F	1.5 Hg vacuum (liquid)	5.6 Hg vacuum (liquid)	126 psig pressure (gas)	74 psig pressure (gas)

Regarding the refrigeration cycle, all refrigeration machines, from refrigerators to car air conditioners and 1000-ton industrial chillers, have four components in common. These four components are as follows:

- Compressor
- Condenser
- Metering device
- Evaporator

Figures 18.4.10 and 18.4.11 are diagrams of a typical refrigeration cycle for a two-stage and three-stage centrifugal chiller, along with its corresponding P-H (pressure vs. enthalpy) diagram. In the next few pages, we will describe the various styles of the aforementioned four components with respect to centrifugal chillers, including their function and any potential problems that an operator may experience. The solution of these problems will be addressed in the reliability section.

18.4.12.2 Compressor

The heart of a centrifugal chiller is the *compressor*. There are several different styles of compressors and each type comprises several components. In the following paragraphs, we describe the most common types of compressors and their major components.

The main function of the compressor is to take the low-temperature, low-pressure gas from the evaporator, raise its temperature and pressure, and send a high-temperature high-pressure gas to the condenser. The *centrifugal* compressor creates this pressure differential between the evaporator and condenser. This pressure differential is referred to as *lift*. A centrifugal compressor is designed with a specific lift that at a specific suction pressure, it will deliver a specific discharge pressure. Unlike a screw compressor, which is a positive displacement device, a centrifugal compressor cannot deliver a certain discharge pressure at a given suction pressure. This means that a centrifugal compressor will surge at too high condensing pressure or too low evaporator pressure. Both types of surging (low side and high side) can damage a centrifugal compressor.

One of the main components in the compressor is the *motor*. On chillers with *hermetic motors*, the refrigerant is sprayed directly on the motor windings to cool the motor.

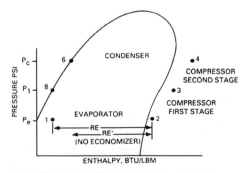

FIGURE 18.4.10 Two-stage chiller, P/E diagram. (*Courtesy of The Trane Company.*)

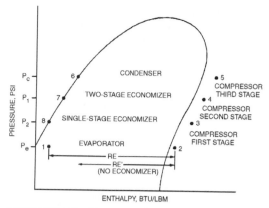

FIGURE 18.4.11   Three-stage chiller, P/E diagram. (*Courtesy of The Trane Company.*)

Moisture and acid levels are important to prevent motor burnout, because contaminants can erode the insulation on the motor windings. On chillers with an *open-drive motor*, the motor is located outside of the refrigeration circuit and is cooled by ambient air. Although the risk of motor failure is reduced, the motor must be kept clean. Dust and dirt in the equipment room can accumulate on the windings and cause overheating and, potentially, even motor failure, although this is very rare, according to my contacts at a chiller manufacturer. Also, the hot air coming off of the open-drive motor must be directed away from sensitive electronic components in the starter and control panel to prevent component failure.

In the United States, the most common voltage for these motors is 460/480 V. There are also some 208/230 V motors, and some 4160 V motors. These motors normally operate at 60 Hz power, although there are some chillers that utilize a variable frequency drive which varies the frequency of the power to adjust motor speed according to the load. In Europe, 50 Hz is standard with several different voltages. Canada, with 60 Hz power, often has non-standard voltages.

Almost all of the motors used in centrifugal chillers are three-phase induction motors. The two main components of these motors are the *stator* and the *rotor*. The stator is a group of windings which are coiled, insulated wire, wrapped around a core of iron laminates. These coils are either form wound or random wound. The insulation on these motors is critical, especially on hermetic motors. In the case of hermetic motors, which are sealed inside the refrigeration circuit, the windings are cooled by liquid refrigerant. In the case of open motors, which are exposed to the atmosphere, the windings are cooled by ambient air. The insulation integrity is tested through a resistance test, or *megging* the motor. The term megging refers to measuring the resistance of the stator, and the units are megohms (a megohm is 1,000,000 ohms). When a motor "burns out," typically one set of windings (single phasing) or all three sets of windings (complete burnout) are shorted and go to ground. There can also be a phase to phase short. If the motor failure is severe enough, part of the iron core could be melted and damaged beyond repair, requiring the purchase of a new motor. Motor failure can also be caused by a starter malfunction, such as a contactor welding shut, or (in a star-delta starter) the motor staying in transition too long. Excessive cycling—turning on and off too many times in a certain period—can also cause motor failure.

The rotor fits inside of the stator with a small air gap in between. The rotor shaft is constructed of hardened alloy steel. On the outer diameter of the rotor are aluminum or copper bars. The most common type of rotor failure is an open rotor. This condition exists when one of the bars becomes loose and the current is discontinuous (or open) at that bar as it rotates. If left unrepaired, an open rotor can cause premature bearing wear.

Another cause of rotor failure could occur after a power outage. After power failure, while the shaft is coasting down, the motor acts as a generator. If power is returned during this coast-down period, and the power from the utility is out of phase with the current generated by the motor, the rotor is stopped instantly. This can break the shaft and/or destroy the impeller—not an easy feat considering the shaft is hardened steel 1 to 3 in (25 to 75 mm) in diameter.

One advantage that open-drive motors have over hermetic motors is the reduced risk of motor failure. In the event of motor failure, it is generally less expensive and faster to repair an open-drive motor than a hermetic motor. Another advantage of open-drive centrifugal chillers is that the conversion cost from CFC-11 to HCFC-123 or CFC-12 to HFC-134a is significantly less expensive. One disadvantage of open-drive motors is the *shaft seal*. Shaft seals are typically a mechanical seal around the low-speed compressor shaft. An additional sealant is provided by a thin film of pressurized oil between the shaft and the mechanical shaft seal. The oil continually drips out of the shaft seal. The rate of dripping must be checked and if it increases beyond an acceptable, predetermined rate, the shaft seal may need to be replaced. During the annual inspection, the pressure of this oil can be checked and adjusted.

The motor shaft to impeller connection further segments the different styles of centrifugal compressors. Some chiller manufacturers attach the impeller (or impellers in a multistage compressor) directly on the motor or compressor shaft. These compressors are referred to as *direct drive compressors*. Figure 18.4.12 is a schematic diagram of a hermetic, direct-drive compressor.

Since most of the motors rotate at 3600 rpm, the impeller also rotates at 3600 rpm. Some chiller manufacturers attach a bull gear at the end of the motor or compressor shaft, which turns one or more smaller pinion gears that are attached to the impeller shaft. This increases the speed of the impeller to 5000, 7000, or 10,000 rpm based on the transmission ratio. These compressors are referred to as *gear-driven* or *high-speed compressors*. Figure 18.4.13 is a cross-section of an open-drive, gear-driven, centrifugal compressor.

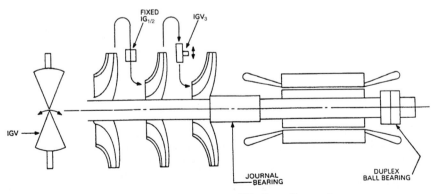

FIGURE 18.4.12   Direct-drive compressor. (*Courtesy of The Trane Company.*)

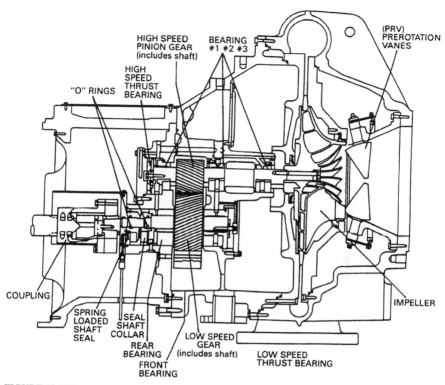

FIGURE 18.4.13   Gear-driven compressor. (*Courtesy of The Trane Company.*)

On *high-speed compressors*, the gears (like the rotor) are made of hardened steel and the teeth are precision machined. Gears rarely fail, but when they do, it is generally due to one of several causes:

- *Open rotor*—The damage from an open rotor was discussed in the rotor section.
- *Halogen fire*—a nonaerobic flame (occurs in the absence of oxygen) caused by the refrigerant burning—can cause aluminum to melt and or burn.
- *Wear from rust*—again the age-old culprit—rust—the enemy of mechanical equipment.
- *Viscosity breakdown*—lubrication fails, usually due to either old oil, acid, moisture, contamination of the lube oil, or less viscosity due to higher temperature.
- *Bearing failure*—this damages the wear surface on the gears.

The *impeller* imparts kinetic energy into the refrigerant gas stream and thereby compresses the refrigerant. Impellers on direct-drive machines are typically larger (2 to 4 ft [60 to 120 cm] diameter) than impellers on high-speed machines (6 to 2 ft [15 to 60 cm]). Also, impellers on direct-drive machines are usually shrouded while the impellers on high-speed machines are often unshrouded, where the blades are exposed. The impeller to volute clearance on unshrouded impellers is smaller and more critical than on direct-drive machines with shrouded impellers. Direct-drive machines are often multistage (2 to 3), while high-speed machines are often a single stage. Figures 18.4.14 and 18.4.15 are P-H diagrams comparing single-stage vs. three-stage compressors. Impellers rarely fail. The few instances of impeller failure are listed below:

- Bearing failure (journal or thrust)
- Metal fragment/liquid refrigerant drop hitting impeller face
- Halogen fire
- Impeller eye seal or rear hub seal wear

Again, since compressor oil lubricates bearings and gears, rust and contaminants can damage these parts. Proper lubrication is another key to reliability and performance. The lubrication system is comprised of an oil pump, oil filters, oil sump(s), oil heater, oil cooler, temperature regulator, and oil lines and channels throughout the compressor. The oil must be clean and maintained at the proper temperature, pressure, and viscosity. It must be kept hot enough to boil off refrigerant, but it must be kept cool enough to maintain its viscosity. Consult the manufacturer's recommendations on the type of oil to use and its optimal temperature range. The manufacturer should also provide information how to monitor the oil and when to change it. Typically, the oil in centrifugal chillers should be changed annually, but more often if contaminants are found in an oil sample.

Internal to the compressor are seals and bearings. The seals keep refrigerant out of the lubrication circuit and keep oil inside the compressor. Often the seals inside the compressor are referred to as *labrynth seals*. Labrynth seals are usually made out of aluminum and rarely wear. When they do wear, it usually is due to improper alignment with respect to the shaft (eccentricity) or severe bearing wear. Wear can be caused by the shaft or the impeller rubbing against the seals.

There are several types of bearings used in centrifugal chillers. The most common type of bearings are *journal bearings*. These bearings support the shaft or gears vertically and allow proper rotation. Another type of bearing is a *thrust bearing*. Thrust bearings allow for horizontal movement of the impeller as the compressor loads and unloads. Thrust bearing adjustments are critical and leave little margin for error. If the thrust is too tight, it can reduce machine capacity since the clearance between the impeller and the volute is too

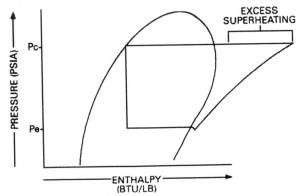

FIGURE 18.4.14 Single-stage without subcooler, P/E diagram. (*Courtesy of The Trane Company.*)

THREE-STAGE WITH TWO-STAGE ECONOMIZER

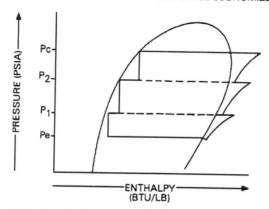

FIGURE 18.4.15 Three-stage compressor with two-stage economizer, P/E diagram. (*Courtesy of The Trane Company.*)

great. If the thrust is too loose, it can cause impeller damage because the clearance between the impeller and the volute is too small. Bearings are *designed to wear* and are made of a soft metal, like babbit (tin and lead) or aluminum. The frequency of bearing replacement may be from every 3 years to 10 to 15 years depending on machine style and manufacturer.

One cause of bearing failure is the loss of lubrication, which could be caused by blockage of flow from debris. Oil pump failure can also cause loss of lubrication, although the low oil pressure cutout should shut the chiller down before wear occurs. Another cause of bearing wear (related to the loss of lubrication) is oil laden with refrigerant. Refrigerant acts as a perfect degreaser and strips the oil from the wear surfaces, which allows contact with the shaft. Also, any particulate matter or corrosion can prematurely wear the bearings.

Steam-driven centrifugal chillers are rare in the western United States, but are more common in the eastern United States. There are even a few engine-driven (either natural gas or diesel) chillers.

The last component in the compressor to be discussed is the vane assembly. There can be multiple vane assemblies in a centrifugal compressor, typically one set of vanes for each stage of compression. Essentially, the vanes perform two functions: one is to provide a means for capacity control and the other is to prerotate the gaseous refrigerant entering the impeller. The vane assembly controls the load on the compressor by regulating how much refrigerant is allowed to enter the compressor. There are several styles of vane assemblies and several designs that have evolved. Common in all designs is that the individual vanes are linked, either through a cable or through a series of mechanical levers or joints.

Vane assemblies rarely fail, but one cause can be moisture (or rust) in the refrigerant circuit. Symptoms of vane failure are improper loading/unloading, cycling, and/or current fluctuation.

18.4.12.3 Condenser

After the low-temperature, low-pressure gaseous refrigerant leaves the compressor as a high-temperature, high-pressure gas, it is pumped into the condenser. While there are a few centrifugal chillers with air-cooled condensers, most centrifugal chillers are water-cooled. Most water-cooled chillers, to diffuse the refrigerant and prevent tube failure, contain a *baffle plate* inside the shell, directly in line with the refrigerant path. The condenser receives the high-temperature, high-pressure gas from the compressor and condenses it into a liquid. When the refrigerant comes in contact with condenser tubes, it condenses. This phase change releases the heat of vaporization picked up in the evaporator along with the heat of compression picked up in the compressor. This heat is transferred from the refrigerant across the tube surface, and into the condenser water, where the heat is then transferred from the condenser water to the atmosphere in the cooling tower.

One useful rule of thumb concerns flow rates through condensers and evaporators. On typical comfort cooling conditions (45°F [8°C] supply and 55°F [13°C] return chilled water temperature, 85°F [25°C] supply and 95°F [35°C] return condenser water), the condenser water flow rate is typically 3 gpm (11 L/m) per ton, and the chilled water flow rate is typically 2.4 gpm (9 L/m) per ton. For example, a 200-ton water cooled centrifugal chiller should have a flow rate through the condenser of about 600 gpm (2200 L/m), and a flow rate through the evaporator of about 480 gpm (1650 L/m). Remember, this is only a *rule of thumb*. Lower condenser water temperatures and system load conditions greatly change this specific rule of thumb.

There are several problems that can occur in the condenser. One relates to leaks. Since the condenser is the high pressure side of the machine, any air or other noncondensibles will end up in the condenser. One of the main problems associated with condensers is *high head pressure* or high condenser pressure, which can be caused by the following:

- Leaks
- Fouled tubes
- Refrigerant stacking
- Low condenser water flow
- High condenser water temperature

Since the condenser and evaporator are so similar (they are both shell and tube heat exchangers with similar construction), we will discuss their individual components in the heat exchanger section.

18.4.12.4   Metering Device

The function of the *metering device* is to create a pressure differential between the high-pressure side of the system (condenser) and the low-pressure side of the system (evaporator). The metering device allows the liquid refrigerant from the condenser to flow into the evaporator. It is usually located on the liquid line between the condenser and the evaporator. On older centrifugal chillers, the metering device is usually a float valve. One common problem associated with the *float valve* is the tendency for refrigerant to stack in the condenser if the float valve sticks and doesn't allow enough refrigerant into the evaporator. On newer centrifugal chillers the metering device is usually an *orifice plate*.

18.4.12.5   Evaporator

The evaporator in a centrifugal chiller receives the liquid refrigerant from the condenser through the metering device. As the liquid refrigerant boils in the evaporator, it absorbs heat from the water in the chilled water loop. Inside the evaporator is a set of screens called *eliminators*. The eliminators are designed to prevent any drops or droplets of liquid refrigerant from entering the compressor. These eliminators are subject to corrosion from acid in the refrigerant circuit. If the eliminators are corroded, it is a symptom of a serious corrosion problem and you've got more to worry about than just eliminators. At the bottom of some evaporators is a liquid distribution system. Its function is to evenly distribute the liquid refrigerant throughout the entire evaporator.

Heat Exchangers. Heat exchangers are just like they sound—surfaces where heat transfer takes place. Since the heat transfer surfaces play such an important role in capacity and efficiency, we will examine those aspects of their function in the performance section. In this section, we are more concerned with reliability.

Both the condenser and evaporator comprise the following components:

• Vessel
• Tubesheets
• Support sheets
• Water box covers
• Tubes

Vessels. Each vessel is cylindrical in shape and constructed of cast iron, usually $\frac{1}{2}$ to 2 in (1.3 to 5 cm) thick. Inside the shell are *tubes* and their *support sheets*. At the end of each heat exchanger are the *tubesheets*. The tubesheets are welded to the ends of the condenser and evaporator. Connecting the water side of the tubes to the chilled water and condenser water pipe is the water box cover. Inside the water box may be no, one, or several divider plates for a one, two, or multiple pass heat exchanger.

In addition to the condenser and evaporator there is one other vessel we haven't mentioned. This other vessel is called an *economizer* and it is only present on multistage centrifugal compressors. Although economizers have no tubes, some have internal features, such as eliminators, orifice plates, and weir devices.

The function of the economizer is to provide additional cooling and subcooling of the liquid refrigerant between the stages. It also may provide motor cooling. It is essentially a liquid receiver that gets liquid refrigerant from the condenser and is vented to the compressor and possibly the motor.

While we're dealing with vessels, now is a good time to discuss refrigerant safety again. Since most of the refrigerant is present in the vessels, as a safety measure, both low-pressure chillers and high-pressure chillers have a pressure relief system. Low-pressure (HCFC-123, CFC-11, CFC-113) machines were designed with a *rupture disk* while high-pressure (HFC-134a, HCFC-22, CFC-12) machines have *relief valves*. Both devices will release refrigerant to relieve pressure in an emergency. The relief valves reseat, while the rupture disks do not. With the increased cost of the low-pressure CFCs (CFC-11 and CFC-113) and the health and safety concerns with HCFC-123, equipment owners have been retrofitting their machines with relief valves. Instead of losing an entire charge with a rupture disk bursting, they have a chance of keeping at least part of the refrigerant charge if the relief valve reseats. Whichever system is used (rupture disk or relief valve), the refrigerant exiting the chiller must be vented outside, per ASHRAE guidelines. This prevents the refrigerant from burning and creating poisonous gases.

Why a pressure relief device? If a boiler explodes, or a fire is generated in the equipment room, or any emergency causes the temperature in the equipment room to rise quickly, the chiller can become a locomotive-sized hand grenade. If the pressure was great enough, it could explode, sending metal fragments through walls. Besides the projectile danger, the burning refrigerant could turn into phosgene gas, or another of several types of deadly gases that could cause harm to building personnel or the general public. To prevent these catastrophes, equipment designers have designed pressure relief systems.

Several other things can cause a pressure relief device to release refrigerant: a heat source in the equipment room—like a fire, or a boiler explosion, or a large steam leak. Also, too high an entering chilled water temperature can cause excessive pressure in the evaporator and cause the rupture disk to rupture. Water flowing through an idle low-pressure chiller can also cause a refrigerant release.

Tubesheets.   The tubesheets secure the tubes at each end and form a seal between refrigerant on the outer diameter of the tube and water on the inner diameter of the tube. On boilers, tubes are welded into tubesheets, but with chillers, the soft copper tubes are mechanically expanded into holes in the tubesheet. The tubesheets are usually 1 to 2 in (2.5 to 5 cm) thick.

Tubesheets rarely fail. When they do, it's usually due to poor water treatment. If the tube sheet is eroded or corroded, water can get in between the tube and tubesheet mechanical seal. To prevent this from happening, you can coat your tube sheets with some type of epoxy or ceramic polymer metal.

Support Sheets.   The support sheets do exactly what they sound like they do—they support the tubes. Since evaporator and condenser tubes can be from 8 to 20 ft (2.4 to 6 m) long, without any support along the way, they would sag in the middle if not supported. They are usually the size of the diameter of the vessel, and $1/4$ to $1/2$ in (6 to 12 mm) thick, with several holes drilled for the tubes. The longer the heat exchanger, the more support sheets. Typically, they are spaced every 2 to 3 ft (0.6 to 0.9 m). In most chillers, each heat exchanger would have 3 to 5 support sheets. The only time that support sheets fail is when the refrigerant circuit is badly contaminated from acid and rust.

Support sheets are especially critical in evaporators, since the turbulent action of the refrigerant boiling can rattle the evaporator tubes. Their function is not as critical in the condenser.

Water Box Covers.   The water box covers cover the entire face of the tube sheet. The water box covers on at least one should be removed annually to allow for inspection of the tubes and tubesheets. There is also a gasket that prevents water from the leaking out. If the tubesheet is corroded, the water box is probably also corroded and should be coated with a protective coating if corrosion is present. Some water boxes (with multiple passes) may have a divider plate that separates one pass from another. If the divider plate leaks, the flow could be short-circuited and cause high condenser or low evaporator pressure, since not enough heat is being transfered between the refrigerant and the water.

Tubes.   The tubes in the condenser and the evaporator are the heat transfer surface in the refrigeration system of almost all centrifugal chillers. Since they are the heat transfer surface, they have a big impact on machine performance. If the tubes are fouled, capacity is reduced, as is efficiency. Maintenance items related to tubes will be addressed in the reliability section, and their effect on capacity and efficiency will be addressed in the performance section.

The tubes in both the condenser and evaporator are almost always copper. Occasionally you find cupronickel, brass, carbon steel, or stainless steel tubes in corrosive environments, process cooling loops, and marine/saltwater condensers. The tubes may be prime surface copper or they may have internal or external enhancements.

Tubes can fail in several ways—almost all of them catastrophic. In considering the cost of a retube, include secondary and tertiary costs. Listed below are the most common tube failures:

FREEZE-UPS.   Freeze-ups are usually caused by either operator error or control failure. Occasionally it is due to a design problem. To freeze up a chiller while it's running, three components have to fail: flow switch, low water temperature cutout, and the low refrigerant cutout (either pressure or temperature). Another way to freeze up a chiller is when a refrigerant charge is lost quickly, and the sudden drop is pressure causes a sudden drop in temperature. Yet one more creative way to freeze up a chiller is to remove the refrigerant charge without operating the chilled water pump to add the heat of vaporization.

One unusual case of a tube freeze-up occured on a low temperature screw chiller application at an ice skating rink. While one chiller was idle, it had a 12°F (−12°C) glycol brine flowing through its evaporator and there was no positive shutoff between the evaporator and condenser. The refrigerant in the evaporator was chilled by the brine, and then thermal migration occurred between the condenser and evaporator through the liquid line. Since the machine was idle, there was no condenser water flow and the low-temperature refrigerant (12°F [−12°C]) took heat out of the water in the condenser tubes until they ruptured.

INNER DIAMETER PITTING, EROSION.   This is caused by incorrect or incomplete water treatment, sand, grit, or another abrasive substance circulating through the condenser loop, or too great a flow rate. This happens more frequently on condensers and open-loop chilled water circuits.

OUTER DIAMETER PITTING.   Outer diameter pitting is usually caused by water in the refrigerant that forms acids and damages the tube surface from the outside. This can occur on either the condenser or evaporator, but would only happen in a machine with a lot of water infiltration over a long period of time. This could only occur under a state of negligence.

SADDLE DAMAGE.   This usually occurs only in the evaporator. In the evaporator, the refrigerant is boiling. In most chillers, the tube is rolled into a support sheet. If the tube is overrolled, the soft copper will be deformed and worn just from overrolling. If it is under-rolled, the tube is too loose. In both cases, the boiling action of the refrigerant will cause the tube to rub against the support sheet and if left unrepaired, can cause tube failure and consequential damage.

MANUFACTURING DEFECTS.   Tube failures can also be caused by any of the following manufacturing defects: zipper cracks from seams, welding slag on tube outer diameters, over-rolling, or pin holes. One large real estate developer I have worked with has every chiller eddy current tested by an independent testing contractor before he accepts it.

Another trend in chiller manufacturing (to squeeze additional efficiency out of the chillers) is to decrease the tube wall thickness and to add enhancements on both the inside diameter and the outside diameter. Internal rifling, similar to rifling in a gun barrel, causes more turbulence and increases the heat transfer. Also, the number of fins per inch have steadily increased from 19 or 26 fins per inch (8 to 10 cm) to 40 or 50 fins per inch (6 to 20 cm). Then, the fins are cross-hatched, or folded over, to get additional surface area and hence, better heat transfer.

Other Chiller Components and Systems

Controls.   There are two types of controls—*operating controls* and *safety controls.* Operating controls start and stop the chiller and allow for capacity control. Safety controls prevent the chiller from destroying itself. Controls can get out of calibration and not perform properly, or they can even outright fail. Below is a partial list of common operating and safety controls:

Operating Controls

- Start/stop/status
- Capacity
  - vanes/VFD (variable frequency drive)
  - load limit
  - current limit

Safety Controls

• Flow switch
• High pressure cutout
• Low pressure cutout
• Low refrigerant cutout
• High/low oil temperature
• Low oil pressure
• High motor temperature
• Low water temperature

This list of operating and safety controls is not all inclusive. With additional automation, new controls and sensors are continually being developed, tested, and manufactured.

Starter.   In older chillers, starters came separate and were usually mounted on a wall near the chiller. Now you can order the starter chiller mounted in the factory, saving installation time. Also, with solid state features, starters have a multitude of additional features offering the following:

• Soft start to reduce start-up current
• Solid state overloads and voltage, current, and frequency sensors to aid in motor protection

If your chiller will sit outdoors, care must be taken to have the proper NEMA-approved enclosure. If the chiller will operate in a classified area like a refinery or chemical plant, further steps may be needed to meet the requirements for the classified area. Typical classified areas are spaces that may contain explosive hydrocarbon gases, like propane or butane. Any spark producing component (e.g., starters and switchgear) must be sealed to prevent gas from entering the device where it can come into contact with a spark and cause an explosion.

Other system components and systems external to a chiller that affect chiller performance are as follows:

• *Chilled water and condenser water piping and pumps*—delivering consistent and unvarying designed water flow at the proper pressure
• *Water treatment*—keeping tubes clean and clear of inorganic and organic fouling
• *Cooling tower*—Rejecting heat from the condenser water loop and delivering design temperature condenser water supply
• *Controls*—Proper sequencing, staging, loading, coordination, and operation of chillers and ancillary equipment
• *Seismic bracing*—requirement for seismically sensitive areas, for example, California and Japan
• *Power*—correct and unvarying voltage, current, and frequency

Quality power is especially important for chillers utilizing microprocessors and SCRs (silicon-controlled rectifiers) in control panels and starters.

Chillers are designed and installed to do one thing—remove heat from something. It's about energy transfer—taking heat from a conditioned space or a process and transferring it to the chilled water loop, which then transfers it to the refrigerant, which is then compressed. In the condenser, the refrigerant condenses and releases the heat picked up in the evaporator and the heat of compression. If the chiller either fails to operate or fails to remove enough heat, either the people, equipment, or substance being cooled will overheat. The cost could be minimal at best, or expensive at worst, and could possibly include the loss of life.

The most common denominator in good chiller performance has been organizations that not only maintained their equipment well, but also took ownership of the entire chilled water plant. These organizations also paid attention to their equipment to prevent things from becoming worse and acted fast when a problem was encountered. Here is a procedure:

- Perform equipment survey and needs analysis
- Make repairs to return equipment to design conditions
- Choose a strategy of either containment, conversion, or replacement
- Chiller optimization—either incrementally or all at once
- Chilled water plant optimization
- Feedback

The first step is to find out where you're at. We perform an equipment survey and a needs analysis, so you know where you stand with your equipment condition. Then you get the chiller plant repaired and performing up to design conditions. Then you choose a strategy of either *containment, conversion*, or *replacement*. Containment involves reducing refrigerant emissions. Conversions involves converting your chiller to operate on one of the new refrigerants. The next step involves optimization. Then you choose the schedule of implementation, either all at once or incrementally. As you install upgrades that generate energy savings, you use those savings to fund your next upgrade. Then you need a monitoring tool for feedback. This ensures that what you did worked and it also justifies the expenditures.

Now that we've looked at the different parts and components in centrifugal chillers and you know most of the ways they can fail, you might ask yourself how to prevent some of these catastrophes from happening? Sometimes you can't—you get a current or voltage surge, or you get an imbalance among the phases, and your motor fails—even though you've done everything humanly possible to prevent it from happening. Some things are out of your control. These are only about $1/_2$ of 1 percent of all failures. The other 99.5 percent can be prevented. In the next section, we'll look at specific methods and tasks to prevent chiller failure.

In the next section we will list specific tasks to keep machines operating on line continually and reliably. *With proper planning, tools, and equipment*, most of the equipment damage and failures can be prevented.

18.4.13   RELIABILITY

In this section we will discuss both the general and the specific along with the philosophical and the pragmatic aspects of maintenance, not only for centrifugal chillers, but as it applies to your entire physical plant. We will also address the solution and the prevention of problems that an operator may encounter.

Preventive maintenance on mechanical equipment is sometimes neglected. In a lot of organizations it is one of the first things to be reduced during budget cuts. When additional capital or operating funds are available it is one of the last things added back into a budget.

In some instances, a temporary "deferral" of maintenance is acceptable. In other instances, such as when downtime is critical, it can be catastrophic. For certain industries, such as semiconductor manufacturing, chemical plants, refineries, and data centers, a single day or sometimes even a single hour of downtime may cost an organization more than an entire year of preventive maintenance.

Here are four other factors that affect machine reliability:

- Design
- Application
- Installation
- Operation

18.4.13.1 Design

Regardless of your specific skills, background, and experience, we recommend hiring a consulting engineer to design or at least help design the mechanical systems. Listed below are some criteria for evaluating different designs:

- Comfort vs. process cooling
- Future needs and growth
- Cost: installed and life cycle
- Proper clearances/access
- Redundancy
- Ease of use
- Automation (BAS/EMS)

18.4.13.2 Application—Machine Selection and Sizing

Even with an excellent design, if equipment is not sized or selected properly with adequate features, problems can arise with the best of maintenance programs.

Whether for new construction or retrofitting an existing building, *machine selection* is the beginning. A large factor related to reliability is the application of the correct equipment for the specific purpose. Depending on the use of the chiller—whether for comfort or process cooling, low or medium temperature, high or low run time, high or low cycling (starting and stopping), several makes and models may work. Or possibly, only one make or only one model will deliver the desired performance. It may require the use of another style of compressor—e.g., screw, reciprocating, or scroll. Due to the availability and price of steam or natural gas, an absorption chiller may be the best choice.

If the chiller is sized improperly, it may not deliver the desired tonnage. If undersized, the chiller may not meet the load or provide redundancy. If oversized, the chiller could fail prematurely due to excessive cycling. Other factors in machine selection:

- Usage
- Budget
- Future growth
- Building purpose

18.4.13.3 Installation

Another factor that impacts reliability is the installation of the equipment. Once the mechanical room is designed, the equipment is selected and ordered. The next step is installation.

Besides the pertinent laws, codes, and regulations, each equipment manufacturer will have specific installation instructions for their whole line of equipment, and possibly various installation requirements for various machine models.

In most comfort cooling applications, the chillers are piped in parallel. Two benefits of this arrangement are redundancy and serviceability. In process cooling applications, there are several possibilities. The chillers may be piped either in parallel or series, depending on several conditions. For example, if the load has a high temperature difference or if it is a batch or erratic process where the heat load can vary, it may be better to pipe machines in series, staging the machines as the load changes. Or if there is a high flow rate, it may be better to pipe machines in parallel, splitting the flow between two chillers. Figures 18.4.16 and 18.4.17 are schematic diagrams of both parallel and series piping configurations.

A chiller in a manufacturing facility in California often tripped out on high head pressure because the water returning from the cooling tower was too hot. The reason was that the piping was not plumb and did not distribute the hot water coming from the condenser evenly in the cooling tower. The chiller and cooling tower operated perfectly, but the installation caused reliability problems. In addition to obeying codes and piping the equipment properly, there is a myriad of other considerations to a proper installation.

18.4.13.4   Operation

Listed below are four approaches to operating your chiller plant:

- In-house
- In-house with service contracts
- Outtasking
- Outsourcing

In-house is just that; operating and maintaining your chillers with in-house labor. Another variation is to do all of the operation in-house and contract for all or part of the maintenance. The third approach is contracting out for an entire discipline—all HVAC, or all fire and life safety maintenance. Outsourcing is contracting for *all* disciplines—both for operations and maintenance.

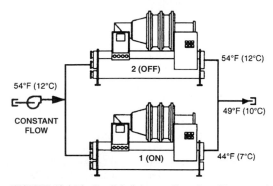

FIGURE 18.4.16   Parallel piping configuration. (*Courtesy of The Trane Company.*)

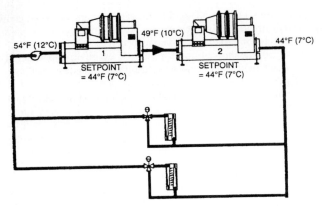

FIGURE 18.4.17   Series piping configuration. (*Courtesy of The Trane Company.*)

After the chilled water plant has been designed and installed, nothing affects machine reliability more than proper operation. Whether the chiller plant is operated manually or through some type of automation or control system, the operation of the equipment is the responsibility of the operators.

Some buildings do not have dedicated operators on site for their centrifugal chillers, but someone must operate the equipment. It can be started and stopped remotely using a building automation system (also known as an energy management system). With full integration, log readings can even be taken automatically. No control system can duplicate what an experienced and trained operator can do—notice a strange sound, notice a drop of oil on the floor or on the machine that wasn't there yesterday. Not only can highly skilled operators save the property owner several times their salary in energy expenses, they can prevent a lot of expensive failures or catch the failures *before* they cause secondary or tertiary damage.

Make it a high priority to get a skilled crew of operators. Operators must be trained. While you may not need a journeyman chiller mechanic on site (you may if you have a critical operation or a large plant—10,000 tons or more), you need operators who understand the refrigeration cycle, basic electricity, some HVAC and/or controls knowledge, plumbing skills, and especially troubleshooting skills. They must know how to take readings and decipher which readings are abnormal or out of range. They also should know what the temperatures and pressures should be in the condenser, evaporator, and compressor. They should be able to spot trends and take corrective action. *If your operators don't know this,* they need training. If it is beyond their skills or if they are limited by time, they should be instructed to call in a specialist for further diagnosis or for repair.

Another skill often overlooked in operators is communication skills. If you depend on outside contractors for chiller repairs, you can save yourself a lot of money if you can communicate effectively with your contractors' personnel. If you have a good relationship with a chiller repair company (either the manufacturer's local service department or a good independent mechanical service contractor), you can often describe the chiller's symptoms to a qualified mechanic and get suggestions on remedial action, saving the expense of a service call. Even if a chiller mechanic must make a service call, the more information you can provide to him or her the more you can save in troubleshooting expenses.

A strong operator crew can do most of your maintenance on your chillers. In the lists of tasks discussed in this chapter, a qualified stationary engineer can do all of them up to the annual inspections. All daily, weekly, monthly, quarterly, and semiannual tasks can be performed with in-house personnel. For annual inspections, it is worth the expense of bringing in an outside specialist like a journeyman centrifugal chiller mechanic. For major repairs or special testing, we recommend contracting out.

As a minimum, the following is a list of what an operator should do:

Operator Duties and Tasks

- Start-up—daily, weekly, seasonally
- Shutdown—daily, weekly, seasonally
- Extended shutdown (seasonal, temporary)
- Start-up after extended shutdown
- Documentation—log readings, federal regulations and codes, EPA refrigerant compliance, OSHA, state, and local documentation
- Inspecting, monitoring, and maintaining chillers—lubrication, minor adjustment, painting and corrosion prevention, insulation repair
- Monitoring and inspecting ancillary equipment
- Ensuring proper flows and temperatures in and out of the chiller on both the condenser and chilled water loop

Although the operation of a centrifugal chiller can sometimes appear to be mundane since it is such a reliable piece of equipment, proper operation is crucial for machine reliability. For example, one important aspect that could be taken for granted is proper flow through the condenser and evaporator tube bundles. Lack of flow or an elevated entering temperature in the condenser water loop can cause high head pressure and even surging. This causes damage to the chiller from possible bearing or impeller failure. Surging could also cause the chiller to use 10 to 20 percent more energy.

If you have a limited operator staff, you may have to contract out for your monthly and quarterly inspections. As a manager or director, there is a trade-off you may have to make. You can trim your staff to save expenses, but the maintenance tasks on your chillers and other equipment and systems don't go away. Although you may save money on overhead and benefits, you could lose some control of your maintenance. If done properly, you can gain additional control through contracting out. Another choice is to outtask or outsource, turning over even the operation of your facility to an outside contractor. Whether you contract out for major repairs, regular maintenance, or all or part of your operations, there are benefits, costs, and risks associated with each choice or combination of choices. Choose carefully.

18.4.14 MAINTENANCE

There are a lot of philosophies and approaches toward maintenance. (See Chap. 12 of this book.) Some organizations put a lot of effort and energy toward maintaining their equipment, while others do little. *Maintenance is the key to reliability.* Overall, there are three main approaches to maintenance listed below:

- Breakdown maintenance (also called "run to failure")
- Preventive maintenance
- Predictive maintenance

Some organizations still have a breakdown maintenance plan. Their reasoning is that even after spending a large amount of money on preventive and predictive maintence, chillers still fail and need major repairs or replacement. To them, a large expenditure is just around the corner, so why waste money on all of this maintenance when mechanical failure is inevitable? Breakdown maintenance is the least expensive method to implement since there is no cost, but is the most expensive, once the life-cycle costs are considered. These life-cycle costs include (but are not limited to) the following:

- Major equipment failure—repair or replace
- Energy
- Downtime
- Risk and liability

There *is* a grain of truth to the philosophy of the breakdown maintenance supporters. Even with performing every feasible preventive and predictive task, chillers do require a major repair sometime in their life, and eventually they are replaced. But breakdown maintenance people are short-sighted, in that a good preventive maintenance program can pay for itself in energy costs alone, while avoiding down-time and providing better control of when repair or replacement dollars are spent. Since I do not recommend a breakdown maintenance philosophy, I will not elaborate on that approach. We will describe preventive and predictive maintenance approaches in more detail.

Preventive maintenance is just that; it is maintaining equipment to prevent secondary and tertiary damage. The result of preventive maintenance is increased reliability, reduced downtime, and reduced loss of energy efficiency.

Predictive maintenance is taking preventive maintenance one step further. The key word is predictive—predicting when chillers need major service. Once up and running, over a period of time it can help you stretch your preventive maintenance dollars so that you don't overspend on maintenance. A predictive maintenance program is most expensive up front, but saves money in long-term secondary and tertiary damage and energy. A thorough predictive maintenance plan also has some beneficial side effects: it reduces downtime further, gives the owner additional control and time for planning, and can reduce preventive teardown inspections.

Whichever philosophy you subscribe to, there are other tools to help plan and implement your maintenance program. One of these new tools is a computerized maintenance management system (CMMS). There are several CMMS programs commercially available, the best of which offer a large database of tasks, labor and material planning, cost tracking, inventory control, and especially refrigerant tracking. Besides a CMMS, there are other types of software for a specific task, such as refrigerant tracking and EPA documentation. See Chap. 14, *Computerized Maintenance*.

Whether you contract out all of your maintenance or do it all with in-house labor, and track it with either a card file or use a sophisticated software program, there are three ways it should be implemented to get the most benefit out of it. It should be a *proactive* tool and used for planning. It should also address and meet your *documentation* needs. Finally, someone should see that it is *standardized* and integrated into your department. If you manage a large facility with several shops or several locations, then standardization is crucial for you.

Being *proactive* is the key to success.

Documentation is getting more important every year—both from a code compliance standpoint and a legal standpoint. Besides protecting yourself and your company, it's also helpful to document your successes and prove that the money invested in your maintenance program is justified. This documentation should reflect uptime, capacity, maintenance, and energy savings. The following information should be included in your documentation:

Start-up Information

- Submittals—specifications and drawings
- Original model and serial numbers
- Design conditions
- Logs

Maintenance History

- All logs
- Preventive and predictive maintenance tasks performed
- Refrigerant tracking
- Monthly/quarterly/annual inspection checklists
- Major service
- Failure history—machine, component, part
- Financial information
- Capacity and efficiency reports

Standardization is crucial for a large facility or an organization with multiple locations; it goes hand-in-glove with organizational quality initiatives such as total quality management (TQM) or any other statistical-based quality process. One of the basic tenets of a quality process is that quality is not the most expensive path—it is the most *cost-effective* path. It is not overspending or underspending. This is true for preventive maintenance. You can also create a "Best Practices" process for your maintenance or facilities department.

Intimately related to standardization and documentation is the subject of logging your machines. It is important to take log readings on your centrifugal chillers. At a bare minimum it must be done once per day, and the more readings the better. Hourly readings would be ideal. With new building automation systems, this logging can be done automatically. You also can design a program to signal an alarm if a reading is too high or too low. The reason for this is when there is a machine problem, it make take a while to manifest itself. Big problems usually start as little problems. Of course, it matters who analyzes the readings. Operators need to know what to look for, and it saves time and money to provide a set of log readings when a contractor shows up.

The following preventive maintenance tasks are recommended for your centrifugal chillers. Since there are many makes and models of centrifugal chillers, these tasks are generic. When they differ from the manufacturer's recommendations of your specific machine, use the manufacturer's recommendations.

Hourly

- Logs, if possible—manually or automatically through building automation system

Daily

- Start/stop
- Load/unload
- Look, listen, and feel for any unusual visual cues, sounds, temperatures, or vibrations

Weekly

- Inspect all sight glasses—record refrigerant and oil levels
- Inspect system piping and pumps on both condenser and chilled water systems

Monthly

- Record hours of run time and the number of starts of machine
- Record minutes of run time on purge unit
- Inspect all refrigerant and oil piping, fittings, flanges, gaskets, o-rings for leaks
- Check operating and safety controls
- Inspect cooling tower
- Inspect lubrication system

Quarterly

- Inspect all components and subsystems
- Adjust and lubricate where necessary
- Inspect all electrical connections
- Leak check (Some people feel that monthly or quarterly leak checks are necessary. On a well running machine with a history of few leaks, it's probably not necessary. If you have had a lot of leaks, you probably should check for leaks quarterly. Some local codes require semi-annual leak checks.)

Annual Tasks

Nothing is more crucial than a thorough annual inspection by a qualified technician. The following tasks are the *minimum* that should be performed *annually* on your centrifugal chillers:

- Pull oil sample and analyze for wear metals and contaminants
- Leak check
- Meg main motor and oil pump motor
- Check and calibrate operating and safety controls
- Service purge unit or service dehydrator and/or pump out vessel
- Pull heads and inspect tubes and tubesheets
- Clean tubes (if necessary); this may need to be performed semi-annually or quarterly depending on water conditions
- Change oil and oil filter
- Cut oil filter in half and inspect for bearing shavings or fragments
- Thoroughly inspect and test lubrication circuit
- Change refrigerant filters and strainers
- Inspect vane linkage
- Inspect and clean starter

- For open motors: inspect coupling alignment
- On gear driven machines: inspect gears for wear
- Inspect shaft seal assembly
- Start-up, test, and run chiller
- Log machine and document all tests and readings

Consult the equipment manufacturer for additional or more specific tasks on your chiller.

Water Treatment.   Quality water is important not only for reliability, but also for energy efficiency. Improperly treated water will foul condenser tubes and increase head pressure. If the head pressure gets too high, the chiller could trip off on the high head pressure control. This increased pressure can also damage the thrust bearing on high-speed machines.

The two main groups of culprits in tube fouling are dissolved solids and microorganisms. Chlorine and ozone (and other biocides) work well on microorganisms, but do nothing for dissolved solids and scale. Depending on the hardness of your water, you may need one or more chemicals to keep the dissolved minerals suspended. Regardless of the quality of your water treatment, you need to make sure the blow down mechanism is functioning on your cooling tower. Since water quality varies, contact one or two water treatment companies, have them test your water, inspect your machines, water boxes, pipes, and cooling towers. They can then recommend a water treatment program. You also can buy chemicals yourself and perform this service in-house. We would recommend an automatic type feeder to add the proper chemicals based upon water hardness and conductivity.

Water treatment is so important for reliability and performance. If your concentration of solids builds up too much, it doesn't take long to scale up condenser tubes. If you *don't* maintain the proper chemical levels, the condenser can scale up in a matter of hours or even minutes. Since the chilled water loop is usually a closed loop, it needs to be treated only as often as you add make-up water.

Multiyear Service.   Even with the best preventive maintenance program, most machines require a major internal inspection on some interval. Some manufacturers have 3- and 6-year major inspections. Others recommend 5-year teardowns. With new manufacturing technologies and closer tolerances, some chillers can go for 8, 10, or maybe 15 years before a major service is required. Follow the manufacturer's recommendations and, as a double check, compare the machine's performance along with the analyses from predictive maintenance tools to decide when a machine needs a major inspection or service.

In addition to internal inspections, you may need some or all of the following major repairs:

- Overhaul
- Retube
- Motor rewind or replacement
- Compressor or driveline replacement

Predictive Maintenance.   We recommend using predictive maintenance tasks. If you don't want to launch into a full blown predictive maintenance plan, one strategy may be to start with the simple, low cost items first, like oil analysis and vibration analysis.

Those tasks can point you in the right direction to allocate money to either repair the worst chillers or prevent the worst failures.

Over a short time frame (less than 1 year), predictive diagnostics can be used to monitor symptoms. Each analysis can give you a snapshot of the machine's condition at one particular point in time. If your machine is experiencing a particular problem, a predictive diagnostic tool or a combination of diagnostics may be used to uncover an acute problem or confirm a suspicion. Predictive tools are not always 100 percent exact or accurate. They should be used in conjunction with each other and with the operators' and chiller mechanics' input. For example, oil analysis and vibration analysis are a good check for each other.

Over a longer time frame (2 to 10 years), predictive diagnostics can be used for *trending* to plan for major inspections or repairs 1, 2, or maybe even 3 years in advance. An entire booklet of snapshots can give you an idea of where a machine's condition is headed by looking where it came from. It may take 2 to 4 readings per year in the first year or two to establish a trend, but after a trend has been established, it may require only one or two analyses per year.

There are two types of predictive maintenance tasks—*nonintrusive* and *intrusive*. Nonintrusive tasks can be performed while the machine is running. Intrusive tasks require the machine to be shut down or in some instances, partially disassembled. The following is a list of predictive maintenance tasks, their purposes, and the recommended frequency:

Nonintrusive Tasks	*Frequency*
Oil analysis	**At least once per year**
Wear metals—tin, lead, aluminum	
Contaminants—water, refrigerant, acid	
Refrigerant analysis	**On suspicion of difficulty**
Water/Moisture	(contamination due to a tube leak,
Acid	air leak, urgent malfunction)
Oil	
Thermography	**Every three years**
Electrical wiring and connections	
Switchgear, starters, control panels	
Motors	
Vibration analysis	**Once or twice per year**
Motors, bearings, shaft	
Impeller	
Gears	
Motor current analysis	**Once per year**
Open rotor bar	
Intrusive tasks	
Megger test	**At least once per year**
Insulation integrity	
Eddy current analysis	**Every three years**
Tube condition	
Special occasions	
Boroscope/fiber-optic inspection	**On suspicion of difficulty**
Metallurgical analysis	**On suspicion of difficulty**

Even with the best operators and the most complete preventive and predictive maintenance program, your chiller could fail. You can always rent a piece of equipment—not only your chiller, but pumps, cooling towers, air handlers, electrical cable and distribution panels, ancillary equipment—even electrical generators to power your entire facility. You can rent until your equipment is repaired or new equipment is installed.

We suggest forming a relationship (maybe even a partnership) with an equipment rental company to provide this equipment on an emergency basis. Listed are criteria important in evaluating a rental equipment partnership:

- Contingency plans
- Experience
- Equipment availability and diversity
- Disaster recovery expertise
- Local, 24-h service
- Age and condition of fleet
- Ability to deliver a complete, turnkey solution for power, cooling, dehumidification, heating
- Equipment with modem capability for remote starting and monitoring
- Support—engineering and managerial

Maintenance Review—19 Questions to Ask Yourself:

1. What per cent of the time is my equipment available for operation?
2. How many minutes of downtime were there this past year?
3. Is our chiller plant meeting the demands of the load?
4. What is the capacity of each chiller? 4a. chilled water plant?
5. What is the efficiency of each chiller? 5a. chilled water plant?
6. What are our annual utility expenses?
7. Has it been increasing or decreasing with time?
8. What is our annual maintenance expense? (Including internal costs like overhead, allocated labor, overtime)
9. Is it increasing or decreasing?
10. What is the condition of our chillers?
11. Is it improving?
12. How many unscheduled breakdowns did we have?
13. How fast did we get the chillers back on line?
14. How much money was spent on corrective actions and repairs vs. preventive maintenance?
15. Do we have enough people?
16. Are we getting the kind of value-added services we need from our vendors?
17. If we're not where we need to be, how are we going to get there?
18. What does 1 day (or 1 h or even 1 min) of downtime cost me or my organization?
19. Why do those Dilbert cartoons remind me so much of my work-place?

Every now and then it helps to take a step back and question some of the assumptions you have made. You must analyze some of the things you've been doing and how you've been working. We also strongly recommend performing a risk analysis. In the refinery and petrochemical businesses, engineers refer to this as *Hazard Operations*, or HazOps for short. In this exercise, everyone plays devil's advocate with each other to come up with every possible scenario of equipment failure, and then a plan to handle the consequences of that failure. It usually starts off with a bunch of what if questions—what if this chiller fails?, what if this pump fails?, what if this valve fails to close or fails to open? Then, when all possible scenarios are examined, contingency plans are created to handle every possible out-come.

This risk analysis can also help justify a preventive maintenance program once the following costs are considered and quantified:

- Downtime
- Rentals
- Overhead

Regardless of what you decide to do, we suggest you have a strategy and a plan to implement that strategy.

Sometimes preventive maintenance is not performed often enough or thoroughly enough. On the other hand, you can perform *too much* preventive maintenance and waste money, or worse yet, degrade the performance of the machine. You want to do enough maintenance, but not love your chiller to death. A good yardstick is to do things that add value your chiller. If a task doesn't extend the useful life of the chiller or if it does not increase the capacity or efficiency of the chiller, don't do it. Spend your money elsewhere.

Energy savings provide additional cost reduction as a collateral benefit of a good maintenance program. The first step in optimizing performance is a good reliable machine. Once a machine is operating properly, we can start to improve its capacity and efficiency. We will investigate this in the next section.

18.4.15 PERFORMANCE

The performance of a chiller is measured in two ways: capacity and efficiency. Capacity is the total refrigeration capability of the machine. When the chiller is started up, it should be operated at its design conditions, to assure that it will perform as expected. Most chiller manufacturers offer a witness test when you purchase a chiller, whereby you can have the chiller tested at factory under controlled conditions.

After your chiller is installed at your plant, and after the flows have been adjusted, you can do a rough check of the chiller capacity. If you don't have flowmeters installed, you can at least install pressure gauges and check the pressure drop across the evaporator and the condensers. You can then interpolate the flow rate from a pressure drop vs. flow rate chart. Once the flow rate is determined, the temperature change, or ΔT must be checked. Measure the difference between entering and leaving chilled water temperatures. Do the same for the condenser water temperatures.

English Units

$$\frac{\Delta T \times \text{flow rate}}{24} = \frac{\text{tons of refrigeration}}{\text{(at 45°F leaving chilled water temp)}}$$

where $\Delta T = \text{temperature}_{in} - \text{temperature}_{out}$ (°F) flow rate is measured in gal/min.

SI Units

$$\Delta T \times \text{flow rate} \times 5.13 = \text{watts of refrigeration}$$

where $\Delta T = \text{temperature}_{in} - \text{temperature}_{out}$ (°C) flow rate is measured in L/s.

$$1 \text{ ton of refrigeration} = 12{,}000 \text{ Btu/h} = 3516.2 \text{ W}$$

The calculation from there is simple if you're at 45°F (9°C) leaving chilled water temperature and 85°F (30°C) entering condenser water temperature. If your temperatures vary from that, you may have to derate your calculation. You take the flow rate (in gpm) and multiply it by the temperature difference. You then divide that number by 24 and that gives you your tons of refrigeration. If you multiply your tonnage by 12,000, you get the total Btu/h. The tonnage through the condenser should be larger than through the evaporator, because the condenser not only has the heat from the load that was transferred in the evaporator, but also the heat of compression. The total heat rejected through the condenser will always be larger than the heat picked up through evaporator.

As far as capacity is concerned, a well running, reliable machine is probably operating at or near capacity. If the tubes are clean, the clearances in the compressor are all within specifications, there is proper flow and pressure drop across the condenser and evaporator, there is no leakage through the divider plate, there is no flow blockage in the piping systems, there is the proper refrigerant charge in the chiller, and maybe a few other factors, the chiller should be operating at or near capacity.

The most common factor affecting capacity occurs at the heat transfer surfaces: the condenser and evaporator tubes. That is why proper water treatment is so important. With fouling factors of less than 1/1000 (25 μm) of an inch, it doesn't take much fouling to affect capacity and efficiency. Fouled tubes are often misdiagnosed when something else is causing high head pressure, much to the dismay of the water treatment companies. It could be caused by fouled tubes, or any of the following conditions:

- Air or other noncondensibles in the machine
- A leaking or broken division plate
- Low condenser flow
- Elevated condenser water temperature
- Condenser water bypassing tubes
- Blocked tubes
- Improper refrigerant charge
- Refrigerant stacking

If you are operating at less than design capacity or efficiency, it could also be caused by any of the following low-side problems:

- Low refrigerant charge
- Dirty evaporator tubes
- Excessive oil in the refrigerant
- Refrigerant "stacking"
- Chilled water bypassing tubes
- Low chilled water flow

If the chiller checks out O.K., it could be any of the following system problems:

- Low flow
- Excessive flow
- Erratic flow
- Cooling tower
- Pumps
- Flow blockage, like a plugged strainer
- Controls/sequencing

Efficiency is measured as the energy output per energy input. Most chiller manufacturers express efficiency as kW/ton of refrigeration. At full load, efficiencies below 0.6 kW/ton are not uncommon, and some chiller manufacturers claim efficiencies below 0.6 kW/ton and some below 0.5 kW/ton. Typically, this is referenced at full load at ARI conditions. What is not typical is chillers running at or near full load. Chillers often run at partial load, so full load efficiency is only part of the story. Some chillers that may be efficient at full load, are not nearly as efficient at partial load, while other chillers maintain their efficiency at partial loads and some others actually increase in efficiency at partial loads. Since chillers operate so often at partial load, the part-load efficiencies should also be considered.

Another fact to consider is that chillers may not operate at or close to design conditions. The machine selection process may have selected a machine that was efficient only across a small band of operating parameters. Once those conditions changed, efficiency dropped.

Once capacity has been determined and efficiency has been calculated, there may be several things to improve efficiency. If you are already operating very efficiently, there may not be much you can do. Here are some ideas to wring out additional efficiencies.

18.4.15.1 Optimization

Controls.   There are several control strategies and sequences of operations for your chilled water plant that you can implement to increase efficiency. With the state-of-the-art controls systems, you can program into your system almost anything you can imagine, often only with software changes. Sometimes you may need to add sensors and possibly valves, but often it's only a change in your sequence of operations.

Chilled Water Reset.   At lower loads, you can let the chilled water supply temperature climb 1° or 2° or more. This saves energy because it takes less energy to make warmer chilled water. Of course, you must check your load to make sure it won't be adversely affected. If you are in a humid area, this may not work because you need a lower chilled water temperature for dehumidification. Also, a process may not tolerate a 1° or 2° change in temperature.

Condenser Water Reset.   Similar to chilled water reset, you can adjust the condenser water temperature to reduce energy. This is also referred to as *floating head pressure* or *floating condenser pressure.*

Compressor Head Reduction.   Similar to floating head pressure—if you can run cooler condenser water temperatures the compressor is pushing against less pressure and this translates into less horsepower, which is less current, which will save you money.

Sweet Spot. Chillers have a point at maximum efficiency, and it's not necessarily at full load. Every chiller has a set of curves that show how efficiency varies with the load percentage. Once the sweet spot for each chiller is found, you can program your controls to keep each chiller at its sweet spot. Some chiller manufacturers iteratively find this sweet spot at all systems parameters and keep the chiller operating as close as possible to these sweet spots, as often as possible, for maximum energy savings. If you have multiple machines, you can locate the sweet spot for your entire chilled water plant and set your controls to sequence your chillers at that sweet spot.

Free Cooling. There are several ways to achieve "free cooling." One way is to run the cooling tower water through the chilled water system. Another way is to cool the chilled water with cooling tower through a heat exchanger. The first way is more effective, but if you're not careful you can contaminate your entire chilled water piping system. With plate and frame heat exchangers, you get approaches close enough to make this a very attractive payback.

VFDs. Variable frequency drives (VFDs) have been put on supply fans, exhaust fans, chillers, pumps, and cooling towers—essentially anything with a motor. As a load decreases, you can slow down a motor by reducing the frequency. As it slows down and operates at slower speeds, it draws less amps and consumes less energy. Since there is a cubic relationship between horsepower and energy, at half speed the motor uses one-eighth the energy of full speed. Listed below are some examples of VFD applications:

Pumping. You can convert your piping system to primary loop through your chillers at a constant flow with a secondary loop through the load at variable flow. Reduced flow means reduced pump horsepower. You vary the flow on the secondary loop only because you don't want to vary the flow through the chiller; this can cause reliability problems and machine damage. Figure 18.4.18 is a typical primary/secondary pumping schematic.

Chiller Driveline. Although more involved than putting a drive on a pump or a fan motor, substantial energy savings can be achieved through applying a VFD on a chiller motor.

Cooling Tower. Putting a VFD on a cooling tower has a very good payback, often under 1 year. It also gives an additional collateral benefit, in that it also increases chiller reliability, because you can gain much better control of your condenser water temperature. Some chillers are susceptible to oil migration at low condenser water temperatures.

Now that we've looked at capacity and efficiency, in the next section we will detail additional upgrades for increased reliability.

18.4.16   IMPROVEMENTS AND UPGRADES

Over the years, the chiller manufacturers, equipment owners, consultants, and independent service companies have dealt with a variety of problems in centrifugal chillers related to reliability and performance. Luckily, you can benefit from this vast knowledge base.

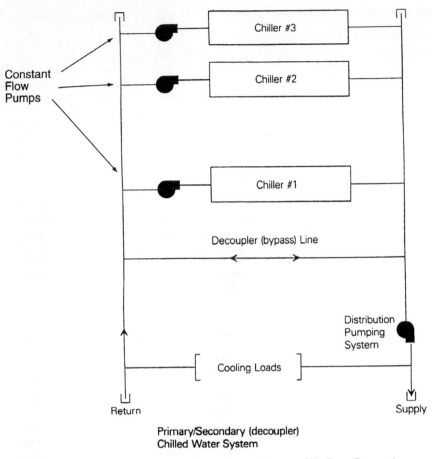

FIGURE 18.4.18   Primary/secondary chilled water system. (*Courtesy of The Trane Company.*)

With a few, rare exceptions most of the problems you will encounter as an operator have been experienced, dealt with, agonized over, and solved by someone else. It is just an issue of information access. We will document a few of the upgrades available for centrifugal chillers. For additional upgrades or more detailed information, consult your chiller manufacturer or your independent mechanical service company.

18.4.16.1   Motor Protection System

Reliability.   Motor failure, although rare, is an expensive event. Not considering downtime and overhead expenses, a motor failure will cost between $20,000 and $150,000 (2003 dollars) in primary and secondary damage. In addition to the electrical distribution system (switchgear, circuit breakers, starter) you can install additional sensors and overloads to protect the motor on your centrifugal chillers.

18.4.16.2   Epoxy Coating Tubesheets

Reliability.   If you have eroded or corroded tubesheets, we recommend sandblasting them to bare dry metal and applying either an epoxy or a ceramic metal coating to prevent tubesheet failure. There are several qualified companies in every major metropolitan area that specialize in this type of work.

18.4.16.3   Additional Refrigerant Filter

Reliability.   If you have either a wet machine or a machine that is full of rust, I recommend installing an additional refrigerant filter consisting of four 1 gal (4 l) filter cores to clean up the moisture and/or rust. This is also commonly referred to as a clean-up kit.

18.4.16.4   Solid-State Starters

Reliability and Energy.   Besides offering additional motor protection, solid-state starters offer energy savings and reliability improvement with the ability of soft starting. Solid state starters also offer additional documentation, as most will have the capability to save and print out information. Figures 18.4.19 and 18.4.20 depict the energy savings due to the reduced current and torque.

Solid-state starters achieve these increases in efficiency through solid-state switching. This is accomplished through three pairs of silicon controlled rectifiers (SCRs), one pair for each phase or motor leg. Figure 18.4.21 is an electrical schematic of a six-SCR configuration. Figure 18.4.22 shows the current waveforms of reduced power consumption at various stages.

By using a solid-state starter, reliability is increased due to the following features:

- Arc-less switching, longer contact life
- Reduced mechanical shock to equipment due to less torque
- Reduction or even removal of current and voltage transients and harmonics

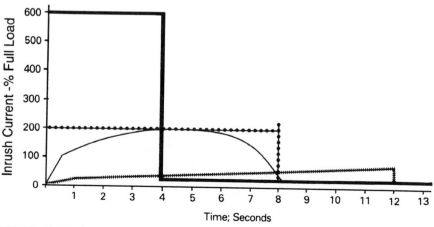

FIGURE 18.4.19   Inrush current versus time. (*Courtesy of The Trane Company.*)

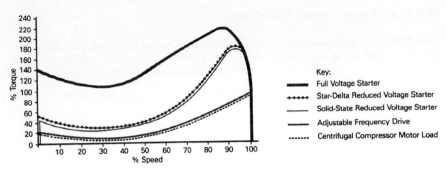

FIGURE 18.4.20   Torque versus speed. (*Courtesy of The Trane Company.*)

Applying a VFD yields similar reliability and efficiency gains, but saves even more energy since chillers often run at partial load.

18.4.16.5   Electronic Chiller Control Panel

Reliability and Energy.   A lot of older centrifugal chillers were built using pneumatic controls for part of the operating and safety controls. The early electric and electronic controls are not as sophisticated as modern electronic control panels. Upgrading your chiller with an electronic control panel will offer you more diagnostics, better control, more reliability, and possible energy savings.

18.4.16.6   Integration with Building Automation System

Reliability and Energy.   To benefit from all the advances in hardware and software, you're not getting the most out of either your chiller plant or your control system if the chillers are not integrated with your building automation system. By integration, we don't mean just on/off, start/stop, and status, but *true integration*. True integration involves getting all of the information from the chiller control panel to the control systems user interface or workstation, along with the ability to monitor and affect changes.

If you still have a chiller that operates on either CFC-11 or CFC-113, with no plans to convert or replace it, we strongly recommend the following items. (If you don't have an CFC-11 or CFC-113 chiller, you can skip to the next section).

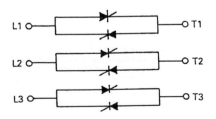

FIGURE   18.4.21 Six-SCR   configuration. (*Courtesy of The Trane Company.*)

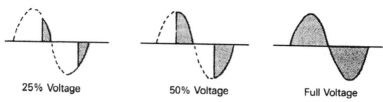

25% Voltage   50% Voltage   Full Voltage

FIGURE 18.4.22   Wave forms. (*Courtesy of The Trane Company.*)

The low pressure chiller you have was designed when CFC-11 cost less than $1.00 per pound. By the late 1980s and early 1990s CFC-11 approached $10 per pound. As you read this, it could be $15, $20, or more per pound (or even less). Refrigerant prices have been unstable and could vary up or down based on demand and supply. If you have lost an entire refrigerant charge the following refrigerant containment devices could pay for themselves many times over:

- High-efficiency purge unit
- Vacuum prevention
- Rupture disk relief valve retrofit
- Leak detection system
- External oil filter
- Isolation valves on oil and refrigerant filters

If you are interested in additional upgrades or if you have a problem or situation that was not addressed in these sections, consult your chiller's manufacturer or a qualified independent mechanical service contractor.

18.4.17 REPAIR VERSUS REPLACEMENT DECISIONS

Deciding whether to repair or replace a centrifugal chiller or any expensive piece of capital equipment can often be a difficult choice. We have assembled the following list of potential criteria used in evaluating repairing or replacing a centrifugal chiller, or any other large expensive piece of capital equipment. This list of criteria is not necessarily all inclusive; based on your specific situation any or all of the following criteria should be considered when faced with these decisions:

Age   Make and model
Type and amount of refrigerant   Downtime costs and risks
Capacity   Efficiency
Present and future load requirements   Hours of service/run time
   comfort vs. process   Primary vs. backup
   redundancy   Safety
Internal rate of return   Reliability

Budget needs, availability of capital

Equipment history

   compressor

   starter

   motor

   bearings

   tubes

Present machine condition

   compressor

   starter

   motor

   bearings

   tubes

Repair parts costs

Repair parts availability

Noise level

Access/space limitations

Cooling tower capacity

Expected operating life of equipment

Codes, laws, and requirements (EPA, etc.)

You may want to add additional criteria for your particular situation. There are more resources to help you in your repair/replace decisions. You can consult the manufacturer of your equipment, a trusted mechanical consultant, internal technical experts, independent mechanical service contractors, mechanical contractors, controls companies, and even your boiler and machinery insurance carrier.

If you choose the repair option, it is straightforward. You just repair the chiller and while you're at it, it may be a good time to convert your chiller to a new refrigerant, or add performance-enhancing components. If you choose the replace option, the choices are more numerous. You can choose an exact replacement, you can redesign your entire chilled water plant, or something in between.

The basis of this technical evaluation is to quantify and justify the costs of each decision. These costs can be divided into two areas: capital expenditures and operating expenses. In most organizations, replacing a piece of equipment would fall under the capital expenditure, while servicing and repairing a piece of equipment would be categorized as an operating expense and come out of a separate budget.

If there is ample capital and operating expenses are tight, an investment in a new chiller could reduce operating expenses because a new chiller comes with a warranty. Repairs will cost very little the first 5 to 10 years. Also, energy expenses will drop because a new chiller is more efficient. Conversely, if capital dollars are tight and operating dollars are easy to come by, you could increase maintenance and add energy saving upgrades to extend equipment life and avoid or delay a capital expenditure. I have several clients that have successfully chosen one path or the other.

There is a third option in addition to the traditional choice between repairing or replacing a chiller: leasing, and for certain clients this may be a viable alternative. If the cooling need is seasonal, capital spending is being reduced, the heat load increases or decreases faster than permanent equipment can be delivered, or there is a lack of maintenance personnel, leasing or renting a chiller may be the most cost-effective path. It may seem ludicrous at first glance, but renting a chiller or even an entire chilled water plant may be a financial tool usable to solve a technical problem.

Funding a chiller replacement is another matter. A creative option that helps with expensive retrofits is a concept called *performance contracting*. In a performance-based contract, a chiller retrofit can be paid for out of the energy savings created by replacing or upgrading the lighting, controls, and other mechanical equipment. Some performance contracts even guarantee the energy savings so there is little or no risk for the owner. Several HVAC controls and chiller manufacturers, energy service companies, and mechanical contractors offer this type of program.

ACKNOWLEDGMENTS

The author acknowledges the help of the following individuals: Wesley Wojdon and Ken Mozek of York International; Dave Bishop and John Bauernfeind of Johnson Controls; and Cristi Johnson and Todd Elmgren of The Trane Company.

18.5 SCREW COMPRESSORS

Kenneth Puetzer
*Chief Engineer, Sullair Refrigeration, Subsidiary
of Sundstrand Corp., Michigan City, Indiana*

18.5.1 INTRODUCTION

18.5.1.1 History

The screw compressor was originally developed in Germany in the late 1800s, but was not commercially applied until the mid-1900s. Modern screw compressors for the refrigeration and air-conditioning industry use an oil-flooded arrangement.

18.5.1.2 Design

The screw compressor is a positive-displacement compressor that uses a rotor driving another rotor (twin) or gaterotors (single) to provide the compression cycle. Current designs consist of both the twin screw (Fig. 18.5.1) and the single screw (Fig. 18.5.2). Both designs use injected fluids to cool the compressed gas, seal the rotor(s), and lubricate the bearings. Compressor designs may incorporate an internal valve for capacity control, economizer ports for improved performance, and an internal valve for variation of internal volume ratio.

A typical screw compressor package consists of the compressor, driver, oil separator and reservoir, oil cooling, control panel, and valving (Fig. 18.5.3).

18.5.2 TWIN-SCREW COMPRESSORS

18.5.2.1 Design

The most common type of screw compressor used today is the twin screw. As the name implies, it uses a double set of rotors (male and female) to accomplish the compression cycle. The male rotor usually has four lobes, while the female rotor consist of six lobes; this is normally referred to as a 4 + 6 arrangement. However, some compressors, especially for air-conditioning applications, are using other variations, such as 5 + 7.

Until the mid-1960s, the rotors were cut using a symmetrical or circular profile. This was replaced by the asymmetrical profile, which is a line-generated profile, and which improved the adiabatic efficiency of the screw compressor.

18.5.2.2 Compression Process

The compression process starts with the rotors meshed at the inlet port of the compressor (Fig. 18.5.4). As the rotors turn, the lobes separate at the inlet port, increasing the volume between the lobes. This increased volume causes a reduction in pressure, thus drawing in

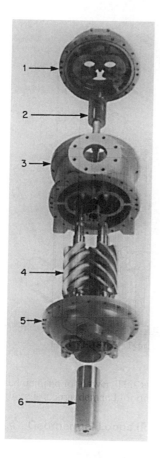

FIGURE 18.5.1 Twin-screw compressor. (1) Discharge housing; (2) slide valve; (3) stator; (4) male and female rotors; (5) inlet housing; (6) hydraulic capacity control cylinder. (*Courtesy of Sullair Refrigeration.*)

the refrigerant gas. The intake cycle is completed when the lobe has turned far enough to be sealed off from the inlet port. As the lobe continues to turn, the volume trapped in the lobe between the meshing point of the rotors, the discharge housing, and the stator and rotors, is continuously decreased. When the rotor turns far enough, the lobe opens to the discharge port, allowing the gas to leave the compressor.

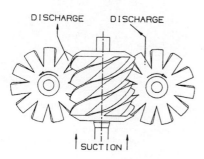

FIGURE 18.5.2 Single-screw compressor. (*Courtesy of Single Screw Compressor, Inc.*)

FIGURE 18.5.3   Screw compressor package. (1) Inlet check valve and strainer; (2) compressor; (3) filter; (4) capacity control actuator; (5) oil cooler; (6) oil separator; (7) discharge connection. (*Courtesy of Sullair Refrigeration.*)

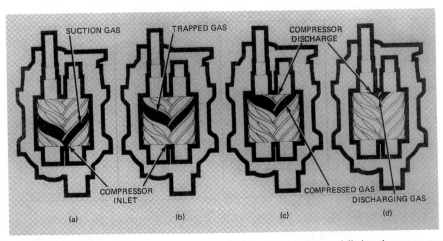

FIGURE 18.5.4   Twin-screw compression cycle. (*a*) Refrigerant gas is drawn axially into the compressor rotors as they turn past the intake port in the housing. (*b*) The rotors have turned past the intake port; gas is trapped in the compressor housing and rotor cavities. (*c*) As the rotors continue to turn, the lobes reduce the volume in the cavities to compress the trapped gas. (*d*) The process is completed as the compressed gas is discharged through the discharge port. (*Courtesy of Sullair Refrigeration.*)

The gas is continuously compressed until the lobes are totally meshed. This eliminates the undesirable condition in reciprocating compressors in which the gas in the clearance volume between the piston and the top of the cylinder reexpands within the cylinder, resulting in reduced volumetric efficiency and an increase in horsepower (wattage).

The ratio of the volume of gas trapped after the intake cycle to the volume of gas trapped just before the lobe opens to the discharge port is known as the *built-in volume ratio VI*. With the injection oil performing a majority of the cooling and with very low pressure differential across each lobe, the screw compressor can reach compression ratios as high as 20:1.

18.5.2.3 Lubrication System

The current refrigeration twin-screw compressor is designed with a relatively large oil flow of up to 100 gal/min (6.3 L/s) for large compressors. This oil flow serves three functions:

- Lubrication of the bearings, gears, and shaft seal
- Cooling of the gas stream
- Formation of a lubricating film between the rotors

The twin-screw compressor requires bearings to handle both the thrust loads and the radial loads. These bearings are usually a combination of either (1) angular contact ball bearings or tapered roller bearings for the thrust loads and (2) sleeve bearings or straight roller bearings for the radial loads. The thrust bearings are designed to carry only the thrust loads in both directions, while the radial loads are carried by the radial bearings only. Some of the smaller compressors use a set of gears to increase the speed of the drive rotor. These gears are also lubricated by this oil.

During the compression cycle, the refrigerant gas picks up heat. The injected oil cools this gas and allows the compressor to operate under the high compression ratios. Even under the high compression ratios, the discharge temperature of a screw compressor seldom exceeds 200°F (93°C), with normal temperatures running around 160 to 180°F (71 to 82°C).

The oil also forms a lubricant film between the two rotors to allow the drive rotor to turn the driven rotor without metal-to-metal contact. The majority of modern compressors use the male rotor to drive the female rotor. However, some of the newer small compressors are using the female rotor to drive the male rotor.

With this type of lubrication system, the compressor package requires a complete lubrication system (Fig. 18.5.5). The system consists of a gas/oil separator and reservoir, an oil-cooling system, an optional oil pump, and an oil distribution system. Since all this oil is injected directly into the gas stream, an efficient oil separation system must be installed.

The oil separation systems are normally designed for oil carryover to the system of less than 20 ppm (20 mg/L) under steady-state operating conditions. However, where the compressor package is built as a complete chiller package, the compressor may be designed for larger carryovers and allow the oil to circulate throughout the system and return to the compressor.

The separator (Fig. 18.5.6) is normally designed to accomplish the separation by several of the following methods:

- Changes in direction of flow
- Reduction in velocity
- Centrifugal or mesh pad devices
- Coalescing materials for removal of fine mist

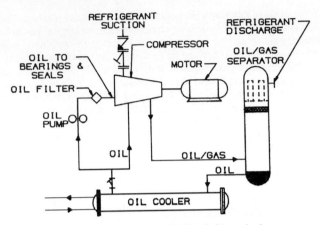

FIGURE 18.5.5   Lubrication system. (*Sullair Refrigeration.*)

The oil pump is used to overcome line and cooler pressure losses and/or to boost the oil pressure higher than the discharge pressure of the compressor. With sleeve bearings, this oil pump is normally required. However, with roller bearings, this pump may be eliminated if there is enough pressure to push the oil into the compressor and if the compressor is designed for the lower oil pressure.

The oil-cooling system is used to remove a major portion of the heat of compression. The amount of heat removed by the oil-cooling system depends on the refrigerant and the compression ratio but can be as high as 40 percent of the total heat rejected from the refrigeration system. Depending on the type of cooling system, this heat may be subtracted from the condenser's heat load.

18.5.2.4   Types of Cooling

The most common forms of cooling the screw compressor are:

- Water or glycol cooling in a heat exchanger
- Refrigerant thermosiphon cooling in a heat exchanger
- Direct injection of refrigerant into the compressor
- Direct-expansion cooling in a heat exchanger

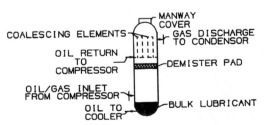

FIGURE 18.5.6   Gas/oil separator—breakdown and part description. (*Sullair Refrigeration.*)

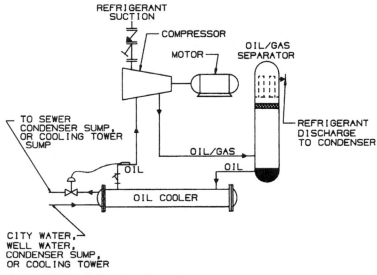

FIGURE 18.5.7 Open type cooling system.

The most common form of oil cooling uses a shell-and-tube heat exchanger with water or glycol in the tubes to remove the heat. The water can be supplied from either an open system (Fig. 18.5.7) or a closed system (Fig. 18.5.8). With the open system, the water can be taken from the normal plant water supply, from the sump of an evaporative condenser, or from a cooling tower. With the closed system, the water or glycol can be taken from an

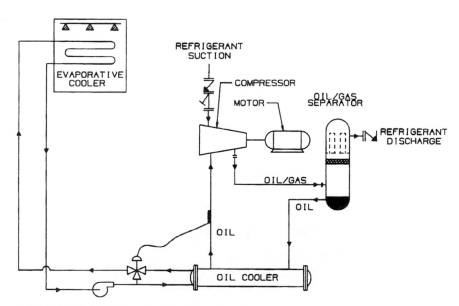

FIGURE 18.5.8 Closed circuit cooling system.

air-cooled heat exchanger, a closed-circuit evaporative cooler, or another heat exchanger that may be used to recover this heat for heating potable water, for boiler makeup, or for building heat. Due to the poor heat transfer, low oil flow, and large oil-temperature drop, the oil cooler is usually designed with outerfin tubes in order to enhance the heat transfer. The design is usually based on a tube-side (water-side) fouling of 0.0005 ft^2 · °F · h/Btu (3.69 cm^2 · °C · s/cal). Therefore, care must be taken to avoid fouling of the oil cooler. The heat removed in the oil cooler can be subtracted from the total refrigeration heat rejection when sizing the condenser.

If a source of clean water is not available or is too expensive, the cooling fluid can be replaced with refrigerant. Some applications have used a direct-expansion cooler and injected the evaporated gas into the compressor at either the suction port or an economizer port (Fig. 18.5.9). This procedure will add to the horsepower (wattage) of the compressor and will decrease the capacity if injected into the suction of the compressor. The condenser must be sized for the entire refrigeration heat of rejection plus the added horsepower (wattage).

The thermosiphon system (Fig. 18.5.10) uses the liquid from the condenser to flood the cooler and return the evaporated gas directly to the condenser. This procedure has no effect on the compressor's performance, but the entire refrigeration heat of rejection must be removed in the condenser.

The direct liquid-refrigerant injection system (Fig. 18.5.11) is used quite often on industrial refrigeration systems using ammonia and packaged freon/screw compressor chillers,

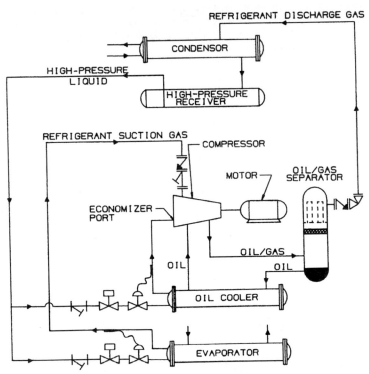

FIGURE 18.5.9   Direct-expansion cooling system. (*Sullair Refrigeration.*)

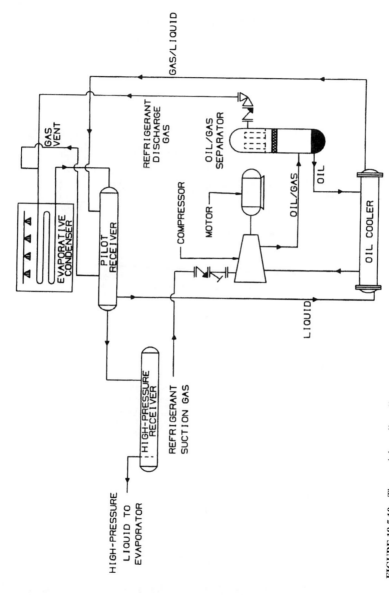

FIGURE 18.5.10 Thermosiphon oil-cooling system. (*Sullair Refrigeration.*)

18.133

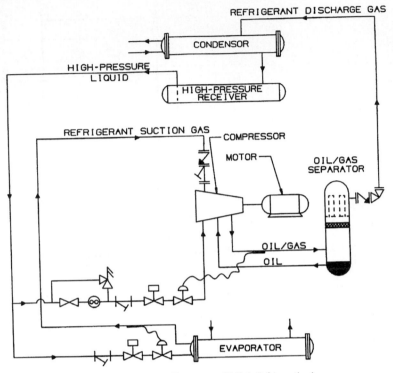

FIGURE 18.5.11   Liquid-injection cooling system. (*Sullair Refrigeration.*)

but has seen very limited use in industrial freon systems. This system injects liquid refrigerant directly into the compressor at a point where the internal pressure is below the discharge pressure. The liquid is boiled off, cooling the gas-and-oil stream in the compressor. The liquid flow is controlled by modulating a liquid valve in response to the discharge temperature. Discharge temperatures are usually either kept around 120°F (49°C) or 20°F (11°C) above the condensing temperature, whichever is higher. This method has a slight effect on compressor capacity and results in a slight increase in compressor horsepower (wattage). The entire refrigeration heat rejection must be removed in the condenser. The main problem with this type of cooling in halocarbon systems is that the halocarbon gas is very soluble in the refrigerant oils; thus, with the low temperatures of the oil and gas and with the high gas solubility, the oil viscosity is usually reduced below the minimum viscosity level required by the compressor.

18.5.2.5 Capacity

The screw compressor has been built with R-22 capacities ranging from 50 to 1600 tons (176 to 5626 kW) when operating at 35°F (1.7°C) evaporating temperature and 105°F (41°C) condensing temperature. Typical performance curves are shown in Fig. 18.5.12. When comparing data from different manufacturers, one must make sure that the ratings are based on

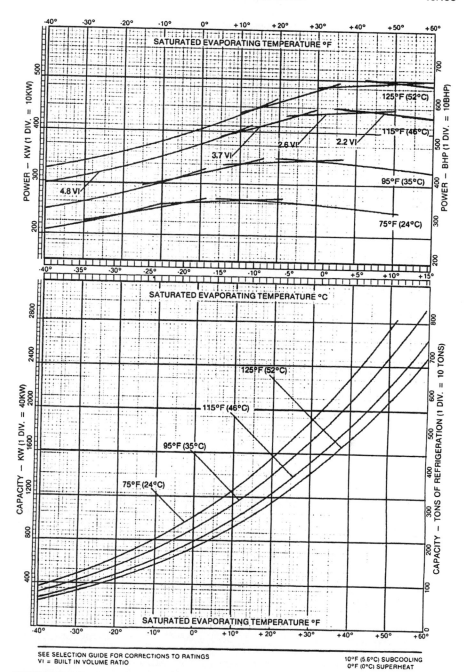

SEE SELECTION GUIDE FOR CORRECTIONS TO RATINGS
VI = BUILT IN VOLUME RATIO

10°F (5.6°C) SUBCOOLING
0°F (0°C) SUPERHEAT

FIGURE 18.5.12 Typical R-22 performance curves.

or corrected to the same rating conditions. Most manufacturers include correction procedures in their manuals. These correction factors consist of one or more of the following:

1. *Degrees of subcooling (C1).* Subcooling of the liquid below the condensing temperature must be accomplished by using a cooling medium other than the refrigerant listed
   in item 2 below.

2. *Degrees of superheating (C2).* The superheat can be either useful (if it is accomplished
   in the evaporator or by subcooling the liquid) or nonuseful (if the superheating occurs
   as a result of heat loss to the atmosphere or if the superheat occurs by cooling some other
   fluid).

3. *Suction pressure drop (C3) and discharge pressure drop (C4).* Most screw compressors include an isolation valve, a check valve, and a strainer in the suction line, and a
   discharge check-valve, an oil separator and a discharge stop valve. These could amount
   to a pressure drop of several psi (kPa). Many compressor manufacturers rate their compressors based on the conditions at the compressor flanges rather than from the suction-
   valve flange to the discharge-valve flange after the separator.

4. *Speed (C5).* For speed variations less than 20 percent of the rated speed, both the capacity and the horsepower (wattage) can be ratioed directly to that speed on which the ratings are based.

Typical correction factors for the above are shown in Table 18.5.1. These corrections
cover the majority of the required corrections, but the manufacturer's manual should be
reviewed for any additional corrections.

As an example of a typical selection, we require a compressor operating with R-22 and
producing 475 tons (1670 kW) when running at 35°F (1.7°C) suction and 115°F (46°C)
condensing with 20°F (11°C) subcooling and 10°F (5.6°C) superheating, which is accomplished in the evaporator. Because of the motor requirements, the compressor will run at
only 3500 rpm. Read up from the 35°F (1.7°C) saturated evaporating temperature line on
the bottom curve of Fig. 18.5.12 to the 115°F (46°C) curve; at the intersection with the
curve, read to the right a capacity of 480 tons (or to the left a capacity of 1688 kW). The
vertical line can then be continued up to the upper curve where it intersects the 115°F
(46°C) horsepower curve; at the intersection with the curve, read to the right a power
requirement of 582 BHP (or to the left a power of 434 kW).

As the design parameters stated, the system will be designed for 20°F (11.1°C) subcooling. From Fig. 18.5.12, the compressor is rated with 10°F (5.6°C) subcooling. From Table
18.5.1, the correction factor is $0.005 \times °F$ ($0.009 \times °C$); therefore, the capacity will be corrected by a factor of $1 + 0.005 \cdot (20°F − 10°F) = 1.05$ [or $1 + 0.09 (11.1°C − 5.6°C) = 1.05$].
The design also calls for a useful superheat of 10°F (5.6°C). Again from Table 18.5.1, the
correction factor is −0.0005/°F (−0.0009/°C); therefore, the capacity will be corrected by a
factor of $1 + (−0.0005 \times 10°F) = 0.995$ [$1 + (0.0009 \times 5.6°C) = 0.995$].

TABLE 18.5.1   Typical R-22 Correction Factors

Description	Value	Applies to
Liquid subcooling (C1)	0.0050/°F (0.009/°C)	Capacity
Suction superheat (C2)		
Useful	−0.0005/°F (−0.0009/°C)	Capacity
Nonuseful	−0.0028/°F (−0.0050/°C)	Capacity
Suction pressure drop (C3)	0.975	Capacity
Discharge pressure drop (C4)	1.02	Capacity and BHP

The curve in Fig. 18.5.12 is based on pressures at the compressor flanges. Therefore, the suction pressure-drop correction of 0.975 and the discharge pressure-drop correction of 1.02 in Table 18.5.1 must be included. The compressor performance will then be as follows:

$$\text{Capacity} = 480 \times 1.05 \times 0.995 \times 0.975 \times 1.02 \times \frac{3500 \text{ rpm}}{3550 \text{ rpm}} = 491.7 \text{ tons}$$

$$(\text{Capacity} = 1688 \times 1.05 \times 0.995 \times 0.975 \times 1.02 \times \frac{3500 \text{ rpm}}{3550 \text{ rpm}} = 1729.1 \text{ kW})$$

$$\text{Power} = 582 \times 1.02 \times \frac{3500 \text{ rpm}}{3550 \text{ rpm}} = 585.3 \text{ BHP}$$

$$(\text{Power} = 434 \times 1.02 \times \frac{3500 \text{ rpm}}{3550 \text{ rpm}} = 436.4 \text{ kW})$$

The horsepower curve shows the horsepower at different operating conditions when using different built-in volume ratios (VI). The curves are drawn using the VI that will give the lowest operating horsepower for a given set of conditions. Since the screw compressor is a fixed-volume-ratio compressor, the adiabatic efficiency will peak when the compression ratio matches the internal built-in volume ratio. Figure 18.5.13 shows a typical R-22 performance curve with adiabatic efficiencies plotted for different VI. The peak adiabatic efficiency occurs at a pressure ratio as calculated in Eq. (18.5.1):

$$PR = (VI)^k \tag{18.5.1}$$

where PR = ideal pressure ratio
VI = built-in volume ratio
k = isentropic exponent

At this point the gas in the compressor will be compressed to a pressure internally (before the lobe opens to the discharge port) that exactly matches the pressure in the discharge line (Fig. 18.5.14, curve 1). As the lobe opens to the discharge port, the compressor will continue

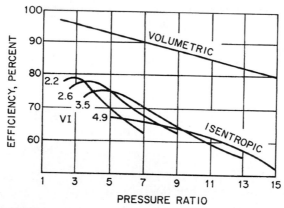

FIGURE 18.5.13 R-22 screw compressor efficiency curves. (*ASHRAE.*)

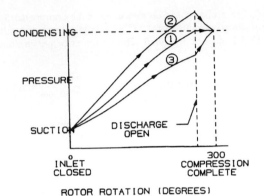

FIGURE 18.5.14 Internal compression curves. (1) Ideal built-in volume ratio (VI); (2) high VI—overcompression; (3) low VI—undercompression. (*Courtesy of Sullair Refrigeration.*)

compression enough to push the gas into the discharge line. When operating off of this peak-efficiency point, the compressor will either be overcompressing before the lobe opens to discharge or undercompressing (Fig. 18.5.14). When overcompressing (curve 2), the gas will be compressed to a pressure internally (before the lobe opens to the discharge line) that is actually higher than the condensing pressure. When this occurs, the gas will create a pulse going into the oil separator when the lobe opens to the discharge line and very little additional compression will occur. Efficiency is lost because the compressor must exert work to compress the gas to a pressure higher than that required by the system. When undercompressing (curve 3), the gas will be compressed to a pressure internally (before the lobe opens to the discharge line) that is actually less than the condensing pressure. When this occurs, the gas will create a pulse going back into the compressor when the lobe opens to the discharge line. This gas, which flows back into the compressor, must be recompressed to the discharge line pressure.

When selecting a compressor with, say, design operating conditions of 35°F (1.7°C) evaporating and 115°F (46°C) condensing, the designer may select a compressor with a built-in volume ratio of 2.6 per the curve in Fig. 18.5.12 in order to give the lowest horsepower at design conditions. However, if the majority of the time the system would be operating at 95°F (35°C) condensing, then a better selection may be a 2.2 VI. This decision must be made by the designer working with the compressor manufacturer.

The last items that must be calculated are the condenser heat load (CHL) and the oilcooler heat load (OCHL). The total heat load (THL) for a screw compressor is calculated by using Eq. (18.5.2):

$$\text{THL} = \text{capacity} \times 12{,}000 + \text{power} \times 2545 \ (\text{Btu/h})$$
$$= \text{capacity} + \text{power} \ (\text{kW})$$

$$(18.5.2)$$

where THL = total heat rejection, tons (kW)
    capacity = compressor capacity, tons (kW)
    power = compressor power, BHP (kW)

If the compressor is using liquid injection or thermosiphon oil cooling, then the CHL is equal to the THL calculated by Eq. (18.5.2).

However, if the compressor is using water or glycol cooling, then OCHL can be deducted from the THL to calculate the CHL. The OCHL is calculated by Eq. (18.5.3):

$$OCHL = THL \times OCHLM \qquad (18.5.3)$$

where OCHL = oil-cooler heat load, Btu/h (kW)
 OCHLM = oil-cooler heat-load multiplier

The OCHLM is dependent on oil flows and on the compressor manufacturer's design parameters; therefore, these multipliers must be obtained from the manufacturer. A typical curve for these multipliers for an R-22 compressor is shown in Fig. 18.5.15. For our example, we would enter the curve at the 35°F (1.7°C) saturated evaporating temperature and draw a straight line up to the 115°F (46°C) saturated condensing temperature curve. At the intersection with this curve, we draw a horizontal line across and read 4.5 percent for the OCHLM. The heat loads are then calculated as follows:

$$THL = 491.7 \cdot 12,000 + 585.3 \times 2545 = 7,389,989 \text{ Btu/h}$$

$$OCHL = 7,389,989 \times 0.045 = 332,550 \text{ Btu/h}$$

$$CHL = 7,389,989 - 332,550 = 7,057,440 \text{ Btu/h}$$

or in kilowatts:

$$THL = 1729.2 + 436.4 = 2165.6 \text{ kW}$$

$$OCHL = 2165.6 \times 0.045 = 97.5 \text{ kW}$$

$$CHL = 2165.6 - 97.5 = 2068.1 \text{ kW}$$

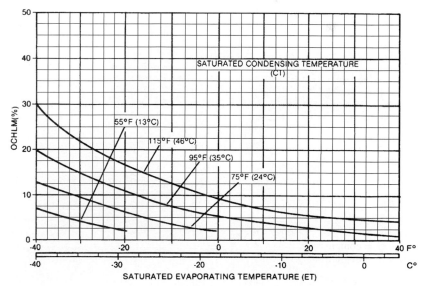

FIGURE 18.5.15   Oil-cooler heat-load multiplier. (*Courtesy of Sullair Refrigeration.*)

18.5.2.6 Capacity Control

An important advantage of screw compressors is the ability to vary compressor capacity as the load varies. This capacity reduction is accomplished in several ways:

1. Internal slide valve
2. Internal turn valve
3. Plug valves
4. Suction throttling
5. Variable speed

The internal slide valve (Fig. 18.5.16) is the most common form of capacity control for the larger compressors. The slide valve is usually constructed by replacing a portion of the stator with a valve shaped identical to the stator (Fig. 18.5.16). When the compressor is first started, the slide valve is located at minimum position. As the compressor loads, the slide valve is moved toward the inlet until it is against the stator when the compressor is at full load.

In order to control the position of this slide valve, a hydraulic piston (Fig. 18.5.17) is used. When high-pressure oil is drained through port B back to suction, the discharge gas will expert a higher pressure on the end of the slide valve than the suction gas exerts on the other end. This pressure differential causes the slide valve to move to the left, closing the gap between the stator and the slide valve and increasing the compressor capacity. To unload the compressor, high-pressure oil is injected behind the piston through port A. The resulting pressure differential across the piston overcomes the differential across the slide valve and drives the slide valve to the right. This allows more gas to flow from the rotors back to suction and less to the discharge port. A typical part-load curve for a screw compressor is shown in Fig. 18.5.18. This curve is based on a constant condensing temperature.

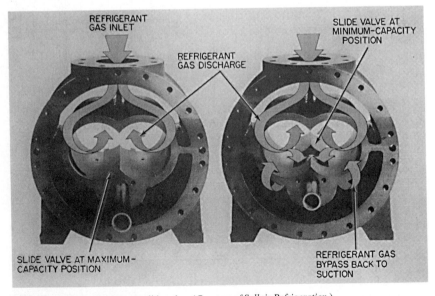

FIGURE 18.5.16   Compressor slide valve. (*Courtesy of Sullair Refrigeration.*)

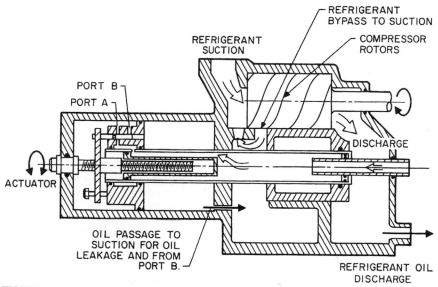

FIGURE 18.5.17 Internal valve and piston. (*Courtesy of Sullair Refrigeration.*)

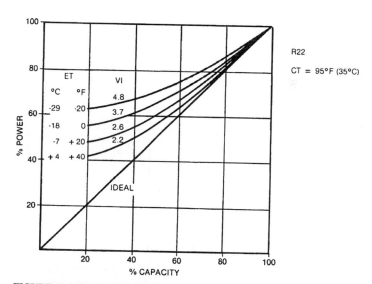

FIGURE 18.5.18 Part-load curves with constant condensing. CT = condensing temperature; ET = evaporating temperature; VI = volume ratio; IT = intermediate temperature. (*Courtesy of Sullair Refrigeration.*)

Figure 18.5.19 shows a part-load curve based on the condensing temperature being reduced in accordance with ARI Standard 550-77.

In order to provide ease of starting, the compressor should be started at minimum position. On a normal shutdown, the compressor will always be at minimum position. However, after an abnormal or manual shutdown, the compressor may not be at minimum position and must be moved to minimum position before restarting the compressor. This is accomplished by using a motor, a spring, or an external source of high oil pressure to drive the slide valve back to minimum position.

As screw compressors became smaller, cheaper means of performing capacity control were required. This introduced the internal turn-valve design. In this design, the stator is machined with slots to allow the gas to bypass back to suction. Also included in an internal turn valve, which is located in the stator just below these slots. The turn valve consists of a hollow cylinder (Fig. 18.5.20), which has slots machined in the surface. As the valve is rotated, the machined-out areas expose more of the slots in the stator and allow the gas to flow back to the suction side of the compressor. The turn valve is rotated by a motor, which can be driven electrically, hydraulically, or pneumatically.

Some of the newer small screw compressors are using a plug valve for capacity control. The plug valve consists of a piston which forms a portion of the stator when loaded

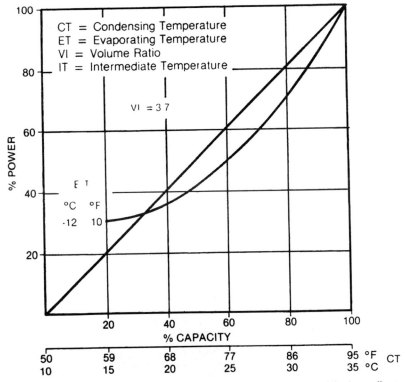

FIGURE 18.5.19 Part-load curves with variable condensing. Variable CT with load according to ARI Standard 550-77-5.1.6. (*Courtesy of Sullair Refrigeration.*)

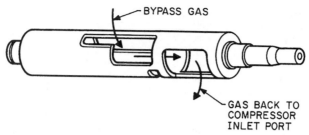

FIGURE 18.5.20 Internal turn-valve unloader. (*Courtesy of Sullair Refrigeration.*)

and is removed to create a hole in the unloaded position. The piston must be machined to the same shape and tolerance as the stator. Most applications use multiple plugs to give step-type unloading. The plugs can be moved either by hydraulics or by electric actuators.

Other screw compressors simply install a butterfly valve in the suction line. Under full load, the valve is wide open and, as the system calls for capacity reduction, the valve is rotated toward the closed position. As the valve closes, it creates a pressure drop, which increases the specific volume of the gas and decreases the mass flow to the compressor.

With the advances in solid-state electrical components, variable-speed drives are becoming less expensive and more efficient. As these advances (along with larger sizes) are made commercially available, they are being applied to the screw compressor as a means of capacity control.

The type of capacity control used is dependent on first cost, operating cost, size of compressor, and system operation. Therefore, the designer should work with the compressor manufacturers to determine the best choice for this application.

18.5.2.7 Variable Volume Ratio

Until the mid-1980s, the screw compressor was built with a fixed built-in volume ratio. This required that the compressor be picked with the correct built-in volume ratio for the design and normal operating conditions. For the industrial market, this created only a couple of percentage points of horsepower losses at full load when operating the compressor at conditions that differed from the ideal pressure ratio, and this could be almost eliminated by proper selection of the built-in volume ratio. The ideal pressure ratio is related to the built-in volume ratio by Eq. (18.5.1).

With increased flexibility required in industrial refrigeration systems, air-cooled screw compressor applications, and ice storage systems, a means to change the built-in volume ratio was required.

The usual method to supply the variable-volume-ratio compressor is to use a split slide valve (Fig. 18.5.21). However, designs using dual slide valves and plug valves have been designed. In order to decrease the built-in volume ratio, both valves are moved toward the inlet end increasing the volume between the lobes at the inlet end, but not changing the volume at the discharge end. When reducing the capacity, the slide valve is moved toward the discharge end while the slide stop remains fixed or is moved towards the inlet end. Under part-load conditions the slide stop and/or the slide valve may be moved in order to change the built-in volume ratio and the compressor capacity. This will be determined by

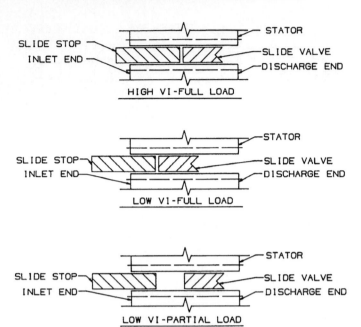

FIGURE 18.5.21   Variable-volume-ratio slide valves. (*Courtesy of Sullair Refrigeration.*)

the manufacturer, depending on the valve lengths and type of control the manufacturer is looking for. In order to know where the build-in volume ratio should be set, the compressor must constantly monitor the suction and discharge pressures and the current built-in volume ratio and capacity along with the refrigerant k value. This type of data monitoring requires the use of a microprocessor for control. If automatic control is not required, the slide stop could be manually set if seasonal changes or mode-of-operation changes are the only changes required.

18.5.2.8   Economizer

With the capability of cooling the gas during compression, the screw compressor can operate at compression ratios as high as 20:1 without having excess discharge temperatures. However, these pressure ratios do reduce the performance considerably. In order to overcome this, the screw compressor has been built with an economizer port. This port allows refrigerant gas to be injected into the compressor after compression has been partially performed. This gas could come from a heat exchanger used to subcool the liquid going to the main evaporator (Fig. 18.5.22), from a flash-type subcooler, or from an external load. A typical curve indicating capacity and power multipliers for an economizer system on an R-22 compressor is shown in Fig. 18.5.23. For an R-22 compressor with a 2.6 VI operating at 30°F (−1°C) evaporating temperature and 115°F (46°C) condensing temperature, the compressor capacity would be multiplied by 1.20 and the power would be multiplied by 1.07 to obtain the performance with an economizer.

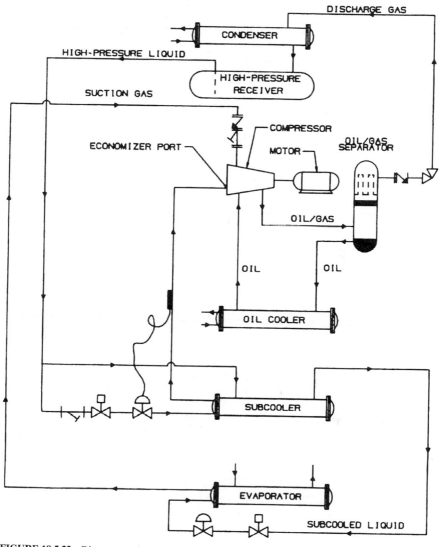

FIGURE 18.5.22 Direct-expansion economizer cycle. (*Courtesy of Sullair Refrigeration.*)

As the compressor unloads to approximately 75 percent, this economizer port pressure will reduce and will equal the suction pressure. Therefore, the system designer must make sure that, if this is a problem, proper precautions are taken, such as using a back-pressure regulator, using a shell-and-tube heat exchanger, or shutting off the economizer port. Deciding what type of system to use will depend on the refrigeration system design, and the designer should work with the compressor manufacturer to decide which is best for the system.

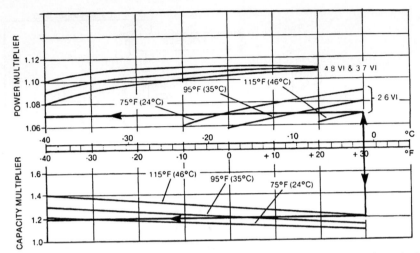

FIGURE 18.5.23   R-22 economizer capcity and power multipliers. (*Courtesy of Sullair Refrigeration.*)

18.5.3 SINGLE-SCREW COMPRESSORS

18.5.3.1 Development

The single screw, or monoscrew, was originally developed in the early 1960s, but was not applied to the refrigeration and air-conditioning industries until 10 years later. Today it is applied mostly to the air-conditioning market, with very little application in industrial refrigeration.

18.5.3.2 Design

As the name implies, the single-screw consists of one main rotor working with a pair of gaterotors (Fig. 18.5.2). The main rotor typically has six helical flutes and a globoid (or hourglass) root profile. The gaterotors each have 11 teeth and are located on opposite sides of the main rotor.

18.5.3.3 Compression Process

The compression process (Fig. 18.5.24) is very similar to the twin screw. As the main rotor rotates, one of the teeth of the gaterotor engages a flute (a) and draws a reduced pressure behind the tooth and within the flute. This reduced pressure draws gas through the suction connection of the compressor.

As the main rotor continues to rotate, it engages a tooth from the gaterotor on the opposite side (b). The gas is now trapped within the flute, the casing wall, and the tooth of the gaterotor. Compression starts and continues until the flute opens to the discharge port (c).

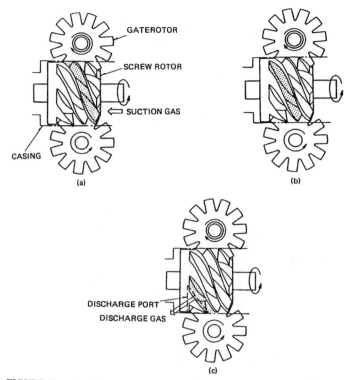

FIGURE 18.5.24 Monoscrew compression cycle. (*a*) Suction; (*b*) compression; (*c*) discharge. (*Courtesy of Single Screw Compressor, Inc.*)

Gas is continuously pushed out of the flute until the volume within the flute and tooth of the gaterotor is zero. As with the twin screw, the volume of gas left in the flute for reexpansion to suction is negligible.

18.5.3.4   Lubrication and Cooling

The lubrication system and means of cooling are identical to the systems outlined under the twin screw.

18.5.3.5   Capacity Control

Capacity control methods for the single screw are basically the same as for the twin screw. However, the geometry of the single screw allows two compression processes to occur simultaneously. The unloading mechanism consists of dual slide valves which are hydraulically activated and which retard the point at which the compression process begins (Fig. 18.5.25).

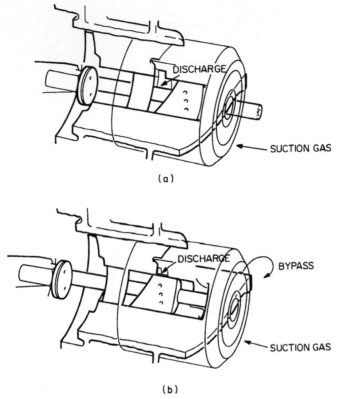

FIGURE 18.5.25   Single-screw axial-type capacity control mechanism. (*a*) Slide valve fully in loaded position; (*b*) slide valve in part-load position. (*Courtesy of Single Screw Compressor, Inc.*)

In addition to the above methods of capacity control, the single screw can also use a rotating ring (Fig. 18.5.26) that is built into the stator housing and allows gas to flow from the compression chamber back to suction through the stator housing.

18.5.3.6 Economizer

Like the twin screw, the single screw can incorporate an economizer cycle. However, the single screw, with dual slide valves, can have one valve operating at minimum position while the second valve is operating at near maximum. This allows the economizer to operate at a lower percent capacity before the port opens up to suction pressure.

18.5.4   SEMIHERMETIC SCREW COMPRESSORS

Both the twin screw and the single screw are available in semihermetic designs (Fig. 18.5.27). With the low discharge temperatures that are inherent with screw compressors, the motor

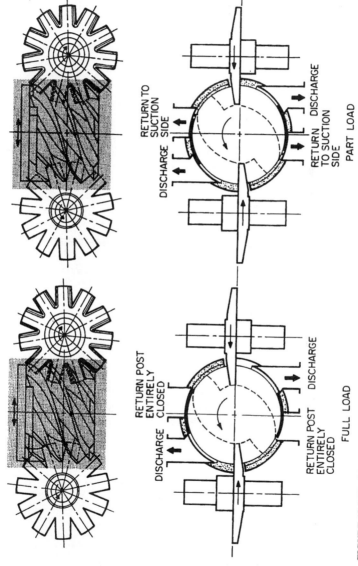

FIGURE 18.5.26 Monoscrew rotary-valve capacity control. (*Courtesy of ASHRAE.*)

RETURN TO SUCTION SIDE

DISCHARGE

RETURN DISCHARGE TO SUCTION SIDE

PART LOAD

RETURN POST ENTIRELY CLOSED

DISCHARGE

RETURN POST ENTIRELY CLOSED

DISCHARGE

FULL LOAD

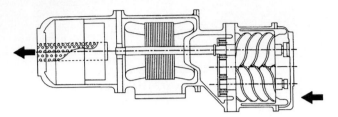

FIGURE 18.5.27   Semihermetic screw compressor.

can be designed to be cooled by suction gas, discharge gas, intermediate gas to the economizer port, or liquid injection. Most semihermetic screw compressors are of the smaller
compressor design, and the oil separator is incorporated with the compressor and motor.

18.6   HEAT PUMPS

Kenton J. Kuffner
WaterFurnace International, Fort Wayne, Indiana

Billy C. Langley, Ed. D, CM
Consulting Engineer, Azle, Texas

PART A: GROUND SOURCE

18.6.1   INTRODUCTION

18.6.1.1   Overall Concept

A ground source heat pump system consists of a water-to-air or water-to-water heat pump, connected to a series of long plastic pipe buried below the earth's surface, or placed in a pond. These systems can also utilize well water in place of the earth loop. As fluid from the earth loop or well water is moved through the unit, the heat pump transfers thermal energy that heats or cools the home or building.

18.6.1.2   History

The first recorded geothermal system was a 1912 Swiss patent. Geothermal systems using well water have been in use since the 1930s. The Edison Electric Institute sponsored closed loop research in the 1940s and 1950s, but the lack of suitable piping materials slowed interest. Swedish researchers began investigating geothermal closed loop systems in the 1970s with the advent of plastic pipe, which was suitable for the application. Installations of the system increased due to the fuel crisis at the time. The International Ground Source Heat Pump Association (IGSHPA) was founded in 1987 at Oklahoma State University to study the applications and establish engineering guidelines. The Geothermal Heat Pump Consortium, a nonprofit group of geothermal equipment manufacturers, electric utilities, and contractors was established in 1994 to promote the use of the technology. The GHPC established the National Earth Comfort Program supported by the U.S. Department of Energy and the Environmental Protection Agency. The number of installations throughout the United States and worldwide have increased significantly due to a heightened awareness of the technology, along with concerns about energy costs and the environment.

18.6.1.3   Energy Source

The ground serves as a giant solar collector, storing heat energy. At depths greater than 30 ft (9 m) deep, the temperature is about the same as the average annual outdoor air temperature for that climate. Air temperatures may fluctuate as much as 50°F (28°C) above and below the average annual temperature. However, only a few feet below the surface, the changes in earth temperatures are much less severe. Earth temperature variations decrease with increasing depth.

During heating, the earth serves as a heat source. During cooling, the earth serves as a heat sink.

18.6.1.4   The Basic Ground Source Heat Pump System

The earth loop is placed in the ground either horizontally or vertically, or it can be placed in a pond. Water (or water and antifreeze) is circulated through the pipe, transporting heat to the heat pump during the heating mode and away from the heat pump during the cooling mode. The heat transfer takes place inside the heat pump in a water-to-refrigerant heat exchanger.

18.6.1.5   Key Topics Discussed in This Chapter

- Geothermal heating and cooling system concept and principles
- Mechanics of the geothermal heat pump
- Geothermal loops and applications
- Design and installation concepts
- System benefits

18.6.1.6   Geothermal

Dictionary definition: "*Relating to the internal heat of the earth.*"

"High temperature" geothermal refers to heat temperatures from typically hundreds of feet deep within the earth, sometimes exceeding 300°F (150°C). These include geysers and other hydrogeothermal reservoirs, and "hot rocks" which have limited locations. Electricity is sometimes generated from these high-temperature geothermal sources. The term "Geothermal Energy" most often refers to this type. This text does not cover high-temperature sources.

"Low temperature" geothermal refers to shallow earth temperatures found anywhere, utilizing a mechanical device (geothermal heat pump) to transfer heat to and from the ground at shallow depths producing heating, cooling, and hot water for buildings and other applications. This is the "geothermal" referred to in this text.

Terms commonly and interchangeably used when referencing geothermal heating and cooling systems:

- Geothermal heat pump (GHP)
- Geothermal comfort system
- Ground source heat pump (GSHP)
- Ground coupled heat pump (GCHP)
- Earth coupled heat pump
- GeoExchange system

The term "water source heat pump" (WSHP) refers to units used in a boiler/cooling tower application without an earth loop. *WSHP applications are not geothermal applications and therefore are not covered in this chapter.*

18.6.2 HEAT PUMP OPERATION

Geothermal heat pump systems consists of four circuits (Fig. 18.6.1):

1. *Air circuit*—The duct system that distributes the air throughout the home or building and returns it to the unit.
2. *Refrigerant circuit*—A sealed and pressurized circuit of refrigerant including compressor, expansion valve, water-to-refrigerant heat exchanger, air coil, reversing valve. The refrigerant is either R-22 or R-410A.
3. *Ground loop circuit*—The piping system buried in the ground or in the pond (or well water), with fluid that is circulated by pumps to and from the geothermal unit.
4. *Hot water circuit*—Domestic water can be heated in a geothermal unit with a device called a desuperheater. A piping connection is made from the geothermal unit to the water heater.

Each of these circuits is closed and sealed from the others—there is no direct mixing. However, heat energy does mix from the refrigeration circuit to the other three circuits.

The air circuit, the earth loop circuit, and the domestic hot water circuit always travel in the same direction. However, the refrigeration circuit will change direction depending on what mode (heating or cooling) the unit is in. (The exception to the change in direction of refrigerant flow is the flow through the compressor. This change of direction is controlled by the reversing valve.)

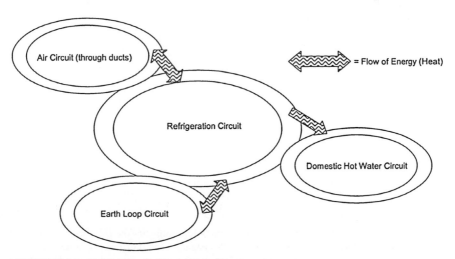

FIGURE 18.6.1 The four circuits in a geothermal heat pump system.

18.6.2.1   Component Comparison: Geothermal Unit vs. Gas and AC

Component	Geothermal	Gas furnace and AC
Compressor	Inside (unless OD split)	Outside
Indoor air coil	Evaporator during cooling	Evaporator
	Condensor during heating	
Outdoor air coil	None	Condensor
Coax (water coil)	Evaporator during heating	None
	Condensor during cooling	
Reversing valve	Yes	No
Expansion valve	Yes	Yes
Line set	Only if split	Yes
Outdoor fan	No	Yes
Blower	Yes	Yes
Duct system	Yes	Yes
Desuperheater	Yes	Rarely
Gas line and flue	Not required	Yes
Loop pumps	Yes	Not required
Thermostats and controls	Standard	Standard

18.6.2.2   Component Comparison: Geothermal Unit vs. Air Source Heat Pump

Component	Geothermal	Air source heat pump
Compressor	Inside (unless OD split)	Outside
Indoor air coil	Evaporator during cooling	Evaporator during cooling
	Condensor during heating	Condensor during heating
Outdoor air coil	None	Condensor during cooling
		Evaporator during heating
Coax (water coil)	Evaporator during heating	None
	Condensor during cooling	
Reversing valve	Yes (inside)	Yes (outside)
Expansion valve	Yes	Yes
Line set	Only if split	Yes
Defrost cycle	No	Yes
Crankcase heater	Only if split	No
Outdoor fan	No	Yes
Blower	Yes	Yes
Duct system	Yes	Yes
Desuperheater	Yes	Rarely
Loop pumps	Yes	Not required
Thermostats and controls	Standard	Standard

18.6.3   PERFORMANCE RATINGS OF GEOTHERMAL AND GROUNDWATER HEAT PUMPS

The Air-Conditioning and Refrigeration Institute (ARI) publishes the *Directory of Certified Applied Air-Conditioning Products* that includes capacities and efficiency ratings for various types of air conditioners, heat pumps, chillers, ventilators, etc., including geothermal and groundwater heat pumps.

Geothermal and groundwater heat pumps, along with water source heat pumps are grouped together in the category of "water-to-air and brine-to-air heat pumps" (WBAHP) and are tested under ISO Standard 13256-1, which replaces the former standards of ARI-330 (ground loop), ARI-325 (groundwater/open loop) and ARI-320 (water loop/boiler-tower). This standard uses different entering water temperatures for heating and cooling and for each type of WBAHP. Listed below are rating points for groundwater and ground loop heat pumps. (Water source/water loop heat pumps are not shown here, as they are not "geothermal" heat pumps.)

	Groundwater Units	**Ground Loop Units**
Single-speed units (and two stage units in high)		
Entering water temperature (cooling)*	59°F (15°C)	77°F (25°C)
Entering water temperature (heating)*	50°F (10°C)	32°F (0°C)
Two-stage units in low		
Entering water temperature (cooling)*	59°F (15°C)	68°F (20°C)
Entering water temperature (heating)*	50°F (10°C)	41°F (5°C)

For manufacturers who participate in the program, units are randomly selected by ARI for testing. Various tests are performed under strict conditions to ensure the units perform without failure and to the manufacturer's published ratings. ARI publishes the performance data for heating and cooling capacity and efficiency.

ARI has designated the efficiency ratings for water-to-air heat pumps as:

EER (for cooling operation)—energy efficiency ratio: EER = BTU output divided by Watt Input (under steady-state test conditions)

COP (for heating operation)—coefficient of performance: COP = BTU output divided by BTU input (under steady-state test conditions)

These efficiency standards differ from those used for air source heat pumps that use seasonal energy efficiency ratio (SEER) and heating season performance factor (HSPF) ratings. The SEER is a measure of the cooling efficiency over the entire heating season in a given climate. HSPF is similar in that it is a measure of heating efficiency over the entire course of the heating season in a given climate. Both SEER and HSPF include a cycling factor in the calculation.

Because EER and COP are based on steady-state conditions, and SEER and HSPF are based on seasonal conditions, the ratings cannot be compared directly. There is no "conversion formula" to go from one to another. The best way to compare efficiencies and resulting energy costs is to use software modeling programs that can calculate seasonal performance and operating costs for a given home in a specific climate.

18.6.4 GEOTHERMAL SYSTEMS AND THE LAWS OF THERMODYNAMICS (PART A)

A basic understanding of the Second Law of Thermodynamics is useful in order to fully understand how a geothermal system works (The First Law of Thermodynamics also applies to geothermal heating and cooling, but it will be reviewed later.)

18.6.4.1 The Second Law of Thermodynamics

The first statement in the Second Law of Thermodynamics is that *energy (heat) flows spontaneously from an area of high concentration (hot body) to an area of low concentration (cold body)*. From hot to cold. Never vice-versa. It is why ice cubes melt on a hot day, and why boiling water cools down after it is removed from the stove. Energy/heat wants to spread out or disperse like electricity in a battery, and it will do so unless hindered by an outside force.

This concept has two applications in geothermal heating and cooling systems.

- *In the refrigeration system*—In the heating mode, heat is absorbed by cool refrigerant in the water-to-refrigerant heat exchanger (coax). And heat is released from the refrigerant into the cooler return air by the air-to-refrigerant heat exchanger (air coil). In the cooling mode, heat is absorbed by the cool refrigerant in the air coil, and the cool loop fluid absorbs heat from the refrigerant in the coax. (This applies in the hot water mode as well, using the desuperheater.)

- *In the water (loop) system*—Heat is constantly moving from the ground to the loop during heating. As heat is removed from the earth loop, the earth loop is cooled to a point where it is cooler than the surrounding soil. Applying the Second Law of Thermodynamics, we know that the warmer earth will release its heat energy into the cooler loop. So in effect, the cool loop "draws" heat from the earth. During cooling, heat is removed from the home and transferred to the loop. Because the loop is warmer than the cool earth, the heat simply moves away from the loop—the area of high concentration, and is drawn into the cool earth—the area of low concentration.

18.6.5   THE REFRIGERATION CYCLE

18.6.5.1   Heating Mode

During heating, a geothermal system absorbs the heat from the ground via the earth loop. The heating cycle starts as cold, liquid refrigerant passes through the water-to-refrigerant heat exchanger (coax, and also the evaporator during heating). The coax is made of copper, and consists of a tube within a tube—water from the loop travels through one tube (the inside tube), refrigerant passes through the other (outer) tube.

As the loop fluid flows through the coax, the heat energy transfers from the loop fluid to the refrigerant through the copper wall separating the two. This heat transfer causes the cold liquid refrigerant to turn into a gas. (Unlike water, refrigerant changes from a liquid into a gas at a very low temperature.) The now gaseous refrigerant is sucked into the compressor where it is compressed. After compression, the refrigerant will be very hot (approx. 165°F) (73°C) and discharged through the reversing valve and into the air coil.

The air coil is a radiatorlike device that has thin aluminum "fins" attached to the copper refrigerant tubing. The refrigerant passes through the air coil (the condenser during heating). As air from the return air duct system passes over the air coil, heat is released from the refrigerant and absorbed by the cooler air. The result is warm air (typically 95°F to 105°F) (34°C to 40°C), which is delivered through the duct system by the blower.

The refrigerant, now cooled again, passes through the expansion valve (which acts as a flow control), returning to the coax where it can accept more heat from the warmer loop fluid.

This process is continuous during the heating mode. The refrigeration heating mode circuit is shown in Fig. 18.6.2.

18.6.5.2   Cooling Mode

During cooling, a geothermal system rejects the heat from the indoor air into the earth loop. The cooling cycle starts as cold, liquid refrigerant passes through the air coil (the evaporator during cooling).

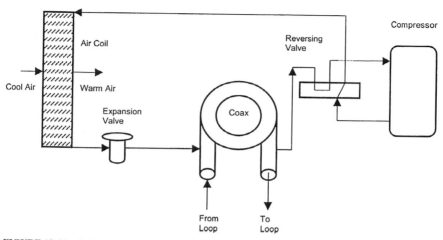

FIGURE 18.6.2   Refrigeration circuit heating mode.

As the refrigerant flows through the air coil, the heat energy transfers from the warm return air to the refrigerant. This heat transfer causes the cold liquid refrigerant to turn into a gas. The compressor draws the refrigerant gas, compresses it, and discharges it through the reversing valve. During cooling, the reversing valve is energized, which changes the openings from one port to another, causing the refrigerant flow to go in the opposite direction that it was in the heating mode. (However, the flow to the compressor does not change direction.)

After compression, the hot refrigerant passes through the coax (the condenser during cooling). In the coax, the hot refrigerant releases its heat energy to the cool loop fluid through the copper walls. Now cooled and liquified, the refrigerant passes through the expansion valve, back to the air coil. Warm air passing over the cool air coil causes the air to be cooled and dehumidified.

This process is continuous during the cooling mode. The refrigeration circuit cooling mode is shown in Fig. 18.6.3.

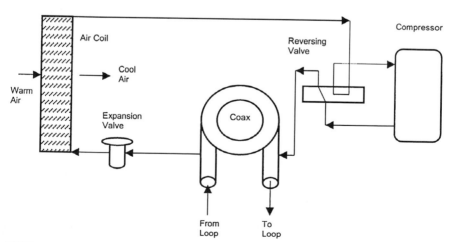

FIGURE 18.6.3   Refrigeration circuit cooling mode.

18.6.5.3 Hot Water Cycle

Many geothermal units installed in homes have an optional feature called a *desuperheater*. This component consists of a refrigerant-to-water heat exchanger installed at the discharge of the compressor. The hot gas at this point is in a "superheated" condition. In the desuperheater, the refrigerant releases some of the heat into the cooler water through the copper wall of the desuperheater heat exchanger. A small circulator moves the water from the water heater to the heat pump and back to the water heater.

This excess hot gas is available in both the heating and cooling modes. However, there is a greater hot water benefit during cooling because some of the heat that is extracted from the air ends up in the superheat, and is transferred to the water. When operating in second or third stage heating, some geothermal units disable the circulator so that all the heat energy is devoted to the air.

The amount of hot water generated is a function of the run time of the unit. On very hot days and cold days, the desupherheater may be able to generate the majority of the hot water required for the home due to the long run times of the unit. On milder days when the unit has short duty cycles, the electric elements in the water heater will maintain the desired temperature so there will always be enough hot water for the homeowner. A safety device (sensor) shuts off the circulator for the desuperheater in the event that the water temperature reaches 130°F (55°C).

The refrigeration circuit relating to the desuperheater is shown in Fig. 18.6.4.

Some geothermal units are able to heat water "on demand" for domestic hot water, radiant floors, fan coils, pools, spas, etc.

Units with desuperheaters are connected to the water heater with fittings that allow the installer to maintain a drain valve at the bottom of the water heater and still deliver hot water into the bottom of the tank (Fig. 18.6.5).

If a larger volume of hot water is required, a second tank may be incorporated into the design (Fig. 18.6.6). The first tank would be installed as shown in Fig. 18.6.5. However, the supply piping coming out of the first tank would lead to the cold water inlet of the second tank. The first tank would serve as the preheat tank. In this scenario, the first tank would not be connected to power. The second tank would be connected to power, but due to the volume of preheated water being supplied by the first tank, the second tank will have

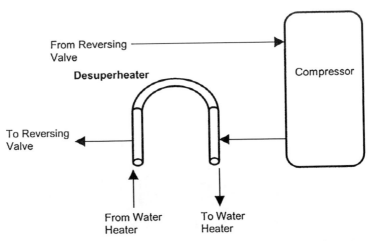

FIGURE 18.6.4   Desuperheater refrigeration circuit.

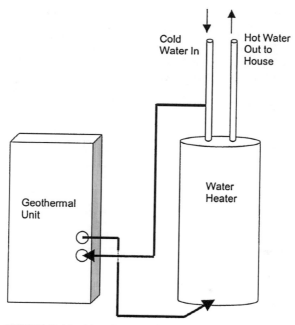

FIGURE 18.6.5　Typical hot water piping.

a reduced demand for power. This arrangement is recommended for added energy savings where there is a large demand for domestic hot water.

18.6.5.4　Geothermal Loops (Figs. 18.6.7*a* to *d*)

Computer programs are used to size the geothermal unit and the loop, based on the heating and cooling loads, climate, soil conditions and depth of the loop.

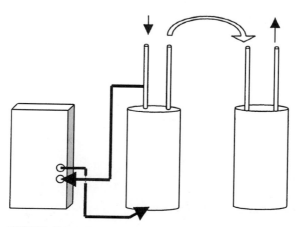

FIGURE 18.6.6　Two-tank system.

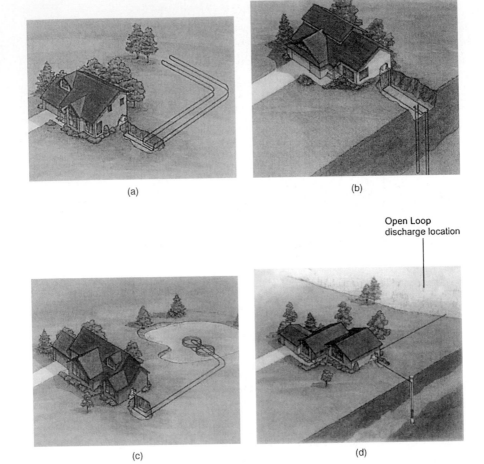

(a)

(b)

Open Loop
discharge location

(c)

(d)

FIGURE 18.6.7   Geothermal loops (*a* to *d*).

Closed Loops

(a) Horizontal
(b) Vertical
(c) Pond/lake

Open Loops

(d) Well water

18.6.5.5 Typical Earth Loop Features

These are shown in Fig. 18.6.8.

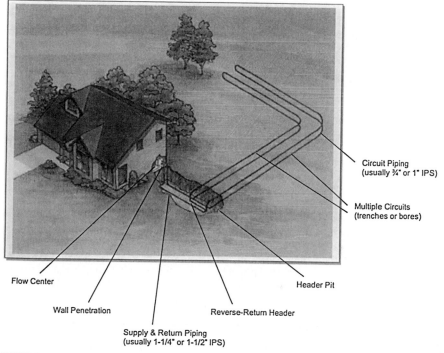

Circuit Piping
(usually ¾" or 1" IPS)

Multiple Circuits
(trenches or bores)

Flow Center   Header Pit

Wall Penetration   Reverse-Return Header

Supply & Return Piping
(usually 1-1/4" or 1-1/2" IPS)

FIGURE 18.6.8   Earth loop features.

Pipe Material.   Geothermal earth loops are constructed using pipe with the following characteristics:

- Polyethylene—designated as High Density PE3408.
- Schedule 40; IPS sizes $^3/_4$, 1, $1^1/_4$, $1^1/_2$, 2 in (larger sizes also available).
- All connections done using heat fusion tools at 500°F.
- Highest rating available for stress crack resistance.
- Retains flexibility down to −180°F.
- Carbon black provides protection from ultraviolet radiation from sun.
- Elongation of six times original length before failure.

Heat Transfer of Polyethylene vs. Copper

Some people question why polyethylene pipe is used instead of copper due to copper's better thermal conductivity. In reality, there is no heat transfer advantage to using copper for earth loops because the thermal conductivity value of the ground is the limiting factor. The heat transfer ability of the pipe is not the limiting factor—the earth is the limiting factor.

(In addition, copper would be much more expensive and subject to corrosion and joint failure.)

18.6.5.6   Reverse-Return Header System

Using a reverse-return header system (Fig. 18.6.9) ensures equal flow through each circuit due to equal pressure drops. If a reverse-return header system is *not* used, the flow rate in each circuit will vary depending on the pressure drop in each circuit, resulting in less than optimal performance.

Headers are typically reducing headers, meaning that larger pipe is reduced to smaller pipe using reducing tees. Using reducing headers facilitates flushing.

18.6.5.7   Horizontal Loops

Horizontal earth loops are used where the space allowed for the loop is not extremely limited. There are various designs of horizontal loops. There is no one type of horizontal loop that is best for every application. The selection of which type to use should be based on system size, space available, soil conditions, and the type of excavating equipment used. Regardless of the type selected, operating costs will not vary substantially.

Cross sections of horizontal loops showing the positioning of the PE pipes are shown in Fig. 18.6.10. The first 2-pipe design shown is dug with a chain trencher. The other 2-pipe, and the 4-pipe and 6-pipe trenches are dug with a backhoe or trackhoe. When trenches are dug with a backhoe, the typical trench width is 2 to 3 ft (0.6 to 0.9 m). Typical minimum separation between trenches is 10 ft (3 m). Loop systems are generally designed to the rule-of-thumb of one circuit per ton. Shorter circuits are used in order to minimize the pressure drop throughout the system.

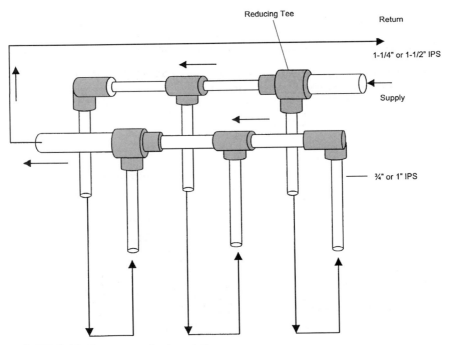

FIGURE 18.6.9   Reverse-return header system.

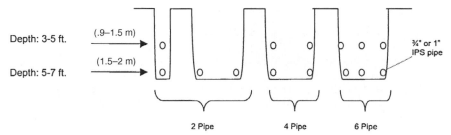

FIGURE 18.6.10   Horizontal loops.

- A trench containing a 2-pipe loop has one circuit.
- A trench containing a 4-pipe loop has two circuits.
- A trench containing a 6-pipe loop has three circuits.

The more piping installed per foot of trench, the shorter the trench becomes. (However, more pipe is required.)

- 2-Pipe systems generally require about 250 ft (75 m) of trench per ton.
- 4-Pipe systems generally require about 180 ft (55 m) of trench per ton.
- 6-Pipe systems generally require about 120 ft (35 m) of trench per ton.

18.6.5.8   Horizontal Slinky

Another type of horizontal loop is called the Slinky. In this installation, a trench is dug with a backhoe 5 or 6 ft (1.2 to 2 m) deep, and about 3 ft wide. The coils are "layed off" and spaced evenly throughout the length of the trench. Slinkys can be designed as "compact" or "extended."

Compact slinkys are generally installed with 800 to 1000 ft (250 to 300 m) of pipe in a 100-ft (30-m)-long trench (per ton). Extended slinkys are generally installed with 600 ft (180 m) of pipe in a 150-ft (45 m) long trench (per ton).

The top view of a slinky loop is shown in Fig. 18.6.11.

18.6.5.9   Horizontal Bore

Where there is adequate space for a horizontal loop, but there is a desire to minimize landscape disruption, the horizontal bore loop may be the preferred solution. (Figs. 18.6.12a,b,c).

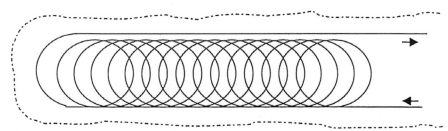

FIGURE 18.6.11   Slinky loop.

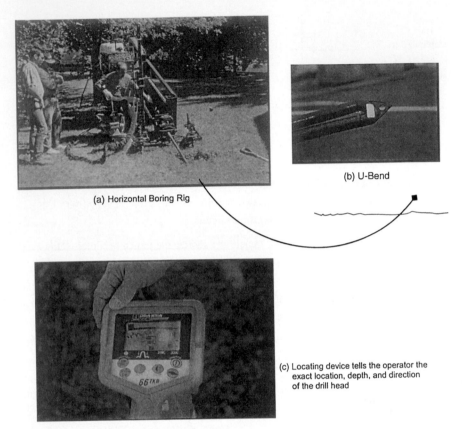

(a) Horizontal Boring Rig

(b) U-Bend

(c) Locating device tells the operator the
exact location, depth, and direction
of the drill head

FIGURE 18.6.12   Horizontal bore operations.

A special horizontal drilling machine is used. This machine has the capability to start at the surface and drill at a slight angle down to a typical depth of 10 to 21 ft, then drill back to the (3 to 3.7 m) surface. Using the right technique, the operator can "steer" the drilling activity to go deeper or shallower, or turn right or left. The drill head emits a radio signal that can be detected by a special device that tells the operator exactly where and how deep it is.

The machine drills the horizontal bore, then causes the drill bit to come back to the surface, typically about 200 ft (60 m) away. At that point, two ends of pipe are attached to the end of the drilling pipe in place of the drill bit, and are pulled back through the hole to the header, and until the u-bend at the end of the pipe is buried.

The only digging required is for the header and the supply and return piping into the house. This type of loop is most often used in a retrofit situation to minimize disruption to the landscape.

18.6.5.10   Vertical Loops

Vertical loops (Fig. 18.6.13) are used where space is limited or where soil conditions are not conducive to horizontal loops. Installing vertical loops requires the use of a drilling rig.

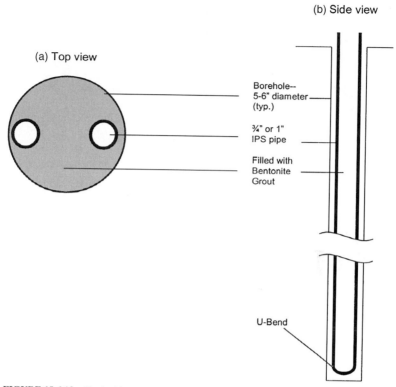

FIGURE 18.6.13   Vertical loop.

Multiple holes are bored. A double pipe connected with a U-bend is inserted into each hole. The hole is thin filled with grout to provide good contact around the pipe and to seal the hole. The vertical pipes are then connected to a header system horizontally a few feet below the surface.

The depth of the holes is dependent upon soil conditions and the size of the system. Different soils and rock types have different thermal conductivity values or "K-values" (Btu/h · ft · °F). These values typically range from 2.0 for dense rock to 0.20 for a light dry soil. Heavy damp soil has a K value of 1.40. Therefore, there is no specific depth that needs to be reached. Capacity is not based on depth; rather how much pipe is in the ground and the overall K-value of the borehole. K-values can be measured using sophisticated equipment in a test bore.

Typically, one hole per ton is used. Vertical loops are generally spaced with a minimum of 10-ft (3 m) separation, and about 150 to 250 ft (45 to 75 m) of borehole per ton. (*Note:* When referencing vertical loops, the term "borehole" is more appropriate than "wells." The term "wells" generally implies the use of groundwater.)

18.6.5.11   Pond Loops

Pond loops (Fig. 18.6.14) are a cost-effective way to install a geothermal system, because trenching is limited to only the supply and return piping from the pond to the house.

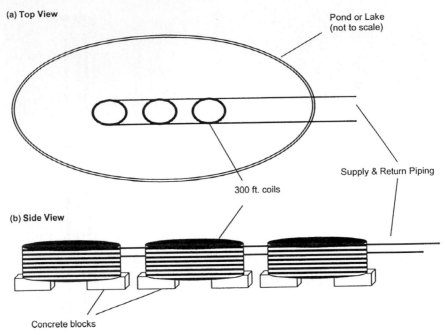

FIGURE 18.6.14   Pond loops.

Pond loops consist of a series of coils connected together and placed at the bottom of the pond. In order for a pond to be suitable for a geothermal application for a typical home, the pond should be at least $\frac{1}{2}$ acre in surface area and at least 8 ft deep, even during a dry spell. Ideally, the pond should be close to the home (less than 200 ft). If the pond is farther from the home, the benefit of using a pond loop is reduced due to added trenching, materials, and pumping costs.

The coils can be connected together on dry land, and then floated into location. Once filled with fluid, they will sink to the bottom. Generally, a 300 ft coil is used for each ton of capacity. This is less pipe than is used in an earth loop because water is a better conductor of heat energy.

Even with an ice cover, the temperature at the bottom of a pond in the winter will generally be about 38°F (5°C)

18.6.6 TYPICAL INSTALLATION CLOSED LOOP SYSTEM

18.6.6.1 Circulating Systems

A module called the Flow Center circulates the fluid in the loop. This device typically contains one or two small pumps (1/6 hp circulators) that are engaged when the unit is heating or cooling. Generally, one pump is used for units up to 3 tons; two pumps are used for units from $3\frac{1}{2}$ to 6 tons. Systems larger than 6 tons may require a third pump, or the use of a single larger pump, or possibly separate loops and pumps for each unit (Fig. 18.6.15).

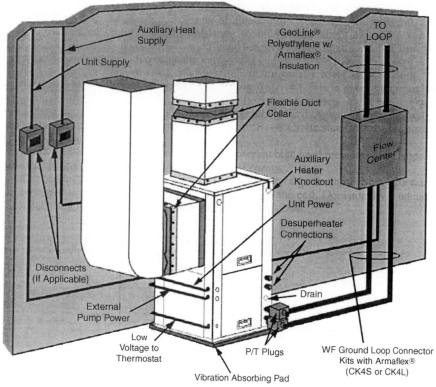

FIGURE 18.6.15 Closed loop system (typical installation).

18.6.7 *FREEZE PROTECTION FOR CLOSED LOOP SYSTEMS*

In climates subject to freezing earth conditions, antifreeze is used for freeze protection. Various types and concentrations of antifreezes are used. Listed below are the most common antifreeze solutions. Each type has advantages and disadvantages.

	Ethanol	Methanol	Propylene glycol	Ethylene glycol
Corrosiveness	Low	Low	Low	Low
Flammability	High if pure	High if pure	Low	Low
Toxicity	Low	High	Low	High
Heat transfer	High	High	Low	Low
Pumping requirements due to viscosity	Low	Low	High	High
Cost	Moderate	Low	High	High

18.6.7.1   Calculating Antifreeze Volume

In order to calculate the volume of antifreeze required for a given closed loop system, it is first necessary to calculate the total volume of fluid. This calculation is made based on the length of the various sized pipes used in the earth loop, inside piping, unit connections, and the unit itself.

Once the total volume of loop fluid has been calculated, determining the specific percentage of antifreeze mixed with water will be a function of the desired level of freeze protection, the type used, and total loop volume.

18.6.8   TYPICAL INSTALLATION OPEN LOOP SYSTEM

18.6.8.1   Open Loop Components (Fig. 18.6.16)

- Electric solenoid water valves open only when the unit is operating.
- Flow regulators are used to control the gpm.
- The expansion tank is sized large enough to provide a minimum of 2 min of run time without the pump.
- Heat exchangers made of a copper-nickel alloy (cupronickel) are recommended instead of copper to reduce scaling.

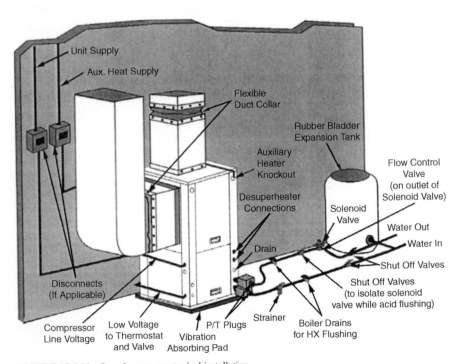

FIGURE 18.6.16   Open loop system typical installation.

18.6.8.2   Key Considerations for Open Loop Applications

1. Adequate water supply (1.5 gpm/ton plus household requirements) [6 liters/min]
2. Quality water (low mineral content)
3. Adequate discharge location (drainage ditch, field tile, pond, etc.)

18.6.9   GEOTHERMAL SYSTEMS AND THE LAWS OF THERMODYNAMICS (PART B)

18.6.9.1   The First Law of Thermodynamics

Otherwise known as the Law of Conservation of Energy—basically, you can't get more energy out than you put in. Then how does a geothermal system have an "efficiency" greater than 100 percent (coefficient of performance greater than 1.0)—generally 3.5 to 4.5 COP? The formula for COP is BTU output divided by BTU input. In this case, the BTU input is just a measure of the electricity used, and does not include the heat input from the loop.

    EXAMPLE   A typical 4-ton geothermal unit at 50°F (10°C) EWT:

KW = 3.3 (electricity input) kW converted to BTU: $3.3 \times 3413 = 11,263$ BTU input
Heat of extraction = 36,400 BTU (input from loop)
Heating capacity = 47,663 BTU output (11,263 + 36,400)
COP = 4.23 (47,663 divided by 11,263)

If you define efficiency in the purest sense—comparing what you get out to what you put in (including the free energy from the loop), then geothermal isn't 350 percent efficient. The formula for COP includes only the input from the electricity, and not the loop. So it would be more accurate to say that a geothermal system produces three to four times the amount of heat energy for every one unit of electrical energy purchased to operate the system.

18.6.9.2   Key Benefits of Geothermal Systems

- Energy savings for heating, cooling, and hot water
- Quiet operation
- No outdoor condensing unit required (some manufacturers do offer splits suitable for outdoor installations, but there is no noisy fan used)
- Less maintenance
- Long life
- Safe and clean—no flames, fumes, flues
- Environmentally friendly
- Decreases peak demands for electricity (benefit to electric utilities)

Interesting Facts

- According to the Environmental Protection Agency (EPA), GeoExchange technology is the most energy-efficient, environmentally clean, and cost-effective space conditioning system available.
- The EPA found that the systems can reduce energy consumption and corresponding emissions by over 40 percent compared to air source heat pumps and by over 70 percent compared to electric resistance heating with standard air-conditioning equipment.
- If one in 12 California homes installed a GeoExchange system, the energy savings would equal nine new power plants.
- Installing a GeoExchange system in a typical home is equal, in greenhouse gas reduction, to planting an acre of trees, or taking two cars off the road.
- Current GeoExchange installations equal 14 million barrels of crude oil saved per year.
- The ground absorbs 47 percent of the sun's energy that reaches the earth. This amount of energy represents 500 times more than mankind needs every year.

18.6.9.3   Other Applications

While most applications of geothermal systems are in residential applications, the technology can also be used for:

- Commercial applications—offices, retail, hotels, apartments/condos, resorts
- Institutional applications—schools, hospitals, nursing homes
- Industrial applications—process water heating and cooling, manufacturing plants
- Pools and spas
- Snow melt/ice melt
- Radiant floor heating
- Fan coils
- And any other application where warm or cold air, or heated or chilled water is required.

GHP units can also be used as water source heat pumps connected to a boiler and cooling tower or chiller. With the extended range of water temperatures these units can operate in, along with higher efficiencies, GHP units can produce considerable savings in operating costs compared to traditional low range/low efficiency units used in boiler/cooling tower applications.

RESOURCES

Organizations

WaterFurnace International
> 9000 Conservation Way, Fort Wayne, IN 46809
> Phone: 1-800-222-5667
> Web site: www.waterfurnace.com

Geothermal Heat Pump Consortium
> 701 Pennsylvania Ave. N.W., Third Floor, Washington, DC 20004-2696
> Phone: 1-888-ALL 4 GEO
> Web site: www.ghpc.org

International Ground Source Heat Pump Association
> 499 Cordell South, Stillwater, OK 74078-8018
> Phone: 1-800-626-4747
> Web site: www.igshpa.okstate.edu

Publications

Closed-Loop/Ground Source Heat Pump Systems Installation Guide, Oklahoma State University, Division of Engineering Technology, ISBN: 0-929974-01-8

Mitchell R. LeClaire, and Michael J. Lafferty, *Geothermal Heat Pump Training Certification Program.*

Soil and Rock Classification for the Design of Ground-Coupled Heat Pump Systems—Field Manual, STS Consultants, ISBN: 0-929974-02-6.

Closed-Loop Geothermal Systems—Slinky Installation Guide, Fred Jones, ISBN: 0-929974-04-02.

Commercial/Institutional Ground Source Heat Pump Engineering Manual, Caneta Research, ISBN: 1-883413-21-4.

Directory of Certified Applied Air-Conditioning Products, Air-Conditioning & Refrigeration Institute

PART B: AIR AND WATER SOURCE

18.6.10   AIR-SOURCE HEAT PUMP BASICS

Electric heat pumps are refrigeration systems that will either heat or cool a space as required by occupants. This is done by removing heat from one place and placing it in another.

This heat movement is possible whether the heat is being moved from a cooler space to a warmer space or from a warmer space to a cooler one. Most heat pumps use the vapor-compression cycle to move this heat.

All heat pumps have the same basic components, whatever their heat source. They contain:

- A fluid (the refrigerant) circulated continuously inside a closed system while it is running;
- A motor-compressor;
- Two heat exchanger coils that alternate their function as the evaporating coil and the condensing coil;
- A reversing valve that changes the direction of refrigerant flow; and
- A flow-control device that controls the pressure drop and temperature of the refrigerant.

Some models, depending on the heat source, require that auxiliary heaters be used; some require accumulators, pumps, and fans.

All heat pumps require certain control and safety devices, like thermostats, pressure controls, and capacity controls, to name a few.

18.6.10.1 Basic Operation

All heat pump systems operate the same way, because of the vapor-compression cycle used.
   Certain characteristics of the heat source will determine the different modes of heat pump operation necessary to meet the needs of the building. For example, a defrost cycle is required only when air-source heat pumps are used, and are required to operate in out-door temperatures below 40°F (5°C).

Cooling Cycle.   When the system is operating in the cooling mode, hot gas is discharged from the compressor through the reversing valve and into the outdoor coil.

- The hot gaseous vapor is condensed into a liquid in the outdoor coil by the cooler air passing through the coil fins.
- The liquid refrigerant then passes through a check valve in the outdoor unit. The check valve operates as a bypass around the outdoor flow control device.
- The liquid then flows to the indoor coil, where it passes through the flow control device and its pressure and temperature are reduced.
- The liquid refrigerant then evaporates in the indoor coil, where it absorbs heat from the warmer air passing through the indoor coil fins.
- The cooled air is then distributed to the conditioned space to cool the occupants (see Fig. 18.6.17).
- The evaporated refrigerant, including any that did not completely evaporate in the indoor coil, is drawn back to the compressor through the reversing valve, and into the compressor, where it is again compressed and the cycle is started over.

   The purpose of the accumulator is to collect and evaporate any liquid refrigerant that reaches this point. It prevents liquid refrigerant entering the compressor where damage could be done.

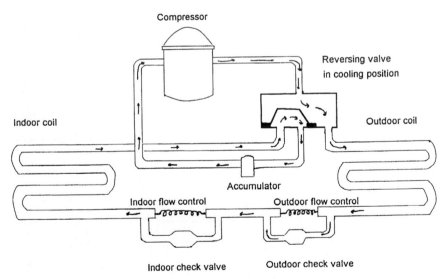

FIGURE 18.6.17   Air-to-air heat pump in cooling cycle.

Heating Cycle.   When the system is operating in the heating mode, the refrigeration cycle is exactly the opposite of the cooling cycle (see Fig. 18.6.18).

- The reversing valve has changed positions so that the hot gaseous refrigerant will flow to the indoor coil rather than the outdoor coil.
- The refrigerant vapor is condensed in the indoor coil, because the air flowing through the coil fins is cooler than the refrigerant vapor.
- As the indoor circulating air condenses the refrigerant vapor, heat is given up to the air. The air is then circulated to the conditioned space.
- The liquefied refrigerant passes through the indoor check valve, which is acting as a bypass.
- The refrigerant then flows to the outdoor unit, where it passes through the outdoor flow control device where its pressure and temperature are lowered.
- It then passes through the outdoor coil, where it is mostly evaporated, and goes on into the reversing valve.
- From the reversing valve it flows into the accumulator, when used, where it is captured and any liquid left is evaporated.
- The refrigerant vapor then flows into the compressor, where it is again compressed and the cycle starts over.

It should be noted that regardless of how cold the outdoor air temperature is, heat can still be removed from it.

However, the refrigerant inside the outdoor coil must be cooler than the ambient air, so that the heat necessary for refrigerant evaporation can be removed from it.

Defrost Cycle.   All air-source heat pumps that operate below 40°F (5°C) must have a defrost cycle. The defrost cycle is needed because the colder coil will freeze moisture from the air and it will collect on the coil surface.

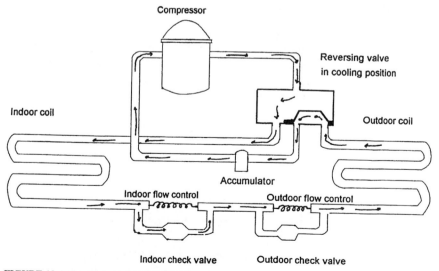

FIGURE 18.6.18   Air-to-air heat pump in heating cycle.

The defrost cycle is especially needed when the outdoor humidity is 60 percent or higher, and the outdoor air temperature is between about 30°F (1°C) and 40°F. When the outdoor air temperature is below 30°F, there is usually less moisture present.

Frost accumulation will depend on the relative humidity, rain fall, wind velocity, outdoor air temperature, and the amount of unit running time.

When the defrost period is needed, the system actually goes into the cooling mode. This changes the outdoor coil into the condensing coil and the indoor coil into the evaporating coil.

The hot gas being discharged into the outdoor coil melts the frost and ice from the coil surface. Usually the outdoor fan motor is stopped to allow the hot gas to defrost the coil much faster.

The indoor fan will remain on to add heat to the refrigerant inside the indoor coil. Usually some type of auxiliary heat is needed during the defrost cycle to add some heat to the conditioned air rather than blowing the cooler air into the space.

The auxiliary heat will cause the heat pump to lose about 3 to 10 percent overall efficiency. This reduction varies depending on the length of the defrost cycle, and the frequency of defrost cycles.

Supplementary Heat Cycle.   During normal operation, supplementary heat is used only during the defrost cycle. Sometimes it is used when the heat pump is operating below its balance point and more heat is needed to heat the space.

The balance point of most air-to-air heat pump systems is from about 15°F (−12°C) to 35°F (2°C). However, heat pump systems that have two-speed compressors have a lower balance point.

Most supplementary heat is in the form of resistance heat strips in the discharge airstream of the indoor unit. They are automatically turned on and work with the heat pump to provide the necessary heat to the space.

They are also used to provide emergency heat if the heat pump system should not be operating properly (or not at all).

Using the supplemental heat is usually very expensive and should be avoided when possible.

Forced-Ventilation Cycle.   Forced ventilation is used during mild weather when neither heating nor cooling are required. The fan is operated to keep the conditioned air moving to provide comfort to the occupants.

To operate the forced-ventilation cycle, turn the fan switch on the thermostat to the "on" position. The heat-cool switch is placed in the "off" position. The fan will then run continuously until the switch is changed to the "auto" position.

18.6.10.2 Performance Ratings

The Air-Conditioning and Refrigeration Institute (ARI) has developed a set of standards by which the heating and cooling cycles of a heat pump can be rate for efficiency, capacity, and estimated annual operating costs.

The performance standards for heat pump equipment below 135,000 Btuh (40.5 kW) are:

- ARI Standard 240–81: Air-Source Unitary Heat Pump Equipment;
- ARI Standard 380–85: Packaged Terminal Heat Pumps;
- ARI Standard 230–86: Water-Source Heat Pumps;
- ARI Standard 325–85: Ground-Source Heat Pumps.

The U.S. Department of Energy has more stringent testing and rating procedures for heat pumps rated less than 65,000 Btuh (19.5 kW) and operating on single-phase electricity. The ARI unitary directory lists these heat pumps as DOE-covered products.

The four different terms used for rating heat pump efficiency are based on two general conditions:

1. Rating is based on the performance of the heat pump at predetermined and specified operating conditions. These are known as *single-point conditions.*
2. Rating is based on the performance over a wide range of operating conditions. These are known as *seasonal conditions.*

For the heating mode, the seasonal rating is known as the heating seasonal performance factor (HSPF). The single-point rating is known as coefficient of performance (COP).

The seasonal rating for the cooling mode is known as the seasonal energy efficiency ratio (SEER). The single-point rating is known as the energy efficiency ratio (EER).

The **HSPF** includes the effects of all operating conditions for a typical heating season in a certain region of the United States.

These conditions include the weather, defrost losses, cycling losses, and the auxiliary heat. To calculate the HSPF use the following formula:

$$HSPF = \frac{\text{total heating during a season in Btu}}{\text{total electric power input in watt-hours}}$$

The higher the HSPF, the greater the efficiency of the unit. A heat pump rating of 9.0 means that the unit will deliver 9 Btu for each 1.0 watt-hour input.

It must be remembered that tables listing HSPFs do not take into account local weather conditions, or the particular characteristics of a building. ARI, in its latest *Directory of Certified Unitary Air-Source Heat Pumps* does, however, include a method for calculating the annual heating costs and adjusting for local electric rates.

Heat pump seasonal heating performance is sometimes expressed as the seasonal heating coefficient of performance, or SCOP. To calculate the SCOP, use the following formula:

$$SCOP = \frac{HSPF}{3.413}$$

The **COP** of a heat pump is the amount of useful heating or cooling that can be obtained from the unit to the energy consumed in delivering this heating or cooling.

The amount of useful output rate includes the circulating fan but not the auxiliary heat. Use the following formula to calculate the COP:

$$COP = \frac{\text{rate of useful energy output in Btuh}}{\text{rate of energy input in watts} \times 3.413 \text{ Btuh/watt}}$$

The procedures used by ARI assume that certain indoor standards are provided. These standards include the indoor air temperature, humidity, and heating sun steady state. They are calculated for both 17° and 47° (−8°C and 10°C) outdoor air temperature.

The *ARI Directory of Certified Unitary Air-Source Heat Pumps* lists the different manufacturers. The higher the COP rating, the more heat the unit is capable of delivering.

The **SEER**, like the HSPF, of a heat pump takes into consideration the performance of the unit over a wide range of time and operating conditions. To calculate the SEER of a heat pump, use the formula:

$$SEER = \frac{\text{total cooling during season in Btu}}{\text{total energy input in watthours}}$$

A higher SEER indicates that the unit is more efficient than the lower ones

Air-source heat pump SEERs range up to about 12.0 or 14.0. This is close to the output of electric cooling units.

The *ARI Directory of Certified Air-Source Heat Pumps* lists these ratings along with the HSPF of the unit in question. The Appliance Efficiency Standards Act of 1986 requires that all air-source heat pump units manufactured after 1992 have a minimum SEER of 10.0.

The **EER** is much like the COP, with the exception that it is expressed in Btuh per watt. Note that the EER, like the COP, is a steady-state rating. To calculate the EER, use the formula:

$$EER = \frac{\text{cooling provided in Btuh}}{\text{electrical input in watts}}$$

For example, we have two units, A and B. Unit A has an EER rating of 14. Unit B has an EER rating of 7. This indicates that unit A is two times as efficient as unit B at the specified rating conditions.

18.6.11   *WATER-SOURCE HEAT PUMPS*

Water-source heat pumps are gaining in popularity in the industry. This is probably because they are more economical to operate and generally cause fewer problems than other types.

The principle on which this system operates is that heat is rejected to the water from the air being cooled during the cooling cycle, and heat is gained from the water by the air during the heating cycle.

There are no defrosting cycles on water-source heat pumps.

During the cooling cycle, humidity is removed from the air because the coil generally operates below the dewpoint temperature of the air. This moisture is disposed of in the form of condensate which drains into the sewer system.

The medium used to transfer this heat between the air and water is the refrigerant inside the system. The basic components used on water-source heat pumps are the compressor and the coaxial heat exchanger.

Just like in the air-to-air system, the change from the heating mode to the cooling mode is accomplished by a reversing valve.

These system are more popular in residential and small commercial applications; however, they are also gaining popularity in some larger applications.

There must be an adequate supply of water for the system at the location. When insufficient water is available, the system will not function properly.

Also, there must be a suitable method of disposing of the used water after it has passed through the condenser.

18.6.11.1   Electrical Controls

As with any type of air-conditioning system, understanding the controls is necessary to properly install, service, start-up, and troubleshoot the system.

The contractor also must be familiar with the controls for the pump and water system, so that complete service can be provided to the customer. Without this knowledge, successful installation and service procedures will be very difficult.

There are two major classifications of heat pump controls: electrical and refrigeration.

The electrical controls are of first interest to the technician, because they usually interface with the water system and any other items that may be installed.

Groundwater heat pump controls, for the most part, operate in a straightforward manner. We will explain the individual components and how they operate together to properly control the system.

However, each manufacturer's equipment must be studied to learn any specifics that may be used in its particular line. Heat pump electrical systems are further divided into two major groups: line and low, or control, voltage.

The *line voltage* is generally used for the operation of motors, compressors, and other devices that require extra power for operation.

The *low voltage* is used in the control circuit that the controls open and close (see Fig. 18.6.19).

However, some controls may be used on either line or low voltage, depending upon the desires of the manufacturer.

When low voltage is used for the control voltage, controls that have a lighter amperage rating can be used. The wiring is also smaller and less expensive than when line voltage is used.

There are rules that the manufacturer must adhere to when using a low voltage control system.

For a control system to have an NEC (National Electrical Code) Class II rating, the control voltage must be no greater than 30 V. The power that the system can carry is limited to 75 VA (volt-amperes). The equipment manufacturer must provide some means of preventing a greater power level if an overload or shorted conduit in the control circuit should occur.

However, there are three possible ways that this can be done: by using a fuse, a circuit breaker, or an energy-limiting transformer. When these rules are followed, the system will have a Class II control circuit that may be connected very quickly and economically.

It must be noted that there is a built-in limitation to how much power can be utilized in this system. In this type of circuit, only small loads, such as magnetic coils, relay coils, and solenoid coils, may be used.

These smaller loads are then used to control larger loads, requiring line voltage for the heat pump unit. At 24 V, 75 VA the maximum current is 3.2 A that would flow through the circuit.

When a small number of current-using controls are used in the circuit, then a correspondingly smaller amount of power is provided for the control circuit. In a residential and small commercial units, 40-VA transformers are generally used.

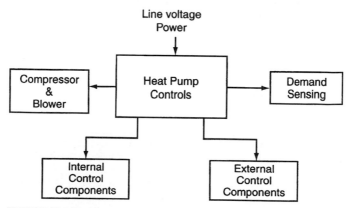

FIGURE 18.6.19   The use of control and line voltage.

These circuits will have a maximum current flow of 1.6 A. Notice that the amount of current supplied is limited by the size of the transformer used.

Room Thermostats.   Thermostats are normally low-voltage components. They are sometimes called the "demand control."

Room thermostats used on modern equipment are complex devices. This complexity is required so the thermostat can properly respond to any changes in room temperatures.

However, the major purpose of the thermostat is to complete an electrical circuit that tells the heat pump when to operate and when not to operate (see Fig. 18.6.20).

The thermostat is simply a switch that opens and closes in response to the room temperature it is controlling.

Magnetic Contactor.   The low-voltage signal from the thermostat is used to control components such as the compressor, fan motors, and pumps by using a magnetic contactor.

The contactor is a means of keeping the line and low voltages separate. They may be connected to one another by either mechanical or magnetic coupling.

In operation, the low voltage passes through the contactor coil, causing a magnetic field inside the coil. This magnetic field attracts the contactor's armature, which carries the contactor movable contacts into contact with the stationary contacts connected to the line voltage circuit.

When in this position, the contactor is said to be "pulled-in." The line voltage load is now energized. This is how low voltage is used to control line voltage while the two are kept separate.

The electrical symbol for a magnetic contactor is shown in Fig. 18.6.21. Notice that the line-voltage and low-voltage terminals are separate to prevent burning and shorting the low-voltage circuit, and that the two line-voltage phases are kept separate from each other to prevent short circuits.

Magnetic Relays.   These types of relays operate much the same as the contactor; however, they are used for controlling other low-power loads, either line or low voltage.

Also, the switched side of the relay can have several poles that may switch separate circuits used for controlling different system components. These poles are caused to switch their load in response to a signal received by the relay coil.

Relays may be equipped with both normally open and normally closed contacts, depending on the needs of the system.

Figure 18.6.22 is a representation of a magnetic relay with the coil and the different contact arrangements. The contacts with a diagonal line through them are normally closed, and will break the circuit when the coil is energized.

Pressure Switches.   Both high- and low-pressure switches are used on water-source heat pumps. They may be considered either electrical or refrigeration controls.

Pressure controls are connected both to the electrical system and to the refrigeration system. Should either of these controls turn the unit off, the system must usually be reset before

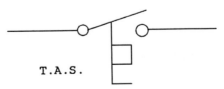

FIGURE 18.6.20   Electrical symbol for a thermostat.
T.A.S. stands for "Temperature Actuated Switch."

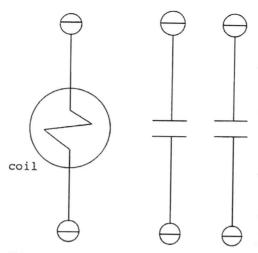

FIGURE 18.6.21 Electrical symbol for a magnetic contractor.

the unit will operate again. This is normally done by turning the thermostat system switch to the off position, or turning off the electrical power to the unit. In either case, it is advisable to wait a few minutes before restarting the unit, to allow system pressures to equalize. They are either on or off, depending upon the refrigerant pressures. Figure 18.6.23 is a symbol for a pressure switch.

Transformer. The control, or low, voltage is supplied by a step-down transformer.

The transformer is usually energized any time line voltage is supplied to the unit. It consumes very little power, and control voltage is available when needed. Figure 18.6.24 is a symbolic representation of a transformer.

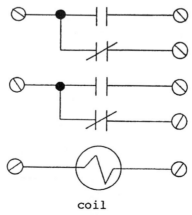

FIGURE 18.6.22 Electrical symbol for a control relay.

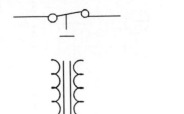

FIGURE 18.6.23   Electric symbol for a normally closed pressure switch.

FIGURE 18.6.24   Transformer electric symbol.

Reversing Valves.   Reversing valves are used to change the direction of refrigerant flow in heat pump systems. They are generally controlled by a control voltage pilot-operated solenoid valve.

The low voltage causes the pilot solenoid to change positions, which in turn allows the difference in refrigerant pressure on both sides of the valve plunger to make the valve change positions (see Fig. 18.6.25).

Reversing valves also may be considered either electrical or a refrigerant control.

18.6.11.2 Ladder Diagrams

When these controls are placed in the system, some means of determining which equipment steps are taken, and when, are readily available on the diagram. Earlier we learned how a thermostat demands equipment operation by signaling the contractor or control relay coil. Occasionally, this signal may be ignored.

It may be interrupted after the system is started, or route the power to another place. Figure 18.6.26 is an indication of a single "rung" of a ladder diagram.

The voltages to be controlled are the sides of the diagram. In this rung are a temperature-actuated switch, a pressure control, and a normally closed set of contacts in the control relay. All are placed in series with the contactor coil. All of the contacts must be closed before the contactor coil will be energized. Should either switch open the circuit, the contractor coil will be de-energized. Note that the purpose of the ladder diagram is to show the logical sequence of operation. The physical locations of the controls are not shown in a ladder diagram. To make a ladder diagram useful, several rungs must be included. They may be

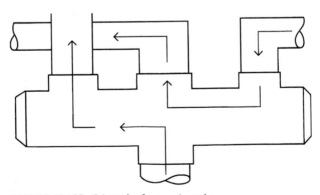

FIGURE 18.6.25   Schematic of a reversing valve.

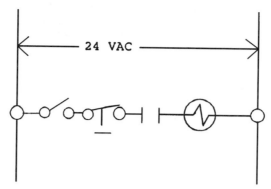

FIGURE 18.6.26 A single "rung" on a ladder electrical diagram.

stacked and interconnected to permit the many functions to occur as needed. Figure 18.6.27 is a typical example of a ladder diagram.

The components below the transformer are in the control voltage section of the diagram. The components found above the transformer operate with line voltage, in this case 240 V.

The control relay (labeled "CR2") has its coil in one rung of the ladder and its contacts in another. Notice that it is a two-pole relay. One set of contacts operates in the control voltage, and the other set operates in the line voltage.

When the relay coil is energized, both contactor 1 and the 240-VAC valve will be energized. When the ladder diagram is analyzed in this manner, the complete operating sequence of the unit can be more fully understood.

18.6.11.3 Refrigerant Controls

The refrigerant expansion device is probably the most important of the refrigerant controls. The TXV regulates the amount of refrigerant entering the evaporating coil. The refrigerant requires heat to change into a vapor inside the evaporating coil. When the unit is operating in the heating mode, the heat source is water. When operating in the cooling mode, the air inside the building is the heat source.

One type of popular expansion device is the **capillary tube**, like those used on some cooling units and on domestic refrigerators and freezers. Another popular expansion device is the **thermostatic expansion valve (TXV)**. Sometimes both are used in the same system, one on each coil.

The purpose of the TXV is to maintain the desired superheat inside the coil.

Refrigerant Check Valves. Some water-source heat pumps use check valves to allow the refrigerant to flow in only one direction through that part of the system. They are usually placed to allow the refrigerant to flow either through or around the expansion device.

Suction Line Accumulator. This device is placed directly ahead of the compressor suction connection. Its purpose is to collect any liquid refrigerant and prevent it from entering the compressor and possibly causing serious damage.

The time when liquid refrigerant would be more likely to be in the suction line is during startup, system overcharging, sudden refrigerant reversal, or some component failure.

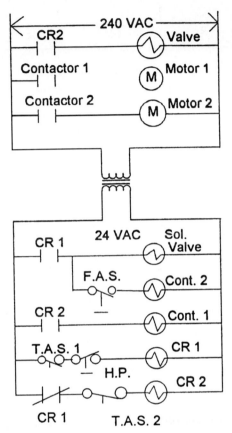

FIGURE 18.6.27   Typical ladder diagram.

Filter-Drier.   Most manufacturers recommend that a liquid line drier be placed in their remote-type systems at the time of installation. Caution must be used to make certain that one especially designed for heat pump use is installed.

The purpose of the filter-drier is to remove moisture and debris from the refrigerant before it can enter and plug any of the other devices.

Water Flow Control Valves.   The same types of water flow control valves are used on water-source heat pumps as with other types of water-cooled equipment.

One valve is used for the heating mode and one for the cooling mode. Each is adjusted separately to provide the desired temperature inside the coil.

The one used for the cooling mode should be set to maintain the desired condensing pressure for the refrigerant used. The one for the heating mode should be set to maintain about a 40°F (5°C) suction temperature. These settings are very important for the efficient and economical operation of the system.

As with any type of adjustment to the refrigeration system, it takes time for the system to respond to the changes. Therefore, time between adjustments is required to properly adjust system operating pressures.

18.7 COILS

Ravi K. Malhotra, Ph.D., P.E.
President, Heatrons Corp., Fenton, Missouri

Billy C. Langley, Ed. D., CM
Consulting Engineer, Azle, Texas

PART A: DESIGN AND APPLICATION

18.7.1 INTRODUCTION

Coils are used for cooling and heating an air stream under natural or forced convection. This chapter will cover design and application under forced convection. Coils are used as components in room air conditioners, central station air-handling units, and various factory-assembled air heating, cooling, and refrigeration units. The applications of each type of coil are limited to the design, code regulation, and materials of construction.

18.7.2 COIL CONSTRUCTION AND ARRANGEMENT

Basically there are two types of coils, one consisting of bare tubes and the other having extended fin surfaces. Bare-tube coils generally are made with copper, steel, stainless steel, and aluminum tube material. Material selection depends on the type of application. Bare-tubed coils are used mainly for evaporative cooling and sprayed coil dehumidifiers. The design and tube arrangement of bare-tube coils vary with the application and manufacturing capabilities. This chapter deals with extended fin surfaces only.

Several designs and arrangements of extended fin surface coils are available. The construction of extended fin surface coils involves the following consideration:

1. *Tube diameter.* Generally $\frac{1}{4}$- to 1-in (6- to 25-mm) diameter tube coils are used in most air-cooling and heating applications.

2. *Tube arrangement.* Coils are made with staggered or in-line tube arrangement. Staggered-tube coils have a higher airside heat-transfer coefficient compared to in-line tube coils. However, staggered-tube coils also have higher air pressure drop.

3. *Fin type.* Fins are generally spiral or continuous plate made out of aluminum, copper, steel, copper-nickel, or stainless steel. Continuous-plate fins are more commonly used in air-conditioning applications. Fins are flat, corrugated, or louvered. Corrugated or louvered fins have a higher airside heat-transfer coefficient and higher air pressure drop. Fins are spaced from 3 to 20 per in (1 to 8 per cm) depending on the application with given load and air pressure drop. Figure 18.7.1 shows some fin-coil arrangements. Spiral fins are wound on the tubes under pressure and then are solder-coated for good fin bond. In plate-fin coils, tubes are mechanically expanded after fins are assembled. Fins have collars to grip the tube for good fin bond. The collar height can be adjusted to space the fins automatically. The inside surface of the tube is usually smooth. To enhance heat transfer, internal fins or turbulators are used; these are either fabricated or extruded.

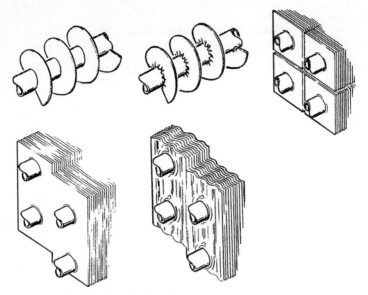

FIGURE 18.7.1   Various fin-tube coil arrangements.

18.7.3   COIL TYPES

18.7.3.1   Water Coils

Water coils are used for air cooling or heating applications. The performance of water coils depends on the air and water velocity, entering air and water temperatures, and mass flow rates. The water velocity inside the tubes usually ranges from 1 to 8 ft/s (0.3 to 2.44 m/s) depending on the design water pressure drop; the optimum water velocity is 3 to 5 ft/s (0.9 to 1.52 m/s). Water coils are provided with various water circuit arrangements. For example a typical eight-tube-high coil can be arranged for 2, 4, 8, and 16 circuits. Figure 18.7.2 shows typical water coil circuiting of an eight-tube-high and four-row-deep coil. Coils are provided with a vent and a drain for proper functioning of the coil. Unless vented, air may be trapped in the coil, possibly causing a reduction in capacity and noise in the system. For job conditions where the waterside may be fouled, a fouling factor is used in designing the coils.

Several types of water coils are used in the field, depending on the application. Following is a description of these coils.

Standard Water Coils.   Figure 18.7.3 shows standard water coils with brazed copper headers and return bends. Full-tube size return bends are used to minimize fluid friction. Headers are correctly sized for each application to provide uniform distribution to all coil circuits. Each header has vent or drain plugs. These coils are constructed of various materials to suit the application. The coils are circuited for draining. However, to ensure complete drainability, coils have tubes pitched in the frame and auxiliary headers are provided for rapid drainage. Figure 18.7.4 shows coils with auxiliary headers for complete and rapid drainage.

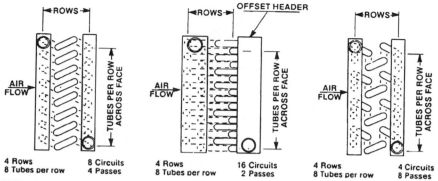

FIGURE 18.7.2 Water coil circuiting of eight-tube-high, four-row-deep coil.

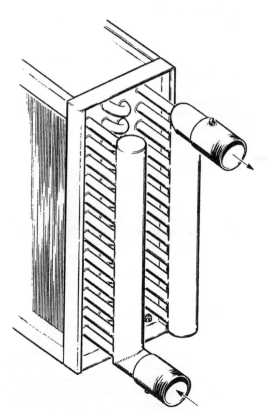

FIGURE 18.7.3 Standard water coil with brazed copper headers.

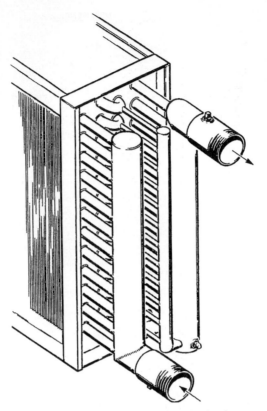

FIGURE 18.7.4   Water coil with auxiliary headers for complete and rapid drainage.

Internally Cleanable Coils.   Figure 18.7.5 shows cleanable plug coils. Coils have plugs on the supply or return connection end of each tube to permit cleaning of the tube interiors. Full-tube-size return bends are used on one end only. Each tube has a special full-size fitting with an easily removable plug with neoprene O-ring to permit insertion of cleaning tools into the tubes. Plugs may be provided on both sides of the coils for cleaning. Headers are sized for each application to provide uniform distribution to all coil circuits. Each header has vent or drain plugs. In general, coils are made of copper tubes; headers and return bends use brass plugs. Special-construction coils are also available for higher pressures and temperatures. Figure 18.7.6 shows a coil with a removable box-type steel header on the supply connection end. This construction offers access to the tube interiors when periodic mechanical cleaning is necessary to remove sediment deposits or scale. Tubes are rolled into a heavy steel tube plate. The box header is baffled for correct circuiting and is gasketed and bolted to the tube sheet. Each connection contains a drain or vent. The opposite end of the coil has full-size return bends so that access to the tube interior exists at only one end. Coils may have removable water box-type steel headers on both ends of the coil. This type of coil provides straight-through mechanical cleaning of the tube interiors. Depending upon the application, these coils may be constructed of other materials.

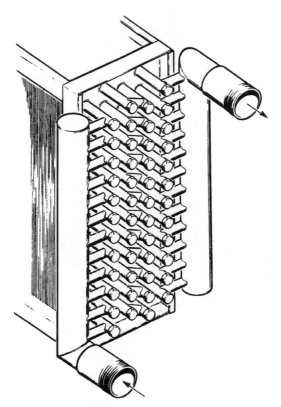

FIGURE 18.7.5   Water coil with cleanable plugs.

18.7.3.2   Direct Expansion Coils

Refrigerant coils are more complex than water coils. Refrigerant distribution and loading are critical to coil performance. These coils are used for recirculation, flooded, and adiabatic expansion systems. Flooded and circulation coils are used mainly for low-temperature applications. Expansion coils are used in air-conditioning applications. Direct expansion coils have either a capillary tube or a thermostatic expansion valve to regulate the flow. The capillary tube system is used in small packaged units such as room air conditioners. The bore and length of capillary tube are sized for full-load design conditions to evaporate liquid refrigerant. Capillary tube systems do not operate efficiently at other than design conditions. Thermostatic expansion valves are used for larger systems. A typical coil with a thermostatic valve assembly is shown in Fig. 18.7.7. The thermostatic expansion valve regulates the flow or refrigerant to the coil in direct proportion to the load. It also maintains the superheat at the coil suction outlet within the predetermined limits. The superheat is generally set at the range of 5 to 10°F (2.8 to 5.6°C).

Direct expansion coils are circuited to provide good heat transfer and oil return at reasonable pressure drop across the circuit. Circuit loading will depend on the tube diameter, circuit length, type of refrigerant, and operating conditions. For multicircuit coils, a distributor is provided to evenly distribute the refrigerant in each circuit. Table 18.7.1 lists the recommended circuit loading for various tube-size coils.

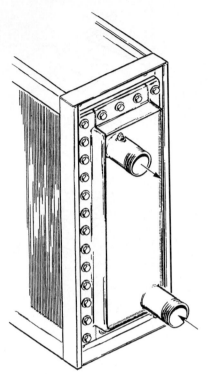

FIGURE 18.7.6 Water coil with removable box-type steel header.

Some coils may have multiple thermostatic expansion valves. If the valves are located on the face of the coil, it is called "face control." If the valves are located in the (rows) depth of the coil, it is called "row control." Figure 18.7.8 shows a coil with both face and row control arrangements. The face of the coil is controlled by two thermostatic expansion valves. Face control is used in coils to provide equal loading on all refrigerant circuits within the coil. Row control is usually designed for special coils, and loading can be divided in several ways. Table 18.7.2 lists the progressive loading per row.

18.7.3.3 Steam Coils

Standard Steam Coils.   Coils are made out of single tubes. Steam enters through one end of the coil, and the condensate comes out the other end. Coils are provided with tubes pitched in the casing for condensate drainage or can be mounted with the tubes vertical. Generally these coils are used with entering air temperatures above freezing. However, vertical tube coils can be used with entering air below freezing. These coils are constructed of copper, steel, and stainless steel tubes with aluminum, copper, steel, and stainless steel fins. Figure 18.7.9 shows a standard steam coil.

Distributing Steam Coils.   Distributing steam coils are constructed with orificed inner steam distributing tubes, centered and supported to provide uniform steam distribution and

FIGURE 18.7.7   Direct expansion coil with thermostatic expansion valve.

TABLE 18.7.1   Recommended Circuit Loading. Btu/h*

Application	Refrigerant	Tube outer diameter		
		$^3/_8$ in (9.5 mm)	$^1/_2$ in (12.7 mm)	$^3/_8$ in (15.9 mm)
Comfort air conditioning	12	6000	12,000	18,000
	22	9000	18,000	24,000
	502	8000	16,000	22,000
	134A	7500	15,000	22,000
	404A	5800	11,500	17,000
Commercial refrigeration	12	2000	4,000	6,000
	22	3000	6,000	9,000
	502	2800	5,400	7,800
	404A	2500	5,000	7,500
		1800	3,600	5,500

*For metric conversion factor, see Appendix B.

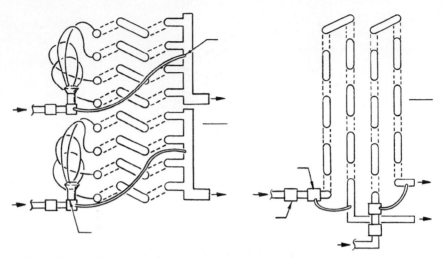

FIGURE 18.7.8   DX coil with face and row control arrangement.

maximum protection against freeze-ups. This design prevents freezing of the condensate, provided a sufficient amount of steam is supplied to the coil and the condensate is removed by proper trapping as fast it is condensed. Coils are provided with tubes pitched in the casing. Figure 18.7.10 shows distributing steam coils. Coils are constructed of copper, steel, and stainless steel tubes with aluminum, copper, steel, and stainless steel fins.

When uniform leaving air temperatures over the entire coil face area are required, coils are designed with two steam inlet connections feeding the alternate tubes. As shown in Figure 18.7.11, this design prevents stratification of heated and unheated air. By keeping both ends of the coil equally warm, maximum protection against freeze-up is provided. These coils have inner steam distributing tubes. Steam supply and return connections are on the same end of the coil, and tubes are pitched in the casing for condensate flow. Headers are sized for each application to provide uniform steam distribution to all coil circuits. Generally, coils are made of copper tubes and aluminum fins. Special construction for higher-pressure applications is also available.

TABLE 18.7.2   Progressive Loading per Row

Rows deep	Approximate percentage of total capacity by rows							
	1	2	3	4	5	6	7	8
2	55	100						
3	40	73	100					
4	33	60	82	100				
5	29	52	71	87	100			
6	26	47	60	78	90	100		
7	24	43	58	70	81	91	100	
8	21	39	53	65	76	85	93	100

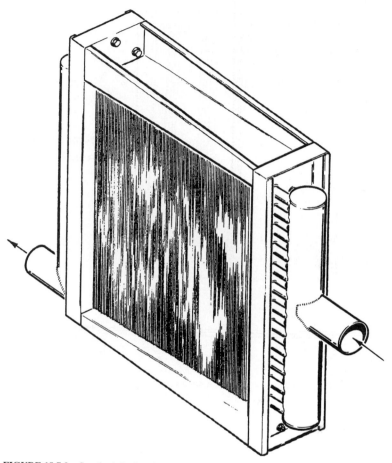

FIGURE 18.7.9 Standard single-tube steam coil.

18.7.4 COIL APPLICATIONS

Coils are used either by original equipment manufacturers or in-field built-up systems. Cooling coils are provided with a drain pan to catch the condensate formed during the cooling cycle. In the case of stacked coils, the condensate trough is provided in between the two coils to prevent flooding of the bottom coil. The drain pan connection should be on the downstream side of the coils and should be big enough for rapid drainage. Drain pans should be insulated to prevent sweating. Generally factory-assembled central station units incorporate these features.

The American Refrigeration Institute (ARI) standard for forced-circulation air-cooling and air-heating coils[2] covers the application range for all types of coils. These are the application and design ranges of various types of coils.

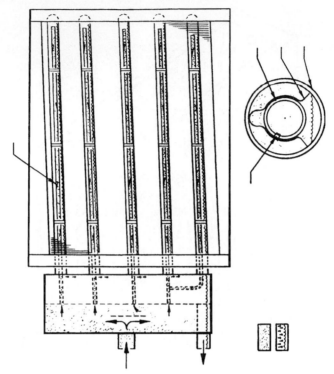

FIGURE 18.7.10   Distributing steam coil with orificed inner steam distributing tube.

For cooling coils:

Entry air, dry bulb	65 to 100°F (18.3 to 27.8°C)
Entry air, wet bulb	60 to 85°F (15.6 to 29.4°C)
Air face velocity	200 to 800 ft/min (61 to 244 m/min)
Fluid velocity	1 to 8 ft/s (0.3 to 2.44 m/s)
Entry fluid temperature	35 to 65°F (1.7 to 18.3°C)
Saturated suction temperature	30 to 55°F (−1.1 to 12.8°C)

For heating coils:

Entry air, dry bulb	0 to 100°F (−17.8 to 37.8°C) for hot-water coils; −20 to 100°F (−28.9 to 37.8°C) for steam
Air face velocity	200 to 1500 ft/min (61 to 457 m/min)
Fluid velocity	0.5 to 8 ft/min (0.15 to 2.44 m/s)
Entry fluid temperature	120 to 250°F (48.9 to 121.1°C)

For special applications such as low-temperature refrigeration (industrial or processing), the above-mentioned ranges may be varied.

In air dehumidifying coil applications, the air face velocity should be kept low to prevent condensed moisture from being blown off the coil. The maximum recommended face

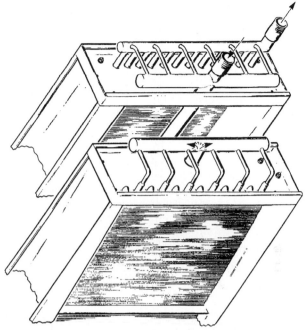

FIGURE 18.7.11 Distributing steam coil feeding alternative tube from both ends to prevent stratifications.

velocity is 550 ft/min (168 m/min). Over 550 ft/min (168 m/min) the condensate will be blown off the coil into the ductwork. In this case eliminators are used to prevent any water carryover. To enhance heat transfer, cooling coils are sprayed with water. In this case the leaving air temperature approaches the saturation temperature. Coils may have air bypass to control the air temperature.

18.7.5 COIL SELECTION

The following variables are to be considered in choosing a coil:

- Total load, Btu/h (W)
- Entering air temperatures, dry and wet bulb
- Available space for the system
- Cooling or heating media flow rate and entering temperature
- Air quantity
- Allowable air and water pressure drop
- Special coil material considerations

The total load is determined based on information available elsewhere in this handbook and on actual room load. Outdoor design temperature data are also available elsewhere in this handbook. With other known variables, a specific coil can be selected from various coil manufacturers' catalogs.

18.7.6 HEAT-TRANSFER CALCULATIONS

For sensible cooling coils, design problems with dry finned-tube heat exchangers require solution of the equation

$$q = U_o A \, \Delta t_m \qquad (18.7.1)$$

where q = rate of heat transfer, Btu/h (J/s)
U_o = overall heat-transfer coefficient, Btu/(h · ft² · °F) [J/(s · m² · °C)]
A = total outside surface, ft² (m²)
Δt_m = log mean temperature difference, °F (°C)

The key coil design parameter in the formula is the overall heat transfer coefficient U_o, which is a function of (1) the metal thermal resistance of external fins and tube wall, (2) inside surface heat-transfer coefficient, and (3) the airside or outside surface heat-transfer coefficient. These are expressed in mathematical terms by the equation

$$U_o = \frac{1}{R_m + B/h_i + 1/h_o} \qquad (18.7.2)$$

where U_o = overall heat-transfer coefficient of dry surface, Btu/(h · ft² °F) [J/(s · m² · °C)]
R_m = metal thermal resistance of external fins and tube wall, ft²/(h · °F · Btu) [m²/(s · °C · J)]
h_i = heat-transfer coefficient of inside surface, Btu/(h · ft² · °F) [J/(s · m² · °C)]
h_o = heat-transfer coefficient of outside surfaces, Btu/(h · ft² · °F) [J/(s · m² · °C)]
B = ratio of outside and inside surfaces = A_o/A_i
A_o = total outside surface area, ft² (m²)
A_i = total inside surface area, ft² (m²)

18.7.7 METAL RESISTANCE OF EXTERNAL FINS AND TUBE WALL

The total metal thermal resistance R_m to heat flow through external fins and the prime tube wall may be calculated as follows:

$$R_m = R_f + R_t \qquad (18.7.3)$$

Here the variable-fin thermal resistance R_f, based on the total surface effectiveness of dry surface, is

$$R_f = \left(\frac{1-\eta}{\eta}\right)\frac{1}{h_o} \qquad \eta = \frac{\phi A_s + A_p}{A_o} \qquad (18.7.4)$$

where η = total surface effectiveness
ϕ = fin efficiency
A_s = secondary fin surface, ft² (m²)
A_p = net primary surface, ft² (m²)

The tubeside resistance R_t is calculated from

$$R_t = \frac{BD_o}{24K_t}\ln\frac{D_o}{D_i} \qquad (18.7.5)$$

where D_o = tube outside diameter, in (mm)
D_i = tube inside diameter, in (mm)
K_t = thermal conductivity of tube material, Btu/(h · ft²) [J/(s · m²)]

Metal thermal resistance is mainly dependent on the tube material, fin material, outside heat-transfer coefficient, and surface effectiveness.

18.7.8 HEAT-TRANSFER COEFFICIENT OF INSIDE SURFACE

The inside heat-transfer coefficient varies due to the tube diameter, mass flow rate, and physical properties of the fluid. For all water coils with smooth internal tube walls, the tube-side heat-transfer coefficient can be calculated from

$$h_i = \frac{150(1+0.011t_{wm})(V_w)^{0.8}}{D_i^{0.2}} \tag{18.7.6}$$

where t_{wm} is the mean water temperature and V_w is the water velocity.
   For other than water, the following equation[1] should be used to calculate the inside heat-transfer coefficient:

$$h_i = 0.023 \left(\frac{K}{D_i} \right) \text{Re}^{0.8} \text{Pr}^{0.33} \tag{18.7.7}$$

where K = thermal conductivity of fluid, Btu/(h · ft² °F) [J/(s · m² · °C)]
   Re = Reynolds number
   Pr = Prandtl number

Due to the two-phase flow characteristics of refrigerant coils, it is not possible to accurately calculate the inside heat-transfer coefficient as a function of flow rates. However, information is available in various handbooks for approximate data. Unit efficiency dictates that the evaporator pressure drop be kept low [perhaps 2 to 4 lb/in² (13.8 to 27.6 kPa)] to maintain as large a log mean temperature difference as possible for maximum capacity. Condenser refrigerant pressure drop should be limited [perhaps 10 to 15 lb/in² (68.9 to 103.4 kPa)], to keep the head pressure low for minimum compressor input.

18.7.9 HEAT-TRANSFER COEFFICIENT OF OUTSIDE SURFACE

This coefficient typically accounts for 50 to 80 percent of the overall coefficient and therefore has been the target of past design work to improve it. Initial development for improved capacity has moved from in-line to staggered-tube arrangement and from flat to corrugated fin surfaces. With every improvement an increase in coil air friction has occurred, so the type of fin surface must be evaluated before a coil is designed. The finside heat-transfer coefficient is generally determined by tests; however, some empirical data are available.

18.7.10 DEHUMIDIFYING COOLING COILS

When the dewpoint of entering air is higher than entering fluid temperature, normally the coil will remove moisture in addition to sensible cooling. In this case, the dewpoint temperature of air leaving a cooling coil is lower than that of the air entering the coil.

In most air-conditioning applications air contains a mixture of water and dry air and enters the coil to lose sensible and latent heat. Latent heat occurs only in those parts of the coil where the temperature of the coil surface is below the dewpoint temperature of the entering air.

ARI (American Refrigeration Institute) Standard 410-81 covers in detail the procedure to rate or select dehumidifying cooling coils. The coils may be fully wet or partially wet depending on the entering air and water temperatures. Most coil manufacturers have published rating tables to select coils. These tables have been derived from test data. The method of testing coils is given in American Society of Heating, Refrigeration, and Air-Conditioning Engineers' (ASHRAE) Standard 33-78.[3] Due to the complexity of calculations, most manufacturers have computer programs to rate or select dehumidifying cooling coils. The dry portion of the coil is rated per dry-bulb temperatures, and the wet portion of the coil is rated per enthalpy of the air and fluid.

PART B: OPERATION AND MAINTENANCE

18.7.11 INTRODUCTION

The two coils used in air conditioning and refrigeration systems are the evaporator and the condenser. In heat pump systems they are called indoor and outdoor coils, respectively.

By knowing the finer points of their operation and maintenance, service technicians will help customers achieve improved operating efficiencies through their HVAC systems.

18.7.12 EVAPORATORS

Evaporator coils are found inside the conditioned space. Their purpose is to absorb heat from the space and products stored inside into the refrigerant as it flows through the system.

Several types of evaporator coils are in use. However, the most popular is the forced-air coil. They are generally more efficient than the other types.

Forced-air coils are further divided into blow-through and draw-through types. The draw-through type is the most efficient of the two. However, the blow-through is used extensively because of the extra space required by the draw-through type.

The draw-through type is more efficient because air flows through most of the coil surface in an equal amount. In the blow-through type, most of the air is forced through the coil directly in front of the blower, leaving the outside edges of the coil operating at less than the designed efficiency.

Other types include the flat-plate coil, and other types of fast-freeze evaporators used in food-processing plants.

The immersed type has the evaporator placed in a fluid bath of some type, with the fluid stored at a given temperature to be used later. This type is popular in some larger a/c applications that take advantage of off-peak electric rates.

18.7.12.1 Watch for Low Load

Efficient operation of any type of evaporator is due to the rate at which heat is absorbed into the refrigerant as it passes through the coil.

Probably the most common cause of reduced evaporator efficiency is low load. This may be caused by several factors.

1. *Dirty evaporator surface*—This dirt, which is carried with the air, contacts the wet coil surface and sticks to it. The dirt acts as an insulator, reducing the amount of heat absorbed. It also prevents the proper amount of airflow through the coil.

To bring the evaporator back to peak efficiency, it must be cleaned. Sometimes this can be done by brushing the fins in the direction of the fins with a wire brush. Sometimes it can be done by spraying the coil with a nonacid coil cleaner and then spraying it with a high-pressure water nozzle in the opposite direction of airflow.

In extreme cases, the coil will need to be steam cleaned to remove the oil and dirt clogging the fins. This cleaning sometimes requires that the coil be removed from the unit.

Be sure to use the proper procedures to remove the refrigerant. Make certain that the tubing is properly sealed to prevent moisture and dirt from entering the system. And protect the motors and electrical equipment from the water or steam to prevent damage.

2. When cleaning the coil, all bent fins should be straightened so that the air can flow freely through the coil. When the airflow is restricted by bent fins, the heat transfer rate is reduced because the air cannot contact the complete coil surface.

3. Frost on the evaporator can also cause the system to be under-loaded. A thickness of ice or frost of $1/_8$ in (3 mm) can reduce the coil's efficiency approximately 25 percent.

Frost or ice on the coil is usually caused by:

a. Low load caused by a dirty coil, dirty filter, low refrigerant charge;

b. Low ambient outdoor temperatures;

c. Dirty blower; or

d. Blower that is running too slow.

A blower that is running too slow is usually indicated by a high temperature drop across the coil. Usually this temperature drop should be about 20°F (−6°C) when 400 Cu ft/min of air per ton of capacity is moving through the coil.

Some higher-efficiency units have a temperature drop more than 20°F (−6°C), and an airflow greater than 400 cfm (11 m^3) per ton. Follow the equipment manufacturer's recommendations.

Sometimes, when proper piping practice has not been followed or when the coil was not installed level, oil will collect in some of the tubes and reduce the amount of coil surface used.

Sometimes when a compressor or condensing unit has been replaced, one that has too much capacity for the evaporator will be used. This causes the suction pressure to be too low, in turn causing frost or ice to build up on the evaporator.

To correct this problem, the proper size compressor or condensing unit must be installed.

18.7.13 CONDENSER COILS

18.7.13.1 Condensers

The purpose of the condenser is to reject heat absorbed by the evaporator, plus the heat of compression created by compressing the refrigerant.

The condenser must be sized to fit the system for the circumstances under which it must operate. Two types of condensers used are as follows:

Air Cooled. Both draw-through and blow-through types. Draw-through types are more efficient than blow-through types, because the air is pulled equally through the complete surface of the condenser coil. (See Fig. 18.7.12.)

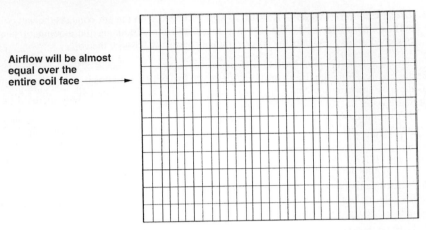

Airflow will be almost
equal over the
entire coil face ——————→

FIGURE 18.7.12   Draw-through coil.

Blow-through coils are less efficient because most of the air flows through the condenser coil directly in front of the fan discharge. Thus, the outer edges of the condenser are not used to their full capacity. (See Fig. 18.7.13.)

In condensing units, the draw-through type is most popular because sufficient room is available for the extra space needed.

Air-cooled condensers are the most popular because they do not require special attention to prevent them from freezing during cold weather, especially when they are installed outdoors.

Air-cooled condensers must be installed away from shrubs, trees, flowers, and other plants whose debris can get into the condenser coil. Air-cooled condensers should not be installed near patios or under bedroom windows, where the noise could interfere with normal use of the adjacent area.

The prevailing wind should also blow in the direction of airflow through the condenser. This will aid in keeping the condenser coil cool, and not overload the condenser fan motor.

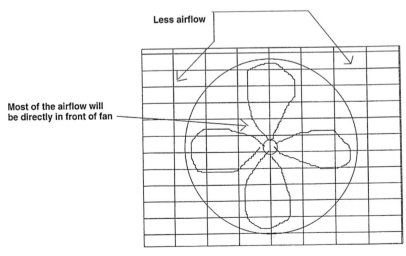

Less airflow

Most of the airflow will
be directly in front of fan ——————

FIGURE 18.7.13   Blow-through coil.

Usually the condenser requires about 1000 cfm (28 m³) of air per ton of capacity to remove the heat absorbed by the evaporator, plus the heat caused by compressing the refrigerant.

The temperature rise through the condenser should be from about 25 to 35°F (−3 to 2°C). Some higher-efficiency units have different temperature rise requirements. Check the equipment manufacturer's recommendations.

The last two or three rows of condenser tubing are usually used for subcooling the refrigerant before it enters the liquid line. (See Fig. 18.7.14.) This is to help increase the efficiency of the unit by reducing flash gas at the flow-control device.

A subcooling of 10 to 15°F (−13 to −10°C) is generally used. Again, on some high-efficiency models the subcooling may be as high as 20°F (−6°C). Be sure to check the equipment manufacturer's recommendations.

Air-cooled condensers must be kept clean to prevent high head pressures and compressor overheating. *Heat is the compressor's greatest enemy.*

The coil may be cleaned with a high-pressure nozzle on a garden hose. Spray the water in the opposite direction of the normal airflow through the coil. Occasionally a nonacid cleaning agent may be needed to remove all the dirt and grease.

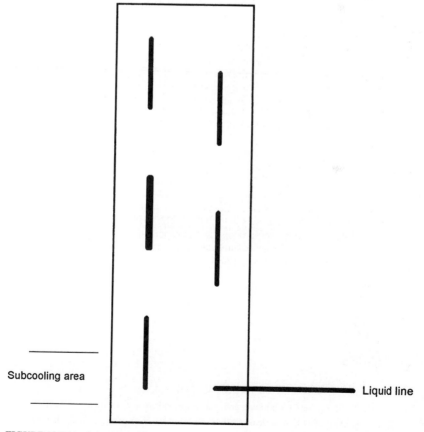

FIGURE 18.7.14   Subcooling area.

Unremoved, the dirt causes the heat transfer to fall below that required for the system. Also, it will prevent the proper amount of airflow through the coil for proper cooling and condensation of the refrigerant.

Always remove any debris left inside the condensing unit housing; it will only get back into the coil.

When flushing the condenser coil with water, protect the condenser fan motor and the electrical compartment to keep water out of them. They should be covered with a piece of plastic.

The fins on the coil must be straightened so that air can flow freely through the coil. When a large portion of the fins are bent over, the proper amount of air cannot flow through the condenser, causing high head pressure. (Fin combs are available at most supply houses that will help in straightening the fins.)

Water-Cooled Condensers.   The type used usually depends upon the equipment manufacturer. However, they all operate on the same basic principle.

Water flows in the opposite direction of the refrigerant. Thus, the cooler entering water contacts the cooler leaving refrigerant first.

Water-cooled condensers require more maintenance than do air-cooled condensers. Maintenance generally depends upon the type of water used for cooling.

Water-cooled condensers will have an accumulation of scale on the tubes that slows the heat transfer from the water to the refrigerant. This scale must be removed so the condenser can operate as designed.

Care must be taken in removing the scale, because an acid-type chemical is used. Too much of the chemical also will remove some of the copper from the tubing inside the condenser shell.

When enough of the tubing has been removed, the tube will give way under the refrigerant pressure inside it. The refrigerant will leak out and the water will leak into the refrigerant system. This causes much trouble for the system and for the service technician who finds it.

Water-cooled condensers are designed to operate properly with 3 gal (11 L) of water/ton/minute flow through the condenser with a 10°F (5°C) rise. A temperature rise lower than this is usually an indication that the condenser is scaled and must be cleaned. A temperature drop higher than this indicates insufficient water flow through the condenser.

Insufficient water flow is usually caused by a plugged pump strainer or a pump that is not running at the proper speed. Also, the spray nozzles on the tower may be plugged.

Reduced water flow will generally cause higher-than-normal head pressure. A scaled condenser coil also causes higher-than-normal head pressure.

The amount of stoppage or scale determines how high the head pressure will rise. Normally, on water-cooled condensers using HCFD-22, the head pressure will be around 215 psig (1482 kPa) when everything is operating properly.

When a water-cooled condenser is subjected to freezing temperatures, it must be protected, usually by draining the water circuit, including the cooling tower, and blowing any water from the condenser tubes.

If the tubes should freeze and burst when the system is again placed in operation, water may be pumped into the refrigerant circuit. This condition requires much work and is very expensive to repair.

REFERENCES

1. W. H. McAdams, *Heat Transmission*, 3rd ed., McGraw-Hill, New York, 1954.
2. ARI Standard 410-91, Air Conditioning and Refrigeration Institute, Arlington, VA, 1991.
3. ASHRAE Standard 33-78, American Society of Heating, Refrigeration and Air-Conditioning Engineers, Atlanta, GA, 1978.

18.8 CONDENSATE CONTROL

Warren C. Trent, M.S., P.E.
CEO, Trent Technologies, Inc., Tyler, Texas

C. Curtis Trent, Ph.D.
President Emeritus, Trent Technologies, Inc., Tyler, Texas

Hudy C. Hewitt, Jr., Ph.D.
*Chairman, Department of Mechanical Engineering,
University of New Orleans, New Orleans, Louisiana*

18.8.1 INTRODUCTION

One of the primary functions of an air-conditioning system is to remove water from the air it circulates. Proper collection and disposal of this water is essential to maintaining a dry, clean, and uncontaminated HVAC system.

Water is both the most useful and most destructive of compounds. In the right places it is invaluable. In the wrong places it can cause disaster.

In a HVAC system the *right* places—that is, the only acceptable places—for water (condensate) are:

1. Surfaces of the cooling coil,
2. A "small and well drained" condensate drain pan, and
3. A free-flowing (nonstagnant) condensate drain line.

The *wrong* places for condensate in a HVAC system are on

1. Internal insulation;
2. Surfaces of walls, top, and floor of air handlers;
3. Large condensate drain pans that cover the floor of air handlers;
4. Surfaces of the fan motor, fan, fan casing, and blades;
5. Internal duct walls;
6. Air supply grilles; and
7. Surrounding structures that support the air handler.

In these places, water is not only destructive to the HVAC system, surrounding property, and building contents, it creates a growth haven for algae, fungi (mold, mildew, etc.), and various forms of bacteria, including *Legionella pneumophilia*.

The consequences of these conditions are excessive and unnecessary costs for building owners-users in terms of service calls, maintenance effort, equipment damage, surrounding property damage, and the threat to human health.

All these costly consequences are avoided by systems designed to confine the spread of condensate and eliminate internal system wetness attributable to the following:

1. Condensate carryover and drips
2. Condensate drain pan
3. Humidity and temperature in the air supply system

4. Position of fan (blower) in the air handler

5. Seal on condensate drain line

6. Condensate drain line sizes and slope

Far too many existing systems in this country exhibit deficiencies in one or more of the above categories. The deficiencies inevitably increase building operating costs and contribute to "sick building syndrome" and "building related illnesses."

How to prevent these deficiencies in new system designs, remedy them in existing systems, and minimize system maintenance is the subject of this chapter.

18.8.2   SYSTEM DESIGN

A successful condensate drain system design must address each of the areas of potential deficiency enumerated above, including consideration of critical design factors, economic factors, and clarity of specifications.

18.8.3   CONDENSATE CARRYOVER AND DRIPS

Condensate carryover and condensate drips are common causes of wet, dirty, and contaminated HVAC systems.

Excessive air velocity through the cooling coil will blow condensate from the coil onto the plenum floor and/or onto other components of the system. Sloped cooling coils and noninsulated coolant (refrigerant or circulated fluid) lines often allow condensate to drip onto internal surfaces, damage equipment, and cause contamination inside the air handler.

18.8.3.1   Critical Design Factors

Condensate Carryover.   The capacity of a particular cooling coil design to resist condensate blow-off (carryover) depends upon a number of factors, including the following:

1. Velocity of the air approaching the cooling coil

2. Diameter of tubes

3. Distance between tubes

4. Number of fins per inch

5. Thickness of fins

6. Amount of condensate present on surfaces of the cooling coil

The velocity of air approaching the cooling coil (coil-face air velocity) is too low to entrain condensate and cause carryover. However, as air passes through the coil, it is accelerated to higher velocities. Whether this velocity reaches a level sufficient to entrain condensate from the coil surfaces depends upon the reduction of flow area effected by tubes, fins, and the condensate held on the coil surfaces.

How these variables interact to affect condensate carryover is illustrated in Fig. 18.8.1, for a typical cooling coil design. For this particular coil type, coil-face velocities above those shown will result in condensate carryover. Other coil designs may exhibit somewhat

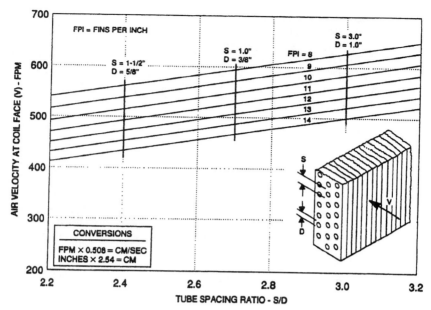

FIGURE 18.8.1   Coil face velocity above which condensate carryover occurs, typical cooling coil.

different characteristics, but the variables and their general relationships remain the same. For a specific coil design, applicable data—in various forms—are usually available from the coil manufacturer.

Cooling coil design arrangements that permit condensate to be carried down stream are unsuitable for HVAC applications. Large drain pans extended to protect the floor from condensate carryover are of little value. The floor still becomes wet, as do fan surfaces and other downstream components, creating serious cost and health problems for building owners-users, see Secs. 18.8.4 and 18.8.7.

Condensate carryover "eliminators," placed at the leaving (trailing) edge of the coil, are unsuitable as a preventive measure. They introduce appreciable air pressure losses that increase energy consumption. In addition, they not only create costly maintenance problems, but their wet surfaces promote the growth of health threatening fungi and bacteria.

Condensate Drips from Sloped Coils.   Sloped cooling coils allow condensate to drip away from the coil surfaces. At low slope angles, surface tension may be adequate to retain the condensate and allow it to drain into the condensate pan. However, foreign deposits on the coil can easily destroy the effects of surface tension and cause dripping to occur.

While the air velocity entering the coil tends to reduce dripping, when the fan is operating, it is not a reliable force to prevent condensate dripping.

Extending the condensate drain pan to catch condensate drips introduces another equally serious condition—large drain pans—discussed in Sec. 18.8.4.

Under no circumstances should an air filter be placed beneath a sloped coil. A wet filter forms an ideal place for the growth of contaminating organisms.

Condensate drips from cooling coils and their detrimental effects are easily avoided, by utilizing only vertically oriented coils.

Condensate Drips from Coolant Lines.   Noninsulated coolant lines that are exposed to conditioned air, inside the cooling air flow path, form condensate that will drip onto the floor and components—damaging equipment and contaminating the HVAC system.

Condensate drips from coolant lines and their detrimental effects are easily avoided by simply applying proper insulation to all exposed lines.

18.8.3.2 Economic Factors

Changes in cooling coil design for the purpose of reducing air velocity and the potential for condensate carryover will almost certainly add some cost to the HVAC unit. That is, reducing the air velocity at the coil face (to reduce condensate carryover) from 600 to 500 ft (150 m) per min can be expected to increase coil cost by about 15 percent. This, however, is a small cost compared to the potential savings.

For example, preventing condensate carryover removes any possible justification for extending the condensate pan more than a few inches downstream of the coil. Thus, the cost of fabricating and installing a large stainless steel condensate pan is eliminated.

Far more significant, however, are the savings that accrue as a result of preventing condensate from being blown onto the HVAC components and duct work, downstream of the cooling coil. Easily recognized by the building owner-user, these savings appear in the form of reduced maintenance, reduced property damage, reduced indoor air contamination, and longer equipment life.

The hardware costs for eliminating sloped coils depend upon the angle of slope involved. Costs for insulating coolant lines are minimal to insignificant. Regardless of initial costs, however, highly sloped coils and noninsulated coolant lines represent unacceptable design compromises.

18.8.3.3 Suggested Statements for Specification

1. The condensate cooling coil shall be designed to accommodate the maximum air flow of the HVAC unit, with a condensing rate of at least 0.013 lb (90 gr) of water per pound of dry air, without allowing condensate carryover.

2. All cooling coils shall be installed vertically, without slope.

3. All coolant lines in the air-handling unit must be insulated unless they pass directly over the condensate drain pan.

18.8.4   CONDENSATE DRAIN PAN

A properly designed drain pan is essential to the successful removal of condensate from a HVAC system. Pans that allow condensate to stand and stagnate form an ideal growth haven for the proliferation of biological and microbial agents—disease causing organisms. When condensate stagnation occurs, the HVAC system becomes a source of air contamination and poses a threat to human health.

The major considerations involved in designing a satisfactory drain pan include the following:

1. Condensate flow rate;

2. Drain ports—position, type, and size;

3. Condensate drainage provisions; and

4. Pan material.

18.8.4.1 Critical Design Factors

Condensate Flow Rate.   The rate of condensate flow from a HVAC unit is determined by the total cooling capacity and how this capacity is divided between latent heat and sensible heat. Sensible heat ratio at which a HVAC system must operate may vary from near 0 to 100 percent of the total cooling capacity, depending upon a number of variables. These variables include the (a) amount of internally generated moisture—human, animal, and equipment; (b) amount of infiltration; (c) amount of outdoor ventilation air; and (d) the absolute humidity (humidity ratio) of the outdoor air.

Figure 18.8.2 shows how sensible (and latent) heat ratios affect the amount of condensate removed from cooling air. To determine the condensate flow rate for a given system, simply enter Fig. 18.8.2 with the operating sensible heat ratio at the nominal rate cooling capacity. For example, a system rated at 60 tons operating at a sensible heat ratio of 0.80 will remove about 0.30 gal (1.1 L) of condensate per min. The same system operating at a sensible heat ratio of 0.30 will remove about 1.05 gal (4 L) of condensate per min.

The shaded areas in Fig. 18.8.2 indicate typical operating ranges of sensible heat ratio for various levels of outdoor ventilation air.

Drain Ports.   Drain port position, type, and size all affect how deep condensate stands in the drain pan. Each of these factors affects the potential for stagnation. Even slight variations can have an appreciable effect on the minimum level at which condensate stands in the pan.

Drain Port Position.   Condensate drain ports may be positioned on a pan wall or in the bottom of the pan. For most effective drainage, ports must be flush with the bottom of the pan.

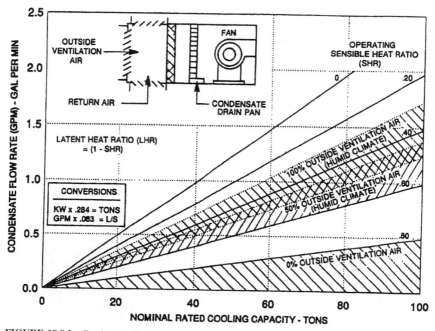

FIGURE 18.8.2   Condensate flow rate as a function of nominal ton rating and operating sensible heat ratio.

Drain Port Type.   Male drain connections are more desirable than female connections. A female connection requires an internal drain connection fitting which reduces the flow area. This restriction blocks condensate flow and catches debris. In addition, in a pan wall drain port, this restriction raises the depth that condensate stands in the pan.

Drain Port Size.   The drain port size depends upon the condensate flow rate, whether the port is in a side wall or in the bottom of the pan, and what condensate depth can be tolerated in the pan.

Condensate Drainage Provisions.   In order for condensate to flow from a particular drain pan, under the force of gravity, it must rise to a predetermined depth. This level depends upon two forces: surface tension of the water and the water level (head) necessary to provide the required flow rate.

Recent test data (Ref. 1) indicate that about $1/8$ in (3.2 mm) of water is required to overcome surface tension and permit condensate flow. Thus, independent of the position, type, and size of the drain port, condensate—when present—will stand in the pan at some finite depth. The total depth, of course, increases in proportion to the condensate flow rate.

Figure 18.8.3 shows how condensate depth in a level pan varies with the condensate flow rate for different port sizes and positions. As shown, a bottom drain port provides better drainage than does a side wall port. Neither drain port arrangement, however, allows for complete drainage from a level pan. That is, condensate will remain in the pan, at some depth, as long as moisture is being removed from the circulated air. If allowed to "back-up" into the pan and stagnate, condensate—at any depth—promotes the growth of biological and microbial agents that contaminate the HVAC system.

Although a bottom drain port affords better drainage than does a wall drain port, either can be used successfully if properly integrated with the condensate pan design.

The importance of proper pan drainage and the avoidance of condensate stagnation has been recognized for years. For example, ASHRAE Standard 62–1999, paragraph 5.11 (Ref. 2) includes the following statement: "Air handling unit condensate pans shall be designed for self-drainage to preclude the buildup of microbial slime" (see Ref. 2). In this context, self-drainage is taken to mean that condensate shall flow from the pan in such a way that no stagnant water remains to support the growth of contaminating organisms (microbial slime).

One means of assisting self-drainage is to slope the condensate drain pan in the direction of the drain port. However, it is generally not practical to add enough slope to prevent condensate from covering a large area of the drain pan, during the cooling operation. *How sloping the pan in one direction by $1/4$-in per ft (20 cm/m)—in combination with other pertinent variables*—affects the amount of pan area covered with condensate is shown in Fig. 18.8.4. Doubling this slope, of course, reduces the condensate coverage by one-half, while reducing the slope by one-half, doubles the area covered by condensate.

Although a sloped pan enhances flow, a portion of the pan will remain covered with condensate, see Fig. 18.8.4. Under these circumstances the pan width—the distance the pan extends downstream of the cooling coil—is critical in preventing stagnation of condensate and precluding the growth of microbial slime.

Figure 18.8.5 illustrates the problems created by wide pans, often designed to catch condensate carryover from cooling coils. In such a pan, it is inevitable that a considerable amount of condensate will stagnate even if the pan is sloped, see Fig. 18.8.6. *The quantity of stagnant condensate in a wide pan can, of course, be reduced by sloping the pan in both directions. But stagnant condensate is only one of the problems posed by wide pans. The primary purpose of a large pan is to catch condensate droplets* blown off the cooling coil. Even if no condensate stands in the pan, condensate droplets, deposited on the floor of the pan, are held in place by surface tension and promote the growth of various unhealthy fungi and bacteria. *Condensate carryover must be eliminated.*

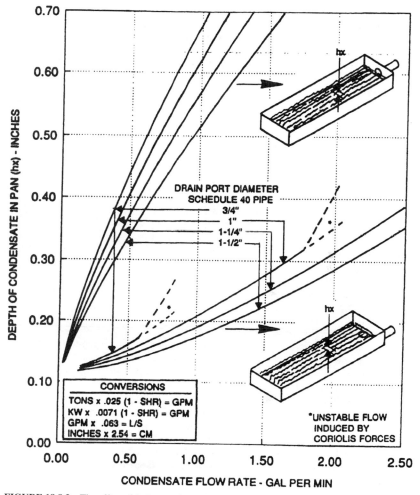

FIGURE 18.8.3 The effect of drain port size and position on condensate flow rate under the force of gravity.

In order to avoid condensate stagnation and droplet deposits, pan width must be limited to ensure that condensate leaving through the drain port will flow continually over the total surface of the floor of the pan.

Figure 18.8.7 suggests pan widths suitable for units with various condensate flow rates, and for port sizes commonly used for side wall drain ports. These same width and drain port size relationships may also be used, conservatively, for selecting pans with bottom floor drain ports.

Greater pan widths and smaller drain port sizes increase the potential for condensate stagnation and system contamination.

Pan Materials. The condensate drain pan may be constructed of either metallic or non-metallic materials, but it is important to avoid iron-based materials that are subject to rapid

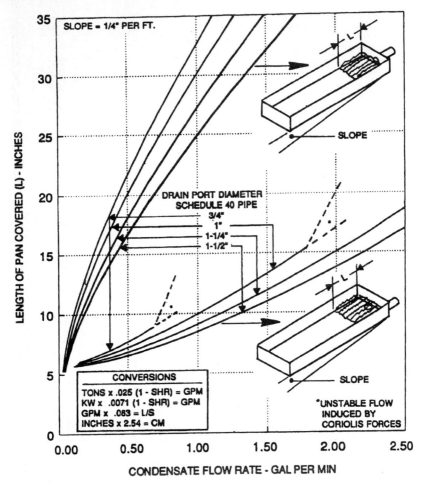

FIGURE 18.8.4   The effect of pan slope on the area covered by condensate.

oxidation in the presence of water and humid air. This is because iron is known to accelerate the growth of *Legionella pneumophilia*, the bacteria that causes Legionnaire's disease. Although *Legionella pneumophilia* has been found in the condensate pans of some HVAC units, no outbreak of Legionnaire's disease has been attributed, officially, to these conditions. Investigators often give two questionable reasons why the HVAC unit is not a threat to the generation and spread of this disease causing bacteria.

First, it is argued that the condensate temperature is too low for sufficient multiplication of the bacteria to occur. Generally speaking this may be the case. However, *Legionella pneumophilia* has been found in water temperatures ranging from 42.3° to 145°F (5.7° to 63°C). Condensate temperatures, typically around 55°F (13°C), are well within this range. Moreover, in the presence of certain iron concentration levels, the multiplication rate of this bacteria increases more than 200 times (Ref. 3, p. 4). In any case, this potential health threat dictates that if an iron-based material is to be used in construction of the condensate pan, it must be stainless steel.

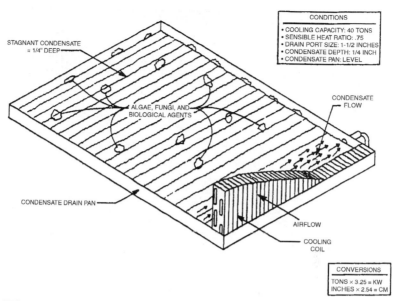

FIGURE 18.8.5   Contamination problems created by wide drain pans.

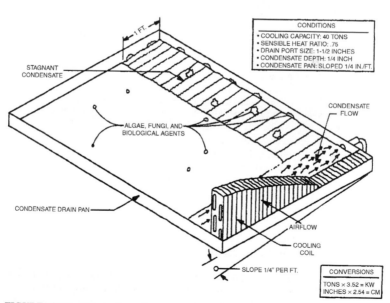

FIGURE 18.8.6   Effects of pan slope on condensate flow rate and contamination.

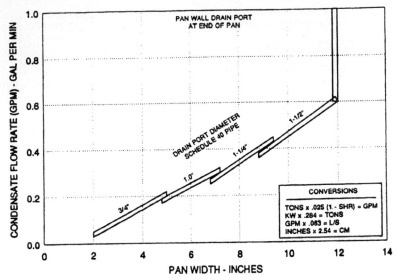

FIGURE 18.8.7   Fan widths suitable for units with various flow rates and for commonly used port sizes.

Second, it is contended that there is no mechanism for aerosolizing the bacteria, a condition essential for its dissemination. Contrary to this contention, as shown in Sec. 18.8.7 of this chapter, a very effective aerosolizing mechanism does exist and frequently operates inside the HVAC unit. This fact emphasizes the importance of utilizing a pan material that avoids high iron concentrations in the condensate.

18.8.4.2   Economic Factors

Incorporating the above design features in the condensate pan has little effect on the initial cost of the HVAC system. In fact, a smaller pan reduces material and fabrication costs.

An internally sloped drain pan necessitates a slightly higher enclosure for the pan and cooling coil; hence, some added cost will be incurred. However, in many systems, the desired pan slope may be achieved by tilting the HVAC unit toward the condensate drain port. This can usually be done without compromising the performance of the system and at little or no cost.

In any case, the added costs of a self-draining stainless steel condensate pan is dwarfed by subsequent savings in terms of reduced maintenance, equipment damage, and fewer human health problems.

18.8.4.3   Suggested Statements for Specifications

1. The condensate drain pan shall be constructed of 18 gauge* stainless steel. It shall have a depth of no more than 2 in* (5 cm), enclose the base of the cooling coil, and extend no more than 10 in* (25 cm) downstream.

---

*Typical values only.

2. A 1-in* (2.5 cm) stainless steel drain pipe, with a male pipe thread shall be connected to the end of the pan wall, flush with the bottom of the pan, and extended 2 in* (5 cm) from the wall.

3. The floor of the condensate pan shall be sloped $1/4$-in per ft (2 cm/m) toward the drain port. The slope may be effected by sloping the pan internally or by externally tilting the entire HVAC unit toward the drain port. External tilting, if employed, shall not affect the performance of the HVAC unit nor the manufacturer's warranty.

18.8.5   HUMIDITY AND TEMPERATURE IN AIR SUPPLY SYSTEM

High relative humidity and low temperature in the HVAC air supply system can result in serious degradation in indoor air quality, and considerable damage to exposed hardware components. Good design practices dictate that these factors be considered carefully in HVAC system design.

18.8.5.1   Critical Design Factors

The conventional HVAC unit provides air cooling and moisture removal by reducing the dry bulb temperature of the air it handles. Passing through the cooling coil, the temperature is reduced to a dewpoint equal to or below that of the indoor air. This process reduces the humidity ratio (moisture in the air) but the average relative humidity is increased typically to about 95 percent. Figure 18.8.8 illustrates the cooling process where the cooling load on

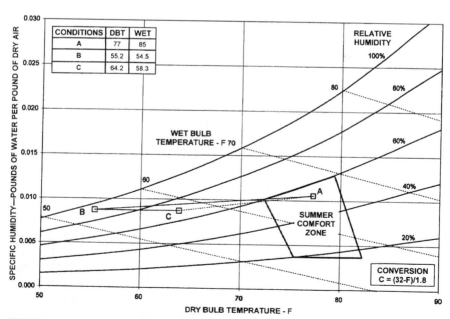

FIGURE 18.8.8   Psychrometric chart showing cooling and reheat process paths.

the system is 25 percent latent heat and 75 percent sensible heat: a sensible heat ratio (SHR) of 0.75. Point A represents the indoor conditions. Point B represents the condition of the air entering the supply system; where the relative humidity is 95 percent and the dry bulb temperature is 55.2°F (13°C). In a tightly sealed and well insulated air supply system, the relative humidity and dry bulb temperature of the air remain essentially constant as it passes through the supply system.

Humidity Conditions.   High relative humidity conditions, as shown in Fig. 18.8.9 (Ref. 4), are conducive to and support the growth of contaminating organisms, including bacteria, viruses, fungi, etc. The relatively low air temperature may suppress rapid growth of organisms and air movement may prevent accumulation of contaminates. However, this condition can degrade indoor air quality, and it represents a potential health hazard which must be carefully considered in HVAC system design.

One simple, but often expensive, way to reduce relative humidity in the supply air is to provide reheat. As illustrated in Fig. 18.8.8, an increase of about 9°F (5°C) in the air supply temperature—path B to C—decreases the relative humidity to 70 percent. At this lower relative humidity the growth of bacteria, viruses, and fungi is greatly reduced, see Fig. 18.8.9.

Temperature Conditions.   Inside the air supply system, low temperature air has no particular adverse effect on indoor air quality. Its effect occurs inside the conditioned space, at the air supply grille. Cold supply air often reduces the temperature of the grille below the dewpoint of the indoor air. When this happens, condensate can form on exposed surfaces of the grille. There, it promotes the growth of contaminating organisms, including mold, mildew, and other fungi. The discoloration frequently observed on supply grilles is often caused by these conditions and not always by dirty air filters, as is sometimes suggested.

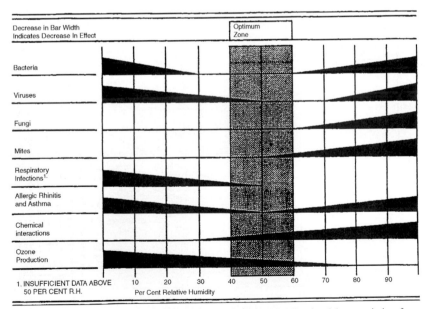

FIGURE 18.8.9 Optimum relative humidity ranges for health. (Reprinted by permission from ASHRAE Transactions, 1985, Vol. 91, Part 1B.)

To minimize this problem, consideration must be given to both supply air temperature and grille design.

The factors that affect air supply temperature are indicated in Fig. 18.8.10, for a typical design condition. Here, the supply air temperature is shown as a function of the SHR of the system, at various values of indoor wet bulb temperature and relative humidity. Also shown, for the various conditions, is how much the supply air temperature is below the dewpoint of the indoor air.

The difference between the indoor air and dewpoint temperatures, shown in Fig. 18.8.10, varies with indoor operating conditions. Lower indoor dry bulb temperatures reduce the differences. Higher indoor dry bulb temperatures and lower wet bulb temperatures increase the differences. For any specific design, these differences can be determined through the use of the psychrometric charts in Appendix A of this handbook.

Low temperature air does not *always* cause condensate to form on air supply grilles. At any particular temperature level, condensate formation depends upon specific features of the grille design. Some of the most significant features include the following:

1. Thermal conductivity of the grille material,

2. Jet velocity of the supply air, and

3. Amount of grille surface area exposed to the indoor air and not washed by the supply air.

Materials applicable to grille fabrication with the lowest and most desirable thermal conductivity characteristic are listed in order as follows: plastics, steel, aluminum, and copper.

High jet velocity causes air supply grilles to cool to a low temperature, which increases the potential for condensate formation. Flush mounted ceiling grilles that provide long

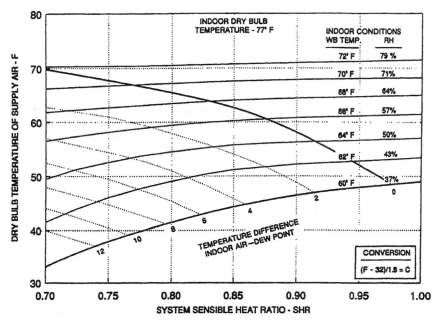

FIGURE 18.8.10 Effects of system sensible heat ratio and indoor wet bulb temperature on temperature of supply air.

throw distances usually operate with high jet velocities. Ceiling mounted concentric supply grilles, with short throw distances, often operate with low jet velocities.

Large grille surface areas exposed to indoor air increase the potential for the formation of condensate. Large unwashed surfaces usually accompany flush mounted ceiling grilles that discharge in one direction only. Ceiling mounted concentric grilles usually exhibit small areas of unwashed surfaces.

The condensate formation characteristics of the various types of air supply grilles are generally not available. However, grille manufacturers are cognizant of this problem. They realize that certain designs are superior to others, and they can aid designers in coping with the subtleties of this problem. In the meantime, designers should take advantage of the grille design fundamentals reviewed above when selecting supply air grilles from production units.

18.8.5.2 Economic Factors

The costs associated with high relative humidity and low temperature inside the air supply system can be appreciable.

The cost of degraded indoor air quality is difficult to establish. In most cases, it cannot be determined with any degree of accuracy. It is real, however, and could result in costly health care and other health related problems.

The cost of equipment damaged by high humidity and low temperature is much more tangible. The effects of high humidity will become evident in damage to air handling equipment and other downstream hardware. The cost of grille maintenance and damage caused by low air temperature will be both visible and definable.

Remedies to the air supply problem may not be expensive. There may be little or no hardware costs associated with selecting a supply grille of superior design.

Reheat, if provided with electricity or fossil fuel, involves appreciable initial costs plus significant and continuing operating costs. For this type of application, a heat pipe system may be attractive. The technology is mature. The initial costs are not exorbitant, about 30 percent of the cost of the basic air handler. There is no significant continuing operating cost, and maintenance costs are minimal.

18.8.5.3 Suggested Statement for Specifications

The HVAC system shall provide conditioned air to the air supply system at a relative humidity no greater than 70 percent and the dry bulb temperature shall be sufficiently high to prevent condensate formation on supply grille surfaces during all operating conditions.

18.8.6 POSITION OF FAN IN AIR HANDLER

The position of the fan in an air handler system can be a significant factor in condensate control. It directly affects internal airflow conditions and can seriously degrade air handler performance.

The flow condition of the air approaching the fan inlet is critical, as is the condition of the air leaving the fan exit.

Air flow conditions at the fan inlet are influenced by (a) the direction at which air enters the fan compartment, (b) hardware located in the airstream that can alter flow direction, and (c) how the approaching air interacts with the rotating fan blades. The airstream may be simply distorted, or it may contain highly harmful vortices. Vortices may be initiated by (a) fan blades and swirling flow at the inlet, (b) flow disturbances caused by obstructing hardware, or by (c) interactions between airstreams of different velocities. These conditions degrade fan efficiency. The interactions are complex, but in some cases their effects can be approximated using information from the Air Movement and Control Association (Ref. 5). The lower fan efficiency caused by these air flow conditions may be acceptable, in some cases. However, any combination of conditions that allows condensate to be entrained and spread into the HVAC system cannot be tolerated.

Air discharged from the fan leaves the exit at a high velocity with a greatly distorted velocity profile. In order to avoid excessive losses in fan system efficiency and achieve a reasonably uniform flow condition, a suitable extension to the exit duct must be provided. Failure to provide a suitable diffuser not only results in reduced efficiency, it can introduce unacceptable local air velocities that entrain condensate and spread it onto components inside the HVAC system.

18.8.6.1 Critical Design Factors

The factors that influence fan efficiency affect all types of air handlers in a similar way. Their effects on condensate problems, however, differ depending upon whether the air handler is a *draw-through* or a *blow-through* type unit. In the draw-through unit, only those conditions ahead of the fan inlet are important to condensate control. In the blow-through unit, the extreme air velocity profile at the fan exit is the primary cause of serious condensate problems.

Draw-Through Air Handler.   In a draw-through air handler, air is drawn through the cooling coil into the compartment that houses the fan and its inlet. Condensate is precipitated by the coil and collected in a pan inside the fan compartment. Thus, air entering the fan compartment is exposed to condensate both in the coil and in the condensate pan. Generally, the average velocity of the air passing into and through this compartment is too low to entrain condensate. However, certain fan and surrounding hardware arrangements can distort flow conditions and create local air velocities sufficiently high to entrain condensate and propel it into the fan inlet.

Published data and analytical techniques available to designers are not adequate to ensure that a selected fan position and adjacent hardware arrangements will be free of condensate entrainment and associated problems. Thus, successful design requires a considerable appreciation of the flow conditions that cause distorted airflow, along with substantial judgment. In some designs, full scale or model testing may be necessary.

Some of the various airflow conditions that create distorted local flow velocities, including vortices, are illustrated in Fig. 18.8.11. Here, a centrifugal fan with the inlet facing the cooling coil is depicted. Changing the position of the fan by 90° presents a different airflow pattern, but the fundamental airflow considerations remain the same.

In this illustration, air enters the fan inlet at 2500 ft/min (762 m/min) and the effect extends upstream from the inlet. The lines of constant velocities shown were computed, based on still air surrounding the fan, assuming no upstream flow distortion (Ref. 6).

These values are somewhat lower than what will be experienced in practice, because the system airflow adds a velocity component. This component increases the peak velocity and distorts the profile, as indicated by the dashed line in Fig. 18.8.11, for velocities of 600 and 1000 ft/min (182 and 305 m/min). Although less than precise, these results provide useful information for the designer. For example, if the cooling coil is placed closer than about 0.80 diameters of the inlet, the coil may be exposed to an air velocity of 600 FPM (182 m/min), which will likely cause condensate entrainment and carryover. On the other hand, at an

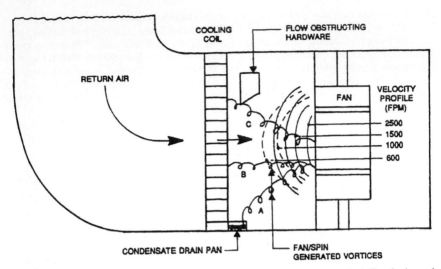

FIGURE 18.8.11   Air flow conditions that create distorted local flow velocities, including detrimental vortices—draw-through unit (Ref. 6).

distance greater than 1.5 diameters, condensate carryover (due to the fan alone) is unlikely to occur.

Vortices can create an even more serious problem. The spiral lines symbolize vortices and show how condensate can be entrained by the air and ingested into the fan inlet. Lines A and B indicate vortices generated by fan blades and externally induced spin or swirl (distortion). Line C indicates how a vortex may be created by obstructing hardware in the airstream. Eliminating airflow distortion, whether caused by obstructing hardware, unequal entering velocities, or other conditions can essentially negate the possibility of damaging vortices.

Blow-Through Air Handler.   The highly distorted velocity profile of air discharged from the fan of a blow-through air handler imposes significant constraints on the position of the fan relative to the cooling coil. The distance between the fan and coil must be adequate to eliminate flow distortion and reduce the air velocity sufficiently to avoid condensate entrainment and carryover. As indicated in Fig. 18.8.12, for a centrifugal fan, the required distance from the coil may be 2 to 5 equivalent duct diameters.

Shorter diffuser duct lengths introduce both undesired duct losses and unfavorable velocity profiles, and cause condensate carryover. Baffles used to reduce flow distortion often introduce unacceptable pressure losses and low fan system efficiencies.

18.8.6.2 Economic Factors

A fan positioned in an air handler such that it entrains condensate and propels it into the HVAC system creates a costly situation for the building owner-user. The cost of service calls, maintenance, shortened equipment life, and property damage can be excessive. Health problems, resulting from contamination caused by condensate entrained and deposited on internal surfaces of the system, add to the owner-user costs.

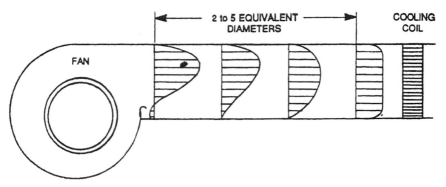

FIGURE 18.8.12   Velocity profiles at fan discharge and at cooling coil entry—blow-through unit.

The extra costs added to ensure that an air handler is free of fan induced condensate problems are due primarily to two factors: Loss in fan system efficiency and the designer's use of precautionary design features. Fan efficiency, for example, may be compromised by adding baffles to reduce flow distortion—particularly in the fan discharge airstream. This adds to operating costs. Also, flow distortion caused by complex interactions of the airstream with internal components cannot be precisely defined. Thus, the designer may add space and distance—walls, floor, ceiling—that may not be needed, in order to ensure that high velocity air and condensate entrainment are avoided. Whatever this cost, however, it is minimal compared to the cost incurred when condensate is entrained by the airstream and deposited on the internal components.

18.8.6.3   Suggested Statement for Specifications

The fan shall be positioned in the air handler so that local air velocities and induced vortices do not entrain condensate and propel it onto components in the HVAC system.

18.8.7   SEAL ON THE CONDENSATE DRAIN LINE (DRAW-THROUGH SYSTEMS)

An effective and reliable seal on the condensate drain line of a draw-through HVAC system is essential for successful condensate removal and for keeping the system dry inside.

In a properly designed HVAC system, all the condensate removed from the circulated air collects in a pan below the cooling coil. From there, it is drained to a selected condensate disposal place.

Achieving suitable drainage, however, can be very difficult, depending upon the type of HVAC unit involved. In this regard, there are two basic types of units: the "blow-through" and "draw-through." These designations stem from the relative positions of the air-circulating fan and the cooling coil.

In the blow-through type unit, the fan is located upstream and blows air through the cooling coil. As a result, the air pressure surrounding the condensate drain pan is positive (above ambient). The positive pressure alleviates somewhat the problems of condensate removal. Even with this advantage, however, the condensate trap is not a suitable device for controlling condensate removal. Many of the inherent problems with a condensate trap remain. Flow blockage and flooding are common, and dry traps at start-up often allow condensate to blow onto floor surfaces creating hazardous and health threatening conditions.

Condensate removal from draw-through units is much more critical. In this type unit, the fan draws air through the cooling coil, creating a negative (below ambient) pressure condition in the drain pan compartment. This condition produces two very adverse effects: (a) it impedes, or may even prevent, the drainage of condensate and (b) it effects ingestion of air and other gases from outside the unit. To avoid the health problems and property damage caused by these conditions, *a seal must be provided in the condensate drain line*, which permits condensate to flow freely and precludes ingestion of outside air or other gases.

Operating a draw-through HVAC unit without an effective seal on the condensate drain line causes serious problems (Refs. 8 and 9) including the following:

1. Negative pressure in the condensate drain pan area impedes the flow of condensate and frequently causes flooding and overflowing that damage the HVAC unit and surrounding property.

2. Inrushing air which in some cases may exceed hurricane velocities creates a geysering effect that propels condensate into the system and keeps it wet inside. The results are property damage and a growth haven for health threatening bacteria, mold, yeast, mildew, and other fungi.

3. The blowing of condensate can also create an aerosol mist—a known mechanism for spreading Legionella pneumophila: The disease that killed 34 Legionnaires at a Philadelphia hotel in 1976. A study of this incident, which strongly implicates the condensate drain system as the most likely source of the disease, is provided in Sec. 18.8.11.

4. Outside air, which may be polluted with carbon monoxide or other contaminates, can be drawn into the system and spread throughout the conditioned space.

5. Condensate flooding due to flow blockage by debris and algae growth.

The consequences of the above problems are illustrated in Fig. 18.8.13. These consequences are clearly unacceptable and dictate that an effective and reliable seal be included on every draw-through HVAC unit. Regarding this matter, ASHRAE Standard 62–2001, Ventilation for Acceptable Indoor Air Quality, paragraph Addendum 62t (Ref. 7), states: *"For configurations that result in negative static pressure at the drain pan relative to the drain outlet (such as a draw-through unit), the drain line shall include a P-trap or other sealing device designed to maintain a water seal while allowing complete drainage of the drain pan, whether the fan is on or off."*

18.8.7.1 Critical Design Considerations

Within the industry, three different types of devices are being used to form condensate drain seals for draw-through HVAC systems: (a) condensate (water) traps, (b) condensate pumps, and (c) a fluidic flow control device—the latter a recent technological development.

Each of these devices exhibits unique physical and operating characteristics and provides a different level of effectiveness and reliability.

Condensate Trap

Description.   The condensate trap is widely used as a seal in condensate drain lines. It is usually mounted outside the HVAC unit, as indicated in Fig. 18.8.14.

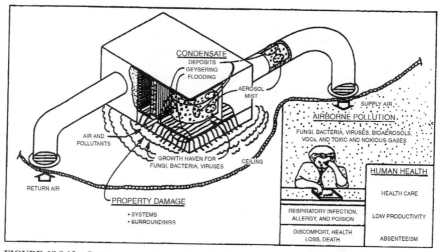

FIGURE 18.8.13   Consequences of no seal or a dysfunctional seal on the drain line of a draw-through HVAC unit.

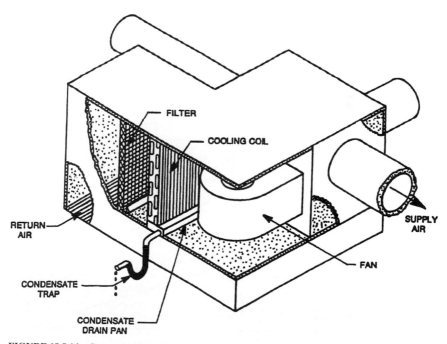

FIGURE 18.8.14   Conventional condensate trap in operation.

The seal is formed by gravitational forces acting on trapped water and a water column. The trap depth and water column depth necessary to form a seal depend upon the pressure inside the drain pan compartment. For any compartment pressure, the required trap depth and water column depth remain fixed. However, the total depth varies with the diameter and thickness of the trap wall. Figure 18.8.15 defines the required seal trap geometry for various drain pan pressures and trap diameters. Trap geometry established in accordance with Fig. 18.8.15 ensures a positive seal, allows drainage, and prevents condensate from standing in the condensate pan due to negative pressure, providing the trap is filled with water. Unfortunately, in practice these geometric dimensions have only academic value,

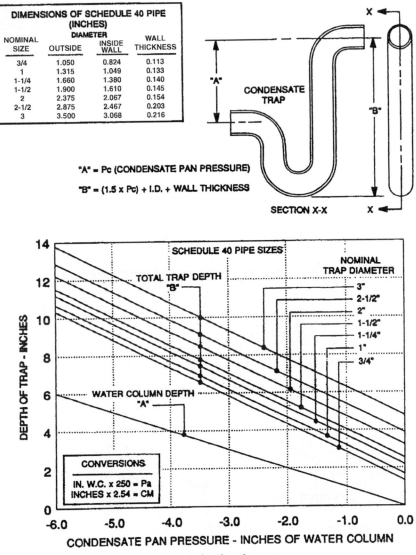

FIGURE 18.8.15 Dimensions of a properly designed condensate trap.

and are not generally applicable to actual operating systems. Any significant reduction in water level destroys the seal. For example, evaporation during periods when there is no flow of condensate (shutdown periods and winter heating) renders traps ineffective. Evaporation rate from traps is typically more than 1 in per month.

To overcome the effects of evaporation and allow an effective seal to be formed, trap depth must be greater than the values shown in Fig. 18.8.15. In some geographic locations, it may be necessary to increase the trap depth several inches over the values shown, in order to prevent so-called "dry trap syndrome." Dry traps permit the ingestion of outside air (possibly contaminated), allow condensate blowing onto internal components, and can cause flooding at start-up of cooling operations. In addition to dry trap syndrome, traps are inherently susceptible to freeze damage in outside locations and to flow blockage and pan overflow caused by trapped debris and algae growth.

Generally, condensate trap seals stocked by supply houses are unsuitable. A photograph of two such devices is shown in Fig. 18.8.16. In this figure, trap (a) provides a seal, but has no depth for water column. Thus, during system operation, it allows water to stand in the drain pan at a depth at least equal to the pan pressure, in inches of water. Trap (b) provides a water column and a seal over a limited range of internal pressures, but its adaptability to any particular application must be assessed by the designer. Under no circumstances should trap selection be left to the installer.

Effectiveness and Reliability. Under *ideal* conditions, the condensate trap can form an effective seal for the condensate drain line. However, a conventional trap exhibits so many failure modes that *its reliability is generally unacceptable for use as a drain line seal.* This situation has been recognized for years. Indeed, in 1996 the informative language in BRS/ASHRAE 62-1989R (Ref. 10) contained the following statement:

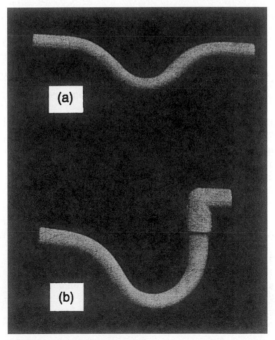

FIGURE 18.8.16   Condensate traps typically available at supply houses.

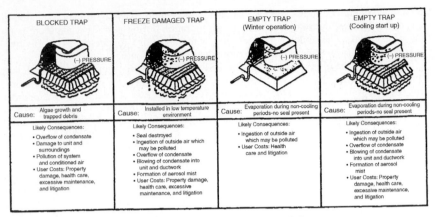

BLOCKED TRAP	FREEZE DAMAGED TRAP	EMPTY TRAP (Winter operation)	EMPTY TRAP (Cooling start up)
Cause: Algae growth and trapped debris	Cause: Installed in low temperature environment	Cause: Evaporation during non-cooling periods-no seal present	Cause: Evaporation during non-cooling periods-no seal present
Likely Consequences: • Overflow of condensate • Damage to unit and surroundings • Pollution of system and conditioned air • User Costs: Property damage, health care, excessive maintenance, and litigation	Likely Consequences: • Seal destroyed • Ingestion of outside air which may be polluted • Overflow of condensate • Blowing of condensate into unit and ductwork • Formation of aerosol mist • User Costs: Property damage, health care, excessive maintenance, and litigation	Likely Consequences: • Ingestion of outside air which may be polluted • User Costs: Health care and litigation	Likely Consequences: • Ingestion of outside air which may be polluted • Overflow of condensate • Blowing of condensate into unit and ductwork • Formation of aerosol mist • User Costs: Property damage, health care, excessive maintenance, and litigation

FIGURE 18.8.17   Some trap failures, causes, and consequences (Ref. 8).

"*Condensate traps exhibit many failure modes that can impact on indoor air quality* [and property damage*]. *Trap failures due to freeze-up, drying out, breakage, blockage, and/or improper installation can compromise the seal against air ingestion through the condensate drain line. Traps with insufficient height between the inlet and outlet* [and other trap design deficiencies*] *on draw-through systems can cause the drain to back up when the fan is on, possibly causing drain pan overflow or water droplet carryover into the duct system. The resulting moist surfaces can become sources of biological contamination* [and associated property damage*]. *Seasonal variations, such as very dry or cold weather, may adversely affect trap operation and condensate removal.*" Many of the most common failure modes, which are inherent in the condensate trap, are discussed in Ref. 11. A few of the more serious causes of these failures, and their consequences are summarized in Fig. 18.8.17.

Some of the most critical failure modes of the condensate trap can be alleviated by incorporating the following design features and maintenance procedures:

1. A water replenishing system to ensure that the trap provides a seal under all operating conditions;

2. Heating provisions, to prevent water in the trap from freezing during winter operations, in outside locations (The use of freeze plugs for protection against trap damage caused by freezing temperatures should be avoided. The maintenance effort for replacing plugs after freeze-ups is impractical to implement effectively); and

3. Mandatory maintenance procedures to ensure that traps are (a) inspected frequently for flow blockage, (b) filled with water prior to each cooling start-up, and (c) thoroughly cleaned or replaced at least annually (and filled with water after each action).

These added trap design features and the defined maintenance procedures will alleviate significant trap failure modes. But they introduce others. Both the water replenishing and water heating systems are subject to failure. And, even "mandatory" maintenance procedures are difficult to enforce. The designer must evaluate all these factors when considering the trap as condensate drain seal.

---

*Added by the authors.

Condensate Pump

Description. There are two basic types of condensate pumps: positive displacement and centrifugal. They require external power (usually electrical) to provide condensate removal. Pumps are used primarily where there is insufficient depth for the installation of a gravity dependent drain system or where condensate must be discharged to a level above the drain pan. When used for condensate removal, pumps are usually placed inside the condensate drain pan.

The configuration of condensate pumps differs greatly among types and manufacturers. Some designs fit better into condensate drain pans than others. In most instances, however, the condensate pan must be equipped with a suitable water sump.

The physical dimensions of a particular pump, of course, depend primarily upon the manufacturer and the volume of condensate it must handle. The maximum condensate flow rates from various size HVAC units, operating under specific cooling conditions, are shown in Fig. 18.8.2.

In operation, the rate of condensate removed by a HVAC unit varies from some maximum value to zero. To accommodate this variation, an on-off switch—which usually operates with a float—must be provided.

Effectiveness and Reliability. The condensate pump provides an effective and positive way to remove condensate from the drain pan of a HVAC unit. The positive displacement pump provides a firm seal and prevents the ingestion of outside air under all operating conditions. The centrifugal pump does not provide a firm seal when the pump is not operating. However, depending upon the particular design and installation, it may restrict air ingestion to an acceptably low level.

Condensate pumps are well developed and have been used successfully in certain applications. Their long term reliability, however, must be questioned. They depend upon moving parts that must operate in a cold and humid environment. The pump must handle condensate that often carries significant quantities of debris, which can interrupt pumping action and prevent flow. The on-off switch required to control condensate flow is exposed to the same hostile environment, and is subject to frequent failures.

Fluidic Flow Control Device

Description. The fluidic flow control device was developed primarily to provide a seal on the condensate drain line of draw-through HVAC units. However, a condensate flow control device for blow-through units, based on the same operating principles, is available. It is discussed separately in Sec. 18.8.12.

The fluidic flow control device is simple and has no moving parts. In draw-through system applications the desired seal is formed by a unique combination of hydraulic and pneumatic forces, readily available in the HVAC unit. One key feature of the device is that it uses air as a seal instead of water. Thus, it negates the problems associated with a water seal (Ref. 12).

The operating principles of the fluidic flow control device are illustrated in Fig. 18.8.18.

During both heating and cooling operations, the air seal is formed as follows: Fresh air from the fan discharge is supplied to point (a) at a pressure slightly above atmospheric. Some of the air flows away from the HVAC unit, thus preventing ingestion of outside air. A portion of the fresh air returns to the HVAC unit, passing through points (b) and (c). The quantity of air returning to the unit is minimized by the high pressure loss in the mitered elbows. This pressure loss plus the air flowing through the bypass connected at point (c), ensures that the air entering the condensate drip pan will not produce blowing and geysering and an aerosol mist.

Condensate flows through the device without being trapped. At the same time, the counterflow of condensate and air creates a pulsing action that ensures free passage of debris. Hence, the potential for freeze-up and flow blockage—common problems with traps—are nil.

A typical field installation of the fluidic flow control device is shown in Fig. 18.8.19.

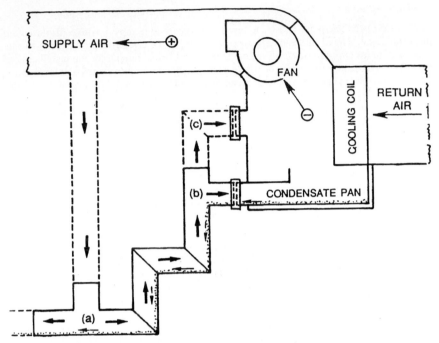

FIGURE 18.8.18   Operating principles of the fluidic flow control device.

The fluidic flow control device is a patented product manufactured and marketed by Trent Technologies, Inc., under the trade name CostGard Condensate Control Device. Detailed performance and installation data are available from the manufacturer.

The product, made from polyvinyl chloride (PVC) material, is available in sizes suitable for HVAC units up to 100 tons of cooling capacity and −5.0 in (127 mm) water column at the condensate drain outlet. Figure 18.8.20 shows how the physical dimensions of the device vary with system cooling capacity and drain pan pressure.

Effectiveness and Reliability.   The fluidic flow control device provides an effective and reliable seal for condensate drain lines during all cooling and heating conditions. Its effectiveness and reliability (see Refs. 8, 11, and 12) stem from unique characteristics and features incorporated in the design which:

1. Allow condensate to flow freely and unimpeded from the air-conditioning unit;
2. Prevent air (which may be contaminated) from being drawn into the system through the condensate drain pipe during heating and cooling start-up operations (when p-traps are usually empty).
3. Prevent condensate in the drain pan from being blown into the air conditioning unit and the duct work (during both normal and start-up operations);
4. Remove the condensate drain system as a source of an aerosol mist;
5. Eliminate condensate overflow caused by trap blockage and negative pressure inside the system;
6. Are not affected by algae growth;
7. Are not affected by condensate evaporation (as are traps);

FIGURE 18.8.19 Typical field installation of the fluidic flow control device on a York Predator unit.

8. Preclude damage from freezing temperatures;

9. Include no moving parts; and

10. Are self-cleaning and self-regulating.

The device is essentially free of failure modes, has been in use for many years with no failures.

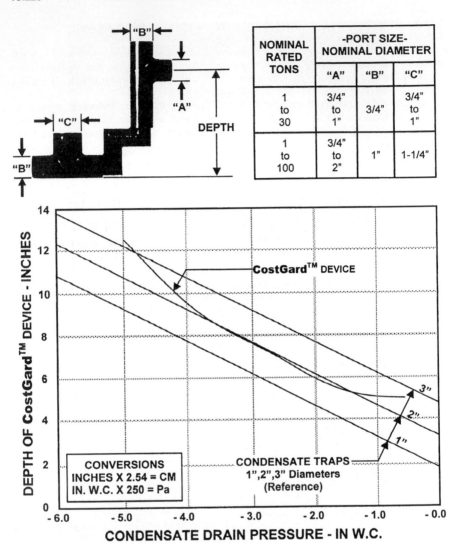

NOMINAL RATED TONS	-PORT SIZE- NOMINAL DIAMETER		
	"A"	"B"	"C"
1 to 30	3/4" to 1"	3/4"	3/4" to 1"
1 to 100	3/4" to 2"	1"	1-1/4"

CONVERSIONS
INCHES X 2.54 = CM
IN. W.C. X 250 = Pa

CONDENSATE TRAPS
1",2",3" Diameters
(Reference)

FIGURE 18.8.20   Physical dimensions of the **CostGard** condensate control device.

18.8.7.2 Economic Factors

Economic factors are paramount in the selection of a seal for condensate drain lines. Both the initial costs and operating costs are important.

Of the three types of seals discussed, the initial cost of the conventional condensate trap is lowest, and the condensate pump is the highest. The initial cost of the fluidic flow control device falls in between. However, total costs, which include both initial and operating costs are not at all related to the initial costs.

Despite the low initial cost of the conventional condensate trap, its operating and total costs are extremely high. This is because of excessive service calls, expensive maintenance effort, damage to surrounding property, and shortened equipment life. Although less visible and definable than others, the effect of the condensate trap on indoor air quality and health related costs may far exceed other operating costs.

Adding a water replenishing system and a heating device addresses two major failure modes of the condensate trap—empty traps and freeze damaged traps. But they introduce other failure modes, add to the initial costs, and contribute little, if anything, to increasing reliability and reducing system operating costs.

The cost of a condensate pump installed in a HVAC drain system, including special condensate drain pan designs, is greater than the cost associated with either of the other drain line seals. Years of field experience, however, indicates that with the fluidic control device, these costs are nil.

18.8.7.3 Suggested Statements for Specification

The seal on the condensate drain line of draw-through HVAC systems shall be designed to:

1. Allow condensate to flow freely and unimpeded from the air-conditioning unit;
2. Prevent air (which may be contaminated) from being drawn into the system through the condensate drain pipe during both heating and cooling start-up operations (when p-traps are usually empty).
3. Prevent condensate in the drain pan from being blown into the air-conditioning unit and the duct work during both normal and start-up operations;
4. Remove the condensate drain system as a source of an aerosol mist;
5. Eliminate condensate overflow caused by trap blockage and negative pressure inside the system;
6. Not be affected by algae growth;
7. Not be affected by condensate evaporation (as are traps);
8. Preclude damage from freezing temperatures;
9. Have no moving parts; and
10. Be self-cleaning and self-regulating.

18.8.8 CONDENSATE DRAIN LINES

The condensate drain line, which extends from the drain seal to the condensate disposal place, is the last link and a critical component in the condensate removal system. It can be, and frequently is, the source of serious condensate problems.

For successful drainage, this line must be carefully defined and clearly specified by the system designer. Under no circumstances should this responsibility be left to the installation contractor.

18.8.8.1 Critical Design Factors

An acceptable drain line is simply one that has adequate flow capacity and offers minimum potential for flow blockage. The following factors are critical to the design: diameter, length, routing, drainage slope, supports, and materials.

The diameter of the drain line must be equal to or greater than the exit diameter of the drain seal device. The line length should be the minimum possible, following the shortest path to the condensate disposal place (the shorter, the better). It should include the least possible number of elbows.

The line must be sloped away from the drain seal at a rate of *no less* than $1/_8$-in/ft (1 cm/m). Solidly fixed drain line supports must be provided at intervals which ensure that a uniform slope is maintained to avoid dips in the line that can trap condensate and debris.

Drain lines may be constructed of PVC, copper, or steel piping. PVC and steel pipe must be Schedule 40 or heavier. Copper pipe must be Type L or heavier.

To avoid excessive dips in the drain line and prevent shifting of position, fixed supports should be provided as follows: PVC, 2 to 3 ft (0.6 to 0.9 m); copper, 4 to 6 ft (1.2 to 1.8 m), and steel 8 to 10 ft (2 to 3 m) intervals.

Clearly, a long and meandering condensate drain line requires careful engineering design and proper installation which adds appreciable cost to the system. Where possible, consideration should be given to eliminating the drain line entirely. For example, in rooftop installations, condensate flowing from the drain seal may be dumped onto the roof. Any potential roof problems caused by this arrangement should be assessed against the cost of installing the drain line and the probable condensate drainage problems caused by the use of a drain line.

18.8.8.2   Economic Factors

The cost of the condensate drain line depends primarily on the type of pipe used. The unit cost of the pipe and the cost of installation are both important.

The unit cost of pipes differ substantially. The cost of PVC pipe is one-half to one-third that of steel. The cost of steel pipe is one-half to one-third that of copper. It follows then that copper pipe is four to six times more costly than PVC.

Installation costs consists of labor plus the material required for supporting the drain line. The cost of labor involved in assembling PVC pipe is minimal. It consists of connecting slip joints with PVC cement, for which little labor is involved. The closely spaced pipe supports required, however, add to both labor and material costs. Even so, a condensate drain line of PVC costs considerably less than either steel or copper.

The cost of labor required to assemble a copper drain line far exceeds that of one constructed of PVC. Each connection must be soldered and the necessary equipment must be transported to the installation site. However, the costs of material for pipe supports and the cost of installing them are only about one-half that for PVC pipe.

The cost of labor necessary for assembling a steel drain line is even greater than for copper. Each connection must be threaded and screwed together. Equipment for cutting threads must be transported to the installation site. These factors plus heavy pipe impose an extraordinary amount of labor on the installer. The cost of pipe supports and material are minimal.

In summary, when both pipe and installation costs are included, a drain line system with PVC pipe is the lowest of the three. The cost of systems constructed with steel and copper pipe are about the same. The installation of copper is less, but the higher cost of copper pipe tends to offset the higher installation costs of steel pipe. In any particular application the cost difference between copper and steel systems may vary, depending upon the experience of the contractors involved.

Building owners-users sometimes prefer one material over others. Also, in some instances, local codes may prohibit the use of PVC. However, all three of the mechanical codes used in the United States permit the use of PVC.

18.8.8.3   Suggested Statements for Specifications

1. The diameter of the drain line must be equal to or greater than the exit diameter of the drain seal device. The line length should be the minimum possible—following the shortest path to the condensate disposal place. It should include the least possible number of elbows.

2. The line must be sloped away from the drain seal at a rate of no less than $\frac{1}{8}$-in/ft (1 cm/m).

3. Drain line supports must be fixed solidly in place and provided at intervals which ensure that a uniform slope is maintained and that any dips formed in the line do not trap condensate and debris.

4. Drain lines may be constructed of PVC, copper, or steel piping. PVC and steel pipe must be Schedule 40 or heavier. Copper pipe must be Type L or heavier.

5. To avoid excessive dips in the drain line and prevent shifting of position, fixed supports should be provided as follows: PVC, 2 to 3 ft (0.6 to 0.9 m); copper, 4 to 6 ft (1.2 to 1.8 m), and steel 8 to 10 ft (2 to 3 m) intervals.

18.8.9 SYSTEM REMEDIATION

Suitable condensate control is achieved and an acceptable HVAC maintenance program can be implemented only when condensate is confined to (a) surfaces of the cooling coil, (b) a small and properly sloped condensate drain pan, and (c) a well-drained system through which condensate flows freely and never stands nor stagnates.

Confining condensate to these three areas allows the system to operate virtually free of excessive maintenance, property damage, and health-threatening biological growth. When condensate is confined in this manner, the required system maintenance consists of rather simple periodic scheduled procedures: inspecting, cleaning, and flushing the drain system (pan, seals, and lines).

Unfortunately, far too many systems now in operation are not designed to restrict the spread of condensate, and are not amenable to reasonable maintenance.

A successful maintenance program for these systems must begin with an assessment of the capacity of each to *confine condensate to the cooling coil, drain pan, and drain system.* Anytime condensate spreads beyond these areas, system remediation is necessary to ensure that condensate is properly confined, under all operating conditions.

System deficiencies that allow the spread of condensate beyond the cooling coil and the drain system include the following:

- Condensate carryover from cooling coils
- Condensate drips onto internal HVAC system components
- Unsuitable drain pan designs
- Very low supply air temperatures
- Improper fan position inside the air handlers
- Ineffective seals on condensate drain lines
- Unsuitable installation of condensate drain lines

When these deficiencies are present, no amount of system maintenance can prevent equipment damage, surrounding property damage, and health-threatening biological growth. *Scheduled maintenance has only limited value.* It occurs after property damage has been done and biological growth has had its effect. Moreover, the damaging effects begin all over as soon as system operation is resumed. The design considerations necessary to avoid these conditions in future systems are provided in Sec. 18.8.2.

Whenever the above deficiencies appear in any system, they must be remedied before a meaningful maintenance schedule can be defined and implemented. The following paragraphs suggest suitable remedies and define a drain system that can be maintained with reasonable routine and preventive maintenance programs.

Specific system deficiencies that preclude the implementation of a feasible maintenance program, along with remedies to these deficiencies are reviewed below.

18.8.9.1 Condensate Carryover from Cooling Coil

Condensate carryover in any observable quantity is incompatible with a practical and acceptable system maintenance program. Damage and contamination begin when carryover occurs, neither will wait for the next scheduled maintenance action.

Condensate carryover occurs when the velocity of the air passing through the cooling coil is sufficient to entrain condensate and blow it off the coil. Any time the system components or other surfaces downstream of the cooling coil become wet, condensate carryover is a possible cause that must be assessed. The presence of carryover can best be established by visual observation (portholes, fiber optics, etc.) downstream of the coil, during the cooling operation when the latent heat load (water removal) is high. The entire surface of the coil must be viewed in order to determine the cause and location of the deficiency. Uniform carryover indicates one deficiency, carryover in local areas indicates other deficiencies. The three most common causes of condensate carryover are (a) unsatisfactory coil design, (b) dirty cooling coils, and (c) distorted air velocity profile entering the coil.

Unsatisfactory Coil Design.   Coil design is unsatisfactory when condensate carryover appears somewhat uniformly over the entire face of a *clean* cooling coil. The following design parameters determine the cooling coil condensate carryover characteristics: airflow of the air handler; height and width of the cooling coil; size and spacing of the coil tubes; and thickness and spacing of fins on the tubes. Figure 18.8.1 illustrates, for a typical coil design, the relationship among these parameters. The air velocity at the coil face, shown in this figure, is determined by dividing the air handler airflow by the face area of the coil.

Problem Definition.   When condensate carryover, due to coil design, occurs in a particular system, it can be remedied only by changing one or more of the parameters included in Fig. 18.8.1. Generally, in existing systems, it is not practical to make significant changes in coil geometry. Thus, the most practical way to eliminate condensate carryover is to reduce air velocity at the coil face; that is, reduce airflow.

Remedies.   Reduced airflow can be achieved most effectively by changing fan speed, which involves either a change in size of pulleys or a change in the motor speed, if the motor speed is variable.

It is often possible to reduce airflow without causing system problems. Many times HVAC systems are oversized. In such cases, reduced fan speed introduces no penalty in cooling performance. Moreover, total cooling capacity is relatively insensitive to airflow. For example, a reduction in airflow of 20 percent typically reduces total cooling capacity by about 5 percent at the rated point. Sensible cooling capacity is reduced more, but latent cooling capacity is increased. Even in those instances where the total cooling capacity is compromised by reduced airflow, the best choice may be to accept this compromise and eliminate the serious property damage and health problems associated with condensate carryover.

In installations where the coil design is similar to that shown in Fig. 18.8.1, the reduction in air velocity needed to eliminate carryover can be approximated as follows: Compute air velocity at the coil face in feet per minute by dividing the airflow (CFM) by the face area (square feet—height times width) of the cooling coil. With the coil face velocity, enter Fig. 18.8.1 at the spacing ratio and fin spacing determined from coil measurements. The difference between this point and the point where carryover is indicated by Fig. 18.8.1, indicates the reduction in velocity and, therefore, the airflow reduction required to eliminate carryover. At best, however, this process provides only a starting point. The proper airflow reduction is that which eliminates condensate carryover, a condition that must be determined by visual observation or other suitable means.

The installation of moisture "eliminators" downstream of the coil is not a viable method for preventing carryover. Eliminators add significantly to the pressure loss in the system and, therefore, reduce airflow. In addition, they introduce another potential growth place for biological contaminates.

Dirty Cooling Coils. Condensate carryover may sometimes be observed in systems where the cooling coil design is entirely satisfactory. The cause may be dirty coil surfaces. Carryover may occur in limited local areas of the coil because dirt does not always collect uniformly on the cooling coil.

Problem Definition. Dirt and other foreign material deposited on the surfaces of cooling coils can reduce the area for airflow and increase the air velocity sufficiently to effect condensate carryover. Unrelated to condensate carryover, dirty coils have other adverse consequences. They reduce heat transfer and decrease system efficiency.

Remedies. Within the industry, the most widely endorsed solution to dirty coil conditions is maintenance program that involves periodic coil cleaning. When properly defined and performed regularly, coil cleaning can be adequate to avoid carryover resulting from dirty coils. But, coil cleaning is a costly and time-consuming process. Furthermore, in many existing systems, the cooling coils are so inaccessible and cleaning is so difficult that it is often deferred until a major problem arises.

Probably the most dependable and cost-effective way to maintain coils in an acceptably clean condition is to utilize filters with adequate capacity to remove particles that accumulate on the coil; thereby, avoiding the need to perform frequent cleaning. Available data (Ref. 13) indicate that filters with dust spot efficiency ratings (as defined in Ref. 14) of 25 percent (or AHSRAE MIRV 7 Filter) or higher can virtually eliminate dirty coils and the airflow problems they cause. Of course, to be effective a filter must be placed in an essentially air tight holder, otherwise bypassed particles can reach the coil and be deposited on the surfaces.

Distorted Air Velocity Profile Entering the Cooling Coil. Air entering the cooling coil with a nonuniform velocity profile can cause carryover even when the average face velocity is below where carryover would occur, as indicated in Fig. 18.8.1. Carryover caused by this condition can be defined by visual observation. And, the location and source of the distorted airflow can be identified and corrected.

Problem Definition. In draw-through systems, distorted airflow can be generated; for example, when air enters a short-coupled return plenum at right angles to the coil. In blow-through systems the coil may be subjected to a very adverse velocity profile created at the fan discharge. Carryover of this type is evidenced by concentration in local areas of the coil, where airflow distortion occurs.

Remedies. Carryover caused by distorted airflow may be remedied by installing longer plenums and/or turning vanes. Properly installed turning vanes and longer, more efficient diffusers not only improve the velocity profile, they can significantly reduce pressure losses. Turning vanes and longer plenums require more space, additional hardware, and added costs. Nevertheless, in certain cases, major changes may be necessary to eliminate the detrimental spread of condensate if an effective maintenance program is to be achieved.

18.8.9.2 Condensate Drips onto Internal HVAC System Conponents

Any HVAC system that allows condensate to drip onto internal surfaces and components is subjected to internal damage and the growth of contaminating organisms. Sloped cooling coils and noninsulated coolant lines (refrigerants or water) are often the source of condensate drips. Systems that exhibit these qualities must be modified, because routine scheduled maintenance does not protect against these conditions, nor does it remedy the causes.

Drips from Sloped Cooling Coils.   Sloped cooling coils included in some HVAC systems are prone to drip condensate onto surfaces outside the condensate drain pan.

    Problem Definition.   Condensate that drips from a slanted coil onto surfaces outside the drain pan creates destructive and contaminating conditions. At small slope angles, surface tension may be adequate to retain the condensate and allow it to drain into the condensate pan. However, foreign deposits on the coil can easily destroy the effects of surface tension and cause dripping to occur.

    Air velocity entering the coil tends to reduce dripping from a coil that is sloped rearwardly (from the top), when the fan is operating. But it is not a reliable force for preventing condensate dripping.

    Remedies.   Extending the condensate drain pan to catch condensate drips is not a viable solution. It introduces another equally serious condition, large drain pans, discussed in Sec. 18.8.9.3. ("Unsuitable drain pan designs"). In order to ensure adequate condensate control and a dry system that is free of harmful contaminates, sloped coils must be kept clean at all times and the fan should not be turned off during the cooling period. The air filter must never be placed below the cooling coil, where dripping condensate creates conditions conducive to the growth of health threatening organisms. Filters mounted below the coil must be moved upstream to eliminate a major source of system contamination. Slanted coils increase the susceptibility of systems to internal wetness and contamination; therefore as a minimum, a more stringent and high frequency maintenance program is necessary for these systems.

Drips from Noninsulated Coolant Lines.   All too frequently, existing HVAC units have noninsulated coolant lines that pass through the cooling airflow passage, causing condensate drips along their paths.

    Problem Definition.   Condensate that drips on surfaces outside the drain pan causes damage to the HVAC system and creates conditions conducive to health-threatening biological growth.

    Remedies.   This condition, much too common in current systems, is simple to remedy. It can and must be eliminated by applying suitable insulation to all bare coolant lines.

18.8.9.3   Unsuitable Drain Pan Designs

Condensate drain pans in many HVAC systems now in the field are so configured that the necessary maintenance effort varies between very difficult and impractical to perform. Among the most troublesome features are (a) large drain pans, (b) primary drain port location, and (c) internal baffling. Systems that exhibit these characteristics must be modified before a reasonable maintenance program can be implemented.

    Large Drain Pans.   Systems that make use of large condensate drain pans (often extended downstream of the cooling coil to protect against condensate carryover)—cannot confine the spread of condensate within the boundaries necessary to permit successful maintenance.

    Eliminating condensate carryover, as discussed earlier, will not keep large condensate pans dry and free of damage and biological growth.

    Problem Definition.   As condensate forms and drains into the pan, it will stand there at some finite depth. If the drain pan is level—as is common for systems now in the field—condensate will cover the entire pan. The precise depth at which condensate stands depends upon the pan geometry, and the rate at which condensate is drained from the pan. Typically, the depth varies between about $1/_8$ in (3 mm) and 1/2 in (12 mm) (or greater). See Fig. 18.8.3.

    During system operation, condensate in the drain pan will flow from the area below the cooling coil to the drain port, leaving the remainder—in fact most—of the condensate in a stagnant state. There it becomes a growth haven for contaminating organisms, as illustrated in Fig. 18.8.5. A photograph of one such result is shown in Fig. 18.8.21.

FIGURE 18.8.21 Photo showing contamination in wide drain pan.

Remedies. For HVAC units now in the field, sloping the drain pan by tilting the HVAC unit is one way to reduce the pan area covered with stagnant condensate. Figure 18.8.6 illustrates the effect of sloping a drain pan in one direction. Sloping the pan in both directions, of course, further reduces the area of stagnant condensate. Hence, with sufficient slope in two directions, it is possible to virtually eliminate stagnate condensate in the pan.

Sloping the pan, however, by no means makes a large pan an acceptable remedy for condensate carryover. Carryover droplets deposited on the pan will not drain readily to the drain port. Instead, they will be held in place by surface tension providing another potential source of biological contamination. In reality, the large drain pan serves no useful purpose. It is not a solution to condensate carryover, a condition that must be prevented, as discussed earlier.

The equipment damage and contamination problems caused by large drain pans now in the field can be remedied by simply reducing the pan size to that required to catch the condensate and accommodate the flow in the drain pan. The length of the pan must be sufficient to cover the base of the cooling coil. The pan area is then fixed by pan width—the distance the pan extends away from the cooling coil. The width of the drain pan must be sufficient to accommodate the maximum condensate flow rate, yet not so wide as to allow condensate to stagnate. Pan widths considered acceptable for systems with various cooling capacities and condensate drain sizes are shown in Fig. 18.8.7.

The most effective way to reduce large pans to a suitable size depends upon the specific system involved. Where possible, the most desirable pan is one constructed of durable non-metallic material or stainless steel. Often, however, the most practical way to effect pan size

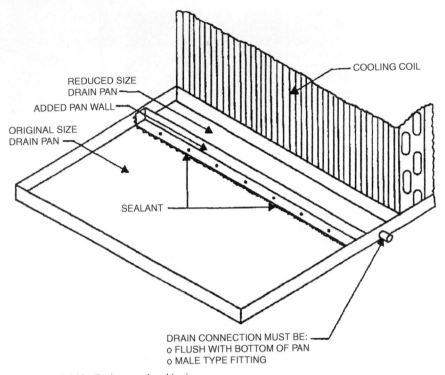

FIGURE 18.8.22   Drain pan reduced in size.

reduction is to install a wall inside the current drain pan, as illustrated in Fig. 18.8.22. In some cases, it may be necessary to relocate the drain port and place it at the end of the pan as indicated.

If the pan and attachments already in place are constructed of ferric metals, they must be replaced or treated with durable (long-life) protective coatings. This is because under some conditions the presence of iron accelerates the growth of certain harmful bacteria. (See Ref. 3.)

Condensate drainage can be enhanced, and protection against the formation of water puddles can be realized by tilting the HVAC unit toward the drain port, as discussed above. A slope of $\frac{1}{4}$-in/ft (2 cm/m) is usually adequate. However, the unit should be tilted only with the approval of the equipment manufacturer.

Large pans modified as defined above will permit condensate to be confined within boundaries that allow practical maintenance and ensure a minimum of property damage and biological contamination.

Primary Drain Port Location.   Proper drain port location is essential to adequate drainage and condensate control.

Problem Definition.   Drain ports located above the bottom of the drain pan inherently prevent complete drainage of condensate. Water standing in the pan below the drain port level collects and retains debris, supports the growth of biological agents, and contaminates the system. Frequent cleaning and maintenance are required to prevent serious property damage and human health problems. Moreover, location of the drain port above pan floor level makes cleaning and scrubbing unduly difficult and time consuming. Maintaining such a pan free of contaminates is not realistically feasibile.

Remedies. An acceptable maintenance program dictates that drain ports be flush with the bottom of the drain pan. Whenever feasible, the drain port should be placed in the bottom floor of the drain pan, to further improve drainage. See Figs. 18.8.3 and 18.8.4.

Baffles in Condensate Drain Pans. The drain pans in some draw-through HVAC packaged units now in the field are equipped with internal baffles. Evidently, these are intended to prevent condensate from blowing into the system where it can cause damage to internal components and promote health-threatening biological growth. In most draw-through systems—in the field today—condensate blowing is a serious problem. The problem arises when no seal has been installed or when the seal depends upon a trap that is dry; for example, during initial system start-up, or start-up for summer cooling. (Traps become dry in winter due to evaporation and/or freeze-plug expulsion.)

Problem Definition. During the above operating conditions, baffles can reduce, although they rarely eliminate, condensate blowing when no drain seal is present. But they present a significant maintenance problem. More surface area is exposed to condensate and the potential for system contamination is magnified. And because of the small air and water passages in baffles, condensate pans seldom drain well and condensate flow is often blocked by debris. Frequent cleaning of baffled systems is therefore imperative. Yet, the interior of baffle arrangements are so inaccessible that reasonable maintenance procedures are usually not feasible.

Remedies. The implementation of a practical maintenance program requires the removal of troublesome baffles and the use of a reliable and effective drain seal that eliminates the condensate blowing, for which the baffles were initially installed. See Sec. 18.8.9.6 "Ineffective seals on condensate drain lines" in the following pages.

18.8.9.4 Very Low Supply Air Temperatures

Low supply air temperatures present a special problem in condensate control, which occurs too frequently to be ignored. The problem appears in the form of damage to and biological contamination of air supply grilles.

Air Supply Grilles. In buildings where the latent heat load is high, the dry bulb temperature of the air leaving the cooling coil is usually below the dewpoint temperature of air in the conditioned room.

Problem Definition. Whenever supply air reduces the supply grille temperature to the dewpoint temperature inside the room, condensate can form on the surfaces of the grille. Deposited there, it causes the formation of molds and other fungi that discolor, damage, and contaminate surfaces of the grille. Systems that allow formation of condensate on supply grille surfaces require constant attention and must be modified to be amenable to a realistic maintenance program.

Remedies. There are three possible solutions to this problem: (a) change in grille design, (b) increase in cooling airflow rate, and (c) addition of cooling air reheat.

The internal and external geometry as well as the material of construction determine how resistant a supply grille is to condensate formation. Some grille manufactures are aware of the problem and can be helpful in selecting designs that are the least susceptible to condensate formation.

An increase in airflow rate will increase the supply air temperature and reduce moisture formation on supply grilles. Many times it may be possible to increase airflow sufficiently to avoid this problem and still provide adequate system performance. This possibility should always be considered, anytime condensate is found on air supply grilles. (The effect of increasing airflow on condensate carryover from the cooling coil must be evaluated when increased airflow is considered.)

Providing reheat to cooling air is an effective way to avoid condensate formation on supply air grilles, but it is usually an inefficient and a costly process. There are, however, other reasons for using reheat. If needed for other purposes, adding reheat may offer an acceptable method of eliminating condensate on supply grilles.

More details on what can be done to avoid condensate formation on supply grilles are provided in Sec. 18.8.5.

18.8.9.5   Improper Fan Position Inside Air Handlers

Fans improperly positioned within an air handler often cause condensate problems that cannot be remedied by routine maintenance. Poorly located fans can (a) cause condensate to be entrained directly from surfaces of the cooling coil or (b) generate vortices that ingest condensate from either the drain pan or the cooling coil.

Condensate Entrainment.   Air entering the fan inlet is accelerated to a velocity much higher than that in the coil plenum. This condition generates a high velocity core that extends well beyond the fan inlet.

Problem Definition.   If fans are placed too close to cooling coils, the high velocity core will extend into the coil, entrain condensate, and blow it onto downstream components.

Remedies.   In instances where the fan draws condensate directly from the coil, the separation distance must be increased in order to alleviate the problem. Section 18.8.6 provides information on the separation distance necessary to avoid condensate entrainment.

Vortex Ingestion.   Airflow distortion developed upstream of fan inlets can interact with fan blades and generate troublesome vortices. The presence of vortices is most readily confirmed by visual observations.

Problem Definition.   Vortices touching the coil surfaces or condensate in the drain pan can carry streams of condensate into the fan. From there, it is spread onto downstream components causing damage to equipment and promoting the growth of biological contaminates.

Remedies.   Vortices can usually be eliminated by installing properly designed turning vanes.

18.8.9.6   Ineffective Seals on Condensate Drain Lines

Ineffective seals on condensate drain lines of draw-through systems is a major cause of equipment damage, surrounding property damage, and system contamination. In fact, dysfunctional drain seals appear to be the primary cause of biological growth in HVAC systems.

An ineffective drain seal can be identified in a number of ways (Table 18.8.1).

The consequences of operating a draw-through HVAC system with an ineffective drain seal are summarized in Fig. 18.8.13. These consequences are clearly unacceptable and dictate that every draw-through HVAC unit be equipped with an effective and reliable condensate drain seal.

Within the industry, three different types of devices are being used to form condensate drain seals for draw-through HVAC systems: (a) condensate (water) traps, (b) condensate pumps, and (c) a fluidic flow control device—the latter a recent technological development.

Each of these devices exhibits unique physical and operating characteristics and provides a different level of effectiveness and reliability.

Condensate traps.   The condensate trap is widely used as a seal in condensate drain lines. It is usually mounted outside the HVAC unit, as indicated in Fig. 18.8.14. The seal is formed by gravitational forces acting on trapped water and a water column.

TABLE 18.8.1 Determining Seal Effectiveness

Symptom	To determine seal effectiveness
Outside air is drawn into HVAC unit during normal operation.	Place soap bubbles or a small flame at the drain exit, or at any open port (downstream of the drain seal). If either the bubbles or flame is drawn into the system, the seal is ineffective and unsatisfactory.
Outside air is drawn into HVAC unit when the fan is operating with the drain pan, trap,* and drain system dry.	Place soap bubbles or a small flame at the drain exit, or at any open port (downstream of the drain deal). If either the bubbles or flame is drawn into the system, the seal is ineffective and unsatisfactory.
Condensate stands in the drain pan during the cooling operation.	Determine the depth of condensate in the drain pan. If it exceeds the level shown in Fig. 18.8.3 the seal is ineffective and unsatisfactory.
Condensate is blown from the drain pan during cooling system start-up, or normal operation, and condensate is spread onto internal surfaces.	View conditions of the drain pan and cooling coil through port holes or other means.

*Unless the seal is equipped with a water replenishing system.

Problem Definition. Properly configured, as described in Sec. 18.8.7, the conventional condensate trap can provide a seal under *ideal* circumstances. Even when configured properly, however, the trap exhibits numerous failure modes and is so unreliable that it is ineffective and is unsuitable for use on a draw-through HVAC system.

Some failure modes are inherent when using the condensate trap: (a) flow blockage—pan overflows; (b) trap freeze damage (in outside locations)—seal destroyed; (c) evaporation of condensate—seal destroyed. The causes of these failures and the destructive consequences are summarized in Fig. 18.8.17.

In addition to inherent deficiencies, traps are susceptible to design deficiencies and unwise field practices, which add failure modes and further decrease trap reliability and effectiveness.

Reference 10 presents a comprehensive summary of failure modes that plague the condensate trap, and torment building owners/users.

Remedies. The methods available for improving trap effectiveness and reliability are limited.

Trap flow blockage is virtually impossible to avoid. Because traps trap water, they likewise trap debris, which eventually blocks flow. In addition, the cool water in the trap promotes the growth of algae, a condition that almost ensures periodic flow blockage. Annual cleaning of traps is one possible way to overcome flow blockage. However, because of prolific growth of algae and the drying-out and hardening that takes place during winter months, trap cleaning is not always effective. The best approach is to replace traps annually.

Trap freeze-up and seal destruction, in outside locations, can be avoided by applying heat to the condensate during freezing periods. Although such a system requires a heating element and a sensor device, it can be effective and reasonably reliable, but not cost free. The use of freeze plugs, often employed to prevent trap damage, is an unsuitable choice because when expelled they destroy the seal and render traps ineffective until the plugs are replaced. Further, in many applications, it is not feasible for maintenance personnel to replace freeze plugs after each freeze-up. Thus, after thaw-out the seal is lost until plugs are replaced and traps are filled, often months later. Evaporation of condensate from traps, which begins as soon as the cooling operation and condensate-flow ceases, can be overcome by using a water replenishing system. The control of water flow may be affected by a continual drip arrangement, an on/off timer, or level sensor. In any case, a water replenishing system adds cost and decreases reliability. For successful operation in outside locations (where freezing occurs), a water replenishing system must be equipped with methods for keeping the water above freezing temperature.

Condensate traps should be used as drain seals only as a last resort. Even when water replenishing and condensate heating systems are employed, systems must be monitored carefully—a very expensive procedure.

Condensate Pumps.   Condensate pumps are usually installed in the condensate drain pan, inside a suitable water sump. They are used primarily where there is insufficient depth for the installation of a gravity dependent drain system or where condensate must be discharged to a level above the drain pan. Pumps provide a positive and effective seal during all phases of operation. External power (usually electrical) is required. To accommodate variations in condensate flow, an on-off switch, which is usually operated by a float, must be provided.

Problem Definition.   Although they are well developed and have been used successfully in special applications, the long-term reliability of condensate pumps must be questioned. They depend upon moving parts that operate in a cool and humid environment. Pumps must handle condensate that often carries significant quantities of debris, which can interrupt pumping action and cause flooding. The on-off switch used to control pump operation is exposed to a somewhat hostile environment, and is subject to frequent failures.

Condensate pumps may be damaged when exposed to freezing conditions. Freezing, generally, does not destroy the seal, but it can result in pump failure and condensate overflow whenever cooling operation begins. Because they usually require considerable space inside the drain pan, pumps are not often used as replacements for condensate traps.

Remedies.   Acceptable pump reliability can be achieved by implementing an aggressive maintenance and monitoring program. It must include periodic cleaning of debris from the sump, and replacement of aging and deteriorating electrical and mechanical components. In outside locations, freeze damage can be avoided by providing a heating system to maintain water in the sump at a temperature above freezing.

Fluidic Flow Control Device.   The fluidic flow control device is a recent advancement in drain seal design. It was developed specifically for use on the drain lines of draw-through HVAC units, to be free of the failure modes common to traps and pumps. The device is connected to the condensate drain port much like a conventional condensate trap. The desired seal is formed by a unique combination of hydraulic and pneumatic forces, available in the HVAC unit. It has no moving parts. A key feature of the device is that it uses air as a seal instead of water. Thus, it negates the problems associated with a water seal. The operating principles of the fluidic flow control device are illustrated in Fig. 18.8.18 and described in Sec. 18.8.7.

Problem Definition.   Installation procedures for the device are new to HVAC service personnel, and the effort required is slightly more than that required to install a conventional condensate trap (although, less than that required for installing a trap with a condensate heater and a water replenishing system). Not all existing systems are adaptable to installation of the new device. For example, units that are too close to the floor to allow installation of a trap, cannot accommodate the fluidic control device either, without major changes.

Remedies.   Detailed step-by-step installation procedures, including a pictorial guide, are available to assist service personnel with their first installation. Subsequent installations become routine. Once installed, the device operates virtually void of maintenance effort and free of the many problems caused by the condensate trap, in that it:

1. Allows condensate to flow freely and unimpeded from the HVAC unit;
2. Prevents air (which may be contaminated) from being drawn into the system through the condensate drain pipe during heating operations, cooling operations, and cooling system start-up operations (when p-traps are usually empty);
3. Prevents condensate in the drain pan from being blown into the air-conditioning unit and the duct work (during both normal and start-up operations);
4. Removes the condensate drain system as a source of an aerosol mist;
5. Eliminates condensate overflow caused by trap blockage and negative pressure inside the system;
6. Is not affected by algae growth;
7. Is not affected by condensate evaporation (as are traps);
8. Precludes damage from freezing temperatures;
9. Includes no moving parts; and
10. Is self-cleaning and self-regulating.

18.8.9.7 Unsuitable Installation of Condensate Drain Lines

Condensate drain lines, which extend from the drain seal to the condensate disposal place, can be—and frequently are—the source of serious condensate problems. Drain lines receive less design, installation, and maintenance attention than any component of the HVAC system. As a result, line failures are significant contributors to property damage, equipment damage, and system contamination.

An acceptable drain line is simply one that has adequate flow capacity and offers minimum potential for flow blockage. The critical design factors are basic geometry and the support system for the drain line.

Drain Line Geometry. Unsatisfactory drain line geometry is a common cause of drain line flow blockage that results in condensate overflow problems.

Problem Definition. Small diameter, long and meandering, non-sloped and deflected drain lines are major causes of drain line blockage. Lines that are too small are easily blocked by debris and algae growth. Long lines are more susceptible to flow blockage, because there are simply more places for blockage to occur. Meandering lines with elbows are highly prone to flow blockage. Non-sloped drain lines retain condensate along with debris. Retained or standing condensate supports the growth of algae, which restricts and blocks flow. The effect of drain line blockage, which occurs all too often, is flooding and the overflow of the condensate pan. The consequences are damage to equipment and surrounding property plus contamination of the HVAC unit and the surroundings.

Remedies. The diameter of the drain line must be equal to or greater than the exit diameter of the drain seal device. When more than one HVAC unit drains to a common line, the area of the common line must be increased proportionally, at the downstream point where the additional HVAC drain line enters. The line length should be the minimum possible, following the shortest path to the condensate disposal place (the shorter, the better). And it should include the least possible number of elbows. The line must be sloped away from the drain seal at a rate of *no less* than $1/_8$ in per ft (1 cm/m).

Drain Line Support. A firm fixed drain line support system is essential to ensure satisfactory condensate drainage.

Problem Definition.   Condensate drain lines are often located in areas of high maintenance activity; for example, building roof tops. In this environment, if lines are not securely fixed they can be broken or damaged, by careless personnel, and permit condensate to be drained to unwanted places. Damage to drain lines frequently destroys the condensate drain seal (a very common occurrence). A destroyed condensate drain seal results in all the property damage and contamination problems discussed in Sec. 18.8.9.6.

A fixed support system is also required to ensure that the drain line maintains a satisfactory slope and prevents line defections sufficient to create harmful secondary traps. Any amount of line defection allows condensate to collect at the low point, where it promotes the growth of algae and increases the potential for flow blockage. Deflections greater than one diameter of the drain line create a much more detrimental situation. In systems with a conventional condensate trap, a trap formed by a dip in the drain lines forms an airlock, which will block condensate flow and cause the pan to flood and overflow.

Remedies.   To avoid excessive deflection of the drain line, prevent shifting of position, and ensure a satisfactory slope, fixed supports should be provided at distances no less than the following: PVC (Schedule 40 pipe), 2 to 3 ft (0.6 to 0.9 m); copper, 4 to 6 ft (1.2 to 1.8 m), and steel 8 to 10 ft (2.4 to 3 m) intervals.

18.8.10   *SYSTEM MAINTENANCE*

HVAC systems that are wet, dirty, and contaminated inside are often attributed to poor maintenance. Most often this is not the case. Systems with one or more of the deficiencies defined in Sec. 18.8.9 are commonplace. These deficiencies allow condensate to penetrate into places where successful maintenance is at best impractical.

In new system designs, these deficiencies can be avoided by applying the design procedures provided in Sec. 18.8.2.

In existing systems the remediation procedures outlined in Sec. 18.8.9, System remediation, can rid these systems of the deficiencies that prevent successful maintenance.

In reality, however, it is not practical to design around or remedy the inherent deficiencies and the nonmaintainability characteristics of the condensate trap. Table 18.8.2, prepared in the form of a routine and preventive maintenance program, shows why it is not possible for the conventional condensate trap to provide an acceptable drain seal. In summary, this is because the trap:

1. Requires frequent periodic cleaning to remove algae and debris, in order to prevent condensate flow blockage and overflow, resulting in system contamination and property damage
2. Must be filled with water frequently,
    a. During noncooling periods—e.g., winter—to prevent the ingestion of potentially toxic and noxious gases, and
    b. Prior to each cooling system start-up, to prevent drenching and/or flooding that contaminate and damage the system interior and surroundings
3. Is not suitable for use during winter cooling in outdoor locations where the temperature is below freezing, because of flow blockage and trap destruction.

Although it is possible by routine and preventive maintenance to minimize the damage caused by trap blockage, the procedure is costly, because traps must be cleaned or replaced frequently, requiring a significant maintenance effort.

Overcoming evaporation, which destroys the drain seal, during non-cooling periods and at system start-up time for cooling, places an enormous burden on maintenance personnel. Filling each condensate trap with water, frequently, to avoid gas ingestion, drench

TABLE 18.8.2   Routine and Preventive Maintenance Program for Conventional Condensate Trap

---

Trap located indoors or outdoors, with outdoor temperatures above freezing

1. Frequency and time of inspection and service:
   (a) For systems that provide summer cooling and winter heating during cooling operation:
      - Annually—at initial system start-up for cooling
      - Semiannually—at initial system start-up and at second system start-up if facility is shut down annually for a week or more, e.g., schools
      During Heating operation:
      - Biweekly, between cooling system shutdown and the beginning of winter heating
   (b) For systems that provide summer cooling and winter cooling
      - Semiannually—at 6-mo: intervals (one inspection must be made at system start-up, following an annual shutdown of facility for a week or more, e.g., schools
2. Maintenance effort required:
   (a) At each annual inspection (and semiannually if need is indicated)
      - Physically remove flow-blocking algae and/or debris, or replace trap
      - Flush with water
      - Treat with EPA approved biocide and
      - Fill trap with water and add biocide tablets
   (b) At each biweekly inspection
      - Fill with water and add biocide tablets if need is indicated.
3. Equipment and material needed:
   (a) Internal pipe scraper
   (b) New trap
   (c) Water hose
   (d) Biocide
4. Estimated time required:
   (a) Annually and semiannually:
      - 5 min per inspection + (25 min travel time to and from maintenance shop and system site)
      - 0 to 60 min per time serviced + (25 min travel time to and from maintenance shop and system site)
   (b) Biweekly:
      - 5 min, per time serviced + (25 min travel time to and from maintenance shop and system site)

Trap located outdoors, with outdoor temperatures below freezing

1. Frequency and time of inspection and service:
   (a) For systems that provide summer and winter cooling and winter heating during cooling operation:
      - Not possible to maintain drain seal with a trap during winter cooling under these conditions—flowing condensate will freeze in trap, block flow, and damage trap
   (b) During heating operation:
      - Not possible to maintain drain seal with a trap during winter heating under these conditions—unless the trap is filled with water, it will not hold a seal and when filled, water will freeze and block condensate flow

---

problems, and condensate flooding places impractical demands on maintenance organizations, even in indoor locations.

For example, using information in Table 18.8.2, it has been estimated that the maintainance effort for a typical system located indoors—used for both heating and cooling—is about 6 h per year per unit.

In outdoor locations, where outdoor temperatures are below freezing, it is virtually impossible to implement an effective trap filling program. Freezing condensate will destroy the trap, rendering it ineffective until it is replaced and refilled with water. Freeze plugs are of little value; even if they protect the trap, they destroy the seal.

TABLE 18.8.3   Routine and Preventive Maintenance Program for CostGard Condensate Drain Seal located indoors or outdoors, temperatures below or above freezing

---

1. Frequency and time of inspection and service:
   (a) For systems that provide summer cooling, winter heating, and cooling
       • Annually—during cooling operation, when condensate is flowing
2. Maintenance effort required:
   (a) If condensate is not flowing freely during cooling operation and/or condensate is standing in the pan more than the operating level indicated in Fig. 18.8.3, at drain outlet:
       • Check for debris inside the device. If present, physically remove and flush inside with water
       • Check operating pressures per manufacturer's instructions
   (b) Otherwise, no effort is required.
3. Equipment and Material Needed:
   (a) Water hose
   (b) Pressure gauge
4. Estimated time required:
   (a) Less than 5 min per inspection + (25 min travel time to and from maintenance shop and system site)
   (b) 0 to 30 min per time serviced + (25 min travel time to and from maintenance shop and system site)

The maintainability of the condensate trap can be made feasible by utilizing a water replenishing system and a condensate heating system, to ensure a water-filled trap and prevent freezing in outside locations. This does not, however, eliminate all the maintenance problems of the conventional trap and it introduces others. The system is subject to the same trap blockage from algae growth and debris, as is the conventional trap. In addition, both the water replenishing and the condensate heating systems require some type of control and usually utilize moving parts. Such systems require appreciable maintenance, which is often much too great for a practical drain seal.

The fluidic flow control device identified under Sec. 18.8.7 provides an effective and reliable drain seal, which eliminates all the problems caused by the condensate trap, and is virtually maintenance free.

Table 18.8.3 shows a routine and preventive maintenance program for the CostGard condensate drain seal. As shown, the projected maintenance effort is small. The only effort for which a finite time is stated is that for scheduled inspections. Experience shows that the time required for servicing these systems is minimal. The estimated time for maintenance service varies from zero to a few minutes. Accordingly, in any particular installation, maintenance experience may indicate that the inspection and service frequencies stated in the table can be decreased significantly, thus reducing the effort and cost.

18.8.11   LEGIONNAIRES' DISEASE: PHILADELPHIA REVISITED

18.8.11.1   Background

More than a quarter century has passed since 34 persons died from Legionnaires' disease allegedly contracted in a Philadelphia hotel. In an effort to find the cause of this catastrophe, federal, state, and local agencies launched one of the most extensive investigations in medical history. The best public documentation of what went on during the investigation is provided in the records of U.S. Senate and U.S. House of Representatives Hearings (Refs. 15, 16) on Legionnaire's disease.

Somewhere between 100 and 200 medical professionals participated in this investigation, for a period of several months. Because most of the investigative effort went into searching for what infected the victims, that is what was found. About five months after the outbreak in Philadelphia, the culprit was identified. It was a very tiny bacterium discovered, in pathologic specimens taken from victims, by Dr. Joseph McDade of the U.S. Center for Disease Control (CDC), Atlanta. The bacterium was named *Legionella* and legionellosis became the medical name of the disease.

It was also determined that *Legionella* causes illness only when it penetrates to the deep portion of the human lungs. For this to happen, it was concluded that the bacteria must be airborne in an aerosol, likely in very small water droplets.

From the viewpoint of medical science, the discovery of *Legionella* was clearly a brilliant achievement and a major success for the medical profession. Unfortunately, in the Philadelphia investigation, **the source of the Legionella was never determined and the spreading mechanism was not identified**.

Thus, a remarkable research effort and discovery has done little in terms of human health benefits and lives saved. The following statement from the January 1997 issue of the *ASHRAE Journal* (Ref. 17, p. 27) well summarizes the current situation:

"Despite two decades of ever increasing information about many aspects of legionellosis, the disease seems to be as common as ever producing tens of thousands of cases and thousands of deaths each year."

In outbreaks of legionellosis—following the discovery of *Legionella*—of which there have been many, the bacteria have been found somewhere at most of the affected sites. Cooling towers and potable water systems, because they are frequently contaminated with legionellae, are often alleged to be the source of the bacteria. However, how the bacteria are spread from these sources to the victim's lungs is not at all clear. In some instances, it has been postulated that *Legionella* aerosolized in cooling towers was airborne for several hundred meters, where it resulted in human illnesses and deaths. Still, other observers contend that the aspiration of water from potable water systems is the primary mode by which legionella is spread. Neither of these arguments is very convincing.

18.8.11.2   Missing Information

In the Philadelphia investigation, little attention was given to the role of the air handlers, as the possible source of the bacteria and as the mechanism for spreading them. Reports of the investigation defined only the location of the air handlers, type of refrigerant used in the systems, type of air filters in each unit, location of the water chilling equipment, and the source and approximate percentage of outside air used by each unit.

The details necessary for assessing how the air handlers could contribute to the growth and spread of the disease causing agents were not sufficiently defined, or at least not in public documents. For example, nothing was reported regarding the following important factors: Type and size of air handlers; geometry of the cooling coils; the geometry, condition, and contents of the drain pans; and the type of drain traps (seals), if any. Although samples of water for testing were taken from the chiller system, potable water system, and cooling towers, there is no evidence that condensate samples were taken from the drain pans of the air handler.

18.8.11.3   Hindering Myths

Despite the obvious potential for contamination inside an air handler, two industry myths have hindered critical investigation of air handlers as the source of legionellosis outbreaks.

These myths are (1) the conditions inside air handlers, including the condensate drain pan, will not support the growth and proliferation of *Legionella* and (2) there is no mechanism for creating the aerosol necessary for transporting the bacteria. Contrary to these myths, examination of available information suggests that air handlers were indeed the likely source of legionellosis outbreaks in Philadelphia, as well as at others locations.

Myth 1.   Legionella cannot grow and proliferate in condensate drain pans.   The first myth is that *Legionella* cannot grow and multiply in a condensate drain pan because the water temperature is too low. While the growth rate of *Legionella* is found to be greatest in water where the temperature is near 98.6°F (37°C) (Ref. 18), it survives at much lower temperatures. In fact, legionella has been found, in water at temperatures varying between 42°F and 145°F (5 to 70°C) (Ref. 3, p. 168) and it will grow and multiply at temperatures as low as 60°F (7°C) (Ref. 3, p. 300).

Legionella can be found almost anywhere there is water and suitable nutrients (e.g., dirt and biological growth). Its presence has been reported in numerous places, including the following: cooling towers, humidifiers, evaporative condensers, evaporative coolers, condensate drain pans, water fountains, spas, potable water systems, outdoor ponds, ice makers, and many other places (Ref. 18, p. 19).

In a typical air handler, operating at design conditions, the temperature of the air leaving the cooling coil is near 55°F (14°C). Under these conditions, the temperature of condensate in a free flowing drain pan is about 60°F (15°C). However, under part-load conditions, condensate temperatures change markedly. When coolant flow is reduced in response to reduced cooling loads, the temperature of the condensate will increase. Under such conditions, condensate temperatures of 70°F have been measured. *Legionella* is said to proliferate between 68°F and 113°F (Ref. 20).

The growth-rate of the *Legionella* bacteria is relatively low in water at temperatures of 60°F to 65°F. However, in the presence of iron (e.g., rusty pans) the growth-rate may increase more than 100 times (Ref. 3, p. 4). Accordingly, even at relatively low temperatures, there is a potential for high growth-rates and great concentrations of *Legionella*, in drain pans of air handlers. Thus, among the millions of HVAC systems in this country, many are undoubtedly infested with high concentrations of legionellae.

Myth 2.   There is no mechanism present to aerosolize and spread **Legionella.**   The second myth stems from the assumption that the only mechanism present for generating an aerosol inside the air handler is the conditioned air flowing over the surface of the condensate in the drain pan. Typically, this velocity is no greater than 600 ft per minute (fpm). According to the Kelvin-Hemholtz instability limit (Ref. 3, p. 51), water will not be aerosolized from the surface until the velocity reaches 1400 fpm (7 m/s). Thus, it can be correctly concluded that the 600 fpm is too low to create an aerosol.

There is, however, another means, not widely recognized, whereby condensate in the drain pan can be and frequently is aerosolized. It is common to draw-through air handlers—the most common type used in public, commercial, and industrial applications. In this type system the static pressure inside the drain pan is always negative. Unless the condensate drain ports are equipped with seals, air will be ingested. And during operation, the level of the condensate will rise in the pan and the velocity of the entering air can entrain and aerosolize the condensate. The negative pressure in condensate drain pans varies among systems. Typically, these pressures range from about −0.50 to −5.00 in of water column (in wc). Without a seal on the condensate drain line, water will stand in the pan at about 0.50 and 5.00 in, respectively. For these conditions, the velocity of the air entering the drain port will vary from about 2000 fpm (23 mph) to about 6200 fpm (70 mph), respectively.

Figure 18.8.23 shows how the negative pressure inside the drain pan compartment affects the velocity of the entering air, when no seal is present or when the seal (usually a

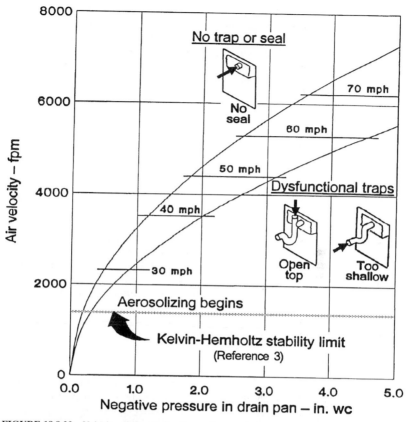

FIGURE 18.8.23   Velocity of air entering the condensate drain pan through a missing or dysfunctional trap.

condensate trap) is dysfunctional. As shown here, for a range of operating pressures, the air velocity in both cases is well above that where aerosolizing begins. Systems operating without adequate drain seals are common, nationwide.

18.8.11.4   Aerosol Generation

Condensate in a drain pan of a draw-through HVAC system will usually be aerosolized whenever the system is operated without a drain seal or with a dysfunctional trap. The seriousness of the condensate trap problem has been known to qualified professionals for years. In fact, in 1999 a statement to this effect was included in the informative language in BSR/ASHRAE Standard 62–1989R (Ref. 10). It reads as follows:

"Condensate traps exhibit many failure modes that can impact on indoor air quality. Trap failures due to freeze-up, drying out, breakage, blockage, and/or improper installation can compromise the seal against air ingestion through the condensate drain line. Traps with insufficient height between the inlet and outlet [poor design] *on draw-through*

systems can cause the drain to back-up when the fan is on, possibly causing drain pan over-flow or water droplet carryover into the duct system. The resulting moist surfaces can become sources of biological contamination. Seasonal variations, such as very dry or cold weather, may adversely affect trap operation and condensate removal."

These hazardous conditions are readily observable by anyone interested enough to visit and inspect the air handlers in almost any public, commercial, or industrial facility. Examples of what to expect during such a visit are depicted in Figure 18.8.24. Generally, these air handlers will be of the draw-through type, operating without adequate drain seals—no traps or dysfunctional traps—on the condensate drain lines.

During the past ten years we have visited scores of public, commercial, and industrial facilities in the central and southern parts of this country. And we have inspected hundreds of HVAC systems during the cooling seasons. With very few exceptions, the systems were of the draw-through type. Most were wet, dirty, and visually contaminated inside with various forms of biological growth, due to no seal or a dysfunctional trap. The wetness of internal walls and air supply ducts, which was frequently observed, attests to the fact that air entering the drain port was blowing and aerosolizing the condensate.

Others have reported similar conditions. For example, one report of a NIOSH investigation included the following comment: "*In a building where the air handling units' fans were down-stream from 'chiller' decks,* [draw-through systems] *stagnant water had spawned thick layers of microbial slime* (Ref. 20)." Almost certainly some of the air handlers in the Philadelphia hotel were in this condition.

Contrary to what many in the industry contend, these conditions are not the result of poor maintenance. They are, instead, the result of system design deficiencies—primarily the use of the condensate trap. Under the most favorable conditions, satisfactory maintenance of a conventional condensate trap is neither realistically feasible nor practical. Under other conditions, satisfactory maintenance is virtually impossible. The maintenance effort

(a) Drain pan - large air handler   (b) Drain pan - packaged hvac unit

FIGURE 18.8.24   Common conditions of drain pans—rusty and contaminated.

required for maintaining a condensate trap is summarized in Sec. 18.8.10. Evidently, no such maintenance program was in place at the Philadelphia hotel.

In view of the deplorable conditions of air handler drain systems, which have prevailed in the field for years, the potential for the widespread growth and aerosolizing Legionellae is obvious. Under such conditions, the illnesses and deaths of thousands from legionellosis should surprise no one.

18.8.11.5 The Philadelphia Story

Despite the discovery of *Legionella*, the results from the extensive Philadelphia investigation provides little help in determining the source and in reducing the number of persons affected by legionellosis. However, the evidence is strong and consensus is that exposure to the bacteria occurred in the hotel lobby and that transmission was by air. Hence, the air handler serving the center of the lobby is strongly implicated.

Comments reported in the senate hearings (Ref. 16) support this observation. For example, hotel staff members reported periods of *"poor flow of cooling air"* during the investigation. Also, in the same document, it was reported that, *"The dysfunction of the air handling system that serves the center of the lobby two weeks after the convention might indicate that the system was working imperfectly earlier."*

In addition to the above comments, there are other factors that point to and implicate the air handler serving the lobby. This air handler had been in operation for 22 years. Like air handlers of this vintage, it likely exhibited several characteristics that are conducive to the growth and spreading of legionellae. That is, it almost certainly (1) was the draw-through type; (2) depended upon a condensate trap to prevent air ingestion and permit condensate drainage; and (3) incorporated a large, deep drain pan, constructed of common steel.

Draw-through systems inherently create a negative pressure in the drain pan area. As stated in ASHRAE Standard 62-89R, quoted above, the condensate trap is subject to numerous failure modes, which frequently destroy the drain seal. When there is no seal on the drain line, the negative pressure holds backs the condensate and causes it to stand in the pan. The air rushing through the dysfunctional trap at high velocity entrains and aerosolizes the condensate. The standing and stagnant condensate enriched with iron from a rusty steel drain pan form a fertile place for the growth and proliferation of legionellae. Once the contaminated condensate is aerosolized inside the air handler, the bacteria are swept quickly into the conditioned space. The transmission is thorough and complete, since all the air in a conditioned space, typically, passes through the air handler 5 to 10 times per hour. All these conditions were likely present in the air handler serving the central lobby of the Philadelphia hotel.

Figure 18.8.25 shows a sketch of the envisioned air handler configuration, and its location relative to the hotel lobby. It also summarizes the likely conditions of the air handler and the conditioned space. The proximity of the air handler to the lobby ensures that contaminates from the air handler enter the lobby in high concentrations. One could hardly visualize more favorable conditions for spreading the legionellae bacteria.

18.8.11.6 Eliminating the Source of Legionellosis

Legionellosis is a serious life threatening disease. The discovery of *Legionella* has done little, or nothing, to reduce the incidence of this terrible malady. It has been estimated that about 8000 to 9000 persons in the United States die annually from this disease (Ref. 17, p 22). Efforts to reduce the number of victims have been remarkably unsuccessfully, ostensibly, because the real sources of legionellosis have not been identified. Even so, there is

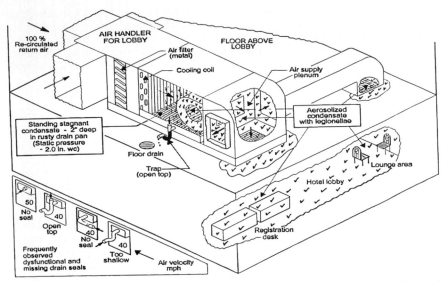

FIGURE 18.8.25   Envisioned features and condition of the air handler serving the central area of the hotel lobby.

no concerted effort, within the industry, to find the sources of the disease. An investigative effort comparable to that being devoted to leukemia (about 11,000 deaths annually) or that devoted to investigating airline crashes (an average of much less than 200 casualities annually) could well identify the source and virtually eliminate these deaths in a short period of time. A comprehensive engineering program plan for defining and eliminating sources of legionellosis is long overdue.

In the meantime immediate action should be taken, on both new system designs and on systems now in operation, using what is already known. For example, we know that *Legionellae* are everywhere. They are often present in condensate drain pans. They thrive in standing contaminated water, grow in condensate at temperatures common to drain pans, and their growth is accelerated many fold in the presence of iron (rusty pans).

Drain pans constructed of stainless steel or of other non-corrosive material can virtually eliminate iron as a factor. Sloped drain pans can prevent condensate puddles and local stagnation. But neither is a remedy for standing, stagnant condensate and the aerosolizing action caused by dysfunctional condensate traps.

Aerosolizing of condensate from drain pans can be prevented in different ways. In the design phase, one way is to select air handlers in which the drain pan is placed under positive pressure, instead of negative pressure (draw-through systems). This, of course, is not possible for the millions of draw-through systems now in service. For these and newly designed draw-through systems, the obvious remedy is an effective and reliable condensate drain seal. The commonly used condensate trap meets neither of these criteria. In fact, the condensate trap is the current problem, not the remedy. In some applications, the condensate pump can effect suitable condensate removal and provide a seal against air ingestion. However, it also exhibits many failure modes, which render it unreliable.

An effective drain seal, for draw-through systems, is essential to preventing stagnant condensate and to eliminating aerosolizing as a source of legionellosis. But an effective drain seal does much more for indoor air quality. It precludes the ingestion of outside air or other gases, when the trap is dry, and therefore prevents contaminates such as sewer gas and carbon monoxide from being drawn into the system. It also prevents flooding at start-up for cooling, when the trap is empty. In addition, by preventing the aerosolizing of condensate, an effective drain seal removes the primary source of internal system wetness. In so doing, it eliminates the accompanying biological growth, which contaminates HVAC components and supply air ducts. This virtually removes the draw-through HVAC system as a contributor to Sick Building Syndrome and Building Related Illness.

In the interest of their clients and building occupants as well as themselves, system designers would be well advised to place increased emphasis on finding an effective and reliable drain seal for draw-through HVAC systems. Clearly, providing an effective drain seal is as important, if not more important, as is providing suitable air ventilation and acceptable humidity control. It therefore warrants equal consideration by the designer.

The mechanics of fluid flow associated with draining condensate from the drain pan of a draw-through air handler are not simple. Nevertheless, system designers should have the basic technical understanding needed to evolve or select a suitable drain seal for the draw-through systems they design. The conventional condensate trap is simply not an acceptable option.

The design and selection of condensate drain seals is treated in Sec. 18.8.2. It reviews traps, condensate pumps, and defines one new drain seal, which is effective, reliable, and suitable for draw-through systems. Ingenious designers may, of course, evolve other equally suitable drain seals.

18.8.11.7 Conclusions

There is a critical need for effective and reliable condensate drain seals for draw-through HVAC systems. In fact, it is safe to say that the spread of legionellosis will not be diminished, nor will Sick Building Syndrome and Building Related Illnesses be reduced appreciably, until draw-through HVAC systems are equipped with effective and reliable drain seals. Moreover, until then, we cannot be assured that another outbreak of legionellosis, like the one in Philadelphia, will not occur.

18.8.12   FLUIDIC FLOW CONTROL— BLOW-THROUGH DRAIN SYSTEMS

In blow-through type HVAC systems, unlike draw-through systems, the fan blows air through cooling coils, creating a positive pressure (above ambient) in the drain pan. The positive pan pressure is favorable to condensate removal, and ingestion of outside air through the drain line is not possible. But, the control of condensate flow is essential. And, as with draw-through systems, a condensate trap is unsuitable for this purpose. This is because it is subject to dry trap syndrome, trap flow blockage, and freeze-up in outside locations.

During noncooling periods when the trap is dry (dry trap syndrome), a relatively large quantity of air may be discharged through the trap. Depending upon the drain size, this could compromise somewhat the overall efficiency of the HVAC system.

In addition, during start-up for cooling—with an empty trap—the discharged air often reaches velocities sufficient to entrain condensate in droplet form and spread it to unwanted places. The velocity at which condensate begins to entrain is about 1400 ft per min (fpm) (430 m/min). The velocity of the air discharged from an empty trap is usually much above that value. For example, at a pan pressure of 1 in (25 mm) of water, (near a minimum value found in practice) the air velocity of discharge is about 2500 fpm (780 m/min). At 5 in (12.5 cm) of water, the velocity approached 6000 fpm (1800 m/min) (near hurricane velocity). At best, the resulting wetness creates a nuisance and at worst it may cause wet floor accidents, property damage, and contamination of local surfaces.

Like a trap on a draw-through system, the trap also collects debris and supports algae growth, which cause frequent flow blockage, condensate pan overflow, and associated damage.

In cold climates, traps placed in outside locations can be damaged by freezing temperatures and their effectiveness destroyed, thus producing an empty trap and all the problems discussed above.

The fluidic flow control device negates or eliminates the problems created by the condensate trap.

An isometric drawing, a photograph, and the characteristics of the fluidic flow control device for blow-through systems are shown in Fig. 18.8.26.

Referring to Fig. 18.8.26, the device operates as follows: Condensate and air (two-phase flow) leaving the drain pan enters part A of the fluidic flow control device. Both fluids then pass through the mitered elbow array, into part B. From there, the condensate and a portion of the air pass into part E and on to the condensate disposal place. The remainder of the air passes into part D and out through the vent.

As the fluids pass through the mitered elbows, there is little resistance to condensate flow. Indeed, condensate flow is accelerated by the air flow and flows freely through the device. At the same time these elbows restrict airflow such that the velocity leaving the unit is far too low to cause entrainment and blowing of condensate. In addition, the air turbulence in the mitered elbows creates a scrubbing effect, which prevents blockage by debris and algae growth.

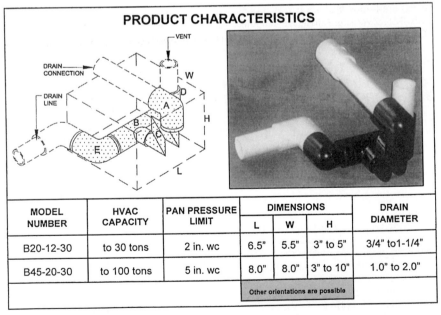

MODEL NUMBER	HVAC CAPACITY	PAN PRESSURE LIMIT	DIMENSIONS			DRAIN DIAMETER
			L	W	H	
B20-12-30	to 30 tons	2 in. wc	6.5"	5.5"	3" to 5"	3/4" to1-1/4"
B45-20-30	to 100 tons	5 in. wc	8.0"	8.0"	3" to 10"	1.0" to 2.0"
		Other orientations are possible				

FIGURE 18.8.26   Fluidic flow control device for blow-through systems.

Two models of the fluidic flow control devices are available. Together, they can accommodate systems with cooling capacities up to 100 tons, and with positive pressures up to 5.0 in (12.5 cm) of water column. The table in Fig. 18.8.3 summarizes the performance and geometric characteristics of the fluidic flow control device.

REFERENCES

1. Unpublished test data from the Department of Mechanical Engineering, University of New Orleans and Trent Technologies, Inc. laboratory, July, 1996.

2. 1999 ASHRAE 62-1999, Ventilation for Acceptable Indoor Air Quality, © American Society of Heating, Refrigerating, and Air Conditioning Engineers, Inc., www.ashrae.org, Para. 5.4.

3. G. W. Brundrett, *Legionella and Building Services*, Butterworth-Heinemann Ltd., London, 1992.

4. E. M. Sterling, A. Arundel, and T. D. Sterling, "Criteria for Human Exposure to Humidity in Occupied Buildings," *ASHRAE Transactions*, Vol. 91, Part B, 1985, pp. 611–622.

5. *Fans and Systems*, AMCA publication 201-90, Air Movement and Control Association, Arlington Heights, IL, 1990, p. 47.

6. "Industrial Ventilation," *A Manual of Recommended Practice, 22nd ed.*, American Conference of Governmental Industrial Hygienists, USA, 1995, pp. 3-6–3-8.

7. 2001 ASHRAE 62-2001, Ventilation for Acceptable Indoor Air Quality, Addendum 62t, © American Society of Heating Air Conditioning Engineers, Inc., www.ashrae.org, Para. 5.11.3.

8. W. Trent and C. Trent, "Indoor Air Quality and the Condensate Trap," paper presented at the Ninth Symposium on Improving Indoor Air Quality in Hot and Humid Climates, Arlington, Texas, May 19–20, 1994.

9. J. Cummings et al., "Uncontrolled Air Flow in Non-Residential Buildings," Final Report FSEC-CR-878-96, Florida Solar Energy Center, Cocoa, FL, April 15, 1996, pp. 14–15.

10. 1996 BSR/ASHRAE 62-1989R, Ventilation for Acceptable Indoor Air Quality, © American Society of Heating, Refrigerating, and Air Conditioning Engineers, Inc., www.ashrae.org, Para 5.6.

11. W. C. Trent, "The Condensate Trap: A Costly Failure," *Air Conditioning, Heating, and Refrigeration News*, Feb. 21, 1994, p. 3.

12. W. Trent and C. Trent, "Consideration in Designing Drier, Cleaner HVAC Systems," *Engineered Systems*, Aug. 1995, p. 38.

13. T. C. Ottney, "Particle Management of HVAC Systems," *ASHRAE Journal, 35*, American Society of Heating, Refrigerating, and Air Conditioning Engineers, Inc., Atlanta, GA, July 1993, pp. 26–34.

14. 2000 ASHRAE Handbook-HVAC Systems and Equipment. © American Society of Heating, Refrigerating, and Air Conditioning Engineers, Inc., www.ashrae.org, p. 24.11, Table 3.

15. U.S. Senate, *Hearings before the Subcommittee on Health and Scientific Research of the Committee on Human Resources, 95th Congress, 1st Session on Follow-Up* Examinations on Legionnaire's Disease, November 9, 1997.

16. U.S. House of Representatives, *Hearings before the Subcommittee on Consumer Protection and Finance of the Committee on Interstate and Foreign Commerce, 94th Congress, 2nd Session*, November 23 and 24, 1976.

17. J. Millar, G. Morris, and B. Shelton, "Legionnaire's Disease: Seeking Effective Prevention," *ASHRAE Journal*, Jan. 1997.

18. ASHRAE, *Legionellosis: Position Paper*, ASHRAE, Atlanta, GA, June 25, 1998.

19. J. Coleman, "Considerations to Prevent Growth and Spread of Legionella in HVAC Systems," *American School and Hospital*, June 1999.

20. V. Bishop, D. Custer, and R. Vogel, "The Sick Building Syndrome: What It Is and How to Prevent It," *National Safety and Health News*, Dec. 1985.

18.9 DESICCANT DEHUMIDIFICATION

Lew Harriman
Mason-Grant Company
Portsmouth, New Hampshire

Douglas Kosar
Senior Project Manager, Gas Research Institute
Chicago, Illinois

18.9.1 INTRODUCTION

Desiccants exhibit such a strong affinity for moisture that they draw water vapor directly from the surrounding air. That affinity can be regenerated repeatedly by applying heat to the material to drive off the collected moisture. Desiccants are placed in dehumidifiers that have been traditionally used in tandem with mechanical refrigeration in specialty air-conditioning systems. The systems have been most commonly applied when air-conditioning systems have large dehumidification load fractions. This situation is common when industrial operations require low humidity. Humidity levels below those necessary for comfort are costly to achieve with mechanical refrigeration and reheat, so systems that use desiccants for this purpose do have an economic advantage.

In the 1930s and 1940s, desiccant dehumidification was also integrated with mechanical refrigeration in air-conditioning approaches for comfort in businesses and homes. Later, the advent of abundant and inexpensive electricity and mass production in factory packaging made the use of electric motor driven vapor compression refrigeration the market choice for cost-effective commercial and residential air-conditioning systems.

From the 1930s through the present, however, desiccant dehumidification has held its economic advantage over dehumidification by mechanical refrigeration and reheat in specialty applications. In industrial air conditioning, numerous moisture sensitive manufacturing and storage applications utilize desiccant dehumidification. The dramatic decrease in product rejection and the direct increase in profitability of the product yields a quick payback on the initial investment in the depressed humidity control equipment. In storage applications, controlling humidity independent of temperature often avoids the need to heat the storage area during cold weather, and reduced losses in storage further improve the economic advantage of desiccant systems.

Since the middle 1980s there has been renewed interest in thermally driven desiccant dehumidification to cost-effectively remove humidity loads introduced by ventilation air in commercial and institutional buildings, and to offset rising electricity costs for air conditioning. Low cost thermal energy, including natural gas, waste heat, solar energy, and other sources, is substituted for electric energy to remove the humidity portion of the air conditioning. Industrial desiccant dehumidification equipment had been considered too expensive compared to assembly line mechanical refrigeration equipment for application beyond the industrial field of use. But today, commercial desiccant dehumidification technology, supported by ongoing research and development, has proven to be cost-effective, particularly for fresh air ventilation for systems with similarly high moisture loads. Beyond those applications, thermally driven desiccant technology is used in commercial buildings to reduce electrical power use, which lowers peak power demand charges.

18.9.2 PSYCHROMETRICS OF AIR-CONDITIONING LOADS

In air-conditioning applications, air must be supplied (to a conditioned space) that is cool enough, dry enough, and in sufficient quantity to remove thermal or sensible heat loads, and the humidity or latent heat loads in the space. To understand these air-conditioning processes, it is essential to discuss the properties of air-water mixtures. These can be understood more clearly through the use of a psychometric chart. Figure 18.9.1 shows such a chart, which is a graphic representation of the condition of air at any given point in an air-conditioning process. The chart relates dry bulb temperature (the air temperature read by a common thermometer or thermostat) on the horizontal scale to absolute moisture content of the air on the vertical scale. If the dry bulb temperature of a given air sample is reduced to the temperature at which it's moisture will condense, it is described as "saturated." This saturation point is the dewpoint temperature. Saturation temperatures are shown by the curved boundary at the left of the chart, also called the 100 percent relative humidity curve.

Sources of the space air-conditioning load are *internal*, including sensible and latent heat gains from people, lights, and equipment; and *external*, including sensible and latent heat gains from transmission, conduction, diffusion, and infiltration. In industrial, institutional, and commercial structures, ventilation air is a major additional sensible and latent load. The amount of fresh air, as determined by state and local building codes, must be treated by the air-conditioning system and introduced to the space to maintain air quality.

As a result of the sick building syndrome and building-related illness of the past decade, the ASHRAE Standard 62 "Ventilation for Acceptable Indoor Air Quality" has dramatically increased outside air requirements from its 1981 to 1989 issue.[1,2] Depending on the climatic conditions, applicable codes, and the operations in the space, a range of sensible and latent heat loads can be encountered when maintaining temperature and humidity setpoints in the space. The fraction of the cooling load, which is sensible, is the sensible heat fraction (SHF) and, likewise, the fraction of the air-conditioning load, which is latent, is the latent heat fraction (LHF).

Standard design practice in the past utilized extreme dry bulb temperature to quantify a building's peak cooling load. For example, four major U.S. cities are evaluated in

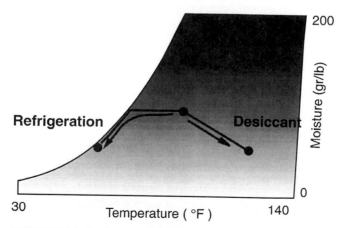

FIGURE 18.9.1 Psychrometric chart with dehumidification process paths.

TABLE 18.9.1   Sensible and Latent Ventilation Air Treatment Loads at 2.5% Dry Bulb (DB) and Mean Coincident Dew Point (MCDP) Design Condition

| (W · s/L) City | 2.5% DB/MCDP °F (°C) | Design load Btu/h/SCFM | | |
		Sensible/SHF percent	Latent/LHF percent	Total
Chicago	89.0/67.8 (31.7/19.9)	15.1 (9.38)/37%	25.6 (15.9)/63%	40.7 (25.3)
Houston	94.0/70.6 (34.4/21.4)	21.6 (13.4)/40%	32.8 (20.4)/60%	54.4 (33.8)
Los Angeles	80.0/62.1 (26.7/16.7)	5.40 (3.35)/30%	12.8 (7.94)/70%	18.2 (11.3)
Phoenix	107.0/51.3 (41.7/10.7)	34.6 (21.5)/100%	NA (NA)/0%	34.6 (21.5)

Note: Loads are calculated compared to a neutral indoor condition of 75°F (24°C) dry bulb temperature and 50% relative humidity. Design dry bulb data per Ref. 3.

Tables 18.9.1 and 18.9.2 at dry bulb and dewpoint temperature extremes to determine the magnitudes of peak sensible and latent loads resulting from outside air. Table 18.9.1 shows that the resulting LHF from ventilation loads is generally high regardless of location (except for arid climates like Phoenix) at extreme dry bulb conditions. However, quantifying loads at extreme dewpoint conditions, as in Table 18.9.2, clearly demonstrates an even greater LHF. In fact, the largest total cooling loads occur at the peak latent load conditions, rather than at the peak sensible load conditions.

In a cooling-only air-conditioning system, air is dehumidified by over-chilling the air with a cooling coil that has a surface temperature below the dewpoint of the air. As the air cools below its dewpoint, moisture condenses on the coil surface and is drained away. The amount of cooling (the sensible heat fraction, SHF) compared to the amount of dehumidification (the latent heat fraction, LHF) accomplished by a such a system will depend on the surface temperature of the coil, and the temperature and dewpoint of the entering air. When the air temperature is high and the coil temperature is low, the system will remove more sensible heat. When the dewpoint temperature of the air is high and the coil surface temperature is low, the system will remove a large amount of moisture.

To maintain the temperature and humidity setpoints in the space, the SHF and LHF of the system cooling capacity must equal the SHF and LHF of the air-conditioning load. In

TABLE 18.9.2   Sensible and Latent Ventilation Air Treatment Loads at 2.5% Dew Point (DP) and Mean Coincident Dry Bulb Design Condition

| City | 2.5% DB/MCDP °F (°C) | Design load Btu/h/SCFM (W · s/L) | | |
		Sensible/SHF percent	Latent/LHF percent	Total
Chicago	72.7/82.7 (22.6/28.2)	8.32 (5.16)/18%	38.6 (24.0)/82%	46.9 (29.1)
Houston	77.0/83.0 (25.0/28.4)	8.64 (5.37)/14%	51.8 (32.2)/86%	60.4 (37.5)
Los Angeles	65.7/73.4 (18.7/23.0)	NA (NA)/0%	20.6 (12.8)/100%	20.6 (12.8)
Phoenix	69.6/83.2 (20.9/28.4)	8.86 (5.50)/23%	30.1 (18.7)/77%	39.0 (24.2)

Note: Loads calculate to space neutral condition of 75°F (24°C) dry bulb temperature and 50% relative humidity. Design dew point data per Ref. 4.

TABLE 18.9.3 Sensible and Latent Cooling by Dewpoint Depression at Saturation Conditions

Dewpoint temperature °F (°C)	Sensible heat removal Btu/hr/SCFM (W · s/L)	Latent heat removal Btu/h/SCFM (W · s/L)
60 to 59 (15.6 to 15)	1.08 (0.671)	1.86 (1.16)
50 to 49 (10 to 9.4)	1.08 (0.671)	1.36 (0.845)
40 to 39 (4.4 to 3.9)	1.08 (0.671)	0.97 (0.602)

certain air-conditioning situations, the LHF of the system cooling capacity is not high enough (or the SHF of the system cooling capacity low enough) to meet the latent load and maintain the space humidity setpoint. To obtain the necessary latent cooling capacity, extra sensible cooling capacity must be designed-in. Either the original amount of air must be sensibly cooled further below the dewpoint temperature, or additional amounts of air must be sensibly cooled down past the dewpoint temperature, to increase latent cooling capacity. The cooler air temperature or the large amount of cooled air is then greater than necessary to meet the sensible load, so reheating equipment must be added to keep the space warm enough for comfort. But standard cooling equipment has limits based on the depth of their cooling coils, and the spacing of fins across the face of the coils.

Such limitations typically restrict flow rates to under 500 SCFM per ton (67 L/s per kW) of cooling. Decreasing the coil temperature to increase latent cooling has its practical limitations also. Although sensible heat removal remains essentially constant per degree of dewpoint temperature reduction at saturation, the amount of latent heat removal per-degree-of-dewpoint-reduction drops off considerably at lower temperatures. Table 18.9.3 illustrates this phenomena by comparing moisture removal for a series of dewpoint depressions.

If sensible cooling loads have already met at the above dew points, then for each additional Btu/h/SCFM (0.621 W · s/L) of latent cooling, that dewpoint must be lowered by the excess sensible cooling and reheating in Table 18.9.4.

When cooling coil surfaces must be held at 32°F (0°C) or lower to produce resulting air dew points of 40°F (4.4°C) or lower, condensed moisture will freeze on the coils and can't be drained away while the system is operating. Defrosting equipment must be added to the system to eliminate ice buildup on the cooling coils. Cooling will be interrupted periodically to defrost the coils before heat transfer, and consequently capacity, are reduced substantially. Dual cooling coils, at a minimum, must be added to the equipment to produce continuous cooling. Finally, all heat (amount depends on defrost method) added to defrost the coils (except that drained away with the melted condensate) will increase the air-conditioning load in the space.

This "over-designed" process, shown in Fig. 18.9.1, yields inefficient operation as unnecessary sensible cooling, defrosting, and reheating increases energy costs when

TABLE 18.9.4 Sensible Overcooling and Reheating for Latent Cooling

Dewpoint temp °F (°C)	Lowered dew-point °F (°C)	Sensible overcooling	Sensible reheating Btu/h/SCFM (W · s/L)	Total excess sensible energy
60 (15.6)	59.46 (15.3)	0.58 (0.36)	0.58 (0.36)	1.16 (0.72)
50 (10)	49.27 (9.6)	0.79 (0.49)	0.79 (0.49)	1.58 (0.98)
40 (4.4)	38.97 (3.9)	1.11 (0.69)	1.11 (0.69)	2.22 (1.38)

meeting high design LHF loads. Equipment first costs increase too as additional unnecessary sensible cooling, defrost, and reheat capacity is purchased to acquire the design latent
cooling capacity. At off-design conditions, when the sensible cooling requirement is lower,
an even higher LHF load can exist, especially in more humid climates. In more humid climates, where the outdoor dewpoint is generally high and close to the dry bulb temperature,
the part load condition has a larger fraction of moisture and a correspondingly smaller fraction of sensible heat. At these part load conditions the operation to control moisture
becomes increasingly inefficient and may be unable to satisfy the humidity setpoint in the
conditioned space.

An alternative to this mismatch in air-conditioning load and system capacity is the
use of a desiccant dehumidifier to meet the latent cooling need. A desiccant dehumidifier utilizes a material which sorbs water vapor directly from the air to its surface or into
a chemical solution. The moisture releases heat as it is sorbed, raising the temperature
of the desiccant, which in turn raises the temperature of the dehumidified air. A desiccant dehumidifier is typically coupled with an aftercooling heat exchanger to reduce the
dry bulb temperature of warm, dry air leaving the desiccant bed. The desiccant bed and
heat exchanger can then be integrated with smaller conventional cooling equipment for
further aftercooling to meet the remaining sensible load in the space. With this "hybrid"
operation shown in Fig. 18.9.1, temperature and humidity can be controlled independently and without excess sensible cooling capacity and reheating and defrosting
requirements.

18.9.3   BEHAVIOR OF DESICCANT MATERIALS

Desiccants are solid or liquid materials. Familiar solid desiccants are silica gel or molecular sieve. Common liquid desiccants include water solutions of lithium chloride or glycol.
The term desiccant is applied to these and other materials having a large capacity for moisture relative to their own weight. Some desiccants also collect other gases or vapors in addition to moisture. Generally known as sorbents in this role; desiccants can remove various
airborne pollutants from air streams. Also, some desiccants such as lithium chloride have
biocidal effects; killing bacteria and viruses through their powerful surface-contact desic-
cation effects.

The weight of water uptake in a desiccant depends on the equilibrium between partial pressures exerted by the moisture in the sorbent compared to the partial pressure of
moisture in the surrounding air. The partial pressure of moisture is termed *vapor pressure*. Vapor pressure in air is directly related to dewpoint temperature. Very humid air
with high dewpoint temperature has a high vapor pressure. Lower vapor pressure at the
surface of a desiccant means it can attract moisture from air which has a higher vapor
pressure.

The equilibrium water capacity of a desiccant at a given vapor pressure decreases as desiccant temperature increases. In other words, as a desiccant is heated, it loses it's ability to
attract and hold moisture. Combining the variables of dry bulb temperature with vapor pressure (or dewpoint temperature) into relative humidity produces a single equilibrium water
capacity curve for 0 to 100 percent relative humidity, which is useful for general comparisons of moisture uptake potentials at various surrounding air relative humidities.
These water vapor equilibrium curves for a few common desiccants are shown in Fig.
18.9.2 (Ref. 3).

The sorption of water vapor in a desiccant raises the temperature of the desiccant itself,
which in turn heats the surrounding air. This temperature rise is caused primarily by the
conversion of latent energy to sensible heat through the release of the heat of condensation

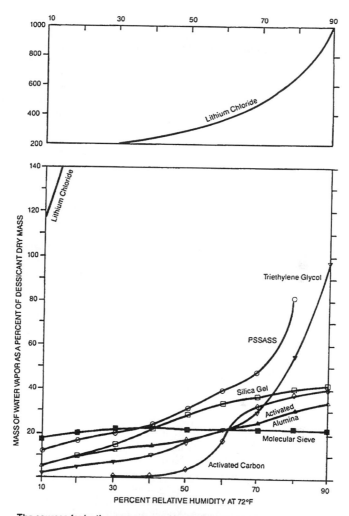

The sources for isotherms presented in the figure include :
PSSASS: Solar Energy Research Insitute Report No. PR-255-3308
Lithium chloride: Cargocaire Engineering Division of Munters USA and Kathabar Systems Division of Sommerset Technologies
Triethylene glycol: Dow Chemical Corporation
Silica gel: Davison Chemical Division of W.R. Grace Co. Inc.
Activated carbon: Calgon Corporation
Activated alumina: LaRoche Chemical Company
Molecular sieve: Davison Chemical Division of W.R. Grace Co. Inc.

FIGURE 18.9.2 Desiccant equilibrium curves.

from the sorbed moisture. However, an additional temperature rise comes from physical, electrical, or chemical interactions between the desiccant and the sorbed moisture. This additional heat can vary from a fraction of, to several times, the heat of condensation depending on the desiccant. In practice, however, dehumidifiers exhibit process air dry bulb temperature increases that typically range from slightly over 1–2 times that associated with heat of condensation alone. The entire heat energy released, including the heat of condensation, is termed the heat of sorption.

The most useful feature of the sorption process is its reversibility through the application of additional heat. Moisture is released by raising the desiccant temperature above the temperature it had during the sorption process. Since the desiccant is hot, it's surface vapor pressure is higher than the vapor pressure in the surrounding air, so water leaves the desiccant in favor of the air. When the desiccant is cooled, the dry material is ready to attract and hold water vapor once again. Heating the desiccant is called "reactivation" or "regeneration." Dehumidifiers are usually regenerated with air at temperatures between 130 and 300°F (54.4 to 148.9°C).

Over time, desiccants can lose a portion of their moisture sorption capacity. The capacity loss depends on the desiccant, the duty cycle, and the contaminant loading in the air it dehumidifies. Thermal and mechanical degradation can cause minor loss in capacity for solid desiccants (5 to 15 percent over five years). A more substantial capacity reduction can occur in both solid and liquid desiccants through contamination with airborne materials or reactive substances. Incoming air must be filtered to remove dust. A 35 percent spot efficiency filter is generally sufficient to prevent measurable capacity reduction. Depending on the desiccant, an application might need an initial assessment for other contaminants as well, but these are usually heavy industrial application issues. In air conditioning applications in commercial buildings, contamination of desiccants is almost never a concern. Where necessary, most equipment manufacturers offer desiccant analysis services for their dehumidifiers.

Although the percent of water sorbed in some desiccants is significantly greater than others, the minimum dewpoint temperature produced by a sorption/desorption process is not determined by the equilibrium capacity [5]. Equilibrium capacity does determine the amount of desiccant that air must contact to achieve a moisture removal rate, i.e., pounds of water per hour (kilograms/hour). Desiccants can readily provide minimum dewpoint temperatures well below freezing (32°F or 0°C) at room air dry bulb temperatures (75°F or 23.9°C).

Ability to maintain this minimum dewpoint temperature at the dehumidifier outlet is limited though, by practical equipment design and operational considerations. Economical sizing for dehumidification capacity typically does not allow equipment to provide air at the minimum dewpoint possible for the desiccant. Also, higher regeneration temperatures lower the dewpoint temperature achievable with some desiccants, but material tolerances and less optimal operating efficiencies may restrict their use in dehumidifiers. As an example, in certain liquid systems, regeneration temperatures are restricted to below levels at which concentrations created would result in crystallization of solids in solution. Available desiccant dehumidifiers attain dewpoint temperatures that generally range from 10°F to −55°F (−12.2°C to −48.3°C) (Ref. 6).

18.9.4   DESICCANT DEHUMIDIFIERS

Granular desiccants are placed in some form of bed to dehumidify air. The older and simpler dehumidifier designs utilize desiccant-filled shallow discs. The perforated disc rotates slowly to expose one portion of the desiccant bed to the process air stream while the other portion simultaneously passes through the regeneration air stream. A partition and flexible seals separate the process and regeneration air streams in the dehumidifier. Moist air enters the process side and passes through the desiccant filled bed and is dehumidified. Regeneration occurs on the other side of the partition where heated air enters from the same or opposite direction, passes through the bed and, laden, with moisture is exhausted from the dehumidifier.

Alternatively, the granular desiccant in some dehumidifiers is contained between inner and outer closely spaced cylinders. A further variation of this design places desiccant in a circular arrangement of compartmentalized vertical beds.

A more recent dehumidifier design employs a parallel passageway wheel to minimize pressure drop. The desiccant is impregnated in a substrate which is typically a sinusoidally corrugated or hexagonal matrix forming channels for air flow.

In all these designs, air flows through the solid which rotates continuously (or step wise) from process to regeneration and back, as shown in Fig. 18.9.3. Often a small amount of process air is used to cool, or purge the heat from the desiccant immediately following regeneration. The purge air stream is then ducted to the entering regeneration air stream. A heat exchanger may be provided that transfers energy from the exiting regeneration air stream to the entering regeneration air stream. Alternatively a heat exchanger may cool the process air stream exiting the dehumidifier and preheat the incoming regeneration air stream in turn. Additional sensible cooling of the process air stream will be required to control dry bulb temperature for comfort in the building.

Liquid desiccants are enclosed in chambers to dehumidify air. The desiccant is distributed in one chamber where it contacts the passing process air stream. As the moisture is

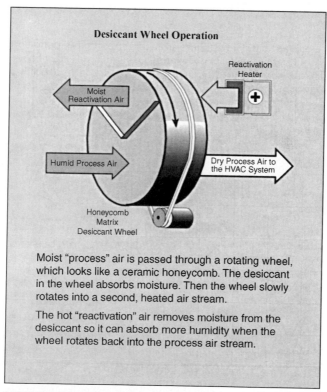

FIGURE 18.9.3 Solid desiccant dehumidifier.

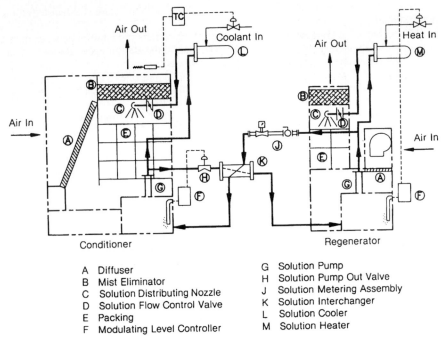

FIGURE 18.9.4   Liquid desiccant dehumidifier.

A Diffuser	G Solution Pump
B Mist Eliminator	H Solution Pump Out Valve
C Solution Distributing Nozzle	J Solution Metering Assembly
D Solution Flow Control Valve	K Solution Interchanger
E Packing	L Solution Cooler
F Modulating Level Controller	M Solution Heater

sorbed by the desiccant, heat is released. A cooling coil in the chamber (or chilled liquid desiccant itself) removes the heat of sorption: creating a quasi-isothermal process (the simultaneous desiccant dehumidification and aftercooling generally follows the dotted line path shown in Fig. 18.9.4. The moisture laden desiccant is then pumped to the other chamber where a heating coil (or "boiling" of the liquid desiccant directly) drives off the water into an exhaust air stream. The desiccant can now be pumped from the regeneration chamber, called the regenerator, to be redistributed in the dehumidification chamber, called the conditioner. An interchanger is often used to cool the warmer desiccant leaving the regenerator by exchanging heat with the cooler desiccant from the conditioner. The liquid desiccant dehumidifier configuration is shown in Fig. 18.9.4. As with other dehumidifiers, additional process air sensible cooling may be required to provide comfortable space dry bulb temperatures.

Dehumidifiers range in size from 25 to 100,000 SCFM (11.8 to 47,200 L/s) with moisture removal capacities from about 0.5 to 10,000 lb/h (0.23 to 4540 kg/h) or 500 to 10,500,000 Btu/h (0.15 to 3078 kW) of latent cooling[7-12]. Individual solid systems are designed from 25 SCFM (11.8 L/s) to 62,500 SCFM (29,500 L/s) and liquid systems are designed from 1000 SCFM (472 L/s) to 100,000 SCFM (47,200 L/s). Figures 18.9.5 and 18.9.6 show representative dehumidifier performance characteristics for solid and liquid systems, respectively. Coefficients of performance (COPs) range from as low as 0.35 to 0.8 or more. COP is based on the latent energy of the moisture removed divided by the heat energy necessary for regeneration. Waste heat utilization in a system design can reduce, and possibly eliminate, the purchase of thermal energy for regeneration.

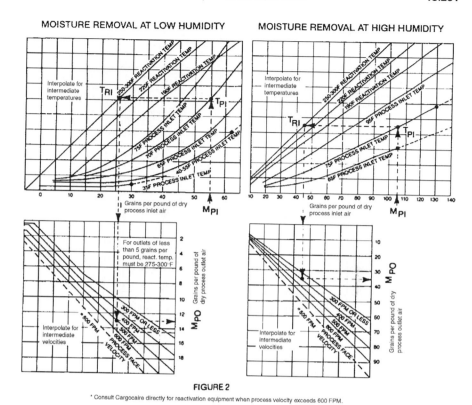

MOISTURE REMOVAL AT LOW HUMIDITY   MOISTURE REMOVAL AT HIGH HUMIDITY

FIGURE 2

* Consult Cargocaire directly for reactivation equipment when process velocity exceeds 600 FPM.

FIGURE 18.9.5   Performance map of solid desiccant dehumidifier.

Manufacturers should be contacted for complete dehumidifier performance data and design specifications.

18.9.5   APPLICATIONS FOR DESICCANT SYSTEMS

In numerous industrial environments, desiccant dehumidification removes unwanted humidity from the air to promote drying and prevent condensation or moisture regain. This allows production processes to proceed efficiently with minimum quality problems. It also allows storage of raw materials and finished goods to be accomplished without degradation problems. To manufacture pharmaceuticals for example, relative humidities down to 10 percent at room temperature are required for granule/powder handling proper compounding, tableting, and packaging for a long product shelf life.

Other industries, including food and beverage production, electronics manufacturing, chemicals preparation, rubber and plastic fabrication, metals treatments, photographic film manufacturing, wood machining, cosmetics manufacturing, printing/painting operations, and glass lamination also use desiccant dehumidification to maintain low humidity levels.

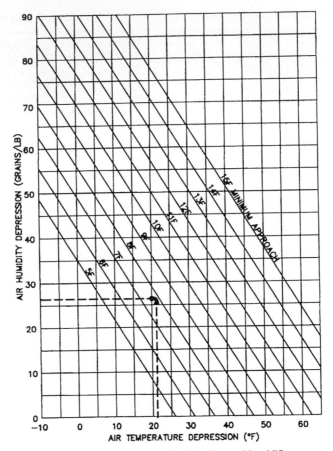

FIGURE 18.9.6 Performance map of solid desiccant dehumidifier.

Many of these same industries also use desiccant dehumidification to reduce humidity levels for transporting and warehousing to prevent corrosion, mold/mildew formation, and other deterioration. A listing of several industries and their temperature and humidity controlled processes is shown in Table 18.9.5 (Ref. 13).

The often dramatic decrease in product rejection and direct increase in profitability of the product is the primary incentive for the investment in low-level humidity control by desiccant dehumidification in the industrial sector. These humidity levels are typically unattainable cost effectively with cooling-based technologies. Small volume and the cost of application engineering support from manufacturers generally places a relatively high initial cost on industrial desiccant dehumidification equipment compared to conventional air-conditioning equipment.

The heavy duty equipment used in industry has, in recent years, been redesigned for cost effective application in many commercial and institutional building types. In many cases, economic benefits to the building operations, in addition to lower air conditioning costs, justify a higher initial investment. In other cases, the desiccant component may reduce the initial cost of the system compared to conventional cooling-only systems.

TABLE 18.9.5 Temperature and Humidity Setpoints for Industrial Air Conditioning

Process	(°C) Dry Bulb (°F)		
ABRASIVE			
Manufacture	79		50
CERAMICS			
Refractory	43 to 65	110 to 150	50 to 90
Molding room	2T	80	60 to 70
Clay storage	16 to 27	60 to 80	35 to 65
Decalcomania production	24 to 27	75 to 80	48
Decorating room	24 to 27	75 to 80	48
DISTILLING			
General manufacturing	16 to 24	60 to 75	45 to 60
Aging	19 to 22	65 to 72	50 to 60
FOUNDRIES*			
Core making	16 to 21	60 to 70	
Mold making			
Bench work	16 to 21	60 to 70	
Floor work	13 to 19	55 to 65	
Pouring	5	40	
Shakeout	5 to 11	40 to 50	
Cleaning room	13 to 19	55 to 65	
ELECTRICAL PRODUCTS			
Electronics and x-ray			
Coil and transformer winding	22	72	15
Semiconductor assembly	20	68	40 to 50
Electrical instruments			
Manufacture and laboratory	21	70	50 to 55
Thermostat assembly and calibration	24	75	50 to 55
Humidistat assembly and calibration	24	75	50 to 55
Small mechanisms			
Close tolerance assembly	22	72*	40 to 45
Meter assembly and test	24	75	60 to 63
Switchgear			
Fuse and cutout assembly	23	73	50
Capacitor winding	23	73	50
Paper storage	23	73	50
Conductor wrapping with yarn	24	75	65 to 70

*Winter dressing room temperatures. Spot coolers are sometimes used in larger installations.

TABLE 18.9.5 Temperature and Humidity Setpoints for Industrial Air Conditioning *(Continued)*

Lightning arrester assembly	20	68	20 to 40
Thermal circuit breakers assembly and test	24	75	30 to 60
High-voltage transformer repair	26	79	5
Water wheel generators			
Thrust runner lapping	21	70	30 to 50
Rectifiers			
Processing selenium and copper oxide plates	23	73	30 to 40

FLOOR COVERING	
Linoleum	32 to 38
Mechanical oxidizing of linseed oil†	90 to 100
Printing	27
Stove process	71 to 121

*Temperature to be held constant.
†Precise temperature control required.

18.9.6 EVALUATING APPLICATIONS FOR DESICCANT SYSTEMS

Deciding whether desiccant systems apply to a given building requires an understanding of the differing needs of each application, combined with the energy economics specific to the site. But there are six circumstances in which desiccant systems may have advantages over competing technology. When two or more of these circumstances apply, the owner would be well-advised to investigate desiccants[17].

18.9.6.1 Economic Benefit to Low Humidity

Desiccants are exceptionally effective at maintaining low humidity. So when there is an economic benefit to such humidity control, desiccants are likely to be the best choice.

The classic extreme example from industry is lithium batter production. Above 6 percent rh, pure lithium can generate enough hydrogen gas to ignite, producing an explosion. Clearly, if a manufacturing company is making lithium batteries, there is a strong economic incentive to maintaining low humidity to avoid explosions. A more common and less extreme example is the supermarket, which avoids excess energy cost by avoiding frost formation on refrigerated display cases and on frozen food. At typical store operating margins of 1.5 percent of sales, the $25,000 annual energy cost reduction produced by a desiccant system is the bottom-line equivalent of selling an additional $1.7 million worth of groceries each year.

A more subtle example is the loading dock of a refrigerated warehouse. Worker's compensation insurance costs are high because of injuries from slips and falls. Keeping humidity low avoids ice and condensation on floors. That is very advantageous economically, as well as being the right thing to do from a purely human perspective.

Other examples of economic benefits to low humidity are as diverse as the last half-century of desiccant technology. For a quick rule-of-thumb, desiccants are worth evaluating if the required humidity control level is below a 50°F (24°C) dewpoint (below 55 gr/lb). Stated in another way: when a problem occurs during the summer, but not during the winter, humidity may be contributing factor, so desiccants may be economically advantageous.

18.9.6.2 High Moisture Loads with Low Sensible Heat Loads

Because desiccants convert a moisture load to a sensible heat load, they do well where the moisture loads are high and the sensible heat loads are small. Supermarkets provide a good example. A great deal of moisture must be removed from the store to avoid frosting on low-temperature coils and frozen product. At the same time, so much excess cold air spills out of frozen food display cases, so the store must be heated to maintain a comfortable shopping environment. In other words, the moisture load is high while the cooling load is low or even non-existent.

Another example is a movie theater. There are no lights, computers, or machinery to generate sensible heat, the walls are well-insulated and there are no windows. The patrons are sedentary, keeping their heat production to an absolute minimum. But each person breathes in and out, releasing large amounts of water vapor into the air, and the fresh air needed to ventilate the theater carries very large amounts of water vapor—particularly in the evening, when theaters are most active. Once again, the moisture loads are high compared to the sensible cooling loads.

One simple way to compare the moisture load to the sensible head load is to calculate the sensible heat ratio (SHR) for the proposed system. The SHR is equal to the sensible heat load divided by the total heat load. For example, if the sensible heat load totals 700 Btu/h; and the total load consists of that 700 Btu/h (21 W)/21 of sensible heat, plus 300 Btu/h (9 W) or latent heat (moisture), the sensible heat ratio is:

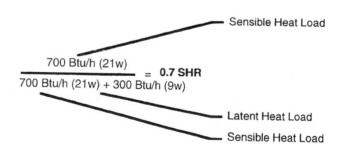

In other words 70 percent of the total load is sensible heat and 30 percent of the total load is latent heat (moisture). For a quick rule of thumb: when a system or a building has a SHR of less than 0.8, the application may or may not benefit from desiccants. But if the SHR is below 0.7, the application is almost certain to gain from desiccants.

18.9.6.3 Need for More Fresh Air

ASHRAE Standard 62 now calls for 15 cfm (0.42 m³) of fresh air per person in most buildings. That standard has been incorporated into all three of the model building codes used in the United States, so it begins to have the force of law.

So many buildings need much more ventilation air than in the past. Retail buildings are especially affected. Where a 100,00 ft² (9300 m²) store would formerly require only 4000 cfm

(112 m^3) of outside air, codes based on ASHRAE Standard 62 now demand that a store be provided with 0.3 cfm (0.08) m^3/m per m per square foot. Ventilation air flow jumps from 4000 to 30,000 cfm (112 to 840 m^3/m) to meet new codes. When such large amounts of fresh air are needed, conventional equipment is simply not designed to deal with the increased load. For example, major manufacturers of rooftop packaged air conditioning equipment estimate that when the proportion of outside air goes above 15 percent of the total supply air, conventional equipment has problems. When the fresh air goes above 20 percent of the total, equipment will regularly fail to maintain the specified temperature levels in a traditional, low-rise commercial building.

For this reason, engineers sometimes combine a desiccant-based make-up air unit with conventional equipment on the same building. The desiccant system handles the majority of the total fresh air requirement, and the conventional units can then operate quite comfortably with less than 10 percent fresh air without violating ventilation codes.

Similarly, engineers often use a desiccant system to add more fresh air to an existing building. The desiccant system satisfies the larger requirement for fresh air as needs of the occupants change over time. Adding a desiccant system is often much less costly and troublesome than reworking all existing HVAC systems to provide additional fresh air.

As a rule of thumb, desiccant systems begin to have advantages over conventional systems when the proportion of fresh air in a given system goes above 15 percent.

18.9.6.4 Exhaust Air Available for Postcooling and Winter Heating

The cost of postcooling the air as it leaves the wheel can favor or limit the applicability of desiccant systems. When exhaust air can be brought back to the desiccant system at little or no incremental cost, the desiccant alternative gains an advantage over conventional systems.

For example, a hospital HVAC system may use 100 percent outside air to meet local ventilation codes. If the exhaust air can be evaporatively cooled and reused through a heat exchanger to cool the warm process air, there may be no need for any conventional postcooling equipment. The overall installed cost of such a system might well be less expensive than the conventional alternative, and the operating cost would be considerably lower. In cooler climates, the advantage is even greater, because the exhaust air can heat the incoming fresh air during winter months, as well as cooling the fresh, dry air during humid months.

In many buildings, returning the air to the desiccant unit is not practical, either because there is little exhaust air, or because the system operates for very few hours a day, or because air must be exhausted far away from the desiccant system. But with the increased amounts of ventilation air required by building codes which adopt ASHRAE Standard 62–89, it is more likely the air can be exhausted through the desiccant system. That is because it is less likely that all the air can be exhausted from any other point, and because standard commercial HVAC design practice usually includes return duct work to bring air back to recirculate through the unit. In such cases, there is no incremental cost to exhausting air at the desiccant system.

As a quick rule of thumb—if the system design already includes return air duct work, the situation is neutral; neither desiccant nor conventional systems has an advantage. But if recovering energy from the exhaust can reduce the installation cost by downsizing equipment, then desiccant systems have an advantage. Conversely, if the desiccant system costs more than conventional, and if the payback time is over 3 to 5 years, then conventional systems have a decided advantage.

In many ways, this criterion is linked to the need for fresh air. If fresh air is a large proportion of the supply air, chances are good that recovering energy from exhaust air will pay off quickly, and may even reduce the installed cost of a desiccant system to less than a conventional system.

18.9.6.5 Low Thermal Energy Cost Plus High Electrical Demand Charges

Geography plays an important role for desiccant technology in unexpected ways. One might expect that desiccants, with their ability to remove moisture efficiently, would be especially advantageous in humid Miami, and less so in dry Detroit. But that is not the case. In Miami, electrical costs are fairly low compared to Detroit, where electrical demand charges are rather high. Also sensible heat loads are high and continuous in Miami, where on an annual basis, sensible heat loads are lower in Detroit.

Clearly, "relatively low thermal energy cost" and "rather high electrical demand charges" are nebulous. It is difficult to establish better, more specific numerical guidelines for this criterion. For example, if a hospital pays $0.90/therm for gas, one would not assume the thermal energy cost is particularly cheap. But if that hospital must operate boilers during the summer for sterilization, then the excess boiler capacity is essentially free when used for desiccant reactivation. In another example, an existing building may have a very low power cost. But if adding more outside air to meet IAQ needs requires more fans and refrigerant compressors, the increased demand may lift the total power cost into another rate category. Also, adding power is not always easy or cheap, and may require changes in the basic service and electrical distribution system. Such changes may turn a low-cost location into a high-cost location because of expenses other than power usage charges.

For a necessarily simplified rule of thumb, inexpensive thermal energy might be defined as a summertime cost of less than $0.60 per therm (100,000 Btu). An electrical demand charge over $9.00/kW would usually be considered a "high" demand charge.

18.9.6.6 Economic Benefit to Dry Duct Work

Bacteria are present throughout all indoor environments, and are especially prevalent in air distribution duct work, where high humidity and accumulated dust are common. But in many cases, the rate of such microbial growth is very low.

In other cases where growth is more rapid and extensive, it may not matter. For example, in many air-conditioned structures in humid climates, a distinctive musty odor is common throughout the building. But until occupants complain, there is no economic incentive for owners to change the system.

In other buildings, the economic consequences of microbial growth are too important to ignore. Hospital operating rooms, for example, place open wounds directly under air discharge ducts. In such cases, complying with the recommendations of ASHRAE Standard 62 to maintain relative humidity in duct work below 70 percent is much more important than it might be for a local pizza parlor or a post office. Consequently, for buildings like hospitals, nursing homes, medical office buildings, and schools, the economic consequences of fungal infection of duct work are focusing more attention on using desiccant system, to keep duct work dry to minimize microbial growth in the HVAC system. A rule of thumb is difficult to formulate in this area, because there are so many unknowns and a wide variety of opinions among experts. But one might suppose that if an owner or engineer believes that amount at risk in future legal costs is higher than the incremental cost of a desiccant-assisted system, it might be prudent to evaluate the hybrid desiccant option in more detail.

Evaluation Example. Figures 18.9.7 to 18.9.9 provide examples of how these criteria can help decide whether desiccants deserve more study. Unfortunately, no such simple analysis

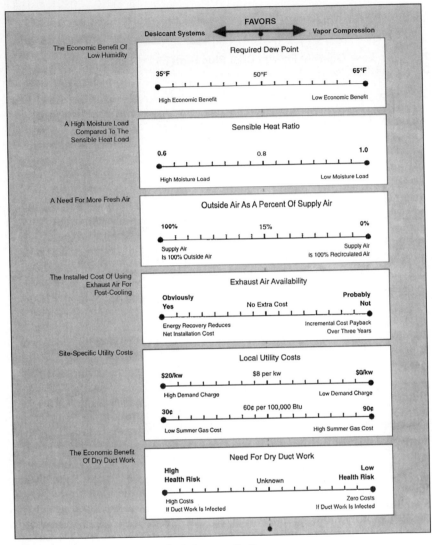

FIGURE 18.9.7   Graphic evaluator for desiccant applications.

can provide instant, comprehensive answers. But by looking briefly at these six key variables, an engineer can make a better prediction of what answers are likely to result from a more rigorous analysis. Some examples are discussed below.

Figure 18.9.8 applies to an operating room. There is a need for a low dewpoint, and failing to provide it may mean the hospital loses revenue as surgeons take patients elsewhere for more complex procedures. Because the operating room often has an all-outside air system, the latent loads are quite high. Also, the need for fresh air is extreme. Exhaust air is generally not available for desiccant postcooling in existing buildings, but in new

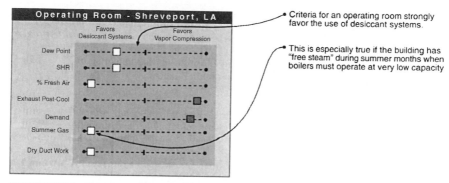

FIGURE 18.9.8   Graphic evaluator for hospital operating room.

construction, it would usually be accessible at low incremental cost. The availability of cheap steam heat during summer months improves economics, and the need to keep duct work dry is typical of a hospital environment. So the desiccant alternative is strongly favored.

Speculative Office Building—Houston, Texas.   Figure 18.9.9 illustrates the results of a graphic assessment for a low-budget, spec office building to be built in Houston, Texas. In this case, desiccants do not look promising.

There is no apparent economic benefit to the developer from maintaining a low humidity level. The building contains office equipment and is only two stories high, so the internal heat generation and sensible heat loads through the building envelope are quite high, and the moisture loads are rather low by comparison. There is no need for a great deal of fresh air, and there is very little exhaust air available for desiccant post-cooling. Utility costs do not especially favor desiccants, and there is normally no need for dry duct work.

Consequently, a desiccant system is not likely to be advantageous in this case, unless the building was experiencing problems with IAQ and needed to keep duct work dry, or to add fresh air.

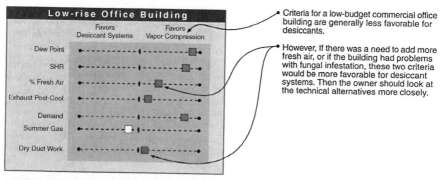

FIGURE 18.9.9   Graphic evaluator for office building.

Supermarkets in Detroit and Miami.   Figure 18.9.10 compares supermarkets in two locations. In both cases, desiccants make sense economically, but the benefits are greater in Detroit, because of local utility cost differences.

Limitations of the Graphic Evaluation.   This graphic technique has limited utility, because different owners weight the importance of each variable differently. For example, if a hospital has had difficulty with contaminated duct work, it may not matter what other factors favor a conventional system; the desiccant system will be chosen regardless of other considerations.

Likewise, in the supermarket industry, some owners have decided that regardless of site utility cost, they will use desiccant systems because of harder-to-quantify benefits such as customer time-in-the-store, since modern desiccant systems for supermarkets are essentially the same cost as conventional equipment. In the opinion of some owners in that industry, the cost of analysis exceeds any cost saving that might result from a site-by-site decision process. And in the case of an ice arena, the need for a low dewpoint makes all economics very clear. Regardless of what other considerations might exist at a given site, if the rink must operate during summer months, then the desiccant alternative is an obvious choice.

Given those all-important weighting considerations, design engineers and building owners will probably find that such graphic evaluation is most useful as a "first cut" rather than as the final decision maker.

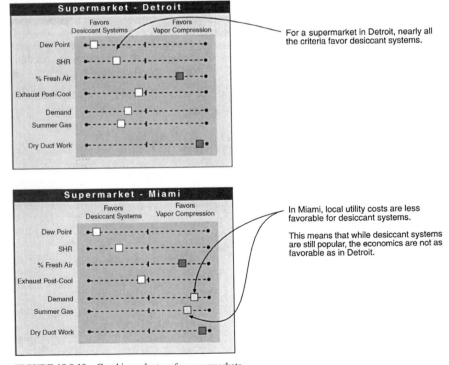

FIGURE 18.9.10   Graphic evaluators for supermarkets.

18.9.7 CONTROLS FOR DESICCANT SYSTEMS

As in conventional systems, controls for desiccant systems are a mixture of components provided by the manufacturer and the control subcontractor. We will divide the discussion into two sections—controls needed for internal system operation, and external controls needed to instruct the equipment to provide a particular supply air temperature and humidity.[18]

18.9.7.1 Internal Controls

Inside the system, the equipment manufacturer generally provides controls to ensure, as a minimum, that:

- The reactivation burner does not ignite if there is no reactivation air flow.
- The reactivation fan continues to operate for a short period after the reactivation burner has turned off to allow the burner to cool down.
- The reactivation burner is shut down if the air temperature entering or leaving the desiccant wheel is excessive, and a fault indicator tripped accordingly.
- An alarm or fault indicator is tripped if the desiccant wheel is not rotating.

Also, manufacturers often provide controls inside the system to modulate reactivation energy in response to changes in moisture load, as previously described in the reactivation heater section.

Manufacturers generally believe it would be unwise and uneconomical for the design engineer to perform these functions in an external control system or in a central building automation system.

18.9.7.2 Dehumidification Capacity Controls

Controlling dehumidification capacity is often a shared responsibility between equipment manufacturer and the control subcontractor. There are three common methods of controlling dehumidification capacity:

- Variable-air bypass (close tolerances)
- Reactivation energy modulation (moderate tolerances)
- On-off reactivation control (loss tolerances)

Each of these methods is effective, depending on the degree of precision needed for the humidity control level in the building.

Variable Process Air Bypass.   The most precise method of controlling humidity requires a bypass air duct and variable-position dampers for the face of the desiccant wheel and for the bypass duct. In this arrangement, when the humidity in the building rises above the setpoint, the bypass damper closes and the face damper opens, allowing more air to flow through the desiccant wheel. Then when the humidity control level is satisfied, the bypass damper opens to allow more air to flow past the desiccant wheel so the building is not over-dried (Fig. 18.9.11).

This method is preferred for industrial process applications, where control within ±1 or 2 percent rh (relative humidity) is essential. Bypass control is used less frequently in commercial buildings, partly because it costs more and partly because the dampers and linkages

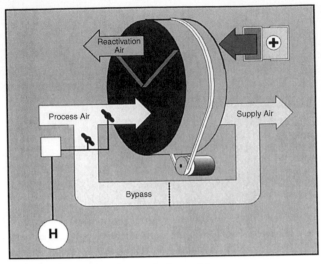

FIGURE 18.9.11   Capacity control using an air bypass.

require some degree of maintenance attention. Finally, many commercial buildings do not need precision control. When control within ±5 or 7 percent rh is sufficient to meet the needs of the building and its internal operations, other methods of control are quite adequate.

Reactivation Energy Modulation.   In this scheme, a humidistat varies the amount of heat produced by the reactivation heaters. When the building rises above setpoint, the controller adds heat to the reactivation air stream, which provides more drying power for the desiccant. When the humidity in the building falls below setpoint, the reactivation heat is reduced, so the desiccant does not remove as much moisture.

This scheme has the advantage of low cost and mechanical simplicity. On the other hand, the control system must have lag-time built in so that the heater does not over-respond to a rise in humidity. For example, if the humidistat in the building calls for more dehumidification, but does not immediately see a fall in humidity, it may continue to call for more dehumidification long after the heater input has been increased. Then the dehumidifier may "overshoot" the desired control level because it is still being fed with extra reactivation heat (Fig. 18.9.12).

These problems can be avoided by locating the humidistat near the supply air outlet of the system, so that changes do not take long to be sensed by the humidistat. Also, a step-function can be built into the control, so that the humidistat calls for a small increment of extra heat, then waits for one wheel rotation before calling for additional heat.

Winter operations add one additional consideration to reactivation modulation as a means of control. When supply air flows through the wheel without reactivation, moisture will build up in the desiccant wheel. Excess moisture supports fungal growth within the core material. In springtime, as the humidistat calls for dehumidification, the warm, humid wheel can give off odors reminiscent of dirty laundry until the heat kills microbial growth accumulated through the winter. But this problem can be easily avoided by periodically rotating and reactivating the desiccant wheel through any season when humidity is too low for the humidistat to call for dehumidification. If this wheel rotation and reactivation is performed

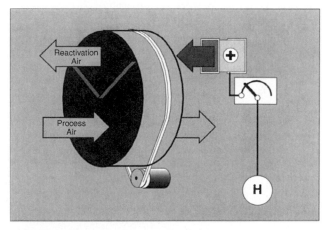

FIGURE 18.9.12   Capacity control using reactivation heat modulation.

once every 8 h for 10 min, the problem will not occur in the spring. Manufacturers may or may not include this feature in their internal controls. So the owner and design engineer should be aware of the issue when dehumidifier capacity control is accomplished by a central energy management or by a building automation system. Finally, this method of control cannot be used for wheels that use lithium chloride as a desiccant, as that highly deliquescent salt would liquefy and migrate out of the wheel structure over time. Bypass control is the only appropriate method of modulating control for lithium chloride-based wheels.

On-Off Reactivation Control.   This method is a variation on the modulating reactivation heat scheme. Instead of modulating the reactivation heat according to the moisture load, the heaters are simply turned on or off according to control signals from a humidistat mounted in the building (Fig. 18.9.13).

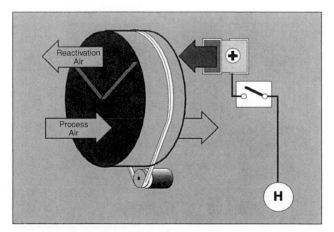

FIGURE 18.9.13   Capacity control using on-off reactivation heat.

The on-off method is even lower in cost than the modulated reactivation method, and gives good results in many situations. For example, when the desiccant system is mounted on the incoming ventilation air, the large mass of return air buffers the effect of changes in outside humidity. On-off control is also used in storage application, where it is seldom necessary to control to tighter tolerances than ±5 percent rh.

The results can be perfectly adequate, for much less cost and complexity than either modulating controls or face and bypass dampers. However, the same cautions as used in the reactivation modulation method apply: if air flows through the unit continuously, the wheel should be rotated and reactivated several times a day, even in periods of low humidity, to avoid microbial growth during inactive seasons.

Additionally, when lithium chloride wheels are used, on-off control is appropriate, provided that air does not flow through the unit when the reactivation energy is off. Humid air flowing through a nonreactivated lithium chloride wheel can saturate the desiccant over time, which would reduce capacity. In those units, the process fan is interlocked with the reactivation heater, so that air only flows through the unit when the reactivation circuit is functioning.

18.9.7.3   Temperature Controls

Many desiccant systems cool the air with conventional vapor compression cooling systems, or with gas cooling systems. Such systems are well understood, as are the methods through which these are controlled. This discussion will review methods of controlling direct evaporative and indirect evaporative coolers, which may be less familiar to owners and design engineers.

As described earlier, air leaves a desiccant wheel warm and dry. In most cases, the air is too warm to send directly to the building, so some sensible heat must be removed. Then, if sensible heat is to be removed from the building as well, the dry air must be cooled below the control condition in the occupied space. There are three stages of cooling available with evaporative cooling and heat exchangers.

18.9.7.4   Stage One—Heat Exchanger Alone

Part of this cooling can be accomplished with the heat exchanger which, in many cases, follows the desiccant wheel. In most desiccant systems, this heat exchanger is not controlled. This is to say, it cools the air as much as possible, regardless of the thermostat setting in the building, because there is always a need to remove heat. However, if the desiccant system is mounted on the ventilation air, there may be a need for winter heat, and with that need, there will be a need to control the heating effect produced by the desiccant wheel. In those cases, the warm air from the desiccant wheel can bypass the postcooling heat exchanger when the space needs heat. Or, if the heat exchanger is a heat wheel, the wheel speed can be modulated in proportion to the heating requirement. If the wheel rotates slowly, most of the heat remains in the supply air. When the wheel rotates more quickly, the supply air is cooled.

18.9.7.5   Stage Two—Indirect Evaporative Cooling

When the heat exchanger by itself does not cool the supply air sufficiently, the next stage of cooling is an evaporative cooling pad located on the cold side (i.e, the reactivation side) of the heat exchanger. As the temperature in the building rises above set point, the control system turns on the water feed to the evaporative cooling pad. The cooling air is reduced in temperature, so more heat can be removed from the supply air.

18.9.7.6 Stage Three—Direct Evaporative Cooling

When the building does not require humidity control, or when the building must be kept cool and humid, as in the case of greenhouses, livestock barns, or vegetable storage, additional sensible heat can be removed by direct evaporative cooling after the heat exchanger. The warm air is passed through an evaporative cooling pad, where it picks up moisture and, therefore, becomes cooler.

This can bring the temperature of the supply air low enough to remove sensible heat from the building, rather than simply bring the air to the same temperature as the building. Direct evaporative cooling has the advantages of low first cost and low operating cost. In return for those advantages, the water supply system and the evaporative pad and drain pan must be maintained regularly, and the supply air is essentially saturated. Consequently the system is no longer a dehumidification system, although such systems can be very useful for maintaining low temperatures and high humidities as described above. When the supply air must remain dry, the third stage of cooling can be accomplished by either conventional vapor-compression or gas cooling systems.

18.9.7.7 Final Cooling—Vapor Compression or Gas Cooling

Designing conventional cooling systems to follow desiccant wheels is quite straight-forward, and in many cases, a desiccant system will include such systems on-board, so the package is a complete comfort-conditioning system. The third stage of cooling will operate for relatively few hours during the year and will be operating at only partial capacity for much of the rest of the year. So, if the postcooler is not equipped with capacity modulation, the building could suffer from poor temperature control and spikes in electrical power demand, as the postcooler turned on and off rapidly when outdoor temperatures and moisture levels are moderate.

When postcooling is provided by the desiccant system manufacturer, one can usually assume that the postcooling will include capacity control. But the point is worth checking, and definitely requires attention when the controls contractor supplies temperature and humidity controls, rather than the manufacturer of the desiccant system.

18.9.8 CONTROLLING LIQUID DESICCANT SYSTEMS

Liquid desiccant dehumidifiers maintain space humidity by controlling solution concentration and temperature to determine moisture removal. Upon sufficient moisture collection, a liquid level controller, usually in the conditioner sump, activates the regenerator. As the dehumidification load lessens, the conditioner solution level lowers. In response, the regenerator temperature controller modulates heat input to return an intended operating solution concentration to the conditioner. Modulating conditioner coil coolant flow, usually based on exiting air dry bulb temperature, regulates the solution temperature.

18.9.9 COMMERCIAL DESICCANT SYSTEMS

Figure 18.9.14 shows a typical example of a desiccant system designed for a commercial building. It includes a desiccant wheel for humidity control and a conventional vapor compression cooling system for temperature control.[19] Such designs combine the best of both

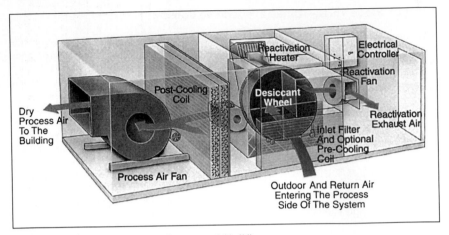

FIGURE 18.9.14   Desiccant system for commercial buildings.

technologies and point to one of the principal advantages of desiccant-assisted HVAC systems, namely, that they can control humidity independently of temperature. The desiccant subsystem is controlled by a humidistat and the cooling coil is controlled by a thermostat. This allows humidity control regardless of what the space may or may not need for heat removal.

But fundamentally, there are two different types of desiccant systems: those which use only desiccants for all cooling and humidity control, and those like the system in Fig. 18.9.7, which combine desiccants with conventional components.

18.9.9.1   All-Desiccant Systems

Figure 18.9.15 shows how a desiccant wheel can be combined with a rotary heat exchanger to form a complete air-conditioning system. Air is dried by the desiccant wheel and then cooled by the heat exchanger.

This configuration has useful advantages when large amounts of fresh air are needed, and when the exhaust air can be evaporatively cooled and used for postcooling the air leaving the desiccant wheel. Under those circumstances, an all-desiccant system is the same physical size as conventional alternatives, because the ventilation air required for the building defines the overall system's air flow.

The system also uses very little electrical power, so it has advantages when electrical demand charges are high. When these two circumstances combine, such as when large amounts of ventilation air must be added to an existing building in an area with high peak demand charges, the all-desiccant system will reduce both energy and first cost compared to other ways of adding the increased fresh air.

The disadvantage of the all-desiccant system is that, at peak design temperatures, it delivers supply air at temperatures above 70°F (21°C). The only exceptions are in far-north and high-altitude climates, where the ambient moisture is so low that evaporative cooling can provide lower air temperatures.

In most climates, if the building does not need a large percentage of ventilation air, and when the exhaust air cannot be collected and brought back to the unit for postcooling, the all-desiccant system has disadvantages compared to a hybrid system. Since it cannot cool

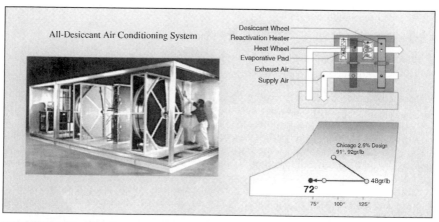

FIGURE 18.9.15 All-desiccant system.

air below 70°F (21°C) on a "design day," the all desiccant system must use large amounts of air to remove a given heat load. Such systems are physically much larger than an equivalent conventional cooling system. The conventional system would supply 55°F (13°C) air, and therefore remove the same internal sensible heat load using less air. For example, consider a small office building maintained at 75°F (24°C) with an internal sensible heat load of 180,000 Btu/h (15 tons). If the supply air can be cooled to 55°F, the system will have to supply 8333 cfm:

$$\text{Btu/h} = \text{cfm} \times 1.08 \times \Delta t$$
$$\text{cfm} = (180,000) \div [1.08 \times (75 - 55)]$$
$$\text{cfm required @ } 55°F = 8333$$

But if the supply air can only be cooled to 70°C, (21°C) the temperature difference between supply and return is only 5°F (28°C), so the air flow needed to remove the load is much greater:

$$\text{Btu/h} = \text{cfm} \times 1.08 \times \Delta t$$
$$\text{cfm required @ } 70°F = (180,000) \div [1.08 \times (75 - 70)]$$
$$\text{cfm} = 33,333$$

However, if that office building needs a great deal of outside air, the all-desiccant system could handle the ventilation load, and a separate system arranged to handle the internal heat load. In that circumstance, the all-desiccant system has advantages over a conventional system.

The desiccant system's 70°F (21°C) delivered air removes some internal load, since the space is being maintained at 75°F (24°C). The heat exchanger in the desiccant system can operate during cooler months, to recover waste heat from the building exhaust. Since the system size is governed by the required outside air quantity and not by the internal load, the all-desiccant system is the same size as a conventional alternative. Installed cost is close to the same, and the desiccant system costs much less to operate because it uses so little electrical power.

In summary, an all-desiccant system is generally attractive when:

- Large amounts of air must be exhausted from the building.
- The exhaust air can be brought back to where the make-up air enters the building.
- Electrical demand charges are high.
- Supplying outside air at 70°F (21°C) is adequate for the application.

In other circumstances, the engineer may wish to consider a hybrid desiccant system.

Hybrid Desiccant Systems.   Figure 18.9.16 shows the wide variety of components that can form hybrid desiccant systems, i.e., systems that include a desiccant component along with gas cooling or conventional coils.

Figure 18.9.17 shows the psychometric behavior of different system alternatives. Note especially the dry bulb temperature leaving the system. To a great extent, the leaving air temperature determines which applications are economically practical for each system alternative.

Some applications, such as hotel corridors and ice rinks, are not sensitive to a leaving air temperature of 78 to 85°F (26 to 30°C) on a design day during the summer. So an indirect evaporative postcooler is the best cooling option because it is quite economical to install and operate. Other applications such as hospital operating rooms, must have air at relative humidities below 50 percent rh and temperatures below 65°F (21°C). In those cases, gas cooling or conventional cooling coils will be required downstream of the desiccant wheel.

To understand each equipment alternative, we will track the process air as it moves through the system. The diagrams in Fig. 18.9.17 assume the system is arranged to handle 100 percent outdoor air on the process side and 100 percent outdoor air for reactivation.

1. Air enters the process side of the desiccant wheel from outside the building. It is hot and humid.

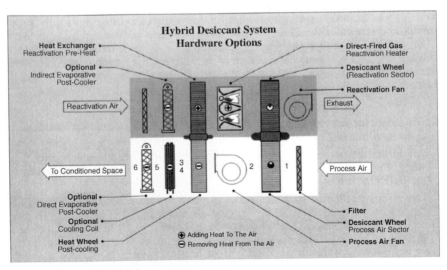

FIGURE 18.9.16   Hybrid desiccant system.

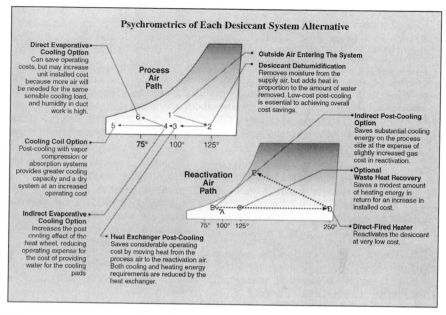

FIGURE 18.9.17 Psychrometrics of hybrid desiccant systems.

2. Air leaves the process side of the desiccant wheel hotter and much drier than when it entered the system. In most cases, this air is too hot to send directly to the building. It must be cooled.

3. Dry air leaves the first stage of postcooling at a lower temperature. The sensible heat has been removed from the process air and transferred to the reactivation air by a heat exchanger. The schematic here shows a rotary heat wheel, but heat pipes and plate-type heat exchangers are used by many system suppliers instead of heat wheels. Regardless of the type of heat exchanger, it provides a double benefit: the process air is cooled using only the energy needed to push the air through the exchanger. So the operating cost of the cooling is very low. Second, the heat from process is used to preheat incoming reactivation air, which saves slightly on the cost of thermal energy.

4. Point 4 represents the additional cooling which can be accomplished by the heat exchanger if the air on the other side of the heat exchanger is evaporatively cooled. In this option, the incoming reactivation air is cooled by an evaporative pad before it enters the heat exchanger. Since the air on the reactivation side of the exchanger is cooler, more heat can be removed from the process side. This diagram shows roughly what happens on a design day, so the evaporative cooling effect is not very large. But when outside air temperature and moisture is lower—99 percent of the time during the year—the cooling effect of the evaporative pad will be substantial. This reduces the need for any subsequent postcooling.

5. Point 5 shows the temperature and moisture leaving the system when a gas cooling or conventional cooling system follows the heat exchanger. Air is sent to the building at a very cool and very dry condition. This configuration is popular because it keeps air distribution ducts and filters dry and free of microbial growth. Low temperature, dry supply air allows the system to do a great deal of cooling and dehumidification with less air than conventional cooling systems.

6. An alternative to conventional cooling coils is a second evaporative cooling pad, this time on the process side. Direct evaporative cooling seldom chills air as deeply as a conventional coil. Also, the supply air is saturated at a comparatively high temperature [73 to 78°F (23 to 26°C) on a design day]. So such systems cannot be used to control humidity unless a relatively warm, highly humid environment is needed, as in a greenhouse.

The evaporative cooling option (point 6) is less expensive to install and uses very little energy compared to conventional postcooling alternatives. So this option has advantages when electrical power cost reduction is the principal goal of a project, rather than humidity control.

To date, hybrid systems have been popular, perhaps because they combine the best characteristics of each technology: desiccants for moisture removal and conventional cooling for sensible heat removal. Hybrid systems are nearly always smaller than all-desiccant systems, because they can provide air at low levels of both temperature and humidity. So smaller hybrid systems can do the same work as larger all-desiccant or all-cooling units.

18.9.10   SUMMARY

Desiccant systems have emerged from a distinguished history in industrial applications to take an increasingly large role in commercial air-conditioning applications. The engineer would be well-advised to consider desiccant technology whenever an application would benefit from a lower-than-usual humidity level, when the application demands a large proportion of ventilation air, or when the cost of electrical power demand charges is higher than usual. Under any of those circumstances, desiccant components can be a useful addition to an air-conditioning system.

REFERENCES

1. *Ventilation for Acceptable Indoor Air Quality*, ANSI/ASHRAE Standard 62-1981, ASHRAE, Atlanta, GA, 1981.

2. *Ventilation for Acceptable Indoor Air Quality*, ANSI/ASHRAE Standard 62-1989 ASHRAE, Atlanta, GA, 1989.

3. *1993 ASHRAE Handbook of Fundamentals*, ASHRAE, Atlanta, GA, 1993, Chapters 19 and 24.

5. Collier, R. K., et al., *Advanced Desiccant Materials Assessment*, Gas Research Institute (National Technical Information Service Document #PB 87-172805), Chicago, IL, 1986.

6. *The Dehumidification Handbook*, Munters Corporation—Cargocaire Division, Amesbury, MA, 1990.

7. Assorted product literature, available from Munters Corporation—Cargocaire Division, 79 Monroe Street, P.O. Box 640, Amesbury, MA, 01913, Telephone: 508-388-0600.

8. Assorted product literature, available from Kathabar Systems Division, P.O. Box 791, New Brunswick, NJ, 08903, Telephone: 909-356-6000.

9. Assorted product literature, available from Air Technology Systems, 1572 Tilco Drive, Frederick, MD, 21701, Telephone: 301-620-2033.

10. Assorted product literature, available from Bry-Air, Inc., P.O. Box 269, Sunbury, OH, 43074, Telephone: 614-965-2974.

11. Assorted product literature available from Airflow Company—Dryomatic Division, 295 Bailes Lane, Frederick, MD, 21701, Telephone: 301-695-6500.

12. Assorted product literature, available from Niagara Blower Company, 673 Ontario Street, Buffalo, NY, 14207, Telephone: 716-875-2000.

13. *2001 ASHRAE Handbook of HVAC Applications*, ASHRAE, Atlanta, GA, 2001, Chapter 11.

14. Assorted product literature, available from Munters Corporation—DryCool Division, 16900 Jordan, Selma, TX, 78154, Telephone: 210-651-5018.

15. Assorted product literature, available from Engelhard/ICC, 441 North Fifth Street, Philadelphia, PA, 19123, Telephone: 215-625-0700.

16. Assorted product literature, available from Seasons 4, 4500 Industrial Access, Douglasville, GA, 30134, Telephone: 404-489-0716.

17. *Application Engineering Manual for Desiccant Systems*, 1996, Chapter 8, The American Gas Cooling Center, 1515 Wilson Blvd, Arlington, VA 22209, Telephone: 703-841-8542.

18. *Application Engineering Manual for Desiccant Systems*, 1996, Chapter 5, The American Gas Cooling Center, 1515 Wilson Blvd, Arlington, VA 22209, Telephone: 703-841-8542.

19. *Application Engineering Manual for Desiccant Systems*, 1996, Chapter 3, The American Gas Cooling Center, 1515 Wilson Blvd, Arlington, VA 22209, Telephone: 703-841-8542.

BIBLIOGRAPHY

American Hotel & Motel Association, *Mold & Mildew in Hotels and Motels*, 1991, AHMA 1201 New York Avenue, Washington, DC 20005-3931, Telephone: 202-289-3196.

Bayer, Charlene W., Ph.D., Crow, Sidney A. Ph. D., and Nobel, Judy A., Ph.D., *Production of Volatile Organic Emissions by Fungi*, pp. 101–109, Proceedings of IAQ '95 Conference, October, 1995, Denver, CO, American Society of Heating, Air Conditioning and Refrigerating Engineers 1791 Tullie Circle, Atlanta, GA 30329.

Brundrett, Geoffrey W., *Criteria for Moisture Control*, 1990, Butterworths, London, England ISBN 0-02374-0.

Colliver, Donald G., Zheng, Hanzhong, Gates, Richard S., and Priddy, K. Tom, *Determination of the 1%, 2.5% and 5% occurrences of extreme dewpoint temperatures with mean coincident dry bulb temperatures*, 1995. Proceeding of the American Society of Heating, Refrigerating and Air Conditioning Engineers, Vol. 101, pt. 2, paper no. 3904 (RP-754) ASHRAE, 1791 Tullie Circle, NE, Atlanta, GA, Telephone: 404-636-8400.

Cook, Edward M., and DuMont, Harmon D., *Process Drying Practice*, 1991, McGraw-Hill, New York, NY, ISBN 0-07-012462-0.

Flannigan, Brian, and Miller, J. David, *Humidity and fungal contaminants*, pp. 43–50, Proceedings of the Bugs, Mold 7 Rot II Workshop, November 1993, National Institute of Building Sciences, 1201 L Street, Washington, DC 20005, Telephone 202-289-7800.

Gutoff, Edgar B., and Cohen, Edward D., *Coating and Drying Defects*, 1995, John Wiley & Sons, 605 Third Avenue, New York, NY 10158, ISBN 0-471-59810-0.

Harriman, Lewis G. III (Ed.), *The Dehumidification Handbook*, 2nd ed., 1990, Munters Cargocaire 79 Monroe St. Amesbury, MA 01913, Telephone (508) 388-0600.

Harriman, Lewis G. III (Ed.), *Desiccant Cooling and Dehumidification*, 1992. American Society of Heating, Refrigerating and Air Conditioning Engineers, 1791 Tullie Circle, NE, Atlanta, GA, Telephone: 404-636-8400. ISBN 0-910110-90-5.

Harriman, Lewis G. II, *Application Engineering Manual for Desiccant Systems*, 1996, The American Gas Cooling Center, 1515 Wilson Blvd. Arlington, VA 22209, Telephone: 703-841-8542.

Iglesias, Héctor A., and Chirife, Jorge, *Handbook of Food Isotherms: Water Sorption Parameters for Food and Food Components*, 1982, Academic Press, 111 Fifth Avenue, New York, NY, ISBN 0-12-370380-8.

Lstiburek, Joseph, and Carmody, John, *Moisture Control Handbook*, 1991, U.S. Department of Energy, Conservation & Renewable Energy, Office of Buildings & Community Systems, Building Systems Division, Oak Ridge National Laboratory, Oak Ridge. TN ORNL/Sub/89-SD350/1.

Martin, Phillip C., and Verschoor, Jack D., *Cyclical Moisture Desorption/Absorption by Building Construction and Furnishing Materials*, 1986, pp. 59–69, Proceedings of Symposium on air filtration, ventilation, and moisture transfer, Ft. Worth, TX, Building Thermal Envelope Coordinating Council, Box 28029, Washington, DC 20038.

Odom, J. David, III. and DuBose, George, P.E., *Preventing Indoor Air Quality Problems in Hot and Humid Climates*, 1996, CH2M/Hill, 25 East Robinson St., Suite 405, Orlando, FL 32801, Telephone: 407-423-0030.

Oscik, Jaroslaw, *Adsorption*, 1982, Ellis Horwood Ltd., Market Cross House, Cooper St., Chichester, West Sussex, P019 1EB, England, ISBN 0-85312-166-4.

Treschel, Heinz R. (Ed.), *Moisture Control in Buildings*, 1994, ASTM, 1916 Race St. Philadelphia, PA 19103, ASTM Manual MNL 18, ISBN 0-8031-2051-6.

Treschel, Heinz R., and Bomberg, Mark (Eds.), *Water Vapor Transmission Through Building Materials and Systems: Mechanisms and Measurements*, 1989, ASTM, 1916 Race St. Philadelphia, PA 19103, ASTM STP 1039, ISBN 0-8031-1254-8.

Valenzuela, Diego P., and Myers, Alan L., *Adsorption Equilibrium Data Handbook*, 1989, Prentice Hall, Englewood Cliffs, NJ, ISBN 0-13-008815-3.

CHAPTER 19
COOLING DISTRIBUTION SYSTEMS

19.1   CHILLED WATER AND BRINE*

Gary M. Bireta, P.E.
Project Engineer, Mechanical Engineering,
Arcadis Giffels, Inc., Southfield, Michigan

Ernest H. Graf, P.E.
Assistant Director, Mechanical Engineering,
Arcadis Giffels, Inc., Southfield, Michigan

19.1.1   INTRODUCTION

Water systems are used in air-conditioning applications for heat removal and dehumidification. The two most common systems use *chilled water* and *brine*. Chilled water is plain water at a temperature from 40 to 55°F (4 to 13°C). Brine is a water/antifreeze solution at a temperature below 40°F (4°C). Here we describe the basic principles and considerations for chilled water. Additional considerations for brine follow.

19.1.2   SYSTEM DESCRIPTION

A chilled-water system works in conjunction with air-handling units or process equipment to remove the heat generated within a conditioned space or process. The terminal unit cooling coil(s) collect(s) the heat and then transfers it by conduction and convection to the water, which is conveyed through connecting piping to the evaporator side of a chiller. The chiller, which is a packaged refrigeration machine, internally transfers the heat from the evaporator to the condenser, where heat is discharged to the atmosphere by the condenser system.

---

*Updated by James Fox, David H. Helwig, William C. Newman, Wayne M. Lawton, James D. Robinson, PE, James Runski, Paul D. Bohn.

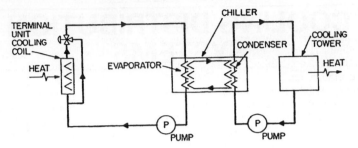

FIGURE 19.1.1   Schematic diagram of chilled-water system.

The chilled water leaving the evaporator is circulated back to the coils, where the heat-removal process repeats again. Figure 19.1.1 is a schematic diagram of the chilled-water system. In this chapter we discuss the basic principles and details regarding the chilled-water or evaporator side of the system.

19.1.3 WHERE USED

Chilled-water systems are applicable when:

- The design considerations of a proposed air-conditioned facility require numerous separated cooling coils plus the restriction that the refrigeration system(s) of the facility be located in a single area.
- There is a need for close control of coil leaving air temperature or humidity control. Control can be very smooth and exact because of the infinite modulating capability of the chilled-water valve.
- Future expansion will require additional cooling capacity. The additional capacity might be merely a matter of new terminal units and branch piping from the chilled-water mains to the coils, although this is limited by the unused capacity of the chiller, pumps, and water distribution system.
- The coil leaving air temperature desired is 45°F (7°C) or higher. The leaving air will be at least 5°F (3°C) warmer than the coil entering water temperature.

19.1.4 SYSTEM ARRANGEMENT

The system designer must consider the cooling loads involved and the type and arrangement of the facility during the conceptual phase of a chilled-water system design. During initial design development, the designer should consider the impact of future system loads. *System expansion costs can be reduced if space for additional equipment and the flow rates are planned for during the initial design.* The module design concept adapts well for planning for future expansion.

Large facilities commonly consist of terminal units located near the area they serve. The total combined loads of the facility result in a large peak demand with a wide operating range that is beyond the capability of a single chiller. A chilled-water arrangement for a large installation would commonly consist of multiple chillers centrally located with multiple cooling towers of the condenser system situated nearby outside. Figure 19.1.2 shows the evaporator side of a multiple-chiller arrangement. Installations of this type are typically arranged in modules with a chiller, cooling tower, and associated pumps dedicated to part of the peak load. A single distribution system transports the chilled water to the various areas and terminal units.

Two-way modulating control valves are used to vary the flow to the terminal units based on a signal from the conditioned space room thermostats. Two-way valves are preferred to three-way valves because the total system pumping cost is reduced during part-load conditions. Chiller manufacturers, however, typically demand a constant flow through the chillers for stable refrigeration control. In this situation a pump arrangement is needed that allows variable flow to the terminals while maintaining a constant flow through the chillers. This problem is solved by installing two sets of pumps in a primary/secondary arrangement. The secondary pumps can be controlled to match the demand of the terminals, while the primary pumps maintain a constant flow through the chillers. The system bypass decouples the two pump sets, which allows them to operate pressure independent of each other. Reference 1 provides further explanation of primary/secondary pumping.

A small installation for an individual building or process may consist of a single chiller, cooling tower, pump, and small distribution system connected to nearby terminal units. The condenser system cooling tower would typically be located on the building roof or nearby, outside the building.

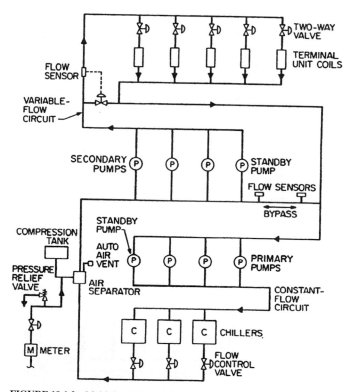

FIGURE 19.1.2 Multiple-chiller arrangement.

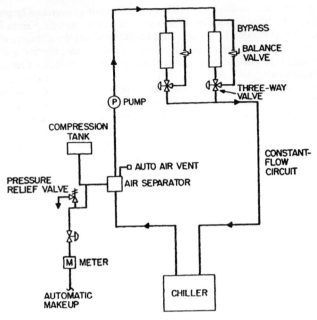

FIGURE 19.1.3   Single-chiller arrangement.

A three-way mixing valve modulates the terminal unit flow based on the cooling load demand while maintaining a constant flow through the chiller and pump. For small installations, the increased pumping cost is offset by the savings realized from fewer pumps and less complicated controls. Figure 19.1.3 shows the evaporator side of a single-chiller arrangement.

19.1.5   DISTRIBUTION SYSTEMS

There are two basic distribution systems for chilled water: the two-pipe reverse-return and the two-pipe direct-return arrangements. Figure 19.1.4 illustrates the direct-return and reverse-return configurations.

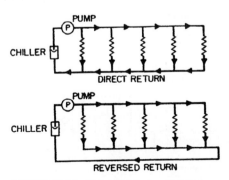

FIGURE 19.1.4   Piping distribution systems.

The *reverse-return* system is preferable from a control and balancing point of view, since it provides very close equivalent lengths to all terminals, resulting in closely balanced flow rates. In large installations, however, the additional piping for a reverse-return system is usually not economical.

The *direct-return* system is more commonly used. The system must be carefully analyzed to avoid flow-balancing problems. Balancing valves and flow meters should be provided at each branch takeoff and terminal unit. Control valves with high head loss are recommended and must be analyzed for varying "shutoff" heads through the system. In large systems, it is sometimes desirable to use a combination of a direct-return system for the mains with a reverse-return system for branch piping to sets of terminal units. This combination provides an economical main distribution segment with easier balancing within the branches. The overall system should be analyzed to determine the most economical distribution system for the application.

19.1.6 DESIGN CONSIDERATIONS

19.1.6.1 Design Temperatures

The chilled-water design temperatures must be established before the terminal unit flow rates can be ascertained. Chilled-water supply temperatures range from 40 to 55°F (4 to 13°C), but temperatures from 42 to 46°F (6 to 8°C) are most common. Temperature differentials between supply and return are in the 7 to 11°F (4 to 6°C) range for small buildings and the 12 to 16°F (7 to 9°C) range for conventional systems. Higher temperature differentials are preferred since they reduce the system flow rates, resulting in smaller piping and pumps, less pumping energy, and increased chiller efficiency. Before the design temperatures are finalized, the designer should ensure that the design temperature selected will result in terminal devices properly sized for their applications.

For large distribution systems, a terminal supply temperature approximately 1°F (0.5°C) higher than the leaving chiller temperature is sometimes assumed. The 1°F (0.5°C) increase accounts for pump and pipe heat gains between the chiller and terminal units. The additional load from these sources must be included during the sizing of the chiller(s).

If the system is subject to freezing, a water/glycol solution (brine) may be required. Refer to the discussion of brine for additional design considerations.

After the terminal units have been located and the flow rates established, a system flow diagram should be created. Several chiller manufacturers should be consulted for sizes, types, and operating ranges available. The designer must analyze the facility for an appropriate distribution system, pump arrangement, and control sequence for all components.

If continuous system operation is required, standby pumps are recommended to ensure system operation in the event of a pump failure. Standby chillers are not usually included because of their high initial cost and rare failure.

19.1.6.2 Piping

Figure 19.1.5 is provided as a general guide for selecting pipe sizes once the system flow diagram has been established. The shaded area provides economical pipe sizes as a function of flow, velocity, and friction loss. In situations where two pipe sizes are capable of handling the design flow, the larger of the two should be selected, in case of an unexpected increase in the flow rate. The system designer should carefully size the inlet and outlet connections to terminal units. If they were left unsized, these branches could be installed, the

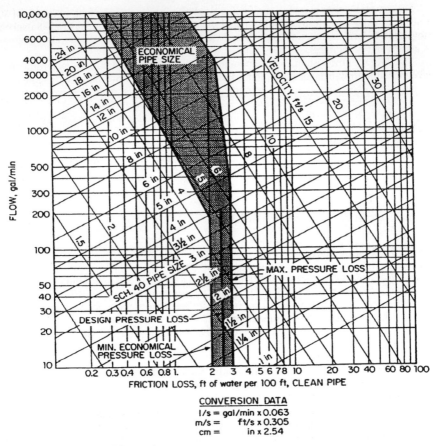

CONVERSION DATA
l/s = gal/min x 0.063
m/s =   ft/s x 0.305
cm =   in x 2.54

FIGURE 19.1.5   Pipe-sizing graph.

same size as the terminal unit connection, which might result in abnormally high pressure losses.

Corrosion inhibitors are commonly added to chilled-water systems to reduce corrosion and scale. Refer to Chap. 8 for recommended water conditioning. With proper water treatment and a closed system, the pipe interior should remain relatively free of scale and corrosion. The calculated pump head can be based on relatively clean pipe, although it is prudent to assume a minimum fouling factor. A 25 percent fouling factor is equivalent to using $C = 130$ in the Williams and Hazen formula for steel pipe.

19.1.7   INSTALLATION CONSIDERATIONS

19.1.7.1   System Volume

Automatic makeup water with a positive-displacement meter is recommended for filling, monitoring volume use, and maintaining system volume. The systems should also be attached to an adequately sized compression or expansion tank and to a pressure relief valve. If the tank is not provided and the system has automatic makeup water, the relief valve will discharge with each rise in average water temperature, wasting water and chemicals.

19.1.7.2 Air Control

Manual or automatic air vents must be installed at system high points to vent air when the system is filled. Automatic vents should be provided with an isolation valve to enable replacement. Vent blowoff lines should be piped down to the closest waste drain. Air, in horizontal mains, can generally be kept out of the branch piping when the branch connections are in the bottom 90° arc of the main. A branch pipe with vertical downflow that connects to the bottom of a main can accumulate pipe scale and similar debris. Dirt legs, such as for steam drips, or strainers may be useful in these branches, especially for 2-in (51-mm) and smaller piping. Air in vertical piping will flow *down* with the water at 2 ft/s (0.6 m/s) or greater water velocity.

19.1.7.3 System Isolation

All equipment requiring maintenance and branch piping should be provided with manual isolation valves. Chain-wheel operators are recommended for frequently used valves located out of the operator's reach. Drain connections should be provided at low points to allow partial system drainage of isolated sections.

19.1.7.4 Coil Control

The coil capacity or degree to which an airstream is cooled or dehumidified as it passes through a chilled-water coil is generally controlled by varying the quantity of water flowing through the coil. (Capacity can also be controlled by varying the water temperature or by varying a portion of air that bypasses the coil and remixing it with the portion passed through the coil.) Ideally the control valve will vary the water flow continuously and uniformly as the valve strokes from full open to full closed; that is, 10 percent of stroke should cause a 10 percent change in flow. However, if a line-size butterfly valve is installed, it might pass 80 percent of full flow when only 20 percent open, which means the valve is trying to control 0 to 100 percent flow while 0 to 25 percent "open." The flow tends to be unstable in these situations, especially at low flow where the valve is apt to wire-draw, chatter, and hunt. Valve size is very important. In the interest of satisfactory control stability, the valves are frequently less than line size because the pressure loss through the fully open valve should be at least 33 percent of the total pressure loss of the circuit being controlled. For example, if the total coil circuit loss is 7 lb/in^2 (48 kPa), a valve sized for a drop of 3 to 5 lb/in^2 (21 to 34 kPa) would be satisfactory. Strainers are recommended upstream of control valves and any other piping elements that require protection against pipe scale and debris within the system.

19.1.8 SYSTEM MONITORING

Pressure gauges permanently installed in a system deteriorate over time from constant vibration. Gauges should be installed only at points requiring periodic monitoring. At points where infrequent indication is required, gauge cocks should be installed with a set of spare gauges provided to the operator. Thermometers are recommended at all terminal units and chillers.

For additional explanation and considerations for the design of a chilled-water system, see Ref. 2.

19.1.9 BRINE

The term "water" is used throughout this chapter for convenience, whereas it could be plain water or a brine. The term "brine" includes a water/glycol solution, a proprietary heat-transfer liquid, water and calcium or sodium chloride solution, or a refrigerant. The best choice of brine will depend on the parameters of the system, but plain water or a water/glycol solution is the overwhelming choice for comfort air-conditioning chilled-water systems. Propylene glycol is the least toxic of the glycols and should be used if there is any possibility (e.g., piping leaks) of contact with a food or beverages. Glycol typically requires system components to be provided with secondary containment.

19.1.9.1 Where Used

An antifreeze must be added to the water, or a nonfreezing liquid must be used, whenever any portion of the water is subject to less than 33°F (0.6°C). Be sure to check the temperature of the refrigerant, or cooling medium, in the chiller. If it operates below 33°F (0.6°C), freeze protection is needed. In systems where the refrigerant is only a few degrees below 33°F (0.6°C), chiller freezeup can be precluded by maintaining chilled-water flow through the chiller for some period after the chiller has been shut off. This water flow will "boil" the refrigerant remaining in the evaporator.

19.1.9.2 Design Considerations

Adding an antifreeze to water will generally reduce the specific heat and the conductivity, and increase the viscosity of the solution. These, in turn, generally necessitate increased heat-transfer surface in the chiller and cooling coils, increased chilled-water flow, and increased pump head. See Fig. 19.1.6. For example, suppose a plain water system involves 8 in (20.3 cm) pipe, 1000 gal/min (63.1 L/s), 50 lb/in² (345-kPa) pressure drop for pipe friction, and 29 hp (21.6 kW) to overcome pipe friction. Then

- If 10 percent glycol is added, the parameters become 1010 gal/min (63.7 L/s), 53 lb/in² (365 kPa), and 31 hp (23.1 kW).
- If 40 percent glycol is added, the parameters become 1150 gal/min (72.6 L/s), 75 lb/in² (516 kPa), and 53 hp (39.5 kW).

The piping size and solution temperature rise are assumed the same in the plain water and glycol systems.

The curves drawn in Fig. 19.1.7 show how the pressure loss caused by pipe friction is affected by the solution temperature and ethylene glycol concentration for various pipe inside diameters. Note that the curves are specific for a solution velocity of 6 fps (1.83 m/c) because this was used in the formula to determine Reynold's Number which in turn was used to establish the curves.

The freezing point of aqueous glycol solution can be found from charts similar to Fig. 19.1.8. Note that 40 percent glycol is needed to lower the freezing point to −10°F

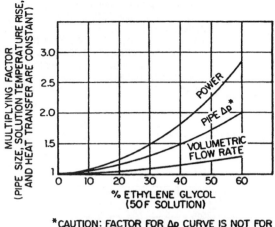

*CAUTION: FACTOR FOR Δp CURVE IS <u>NOT</u> FOR
EQUAL VOL. FLOW RATES OF PLAIN WATER &
E.G. SOLUTION

FIGURE 19.1.6   Flow, pressure drop, and power consumption factors for ethylene glycol solutions versus plain water. (See para. 9, *design considerations.*)

(−23°C). By definition, the freezing point is that at which the first ice crystal forms. Chilled-water piping has been protected from freeze damage with as little as 10 percent or even 5 percent glycol in 0°F (−18°C) weather. This is referred to as "burst protection." Ice crystals may form at 25 to 29°F (−4 to −2°C) in the 10 percent or 5 percent water/glycol solution, but the solution merely forms a slush and does not freeze solid. The slush must be permitted to expand. If it is trapped between shutoff valves, check valves, automatic valves, etc., pipe rupture may occur. The slush must be permitted to melt before the chilled-water pumps are started.

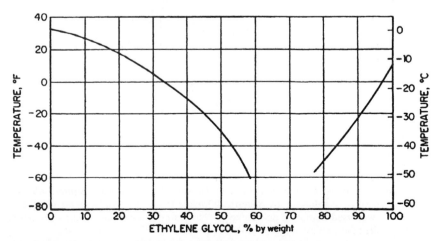

FIGURE 19.1.7   Freezing points of aqueous ethylene glycol solutions. (*From Union Carbide Corp., Ethylene Glycol Product Information Bulletin, Document F 49193 10/83-2M.*)

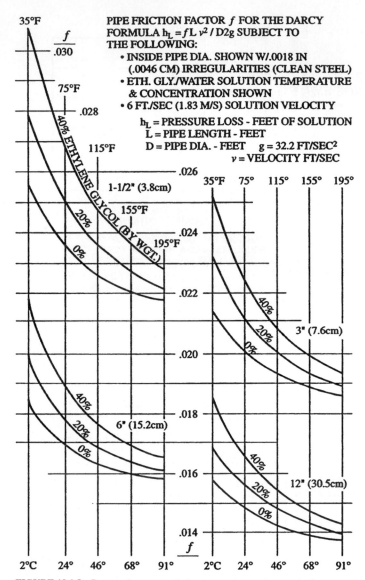

FIGURE 19.1.8   Pressure losses vs. solution temperature and concentration.

Automotive antifreeze solutions should not be used. The corrosion inhibitors added to automotive antifreeze solutions are specifically made for the materials encountered in an automobile engine. Automotive antifreeze is not meant for long life, whereas industrial heat-transfer fluids may last 15 years with proper care.

A chemical analysis of the makeup water must be checked for compatibility with the proposed chiller, pump, piping, and coil materials, chemical treatment, antifreeze, corrosion inhibitors, etc., to preclude the formation of scale, sludge, and corrosion. The water should be checked regularly for depletion of any components.

19.1.10 STRATIFIED CHILLED-WATER STORAGE SYSTEM

The chilled-water storage system is a conventional chilled-water system with the addition of a thermally insulated storage tank. Figure 19.1.9 shows a typical chilled-water storage arrangement. During the daily cooling cycle, the chillers operate to maintain cooling until the load exceeds the capacity of the system. At that point, the chillers and tank work in conjunction to handle the peak demand. As the load falls below the chiller's capacity, the chillers continue to operate to recharge the tank for the next day's demand.

The advantage of this arrangement is that a portion of the equipment required for a conventional system to handle peak loss can be replaced by a less expensive storage tank. In addition, the owner's electric power rates are reduced since the tank has shaved the monthly peak power demand.

The system is classified as a stratified storage system because warm water and cold water within the storage tank remain separated by stratification. During operation, as a portion of chilled water is removed from the tank bottom for cooling, the identical portion of warm return water is discharged back into the tank at the top. A thermal boundary forms with the warmer, less dense water stratifying at the top and the denser, colder water remaining below. During periods of reduced load, the tank is recharged by removing the warm stratified water from the tank top, chilling it, returning it to the tank bottom. During a daily cycle, the thermal boundary moves up and down within the tank, but the total water quantity remains unchanged.

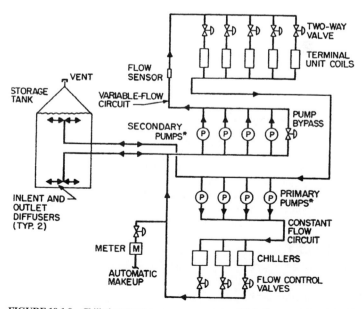

FIGURE 19.1.9   Chilled-water storage arrangement. Asterisk indicates standby pumps.

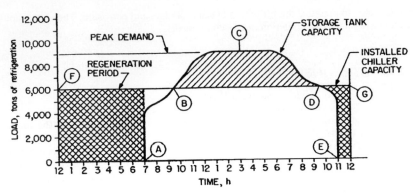

FIGURE 19.1.10   Cooling-load profile. See text explanation of letter symbols.

19.1.10.1   Load Profile

Figure 19.1.10 shows a typical building cooling load profile utilized for storage applications. Curve *ABCDE* represents the cooling load profile during the day. Point *C* represents the maximum instantaneous peak line *FG* represents the installed chiller capacity. The area within *ABDE* represents the portion of cooling provided by the chillers, and area *BCD* represents the portion of cooling provided by the tank. The remaining areas, *FBA* and *DGE*, represent the chiller capacity available to recharge the tank. For storage applications, units of ton-hours (kWh) are used to determine the cooling load and storage requirements of the system.

19.1.10.2   Where Used

Chilled-water storage systems generally become economical in systems in excess of 400 tons (1407 kW) of refrigeration. In all cases, the economics of the installation should be the deciding factor for choosing a storage system.

19.1.10.3   Design Considerations

The design engineer must analyze the operating parameters of the facility in order to accurately predict the load-cycle hours for a given day. The number of hours required for cooling and available for tank regeneration must be determined. Once established, the daily cooling load can be calculated. If available, a computer cooling-load program capable of providing an hour-by-hour analysis is recommended for predicting the cooling load profile. Example 7.1.1 demonstrates the method for creating a load profile diagram and determining the refrigeration and tank capacity for a storage installation.

EXAMPLE 19.1.1

Find:

1. Total refrigeration load
2. Tank capacity

Given:

Cooling period, 8 a.m. to 8 p.m. (12 h)
Regeneration period, 8 p.m. to 8 a.m. (12 h)
Note: Hourly loads provided are not from Fig. 19.1.10 load profile

Daily Cooling Load, Hour-by-Hour Analysis

Time		Total load
8 a.m.	30,084 MBtu	$(7.58 \times 10^6$ kcal)
9	36,972	(9.31×10^6)
10	45,144	(1.14×10^7)
11	59,268	(1.49×10^7)
12	72,168	(1.82×10^7)
1 p.m.	80,952	(2.04×10^7)
2	82,212	(2.07×10^7)
3	83,100	(2.09×10^7)
4	85,392	(2.15×10^7)
5	85,320	(2.15×10^7)
6	83,436	(2.10×10^7)
7	81,900	(2.06×10^7)
Total	825,948 MBtu	$(2.08 \times 10^8$ kcal)

Total daily load (ton/h)

$$= 825{,}948 \text{ MBtu}/[12 \text{ MBtu}/(\text{ton} \cdot \text{h})]$$
$$= 68{,}829 \text{ ton/h}$$
$$= (2.08 \times 10^8 \text{ kcal}/3022 \text{ kcal/ton} \cdot \text{h})$$
$$= (68{,}829 \text{ ton/h})$$

Solution

1. Total chiller capacity

$$68{,}829 \text{ ton/h}/24 \text{ h} = 2868 \text{ tons}$$

(Select two 1500-ton chillers)

2. Storage tank capacity

$$68{,}829 \text{ ton/h} - (3000 \text{ ton} - 12 \text{ h})$$
$$= 32{,}829 \text{ ton/h}$$

Based on 20°F temp. diff (11.1 K)

$$\frac{(7.48 \text{ gal/ft}^3)\,(32{,}829 \text{ ton/h})\,[12{,}000 \text{ Btu}/(\text{h}\cdot\text{ton})]}{(62.4 \text{ lb/ft}^3)\,(20°\text{F})\,[1.0 \text{ Btu}/(\text{lb}\cdot°\text{F})]}$$

$$\left(\frac{(1000 \text{ L/m}^3)\,(32{,}829 \text{ ton/h})\,[3022 \text{ kcal/ton}\cdot\text{h})]}{(1000 \text{ kg/m}^3)\,(11.1\text{K})\,[0.998 \text{ kcal}/(\text{kg}\cdot\text{K})]}\right)$$

capacity = 2,361,162 gal (8,955,680 L)
Increasing tank volume 20 percent for mixed zone and internal piping:
Total capacity = 2,833,395 gal (10,746,817 L)

19.1.10.4   Design Temperatures

As previously described for the conventional chilled-water system, the design temperatures of the system must be established. Temperature differentials for storage applications typically range from 16 to 22°F (−10 to −6°C). A higher temperature differential is preferred since it will reduce the size of the tank and system flows. Before the design temperatures are finalized, the terminal unit coils must be checked to ensure that they can be sized for proper operation. High differential temperature can result in low water velocities within the coil tubes, which may lead to poor part-load performance. Tube water velocities between 2 and 5 ft/s (0.6 and 1.5 m/s) are preferred for efficient heat transfer.

19.1.10.5   Tank Sizing

After the design temperatures have been established, the storage tank capacity can be determined. The capacity previously calculated in ton-hours can be converted to gallons, as illustrated in Example 19.1.1. The tank sizing must also allow both for unused space due to piping and related apparatus and for the mixed thermal boundary between warm and cold fluids. ASHRAE standard ISO-2000 "method of testing the performance of cool storage systems" may be used to determine the actual performance of a new or existing storage tank system.

19.1.10.6   Installation Considerations

Tank.   Tanks for storage applications are field-fabricated of steel or concrete. Steel is generally preferred over concrete for stratified storage because steel readily absorbs and rejects the changes in water temperature without disturbing the thermal boundary. For example, when a concrete tank is recharged, the rising chilled water is warmed by the heat stored in the concrete.

The tank should have a roof, to keep out unwanted debris. The exterior of the tank must be insulated; spray foam or rigid board insulation is used for this type of installation. The tank should be equipped with provisions for access, filling and draining, venting, and overflow, with associated controls for temperature and level monitoring.

Diffuser.   The size, number, and orientation of the diffusers within the tank depend on the design parameters of the system. The intent of the diffusers is to allow removal and replacement of the tank water without disrupting the stratified thermal boundary. The tank should be installed in parallel with the chillers. This arrangement will result in two diffusers, one at the top and one at the bottom of the tank.

Radial-type diffusers have been used with success for cylindrically shaped tanks. The diffuser consists of two steel plates with the inlet pipe located at the center. As the water enters the diffuser, the flow is distributed in all directions toward the outside wall. The diffuser should be designed for low outlet velocity. This is determined by designing for a Froude number of 1 or below. Example 19.1.2 demonstrates the method for sizing a radial diffuser.

EXAMPLE 19.1.2

Find: Radial diffuser height between plates
Given:

Maximum flow = 3500 gal/min (220 L/s)
Diffuser diameter = 10 ft (3.048 m)
Water temperatures = 44°F (6.6°C), 65°F (17.7°C)

Solution The Froude number is dimensionless and defined by the following equation:

$$F = \frac{Q}{\sqrt{g(\Delta p/p)h^3}}$$

where F = Froude number = 1
  Q = outlet/inlet flow ratio, defined as ft³/s (m³/s) flow divided by outlet perimeter in ft (m)
  g = gravitational effect, ft/s² (m/s²)
  p = water density, lb/ft³ (kg/m³)
  Δp = difference in water density between inlet and outlet, lb/ft³ (kg/m³)
  h = distance between plates, ft (m)

$$1 = \frac{\dfrac{3500 \text{ gal/min}}{7.48 \text{ (gal·min)}/\text{ft}^3 \times (60 \text{ s/min}) \times 10 \text{ ft} \times 3.14}}{\sqrt{32.2[(62.40-62.31)/62.40]h^3}}$$

or

$$1 = \frac{\dfrac{220 \text{ L/s}}{1000 \text{ L/m}^3 \times 3.048 \text{ m} \times 3.14}}{\sqrt{9.8 \text{ m/s}^2[(999.7-998.3)/999.7]h^3}}$$

Solving for h gives 1.13 ft (0.34 m).

Temperature Monitoring. The tank should be equipped with temperature controls to monitor the temperature gradient within the tank. Thermocouples are typically used and installed vertically along the tank wall. The spacing should be adequate to identify the location of the thermal layer at all times. Generally, spacing of 1 to 2 ft (0.3 to 0.6 m) should be adequate. However, judgment is required depending on the tank's dimension.

REFERENCES

1. *Primary Secondary Pumping Application Manual*, International Telephone and Telegraph Corp., Bulletin TEH-775, 1968.

2. The American Society of Heating, Refrigeration, and Air-Conditioning Engineers, Inc., 2000 *ASHRAE Handbook, Systems and Equipment*, ASHRAE, Atlanta, GA, Chaps. 12 and 13.

3. The American Society of Heating, Refrigeration, and Air-Conditioning Engineers, Inc., 2003 *ASHRAE Handbook, Applications*, ASHRAE, Atlanta, GA, Chap. 34.

19.2 ALL-AIR SYSTEMS*

Ernest H. Graf, P.E.

*Assistant Director, Mechanical Engineering,
Arcadis Giffels, Inc., Southfield, Michigan*

Melvin S. Lee

*Senior Project Designer, Arcadis Giffels, Inc.,
Southfield, Michigan*

19.2.1 SINGLE-ZONE CONSTANT-VOLUME SYSTEM

The single-zone constant-volume system is the basic all-air system used in a temperature-controlled area (Fig. 19.2.1). This system will maintain comfort conditions in an area where the heating or cooling load is fairly uniform throughout the space.

The method of controlling the temperature in the area can vary based on the functions performed in the space. The basic general-use area will require only modulating the cooling or heating medium at the air-handling equipment to maintain desired space conditions. Areas requiring a closer temperature-humidity control will need a unit that can cool and reheat at the same time, to maintain a close temperature range.

19.2.1.1 Central Equipment

A single-zone unit, shown in Fig. 19.2.1, consists of a supply fan, cooling coil, heating coil, filter section, and return-air/outdoor-air mixing plenum.

The combination of these components in the unit provides a system that can maintain a basic temperature-controlled environment with a change in either heating or cooling loads.

Variations and additions to these components can provide a system that can maintain a closely controlled temperature and humidity environment throughout the year. With the addition of a return-air fan to systems having long return air ducts, outdoor air can provide the required cooling medium during certain periods of the year. Adding a humidifier to the unit provides means to control and maintain a precise level of humidity to match the area function.

19.2.1.2 Ductwork System

The layout of the single supply duct to the conditioned space should be routed with a minimum of abrupt directional and size configuration changes.

The supply ductwork should be sized by using the static regain method to assist in a balanced air distribution in the duct system. The branch or zone mains should be provided with a balancing damper at the point of connection to the supply main. This will enable fine adjustments to be made to the air distribution system within the zone.

The supply air terminals in the space should be selected, sized, and located to provide even distribution throughout the space without creating drafts or excessive noise. Each terminal should have a volume damper to permit individual air balancing.

---

*Updated for this book by James R. Fox, P.E., David H. Helwig, William C. Newman, Joseph D. Robinson, P.E., James Runski, P.E., and Paul D. Bohn.

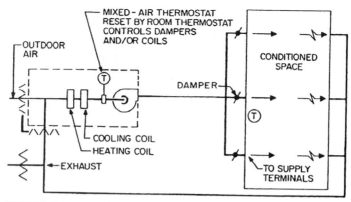

FIGURE 19.2.1 Single-zone system.

The return-air ductwork should be sized by using the equal-friction method from the space return registers back to the central equipment. The same ductwork configuration considerations and accessories should be used in laying out the return ductwork as listed for the supply duct system.

The location of the central equipment relative to the conditioned space should be considered when one is evaluating the need for acoustically lined ductwork or sound traps at the central equipment, to prevent transmission of noise through the duct system to the space.

19.2.1.3 Applications

The single-zone systems are generally used for small offices, classrooms, and stores. The single-package type of individual air-handling unit, complete with refrigeration and heating capabilities, can be roof-mounted or located in a mechanical space adjacent to or remote from the conditioned space.

19.2.2 SINGLE-ZONE CONSTANT-VOLUME SYSTEM WITH REHEAT

A single-zone constant-volume system with reheat has the same equipment and operating characteristics as the single-zone system, but has the advantage of being able to control temperatures in a number of zones with varying load conditions. See Fig. 19.2.2.

Areas made up of zones with varying loads can be supplied by a single supply air system of constant volume and temperature. The air quantity and air temperature are based on the maximum load and comfort conditions established for the area. The individual rooms or zones within the area can be temperature-controlled with the addition of a reheat coil to the branch supply duct.

This system does not comply with current energy codes.

19.2.2.1 Central Equipment

The same equipment as described for the single-zone constant-volume system is used, except for the addition of reheat coils in the branch ducts serving zones with changing heating and cooling loads.

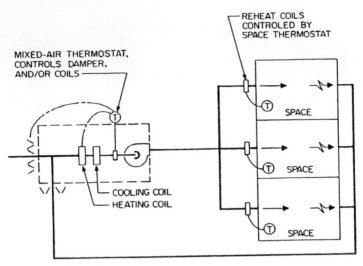

FIGURE 19.2.2   Reheat system.

The heating coil can be electric, hot water, or steam. A space thermostat modulates the reheat coil to maintain the desired space temperature.

19.2.2.2 Ductwork System

The ductwork for the constant-volume system with reheat requires the same considerations as listed for the single-zone constant-volume system.

The addition of a reheat coil will require ductwork enlargement transition before the coil and a reducing fitting after the coil, to ensure proper air flow over the entire face area of the coil. Access doors should be provided in the ductwork on the entering the leaving sides of the coil for cleaning and inspection.

19.2.2.3 Applications

The single-zone system with reheat coils is used for small commercial facilities which may be divided into a number of areas and/or offices with varying internal and perimeter loads. These systems, which use reheat to maintain comfort, should be provided with controls to automatically reset the system cold-air supply to the highest temperature level that will satisfy the zone requiring the coolest air.

The leaving air temperature of a reheat coil depends on several factors:

* The design space heating temperature
* Whether there is a supplementary heating system along the exterior perimeter of the building (such as fin pipe convectors, fan coil units, etc.) for the zone served by the reheat coil
* Whether there is a space equipment cooling load during the heating season

For instance, one of the following conditions can determine the leaving air temperature of a reheat coil:

- *Condition 1:* When the space or zone does not have an exterior exposure or has a supplementary perimeter heating system and there is no equipment cooling load during the heating season, the reheat coil leaving air temperature should nearly equal the space design temperature.

- *Condition 2:* When the space or zone has an exterior exposure without a supplementary perimeter heating system and there is no equipment cooling load during the heating season, the reheat coil leaving air temperature should equal the space design temperature plus the temperature difference calculated to offset the space or zone heating loss from the exterior exposure.

- *Condition 3:* The space or zone is the same as for condition 1 except there is an equipment cooling load requirement during the heating season. Then the reheat coil leaving air temperature should be equal to the space design temperature minus the temperature difference calculated to offset the space equipment cooling load.

- *Condition 4:* The space or zone is the same as for condition 2 except there is an equipment cooling load requirement during the heating season. Then the reheat coil leaving air temperature should be equal to the space design temperature plus the temperature difference calculated to offset the space or zone heat loss from the exterior exposure minus the temperature difference calculated to offset the space equipment cooling load.

19.2.3 MULTIZONE SYSTEM

This type of system (Fig. 19.2.3) is used when the area being served is made up of rooms or zones with varying loads. Each room or zone is supplied by means of a single duct from a common central air-handling unit.

The central air-handling unit consists of a hot-air plenum and cold-air plenum with individual modulating zone dampers mixing hot and cold air streams and supplying the mixture through a dedicated duct to the space. A thermostat located in the occupied space modulates the zone dampers at the unit to achieve the desired temperature conditions.

This system does not comply with current energy codes.

19.2.3.1 Central Equipment

The multizone unit shown in Fig. 19.2.3 may be a factory-assembled package unit consisting of a mixing plenum, filter section, supply fan, heating coil, cooling coil, and damper assemblies on the discharge side of coils. A humidifier can be added to the unit to maintain a winter humidity level.

19.2.3.2 Ductwork System

The supply ductwork for the multizone system originates at the central unit dampered discharge outlet from the hot and cold deck. Each zone will be supplied by a single duct with a number of supply air terminals. The supply ductwork should be sized by using the static region method, to assist in a balanced air distribution in the duct system. Branch duct take-offs from the duct mains should be provided with balancing dampers to permit fine adjustments to the air distribution system within the individual zones.

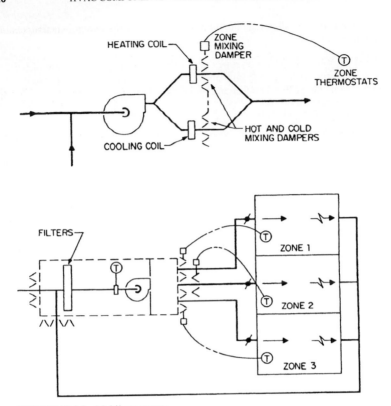

FIGURE 19.2.3   Multizone system.

The supply air terminals in the space should be selected, sized, and located to provide even distribution throughout the zones without creating drafts or excessive noise. The supply air terminal should be provided with a volume damper to balance air quantities at the individual outlets.

The return-air ductwork should be sized by using the equal-friction method from the space return registers back to the central equipment. The same ductwork configuration considerations and accessories shall be used in laying out the return duct as was used in designing the supply duct system.

The location of the central equipment relative to the conditioned space should be considered when one is evaluating the need for acoustically lined ductwork or sound traps at the central equipment, to prevent transmission of noise through the duct system to the space.

19.2.3.3   Application

The multizone type of system is considered for office buildings, schools, or buildings with a number of floors and interior zones with varying loads.

The multizone system and dual-duct system, to an extent, will give similar performances inasmuch as the dual-duct system is sometimes described merely as a multizone system with extended hot and cold decks. However, the following real differences do exist:

- Packaged multizone air handlers are available with up to 14 zones, whereas dual-duct systems have virtually no limit as to zones.

- Building configuration may be better suited to the numerous small ducts from a multizone system than to the two large ducts off a dual-duct air handler.

- The small zone off a multizone which also has large zones will have erratic air flow when the large zone dampers are modulating. The pressure-independent mixing boxes of a dual-duct system preclude this.

- The damper leakage at "economy" multizone units can be excessive, especially when maintenance is poor.

- It is undoubtedly more costly and cumbersome to add a zone to an existing multizone system than to use a dual-duct system.

- Packaged multizone systems are suitable for small systems and as such may include direct expansion cooling and gas-fired heating equipment. The step capacity control included with this equipment can result in noticeable cycling of space temperatures. The larger cooling and heating equipment generally accompanying dual-duct systems includes modulating capacity controls, and this precludes the space temperature cycling.

- The air in the short hot and cold plenums of multizone units can experience the same temperature gradient as that of a heating coil which has a "hot end," especially during low loads (and similarly for cooling coils). This temperature gradient can result in improper hot (or cold) air entering the zone duct. The long hot and cold ducts of the dual-duct system permit thorough mixing of air off the coils and eliminate the gradient.

19.2.4   INDUCTION UNIT SYSTEM

The induction unit system is used for the perimeter rooms in multistory buildings such as office buildings, hotels, hospital patient rooms, and apartments. See Fig. 19.2.4.

This system does not comply with current energy codes.

The system consists of a central air-handling unit which supplies primary air, heated or cooled, to offset the building transmission loss or gain; a high-velocity duct system for conveying the primary air to the induction units; an induction unit with a coil for each room or office; and a secondary water system, which is supplied from central equipment. The secondary water system is heated or cooled depending on the time of year and the requirements of the space being served.

A constant volume of primary air is supplied from the central air-handling unit through a high-pressure duct system to induction units located in the rooms. The air is introduced to the room through the high-pressure nozzles located within the unit that cause the room air to be drawn over the unit coil. The induced air is heated or cooled depending on the secondary water temperature and is discharged into the room.

19.2.4.1   Central Equipment

The primary air supply unit for the induction system generally includes a filter, humidifier, cooling coil, heating coil, and fan. A preheat coil is also included when the unit handles large quantities of outdoor air which is less than 32°F (0°C). The heating coil may be in the form of zone reheat coils when the unit supplies induction units on more than one exposure (north, east, south, or west).

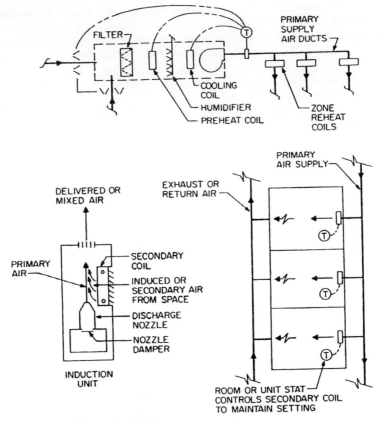

FIGURE 19.2.4   Induction unit system.

The supply fan is a high-static unit sized to provide the primary air requirements for each induction unit. The chilled water or refrigerant cooling coil dehumidifies and cools the primary air during the summer months. Primary air is supplied at a constant rate to the induction units and is generally 40 to 50°F (4 to 10°C) year-round. The final room temperature is maintained by the secondary coil.

19.2.4.2 Ductwork System

The air supply to the induction units originates from a central air-handling unit. The supply header ductwork should be routed around the perimeter of the building, with individual risers routed up through the floors supplying primary air to the induction units.

Limited available duct space frequently dictates that velocities in the risers be maintained at 4000 to 5000 ft/min (20 to 26 m/s). Rigid spiral ductwork is used; elbows and take-offs are of welded construction. Close attention must be paid to prevent noisy air leakage in the duct system.

A sound-absorbing section of ductwork should be provided at the discharge of the central air handler to absorb noise generated by the high-pressure fan.

The supply header and risers should be thermally insulated to prevent heat gain and sweating during summer operation and heat loss during winter operation.

The supply ductwork system should be sized by using the static regain method.

19.2.4.3 Application

The induction unit system is well suited to the multistory, multiroom buildings with perimeter rooms that require individual temperature selection.

The benefits in using the induction system in these types of buildings is in the reduced amount of space required for air distribution and equipment. The secondary coil in the induction unit is frequently connected to a two-pipe dual-temperature system which provides the coil with hot water during the winter and chilled water during the remaining seasons. The thermostat modulates water flow and therefore varies the temperature of the delivered or mixed air to compensate for the room heat loss or heat gain.*

19.2.5 VARIABLE-AIR-VOLUME SYSTEM

This system is used primarily when a cooling load exists throughout the year, such as the interior zone of office buildings. This air supply system uses varying amounts of constant-temperature air induced to a space to offset cooling loads and to maintain comfort conditions. The system operates equally well at exterior zones during the cooling season, but during the heating season supplementary equipment such as reheat coils, finned radiation, etc., must be provided at all spaces with an exterior exposure. When a reheat coil is added to a variable-air-volume (VAV) box, the temperature controls should reduce air flow to the minimum acceptable level for room air motion and makeup air and then activate the reheat coil (Fig. 19.2.5).

The system typically consists of a central air-handling unit with heating and cooling coil, single-duct supply system, VAV box, supply duct with air diffuser, return air duct, and return air fan.

Constant-temperature air is provided from the central air-handling unit through a single-supply air duct to the individual VAV box which regulates supply air to zone to offset cooling load requirements.

19.2.5.1 Central Equipment

The air supply unit consists of a supply fan with variable inlet vanes, variable speed control or discharge dampers, cooling coils using refrigerant or chilled water, filters, heating coil using steam or hot water for morning warmup, return air fan which is modulated through controls to match supply fan demands, and mixed air plenum to provide outdoor air requirements to the system.

The supply fan should be selected for the calculated load and system static pressure. During system operation, the supply air demand varies with space load requirements. To meet this demand, the supply and return fans' discharge air quantities must be modulated in unison with variable inlet vanes, variable speed control, or discharge dampers.

---

*For more details see *2000 ASHRAE Handbook: HVAC Systems and Applications*, American Society of Heating, Refrigerating, and Air-Conditioning Engineers, Atlanta, GA, 2000.

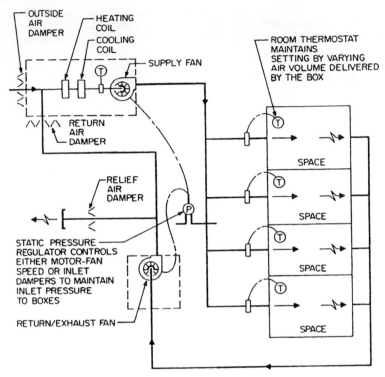

FIGURE 19.2.5   Variable-volume system.

The cooling coil, being either a direct-expansion or chilled-water type, automatically controls the discharge air temperature for the unit. During the winter cycle, the mixed air damper and return air damper are modulated to maintain the discharge air temperature.

The following items may need special consideration when VAV systems are designed:

Minimum Outdoor Air.   The outdoor air drawn into a building will tend to reduce as the supply fan volume reduces. This can become detrimental when the supply fan has a large turndown from its maximum flow and/or when the VAV system provides makeup air for constant-flow exhaust systems. The minimum outdoor air can be maintained by providing a short duct with a flow sensor downstream of the minimum outdoor air damper. The flow sensor modulates the outdoor and return air dampers to maintain the required minimum.

Building Static Pressure.   It is important that return fan volume be properly reduced when the supply fan volume reduces. The return should reduce at a greater rate so as to leave a fixed flow rate for the constant-flow exhaust systems and building pressurization. Flow sensors at the supply and return fans can monitor and maintain a constant *difference* between supply and return air by modulating the return air and exhaust air flow.

Room Air Motion.   Select air diffusion devices for proper performance at minimum as well as maximum flow to preclude "dumping" of air.

Building Heating. Calculations frequently show that the internal heat gain (lights, equipment, people) during occupied hours of basement, interior, and sometimes perimeter spaces is more than sufficient to keep these spaces warm. So it may appear that a "mechanical" heat source is not required. But these heat gains might not exist during unoccupied nights, weekends, and shutdown periods, and the spaces will cool down even when the only exposure is a well-insulated wall or roof. The central equipment of VAV systems is sometimes designed without heating coils and in itself cannot heat the building (it "heats" by providing less cooling). Unit heaters, radiation elements, convectors, or heating coils including controls coordinated with the VAV system at zero outdoor air are required for a timely morning warmup and heat when the space is unoccupied.

Calculations for winter usually show a need for heat at perimeter spaces. If the VAV boxes have "stops" for minimum air supply, there must be sufficient heat to warm this minimum air [usually 55 to 60°F (13 to 15°C)] in addition to that required for transmission losses.

19.2.5.2 Ductwork System

The supply air mains and branch ducts can be sized for either low or high velocity depending on the space available in the ceiling. A low-velocity design will result in lower operating costs.

Supply ducts should have pressure relief doors, or be constructed to withstand full fan pressure in the event of a static pressure-regulator failure concurrent with closed VAV boxes.

19.2.5.3 Application

The variable-air-volume system is considered for a building with a large interior zone requiring cooling all year. The varying amounts of cool air match the varying internal loads, as people and lighting loads change throughout the day and night. When used in conjunction with perimeter heating systems, the VAV system can be used for the perimeter zone of a building also. These conditions describe the typical operations of office buildings, schools, and department stores, which are the prime users of this type of system.

19.2.6 DUAL-DUCT SYSTEM

This type of system is used when the area served is made up of rooms or zones with varying loads, with the entire area being supplied from a central air-handling unit. See Fig. 19.2.6. The central unit supplies both cold and hot air through separate duct mains to a mixing box at each zone. The zone box controlled by the space thermostat mixes the two air streams to control the temperature conditions within the zone.

19.2.6.1 Central Equipment

The dual-duct unit may be a factory-assembled packaged unit consisting of a mixing plenum, filter section, supply fan, heating coil, cooling coil, and discharge plenums. A humidifier can be added to maintain winter humidity level.

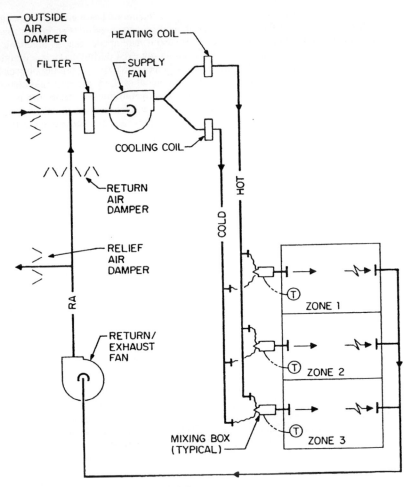

FIGURE 19.2.6   Constant-volume dual-duct system.

19.2.6.2 Ductwork System

The hot and cold supply ductwork headers originate at the unit discharges and run parallel throughout the building, connecting to the individual mixing boxes which supply the zone ductwork.

The cold duct header is sized to carry the peak air volumes of all zones. The hot duct header is sized to carry a certain percentage of the cold air, usually 70 percent.

The zone mixing box responds to a space thermostat and modulates and mixes quantities of cold and hot air and delivers a constant volume of air to the space to maintain desired temperature levels. The size of the mixing box is based on the peak air volume of the zone or room.

The mixed air from the mixing box is discharged through a single duct terminating with a number of supply air diffusers, chosen and sized to provide even distribution throughout the zone.

Supply ductwork may be sized by the static regain or equal-friction method with aerodynamically smooth fittings and velocities not exceeding 3000 ft/min (15 m/s).

The return-air ductwork is often sized by the equal-friction method, but it does not exceed 1500 ft/min (7.6 m/s). The routing and configuration of the supply headers must satisfy the space limitations in ceiling and shaft areas. Access and space requirements for the mixing box should also be considered when the routing of the duct system is laid out.

The location of the central equipment relative to the conditioned space should be considered when deciding the need for acoustically lined ductwork or sound traps at the central equipment, to prevent transmission of noise through the duct system to the space.

19.2.6.3 Application

The dual-duct type system is considered for office buildings, schools, or buildings with a number of floors and zones with varying loads. Generally, however, this system has been "replaced" by the VAV system because of higher operating and first costs and increased duct space requirements.

BIBLIOGRAPHY

The American Society of Heating, Refrigerating, and Air-Conditioning Engineers, Inc., 2000. *ASHRAE Handbook: HVAC Systems and Equipment*, ASHRAE, Atlanta, GA, Chapter 2.

19.3 DIRECT EXPANSION SYSTEMS

Simo Milosevic, P.E.*†

Project Engineer, Mechanical Engineering,
Arcadis Giffels Inc., Southfield, Michigan

19.3.1 SYSTEM DESCRIPTION

To air-condition the interior space of a building, recirculating and makeup air is commonly cooled and simultaneously dehumidified while passing across a cooling coil. This coil could be a direct expansion (DX) type.

DX refrigeration utilizes refrigerant (working fluid of the cycle) fluid temperature, pressure, and latent heat of vaporization to cool the air. To evaporate liquid refrigerant to a vapor, the latent heat of vaporization has to be applied to the liquid. The quantity of heat necessary to evaporate 1 lb (0.45 kg) of liquid refrigerant varies with the thermal characteristics of different refrigerants. The boiling point of an ideal refrigerant has to be below the supply air temperature and above 32°F (0°C), so as to not freeze the moisture condensed from the building supply air. Most refrigerants in use today have a relatively low boiling point and nonirritant, nontoxic, nonexplosive, nonflammable, and noncorrosive characteristics for use in commercially available piping materials.

Early air-conditioning and refrigeration systems used toxic and hazardous substances such as ether, chloroform, ammonia, carbon dioxide, sulfur dioxide, butane, and propane as refrigerants. The most common refrigerants in use today are numbers 22, 23, and 134A.* There is concern that these refrigerants may be damaging the earth's ozone layer and alternative refrigerants are being offered for use in new equipment. Refrigerant 22 is still in use, but efforts are underway to find a suitable replacement. Production of refrigerant 11 and 12 ended in 1995. The leading alternatives for R-22 are R-407C and R-401A.

A simple refrigeration system is illustrated in Fig. 19.3.1. Basic elements of the system are the expansion valve (pressure-reducing valve), evaporator (cooling coil or DX coil), compressor, condenser, and interconnecting piping. The compressor and expansion valve are points of the system at which the refrigerant pressure changes. The compressor maintains a difference in pressure between the suction and discharge sides of the system (between the DX coil and condenser), and the expansion valve separates high- and low-pressure sides of the system. The function of the expansion valve is to meter the refrigerant from the high-pressure side (where it acts as a pressure-reducing valve) to the low-pressure side (where it undergoes a phase change from a liquid to a vapor during the process of heat absorption).

The compressor draws vaporized refrigerant from the evaporator through a suction line *A*. In the compressor, the refrigerant pressure is raised from evaporation temperature and pressure to a much higher discharge pressure and temperature. In the discharge line *B*, refrigerant is still in the vapor state at high temperature, usually between 105 and 115°F (40 and 60°C). A relatively warm cooling medium (water or air) can be used to condense and subcool the hot vapor. In the condenser, heat of vaporization and of compression is transferred from the hot refrigerant gas to the cooling medium through the walls of the condenser

---

*These refrigerants can be found on the market under the various trade names of Freon (registered trademark of E. I. du Pont de Nemours Co.), Genetron (registered trademark of Allied Chemical Corp.), and Isotron (registered trademark of the Pennsalt Chemicals Corp.).

†Updated by James Runski, P.E., Wayne M. Lawton, Arcadis Giffels.

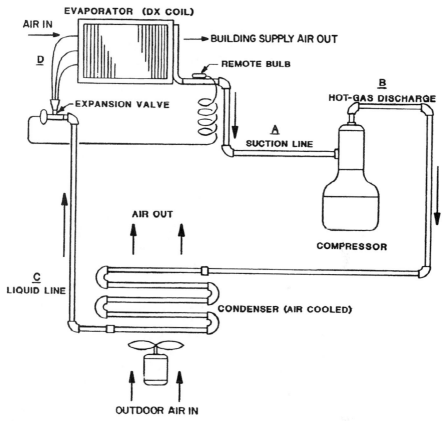

FIGURE 19.3.1 Mechanical refrigeration system.

heat-exchange surfaces while the gas becomes liquefied at or below the corresponding compressor discharge pressure *C*.

The expansion valve separates the high-pressure or condenser side of the system from the low-pressure or evaporator side. The purpose of the expansion valve is to control the amount of liquid entering the evaporator, such that there is a sufficient amount to evaporate, but not flood, the evaporator *D*.

In the evaporator, liquid refrigerant is entirely vaporized by the heat of the building supply air. Heat equivalent to the latent heat of vaporization has been transferred from building air through the walls of the evaporator to the low-temperature refrigerant. Thus the building supply air is cooled and dehumidified. The boiling point (temperature of evaporation) at the evaporator pressure is usually between −41 and 82°F (1 and 7°C) for refrigerants 22, 123, and 134A. It is even lower for refrigerant 502.

From the evaporator, vaporized refrigerant is drawn through suction piping to the compressor, and the cycle is repeated.

All refrigerants have different physical and thermal characteristics. Depending on the available condensing temperature, required evaporation temperature, and cooling capacity, different refrigerants are used for different applications.

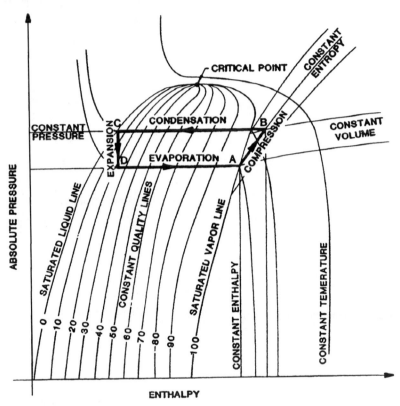

TEMPERATURE, °F (°C)
SP. VOLUME, FT³/lb (m³/kg)
ENTHALPY, BTU/lb (kJ/kg)
ENTROPY, BTU/lb•R (kJ/kg•K)
PRESSURE, lb/in² (kPa)
QUALITY, %

FIGURE 19.3.2 Mollier diagram.

Figure 19.3.2 illustrates the theoretical refrigeration cycle (without pressure losses in the system and without subcooling of the liquid or superheating of the vapor) shown on a Mollier or pressure-enthalpy diagram.

19.3.2 EQUIPMENT

The purpose of this section is to describe the components of a DX refrigeration system and their functions. In addition to the basic elements already mentioned (compressor, condenser, expansion valve, evaporator, and refrigerant piping), the typical refrigeration system incorporates other components or accessories for various purposes. Figure 19.3.3 illustrates a simple DX refrigeration system with an air-cooled condenser designed for operation in cold weather.

19.3.2.1 Compressor

This is a vapor-phase fluid pump which maintains a difference in refrigerant gas pressure between the DX coil (low-pressure or suction side) and the condenser (high-pressure or discharge side) of the system. Compressors can be categorized as to construction, i.e., hermetic, semihermetic, and open (direct- or belt-driven). They also can be categorized by the type of machine, i.e., reciprocating, centrifugal, and screw. More about compressors can be found in Secs. 18.2 to 18.5.

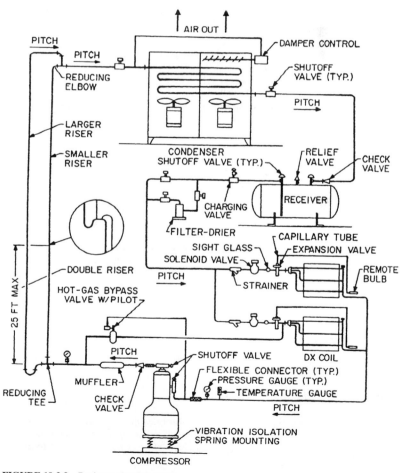

FIGURE 19.3.3 Reciprocating DX system.

19.3.2.2 Condenser

This is a heat-exchange device in which heat of vaporization and compression is transferred from hot refrigerant gas to the cooling medium, in order to change the refrigerant from a superheated vapor to a liquid state and sometimes to subcool the refrigerant. Condensers can be air-cooled, where outdoor air is used to condense and subcool the refrigerant, or water-cooled, where city water or cooling tower water is used as the cooling medium. Evaporative condensers use both water and air to condense and subcool refrigerant: recirculating water is sprayed over tubes containing hot refrigerant and is evaporated by moving outdoor air, thus removing heat from the refrigerant.

Air-cooled condensers can be single- or multifan types. Axial fans are most commonly used because axial fans economically handle large air volumes at low static pressure. Centrifugal fans, which are capable of generating higher static pressures, are used in certain applications.

Condensers can also be categorized as single- and multicircuit, according to whether they are connected to one or multiple compressors.

19.3.2.3 Expansion Valve

This is a throttling or metering device with a diaphragm operator. The space above the diaphragm is connected to a remote sensor bulb with capillary tubing and filled with the same refrigerant as is used in the system. The valve controls flow of fluid refrigerant to maintain a set-point pressure in the evaporator.

The remote temperature-sensing bulb is normally strapped or soldered to the suction line (leaving evaporator) for maximum surface contact. An increase in heat load on the evaporator is sensed by the bulb, causing a corresponding increase in vapor pressure within the bulb, capillary tube, and space above the diaphragm. This pressure, transmitted by the diaphragm, moves the valve off its seat, to admit more liquid refrigerant into the evaporator, for evaporation by the increased heat load.

When the cooling requirements are satisfied, the process reverses.

19.3.2.4 Evaporator (DX Coil)

This is an extended-surface (finned tube) device in which heat exchange occurs between building supply air and the liquid refrigerant in the coil tubes, causing the refrigerant to vaporize.

DX coils are either dry or flooded (with refrigerant liquid). The coils can have 20 or more parallel circuits and are of one- or multirow construction.

19.3.2.5 Refrigerant Piping

Typically, Type L copper tubing is used for handling refrigerants, discussed earlier.

19.3.2.6 Hot-Gas Bypass Control

This control is a way to maintain a reasonably stable evaporator suction pressure when a refrigeration system is operating at minimum load. Two hot-gas bypass methods are in use today on direct expansion systems: hot-gas bypass to the evaporator inlet and to the suction line.

Hot-gas bypass to the evaporator inlet introduces compressor discharge vapor to the DX coil after the expansion valve (see Fig. 19.3.3). This acts as an artificial heat load on the DX coil and raises the temperature at the coil outlet. The remote bulb of the expansion valve senses this temperature rise and opens the valve to increase the flow of refrigerant through the coil, resulting in a rise of the suction pressure and stabilization thereof.

The effectiveness of this method is a function of the distance between the (compressor) and the DX coil. This method should not be used when the distance is greater than 50 ft (15 m) for hermetic compressors (hot gas might start condensing, resulting in oil "holdup" and compressor lubrication problems).

Hot-gas bypass to the suction line introduces hot vapor from the compressor discharge to the inlet (suction) side of the compressor (see Fig. 19.3.4). This method requires an additional liquid line solenoid valve and expansion valve. When low suction pressure is sensed because of reduced heat load, hot gas is introduced to the compressor inlet through the hot-gas bypass valve. This causes the supplementary expansion valve to open and to introduce liquid refrigerant into the hot gas. The liquid refrigerant evaporates and increases compressor suction pressure to stabilize operation.

Disadvantages of this method include the additional cost of expansion and solenoid valves, oil trapping in the DX coil, and the possibility of liquid slugs entering the compressor.

19.3.2.7 Suction and Hot-Gas Double Riser

This is a pipe assembly which promotes oil movement to the compressor on the suction side and from the compressor on the hot-gas side. A double-riser arrangement is used in a vertical piping layout when the compressor is below the condenser and/or when the compressor is above the DX coil. Figure 19.3.3 illustrates the double riser on the hot-gas side. A similar setup would be provided on the suction side if the compressor were located above the DX coil.

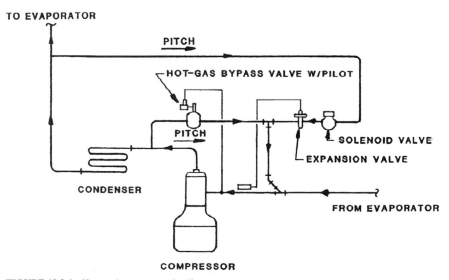

FIGURE 19.3.4 Hot-gas bypass to suction line.

Operation at minimum compressor capacity (with the compressor unloaded or one compressor running in multiple-compressor installations) reduces oil conveying velocities in the system, which causes compressor oil to fill the trap, thus directing gas flow to the smaller riser. This riser is sized to produce a velocity of 1000 ft/min (5 m/s) which is sufficient to convey oil upward for return to the compressor. When full-load capacity is restored, pressure clears the trap and flow is established through both risers. The larger riser is sized for velocities between 1000 and 4000 ft/min (5 and 20 m/s) at full compressor load.

The maximum vertical rise of a double riser should not exceed 25 ft (8 m). If greater than 25-ft (8 m) rise is required, an intermediate trap should be incorporated every 25 ft (8 m) of rise (see Fig. 19.3.3).

19.3.2.8 Filter-Drier

This is usually installed in the liquid line to protect the expansion valve from dirt or moisture that may freeze in the expansion valve and to protect motor windings from moisture. The filter-drier core has an affinity fore and retains water while simultaneously removing foreign particles from the liquid refrigerant (see Fig. 19.3.3).

19.3.2.9 Condenser Pressure Control

This control is necessary with lower outdoor air temperatures when the air-cooled condenser capacity increases and system load decreases, causing low condenser pressure. This is controlled by modulating air flow through the condenser with an outlet damper whose operator is driven by condensing pressure. In multifan condensers, cycling of the fans by outdoor air temperature thermostat provides step control of the air flow through the condenser. The last operating fan might have an outlet damper operated by condensing pressure.

In water-cooled condensers, water flow is usually controlled by a flow-modulating valve controlled by condensing pressure.

19.3.2.10 Hot-Gas Muffler

This muffler is usually installed on the discharge side of reciprocating compressors with long piping systems. This reduces gas pulsation and noise produced by reciprocating equipment.

19.3.2.11 Solenoid Valves

These valves are electrically operated two-position valves. They permit isolation of coil circuits to reduce the cooling produced and pumpdown of the low pressure side for eventual compressor shutoff when heat load is zero.

19.3.2.12 Sight Glasses

These should be installed in every system in front of the expansion valve. The operator can verify the flow of liquid and absence of gases or vapors upstream of cooling coils.

19.3.2.13 Shutoff Valves

These valves are usually the capped, packed, angle type mounted directly on the compressor or liquid receiver. The purpose of shutoff valves is to isolate portions of the refrigeration circuit to enable maintenance or repair.

19.3.2.14 Charging Valve

This is the point at which refrigerant is introduced (charged) into the system. Normally the charging valve is installed in the liquid line after the condenser or after the liquid receiver, if one is used.

19.3.2.15 Relief Valves and Fusible Plugs

These devices protect the refrigeration system from excessive pressure buildup. In the case of fusible plug activation, all refrigerant charge is released when the plug melts because of excessive temperature.

19.3.2.16 Check Valves

These are usually used in front of the liquid receiver and after the compressor, to prevent vapor migration from the receiver to the condenser or liquid migration from the condenser to the discharge of the compressor, after the system shutdown. This is especially important in systems where the receiver is located in a hot space or the compressor is located in a space cooler than the condenser.

19.3.2.17 Strainers

These are installed in liquid lines to protect solenoid and expansion valves from dirt.

19.3.2.18 Liquid Receiver

When condensers (evaporative, air-cooled) which inherently have a small refrigerant storage volume are used and the system is sufficiently large, a liquid receiver is installed after the condenser to collect and hold the system liquid refrigerant until it is required by the peak load.

19.3.2.19 Pressure and Temperature Gauges

These are used to indicate suction and discharge compressor pressures and temperatures, condenser water temperature, and compressor lubricating oil pressure.

19.3.3 APPLICATIONS

Direct expansion air-conditioning systems are available as window-type units with capacities from less than $^1/_2$ ton (1.8 kW) to large packaged or built-up units of over 100 tons (352 kW) of refrigeration. They can be of self-contained construction (rooftop) where all elements, including the controls, are built in one cabinet, or they can be a split system where the condenser and compressor are located in one cabinet outside the building (condensing unit) and the DX coil and expansion valve are located inside the building in the air-handling unit. In the latter case, interconnecting refrigerant piping (liquid and suction lines) has to be field-installed and insulated.

Capacities of split and self-contained air-conditioning systems (excluding window air conditioners) normally start at 2 tons (7 kW) and for self-contained systems range to 100 tons (352 kW), while split systems range to 120 tons (422 kW) of refrigeration.

Split and self-contained DX systems are normally used in situations that require individual temperature and humidity control for numerous small spaces within a large building. Typical applications are apartment buildings, condominiums, small shopping malls (strip stores), office buildings with multiple tenants, medical buildings (doctors' and dentists' offices), and various departments within a manufacturing building. These systems are also used in nonair-conditioned plants where it is necessary to air-condition in-plant offices and spaces. In this case, packaged systems are frequently located on the roof of the in-plant space with heat rejection to plant space. These air-conditioning systems can be individually shut off when not needed, thereby saving energy. Also the systems can be individually metered to facilitate allocating the operating cost directly to the tenants or departments. In case of failure in one system, only the space being served will be affected whereas failure of a large, central built-up system would affect the entire building or buildings.

Typical use of DX system air conditioning is found in churches and restaurants where zoning of different spaces is important (different temperatures or times of use of different spaces such as bar, kitchen, dining area, recreation hall, and church area).

The most common use of small DX systems is in residential spaces. Capacities of these systems start at 2 tons (7 kW) of refrigeration, which is enough for a small home. When two or more small DX systems are installed in larger homes, separate zones with independent temperature control and operating periods are established, e.g., sleeping areas which are cooled only at night, living quarters cooled only during the day, rooms with west exposure cooled only in late afternoon or evening, etc.

Special fields of use of DX cooling systems include computer rooms and vehicles. Self-contained, water-cooled condenser units are often used for computer rooms. They are normally designed for recirculation air only with bottom (under floor) discharge and sized to handle large sensible loads. However, computer rooms which have uniform heat release throughout the room can be conditioned by ceiling air distribution systems. For transportation vehicles (subway cars, public buses, and cars), modular systems with air-cooled condensers are used. All three major parts of the system (condenser, evaporator, and compressor) are in different locations in the vehicle and are connected with insulated piping.

Split systems are applicable as a retrofit or an option to standard air-handling units. A typical case is a residential furnace where space for a future DX coil is provided.

19.3.4   *DESIGN CONSIDERATIONS*

During initial design development, the designer must consider the type and function of facility to be air-conditioned, cooling loads involved, building layout, provisions for future expansions, and degree of required temperature and humidity control. If the designer decides that a DX system is suitable for the project, the next steps include evaluation of available condenser cooling media, type of system to be used, and location of the condenser and air handler if the system comprises multipackage units.

The simplest approach is to provide an air-cooled, single-package, rooftop-mounted air-conditioning unit. This system is completely self-contained including controls, so that the designer has only to connect ductwork to the unit and to bring in electric power and thermostat wiring. With restrictions on water use and the high cost of water in many areas of the country, air-cooled condensers have long been popular.

In general, air-cooled condensers have lower initial cost, they are lighter, maintenance is easier, and there is no liquid disposal problem. However, there are certain disadvantages and design considerations that the designer has to recognize before choosing a type of condenser. Air-cooled condensers require large amounts of relatively cool air, which could be a problem, especially with a condenser located indoors. Axial-flow fan condensers can be noisy. They require relatively clean air (condenser plugging problem). Startup difficulties at low outdoor air temperatures, capacity reduction on high outdoor temperatures, and operating problems at part load are common problems. Air-cooled condensers require locations free of any obstructions on both inlet and outlet sides. Usually clearance of 1.5 times the condenser height is required around the condenser. If a possibility of air short-circuiting (recirculation of hot air) occurs, the designer should consider condenser fan discharge stacks. Since the north side of the building is cooler and is in shade for most of the day, the condenser should be located in this area, if possible.

When a system operates for a longer period on minimum load, the suction pressure drops, as does the corresponding temperature. This can result in frost or ice on the cooling coil, restricting air flow through it. Also reduced refrigerant flow through the system may cause compressor lubrication problems and motor cooling problems in hermetic compressors.

In general, capacity control in a reciprocating compressor DX system is a problem. Control is achieved in steps, either with multiple-compressor arrangements or by compressor valve control (unloading compressor cylinders). In any case, this is step (nor modulating) control, therefore, precise temperature control cannot be expected from DX systems. For more precise capacity control, multispeed and variable-speed motors are usually considered.

Temperature and humidity control can be achieved with parallel- and series-arrangement DX coils. A parallel coil arrangement is less expensive and provides better humidity control, but maintenance of constant leaving air temperature is difficult. Therefore, parallel coil arrangements are not recommended for reheat air distribution systems where a constant air temperature in front of reheat coils is important. Coils arranged in series are usually split to carry half the capacity each and are connected to separate compressors of the same capacity (two circuits). This is done so that the first coil has one-third of the total number of rows and the second coil has the remaining two-thirds, because the first coil has greater air temperature differences and will still carry one-half of the total cooling load. The disadvantage in this arrangement is that one compressor (the one connected to the upstream coil) is always leading on load demand and is therefore wearing faster.

Air velocity through the cooling coil is limited to 550 ft/min (2.8 m/s) maximum because of condensate moisture carryover from coil fins.

Part-load system operation can increase lubricating oil migration problems. On long vertical piping runs, this is solved with double-riser piping arrangements, as discussed earlier.

If a split system is selected, the designer must consider the distance between the condensing unit and DX coil. This distance is limited to 50 ft (15 m) total length of piping for hermetic compressors of 20 tons (70 kW) of refrigeration capacity, and under and up to 150 ft (46 m) for semihermetic compressors with capacity of over 20 tons (70 kW) of refrigeration.

When modular systems are used, compressor vibration and noise factors must be recognized. These disadvantages can be mitigated by installing vibration isolators under the compressor and by providing muffler and flexible connectors at the compressor. Piping flexibility can be improved by using two or three 90° elbows in the piping near the compressor.

If the air-conditioning unit is not easily accessible, remote panel indication of air filter status, different pressures, and temperatures should be considered. All equipment requiring maintenance should be provided with manual shutoff valves. Some municipalities require licensed operators for compressor motors above certain sizes. This can be avoided by use of multiple compressors of smaller size.

The DX systems are found to be the most expensive when analyzed over a building's life cycle. This is mainly due to energy consumption, maintenance, and replacement cost based on life expectancy.

REFERENCE

ASHRAE Handbook, HVAC Systems and Equipment, 2000, Ch. 5, 21, 34, 35, The American Society of Heating, Refrigeration, and Air-Conditioning Engineers, Inc., Atlanta, GA.

19.4   COOLING TOWERS*

John C. Hensley
*Marketing Services Manager, The Marley Cooling Tower
Company, Overland Park, Kansas*

Joe Gosmano†
*Gulf Coast General Manager,
Marley Cooling Technologies, Houston, Texas*

19.4.1   INTRODUCTION

As a general rule, air-conditioning and refrigeration systems in excess of 150 to 200-ton (528- to 704-kW) capacities make use of water as the medium for heat rejection, and the majority of such installations utilize cooling towers for the ultimate rejection of this heat to the atmosphere. In smaller systems, air-cooled heat exchangers and evaporative condensers are increasing in use, but cooling towers continue to be the method of choice where limiting the energy usage is a primary consideration.

This chapter defines the various types and configurations of cooling towers utilized for air-conditioning and refrigeration service and explains the principles by which they operate. It also discusses the environmental and energy-consuming aspects of towers and the external factors that can adversely affect their thermal performance capability.

Within reasonable limits, a cooling tower will dissipate whatever heat load is imposed on it, regardless of its size and efficiency. The tower *does not* establish the load—it merely reacts to it. However, the size and capability of the tower *does* establish the temperature level at which the heat load is dissipated, and this level, in turn, determines the operating efficiency of the system as a whole. Therefore, the student of air-conditioning is well advised to become familiar with the application of cooling towers—and with their operating characteristics.

19.4.2   TOWER TYPES AND CONFIGURATIONS

Cooling towers are designed and manufactured in several types and materials of construction, with numerous sizes (models) available in each type. They range in size from the very small [as little as 10 gal/min (63×10^{-5} m³/s)] to the very large 250,000 gal/min (15 m³/s) and larger]. Those which are normally utilized in air-conditioning service are described in this section.

Understanding the primary types and configurations, along with their advantages and limitations, can be of vital importance to the specifier and user and is essential to the full understanding of this chapter.

---

*The contents of Sections 19.4.1 to 19.4.11 were adapted by John C. Hensley from *Cooling Tower Fundamentals* and *Cooling Tower Information Index*, published by The Marley Cooling Tower Company. All art used courtesy of The Marley Cooling Tower Company.
†Sections 19.4.11 to 19.4.19.

19.4.2.1 Atmospheric Type

Atmospheric towers do not utilize a mechanical device (fan) to create air flow through the tower. There are two main types of atmospheric towers: large and small. The large hyperbolic towers are equipped with "fill"; since their primary application is with electric power generation, which is beyond the scope of this book, the reader is referred to T. C. Elliott, ed., *The Standard Handbook for Powerplant Engineering* (McGraw-Hill, New York, 1989) for additional information on hyperbolic towers. The smaller atmospheric towers utilized in the air-conditioning industry are not normally equipped with fill to enhance the transfer of heat.

The smaller atmospheric tower shown in Fig. 19.4.1 derives its air flow from the natural induction (aspiration) provided by a pressure-spray type of water distribution system, and the surface contact between air and water is limited by the pressure at the sprays and by the characteristics of the nozzles.

Although relatively inexpensive, atmospheric towers are usually applied only in very small sizes, and they tend to be energy-intensive because of the high spray pressures required. Because they are far more affected by adverse wind conditions than are other types, their use on systems requiring accurate, dependable cold-water temperatures is not recommended.

19.4.2.2 Mechanical-Draft Types

Mechanical-draft towers use either single or multiple fans to provide the flow of a known volume of air through the tower. Thus their thermal performance tends toward greater stability and is affected by fewer psychrometric variables than that of the atmospheric towers. The presence of fans also provides a means of regulating the air flow to compensate for changing atmospheric and load conditions; this is accomplished by fan capacity manipulation and/or cycling, as described in Sec. 19.4.9.

Mechanical-draft towers are categorized as either forced-draft towers (Fig. 19.4.2), wherein the fan is located in the ambient air stream entering the tower and the air is blown through, or induced-draft towers (Fig. 19.4.3), wherein a fan located in the exiting air stream draws air through the tower.

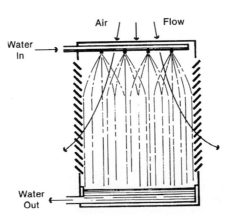

FIGURE 19.4.1 Atmospheric spray tower.

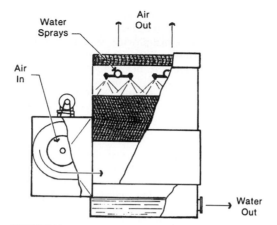

FIGURE 19.4.2 Forced-draft, counterflow, blower-fan tower.

Forced-Draft Towers.   These are characterized by high air-entrance velocities and low exit velocities. Accordingly, they are extremely susceptible to recirculation (see Sec. 19.4.5.1) and are therefore considered to have less performance stability than induced-draft towers. Of equal concern in northern climates is the fact that forced-draft fans located in the cold entering ambient air stream can become subject to severe icing, and the resultant imbalance, when moving air laden with either natural or recirculated moisture.

   Usually, forced-draft towers are equipped with centrifugal blower-type fans, which, although requiring approximately twice the operating horsepower (wattage) of propeller-type

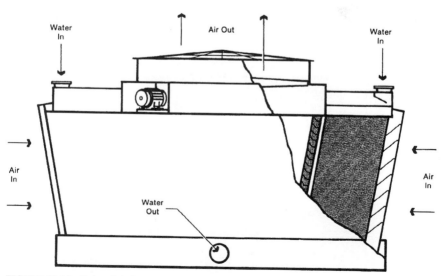

FIGURE 19.4.3 Induced-draft, crossflow, propeller-fan tower.

fans, have the advantage of being able to operate against the high static pressures associated with ductwork. So equipped, they can be installed either indoors (space permitting) or within a specially designed enclosure that provides sufficient separation between the air-intake and -discharge locations to minimize recirculation.

Given the increasing economic pressure dictating reduced energy consumption in cooling towers and the increasing wintertime usage of cooling towers for "free cooling" (see Sec. 19.4.9.2), it is to be anticipated that the use of forced-draft towers, particularly those equipped with blower fans, will soon be relegated to the exceptional situation and that induced-draft configurations will prevail.

Induced-Draft Towers. These have an air-discharge velocity three to four times higher than their air-intake velocity, with the intake velocity approximating that of a 5-mph (0.2-m/s) wind. Therefore, there is little tendency for a reduced-pressure zone to be created at the air inlets by the action of the fan alone (see Sec. 19.4.5.1). The potential for recirculation on an induced-draft tower is not self-initiating, and it can be more easily quantified and compensated for (see Sec. 19.4.6) purely on the basis of ambient wind conditions.

Induced-draft towers are also typically forgiving of wintertime operation. The location of the fan within the warm air stream (the movement of which continues even with the fan off) provides excellent protection against the formation of ice on the mechanical components. As will be seen in Sec. 19.4.10, fans so located actually provide a means by which to facilitate deicing in extreme conditions.

19.4.2.3 Characterization by Air Flow

Cooling towers are also classified in terms of the relative flow relationship of the air to the water within the tower as being either counterflow or crossflow towers.

Counterflow Towers. In counterflow towers (Fig. 19.4.4), the air moves vertically upward through the fill, counter to the downward fall of the water. Historically, because of the need

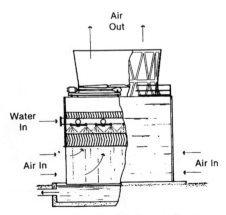

FIGURE 19.4.4   Induced-draft counterflow tower.

for expansive intake and discharge plenums, the use of high-pressure spray systems, and the typically higher air-pressure losses, some of the smaller counterflow towers were physically higher, required greater pump head, and utilized more fan power than their crossflow counterparts. With the advent of more sophisticated fills, fans, and spray systems, however, this operational economic gap is rapidly closing.

Although the enclosed nature of a counterflow tower tends to make it somewhat more difficult to service, it also restricts exposure of the circulating water to direct sunlight, thereby retarding the growth of algae.

Crossflow Towers. Crossflow towers (Figs. 19.4.3 and 19.4.5) have a fill configuration through which the air flows horizontally across the downward fall of the water. The water to be cooled is delivered to hot-water inlet basins located above the fill areas and is distributed to the fill by gravity through metering orifices in the floor of these basins. This obviates the need for a pressure-spray distribution system and places the resultant gravity system in full view for ease of maintenance.

19.4.2.4 Characterization by Construction

Field-Erected Towers. These are assembled primarily at the site of ultimate use. All large towers, as well as many of the smaller towers, are prefabricated, piece-marked, and shipped to the site for assembly. The labor and/or supervision for site assembly is usually provided by the cooling tower's manufacturer.

Factory-Assembled Towers. These are almost completely assembled at their point of manufacture and are shipped to the site in as few sections as the mode of transportation will permit. The relatively small tower shown in Fig. 19.4.6 would ship essentially inact. The larger, multicell towers (Fig. 19.4.7) are assembled as "cells" or "modules" at the factory and are shipped with appropriate hardware for ultimate joining by the user. Factory-assembled towers are sometimes referred to as "packaged" or "unitary" towers.

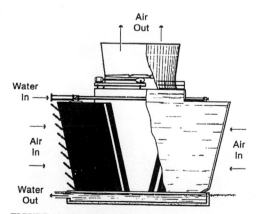

FIGURE 19.4.5 Double-flow crossflow tower.

FIGURE 19.4.6   Small factory-assembled tower.

FIGURE 19.4.7   Multicell factory-assembled tower.

19.4.3 HEAT EXCHANGE CALCULATIONS

A cooling tower is a specialized heat exchanger in which two fluids (air and water) are brought into direct contact with each other to effect the transfer of heat. In the spray-filled tower shown in Fig. 19.4.8, this is accomplished by spraying a flowing mass of water into a rainlike pattern through which an upward-moving mass flow of cool air is induced by the action of the fan.

Ignoring any negligible amount of sensible heat exchange that may occur through the walls (casing) of the tower, the heat gained by the air must equal the heat lost by the water. Within the air stream, the rate of heat gain is identified by the expression $G(h_2 - h_3)$, where:

G = mass flow of dry air through the tower, lb/min (kg/min)

h_1 = enthalpy (total heat content) of entering air, Btu/lb (J/kg) of dry air

h_2 = enthalpy of leaving air, Btu/lb (J/kg) of dry air

Within the water stream, the rate of heat loss wood *appear* to be $L(t_1 - t_2)$, where:

L = mass flow of water entering the tower, lb/min (kg/min)

t_1 = temperature of hot water entering the tower, °F (°C)

t_2 = temperature of cold water leaving the tower, °F (°C)

This derives from the fact that a Btu (calorie) is the amount of heat gain or loss necessary to change the temperature of 1 lb (1 g) of water by 1°F (1°C). However, because of the evaporation that takes place within the tower, the mass flow of water leaving the tower is actually less than that entering it, and a proper heat balance must account for this slight difference. Since the rate of evaporation must equal the rate of change in the humidity ratio (absolute humidity) of the air stream, the rate of heat loss represented by this change in humidity ratio can be expressed as $G(H_2 - H_1)(t_2 - 32)$, where:

H = humidity ratio of entering air, lb (kg) of vapor per lb (kg) of dry air

H_2 = humidity ratio of leaving air, lb (kg) of vapor per lb (kg) of dry air

$(t_2 - 32)$ = an expression of water enthalpy at the cold-water temperature, Btu/lb (J/kg)
　　　　　[the enthalpy of water is considered to be zero at 32°F (0°C)]

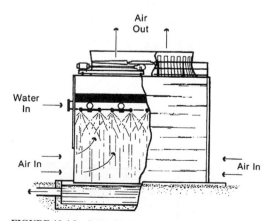

FIGURE 19.4.8　Spray-filled counterflow tower.

Including this loss of heat through evaporation, the total heat balance between the air and the water, expressed as a differential equation, is

$$Gdh = L\,dt + GdH(t_2 - 32)$$
$$[Gdh = L\,dt + GdH(t)]$$

(19.4.1)

19.4.3.1 Heat Load, Range, and Water Flow Rate

The expression $L\,dt$ in Eq. (19.4.1) represents the heat load imposed on the tower by whatever process it is serving. However, because the mass of water per unit time is not easily measured, the heat load is usually expressed as

$$\text{Heat load} = \text{gal/min} \times R \times 8.33 = \text{Btu/min}$$
$$[\text{L/min} \times R(°C) \times 0.998 = \text{kcal/min}]$$

(19.4.2)

where gal/min (L/min) = water flow rate through the system
  R (the range) = difference between the hot- and cold-water temperatures [in °F (°C) (see Fig. 19.4.9)]
  8.33 = pounds per gallon of water (0.998 = kilogram per liter of water)

Note from Eq. (19.4.2) that the heatload establishes *only* an imposed temperature differential in the condenser water circuit and is unconcerned with the actual hot- and cold-water temperatures. Therefore, the mere indication of a heat load is meaningless to the application engineer attempting to size a cooling tower properly. More information of a specific nature is required.

Optimum operation of an air-conditioning system usually occurs within a relatively narrow band of condenser water flow rates and cold-water temperatures. This establishes two of the parameters required to size a cooling tower accurately: the water flow rate and the cold-water temperature.

The total heat to be rejected from the system establishes a third parameter: the hot-water temperature coming to the tower. For example, let us assume that a refrigerant-compression type of air-conditioning unit of 500-ton (1760-kW) nominal capacity rejects heat to the condenser water circuit at a rate of 120,000 Btu/min (30,240 kcal/min) (including heat of

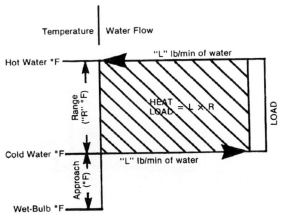

FIGURE 19.4.9   Range-approach diagram.

compression) and performs most efficiently if supplied with 1500 gal/min (5677.5 L/min) of water at 85°F (29.4°C). With a slight transposition of Eq. (19.4.2), the condenser water temperature rise can be determined as

$$R = \frac{120,000}{1500 \times 8.33} = 9.6°F \qquad R = \frac{30,240}{5677.5 \times 0.998} = 5.3°C$$

Therefore, the hot-water temperature coming to the tower would be $85 + 9.6 = 94.6°F$ ($29.4 + 5.3 = 34.7°C$).

19.4.3.2 Wet-Bulb Temperature

Having determined that the cooling tower must be able to cool 1500 gal/min (5677.5 L/min) of water from 94.6°F (34.7°C) to 85°F (29.4°C), what conditions of the entering air must be known?

Equation (19.4.1) would identify enthalpy to be of prime concern, but air enthalpy is not something that is routinely measured and recorded at any geographic location. However, wet-bulb and dry-bulb temperatures are values easily measured, and a glance at Fig. 19.4.10 shows that lines of constant wet-bulb temperature are essentially parallel to lines of constant enthalpy. Therefore, the wet-bulb temperature is the only air parameter needed to size a cooling tower properly, and its relationship to other parameters is as shown in Fig. 19.4.9.

19.4.3.3 Enthalpy Exchange Visualized

To understand the exchange of total heat (including a slight amount of mass exchange) that takes place in a cooling tower, let's assume a tower designed to cool 120 gal/min (1000 lb/min) [454.2 L/min (543.6 kg/min)] of water from 85°F (29.4°C) to 70°F (21.1°C) at a design wet-bulb temperature of 65°F (18.3°C) and (for purposes of illustration only) a coincident dry-bulb temperature of 78°F (25.6°C). (These conditions of the air are defined as point 1 in Fig. 19.4.10.) Let's also assume that air is caused to move through the tower at the rate of 1000 lb/min (approximately 13,500 ft³/min) [543.6 kg/min (382.3 m³/min)].

Since the mass flows of air and water are equal, 1 lb (0.4536 kg) of air can be said to contact 1 lb (0.4536 kg) of water, and the psychrometric path of one such unit of air has been traced in Fig. 19.4.10 as it moves through the tower.

Air enters the tower at condition 1 [65°F (18.3°C) wet bulb and 78°F (25.6°C) dry bulb] and begins to gain enthalpy (total heat) and moisture content in an effort to achieve equilibrium with the water. This pursuit of equilibrium (solid line) continues until the air exits the tower at condition 2. The dashed lines identify the following changes in the psychrometric properties of this pound of air due to its contact with the water:

- The total heat content (enthalpy) increased from 30.1 Btu (7.59 kcal) to 45.1 Btu (11.37 kcal). This enthalpy increase of 15 Btu (3.78 kcal) was gained from the water. Therefore, 1 lb (0.4536 kg) of water was reduced in temperature by the required amount of 15°F (8.3°C).

- The air's moisture content increased from 72 grains (gr) [4.67 grams (g)] to 163 gr (10.56 g) (7000 gr = 1 lb). These 91 gr (5.89 g) of moisture (0.013 lb of water) were evaporated from the mass flow of water at a latent heat of vaporization of about 1000 Btu/lb (556 kcal/kg). This means that 13 of the 15 Btu (3.28 of the 3.78 kcal) removed from the water (86 percent of the total) occurred by virtue of *evaporation*.

Note: Water's latent heat of vaporization varies with temperature from about 1075 Btu/lb at 32°F (597.7 kcal/kg at 0°C) to 970 Btu/lb at 212°F (539.3 kcal/kg at 100°C). Actual values at specific temperatures are tabulated in various thermodynamics manuals.

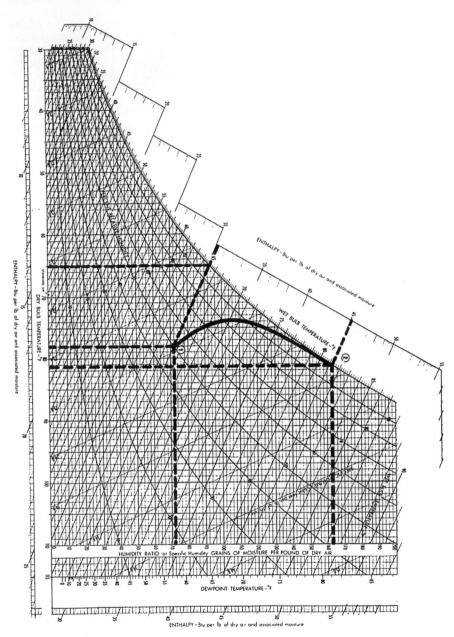

FIGURE 19.4.10 Psychrometric chart.

19.4.3.4 Effects of Variables

Although several parameters are defined in Fig. 19.4.9, each of which will affect the size of a tower, understanding their effect is simplified if one thinks only in terms of (1) heat load, (2) range, (3) approach, and (4) wet-bulb temperature. If three of these parameters are held constant, changing the fourth will affect the tower size as follows:

1. Tower size varies directly and linearly with heat load (Fig. 19.4.11).
2. Tower size varies inversely with range (Fig. 19.4.12). Two primary factors account for this. First, increasing the range (Fig. 19.4.9) also increases the temperature differential between the incoming hot-water temperature and the entering wet-bulb temperature.

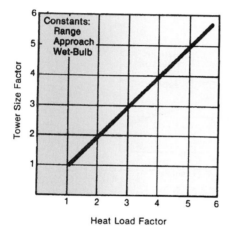

FIGURE 19.4.11 Effect of heat load on tower.

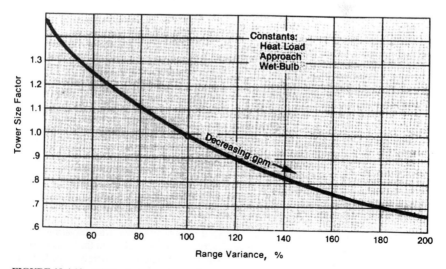

FIGURE 19.4.12 Effect of range on tower size.

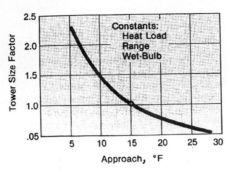

FIGURE 19.4.13   Effect of approach on tower.

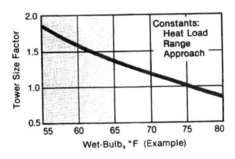

FIGURE 19.4.14   Effect of wet bulb on tower.

Second, as evidenced by Eq. (19.4.2), increasing the range at a constant heat load requires that the water flow rate be decreased. The resultant decrease in internal static pressure increases the flow of air through the tower.

3. Tower size varies inversely with approach (Fig. 19.4.13). A longer approach requires a smaller tower. Conversely, a closer approach requires an increasingly larger tower, and at a 5°F (3°C) approach the effect on tower size begins to become asymptotic. For this reason, *it is not customary in the cooling tower industry to guarantee any approach of less than 5°F (3°C).*

4. Tower size varies inversely with wet-bulb temperature. When the heat-load, range, and approach values are fixed, reducing the design wet-bulb temperature increases the size of the tower (Fig. 19.4.14). This is because *most* of the heat transfer in a cooling tower results from evaporation, and air's ability to absorb moisture is reduced as the temperature is reduced.

19.4.4   COOLING TOWER FILL

Although cooling tower fill is often acceptably referred to as a "heat-transfer surface," such terminology is not true in its strictest sense. The heat-transfer surface in the classic cooling tower is actually the exposed surface of the water itself. The fill is merely a medium that causes more water surface to be exposed to the air (increasing the *rate* of heat transfer) and increases the time of air-water contact by retarding the progress of the water (increasing the *amount* of heat transfer).

At a given rate of air moving through a cooling tower, the extent of heat transfer that can occur depends on the amount of water surface exposed to that air. In the tower shown in Fig. 19.4.8, the total exposure consists of the cumulative surface areas of a multitude of random-size droplets, the size of which depends on the type of nozzle utilized and on the pressure at which the water is sprayed. Higher pressure normally produce a finer spray and greater total surface-area exposure.

However, droplets contact each other readily in the overlapping spray patterns and coalesce into larger droplets, which reduces the net surface-area exposure. Consequently, predicting the thermal performance of a spray-filled tower is difficult at best and is highly dependent on good nozzle design as well as on a constant water pressure.

This relationship between water-surface exposure and heat-transfer rate is intrinsic to all types of towers, regardless of the type of water distribution system utilized. The more water surface exposed to a given flow of air, the greater will be the rate of heat transfer.

For a specific heat-dissipation problem, however, the *rate* at which enthalpy will be exchanged is important, in the sense that it allows the designer to predict a finite *total* exchange of heat, with the total being a function of the period of time that the air and water are in intimate contact. Within psychrometric limits, the longer the contact period, the greater will be the total exchange of heat and the colder will be the cold-water temperature.

In the spray-filled tower, the total time of air-water contact can only be increased by increasing the height of the tower, causing the water to fall through a greater distance. Given a tower of infinite height, the cold-water temperature produced by that tower would equal the incoming air's wet-bulb temperature, and the leaving air temperature would equal the incoming hot-water temperature. (These are the psychrometric limits previously mentioned.)

Obviously, a tower of infinite height would cost an infinite amount of money. More practically, structural limitations would begin to manifest themselves very early in the attempt toward infinite height. Early cooling tower designers quickly discovered these limitations and devised the use of fill as a far better means of increasing not only the rate of heat transfer, but its amount as well.

19.4.4.1 Types of Fill

The two basic types of fill utilized in present-day cooling towers are the splash type (Fig. 19.4.15) and the film type (Fig. 19.4.16). Either type may be used in towers of both counterflow and crossflow configuration, situated within the towers as shown in Figs. 19.4.4 and 19.4.5, respectively.

Both types of fill exhibit advantages in varied operating situations, assuring that neither type is likely to endanger the continued utilization of the other. Offsetting cost comparisons tend to keep the two types competitive, and the operational advantages peculiar to a specific situation are usually what tip the scales of preference. Therefore, specifiers are cautioned against excluding either type unless there are overriding reasons for doing so.

Splash Type. Splash-type fill (Fig. 19.4.17) causes the flowing water to cascade through successive elevations of parallel "splashbars." Equally important is the increased time of air-water contact brought about by repeated interruption of the water's flow progress.

Since the movement of water within a cooling tower is essentially vertical, splash-type fill must obviously be arranged with the wide dimension of the splashbars situated in a horizontal plane; otherwise, maximum retardation and breakup of the water could not be realized. When wood splashbars, for example, are utilized, they are typically $^3/_8$ in (1 cm) thick

FIGURE 19.4.15   Splash fill, installed.

by $1^{1}/_{2}$ in (3.8 cm) wide by about 4 ft (1.2 m) long, and, as situated in the tower, only the $^{3}/_{8}$-in (1-cm) dimension is vertical. Consequently, splash-type fill provides the least opposition to air flow in a horizontal direction, which accounts for the fact that splash-type fill is not routinely applied in counterflow towers.

Because of the water dispersal that takes place within splash-type fill, splash-filled towers are quite forgiving of the poor initial water distribution that can result from clogged or

FIGURE 19.4.16   Film fill, installed.

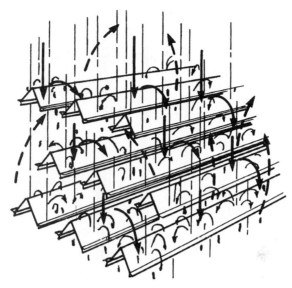

FIGURE 19.4.17 Splash-fill concept.

missing nozzles. The splashing action effectively redistributes the water at each level of splashbars. The relatively wide spacing of splash-type fill renders it ideal for use in a contaminated-water situation, where fills of closer spacing would be subject to clogging.

Splashbars are typically manufactured of polyvinyl chloride (PVC), polypropylene, polyethylene, or treated wood. Occasionally they are made of rolled aluminum or stainless steel to withstand high temperatures or particularly aggressive water conditions. Splashbars are normally supported by fiber-reinforced polyester (FRP) grids.

Film Type. Film-type fill has gained prominence in the cooling tower industry because of its ability to expose greater water surface within a given packed volume. Hence, film-filled towers tend to be somewhat smaller than splash-filled towers of equal performance. Film fill is equally effective in both crossflow and counterflow towers.

As can be seen in Fig. 19.4.18, water flows in a thin "film" over vertically oriented sheets of PVC fill, which are usually spaced approximately $^3/_4$ in (1.9 cm) apart. For purposes of clarity, the fill sheets in Fig. 19.4.18 are shown as if they were flat. In actuality, these sheets are usually molded into corrugated, or "chevron," patterns (Fig. 19.4.16) to create a certain amount of turbulence within the air stream and to extend the exposed water surface.

Unlike splash fill, film-type fill affords no opportunity for the water to redistribute itself during its vertical progress. Consequently, uniformity of the initial water distribution at the top of the fill is of prime importance, as is vigilant maintenance of the water distribution system. Areas deprived of water will become unrestricted paths of useless air flow, and thermal performance will degrade.

Where normally "clean" circulating water is anticipated, film-type fill will usually be the proper choice. However, the narrow passages afforded by the close spacing of fill sheets makes film fill very susceptible to poor water quality. High turbidity, leaves, debris, or the presence of algae, slime, or a fatty acid condition can diminish the passage size and affect the heat-transfer efficiency.

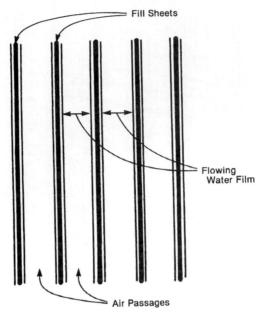

FIGURE 19.4.18   Film fill concept.

19.4.5   *EXTERNAL INFLUENCES*
ON PERFORMANCE

Unlike other components comprising the total air-conditioning system, the cooling tower is normally exposed to the vagaries of wind and weather, which can affect its level of performance. Being a "breathing" device, it can also be affected by abnormalities relating to both the quality and quantity of its air supply.

Although cooling tower manufacturers strive for designs that will minimize the impact of such external forces, it is ultimately incumbent upon the site designer (1) to allocate an adequate, properly configured space for the cooling tower or (2) to adjust the thermal design conditions accordingly. This section discusses the first of these options, and Sec. 19.4.6 discusses the second.

19.4.5.1   Recirculation

The air moving through a cooling tower undergoes considerable increase in wet-bulb temperature, compared to its entering condition. Occasionally a portion of this warm, saturated, leaving air stream will be reintroduced into the tower, causing an artificial elevation of the entering wet-bulb temperature and a resultant increase in the tower's cold-water temperature. This condition is called "recirculation," and the potential for it to occur is directly related to the type of tower utilized and to its orientation with respect to the prevailing wind.

In the induced-draft tower (Fig. 19.4.3), discharge air leaves the tower at a velocity of approximately 2000 ft/min (10 m/s), whereas the intake velocity is usually less than 700 ft/min (3 m/s). This velocity differential essentially precludes self-imposed recirculation. However, the velocity relationships in the forced-draft tower (Fig. 19.4.2) are the reverse of those encountered in the induced-draft tower. Air enters the fan at a velocity exceeding 2000 ft/min (10 m/s) and leaves the tower at velocities normally less than 700 ft/min (3 m/s). The high entrance velocity evacuates a low-pressure zone at the intake, in which a part of the billowing exhaust is often recaptured.

When flowing wind encounters an obstruction, such as a cooling tower, the normal path of the wind is disrupted and a reduced-pressure zone forms on the lee side (downwind) of the obstruction. In trying to fill this "void," the wind will attempt to carry with it the tower's saturated exhaust air, which has been suitably deflected in the direction of the trouble zone. The wind's success or failure in this attempt depends, of course, on the velocity of the air leaving the tower. Higher-velocity discharges obviously suffer less downwind deflection than do lower-velocity discharges.

Whether or not the plume deflection results in recirculation also depends on the nature of the downwind tower face. If that face is an air intake, some degree of recirculation will definitely occur. If it is a cased (solid) face, there will be small likelihood of recirculation. Therefore, proper orientation of the tower with respect to the prevailing wind (coincident with the design wet-bulb temperature) is very important. Normally, seasonal shifts in wind direction are of little concern because they are usually accompanied by reductions in the wet-bulb temperature.

19.4.5.2 Interference

Intervening heat sources of higher enthalpy content located upwind of a cooling tower can also cause an elevation in the wet-bulb temperature of the entering air. These sources, most often, are other cooling towers, whose discharge air (depressed by the wind) "interferes" with the thermal quality of the air supplied to the tower in question. Where location of an intended cooling tower downwind of such a heat source is unavoidable, the design wet-bulb temperature should be adjusted, or higher system operating temperatures should be anticipated.

19.4.5.3 Air Restrictions

In residential, commercial, and small industrial installations, cooling towers are frequently shielded from view with barriers or enclosures. Usually this is done for aesthetic reasons. Quite often these barriers restrict air flow, resulting in low-pressure areas (inviting recirculation) and poor air distribution to the tower's air inlets. Sensible construction and placement of screening barriers will help minimize any negative effect on the thermal performance of the cooling tower.

Screening in the form of shrubbery, fences, or louvered walls should be placed several feet from the air inlet to allow normal air entry into the tower. When a tower is enclosed, it is desirable for the enclosure to have openings opposite each air-intake face and for the net free area of the openings to be at least equal to the gross intake area of that tower face.

Although recommended clearances vary with the type of enclosure under consideration, the basic rule of thumb for an induced-draft tower is to place the enclosure wall no closer to the tower than the gross height of the tower's air-intake opening. For forced-draft towers, this distance should be at least three times the air-intake height. However, since these recommended distances tend to increase with tower length and number of operating cells, screening barriers or enclosures should not be installed without prior consultation with the cooling tower manufacturer.

19.4.6 CHOOSING THE DESIGN WET-BULB TEMPERATURE

Selection of the design wet-bulb temperature must be made on the basis of conditions existing at the site proposed for the cooling tower. It should be that which will result in the maximum acceptable cold-water temperature at, or near, the time of peak load demand.

Performance analyses have shown that most air-conditioning installations based on wet-bulb temperatures that are exceeded in no more than 1 percent of the total hours during a normal summer have given satisfactory results. The hours in which peak wet-bulb temperatures exceed the upper 1 percent level are seldom consecutive hours and usually occur in periods of relatively short duration. The "flywheel" effect of the total water system inventory is usually sufficient to carry through the above-normal periods without detrimental results.

Air temperatures, wet-bulb as well as coincident dry-bulb, are routinely measured and recorded by the U.S. Weather Bureau, worldwide U.S. military installations, airports, and various other organizations to whom anticipated weather patterns and specific air conditions are of vital concern. Compilations of these data exist that are invaluable to both the users and designers of cooling towers.

However, it must be realized that the wet-bulb temperature determined from these publications represents the ambient for a geographic area and does not take into account localized heat sources and the potential for recirculation, which may artificially elevate that temperature at a specific site. Where local effects and anticipated recirculation are known, the design wet-bulb temperature must be increased by an appropriate amount. Where doubt exists, the design wet-bulb temperature should be increased by 1°F (0.5°C) for an induced-draft tower, or 2°F (1°C) for a forced-draft tower. (Readers interested in the derivation of this "rule" are referred to "External Influences on Cooling Tower Performance," *ASHRAE Journal*, January 1983.)

19.4.7 TYPICAL COMPONENTS

Although techniques and technology vary among cooling tower manufacturers, in the realm of mechanical-draft types (which predominate in air-conditioning applications) certain primary and essential components are common to all. Each, for example, will assume a modified box shape designed to contain and control a transient flow of water while, at the same time, permitting the relatively free passage of a continuous flow of air. Each will also be equipped with a fan to provide that amount of air, and virtually all will be equipped with fill to promote intimate contact between air and water.

Beyond these essential concerns, means must be provided to recover the flowing mass of water and replenish that which is lost, to assure mechanical operation within design limits, to promote thermal performance stability, to simplify maintenance, and to reduce adverse environmental impact. Components devoted to these considerations are described in this section. Fill was discussed in Sec. 19.4.4, and fans will be discussed in Sec. 19.4.9.

19.4.7.1 Cold-Water Basin

The cooling tower basin serves the two fundamentally important functions of (1) collecting the cold water following its transit of the tower and (2) acting as the tower's primary foundation. Because it also functions as a collecting point for foreign material washed out of the air by the circulating water, it must be accessible, cleanable, have adequate draining facilities, and be equipped with suitable screening to prevent entry of debris into the suction-side piping.

Ground-level installations, typical of many larger projects, may utilize concrete basins. Occasionally, concrete basins are also used on rooftop installations, with the basin slab either poured integrally with the roof or separated by a waterproof membrane. In all cases, concrete basins are designed by the structural engineer for construction by the general contractor, based on dimensional drawings and load schedules provided by the cooling tower manufacturer. Contiguous sumps (Fig.19.4.19) provide outflow submergence and house the facilities for overflow and cleanout.

Factory-assembled towers (Figs. 19.4.6 and 19.4.7), as well as elevated field-erected towers (Fig. 19.4.20), are normally equipped with basins provided by the cooling tower manufacturer. The materials utilized are compatible with the tower's overall construction; they include wood, steel, stainless steel, and plastic. For towers that come equipped with basins, the cooling tower manufacturer will include drain and overflow fittings, makeup valve(s), sumps and screens (Fig. 19.4.21), and provision for anchorage. The basin of the factory-assembled tower shown in Fig. 19.4.22 merely dumps by gravity into a concrete collection basin, the side curbs of which serve to support the tower.

A grillage of steel or concrete is normally utilized for support of a tower installed over a wood, steel, or plastic basin (Fig. 19.4.20). Grillages must be designed to withstand the total wet operating weight of the tower and connecting piping, plus the dead loads contributed by stairways, catwalks, etc. They must also accept transient loads attributable to winds, earthquakes, and maintenance traffic. The grillage members must be level and of sufficient strength to preclude excess deflection under load.

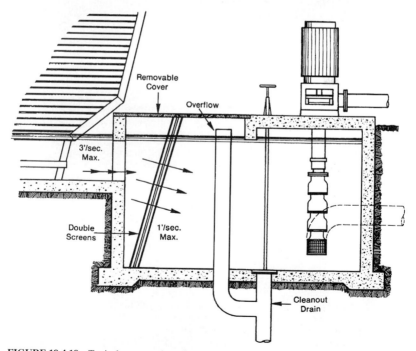

FIGURE 19.4.19   Typical cross-section of concrete sump pit.

FIGURE 19.4.20   Steel grillage supporting tower.

FIGURE 19.4.21   Plywood cold-water basin floor.

FIGURE 19.4.22   Two-cell factory-assembled tower.

19.4.7.2   Louvers and Drift Eliminators

Water management and control are of prime concern to the cooling tower designer. Where air enters or leaves the tower, water also has the opportunity to escape. On the air-intake sides of induced-draft towers, louvers are devised to prevent the escape of random water droplets. Also, because a cooling tower promotes maximum contact between water and relatively high-velocity air [normally 500 to 700 ft/min (2.5 to 3.5 m/s)], water droplets become entrained in the leaving air stream. Collectively, these entrained water droplets are called "drift" (and are not to be confused with the pure water vapor with which the effluent air stream is saturated or with any droplets formed by the condensation of that vapor). The composition and quality of drift is the same as that of the circulating water flowing through the tower. Its potential for nuisance in the spotting of cars, windows, and buildings is considerable. Located upwind of critical areas, a cooling tower producing significant drift can pose an operating hazard.

Drift eliminators remove entrained water from the discharge air by causing it to make sudden changes in direction. The resultant centrifugal force separates the drops of water from the air and deposits them on the eliminator surface, from which they flow back into the tower.

As with fill, PVC has become the dominant material from which to manufacture louvers and drift eliminators. They are normally formed into a honeycomb configuration with labyrinth passages and, for use with crossflow towers, can be molded integrally with the fill sheets (Fig. 19.4.23). Although current industry standards continue to limit allowable drift to 0.2 percent of the circulating water rate, actual drift rates seldom exceed 0.02 percent with present technology.

19.4.7.3   Fan Drive Mechanisms (See also Chap. 23)

The optimum speed of a cooling tower fan seldom coincides with the most efficient speed of the driving motor. Even in the smaller fan sizes, design speeds are usually less than 500 rpm, whereas motors are usually applied at a nominal speed of 1800 rpm. Speed reduction is therefore necessary, and it is usually accomplished either by differential gears of positive engagement or by differential pulleys connected through V-belts.

FIGURE 19.4.23   Film fill.

FIGURE 19.4.24   Fan and drive mechanism mounted on support assembly.

Typically, gear reduction units are of the right-angle type, coupled to the motor through an extended drive shaft (Fig. 19.4.24). This permits the motor to be located outside the fan cylinder where it is more accessible (Fig. 19.4.25) and operates in a less aggressive environment. Gear reduction units are applied throughout a wide range of operating powers, from 250 hp (175 kW) and larger down to as little as 5 hp (3 kW). V-belt drives, on the otherhand, are seldom applied at powers greater than 50 hp (35 kW) in cooling towers. Prudent limiting of belt lengths requires that the motor be located in the hot, humid atmosphere within the tower. In addition to complicating the routine maintenance of the motor, this tends to increase the frequency of necessary belt adjustment to reduce the energy loss associated with belt slippage.

FIGURE 19.4.25   External motor location.

19.4.8 MATERIALS OF CONSTRUCTION

The structures of cooling towers are typically made of wood or steel, depending on one's preference and on local building or fire codes. Casings, basins, and decks are usually of like material, with casings also being manufactured of fire-retardant glass-reinforced polyester plastic. Woods used are Douglas fir and redwood, either of which should be pretreated to prevent fungal attack. Being the single most available and most renewable structural material, wood is expected to continue as one of the predominant cooling tower materials for the foreseeable future.

Because it lends itself well to plant manufacturing techniques, steel is the principal material used in factory-assembled cooling towers. Appropriate grades of carbon steel are utilized for framing, casing, decking, hot- and cold-water basins, etc., and are usually galvanized for protection against corrosion. Minimum protection offered in the industry consists of G-90 galvanizing, ranging to G-210 galvanizing at the premium end of the scale. In addition, coatings are often applied over the galvanizing in an effort to extend its useful service life.

Although some towers in smaller sizes have been produced primarily in plastic, their susceptibility to fire and to ultraviolet deterioration has discouraged their widespread use. With technological advances in the plastics industry, such towers may well become more prevalent in the future.

Currently, and for the foreseeable future, those who seek to obtain a cooling tower offering the longest possible service life typically specify stainless steel. Such towers combine the fire resistance, corrosion immunity, and service longevity sought by the nonspeculative buyer. Furthermore, the increased use of stainless steel, plus economic pressures in the steel market as a whole, have tended to decrease the premium price differential. Therefore, the movement toward cooling towers of higher-quality construction can be expected to increase.

19.4.9 ENERGY MANAGEMENT AND TEMPERATURE CONTROL

Energy is consumed by a cooling tower in the operation of fans for the movement of air. It is also consumed within the system served by the cooling tower in the operation of condenser water pumps. Of these two energy-consuming aspects, fan manipulation offers the most productive means of controlling temperature—and thereby of controlling energy use.

Unless specifically permitted by the cooling tower manufacturer, *pump manipulation to vary water flow over the tower should not be used*. This is because the distribution system in a given tower is designed to produce optimum fill coverage at a specific water flow, and variations in that flow tend to reduce the tower's thermal efficiency.

Before attempting to manage energy use by the manipulation of fans, however, the operator should make sure that such manipulation is not premature—and therefore not self-defeating. Observing that the cold-water temperature produced by a cooling tower reduces as the outside wet-bulb temperature reduces (Fig. 19.4.26), many operators will begin cycling fans without regard to the effect that this may have on the overall system's energy consumption.

Bear in mind that the system compressor horsepower (wattage) is perhaps 20 times greater than that consumed by the cooling tower fans, and some air-conditioning systems reward a reduction in condenser water temperature by a commensurate reduction in compressor horsepower (wattage). Although those systems which do benefit from reduced condenser water temperature usually have some limiting temperature [perhaps 75°F (24°C)]

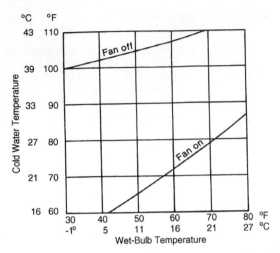

FIGURE 19.4.26 Typical performance curve—single-cell tower with single speed motor.

below which no further reduction in compressor horsepower (wattage) is realized, the operator should allow that cold-water temperature from the tower to be reached before undertaking any fan manipulation.

19.4.9.1 Fan Cycling

The ability to achieve effective air-side control depends primarily on the number of cells comprising the tower and on the speed-change characteristics of the motors driving the fans. Figure 19.4.26 defines the operating modes available with a single-cell tower equipped with a single-speed motor. In this most rudimentary of cases, control of the cold-water temperature can be attempted only by cycling the fan motor on and off, and great care must be exercised to prevent an excessive number of starts from burning out the motor. (As a basic rule of thumb, 30 s/h of starting time should not be exceeded.)

Conversely, the operating characteristics of a three-cell tower equipped with single-speed and two-speed motors are shown in Figs. 19.4.27 and 19.4.28, respectively. The numbers in parentheses represent the approximate percentage of total fan power consumed in each operating mode. Note that the opportunity for both temperature control and energy management is tremendously enhanced by the use of two-speed motors. At any selected cold-water temperature, it can be seen that an increase in the number of fan-and-speed combinations causes the operational mode lines to come closer together. It follows, therefore, that the capability to modulate a fan's speed or capacity (within the range from 0 to 100 percent) would represent the ultimate in temperature control and energy management.

The technology by which to approach this ideal situation currently exists in the form of the automatic variable-pitch (AVP) fans (Fig. 19.4.29) and electrical frequency modulating devices provided by several manufacturers. Figure 19.4.30 indicates the cold-water temperature control and energy reductions that can be achieved with AVP fans. For purposes of the curve, it is assumed that the cold-water temperature (dashed line) has been set to reduce no lower than 75°F (24°C). In a reducing wet-bulb temperature, fan power (solid line) will remain at 100 percent until such time as that cold-water temperature is reached,

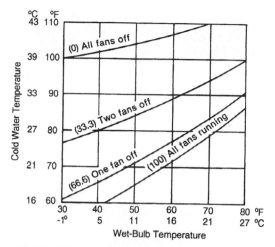

FIGURE 19.4.27 Typical performance curve—three cell tower with single-speed motors.

after which the fan automatically depitches to maintain that cold-water temperature. In this typical case, note that a further reduction of only 20°F (11°C) in wet-bulb temperature results in an 80 percent reduction in fan power.

Were it not for the physical characteristics of standard fans as well as the limitations of their drive mechanisms, results similar to those shown in Fig. 19.4.30 could be achieved with frequency modulation control devices. Most fans have at least one critical speed that occurs between 0 and 100 percent of design rpm, and some have several. Typically, fans are designed such that their critical speeds do not coincide with the rpm's that will be produced by normal motor speed changes. They are also, of course, designed to minimize the

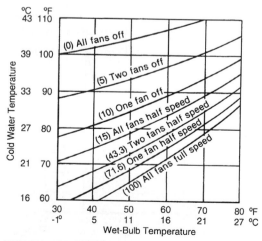

FIGURE 19.4.28 Typical performance curve—three-cell tower with two-speed motors.

FIGURE 19.4.29   Automatic variable-pitch (AVP) fan.

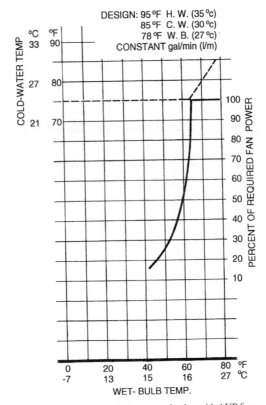

FIGURE 19.4.30   Typical power reduction with AVP fans.

effect of any critical speed. Nevertheless, it would be foolhardy to run the risk of protracted operation at a critical speed; therefore, it becomes necessary to predetermine the critical speeds and to prevent corresponding frequencies from being utilized. Avoiding critical speeds, however, reintroduces some miniature "notches" in an otherwise constant cold-water temperature line and, to avoid "hunting," requires that some appreciable tolerance be designed into the control mechanism.

19.4.9.2 Free Cooling

Because of increased computer loads, process loads, and the need for "core cooling" of larger buildings, air-conditioning systems are being used for longer periods of the year and, in many cases, year-round. The ability of cooling towers to produce water temperatures typical of chilled-water temperatures in colder months has led to an increase in the utilization of "free cooling" as the best means to reduce overall system energy demand.

Direct Free Cooling. The simplest and most thermally effective—*yet least recommended*—arrangement for free cooling is shown in Fig. 19.4.31, where a simple bypass system physically interconnects the chilled-water and condenser water loops into one common water path between the load and the cooling tower. The dashed lines indicate the water flow path during the free-cooling mode of operation.

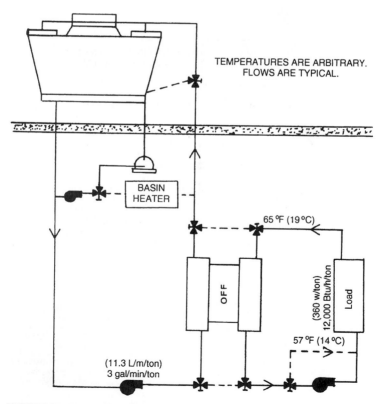

FIGURE 19.4.31 Direct free-cooling system.

The absence of a heat exchanger separating the two water loops precludes the need for a temperature differential, so the load benefits from the cooling tower's full capability.

The reason why this direct system is least recommended is because the intermixing of the two water streams contaminates the "clean" chilled water with "dirty" condenser water, a situation most operators are reluctant to allow. Those who have been most successful in the utilization of this direct-connected system have used a "side stream" filtration arrangement to filter a portion of the total water flow continuously. In the figure, the flow from the small pump would be directed through the filter during tower operation and through an instantaneous heater during periods of shutdown.

Indirect Free Cooling.   By the simple expedient of installing a heat exchanger, piped in a parallel bypass circuit with the chiller, free cooling is accomplished without concern for contamination of the chilled-water loop (Fig. 19.4.32). Because plate heat exchangers can function properly with only a small temperature difference [as little as 2°F (1°C), depending on size], they permit separation of the water loops with minimal sacrifice of free-cooling opportunity.

Tower Selection for Free Cooling.   One of the prime advantages of free cooling is that it can be applied to many existing systems. Most towers are capable of producing the

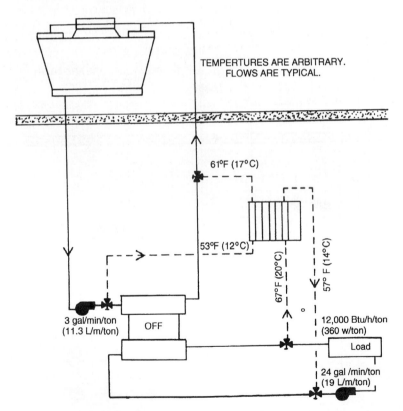

FIGURE 19.4.32   Indirect free-cooling system.

temperatures required at some point in the year. Designers of new systems, however, have the opportunity to take free cooling into account early in the design and maximize free cooling to its fullest extent. Those who do usually choose a tower size that will produce a necessary "chilled"-water temperature at the highest possible wet-bulb temperature. This ensures that compressor energy will be avoided for the maximum number of annual operating hours.

19.4.10 WINTERTIME OPERATION

Although cooling towers have always been used 12 months of the year in process industries, such usage has only recently become prevalent in the air-conditioning industry. Accordingly, operators who have been accustomed to shutting down their tower in October and restarting it in May must now concern themselves with the fact that the tower may become a potential ice-maker in freezing weather.

19.4.10.1 Periods of Shutdown

The most obvious time of concern, of course, is when the system is shut down for an interim period of time (i.e., nights, weekends, etc.). During these shutdown periods, no system heat is being added to the condenser water (the water is stagnant) and freezing is free to occur in the cooling tower basin as well as in all exposed lines. At these times, the operator basically has two possible courses of action: (1) drain the system to the point where all outside components are empty of water or (2) find a way to add heat to the exposed portion of the system.

With proper forethought, system draining can become an automatic process. The "indoor tank" method shown in Fig. 19.4.33 allows water in the tower's cold-water basin to drain continuously into an indoor storage tank, from which the condenser water pumps

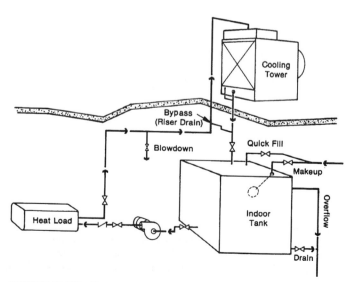

FIGURE 19.4.33 Use of indoor storage tank.

take suction. The small bypass drain line connecting the main supply and return lines, just below the "roof" level, is there to ensure drainage of the tower supply line at the time of pump shutdown. This bypass line may be equipped with an automatic valve designed to open on pump shutdown, or it may merely be a relatively small open line that would allow a limited amount of warm water to bleed continuously into the tank.

Where space does not allow the use of an indoor tank, heat must be added to the exposed system by means of some type of temperature-actuated device. Foremost among the systems used are electric immersion heaters submerged in the water of the cooling tower basin, although steam or hot-water loops are also routinely utilized. Where electric immersion heaters are used, care must be taken to ensure that the heater element is always adequately covered with water. Such heater elements operate at about 1500°F (815°C) in the open air and constitute an extremely high fire risk.

Where such devices are used in the cooling tower basin, the designer must realize that they give no protection to exposed piping. Therefore, all exposed piping must be heat-traced and insulated.

A very effective way to protect both the cooling tower basin and the primary exposed piping is shown in Fig. 19.4.31. Note the presence of a small bypass pump and an instantaneous heater. During shutdown periods, this pump would be temperature-actuated to pump a small amount of water through the heater and back to the cooling tower basin. Note also that a bypass valve has been added to the tower riser to divert this small flow directly into the cooling tower basin; this is to prevent the water from going over the top of the tower, where it could induce freezing on the fill. In the case of this system, the designer must remember that the exposed makeup line to the tower gets no protection from freezing and must make provisions accordingly.

19.4.10.2 Periods of Operation

Manipulation of the air flow (see Sec. 19.4.9) is an invaluable tool, not only in the retardation of ice formation but also in the reduction or elimination of ice already formed. In addition to bringing less cold air into contact with the circulating water, reducing the entering air-flow velocity alters the path of the falling water, allowing it to impinge upon—and melt—ice previously formed by those random droplets which wind gusts or normal splashing may have caused to escape the protection of the relatively warm main stream of water.

The falling-water patterns associated with each type of tower (crossflow and counterflow) have much to do both with the type of ice formed and with its location.

Crossflow Towers.   Crossflow towers equipped with splash-type fill tend to form louver ice, wherein water droplets generated by the splashing action may escape the fill area, impinge upon the louvers, and be frozen almost instantaneously. Usually this ice can be controlled merely by reducing the fan's speed or by turning the fan off for a short period of time. Severe cases may require that the fan be reversed for an interim period. This causes warmed air to flow outward through the louvers to effect deicing. However, since this also causes frigid air to flow downward through the fan, the amount of time in this operating mode must depend on the tendency for ice to form on the fan cylinder and mechanical equipment. Therefore, fan reversal should be limited to the minimum time required to accomplish acceptable deicing and should be monitored by the operator.

Also, the direction of a fan's rotation should not be changed without allowing the fan to come essentially to a stop; otherwise, tremendous stresses will be imposed on the entire drive train. Prudence dictates that the starter be equipped with a 2-min time delay between directional changes.

Used in crossflow configuration, PVC film-type fill also tends to limit the amount of ice formed. This is due to the absence of any splashing action within the fill. Typically, thin ice may tend to form on the lower leading edges of the fill, turn to slush as the air flow becomes blocked, and shortly disappear. This is particularly true of current fill configurations having louvers molded integrally with the fill. Given normal control measures, these towers have proved to be very civilized in their wintertime operation. This can be very important in the low-load, low-temperature situations encountered in free cooling.

Counterflow Towers. Since the fill of a counterflow tower is elevated appreciably above the cold-water basin's level, the generation of random water droplets produced by this free-fall tends to be irrespective of the type of fill utilized. Droplets that splash in an outboard direction may freeze on the louvers and lower structure.

Deicing measures for counterflow towers are similar to those utilized for cross-flow towers but tend to be somewhat less effective. With the fans off, the normally vertical sides of a counterflow tower place the air inlets beyond the reach of the falling-water pattern. In this fans-off mode, only the convective heat from the water promotes deicing, and the process may take considerably longer than with a cross-flow tower. Although fan reversal can also be effectively utilized for deicing of a counterflow tower, many operators are reluctant to reverse fans because of the small amount of water caused to escape the air inlets by the outward flow of air; for this water may produce sufficient ice in the immediate region of the tower to be considered hazardous, requiring separate measures for its control.

Fan Options. Single-speed motors afford the least opportunity for air-flow variation, and towers so equipped require maximum vigilance on the part of the operator to determine the proper cyclic operation of the fans to accomplish the best ice control.

Two-speed fan motors offer significantly greater operating flexibility and should be given maximum consideration in the purchase of towers for use in cold climates. To achieve a balance between cooling effect and ice control, the fans may individually be cycled back and forth between full speed and half speed, as required; this practice is limited only by the maximum allowable motor insulation temperature (which temperature could be exceeded if there were an abnormal number of speed changes per hour). In many cases, it will be found that operating all the tower fans at half speed produces the best combination of cooling effect and ice control.

Operators who wish to be relieved of the burden of selecting the proper air flow for any operating situation may choose such devices as the AVP fan (Fig. 19.4.29) described in Sec. 19.4.9. In most operating cases, by the time the ambient air temperature has dropped low enough to produce ice, the fan will have depitched to a much reduced air-flow capability, making the formation of ice unlikely. By the simple expedient of adjusting the cold-water temperature control point, operators of sensitive systems can easily strike a balance between maximum cooling and minimum icing.

For water treatment and blowdown rates, see Chap. 8.

19.4.11 ROLE OF THE COOLING TOWER IN THE HVAC SYSTEM

In the simplest of terms, a cooling tower is the vehicle that dissipates heat that is generated by equipment to the atmosphere. During the process of working, most air-conditioning and industrial systems produce various amounts of heat. To extract this heat from the exchanger or condenser, water is the most commonly used cooling mechanism.

Figure 19.4.34 is a simple look at the cooling tower and its relationship to the air-conditioning unit.[1] The cooling water flows from the cold well of the tower to the condenser, thus removing heat from the condenser. As the unit works and generates heat, water flows through the unit. The heated water is then returned to the tower for cooling. The water circulates over the top of the unit, and as it falls through packing inside the tower, it is contacted with air to facilitate the removal of the heat.

In the past, cooling was accomplished by using streams of water, either from a well or nearby water source. Once the water collected the heat to be removed, it returned to the stream or river and the process was repeated. This circular process required an unlimited supply of water and quickly became expensive. The role of environmental agencies emerged when concerns arose about the heated water and its return to rivers and streams. The heated water, it was concluded, may have an effect on the wildlife.

Because of these environmental concerns, and most importantly, because the supply of water is limited, the cooling tower was introduced as an efficient means to control a precious resource.

On average, cooling towers reduce the amount of water necessary for cooling a once-through system to less than 5 percent.[2] Towers can also cool to within 5°F (3°C) of the ambient wet bulb temperature. A cooling tower cools by combining water and air. Typically, water is distributed over such things as a heat transfer media, a packing material of wood, or PVC. As it falls through the media, air is drawn across or past that fill material. The water droplets are exposed to the passing air and the transfer of heat is accomplished. The air can be introduced either through fans or by natural draft. Either method requires an adequate amount of air to properly contact the water. When using a motorized fan to introduce the air, issues of energy consumption arise.

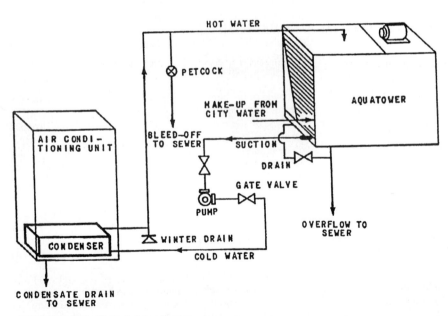

FIGURE 19.4.34   Typical piping diagram.

When water flows and horsepower moves the cooling tower fan, the tower must be clean and properly maintained to be efficient and in the best possible working condition.

19.4.12 IMPACT OF THE COOLING TOWER ON THE SYSTEM

Cooling tower manufacturers design the unit for a given set of performance conditions, such as by the type of chiller used, ambient temperatures, location, and specifications. As the system ages it may lose efficiency. *The cleanliness of the tower and its components become crucial to the success of the system.* If unattended, the cold water temperature will rise, sending warmer water to the chiller. When the chiller kicks out on high head pressure, the system may shut down. Certain precautions can be taken to keep this shutdown from occurring.

Cooling towers are usually remotely located; it becomes necessary to regularly inspect and clean the tower according to the manufacturers' recommendations. The few hours each month spent on inspecting the cooling tower and maintaining it will pay dividends. The life of a tower varies according to materials of construction, location within the system, and the location of the city or country.

Typically, the premium materials of construction are wood, concrete, stainless steel, and fiberglass. These units are expected to last from 20 to 30 years if properly cared for. The less expensive units, made of galvanized steel, will operate for 8 to 20 years. Of course, tower life will vary due to the extremes of weather, number of hours used each year, and type of water treatment. It is sufficient to say that in order to get the most out of the tower, cooling tower manufacturers want to make that tower last as long as possible.

19.4.13 TYPES OF COOLING TOWERS

In the air conditioning world there are four types of cooling towers that play key roles: induced draft crossflow (single flow), induced draft crossflow (double flow), induced draft counterflow, and counterflow forced draft. All serve the same purpose, but meet various specified needs.

The induced draft crossflow (see Figs. 19.4.35a and 19.4.35b) are propeller fan towers. Both are mechanical draft, meaning they use a fan to introduce air to cool the water. With the induced draft design the fan is located in the exiting air stream and air is drawn through the tower. In the crossflow configuration, the water flow in the fill (packing) is perpendicular to the flow of air. Water is delivered to the top of the tower and as it falls, air is drawn across the fill. The *singleflow* has the packing and water flow on one side only, while the *doubleflow* takes advantage of two sides.

The induced draft counterflow (Fig. 19.4.36) also uses the fan in the leaving air stream. In counterflow towers the air moves counter, or vertically upward, through the fill. The air and water make contact as they cross paths. This type of tower is characterized by the high-pressure spray system, taller design, and (compared to the crossflow design) may use much more power.

The counterflow forced draft (Fig. 19.4.37) is the fourth type of conventional air-conditioning tower. It is characterized by the fan located on the side in the airstream, but the air is blown through the tower. These towers are more susceptible to hot discharge air recirculation and may have less performance stability than induced draft towers.

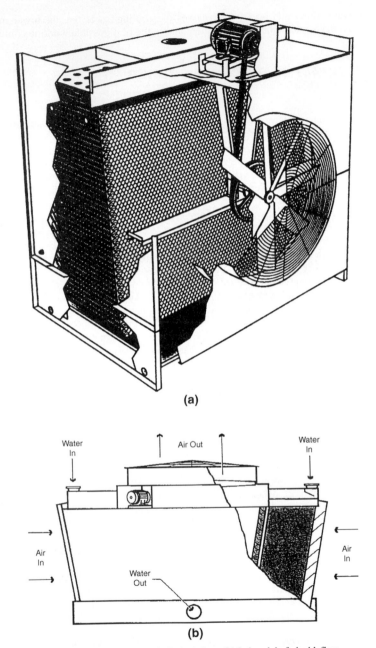

(a)

(b)

FIGURE 19.4.35   (a) Induced draft singleflow, (b) Induced draft doubleflow.

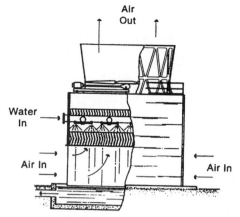

FIGURE 19.4.36 Induced draft counterflow.

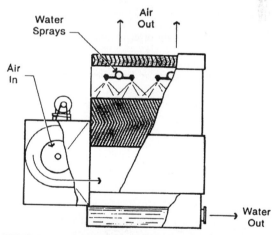

FIGURE 19.4.37 Forced draft counterflow.

Usually these towers are equipped with centrifugal blower type fans which require considerably more horsepower than the propeller fan. These types of towers are often located in smaller enclosures or indoors and ducted to the atmosphere.

19.4.14 MATERIALS OF CONSTRUCTION

Industrywide, there are four types of typical exterior construction materials: *galvanized steel, fiberglass, stainless steel*, and *wood*. All are usually factory assembled and delivered to the jobsite ready for installation. And in some cases, the larger designs are prefabricated and assembled at the location.

For packaged tower designs that range from 10 tons (9 metric tons) of air conditioning to 1000 tons (900 metric tons) in a single unit, steel is the material of choice. Steel's strength makes it economical. In most cases, with normal circulating water conditions, the structure, hardware, sides, and top and bottom pans are made of mill galvanized material. For severe water service many components must be coated in addition to being galvanized. Coatings offer the advantage of added protection, but with that comes the need for added maintenance. Coatings are susceptible to peeling, thereby exposing bare metal to the severe water. Added steps are necessary to achieve long-term protection.

When the process is severe, or when the tower owner wants extended life, stainless steel should be considered as a construction material. Although it is the premium material for cooling towers, stainless steel represents significant added expense. Another reason to consider stainless steel is the tower location. If in an area that is not easily accessible for change-out, such as on the roof of a tall building, stainless steel offers the one-time expense for equipment and installation.

Fiberglass is the latest material to enter the market. The capability to be molded into single parts of complex shapes and dimensions offers distinct advantages. Generally used only for the fan blades, louvers, and external casing, fiberglass has become a successful material for the tower structure. The lighter weight components allow for installation in areas sensitive to high concentrations of load. Typically, fiberglass is less susceptible to abnormal water conditions because it is inherently resistant to microbiological attack, corrosion, and erosion. It is more expensive than the standard galvanized steel version, but the trade-off with extended life and lighter weight offer advantages unavailable when using other construction materials.

For larger installations, the use of wood as the primary building component has existed for years. It is readily available, is relatively low in cost, and extremely durable under a variety of weather and water conditions. Such a structure is delivered ready for field installation. Douglas fir and redwood, because of their external structural capabilities, are the most common types of wood used. Most wood towers are built in the crossflow and counterflow induced draft designs. Regardless of the type of wood, it must be treated for long-term usage and protection against bacteria and fungal attack. Pressure treatment takes the form of chromate copper arsenate or acid copper chromate. This preservative, which offers life to the tower, is injected into the wood under pressure. Although a wood tower is the most expensive, it offers the longest service life and is easiest to repair.

19.4.15 MAINTENANCE PROCEDURES

As the seasons change, the maintenance staff is offered a chance to inspect and take necessary steps to ensure the cooling tower will remain trouble free for many years to come. *Late fall is the best time to start inspections.* Beginning inspections around November gives the staff a chance to make repairs during the early winter months. The milder days allow crews to make repairs before the bad days of winter set in. It also guarantees the unit will be ready for the hot summer months and peak cooling period.

Figure 19.4.38 is the anatomy of a cooling tower. Regardless of size, materials of construction, or manufacturer, all towers have similar parts. Structure, water distribution, heat transfer media, and mechanical components are the mainstay of all units. These are the major areas to inspect, repair, and keep in good operating condition.

Before any attempt is made to begin the cooling tower inspection, safety issues must be addressed. No inspection should be conducted with safety gear that fails to meet local safety codes. If the inspection is conducted with the fan running and water flowing, use caution when opening inspection doors or walking around the top of the unit. Always be aware

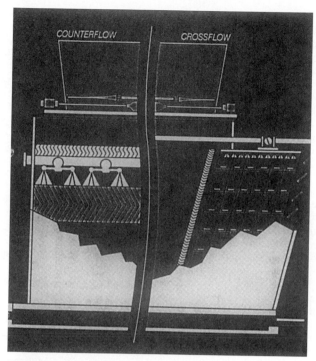

FIGURE 19.4.38   Anatomy of a cooling tower.

of the hazards of being around operating mechanical components. If any preliminary corrections are going to be made or work performed, always shut off the electrical power to the tower fan motor. All electrical switches should be locked out and tagged to prevent others from turning the power back on. This is especially true if the power can be turned on from a remote source, such as the main control panel that may be away from the cooling tower.

19.4.16   TOWER INSPECTION

Walk around the operating cooling tower and visually inspect the structure. Look for signs of leaks, cracks, or deterioration. Notice any separations in the end panels and make certain connections are tight and firmly attached to the structure. In most cases the end panel casing, or skin material, is fiberglass, steel, or wood. In older units it may be an asbestos, cement board or corrugated material. If the budget allows or if corporate policies require, consider replacing it immediately—regardless of condition. Inspect wood casing for signs of rot or plywood separation, commonly referred to as delamination. When inspecting a steel tower, look for corrosion or signs of pitting. If the skin is fiberglass, look for brittleness, cracking, or fiber separation.

Next, conduct a thorough inspection of the structure, making sure that the fan and water are turned off. This allows easy, safe entry, and ease in inspecting. When inspecting a steel tower, look for corrosion and most importantly, loss of metal or major areas of discoloration. In addition spot check tightness of bolts, making certain the size of bolt hole has not created an area where water can leak through. Check any welds for cracks. This will mean cleaning an area near welds to get an up-close look. If possible go under the tower. The cold water collection basin in most towers and the associated structure are the least inspected. They are not easy to access, yet a leak or failure in these locations creates major problems. Like an automobile, anything above the frame can be replaced or repaired, but trying to fix the frame itself is a very expensive proposition. If there are leaks or pin holes, use an oversized rubber-backed washed and sheet metal screw to fix. If the area is large, use a sheet metal panel and silicon-based caulk to cover the affected area. Of course, the area must be cleaned and ultimately recoated with a high grade epoxy coating.

The inspection process for a wood tower is very similar. First, look for signs of rot or decay. Inspect with a small hammer and tap the wood member. A dull low-pitched sound indicates softness; a sharp high-pitched sound indicates good solid wood. If areas of rot exist, probe them carefully with a knife or ice-pick, paying attention to the depth of penetration of the probe. Especially examine the areas around the bolts and where there is contact with steel or cast iron.

When checking the tightness of bolts, look at the indentations in the wood. Be sure they are not excessive. The joint connectors, either steel or fiberglass, must be inspected. Look for deterioration, cracking or corrosion. Some areas in the cooling tower are inaccessible. For example, in the heat transfer media, material must be moved before there is access to the unit. Reach as many bolts as possible and along the way check all wood pieces with the tap of the hammer.

When checking fiberglass or coated steel towers, check the assembled joints for tightness. Look for cracking or signs of distress in the coating. Many minor scratches and scrapes that do not require extraordinary corrosion protection can be touched up with a repair kit.[3] If the extent of damage or protection is uncertain, contact the manufacturer.

The most often overlooked area of the structure is the *top surface or fan deck*. It is always a good idea to get on top and take a firsthand look. On steel towers, look for pitting or corrosion. On wood units watch for softness of the lumber, delamination of the plywood, or holes caused by rot. A lot of information can be gained by lightly jumping up and down. The spring action under your feet is an indication of softness. Remember, it may not be the deck itself, but the supports underneath. They should be accessed from the bottom and ends. Be aware of uneven walking surfaces. The sheets of plywood on wood towers or connections on steel towers can come apart or shift over time, presenting a tripping hazard. In all cases, the connections should be tight and in line.

19.4.17   *WATER DISTRIBUTION SYSTEM*

The heart of the cooling tower is the *water distribution system*. Before the cooling can be accomplished, the water must flow evenly over the heat transfer media (fill). Regardless of the type of tower and internals, a pattern of water fall must be established and maintained throughout the life of the unit. Keeping the components of the distribution system clean and operational is a primary function of maintenance.

Starting at the top of the tower, inspect the hot water distribution basins, also referred to as *water pans*. These metal or wood boxes are located on the top of the tower and hold the water before it cascades over the fill. In steel units check for corrosion and loss of metal. If the basin is wood, usually plywood, look for delamination or wood decay. In both cases check for leaks between joints. Inspect the integrity of all basin support members. Check the tightness of bolted joints in steel or wood basins.

Each tower basin contains nozzles. In crossflow designs, they are plastic holes with an extended stem that hangs off the bottom. From the basin top, all that is visible is the plastic hole. Carefully clean all debris that may be clogging the nozzle opening. Do not use a probe that enters the nozzle to clean away sludge, leaves, and the like. Too long a probe, like a broom handle, could break off the bottom part of the nozzle, reducing the effectiveness of the nozzle.

Counterflow type towers contain a spray-type nozzle. To properly inspect these nozzles, it is necessary to gain access into the tower and remove them. They generally screw into the piping system and can be easily removed. Clean away debris and reinstall. It is always important to have the proper number of nozzles installed. Do not insert a plug in place of a nozzle—the performance of the system could be affected.

On some towers, flow control valves are located on top of these basins. Their purpose is to even the water flow across the entire area. Each valve has a grease fitting, which keeps the valve disc and steam lubricated. Follow the vendor's maintenance instructions and lubricate at least once a year. The flow control valves have a locking bar, which must be properly turned to lock the valve disc in place. Under water pressure the disc tends to shift and, if not locked down, will eventually fail.

The water is brought to the tower by means of a distribution pipe, generally located on top of the tower. Inspect the pipe, usually made of iron, for signs of corrosion or loss of coatings. Examine all supports for structural integrity. Spot check for leaks and check the tightness of all bolts. On fiberglass pipe look for signs of cracking or distress.

Many towers are equipped with a single inlet pipe arrangement, whereby water enters into a chamber before being distributed to the upper basin. This arrangement is used when there are no balancing valves involved. Usually it is equipped with an internal strainer and blow-down connection. The strainer can be easily removed and cleaned. The bottom of the chamber has a plugged connection, allowing for dirt and debris to be collected for removal.[4]

The lower water basin, where the water is collected, accumulates sludge and debris. Periodically, when the tower can be shut down, it should be cleaned. This removal of deposits keeps the sediment from getting corrosive and attacking the basin metal or wood. It also removes a potential breeding ground for harmful bacteria. The sumps and screens should be inspected and cleaned of trash to allow for the free flow of return water.

19.4.18 HEAT TRANSFER MEDIA

Inside every cooling tower are heat transfer media, commonly referred to as "fill." Its only function is to ensure that the water droplets are broken up to allow for easy contact with the incoming air. The evaporation of water provides the cooling. There are various types of fill arrangements and materials, the most common being splash and film, either constructed of wood or PVC. See Figs. 19.4.39a, b.

There are several types of splash fill. Generally they consist of wood or plastic bars supported in wire or fiberglass hangers. Ceramic tile is also a type of splash fill. When inspecting splash fill bars, which can be viewed from the outside of the tower, look for sagging, broken, or decayed bars. These bars can be seen from the outside of the tower. There may also be scale buildup on the bars. Also, pay attention to the pattern of the bars—misplaced bars will not act as efficiently as designed. If water is running over the fill, look at the pattern of water flow, making certain the flow is even.

When inspecting the supporting grids or hangers, see that they are in place and not sagging. Check the coating, if there is any, on steel grids and the condition of the welds on stainless steel grids. If the grids are fiberglass, check the condition and look for any breakage. Any section that shows excessive deterioration should be replaced. Most importantly,

(a)

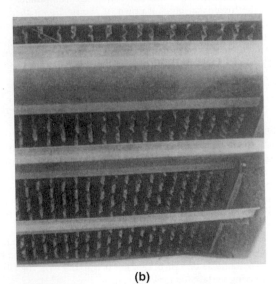

(b)

FIGURE 19.4.39   (*a*) Splash-type fill, (*b*) Film-type fill.

always look at the supporting structure where the grids are attached. Pay particular attention to sagging or damaged pieces.

Film fill, which consists of several sheets of PVC, is a second type of packing. Film fill is either hung from tubes or resting on the bottom of the structure. For this type of packing, look for buildup of algae or scale on the surface and within the gaps of the sheets. Also check for sagging, tears, erosion, or ice damage. Continue to inspect the supporting members just as you would for the splash type.

Some older towers contain fill types that were manufactured from asbestos material. When possible, *consider replacing them immediately*. Besides removing the environmental concerns, the new fills of today are more efficient than the older type. The benefit of increased performance could pay for these repairs.

The next major component is the drift (mist) eliminators, which are located inside the doors at each end of a crossflow tower and above the spray system of a counterflow tower. Usually wood or PVC, a visual inspection can be done from either inside or outside the tower.

To work properly, all air passages must be clear of mud and debris. Look at the arrangement of the eliminator and note any shifting or dislodging of material. Check for visible gaps between the packs or frames. These seals should be caulked, and in some cases, replaced to prevent air and water from exiting the tower. Most eliminators are wood or plastic bars supported in a framework of wood, or PVC cellular packs.

Lastly, go outside and look at the front of the tower. The louvers, which are attached to the structure, are angled pieces that direct the air into and keep water droplets from splashing out of the tower. Usually of wood, steel, or PVC, these pieces should be in place and free of algae or scale buildup. They should be clean and sitting properly in the supports. Check the condition of both the individual louvers and the supporting members. Look for signs of deterioration. In newer units the louvers are PVC and molded into the face of the fill sheets. Inspect and clean as you would the fill.

Older models used asbestos-containing materials. *Consider replacing them immediately*.

19.4.19 MECHANICAL EQUIPMENT

Cooling towers usually have two methods of turning the fan: either through belts or a gearbox (transmission). For belt drive systems, check the pulleys and belts at least monthly. Look for signs of corrosion or loss of metal on the pulleys. On the belts, identify wear patterns. If nicks or gouges exist consider replacing the belt. The procedure for belt tightening depends on the manufacturer. Review the operation and maintenance manuals provided for recommendations and frequency.

The pulleys are driven through a bearing housing or pillow block bearings. Lubricate as instructed by the manuals. Check for loose or damaged bearings. Does the input or output shaft have any play? Look at the support for this housing. Identify signs of wear or corrosion and repair as necessary. The bolted connections should be inspected and tightened.

Because of their design, gearbox driven units have more areas to pay attention to. The primary concerns are the oil level and quality of the oil during operation. Like an automobile, the oil level should be checked when the unit is off and cold. Each manufacturer will provide a minimum time interval between oil changes, but as a minimum, it should be semiannually. Some of the newer gearboxes have a new sealed design that uses synthetic oil, which requires an oil change every 5 years.[5]

When examining the oil, drain a little off and look for water, sludge, or metal shavings. Check for leaks at both the input and output seals. It may be necessary to replace these seals.

Rotate the input shaft and be sure that there is no slippage. The gear teeth should engage without excessive movement. The endplay on the gears can be determined by pulling up and down on the fan blade. When uncertain about the amount of movement, a service contractor or cooling tower manufacturer can help determine the desirable amount of movement.

In observing the gearbox housing, as with any metal part, check for corrosion or loss of metal. It's easy to see if the seals are holding by spotting any oil buildup on the case. The hardware holding the box to the support should be tight. Some boxes are equipped with an external oil fill/drain line and dipstick, which usually extends outside the fan and are easily accessible from the deck. Look at the connection to the back of the gearbox. Is the connection cracked? Is it leaking? This connector hose can be a hydraulic type and over time will crack or break due to exposure to the elements and vibration. A sudden failure of this hose causes loss of oil and the gearbox will cease working.

In direct drive arrangements, that is a motor attached directly to the gearbox, make sure that the fan bushing is tight and in good condition. There are no areas to lubricate in this bushing. It should simply be checked to be sure it fits tight to both the motor and gearbox.

If the unit operates through a driveshaft and coupling, attention should be paid to the connection and alignment. Examine the metal parts, looking for corrosion and tightness of hardware. Some units have flexible metallic elements that allow for minor amounts of vibration or misalignment. The bushings should be checked for cracking or brittleness. The driveshaft itself is either one or two piece. If the shaft has mid-span bearings, they should be lubricated regularly. Newer towers will have a single metal or fiberglass shaft. The alignment of the shaft is critical and should be checked with a dial indicator. All hardware should be properly torqued according to manufacturers' instructions.

The two-piece shafts can easily be converted to a single shaft. Consult with a mechanical contractor or cooling tower manufacturer. Having only one integral shaft reduces maintenance costs and possibility of failures.

The fan attaches to the top shaft of the gearbox or pulleys on a propeller fan arrangement. Fans are either fixed or adjustable pitched. Fixed blades are permanently attached to a hub, while the adjustable type are clamped and can be moved to accommodate changes in design and air flow. Regardless of type, look for metal corrosion. The bolts holding the blades in place should be torqued at least twice a year. Blades must not have excessive buildup, chips, or cracks. Look at the leading edge of each blade for any nicks or separation of material.

When working with adjustable pitched fans, the blade angle can be measured and changed. At any fan speed, the pitch determines the amount of air flow. By loosening the bolts and changing the pitch, more air can be drawn and, in some cases, improve cooling. However, consult your cooling tower manufacturer first.

Each fan operates within the confines of a cylinder or shroud, which can be made of wood, fiberglass, or steel. In each case check the material for loss or corrosion. The blade tracks inside that shroud and must be sufficiently close to minimize air loss but not too close that it hits the sides. Adjust as necessary to accomplish this. The shroud can be realigned by adjusting the bolts and tightening.

For a fixed, centrifugal fan, similar inspections are necessary. Look at the blades for corrosion, nicks, or breaks. Check the welds of the blades to the hub and the tightness of all bolting. Refer to Chap. 23, *Fans and Blowers.*

The entire drive train rests on a mechanical equipment support made of steel. As with all the other components, the metal on the drive train should be looked at for corrosion and tightness. If this support is on a wood tower, look at the connection to the wood and see if iron rot of the wood exists. If the tower is fiberglass, check the bolted connections.

The last member of the mechanical package is the fan motor. Most motors are open drip proof, totally enclosed, or fan cooled. All three types have grease fittings for the lubrication of the bearings. The housing metal should be checked for corrosion, while the bolts that attach it to the framework should be checked for tightness. Listen for sounds that indicate a bad bearing or fan hitting the motor housing.

19.4.20 SAFETY

Most areas of the tower must be inspected for safe working and operating conditions. These items should be inspected yearly and repaired immediately to guarantee the safety of maintenance personnel.

Most towers are equipped with handrails around the top. Inspect the members and the hardware to be sure they are properly connected. If the tower has a ladder or stairway to the top, inspect the members for corrosion or breakage. If corrosion or breakage exists, it should be corrected immediately. Stairway landings should be sturdy and tightly bolted.

The ends of many cooling towers have access through a door. Check the operation of that door and its attachment to the structure. Keep all hardware tight and the door in good working condition. Towers also have a guard, generally made of steel, that protects the fan. Again, check for corrosion and tightness of the hardware. Be sure that individuals cannot stick their fingers into the fan and that flying debris cannot enter the fan.

To protect against failure of the fan due to a loss of blade, the tower can be equipped with a vibration limit switch. There are many types, but to be effective it must be capable of shutting the fan down the moment excessive vibration is noticed. A remote reset should be included, requiring the maintenance person enter the tower and inspect for damage. This switch can be wired to the main control panel, lighting up when a problem exists.

19.4.21 FREQUENCY OF INSPECTION

Recommended frequency of inspections depend on tower location and number of available staff. The following are minimums that should be considered:

Daily: Observe, listen, and walk around the unit. Become accustomed to the tower's appearance, sound, and level of vibration.

Weekly: Look at the motor, drive shaft, gearbox, belts, and fan. Shut off the fan and, if it is equipped with a gearbox, check the oil for proper operating levels.

Monthly: Check for silt buildup in the basin sump, check the operation of the float valve, inspect the upper distribution chambers for cleanliness of the basin and nozzles, and drain gearbox oil sample.

Semiannually: Drain and refill the gearbox. Check the condition of belts and replace, if necessary. Relubricate the fan motor. Check all bolting for tightness in the mechanical region. Visually inspect tower structure; heat transfer media; drift eliminators; and condition of stair, ladder, and handrail.

Annually: Inspect the tower thoroughly, taking advantage of the operation and maintenance manuals as a guide. If the tower is equipped with a protective finish, inspect it. Take the time to make corrective repairs. Ask for an annual inspection by a service contractor or the local cooling tower manufacturer.

The operation and installation manual provided with the equipment should be used as a guide to efficiently operate the tower. Each tower is supplied with a serial number.

That number is the key to maintenance history and will always be the first thing a supplier will ask for when ordering parts. It should be kept handy.

Figure 19.4.40 is a standard troubleshooting guide. Most problems with cooling towers are in the mechanical equipment. This two-page guide offers examples of a problem, possible cause, and remedy. Figure 19.4.41 is an easy-to-use cooling tower inspection checklist. It covers all aspects of the cooling tower and can be used by maintenance personnel to keep a record of cooling tower repair activity. Such records are important to itemize work to be completed and justify those repairs or replacements. It also offers a track record, so multiple occurrences of a problem can be identified and corrective action taken.

Troubleshooting

Trouble	Cause	Remedy
Motor Will Not Start	Power not available at motor terminals	• Check power at starter. Correct any bad connections between the control apparatus and the motor.
		• Check starter contacts and control circuit. Reset overloads, close contacts, reset tripped switches or replace failed control switches.
		• If power is not on all leads at starter, make sure overload and short circuit devices are in proper condition.
	Wrong connections	Check motor and control connections against wiring diagrams.
	Low voltage	Check nameplate voltage against power supply. Check voltage at motor terminals.
	Open circuit in motor winding	Check stator windings for open circuits.
	Motor or fan drive stuck	Disconnect motor from load and check motor and Geareducer for cause of problem.
	Rotor defectve	Look for broken bars or rings.
Unusual Motor Noise	Motor running single-phase	Stop motor and attempt to start it. Motor will not start if single-phased. Check wiring, controls, and motor.
	Motor leads connected incorrectly	Check motor connections against wiring diagram on motor.
	Bad bearings	Check lubrication. Replace bad bearings.
	Electrical unbalance	Check voltages and currents of all three lines. Correct if required.
	Air gap not uniform	Check and correct bracket fits or bearing.
	Rotor unbalance	Rebalance.
	Cooling fan hitting end bell guard	Reinstall or replace fan.
Motor Runs Hot	Wrong voltage or unbalanced voltage	Check voltage and current of all three lines against nameplate values.
	Overload	Check fan blade pitch. See Fan Service Manual. Check for drag in fan drive train as from damaged bearings.
	Wrong motor RPM	Check nameplate against power supply. Check RPM of motor and gear ratio.
	Bearings overgreased	Remove grease reliefs. Run motor up to speed to purge excessive grease.
	Wrong lubricant in bearings	Change to proper lubricant. See motor manufacturer's instructions.
	One phase open	Stop motor and attempt to start it. Motor will not start if single-phased. Check wiring, controls, and motor.
	Poor ventilation	Clean motor and check motor ventilation openings. Allow ample ventilation around motor.
	Winding fault	Check with Ohmmeter.
	Bent motor shaft	Straighten or replace shaft.
	Insufficient grease	Remove plugs and regrease bearings.
	Too frequent starting or speed changes	Limit cumulative acceleration time to a total of 30 seconds/hr. Set on/off or speed change set points farther apart. Consider installing a Marley VFD drive for fine temperature control.
	Deterioration of grease, or foreign material in grease	Flush bearings and relubricate.
	Bearings damaged	Replace bearings.
Motor Does Not Come Up To Speed	Voltage too low at motor terminals because of line drop	Check transformer and setting of taps. Use higher voltage on transformer terminals or reduce loads. Increase wire size or reduce inertia.
	Broken Rotor bars	Look for cracks near the rings. A new rotor may be required. Have motor service person check motor.
Wrong Rotation (Motor)	Wrong sequence of phases	Switch any two of the three motor leads.

FIGURE 19.4.40   Troubleshooting guide.

Troubleshooting

Trouble	Cause	Remedy
Geareducer Noise	Geareducer bearings	If new, see if noise disappears after one week of operation. Drain, flush, and refill Geareducer. See Geareducer Service Manual. If still noisy, replace.
	Gears	Correct tooth engagement. Replace badly worn gears. Replace gears with imperfect tooth spacing or form.
Unusual Fan Drive Vibration	Loose bolts and cap screws	Tighten all bolts and cap screws on all mechanical equipment and supports.
	Unbalanced drive shaft or worn couplings	Make sure motor and Geareducer shafts are in proper alignment and "match marks" properly matched. Repair or replace worn couplings. Rebalance drive shaft by adding or removing weights from balancing cap screws. See Drive Shaft Service Manual.
	Fan	Make certain all blades are as far from center of fan as safety devices permit. All blades must be pitched the same. See Fan Service Manual. Clean off deposit build-up on blades.
	Worn Geareducer bearings	Check fan and pinion shaft endplay. Replace bearings as necessary.
	Unbalanced motor	Disconnect load and operate motor. If motor still vibrates, rebalance rotor.
	Bent Geareducer shaft	Check fan and pinion shaft with dial indicator. Replace if necessary.
Fan Noise	Blade rubbing inside of fan cylinder	Adjust cylinder to provide blade tip clearance.
	Loose bolts in blade clamps	Check and tighten if necessary.
Scale or foreign substance in circulating water system	Insufficient blowdown	See "Water Treatment" section of this manual
	Water treatment deficiency	Consult competent water treating specialist. See "Water Treatment" section of this manual
Cold Water Temperature Too Warm (See "Tower Operation")	Entering wet bulb temp. is above design	Check to see if local heat sources are affecting tower. See if surrounding structures are causing recirculation of tower discharge air. Discuss remedy with Marley representative.
	Design wet bulb temp. was too low	May have to increase tower size. Discuss remedy with Marley representative.
	Actual process load greater than design	May have to increase tower size. Discuss remedy with Marley representative.
	Overpumping	Reduce water flow rate over tower to design conditions.
	Tower starved for air	Check motor current and voltage to be sure of correct contract horsepower. Re-pitch fan blades if necessary. Clean louvers, fill, and eliminators. Check to see if nearby structures or enclosing walls are obstructing normal airflow to tower. Discuss remedy with Marley representative.
Excessive Drift Exiting Tower	Distribution basins overflowing	Reduce water flow rate over tower to design conditions. Be sure hot water basin nozzles are in place and not plugged.
	Faulty drift elimination	Check to see that integral fill, louvers, and eliminators are clean, free of debris, and installed correctly. If drift eliminators are separate from fill, make sure they are correctly installed in place. Clean if necessary. Replace damaged or worn out components.

FIGURE 19.4.40   *(Continued.)*

19.4.22 REPAIR CRITERIA

Given the expected life as discussed above, it is the responsibility of the chief engineer and cooling tower contractor to make the most out of the equipment. The decision to repair or replace becomes a matter of dollars. In most cases the life can be extended by following the operation and maintenance procedures and doing repair work as needed. As with any product, *waiting until the tower is in unusable shape will only hasten the need to replace.* Use the resources available, conduct regular inspections, and keep good records. The actual need to repair or replace will vary with the owner and the contractor.

Common sense and loss of cooling efficiency will dictate when repair work needs to be done. Two additional topics should be discussed to help extend the service life of the tower and protect if from airborne bacteria. The whole issue of water cleanliness and tower life

Cooling Tower Inspection Checklist

Date Inspected: _____   Inspected By: _____
Owner: _____   Location: _____
Owner's Tower Designation: _____
Tower Manufacturer: _____   Model No. _____ Serial No. _____
Process served by tower _____   Operation: Continuous ☐ Intermittent ☐ Seasonal ☐
Design Conditions: GPM _____ , HW _____ °F, CW _____ °F, WB _____ °F
Number of Fan Cells: _____   Tower Type: Crossflow ☐   Counterflow ☐

Condition: 1 — Good, 2 — Keep an eye on it, 3 — Needs immediate attention

	1	2	3	Comments
Structure				
Casing Material _____				
Structural Material _____				
Fan Deck Material _____				
Stairway? _____ Material _____				
Ladder? _____ Material _____				
Handrails? _____ Material _____				
Interior Walkway? _____ Material _____				
Cold Water Basin Material _____				
Water Distribution System				
Open Basin System				
Distribution Basin Material _____				
Inlet Pipe Material _____				
Inlet Manifold Material _____				
Flow Control Valves? _____ (Size _____ inches)				
Nozzles (Orifice diameter _____ inches)				
Spray Type System				
Header Pipe Material _____				
Branch Pipes Material _____				
Spray Nozzles (Orifice diameter _____ inches)				
Up Spray ☐ Down Spray ☐				
Heat Transfer System				
Fill (Type & Mat'l _____)				
Drift Eliminators (Type & Mat'l _____)				
Louvers _____				

Use this space to list specific items needing attention: _____

FIGURE 19.4.41   Inspection checklist.

and efficiency are tied together. A cooling tower will accumulate dirt and debris because of the nature of drawing air into the path of falling water. This buildup of sludge can collect on the condenser, heat exchanger, and other equipment, forming an insulating layer that decreases the heat transfer capacity.[6] A water filtration system can reduce this buildup. Such a piece of equipment not only maintains component efficiency but can reduce the amount of chemicals necessary in the circulating water system. A single important byproduct is the reduced maintenance time to clean the heat exchangers, cooling tower basins, and condenser.

A final issue related to water cleanliness is the *possible breeding ground for bacteria.* The single most difficult issue is stagnant water. System piping should be free of "dead legs," and tower flow should be maintained. When dirt accumulates in the collection basin of a tower, the combination of moisture, oxygen, warm water, and food supply[7] *creates the*

Cooling Tower Inspection Checklist (Page 2)

	1	2	3	Comments
Mechanical Equipment				

Speed Reducer Type: Belt ☐   Gears ☐   Direct Drive ☐

Belt Drive Units

	1	2	3	Comments
Belt (Designation _____)				
Fan Pulley (Designation _____)				
Motor Pulley (Designation _____)				

Gear Drive Units

Manufacturer _____ Model _____ Ratio_____

Oil Level: Full ☐   Add Immediately ☐   Low, check again soon ☐

Oil Condition: Good ☐   Contains Water ☐   Contains Metal ☐   Contains Sludge ☐

Oil Used (Type_____)

	1	2	3	Comments
Seals				
Back Lash				
Fan Shaft End Play				
Any Unusual Noises? No ☐ Yes ☐ Action Required:_____				
Drive Shafts (Mfr. & Mat'l _____)				

Fans

Propeller ☐   Blower ☐   Wheel Diameter_____

Manufacturer_____   Fixed Pitch ☐   Adjustable Pitch ☐

Dia. (_____ feet _____ inches)   Number of Blades_____

	1	2	3	Comments
Blade Material _____				
Hub Material _____				
Hub Cover Material _____				
Blade Assembly Hardware _____				
Blade Tip Clearance_____ " min. _____ " max.				
Vibration Level				
Fan Stacks (Dia.at Fan _____ ft.; Height _____ ft.)				
Mech. Eqpt. Support (Mat'l _____)				
Fan Guards				
Oil Fill & Drain Lines				
Oil Level Sight Glass				
Vibration Limit Switches				
Make-up Valves				
Other Components:_____				

Motor Mfr.:_____ .

Name Plate Data: HP_____ RPM _____ Phase_____ Cycle_____ Volts_____

F.L. Amps _____ Frame_____ S.F. _____ Special Info._____

Last Lubrication (Date_____)

Grease Used (Type_____)

Any Unusual Noise?   No ☐   Yes ☐   Action Required:_____

Any Unusual Vibration?   No ☐   Yes ☐   Action Required:_____

Any Unusual Heat Build-up? No ☐   Yes ☐   Action Required:_____

FIGURE 19.4.41   (*Continued.*)

possibility of Legionella bacteria. These bacteria can be found in water supplies and around rivers or streams. They are contained in water droplets and become airborne, making humans susceptible to breathing the contaminated air. There are no chemicals that can positively eliminate all bacteria from the water supply in a cooling tower. However, evidence suggests that good maintenance along with a comprehensive treatment program will dramatically minimize the risk.[8]

One final comment on cooling tower maintenance: Take frequent walks around the tower and listen to its sounds. Become aware of the way it operates and looks on a regular basis. There is no substitute for common sense and using the knowledge gained on working with equipment of all types. Rely on the cooling tower manufacturer and service contractor as a source of skilled labor and parts. Being in charge of keeping the cooling tower running efficiently does not have to be a difficult task.

REFERENCES

1. *Cooling Tower Fundamentals*, Published by Marley Cooling Tower Company, 2nd ed., 1998.

2. *ASHRAE Handbook*, Systems and Equipment Edition, Atlanta, GA, p. 36.1, 1996.

3. Baltimore Aircoil Company, Operation and Maintenance Instructions, bulletin M24/1-0AB, p. 8, Baltimore, MD.

4. Ibid, p. 5.

5. Marley Cooling Tower Company, Owners Manual OM-NC2 A, Number 92-1327B, p. 20, Overland Park, KS, 2002.

6. Process Efficiency Products, Inc., "Cooling Tower Filtration Systems," p. 3, Mooresville, NC, 2002.

7. L. S. Staples Company, "Legionnaires' Disease-Still With Us," p. 1, 1996.

8. Ibid, p. 3.

AIR DISTRIBUTION MANAGEMENT AND MAINTENANCE

CHAPTER 20
AIR FILTRATION AND AIR POLLUTION CONTROL EQUIPMENT*

Keiron O'Connell, P.E.
AAF International, Louisville, Kentucky

20.1 GAS PURIFICATION EQUIPMENT CATEGORIES

Gas purification devices are found in a very broad array of air- and gas-handling systems. Ultra high efficiency filters remove particles from the air of clean rooms where semiconductor integrated circuits are manufactured; inexpensive fiber filters protect the heat exchangers in residential air conditioners; large-scale big filter, electrostatic precipitators and scrubbers remove fly ash and sulfur dioxide from power plant stack gases. Gas-cleaning technology is usually divided into two categories, based on the purpose of the device considered.

- *Air filters:* These devices remove pollutants entering buildings and industrial processes or recirculated within them. Such air is commonly referred to as àtmospheric air. It is common to use the terms "air cleaner," "air filter," or merely "filter" to refer to this type of device, regardless of whether air or some other gas is the working fluid.

- *Air pollution control equipment:* These devices capture pollutants emitted by industrial processes, to prevent their release into the atmosphere.

Pollutant concentrations for the two categories differ greatly. Air pollution control equipment typically must deal with pollutant concentrations 1000 to 100,000 times those found in air filtration applications. A few devices, such as systems to protect turbines and compressors from desert sandstorms, may operate at contaminant concentrations between the filtration and pollution control categories.

Gas purification devices in the above application areas are further subdivided into devices which capture *particulate* contaminants and those which capture *gaseous* contaminants. The operating principles for capture of the two contaminant types are quite

---

*Updated by author from *HVAC Systems and Components Handbook*, McGraw-Hill, 1998.

different, and in only a few cases does the same device remove both gaseous and particulate contaminants.

Before we discuss the operating principles and design of gas purification equipment, it will be helpful to examine the nature of gasborne contaminants and to define some terms used in gas purification technology.

20.2 PARTICULATE CONTAMINANTS

20.2.1 Size Range of Aerosols

Both solid and liquid particles can be undesirable contaminants in gases. The range of particle sizes which can exist in a flowing gas stream is about 0.001 to 100 μm (1 μm = 1 micrometer = 10^{-6} m). Below 0.001 μm, a *particle* includes so few molecules that it behaves as a gaseous molecule. Particles in the 0.001– to 0.01– μm size range tend to agglomerate rapidly, forming larger particles. Particles larger than 100 μm are so heavy that they do not remain suspended in gases; they can be transported only by repeated bouncing from surfaces. Particles which remain in suspension for extended times are called *aerosols*.

20.2.2 Aerosol Shape

Liquid aerosols in small sizes are very nearly perfect spheres. Larger liquid aerosols (droplets) may be distorted by aerodynamic and gravity forces, but are still nearly spherical. Solid aerosols may be spherical, but are often distorted or ragged in appearance. Some aerosols (e.g., soot particles) are randomly shaped chains or agglomerates of smaller particles. Aerosols which have length-to-diameter ratios greater than about 5 (asbestos, lint) are usually considered fibers and behave somewhat differently from near-spherical aerosols.

20.2.3 Aerosol Size Distribution Statistics

A few aerosols are of nearly uniform and constant size. More typically, however, an aerosol cloud includes a range of particle sizes. Statistical concepts are therefore useful in describing most aerosols and are almost always necessary to calculate the performance of gas-cleaning devices. The first step in defining the statistics of an aerosol cloud is to subdivide it into size groups, each of which contains only a narrow range of sizes. This may be done with an instrument which actually separates the particles by size or one which measures an image or other size-dependent property of individual particles. The number or mass of particles in each narrow size range is determined, and a particle-size histogram (Fig. 20.1) is plotted. Aerosol histograms may have a single peak (as in Figure 20.1) or several peaks (Fig. 20.2). Each peak is called a "*mode*." The terms "unimodal," "bimodal," "trimodal," and "multimodal" are used to label aerosols with one, two, three, and many modes. Where more than one mode is present, the measured aerosol comes from more than one source or is the result of more than one production mechanism.

It is possible to divide the histogram into the components which make up each individual peak. When this is done, the histogram is usually skewed to one side of the peak, as in Fig. 20.1. If the data are replotted on a logarithmic scale for particle diameters, the histogram appears symmetric about the peak (Fig. 20.3). Such a histogram can be fitted quite

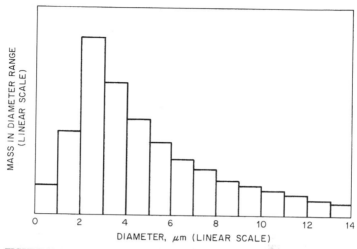

FIGURE 20.1   Particle-size histogram with linear particle-diameter scale and single peak.

closely to a common statistical function, the "normal" or "gaussian" distribution. Such an aerosol is said to be "lognormally distributed." We can use this curve to replace the data array used to plot the histogram with just three parameters: the total count of particles, mean diameter (diameter at the peak of the smooth curve), and standard deviation of the curve (half the width between the inflection points).

The process of fitting a smooth curve to the aerosol size distribution histogram is greatly simplified by plotting the *sum* of counts for all sizes larger than a given size against the logarithm of the size, using a special long-probability graph paper.

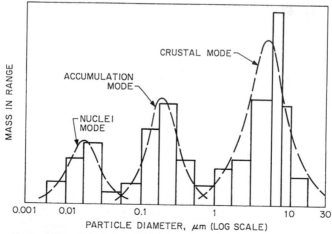

FIGURE 20.2   Particle-size histogram with multiple peaks.

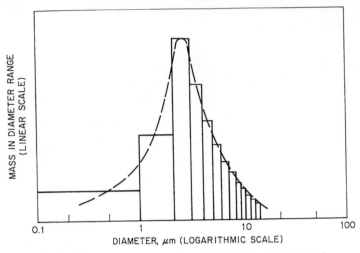

FIGURE 20.3   Particle-size histogram plotted with logarithmic diameter scale.

For aerosols which are lognormally distributed, this form of plot (Fig. 20.4) appears as a straight line. Small deviations in measured size distributions can thus be smoothed out, and the distribution parameters picked off the plot. These parameters are

d_g = geometric mean diameter, above which lies
    50 percent of distribution count

d_{σ^-} = diameter above which lies 84.13 percent of distribution

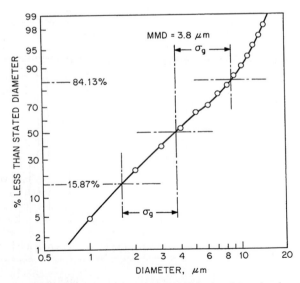

FIGURE 20.4   Plot of cumulative particle-size data using log-probability scales.

$d_{\sigma+}$ = diameter above which lies 15.87 percent of distribution

σ_2 = geometric standard deviation

$$\sigma_g = \left(\frac{d_{\sigma+}}{d_g}\right) = \left(\frac{d_g}{d_{\sigma-}}\right)$$ (20.1)

The value of d_m is considerably larger if we have based the distribution histogram on the masses of particles in each diameter range, rather than the count. This distinction is very important when aerosol data are reported or used. If the distribution is indeed lognormal, the relations between mass and count distribution (geometric) mean sizes can be calculated from the Hatch-Choate equation:

$$\ln d_m = \ln d_g + 3(\ln \sigma_g)^2$$ (20.2)

This conversion is useful when one needs to compute the penetration of a particulate filter or air pollution control device on a mass basis, yet only count data are available for the aerosol; or when penetration on a count basis is wanted, but only mass data are available for the aerosol.

The lognormal plot is not the only way in which aerosol distributions can be presented. A form often used is the Lundgren plot, shown in Fig. 20.5 Here the horizontal axis is the logarithm of the particle diameter, and the vertical axis is the logarithm of the term $\Delta c/\Delta\ln d$,

where Δc = particle count per m^3 for particle size range d_i to d_{i+1}

$\Delta\ln d$ = difference between natural logarithms of d_i and d_{i+1}

Similar plots can also be made with $\Delta w/\Delta\ln d$ or $\Delta s/\Delta\ln d$ on the vertical axis. In these cases, Δw is the total mass of particles per cubic meter in the size range d_i to d_{i+1}, and Δs is the total surface of the particles in the range from d_i to d_{i+1}. This type of plot shows the modal pattern of the aerosol quite clearly. Another form, with the logarithm of particle counts greater than a given diameter plotted against the logarithm of these diameters, is used in determining the "class" of clean rooms (see Fig. 20.6).

It is possible to convert any of these plots to one of the others. Programs for hand calculators and microcomputers exist to help in this rather tedious work.[1] Further discussions of aerosol statistics are found in many texts.[2,3]

20.2.4 Aerosol Concentrations

Aerosol mass concentrations are expressed in mass of particulate per unit volume of gas. Units commonly used are $\mu g/m^3$, g/m^3, mg/m^3, gr/ft^3, gr/ft^3, and $gr/1000$ ft^3. Table 20.1 gives some typical values for industrial-process concentrations. The outdoor mass concentration (total suspended particulates, or TSP) is measured continually at hundreds of stations throughout the world. Average values obtained for 80 U.S. cities are listed in Table 20.2 and for 90 rural stations in Table 20.3. Outdoor concentrations are generally high in arid regions with agricultural activity and in cities with high automotive vehicle populations. The trend in TSP concentrations has been downward since the advent of national pollution control regulations and the shift from coal-fired residential heating. Tables 20.2 and 20.3 give average values; the percentage of each year for which TSP concentrations exceed a given level

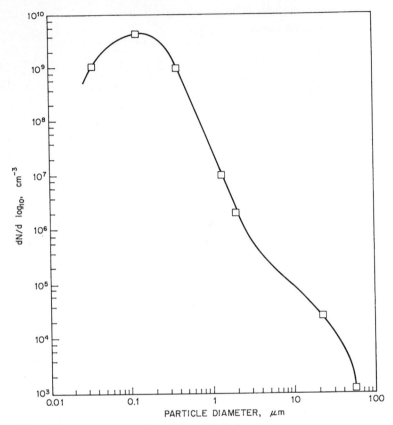

FIGURE 20.5 Lundgren plot of particle-size distribution.

is lognormally distributed in almost all locations. Desert zones have special patterns. There seasonal winds bring sandstorms and duststorms with extremely high concentrations (see Table 20.3).

    Average aerosol concentration has been found[6] to decrease with height above grade, approximately as given by Eq. (20.3):

$$C = C_o e^{-0.008h} \tag{20.3}$$

where C_o = concentration near surface, mg/m^3
      C = concentration at height, h, mg/m^3
      h = height above grade, m

Indoor aerosol concentrations are the result of the combined effects of outdoor concentration, air leakage into and from the space, internal generation and deposition, air-flow patterns (especially the recirculation rate), and filtration. Calculation of indoor dust concentration is detailed in Sec. 20.4.3.

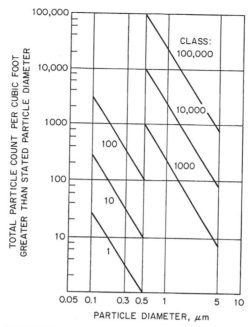

FIGURE 20.6 Particle-size limits for clean-room classes.

20.2.5 Aerosol Sources and Generation Modes

Outdoor aerosol particle-size distributions exhibit three modes:

- *Coarse-particle or crustal mode:* Chiefly soil particles, suspended naturally by wind erosion or by human activity in agriculture, construction, or travel on unpaved roads. Particles are produced by mechanical fracture of solids.
- *Fine-particle or accumulation mode:* Chiefly combustion products, with some production by atmospheric photochemical reactions, agglomeration, and suspension of very fine soil particles.
- *Nuclei mode:* Produced by evaporation, recondensation, and reactions of various compounds, primarily in combustion processes (including gasoline and diesel engines). Some particles in this mode originate from sea spray and radon gas. Nuclei-mode particles are short-lived, coagulating into accumulation-mode particles. Since there is essentially continuous production, a distinct nuclei mode will appear in ambient aerosols.

Indoor aerosols exist in the same three modes, in part because they arise from infiltration of outdoor aerosols into indoor spaces. However, interior generation sources often add to one or more of the modes, thus altering the overall distribution. In the past a major indoor aerosol has been tobacco smoke, which contains both liquid and solid particles. However, as the trend towards the prohibition of or control of smoking in public and commercial buildings has accelerated, the presence of tobacco smoke aerosols is rapidly being eliminated.

Industrial processes can produce aerosols of virtually any size and chemical composition. Concentrations are usually high enough to allow reasonably rapid aerosol concentrations

TABLE 20.1 Industrial Process Dust Characteristics

Service	Operation	Concentration, g/m^3	MMD, μm	σ_g	d, g/cm^3	Code*
Boiler fly ash	Chain grate, coal	1–5	4–20	3	0.7–3	1
	Stoker, coal	5–7	4–20	2–3	0.7–3	2
	Pulverized, coal	10–30	4–20	2–3	0.7–3	3
	Wood-fired	5–20	12	4		4
Ceramics	Frit spraying	1–8	5	2	2.7	5
Chemicals	Crushing, grinding	5–15	5–20	2	2.5	6
	Dryers, kilns	10–20	10	2	2.5	7
	Bin ventilation	1–5	5–10	2	1	8
	Materials handling	1–20	5–20	2	1	9
Coal mining and	Dedusting, cleaning	10–20	10	1.7	2	10
handling	Drying	10–20	5–11	1.7	2	11
Fertilizers	Ammoniators	1–7	5–10	2	2	12
Food	Sugar processing	1–11	5–10	2	1.7	13
	Coffee roasting	1–5	1–2	2	2	14
	Grain drying	1–5	20–30	3	2	15
	Flour handling	7–10	10	2	1.5	16
Foundries	Cupolas	2–10	1–10	2	4	17
	Electric furnaces	0.7–2	0.5–1.5	2	4.5	18
	Molding	7–10	2–10	2	2.5	19
	Shakeouts	3–12	1–10	4	2.5	20
Leatherworking	Buffing	7–11	11	2.5	0.7	21
Metalworking	Grinding	0.5–8	10–40	2.5	3–6	22
	Polishing, buffing	1–5	5–15	2.5	4	23
	Machining	1–8	1–5	2.2	3.7	24
	Welding, cutting	0.02–1.5	0.5–3	1.8	4	25
	Abrasive cleaning	7–30	5–10	2	2.5	26
	Spray painting	0.2–3	15–30	2	1.1	27
Papermaking	Paper cutting	7–10	12	4	1	28
Pharmaceuticals	Coating, packing	5–10	5–10	2	1.7	29
Plastics, rubber	Mixing	7–11	5	2	1.2	30
	Grinding, buffing	7–11	20	2.5	1.2	31
Plating	Acid mist collection	1–30	1.6–2	1.7	1	32
Printing	Press rooms	0.001–0.01	10	3	1.5	33
Rock products	Crushing, screening	10–30	5–10	2.5	2.5	34
	Dryers, kilns	7–20	10–15	2.5	2.5	35
Steelmaking	Open-hearth furnace	1–20	0.5–3	2	5.2	36
	Electric-arc furnaces	1–5	0.5–1	2	5.2	37
	Basic oxygen furnaces	5–30	0.5–1	2	5.2	38
	Scarfing	1–5	1.5	2.5	6	39
	Sintering, pelletizing	10–20	5–12	2	5.4	40
	Strip mills	1–5	5	2.5	7	41
Textiles	Carding, weaving	0.001–2	3–10	2.5	1.2	42
	Synthetic fiber drawing	0.01–2	5	2	1	43
Woodworking	Sawing, planing	7–12	10–1000	5	1	44
	Sanding	7–10	10–30	2.6	1	45
Miscellaneous	Incinerators	1–15	10–30	3	0.7–2	46
	Engine, compressor intakes	0.001–0.03	1–30	2	2.5	47

*Operation codes are used in Table 20.10.

TABLE 20.2   Total Suspended Particulate Concentrations for Major U.S. Population Centers, 1981

Population center	TSP, $\mu g/m^3$	Population center	TSP, $\mu g/m^3$
Birmingham, AL	111	Long Branch, Asbury Park, NJ	62
Phoenix, AZ	178	Newark, NJ	95
Tucson, AZ	53	New Brunswick, NJ	69
Anaheim, Santa Ana, CA	104	Albany, Schenectady, NY	59
Fresno, CA	109	Buffalo, NY	97
Los Angeles, Long Beach, CA	121	Nassau, Suffolk Counties, NY	56
Oxnard, Ventura, CA	90	New York, NY	68
Riverside, San Bernadino, CA	157	Rochester, NY	73
Sacramento, CA	68	Syracuse, NY	76
San Diego, CA	95	Charlotte, Gastonia, NC	67
San Francisco, Oakland, CA	56	Greensboro, Winston-Salem, NC	61
San Jose, CA	64	Raleigh, Durham, NC	53
Denver, Boulder, CO	183	Akron, OH	67
Hartford, CT	47	Cincinnati, OH	84
Wilmington, DE	65	Cleveland, OH	129
Washington, DC	67	Columbia, OH	74
Ft. Lauderdale, FL	69	Dayton, OH	77
Jacksonville, FL	79	Toledo, OH	72
Miami, FL	97	Youngstown, Warren, OH	112
Orlando, FL	67	Oklahoma City, OK	96
Tampa, St. Petersburg, FL	87	Tulsa, OK	99
West Palm Beach, FL	59	Portland, OR	114
Atlanta, GA	79	Allentown, Easton, PA	84
Honolulu, HI	51	Philadelphia, PA	75
Chicago, IL	111	Pittsburgh, PA	100
Gary, Hammond, IN	121	Scranton, Wilkes-Barre, PA	61
Indianapolis, IN	80	San Juan, PR	94
Louisville, KY	92	Providence, Pawtucket, RI	57
New Orleans, LA	82	Greenville, Spartanburg, SC	63
Baltimore, MD	90	Memphis, TN	74
Boston, MA	62	Nashville, TN	74
Springfield, Holyoke, MA	73	Austin, TX	96
Detroit, MI	116	Dallas, Ft. Worth, TX	77
Flint, MI	60	Houston, TX	159
Grand Rapids, MI	58	San Antonio, TX	73
Minneapolis, St. Paul, MN	100	Salt Lake City, UT	67
Kansas City, MO	96	Norfolk, Portsmouth, VA	64
St. Louis, MO	190	Richmond, VA	50
Omaha, NE	63	Seattle, Everett, WA	87
Jersey City, NJ	78	Milwaukee, WI	73

Note: Population center = metropolitan statistical area.
Source: Reference 4.

and particle-size measurements to be made. Table 20.4 gives typical values for MMD (geometric mass-mean diameter), σ_g (geometric standard deviation), and density for these aerosol modes. Table 20.4 also lists models for the percentage of the three modes in aerosols in different geographical situations. These models, coupled with TSP concentration data from the Environmental Protection Agency (EPA), make it possible to characterize the outdoor aerosol with sufficient accuracy for the calculation of air filtration system performance

TABLE 20.3 Rural Aerosol Concentrations in the United States

Site	Concentration, $\mu g/m^3$	Site	Concentration, $\mu g/m^3$
Clanton, AL	54	Kalispell, MT	75
Florence, AL	75	North Platte, NE	66
Kenai, AK	47	Winnemuca, NV	80
Kingman, AZ	56	Northumberland, NH	46
Miami, AZ	51	Glassboro, NJ	48
Yuma Co., AZ	123	Roswell, NM	98
Dumas, AR	98	Zuni Pueblo, NM	41
Russellville, AR	78	Lake Placid, NY	33
Alturas, CA	47	Ulster Co., NY	35
Barstow, CA	152	Warsaw, NY	42
El Centro, CA	112	Brevard, NC	41
Napa, CA	59	Chapel Hill, NC	48
Visalia, CA	167	New Bern, SC	58
Clear Creek, CO	80	Sheridan Co., ND	40
Cortez, CO	49	Findlay, OH	75
La Junta, CO	97	Chillicothe, OH	67
Danbury, CT	64	Coshocton, OH	60
Seaford, DE	57	McAlester, OK	91
Merritt Is., FL	40	Corvallis, OR	35
Naples, FL	39	Bend, OR	64
Okeechobee, FL	27	Erie Co., PA	49
Panama City, FL	47	Johnstown, PA	112
Putnam Co., FL	36	York Co., PA	60
Dalton, GA	51	Sumter, SC	59
Cordele, GA	50	Huron, SD	48
Maui Co., HI	60	Dyersburg, TN	76
Blaine Co., ID	45	McMinville, TN	63
Mt. Vernon, IL	70	Bryan, TX	63
Petersburg, IL	64	Eagle Pass, TX	117
Bloomington, IN	75	Jeff Davis Co., TX	24
Estherville, IA	48	McAllen Co., TX	85
Dodge City, KS	53	Plainview, TX	150
Glasgow, KY	70	San Juan Co., UT	37
Iberville Pat., LA	51	Tooele, UT	58
Millinocket, ME	38	Rutland, VT	56
Acadia National Park, ME	21	Accomack, VA	52
Garrett Co., MD	48	Martinsville, VA	56
Athol, MA	49	Aberdeen, WA	41
Plymouth, MA	41	Moses Lake, WA	75
Albion, MI	44	Mt. Vernon, WA	47
Petosky, MI	57	Weston, WV	77
Hibbing, MN	46	La Crosse, WI	51
Stearns Co., MN	52	Rhinelander, WI	40
Columbia, MO	80	Converse Co., WY	25
Big Horn Co., MT	37	Yellowstone Park, WY	13

Source: EPA-450/2-78/040 (Ref. 5).

TABLE 20.4   Recommended Models for Ambient Aerosols

	Coarse (crustal)	Fine (accumulation)	Nuclei
Aerosol mode			
MMD, μm	18	0.4	0.02
σ_g†	4.8	3.1	2.0
Density ρ, g/cm³	2.5	1.0	1.0
Composition models			
Urban, high automotive, arid, %	13.0	86.1	0.9
Urban, low automotive, arid, %	45.0	54.5	0.5
Urban, high automotive, temperate, %	30.0	69.4	0.6
Urban, low automotive, temperate, %	25.0	74.3	0.7
Rural, arid, %	55.0	44.5	0.5
Rural, temperate, %	20.0	79.4	0.6
Dust storms,‡ %	100.0	—	—
Cigarette smoke (sidestream), %	—	80	20
Radon progeny, %	—	20	80
Street dust, %	100.0	—	—

†$\sigma_g = d_{84.13\%}/d_{50\%}$ (geometric standard deviation), where $d_p\%$ = diameter above which lies p percent of the mass.
‡Concentration is 1 to 200 mg/m³.
Sources: References 7–11.

(see Sec. 20.4.3). Some important aerosols appearing indoors are also characterized in this table.

20.2.6   Liquid Aerosols

The most common liquid aerosols are water: fog and rain. These are nontoxic but may be harmful to air filtration equipment. Electrostatic air filters, e.g., cannot operate when free water covers their high-voltage insulators; some fibrous filter mats collapse when wetted, non-woven filter papers (such as those used in medium and high efficiency pleated filters) will become impervious to airflow and the pores of activated carbon can be plugged and rendered ineffective with liquid water. Hence systems with danger of heavy rain or fog exposure must be protected with louvers and/or coalescer pads (coarse fiber filters) which can capture and drain off water droplets. In severe situations, dehumidification and reheat of the gas stream may also be required. Liquid aerosols sometimes appear in industrial situations (e.g., paint overspray, textile fiber finishes, machine-tool coolants, ink mists). Fiber-bed plugging can be serious if the liquid viscosity is high. Reentrainment of collected liquid into the air stream is a problem if viscosity is low. Special equipment types have been developed which allow collected liquid aerosols to drain from their filter media. Volatile aerosols may be collected by an air filter, evaporate, pass through the filter, condense, and thus reappear downstream of the filter.

20.2.7   Radioactive Aerosols

Some alpha-particle emitters produce recoil effects which can dislodge collected particles from filter fibers. Intense radiation from collected particles may damage filter media, seals, and structures. Reliability requirements for air filtration systems in nuclear applications are very stringent, and the handling and disposal of collected dust and used filters poses special problems. The needs of this technology have been dealt with in extensive literature,[12–15] some of which is applicable to general filtration problems.

Radon gas creates unique air filtration problems. It enters a space as a gas, but then by radioactive decay it is converted to extremely fine (<0.02 μm) "progeny" or "daughter" particles. (See Table 20.4.)

20.2.8 Biological Aerosols

Microorganisms (bacteria, viruses, pollen, spores) have the filtration properties of other aerosols of like size and density. Densities for pollen (~1.0 g/cm^3) are, however, less than densities for soil particles, which range from 1.0 to 20.0 μm in diameter. Under appropriate environmental conditions, viable organisms can multiply after being captured by a filter and grow through it. In recent years concerns about indoor air quality and *Sick Building Syndrome* and the effects that these might have on the health of building occupants has led to renewed interest in the role that filters can play in both causing and preventing such problems. The application of antimicrobial compounds in the form of biocides and biostats to filter media in order to limit the growth of biological organisms is currently one of the more important innovations in filtration technology.

20.2.9 Fibrous Aerosols

Lint, plant parts, hairs, and fibers of polymers, glass, and asbestos all become airborne. With the exception of asbestos and debris from raw cotton and similar vegetable fibers, these are generally nontoxic. They may, however, be very damaging to mechanisms such as computer disk drives or to processes like paint spraying or photography. The lengths of fibrous aerosols make them relatively easy to filter, but they tend to form mats over the surfaces of filter media. This causes cake formation and rapid increase of air-flow resistance. Special filters with very thin, filter webs which are automatically "renewed" are used to collect the high fiber concentration found in textile and printing plants.

20.2.10 Aerosol Charging

A substantial fraction of all aerosols carry electric charges of either + or − sign. Aerosols which are freshly generated, whether by grinding, spraying, or condensation from combustion gases, carry somewhat higher charges than aged aerosols. The maximum charge which can be put on an aerosol particle is a function[2,3,9] of its diameter, its dielectric constant, and the ion density, electric field, and gas temperature in the vicinity of the particle. Charged particles move in electric fields, and create fields. These effects are the basis of electrostatic precipitators and electrically augmented air filters, whose operation is described in Sec. 20.5. Charged aerosol particles may be rendered nearly neutral by flooding them with a cloud of bipolar ions, containing nearly equal numbers of positive and negative ions.

20.3 CONTAMINANT EFFECTS

The control of gaseous contaminants in ventilation air and in stack gases is based on the damage they can inflict on people, processes, materials, or the general environment. The decision to control a given contaminant therefore depends on determination of its damaging properties. For some gaseous contaminants, this determination has been quite thorough, and allowable air concentration levels in workplaces or stack emission rates are mandated by federal and local laws. For others, contaminant control levels must be based on published studies and such poorly defined quantities as odor thresholds.

20.3.1 Gaseous Contaminant Toxicity

Air pollution health effects usually depend on the cumulative dose of contaminant inhaled. However, there are upper limits to permissible concentration even for short exposure periods. Table 20.5 lists a few well-recognized toxic gaseous contaminants along with their physical and toxicological properties. The abbreviations in the table are defined as follows:

$$MW = \text{molecular weight, g/(gmol)}$$
$$BP = \text{boiling point, C at 1-atm (760-mmHg) pressure}$$
$$DEN = \text{contaminant gas density, g/m}^3$$

Maximum allowable concentrations (MACs):

IDLH = concentration immediately dangerous to life and health

AMP = acceptable maximum peak for short-term exposure

ACC = acceptable ceiling concentration, not to be exceeded during any 8-h shift, except for periods defined by AMP

TWA8 = time-weighted average concentration, not to be exceeded in any 8-h period of 40-h week

OTH = odor threshold, concentration below which 50% of observers cannot detect contaminant's presence

Alternative terms for the above items are MAC (maximum acceptable ceiling) = AMP, STEL (short-term exposure limit) = ACC, and TLV (threshold limit value) = TWA8. The AMP, ACC, and TWA8 concentration levels are requirements of OSHA (Occupational Safety and Health Administration). Table 20.6 lists the group of contaminants defined by OSHA as "high-hazard"; these must be controlled in ways specified by OSHA. Reference 19 gives details on the control regulations for these and several hundred other gaseous contaminants. Medical opinion is by no means unanimous that the allowed concentrations are correct or that all significant toxic substances are on the lists.

Air known to contain the substances in Table 20.6 in clearly detectable amounts must not be recirculated. Trace quantities of some compounds in this list have been detected in tobacco smoke, and other highly suspect carcinogens are present in tobacco smoke. In most cases, however, the annoyance level of contaminants—odor, lacrimation, and reduction in visibility—rather than matters of health, will determine control levels. In general, we find that

$$IDLH > AMP > ACC > TWA8 > OTH$$

There are important exceptions to this sequence, however; e.g., carbon dioxide and carbon monoxide have no odor. Such substances give no warning of their presence and so are very dangerous. Air-handling systems must provide enough makeup air to eliminate the possibility of odorless gases reaching dangerous levels.

TABLE 20.5 Gaseous Air Pollutants: Toxicity, Odor, and Physical Properties

Compound	Allowable concentration, mg/m³					BP, °C	MW
	IDLH	AMP	ACC	TWA8	OTH		
Acetone	48,000			2,400	47	56	58
Acrolein	13			0.25	0.35	52	56 T
Acrylonitrile	10			45	50	77	53
Allyl chloride	810		9	3	1.4	44	77
Ammonia	350		35	35	33	−33	17 T
Benzene	10,000		25	5	15	80	78 T
Benzyl chloride	50			5	0.2	179	127
Carbon dioxide	90,000		54,000	9,000	x	−78	44
Carbon disulfide	1,500	300	90	60	0.6	46	76
Carbon monoxide	1,650		220	55	x	313	28 T
Carbon tetrachloride	1,800	1,200	150	60	130	77	154
Chlorine	75		1.5	3	0.007	−34	71
Chloroform	4,800		9.6	240	1.5	124	119
Chloroprene	1,440		3.6	90		120	89
p-Cresol	1,100			22	0.056	305	108 T
Dichlorodifluoromethane	250,000			4,950	5,400	−30	121
Dioxane	720			360	304	101	68
Ethylene Dibromide	3,110	271	233	155		131	188
Ethylene Dichloride	4,100	818	410	205	25	84	99
Ethylene Oxide	1,400		135	90	196	10	44
Formaldehyde	124	12	6	4	1.2	97	30
n-Heptane	17,000			2,000	2.4	98	100
Hydrogen fluoride	13		5	2	2.7	19	20
Hydrogen sulfide	420	70	28	30	0.007	−60	34 T

Substance	IDLH	AMP	ACC	TWA8	OTH	BP	MW
2-Butanone (MEK)	8,850			590	30	79	72 T
Mercury	28			0.1	x	357	201
Methanol	32,500			260	130	64	32 T
Methylene chloride	7,500		3,480	1,740	750	40	85 T
Nitric acid	250			5		84	63
Nitrogen dioxide	90			9		21	46
Ozone	20		1.8	2	51	-112	48
Parathion	20			0.11	0.2	375	291
Phenol	380			19		182	94 T
Phosgene	8		60	0.4	0.18	8	90
Sulfur dioxide	260			13	4	-10	64 T
Sulfuric acid	80		0.8	1	1.2	270	98
Tetrachloroethane	1,050			35	1	146	108
Tetrachloroethylene	3,430	2,060	1,372	686	24	121	166
o-Toluidine	440			22	140	199	107
Toluene	7,600	1,900	1,140	760	24	111	92 T
Toluene diisocyanate	70		0.14	0.14	8	251	174
1,1,1-Trichloroethane	2,250			45	15	113	133
Trichloroethylene	5,413	1,620	1,080	541	1.1	87	131
Vinyl chloride		1,080	0.014	0.003	120	-14	63
p-Xylene	43,500		870	435	2	137	106 T

Notes:

IDLH, AMP, ACC, TWA8, and OTH are defined in the text.

BP = boiling point at 760 mmHg; MW = molecular weight, g/gmol.

The letter T in the rightmost column indicates that this compound has been found in tobacco smoke.

To convert from mg/m³ to ppm, use the following equation: $\text{ppm} = 9.62\, sP/[(MW)(t + 273)]$, where

P = carrier gas pressure, mmHg; t = carrier gas temperature, °C; and s = contaminant concentration, mg/m³.

Source: References 18 through 20.

TABLE 20.6   OSHA-Regulated High-Hazard Toxic Airborne Contaminants

Asbestos	4-Aminodiphenyl
4-Nitrobiphenyl	Ethyleneimine
α-Naphthylamine	β-Propiolactone
Methyl chloromethyl ether	2-Acetylaminofluorene
3,3′-Dichlorobenzidine	4-Dimethylaminoazobenzene
(and its salts)	N-Nitrosodimethylamine
bis-Chloromethyl ether	Vinyl chloride (monomer)
β-Naphthylaminc	Benzene
Benzidine	Coke oven emissions

Note: 4,4-Methylene-bis(2-chloroaniline) (MOCA) was originally on this list, but was removed for legal reasons rather than because of any proof of lower carcinogenicity.
Source: Reference 19, secs. 1910.1001 through 1910.1029.

The allowable levels of gaseous contaminant emissions from stacks are determined first by EPA regulations. Very few stack gas emissions are controlled as of this writing; Table 20.7 lists those that are. Beyond this, odors, corrosive effects, and the desire to live in peace with one's neighbors often determine which contaminants get controlled.

20.3.2 Odors

Individual humans have widely different abilities to detect the odor of any one compound. Perceived odor intensities are approximately proportional to concentration to the nth power, with $0.2 < n < 0.7$. The nose sometimes quickly loses its sensitivity to a compound on exposure; in other cases, an odor can be "learned" and hence becomes easier to detect at lower levels. Odor thresholds, the intensity of odors in a space, or the effectiveness of an odor control device must be determined by panels of observers making "blind" choices in

TABLE 20.7   National Ambient Air-Quality Standards, 1985

Pollutant	Averaging time*	Primary standard,† $\mu g/m^3$	Secondary standard,‡ $\mu g/m^3$
Total suspended particulate matter	Annual	75	60
	24 h	260	150
Sulfur oxides	Annual	80	—
	24 h	365	—
	3 h	—	1,300
Carbon monoxide	8 h	10,000	10,000
	1 h	40,000	40,000
Nitrogen oxides (NO_2)	Annual	100	100
Ozone	1 h	240	240
Nonmethane hydrocarbons	3 h	160	160
Lead	3 months	1.5	1.5

*Not to be exceeded more than once a year.
†This standard is intended to protect public health.
‡This standard is intended to protect the general environment from damage and to avoid problems with visibility, comfort, etc.
Source: Reference 20.

comparison with standards. Testimonials without such experimental controls are useless. The OTH values listed in Table 20.5 have been gathered from generally respected sources, but should be taken as order-of-magnitude guidelines rather than absolute values.

Stack emissions cause no odor problems if they are diluted to the point where odorous compounds are below threshold concentrations at ground levels. Determining what emission level is permissible before control equipment is installed requires a calculation of the diffusion of the stack plume over the surrounding countryside. Standard methods have been established for doing these calculations.[21]

20.3.3   Special Contaminants: Ozone and Radon

Ozone (O_3) is a highly reactive, unstable form of oxygen (O_2). Ozone quickly disappears in a confined space unless it is continuously resupplied to the space. Office copiers, welding, electric-motor commutators, and incorrectly designed electronic air cleaners (including desktop units) all produce ozone. Ozone has a strong odor, and persons exposed to it will ordinarily complain of its presence before concentrations reach hazardous levels—but not necessarily before they develop headaches or eye irritation. *Ozone is very corrosive to rubber and plastics.*

Radon is present in soils, groundwater, stone, concrete, indeed virtually anything that comes out of the ground. It is a harmful gas when inhaled, but by radioactive decay it generates far more toxic "progeny." These are condensation aerosols of polonium, bismuth, and lead, with diameters from 0.01 to 0.1 μm; they are trapped deep within the human lung, where they emit beta and alpha particles. There is strong evidence that radon progeny form a significant source of lung cancer. Reference 22 reviews the current knowledge of the radon problem.

20.3.4   Material and Product Damage

Both gaseous and particulate contaminants damage structures and process materials. A common example is wall discoloration and streaking. The dark patterns on walls are the result of particle deposition under combined turbulence, thermal, and electrostatic effects. Overall discoloration of walls can be due to deposition of tar aerosols (from tobacco smoke) or reactions with gaseous contaminants, such as sulfur dioxide. Fabrics, silver, artworks, paper, photographic film, plastics, and elastomers are vulnerable to degradation by gaseous contaminants. Ozone, oxides of nitrogen, and sulfides and sulfates are especially prevalent and troublesome.

The only sure way to decrease the rate of wall discoloration is to remove the contaminants causing it from ventilation air. The ASHRAE dust spot efficiency test for particulate air filters[23] is in essence a measure of wall blackening. Measurement of other damaging effects is more difficult, for one must usually measure the long-term cumulative exposure from very low concentrations.

Some industrial operations require careful filtration of process gases and the air in the work environment. Paint spraying and high-quality printing are examples of operations where dust causes product defects. The most demanding area for this type of control is semiconductor microcircuit manufacture. Here particles with as small as a 0.01-μm diameter can produce defects and reduce yields significantly. To cope with this cleanliness requirement, "clean rooms" are constructed. A typical configuration is shown in Fig. 20.7. The activities in the room are continuously washed with a downflow of air that is essentially particle-free. The air leaving through the floor grills carries off internally generated particles which might otherwise diffuse onto the product. Similar arrangements are used in

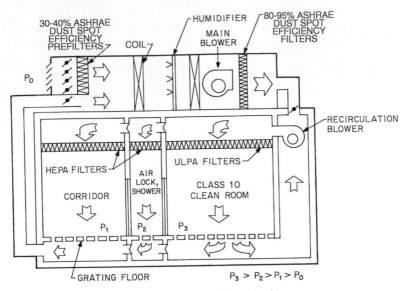

FIGURE 20.7   Typical downflow clean-room configuration, classes.

pharmaceutical production to maintain sterility and chemical purity. Figure 20.6 defines the clean-room classes established under Federal Standard 209.[24] For a room to qualify for a given class, the per-cubic-foot count of particles larger than each size must fall below the line labeled for that class. Standard 209 details procedures for the counting process.

20.3.5   Particulate Contaminant Toxicity

Reference 19 lists OSHA-mandated TWA8 and ACC values for both particulate and gaseous contaminants. Among those listed are heavy metals; organotin compounds; beryllium and cadmium; silica, talc, and some other minerals; several pesticides and herbicides; and raw cotton dust. Asbestos is singled out along with some other carcinogens for prescriptive control measures, rather than allowable concentrations. *This list does not include all materials which have been implicated by toxicity studies.*

References 16, 17, 19, and 25 provide additional guidance with regard to allowable concentrations for particulate contaminants, and the journals listed in the Bibliography provide periodic reviews of toxicity knowledge as well as reports of studies on specific materials. Allowable airborne levels of radioactive substances are controlled by a different set of regulations, promulgated in most of the U.S. state agencies acting on behalf of the U.S. Nuclear Regulatory Commission. Reference 20 defines these limits, which are unique in that the age of the exposed subjects is an important factor.

20.3.6   Environmental Damage

Pollutants emitted into the atmosphere are windborne away from the emission point. Both aerosols and gaseous contaminants can travel immense distances; *dusts from the Sahara have been detected in North America.* The major impact of industrial pollutants is relatively close to the emission point since concentrations diminish as we move away from the source. Concentration at a distance from a source is a function of wind speed, stack height stack velocity, meteorological parameters, and observer location. (See Sec. 20.4.2 for details).

The damage caused to the environment may be due to the contaminant actually emitted, e.g., sulfur dioxide destroying plant growth near a copper smelter, or the damage may be due to some sort of interaction between the pollutant and other compounds. The creation of "smog" or "brown clouds," e.g., is a complex chemical reaction involving organic vapors from vehicles, combustion products, and energy supplied by sunlight. *Direct* damage may occur when the pollutant finally deposits on the earth's surface, as in the case of acid rain. *Indirect* damage may result from the increased incidence of ultraviolet radiation at the earth's surface resulting from impairment of the atmosphere's ozone layer due to the effects of various pollutants. The phenomena involved are complex, and data interpretation is frequently controversial.

Engineers concerned with air pollution control must meet at least the standards set by the EPA for emission levels. These standards appear in the Federal Register and are gathered in Ref. 26. Local standards may be more stringent. In general, emission standards limit both concentration and total emission, with stricter concentration limits on larger installations. Sometimes (especially in the case of radioactive emissions) the concentration at the boundary of the owner's site is controlled. Where this is the case, calculation of dispersion from the emission point is required.

20.3.7 Visibility Problems

Aerosols reduce visibility by simultaneously absorbing and scattering light. The percentage of light transmitted by a column of polluted air is

$$T = 100e^{-NC_s x} \qquad (20.4)$$

where T = transmittance, percent
N = no, particles per volume of air, m^{-3}
C_s = scattering cross section for particle, m^2
x = length of air column, m

Here C_s is a complicated function of light wavelength and the diameter and index of refraction of the particles. In general, it has a maximum when the particles radius is near the wavelength of the light (about 0.4 to 0.7 μm for the visible spectrum). Since N is usually larger for small particles, aerosols with diameters near 0.4 μm are the most troublesome from the standpoint of visibility. Formerly the "blue haze" observed in sports arenas was the result of absorption and scattering of light by tobacco smoke aerosols, which are typically in this size range. Other examples from industry are welding fumes from welding operations and oil mist from machining operations.

The visibility of smoke plumes emerging from a stack was at one time a measure of the acceptability of the discharge. Present-day air pollution control equipment reduces discharges to near invisibility, except for harmless water droplets, which evaporate and become invisible almost instantly. *The classic Ringlemann chart method of smoke shade measurements has been shown to be very subjective and is rarely specified.* Instruments which measure the light transmission across a stack by lamp or photocell systems are sometimes used to monitor the performance of an air pollution control system, but cannot measure the actual mass of aerosols emitted.

20.3.8 Nuisances

Dust kittens, cobwebs, minor eye irritation, nontoxic fallout, and temporary stimulation of coughing and sneezing may all be considered nuisances caused by air contaminants. These matters alone may dictate the quality of gas cleaning in an installation, but they are virtually impossible to quantify.

20.3.9 Flame and Explosion Hazards

Combustible gaseous contaminants can, under certain conditions, burn or explode. The minimum vapor concentration in air necessary to support combustion, called the "lower flammable limit" or "lower explosive limit" (LEL), ranges from about 1 to 20 percent by volume. Safety engineers recommend holding calculated concentrations to below one-fourth the LEL, to allow for imperfect mixing with ventilation air. Even those concentrations will usually be many times toxic or odorous levels, hence not likely to exist in spaces ventilated for human occupancy. Nevertheless, the design process should examine the possibility of reaching one-fourth the LEL and should provide sufficient ventilation or dilution air in a reliable manner.

Combustible dusts are also explosive. This is a serious problem in pollution control systems, where even if average concentrations are below the LEL, there are many opportunities for locally higher concentrations. It is important to keep runs to pollution control equipment short and clean, to prevent dust buildup. In many cases, vents are desirable to allow pressure relief when an explosion does occur. The National Fire Protection Association (NFPA) provides codes for appropriate vent designs and tables of LEL values for both gases and dusts.[27,28]

20.4 AIR QUALITY

20.4.1 Outdoor Air Quality

In the United States, outdoor air quality is regulated on a national scale by several acts of Congress from 1955 to 1977. During this period, substantial research was undertaken by the Environmental Protection Agency to determine existing levels of pollution, detrimental effects of pollution, and the technical feasibility and cost of air pollution. By 1980, it became apparent that the standards mandated by Congress were not being met. Thereafter, regulatory effort was directed in the following manner:

• Strict control of airborne carcinogens and other pollutants of verified, high toxic risk (see Table 20.6).
• Use of best-available control technology (BACT) for air pollution from new plants in specific industries [new source performance standards (NSPS)].
• Maintenance of areas of the nation with high-quality air as close as possible to their present condition by prevention of significant deterioration (PSD) rules.
• Use of "offsets" and "bubbles," allowing the owner of plants in a given region to emit excess pollution from a new source, provided a greater compensating reduction is made on emissions from a existing source.

This regulatory procedure makes for considerable complexity, but also allows considerable exercise of engineering skill to minimize pollution. For large facilities at least, extensive environmental impact analyses are required before an EPA construction license can be obtained. References 29 and 30 give details on these regulations and their impact on pollution control equipment design. The fact that air-quality levels and emission standards have been promulgated for very few substances does not exempt the designer of pollution control systems from paying close attention to the hazards possible from the release of other substances. References 3, 7, 11, 16, 19, 25, 29, 30, and 31 all provide guidance on the current understanding of the toxicity of air pollutants, as do the sources listed in the Bibliography.

The goals for outdoor air quality remain those established by Congress. However, only a few such goals—national ambient air-quality standards (NAAQS)—have been promulgated (See Table 20.7). Reports[4] are made periodically showing the level of attainment of these standards throughout the United States. It is these reports which provide the estimates of existing concentrations, used in Sec. 20.2.

20.4.2 Dispersion of Air Contaminants

Pollutants emitted outdoors from a point source are carried away by winds. As a plume moves with the wind, air turbulence causes it to spread, both horizontally and vertically. The pollutant concentration in the plume thus (on average) decreases with the distance from the source. Plume concentrations are predictable with reasonable accuracy, provided that the meteorological parameters which define the wind and turbulence structure are known and, of course, that the physical properties of the contaminant and such items as height of release point, stack exit velocity, gas temperatures, etc., are known. Reference 21 provides a review of these relationships. With a reasonable amount of calculation one can estimate the average and peak concentrations to be expected near pollutant-emitting stacks and thus the potential for health hazard, odors, or nuisance.

Not all air pollutants are emitted from stacks. So-called fugitive emissions, which escape directly from processes without being captured by a hood, may contribute greatly to the surrounding pollution. Reference 32 shows that total material escaping from a process equipped with a hood and an air pollution control system is

$$G_t = (1 - 10^{-4} E_h E_e)G \qquad (20.5)$$

where G_t = total emission to surroundings, g/s
$\quad E_h$ = hood capture efficiency, percent
$\quad E_e$ = air pollution control equipment efficiency, percent
$\quad G$ = process pollutant emission rate, g/s

Clearly, it makes no sense to invest in high-efficiency pollution control equipment and then couple it with a hood system of low efficiency. Reference 33 is a guide to proven hood design principles.

Dispersion of contaminants indoors is more difficult to predict than outdoors, for there are ordinarily many sources, many cross currents of air flow, and even the motion of people within the space affects dispersion. Most investigation of indoor contaminant dispersion has been done in connection with clean rooms. These usually have a relatively low uniform air velocity directed downward from a ceiling completely covered with air filters. Pollutants in these "laminar-flow rooms" are washed downward and away from the workspaces in the room, leaving either through low sidewall vents or a perforated floor. In more conventional rooms, the most common assumption is that pollutants are completely mixed with ventilation air.

20.4.3 Indoor Air Quality

Ventilation.   Most city dwellers spend the majority of their lives indoors; indoor air quality is therefore of equal or greater importance to them than outdoor air quality. At present,
there are no regulations on indoor air quality except in industrial workplaces (see Sec. 20.3.1
and Ref. 19) and implicitly in the ventilation requirements of building codes. Indoor air
quality can be maintained by ventilation alone, provided outdoor air for ventilation is itself
acceptable. A minimum "fresh" air supply of about 5 ft³/min (8.5 m³/h) is needed for each
occupant of a space, just to carry away carbon dioxide and supply oxygen. This cannot be
made up by recirculation without the very complicated systems used in spacecraft and
submersibles. Such minimal ventilation is not, however, adequate to eliminate odors,
internally generated contaminants such as cigarette smoke, or in some cases radon infiltration. ASHRAE Standard 62-89 establishes higher ventilation rates, based on space use
and number or occupants in the space.[35] A proposed revision 62-89R, is now being
reviewed by ASHRAE. Obtain the final version before proceeding with any operations.
ASHRAE 55-1992 "Thermal Environmental Conditions for Human Occupancy" should
also be consulted. Ventilation rates as high as seven times the above minimum are recommended for some situations. The standard allows fresh air above the 5 ft³/min (8.5 m³/h)
minimum to be obtained by adequate filtration processes and sets standards for the allowable contaminant content after filtration. The specified levels for particulates are fairly
easy to obtain; specified gaseous contaminant levels are very difficult to obtain with the
present state of the art.

Calculation of Indoor Air Quality.   Both particulate and gaseous contaminant concentrations can be calculated with reasonable accuracy by use of system models, such as shown
in Fig. 20.8. The steady-state concentration in the space, assuming complete mixing, is

$$C_i = \frac{G_i + 0.01(P_iQ_i + 0.01P_2P_1Q_m)C_o}{Q_{v2} + Q_x + K_dA_d + Q_r(1 - 0.01P_2)} \qquad (20.6)$$

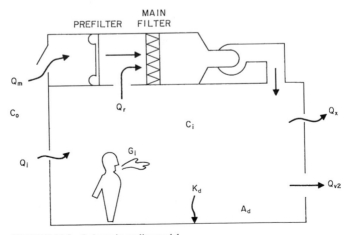

FIGURE 20.8   Indoor air-quality model.

where C_i = steady-state indoor concentration, mg/m^3
$\quad\quad C_o$ = outdoor concentration, mg/m^3
$\quad\quad G_i$ = rate of indoor generation, mg/min
$\quad\quad Q_i$ = air leakage through cracks into space, m^3/min
$\quad\quad Q_m$ = makeup air flow through filtration system, m^3/min
$\quad\quad Q_x$ = air leakage from space, m^3/min
$\quad\quad Q_{v2}$ = exhaust air, m^3/min
$\quad\quad Q_r$ = recirculation air flow, m^3/min
$\quad\quad P_1$ = penetration of contaminant through prefilter, percent
$\quad\quad P_2$ = penetration of contaminant through main filter, percent
$\quad\quad P_i$ = penetration of outdoor contaminant through cracks, percent
$\quad\quad K_d$ = indoor dust deposition velocity, m/min
$\quad\quad A_d$ = indoor dust deposition area, m^2

Equation (20.6) applies to each contaminant, i.e., to each gaseous contaminant compound, and to each particle size, in the case of particulate contaminants. This means that, in the particulate case, it is necessary to divide the aerosol spectrum into narrow bands and make a calculation for that narrow size range using the filter penetrations appropriate to that size. Hand calculation is extremely tedious, but computer programs are available to do the work. Reference 35 details these calculations and includes procedures for more complicated recirculation patterns.

Infiltration and Exfiltration. Equation 20.6 assumes that the amount of air entering the space through cracks and leaving, either through cracks or intentional exhausts, is known. The *ASHRAE Handbook, Fundamentals,*[36] provides considerable guidance on this subject. For spaces where infiltration is not allowable (such as clean rooms), the protected space must be pressurized to prevent air entry through any path other than the filter system. Pressurization of about 25 Pa (0.1 in WG) is generally adequate for this purpose.

20.5 PARTICULATE AIR FILTERS

20.5.1 Operating Principles

Fibrous Filters. Fibrous air filters have four general configurations:

1. *Panel filters*, where the filter medium is very open, with low air-flow resistance. Medium is held in relatively thin panels oriented perpendicular to the main air-flow direction so that the medium velocity is equal to the duct velocity.

2. *Pleated-medium filters*, using denser, high-resistance medium pleated in zig-zag fashion so that considerable medium area is available and air velocity through the medium is low. The filter medium is held more or less rigidly in parallel planes by spacers or separators, which can be corrugated paper or aluminum inserts, or ribbons, threads, or glue-beads attached to the surface of the filter media, or corrugations of the media itself.

3. *Tubular pocket or bag filters*, where the medium is formed into tubes or pockets which are closed at one end and sealed at the opposite end into a header plate or frame. The aim of this construction is the same as in pleating—to expand the medium area, reduce medium velocity and hence pressure drop and increase efficiency. These filters are usually collapsible and are held in extended form by their own air-flow resistance or by wire-frame supports.

4. *Roll filters*, where a flexible, compressible panel-type medium of glass or polymer fiber is supplied in the same manner as camera roll film. Clean filter medium is usually

advanced into the air-filtering zone by an automatic transport mechanism, activated by a clock or sensor of resistance or medium dust load.

Fibrous filter media are nonwoven webs or blankets of glass, polymeric, metal, ceramic, or natural fibers. Fiber diameters range from about 0.2 to 100 μm. Media fibers are normally bonded together to give dimensional stability and resist compression. Practical media are made with as much open space as possible, to allow air passage at low pressure drop. Such media are often manufactured to progressively increase in density and have decreasing fiber diameter from front to back of the media with the less dense media and higher diameter fibers on the inlet air side. The purpose of this is to capture varying size particulate at different depths within the media, more fully utilizing all of the media and increasing the media dust loading capacity. They often contain no more than 1 percent solids.

Fibrous air filters capture particles by a combination of the following effects:

Aerodynamic Capture.   This occurs because air flow must change direction to pass around filter fibers. Particles have greater inertia than gas molecules and thus tend to cross flow streamlines as these streamlines bend. Particles which pass close enough to a filter fiber to strike it will usually adhere to the fiber (Fig 20.9a). This type of capture, the result of aerodynamic effects, is also called *impaction, impingement*, or *interception*. The effect increases capture probability as particle mass, density, and velocity increase.

Diffusional Capture.   This is the result of random thermal motion of the molecules of the gas carrying the particles. Individual gas molecules impact the particle from all directions. If the particle is small enough, the number of impacts in a small time interval may be predominantly from one direction. This drives the particle in erratic fashion out of its normal path (Fig. 20.9b). In effect, this erratic motion greatly increases the width of region in which a particle-fiber collision can occur. The effect increases as the particle's size decreases and as the gas velocity decreases.

Electrostatic Effects.   These are the result of unbalanced electric charges on aerosol particles and filter medium fibers or both. Some fraction of the particles in an aerosol cloud are always charged. When such a charged particle approaches a fiber, the particle induces an opposite charge on the fiber surface. These two charges produce an electric force which drives the particle toward the fiber, and they increase as the gas velocity decreases and particle size increases. The effect can be intensified by intentionally charging the aerosol particles and by impressing electric fields on the filter media (Fig. 20.9c through 20.9e)

Straining.   This occurs when particles are larger than the spaces between fibers. The effect is of little importance for ventilation-type filters, except in the collection of lint and other fibrous particles. Most aerosols are much smaller than the passages through ventilation-type filter media. Membranes, however, may have openings which are smaller than many aerosols, and straining becomes a significant capture mechanism (Fig. 20.9f).

Membrane Filters.   Membrane filters are made from very thin polymer sheets pierced by extremely small air-flow passages. They are available in several forms:

1. Collimated pore membranes, whose air passages are micrometer-diameter holes, all of nearly uniform diameter, running straight through the membrane. These membranes have a high solid content and high air-flow resistance, but can offer absolute removal of particles larger than their tube diameters.

2. Granular symmetric membranes. These are formed like sand beds or sponges on a microscopic scale.

3. Granular asymmetric membranes. These have a relatively open, spongelike structure which acts as support for a very thin face layer containing ultrafine pores. This thin layer is the actual filter.

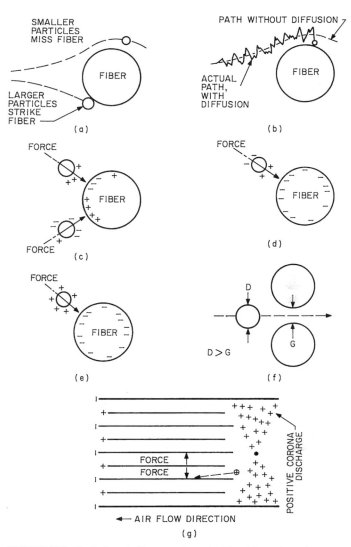

FIGURE 20.9 Particulate air filter operating principles: (*a*) aerodynamic capture; (*b*) diffusional capture; (*c*) charged-particle-induced electric force; (*d*) dielectric force with charged filter, uncharged particle; (*e*) coulomb force with charged filter, charged particle; (*f*) straining; (*g*) two-stage electrostatic air filter.

4. Fiber membranes, which are similar in structure to the fibrous media described above, but have considerably smaller-scale geometries.

   Membranes are usually pleated and formed into panels or cylindrical cartridge filters for gas filtration use. The rather fragile nature of membranes, their cost, and the relatively high pressure drop for a given flow have, to date, limited their application to the production of ultrapure gases, to air sampling, and to fabric air pollution control equipment.

Future developments might make them practical for ventilation applications. Membranes differ from usual gas filtration media in that they often have more media volume filled with solids than open space and hence provide very little space for dust storage.

Electrostatic Filters.   Electrostatic air filters for ventilation service have the form shown in Fig. 20.9g. The filter has two stages: an ionizer to charge the aerosol and a plate package to capture it. In the ionizer stage, fine tungsten wires are suspended parallel to the grounded struts, midway between them. Each wire is maintained at a high voltage (+6 to +15 kV, depending on the strut spacing). The intense electric field near the wire ionizes oxygen molecules in that area. Positive ions of each ion pair formed are driven across the space between the ionizer wire and struts. During the transit, some ions attach themselves to passing aerosol particles. These charged particles then enter the plate package. The adjacent plates of the plate package are alternately grounded and charged, so that fields perpendicular to the air flow exist between the plates. These fields drive the charged particles to the plates. Here the particles are retained, sometimes with the help of a low-flammability viscous liquid adhesive on the plates. Electrostatic air cleaners can readily remove about 90 percent of atmospheric aerosols from the air flow, with an overall pressure drop of no more than 3 mmWG. This pressure drop does not increase appreciably as dust builds up on the plates. Eventually, however, the plates must be washed clean and recoated with adhesive.

In an alternative system, the plates are left dry, and collected aerosols are allowed to build up on the plates until they blow off. Released particles are agglomerates much larger than the entering aerosols and can be collected in a roll or fine-fiber-medium filter. The rate of resistance rise is much slower for electrostatically agglomerated aerosols; hence these systems have favorable maintenance characteristics.

Combined Forms.   Electric fields can be maintained in fiber beds, provided the fibers do not become too conductive. Such fields can aid the collection of both charged and uncharged aerosols; the effects are much stronger for charged aerosols than for uncharged. However, when the medium becomes conductive—from humidity usually but also because ambient aerosols are conductive—the collection effectiveness drops to the level of the fiber bed alone, without electrostatic augmentation. Electrets, or "permanent" electric fields, have been impressed on fibers in air filters. This gives the same effect as an externally supplied electric field. However, once the electrets are drained down by conduction, they do not necessarily recover; hence this type of medium is very humidity-sensitive. Examples of air filter configurations are shown in Fig. 20.10.

20.5.2 Performance Characteristics

Penetration and Efficiency.   Figure 20.11 shows the penetration and efficiency of a range of air filters as a function of particle diameter. The scales used in the figure (logarithmic for diameter, probability for penetration and efficiency) allow presentation of a wide range of data in a single chart. The curves were developed from test data, rather than theoretical calculations; manufacturers can supply similar curves for specific filters operating at chosen velocities. Such data are essential to the air-quality calculation described in Sec. 20.4.1. Note the penetration maxima in the region below 0.3 μm. This is the result of the competing effects of diffusion and aerodynamic forces. For low-efficiency (high-penetration) filters, another penetration peak occurs at large particle diameters, because of the poor adhesion of large particles to the filter fibers.

Pressure Drop and Service Life.   The pressure drop across an air filter (usually called "filter resistance") is dependent on the quantity of gas flow through the filter, on the

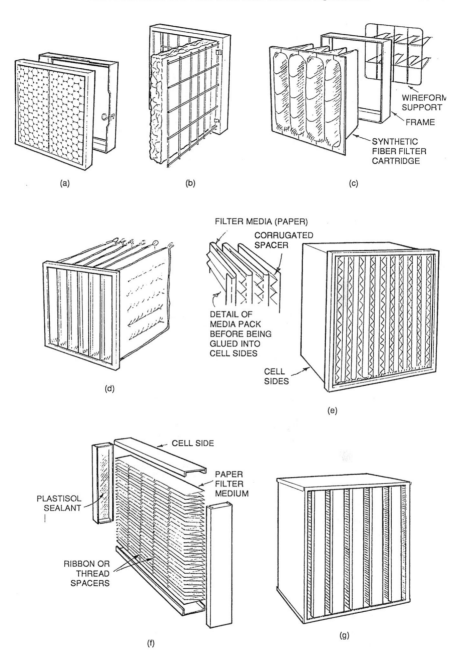

FIGURE 20.10 Particulate air filter configurations: (*a*) throwaway viscous impingement panel; (*b*) replaceable medium panel; (*c*) replaceable medium extended surface, supported; (*d*) self-supporting "soft" cartridge, extended surface; (*e*) rigid cartridge pleated medium, with separators; (*f*) rigid cartridge pleated medium, no separators; (*g*) zigzag cartridge, no separators; (*h*) roll filter; (*i*)self-cleaning viscous impingement curtain; (*j*) two-stage electrostatic air filter.

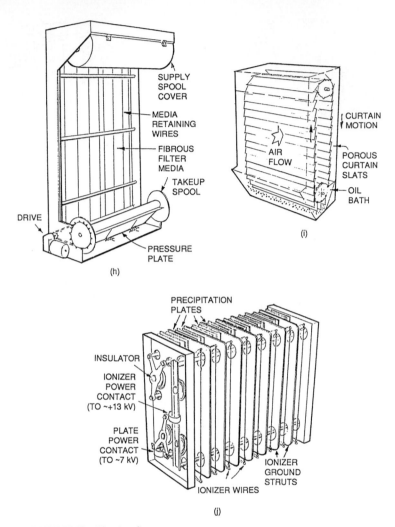

FIGURE 20.10   *(Continued)*

physical properties of the gas (density and viscosity), on the geometry of the filter, and, in the case of fibrous filters, on the amount, size, and shape of aerosol stored in the filter media. It is customary to express pressure drop as a function of the air filter average face velocity, which is air flow divided by the face area of the filter. This allows the same data to be applied to like filters of any flow capacity. Gas in passing through a filter can experience any of three flow regimes: laminar, turbulent, or an unstable, transitional regime between these two. For fine-fiber-medium filters, the flow in the fiber bed is most likely to be laminar, while the flow through entrance and exit passages is probably turbulent. Even filters which operate at duct velocity (roll filters and panel filters) will exhibit both flow regimes, but in that case the turbulent-flow term is dominant. Filter manufacturers supply curves of filter resistance versus air flow, made in accordance with the test procedures of

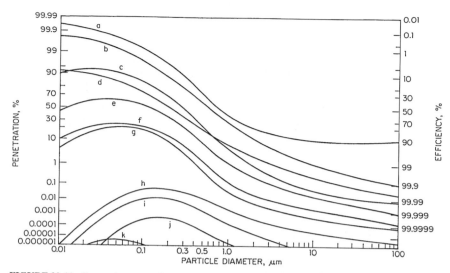

FIGURE 20.11 Panetration vs. particle size for typical air filters: (*a*) roll filter, 20% ASHRAE 52-76 dust spot efficiency (ADSE); (*b*) pleated-medium pocket filter, 45% ADSE; (*c*) pleated-medium pocket filter, 85% ADSE; (*d*) 85% ADSE electrostatic air cleaner; (*e*) pleated-medium pocket filter, 95% ADSE; (*f*) 95% ADSE pocket filter with electrostatic agglomerator; (*g*) pleated glass-fiber paper medium, 5% DOP penetration (DOPP); (*h*) standard HEPA filter, 0.03% DOPP; (*i*)HEPA filter, 0.01% DOPP; (*j*) ULPA filter, 0.001; DOPP; (*k*) membrane filter, 0.2-μm pore diameter.

ASHRAE Standard 52-76.[23] Unfortunately, the standard requires this "resistance traverse" for a clean filter only; designers actually need velocity-resistance curves for aerosol-loaded filters as well. Where traverse data are available, we can solve for K_T and K_L in the general resistance expression

$$R = K_T\rho V^2 + K_L\mu V \frac{A_f}{A_m}$$ (20.7)

where R = device resistance, Pa
$\quad V$ = device face velocity, m/s
$\quad \rho$ = gas density, kg/m^3
$\quad \mu$ = gas viscosity, poise (P)
$\quad K_T$ = constant for device or its turbulent-flow elements (dimensionless)
$\quad K_L$ = constant for device or its laminar-flow elements, 1/m
$\quad A_f$ = device face area, m^2 (ft^2)
$\quad A_m$ = device medium area, m^2 (ft^2)

(If units of inches per second are used, the numerical values obtained for K_T and K_L will be different from those obtained using SI units.)

Some filtration devices (membrane filters, HEPA, and ULPA filters) operate at such low velocities that they have essentially zero values of K_T. Others (cyclones, louvers, electrostatic precipitators, and, of course, ductwork and orifices) have only turbulent elements, hence for them $K_L = 0$. Another group (fabrics filters, extended-medium filters, adsorbers, panel filters, roll filters) have both laminar and turbulent elements ($K_L > 0$, $K_T > 0$). The value of K_T is constant for purely turbulent-flow devices, there being no dust buildup. The turbulent-flow term in the general resistance equation comes from the turning, converging, and diverging of gas flow in the filter structure, and this does not change greatly as the filter medium loads with dirt. The value of K_L is, however, strongly dependent on the amount and type of dust loaded into the device. A given mass of fine dust captured by a piece of filter medium produces a greater increase in K_L than does the same mass of coarser dust. No very satisfactory theory relating dust size and filter medium geometry has been developed.

Most designers will be content to take manufacturers' estimates of filter service life for their operating conditions. However, those who perform the indoor air-quality calculations of Sec. 20.4.1 and who want to know the service life of filters predicted by that model will need life data on the type of duct actually reaching the device. Dust-holding capacity by the ASHRAE Standard 52-76 test cannot be translated directly to dust-holding capacity in service; the ASHRAE dust-holding capacity can be taken only as a rough comparative value between filters.

Maintenance and Energy Consumption. Most air filters are disposable items; they are thrown away when the filter resistance exceeds the design value. Some, however, are cleaned and reused. Renewable filters (including electrostatic air cleaners) often have a coating of a liquid adhesive to bind captured dust to the filter; this must be replaced after the filter is cleaned.

Systems are usually supplied with draft gauges measuring filter resistance (the static pressure drop between upstream and downstream filter faces) to indicate when replacement or renewal should occur.

The energy use by a filter is given by

$$P = \frac{0.001QR}{E_m E_d E_b} \tag{20.8}$$

where P = power consumed by system, kW
  Q = air flow, m³/s
E_m, E_d, E_b = fractional efficiencies of motor, drive, and blower at flow Q
  R = filter resistance, Pa

(If Q is in ft³/min and R is in WG, the factor 0.001 should be replaced by 0.0001175.)

If air flow Q changes, the energy consumption drops—but only if this change is produced by a change in blower speed as might be obtained by the use of a variable speed motor. If flow is controlled by dampers, these eat up the energy not used by the filter bank. Thus a low-resistance filter is useful only if the original design of the system contemplated this low resistance and a blower appropriate to the overall system resistance was chosen. The blower must be capable of supplying the terminal resistance of the filter at design air flow. Controls to regulate flow when the filter is below terminal resistance can save energy, for a low *initial* filter resistance will increase air flow and fan power, rather than reducing energy consumption.

For electrostatic filters, there will be some energy input for high-voltage power supplies. The energy used by any auxiliary blowers or motors in a system must also be included in total consumption.

Operating Limitations and Hazards. The threats to reliable air filter operation are high temperatures, excessive air velocity and filter resistance, wetting, vibration, and corrosion. Local or temporary excursions of any of these factors can cause catastrophic or gradual failures.

It is therefore important to estimate the ranges of temperature and air velocity expected and to check the design of the filter for material and structural compatibility with these conditions. Air-flow distribution over filter banks should be as nearly uniform as possible. Filters that may be exposed to rain and snow should be protected by inlet louvers: snow hoods, air washers, and cooling coils ahead of filters must never operate at velocities which allow carryover of sprays and condensate droplets. The air flow through filters provides immense potential for oxidation; construction materials must therefore be polymers, coated metals, aluminum, or stainless steel, depending on service requirements.

Filters are tested and rated under idealized conditions. However, it must be realized that in practice any filter will only be as effective as the method used to support it within the airstream. Individual filters are normally supported in filter holding frames. Gaskets are often utilized to seal between the filter and the holding frame and various attaching devices are used to hold the filter in position while providing a compressive force against the gasket. Individual holding frames should be caulked or welded at all connecting points to prevent by-pass of contaminated air around the filter. In general, the higher the efficiency of the filter the more important it is to seal against by-pass, with a zero tolerance being the norm for HEPA filter systems.

It is desirable for air filters to be nonflammable or at least to contribute minimally to fire and smoke hazards. For this reason, filters must usually meet one of the flammability categories specified in UL 900.[37] Class 1 filters do not contribute fuel when attacked by flame and emit only negligible amounts of smoke; class 2 filters may burn moderately or emit moderate amounts of smoke or both. UL 900 species tests for qualification in these classes. Tests are performed on clean filters, and hence, may underestimate the hazard present if dust of high flammability or smoke is trapped in a filter. Electrostatic filters are powered by high-voltage suppliers (from 6 to 20 kV), some of which are capable of supplying up to 30 mA of direct current. These are quite lethal levels. High-voltage cables must be insulated with ample safety factors and be run through metal conduit. All doors to plenums accessing electrostatic filters must be provided with closures which turn off line power to power supplies, then force a delay of 15 s or so to allow the power supply capacitors to bleed to about 10 V before anyone can enter the plenum. The power supplies themselves must be able to withstand essentially steady arc-overs within the filter elements.

20.6 GASEOUS CONTAMINANT AIR FILTERS

20.6.1 Operating Principles

Removal of an unwanted gaseous contaminant from air or other carrier gas is a mass-transfer process. The contaminant molecules must be driven across as interface of such a nature that they cannot easily return. Some porous membranes have this asymmetric permeability for specified gases, and development continues to widen the number of contaminant types that can be separated by membranes.

More common, however, for air filtration applications are adsorption materials. In these the interface is a solid surface covered with vast numbers of microscopic pores which trap contaminant molecules. If the trapping process at the surface involves a chemical reaction, the process is called "chemisorption." For air pollution control, the interface is often the surface of a liquid into which the contaminant molecules dissolve. (This is called "absorption.") Ventilation air which passes through a liquid scrubber emerges very nearly saturated with the liquid vapor; even if the vapor is water, most of it must be removed, at considerable expense and complexity. Catalysts find limited use in air filtration systems, because only a few are available which operate at ambient temperatures; the cost of heating and then cooling air passing through a catalytic bed ordinarily prohibits their use. Dry adsorbers, chemisorbers, and low-temperature catalysts currently available for air filtration applications are listed in Table 20.8, with the contaminant categories for which they are suited.

Adsorption. In general, high-boiling-point contaminants condense most easily and are therefore easiest to adsorb. Molecular size and structure influence the process, however; adsorption is generally easier for large molecules than for small, at least at the low contaminant concentrations of concern in air filtration. Low temperatures aid absorption; some substances, such as the noble gases, cannot be absorbed effectively above cryogenic temperatures. Contaminants which are readily absorbed will tend to block and absorption of those which are more difficult to adsorb, and may even replace them. Water vapor (humidity) is the most common source of this interference. If the temperature of an adsorber is increased, adsorbed contaminant vapor may be driven out of pores and returned to the carrier gas stream. This process is called "desorption" or "elution."

The most important requirement for an absorber is that it have a very large surface area covered with large numbers of pores of dimensions appropriate to the contaminant to be absorbed. Surface areas of 50 to 2000 m^2/g are available. Beyond this, the adsorber

TABLE 20.8 Low-Temperature Adsorbers, Chemisorbers, and Catalysts

Material	Impregnant	Vapors or gases captured
	Physical adsorbers	
Activated carbon	None	Organic vapors; SO_2, H_2S, acid gases, NO_2
Activated alumina	None	Polar organic compounds*
Activated bauxite	None	Polar organic compounds
Silica gel	None	Water, polar organic compounds
Molecular sieves (zeolites)	None	Carbon dioxide, iodine
Porous polymers	None	Various organic vapors
	Chemisorbers	
Activated alumina	KNO_3	Hydrogen sulfide, sulfer dioxide
Activated carbon	I_2, Ag, S	Mercury vapor
Activated carbon†	I_2, KI_3, amines	Radioiodine, organic iodides
Activated carbon	$NaHCO_3$	Nitrogen dioxide
LiO_3, NaO_3, KO_3	—	Carbon dioxide
LiO_2, NaO_2, KO_2, $Ca(O_2)_2$	—	Carbon dioxide
Li_2O_2, Na_2O_2	—	Carbon dioxide
LiOH	—	Carbon dioxide
$NaOH + Ca(OH)_2$	—	Acid gases
Activated alumina, activated carbon	(Various, some proprietary)	Formaldehyde, mercury vapor
	Catalysts	
Activated carbon	None	Ozone
Activated carbon‡	Cu, Cr, Ag, NH_4	Acid gases, chemical warfare agents
Activated alumina	Pt, Rh oxides	Carbon monoxide

*Polar organics are alcohols, phenols, aliphatic and aromatic amines, etc.
†Mechanism may be isotopic exchange as well as chemisorption.
‡"ASC Whetlerite"

is more effective if it exhibits the special pore size and binding properties mentioned above.

Adsorbers are made into granules, pellets, and fibers, all with the intent of allowing the formation of porous beds through which air can pass while it comes into contact with the adsorber surface. These granules must be rigid, so that the air passages are kept open, and must not break up into powder when the bed is vibrated.

Chemisorption.   Most chemisorbers are adsorbers whose surfaces have been coated or "impregnated" with a chemical compound which reacts with the contaminant to be captured. In some cases, the body of the chemisorber is itself a substance which reacts with the contaminant. The reaction must convert the contaminant to a substance which is nonvolatile or of low volatility. In addition, the reaction must not yield any gaseous product which is itself unwanted. Because many chemical reactions are ionic, the presence of some condensed water is essential to the performance of some chemisorbers. Chemisorbers generally are more effective as temperatures rise, because chemical reactions are more rapid at higher temperatures. The reactant material must, of course, not react appreciably with the carrier gas or decompose or evaporate. Mechanical requirements are the same as those for adsorbers.

Catalysis.   For ventilation applications, oxidation is about the only catalytic process likely to be useful. One must be careful that the reaction product is innocuous and nonodorous, or else the effluent may be more troublesome than the contaminant being controlled. Catalysts are sensitive in difficult-to-predict ways to surface adsorption of extraneous materials, "poisons." Considerable testing is necessary to determine their level of penetration and life as a function of temperature and contaminant concentration.

20.6.2   Adsorption Filter Configurations

It is most important that the absorber prohibit air passage without contact with the adsorption medium. There must be no thin spaces or pockets in the medium body. Air channels either through the medium body or around its edges will increase penetration substantially. To prevent these failures, the edges of each medium-holding cell are covered with "baffles" to force air flow through the main medium body. Medium granules are packed tightly into the cells, by allowing the granules to fall into place rather slowly, as individual grains, usually with additional gentle vibration of the cell. For high-reliability cells, the perforated faces of each cell are drawn together after filling, to compress the medium fill. Elastomeric "springs" are sometimes inserted in cell ends to maintain pressure on the medium granules. Regardless of how deep the bed is, the granules cannot be unrestrained, for air flow will "mine" wide passages through the granules. Units for toxic contaminant capture have all metal-to-metal joints sealed by seam welding, gaskets, or sealants.

Three types of adsorption units are in general use for air filtration systems. The first, for commercial odor control applications (Fig. 20.12a), is made of perforated panels 13 to 25 mm in depth arranged in zigzag fashion to give bed velocities about 15 percent of duct velocities. The panels are removable so that the carbon they hold can be removed, regenerated, and put back into service. The second type, illustrated in Fig. 20.12b, is of sinuous form, with edge baffles to reduce by passing. Beds are usually 50 mm deep. The third configuration, shown in Fig. 20.12c, has horizontal trays with baffles and controlled spacing of perforated sheets. Bed depths are from 50 to 300 mm. The perforated sheets can be lined with cloth to allow finer granule adsorption medium (down to 30 mesh) to be held. Both sinuous and horizontal panel units can be reloaded with adsorption medium by removing the entire unit from the system.

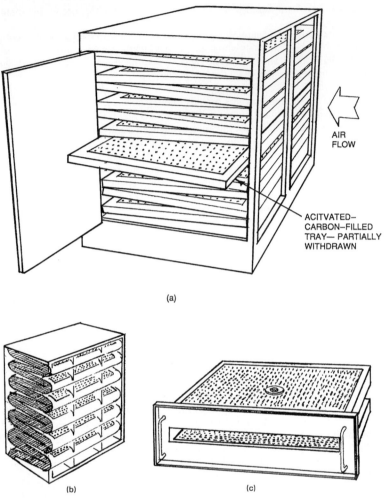

AIR
FLOW

ACITVATED–
CARBON–FILLED
TRAY— PARTIALLY
WITHDRAWN

(a)

(b)   (c)

FIGURE 20.12   Adsorber configurations for air filtration: (*a*)removable panel odor filter; (*b*) sinuous-bed filter; (*c*) flat panel nuclear or toxic contaminant filter.

20.6.3   Performance Characteristics

Penetration and Breakthrough.   The penetration of adsorbates through any adsorber or chemisorber bed increases as contaminants are trapped by the bed. If the bed is deep enough and if adsorption is vigorous, the unpolluted bed adsorbs essentially all the adsorbate. Penetration is very low until nearly all active adsorption pores are used up. Thereafter very little adsorbate is captured, and penetration rises to 100 percent. Because the change in penetration is often sudden, it is usually called "breakthrough." For shallower beds with low inlet concentrations and weaker adsorption, this breakthrough is less sharp. Figure 20.13 shows the typical form of the passage of an adsorption wave through an adsorber bed.

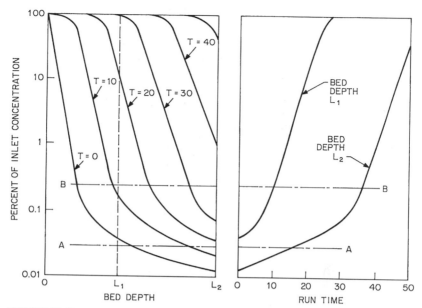

FIGURE 20.13 Penetration of gaseous contaminant through an adsorption bed.

For the fresh bed ($T = 0$) the adsorbate concentration in the flowing gas decreases nearly exponentially with bed depth until the concentration in the bulk gas stream begins to equal the concentration at the adsorber surface. The force driving adsorbate into the pores is now very small, and the adsorbate concentration in the stream decreases more slowly for a given amount of bed depth. Finally the initial portions of the bed become filled with adsorbate and useless; the concentration curve essentially starts anew at the boundary of this spent bed. (Note curves for $T = 10$ to $T = 40$ h in Fig. 20.13.)

It is not customary for suppliers of adsorbers and chemisorbers to supply this type of data for their products. They appear in published papers (e.g., Nelson and Correia)[37] and can be developed with reasonable accuracy from theory. The following must all be known to evaluate penetration test data for an adsorber:

- Carrier gas temperature, relative humidity, pressure, and velocity
- Adsorbate physical properties and concentration at bed inlet
- Adsorber bulk density, particle-size distribution, activity, and retentivity for adsorbates of interest
- Bed depth and pressure drop

Service Life.   The operating parameters just listed plus the allowable bed penetration determine the service life of an adsorber bed. Two breakthrough levels are indicated in Fig. 20.13. For the bed with depth L_1, breakthrough level A is reached instantly; level B requires about 10-h exposure. For a bed of depth L_2, breakthrough at level A require 16 h and at level B, 36 h. (Times given are merely for illustration and do not apply to any specific adsorber.) Detailed calculation of such breakthrough curves is time-consuming. A preliminary best-case analysis will frequently determine whether detailed calculation or, better, testing of the adsorber-adsorbate system is worthwhile. This calculation proceeds as follows:

If the adsorber is assumed 100 percent efficient on the adsorbate of interest, a simple mass-balance calculation shows that the maximum life of the adsorber bed is

$$t = \frac{360 f \rho_u L}{cv} \tag{20.9}$$

where t = life, h
ρ_u = adsorber bulk density, g/cm^3
L = bed depth, cm
c = adsorbate concentration upstream of bed, mg/m^3
v = carrier gas velocity through actual bed, m/s
f = ratio of adsorbate mass in bed to adsorber mass, at saturation

The term f is called the "fractional retentivity," and it is usually tabulated as a percentage. The manufacturers' literature for activated carbons give values from 2 to 40 percent for retentivity with values of 15 percent being typical. Carbon bulk densities are about 0.5 g/cm^3, bed depths are 2.5 cm, and velocities are about 0.2 m/s. Given these values, we get

$$t = \frac{(3170)(0.15)(0.5)(2.50)}{(c)(0.2)} = \frac{2972}{c} \text{ h}$$

A change-out period of 6 months (4380 h) might be acceptable from an economic standpoint. In that case.

$$c_{max} = \frac{2972}{4380} = 0.68 \text{ mg/m}^3$$

This is the maximum inlet concentration tolerable for these assumptions. Table 20.5 shows that this is in the order of magnitude of some odor thresholds, but considerably below most TWA8 values. Thus fixed activated carbon beds are likely to require replacement far more often than every 6 months in industrial situations where concentrations are near TWA8 levels. This is especially true where the assumptions made for this best-case analysis are unrealistically favorable.

A different limitation will exist for highly odorous or toxic adsorbates, where allowable concentrations may be quite low. In these cases, the adsorbate concentration downstream of the adsorber must be substantially lower than that upstream, or else there will be little justification of the air purification system. Reduction of the mass concentration of an odorous adsorbate by 75 percent, e.g., will be barely noticeable by most observers. Toxic adsorbates must be removed with substantial safety margins. Thus penetrations in the order of 1 to 10 percent must be achieved for toxic or odorous adsorbates throughout the useful life of the bed. Such penetrations are difficult to achieve at the low concentrations listed in the OTH column of Table 20.5 unless deep adsorber beds and long stay times are employed. This does not tend to make economic systems.

Detailed Life Calculations.   Economic comparisons and, indeed, decisions on whether to use adsorption at all (instead of ventilation) depend on knowing the service life of the adsorber more accurately than through best-case calculations. The theory of adsorption is poorly developed for low concentrations and multiple adsorbates, which is usually the situation for air filtration. Some procedures require that the user know contaminant physical properties which are not generally available. A fairly simply semiempirical procedure has been proposed by Nelson and Correia.[37] They obtain this expression for the time required for penetration of a single contaminant to reach 10 percent:

$$t_{10} = \frac{280Wf}{C^{2/3}M^{1/3}Q}$$ (20.10)

where W = carbon mass, g
  C = inlet concentration, mg/m^3
  M = molecular weight, g/(gmol)
  Q = air flow, m^3/min
  f = ratio of mass of contaminant adsorbed to mass of carbon at 10 percent penetration breakpoint, percent

Nelson and Correia (Ref. 37) give a means of calculating f based on the boiling point of the adsorbate. Adsorber suppliers often provide f, based on test data. In addition, Ref. 39 indicates that interferences between contaminants are not too severe at low loadings. That paper gives a procedure for handling multiple adsorbates.

Life of Chemisorbers.   Similar procedures are applicable to chemisorptive beds as well as those which operate on purely physical adsorption. For chemisorbers, the best-case analysis is even simpler, for once all the impregnant has reacted with the adsorbate, the bed is spent, or at least reduced to the performance of an un-impregnated bed. If the carrier for the impregnant is a poor physical adsorber (which is the case for activated alumina with most contaminants, e.g.), the bed is a useless energy consumer. Truly catalytic beds, of course, suffer less from these disabilities, but even they eventually become "poisoned" and must be replaced.

Inlet Concentrations.   For any single gaseous contaminant, the concentration reaching the adsorber inlet can be calculated by the procedures described in Sec. 20.4.3. Indoor contaminant source terms may be derived from the data in Table 20.9. EPA annual air-quality reports, such as Refs. 4 and 5, give data for outdoor levels of the priority pollutants (Table 20.7). Others can be obtained from the periodicals listed in the Bibliography. Intakes are often located well above ground, where gaseous contaminant concentrations may actually be *higher* than at grade, where EPA samples are often located. The following expression fits typical vertical distribution patterns for gaseous contaminants (39):

$$C = C_2 \exp[2.9 \times 10^{-5}(h^2 - 274h + 545)]$$ (20.11)

where C = concentration at height h above grade
  C_2 = concentration at local sampling station (assumed 2 m above grade)
  h = height above grade, m

Regeneration.   Activated carbon can be restored to nearly new performance by passing air, nitrogen, or steam at elevated temperatures through it. Steam is generally preferred, because it is cheaper than nitrogen, yet does not support combustion. Collected contaminants are desorbed or eluted from the carbon, leaving cleaned pores available for another adsorption cycle. The desorbed contaminants, of course, appear in the steam; unless some means of detoxification of the contaminant is provided, regeneration merely moves a contaminant from one place to another. The usual detoxification process is incineration, which may be economically feasible if the desorption flow is low enough. The cost of handling and transport to and from the regenerator must be added to the regeneration costs to determine whether regeneration is practical. The alternative to regeneration is to bury or burn

TABLE 20.9 Indoor Gaseous Contaminant Generation

	Activities									
	Theaters, concert halls	Meeting rooms	Trade shows	Quality restaurants	Cafeterias	Bars	Enclosed sports arenas	Corridors, foyers	Office space	Units
Occupancy factors										
Population*	1.5	0.8	0.3	0.5	0.8	1.5	1.5†	0.5	0.2	occupant/m²
Smoking	0.0	0.5	0.2	1.2	1.5	2.5	2.0	0.2	0.7	cigarette/(h · occupant)
Human emissions										
Carbon dioxide	24	33	48	40	62	55	55	21	35	g/(h · occupant)
Methane	2	2	2	2	2	2	2.7	1	2	mg/(h · occupant)
Hydrogen	1	1	1	1	1	1	1.3	0.5	1	mg/(h · occupant)
Hydrogen sulfide	0.015	0.015	0.015	0.015	0.015	0.015	0.02	0.007	0.015	mg/(h · occupant)
Ammonia	9	18	30	18	40	35	35	7	20	mg/(h · occupant)
Smoking										
Carbon dioxide	—	0.7	0.7	0.7	0.7	0.7	0.7	0.7	0.7	g/cigarette
Carbon monoxide	—	80	80	80	80	80	80	80	80	mg/cigarette
Organics	—	15	15	15	15	15	15	15	15	mg/cigarette
Building emissions										
Formaldehyde*	0.003	0.003	0.003	0.003	0.003	0.003	0.003	0.003	0.003	mg/(min · m²)
Ozone	—	—	0.1	—	—	—	—	—	0.1	mg/photocopy

*Per m² of floor area.
†Seating areas only.

the spent carbon and replace it. Adsorbents for critical service should always be new material, not regenerated.

Pressure Drop and Energy Consumption.   Bed pressure drop can be calculated approximately if the bed depth granule diameter, percentage of solids, gas velocity, and gas physical properties are known. Suppliers of granular adsorbers provide pressure drop test data for air on their various grades. Overall pressure drop is the drop through the granular bed plus the drop through any structure used to hold the granules. Structure resistance may not be negligible; it is best to obtain actual test data on the complete adsorber unit. Pressure drop through an adsorber does not change if the adsorber is protected from dust buildup, which can be done by placing particulate filters upstream of the adsorber.

Environmental Considerations.   Humidity is always a factor in gaseous contaminant air filtration. Physical adsorbers generally work best at low humidities and may fail entirely if they absorb too much water. Chemisorbers, however, usually cannot operate at extremely low humidities. Chemical reactions tend to work better at high temperatures, hence these favor chemisorbers and catalysts. Temperature excursions above specified operating conditions, however, may drive adsorbates out of physical adsorbers and may evaporate or decompose impregnants, chemisorptive compounds, or catalysts. Activated carbon is flammable and a poor heat conductor; adsorption is a process which generates heat. Spontaneous ignition is possible when adsorbate concentrations are high. Where unusual conditions are expected, the adsorber/adsorbate system should be tested at those conditions.

Vibration of the adsorber bed can cause attrition of adsorber granules. Standard tests for attrition resistance, ignition point, and other adsorber properties have been established by the American Society of Testing and Materials (ASTM).[41]

20.7 PARTICULATE AIR POLLUTION CONTROL EQUIPMENT

Table 20.10 gives characteristics and typical application areas for various types of particulate air pollution control equipment; Figs. 20.14 and 20.15 give schematic illustrations of their operating principles. References 3, 40, 42, and 43 provide further design advice. In general, the higher the energy input to the device, the lower the penetration through it. Devices which circumvent this rule (e.g., electrostatic precipitators) will probably require high initial investment and more extensive control systems and maintenance. Equipment choices are usually based, first, on which system will provide the needed penetration level and, second, which will function reliably in the operating environment.

20.7.1 Operating Principles

Inertial (Aerodynamic) Collectors

Large-Scale Cyclones (Fig. 20.14a).   Gas enters tangentially at high velocity at the top of the cyclone, changes direction at the bottom, and leaves through the central cylinder. Dust is centrifuged to the cyclone walls and falls into the hopper at the gas turnaround zone. Higher velocities—up to about 30 m/s—produce lower penetrations, but these must be considered large-particle control devices, or precleaners for more effective devices.

Small-Scale Cyclones (Fig. 20.14b).   Sometimes these are simply small-scale versions of Fig. 20.14a, with many units mounted in parallel. Spinning can be imparted to the gas

TABLE 20.10 Particulate Air Pollution Control Equipment: Characteristics and Application Areas

Control equipment type	Pressure drop,* Pa	Temperature range,† °C		Application areas (see Table 36.1 for codes)
		Low	High	
Large-scale cyclones	125–800	−40	550	4, 5, 9, 10, 15, 16, 22, 28, 31, 35, 42, 44
Small, multiple cyclones	500–1500	−40	550	1–5, 7, 9, 10, 15, 16, 22, 26, 31, 34, 35, 42
Louvers	125–500	−40	350	47
Dry dynamic collectors‡		−30	370	5–7, 10, 11, 16, 22, 23, 26, 29, 31, 34, 44, 45
Wet dynamic collectors‡		5	70	5, 7, 9, 13, 19–21, 29, 31, 45
Centrifugal mist collectors‡		5	110	24
Shaker baghouse	500–1500	−30	300	6–10, 17, 26, 35, 42, 44, 45
Collapse-cycle baghouse	500–1500	−30	300	1–10, 13, 15–17, 20–22, 26, 29, 30, 40, 44, 45
Reverse-pulse baghouse	500–1500	−30	300	5, 8, 9, 13, 22, 24, 25, 29-31, 45, 47
Pulsed cartridge collectors	300–1000	−20	110	46
Granular beds		−40	800	13, 28, 29, 33, 42
Lint arrestors	180–400	−20	65	14, 22, 24, 25, 31, 43
Low-voltage, two-stage electrostatic precipitators	125–375	−30	60	
High-voltage, one-stage electrostatic precipitators	125–375	−20	400	1–3, 17, 18, 35–41, 46
Venturi scrubbers	2000–20,000	5	90	7, 11, 12, 17, 18, 35–39, 46
Packed-bed scrubbers	250–2500	5	90	32
Impingement scrubbers	500–2000	5	90	6, 7, 9, 19–26, 29, 31, 32, 34, 35, 39, 41
Submerged-nozzle scrubbers	500–1500	5	90	6, 7, 9, 19–27, 29, 31, 32, 34, 35, 39, 41

*249 Pa = 1 inWG.
†Temperature range is determined by materials of construction, except for wet scrubbers. Values given are typical for commercially available devices. To convert to USCS units: °F = 1.8 × °C + 32.
‡These devices combine blower and collector; they usually provide from 500 to 2500 Pa external static pressure. Motors may require protection to meet listed temperatures.

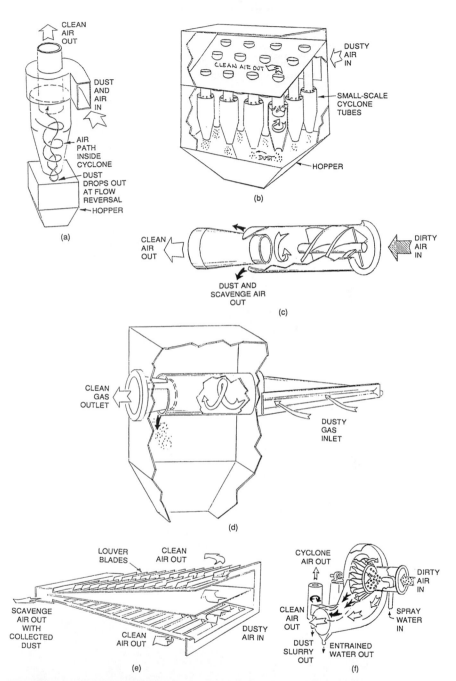

FIGURE 20.14 Particulate air pollution control equipment: (*a*) large-scale cyclone: (*b*) multiple bank of small-scale cyclones; (*c*) helical-vane cyclone; (*d*) straight-through cyclone; (*e*) louver; (*f*) dynamic blower-collector; (*g*) centrifugal mist collector; (*h*) shaker baghouse; (*i*) collapse-cycle baghouse; (*j*) reverse-pulse baghouse; (*k*) pulsed panel cartridge collector; (*l*) pulsed cylindrical element cartridge collector; (*m*) vacuum-type lint collector; (*n*) paper web lint/ink mist collector.

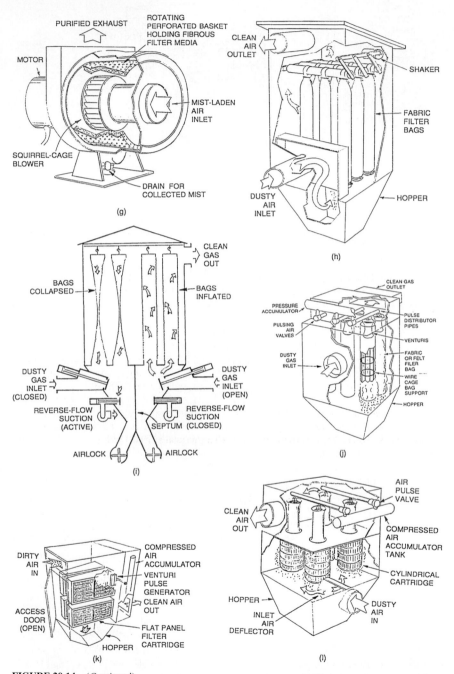

FIGURE 20.14   (*Continued*)

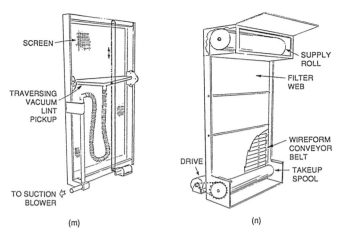

FIGURE 20.14   (*Continued*)

by "twirlers" helical vanes at the gas inlet to each cyclone (Fig. 20.14c); in this case, gas entry is axial instead of tangential. In another version (Fig. 20.14d) the clean air passes out through a central tube, with no flow reversal. A secondary air flow (for the dust-heavy gas) greatly improves performance.

Louvers (Fig. 20.14e).   Gas is forced to reverse direction by louver blades. Dust-heavy gas is drawn out a slot at the downstream end of the device.

Dynamic Collectors (Fig. 20.14f).   These combine a fan and a particulate separator. The specially shaped fan blades centrifuge particles to the outside of the fan spiral, where they are skimmed off into a hopper. The device can be assisted with water sprays in the inlet.

Centrifugal Mist Collector (Fig. 20.14g).   In this device, a fibrous cylindrical air filter is rotated rapidly. Air entrained by the filter exhausts radially. Entrained droplets, however, impinge on the filter fibers and drain off through a perimeter slot. The device is used to capture coolant-oil mists from machine tools.

Fabric Collectors

Shaker Baghouses (Fig. 20.14h).   These operate on the same principle as a household vacuum cleaner. Gas passes through the fabric, where a cake of dust builds up, greatly improving filtration efficiency. At some point, groups of bags are shaken to dislodge the collected cake, and the cycle starts again. Units with flat fabric panels are available. Fabrics used include woven and felted natural synthetic and glass fibers as well as laminations of these and membrane materials to improve base collection efficiency and cake-release properties.

Collapse-Cycle Baghouses (Fig. 20.14i).   Here the particulate cake is removed by a complete reversal of gas flow, which collapses each tube.

Reverse-Pulse Baghouses (Fig. 20.14j).   In these devices, particles are collected on the outside of the fabric tubes, which are kept from collapsing by an internal wireform. When the dust cake builds to the maximum desired level, a pulse of compressed air is injected through a venturi at the top of each of a group of bags. This pulse induces gas flow into the bag, causing a momentry flow reversal to ripple down each pulsed bag and break the cake loose. The system has the advantage of having a minimum of moving parts in the gas system.

Pulsed-Cartridge Collectors (Fig. 20.14k and l).   These are similar in principle to reverse-pulse baghouses, except that the collection medium is a pleated paper web instead of fabric. Both cylindrical and flat-panel configurations for the cartridges are available. The horizontal pulsed cartridge is now the industry norm.

Lint Arrestors.   Collectors for lint in textile and printing plants have two forms. One form makes use of an endless wire-screen band (Fig. 20.14m). Linty air passes through the band twice, with almost all the lint depositing on the screen during the first pass. The band moves slowly, so that the collected lint is steadily fed into the nozzle of a vacuum cleaner for transfer to a fabric collector of much smaller capacity. The second type of lint collector (Fig. 20.14n) makes use of an endless wire-mesh band, but this is merely a support for a very thin, porous paper web. Lint builds up on this to make an effective filter. The web and its collected lint are rolled up to form a disposal package. This form of lint collector is also effective in capturing the ink mist which often accompanies lint in printing operations.

Electrostatic Precipitators

Low-Voltage Two-Stage Precipitators.   A few industrial applications (e.g., welding-fume collection) use the same type of two-stage collector described for air filtration use in Sec. 20.5. These will be made somewhat more rugged for this service and may include a deionizing stage downstream to eliminate the space-charge effects which occur when dust concentrations are high.

High-Voltage, Single-Stage Precipitators (Fig. 20.15a).   These are widely used for stack gas cleaning, especially in fossil-fuel power plants. They consist of parallel grounded plates, with ionizer electrodes suspended midway between these, separated from each other by a distance about equal to the plate-to-plate spacing. The ionizer electrodes may be wires, but are often tubes which carry barbs of various forms; the ions are created at these barbs. Plate-to-plate spacings are large—typically 30 cm—and voltages high—typically 70 kV. Dust collected on the plates is removed by rapping the plates. Voltage is adjusted so that there is a controlled arcing rate between electrodes.

Wet Scrubbers

Venturi Scrubbers.   A venturi scrubber (Fig. 20.15b) is a duct with a short converging entrance and a longer diverging exit. This shape allows the gas to accelerate to a high velocity at the smallest cross section (the throat) but to drop back to the normal duct velocity without excessive pressure losses. Water is injected into the venturi throat, through very crude, difficult-to-plug nozzles or slots. The very large difference between the water and gas velocities tears the water into a cloud of fine droplets, which, however, are not travelling as fast as the aerosol particles in the gas. The particles therefore collide and combine with the water droplets, which are subsequently removed by a cyclone or other low-energy scrubber. Penetrations can be very low through venturi scrubbers and reliability high—but at the cost of high energy consumption.

Packed-Bed Scrubbers (Fig. 20.15c to e).   In these scrubbers the gas is passed through beds of packings (marbles, rings, saddles, impingement plates, or fibers, which are continuously irrigated with a liquid spray). Particles are captured by impaction on both the packing materials and droplets of the liquid and are continuously removed in the flowing liquid. The liquid is filtered externally and recirculated.

Submerged-Nozzle Scrubbers (Fig. 20.15f).   In this device, liquid-particle contact is made in a sort of upside-down waterfall formed in a half-submerged, sinuous passage. The device has neither spray nozzles nor liquid filters to maintain; the collected dust settles out in the liquid tank, from which it is periodically removed by scraping. Another form uses a submerged jet (Fig. 20.15g).

Granular Bed Collectors (Fig. 20.15h).   In these, the gas passes through a rather deep bed made of granules with progressively smaller diameters as it moves from the inlet to outlet face of the collector. (Some reverse gradation near the outlet face may be necessary to

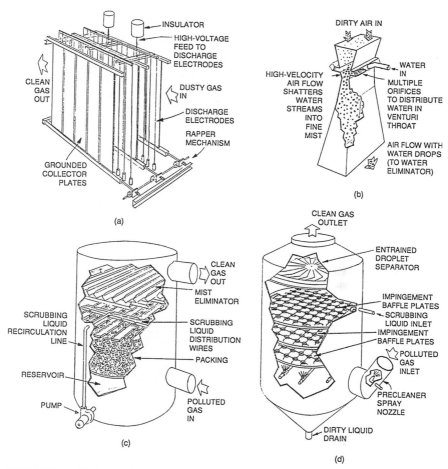

FIGURE 20.15 Particulate air pollution control equipment: (*a*) single-stage electrostatic precipitator; (*b*) venturi scrubber; (*c*) packed tower scrubber; (*d*) impingement baffle scrubber; (*e*) fiberbed scrubber; (*f*) sinuous submerged-nozzle scrubber; (*g*) submerged jet scrubber; (*h*) granular bed-collector.

keep the finer granules from entraining into the gas stream.) Granules are usually sand and pebbles. Collection is by impaction, sedimentation, and diffusion of dust onto the granules. This concept has been used in high-temperature systems and in highly radioactive streams.

20.7.2 Performance Characteristics

Penetration and Efficiency.   Typical penetration-particle diameter curves for selected air pollution control devices are shown in Fig. 20.16. Detailed data for specific devices are usually available from device manufacturers. Such data must be quite specific with regard to operating conditions during the test, giving gas constituents, temperature and pressure,

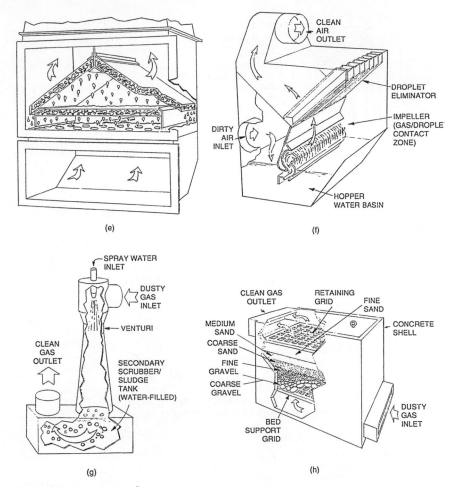

FIGURE 20.15   (*Continued*)

and particle density. A common problem in pollution control equipment selection is to estimate the device penetration at some gas condition entirely different from the test, or for particles with density different from that used in the test.

A reasonable approximation to this can be obtained by converting the diameter scale of the diameter-penetration curve to an "interception parameter" scale. This interception parameter is, for a given device,

$$N_i = \frac{\rho_p d^2 V}{\mu} \qquad (20.12)$$

where μ = gas viscosity, g/(cm · s)
ρ_p = particle density, g · s/cm

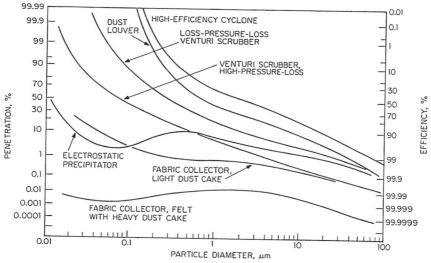

FIGURE 20.16 Penetration vs. particle diameter for typical air pollution control devices.

d = particle diameter, cm
V = gas velocity through device, cm/s
N_i = interception parameter

This same parameter can then be calculated for the operating conditions for which penetration data are desired, using diameters over the range of interest. One can then pick off penetrations for the values of N_i which correspond to the chosen diameters and plot a new diameter-penetration curve. The units used are immaterial but must, of course, be kept the same for all conditions examined.

Pressure Drop and Service Life.   Pressure drops through pollution control devices are obtainable from manufacturers of the devices, with proper consideration of gas temperature and density and liquid flows, if any. The pressure drop through fabric collectors and pulsed cartridge collectors rises to a nearly steady-state level, which is a function of aerosol concentration, medium velocity, and frequency and vigor of the cleaning cycle. The other particulate pollution control devices listed above have essentially steady pressure drop, for they do not store dust on medium which remains continuously in the air stream. The media in fabric collectors and pulsed cartridge units do have finite service lives and must be replaced when they begin to lose strength or become clogged with dirt. For lint collectors which use replaceable filter webs, media must be replaced when the roll has run out. All pollution control equipment is subject to corrosion and abrasive wear.

Disposal of Captured Dust.   Dust captured by dry collectors is usually deposited in a pyramidal hopper. Rotary "air locks" allow dust to be removed without inrush or outrush of gas. The dust removed may be bagged or dropped into containers, but care must be taken to avoid loss during this transfer. Wet collectors may transfer innocuous wastes directly to sewers, but it is more common to allow the collected material to settle in a pond and recirculate the scrubbing liquid; even water is sometimes precious. Whether the collected dust is dry, in a slurry, sludge, or pelletized form, final disposal may be simple or difficult.

Innocuous dusts may be piled or buried, but there are often regulations with regard to disposal of hazardous materials. Many dusts have recovery value; this should be investigated.

Operating Limitations and Maintenance.   Air pollution control equipment must be designed to withstand the temperature and pressure excursions normal to the process involved. If the gases involved are known to be corrosive, the materials chosen must withstand this corrosion for an economic equipment lifespan. Wet collectors especially require operation within fairly narrow flow limits. Fabric collectors need to be inspected internally at regular intervals to check on failed bags; regular preventive maintenance schedules should be followed on all equipment. Support structures and foundations must meet the usual American Institute of Steel Construction (AISC) requirements, and maintenance ladders and catwalks must be well thought out and designed to OSHA standards. Safety interlocks are essential on all electrostatic precipitators.

20.8 GASEOUS CONTAMINANT AIR POLLUTION CONTROL EQUIPMENT

20.8.1 Operating Principles

Equipment for control of gaseous air pollution emissions uses many of the same principles described in Sec. 20.6 for air filtration systems. The differences are matters of scale, maintenance, and operating environment. In addition, the emission from a pollution control system can sometimes be saturated with water vapor without causing problems; stack emissions need not be at air-conditioning temperature; reaction products which might not be tolerable in air for breathing may be at a low enough level to be emitted into the atmosphere and diluted. Incineration is often practical, and condensation, even at cryogenic temperatures, is sometimes used.

Air pollution control systems for control of gaseous emissions are basically chemical engineering problems. The literature on the subject is immense. References 3 and 42 through 44 provide design guidance beyond the brief outline given below, and periodicals listed in the Bibliography, give current developments.

Adsorption.   The fundamentals of adsorption for stack gas cleaning are the same as for air filtration service. (See Sec. 20.6.) Dry adsorbers are less used for pollution control than for air cleaning, because temperatures are often high. Where they are used, the normal maintenance arrangement is in situ regeneration. Highly practical solvent recovery systems are built in this way. Dry adsorbers (e.g., activated carbon sulfur dioxide capture) are sometimes used as a device to concentrate a contaminant, which is subsequently immobilized by a second system when the adsorber is regenerated. Contaminant concentrations in air pollution control systems are usually much greater than those in air filtration, regenerative adsorption systems are therefore more likely to be practical.

Adsorption and Chemisorption.   Dry adsorbers impregnated with various reactants are used to capture mercury, radioiodine and its organic compounds, hydrogen sulfide, and acid gases (especially those in the highly toxic chemical-warfare-agent class).

Liquid scrubbers, both with and without added reagents, are commonplace in gaseous contaminant pollution control. Scrubbing liquids are sprayed into chambers which may be packed with rings, saddles, marbles, fibers perforated plates, etc., to provide large liquid-gas interface areas (Fig. 20.15c to e). Bed length and flow velocities must be adjusted to give adequate contact time, yet not allow entrainment of the scrubbing liquid into the gas stream.

The reaction used must not create solid precipitates which will fill up the passages through the packing. Final disposal of the reacted contaminant usually involves a secondary cycle outside the scrubber, where precipitates can be filtered, centrifuged, or settled out of the scrubber liquid. The mass-transfer area and contact time for particulate-type wet scrubbers are usually inadequate for gaseous contaminant collection.

Condensation.   If the contaminant has a higher boiling point than the carrier gas, it is possible to condense it out of the carrier gas on a cool surface. This technique is often used in equipment for stack gas sampling, where flows are small and one wants to capture all condensables. The condensation of radioactive xenon, krypton, and radon in activated carbon at cryogenic temperatures is one of the few practical ways to capture these gaseous contaminants.

Catalysis.   Catalysis is generally more useful in air pollution control than in air filtration because the control targets are usually a few known contaminants and higher temperatures are available. Reaction heats add to the energy available to sustain catalytic actions.

Catalytic Incineration.   Catalytic incineration is sometimes a practical alternative to thermal incineration or regenerative adsorption systems. With a catalyst present, oxidation of the contaminant can be achieved at lower temperatures (300 to 600°C) than with thermal incineration (>650°C). The entire carrier gas stream must be held at these temperatures, in either case. Recuperative heat exchangers downstream of the catalyst bed or incinerator return energy from the exhaust stream for use in preheating the carrier gas or some other process. A problem with catalytic incineration is that combustion of the contaminant is sometimes incomplete and so the resulting products may be odorous or toxic.

Thermal Incineration.   In thermal incineration, the gas stream is heated, usually with direct-fired natural-gas burners, to the temperature at which the contaminant oxidizes. This is a form of combustion, and once the process starts, it may be possible to reduce the fuel input—or, with very high concentrations, the reaction may be self-sustaining. The method is simple, nonspecific, generally reliable, but expensive, for the gas temperature must be raised substantially.

20.8.2   Performance Characteristics

System Penetration.   Higher device penetrations are usually allowable for gaseous contaminants in air pollution control systems than in air filtration, because the chief aim is to bring the effluent down to some safe, legislated or aesthetically acceptable control level. Atmospheric dispersion from tall stacks tends to increase safety and aesthetic acceptability. Concentrations of contaminant and carrier gas constituents at the control equipment inlet must be known, along with gas temperature and pressure and many gas properties. Mass transfer, adsorption, catalysis, and combustion theory are comparatively well developed at the concentration levels met in air pollution control, and traditional chemical engineering techniques can be applied. Much engineering effort is needed for accurate prediction of performance. In general, performance for well-designed systems remains stable for a given inlet condition; the exception to this is the catalytic incinerator. These decline in performance as the catalyst gradually becomes poisoned. Systems with catalysts must operate at some minimum temperature, while absorbers fail miserably at high temperatures.

Environmental Limitations.   Temperatures, gas and liquid flows, and reagent concentrations must be accurately controlled if systems are to operate as designed. Buildup of precipitates in liquid scrubbers is not necessarily reversible; corrosion is a major problem in all systems, especially where sulfuric acid can condense out of the gas stream. The combination of heat, free water, and catalysts is inherently destructive to metals; stainless steel, monel, titanium, glass, and plastics (such as polyester/glass-fiber composites) are often needed for materials of construction. The disposal of liquid and solid wastes from control processes is a serious problem, subject to controls as stringent as those for the gasborne pollutants themselves.

Pressure Drop and Energy Consumption.   Suppliers provide data on system pressure drops at operating conditions (which include liquid-flow rates for scrubber systems). Overall energy consumption calculations must include the cost of air horsepower as well as liquid pumping costs and fossil-fuel consumption. Reagent costs, catalyst or adsorber replacement, regeneration steam, heat-exchanger fan power, waste disposal, and amortization of capital are among overall life-cycle system costs.

20.9 GAS PURIFICATION EQUIPMENT: PERFORMANCE TESTING

Purchasers of pollution control equipment often require a performance guarantee, to be proved by tests on the installed equipment. The only logical test is the percentage penetration for the equipment at specified operating conditions. It is not logical to specify concentrations downstream of the equipment, for this is always a function of upstream concentration, which is beyond the control of the pollution control equipment supplier. Thus performance evaluation of air pollution control equipment requires measurement of pollutant concentrations both upstream and downstream of the tested device. This validation process is usually a small part of the cost of the installed equipment; it can be done relatively quickly and may be required by regulatory agencies. If the device performance is sensitive to changes in operating velocity, temperature, etc., tests should be run over the expected range of these parameters. A useful survey of test methods is given in Ref. 45.

Testing of installed air filtration systems has been rare, except in critical cases. Customers usually rely on tests of filter samples. These tests should be made according to industry standard procedures to allow comparison between suppliers. (See "Laboratory Tests" below.) Field testing is possible but expensive in relation to equipment cost, and there are almost no standard procedures available, except for clean rooms and nuclear safety systems.

20.9.1 Aerosol Contaminant Testing

Traditional aerosol testing methods make use of integrated samples. Known sampling flows are drawn simultaneously upstream and downstream of the gas cleaner. Filters having extremely low penetrations for the aerosol present are placed in the sampling lines, to gather as much contaminant as possible. Enough sample must be collected to allow reliable measurement of the masses of aerosol captured.

Sampling filters are never perfectly stable; some gain or loss in filter mass is to be expected when no aerosol at all is collected. If we want the aerosol mass to be at least 20 times the expected gain or loss in the sampling filter, then the sampling time t (in minutes) must be

$$t = \frac{333 f W}{PCQ} \tag{20.13}$$

where f = expected sampling filter gain or loss, percent (due to moisture adsorption, etc., *not* aerosol accumulation)

W = sampling filter mass, g

P = air pollution control system penetration, percent

C = upstream aerosol concentration, mg/m^3

Q = sampler flow, m^3/s

Substitution of reasonable values in Eq. (20.13) shows that even for high upstream concentrations, the sampling filter must be very stable (low f) and the mass of the sampling filter kept low.

The EPA has mandated a method for particulate measurement in stacks (method 5, Ref. 26, part 60, App. A). This procedure is usually specified for validation of stack gas particulate emissions.

Equation (20.13) applies to air filtration situations as well. In this, concentrations are so low that sampling times must be on the order of days, not minutes. Sensitive detectors have been developed to overcome some of these problems. Piezoelectric and beta-ray mass sensors are available to determine mass concentrations within a few minutes, with reasonable reliability. Optical particle counters take an entirely different approach. These measure the amount of light scattered by single particles passing through the instrument. This scattered light intensity is chiefly a function of the diameter of the particles. By counting the number of light pulses in a range of size bands, the instrument constructs a histogram of the particle sizes in the aerosol, as in Fig. 20.1. Calculations from the histogram data enable determination of particulate mass concentrations, provided the aerosol density is known. Users of optical particle counters should be cautious in interpreting data from these instruments. The particle sizes they give are not necessarily the true, "geometric" diameters of the particles observed, nor do these counters directly measure the filtration properties or mass of individual particles. Particles of several different diameters may appear of equal size in an optical particle counter. They are, however, widely applied in clean rooms, where their high sensitivity is necessary and the operating environment is gentle enough for their use. References 44 and 45 examine the capabilities and limitations of aerosol detection instruments in detail.

Laboratory Tests for Air Filters.   The test standards most used in the United States for air filters are ASHRAE 52-76[23] and MIL-STD-282.[47] ASHRAE 52-76 prescribes four tests for air filters for general ventilation service:

Resistance, or the pressure drop across the filter as a function of air flow through it

Dust-holding capacity, or the increase in resistance as a function of the amount of ASHRAE standard air filter test dust loaded on it

Arrestance, or the percentage (by mass) of this dust captured by the filter at sequential dust loadings

Dust spot efficiency, or a measure of the ability of the filter to remove dust which will blacken walls

An additional test standard, ASHRAE 52.2, is currently in development. This will prescribe a methodology for measuring efficiency by particle size. It is expected that this standard will eventually make the use of the AHSRAE 52.1 *Dust Spot Efficiency test* obsolete.

The dust spot test is made by sampling air both upstream and downstream of the tested filter. These samples are passed through glass-filter "targets," which become darker as aerosol builds up on them. The rate of change of light transmission of these dust spots is used to compute the dust spot efficiency.

MIL-STD-282 is used to evaluate HEPA filters. An aerosol of DOP (dioctyl phthalate, or diethylehexyl phthalate) having a mass-mean diameter near 0.3 m is fed to the filter. The total light scattered by the aerosol cloud is measured upstream and downstream of the filter. Penetrations as low as 0.002 percent are routinely measured by this method. For lower penetrations, particle count techniques[24,46] must be used.

20.9.2 Gaseous Contaminant Testing

Many detection systems are available for both integrated and instantaneous measurement of gaseous contaminant concentrations and thus the performance of gaseous contaminant air pollution control and air filtration devices. Table 20.11 lists the major types of detectors available. Their operating principles are, briefly:

Gas washers (also called "bubblers" or "impingers"): These are miniature liquid scrubbers, frequently used in series to provide adequate removal efficiency. Reagents are often used to make them efficient and specific to a chosen contaminant.

Gas adsorber tubes: These are small tubes filled with adsorber granules (usually impregnated to give low penetration of a specific gas). After the sample is captured by drawing polluted air through the tube, the contaminant is desorbed by flushing with an appropriate liquid. Various liquid analysis techniques can be used to detect the amount of contaminant desorbed.

Gas detector tubes: These are small tubes filled with a chemisorber that shows a color change with a chosen contaminant. The progress of the adsorption wave down the tube can be seen and measured.

Chemiluminescent analyzers: These have a lightproof chamber through which the polluted gas passes. Another gas, selected to react with the chosen pollutant, is mixed into

TABLE 20.11 Instruments for Measuring Gaseous Contaminant Concentration

Instrument type	Sample*	Contaminants detected
Gas washers (bubblers, impingers)	IM	Acid gases, ozone, organics
Gas adsorption tubes.	IM	Many, with specific tube fill
Gas detector tubes	IM	Many, with specific tube fill
Chemiluminescence	RT	Ozone, ethylene oxide, organics
Infrared spectrometry	RT	Ozone, organics, acid gases
Laser spectrometry (LIDAR)	RT	Many gaseous; also aerosols
Gas chromatography (GC)	RT	Many, with proper column and detector
Mass spectroscopy (MS) (also GC/MS)	RT	Almost all compounds
Flame ionization	RT	Many, especially with GC
Atomic adsorption	RT	Many, with proper light wavelength
Passive detector badges	IM	Growing number
Piezoelectric	RT	Limited; based on dry chemisorbers
Freeze-trapping	IM	All condensable gases
Thermal conductivity	RT	Gases with thermal properties different from those of air

*Sample codes: RT = real time, instantaneous or nearly so; IM = sample is integrated over extended period.

this flow. The reaction is of a type that emits photons, which are observed by a sensitive photocell.

Infrared spectrometry: The transmission of infrared light is measured through a very long path containing the polluted gas. The path is folded with mirrors to make a compact instrument.

Laser spectrometry: A pulse of narrow-beam laser light is directed through a region containing the contaminant. Contaminant molecules scatter the laser light (in proportion to their numbers) in various directions, including back to the source. The intensity of the returned pulse is a measure of contaminant concentration. The method is unique in its ability to measure concentrations at a distance from the instrument.

Gas chromatography: A small grab sample (of microliter size) is mixed with an inert carrier gas and drawn through a small-diameter adsorber column. Individual components of the polluted gas sample migrate down the column at different, but predictable, rates. The mass of each component present is measured as it comes out the end of the column, using one or more detection devices.

Mass spectroscopy: Individual molecules of the contaminant are injected through electrostatic and/or magnetic fields in a vacuum. The flight path of the molecule is bent, with the path radius being a function of the molecule mass. The numbers of molecules of various masses are measured electronically. The method is sometimes coupled with gas chromatography.

Flame ionization: The contaminant is passed through a colorless flame, where it is ionized and changes the conductivity of the flame region; this change can be detected electronically.

Atomic adsorption: The contaminant is passed through a colorless flame, where it is ionized. A light beam from a narrow-wavelength-range source also passes through the flame. The ionized contaminant reduces the transmission of light through the flame; this change measured contaminant concentration. There are also flameless techniques for exciting the contaminant atoms.

Passive detector badges: These diffusively chemisorb specific contaminants, which are later desorbed for analysis. No pumps are needed, hence the badges are easily worn by individuals to measure personal exposure.

Piezoelectric: A quartz crystal is coated with a chemisorber specific to the chosen contaminant. As contaminant mass is added to the chemisorber, the oscillation frequency of the crystal changes.

Freeze-trapping: A tube is immersed in a liquid-nitrogen bath. Any gas which can condense at the tube (trap) temperature will condense out when passed through it. The condensed liquids can be distilled off later for individual contaminant measurement.

Thermal conductivity: The pressure of contaminant molecules can alter the thermal conductivity of a gas, roughly in proportion to the number of contaminant molecules present. This method is not highly specific, hence it can be used only where a single contaminant is present or in conjunction with gas chromatography.

Gaseous contaminant detectors are quite sensitive to environmental conditions, and much care is needed to avoid spurious readings. The low concentrations present in air filtration situations (often parts per billion) frequently demand special procedures, such as accumulation of the contaminant on an adsorber, with subsequent desorption at higher concentration and higher temperature. (See Ref. 48).

NIOSH[31] has validated test procedures for many gaseous contaminants. Presumably these will be updated as new techniques are shown to be reliable.

REFERENCES*

Many references are constantly updated. Please refer to latest edition of reference cited.

1. J. W. Johnson et al., *A Computer-Based Cascade Impactor Data Reduction System*, EPA-600/7-78/042, NTIS, 1978.

2. H. L. Green and W. R. Lane, *Particulate Clouds*, 2d ed., D. Van Nostrand, Princeton, NJ, 1964.

3. A. C. Stern (ed.), *Air Pollution*, 3rd ed., Academic Press, New York, 1977.

4. *National Air Quality and Emissions Trends Report, 1982*, EPA-450/4-84/002, NTIS, 1984.

5. *Air Quality Data—1977 Annual Statistics*, EPA-450/2-78/040, NTIS, 1778.

6. P. J. Lioy, R. P. Mallon, and T. J. Kneip, "Long-Term Trends in Total Suspended Particulates, Vanadium, Manganese and Lead at near Street Level and Elevated Sites in New York City," *Journal of Air Pollution Control Association*, vol. 30, 1980, p. 153.

7. National Research Council Subcommittee on Airborne Particles, *Airborne Particles*, University Park Press, Baltimore, MD, 1981.

8. International Symposium on Indoor Air Pollution, *Health and Energy Conservation*, CONF-811048-SUMS (DE84004447), NTIS, 1981.

9. B. Y. H. Liu, D. Y. H. Pui, and H. J. Fissan (eds.), *Aerosols: Proceedings of the 1st International Conference*, Elsevier, New York, 1984.

10. C. S. Dudney and P. J. Walsh, *Report of Ad Hoc Task Force on Indoor Air Pollution*, ORNL/TM-7679, NTIS, 1981.

11. National Research Council, *Indoor Pollutants*, EPA-600/6-82/001, PB82180563, NTIS, 1982.

12. M. J. First (ed.), *Proceedings 16th DOE Nuclear Air Cleaning Conference*, CONF-801038, NTIS, 1980 (vol. 3 is index to all previous conference proceedings).

13. M. J. First (ed.), *Proceedings 17th DOE Nuclear Air Cleaning Conference*, DE83009767 and DE83009768, NTIS, 1983.

14. M. J. First (ed.), *Proceedings 18th DOE Nuclear Air Cleaning Conference*, CONF-840806, NTIS, 1984.

15. C. A. Burchsted, A. B. Fuller, and J. E. Kahn, *Nuclear Air Cleaning Handbook*, 2d ed., ERDA 76-21, NTIS, Springfield, VA, 1976.

16. F. W. Makison, R. S. Stricoff, and L. J. Partridge, *NIOSH/OSHA Pocket Guide to Chemical Hazards*, DHEW-NIOSH Pub. 78-210, USGSD, Washington, D.C., 1978. See also comment, *American Industrial Hygiene Association Journal*, vol. 44, Sept. 1983, p. A-6.

17. A. Dravnieks and B. K. Krotoszynski, "Systematization of Analytical and Odor Data on Odorous Air," in *ASHRAE Symposium, Odors and Odorants: The Engineering View*, ASHRAE, Atlanta, GA, 1969.

18. C. E. Billings and L. C. Jonas, "Odor Thresholds as Compared to Threshold Limit Values," *American Industrial Hygiene Association Journal*, vol. 42, 1981, p. 479.

19. *Code of Federal Regulations. Title 29 (Labor)*, Part 1900.1000ff, USGSD, Washington, DC, annual.

20. *Code of Federal Regulations, Title 10 (Energy)*, Part 20, USGSD, Washington, DC, annual.

21. *UNAMAP (Version 5) User's Network for Applied Modeling of Air Pollution*, EPA-DF-83/007, PB83-244368, NTIS, Springfield, VA, 1983.

22. R. G. Sextro, "Control of Indoor Radon and Radon Progeny Concentrations," symposium paper HI85-39-3. ASHRAE. Atlanta, GA, 1985.

23. American Society of Heating, Refrigerating, and Air-Conditioning Engineers, *Methods of Testing Air-Cleaning Devices Used in General Ventilation for Removing Particulate Matter*, Standard 52-76, ASHRAE, Atlanta, GA, 1976.

---

*Note: NTIS = National Technical Information Service, Springfield, VA 22161; USGSD = Superintendent of Documents, U.S. Government Printing Office, Washington, DC 20402.

24. *Clean Room and Work Station Requirements, Controlled Environments*, Federal Standard 209d. General Services Administration, Washington, DC, 1988.

25. *Threshold Limit Values for Chemical Substances in the Work Environment*, American Conference of Governmental Industrial Hygienists, Cincinnati, OH, 1984.

26. *Code of Federal Regulations, Title 40 (Environment)*, USGSD, Washington, DC, annual.

27. *Guide for Explosion Venting* (NFPA 68-1978) National Fire Protection Association, Quincy, MA, 1978.

28. *Fire Hazard Properties of Flammable Liquids, Gases and Volatile Solids* (NFPA 325M-1984), National Fire Protection Association, Quincy, MA, 1984.

29. *Control Techniques for Particulate Emission from Stationary Sources*, vol 1: EPA 450/3-81-005A, PB83-127498; vol. 2: EPA 450/3-81-005B, PB83-127480; NTIS, Springfield, VA, 1981, 1983.

30. *Control Technology for Volatile Organic Emissions from Stationary Sources*, EPA 450/2-78-022, PB284804, NTIS, Springfield, VA, 1978.

31. P. M. Eller (ed.), *NIOSH Manual of Analytical Methods*, 3d ed., vols. 1, 2, PB8J-179018/GAR, NTIS, Springfield, VA.

32. M. J. Ellenbecker, "Evaluation of Total Airborne Emissions from Industrial Processes," *Journal of Air Pollution Control Association*, vol. 33, 1983, p. 884.

33. Committee on Industrial Ventilation, *Industrial Ventilation*, 18th ed., American Conference of Governmental Industrial Hygienists, Cincinnati, OH, 1984.

34. ASHRAE, *Ventilation for Acceptable Indoor Air Quality*, Standard 62-89, ASHRAE, Atlanta, GA, 1989.

35. R. D. Rivers, "Predicting Particulate Air Quality in Recirculatory Ventilation Systems," *ASHRAE Transactions*, vol. 88, part 1, 1982.

36. ASHRAE, *1985 ASHRAE Handbook, Fundamentals*, ASHRAE, Atlanta, GA. 1985, chap. 22.

37. *Standard for Test Performance of Filter Units* (UL 900), 4th ed., Underwriters' Laboratories, Northbrook, IL, 1983.

38. G. O. Nelson and N. Correia, "Respirator Cartridge Efficiency Studies," *American Industrial Hygiene Association Journal*, vol. 37, 1976, p. 514.

39. L. A. Jones, E. B. Sansone, and T. S. Farris, "Prediction of Activated Carbon Performance in Binary Vapor Mixtures," *American Industrial Hygiene Association Journal*, vol. 44, 1983, p. 716.

40. C. I. Harding and T. R. Kelley, "Horizontal and Vertical Distribution of Corrosion Rates in an Industrialized Seacoast City," *Journal Air Pollution Control Association*, vol. 17, 1966, p. 545.

41. *1985 Annual Book of ASTM Standards*, sec. 15.01 (Standards D2652, D2854, D2862, D2866, D2867, D3466, D3467, D3802, D3803, D3838), American Society of Testing and Materials, Philadelphia, PA, 1985.

42. H. W. Parker, *Air Pollution*, Prentice-Hall, Englewood Cliffs, NJ, 1977.

43. P. N. Cheremisinoff and R. A. Young, *Air Pollution Control Design Handbook*, Marcel Dekker, New York, 1977.

44. J. A. Danielson (ed.), *Air Pollution Engineering Manual*, 2d ed., EPA Publication AP-40, NTIS, 1973.

45. R. R. Wilson et al., *Guidelines for Particulate Sampling in Gaseous Effluents from Industrial Processes*, EPA/600/7-79/028, PB290899/4, NTIS, Springfield, VA, 1979.

46. *Military Standard; Filter Units, Protective Clothing, Gas Mask Components and Related Products, Performance Test Standards*, MIL-STD-282, Naval Publications and Forms Center, Philadelphia, PA, 1956.

47. D. A. Lundgren et al. (eds.), *Aerosol Measurement*, University Presses of Florida, Gainesville, FL, 1979.

48. J. F. Walling et al., *Sampling Air for Gaseous Organic Chemicals Using Solid Adsorbents: Applications to Tenax*, EPA/600/4-82/059, PB82-262189, NTIS, Springfield, VA, 1982.

BIBLIOGRAPHY

Books and Monographs

Hidy, G. M.: *Aerosols: An Industrial and Environmental Science*, Academic Press, Orlando, FL, 1983.

Party's Industrial Hygiene and Toxicology, 3d ed., Wiley-Interscience, New York, 1985; vol. 1: G. D. Clayton and F. E. Clayton (eds.), *General Principles*; vol. 2*a, b, c:* G. D. Clayton and F. E. Clayton (eds.), *Toxicology*; vol. 3*a, b:* L. J. Cralley and L. V. Cralley (eds.), *The Work Environment.*

Registry of Toxic Effects of Chemical Substances, vols. 1, 2, U.S. Dept. of Health and Human Services, HE 20.7112:980, USGSD, Washington, DC, 1980.

Sax, N. I.: *Dangerous Properties of Industrial Materials*, 7th ed., Van Nostrand-Reinhold, Florence, KY, 1988.

Periodicals

ASHRAE Journal, American Society Heating, Refrigerating, and Air-Conditioning Engineers, Atlanta.

Atmospheric Environment, Pergamon Press, Elmsford, NY.

American Industrial Hygiene Association Journal, AIHA, Akron, OH.

Aerosol Science and Technology, Elsevier Science Publishing Co., New York, NY.

Journal of Aerosol Science, Pergamon Press, Elmsford, NY.

Journal of the Air Pollution & Waste Management Association, A&WMA, Pittsburgh, PA.

Environmental Science and Technology, ACS, Washington, D.C.

Filtration and Separation, Uplands Press Ltd., London, England.

Chemical Engineering, McGraw-Hill, New York, NY.

NTIS Published Searches (annually, NTIS), Springfield, VA.

U.S. Government Research and Development Reports (biweekly, NTIS).

CHAPTER 21
INDOOR AIR QUALITY

William A. Turner, M.S., P.E.
President, Turner Building Science, LLC,
The H.L. Turner Group Inc., Concord, New Hampshire

21.1   THE ISSUE OF INDOOR AIR QUALITY

Concerns over Indoor Air Quality (IAQ) have propelled it into the forefront of the workplace and educational environment, and in many cases it has become a dominant issue confronting facility management and maintenance personnel. As the public has become more aware of the health and comfort implications of IAQ, pressure has increased on all types of facilities to maintain an acceptable level of air quality. In most facilities, IAQ concerns are now regarded as health and safety concerns, as they likely should be.

Like many other building related issues, unresolved IAQ problems often result in tensions between the occupants affected and those with the responsibility for the building's management/operation. We have learned that successfully resolving IAQ problems is based as much on a person's understanding that something is being done to resolve the problem, as it is on the actual expenditure of resources leading to the solution. Consequently, a rapid and well thought-out response to calls for assistance begins to build the foundation for a positive relationship (educating both parties regarding the concerns and what is being done about it) and the potential for eventual resolution.

Before discussing the issue of maintenance of mechanical systems and its effects on indoor air quality, it is important that the reader understand the variety and range of factors that often affect indoor quality. This chapter will discuss the types of sources that can create a concern for indoor air quality, how pollutants are distributed (transported), the general professional guidelines regarding IAQ, and ultimately, what can be done to control the pollutants. In addition, we will discuss the positive or negative impact that building occupants can have on indoor air quality. A table with suggested preventive maintenance scheduling is included.

21.2   THE IMPACT OF INDOOR AIR QUALITY

Much information has been learned about the relationship of indoor air quality and the maintenance of mechanical systems in commercial and educational (school) buildings. Although it is now widely recognized by IAQ consultants, property owners and managers are only recently beginning to accept that *the expense and effort required to address indoor air quality in a pro-active manner is almost always less than the expense and required to address a problem after it has occurred.*

Many organizations have learned that with proper training, indoor air quality problems *can be prevented*, or the likelihood of them happening reduced. This training often begins by making sure that the facility's management staff and operations personnel understand the many factors that can create IAQ problems. In addition, the facilities personnel must be well-trained in the operations of their specific HVAC equipment and must have budgets and support staff to allow them to remain diligent in year-round preventive maintenance activities.

Many indoor air quality problems that occur in high-rise buildings are very similar, in nature and cause, to those problems that occur in one-or two-story strip mall centers, low rise educational buildings, and even homes. Consequently, the design and use of the building does assume the same level of importance as the building's maintenance and operations activities. Too often, in a fast track value engineering (cost cutting) construction environment, factors that will govern the air quality a building will ultimately be capable of, are overlooked.

21.3   FACTORS AFFECTING IAQ

The air quality found inside a building is the result of many factors, some of which have to do with the building design, but many of which have to do with daily use patterns and the occupants' activities on a given day. There are routine factors that can change rather quickly, such as occupant loading or a change in control over space conditioning. There are often unplanned events, such as odor or contaminant releases, that can occur in an uncontrolled manner.

In addition, there are other factors which can develop into air quality problems over extended periods, such as the long-term effect of moisture intrusion within a wall system. Problems of this nature include the development of biological growths or the deterioration of building components. Another factor that may vary in the long term are gradual modifications to the tenant makeup, affecting how and when the building and its mechanical systems are being used.

Experience has shown, however, that for an indoor air quality problem to occur there need to be four specific factors present. Those factors are:

- A contaminant source (pollutant)
- A pollutant pathway (hole, tunnel, duct, hollow wall)
- HVAC systems (transport mechanism)
- Building occupants (the user)

While any of these factors can play an important part in an overall indoor air quality concern, all four must be present for an occupant complaint to occur.

Take, for example, a source such as an odor present in a building, away from any occupied spaces. The building's occupants, however, are located at the other end of the building, away from the source. In this instance, with the odor occurring and the building occupied, if there is no transport mechanism present (HVAC system) the chance of an indoor air quality complaint to occur is not likely.

Given a second example in a building that clearly has a contaminant source and a fully functional HVAC system, but no building occupants. In this instance the chance for an indoor air quality complaint to occur is also limited. However, in any instance where an

HVAC system, contaminant source, and building occupants are present concurrently, the chance for an indoor air quality complaint to occur increases.

21.4 CONTAMINANT SOURCES

The "source" is considered the area or product, which is creating the indoor air quality contaminant in the air. This source can be an odor or fume brought into the building through the outside air supply. This source can also originate from within a building from deteriorating components, off-gassing of fixtures or furnishings, or can be generated by a building occupant activity (such as photocopying, laminating, gluing, soldering, etc.).

As previously discussed, indoor air quality sources can occur *outside* of the building or can be generated *inside* of a building. Whenever these sources are uncontrolled or not planned for, the potential for an indoor air quality problem to occur is increased dramatically. Indoor air quality training programs developed by the U.S. EPA and others have identified a variety of areas in which sources can occur.

21.4.1 Sources Outside of the Building

Sources found outside of the building can include outdoor air, which has been contaminated by a variety of things both naturally occurring and man-made. The naturally occurring ones are usually anticipated in the forms of pollen, dust, etc., which are generated by trees, grasses, fungi, and other living things. Man-made pollutants can include gasses, dusts, and fumes from industrial activities.

In addition, outside air fans can bring in other outside sources, such as exhaust from a boiler stack, lab hood fan, or sanitary vent, or exhaust from vehicles parked at building loading docks, or even from idling traffic (buses, trucks, cars) in parking lots and nearby congested freeways.

Sources can also be brought in with soil gas from the earth. Examples include Radon and Volatile Organic Compounds (VOC) from spills or underground storage tank leaks.

21.4.2 Mechanical Equipment Sources

Sources of indoor air quality contaminants related to the HVAC system can include dirt and dust, which is found in ductwork or on other components deposited from historic use of poor air filters. Most often seen is microbiological growth from condensate drain pans (that do not drain completely under all operating conditions), poorly-maintained humidifiers, leaking duct coils, or poorly-insulated ducts which allow portions of the system to become wet or damp.

In addition, indoor air quality investigations often turn up by-products of combustion from improper venting of combustion products such as carbon monoxide from water heaters, or fine particles from back-puffing oil fired boilers, or improper gasket installations, or poorly-tuned gas-fired appliances.

21.4.3 Building Occupant Activities

Activities that can create contaminant sources in buildings come from a variety of areas. Building management firms must be apprised of the fact that housekeeping activities as well as maintenance activities can create potential indoor air quality sources. In addition, occupant activities have been shown as a source for a variety of air quality contaminants. These contaminants include the obvious, such as tobacco smoking, and may include cooking, cosmetic odors, and even the use of certain office supplies that may affect more allergic individuals.

Building housekeeping and maintenance crews must also be made aware of the fact that the materials used for cleaning, pest control activities, maintenance, and lubricating supplies can also generate odors, vapors, and fumes, which can be considered potential indoor air quality sources.

Under many laws in some states, the occupants have a right to know if they are being exposed to some chemical that is added to the air as it is utilized.

21.4.4 Building Components and Furnishings

Building components and furnishings can create indoor air quality problems. Such products can act as collectors for dust or fallen airborne fibers and other particles. These furnishings include actual hard-surface fixtures such as cabinets, desks, bookshelves, etc., and soft surfaces such as books and reports.

In addition, carpeting has been found to be one of the largest collectors (storers) of dust and other indoor air contaminants. This problem increases as most carpeting ages since the buildup of dirt and other contaminants can never fully be removed through cleaning processes in most carpets. One exception to that is impervious backed (non-flow-through) carpets with short piles such as the Powerbond product made by Collins and Aikman, and a few other competing manufacturers. These types of completely waterproof backed materials can often be "effectively restoratively cleaned" (cleaned to like new condition) with chemical pretreatment, aggressive high-temperature extraction, and rapid drying.

Other types of unsanitary conditions can be created as a result of water damage that may occur in occupied space. Water leaks or spills which come in contact with porous materials such as cellulose ceiling tiles, paper covered gypsum wallboard and even the aforementioned carpeting, present a tremendous potential for the growth of microbiological organisms if they are not dried completely in 24 to 48 h after wetting.

When chemically-blended products, fixtures, and furnishings are new, the potential for indoor air quality concerns also exists. Research has shown that dyes and glues used in the manufacture and installation of items such as wallpaper, carpeting, and even drapery systems can off-gas irritating chemicals into a building environment. A method for "forcing" the premature off-gassing of these products has been developed, and is referred to as "bake-out" or "flush out." In general a bake-out process has not been shown to work well in most environments and may damage materials or drive VOCs out of some materials and into others, so it is not recommended. Flushing out a new or renovated facility with as much outdoor air-conditioned outdoor air as the mechanical system can provide has been shown to work in certain environments. Flush-outs should only be attempted under the proper guidance of an experienced mechanical engineer and environmental consultant who understands the limits of the HVAC system.

21.4.5 Other Unusual Sources

There are many other sources of indoor air quality contaminants which are often considered unusual, yet can be found in most buildings. The most common of these is a small kitchenette area, or a high-use photocopier, both found in many office suites. Odors are often the primary cause of IAQ complaints where these are found. In addition, odors from a building or tenant cafeteria can create indoor air quality complaints.

Other areas which can generate unique, though not unusual, problems include laboratory areas, copying or printing shops, gymnasium or health spa areas, pools, beauty salons, smoking lounges, and general office areas with high volume copy machines. High-use photocopies are often a concern because they bond a mixture of carbon black, iron filings, and styrene monomer to paper using electrostatic charge and high temperatures. Thus, given the process, the by-products of a high use photocopier are most often heat, chemical odors, and fugitive dusts. In general, it is most cost effective to exhaust specific known pollutants such as high use photocopier emissions from a facility via local exhaust or tying the area into a general exhaust vs. recycling the by-products into the occupants' breathing air by returning the air to the HVAC system.

21.5 POLLUTANT PATHWAYS

The way in which air flows over, around, and at times under a building, can create pollutant pathways. These pathways, created by air pressure differences or air movement, can act as a carrier or transport mechanism for the pollutant, taking it from the source to previously unaffected areas of the building. Specific examples of this type of situation are utility trenches that are common in many facilities, or rooftop reentrainment (from exhausts into air intakes).

A fourth and often-critical factor affecting indoor air quality involves pollutant pathways and naturally occurring driving forces. Pollutant pathways can most commonly be referred to as small pressure differentials that occur throughout the inside of a building.

These pressure differentials can be caused from an oversupply of conditioned air being provided to one room, while limited quantities of return air are being allowed. When this occurs, a positive pressure differential happens. If this room as described were a laboratory, high-use copying area, kitchen, or other source of potential contaminants, those contaminants can then be delivered through the use of an unplanned pollutant pathway to other occupied areas in the building.

Driving forces describes the activity that generally occurs through natural forces exerted on the outside of a building such as the wind or the sun, or internal forces generated by powerful fans. These forces can include high winds on the north side creating a low-pressure area on the south side of the building, or the phenomenon known as *stack effect,* which can occur most dramatically during winter months. During stack effect, warm air rises inside of a building causing an accumulation of warm air into the upper floors, while at the same time generating an obvious lack of warm air in the lower floors. As heated air rises, it is replaced by cold air being "sucked" into the building at lower floors.

Even when a building is designed and maintained under a positive pressure, there is always an area or two that will occur to be periodically under a negative pressure. As this occurs so does the potential for a source to affect the low-pressure areas. The interaction between pollutant pathways and intermittent or variable driving forces can lead to the problem where a single source is causing indoor air quality complaints in various and sometimes distant, areas within a building. This is especially true if a building is operated under extreme negative pressures [such as 0.10 to 0.20 in (2.5–5.0 cm) water column negative] where sources can be pulled large distances through a facility. Diagnosing and mitigating this type of a problem is often complicated and requires a good understanding of how to make low pressure difference measurements with a digital micromanometer, and the ability to understand how the HVAC systems operate.

21.6 HVAC SYSTEM

Although the HVAC system may not always be able to control air contaminants that are being generated at the source (via local exhaust), it can often be used to affect internal pressure relationships and even dilute (through the ingestion of outside air) indoor air quality contaminants. For the sake of our discussions here it is important to note that while the indoor air quality source can in fact be the HVAC system, it is most commonly found as the carrier (or transporter) of the source contaminants to the building occupants. In general it is always most cost-effective to use the exhaust component of an HVAC system to capture locally generated pollutants and take them out of the building.

21.7 BUILDING OCCUPANTS

Without a person or persons who are impacted by the presence of a contaminant source, there would be no indoor air quality complaints. Facility managers must keep in mind that building occupants may be very healthy or in fact may be immune system-compromised due to various health situations such as chemotherapy, an organ transplant, or disease. Although indoor air quality concerns in nonindustrial facilities seldom deal with situations that would violate OSHA air pollution standards, complaints are often driven by low-level exposures to a certain pollutant or a mixture of materials. Finally it is important for the facilities staff to educate the building occupants as to what types of things they should and should not do that would contribute to indoor air quality problems.

21.8 STANDARDS GOVERNING INDOOR AIR QUALITY

Although at present there is no federal legislation that sets standards for acceptable indoor air quality, there are a number of states that have enacted legislation requiring that. Many states are currently considering laws that would govern how mold is diagnosed and remediated. Some of these laws will be aimed at avoiding conflict-of-interest situations between the diagnostician and the mitigator.

The scope and nature of indoor air quality regulation in those affected states varies on a state-by-state basis. However, a good rule of thumb is to include what legal experts refer to as "providing standard practice of care" when determining the level of indoor air quality or HVAC system hygiene that is acceptable and the behavior is warranted. Additionally both certified industrial hygienists and professional engineers follow a code of ethics that requires them to protect the public good.

During the design, construction, and even retrofit of commercial buildings there are recommended trade association standards, which should be clearly understood (as to their basic intent) and followed. The standards written by the American Society of Heating, Refrigerating, and Air Conditioning Engineers (ASHRAE) were created to cover several different aspects of commercial building design and operation/management which can affect indoor air quality. Standards written by Sheet Metal Associated Contractors of North America (SMACNA) offer not only standards for the construction of ductwork, but also include IAQ guidelines for buildings under construction. The American Conference of Government Industrial Hygienists (ACGIH) and U.S. EPA also offer guidelines for dealing with bioaerosols, and mold remediation in schools and commercial buildings.

When reviewing observed data in a facility and comparing it to guidance, it is important to remember, for example, that in a nonindustrial setting, the fact that there is carbon monoxide (CO) at levels higher inside a building than outside, indicates a likely source, whether the level is higher than the guideline of 9 ppm or not. In addition to understanding the level of pollutants, it is very important to understand why a pollutant is present, what the source is and what is controlling it, and to be able to predict whether it can rise to a life threatening or hazardous situation.

21.8.1 ASHRAE standard 62-2001

The most common standard recommending guidelines for acceptable indoor air quality is referred to as standard 62-2001: "Ventilation for Acceptable Air Quality".

ASHRAE 62-2001 is intended to assist professionals in the proper design and operation of ventilation systems for buildings. This standard presents two possible procedures for ventilation design: A "ventilation rate" procedure and an "air quality" procedure.

Whichever procedure is used during the design of mechanical systems, the standard states that the design criteria in assumption shall be documented and made available to those responsible for the operation and maintenance of the system. Some of the important features of ASHRAE 62-1989 include the following:

- A definition of acceptable air quality
- A discussion of ventilation effectiveness
- The recommendation of the use of source control of contaminants
- Recommendations on the use of heat recovery ventilators
- A guideline for acceptable carbon monoxide levels
- Guidelines for exhausting tobacco smoke out of buildings
- Guidelines for construction and system start-up
- Guidelines for operation and maintenance
- Appendices listing suggested possible guidelines for common indoor air pollutants

The latest revision, ASHRAE 62-2001, is currently available, along with several addenda that clarify or elaborate information. Readers are urged to obtain a copy from ASHRAE, 1791 Tullie Circle, N.E., Atlanta, GA 30329-2305, www.ASHRAE.org.

21.8.2 ASHRAE Standard 55-1992

The second of the two ASHRAE standards, which can have an affect on indoor air quality, is referred to as standard 55-1992: "Thermal Environmental Conditions for Human Occupancy."

ASHRAE 55-1992 covers several environmental perimeters including the temperature, radiation, humidity, and air movement designed within a building's structure. For example, the standard conveys what air velocities would be expected to contribute to complaints of cold drafts.

Since the perception of being too warm or cold, too humid or dry, can often be misinterpreted as an indoor air quality issue (or at best will confound the issue), thermal comfort within the occupied space becomes important.

Some of the important features of ASHRAE 55-1992 include:

- A definition of acceptable thermal comfort
- Recommendations for summer comfort zones and winter comfort zones clearly defined in graphs
- Guidelines for making adjustments in air delivery depending on occupant activity levels

21.9   IAQ PROGRAM MANAGEMENT

Once the construction of a building has been completed (and the HVAC system has been commissioned and fine-tuned to work as designed) the responsibility for preserving good IAQ rests with the building operations maintenance functions. A comprehensive facilities program should exist within each occupying firm or building manager. This program will be used to manage and monitor acceptable indoor air quality. Elements of the plan can include limited scheduled proactive testing and analysis, identifications of a need for repairs based on observations or testing, as well as to guide the reaction to unscheduled breakdown repairs and maintenance. There must also be provisions in the plan for evaluating and documenting changes in the building use and/or alterations to the HVAC systems.

To implement an IAQ program, three elements must be provided:

- Management support and commitment from all levels of the organization
- Staffing by trained and competent personnel
- Budgetary allocations to provide the fiscal resources necessary to perform the comprehensive preventive care services required to maintain good indoor air quality

The following pages cover areas that are shown to have an effect on the indoor air quality of a building.

21.9.1   Facility Inventory

An audit of the condition of all buildings should be conducted as early as possible to establish a baseline for the physical plant. In addition to anticipating potential IAQ problems, an audit has value to all other aspects of operations and maintenance. Results of an effective and periodic audit can also be used as due diligence groundwork in instances of lawsuits related to building IAQ. Such audits can be performed by properly prepared and knowledgeable mechanical/environmental engineers.

The major building components that should be considered for evaluation include:

- Roofing system (including roof drains, scuppers, and any surface drains)
- HVAC system (including local exhaust systems in specialty areas)
- Plumbing system (including traps and vents)
- Electrical (especially as it pertains to HVAC controls)
- Sub-floor/crawl space areas (including basements and utility trenches)
- Other areas: Loading docks, food services, duplicating and copy rooms, laboratories, storage facilities, industrial and manufacturing areas, and so forth.

21.9.2   Systems Descriptions

Providing a description of each system is a particularly important but often overlooked program element. It is important that both existing and new staff in the building understand the basics of what the design intent is, and the capacities and capabilities of the systems to assure that the building is performing to its optimum.

Review your existing equipment lists and mechanical plans. If you have them, compare the as-built drawings to the equipment installed. If you do not have as-built drawings or the drawings have not been kept up to date, an effort should be made to accomplish this. This helps to assure that components are receiving regular attention. In addition, equipment that has been installed in inaccessible locations is often overlooked during routine maintenance. By verifying all equipment noted on the mechanical drawings and what it originally served, you are forced to confirm its presence and condition, and the current use for the area it serves.

21.9.3 Operations Plan

An operations manual should describe the basic design intent of the HVAC systems and how to run the building systems. The plan should provide a detailed systematic procedure covering operation from start up to shutdown. This document, an essential reference for day-to-day activities, will be a useful training document for new employees. A vital part of this plan should be a compilation of all available manufacturers' literature and manuals.

21.9.4 *COMPREHENSIVE PREVENTIVE MONITORING PLAN*

Preventive monitoring is incorporated into all building IAQ programs. The objective of this monitoring is to prevent small deficiencies from blossoming into major costly outages or repairs. For example, regularly oiling a bearing on a fan system will extend the useful lifetime of the unit, and prevent the potential loss of make-up or exhaust air that is necessary for proper IAQ. The table of preventive monitoring tasks (Table 21.1) lists the typical tasks that are necessary in any program related to IAQ.

The management staff can either self-educate to a point where they periodically check basic IAQ parameters or they can schedule periodic consultant audits for a determination of indoor air quality. Both frequency and timing are important. For example, biological growths of fungi and bacterial organisms have often been found in condensation pans. In some environmental studies, these organisms have been shown to cause disease. HVAC systems run throughout the year. If the condensate drain pans do not drain completely under all operating conditions, (see Sec. 18.8) they will very likely accumulate biological growth as the year progresses and may disseminate biologically active aerosols and odors to the building's occupants all of the time. *If building management waits until a chosen annual date to clean and disinfect these units, they may pose a greater risk to building occupants than if they were inspected on a regular schedule and cleaned/disinfected as problems begin to develop.*

Scheduled maintenance and any preventive monitoring must also be tied to a building's utilization schedule. Painting, coil cleaning, pest control, or other projects involving the use of volatile organic chemicals should be scheduled when there are few, if any, occupants in the immediate vicinity. The work area should be well ventilated, using fans or supplying large quantities of outside air wherever possible. In contrast, the monitoring phase must be scheduled when the building is occupied to allow for retrieval of "occupied" and/or potential exposure analysis figures.

21.9.5 Managing Buildings During High Dew Point Events

Managing buildings during high dew point events is likely one of the most difficult and overlooked tasks. For example, when the outdoor air dew point is above 55°F(11°C), the outdoor air has very little drying potential when it is brought indoors. When it is above 65°F (21°C) it has no drying potential and most often will wet many materials if the air is brought in

TABLE 21.1   Preventive Monitoring Tasks and Recommended Frequency

Air-handling system	
Heat/cool/vent	
Supply and return fans	
Clean or replace air filters (minimum MERV 7)	Quarterly or as needed
Clean coils & spray with approved disinfectant	Annually or as needed
Clean & flush condensate pans with approved disinfectant as needed	Quarterly or as needed
Check filters. Replace if needed	Monthly
Check condensate pan drains	Monthly
Check & clean air intake louvers and damper	Monthly
Check air distribution boxes, reheat coils, and overall duct system for cleanliness, leaks, collapse	Quarterly
Clean heat & cool coils	Annually
Check & clean fan/blower blades of dirt & trash build-up	Quarterly or as needed
Unit heaters	
Check thermostat & clean coil & fan blades	Quarterly or as needed
Fan cooling units	
Clean & flush condensate pans with approved disinfectant as needed	Quarterly
Check & clean filters as needed	Quarterly
Check & clean coil with spray and disinfectant	Annually
Ceiling fans	
Check & clean fan blades	Monthly
Roof top units	
Heat/cool/vent	
Check operation of outdoor air damper	Monthly
Check or replace air filters	Quarterly or as needed
Clean coils & spray with disinfectant	Quarterly or as needed
Clean & flush condensate pans with disinfectant as needed	Quarterly or as needed
Clean all coils	Annually or as needed
Check to see roof near air intake is white or silver color, not black	Annually
Drainage systems	
Roof drains	
Check & clean roof or floor drains in area	Weekly
Check to ensure no standing water at air intake	Quarterly
Scuppers, check for drainage not hitting building	Quarterly
Roof drip lines, check for proper drainage away from the facility	Yearly
Walls	
Check for signs of moisture intrusion after heavy rain	Quarterly
IAQ audit	
In house or independent testing & analysis of mechanical systems and tenant areas for acceptable air quality levels	Quarterly, or twice yearly

without drying it by passing it over a cold cooling coil (less than 45°F (7°C)) or passing it through a dehumidification unit. In most northern climates, if high dew point air (greater than 65°F (21°C)) is brought directly into a crawlspace or basement, it will condense and lead to mold problems. If a building does not have the ability to dry out the air within it during high dew point weather, carpets or other porous materials should never be purposely wet i.e., shampooing or wet extraction, as mold growth will likely result within the material itself.

Mechanical Systems. Pulleys, belts, bearings, dampers, heating and cooling coils, and other mechanical systems must be checked periodically. These are included in Table 21.1; consequently it is unlikely that these should be the source of an indoor air quality problem if the staff is following the table schedules.

Pulleys and belts should be tightened as needed and changed prior to failure, a belt tension meter should be used on any belt that is not lasting for a year or more to begin to figure out why the shortened life is occurring. Bearings should be lubricated or replaced to prevent major failure of vital system components. Air dampers and baffles should be cleaned and cleared of debris periodically. Failure to perform these activities will result in an increase in resistance, which causes a decrease in air supply.

The air conveyance system (ductwork) should have access ports available to allow for periodic observations or periodic air sampling at various points in the system. Proactive air sampling data, when retrieved, can offer accurate information on the unexpected elevated levels of mold, bacteria, dust, etc., allowing building management to address a problem before any concerns are raised. See Chapter 26: "Ductwork."

Building ventilation networks are systems that serve multiple locations. The practice of arbitrarily adjusting the dampers or baffles to accommodate complaints from one area must be avoided. By changing the airflow in one area, the system balance is shifted and distribution throughout the entire network is affected. If there are distribution complaints, a test of the building's air-handling system should be performed to confirm that the HVAC system and distribution network are in balance and are adequate for the current use of the space that is served.

Condensate Pans. Components that are exposed to water, such as condensation drain pans, require scrupulous maintenance to prevent biological growth and the entry of undesired biological contaminants into the indoor airstream. This is an item in Table 21.1. In addition, this has been proven to be extremely important. Pans should be designed to drain completely under all operating conditions. High levels of UV light are considered to be another means of caring for drain pans. In general, UV light should only be relied upon as a secondary defense. Proper hygiene of the pan should always be the first line of attack to prevent or minimize growth. Therefore, if a pan is found to be the source of an IAQ problem it needs either to be redesigned or better maintenance. The need to use an approved biocide or algaecide is an indication of inadequate design or inadequate maintenance. Refer to Sec. 18.8 "Condensate Control."

Air Filters. Mechanical equipment for ventilation, heating, and air conditioning contain air filter media or in some cases screens that minimize particulate matter from collecting on the coils, fans, and interior housings and work. Originally, the primary consideration was to protect the system from loss of efficiency from clogging resulting in equipment down and extensive cleaning. Recognition is now given to the role that filters can play in improving indoor air quality and protecting the occupants from either outdoor origin materials or the recirculation of indoor origin materials.

Building systems are primarily concerned with mechanical filtration filters. Commonly used are replaceable fiberglass filter media, mechanical screen media that requires cleaning and recoating, and bag-type filters. Air filters are rated by efficiency. The method of particle trapment is by impingement, which locks the particulates within the filters. Filters have a life expectancy rated in hours when used for a given air flow with expected contaminants. Severe dry spells that blow excessive dirt, or interior dust producing activities, or construction at a site nearby, which sends dust toward the building's outside air intake, can increase contaminants reaching the filter.

Independent investigations have found that most often "permanent" metal filters or screens are not being cleaned properly or often enough. Improper maintenance of these filters can have the following effects:

- Equipment efficiency is reduced and more energy is required to push the same quantity of air through the filter.
- Some contaminants pass through the overloaded filter, clog the coils, and enter the occupied spaces.
- Dirt collection on the coils diminishes the thermal transfer efficiency of the unit resulting in higher energy consumption.
- Contaminants on the coils become a breeding ground for bacterial and fungal growth.

Based on considerable experience, it is recommended that all "permanent" filters be replaced with at least medium efficiency disposable type filters with a minimum of a MERV 7 rating. The recommended period for checking and changing air filters is covered in the list of preventive monitoring tasks, Table 21.1.

Vacuuming to Clean Surfaces.   Normal industrial vacuums emit particles and fibers in their exhaust. An improvement in performance can be obtained if the vacuum can be fitted with a high-efficiency particulate air filter (HEPA). HEPA vacuums should be used in areas that might have spores or microorganisms. HEPA vacuum filtration ensures that potentially toxic or harmful aerosols are not dispersed while responding to a problem. Most manufacturers offer a HEPA filter for units. Dry sweeping in problem areas should always be avoided. Poor floor cleaning maintenance and general poor housekeeping has been shown to increase IAQ complaints. Conversely, deep cleaning of a given floor has been shown to decrease most symptoms.

Roofing.   Poorly designed or improperly drained roofs may be a potential source of poor IAQ. Flat roofs will invariably collect water and cause pooling. Stagnant, standing water on roofs can support the growth of algae and bacteria that can be drawn into building air systems. Leaks in roofs result in water damage or accumulation in ceiling tile, carpeting, or internal wall spaces. Fungi and bacteria that develop in this moisture have been found to be responsible for allergies and respiratory disease. Consequently, when roofs are sloped inadequately or roof repairs are postponed, IAQ can easily be compromised. Water damage materials must be rapidly dried or removed and replaced in a timely fashion (less than 24 to 48 h) before they serve as a breeding ground for biological growth. Roofs that allow the wall system or foundation system to be charged with rainwater often end up harboring hidden microbial growth within the wall or foundation system that is being wet. Roofs should always drain water away from the facility.

Pipe Leaks.   Pipe leaks can occur through normal corrosion, microbial-induced corrosion (MIC), mechanical failure, or because of the age of the facility. In any case of leakage, repair or replacement of the damaged pipe section must be performed immediately and the contaminated water quickly removed and disposed of by pouring it into an appropriate drain. It is prudent to have wet vacuums, submersible pumps, squeeze brooms, and mops available to handle water emergencies. Water-damaged ceiling tiles, rugs, insulation, or walls must also be moved and replaced in a timely fashion to prevent mold from growing. Following storms, it is good practice to inspect the building for discolored ceiling tiles or leaks as signs of water problems. Pipes that may be subjected to situations causing condensation should be insulated to prevent the problem. It is always more cost-effective to deal with wet materials quickly to dry them out than to have to deal with mold-contaminated materials because items were allowed to stay wet too long and mold grew on them.

Drains.   The antisiphon "P" traps in sinks must contain water to prevent noxious odors from the sanitary sewer lines from migrating back into the indoor air spaces. Sinks and drains that are used infrequently can dry out, allowing a path for sewer gases to enter. Cupsinks in laboratory fume hoods and on benches frequently dry out and have often been found to be the sources of odors. This problem can be resolved and prevented by periodically running water in these drains, plugging unused drains with a rubber stopper, or using a nontoxic liquid with a low vapor pressure to fill the P trap. Drains located in mechanical or air-handling rooms are often designed for condensate pan runoff. If no moisture is evident, these lines can often dry out (this occurs most frequently in dry climates and during the winter months in humid climates). Dry P traps allow a path for noxious odors from sewer lines to enter. These odors can then be unknowingly distributed throughout the building by the air distribution system. Drains in laboratories must be kept clear and in working order. Sediment in drain traps can provide conditions that support growth and accumulation of biological organisms.

Sumps and Grease Traps.   Some plumbing systems have sumps for art rooms or kitchen grease to prevent materials from entering a sewer or septic system. It is important for these sumps to be kept odor tight and in some cases to be properly vented to the outdoors as part of the sanitary vent system. If mysterious foul odors occur in any area with a sump, it should be high priority for checking for both proper hygiene and venting.

21.9.6   Indoor Liquid and Nonsolid Materials Management Program

It is important to understand that most chemicals used in housekeeping, maintenance, operations, pest control, and cafeteria service can affect indoor air quality. Their use in the building should be properly managed. In most states all pesticides and biocides must be used by trained individuals. Material safety data sheets (MSDS) are obtained from the manufacturers or distributors and should be on file in the building where any potentially hazardous material is to be stored or used. MSDS on all chemical products used should be kept in an easily accessible area for employee access. A master file should also be kept in the offices of personnel responsible for the chemical's use. A building's population should not be subjected to unknown and potentially harmful effects of any material because of the absence of an MSDS.

The chemical composition of hazardous ingredients must be evaluated along with the hazardous reaction potential. Wherever possible, materials, chemicals, and reagents that present the lowest toxic potential should be used. There are many authoritative sources available to help in this determination, including vendor salesperson, vendor chemists, and even the local or state health department.

Less toxic materials should be substituted for more toxic materials. In general, water soluble materials should be given preference to organic solvent systems. Often, water soluble materials will still have some level of preservative chemical contained in the product to keep it from growing materials in its own storage container. Materials that are higher in flash point (ignition) and/or have a lower vapor pressure are also preferred.

Building Use Changes.   Special care must be exercised when building space utilization is changed. This is especially true when occupant density is increased, or heat sources such as computers are added. Renovation, redesign, or changes in building use can create situations that may lead to compromises in IAQ. For example, if a high-use copy machine is installed in a small closet or other unventilated space, chemical emissions such as ozone or carbon black dust or ammonia or styrene may become problems. Further, the addition of blueprint processes, paper bursting activities, or even microwave cooking can add unexpected odors, dusts, or gases to an unprepared or improperly engineered area. The effect on heat and noise level must also be anticipated if new equipment is added to an area.

A common renovation problem arises when additional personnel need to be accommodated in a space. Office or instructional spaces are often partitioned and additional furniture and equipment is installed. Managers need to anticipate the need to modify the air distribution in these situations. Conversely, when partitions are removed, creating new spaces, the ventilation distribution and balance must be revised. Care must be taken to ensure that, in the final layout, air supply grills are located far enough away from the returns so that complete air balance/mixing does occur. Otherwise, stagnant areas will develop. This is especially true if the air system has to provide both heat and cooling to an area.

Evaluation of Building Materials (Prior to Construction or Renovation).   New materials used during construction or renovations present the potential for occupant exposure to emissions (off-gassing) from those materials. Many IAQ problems can be avoided by selecting building materials that are less likely to pollute, and these will not smell as much as other products for the same use. In cases where this is not possible, the responsibility to address the issue of off-gassing products falls to building management. The most likely avenue to pursue is dilution of the concentrations caused by the source. Chemical content, chemical emission potential, and the potential for toxicity and irritation are all issues that should be considered in construction or renovation.

Most manufacturers are now supplying MSDS with the delivery of their products. If the product is a carpet, wallboard, paneling, or material, some vendors can now supply product "emission" statistics on these items. In cases where the product emission rates are not available, most vendors will employ staff or consultants who can relate information about the products' effects on IAQ.

The simplest approach to material evaluation is to identify the potential use of materials that can raise IAQ concerns, and select the lowest-emitting and least toxic products. In general, blended, mixed, or composite materials are more likely to off-gas than materials with a purely natural origin. Most materials used in buildings today fall into the former category, which is why specifying low emitting or low odor materials is appropriate in most instances.

Ventilation and Isolation During Construction or Renovation.   During renovation projects, special care must be taken to isolate the area being remodeled from all occupied areas. The contaminants can spread via the HVAC system, by airflow through corridors, up and down elevator shafts, and through ceiling plenum areas. Contamination of occupied areas during remodeling can lead to tenant complaints ranging from dry eyes, itchy skin, and burning throats to nausea and even vomiting. Individual effects almost always include lower productivity, which is used to support claims of tenant move-out.

There should be an effort to supply sufficient exhaust-only ventilation to areas that are involved in this construction or renovation process. Further, adjoining (occupied) areas should be kept under a slight "positive" pressure to avoid cross-contamination. This positive pressure in occupied areas, along with exhaust ventilation at the source of the contaminants, will help isolate and remove most contaminants. These principles are covered in the document provided by Sheet Metal Associated Contractors of North America (SMACNA) entitled *IAQ Guidelines for Buildings Under Construction.*

After the construction or renovation project has been completed, other strategies fall into play that can have a major impact on the IAQ of the facility. Much has been written about "bake-out" or "flush-out" procedures. *Bake-out* is a process of overheating a building or space to artificially age the materials that are sources of contaminants. In general it has been shown that this approach is not very successful and may drive VOCs into porous materials that did not contain them in the first place. Research has shown that *building flush-outs* may be far more useful and cause less damage. However, buildings should not be operated beyond their intended design parameters or damage may occur.

Pesticides. Building management should be committed to providing the building with a pest-managed environment through the implementation of preventive hygienic methods and chemical strategies when necessary. Integrated Pest Management (IPM) emphasizes the use of nonchemical techniques for the management of pests, relying on the use of pesticides only when nonchemical strategies are not effective.

Each building should receive a scheduled inspection for the purpose of identifying existing or potential problems that may contribute to harboring, feeding, or population growth of pests. Inspectors should recommend nonchemical pest control measures whenever possible. Finally, the inspector will list options for chemical treatments should the nonchemical measures prove to be unsuccessful.

Chemical control treatment should be applied after building hours with the exception of emergency situations. All contract personnel should be required to possess a valid pest control applicators license, and the license must be on file in the building management office.

Building management has a responsibility to notify all tenants about the pending treatment for pests. Management should also assign personnel to accompany the application technician, and to monitor the use and location of all pesticides.

21.9.7 Record Keeping

While there is a clear record keeping requirement for asbestos, other potential contaminants have no set procedure in most states. Therefore, it is recommended that records on issues related to IAQ be kept on a scheduled periodic basis. These records should contain information related to the scheduled PM program, as well as engineering reports showing the IAQ health of the building. This IAQ survey will involve testing for specific contaminants related to indoor air quality and should be performed every 3 to 6 months. If no IAQ problems are encountered during the first year, consideration can be given to extending the schedule to a 6-month basis.

The minimum testing required should include review of the operations of the HVAC system, testing and evaluations for the presence of CO (carbon monoxide), CO_2 (carbon dioxide), temperature and humidity (as a comfort indicator), pressure differential of various rooms (where contaminants are used) and A/C zones, and in some instances samples for the presence of excessive levels of fungi, bacteria, and breathable dusts.

This record keeping should include records developed during an initial baseline audit, laboratory field tests (if any), reinspections (every 3 to 6 months), and any other additional testing for specific materials that may be in use. Action plans and any maintenance or operation activities that have been used to mitigate problems and results of work performed by any outside consultants or contractors that were used should be kept on file as well.

21.9.8 Preventive Monitoring Tasks Guidelines

The preventive monitoring tasks guidelines (Table 21.1) are designed to offer guidance on daily, weekly, monthly, quarterly, and annual procedures that can have a dramatic effect on indoor air quality. These items are offered for guidance only, and should not be treated as suggested requirements. Specific information for each building depends on a number of factors, including occupancy levels, location, number of heating days vs. cooling days, ambient air conditions, type of HVAC system, recommended manufacturer maintenance guidelines, and other unique factors not addressed in this chapter.

21.10 SUMMARY

Reducing the likelihood of having and indoor air quality problem in a facility has been best summarized in Table 21.2.

The purpose of the guidance offered in this chapter is to begin the thought-process regarding items that can exhibit the potential for the development of an indoor air quality problem. *Being aware of what can happen can go a long way to reducing the likelihood of it happening.* When an IAQ event does occur, responding to it in a rapid, well thought out response begins to build the foundation for a positive relationship with all parties involved, increasing the potential for eventual resolution.

TABLE 21.2 Preventing and Solving IAQ Problems*

1.	Understand and educate people
2.	Keep the building dry and clean
3.	Reduce potential sources
4.	Provide exhaust ventilation for unavoidable stationary sources
5.	Provide dilution ventilation for unavoidable mobile or large area sources
6.	Reduce unplanned airflows

*This table contributed by Terry Brennan, Camroden Associates, Rome, N.Y.

CHAPTER 22
AEROBIOLOGICAL CONTROLS

W.J. Kowalski, P.E., Ph.D.
The Pennsylvania State University,
Department of Architectural Engineering,
University Park, Pennsylvania

W.P. Bahnfleth, P.E., Ph.D.
The Pennsylvania State University,
Department of Architectural Engineering,
University Park, Pennsylvania

22.1   INTRODUCTION

Buildings are designed to provide a comfortable living and working environment for humans, but they simultaneously provide an environment in which a wide variety of microorganisms can survive long enough to transmit between hosts. Some microbes find niches indoors that are so favorable to growth that they proliferate as if buildings were giant incubators. In order to counteract the incubation effect of indoor environments on microbes, it is necessary to design heating, ventilation, and air-conditioning systems to be inhospitable to microbes. Primarily, this involves controlling indoor environments through ventilation, air cleaning, and air disinfection. The purpose of this chapter is to provide a general introduction to the use of air cleaning technologies to control the aerobiology of indoor air and to provide basic engineering information to assist engineers in this objective.

22.2   THE AEROBIOLOGY OF INDOOR AIR

Aerobiology is the science and study of airborne microbes in the environment. The microbial content of outdoor air consists primarily of fungal spores, bacterial spores, vegetative bacteria, pollen, and various microbial by-products, including other allergens. The microbial composition of outdoor air varies seasonally, geographically, and even by altitude. In general, the aerobiology of outdoor air is healthy for humans, with one notable exception being the seasonal sufferings of allergic individuals.

The aerobiology of indoor air is affected both by the microbial content of outdoor air and by indoor sources such as humans, animals, plants, building materials, and furnishings.

Indoor air temperatures, relative humidity, the presence of moisture, and even the level of sunlight or artificial light also directly affect indoor aerobiology. As a result of these factors, the microbial composition of indoor air can vary considerably from that of outdoor air, even for buildings with natural ventilation or 100 percent outside air. Indoor air may consist largely of harmless microorganisms, but it is harmful microbes, the pathogens and allergens, on which most attention must be focused in order to engineer healthy indoor environments.

Pathogens are disease causing microbes such as cold viruses, influenza, tuberculosis bacilli, and anthrax spores.[1] Allergens are microbes, parts of microbes, microbial byproducts, or pollen that cause allergic reactions in atopic, or susceptible, individuals.[2] Well over one hundred airborne pathogens and allergens can occur in normal indoor environments. Dozens of other disease-causing microbes can cause inhalation hazards under special conditions such as in laboratories, animal facilities, or when weaponized as biological warfare agents. Detailed lists and descriptions of these microbes are available in the literature and these will not be repeated here, but a general review of these pathogens with select examples will provide insight into the nature of the airborne hazards. Consult the references for more extensive listings of pathogens and allergens.[1,3–5]

Viruses are submicroscopic intracellular parasites and could be considered as either very simple microbes or very complex molecules. The ACGIH defines viruses as living, which makes them the smallest living organisms. Viruses cannot reproduce by themselves; they can replicate only by commandeering a host cell and utilizing its resources. They contain DNA or RNA, but not both. In fact, almost all respiratory viruses are RNA-bases. Only *Poxvirus* and *Adenovirus* contain DNA. The most common viruses are respiratory viruses, which cause the common cold, the flu, most pneumonia, and hemorrhagic fever. They are transmitted by inhalation of droplets aerosolized by coughing, sneezing, and talking, or by direct contact.[6–8]

Bacteria are prokaryotic, or single-celled, organisms in the 0.1 to 10 μm size range. They have a cell membrane, DNA, and some subcellular components. they are ubiquitous in both indoor and outdoor air, although most respiratory bacteria pathogenic for humans are found almost exclusively indoors.[9,10] Some of the exceptions, such as *Pseudomonas aeruginosa*, are found in association with sewage or waste, while some others, such as *Bacillus anthracis* and *Francisella tularensis*, are associated with animals, but can transmit to humans by close association or direct contact.[1]

Fungi are plant-like microbes that are abundant in the environment, where they tend to live off dead organic matter. Fungal spores are seed-like microbes in the 1 to 100 μm size range that produce new fungal growth. Fungi may reach concentrations in the outdoor air of 100 to 1000 cfu/m^3 typically (cfu = colony forming unit), with seasonal variations.[11–13] Many guidelines suggest that levels above 1000 cfu/m^3 represent contamination, although the actual health threat depends on the species.[13,14] Inhalation of fungal spores by those with compromised immune systems can be fatal. Low levels of airborne fungi can contribute to Sick Building Syndrome (SBS) and poor indoor air quality (IAQ).[2,15]

Fungi do not cause contagious infections; only the person inhaling the spores is at risk. Fungi can enter buildings through air intakes, via infiltration, or they can be brought in with materials, carpets, furniture, people, or plants.[11] *Fungal infections inevitably result from fungi being in the wrong place*, often as the result of excessive indoor moisture, poor cleanliness, or improper design of ventilation system components.[15,16] The fungi that may cause problems in buildings are usually certain of the ascomycetes.[11]

Pollen are plant seedlings that occur naturally in the environment. They cause no disease but can produce allergic reactions or respiratory irritation in susceptible individuals. Because of their size, typically about 100 μm they are easily removed by dust filters and are rarely a problem in buildings with forced ventilation.

Dust mites and other insects such as cockroaches may cause allergic reactions as the result of their waste products and parts of these creatures that may be small enough to become airborne. Dander consists of skin cells or other matter from cats, dogs, or other animals like mice. Dander can cause allergic reactions in sensitive individuals. The size of such particles can vary from 1 to 100 μm and so dust filters alone my not always be sufficient for their control.

Volatile organic compounds (VOCs) and other gases can be produced by molds, bacteria, or organic materials. Certain VOCs can irritate nasal passages, the lungs, the eyes, or the skin. Gaseous odors can create unpleasantness and even nausea. VOCs can only be removed or destroyed by gas phase filtration technologies like carbon adsorption.[2]

22.3 SOURCES OF MICROBES

An understanding of the source of microbial contamination is essential for designing systems for their removal. Environmental sources, including the air, water, and soil, produce an endless variety of microorganisms that may find their way into buildings. These environmental microbes pose no threat to healthy people except for allergic individuals. There are no human viruses in outdoor air, only bacteria and fungal spores, and almost no true pathogens. Certain microbes, such as *Legionella pneumophila*, can come from natural waters and cause problems when they get into cooling towers. Microbes produced or released into the air from human sewage may sometimes include pathogens such as *Mycobacteria* and even viruses like *Adenovirus*. Microbes from agricultural sources may also include dangerous pathogens such as *Francisella tularensis* and *Bacillus anthracis*.

In indoor environments, the most dangerous pathogens come from human sources. Viruses and contagious bacterial diseases are routinely aerosolized by infectious individuals and these aerosols may persist for minutes or even hours in recirculated indoor air. Fomites, particles left on surfaces that contain microbes, may persist even longer, and it has been reported that *Mycobacterium tuberculosis* bacilli may survive for days on indoor surfaces.

Animals can be a source of certain hazardous microbes. Cats, dogs, birds, mice, and agricultural animals can produce various airborne pathogens or aerosolizable feces containing pathogens that are also human pathogens or that can cause noncontagious infections in exposed individuals. No plant diseases directly threaten humans, but a number of fungi may grow on plants or in the soil for plants. The spores of fungal allergens may be produced in the soil of indoor plants and become aerosolized indoors.

Buildings can be a sources of microbial contaminants when they either contain microbes or provide an environment in which microbes can proliferate. Figure 22.1 illustrates classifications of buildings according to the levels of indoor microbes and the degree to which

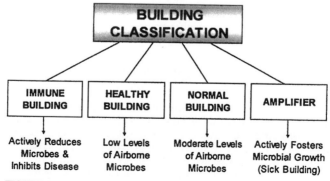

FIGURE 22.1 Building aerobiological classifications.

they promote or suppress disease. Some explanations of these classifications and more specific definitions follow, although these definitions should not be considered absolute since no standards exist to define acceptable levels of indoor microbial contamination. Note that normal buildings fall into a category below that of healthy buildings, since normal buildings tend to have microbial concentrations higher than outdoor air.

Healthy buildings can be loosely defined as those that do not foster the growth of microorganisms and do not directly contribute to or cause airborne infections or exacerbate allergies. Healthy buildings do not necessarily include any air cleaning systems or even forced ventilation. A facility can only really be classified as a healthy building based on epidemiological data, indoor air concentrations of microbes, or comparison with other buildings. The California Healthy Buildings (CHB) Study suggests that levels of mixed airborne spores at or below 100 cfu/m^3 characterized healthy buildings.[2,17] However, the exact composition of spores, or bacteria, must figure into the health hazard of any microbial concentrations. A healthy building should cause no more respiratory problems than would occur from breathing natural outdoor air. It could be argued, however, that indoor levels in a healthy building should be sufficiently below outdoor levels that they would protect atopic individuals against allergic reactions.

Immune buildings actively or passively inhibit the transmission of airborne disease or allergies through the removal or disinfection of airborne microbial contaminants. The definition can be further extended to include the removal of chemical agents and the disinfection of internal building surfaces. Levels of all microbes, pathogenic or otherwise, could be expected to be lower than those of the outdoor air. Although no criteria for indoor levels in an immune building currently exist, it could be assumed that they would be similar to levels characterizing health care facilities and operating rooms. Airborne microbial concentrations in the range of 10 to 50 cfu/m^3 might be expected in a well-designed immune building, although some consideration must be given to the composition as well since the infectious dose for some pathogens can be in the single digits range (i.e., TB). Ideally, an immune building would have zero levels of hazardous microbes, but this is unachievable for any building occupied by humans, who continuously release microbes of various sorts.

A building is classified as an amplifier if it promotes microbial growth or accumulates harmful airborne microbes. In general, an amplifier increases airborne concentrations of mold spores well above the levels in outdoor air. The composition of the air in amplifiers varies considerably from the composition of outdoor air and often includes specific molds that thrive on building materials such as *Aspergillus, Penicillium,* and *Cladosporium.* The fungi that thrive in indoor environments are primarily those that produce toxins. Although amplifiers associated with illness may be classified as cases of SBS or building related illness (BRI), only some 15 to 30 percent of cases of SBS or BRI have been associated with molds.[2,11,15] The remaining cases of SBS or BRI are often associated with chemical or material contamination, vibration, noise, or other factors.

22.4   EPIDEMIOLOGY AND PATHOLOGY

Some airborne diseases are contagious, like tuberculosis, while others do not transmit beyond the first host, like inhalation anthrax. Most airborne pathogens cause respiratory infections, but many cause other infections such as diseases of the eyes, ears, and skin. There are two aspects of medical science that are important to an understanding of the limitations and applications of air disinfection technology; the epidemiology of airborne disease, which is the study of how diseases spread in a population, and the pathology of airborne disease, which is the study of how a disease develops or progresses.

Most airborne pathogens can transmit by the airborne route, by direct person-to-person contact, or by contact with contaminated surfaces. The degree to which any transmission mechanism predominates is not well understood, but it is thought that perhaps one-half or more of cold viruses are transmitted via direct contact.[6] Some data suggest that most cold and flu transmission occurs as the result of hand-to-hand contact or by contact between hands and fomites.[18–20]

Some sources claim that ventilation systems do not transmit disease but this is at best a strained interpretation of limited data. Proximity to an infected individual plays a major role in contagious airborne disease and one study suggests that the probability of airborne disease transmission is greatly increased when individuals are within 5 to 6 ft of each other for extended periods.[10,21,22] However, evidence from Zeterberg[23] demonstrated that infections can be caused by TB bacilli at points in a building that are connected only by ventilation duct and considerable evidence for the transmission of airborne disease via ventilation systems comes from laboratory studies on the viability and settling time of microorganisms suspended in air.[24–26]

An epidemic occurs when a contagious disease spreads naturally through a susceptible population. Like a forest fire, which spreads until the density of combustibles is insufficient to maintain the fire, the disease will spread until either everyone is infected or the density of susceptible individuals is insufficient to sustain further propagation.[1]

Following exposure to an infectious dose of an agent, a disease will generally progress through a series of stages that include incubation, an infectious period, and a recovery period. During incubation period the microbe will multiply and spread during the course of a few hours, days, or weeks. The incubation period may produce symptoms or else it may be asymptomatic. Treatment during the infectious period may limit or prevent the disease. Following incubation there will be a period of infectiousness for contagious diseases. Ultimately the disease may resolve in the recovery period, with or without treatment, as the body recovers or develops antibodies to eliminate the pathogen. Some diseases, like tuberculosis, may persist indefinitely unless treated.[1]

Nosocomial infections are a special class of diseases that result from exposure to opportunistic pathogens in hospital or health care environments. In intensive care units, almost a third of nosocomial infections are respiratory diseases, but not all of these are airborne.[27] Nosocomial infections can be nonrespiratory, such as when common microbes like *Staphylococcus aureus* settle on open wounds, burns, or medical equipment.[28,29]

Patients whose natural defenses have been compromised may be susceptible to common or endogenous microbes that become pathogenic when provided with such opportunities. In theory, the best solution to the increased susceptibility of the patient is to provide decreased concentrations of indoor microorganisms. The reduction of microbial contaminants far below ambient levels is accomplished through the use of isolation rooms, filtration equipment, ultraviolet germicidal irradiation (UVGI), and strict hygiene procedures.[30–32]

22.5 AIR DISINFECTION FUNDAMENTALS

A common characteristic of all microbes is that their populations will decay naturally or under exposure to any biocidal factor such as heating or irradiation. In the natural environment death occurs rapidly for airborne human respiratory pathogens. These parasites have become so adapted to indoor transmission that, with few exceptions, they cannot tolerate even short exposures to environmental extremes.

Various environmental factors destroy airborne microbes. Direct sunlight contains low levels of ultraviolet radiation that are lethal in the long term. Dehydration renders most microbes inactive, although many spores may survive indefinitely and return to activity.

High temperatures will inactivate all pathogens, some more rapidly than others. Freezing will destroy most pathogens, except that some, especially spores, may become preserved. Pollution can be fatal to microorganisms. Plate-out, or adsorption, occurs on all indoor surfaces, and microbes entrained on such surfaces may die out naturally or be reaerosolized. However, adsorbed bacteria and spores may grow and multiply if conditions are suitable (see the previous section on amplifiers).

Any biocidal factor, environmental or otherwise, will tend to reduce pathogen populations according to the following general equation:

$$S(t) = e^{-kIt} \qquad\qquad (22.1)$$

where S = population fraction at time t
$\quad k$ = rate constant for process
$\quad I$ = Intensity of exposure to biocidal factor

The resulting exponential decay curve is known as a survival curve, or death curve. Equation (22.1) is a simplistic model that is adequate for most applications where survival rates are not extreme. The kill rate of an air disinfection system is simply the complement of the survival, or $1 - S$.

Disinfection is defined as the reduction of any microbial population. The rate of disinfection is defined by the rate constant (k in the above equations) which is unique to every species. Sterilization is defined as the complete destruction of a microbial population and the common technical definition of sterilization is a reduction of microbial populations by six logs, or a factor of 1 million. In percentage terms, sterilization is a population reduction from 100 to 0.0001 percent.

22.6 AIR CLEANING AND DISINFECTION TECHNOLOGIES

A variety of technologies exist for the purpose of cleaning indoor air but few of these are sufficiently developed that accurate sizing and prediction of performance are possible. Applications of dilution ventilation, filtration, UVGI, and carbon adsorption are common today for controlling IAQ. Other technologies are available commercially or are in development that may have potential, but most of these are not as well understood. Figure 22.2

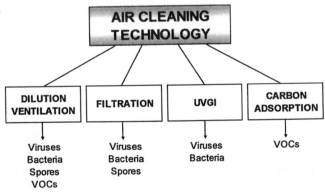

FIGURE 22.2   Commonly applied air cleaning and disinfection technologies.

shows a breakdown of these technologies and the microbiological contaminants against which they are most effective.

22.6.1 Dilution Ventilation

Dilution ventilation is the control of contaminants by supplying clean outside air to a space to replace contaminated air. It removes all contaminants at roughly equal rates. Ventilation may be natural or forced. Natural ventilation depends on air flow through openings in the building envelope driven by wind pressure, stack effect, and diffusion. Prediction of the actual ventilation rate in a naturally ventilated space is not straightforward. Forced ventilation is driven by exhaust or HVAC system fans. Forced ventilation systems provide specific ventilation flow rates by design that can be verified postconstruction.

Ventilation rate requirements for many buildings are determined by ASHRAE Standard 62-1999.[33] For most applications, requirements depend on either the number of occupants or the floor area of a ventilated space. However, the standard also permits demand controlled ventilation based on measured levels of indoor carbon dioxide or other specific contaminants. Refer to Chaps. 20 and 21 for additional information on IAQ and dilution ventilation.

For typical constant air volume (CAV) and variable air volume (VAV) forced ventilation systems complying with ASHRAE Standard 62, the ventilation rate at design conditions is roughly 15 to 25 percent or the supply air flow.[34] Other standards apply to specialized applications and may require up to 100 percent outside air. Dual path systems designed to comply with ASHRAE Standard 62 have separate ventilation systems that also provide 100 percent outside air. The *HVAC Systems and Equipment* volume of the *ASHRAE Handbook* gives detailed descriptions of the various ventilation system types as well as other sources.[2,34]

Systems that do not recirculate air have an apparent advantage over systems that do because they do not redistribute contaminants within the building. However, a 100 percent outside air system supplying the ASHRAE Standard 62 flow rate may also be at a disadvantage relative to a CAV or VAV system with a higher design supply air flow rate because fewer air changes pass through filters and other air-cleaning elements.

At design, the ventilation characteristics of all system types complying with the pertinent standards are similar. Off-design performance, however, may vary considerably from system to system. Systems with design outside air fractions less than 100 percent may bring in up to 100 percent outside air if they have air-side economizer controls. VAV systems with uncontrolled outside air will tend to provide the same outdoor air quantity as a fraction of supply flow rate as supply flow reduces.

Most forced ventilation systems are mixing ventilation systems designed to mix room air with ventilation air as completely as possible. To the extent that this does not occur, ventilation effectiveness is degraded. Displacement ventilation systems create stratified contaminant distributions and, at least in theory, remove contaminants more efficiently than mixing systems. Refer to Part F for more detailed information on types of ventilation systems.

22.6.2 Filtration

Filtration has been used for controlling indoor airborne contaminants for much of the past century.[2] Filtration is treated in detail in Chap. 20. This section addresses only the filtration of airborne microbes.

Most microbes are roughly spherical or ovoid shaped particles in the 0.01 to 10 μ size range and each species is characterized by a range of sizes that is distributed lognormally.[35] Refer to Chap. 20 for additional information on the size distribution of micron-sized particles. Filtration efficiencies depend on the filter type, the air velocity, and the microbial size distribution. The use of the logmean diameter will provide a good approximation of the overall filter removal rate of any microbe. Table 22.1 lists the logmean sizes for over 100 airborne pathogens and allergens. The Type column identifies the microbe as a virus (V), bacteria (B), bacterial spore (BS), or fungal spore (FS).

Using the sizes in Table 22.1, removal efficiencies of microbial contaminants can be predicted from filter performance curves. Unfortunately, performance curves do not extend into the submicron size range that includes many bacteria and all viruses. In order to extend filter performance curves into the submicron size range it is necessary to use a filter model based on physical or empirical data for each filter type.[35] Test data on filter performance is often available in the form of the minimum efficiency reporting value (MERV) test results that are compiled in accordance with the ASHRAE 52.2-1999 standard.[36] Figure 22.3 shows a set of filter performance curves that have been extended down into the virus size range through filter modeling based on MERV test results, which are also shown in the graphs.[37] These curves represent particular manufacturer's filter models and may not be representative of other filter models, even with the same MERV ratings. Actual manufacturer's curves should be used whenever they are available.

22.6.3   Ultraviolet Germicidal Irradiation (UVGI)

UVGI has been around for over a century, but has not developed into a reliable disinfection technology until recently. The problem with UVGI applications in the past is that no special attention was paid to the precise design and operating conditions and effectiveness often fell far short of performance in laboratory settings. Information on methods for accurately sizing, designing, and operating UVGI systems is available today but old notions persist to the point that some still refer to it as a "snake oil" technology.

UVGI can be extremely effective against viruses, which are fragile and highly susceptible to the genetic damage, resulting from irradiation in the 200 to 300 nm wavelength range.

TABLE 22.1   Sizes of Airborne Pathogens and Allergens

Microbe	Type	Logmean size, μm	Microbe	Type	Logmean size, μm
Rhinovirus	V	0.023	Yersinia pestis	B	0.707
Rubella virus	V	0.061	Neisseria meningitidis	B	0.775
Hantavirus	V	0.096	Staphylococcus aureus	B	0.866
Influenza A virus	V	0.098	Streptococcus pyogenes	B	0.894
Coronavirus	V	0.110	Bacillus anthracis	BS	1.118
Measles virus	V	0.158	Nocardia asteroides	BS	1.118
Mumps virus	V	0.164	Serratia marcescens	B	1.225
Varicella-zoster virus	V	0.173	Histoplasma capsulatum	FS	2.236
Mycoplasma pneumoniae	B	0.177	Penicillium	FS	3.262
Respiratory syncytial virus	V	0.190	Aspergillus	FS	3.354
Francisella tularensis	B	0.200	Coccidioides immitis	FS	3.464
Bordetella pertussis	B	0.245	Paracoccidioides	FS	4.472
Coxiella burnetii	B	0.283	Aureobasidium pullulans	FS	4.899
Legionella pneumophila	B	0.520	Cryptococcus neoformans	FS	5.477
Brucella	B	0.566	Stachybotrys chartarum	FS	5.623
Mycobacterium tuberculosis	B	0.637	Rhizopus stolonifer	FS	6.928
Klebsiella pneumoniae	B	0.671	Cladosporium	FS	8.062
Corynebacterium diphtheriae	B	0.698	Altemaria altemata	FS	11.225

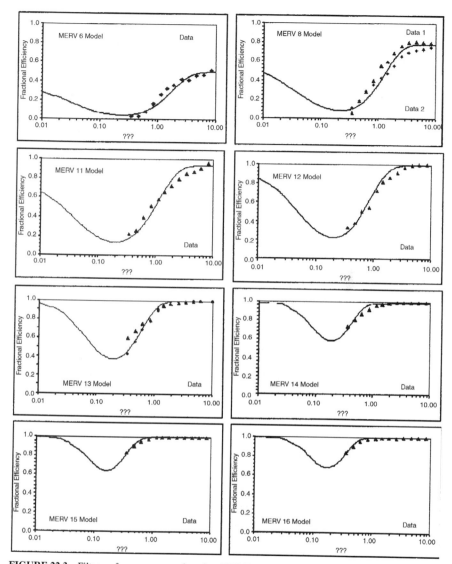

FIGURE 22.3 Filter performance curves based on MERV test results.

The effect of UVGI on bacteria is species dependent; some bacteria are highly susceptible to UVGI and some are highly resistant. Fungal and bacterial spores are as resistant to UVGI as they are to any biocidal factor such as heating or chemical disinfectants.[38]

Figure 22.4 shows a typical UVGI system inside a rectangular duct in which an externally mounted ballast powers an internally mounted lamp.

The dose received by an airborne microbe passing through an enclosure depends on the lamp UV power output, the duct dimensions, and the reflectivity. If the air is well mixed,

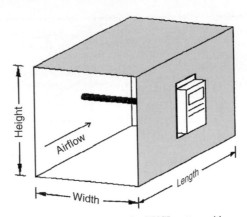

FIGURE 22.4 Diagram of a UVGI system with an externally mounted ballast.

then the dose is computed by multiplying the average intensity in the enclosure by the exposure time. Equation (22.1) can then be used to estimate the kill rate for any microbe rate constant. The methods for calculating the average intensity with or without reflectivity have been detailed elsewhere, and although these methods are complicated, the methods for selecting appropriate system sizes can be as simple as those for selecting filters.[39] Table 22.2 summarizes a proposed standard for rating UVGI systems, which is based on a UVGI rating value (URV).[40] The URV system quantifies 20 levels of UVGI average intensity that cover a broad range of typical UVGI system sizes. The URV represents an average intensity from which may be computed the dose and kill rates for any microbe. Examples of kill rates are shown in Table 22.2 for TB bacilli and smallpox exposed for 0.5 s, and for *Rhizopus* spores exposed for 2 s. These kill rates were computed per Eq. (22.2) using published rate constants.[38,41–43]

The URV rating system is designed to be analogous to the MERV rating system for filters and a URV system can be coupled with a filter of the same MERV rating to produce a balanced reduction of airborne microbial pathogens for exposure times of approximately 0.5 s.

TABLE 22.2 URV Average Intensities and Kill Rates Examples

UVGI rating value (URV)	Average intensity (μW/cm^2)	DOSE at $t = 0.5$ s (μW · s/cm^2)	TB kill rate (%)	Influenza kill rate (%)	DOSE at $t = 2$ s (μW · s/cm^2)	Rhizopus kill rate (%)
6	75	37.5	8	4	150	1
7	100	50	10	6	200	1
8	150	75	15	9	300	1
9	250	125	23	14	500	2
10	500	250	41	26	1000	4
11	1000	500	66	45	2000	8
12	1500	750	80	59	3000	12
13	2000	1000	88	69	4000	16
14	3000	1500	96	83	6000	23
15	4000	2000	99	91	8000	29
16	5000	2500	100	95	10000	35

If, for example, a MERV 11 filter is installed for aerobiological purposes, it can be coupled with a URV 11 system when the exposure time is 0.5 s. This approach offers an extremely convenient way of sizing UVGI systems because once the filter is sized and selected, the UVGI size becomes implicit.

If the exposure time is less than or greater than 0.5 s, then an equivalent URV rating can be estimated by multiplying the actual average intensity by the ratio of the actual exposure time to 0.5 s. In equation form:

$$\bar{I}_{eq} = \bar{I}\left(\frac{E_t}{0.5}\right) \tag{22.2}$$

where $\bar{I}_{eq}$ = equivalent average intensity, cm^2/μW
$\bar{I}$ = average intensity, cm^2/μW
E_t = exposure time, s

Table 22.3 provides a summary of UVGI rate constants for airborne microbes that may be used in conjunction with Eq. (22.2) and Table 22.2 to compute UVGI kill rates. Additional rate constants are available from other sources, but many of these are not for airborne microbes, but come from plate-based or water-based studies and these are, in general, conservative to use for airborne applications.[38]

Filters should always be added in concert with UVGI systems to protect the lamps against dust. Prefilters alone may not be sufficient for this purpose. Attention should also be paid to operating air velocity and air temperature, as these factors can impact UV lamp output. Some attention may also need to be given to relative humidity, as this has been shown to affect the UV rate constant and also impacts photoreactivation, which can occur when visible light exposure causes the repair of UV-damaged microorganisms.[45–47]

22.6.4 Alternative Technologies

A variety of alternative technologies are available or under development today that offer solutions for special applications, but limited data are available on their performance.

TABLE 22.3 UVGI Rate Constants in Air

Microorganism	Reference	Rate constant k cm^2/μW · s
Adenovirus	(38,43)	0.000546
Vaccinia	(38,43)	0.001528
Coxsackievirus B-1	(38,43)	0.001108
Influenza A	(38,43)	0.001187
Staphylococcus aureus	(38,44)	0.003476
Mycobacterium tuberculosis	(38,41)	0.002132
Pseudomonas aeruginosa	(38,41)	0.002555
Serratia marcescens	(38,41)	0.002208
Rhizopus nigricans spores	(38,42)	0.000086
Cladosporium herbarum	(38,42)	0.000037
Scopulariopsis brevicaulis	(38,42)	0.000034
Mucor mucedo	(38,42)	0.000040
Penicillium chrysogenum	(38,42)	0.000043
Aspergillus amstelodami	(38,42)	0.000344

Technologies such as photocatalytic oxidation and pulsed light have the capacity to disinfect airstreams but applications may be limited by their expense.[48,49] Ozone has proven to have great effectiveness in disinfecting water and has considerable potential to do the same in air, but it remains a developmental technology.[50]

22.6.5 Surface Disinfection Technologies

Air disinfection alone may not be able to deal with problems caused by aerobiological contamination, since microbes may settle on surfaces such as rugs, furniture, and equipment. Fomites may also be placed on surfaces such as doorknobs and table tops, and although they may not result from aerosols they could be a major factor in respiratory disease. Any complete solution to indoor aerobiological problems may need to address related nonairborne microbial contamination. Surface model growth on air-handling units, cooling coils, or filters can be controlled by continuous UV exposure and, in the case of filters, antimicrobials may be able to prevent growth on filters.[2,51,52]

Antimicrobials may have other applications in controlling contamination on painted surfaces, carpets, and other indoor materials. Titanium dioxide, which is used in photocatalytic oxidation processes, may have some use in disinfecting fomites on surfaces such as doorknobs. Finally, the value of passive solar exposure for the disinfection of indoor surfaces is an underexplored area, which can offer a simple means of keeping exposed building interiors germ-free.

22.6.6 Air Sampling and Detection

Air sampling can be a valuable tool in identifying and resolving indoor aerobiological contamination problems.[2] Such problems are predominantly caused by fungi, with bacteria a distant secondary cause. Air sampling can be performed by professionals to determine whether indoor airborne concentrations of fungi or bacteria are above normal levels, and they can determine if any dangerous agents, such as *Stachybotris*, are present.

Air samplers typically draw a measured volume of air into or across the surface of a petri dish. This plate is then cultured overnight in a lab and analyzed. The colony forming unit (cfu) counts can be used to estimate the airborne concentrations of fungi and bacteria. Analysis of the colonies can usually establish exactly which fungi or bacteria are present in the indoor air.

22.7 APPLICATIONS

Typical commercial buildings often have nothing more than dust filters in the outside air or supply air ventilation system. Such prefilters rarely have the capacity to control indoor aerobiology and the retrofit of higher rated MERV filters is often the only solution. Retrofitting of filters will invariably have some impact on the air-handling unit. The addition of a filter and the associated pressure loss it adds to the system may cause an increase in fan motor power consumption or a decrease in total system airflow, depending on what type of fan it is (i.e., centrifugal or axial), and where the fan is currently operating on its performance curve. The retrofit of a UVGI system is much simpler since the pressure loss associated with UVGI lamps is typically, but not always, negligible.

Health care facilities, laboratories, and animal facilities are often designed to deal with microbial contamination and provide high levels of aerobiological protection. As a result, retrofitting equipment in these facilities may not always be necessary unless there is an existing problem. However, most health care facilities and laboratories do not have UVGI systems installed (due mainly to existing guidelines which pan UVGI), and there may be some benefit to be gained from retrofitting UVGI systems in facilities where aerobiological problems or concerns exist.[4,39,53]

Residential buildings often have either natural ventilation or small forced air systems with only simple prefilters. The use of filtration and UVGI to control residential aerobiology is uncommon but growing at a rapid rate. It is estimated that 18 percent or more of respiratory infections are contracted at home, and although many of these may be direct-contact related, there is likely to be some benefit from disinfecting residential air.[20] Fungal spores commonly contaminate indoor residential environments and these may be controlled by filtration and by using UVGI on internal duct and cooling coils.

Bioterrorism defense is an area of heightened concern recently, and many building owners have been investigating alternatives for protecting building occupants. Of course, there is little difference between microbes released from a sneeze and those deliberately released into building air except that intentional releases are likely to involve larger quantities of more virulent agents. The technologies for dealing with the problem are virtually identical, but the level of concern about bioterrorism often dictates that the solutions provide maximum protection to building occupants. The installation of MERV filters combined with UVGI systems offers what is possibly the most practical, reliable, and cost-effective solution to bioterrorism defense for typical commercial buildings, but there are many types of buildings in which solutions are not so simple. Each case must be evaluated separately and decisions made regarding the level of safety that can be affordably provided. Bioterrorism defense is a complicated topic that is more adequately addressed elsewhere and these sources should be consulted for further information.[5,54,55]

22.8 AEROBIOLOGICAL MODELING OF SYSTEMS

The physical characteristics that define the removal or disinfection of airborne microbes have been addressed previously. These characteristics can be used to model the removal rates of microbes by dilution ventilation, filtration, and UVGI. Modeling of dilution ventilation to determine removal rates is possible using various methods including calculus, system models, computational methods, multizone modeling software, and computational fluid dynamics (CFD).

Computational methods using spreadsheets are fairly easy to develop when a limited number of well-mixed zones are being analyzed. This approach involves setting up a minute-by-minute analysis of the contaminant concentrations in each zone, the supply and exhaust air quantities, and the outside air flowrates. The contaminant source may be a continuous release, a sudden release, or a measured release at different points in time. In such models it is prudent to include any outdoor airborne spores or environmental bacteria that are added to the building control volume.

Multizone modeling software can be more effective to use than spreadsheets for complex building models, and they often have built-in options for factors such as stack effects, wind, wall, and door leakage, and plate-out on building internal surfaces. Multizone modeling methods assume that air is well-mixed in each zone. For situations in which it is necessary to model or to understand the actual airflow currents inside a single or multizone building model, it is necessary to use CFD software. A variety of such general and special purpose software packages are available today, such as CONTAMW and COMIS.[2,56,57]

As an example of sizing a filtration/UVGI system for an aerobiological application, a system model is adapted here for predicting steady state microbial concentrations in a building modeled as a single zone. The system model presented in Chap. 20 is restated here in aerobiological terms and assumes only one filter and a UVGI system is present:

$$C_i = \frac{G_i + 0.01(P_i Q_i + 0.01 P_f P_u Q_m)C_o}{Q_{v2} + K_d A_d + Q(1 - P_f P_u / 10,000)} \qquad (22.3)$$

where C_i = steady state indoor concentration, cfu/m^3
$\quad\quad C_o$ = outdoor concentration, cfu/m^3
$\quad\quad G_i$ = indoor generation or release rate, cfu/min
$\quad\quad P_f$ = penetration through filter, percent
$\quad\quad P_u$ = penetration through UVGI system, percent

In Eq. (22.3), the units of cfu may be used to represent bacteria, fungal spores, or viruses, even though viruses are more properly defined by plaque-forming units (pfu). The filter penetration of any microbe may be determined by using the logmean diameter from Table 22.1 and the representative filter performance curves in Fig. 22.3. If the actual filter performance curve is available, this should be preferentially used. The UVGI kill rate can be determined by assuming a URV to match the filter MERV rating, and then computing the kill rate (removal rate) based on Eq. 22.2 and the UVGI rate constants provided in Table 22.3.

Following is an example in which concentrations of TB bacilli and *Aspergillus* spores are removed from indoor air by various levels of filtration and UVGI (MERV/URV). The TB bacilli are assumed to be generated internally at 1000 cfu/min and the *Aspergillus* spores are assumed to exist in the outdoor air at 1000 cfu/m^3. In this model single zone building, the total airflow is 2113 m^3/min and the outdoor airflow rate is 317 m^3/min. For simplicity, no leakage or plate-out effects are assumed. The penetration rates of *Aspergillus* spores and TB through the filters and UVGI systems are summarized in Table 22.4. The microbe logmean sizes and UVGI rate constants used are those presented earlier in this chapter. UVGI exposure time is assumed to be 0.5 s.

Figure 22.5 summarizes the results of the previous analysis in terms of the percentage of indoor concentrations when no filter or UVGI system is present. It can be observed that as the filtration removal efficiency and UVGI removal efficiency are increased, the steady-state concentrations level off at approximately MERV 13/URV 13. The implication is that in this model no increases in filtration above this level will provide any significant decrease in indoor concentrations. A simple explanation for this result is that an air-cleaning system has an ideal size for any particular building that is primarily a function of building volume and design airflow rate. It is not certain, however, that this applies to all possible buildings and contamination situations, but it does strongly suggest that cost-effective solutions and simple design principles exist that will enable any building with a CAV or VAV system to be adequately immunized against airborne microbial contamination.

22.9 CONCLUSIONS

The field of aerobiological engineering is too vast to be adequately addressed in this single chapter, but this synopsis has introduced the key concepts and the technologies that can be used to seek solutions to microbial IAQ problems. Design information has been provided on filtration and UVGI that should assist the engineer in developing workable solutions to most aerobiological problems. Additional information should be sought on these and other related topics in the references or from other appropriate resources.

TABLE 22.4 MERV/URV System Evaluation Input and Steady State (SS) Results

TB Bacilli	None	6/6	7/7	8/8	10/10	11/11	12/12	13/13	15/15	16/16	
Filter penetration	100	91	90	79	78	67	51	21	2	2	%
UVGI penetration	100	92	90	85	59	34	20	12	1	0	%
Total penetration	100	84	81	68	46	23	10	3	0	0	%
SS concentration	3.15	1.64	1.52	1.11	0.77	0.59	0.52	0.48	0.47	0.47	cfu/m^3
Aspergillus											
Filter penetration	100	55	47	25	15	8	1	0	0	0	%
UVGI penetration	100	100	100	100	99	98	97	97	93	92	%
Total penetration	100	55	47	24	15	8	1	0	0	0	%
SS concentration	551.3	155.6	137.4	104.4	94.5	88.8	83.4	82.7	82.7	82.7	cfu/m^3

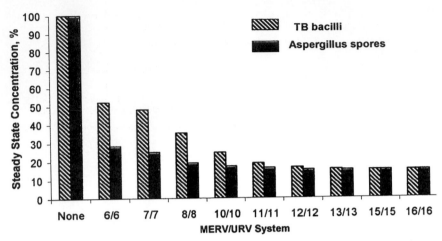

FIGURE 22.5   Reduction in concentrations with increasing levels of air cleaning in example problem.

REFERENCES

1. Ryan, K. J. editor (1994). *Sherris Medical Microbiology.* Appleton & Lange, Norwalk.

2. Spengler, J. D., Samet, J. M., and McCarthy, J. F. (2001). *Indoor Air Quality Handbook.* McGraw-Hill, New York.

3. Kowalski, W., and Bahnfleth, W. P. (1998). "Airborne respiratory diseases and technologies for control of microbes." *HPAC* 70(6).

4. Kowalski, W. J., Bahnfleth, W. P., and Carey, D. D. (2002). "Engineering control of airborne disease transmission in animal research laboratories." *Contemp. Topics Lab. Animal Sci.* 41(3), 9–17.

5. Kowalski, W. J. (2002). *Immune Building Systems Technologies.* McGraw-Hill, New York.

6. Buckland, F. E., Bynoe, M. L., and Tyrrell, D. A. J. (1965). "Experiments on the spread of colds." *J. Hygiene* 63(3), 327–343.

7. Riley, R. R. (1960). "Air-borne infections." *Am. J. Nurs.* 60, 1246–1248.

8. Nardell, E. A., Keegan, J., Cheney, S. A., and Etkind, S. C. (1991). "Airborne infection: Theoretical limits of protection achievable by building ventilation." *Am. Rev. Resp. Dis.* 144, 302– 306.

9. Mitscherlich, E., and Marth, E. H. (1984). *Microbial Survival in the Environment.* Springer-Verlag, Berlin.

10. Baker, S. A. (1995). "Airborne Transmission of Respiratory Diseases." *J. Clin. Eng.* 20(5), 401– 406.

11. Samson, R. A., editor (1994). *Health Implications of Fungi in Indoor Environments.* Elsevier, Amsterdam.

12. Rao, C. Y., and Burge, H. A. (1996). "Review of quantitative standards and guidelines for fungi in indoor air." *J. Air Waste Manage. Assoc.* 46(Sep), 899–908.

13. Burge, H. (1990). "Bioaerosols: Prevalence and health effects in the indoor environment." *J. Allerg. Clin. Immunol.* 86(5), 687–781.

14. Howard, D. H., and Howard, L. F. (1983). *Fungi Pathogenic for Humans and Animals.* Marcel Dekker, New York.

15. Godish, T. (1995). *Sick Buildings: Definition, Diagnosis and Mitigation.* C. Press Lewis Publishers, Boca Raton.

16. Hyvarinen, A., O'Rourke, M. K., Meldrum, J., Stetzenbach, L., and Reid, H. (1995). "Influence of cooling type on airborne viable fungi." *J. Aerosol Sci.* 26(S1), s887–s888.

17. Fisk, W. (1994). "The California healthy buildings study." *Center for Building Science News* Spring 7, 13.

18. Gwaltney, J. M., Moskalski, P. B., and Hendley, J. O. (1978). "Hand-to-hand transmission of rhinovirus colds." *Ann. Int. Med.* 88, 463–467.

19. Lovelock, J. E., Porterfield, J. S., Roden, A. T., Sommerville, T., and Andrews, C. H. (1952). "Further studies on the natural transmission of the common cold." *Lancet* 1952, 657–660.

20. Lidwell, O. M., and Williams, R. E. O. (1961). "The epidemiology of the common cold." *J. Hygiene* 59, 309–334.

21. Hers, J. F. P., Winkler, K. C., and editors (1973). *Airborne Transmission and Airborne Infection.* Tech. Univ. at Enschede Oosthoek Publishing Company, The Netherlands.

22. Riley, R. L., and O'Grady, F. (1961). *Airborne Infection.* Macmillan, New York.

23. Zeterberg, J. M. (1973). "A review of respiratory virology and the spread of virulent and possibly antigenic viruses via air conditioning systems."*Ann. Allergy* 31, 228–299.

24. Lidwell, O. M. (1960). "The evaluation of ventilation." *J. Hygiene* 58, 297–305.

25. Rubbo, S. D., and Benjamin, M. (1953). "Transmission of haemolytic streptococci." *J. Hygiene* 51, 278–292.

26. Goodlow, R. G., and Leonard, F. A. (1961). "Viability and infectivity of microorganisms in experimental airborne infection." *Bacteriol. Rev.* 25, 182–187.

27. Groschel, D. (1979). *Hospital-Associated Infections in the General Hospital Population and Specific Measures of Control.* Marcel Dekker, New York.

28. Weinstein, R. A. (1991). "Epidemiology and control of nosocomial infections in adult intensive care units." *Am. J. Med.* 91 (Suppl 3B), 179S–184S.

29. Tablan, O. C., Anderson, L. J., Arden, N. H., Beiman, R. F., Butler, J. C., MacNeil, M. M. and HICPAC (1994). "Guideline for the prevention of nosocomial pneumonia." *Am. J. Infect Cont.* 22, 247–292.

30. Galson, E., and Guisbond, J. (1995). "Hospital sepsis control and TB transmission." *ASHRAE* May.

31. AIA (1993). *Guidelines for Construction and Equipment of Hospital and Medical Facilities.* Mechanical Standards American Inst. of Architects, Washington, D.C.

32. Gill, K. E. (1997). "Tuberculosis isolation room design: Using CDC guidelines." *HPAC* 69(9), 69–73.

33. ASHRAE (2000). *Handbook of Systems and Equipment.* ASHRAE, Atlanta.

34. ASHRAE Standard 62-1999(1999). *Ventilation for Acceptable Indoor Air Quality,* ASHRAE, Atlanta.

35. Kowalski, W. J., Bahnfleth, W. P., and Whittam, T. S. (1999). "Filtration of airborne microorganisms: modeling and prediction." *ASHRAE Transac.* 105(2), 4–17.

36. ASHRAE (1999). *ASHRAE Standard 52.2-1999.* American Society of Heating Refrigerating, and Air Conditioning Engineers (ASHRAE), Atlanta.

37. Kowalski, W. J., and Bahnfleth, W. P. (2002). "MERV filter models for aerobiological applications." *Air Media* Summer, 13–17.

38. Kowalski, W. J., Bahnfleth, W. P., Witham, D., Severin, B. F., and Whittam, T. S. (2000). "Mathematical modeling of UVGI for air disinfection." *Quantit. Microbiol.* 2(3), 249–270.

39. Kowalski, W. J., and Bahnfleth, W. P. (2000). "Effective UVGI system design through improved modeling." *ASHRAE Transac.* 106(2), 4–15.

40. Kowalski, W. J., and Dunn, C. E. (2002). "Current trends in UVGI air and surface disinfection." *INvironment Professional* 8(6), 4–6.

41. Collins, F. M. (1971). "Relative susceptibility of acid-fast and non-acid fast bacteria to ultraviolet light." *Appl. Microbiol.* 21, 411–413.

42. Luckiesh, M. (1946). *Applications of Germicidal, Erythemal, and Infrared Energy.* D. Van Nostrand, New York.

43. Jensen, M. M. (1964). "Inactivation of airborne viruses by ultraviolet irradiation." *Appl. Microbiol.* 12(5), 418–420.

44. Sharp, G. (1940). "The effects of ultraviolet light on bacteria suspended in air." *J. Bact.* 38, 535–547.

45. Peccia, J., Werth, H. M., Miller, S., and Hernandez, M. (2001). "Effects of relative humidity on the ultraviolet induced inactivation of airborne bacteria." *Aerosol Sci. Technol.* 35, 728–740.

46. Peccia, J., and Hernandez, M. (2001). "Photoreactivation in airborne mycobacterium parafortuitum." *Appl. Environ. Microbiol.* 67, 2001.

47. Setlow, J. K. (1966). "Photoreactivation." *Radiaf. Res. Suppl.* 6, 141–155.

48. Jacoby, W. A., Blake, D. M., Fennell, J. A., Boulter, J. E., and Vargo, L. M. (1996). "Heterogeneous photocatalysis for control of volatile organic compounds in indoor air." *J. Air Waste Mgt.* 46, 891–898.

49. Wekhof, A. (2000). "Disinfection with flashlamps." *PDA J. Pharma. Sci. Technol.* 54(3), 264–267.

50. Kowalski, W. J., Bahnfleth, W. P., and Whittman, T. S. (1998). "Bactericidal effects of high airborne ozone concentrations on *Escherichia coli* and *Staphylococcus aureus.*" *Ozone Sci. Eng.* 20(3), 205–221.

51. Shaughnessy, R., Levetin, E., and Rogers, C. (1999). "The effects of UV-C on biological contamination of AHUs in a commercial office building: Preliminary results." *Indoor Environment '99*, 195–202.

52. Foarde, K. K., Hanley, J. T., and Veeck, A. C. (2000). "Efficacy of antimicrobial filter treatments." *ASHRAE J.* Dec, 52–58.

53. Luciano, J. R. (1977). *Air Contamination Control in Hospitals.* Plenum Press, New York.

54. ASHRAE (2002). *Risk Management Guidance for Health and Safety under Extraordinary Incidents.* ASHRAE, Atlanta.

55. USACE (2001). "Protecting buildings and their occupants from airborne hazards." *TI 853-01*, U.S. Army Corps of Engineers, Washington, D.C.

56. NIST (2002). *CONTAMW: Multizone Airflow and Contaminant Transport Analysis Software.* National Institute of Standards and Technology, Gaithersburg, MD.

57. Zhao, Y., Yoshino, H., and Okuyama, H. (1998). "Evaluation of the COMIS model by comparing simulation and measurement of airflow and pollutant concentration." *Indoor Air* 8, 123–130.

CHAPTER 23
FANS AND BLOWERS*

23.1 SELECTION AND APPLICATION

Robert Jorgensen, P.E.
Retired Vice-President-Engineering,
Buffalo Forge Company, Buffalo, New York

23.1.1 FAN REQUIREMENTS

Fans provide the energy to move the air through the ducts and other apparatus that form the air side of a heating, ventilating, and air-conditioning (HVAC) system. Each system may be served by one or more fans. Sometimes fans are an integral part of a unit containing coils, filters, and other devices. At other times, freestanding fans are used.

23.1.1.1 Pressure

The energy transmitted to the air by the fan must equal exactly the energy lost by the air in moving through the system. Fan requirements and system losses are usually expressed in terms of energy per unit volume of gas flowing, which is known as pressure.

In fan engineering it is necessary to distinguish various pressures according to the method by which they can be measured or by the kind of energy with which they can be identified as follows:

1. The total pressure p_T at a point in a gas stream is the force per unit area which can be measured by a manometer connected to an impact tube that points directly upstream. It is equivalent to the sum of the pressure energy† and kinetic energy of a unit volume of gas and exists by virtue of the gas density, velocity, and degree of compression.

2. The static pressure p_S at a point in a gas stream is the force per unit area which can be measured by a manometer connected to a small hole in the duct wall or other boundary,

---

*All the illustrations and tables in this chapter appear with the permission of the Buffalo Forge Company, Buffalo, NY. The text is condensed from Robert Jorgensen (ed.). *Fan Engineering*, 8th ed., Buffalo Forge Co., Buffalo, 1983, and is also used with permission.

†The concept of pressure energy is a convenience in fan engineering. It derives from the flow work term in the general energy equation.

the surface of which must be parallel to the path of the stream. It can be considered equivalent to the pressure energy of a unit volume of gas and exists by virtue of the gas density and degree of compression.

3. The velocity pressure p_V at a point in a gas stream is the force per unit area which can be measured by a manometer, one leg of which is connected to an impact tube pointing directly upstream and the other leg connected to a small hole in the duct wall (or its equivalent). It is equal to the kinetic energy of a unit volume of gas and exists by virtue of the gas density and velocity.

Pressures in fan systems are usually measured with some form of water column gauge (WG) such as the vertical U-tube or the inclined manometer. The unit of measurement is most commonly the inch water gauge (kilopascal), abbreviated as inWG (kPa).

The Air Movement and Control Association (AMCA) and the American Society of Heating, Refrigerating, and Air-Conditioning Engineers (ASHRAE) jointly published a test code, AMCA 210-97/ASHRAE 51-97, which is the standard of the industry. In it fan pressures are defined as follows:

1. The fan total pressure p_{FT} is the difference between the total pressure at the outlet of the fan p_{T2} and the total pressure at the inlet of the fan P_{T1}. If there is no inlet connection, p_{T1} must be assumed equal to zero.

$$p_{FT} = p_{T2} - p_{T1} \qquad (23.1.1)$$

2. The fan velocity pressure p_{FV} is the velocity pressure corresponding to the average velocity through the fan outlet p_{V2}.

$$p_{FV} = p_{V2} \qquad (23.1.2)$$

3. The fan static pressure p_{FS} is the difference between the fan total pressure and the fan velocity pressure:

$$p_{FS} = p_{FT} - p_{FV} \qquad (23.1.3)$$

The symbols used above are not always those encountered in fan engineering. Frequently *FTP*, *FVP*, and *FSP* are substituted for their obvious counterparts. The test standard noted above does not use either nomenclature, choosing instead to use a capital *P* for pressure and omitting the subscript *F*. The reader should have no trouble with any of these systems provided the distinction between fan pressures and pressures at a point or across a plane are carefully recognized.

23.1.1.2 Capacity

In the standard test code, "fan capacity" is defined as the volume rate of flow measured at the fan inlet conditions. Capacities in fan systems are determined from pressure measurements. These may be velocity pressure traverses of the duct itself or total pressure or static pressure measurements associated with an orifice or nozzle. Such measurements may be converted to a volume rate of flow by proper consideration of air density and duct or nozzle geometry. The unit of measurement for volume rate of flow is most commonly cubic foot per minute (cubic meter per second), abbreviated ft^3/min (m^3/s). Flow rate is conventionally designated by the symbol Q, and fan flow rate can be designated as either Q_F or Q_1. The latter reflects the standard definition. It is not unusual to refer to the fan flow rate as the cubic feet per minute (CFM) of the fan.

23.1.1.3 Density

Air density may be determined by measuring dry-bulb temperature, wet-bulb temperature, and barometric pressure and referring to a psychrometric density chart, such as that illustrated in Fig. 23.1.1

Most performance data are published for the conditions that would be obtained if the fan were handling air at the standard density of 0.075 lbm/ft³ (1.2 kg/m³). This is substantially the density of dry air at 70°F (20°C) and 14.7 psia (101.3 kPa). If other conditions prevail, corrections must be made according to the fan laws.

23.1.1.4 Power Formulas

The formula for power H in terms of the total efficiency η_T, total pressures at inlet and outlet p_{T1} and p_{T2}, respectively, in inWG, inlet capacity Q_1 in cubic feet per minute is

$$H = \frac{Q_1(p_{T2} - p_{T1})K_p}{6362\eta_T} \qquad (23.1.4)$$

The power in terms of the static efficiency η_S, static pressure at the outlet p_{T2}, and total pressure at the inlet p_{T1} is

$$H = \frac{Q_1(p_{S2} - p_{T1})K_p}{6362\eta_S} \qquad (23.1.5)$$

The compressibility coefficient K_p generally can be taken as unity for fans, particularly those that generate less than 10 inWG (2.5 kPa).

The above equations for power are based on U.S. Customary System (USCS) units and yield the power in horsepower. (The standard test code uses the symbol H to designate this quantity and the compound symbol HP is frequently found.) *If Q_1 is measured in cubic meters per second and pressures are measured in kilopascals, then power will be in kilowatts when the factor 6362 is changed to 1.0.*

23.1.2 FAN TYPES

The various aerodynamic types of fans can be distinguished by the direction in which the air flows through the impeller when the energy is being transmitted by the blades or working surfaces. The principal types are:

1. *Axial-flow fans*, including propeller fans, through which the air flows substantially parallel to the shaft
2. *Centrifugal fans*, which might be called radial-flow fans because the air flows in a radial direction relative to the shaft

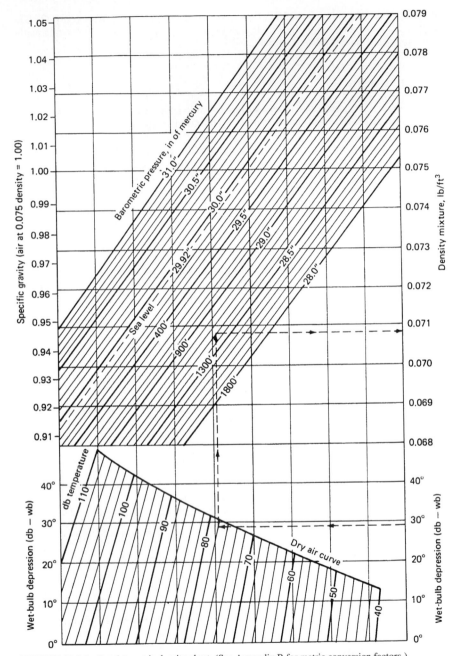

FIGURE 23.1.1   Psychrometric density chart. (See Appendix B for metric conversion factors.)

3. *Mixed-flow fans*, through which the air flows in a combined axial and radial directions

4. *Cross-flow fans*, through which the air flows in an inward radial direction and then in an outward radial direction

23.1.2.1 Axial-Flow Fans

There are numerous aerodynamic types of fans within the axial-flow category. One of the most important distinguishing features is the *hub ratio*, i.e., the ratio between the diameter of the hub (or root of the blade) and the tip of the blade. Generally, the higher the hub ratio, the higher the inherent pressure capability of the impeller. Similarly, the higher the pressure requirements of the fan, the more complex the other parts of the fan must be, to achieve acceptable efficiencies.

Propeller fans are perhaps the simplest and best known of all fan types. For all practical purposes, they have a zero hub ratio, having only enough hub to satisfy the mechanical requirements for driving the fan. Even those fans with the most elaborately contoured blades are only capable of efficiently developing total pressures up to approximately 1 inWG (2.49 kPa). Most propeller fans are used for static pressures close to zero.

Propeller fans may be either direct- or belt-driven. They may be free-standing, wall- or ceiling-mounted, or incorporated in a roof ventilator (see Figs. 23.1.2 through 23.1.5).

Accessories such as bird screens and louvers may be employed. Motors and drives are usually sized for the power requirements at the operating point which will be at or near the best efficiency point. If this is the case, operation at shutoff or low flows should be prevented. Shutters should not be allowed to freeze closed; filters should not be allowed to get too dirty. Figure 23.1.6 shows the performance of a typical propeller fan.

Axial-flow fans are generally built with hub ratios ranging from 0.25 up. It is usually desirable to house the impeller in a cylinder together with a set of guide vanes, located on either the upstream or downstream side of the impeller. When this is done, the fan is generally described as a *vane axial fan*. When lower hub ratios are used, the vanes can be omitted with only a limited sacrifice in efficiency. The resulting fan is usually called a *tube axial fan*.

Axial-flow fans may be constructed for either direct or belt drive, as seen in Figs. 23.1.7 and 23.1.8. Variable-inlet vanes, suction boxes, inlet bells, discharge cones, sound attenuators,

FIGURE 23.1.2 Free-standing, direct-drive propeller fan.

FIGURE 23.1.3   Wall-mounted, direct-drive, heavy-duty propeller fan.

FIGURE 23.1.4   Ceiling-mounted belt-drive propeller fan.

FIGURE 23.1.5   Direct-drive propeller fan in a roof ventilator.

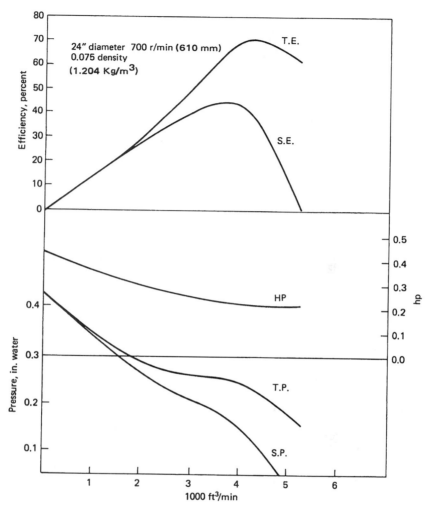

FIGURE 23.1.6   Performance for typical propeller fan. TE = total efficiency; SE = static efficiency; HP = horsepower; TP = total pressure, inWG; SP = static pressure, inWG. (See Appendix B for metric conversion factors.)

etc., may be used. Figures 23.1.9 and 23.1.10 illustrates some special features available for vane axial fans. The impeller may have fixed or adjustable blades. Blade adjustment may be possible when the fan is shut down or while the fan is in operation. The latter is called a variable-pitch or controllable-pitch fan. Depending on the hub ratio, speed, and other design factors, total pressures up to 10 to 20 inWG (2.55 to 5.0 kPa) can be developed in one stage. Multistage fans and fans in series can be used for higher pressures.

Most axial-flow fans require more power at low flows than at high flows at the same rotative speed. Motors need not be sized for the maximum power, if provisions are made to avoid operation at or near the shutoff condition. Figure 23.1.11 gives the performance of a typical vane axial fan.

FIGURE 23.1.7 Adjustable-blade, direct-drive vane axial fan.

FIGURE 23.1.8 Fixed-blade belt-drive vane axial fan.

FIGURE 23.1.9 Vane axial fan with clamshell-type access doors.

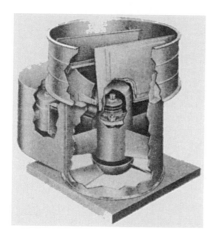

FIGURE 23.1.10 Belt-drive vane axial fan in a roof ventilator.

23.1.2.2 Centrifugal Fans

Within the centrifugal or radial-flow category, there are numerous aerodynamic types. Among the various distinguishing features are blade shape and blade depth (see Fig. 23.1.12). Generally, all centrifugal-type impellers should have blades curved at the heel, but this may be omitted on the deepest blade designs with very little sacrifice in efficiency. When curved, the heel should be curved forward in the direction of rotation. The blade tips may be curved in either a backward or forward direction, or they may be radial. In general, the backward-curved types are the most efficient and most stable.

A typical backward-curved fan and its performance are shown in Figs. 23.1.13 and 23.1.14. Greater efficiencies can be obtained by employing airfoil-shaped, backward-curved blades. Performance for a typical airfoil-blade fan is shown in Fig. 23.1.15. Fans with forward-curved blades are generally smaller for a given duty than the other types. A small vent set with forward-curved blades is shown in Fig. 23.1.16. The performance of a larger forward-curved blade fan is shown in Fig. 23.1.17. Radial blades are usually used in industrial applications where erosion or corrosion is a consideration. The deeper the blade, the more pressure the impeller will be capable of developing. Figure 23.1.18 shows a typical exhauster which can be equipped with various impeller types. Figure 23.1.19 shows a typical pressure blower, and Fig. 23.1.20 illustrates the performance of a larger radial-blade fan.

Centrifugal fans at constant speed and density have their lowest power requirements at the shutoff condition and so are often started with dampers closed. Both radial and forward-curved blades require considerably more power at high flows for a given speed than at the best efficiency point. Accordingly, some protection should be provided to prevent the system resistance from falling below the design value, unless motors are oversized. Overloads can occur if the system resistance is overestimated. By distinction, the backward-curved types exhibit a horsepower characteristic such that if a fan is rated at or near the best efficiency point, no other point of operation at the same speed and density will require very much more power. This characteristic is variously described as nonoverloading, power-limiting, or Limit-Load.*

---

*Registered trademark of Buffalo Forge Co.

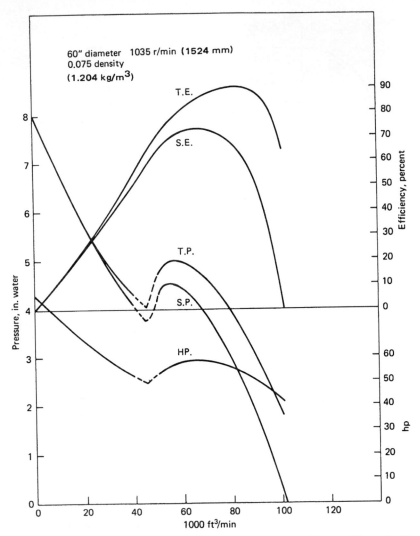

FIGURE 23.1.11 Performance of a typical vane axial fan. TE = total efficiency; SE = static efficiency; HP = horsepower; TP = total pressure, inWG; SP = static pressure, inWG. (See Appendix B
for metric conversion factors.)

Centrifugal fans may be built for either belt or direct drive. Variable-inlet vanes, inlet
boxes, inlet box dampers, outlet dampers, sound attenuators, etc., may be used.

23.1.3   *FAN SYSTEMS*

Fan systems may consist of any combination of fans, duct elements, heat exchangers, air
cleaners, or other equipment through which all or part of the total flow must pass.

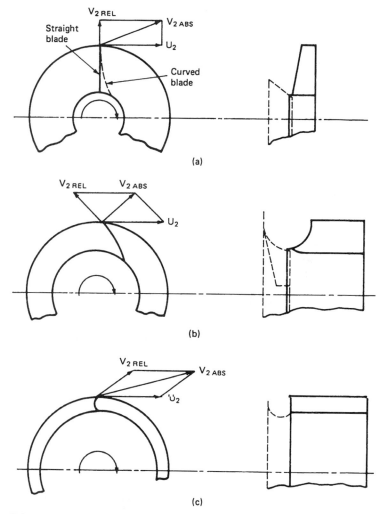

FIGURE 23.1.12 Blade designs: (*a*) radial discharge; (*b*) backward-curved discharge; (*c*) forward-curved discharge.

23.1.3.1 Fan Location

The fan location in any system is dictated by the direction of flow and desired pressure relations. That is, a supply fan may be used to pump air into a space, or an exhaust fan may be used to draw air out of a space. The same through-flow conditions are obtained, but the pressure relations are different. In the first case there is a buildup of pressure in the space, and in the second case a reduction occurs in the space pressure. Both supply and exhaust fans may be used. Then the space pressure will depend on the relative amounts of air handled by each fan; i.e., space pressure will be positive if there is an excess of supply over exhaust and negative if there is an excess of exhaust over supply. Assuming the same capacities and end pressures for all cases, the total energy delivered by the fan or fans to

FIGURE 23.1.13   Centrifugal fan with inlet vanes and backward-curved impeller.

the air passing through a given system is fixed, whether a supply fan, an exhaust fan, or both are used.

Obviously, if one fan is to supply air to several spaces, it must be located upstream of each space. A downstream location relative to each space is required if a fan is to exhaust air from several spaces.

When there must be several branches, the fan should be located as centrally as possible so that each particle of air will require approximately the same amount of energy for transport. Only one pressure can exist at a single point in space or in a duct system. This applies

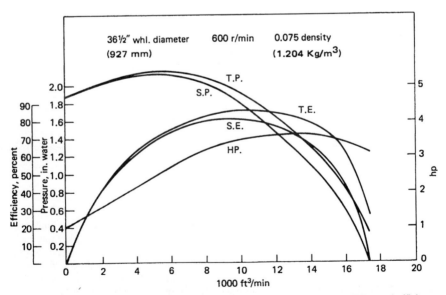

FIGURE 23.1.14   Performance of typical backward-curved-blade centrifugal fan. TE = total efficiency; SE = static efficiency; HP = horsepower; TP = total pressure, inWG; SP = static pressure, inWG. (See Appendix B for metric conversion factors.)

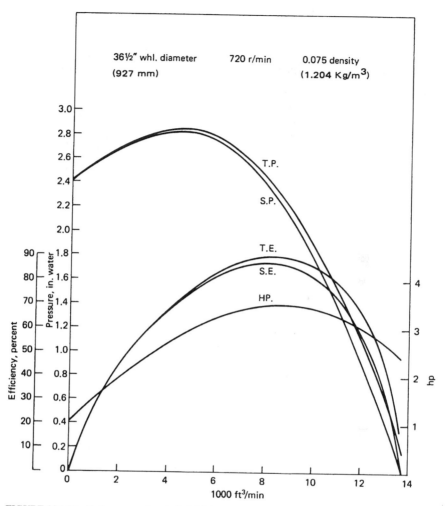

FIGURE 23.1.15 Performance of typical airfoil-blade centrifugal fan. TE = total efficiency; SE = static efficiency; HP = horsepower; TP = total pressure, inWG; SP = static pressure, inWG. (See Appendix B for metric conversion factors.)

at the junction of any two branches regardless of differences in branch size, length, or configuration. Accordingly, the pressure drop along one branch must equal that along the other. If the branches are not designed to provide equal pressure drops at the required flow rates, the flow rates will differ from the design values. When the available pressure exceeds that required, the flow through a branch can be reduced to the design value by dampering. This is a waste of energy, which sometimes cannot be avoided. In many cases, the size of the duct can be reduced to balance the pressure drops.

The fan must be chosen for a pressure sufficient to overcome the total losses based on the flow through the longest run. In other words, the same pressure must be dissipated by

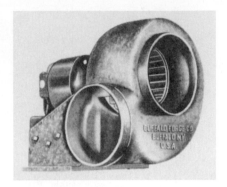

FIGURE 23.1.16   Cast-iron-housed vent set with forward-curved impeller.

each portion of the air flowing, regardless of the lengths of the runs. Balanced design may be accomplished by using balancing dampers or by appropriate duct sizes in all branches.

23.1.3.2   Fan and System Matching

Fan performance must match system requirements. The only possible operating points are those where the system characteristic intersects the fan characteristic. At such points the pressure developed by the fan exactly matches the system resistance and the flow through the system equals the fan capacity. If the flow rate is not equal to specifications, either the

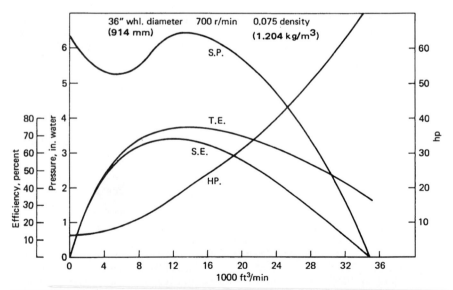

FIGURE 23.1.17   Performance of typical forward-blade centrifugal fan. TE = total efficiency; SE = static efficiency; HP = horsepower; TP = total pressure, inWG; SP = static pressure, inWG. (See Appendix B for metric conversion factors.)

FIGURE 23.1.18   Arrangement 9: industrial exhauster with air, material, and open impellers.

fan characteristic must be altered (by a change in speed, size, or whirl) or the system characteristic must be changed (by altering components or damper settings).

If more than one fan serves a system, their combined characteristics must be used to determine the capacity and performance characteristics of the system.

In any case, fan characteristics should be based on the gas density at the proposed fan location. System characteristics must be based on the individual densities in each of the system components. The overall system resistance is equal to the sum of the pressure losses for the individual components along any one flow path.

FIGURE 23.1.19   (*a*) Arrangement 4: pressure (*b*) radial-blade impeller.

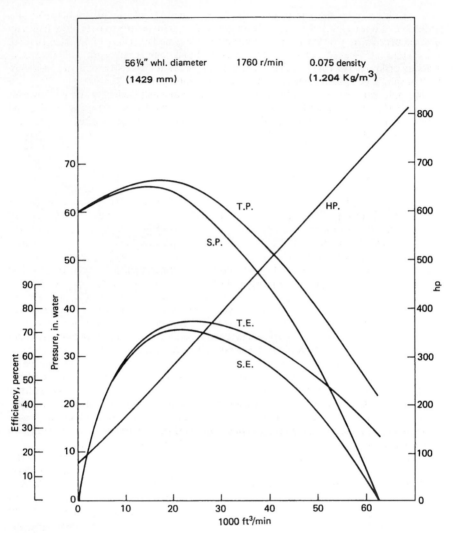

FIGURE 23.1.20   Performance of typical radial-blade centrifugal fan. TE = total efficiency; SE = static efficiency; HP = horsepower; TP = total pressure, inWG; SP = static pressure, inWG. (See Appendix B for metric conversion factors.)

23.1.3.3   Two-Fan Systems

More than one fan may be required on an HVAC system. Supply and exhaust (return) fans are required to avoid excessive pressure buildup in the spaces being served. This combination may be considered a series arrangement in that both fans handle the same air. In a true multistage arrangement, both stages handle the same mass flow rate of air. However, as noted above, supply and exhaust rates may be adjusted to control the pressure in the space being served. The pressure requirements for each fan depend on the pressure in the space as well as the ductwork between the space and the fan.

Fans may also be installed in a parallel arrangement. Double-width, double-inlet fans are essentially two fans in parallel in a common housing. Sometimes two separate fans may be used because they fit the available space better than the larger fan. When such fans are used without individual ductwork and discharge into a common plenum, their individual velocity pressures are lost since they will merge into a common plenum pressure. Therefore the fan should be selected to produce the same fan static pressures.

Multiple-fan operation can be troublesome on some complicated systems. However, most HVAC systems are relatively uncomplicated, and most fans will come on-line and operate satisfactorily without special precautions. Fans that are grossly misapplied may stall or cause other problems.

23.1.3.4 Systems with Mass or Heat Exchange

If the air passing through a system is heated or cooled, its density will be decreased or increased, respectively, and assuming no change of mass, its volume rate will be increased or decreased accordingly. The resistance of any component should be calculated for the actual gas density, volume, and velocity through it. The overall system resistance is the sum of these individual resistances along any one flow path.

The mass rate of flow may be different at various locations. Multiple intakes or outlets have been discussed earlier. The mass rate may also vary due to a change of state, such as when water is evaporated into a gas stream. Conversely, water vapor may be condensed out of an air/vapor mixture, and in any combustion process additional gas may be generated. In any case, it is necessary to determine the rate of mass gain or loss from the appropriate relation for the physical or chemical process involved. The actual densities, volumes, and velocities should be used to determine individual resistances; then overall resistances can be determined by simple addition.

The fan for any system, including one with mass or heat exchange, must develop enough pressure to overcome the system resistance. This pressure requirement is the same regardless of the fan location. The best fan location based on power requirements is at the point where the volume entering the fan is minimized. Naturally, the fan will have to be downstream of all intakes on exhaust systems and upstream of all discharge openings on supply systems. Volume flow rate varies inversely with density when the mass flow rate is constant, and volume flow rate is proportional to weight flow when the density is constant. Both density and weight flow may vary, in which case volume flow will vary accordingly.

The difference in power requirements for various fan locations can be calculated from Eq. (23.1.4). For constant system resistance, the pressure required of the fan $(p_{T2} - p_{T1})$ will be constant regardless of its location. The volume flow Q will vary with position, as noted above. Assuming constant efficiency, the fan location requiring the least power is that place where the density is greatest. Density changes are not usually very great in HVAC systems. It would be appropriate to take advantage of the high-density location, but many other factors influence the choice of where the fan(s) will be located in the system.

23.1.3.5 Mutual Influence of Fan and System

Fan performance data are generally based on tests in which the air approaches the inlet with a uniform velocity, free of whirl. Duct element losses, except where noted otherwise, are based on similar flow conditions.

Elbows, unless provided with adequate turning vanes or splitters, produce uneven velocity patterns that may persist for considerable distances in subsequent straight ducts. Nonuniform inlet velocities may alter fan performance since different portions of the impeller will be loaded differently. Nonuniform velocities may also produce whirls in the inlet flow that affect fan performance. Every reasonable precaution should be taken to ensure uniform flow from all elbows.

The velocity pattern at fan discharge will vary with the design. Performance data are usually based on tests with a straight discharge duct. If an elbow is located close to the discharge, there may be some loss in fan performance due to a reduction in static-pressure regain. If the velocity along the inside radius of the elbow is higher than that along the outside radius, the loss will be higher than normal; but it will be less than normal if the velocity is higher along the outside radius than along the inside radius.

The effects of air-flow conditions different from those that would exit during a standard laboratory test are known as *system effects*. System-effect factors have been developed by various fan manufacturers and by the Air Movement and Control Association.[1] The former should be used in preference to the latter wherever possible because the system effects will vary with individual fan designs. Nevertheless, the AMCA data can be very useful, particularly if the system designer avoids the conditions where high system-effect factors tend to prevail. The most common reason for systems' failing to perform as projected is poor duct conditions near the fan.

23.1.3.6 Capacity Control

If a fan system must operate over a range of capacities, some means of adjusting either the fan characteristics or the system characteristics must be provided.

System characteristics may be altered by inserting additional resistance or by providing an additional flow path. Gradual adjustment can be provided with moveable dampers either in the main passage or in the bypass.

Fan characteristics may be modified by altering blade position, rotative speed, or inlet whirl. Gradual adjustment is usually possible within the limits of the device used, whether it be a variable-pitch rotor, a variable-speed motor or transmission, or variable-inlet vane. Step-by-step adjustment is possible with multispeed motors and multiple-fan arrangements.

The choice of a specific control method should be based on an economic evaluation. Before such an evaluation can be made, it is necessary to estimate the operating times at various capacities. Each of the methods entails a loss in efficiency over most of its operating range. The most inefficient methods are usually the least expensive to install originally. Only a complete evaluation on the basis of both first cost and operating cost will reveal the best method.

The effects of various methods of capacity control on fan operation are illustrated in Fig. 23.1.21. The first three sets of curves are drawn for the same fan (a backward-curved centrifugal) operating at the same full-load speed. The system resistance varies as the square of the capacity.

Figure 23.1.21*a* is drawn for damper control. The pressure required of the fan is higher at reduced ratings than at design values because the damper increases system resistance. The dotted system curves include this increase, which also varies as the square of the capacity. The powers at one-half and three-fourths load are less than at full load because of the nature of the fan characteristics. This would not be true if the characteristics of an axial fan were used instead of those for a centrifugal fan.

Figure 23.1.21*b* is drawn for variable-inlet-vane (VIV) control. Three fan curves are drawn for three different positions or settings of the variable-inlet vanes. The powers required by the fan at reduced capacities are lower than those for damper control.

Figure 23.1.21*c* is drawn for speed control. Instead of three system curves, three fan curves are drawn, one each at full, three-fourths, and one-half speed. There is also a

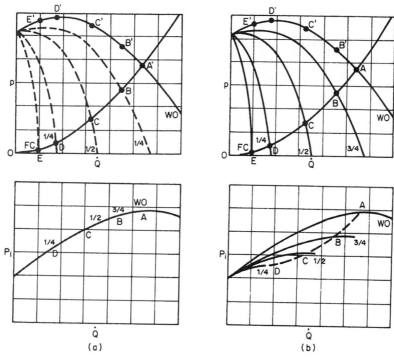

FIGURE 23.1.21 Fan performance and power savings with (*a*) outlet dampers, (*b*) variable-inlet vanes, (*c*) variable pitch, (*d*) variable speed, and (*e*) comparison of power savings. Details for inlet box dampers are not shown since they are quite similar to those for variable-inlet vanes. OD = outlet damper; IBD = inlet box damper; VIV = variable-inlet vanes; VS = input to variable-speed device; VS$_o$= output from variable-speed device; VP = variable pitch. (See Appendix B for metric conversion factors.)

power curve for each speed. The powers at reduced capacities are less than the corresponding powers for either damper or VIV control, and no elements add resistance to the system.

Figure 23.1.21*d* is drawn for variable-pitch control. Three fan curves are drawn for three different blade positions or pitch settings. The power requirement is very close to that for variable speed.

Figure 23.1.21*e* shows the power relations for all four methods. The curves for damper and vane control are exactly as in Fig. 23.1.21*a* and *b*; however, two curves are shown for speed control. The extra curve 2 includes the efficiency of a hydraulic coupling used as a variable-speed transmission. Curve 2 therefore represents the input to the transmission rather than to the fan as for all other curves. This figure indicates that there is a range of capacities near full load where inlet vane control is superior to speed control on a power input basis. The simplicity of a duct damper gives it the advantage in first cost. Variable-inlet vanes are considerably less expensive than most methods of speed control. Speed control does have additional advantages when considerable operation at less than maximum speed is expected. The accompanying reduction in noise and increased life may also be appreciable. As can be seen from the typical variable-pitch-control curve, power savings approximate those of variable-speed control very closely.

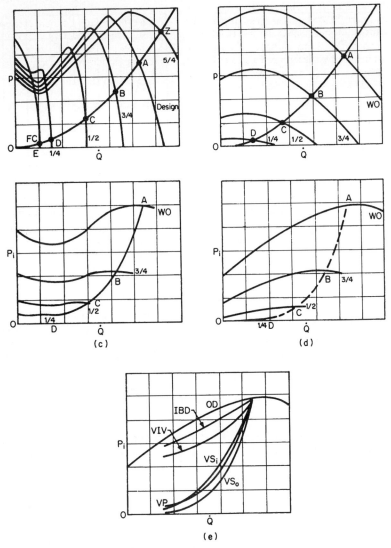

FIGURE 23.1.21   (*Continued*)

23.1.4   FAN LAWS

The fan laws relate the performance variables for any homologous series of fans. The variables involved are fan size D, rotative speed N, gas density P, capacity Q, pressure p, power H, sound power level L_w, and efficiency η.

The fan laws are the mathematical expression of the fact that when two fans are both members of a homologous series, their performance curves are homologous. At the same point of rating, i.e., at similarly situated points of operation, efficiencies are equal. The ratios of all the other variables are interrelated.

The fan laws can be expressed in terms of various combinations of dependent and independent variables. The most frequently used combination is fan law 1 where size, speed, and density are considered the independent variables and capacity, pressure, power, and sound power level are considered the dependent variables. Accordingly, the effects of change in speed or density can be calculated for a fan that has already been installed and such changes need to be evaluated. Table 23.1.1 gives fan law 1. The subscript b indicates that the variable is for the tested fan.

23.1.4.1 Fan Law Restrictions

It cannot be emphasized too strongly that the fan laws should be used only under very special circumstances. The most frequent error made in using the fan laws is trying to predict performance after changing some physical aspect of the system to which the fan is attached or even a damper setting.

In fan engineering terminology, the fan laws are only applicable to exactly similar points of rating. When a system is changed by adding a coil or some other means, the point of rating has to change. Of course, there are ways to predict the effects of these kinds of changes, but they involve the calculation of more than one point on the performance curve.

23.1.4.2 Equivalent Static Pressure

The use of the fan laws can sometimes be simplified by employing a device called *equivalent static pressure (ESP)*. The "equivalent static pressure" can be defined as the pressure that would be developed by a fan operating at standard air density instead of the actual pressure developed when operating at actual density. This is a useful tool when one is selecting a fan from published data, which are based on operation at standard density. Example 23.1.1 will illustrate this.

$$ESP = p_{FS} \frac{\rho_{std}}{\rho_{act}} \qquad (23.1.6)$$

In fact, the concept of equivalent static pressure can be extended to any pair of densities. Since most published data for a fan are based on standard air density, we do not explore the general case any further.

23.1.5 FAN NOISE*

Fan noise consists of a series of discrete tones superimposed on a broadband component. The former, which may be called the *rotational component*, may be traced to the process of energy transfer that also leads to the development of head. The latter, which may be called the *vortex component*, may be traced to the formation of turbulent eddies of one kind or another that usually lead to losses of head.

23.1.5.1 Centrifugal Fan Noise

The predominant tone of the rotational component of fan noise in centrifugal apparatus is usually that at the blade-passing frequency. In very narrow blade designs, the higher

---

*See also Chap. 10, "Noise Control."

TABLE 23.1.1 Fan Laws

For all fan laws: $\eta_a = \eta_b$ *and (point of rating)$_a$ = (point of rating)$_b$*

No.	Dependent variables		Independent variables		
1a	CFM_a	$= CFM_b \times$	$\left(\dfrac{SIZE_a}{SIZE_b}\right)^3 \times$	$\left(\dfrac{RPM_a}{RPM_b}\right)^1 \times$	$\left(\dfrac{\delta_a}{\delta_b}\right)$ (1)
1b	$PRESS_a$	$= PRESS_b \times$	$\left(\dfrac{SIZE_a}{SIZE_b}\right)^2 \times$	$\left(\dfrac{RPM_a}{RPM_b}\right)^2 \times$	$\left(\dfrac{\delta_a}{\delta_b}\right)$
1c	HP_a	$= HP_b \times$	$\left(\dfrac{SIZE_a}{SIZE_b}\right)^5 \times$	$\left(\dfrac{RPM_a}{RPM_b}\right)^3 \times$	$\left(\dfrac{\delta_a}{\delta_b}\right)$
1d	PWL_a	$= PWL_b + 70 \log$	$\dfrac{SIZE_a}{SIZE_b} + 50 \log$	$\dfrac{RPM_a}{RPM_b} + 20 \log$	$\dfrac{\delta_a}{\delta_b}$
	Q_a	$= Q_b \times$	$\left(\dfrac{D_a}{D_b}\right)^3 \times$	$\left(\dfrac{N_a}{N_b}\right)^1 \times$	$\left(\dfrac{P_a}{P_b}\right)$ (1)
	P_{FTa}	$= P_{FTb} \times$	$\left(\dfrac{D_a}{D_b}\right)^2 \times$	$\left(\dfrac{N_a}{N_b}\right)^2 \times$	$\left(\dfrac{P_a}{P_b}\right)$
	H_a	$= H_b \times$	$\left(\dfrac{D_a}{D_b}\right)^5 \times$	$\left(\dfrac{N_a}{N_b}\right)^3 \times$	$\left(\dfrac{P_a}{P_b}\right)$
	L_{wg}	$= L_{wb} + 70 \log$	$\dfrac{D_a}{D_b} + 50 \log$	$\dfrac{N_a}{N_b} + 20 \log$	$\dfrac{P_a}{P_b}$

harmonics may be of equal intensity. Widening the blades progressively weakens the higher harmonies.

Vortices can be created at the leading or trailing edges of the blades, along the sides of the blades, or at locations remote from the blades. In general, the size, rate of growth and decay, and point of origin and movement of these vortices will be random and the resulting noise will have a broadband spectrum.

Streamlining the leading edges of the blades minimizes vortex formation at that location. At the design capacity, both thin blades with rounded edges and thick blades with airfoil sections are quite effective in reducing vortex formation. The airfoil-shaped blade may enjoy some advantage, particularly when the leading-edge angle does not match the entering flow angle across the entire width of the blade.

Large eddies may be formed in the blade passages due to flow separation from a boundary. The greatest benefit to be derived from the use of airfoil-shaped blades is that of reducing separation. This is somewhat offset from a noise standpoint by the decrease in the optimum number of blades compared to that for thin blades. The thickness of the blade apparently has very little effect in centrifugal fans.

The speed of sound so greatly exceeds the airspeed in most fans that the noise is propagated upstream and downstream with equal facility. The acoustical impedances of the inlet and outlet openings are so nearly equal that in most cases the sound power radiated through the outlet can safely be assumed to be equal to that radiated through the inlet. The transmission through the easing walls is so small by comparison that when the total sound power output of a fan is measured, the portions radiated through the outlet and inlet are each reported as one-half of that total. The corresponding sound power levels are therefore each 3 dB less than the total sound power level.

The sound power level curve for a backward-curved thin-bladed centrifugal fan is shown in Fig. 23.1.22 together with other performance characteristics. This curve is typical of that for all centrifugal fan types. The overall shape of the sound power level curve indicates that

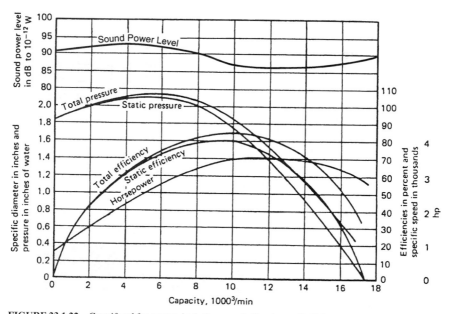

FIGURE 23.1.22   Centrifugal fan curves, including sound. (See Appendix B for metric conversion factors.)

the sound power output of a fan is a function of both capacity and pressure. Tests indicate that sound power outputs are proportional to the capacity ratio multiplied by the square of the pressure ratio, all other conditions being equal. The spectrum for a centrifugal fan may be approximated in most cases by subtracting 8, 4, 8, 9, 11, 16, 20, and 25 dB from the overall level to obtain the levels in the 63-, 125-, 250-, 500-, 1000-, 2000-, 4000-, and 8000-Hz octave bands, respectively, provided the blade-passing frequency falls in any other band.

23.1.5.2 Axial-Flow Fan Noise

The noise characteristics of axial-flow fans are very similar to those of centrifugal fans. The division of fan noise into rotational tones and broadband vortex components applies to both types.

In axial-flow fan apparatus, the predominant tone of the rotational component may be one of the higher harmonics rather than the fundamental blade-passing frequency, if the fan is used to develop appreciable pressure.

Increases in the number of blades generally have a beneficial effect on axial fan noise. The number of blades should differ from the number of guide vanes, to prevent strengthening of the fundamental tone.

The effects of streamlining on leading-edge vortices and side-separation eddies are the same for both axial and centrifugal fans. The effect of trailing-edge thickness is more pronounced in axial fans. There may be a noticeable increase in noise if the wake from one blade is cut by succeeding blades.

The sound power level curve for a vane axial fan is shown in Fig. 23.1.23 together with other performance characteristics. This curve is typical for all axial-flow fan types. The overall shape of the curve is slightly different from that for a centrifugal fan, consistent with the difference in pressure-capacity curves. The spectrum for a high-pressure vane axial fan may be approximated in most cases by subtracting 11, 13, 7, 5, 7, 11, 16, and 19 dB from the overall level to obtain the levels in the 63-, 125-, 250-, 500-, 1000-, 2000-, 4000-, and

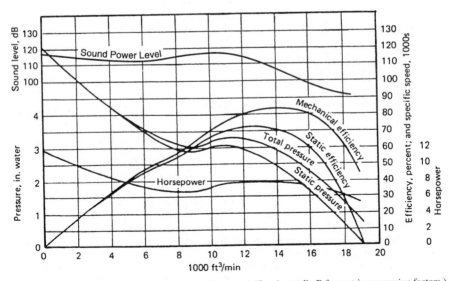

FIGURE 23.1.23   Vane axial fan curves, including sound. (See Appendix B for metric conversion factors.)

8000-Hz octave bands, respectively. The corresponding values for a low-pressure propeller fan are 5, 5, 7, 9, 15, 18, 25, and 30 dB.

23.1.5.3 Sound Power Level and the Fan Laws

The ratio of the sound power levels for two similar fans can be predicted from the fan laws. This characteristic, like the other dependent variables, is predictable only if the fans have the same point of rating and both have good balance, good bearings, etc. Complete test data on one of the fans must be available.

The noise spectra for any two homologous fans at the same point of rating may be considered similar in most cases. The ratio of the sound power levels at any pair of corresponding frequencies will equal the ratio of the overall sound power levels. Corresponding frequencies may be the two blade frequencies or any two harmonics thereof. If the rotative speeds and thus the blade frequencies are equal, the fan laws may be used to predict the sound power levels in each of the standard octave bands, provided the corresponding test data are available. If rotative speeds are not equal, only the overall sound power level can be predicted.

Test data on sound power level will usually be based on measurements of the total noise radiated from the inlet, the outlet, and the casing. If the inlet and outlet have the same size and shape, there is a good chance that the noises radiated in the two directions will be equal. This is approximately true even in centrifugal fans having round inlets and rectangular outlets.

Fan law 1d (see Table 23.1.1) indicates that the ratio of sound power outputs varies as the seventh power of the size ratio, the fifth power of the speed ratio, and the square of the density ratio. These relationships have been verified by Madison and Graham[2] in the Buffalo Forge Company laboratory.

23.1.6 FAN CONSTRUCTION

23.1.6.1 Standard Designations and Arrangements

The fan industry, through the Air Movement and Control Association (AMCA), has devised certain standard designations. Those for rotation, discharge, inlet box position, drive arrangement, and motor position are excerpted here. For a complete set refer to AMCA Publication 99-86.

The method of specifying rotation is to view the fan from the drive side and to indicate whether the rotation is clockwise or counterclockwise. The drive side of a single-inlet centrifugal fan is considered to be the side opposite the inlet, even in those rare cases where the actual drive location may be on the inlet side. It is necessary to specify which of the drives is used for reference on dual-drive arrangements. The rotation of a propeller or axial-flow fan is usually immaterial and a matter of individual design. There is no official designation of drive sides for axial fans, so if it is necessary to specify rotation, the direction from which the fan is viewed should also be specified.

The method of specifying discharge position is indicated in Fig. 23.1.24. If the fan is to be suspended from the ceiling or a side wall, discharge should be specified as if the fan were floor-mounted. The intended counting arrangement should also be given. An angular measure is required for angular positions.

The various drive arrangements have been assigned a number, as indicated in Fig. 23.1.25. Designations for axial fans are consistent with standards for centrifugal fans. The official

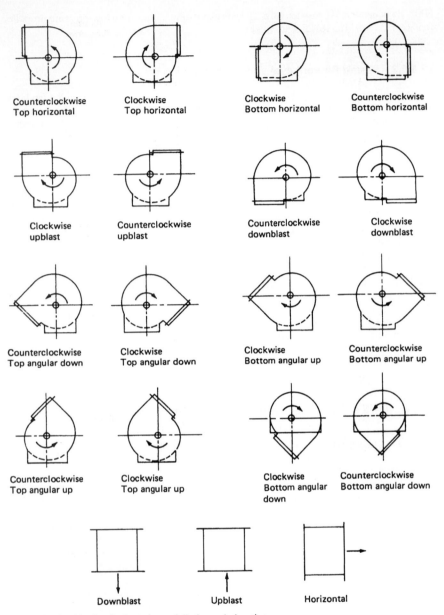

FIGURE 23.1.24 Standard rotation and discharge designations.

arrangement numbers may be used for fans with a bearing on the housing or subbase as shown or for pedestal-mounted bearings. Arrangements involving a bearing in the inlet should be avoided for small fans.

The method of specifying the inlet box position is to view the fan from the drive side (same as for rotation) and indicate the position of the intake opening. Angularity is referred

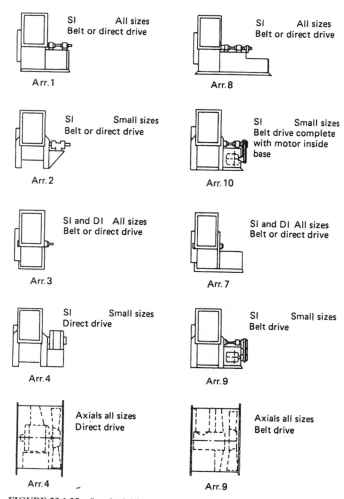

FIGURE 23.1.25   Standard drive arrangements.

to a horizontal centerline, as shown in Fig. 23.1.26. The various motor positions have been assigned letter designations, as indicated in Fig. 23.1.27.

23.1.6.2   Heat-Resistant Materials

Protecting fans in high-temperature gas streams involves both corrosive and structural considerations. Ultimate strengths of steels may improve slightly at moderate temperatures, but eventually decrease rapidly. Yield strengths decrease with temperature even when the ultimate strengths are apparently unaffected. The design criteria may be the ultimate strength, yield strength, creep strength, or rupture strength, depending on the temperature, nature of the stressed part, and service requirements.

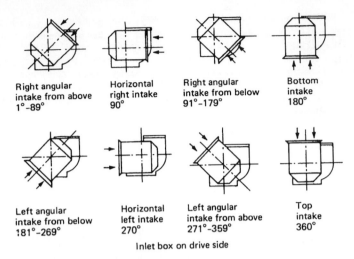

| Right angular intake from above 1°–89° | Horizontal right intake 90° | Right angular intake from below 91°–179° | Bottom intake 180° |

| Left angular intake from below 181°–269° | Horizontal left intake 270° | Left angular intake from above 271°–359° | Top intake 360° |

Inlet box on drive side

FIGURE 23.1.26   Standard inlet box positions.

Mild steel scales rapidly at temperatures above 900°F (470°C) in normal atmospheres. A process involving a metal spray and heat treatment to form a steel-aluminum alloy at the surface may extend this operating range. Heat-resistant paints may provide some protection, but temperatures are limited by the binder used. Various low-alloy steels, stainless steels, and high-nickel alloys provide suitable strength and corrosion resistance in air.

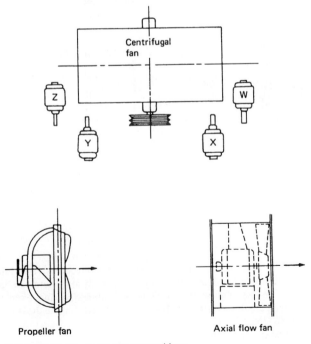

Centrifugal fan

Propeller fan   Axial flow fan

FIGURE 23.1.27   Standard motor positions.

Problems due to expansion and problems of cooling bearings also face the designer of high-temperature fans. In general, bearings should be kept out of the air stream on all but room-temperature applications. Grease lubrication is satisfactory up to 200°F (93°C) on many types of antifriction bearings. Oil lubrication is required for higher temperatures. Above 350°F (175°C) some sort of heat slinger or cooling disk is usually required on the shaft between the fan and bearing. Above 700°F (425°C) the bearing subbase should be physically separated from the fan housing, to prevent the direct conduction of heat. Fans and ductwork should be insulated.

23.1.6.3 Corrosion-Resistant Materials

The choice of a particular material or coating to protect a fan from attack by a corrosive gas is an economic matter. Ordinarily a fan is constructed of the materials that provide the necessary strengths and contours most economically. If the life of such a fan handling a corrosive gas is limited, the extra first cost of a special material or coating may be justified by the increase in life obtained.

Such an evaluation is easily made for any two materials if actual operating experience or test data are available. Uhlig[3] has compiled a list of corrosive agents and materials with various degrees of resistance. The accuracy of any such prediction is limited. The effects of temperature, local concentration, velocity, impurities, and fabrication must be considered when one is translating test results to terms of actual performance.

When corrosion takes place, two types of attack may be involved. Direct chemical attack is generally limited to high temperatures or highly corrosive environments or both. The rapid scaling of steel at temperatures above 900°F (525°C) and the effect of concentrated acids or alkalies are examples. Such reactions may be prevented by controlling the temperature or the concentration of the corrosive substance or by application of inert protective coatings.

Electrochemical attack is much more common. The requirements for such a reaction are that there be discrete anodic and cathodic regions connected by solid material submerged in an electrolyte. Such cells not only are produced by coupling two materials, but also arise from inner variations in stress, surface composition, deposits of dissimilar metals, and condensation of an electrolyte.

Frequent cleaning will remove deposits and thus promote better life for any material. The composition of weld deposits and the heat-affected zones in welding, as well as stress relief and surface treatment, should be evaluated for their effect on corrosion resistance.

The formation of corrosion products may serve to retard or accelerate the reaction. Rust on mild steel tends to promote corrosion, while rust that forms on low-alloy steels may protect the metal from further corrosion. The film that forms on the surface of stainless steel is protective in many corrosive environments, as is that which forms on aluminum.

The protection afforded by many metallic coatings is due to their sacrificial action. Zinc on galvanized products and the anodic covering on aluminum-clad and cadmium-plated materials protect scratches and even sheared edges by this means. However, to avoid undesirable sacrificial action, the formation of galvanic couples by joining dissimilar metals should be avoided.

The protection provided by other coatings such as lead, rubber, or plastics is due primarily to their inertness to corrosive media. The optimum thickness varies with the corrosive medium, but the values in Table 23.1.2 are generally suitable. Temperature and tip-speed limits are shown.

Lead linings may be "burned" on or mechanically attached to the surfaces of fan wheels or housings made of steel. Costs limit use to applications where other means of protection are not particularly suitable.

Rubber coverings, in the thicknesses shown in Table 23.1.2 are sheets adhered to exposed steel surfaces and vulcanized in place. Rubber-coated parts may also be sprayed

TABLE 23.1.2 Protective Coverings

	Thickness	Limitations
Lead linings	$^1/_8$–$^1/_4$ in	12,000 ft/min at 200°F (94°C)
		9,000 ft/min at 400°F (204°C)
Hard rubber coverings	$^1/_8$–$^3/_{16}$ in	6×10^6 RN^{-2},* 150°F (66°C) max
Soft rubber coverings	$^1/_8$–$^3/_{16}$ in	3×10^6 RN^2,* 150°F (66°C) max
Epoxy coatings	5–10 mil	200°F (94°C) max
Phenolic coatings	5–10 mil	350°F (177°C) max
PVC lacquers	5–10 mil	180°F (82°C) max
PVC plastisols	30–10 mil	160°F (71°C) max

*RN^2 = radius in ft × (r/min)^2.
Note: Conversion factors to the metric system are: in × 2.54 × 10^{-6} = mm: mil × 2.54 × 10^{-7}
= mm: ft/min × 5.08 × 10^{-3} = m/s: 6×10^6(RN)^2 [ft · (r/min)^2] = 1.829 × 10^6 RN^2 [m · (r/min)^2];
3×10^6 RN^2] [ft · (r/min)^2] = 0.915 × 10^6 RN^2 [m · (r/min)^2].

or dipped. Various natural and synthetic elastomers may be used including neoprene, Saran, and silicone rubber for increased temperature.

Plastic coatings are generally sprayed on. To obtain the optimum thickness, multiple layers are needed. Phenolics and plastisols require baking at moderate temperatures [375 to 450°F (191 to 232°C)], but vinyl lacquers and epoxies may be air-dried. Only the more flexible types of plastics may be used on fan wheels. Even so, standard fan construction must often be modified to eliminate all gaps and voids. Continuous welds and rounded edges as well as clean surfaces are required. Similar provisions must be made for ceramic and metal sprays.

Fiberglass-reinforced plastic (FRP), both polyesters and epoxies, is used for fan housings and impellers. Figure 23.1.28 shows a vane axial fan made with FRP. The portion of the shaft which will be exposed to the gas is encapsulated. Both manual lay-up and pressure-molding

FIGURE   23.1.28 Fiberglass-reinforced plastic vane axial fan.

techniques are employed. The fiberglass should be covered with a sufficient thickness of a chemical-resistant resin to protect it where the gas might attack glass, as in the case of hydrogen fluoride. Rigid polyvinyl chloride (PVC) is also used to fabricate impellers and housings. It is generally limited to a fraction of the speeds for FRP equipment.

Various metals and alloys have been used to fabricate corrosion-resistant fans. Aluminum-base alloys are not affected by most gases in the absence of water at or near room temperature. Acid gases and SO_2 in the presence of water have some corrosive action on aluminum. All alkalies will attack aluminum. Many industrial fumes and vapors attack aluminum surfaces.

Copper at or near room temperature is not affected by dry halogen gases but absolutely no moisture may be present, even with halogenated refrigerants. Normal and industrial atmospheres do not usually cause corrosion beyond the formation of a protective oxide film.

Copper-zinc, copper-nickel, copper-tin, and copper-silicon alloys are unaffected by most dry gases at ordinary temperatures. The presence of moisture increases the corrosiveness of the halogens, SO_2, and CO_2 considerably. Generally H_2S reacts even at low humidity.

23.1.6.4 Spark-Resistant Construction

The Air Movement and Control Association, in its Standard 99-86,[4] outlines three types of spark-resistant construction:

- *Type A* construction requires that all parts of the fan in contact with the air or gas being handled be made of nonferrous metal.
- *Type B* construction requires that the fan have an entirely nonferrous wheel and a nonferrous ring about the opening through which the shaft passes.
- *Type C* construction specifies that the fan must be so constructed that a shift of the wheel or shaft will not permit two ferrous parts of the fan to rub or strike.

In all three types, bearings should not be placed in the air or gas stream, and the user must electrically ground all fan parts.

Bronze and aluminum are commonly used where nonferrous parts are specified. Stainless steels have been allowed in certain instances, even though they are ferrous.

23.1.7 FAN SELECTION

In the majority of fan applications, it is neither necessary nor desirable to design a completely new fan for the specified job requirements. Standard designs are available in each of the various aerodynamic types. Fan selection is usually a matter of choosing the best size and type from those available.

Selecting a fan begins with the specification of requirements and ends with the evaluation of alternative possibilities. Of the many fans capable of satisfying particular capacity and pressure requirements, the best choice is the one that does the job most economically. First costs, operating costs, and maintenance costs must all be considered.

23.1.7.1 Specification of Requirements

A fan specification should give the fan supplier all the pertinent information regarding performance, service, evaluation, arrangement, etc., so that the best selection can be made.

Most of the important items are listed in Table 23.1.3. Explanatory notes are given for many of these items in the following paragraphs.

The number of fans and their aerodynamic type are items which should be specified only after the various possibilities have been compared. The type of service for which the fan is intended should be specified, to warn the supplier of any unusual conditions. In some instances, a duct layout should be included with the specifications. In any case, the sizes of the connecting ductwork should be indicated.

The capacity of the fan must be specified by the system designer. Due to the nature of the application, the designer may find it convenient to calculate the capacity as a weight rate of flow. Fan capacity should be expressed as a volume rate of flow and specified in cubic feet per minute (cubic meters per second) at inlet conditions, so to determine the

TABLE 23.1.3   General Fan Specifications

General: _____
  Number of fans _____
  Aerodynamic type _____
  Size of connecting ductwork _____
  Service _____
Capacity of each fan—Specify maximum and reduced ratings _____
  ft^3/min (m^3/s) at inlet or lb/h (mg/h) _____
Fan pressure at each capacity: _____
  Inches water gauge (kilopascals) _____
  State whether static or total _____
  Indicate distribution between inlet and outlet _____
Gas composition and conditions at each capacity: _____
  Name or gas analysis _____
  Molecular weight or specific gravity referred to air _____
  Ambient barometer in inHg (kPa) or elevation in ft (m) _____
  Temperature at inlet in °F (°C) _____
  Relative humidity in % _____
  Dust loading through fan _____
Power evaluation factors: _____
  Expected life in years _____
  Expected operation at each rating in h/year _____
  Power rate in $/kWh _____
  Demand charge in $/kW or $/hp _____
Physical data: _____
  Number of inlets _____
  Type of drive _____
  Arrangement number _____
  Direction of entry for inlet boxes _____
  Rotation, discharge, and motor position _____
Construction details: _____
  Appurtenances _____
  Special materials _____
  Type and mounting of bearings _____
Motor data: _____
  Electrical characteristics _____
  Type and enclosure _____

appropriate fan capacity in ft³/min (m³/s), divide the weight rate of flow in pounds per minute by the inlet density in pounds per cubic foot:

$$\frac{ft^3}{min} = \frac{lb/min \; flowing}{lb/ft^3 \; at \; inlet} \left(\frac{m^3}{s} = \frac{kg/s}{kg/m^3} \right)$$ (23.1.7)

If an adjustable flow rate is contemplated, the minimum capacity and any intermediate capacities at which power requirements are to be evaluated should be specified.

The fan must provide the air or gas with sufficient energy to overcome the losses encountered in passing through the system. The usual method of specifying this energy requirement is to stipulate the static or total pressure in inches water gauge (kilopascals) which the fan must develop at each capacity. In any case, the distribution of pressure between suction and discharge should be specified.

When a fan is located at the downstream end of the system, i.e., exhausting, the static pressure required is equal to the sum of the total-pressure losses in all the system components plus any difference in barometric pressure between the two ends of the system. For the duct system calculations, see Chap. 26.1. When the fan is located at any position other than the downstream end, i.e., blowing or boosting, an amount equal to the fan outlet velocity pressure should be subtracted from the requirement as stated above. The fan static pressure p_{FS} may thus be expressed in terms of the sum of the total-pressure losses $\Sigma p_{T_{loss}}$, *the difference between the barometric pressure at the system exit* p_{Rx}, and the barometric pressure at the system entrance P_{Be}, and the fan velocity pressure p_{FV} as

$$p_{FS} = \Sigma p_{T_{loss}} + p_{Bx} - p_{Be} - p_{FV}$$ (23.1.8)

with the proviso that the p_{FV} term is zero when the fan is located at the down-stream end of the system.

Fan velocity pressure depends on the fan size, and since this is not usually known in advance, many designers ignore the p_{FV} term regardless of the fan location. This is usually justified as constituting a factor of safety. Better accuracy, when warranted, can be obtained by using p_{FV} for a trial size and correcting to the actual value in the final selection.

The sum of the total-pressure losses through the system should include an allowance for any elements required to connect the fan to the system. This will be a small amount unless the size of the fan opening differs greatly from the size of the connected ductwork.

In the usual case there is no difference in barometric pressure between the entrance and exit of the system. The exceptions occur whenever there is more than one device supplying energy to the air.

The total pressure required of the fan is equal to the sum of the total-pressure losses plus any difference in barometric pressure (as noted above):

$$p_{FT} = \Sigma p_{T_{loss}} + p_{Bx} - p_{Be}$$ (23.1.9)

The performance of a fan is a function of the density of the air or gas at the fan inlet. The inlet density determines not only the volumetric capacity for a specified weight rate of flow but also the pressure which the fan is able to develop. The factors which affect density (and should therefore be specified) are the barometric pressure, temperature, and relative humidity at the inlet, as well as the name or composition of the gas, the ambient barometer and the gauge pressure at the fan inlet may be specified in lieu of the inlet barometer.

The composition of the gas and information about any entrained material (dust loading, etc.) should be specified so that the fan supplier can offer the best selection based on any previous experience.

Whenever the gas composition and conditions are not specified, the fan supplier usually assumes air at standard conditions. Standard conditions for the fan industry are dry air at 70°F and 29.92 inHg barometer (20°C and 101.3 kPa). The density corresponding to these conditions is 0.075 lbm/ft^3 (1.2 kg/m^3). If the fan requirements are specified in terms of other standards, both the actual and the standard conditions should be given in detail. If the weight rate of flow corresponding to a certain number of standard cubic feet per minute (cubic meters per second) is required, sufficient information to determine both the actual and standard density should be listed.

Power requirements should be evaluated to obtain the best selection. The fan supplier should be advised of the method of evaluation. The usual method is to reduce the expected cost of power during the useful life of the equipment to the present value of an annuity sufficient to yield the annual expenditure. The annual expenditure will depend on the power rate and the expected operating schedule during the life of the equipment as well as the power requirements at the various operating conditions. The size of the hypothetical investment will depend on the expected life and rate of interest which could be obtained. Accordingly, the cost of the expected operation reduced to present value ($\$_{\mathrm{RPV}}$) may be determined from the power rate ($\$/kWh) and the expected annual power consumption (kWh/year):

$$\$_{\mathrm{RVP}} = \frac{\$}{\mathrm{kWh}} \ \frac{\mathrm{kWh}}{\mathrm{year}} \left[\frac{1-(1+i)^{-n}}{i} \right] \qquad (23.1.10)$$

The bracketed term, which is the present value of an annuity to yield \$1 annually for n years if invested at rate of interest i, may be determined from any standard interest table.

Inefficiency may be penalized by charging the fan a flat amount for each horsepower or kilowatt required. In the power generation industry such a demand charge, which reflects the loss of power available for sale at peak load, may be applied in addition to any operating charge based on the cost of fuel.

Operating costs, as determined by Eq. (23.1.10) or any other method, should be added to the first costs and expected maintenance costs for each possible selection. The best selection is the fan having the lowest total cost. Maintenance costs are not determined as readily as first and operating costs. In many cases, maintenance costs can be assumed equal for alternative selections. All pertinent engineering factors should be examined to justify any such assumptions.

The items listed in Table 23.1.3 under "Physical data" and "Construction details" should be specified because the user is usually in a better position than the supplier to decide such issues. Certain of these items, such as rotation and discharge, have no influence on cost. Other items, such as arrangement number and appurtenances, may have great influence on first cost but no effect on the size and type of fan which should be selected. Still other items, such as the number of inlets and type of drive, largely determine the size and type of fan which should be selected.

If there is no connecting ductwork on the inlet side of the fan, either a single- or double-inlet fan can be used in many cases. The first cost is generally lowest for the double-inlet fan, particularly when relatively large capacities and low pressures are involved. Double-inlet fans require less head room but more flow space than a single-inlet fan for the same rating. Single-inlet fans are generally favored for high-pressure, low-capacity ratings. The advantages of providing only one inlet connection rather than two are obvious.

Direct-drive specifications limit the fan speeds to available motor speeds, and this in turn limits the number of possible fan selections, except where adjustable-blade fans are considered. It is quite unlikely that a standard-size fan will be able to satisfy performance requirements exactly at a direct-connected speed. Accordingly, either requirements must be relaxed or a nonstandard fan must be used. The performance of a standard fan can sometimes be changed sufficiently by modifying the wheel diameter or width. At other times an odd-sized fan may be furnished. When warranted, an entirely new design may be furnished. Direct drives generally require less maintenance and involve less power transmission loss than belt drives.

Adjustable-blade and belt-drive specifications make a large number of fan selections possible, and standard-size fans may be used. Should requirements be altered slightly after installation, usually it is a comparatively easy and inexpensive matter to change the blade setting or belt drive.

The standard designations for arrangement number, direction of entry for inlet boxes, rotation, discharge, and motor position were given earlier. Arrangements with the impeller mounted between bearings are generally less expensive than those with overhung impellers. In the smaller-size fans, arrangements with bearings in the inlet are generally avoided because a closely situated bearing may block an appreciable portion of the inlet. Overhung impeller arrangements are frequently used to protect the bearings whenever the fan must handle hot, dirty, or corrosive gas. The alternative is to use inlet boxes and increase the center distance between bearings accordingly. Overhung pulleys or sheaves are preferred for easy maintenance, but jack shafts may be required on some larger drives.

Various appurtenances may be required, including vibration isolation bases, belt guards, drains, access or inspection doors, flanged connections, inlet screens, stack bracing, stuffing boxes, shaft seals, heat slingers, outlet dampers, and variable-inlet vanes.

Standard fans are usually steel-plate products. However, certain standard lines are housed in cast iron. Under certain conditions, special materials or special methods of construction may be justified. When corrosion resistance is required, the materials of construction should be specified by the user, if possible. Several classes of spark-resistant construction are generally available. Abrasion resistance is very difficult to achieve in a fan. Additional thicknesses of material, or special materials, or both may be specified. Centerline support may be required to maintain alignment when high-temperature gases are handled. Special materials may be required to prevent rapid oxidation at elevated temperatures.

Certain fan lines are normally furnished with antifriction bearings, others with sleeve bearings. Any preference should be specified together with details on lubrication system, cooling media, and type of mounting.

The type of motor, its enclosure, etc., should be specified, particularly if furnished by the user. In any case, the electrical characteristics should be listed.

23.1.7.2   Selecting the Proper Size and Type of Fan

Theoretically, almost any size fan of any type could be used to satisfy the maximum requirements of a particular job. Practical engineering and economic considerations reduce the possibilities to a relatively narrow range of sizes and to a few types.

Certain types of fans are designated according to their usual field of application. There are ventilating fans, mechanical draft fans, industrial exhausters, and pressure blowers, with subclassifications in each case.

Ventilating fans are designed for clean-air service at normal temperatures. Some heavy-duty ventilating fans may be used for more severe conditions, both centrifugal and axial designs are available. Centrifugal types may have either backward or forward-curved blades. Maximum efficiencies are obtained with the former, particularly when airfoil-shaped blades are used. Forward-curved blade types are used when space and price are more important than efficiency. Belt drives are normally used so that any rating can be obtained with a standard-size fan. Propeller fans are usually designed for free delivery operation, but may be used up to 1 inWG (0.25 kPa) static pressure in some cases.

After the type or types of fans suitable for an application are determined, it is necessary to choose the best size fan in each type. There is only one size fan in each type that will operate at the point of maximum efficiency for any given rating. This optimum-size fan must be operated at a certain speed to produce the required rating. A smaller-size fan could be selected that would have to operate at higher speed, or a larger-size fan could be selected that would have to operate at a lower speed. In either case efficiency would be lower than that for the optimum size.

Fans that rate to the right of peak efficiency may be called *undersized fans*, and those which rate to the left of peak efficiency may be called *oversized fans*. To the left means lower capacity, and to the right means higher capacity on the base curve. Oversized fans are hard to justify unless future increases in capacity are envisioned. Occasionally the required operating speed of an oversized fan will match a motor speed. Slightly undersized fans are usually chosen because optimum sizes are rarely standard sizes. Ratings slightly to the right of peak efficiency usually have more stable operating characteristics, i.e., have steeper slopes than those at or to the left of peak efficiency. Sometimes a fan that is considerably undersized is the best choice. In such cases, the additional operating costs occasioned by the lower efficiency must be offset by the savings in first cost or some other engineering or economic factor.

23.1.7.3 Rating Fans from Published Data

Rating data are generally published in the form of tables or charts for each size fan of a given type. The average user finds such presentations convenient. Two typical methods of presenting rating data and examples of their use are illustrated below.

Multirating tables are probably the most common type of published data. Portions of three pages from a typical multirating table are illustrated in Fig. 23.1.29. Such tabulations are almost always based on standard air. To use such a table, enter with the required CFM and ESP and read the RPM and BHP on the appropriate line in the appropriate column. If the requirements do not match the listed values of CFM or ESP exactly, linear interpolations will give accurate results. The table value of RPM is the required operating speed. The table value of BHP must be multiplied by the ratio of actual density to standard density, to obtain the required operating horsepower. Example 23.1.1 illustrates the use of such tables.

$$ESP = 5 \left(\frac{0.075}{0.0693} \right) = 5.41 \text{ inWG, from Eq. (23.1.6)}$$

To select a trial size and determine rating, examine the table for size 805. Note that interpolation is required between 5 and $5\frac{1}{2}$ inWG and between 29,824 and 30,756 ft^3/min.

$$1076 + \frac{5.41 - 5.00}{5.50 - 5.00} (1102 - 1076) = 1086 \text{ r/min}$$

$$1095 + \frac{5.41 - 5.00}{5.50 - 5.00} (1120 - 1095) = 1115 \text{ r/min}$$

$$1097 + \frac{30,000 - 29,824}{30,756 - 29,824} (1115 - 1097) = 1100 \text{ r/min}$$

$$33.3 + \frac{5.41 - 5.00}{5.50 - 5.00} (35.8 - 33.3) = 35.4 \text{ hp}$$

$$34.9 + \frac{5.41 - 5.00}{5.50 - 5.00} (37.3 - 34.9) = 36.9 \text{ hp}$$

$$35.4 + \frac{30,000 - 29,824}{30,756 - 29,824} (36.9 - 35.4) = 35.7 \text{ hp}$$

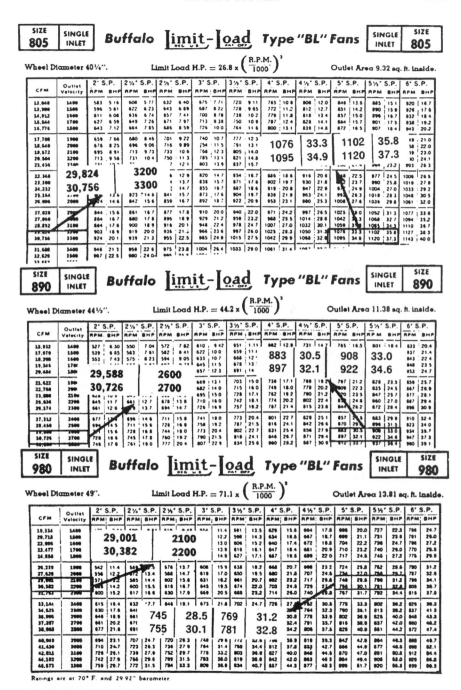

FIGURE 23.1.29 Typical rating table for a centrifugal ventilating fan.

The required operating power is less than the table value since the operating density is less than the standard density for which the table was prepared.

$$3.7 \left(\frac{0.0693}{0.075} \right) = 33.0 \text{ hp}$$

Performing similar interpolations for the other two sizes shown in Fig. 23.1.29 and tabulating will yield the following:

Size 805	Size 890	Size 980
1100 r/min	909 r/min	773 r/min
35.7 hp at 0.075	33.2 hp at 0.075	31.9 hp at 0.075
33.0 hp at 0.0693	30.7 hp at 0.0693	29.5 hp at 0.0693

Note that the larger fan must run slower and the smaller fan faster than the original trial size. The power requirements are all fairly close to 30 hp (22 kW). Even allowing for a 3 percent power loss in the belt drive, the size 805 fan could be driven by a 30-hp (20 kW) motor without exceeding the normal service factor (1.15) of an open motor (1.03 × 33.0 = 34.0 and 1.15 × 30.0 = 34.5). The cost of the extra power and the reduction in motor life should be weighed against the savings in first cost for the smaller fan.

Assuming that the fan is expected to operate 7500 h/year for 20 years and that power costs will average 0.04 $/kWh while interest rates average 15 percent, the present value of the difference in cost for power to drive the 33.0- and 30.7-hp fans can be determined. Assuming equal motor and drive and efficiencies, the power difference is 0.746(33.0 –30.7), or 1.72 kW. The cost of the extra power for 1 year is 0.04(1.72)(7500), or $516. The present value of an annuity which would yield this amount for 20 years if invested at 15 percent may be determined from interest tables or Eq. (23.1.10) as $516(6.26), or $3229.82. This will more than pay for the difference in first cost of the fans without even considering the reduction in motor life.

There are numerous other methods of presenting rating data in use. Figure 23.1.30 illustrates a method of showing the performance of an adjustable-blade fan. As usual, the sample zone curves are drawn for standard air. To use such a chart, enter with the required CFM and ETP and read blade position, efficiency, and horsepower by interpolating between appropriate lines. If nonstandard air is to be handled, the procedures used in Example 23.1.1 must be used to determine ETP and BHP.

Example: Zone Curve Rating

Given: Zone curves for a specific design according to Fig. 23.1.30.
Required: Determine operating characteristics of a 38-A fan to deliver 25,000 ft³/min of air against 1.5-inWG total pressure for standard air.

Solution Enter the chart at bottom of 25,000 ft³/min. Note outlet velocity pressure of 3200 ft/min at top. Velocity pressure is 0.62 inWG at top. Enter chart at right at 1.5 inWG total pressure. Note intersection with capacity line. Blade position is 50. Note horsepower by interpolation is 7.3 hp. Note total efficiency is 80.5 percent.

23.1.7.4 Electronic Catalogs

Many fan manufacturers are now publishing an *electronic catalog* (software). Most of them lead the user through a step-by-step process to facilitate selection of the best size and type for the application. Corporate Websites also offer convenient sources of information.

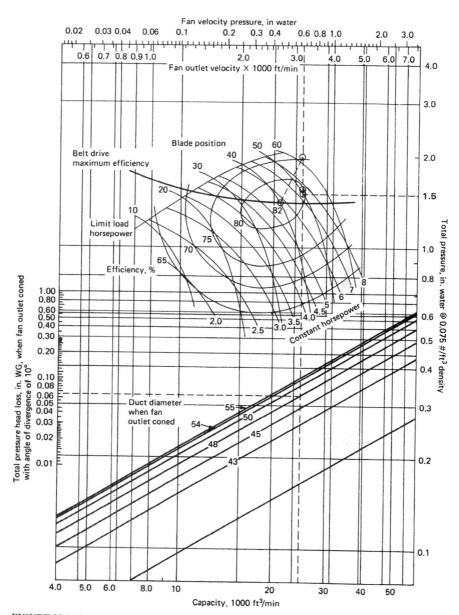

FIGURE 23.1.30 Typical rating chart for adjustable-blade vane axial fan. (See Appendix B for metric conversion factors.)

REFERENCES

1. *Fans and Systems*, AMCA publication 201, Air Movement and Control Association, Arlington Heights, IL, 1990.
2. R. D. Madison and J. B. Graham, "Fan Noise Variation with Changing Fan Operation," *Transactions of the ASHRAE*, vol. 64, 1958, pp. 319–340.
3. H. H. Uhlig, *Corrosion Handbook*, John Wiley & Sons, New York, 1948, pp. 747–799.
4. Standard 99-86, Air Movement and Control Association, Arlington Heights, IL, 1986.

23.2 INSTALLATION AND MAINTENANCE*

Bayley Fan
LAU Industries,
Lebanon, Indiana

23.2.1 SAFETY PRECAUTIONS

Do....

1. Make sure unit is stopped and electrical power is locked out before putting hands into the inlet or outlet opening or near belt drive. We suggest a *LOCK-OUT* and a warning sign on the start switch cautioning not to start the unit.
2. Follow maintenance instructions.
3. Make sure all drive guards are installed at all times fan is in operation. If the inlet or outlet is exposed, a suitable guard should also be installed.
4. Take special care not to open any fan or system access panels while the system is under pressure (negative or positive).
5. Never allow untrained or unauthorized persons to work on equipment.
6. Take special care when working near electricity. Also insure the power is off and cannot be turned on while servicing the fan.
7. Keep area near equipment clean.

Caution....

1. **Do not** put hands near or allow loose or hanging clothing to be near belts or sheaves while the unit is running.
2. **Do not** put hands into inlet or outlet while unit is running. *It is sometimes difficult to tell whether or not it is running* ... be sure it is not running and cannot be operated before doing any inspection or maintenance.
3. **Do not** operate fan with guards removed.
4. **Do not** take chances.

23.2.2 FOUNDATIONS

A rigid level foundation is a must for every fan. It assures permanent alignment of fan and driving equipment and freedom from excess vibration, minimizing maintenance costs. The foundation must be cast separate from any adjacent floor structure and separated around edges by at least $^3/_4$ in (18 mm) tar felt to prevent transmission of vibration in either direction. The subfoundation (soil, stone, rock, etc.) should be stable enough to prevent uneven settling of fan foundation. Foundation bolt locations are found on the certified drawings.

---

*Adapted from Lau Service Manual #0696 C-141.

The natural frequencies of the foundation must be sufficiently removed from the rotational frequency of the fan to avoid resonant conditions.

23.2.2.1 Poured Concrete Foundations (Recommended)

Poured concrete under the fan and all drive components is the best fan foundation. A generally accepted rule of thumb is that the weight of the concrete foundation be at least three times the total weight of the equipment it will support. This weight acts as an inertia block to stabilize the foundation. Where the ground is soft the foundation should be flared or the footing course increased in size to resist settling. The top should extend at least 6 in (15 mm) outside the outline of the fan base and should be beveled on the edges to prevent chipping.

Anchor bolts in concrete should be L or T shaped. They should be placed in pipe or sheet metal sleeves approximately 2 in (7 mm) larger in diameter than the anchor bolts to allow for adjustment in bolt location after the concrete is set. In estimating the length of the bolts, allow for the thickness of the nut and washers, thickness of the fan base, thickness of the shim pack, if required, and extra threads for drawdown.

Seating area for washers and nuts must be clean and threaded area must be clean and lubricated.

23.2.2.2 Equipment-Mounted Fans

If the fan is mounted on equipment having parts that cause vibration, *it is very important that the fan support is rigid enough to prevent such vibration from being carried to the fan.* The resonant frequency of the support should avoid the fan running speed by at least 20 percent. It may be advisable to use vibration isolators under the fan.

23.2.2.3 Duct-Mounted Fans

If the fan is mounted in ductwork, horizontally or vertically, indoors or outdoors, the weight of the fan should be sufficiently supported to assure permanent alignment of the fan assembly. Forces imposed on the fan that may cause misalignment must be isolated to prevent serious damage. Horizontally installed fans should be supported with vibration isolators and flexible duct connections. Vertically installed fans should be supported with guy rods and guy wires. The weight of vertical ductwork supported by the fan should not exceed the weight of the fan without additional external duct support.

23.2.2.4 Structural steel foundation

When a structural steel foundation is necessary, it should be sufficiently rigid to ensure permanent alignment. It must be designed to carry, with minimum deflection, the weight of the equipment plus loads imposed by the centrifugal forces set up by the rotating elements. We recommend welded, riveted, or suitably locked structural bolted construction to best resist vibration. In certain applications, it is recommended that vibration isolators, selected specifically for weight and span conditions, be installed.

Fans installed above ground level should be located near to or above a rigid wall or heavy column. An overhead platform or support must be rigidly constructed, level and sturdily braced in all directions.

23.2.3 *ASSEMBLY AND INSTALLATION*

Most fans and blowers are shipped completely assembled, however, a few are shipped partially knocked-down. Partially knocked-down fans are partially disassembled and the remaining parts match marked for easy field assembly. When installing units allow ample space for removal of impeller, lubrication of bearings, adjustment of motor base and inspection or servicing of complete units.

For units shipped completely assembled:

- Remove skids and any protective coverings over bearings or housing.
- Move fan to its rigid support. Install vibration isolation and secure fan to the mounting bolts.
- Level unit with spirit level by adjusting vibration isolators or shimming as required.
- Tighten all mounting hardware.

For units shipped partially assembled: Fans and blowers shipped partially knocked-down may have parts shipped as follows:

- Fan section with housing, impeller, bearings, and shaft assembled.
- Stack cap with integral gravity back draft damper.
- Motor or V-belt drive separate.
- Variable inlet vanes, screens, outlet damper, etc., separate.

To assemble

- Remove skids and any protective coverings over bearings or housing.
- Move fan to its rigid support. Install vibration isolation and secure fan to the mounting bolts.
- Level unit with spirit level by adjusting vibration isolators or shimming as required.
- Tighten all mounting hardware.
- If fan is V-belt driven and drive is not mounted, assemble sheaves on their shafts and line up V-belts with proper tension. *See V-belt drive section.*
- If fan was shipped separate from a stack cap and roof curb cap, caulk the flanges to be joined with a generous $1/4$ in (6 mm) caulk being careful to caulk completely around all mounting holes. Proper caulking stops rain water leaks. Without proper caulking, leaks will occur because the air passing the flange joints will siphon the rain water into the air stream where it can leak back into the building.
- If variable inlet vane, inlet bell or inlet screen is required, bolt these in position on the fan.
- The fan is now ready to be connected into the system.
- Pay attention to correct direction of rotation.

23.2.4 *ALIGNMENT*

- Level the fan housing on the foundation by shimming where necessary. Use a spirit level on the shaft or on a horizontal portion of the housing. Tighten the nuts on all mounting bolts.
- If fan is received with motor mounted, check tightness of motor hold down bolts. Also check V-belt drive alignment. (Refer to V-belt drive section).

- Spin impeller slowly by hand to see that it clears the fan inlets. Due to transit and handling damage, it may be necessary to loosen the bearing bolts and shift the impeller and the shaft to locate it properly with the fan inlet.
- If fan is on concrete foundation, you are ready to pour to grout. Anchor bolts should be tight. After grout hardens, recheck for final level and alignment of all components.

After motor has been mounted, align and bolt down, wire power supply through a disconnect switch, short-circuit protection, and suitable magnetic starter with overload protection. All motors should be connected as shown on nameplate. Install wiring and fusing in accordance with the national Electrical Code and local requirements.

Be sure power supply (voltage, frequency, and current carrying capacity of wires) is in accordance with the motor nameplate.

If fan is on a steel foundation or mounted in duct work, you are ready for operation.

23.2.5 FIBERGLASS FANS

Fiberglass fans require care similar to steel fans except as noted herein.

- Be sure impeller is not striking the inlet or housing and rotating in the proper direction. Fiberglass may break on impact or fail quickly due to stress caused by improper rotation.
- Do not operate a fiberglass fan in abrasive atmospheres. Abrasives will erode the resin rich surfaces of the FRP material and destroy the corrosion resistance of the fan.
- Do not operate a fiberglass fan in temperatures above 150°F (65°C) without a specific resin and a fan design approved by the fan manufacturer.
- Never attempt to support the fan by one flange. Use mounting brackets, hangers, etc. and appropriate vibration isolators if required. Use *both* flanges if mounted in ductwork.
- Never allow the fan to operate with a vibration problem. The stress caused by vibration will quickly destroy a fiberglass fan.

23.2.6 MOTORS

Warning: Always disconnect or shut off electrical power before attempting to service fan and/or motor.

All A-C induction motors will perform satisfactorily with a 10 percent variation in voltage, a 5 percent variation in frequency or a combination voltage-frequency variation of 10 percent. For motors rated 208 to 220 V, the above limits apply only to 220 V rating. To select control for 208 to 220 V motors, use same amps for either 208 or 220 V.

Motors are received with bearings lubricated and require no lubrication for some time depending on operating conditions (see maintenance section on motor bearings).

23.2.6.1 To Reverse Direction of Rotation Single-Phase Motors

Shaded pole   Rotation cannot be reversed unless motor is constructed so that the shading coil on one half of the stator pole can be shifted to the other half of the stator pole.

Split phase	Interchange connections to supply of either main or auxiliary winding.
Capacitor	All types of capacitor motors are reversed in rotation by interchanging connections to supply to either main or auxiliary winding.
Repulsion	Remove plate on motor end bracket and turn bracket (holding brushes) in direction *opposite* to direction of existing rotation.

Note: It is suggested that rotation change be made for single phase motors by the manufacturer's approved repair shop.

Three-Phase Motors. To reverse rotation, interchange any two line leads.

Normal operation of motors results in temperature rise. Permitted temperature rise depends on the type of motor installation. The total motor operating temperature is the ambient temperature + motor temperature rise. The motor temperature rise includes nameplate temperature rise, service factor allowance, and hot spot allowance.

Important Note: Motors are warranted by the motor manufacturer, the fan manufacturer will assist in locating a local vendor approved repair shop if required.

23.2.7 MOTOR LUBRICATION AND MAINTENANCE

Regrease or lubricate motor bearings according to manufacturer's recommendations. *Do not over-lubricate.* Motor manufacturer's lubrication recommendations are printed on tags attached to motor. Should these tags be missing, the following will apply:

23.2.7.1 Fractional Horsepower Ball Bearing Motors

Under normal conditions, ball bearing motors will operate for five years without relubrication. Under continuous operation at higher temperatures (but not to exceed 140°F (60°C) ambient) relubricate after one year.

23.2.7.2 Integral Horsepower Ball Bearing Motors

- Motors having pipe plugs or grease fittings should be relubricated while warm and at a stand still. Replace one pipe plug on each end shield with grease fitting. Remove other plug for grease relief. Use low pressure grease gun and lubricate until grease appears at grease relief. Allow motor to run for 10 min to expel excess grease. Replace pipe plugs.
- Recommended relubrication intervals (Table 23.2.1)—this is a general guide only.
- These ball bearing greases or their equivalents are satisfactory for ambients from −20°F to 200°F (−29°C to 93°C).

Chevron SRIU #2	(Standard Oil of California)
Chevron BRB #2	(Standard Oil of California)
Premium RB	(Texaco, Inc.)
Alvania No. 2	(Shell Oil Company)

TABLE 23.2.1 Recommended Lubrication Intervals

Power range		Standard duty 8 h/day	Severe duty 24 h/day dirty, dusty	Extreme duty—very dirty high ambients
hp*	kW			
$1^{1}/_{2}$–$7^{1}/_{2}$	2–10	5 yr	3 yr	9 mo
10–40	13–54	3 yr	1 yr	4 mo
50–150+	67–200+	1 yr	9 mo	4 mo

*Conversoin: 1 hp = 746 W.

- Make certain motor is not overloaded. Check power draw against nameplate.
- *Keep motors dry.* Where motors are idle for a long time, single phase heaters or small space heaters might be necessary to prevent water condensation in windings.

23.2.8 FAN START-UP

23.2.8.1 Before Start-up, Check

- **Fastenings**—It is recommended that all foundation bolts, impeller hub set screws and bearing set screws be checked for tightness before start-up.
- **Access doors**—should be tight and sealed.
- The fan bearings, whether pillow block or flange mounted, are pre-lubricated and should not require additional grease for start-up.
- "Bump" the motor to check for proper impeller rotation. The motor should be started in accordance with manufacturer's recommendations.
- **Bearings**—check bearing alignment and make certain they are properly locked to shaft.
- **Impeller**—turn over rotating assembly by hand to see that it runs free and does not bind or strike fan housing. If impeller strikes housing, it may have to be moved on the shaft or bearing pillow blocks moved and reshimmed. Check location of impeller in relation to fan inlets. Be sure fan housing is not distorted. See *Alignment* section.
- **Driver**—check electrical wiring to the motor. See *Motor* section.
- **Guards**—make certain all safety guards are installed properly.
- **V-Belt Drive**—must be in alignment, with belts at proper tension. See *V-belt Drive* section.
- **Duct connections**—from fan to ductwork must not be distorted. Ducts should never be supported by the fan. Expansion joints between duct connections should be usd where expansion is likely to occur or where the fan is mounted on vibration isolators. All joints should be sealed to prevent air leaks and all debris removed from ductwork and fan.

Dampers and variable intake ventilators VIVs should operate freely and blades close tightly. All dampers and VIVs should be partially closed during starting periods to reduce power requirements. This is particularly important for a fan designed for high temperature operation being "run in" at room temperature or at appreciably less than design temperature.

When air is up to temperature, the damper or VIV may be opened. Completely closing dampers could cause the fan to run rough.

• Fan may now be brought up to speed. Watch for anything unusual such as vibration, overheating of bearings and motors, etc. Multispeed motors should be started at lowest speed and run at high speed only after satisfactory low speed operation. Check fan speed on V-belt driven units and adjust motor sheave to give the desired RPM. Balance system by adjusting damper VIVs.

• **At first indication of trouble or vibration, shut down and check for difficulty**.

23.2.9 MAINTENANCE POINTERS

• **Always disconnect or shut off fan before attempting any maintenance**.

• A definite time schedule for inspecting all rotating parts should be established. The frequency of inspection depends on the severity of operation and the location of the equipment.

• Fan bearing alignment should be checked at regular intervals. Misalignment can cause overheating, wear to bearing dust seals, bearing failure and unbalance.

• Fan bearings should be lubricated at regular intervals. Periodic inspection will be necessary. If grease is found to be breaking down, replenish grease by pumping new grease into bearing until all the old grease has been evacuated. See section on Bearing Lubrication.

• Bearings on high speed fans tend to run hot, 75° to 105°F (41° to 58°C) above ambient. Do not replace a bearing because it feels too hot to touch. Place a contact thermometer against the bearing pillow block and check the temperature. Before you investigate high temperature, realize that ball or roller bearing pillow blocks can have a total temperature of 225°F (107°C). High-temperature bearings may be rated at 425°F (218°C).

• Foundation bolts and all set screws should be inspected for tightness.

• Fans should be inspected for wear and dirt periodically. Any dirt accumulated in housing should be removed. Severely worn spots on components other than the impeller may sometimes be built up with weld material, using care to prevent heat distortion. The impeller may have to be cleaned. A wash down with steam or water jet is usually sufficient, covering bearings so water will not enter the pillow blocks. Impellers having worn blades should be replaced. Impellers require careful rebalancing before being returned to service. Replacement impellers should have the balance checked upon start-up and corrected as required to operate properly in it's specific application.

• Repairing of exterior and interior parts of fans and ducts will extend the service life of the installation. Select a paint that will withstand the operating temperatures. For normal temperatures a good machinery paint may be used. Corrosive fumes require all internal parts to be wire brushed, scraped clean and repainted with an acid resisting paint. Competent advice should be sought when corrosive fumes are present.

• Blow out open type motor windings with low pressure air to remove dust or dirt. Air pressure above 50 psi (345 kPa) should not be used as high pressure may damage insulation and blow dirt under loosened tape. Dust can cause excessive insulation temperatures. **Do not exceed** OSHA air pressure requirements.

• Excessive vibration will shorten the life of any mechanical device. Correct any imbalance situation before returning fan to service.

23.2.10 LUBRICATION OF ANTIFRICTION BEARINGS

Bearings on assembled fans receive their initial lubrication from the bearing manufacturer. Bearings shipped separate from the fan or as a replacement may not be lubricated before shipment. When there is the slightest doubt, the safe practice is to assume that the bearing has not been lubricated. Always turn fan off before lubricating.

For grease lubricated ball or roller bearing pillow block, a good grade of grease free from chemically or mechanically active material should be used. These greases are a mixture of lubricating oil and a soap base to keep the oil in suspension. They have an upper temperature limit where oil and soap base oxidize and thermally decompose into a gummy sludge.

Mixing of different lubricants is not recommended. If it is necessary to change to a different grade, make or type of lubricant, flush bearing thoroughly before changing. Regreasing will vary from 3 months to a year depending on the hours of operation, temperature and surrounding conditions. Special greases may be required for dirty or wet atmosphere (consult your lubricant supplier).

When grease is added, use caution to prevent any dirt from entering the bearing. The pipe plug or grease relief fitting should be open when greasing to allow excess grease to flow out. The pillow block should be about one-third full, as excess grease may cause overheating. Use low pressure gun.

These ball bearing greases or their equivalents are satisfactory for ambients from −20°F to 200°F (−7°C to 93°C).

Chevron SRIU #2   (Standard Oil of California)

Chevron BRB #2   (Standard Oil of California)

Premium RB   (Texaco, Inc.)

Alvania No. 2   (Shell Oil Company)

23.2.10.1 Frequency of Lubrication

The bearings are lubricated at predetermined intervals and the condition of the grease established as it is purged out of the seals or by examination of the grease in the housing. An average installation where the environmental conditions are clean and room temperatures prevails may only require bearing lubrication every 3 to 6 months, while operation in a dirty atmosphere at high temperatures will require much more frequent intervals. Base your particular interval on condition of grease after a specific service period. Table 23.2.2 is optimum for various shaft sizes and operating speeds.

23.2.11 V-BELT DRIVES

Fans shipped completely assembled have had the V-belt drive aligned at the factory. Alignment must be checked before operation.

- Be sure sheaves are locked in position.
- Key should be seated firmly in keyway.
- The motor and fans shafts must be properly aligned, with the centerline of the V-belts at a right angle to the shafts.

TABLE 23.2.2 Recommended Frequency of Lubrication—Fan Bearings

Shaft size in	mm	Operating speed (RPM)									
		500	1000	1500	2000	2500	3000	3500	4000	4500	500
		Lubricating frequency (months)									
0.50–1.00	12–25	6	6	6	6	6	6	4	4	2	2
1.06–1.44	27–37	6	6	6	6	6	6	4	4	2	1
1.50–1.75	38–45	6	6	6	4	4	2	2	2	1	1
1.88–2.19	49–56	6	6	4	4	2	2	1	1	1	
2.25–2.44	57–62	6	4	4	2	2	1	1	1		
2.50–3.00	64–76	6	4	4	2	1	1	1			
3.06–3.50	77–90	6	4	2	1	1	1				
3.56–4.00	90–102	6	4	2	1	1					

- Be sure all safety guards are in place.
- Start the fan. Check for proper rotation of impeller. Run fan at full speed. A slight belt bow should appear on the slack side. Adjust belt tension by adjusting the motor on its adjustable base.
- If belts squeal excessively at startup, they are too loose and should be tightened.
- When belts have had time to seat in the sheave grooves, readjust belt tension. Check belt tension after 8, 24, and 100 h of operation.

23.2.11.1 V-Belt Drive Assembly can be Mounted as Follows

- Clean motor and drive shafts. Be sure they are free from corrosive material. Clean bore of sheaves and coat with heavy oil for ease of shaft entry. Remove oil, grease, rust or burrs from sheaves. Place fan sheave on fan shaft and motor sheave on its shaft. Do not pound on sheave as it may result in damage. Tighten sheaves in place.
- Move motor on base so belts can be placed in grooves of both sheaves without forcing. Do not roll belts or use tool to force belt over grooves.
- Align fan and motor shafts so they are parallel. The belts should be at right angles to the shafts. A straight edge or taut cord placed across the faces of the sheaves will aid in alignment with single grove sheaves. If multiple grooves sheaves are installed, use the centerline of the drive as your alignment point.
- Tighten belts by sliding motor in its base. Correct tension gives the best efficiency. Excessive tension causes undue bearing pressure.
- Be sure all safety guards are in place.
- Start the fan and run at full speed. Adjust belt tension until only a slight bow appears on the slack side of the belts. If slippage occurs, a squeal will be heard at start-up. Eliminate the squeal by tightening the belts.
- Belts require time to become fully seated in the sheave grooves—check belt tension after 8, 24, and 100 h of operation. Allowing belts to operate with improper tension will shorten belt life substantially.

- If the shafts become scratched or marked, carefully remove the sharp edges and high spots such as burrs with fine emery cloth or a honing stone. Avoid getting emery dust in the bearings.

- Do not apply any belt dressing unless it is recommended by the drive manufacturer. V-belts are designed for frictional contact between the grooves and sides of the belts. Dressing will reduce this friction.

- Minimum belt center distances are available from factory upon request.

- Belt tension on an adjustable pitch drive is obtained by moving the motor—*not by changing the pitch* diameter of the adjustable sheaves.

23.2.12   *SUMMARY OF FAN TROUBLES AND CORRECTIONS*

1. Capacity or pressure below rating
    - Total resistance of system higher than anticipated—*system problems.*
    - Speed too low—*adjust drive.*
    - *Dampers or variable inlet vanes improperly adjusted.*
    - Poor fan inlet or outlet conditions—*elbows at or too close to fan.*
    - Air leaks in system—*seal joints—correct damper settings.*
    - Damaged impeller or incorrect direction of rotation—*correct.*

2. Vibration and noise
    - Misalignment of impeller, bearings, couplings, or V-belt drive—*loosen, align, tighten.*
    - Unstable foundation—inferior design—*start over.*
    - Foreign material in fan causing unbalance—*remove.*
    - Worn bearings—*replace bearings and shaft.*
    - Damaged impeller or motor—*check and repair.*
    - Broken or loose bolts or set screws—*replace.*
    - Bent shaft—*replace.*
    - Worn coupling—*replace.*
    - Impeller or driver unbalanced—*balance.*
    - 60/120 cycle magnetic hum due to electrical input. *Check for high or unbalanced voltage.*
    - Fan delivering more than rated capacity—*reduce speed.*
    - Loose dampers on VIVs—*adjust and tighten.*
    - Speed too high or fan rotating in wrong direction—*correct.*
    - Vibration transmitted to fan from some other source—*isolate.*

3. Overheated bearings
    - Too much grease in ball bearings. *Allow run time to purge (24 h).*
    - Poor alignment—*correct.*

- Damaged impeller or drive—*inspect, correct, or replace.*
- Bent shaft—*replace.*
- Abnormal end thrust—*loosen set screws and adjust.*
- Dirt in bearings—replace bearing—*use filtered grease.*
- Excessive belt tension—*adjust.*

4. Overloaded motor (draws too many amps)

- Speed too high—*reduce speed or change HP.*
- *Discharge over capacity due to existing system resistance being lower than original rating.*
- Specific gravity or density of gas above design value—*recalculate and correct.*
- Wrong direction of rotation—*correct.*
- Poor alignment—*correct.*
- Impeller wedging or binding on inlet bell—*loosen and adjust.*
- Bearings improperly lubricated—*see par 10.*
- Motor improperly wired—*verify and correct.*

5. Motor problems

- *Check for low or high voltage from power source.*
- High temperature—*drawing too much current or dirt in windings.*
- Armature unbalance—*vibration and noise.*
- Worn bearings—*armature rubs against stator.*
- *Too much or not enough lubrication in bearings.*
- *Commutator brushes on dc motor worn or not seated under proper tension.*
- Loose hold down bolts—*vibration and noise.*
- Low insulation resistance due to moisture—*Check resistance with a megohm meter ("Megger") or similar instrument employing a 500 V dc potential. Resistance should read at least 1 MΩ.*

23.3 AXIAL VANE FANS

Ben Harstine

Supervisor of Field Service, Joy/Green Fan Division,
New Philadelphia Fan Co.,* New Philadelphia, Ohio

23.3.1 INTRODUCTION

Axial vane fans may be *direct drive* in which the rotor assembly is mounted on the motor shaft or driven by an external motor with a flexible coupling, or *belt driven* with the motor mounted directly on the fan or base mounted. Axial vane fans may also be fixed-pitch, adjustable-pitch, or controllable-pitch.

We will look at general fan maintenance for axial vane fans, then into specifics for each drive and mounting arrangement.

23.3.2 SAFETY PROCEDURES

Always lockout and tagout equipment before removing safety screens or guards. *Never attempt to energize a fan while working on the machine.* Always remove all loose material from the fan room or ducted area before energizing fan.

23.3.3 GENERAL FAN MAINTENANCE

It is very important to keep filters and coils clean to allow the fan to have access to incoming air flow. Dirty or clogged filters and coils can starve fans for air; this will cause excessive negative pressure on the inlet side and significant rise in power draw to the motors. These conditions can cause aerodynamic stall. If aerodynamic stall continues for a period of time, blades can fail due to fatigue. The result is catastrophic failure. Similar cautions should be observed down stream to be sure all dampers and VAV function properly and sequentially timed.

Axial fans should be lubricated as instructed by the fan manufacturer. If maintenance personnel are not sure of the proper type and quantity of lubricant or at what intervals to lubricate, the maintenance person should contact the fan manufacturer for specific instructions. Incompatible lubricants can break down, loosing the lubricating quality. *Too much lubricant can be as damaging as too little.*

Periodic preventive maintenance programs should be established for all machinery including fans. The frequency of such programs will depend on the environment that the fan is operating in. Fan operating in a clean environment, such as a clean room application where air is prefiltered and where the ambient temperature ranges from 50 to 80°F (11 to 27°C) would only require an annual or biannual cleaning, where as fans ventilating diesel

---

*Formerly Joy Technologies, Inc.

exhaust from a tunnel, for example, would require much more frequent cleaning. Typical HVAC fans should receive an annual preventive maintenance program.

The preventive maintenance program should involve cleaning the blades and the stationary vanes or struts on the discharge side of the fan rotor assembly. Keeping the fan clean will maintain the fan efficiency and save energy. The preventive maintenance program should also include vibration analysis. Vibration signatures should be recorded and stored, then compared with past signatures to establish a history and to warn of any early defects so that repairs can be scheduled rather than cause catastrophic expensive and inconvenient emergency repairs.

Vibration can be measured with a vibration analyzer/balancer or a vibration data collector then stored in a computer for comparison records. Vibration is measured in velocity inches per second (cm/s) or in mils of displacement. Velocity measures the speed of the movement while mils measures the distance of movement. See the typical vibration severity chart (Fig. 23.3.1). Indications of one times the operating speed indicates out of balance, two times indicates looseness, and three times indicates coupling alignment problems. Other frequencies are bearing frequencies, blade passing (the combination of speed times number of blades and/or the number of stationary vanes). *Never attempt to balance a mechanical problem or a dirty fan.* A fan, or any rotating machine, will not get out of balance unless some mass is added to or taken away from a balanced rotating component.

Vibration signatures should be taken at several locations on the fan. See Fig. 23.3.2. Locations 1 and 3 measure vertical movement, locations 2 and 4 measure horizontal movement and location 5 measures axial movement. When measuring locations 1 through 4, the transducer or accelerometer must be pointed at the center or rotation. These readings will indicate balance, bearing wear and other frequencies such as blade passing. When measuring at location 5, the transducer at the fan flange is placed opposite to, or in the direction of, the air flow. This measures axial movement of the fan and can detect air flow problems such as turbulence or excessive pressure in the system.

Only take and record as much information as is necessary to maintain a history file or to solve a particular problem. Too much or too detailed data may be more confusing than it is helpful. There are many analyzers on the market for maintenance records. A data collector that gives a printed signature and a log that identifies the frequencies is sufficient for a good preventive maintenance program.

23.3.4   ADJUSTABLE- (AP) AND FIXED-PITCH (FP) DIRECT-DRIVE FANS

Adjustable-pitch fans (Fig. 23.3.3) are fans where the blade angles may be adjusted manually. Fixed-pitch fans are ones which have blades cast or welded to the center hub. These types of fans range in size from a few inches (cm) to several feet (m) in diameter with horsepower ranging from fractional to several hundred. These fans usually have the rotor assembly keyed directly to the motor shaft. This type of fan is usually applied to constant volume systems. When applied to variable air volume VAV systems, air flow may be controlled with the use of a variable frequency drive, but never with dampers.

These fans require the least amount of maintenance. An occasional cleaning, proper lubrication, and vibration analysis will reward the operator many years of dependable service. Always consult the performance curve before making any blade adjustments and always check motor and amp draw after adjustments are made. See typical fan performance curve (Fig. 23.3.4).

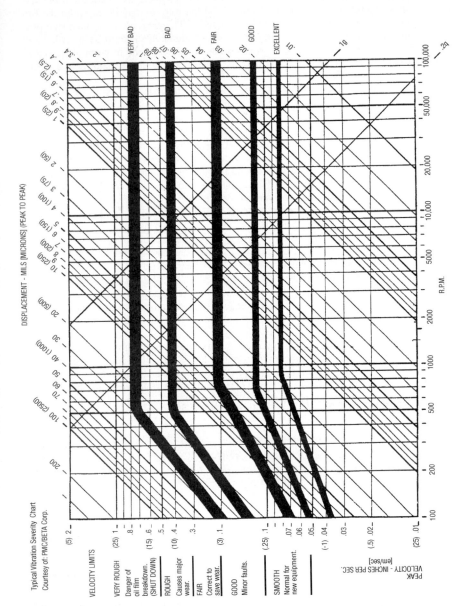

FIGURE 23.3.1 Vibration severity chart.

23.54

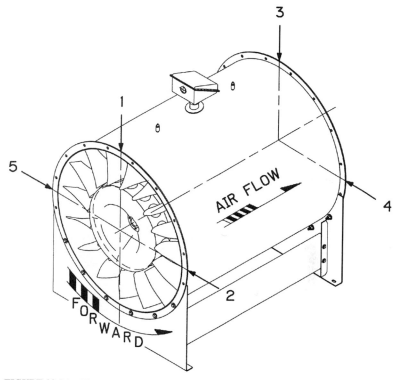

FIGURE 23.3.2   Vibration signature locations.

23.3.5   ADJUSTABLE PITCH BELT DRIVE FANS

This type of fan is similar to the direct drive fan except, of course, the motor is external; which necessitates the use of a fan shaft, pillow block bearings, belts, and sheaves. Belt-driven fans are used when the motor must be outside the air stream or the owner prefers to use a standard motor (Fig. 23.3.5).

When replacing bearings in the belt-driven fan, it is very important to install the bearings such that the bearing closest to the fan wheel is fixed to carry the axial thrust generated by the fan and the bearing closest to the sheave be allowed to float and carry the radial load imposed by the belts. If this procedure is not followed, two situations can result, neither of which is desirable.

- If the axial and radial load are both imposed on the sheave end-bearing, it will overload and fail.
- If the fan end bearing does not carry the thrust load or is permitted to run unloaded, it will skid causing excessive heat and be doomed to failure.

Belts should be inspected regularly for signs of wear, cracks, fraying, or scorching due to slippage. Most belt failures are the result of improper installation and tensioning.

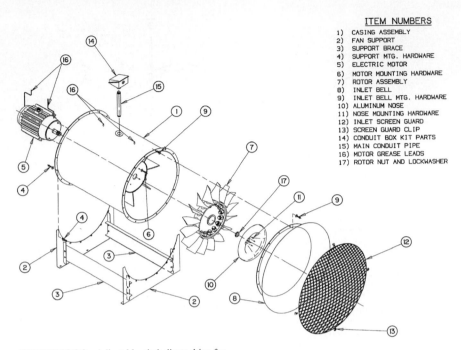

ITEM NUMBERS

1) CASING ASSEMBLY
2) FAN SUPPORT
3) SUPPORT BRACE
4) SUPPORT MTG. HARDWARE
5) ELECTRIC MOTOR
6) MOTOR MOUNTING HARDWARE
7) ROTOR ASSEMBLY
8) INLET BELL
9) INLET BELL MTG. HARDWARE
10) ALUMINUM NOSE
11) NOSE MOUNTING HARDWARE
12) INLET SCREEN GUARD
13) SCREEN GUARD CLIP
14) CONDUIT BOX KIT PARTS
15) MAIN CONDUIT PIPE
16) MOTOR GREASE LEADS
17) ROTOR NUT AND LOCKWASHER

FIGURE 23.3.3   Adjustable-pitch direct-drive fan.

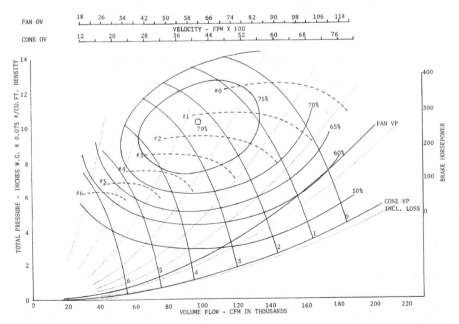

FIGURE 23.3.4   Typical fan performance curve.

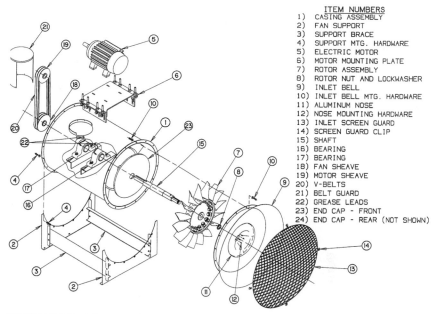

```
                                    ITEM NUMBERS
                                1)  CASING ASSEMBLY
                                2)  FAN SUPPORT
                                3)  SUPPORT BRACE
                                4)  SUPPORT MTG. HARDWARE
                                5)  ELECTRIC MOTOR
                                6)  MOTOR MOUNTING PLATE
                                7)  ROTOR ASSEMBLY
                                8)  ROTOR NUT AND LOCKWASHER
                                9)  INLET BELL
                                10) INLET BELL MTG. HARDWARE
                                11) ALUMINUM NOSE
                                12) NOSE MOUNTING HARDWARE
                                13) INLET SCREEN GUARD
                                14) SCREEN GUARD CLIP
                                15) SHAFT
                                16) BEARING
                                17) BEARING
                                18) FAN SHEAVE
                                19) MOTOR SHEAVE
                                20) V-BELTS
                                21) BELT GUARD
                                22) GREASE LEADS
                                23) END CAP - FRONT
                                24) END CAP - REAR (NOT SHOWN)
```

FIGURE 23.3.5   Adjustable pitch belt drive fan.

When installing replacement belts, always move the motor base toward the fan so that belts may be easily removed and replaced. *Never pry belts over sheave grooves to install or remove.* This will stretch and damage belt fibers and can cause damage to sheaves. Sheave alignment should be accomplished before tensioning. A string alignment or straight edge can be used to align sheaves. Proper belt tension is the least amount of tension required to operate the fan under full load without slippage. Minimal slippage is acceptable upon start-up. When installing new belts allow the fan to operate for at least eight hours then retention the belts as required. Belt manufacturers do not recommend the use of belt dressings or non-slip chemicals.

23.3.6   CONTROLLABLE-PITCH (CP) FANS

Controllable pitch fans may be direct or belt driven; these are fans which have the ability to change pitch while the fan is in operation. Blade angle (pitch) is changed by means of an actuator which may be external or internal of the rotating assembly. External actuated controllable pitch fans are considered to have more reliable hysteresis and blade angle positioning than internal actuated fans and are also applied to heavier duty applications.

23.3.7   EXTERNALLY ACTUATED
CONTROLLABLE-PITCH FANS

Externally actuated CP fans (Fig. 23.3.6) have blade positioning actuators on the outside of the fan casing. Actuators may be pneumatic, electronic, hydraulic, or manual. Others may

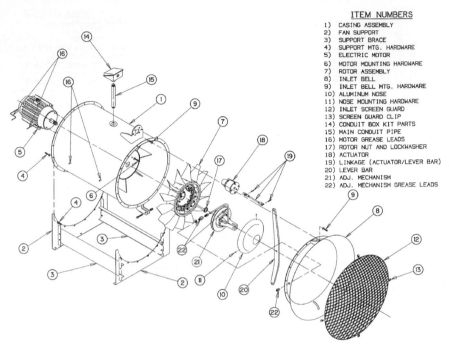

ITEM NUMBERS

1) CASING ASSEMBLY
2) FAN SUPPORT
3) SUPPORT BRACE
4) SUPPORT MTG. HARDWARE
5) ELECTRIC MOTOR
6) MOTOR MOUNTING HARDWARE
7) ROTOR ASSEMBLY
8) INLET BELL
9) INLET BELL MTG. HARDWARE
10) ALUMINUM NOSE
11) NOSE MOUNTING HARDWARE
12) INLET SCREEN GUARD
13) SCREEN GUARD CLIP
14) CONDUIT BOX KIT PARTS
15) MAIN CONDUIT PIPE
16) MOTOR GREASE LEADS
17) ROTOR NUT AND LOCKWASHER
18) ACTUATOR
19) LINKAGE (ACTUATOR/LEVER BAR)
20) LEVER BAR
21) ADJ. MECHANISM
22) ADJ. MECHANISM GREASE LEADS

FIGURE 23.3.6   Externally actuated controllable-pitch fan.

be a combination of electro pneumatic or electro hydraulic. The actuator is connected to a lever or other push-pull device, which is connected to the outer race of a de-spin thrust bearing. The inner race of the bearing is connected by way of a linkage system to each blade.

Maintenance procedures for this type of fan include all of those mentioned for AP and FP (adjustable pitch and fixed pitch) fans. The despin thrust bearing shall receive lubrication which will be similar in quantity as the motor but usually twice the frequency. An annual or bi-annual preventive maintenance program should also include removing the despin thrust bearing, rinsing clean the used lubricant residue, inspecting the bearing, repacking with fresh lubricant and reinstallation. All parts of the rotating assembly should be inspected at this time. Any worn or broken parts should be replaced. Things to check would be worn or sticking rollers, bolts; also check for broken blade clamps. Also check to be sure that the pitch control mechanism slides easily on the driven shaft, and check to be sure all blades are set at the same angle of attack.

23.3.8 INTERNALLY ACTUATED CONTROLLABLE-PITCH FANS

Internally actuated fans (Fig. 23.3.7) are equipped with a positioner on the outside of the fan casing. They also have a high-pressure air line supplying air to a diaphragm through a rotating union. Inflating and deflating the diaphragm causes the pitch control mechanism to move. The pitch control mechanism also has a feedback cable to the positioner to tell the

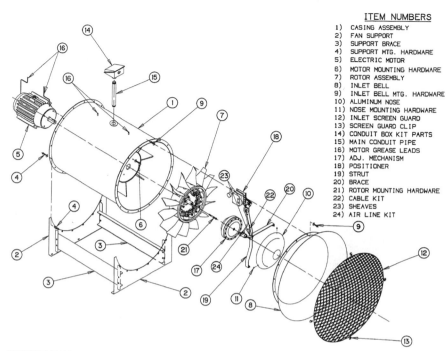

ITEM NUMBERS

1) CASING ASSEMBLY
2) FAN SUPPORT
3) SUPPORT BRACE
4) SUPPORT MTG. HARDWARE
5) ELECTRIC MOTOR
6) MOTOR MOUNTING HARDWARE
7) ROTOR ASSEMBLY
8) INLET BELL
9) INLET BELL MTG. HARDWARE
10) ALUMINUM NOSE
11) NOSE MOUNTING HARDWARE
12) INLET SCREEN GUARD
13) SCREEN GUARD CLIP
14) CONDUIT BOX KIT PARTS
15) MAIN CONDUIT PIPE
16) MOTOR GREASE LEADS
17) ADJ. MECHANISM
18) POSITIONER
19) STRUT
20) BRACE
21) ROTOR MOUNTING HARDWARE
22) CABLE KIT
23) SHEAVES
24) AIR LINE KIT

FIGURE 23.3.7   Internally actuated controllable-pitch fan.

positioner how to function. Inflating the diaphragm increases the angle of attack of the blades. A spring apparatus deflates the diaphragm to reduce the blade angle.

Lubrication and cleaning as described in the previous configurations apply to this type of fan except that there is no despin thrust bearing to maintain.

The annual preventive maintenance program should include replacement of the rotating union and feedback cable to positioner. The internal actuator, supply lines, and fittings should be checked for air leaks. Check the positioner calibration and run a vibration analysis. Some fans of this type have radial antifriction thrust bearing where others have teflon rubbing surfaces for each blade. The antifriction bearing design may require complete tear down to clean and relubricate the radial thrust bearings on each blade. As in the previously discussed types of fans, check to see that all blades are at the proper angle of attack.

Before any preventive maintenance program is put into effect, contact the fan manufacturer for specific instructions and assistance to develop a preventive maintenance program for your fans.

CHAPTER 24
VARIABLE-AIR-VOLUME (VAV) SYSTEMS*

James A. Reese
York International Corp., York, Pennsylvania

24.1 SYSTEM DESIGN

24.1.1 Focus: Variable-Air-Volume All-Air Systems

The experiences of most skilled designers indicates that for most buildings a variation on the basic variable-air-volume (VAV) all-air system will yield the best combination of comfort, first cost, and life cycle cost. Therefore, this discussion focuses on the design and selection of equipment for all-air VAV systems and deals lightly with alternative systems. (See also Sec. 19.2, par. 5)

24.1.2 Comfort

With all the written and verbal discussions concerning energy, controls, building automation, environmental impact, etc., we must remember that the original objective of the building's design, and of the heating, ventilating, and air-conditioning (HVAC) system in particular, is *the comfort of the building's occupants*. It is the task of the designer to provide environmental comfort and acceptable indoor air quality (temperature, humidity, ventilation, and noise level) consistent with meeting the specific project requirements; intent of local, state, and federal codes; and first cost, life cycle cost, and energy budgets.

Temperature Control. The design goal for VAV systems is to provide inexpensive temperature control in large numbers of zones by modulating the flow of constant-temperature air in response to a local thermostat.

Comparable temperature control can be achieved at higher first cost and maintenance cost with multiple-fan coil and hydronic heat pump systems and at higher energy requirements due to mixing and/or reheat in multizone, dual-duct, and reheat systems.

---

*All art reproduced in this chapter is courtesy of York International Corp.

Humidity Control.   Precise humidity control requires reheat with constant-volume air flow or a supply air temperature reset with variable volume. However, one of the major advantages of a VAV system is that comfortable humidity levels are easily maintained at part-load conditions—especially important with higher summer interior design temperatures. Outside air intake must be designed and controlled to provide the required amount regardless of the building thermal load. This means the percentage of outdoor to total air will vary considerably and therefore vary the distribution of outside air as thermal load varies.

Air Movement.   In perimeter zones with nonair heating or with VAV heating, loss of air circulation may occur at the changeover point. Reheating or mixing to maintain air flow should be avoided for energy cost reasons. Maintenance of minimum circulation in perimeter zones can be best accomplished by the use of central fan perimeter heating/cooling or heating-only systems or the use of appropriately selected variable-volume fan terminals, which recirculate unheated air from the core of the building at light-load conditions. Recirculated air from other interior areas must conform to indoor air quality codes in existence, such as those based on ASHRAE Ventilation Standard 62. Suggested systems are described in Sec. 24.2.

In interior zones, modern buildings with low lighting loads sometimes require very few cubic feet per minute of primary air per square foot. A properly designed variable-volume system using high induction slot diffusers can handle interior zone loads as low as 0.3 ft^3/(min·ft^2) [0.5 m^3/(h·m^2)] and still provide adequate air circulation in the space. For lower primary air requirements or for systems using non-coanda (surface-effect) slot diffusers (e.g., perforated metal-covered diffusers and light troughers), parallel fan-powered terminals can be used to mix ceiling plenum air with primary air. Additional information on surface effect may be found in the American Society of Heating, Refrigeration, and Air-Conditioning Engineers' (ASHRAE) chapter on air-diffusing equipment.* Surface effect refers to the ability of air to hug the ceiling and not drop. This will provide constant air circulation and predictable diffuser performance in the interior spaces. For diffuser application rules and selection procedures, see diffuser manufacturers' catalogs and/or the ASHRAE handbooks on fundamentals* and equipment.†

For example, in cold climates where the air flow in perimeter spaces required for comfort cooling and distribution of ventilation may approach zero, the interior spaces requiring year-round cooling would likely be cooled with outside air below about 55°F (13°C). Thus the ceiling air for the perimeter heating fan terminal or dual duct perimeter air should be located above the interior spaces where the return air is "unvitiated" (useful outside air content still available).

Ventilation.   Ventilation requirements were diminished considerably with the press for energy conservation. However, concern over indoor air quality has resulted in a reversal of this practice. Lower temperature supply air allows humidity to be maintained at acceptable levels over a wide range of full load and part load conditions. With a VAV system and other systems that are designed for free cooling (up to 100 percent outside air), any percentage of outside air can be easily provided. Ventilation in the interior-cooling-only VAV system can be accomplished without the energy penalty of heating the ventilation air for central air systems serving both perimeter and interior core areas. It has been generally accepted that if the correct amount of ventilation air is provided, regardless of the total amount of air being circulated, it will be distributed adequately, since nearly all buildings require cooling year-round. When this is not the case (or when codes are more specific) special provisions

---

*2001 ASHRAE Handbook, Fundamentals, ASHRAE, Atlanta, GA, 2001.
†2000 ASHRAE Handbook, Systems and Equipment, ASHRAE, Atlanta, GA, 2000.

may be necessary to supply ventilation air directly to the required areas. This may require transfer air from unvitiated spaces (with excess ventilation) or pressure-independent minimum flow controls with reheat to keep zone temperatures under control. For buildings with a wide difference in cooling load between perimeter and interior spaces, use of parallel fan terminals is recommended for interior zones to transfer "unvitiated" return air flowing from over-ventilated spaces on the perimeter. CO_2 monitoring can be used to vary the fan speed of the parallel fan terminal. CO_2 monitoring can also be used to independently control the amount of recirculated and unvitiated return air supplied through the ventilation damper in a split fan dual duct terminal.

Noise. The proper design of air distribution ductwork and the proper selection of fans, terminals, and diffusers are essential in eliminating noise problems in VAV systems. Excessive noise can make a space unusable for its original purpose. This is a common problem in poorly designed VAV systems. A solution is to use computer duct design programs combining static regain duct sizing with fitting selection, VAV unit sizing, and fan and attenuation selection to achieve supply-side noise criteria requirements. See Chap. 10 of this book for noise control.

Indoor Air Quality. (See also Chap. 21). Double wall or external insulation is preferred over internal lining with exposed fiberglass to prevent fiber migration and absorption of dust and moisture. Where ducts and equipment are lined with fiberglass, it should be covered with non-porous lining downstream of dehumidifying coils and all exposed edges and nosings should be coated. Drain pans should be sloped to provide complete drainage and prevent accumulation of stagnant water. Relative humidity should be kept below 60 percent in the space to provide optimum comfort and minimize growth of microbes. Particular attention should be given to part load controls to insure adequate dehumidification. Good filtration is very important to minimize accumulation of dust in ductwork and equipment. Where possible with densely populated space, select filters to remove respirable size particles. Where there is considerable variation in the amount of outside air required for good indoor air quality, several options should be considered to reduce the cost of energy consumption as follows:

1. Use separate systems for interior and perimeter spaces.
2. Use a 2- or 3-duct dual fan system or transfer fans to add air to the critical spaces from areas with an oversupply of ventilation air.
3. CO_2 demand control.

Note: CO_2 sensors can be used to monitor ventilation requirements at partial occupancy. There are many other control options that should be considered to provide optimum air quality and energy efficiency (please refer to Chap. 7 on controls). Good indoor air quality is dependant on proper commissioning and maintenance of the equipment and controls. Proper documentation of system design, start-up, and service requirements is highly recommended and may be required by code.

See Sec. 24.8 of this chapter for a *design checklist for good indoor air quality*.

Control of Downdraft. Control of downdraft is a major consideration in perimeter zones. This draft control can be achieved with all-air systems by the introduction of heating air as follows:

Skin loss greater than 400 Btu/(h·lin ft) (385 W/lin m) (Fig. 24.1). High induction downblow slot diffusers, located over windows, discharging a constant volume of skin heating air.

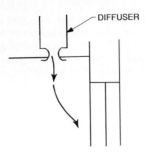

FIGURE 24.1   Skin loss greater than 400 Btu/(h·lin ft) (385 W/lin m). Vertical high-induction slot diffusers located over windows with constant volume of skin heating air. Maximum distance from wall is 12 in (33 cm).

Skin loss of 250 to 400 Btu/(h·lin ft) (240 to 385 W/lin ft) (Fig. 24.2). Slot diffusers next to the outside wall, discharging inward, with a constant or variable volume of skin heating air. Air temperature is reset to approximate the skin load of the building. Variable volume can be used to trim individual zones to minimize energy waste.

Skin loss less than 250 Btu/(h·lin ft) (240 W/lin m). Downdrafts are not experienced in this skin loss range. Therefore, slot diffusers at any location can provide constant-volume or variable-volume heating air.

The current trend toward building skins with lower winter skin heating loss [<250 Btu/(h·lin ft) (240 W/lin m)] is increasing the use of systems which do not separate the heating from the cooling.

24.1.3 Budget

Attention is gradually shifting from concern with first cost only to concern with life cycle cost, or the total cost over a period of owning and operating the building. The design professional now has computer analysis tools available to make intelligent recommendations to the architect and owner:

- To influence the building envelope design (*U* factor, glass, etc.) to minimize total building owning cost and meet energy efficiency requirements.

- To compare various system concepts and equipment selections to choose the best from a comfort and owning-cost standpoint. Important cost factors, trade-offs, and life cycle costing techniques must be well understood and applied intelligently. Computer analysis can do this in the early design stages on a cost-effective basis without demanding an overwhelming and costly engineering effort.

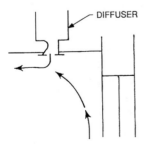

FIGURE 24.2   Skin loss 250 to 400 Btu/(h·lin ft) (240 to 385 W/lin m). Slot diffusers next to outside wall directed inward with constant volume of skin heating air.

It is most important to establish the overall budget properly and early so that inexperienced clients will not expect more than their budgets can provide. First-cost estimates must be developed from experience and will vary from area to area and project to project, depending on the mix and cost of piping, sheet metal, and electrical labor and material.

24.1.4 Energy and Safety Codes

Energy.   Reference 1 relating to air consumption in new buildings offers two alternatives:

1. Follow carefully prescribed dos and don'ts which allow very little originality or trade-offs, particularly where substantial first-cost advantage might be obtained. This is the most straightforward.

2. Demonstrate through computer simulation that the proposed design is as efficient as the design prescribed by the ASHRAE standard. This approach will generally lead to the best system.

Many energy codes follow this pattern. An alternative approach gaining popularity is the use of an energy budget. It permits any design which leads to an annual per-square-foot energy consumption of less than a prescribed amount.

Safety.   There are fire, electrical, and ventilation codes to satisfy. The type of controls, e.g., might be influenced by whether low-voltage control wiring must be enclosed in conduit, whether pneumatic lines must be copper rather than plastic, etc. Fire codes can play an important role in determining the best system, type of controls, and location of air supply units.

24.1.5 Building Specifics

Building Configuration.   Building configuration can have a substantial effect on load, energy consumption, and proper system design. The size, shape, and orientation; equipment room location and space; trunk duct space; and amount of glass may change the best system choice from one type to another.

Space Utilization.   Use of space is another big factor. If the space is owner-occupied and can be prezoned, a system might be used that has limited flexibility but may be very energy-efficient and/or low in first cost. A speculative building, however, needs the combination of flexibility and low tenant completion costs. Special-purpose buildings, such as medical offices, influence choice, as does average zone size. For example, many medical buildings require a system that can provide heating and cooling year-round to all spaces. This rules out cooling-only interior systems or locking out heating in summer.

Ventilation.   Many codes now require that the proper amount of outside air be traceable to every zone in the building under all operating conditions from full load to minimum load. Wide variations in the percentage of outside air required to satisfy ventilation and thermal requirements may favor one system over another or separate systems for various parts of the building. For example the critical zone in a cold climate is the perimeter area in winter and is the interior area in summer. The space requiring the greatest percent of outside ventilation air dictates the requirement for the entire building which may significantly increase energy consumption and operating cost. This can be mitigated by using reheat when the interior space is a relatively small part of the total or with fan terminals or double duct terminals served by a dual fan system to transfer air from over ventilated spaces. Another alternative is separate systems for each major ventilation area.

Zone Size.   Zone size can have quite an effect on system cost. A rectangular building should normally have a minimum of nine zones per floor. The number of additional zones depends on building application and tenant requirements. The finished cost per ft^3/min (m^3/h) of 250 ft^3/min (425 m^3/h) zones is often $2^1/_2$ times the finished cost per ft^3/min (m^3/h) of 2000 ft^3/min zones (3400 m^3/h).

Energy Development and Cost.   The availability and relative cost of different forms of energy (and local expectations of changes in cost and availability) will often determine the type of heating and cooling equipment to be used.

24.1.6   Design Checklist for Energy Efficiency

Given the design suggestions above as well as first-cost and building constraints, the crux of the HVAC system and equipment selection is the energy comparison. A well-written, sophisticated, and flexible computerized energy analysis program is a valuable tool. System energy requirements can be minimized by following this checklist:

1. Take advantage of heat produced by the sun, lights, and people in control of the perimeter heating system.
2. Allow conditioned primary air volume to vary down to minimum supply air ventilation in both interior and perimeter spaces at part-load conditions. Accomplish minimum air circulation in the perimeter with plenum air, provided the minimum required ventilation air requirement is satisfied. Use transfer air when possible to satisfy minimum ventilation requirements.
3. Mixing or reheat should be avoided, if possible. Control systems should be selected so that control miscalibration will not lead to mixing or reheat.
4. If reheat must be used, primary air temperature should be reset to minimize the reheating penalty at part-load conditions. With primary air temperature reset, interior cooling-only terminals and diffusers will have to be oversized.
5. Take return air for perimeter heating from as close to the core of the building as possible, to maximize pickup of heat and unvitiated ventilation from interior loads.
6. Do not condition outside air if it can be avoided. Ventilation air should be distributed to the interior cooling-only zones of the building when codes and building design permit. Keep ventilation air off when the building is unoccupied and during warmup cycles, except as required by code.
7. Optimize the supply air temperature design to minimize system energy consumption as the supply air temperature is reduced, fan energy decreases, compressor energy increases, and outside air load increases.
8. Minimize face velocities on coils and filters as well as terminal and diffuser pressure drop. Use static regain computerized duct designs to reduce fan static pressure and resultant horsepower.
9. Utilize refrigeration, heating, and air-handling controls, which provide maximum free cooling without heating penalty and unload and reset refrigeration and heating equipment for maximum part-load efficiency.
10. Use localized heating with deep night setbacks when the building will experience substantial periods of nonoccupancy. Turn off interior fans during unoccupied periods

unless filtering or ventilation is required to prevent the buildup of air pollutants due to continuous off-gassing.

11. Maximize the efficiency of refrigeration equipment by utilizing low head and high suction pressures. Evaporative-type air-cooled or water-cooled centrifugal or reciprocating equipment is normally the most efficient. Condenser water must be treated to prevent microbial contamination such as Legionella through regular maintenance. Do not forget the energy impact of pumps and other refrigeration auxiliaries, however. For air-cooled machines, consider the effect of dirt on the condenser coils and the resultant degradation of efficiency.

12. Utilize energy transfer and heat recovery whenever possible. Make sure that the net heat saved is more than the energy expended to save it. Heat must be available when it is needed unless storage is provided. If you are considering heat pumps, be sure to analyze the coincidence of the amount and availability of heat when required (include the effect and cost of storage, if necessary). Pay particular attention to the effects of diversity (requirement for instantaneous peak capacity) in determining installed tonnage, cubic feet per minute (cubic meters per hour), and connected load.

13. If thermal storage is used, take advantage of the opportunity to lower the supply air cooling temperature and reduce the size and operating cost of the air distribution system.

14. Use central building automation to schedule equipment operation, provide optimized summer and winter early startup, control equipment performance at maximum efficiency, and control electrical demand.

24.1.7 Use an Energy Analysis Program

Utilizing an energy analysis program enables the professional designer to make quick and accurate estimates of the annual energy consumption of each of the HVAC systems being considered. Trade-offs can then be made and systems compared. For example, a very energy-efficient system may cost more than a less efficient system but have a desirably short payback on its cost premium. Maintenance difficulty should also be reviewed. First-cost or energy cost savings can be negated by a system that is difficult and costly to maintain.

24.2 TYPICAL SYSTEM DESIGNS

To assist the designer in choosing the best system type, control method, and equipment, various possibilities are described and illustrated in the following pages. Each of these systems can be simulated on computerized energy analysis programs and designed with the assistance of a computerized duct design program. Financial analysis is also available through computerized programs.

24.2.1 Series Fan Terminal, Skin VAV Cooling Only-Interior (Fig. 24.3)

Advantages

1. Low first cost with electric heat.
2. Continuous air circulation at all times with excellent downdraft control.
3. Minimal simultaneous heating and cooling, if properly controlled.

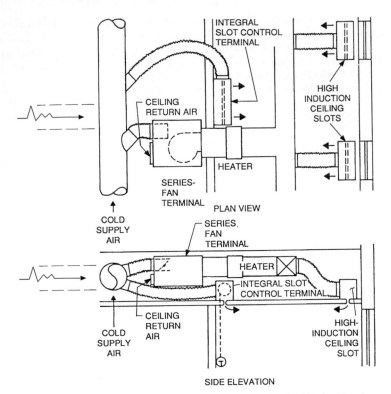

FIGURE 24.3   Series fan terminal skin VAV cooling only—interior. Skin load less than 400 Btu/(h·lin ft) (385 W/lin m). (Above these limits use a downblow diffuser above windows.) *Note:* Make runouts straight for four diameters upstream from box inlet. The inlet for skin heating unit should be located in the interior space where unvitiated air can be transferred when required for ventilation purposes to the perimeter. Relocate unit or extend return air duct as shown.

4. Possibility of free heat and ventilation air transfer from interior to perimeter spaces, if fan terminal units are located near central core.

5. Excellent way to use VAV primary air and maintain any design minimum air-flow circulation rate desired in perimeter spaces.

Application Guidelines.   Install fan terminal units as close as possible to the center core, to minimize wiring or hot-water piping, to transfer unvitiated air from interior spaces, and to reduce noise in occupied spaces.

The System.   Fan terminal units provide constant air circulation year-round. Cold primary air is varied, and the unit fan pulls in return air to make up the difference. Heating is not energized until the fan terminal's cooling damper is closed or reaches a minimum air flow, if required for ventilation codes.

   Fan terminal units handle the skin heating and cooling load and provide downdraft protection where needed. A minimum of one unit per building face per floor is recommended to minimize simultaneous heating and cooling in perimeter spaces in winter. If the lineal wall exceeds 100 ft (30.5 m), additional units should be considered.

Where the heating transmission load exceeds 400 Btu/(h · lin ft) (385 W/lin m) including infiltration, downblow induction diffusers should be used to avoid downdraft. When this load is less than 400 Btu/(h · lin ft) (385 W/lin m), a ceiling high-induction slot diffuser, located at, but blowing away from, the outside wall, can be used. When this load is below 250 Btu/(h · lin ft) (240 W/lin m), any kind of diffuser is satisfactory.

Lights, people, and equipment loads in interior spaces and lights, people, and solar loads in perimeter spaces can be handled for zones up to approximately 400 ft³/min (680 m³/h) by integral diffuser control terminals. Conventional single-duct control terminals supplying one or more ceiling slot diffusers can be used for larger zones. Fan terminal units can be employed to increase air circulation in very low-load interior spaces and provide the required ventilation air.

Variable-volume constant-temperature cold air is used as the primary air for this system.

An efficient heat pump system can be made by using a central double-bundle condenser and hot-water heating coils in the fan terminal units. Thermal storage can also be used with this system but thermal insulation and isolation must be provided if the resulting supply air temperature is below 50°F (10°C).

Controls. Pneumatic, electric, or electronic or direct digital controls (DDC) can be used by selecting a particular zone or using the outdoor or exterior wall temperature to control the series-type skin fan terminal unit. A wide variety of optional velocity limits and dead-band adjustments are available. Networked pressure independent controls can maintain minimum air flows to meet IAQ requirements as well as provide communication to the air-handling unit and maintain duct static pressure at the minimum level.

VAV terminals in perimeter spaces should be interlocked with the fan terminal unit serving the same spaces, and the control sequencing should be arranged to minimize simultaneous heating and cooling.

Central readout and reset of zone temperatures and central night setback with an electronic building automation system should be considered. Night temperature control is obtained by intermittent operation of the fan terminal and its heater without operating the central air-handling unit.

24.2.2 Series Fan Terminal Perimeter, VAV Interior (Figs. 24.4 and 24.5)

Advantages

1. Minimum installed ductwork and low first cost in buildings with large zone sizes.
2. Continuous air circulation at all times with excellent downdraft control.
3. No danger of simultaneous heating and cooling.
4. Low first cost with electric heat.

The System. Series fan terminal units provide constant total air circulation year-round for perimeter spaces and handle the entire cooling/heating load. Primary cold air is varied, and the unit fan pulls in return air to make up the difference. Heating is not energized until the unit cooling damper is closed.

Use one fan terminal unit per zone and select the unit for any ratio of primary to total air less than unity. (Where downdraft may be a problem, high-induction downblow diffusers can be used to blanket the wall with ceiling slot diffusers for the remainder of the air.) Locate perimeter terminal units over interior spaces or extend duct intake to the fan terminal from interior space to provide minimum ventilation in winter from unvitiated interior air.

Where the heating transmission load exceeds 400 Btu/(h·lin ft) (385 W/lin m) including infiltration, high-induction downblow diffusers should be used to avoid downdraft. When this

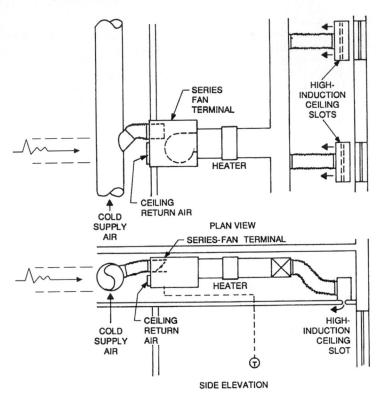

FIGURE 24.4   Series fan terminal perimeter. Skin load less than 400 Btu/(h·lin ft) (385 W/lin m). (Above these limits use a downblow diffuser above windows.) *Notes:* Make runouts straight for four diameters upstream from box inlet. The inlet for skin heating unit should be located in the interior space where unvitiated air can be transferred when required for ventilation purposes to the perimeter. Relocate unit or extend return air duct as shown.

load is less than 400 Btu/(h · lin ft) (385 W/lin m), a ceiling slot diffuser located at, but blowing away from, the outside wall can be used. When this load is below 250 Btu/(h·lin ft) (240 W/lin m), downdraft is not a problem.

Lights and people loads in interior spaces may be handled by integral diffuser control terminals for zones up to approximately 400 ft³/min (680 m³/h) and control terminals with one or more ceiling slot diffusers for larger zones parallel. Fan terminal units can be employed to increase air circulation in very low-load interior spaces to meet IAQ requirements.

Variable-volume constant-temperature cold air is used as the primary air for this system.

This arrangement can also be used for interior spaces such as conference rooms to increase air circulation rates, to transfer unvitiated air from adjacent spaces, and/or to provide local air filtering in conjunction with VAV primary air control (Fig. 24.5). It can also be used with low-temperature VAV primary air to achieve any desired supply air temperature to the room. Caution should be exercised to ensure that modifications are made in the terminal design to accommodate very low-temperature supply air.

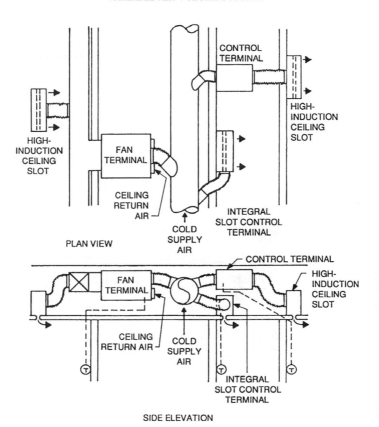

PLAN VIEW

SIDE ELEVATION

FIGURE 24.5 Fan terminal for interior space. *Note:* Make runouts straight for four diameters upstream from box inlet.

Controls. Controls can be pneumatic, electric, or electronic or direct digital control (DDC) with a wide variety of optional velocity limits and dead-band adjustments.

Central readout and reset of zone temperatures and central night setback are available. Night temperature control is obtained by intermittent operation of fan terminals and their heaters without operating the central air handling unit.

24.2.3 Parallel Fan Terminal Perimeter, VAV Interior (Fig. 24.6)

Advantages

1. Minimum installed ductwork and low first cost in buildings with large zone sizes.

2. No danger of simultaneous heating and cooling.

3. Minimum fan energy because smaller fan terminal fans operate only upon a call for heat.

Application Guidelines. This is not recommended for applications with a high winter heating load because of low diffuser velocity in the heating season and marginal downdraft control.

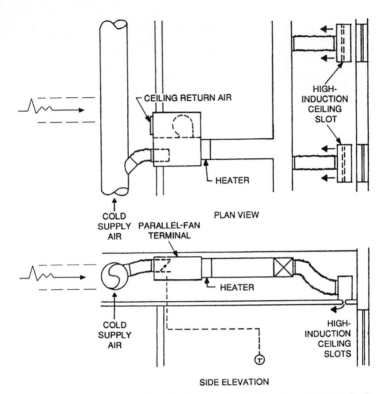

FIGURE 24.6   Parallel-fan terminal perimeter. Skin load less than 400 Btu/(h · lin ft) (385 W/lin m). *Notes:* Make runouts straight for four diameters upstream from box inlet. The inlet for skin heating unit should be located in the interior space where unvitiated air can be transferred when required for ventilation purposes to the perimeter. Relocate unit or extend return air duct as shown.

However, operating the system so the total air quantity (primary air plus transfer air) is relatively constant will mitigate this problem. Fan terminal units should be located over interior spaces or extend return air duct intake to capture unvitiated interior air.

The System.   The parallel fan terminal unit has a unit-mounted fan installed in parallel with a primary air damper. In the cooling season, the unit fan is off and cold primary air bypasses it. Heating is energized only after the primary air damper is closed and the unit fan is up to full speed.

The parallel fan terminal perimeter system is best suited for electric heat applications for medium-size and large zones. One unit can meet the entire heating and cooling requirement if the winter heating load is relatively light. This system is not recommended where winter transmission load plus infiltration exceeds 400 Btu/(h · lin ft) (385 W/lin m). When this load is between 250 and 400 Btu/(h · lin ft) (240 and 385 W/lin m), it can be employed with a ceiling high-induction slot diffuser located at, but blowing away from, the outside wall, provided that the total air quality is held nearly constant throughout the year (select design heating cfm equal to design cooling cfm in perimeter spaces). When this load is below 250 Btu/(h · lin ft) (240 W/lin m), any diffuser and cooling-heating air-flow ratio are satisfactory.

Lights and people loads in interior spaces are handled by single integral slot control terminal units for zones up to approximately 400 ft³/min (680 m³/h) and by control terminals with one or more ceiling slot diffusers for larger zones. Parallel fan terminals can be employed to increase air circulation and/or minimum ventilation in very low-load interior spaces.

Variable-volume constant-temperature cold air is used as the primary air for this system.

Controls. Controls can be pneumatic, electric, or DDC, with a wide variety of optional velocity limits and dead-band adjustments.

Central readout and reset of zone temperatures and central night setback are available. Night temperature control is obtained by intermittent operation of fan terminal fans and heaters.

24.2.4 Dual-Fan Variable-Volume, Double-Duct Perimeter, VAV, or Double Duct Interior (Fig. 24.7)

Advantages

1. Low first cost when start-up occupancy is low because both heating and cooling means can be installed as the tenant finish is completed.

2. Minimum piping with hot-water heating.

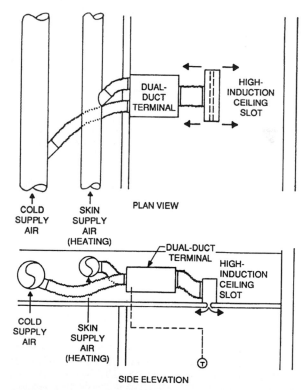

FIGURE 24.7 Variable-volume double-duct perimeter. Skin load less than 250 Btu/(h·lin ft) (240 W/lin m). Single box and diffuser for both systems. *Note:* Make runouts straight for four diameters upstream from box inlet.

3. Simultaneous cooling and heating when required.

4. Reduced fan energy at light-load conditions.

5. Heating duct can be used for providing unheated transfer air to interior spaces in summer and to perimeter spaces in winter to meet minimum ventilation requirements.

Application Guidelines. First, the system must be properly designed and controlled to enjoy its advantages. This system is not recommended for applications with a high winter heating load because of low diffuser velocity in the heating season and poor downdraft control. It can be improved by using split diffusers to maintain higher slot velocity in winter.

The System. A skin heating-only or heating/cooling variable-volume supply air-handling unit and a perimeter and interior space cooling-only variable-volume supply air-handling unit both feed common perimeter double-duct terminal units. Diffusers, boxes, and ductwork do not have to be installed in initially unoccupied spaces.

To minimize reheating of cold "economizer" air in the winter, two separate supply fans are used and outdoor air from economizer dampers is provided only on the interior cooling fan. Both fans must have inlet vane controls or variable-speed controls and properly located duct static pressure sensors. The supply duct pressure sensor should be located as far away from the fan as possible, to minimize overpressurization of ducts but still provide the required pressure to all ducts. Care must be taken in parallel duct systems where the solution involves using multiple sensors controlled to satisfy the one with the greater need. Resetting of the hot duct temperature minimizes hot duct pressure variation and increases air circulation in perimeter spaces at light-load conditions.

The variable-volume two-fan perimeter system is not recommended where the winter transmission load plus infiltration exceeds 400 Btu/(h·lin ft) (385 W/lin m). When this load is between 250 and 400 Btu/(h·lin ft) (240 and 385 W/lin m), it can be employed with split diffusers (heating separate from cooling to maintain high slot velocity at full heating) located at, but blowing away from, the outside wall, provided that the heating diffuser section is properly matched to the heating transmission plus infiltration air flow and that the hot air temperature is reset. When this load is below 250 Btu/(h·lin ft) (240 W/lin m), any high-induction slot diffuser and cooling/heating air-flow ratio are satisfactory.

Lights and people loads in interior spaces are handled by single integral diffuser control terminal units for zones up to approximately 400 ft³/min (680 m³/h) and by control terminal units with one or more high-induction ceiling slot diffusers for larger zones. Series fan terminal units or double duct boxes can be employed to increase air circulation and minimum ventilation requirements in very low-load interior spaces.

Variable-volume constant-temperature cold air is used to offset the interior space lights and people loads in both interior and perimeter spaces. Variable-volume reset temperature air is used for the skin system handling heating transmission and infiltration loads. Cooling transmission and infiltration and solar loads may be assigned to either air distribution system; both ducts can provide cooling in summer to reduce the size of the cooling only duct.

An efficient central heat pump system can be made by using a double-bundle condenser and hot-water heating coils for the skin air distribution system. Thermal storage can be included.

Controls. Controls can be pneumatic, electric, or electronic, or direct digital control (DDC) with a wide variety of optional velocity limits and dead-band adjustments.

The simplest controls are obtained with zero minimum air flow, a fixed small dead band, and heating-only skin system air distribution. This forces the interior air distribution system to handle all cooling loads. This works well in buildings where minimum heating is required and the space requires cooling year-round except for morning warmup.

If the skin air system also has a cooling capability, interior system duct sizes are reduced, but additional controls are required to accomplish changeover.

If a relatively high minimum air quantity is required, the inlet velocity on one or both heating and cooling inlets must be measured and controlled, to avoid the possibility of prohibitive energy consumption. In such cases, it is important to reset at least the hot duct temperature to the greatest possible extent.

Central readout and reset of zone temperatures and central night setback are available. Winter temperature control is obtained by operation of the skin air system only with appropriate temperature resetting.

24.2.5 Central Constant-Volume Skin System, VAV Interior Cooling Only (Fig. 24.8)

Advantages

1. Maximum zone relocation flexibility and relatively low first cost in buildings with many small zones and high occupancy at startup.
2. Minimum piping with hot-water heating.
3. Continuous air circulation at all times with excellent downdraft control.

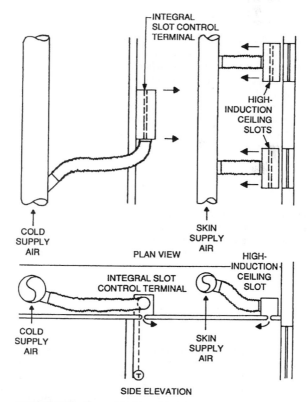

FIGURE 24.8 Constant-volume skin VAV cooling only—interior. Skin load less than 400 Btu/(h·lin ft) (385 W/lin m) (Above these limits use a downblow diffuser above windows.) *Note:* Make runouts straight for four diameters upstream from box inlet.

4. Quiet, reliable, and easy to maintain.

5. Heating duct can be used for providing unheated transfer air to interior spaces in summer and to perimeter spaces in winter to meet minimum ventilation requirements.

Application Guidelines.   Heating reset schedule should be set carefully to avoid unnecessary winter heating.

The System.   A skin heating-only or heating/cooling constant-volume supply unit handles the heating or heating and cooling skin transmission and infiltration load. A cooling-only variable-volume supply unit handles all lights, people, and solar loads, and may also handle the skin cooling transmission and infiltration load.

To minimize reheating of cold economizer air in the winter, economizer dampers are put only on the interior variable-volume supply fan. The constant-volume perimeter system maintains air circulation and ventilation with unvitiated air in perimeter spaces even when variable-volume boxes serving these spaces are closed.

Where the heating transmission load exceeds 400 Btu/(h · lin ft) (385 W/lin m) including infiltration, high-induction downblow slot diffusers should be used to avoid downdraft. When this load is less than 400 Btu/(h · lin ft) (385 W/lin m), a high-induction ceiling slot diffuser located at, but blowing away from, the outside wall can be used. When this load is below 250 Btu/(h · lin ft) (240 W/lin m), any kind of diffuser is satisfactory.

Lights and people loads in interior spaces and lights, people, and solar loads in perimeter spaces are handled by single integral slot control terminal units for zones up to approximately 400 ft^3/min (680 m^3/h) and control terminal units with one or more high-induction ceiling slot diffusers for larger zones. Series fan terminals and parallel fan terminal boxes can be employed to increase air circulation and minimum ventilation requirements in very low-load interior spaces.

When the skin is heating-only, a very efficient central heat pump system can be made by using an auxiliary air-cooled condenser in the main refrigeration circuit as the heating coil for the skin system. When the skin system is a heating/cooling system, a double-bundle condenser and hot-water heating coil are employed. Thermal storage can be included in either case.

Controls.   VAV box controls can be pneumatic, self-powered, electric, or electronic or DDC with velocity-resetting options. The skin system is reset off the outside temperature to handle just the heating transmission plus infiltration load in winter and cooling transmission plus infiltration in summer, if cooling capability is included.

Heating energy is conserved by setting the heating reset schedule to maintain a relatively low inside temperature when lights are off. Heat from lights then brings the winter temperatures up to a comfortable level.

Central readout and reset of zone temperatures and central night setback are available. Night temperature control is obtained by intermittent or slow-speed operation of the skin fan only with appropriate temperature resetting.

24.2.6 Furnace Fan Heating-Only Skin, VAV Cooling Only, Interior (Fig. 24.9)

Advantages

1. Low first cost with electric heat.

2. Good downdraft control.

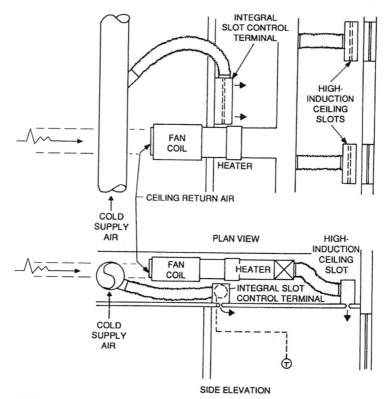

INTEGRAL
SLOT CONTROL
TERMINAL

HIGH-
INDUCTION
CEILING
SLOTS

FAN
COIL

HEATER

CEILING RETURN AIR

COLD
SUPPLY
AIR

PLAN VIEW

HIGH-
INDUCTION
CEILING
SLOT

FAN
COIL

HEATER

INTEGRAL SLOT
CONTROL TERMINAL

COLD
SUPPLY
AIR

SIDE ELEVATION

FIGURE 24.9 Fan-coil heating only, skin VAV cooling only—interior. Skin load less than 400 Btu/(h · lin ft) (385 W/lin m). (Above these limits use a downblow diffuser above windows.) *Notes:* Make runouts straight for four diameters upstream from box inlet. The inlet for skin heating unit should be located in the interior space where unvitiated air can be transferred when required for ventilation purposes to the perimeter. Relocate unit or extend return air duct as shown.

3. Minimal simultaneous heating and cooling, if properly controlled.

4. Reduced fan energy because smaller fans operate only on a call for heat.

5. Possibility of free heat transfer to perimeter spaces if heating fan units are located near central core.

Application Guidelines. Install furnace fan heating units as close as possible to the central core to minimize wiring or hot-water piping and reduce noise in occupied spaces, and transfer unvitiated air from interior spaces for ventilation.

The System. Separate heating-only furnace heating fan coils per building face or per face per floor supply for constant-volume skin systems. With only one central cooling-only VAV system, this arrangement permits zoning of the skin to take advantage of solar and internally generated heat.

Where heating transmission load exceeds 400 Btu/(h · lin ft) (385 W/lin m) including infiltration, high-induction downblow slot diffusers should be used to avoid downdraft.

When this load is less than 400 Btu/(h · lin ft) (385 W/lin m), a high-induction ceiling slot diffuser located at, but blowing away from, the outside wall can be used. When this load is less than 250 Btu/(h · lin ft) (240 W/lin m), any kind of diffuser is satisfactory.

All cooling loads in perimeter and interior spaces are handled by single integral slot control terminal units for zones up to 400 ft³/min (680 m³/h) and by control terminal units with one or more high-induction ceiling slot diffusers for larger zones. Parallel fan terminal boxes can be employed to increase air circulation in very low-load interior spaces.

Variable-volume constant-temperature air is used as the primary cooling air for this system. Heating air is at constant volume during the heating season with reset temperatures.

An efficient heat pump system can be made by using a central double-bundle condenser and hot-water heating coils in the fan coil units. Thermal storage can be included.

Controls.   Pneumatic, self-powered, electric, electronic controls, or DDC can be used by selecting a particular control zone or using the outdoor or exterior wall temperature to control the heating unit. A wide variety of optional velocity limits and dead-band adjustments are available.

With electronic controls there is gradual fan turn-on, and each fan coil can be interlocked with one or more of the perimeter cooling-only VAV boxes. The heating fans are off during the cooling season. In midseason and in winter they operate to provide minimum air circulation and/or minimum ventilation and heating.

Central readout and reset of zone temperatures and central night setback are available. Night temperature control is accomplished by intermittent operation of furnace heating fans and heaters.

24.2.7   VAV Reheat Perimeter (with Cooling or Reheat Interior) (Fig. 24.10)

Advantages.   Minimum installed ductwork and low first cost are the benefits of this system; reheat can be used to increase air circulation and/or minimum ventilation requirements.

Application Guidelines.   The system must be properly designed, controlled, and restricted to light-load heating applications to avoid substantial energy waste. It is not recommended for applications with high winter heating load because of low diffuser velocity in the heating season, high reheat energy penalty, and marginal downdraft control.

The System.   Perimeter spaces are served by a variable-volume terminal unit which is controlled to maintain a certain minimum air flow when a downstream heating coil is on.

A VAV reheat system is not recommended where winter transmission load plus infiltration exceeds 250 Btu/(h · lin ft) (240 W/lin m). When the load is below this value, any diffuser is satisfactory.

Lights and people loads in interior spaces are handled by single induction slot control terminal units for zones up to approximately 400 ft³/min (680 m³/h) and single-duct terminal units with one or more high-induction ceiling slot diffusers for larger zones. Parallel fan terminal boxes can be employed to increase air circulation and/or minimum ventilation requirements in very low-load interior spaces.

Variable-volume temperature-reset cold air is recommended for this system. To be able to achieve substantial upward resetting of the cold supply air temperature without losing control of interior zones in winter, all boxes and diffusers serving pure interior spaces should be oversized. The amount of oversize depends on the ratio of interior to perimeter space,

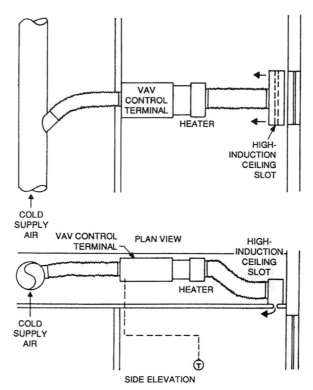

FIGURE 24.10 VAV reheat perimeter. Skin load less than 250 Btu/(h·lin ft) (240 W/lin m). *Note:* Make runouts straight for four diameters upstream from box inlet.

with the greatest amount of oversizing used where the interior is small compared with the perimeter.

Controls. Controls can be pneumatic, electric, or electronic or DDC with a wide variety of optional velocity limits and dead-band adjustments. Unless the duct static pressure is well controlled and uniform throughout the building, special controls are needed to ensure desired air flow while reheat is maintained at the design value for reheating. Duct static pressure control is discussed in Sec. 24.5. If this value is relatively low and the cold air temperature resetting is substantial, the energy wasted in reheating can be kept within reasonable bounds. Humidity control should be considered to over-ride the reset temperature schedule if relative humidity exceeds desired maximum.

Electronic controls enjoy a special advantage in that they permit the air flow to decrease to a cooling minimum air flow which is below the heating minimum air flow. This avoids overcooling and subsequent reheating at times when the net space load is very small. In many mild-climate applications, this additional feature can almost completely eliminate the energy waste associated with reheating.

Central readout and reset of zone temperatures and central night setback are available. Night temperature control is obtained by intermittent operation of the main fan.

24.2.8   Baseboard Heating, Skin VAV Cooling-Only Interior (Fig. 24.11)

Advantages

1. Maximum zone relocation flexibility and relatively low first cost in buildings with many small zones and high occupancy at startup.
2. Good downdraft control when excess heat is provided.
3. Quiet operation.

Application Guidelines.   To maintain perimeter space ventilation in winter, it is usually necessary to overheat with the baseboard system and thereby force the perimeter space cooling boxes to recool with outside air. Alternatively, VAV boxes can be provided with a minimum air setting and the baseboard controlled to maintain space temperature, but, with the exception of systems already utilizing DDC control, this generally has higher first cost.

The System.   An electric or hot-water baseboard skin radiation heating system handles the heating skin transmission and infiltration load. A cooling-only variable-volume supply unit handles all cooling and ventilation air loads. The cooling system often includes an outside air economizer.

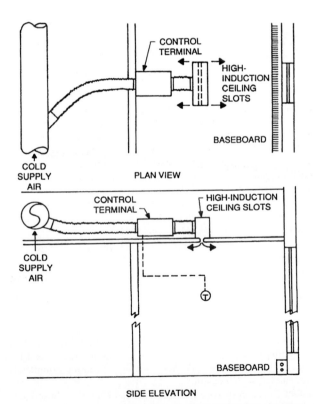

FIGURE 24.11   Baseboard heating, skin VAV cooling only—interior.
Note: Make runouts straight for four diameters upstream from box inlet.

All cooling loads are handled by single-type integral slot control terminal units for zones up to approximately 400 ft³/min (680 m³/h) and by control terminal units with one or more high-induction ceiling slot diffusers for larger zones. Parallel fan terminal boxes can be employed to increase air circulation and/or minimum ventilation requirements in very low-load interior spaces.

Variable-volume constant-temperature cold air is used for the air distribution system.

Controls. VAV box controls can be pneumatic, self-powered, electric, or electronic, or DDC, with velocity resetting options. A hot-water baseboard skin heating system is usually reset from the outside air temperature to provide some overheating for the sake of ventilation. Electric baseboard systems are usually broken into relatively small zones with each zone sequenced to come on as the nearest cooling box approaches some minimum position.

Central readout and reset of zone temperatures and central night setback are available. Night temperature control is accomplished by resetting the hot-water temperature or by intermittent operation of electric baseboard heaters.

24.2.9 System Fans

(See Chap. 23 for additional information on fans.)
Three factors can cause unsatisfactory operation in fans:

1. Surge, which can occur in all fan types.
2. Paralleling, which may result from multiple fans with a characteristic dip in the curve, common with forward-curved (FC) and vane axial types.
3. Resonance, which is a beat frequency produced by multiple fans operating at slightly different speeds.

Fan surge (Fig. 24.12), as it is normally referred to in the industry, is a result of stall as air passes over the fan blades. This produces a static pressure and noise level fluctuation. The magnitude is on the order of 10 percent of block-tight static pressure for an airfoil centrifugal fan. While this pulsation is typically less for FC and vane axial fans, it still results in unsatisfactory operation in any type of fan. Do not choose a fan or allow a fan to operate in this area of its performance curve. Most manufacturers allow some margin of safety between their catalog cutoff limit and the point where tests indicate that surge begins.

With multiple fans (of the same zone) having a dip in their curve, paralleling can occur (Fig. 24.13). This is caused by one fan operating on one side of its peak and the other fan operating on the opposite side of the peak. This results in unstable operation and static pulsation,

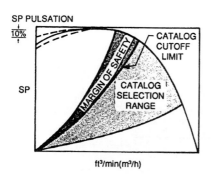

FIGURE 24.12 Fan surge.

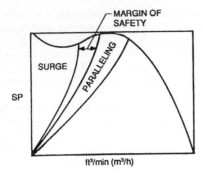

FIGURE 24.13 Multiple fans, paralleling and resonance.

increased horsepower, and increased noise level. The symptoms of paralleling are the same as those for surge, and with multiple fans of this type, it is often difficult to tell which is the cause. It is obvious that a fan cannot be allowed, and should not be selected, to operate under either of these conditions. Resonance may also result from multiple fans. The solution is to operate them at the exact same speed or speeds far enough apart to avoid a beat frequency.

24.3 FAN MODULATION METHODS

The following types of fan capacity modulation can be used:

- Static pressure modulation with inlet dampers and discharge dampers
- Scroll volume
- Runaround
- Speed variation
- Controllable pitch
- Inlet vanes

Inlet dampers and discharge dampers provide load modulation by adding resistance to the system, either upstream or downstream of the fan. Figure 24.14 shows the characteristic

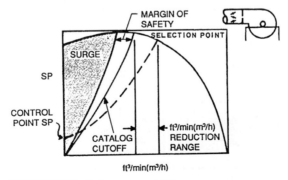

FIGURE 24.14   Static pressure control.

curves of an FC fan with the catalog cutoff limit and surge range indicated. A variable-volume system will require air volume and static pressure as indicated by the dashed line between the selection point and control point. As resistance is added to the system through dampers, the operating point moves up on the fan curve and the load is reduced. Note that the load reduction is limited to where the fan reaches the catalog cutoff limit because of fan surge. This modulation method is used widely on single FC fans in systems requiring relatively low static pressure.

Scroll volume control (Fig. 24.15) is achieved in FC fans by attaching a flat sheet inside the top of the fan housing, which in effect changes the scroll's shape. The result is a variety of performance curves, which will result in a capacity reduction. However, there is little or no horsepower savings and it does not work on BI or AF types. Its primary application is on low-pressure class 1 FC fans or to correct paralleling on multiple FC fans.

The next method used is the runaround (Fig. 24.16). This is a bypass duct which diverts some of the air back from the discharge of the fan to the inlet and prevents it from operating in the unstable range. Keep in mind that while the runaround prevents unstable fan operation, the duct static pressure is not being reduced and an additional set of dampers downstream from the runaround should be used to prevent excessive duct static pressure. Because it works well and is relatively inexpensive, the runaround method is frequently used to correct an existing problem, although there is little horsepower savings and it is rather space-consuming.

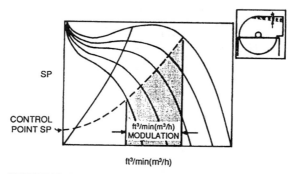

FIGURE 24.15 Scroll volume-control.

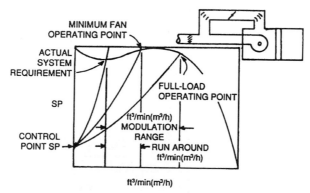

FIGURE 24.16 Runaround bypass.

Speed variation (Fig. 24.17) is another means of controlling fan capacity. It is available through fluid or mechanical speed reducers or solid-state voltage rectifiers. This affords the greatest opportunity for horsepower reduction, providing the power consumed by the speed reducer does not offset this reduction. With careful system design and fan selection, speed variation with direct drive coupled motors is becoming more popular. Sound level is also reduced with this method of control. At present it is normally higher in first cost than other options and must be justified on its operating-cost savings. However, the cost has come down enough to be typically justified within acceptable payback periods. It is also possible to use multispeed motors which will provide two or more steps of reduction. However, the added first cost, plus control complexity, is the main reason for its not being more popular.

Controllable-pitch axial fans (Fig. 24.18) achieve capacity reduction by changing the blade pitch. This produces a family of fan curves which provide capacity reduction. While this method of control may achieve horsepower reduction approaching that of variable speeds, the sound level is not necessarily reduced and may, in fact, be increased. This, coupled with the high sound level generated by this type of fan, requires considerable attention to acoustical treatment downstream of the fan. It is also high in first cost when attenuation requirements are considered.

Inlet vanes (Fig. 24.19) are readily available from almost all manufacturers of centrifugal fans, except in smaller sizes. They achieve speed reduction by creating spin in the direction of rotation of the fan, which reduces the load, static pressure, and brake horsepower (bhp). This spin changes the surge characteristics of the fan slightly, as indicated by the bump in the

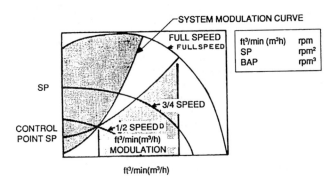

FIGURE 24.17   Speed control.

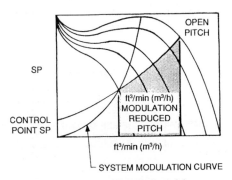

FIGURE 24.18   Controllable pitch—axial.

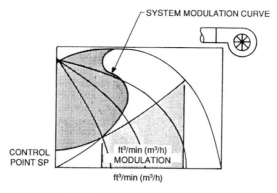

CONTROL POINT SP

- REDUCED HP
- LOW FIRST COST
- INCREASED SOUND LEVEL

FIGURE 24.19   Inlet vanes.

surge curve. However, inlet vanes do not eliminate surge, as is sometimes assumed. Horsepower reductions are achieved, although not to the extent of variable-speed or control-lable-pitch fans. Since the centrifugal fan is typically more efficient than vane axial fans, the centrifugal fan with inlet vanes frequently requires lower horsepower than the controllable-pitch vane axial. Inlet vanes may increase the sound level of centrifugal fans on the order of 5 dB, although centrifugal fan sound levels are quite low for the duty performed. Inlet vanes are relatively low in first cost, typically ranging from 20 to 50 percent of the cost of the fan, depending on fan size. Inlet guide vanes are very non-linear control devices in that flow reduction is not linear with blade rotation. This may cause control instability unless "pro-portional plus integral" (PIP) control is used.

Variable-volume systems without fan control are not recommended. Even though the fan can be undersized to prevent surge at the minimum load requirement, the excess pressure buildup increases noise and leakage problems.

24.4   FAN DEVIATION FROM CATALOG RATINGS

Fans are tested, rated, and cataloged under more or less ideal conditions. The standard Air Moving and Conditioning Association (AMCA) test procedure used is shown in Fig. 24.20. Note that 10 diameters of duct are used and performance is measured $8\frac{1}{2}$ diameters from the discharge. A flow straightener is also used.

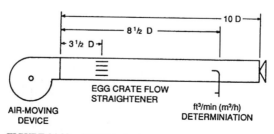

FIGURE 24.20   AMCA test setup.

Most fans do not have a uniform discharge velocity profile (see Fig. 24.21). Part of the high-velocity energy is recaptured as static regain and is included in catalog ratings. However, when a minimum of five fan diameters of duct is not provided in a given application, such as a blowthrough installation, this energy is lost and cataloged ratings must be adjusted accordingly. The typical static regain included in manufacturers' ratings which must be compensated for when not installed in the same manner as rated, is shown in Fig. 24.22. For example, on blowthrough applications, multiply the factor by the outlet velocity pressure and add it to the static pressure required to make the fan selection.

Be sure to consider these factors in any fan selection:

1. Whether full static regain is achieved.
2. Plenum wall within three-fourths wheel diameter of inlet.
3. Effect of drive sheaves and belt guards close to fan inlet.
4. Inlet spin.
5. Nonuniform air flow to inlet.
6. Inlet vane resistance.

24.4.1   Example of Static Regain Effect

To illustrate the effect of static regain and how to compensate for its absence in fan selection, consider the following example (Fig. 24.23). (*Note:* For metric equivalents, BHP × 0.746 = kW, FPM (or ft/min) × 0.0058 = m/s, and inWG × 249 = Pa.)

This is required at entrance to supply duct system: 2830 ft/min velocity, 3 inWG static pressure (all at discharge). Assume a 3 percent speed increase and 13 percent horsepower increase for inlet vanes, and a 2 percent speed increase and 6 percent horsepower increase for drive losses.

Case 1.   Ducted discharge is same size as fan outlet (duct exceeds eight fan diameters). The fan is applied in same manner tested, so catalog ratings may be used to determine the required speed. From the catalog, RPM (or r/min) = 900.

$$\text{RPM required} = \text{catalog RPM}$$
$$\times \begin{pmatrix} \text{assumed increase} \\ \text{due to inlet vanes} \end{pmatrix} \times \begin{pmatrix} \text{assumed increase due} \\ \text{to drive losses} \end{pmatrix}$$
$$= 900 \times 1.03 \times 1.02 = 946$$

TYPICAL CENTRIFUGAL FAN VELOCITY PROFILE

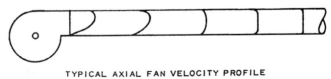

TYPICAL AXIAL FAN VELOCITY PROFILE

FIGURE 24.21   Fan velocity profiles.

FAN TO CLOSE
TO WALL

WITHIN 3/WHEEL DIAMETER - NO EFFECT
WITHIN 1/2 WHEEL DIAMETER - ADD 0.5 VP TO SP
WITHIN 1/3 WHEEL DIAMETER - ADD 1.5 VP TO SP

BLOW THROUGH
APPLICATION

ADD THE FOLLOWING TO CATALOG SP:
FC   1.8 X VP (OUTLET VELOCITY PRESSURE)
BI   1.5 X VP
AF   1.0 X VP
VANEAXIAL   1.6 X VP
TUBULAR CENTRIFUGAL 2.4 X VP

RIGHT ANGLE TURN
INTO FAN

ADD 1.0 X VP TO SP WHEN D IS LESS THAN 1 DIA.

1 DIA. MIN.
3 DIA. REC.

DUCT TURN
CLOSE TO FAN
DISCHARGE

NO STATIC REGAIN PLUS HIGHER FRICTION LOSSES
IN TURN IS 1 DIA. OR LESS INCREASE SP FOR SELECTION:
AF 1.5 X VP, BI 2.0 X VP, FC 2.3 X VP

NOTE:

$$VP = \frac{FAN\ OUTLET\ VELOCITY\ IN\ ft/min}{4005}$$

IF COUNTER TO FAN WHEEL ROTATION:
    SPIN INCREASES ft^3/min
        INCREASE BHP
        REDUCES EFFICIENCY
IF SAME DIRECTION AS FAN WHEEL ROTATION:
    SPIN DECREASES ft^3/min
        INCREASES BHP
        REDUCES EFFICIENCY
THE MAGNITUDE OF THIS EFFECT IS A FUNCTION OF THE DEGREE OF
SPIN INSTALL ANTI—SPIN DEVICES IF PROBLEM OCCURS.

ANTI SPIN
DEVICE

FIGURE 24.22   Performance effect of fan installation.

From catalog, BHP = 23.0.

$$BHP(kW)\ required = catalog\ BHP \times kW$$

$$\times \left(\begin{array}{c} assumed\ increase \\ due\ to\ inlet\ vanes \end{array} \right) \times \left(\begin{array}{c} assumed\ increase\ due \\ to\ drive\ losses \end{array} \right)$$

$$BHP\ required = 23.0 \times 1.13 \times 1.06 = 27.6$$

Case 2: Discharge with Gradual Expansion and Gradual Contraction Transitions.
There is no difference between this application and case 1 since no dynamic losses (ignore friction in this example) occur and the energy is transferred from velocity (leaving the fan) to static (in the plenum) and back to velocity (at the entrance to the main duct).

Case 3: Ducted Discharge to Plenum with Minimum of Three Fan Wheel Diameters of Duct.   In this case, one assumes the velocity pressure is lost by the sudden expansion into the plenum. Therefore, to obtain 3 inWG static pressure + $^1/_2$ inWG velocity pressure (2830 ft/min velocity) ($3^1/_2$ inWG total pressure), we must select the fan to provide $3^1/_2$ in WG

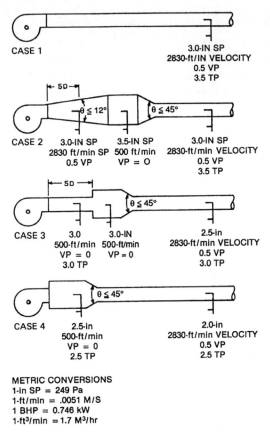

CASE 1
3.0-IN SP
2830-ft/IN VELOCITY
0.5 VP
3.5 TP

CASE 2   3.0-IN SP   3.5-IN SP   3.0-IN SP
         2830 ft/min SP   500 ft/min   2830-ft/min VELOCITY
         0.5 VP   VP = 0   0.5 VP
                            3.5 TP

CASE 3   3.0   3.0-IN   2.5-in
         500-ft/min   500-ft/min   2830-ft/min VELOCITY
         VP = 0   VP = 0   0.5 VP
         3.0 TP   3.0 TP

CASE 4   2.5-in   2.0-in
         500-ft/min   2830-ft/min VELOCITY
         VP = 0   0.5 VP
         2.5 TP   2.5 TP

METRIC CONVERSIONS
1-in SP = 249 Pa
1-ft/min = .0051 M/S
1 BHP = 0.746 kW
1-ft³/min = 1.7 M³/hr

FIGURE 24.23   Effect of static regain at constant fan speed.

total pressure in the plenum ($3\frac{1}{2}$ inWG static pressure since velocity pressure is nearly zero in the plenum).

Select fan for $3\frac{1}{2}$ inWG static pressure with 2830 ft/min outlet velocity (assume airfoil is centrifugal).

From catalog, RPM = 935. So

$$\text{RPM required} = 935 \times 1.03 \times 1.02 = 982$$

From catalog, BHP = 26.0, so

$$\text{BHP required} = 26.0 \times 1.13 \times 1.06 = 31.1$$

If the fan is operated at the same speed as the fan in case 1, the static pressure at entrance to duct system would be only $2\frac{1}{2}$ inWG.

Case 4: Blowthrough with Fan Discharging Directly into Plenum.   This is the same as case 3 except no static regain occurs. Therefore, to achieve 3 inWG static pressure at 2830 ft/min in the downstream duct, assuming an airfoil centrifugal fan (regain effect of $1.0 \times$ velocity pressure, from Fig. 24.22) and loss of fan discharge velocity pressure, we would select the fan for:

Actual static required in discharge plenum	3.0 inWG static pressure (747 Pa)
Absence of duct to achieve regain	0.5 inWG static pressure (125 Pa)
Discharge loss (sudden expansion)	0.5 inWG static pressure (125 Pa)
Fan selection	4.0 inWG (997 Pa)

From catalog, RPM = 970, so

$$\text{RPM required} = 970 \times 1.03 \times 1.02 = 1019$$

From catalog, BHP = 29.4, so

$$\text{BHP required} = 29.4 \times 1.13 \times 1.06 = 35.2$$

If this fan is operated at the same speed as in case 1 (900 r/min), the static pressure at entrance to the duct system would be only 2 inWG, as shown in Fig. 24.23.

This example illustrates the importance of making proper allowances for the difference between catalog ratings and the way that a fan is applied. In this case the horsepower actually required is 53 percent greater than that indicated in the catalog.

24.5 FAN CONTROL SENSOR LOCATION

A very important consideration in the design of the duct and fan system is the location of the sensor which controls fan modulation. Many possibilities exist in any given system with widely ranging results. The following considerations are important:

- Duct system configuration
- Parallel risers and trunks
- Possible use of the space
- Fan selection and surge characteristics
- Duct consideration and tightness
- Sound level characteristics of the terminal unit

Now let

TP = total pressure
SP = static pressure
VP = velocity pressure
HL = head losses due to friction and air turbulence at duct fittings
VEL = velocity
u = upstream
d = downstream
ΣHL = sum of head losses from fan to a given point

Then

$$TP = SP + VP$$

$$TP_u = TP_d + HL$$

$$SP_u + VP_d = SP_d + VP_d + HL$$

$$VEL = \frac{\text{air flow}}{\text{duct area}}$$

$$VP,\ \text{inWG} = \left(\frac{VEL}{4005}\right)^2 \quad VEL \text{ in ft/min}$$

To illustrate, consider the high-velocity static regain design in Fig. 24.24. This is a typical result of computer-optimized design.

The fan is sized to handle 99,000 ft³/min (168,201 m³/h), serving seven floors of a commercial office building. The upper six floors consist of office space which is expected to have very limited use at night and on weekends. However, the bottom floor is a commercial store which may be in full operation at times when the rest of the building is not. With this design and space utilization, the three operating conditions which cover the capacity modulation range and encompass the worst situation that would probably exist are

1. Full load.
2. 50 percent capacity evenly distributed.
3. Only the bottom floor, with 14 percent of load in operation.

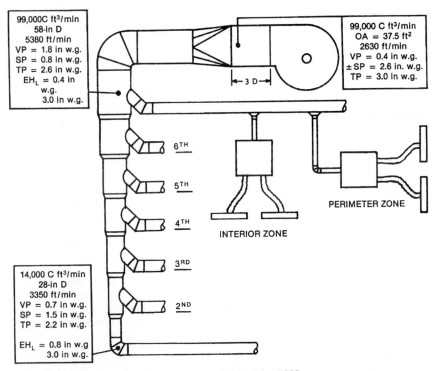

FIGURE 24.24   Fan control—full load. 1 inWG = 249 Pa; 1 ft = 0.305 m.

First, the building is analyzed at full-load conditions at the fan in the main riser just prior to the first branch takeoff and at the bottom of the main riser just prior to the last branch takeoff.

At full load the total pressure at the fan consists of a velocity pressure of 0.4 inWG (100 Pa) and a static pressure of 2.6 inWG (647 Pa). Just downstream in the main riser, the duct size is reduced, the velocity pressure equals 1.8 inWG (448 Pa), and the static pressure equals 0.8 inWG (200 Pa), for a total pressure of 2.6 inWG (647 Pa). The difference between the total pressure at this point and that at the fan is the head losses of 0.4 inWG (100 Pa) due to dynamic losses in the transition and elbow and the friction loss in that length of duct.

At the bottom of the riser, only 14,000 ft³/min (23,786 m³/h) is left in the duct with a velocity pressure of 0.7 inWG (174 Pa), a static pressure of 1.5 inWG (374 Pa), and a total pressure of 2.2 inWG (548 Pa). The difference in total pressure between this and 3.0 inWG (747 Pa) at the fan is the sum of the head losses, equal to 0.8 inWG (200 Pa). The bottom of the main riser in this example would be the proper place to locate the fan capacity control signal set at 1.56 inWG (374 Pa) static pressure. This location provides for significant capacity reduction at the fan without short-circuiting any trunk ducts at the bottom of the main riser.

At each branch takeoff, static regain occurs which increases the static pressure. However, dynamic loss occurs at the takeoff which reduces static pressure. And velocity pressure is increased through the reducer which also reduces static pressure. The objective in high-velocity static regain design is to attempt to maintain, as closely as possible, a constant static pressure throughout the main riser and trunk duct system. This means that the net reduction in static pressure due to friction and dynamic losses, plus a reduction due to a change in velocity pressure through reducers, must be offset by the static regain achieved at each takeoff.

Next consider what happens when the system is at 50 percent capacity, evenly distributed to reach each floor (Fig. 24.25). This situation could exist, e.g., in a building with a lot of glass if the sun went under a cloud. It would be a relatively common occurrence. Under these conditions, the head losses and velocity pressure regain are reduced by the square of the air-flow change. The fan capacity control is set at 1.5 inWG (374 Pa) static pressure (the value required at maximum load), even though the actual requirement on the bottom floor may be less. The net effect in this case is a total pressure reduction at the fan of 1.88 inWG (468 Pa). Note that at the top and bottom of the riser the total pressure, plus the sum of the head losses, is equal to that at the fan.

Finally, consider the situation when only the first floor is in operation (Fig. 24.26). This is not the same "system" as in the previous two situations, because full flow still exists in the last section of the main riser between the first and second floors. In the next section up between the second and third floors, only half the design air flow exists. This is reduced in succeeding sections to $\frac{1}{3}$, $\frac{1}{4}$, $\frac{1}{5}$, $\frac{1}{6}$, and $\frac{1}{7}$ of the design air flow. The resulting head losses in each section are reduced by the square of the load change, and since none of the other floors is taking any air, no static regain occurs.

Enough total pressure must be produced at the fan to provide the required static pressure and velocity pressure at the bottom of the main riser (in this example) as well as offset all the head losses to that point.

The resulting velocity pressure, static pressure, and total pressure are shown at the top of the main riser and at the fan. Note that the velocity pressure now is practically zero at these points, and thus the static pressure component is almost all the total pressure.

Under this condition, if a person came into an office on the top floor, there would be about 2.29 inWG (570 Pa) of static pressure in the entire top-floor trunk duct system. Assuming an internal office so that the terminal unit requires full load as soon as the lights are on and the people load exists, 1.86 inWG (463 Pa) of static pressure would exist at the terminal after the losses in the takeoff fitting and runout ductwork are deducted. The sound

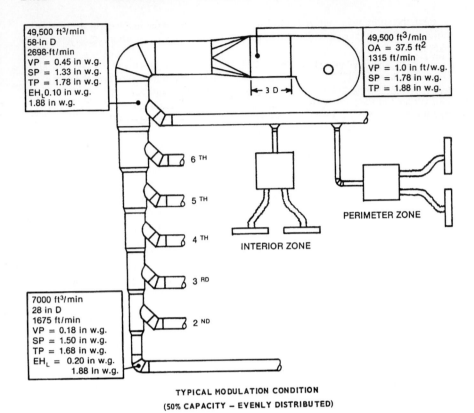

49,500 ft³/min
58-in D
2698-ft/min
VP = 0.45 in w.g.
SP = 1.33 in w.g.
TP = 1.78 in w.g.
EH$_L$0.10 in w.g.
    1.88 in w.g.

49,500 ft³/min
OA = 37.5 ft²
1315 ft/min
VP = 1.0 in ft/w.g.
SP = 1.78 in w.g.
TP = 1.88 in w.g.

← 3 D →

6 TH

5 TH

4 TH

3 RD

2 ND

PERIMETER ZONE

INTERIOR ZONE

7000 ft³/min
28 in D
1675 ft/min
VP = 0.18 in w.g.
SP = 1.50 in w.g.
TP = 1.68 in w.g.
EH$_L$ = 0.20 in w.g.
         1.88 in w.g.

TYPICAL MODULATION CONDITION
(50% CAPACITY – EVENLY DISTRIBUTED)

FIGURE 24.25   Fan control—50 percent of capacity. 1 inWG = 249 Pa; 1 ft = 0.305 m.

level, duct construction, and other considerations would have to be evaluated on the basis of this maximum static pressure condition. Also shown is the same situation involving a perimeter office with only half of the design terminal air flow in the absence of any solar load. Since lower losses exist in the takeoff fitting and runout, the static pressure would be higher at this terminal and the sound level would have to be evaluated on the basis of the reduced load and higher static pressure.

Figure 24.27 indicates the static pressure requirements for each of these conditions. Remember, the static pressure which the sensor is set to control must satisfy the highest static pressure requirement at that point. Thus, if the sensor is located at the fan, it must be set to maintain full-load discharge static pressure of 2.6 inWG (647 Pa) at all times (in this example). This obviously would result in considerable excess pressure at all other conditions and therefore is not the most desirable control point. Locating the sensor at the top of the main riser in the high-velocity ductwork would still require a rather high static pressure at the worst condition (only the bottom floor in operation). Thus, this is obviously not a good location. Locating the sensor at the bottom of the main shaft and setting it for 1.5 inWG (374 Pa) is probably the best location for this example. If a complete DDC system is used with air flow sensors at each VAV terminal, the fan control can be optimized by resetting the duct static pressure lower until one (or more) terminal damper is fully open and cannot maintain desired airflow.

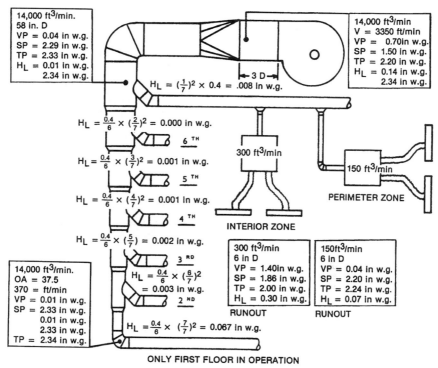

FIGURE 24.26 Fan control—14 percent of full load. 1 inWG = 249 Pa; 1 ft = 0.305 m.

24.6 FAN SELECTION

Now, consider the selection of fans for various applications. Consideration should be given to the varying system resistance in multizone and particularly double-duct systems and its effect on load, fan horsepower, and duct static pressure (see Fig. 24.28).

For example, when the average of all zone requirements is 50 percent hot deck and 50 percent cold deck, the friction loss through the parallel components (i.e., coils, dampers, and supply ducts for double duct) is reduced by the square of the air-flow change. This means that the operating point on the fan is changed, resulting in increased air flow.

On double-duct systems, since the same static pressure exists at the entrance to both ducts, considerable excess pressure may exist in the duct requiring less air. Dampers are sometimes installed and controlled from a remote point in the duct to prevent excessive static pressure buildup (see Fig. 24.29). The same rules for locating the sensor apply as described in Sec. 24.5.

With constant-volume systems, the rule of thumb has typically been to select the largest fan possible, which would also be the most efficient and quietest. But while the largest fan may be the most efficient, it is not the correct fan for a variable-volume system. The reason is that little load variation is possible without putting the fan into surge. This is true regardless of the sensor's location in the system. For that reason, a fan one size, or possibly even

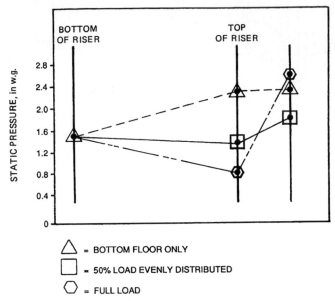

FIGURE 24.27   Variable volume—static pressure required.

two sizes, smaller (Fig. 24.30) is required to provide the necessary load variation. While the smaller fan may be less efficient at its full-load operating point, there may be very little horsepower variation, compared with a large fan, at the typical modulated condition where the fan will operate a majority of the time.

The following illustrates fan selection for a variable-volume system. Figure 24.31 shows a 60-in (1.52-m) double-width fan, which is the second largest size that could be used for the selection. It results in 71 percent static efficiency at full load and 77 percent total efficiency. The minimum system resistance curve is shown as well as the situation that could exist at the fan when only the bottom floor is in operation. These two curves, then, determine the maximum amount of excess pressure that can exist ahead of any terminal. So this gives an easy means of evaluating acceptability of the resulting sound level and duct construction criteria. Note, however, that this fan is entering the surge region at about 50,000 ft³/min (84,950 m³/h) with a static pressure of approximately 3.0 inWG (747 Pa) at the fan at the worst condition. The surge condition in this static pressure range is about the limit of what we could reasonably expect to get by with. If this situation could exist for extended periods with only the bottom floor in operation, a runaround duct, in conjunction, with the inlet vanes, might be considered.

Figure 24.32 shows the situation with a fan two sizes smaller. This 49-in (1.245-m) double-width fan has a considerably reduced static efficiency of 54 percent, but the total efficiency of 67 percent has only dropped off by 10 points. Since total efficiency is the only true criterion in a properly designed drawthrough fan system, this selection looks better than if static efficiency had mistakenly been used for the comparison. But static efficiency is the proper criterion for comparing blowthrough applications where the velocity head is dissipated. This smaller fan allows reduction in load down to approximately 20,000 ft³/min (33,980 m³/h) before entering the surge range. At this operating point, the fan static pressure with inlet vane or variable-speed control is only $1\frac{1}{2}$ to $2\frac{1}{2}$ inWG (8 to 18 Pa).

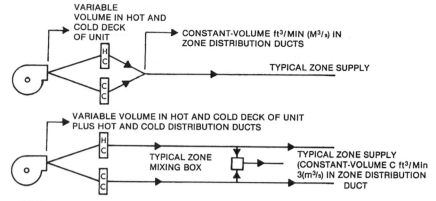

VARIABLE
VOLUME IN HOT AND
COLD DECK
OF UNIT → CONSTANT-VOLUME ft³/MIN (M³/s) IN
ZONE DISTRIBUTION DUCTS

TYPICAL ZONE SUPPLY

VARIABLE VOLUME IN HOT AND COLD DECK OF UNIT
PLUS HOT AND COLD DISTRIBUTION DUCTS

TYPICAL ZONE
MIXING BOX

TYPICAL ZONE SUPPLY
(CONSTANT-VOLUME C ft³/Min
3(m³/s) IN ZONE DISTRIBUTION
DUCT

ON MULTI ZONE (M2): IF COIL PRESSURE DROP IS 1-in w.g. (249-Pa) SP AT MAXIMUM-LOAD ft³/min(m³/s), THEN COIL SP AT 50% LOAD = ¼ in w.g. (62 Pa) [SP AT MAXIMUM LOAD × (½)²], OR ¾-in w.g. (187-Pa) SP VARIATION.

ON DUAL DUCT (DD): IF COIL & TRUNK DUCT PRESSURE DROP IS 4-in w.g. (996-Pa) SP AT MAXIMUM-LOAD ft³/min (m³/s), THEN SP AT 50% LOAD = 1 in w.g. (249 Pa) [SP AT MAXIMUM LOAD × (½)²], or 3-in w.g. (747-Pa) VARIATION.

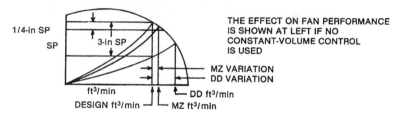

1/4-in SP

SP

3-in SP

THE EFFECT ON FAN PERFORMANCE
IS SHOWN AT LEFT IF NO
CONSTANT-VOLUME CONTROL
IS USED

MZ VARIATION
DD VARIATION

ft³/min — DD ft³/min
DESIGN ft³/min — MZ ft³/min

FIGURE 24.28 Fan selection and system resistance. 1 inWG = 249 Pa; 1 ft = 0.305 m.

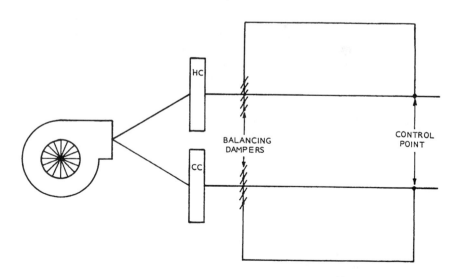

HC

BALANCING
DAMPERS

CONTROL
POINT

CC

STATIC PRESSURE CONTROL IN DOUBLE DUCT SYSTEM

FIGURE 24.29 Static pressure control in double-duct system.

24.35

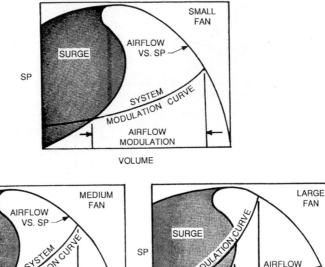

FIGURE 24.30   Fan selection for variable volume.

The resulting increase in noise or pulsation would probably be acceptable at these static pressures. The system designer must evaluate the potential minimum load for the system and the way in which it would be distributed in the building. She or he must then compare this value with the fan operating characteristics to determine the required method of control.

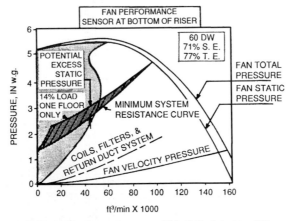

FIGURE 24.31   Fan performance—60-in (1.5-m) double width.

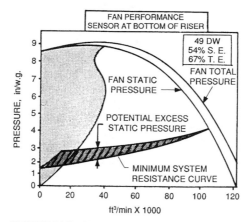

FIGURE 24.32   Fan performance—49-in (1.2-m) double width.

24.7 RETURN-AIR FANS

Use of a return-air fan should be considered whenever high return air duct losses are anticipated. In these cases, a return air fan may prove useful by providing a positive means for exhausting air. It also has the advantage of reducing the negative pressure on the suction side of the supply fan, permitting smaller return duct sizing.

Power relief fans should be used in lieu of return fans if the consideration is solely to provide sufficient building exhaust. Generally, when the return-air path pressure drop is between 0.15 and 0.50 inWG (37 and 125 Pa), power relief fans give good results.

Return-fan volume is calculated by subtracting the amount of air necessary to pressurize the building from the supply fan volume. Typically, this is about 10 percent of supply fan volume. Any forced exhaust should also be deducted from the return air volume.

If the return-air fan does not follow the supply fan (to maintain the design load differential), it could overpower the system. This would create a negative static pressure in the building and a positive static pressure at the outside air entrance (Fig. 24.33). One means of preventing this is to control the return-fan from a differential-pressure controller sensing outdoor static and building pressure on the ground floor. A low limit control must be installed in the return/exhaust plenum to maintain a minimum positive pressure to prevent damage to ductwork on system startup. Control of the return-fan from the same sensor as the supply fan has been used on small systems, but the fan and system characteristics are different, so building pressure control cannot be ensured. Fan tracking systems are available to attempt to control the air flow from the return system to match the volume of the supply system. However, these systems do not sense or control building pressure directly, and the differential between the two fans cannot be controlled accurately enough to assure minimum ventilation airflow meets codes. Volumetric fan tracking is not recommended.

To summarize, remember these points:

1. Use static regain duct design for best results in large systems.
2. Understand the fan characteristics to prevent unsatisfactory operation.
3. Carefully evaluate the use of the building and the expected capacity under various anticipated operating conditions to determine fan control.

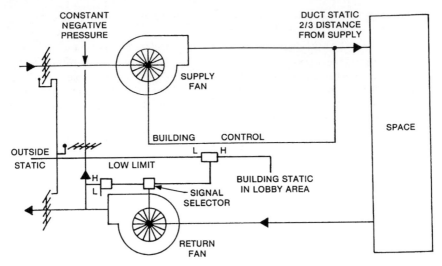

FIGURE 24.33   Return fan control.

4. Consider a smaller fan for variable-volume than for constant-volume systems.

5. Locate the sensor for minimum static pressure variation in the system and to ensure proper static pressure under all operating conditions.

6. Make sure that the combination of the fan, system, and control points is compatible so the sound level and duct static pressure stay within acceptable limits.

7. Control return fans from a building pressure sensor.

8. Consider use of relief fans in lieu of return fans.

24.8   DESIGN CHECK LIST FOR GOOD INDOOR AIR QUALITY (IAQ)*

1. Carefully document all criteria for design, installation, and maintenance related to IAQ.

2. Maintain proper distances between all intake and exhaust openings to minimize reentrainment of contaminants.

3. All air intakes should be designed to prevent moisture entrainment.

4. All drain-pans to be sloped and trapped to leave no standing water.

5. Humidifiers, air washers, and direct evaporative coolers to have potable makeup water supply and no chemicals or additives emitted into the air unless government approved.

6. Dehumidifying coils, air washers, and humidifiers to have no exposed insulation and smooth surfaces downstream that will not absorb moisture within the absorption distance recommended by the manufacturer.

7. Provide access for inspection, cleaning, and maintenance for all ventilation system components including equipment and ductwork.

---

*See also Chap. 21.

8. Duct lining in the air stream should be coated or have metal nosings on all exposed edges and not be subject to "fiber migration."

9. External insulation should have sealed vapor barrier to prevent condensation.

10. Good filtration is extremely important. All filters should be capable of efficient capture of particles down to 3 μ in size and where serving a high density of people, filters capable of capturing particles in the respirable range are recommended. Filter racks should minimize bypass of air. Extra care should be taken during construction to keep equipment and duct systems clean.

11. The air distribution system should be capable of providing the required amount of ventilation air to all spaces as specified by code.

12. VAV systems should be capable of measuring and controlling the required out door airflow at any load condition. Use transfer of unvitiated air from adjacent over-ventilated spaces wherever possible instead of reheat to meet minimum ventilation space requirements.

13. Provide a minimum airflow as specified by code to all spaces for the removal of air contaminants. Use transfer air from adjacent spaces if necessary.

14. Humidity levels should be kept below 60 percent RH to prevent the growth of microorganisms.

15. Buildings should be pressurized to prevent the infiltration of humid air.

16. Regular maintenance to prevent growth of harmful microorganisms and insure proper ventilation is extremely important.

REFERENCE

1. *Energy Consumption in New Buildings*, Standard 90.15, American Society for Heating, Ventilating, and Air-Conditioning Engineers (ASHRAE), Atlanta, GA, 1989.

PIPING AND DUCTWORK— SELECTION AND MAINTENANCE

CHAPTER 25
PIPE SYSTEMS

25.1 PIPING INSTALLATION AND MAINTENANCE

The Mechanical Contracting Foundation
Rockville, Maryland

25.1.1 SYSTEM AND COMPONENT GUIDELINES

Piping systems interconnect equipment and piping components of heating, ventilating, and air-conditioning (HVAC) systems in order to transmit liquids and gases to the equipment for heat transfer processes. Pipe failures in the HVAC system are remote when properly installed; however, the loss of media from the pipe at joints or system components does occur when connections are not properly made by the installer.

Very minor leaks at joints will not affect system operation because most hydronic systems have automatic makeup of media. Systems without automatic makeup will have loss of efficiency and equipment "lock out" when leaks occur; therefore, repair procedures should be initiated promptly.

When selecting piping configurations and equipment locations, proper piping practices must be followed. An improperly—or poorly—installed piping system can later prove to be a major service problem. In addition, it is possible to have a piping configuration that is functionally correct, but it is configured in such a manner that servicing the equipment is difficult, and in some instances impossible.

Never install lines in a position where they will block equipment access panels. Never install piping systems containing liquids where condensation or leaks could occur directly over electrical panels or electrical equipment without proper protection. Electrical access panels must have at least 36 in (92 cm) of clear access.

Piping should be installed to provide clear access for servicing equipment. Piping systems should be properly cleaned and flushed to avoid original-installation debris from accumulating in the system or in connected equipment.

The position of the equipment to be installed should be given initial consideration. The position selected should make the installation less complicated, while creating a safe and workable atmosphere for future service and maintenance work. To address the serviceability of piping, it is necessary to address the types of equipment and pipe accessories that are available.

Maintenance features should address pipe surface temperatures, expansion and contraction, corrosion, vibrations, system pressures, pipe joints, and connections to system components. Maintenance is necessary because of the broad range of operating temperatures, pressures, and many different connected components.

Piping systems discussed are: hot water—low, medium, and high temperature; chilled water; condensing water; steam; condensate; refrigerant; natural gas; and oil.

A. *Hot water piping systems* are normally constructed of copper or black steel pipe. Systems are insulated to retain heat and protect individuals from surface temperatures. Corrosion at joints is an indication of a system leak; the joint should be repaired. Medium and high temperature systems are subjected to considerable expansion and contraction of the pipe materials. Attention should be given to expansion joints, anchors, and pipe guides, which are installed to control the movement of pipe as a result of expansion. Excessive pipe movement can cause damage to insulation, piping components, pipe joints, or surrounding equipment.

B. *Chilled water piping systems* are normally constructed of copper or black steel pipe. Insulation of these systems must maintain external surface temperature above the surrounding dew point and maintain a sealed vapor barrier. Condensation will collect on pipe surface if insulation is defective. This condition shall cause the pipe to rust and components to corrode. Defective insulation must be repaired immediately.

Hot and chilled water pipe systems must be *totally filled* with system fluid because a shortage of fluid will permit air to be retained in the system. This condition will restrict flow to some system components. The system must be vented to obtain full operating conditions. If air is retained in the system, internal corrosion can occur.

C. *Condensing water piping systems* are normally constructed of galvanized, black steel, copper, or plastic pipe. Internal corrosion is a major maintenance problem because the system is open to the atmosphere. Periodic inspections should be regularly scheduled to control this condition. Chemical treatment and cleaning are necessary to maintain clean piping and system components. System bleed-off must be performed to reduce the quantity of solids that could collect in the water system and to prevent excessive corrosion. Closed circuit systems should be maintained as other water systems.

D. *Steam and condensate pipe systems* are normally constructed of black steel pipe. Fiberglass reinforced plastic (FRP) is frequently used for condensate systems. Maintenance conditions are the same as those for the hot water piping systems.

E. *Refrigerant pipe systems* are constructed of copper pipe for freons and black steel for ammonia. Internal pipe surfaces must be clean and free of debris to prevent damage to system components; therefore, system piping maintenance procedures must be followed. Oil slicks at joints are an indication of leakage and should be repaired. Vibration of piping as a result of compressor operating should be controlled by use of vibration eliminators and pipe anchoring.

F. *Natural gas piping systems* are normally constructed of black steel or plastic pipe. These systems are usually maintenance-free when properly installed.

G. *Oil piping systems* are normally constructed of copper, black steel, or plastic pipe. Oil slicks at joints are an indication of leaks and must be repaired.

25.1.2 INSTALLATION GUIDELINES

25.1.2.1 Preassembly Procedures

Piping should be cut accurately to establish measurements and then neatly installed either parallel to or at right angles to walls or floors. Materials should be worked into place without springing or forcing. Sufficient head-room should be provided to clear lighting fix-tures, ductwork, sprinklers, aisles, passageways, windows, doors, and other openings. Materials should not interfere with access to other equipment.

Materials should be clean (free of cuttings and foreign matter inside) and exposed ends of piping should be covered during site storage and installation. Split, bent, flattened, con-taminated, or damaged pipe or tubing should not be used.

Sufficient clearance should be provided from walls, ceilings, and floors to permit weld-ing, soldering, or connecting of joints and valves. A minimum 6- to 10-in (15 to 25 cm) clearance should be provided.

Installation of material over electrical equipment such as switchboards should be avoided. Piping systems should not interfere with safety valves or safety relief valves.

25.1.2.2 Assembly Procedures

A means of draining the piping systems should be provided. A $^1/_2$- or $^3/_4$-in (12 to 18 mm) hose bib (provided with a threaded end) should be placed at the lowest point of the water piping system for draining. Constant grades should be maintained for proper drainage, and piping systems should be free of pockets or traps due to changes in elevation.

Unions should be installed in an accessible location in the piping system to permit dis-mantling of piping or removal of equipment components. Clearance should be provided for using wrenches of sufficient size to break unions after many years of service. Union joints should be clean of system debris prior to installation to permit proper sealing.

Flanges should be installed in an accessible location in the piping system to permit dis-mantling of piping or removal of equipment components. Clearance should be provided for using wrenches of sufficient sizes to break flange bolts after many years of service.

Mechanical joints, when used for dismantling or removal of piping components, should be installed in an accessible location in the piping system. Clearance should be available for using wrenches of sufficient size.

When pipe threads are cut too long, it allows the pipe to enter too far into the fitting or system component. For example, on a valve installation, the pipe end would hit against the valve seat area distorting the seat and body, and this could prevent a tight shutoff when the valve is closed (Fig. 25.1.1).

When a system component is installed on threaded pipe, it is important to use a wrench on the end of the component nearest the joint. Never use a wrench on the component end opposite the end being joined to the pipe, as this could distort the component body or threaded end of the pipe (Fig. 25.1.2).

When pipe joint compound is applied to internal (female) threads, the compound may be forced into a valve seating area, contaminating the fluid in the pipeline. Pipe joint com-pound must be applied to external (male) threads only. The same rule applies when using Teflon tape.

Cracked flanges are usually the result of unequal stress due to improper makeup or poor bolt-tightening procedures. Flanges must be properly aligned before bolts are tightened (Fig. 25.1.3). There should be no gaps between the flanges. Flange bolts should not be used to correct alignment problems.

There are at least three stages of bolt tightening that shall be followed. First, prop-erly align the flanges and gaskets, install the bolts, and hand tighten the nuts. Second, snug up the bolts *but to less than the specified torque*. Depending on the specific requirements of the component being installed, more than one intermediate level of torque may be prudent before applying the final/specified torque. The bolt-tightening sequence demonstrated in Fig. 25.1.4 shall *always* be followed. Third, apply final/spec-ified torque.

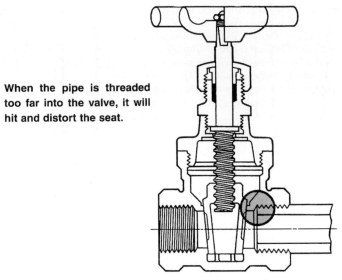

When the pipe is threaded too far into the valve, it will hit and distort the seat.

FIGURE 25.1.1   Threading pipe into a valve. (*Courtesy of Grinnell Corp.*)

Flange gasket selection is also very important. Most flanges have a raised face or a flat face that is the mating surface where the flanges fit together. Ring gaskets should be used with raised-face flanges, and full-face gaskets should be used with flat-face flanges.

The most common gasket material, of the wide variety available, is red rubber. Several asbestos substitutes are being developed for applications where asbestos is no longer approved.

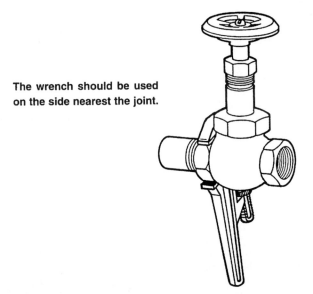

The wrench should be used on the side nearest the joint.

FIGURE 25.1.2   Using a wrench on a valve. (*Courtesy of Grinnell Corp.*)

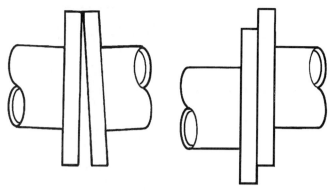

FIGURE 25.1.3 Examples of improper flange alignment and fit-up. (*Courtesy of Grinnell Corp.*)

   The art of soldering joints is well detailed in other manuals, therefore it is not covered here. In general, the most common problem with solder joint connections is when excessive heat is applied while soldering the joint. Resilient seats become scorched, and metal seat rings become distorted.
   Valves should be fully open during the soldering process. In addition, most of the heat should be applied to the tubing, not the component.
   Debris in piping lines is a major cause of contamination. Installation of new piping, in particular, must always be cleared of any foreign material that could penetrate piping.
   Pipe lines must be flushed or blown out (1) to eliminate weld spatter, or small pieces of metal that are created when butt welds are made in piping, (2) to eliminate pipe chips, the metal chips created when cutting pipe threads and drilling holes in piping, (3) to eliminate pipe scale, the scale-like chips in steel pipe caused by severe rusting, (4) to eliminate any internal debris.
   Refrigerant piping systems should be purged with dry nitrogen when brazing to eliminate internal debris.
   A strainer should be installed at unfiltered water sources, pump intakes, and control and metering devices to flush debris from the piping system.

25.1.2.3 Storage Procedures

The proper storage of material requires a reasonably clean and preferably dry storage area for piping system components, an orderly placement of the various material categories, and adequate access for material removal and/or added deliveries.

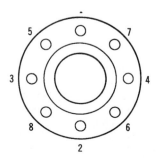

FIGURE 25.1.4 Flange bolts tightening sequence. (*Courtesy of Grinnell Corp.*)

Material that is improperly stored is usually subject to exposure to foreign substances, such as mud, dirt, sand, or other debris. During system erection, most of these elements find a way to stay within the system so that during system start-up operations, costly and time-consuming problems generally develop. Most problems can be avoided by using proper storage techniques.

Piping should be stored on racks or dunnage, off the ground, and protected from the elements. Where practical, pipe fittings should be placed in bins. Larger sizes should be stored on pallets to ensure that clean, dry, and easy handling conditions prevail.

Valves and specialities should be stored and protected in a similar manner. In addition, in almost all cases, they should at least be protected from the elements by waterproof tarpaulins or an enclosed storage area. Gasket materials must always be protected from exposure to ultraviolet light and ozone.

When purchasing piping materials, it is recommended that they be ordered for delivery in such a manner that materials and sized parts are delivered in a prepackaged separate condition [i.e., $1/_2$-in (12-mm) copper elbows separate from $1/_2$-in (12-mm) copper tees]. This will facilitate the establishment of an orderly warehouse arrangement where materials are binned in separate compartments or containers.

The orderly arrangement of material by size and type permits the easy location of needed material during system erection, as well as recognition of potential depletion of stocked material prior to running out of an important system component.

The ability to get to stored material is vital for an efficient storeroom operation. Larger fittings and valves (where mechanical apparatus is needed for proper handling) should have adequate aisle space for equipment (e.g., forklift). Smaller material that can be manually handled requires only personnel access space.

25.1.3   SYSTEM COMPONENTS

Many components are common to various pipe systems. Components requiring maintenance are: pumps, valves, metering devices, relief valves, strainers, and firestops. Other components, such as unions, hangers, and fittings, should be inspected visually. (See also Secs. 25.4 and 25.5).

25.1.3.1   Valves—Common Usage Types

Proper installation is necessary if valves are to perform as expected. Some of the more common but sometimes overlooked installation and service problems are described below. Valves should be installed in a position that allows servicing of units. Consideration should be given to the removal of bonnets and of internal operating parts, and to the ability to pack units.

Isolation valves should be installed in a position that will permit the removal of the piping system or equipment. Isolation valves should be rated at a pressure differential high enough to accommodate full shut-off, with one side open to atmospheric pressure.

Drain valves should be installed at the bottom of the piping system risers in a practical and serviceable area.

The correct valve position for gate, globe, and check valves is an upright position whenever possible. *Exception: A valve in chilled water systems should have its operating stem in a horizontal position to prevent condensation from damaging insulation.* Gate, globe, and check valves are designed to be installed upright. Positioning valves otherwise invites problems because sediment in the line will settle in the valve's working parts and bonnet cavities. Swing check valves will not function properly in the upside-down position under any circumstances.

If valves do not seat properly when first used, flush foreign matter off of the seat by opening a valve slightly, then closing and repeating this procedure as necessary. Since the valve seat may have debris preventing normal closure, it is a mistake to apply excessive force on the handwheel or handle. Abnormal force imposed on valve stems, seats, etc., will distort or break them. Cheater bars or any devices that put more-than-normal force on valve handles should never be used.

Most valve leakage problems fall into one or more of these three categories: *through-the-valve, outside-the-valve, and outside-to-inside-the-valve.*

1. *Through-the-valve* leakage occurs when flow penetrates beyond the closing member of the downstream side of the valve.

2. *Outside-the-valve* leakage is characterized by flow escaping through valve seals, or packing, to outside the valve.

3. *Outside-to-inside* leakage occurs in vacuum service where air is drawn in through worn or damaged packing or seals.

As previously stated, when a valve fails to shut off completely, do not use extra force to close it. Debris on the seat area is often the problem and is handled by opening the valve partially. The resulting turbulent flow will usually flush the debris.

Always open and close one-quarter-turn valves slowly. Operating them too rapidly may result in water hammer and damage to the valve of piping.

25.1.3.2 Balancing Valves

These valves are used in hydronic systems to adjust and proportion the flow through various branches and mains to various portions of the system. If balancing valves are omitted from the piping system, the flow will take the path of least resistance, which may result in some sections of the system not receiving any flow. As an example, the use of a balancing valve on a warm-water heating coil will limit the flow through that coil, limiting the amount of energy used to heat the air passing across the coil.

The three most common balancing valve configurations are: (1) calibrated manual balancing valve, (2) flow-measuring venturi or orifice plate with manual ball or butterfly valve, and (3) automatic flow-limiting (or automatic flow control) valve.

25.1.3.3 Steam Traps*

Soon after steam leaves a boiler, it begins to lose some of its heat to any surface with a lower temperature, and it begins to condense into water. Clearly, it is desirable to separate and remove the accumulated water (condensate) from the steam supply system by sending it back to the boiler through a piped-vacuum or gravity-return system, while retaining the heat or energy that produces steam within the supply steam. This energy-saving function is fulfilled by means of steam traps.

The entire steam distribution system must remain free of air and condensate. Failure to maintain the steam distribution system free of air and condensate often leads to water hammer and slugs of condensate that could damage equipment, system control valves, and steam traps.

---

*See also Sec. 17.1.

Common to all steam distribution piping systems is the requirement for strategically located drip legs. Drip legs permit condensate to escape by gravity from rapidly moving steam within the system. In addition, drip legs store the condensate until it can be discharged by the pressure differential through the steam trap.

The following installation considerations should be considered when installing steam traps.

1. Service, repair, and replacement are simplified by making steam trap configurations identical for each given size and type.

2. Isolation valves are recommended in order to permit the removal of steam traps without shutting down the operating system.

3. When only one union is used, it should be placed on the discharge side of the steam trap. Avoid in-line (horizontal or vertical) installation of more than one union. It is best to install two unions at right angles to each other (one in horizontal run, one in vertical run).

Dirt legs are installed upstream of a trap assembly to prevent scale and dirt from entering the steam trap. Dirt legs should be cleaned periodically by using a blowdown valve or by removing the end cap and free blowing until clear.

Install a strainer upstream of a steam trap, especially where pipe line dirt conditions are present. Some steam traps are made with built-in strainers.

25.1.3.4 Strainers/Filters-Driers

Strainers should be installed in the piping system at locations that will allow removal of the screens for cleaning. Strainers should be used on the inlet side of system components to protect critical operating parts from foreign particles and debris.

The strainer screen must be clean to permit full-flow. The strainer must be cleaned after the piping system is repaired; flush strainers after the system is activated as a result of repair procedures.

Strainer screens are available with various opening sizes and materials. The screen specification must be used to maintain a clean system and to resist corrosion from system media. A fine-mesh type screen may be used for system cleaning after repair; however, the specified screen should be used during normal system operations.

Refrigerant systems normally use filter-driers that function as a strainer and a moisture collector. A three-valve bypass assembly permits ease of removal and replacement of a clogged unit.

25.1.3.5 Flexible Connections

The installation of these connections in the piping system prevents the transmission of vibrations from equipment to piping in other parts of the building.

Piping system components should be supported independently on each side of a flexible connection so that the connector can be easily removed when a piping system is disconnected.

25.1.3.6 Air Vents

These vents should be installed in the high points of any piping system where air could accumulate, and they should be installed wherever the water stream reduces velocity,

changes direction, or is heated. Any such areas should have either a manual or automatic air vent installed. Automatic air vents should *not* be installed on the suction side of pumps.

If the vent port of an automatic air vent discharges in an area where there is danger of water damage, it should be piped to a drain. Automatic air vents should be installed with a manual isolation valve so that the air vent can be replaced or repaired. If automatic air vents are to be left open to the system, it is necessary to install an automatic makeup water system to compensate for the leakage at the air vent. The air vent should be installed in an accessible area for use and service.

25.1.3.7 Thermometers (or Thermometer Wells)

These should be installed on the inlet and outlet of each heat transfer device. Consideration should be given to locations that will permit easy reading of the thermometer. Thermometers should be installed in the piping system at points that protect them from damage during normal service activities around the equipment.

25.1.3.8 Gauges and Gauge Ports with Shut-off Valves

These gauges and gauge ports should be installed on the inlet and outlet of each piece of equipment that reflects pressure differential in a piping system. One water pressure gauge can be used for connection between the inlet and outlet of a component by manifolding so an accurate pressure-differential reading can be obtained. Pressure gauges permanently installed in a piping system should have a pigtail, or where applicable, a snubbing device, between the gauge and the system shut-off valve. The shut-off valve can be used as a snubber while reading system pressure, and it should be shut off when not reading system pressure.

25.1.3.9 Flow Switches

When pressure-differential flow switches are required, they should be installed in the piping system with a minimum of 10 pipe diameters of straight pipe on each side of the pressure-sensing point. Pressure-differential flow switches have proven to be a service problem due to the small pressure drop across the system components.

If the piping-system line size is less than 1-in (25-mm), a piston-type flow switch should be used. If the piping-system line size is greater than 1-in (25-mm), a paddle-type flow switch should be used. Paddle-type flow switches require at least a 1-in (25-mm) pipe diameter spacing and operate best with 10 gpm (38 L/min) or more of water flow.

25.1.3.10 Anchors and Guides

These anchors and guides are installed in piping systems that are subjected to a wide temperature range or pressure shock. Expansion joints or expansion loops are installed to permit system expansion and contraction. Pipe alignment guides and anchors are installed to control direction of movement. Components should be inspected occasionally to verify design performance.

25.1.3.11 Pipe Sleeves and Firestops

The inspection of pipe sleeves is necessary because damage to the packing materials may eliminate fire protection qualities.

Sleeve packing materials are installed to prevent the intrusion of dirt, water, air, rodents, and foreign objects. Packing materials add air tightness, cathodic protection, corrosion resistance, anchorage support, and shock (sound and vibration) absorption. Common packing materials are lead-and-oakum joints, mastics, casing boots, and link seals.

Project specifications determine the types of materials required to seal the annular space between the pipe and sleeve.

To penetrate fire-rated floors and walls with items such as pipes, through-penetration firestop is required. This firestop consists of specific field-installed materials designed to prevent the spread of fire through fire-rated openings.

The firestop codes require that any penetration through a fire-rated wall or floor must be firestopped to a fire rating that is equal to that of the wall or floor before the opening was created. In some cases, state law may require that a building meet NFPA Life Safety Code, Section 101, in addition to local requirements.

The following listings present the materials and hardware associated with firestop installations.

1. *Firestop dam-forming material.* Some form of damming material, such as mineral wool, backer rods, or insulation must be placed in the annular space to help support the firestopping caulk or sealant.

2. *Firestop caulk or sealant.* The caulk or sealant in a firestop is the material that actually stops the fire. Firestopping caulk or sealant is available in four chemical compositions to satisfy different applications.

   • Intumescent caulk or sealant expands when exposed to heat.

   • Silicone caulk or sealant is very flexible and allows movement of the pipe.

   • Endothermic caulk or sealant expels moisture in a fire to protect the seal.

   • A water-based material is the fourth kind of caulk or sealant.

3. *Firestop retaining hardware.* Retaining collars, clips, or wire mesh may be required in some firestop systems to support the firestop system, or to direct the expansion of intumescent materials.

4. *Firestop intumescent wrap strips, firestop kits/devices.* Strips of intumescent wrap (which may expand more than sealants when exposed to heat) are typically used around combustible penetrants. These materials may be used in conjunction with retaining hardware, fasteners, etc.

25.1.4 SAFETY CONSIDERATIONS

Whenever a system must be serviced or repaired, or when a piece of pipe must be removed, it is necessary to plan the service or repair procedure carefully to avoid personal injury and damage to valves or piping. As a minimum, the piping system must always be depressurized before attempting the removal or repair of any piping component.

Special caution must be taken to relieve pressure from any piping system before service or removal. The operator is responsible for knowing the type of fluid, gas, or other medium in the piping system. The process for disposal or recovery of the medium also must be known. Relieving the system pressure must be accomplished in accordance with requirements for the specific medium involved.

The depressurization of any refrigerant system must be accomplished through the normal established recovery process. Depressurization should be handled in accordance with Environmental Protection Agency (EPA) requirements.

Unknowingly, pressure can remain in a piping system even though all of the valves are shut off and the system has supposedly been drained. Even after a section of piping has been sealed by isolating valves, the medium within the pipe can remain under enough pressure or temperature to cause injury. Gaskets may have been sealed to both flange faces by heat, allowing some flange joints to hold pressure, even though the bolts are removed. Ball and plug valves can trap fluid inside and retain it there, even after removal from the piping.

When inspecting valves, always hold them in a manner that prevents fluid from being spilled onto personnel after removal. When dealing with hazardous or corrosive materials, special care must be taken to prevent personal injury and damage to surrounding materials and surfaces.

Opening a drain valves does not always make a piping system safe to work on, as leaking valves can often result in unexpected pressure buildup.

When a flanged connection is being dismantled, the safe practice is to make it leak to relieve pressure before removing all the bolts. Leaking a flanged connection to relieve pressure should be accomplished with any system that is known to be or is suspected to be under pressure; do not depend solely on a pressure gauge to assure a safe condition. *Never trust a pressure gauge; double check line pressure whenever possible!*

25.1.5 SYSTEM PRESSURE TESTING

Piping systems are usually tested at 1.5 times the working pressure of the system. The system should not be buried, concealed, or insulated until it has been inspected, tested, and approved. All defective joints should be repaired, and all defective materials should be replaced. The following procedures briefly summarize a typical pressure test.

1. Examine the system to ensure proper isolation of equipment and system parts that cannot withstand the test pressures. Examine the test equipment to ensure that it is leak-free and that low pressure filling lines are disconnected.

2. Provide temporary restraints for those expansion joints that cannot sustain reactions resulting from test pressure. If temporary restraints are not practical, isolate expansion joints from testing.

3. Isolate equipment that is not to be subjected to the piping system test pressure. If a valve is used to isolate the equipment, its closure shall be capable of sealing against the test pressure without damage to the valve. Flanged joints, at which blinds are inserted to isolate equipment, need not be tested.

25.1.5.1 Water and Steam Systems

The following procedures summarize a typical pressure test:

1. Flush with clean water, clean all strainers. Use ambient temperature water as the testing medium, except where there is a risk of damage due to freezing. Other liquids may be used if safe for workers and compatible with the piping system components. The engineer should be consulted if the system is tested with a liquid other than water to ensure compatibility with other system components.

2. Use the manual air vents that have been installed at high points in the system to release trapped air while filling the system.

3. Subject the piping system to a hydrostatic test pressure, which at every point in the system is not less than 1.5 times the design pressure (high-temperature water piping shall not be tested at more than 500 psig [3500 kPa]). The test pressure *shall not exceed* the maximum pressure for any vessel, pump, valve, or other component in the system under test.

4. After the hydrostatic test pressure has been applied for at least four hours, examine the system for leakage. Eliminate leaks by tightening, repairing, or replacing components as appropriate, and repeat the hydrostatic test until there are no leaks.

25.1.5.2   Refrigerant Systems*

Mechanics must be proficient in testing, evacuation, and dehydration techniques of refrigerant systems to ensure a tight and dry system.

1. Pressurize the complete system with dry nitrogen to 50 psig (350 kPa), and observe over a period of time for any loss in pressure. Joints, fittings, etc., can be leak tested using an approved bubble solution.

2. After the system has been proven leak-tight with dry nitrogen, pressurize with a mixture of dry nitrogen and HCFC-22 (as a leak test gas) to near system-operating pressure or per the applicable local building code. Then test joints, fittings, etc. with an electronic leak detector or other suitable leak detection device.

   Caution: Enough time should be allowed for the refrigerant to mix with the dry nitrogen.

3. Evacuation and dehydration. After determining that there are no refrigerant leaks when the system is pressurized, the system must be evacuated and dehydrated to remove moisture and noncondensables. Evacuation of the system should be accomplished with a vacuum pump to lower the absolute pressure of the system to 500 μ or less. This procedure will reduce the internal pressure of the system below the boiling point of water. External heat may be required to vaporize the water in the system.

A standing vacuum test should then be performed to ensure that the system can hold a vacuum, and that the rate of rise is less than 50 μ per hour. If this test fails, and the system equalizes at the vapor pressure of the ambient temperature, then moisture is still present, and further dehydration must be performed. If the system pressure continues to rise, it will be necessary to check for leaks again, using the procedures listed above. It is advisable to hold the vacuum for 24 h.

When pulling a vacuum on a system, care should be taken to avoid evacuating the system so rapidly that moisture in the system freezes. Moisture in the form of ice can remain in the system and cause internal damage. The vacuum on the system should be broken with the refrigerant specified for the system.

25.1.6   SERVICING PIPING SYSTEMS

The most common defects, cause, and repair/maintenance procedures of piping systems or components are listed in Table 25.1.1. A visual inspection of the systems periodically (quarterly) should reveal defects. Manual valves should be turned a quarter to a half turn to provide freedom of movement of parts. Blow down valves should be opened to flush out

---

*See also Chap. 18.

TABLE 25.1.1 Piping System Troubleshooting Chart

Component	Defect	Cause	Repair and maintenance procedure
Air vents	Leak	Dirt	Remove, flush, or replace (Depress stem to verify shut-off)
	Plugged	Sediment	
Dielectric unions	Leak	Gasket failure resulting from expansion/contraction	Clean joint—replace gasket. (Consider anchors and guides to control movement of pipe.)
		Overtightening at time of installation damaged component	Replace fitting.
Gauges	False reading	Dirt in gauge line Vibration or pulsation	Clean line and fittings. Use snubber valve and shut-off valve. Open shut-off valve only when reading.
Hangers	Loose	Vibration/Movement	Adjust rod length and lock nuts. Add units at equipment to support pipe free of components.
	Noise	Vibration	Consider use of spring or fiberglass hanger isolator.
Insulation	Deterioration	Condensation	Replace—Apply proper vapor barrier Metal jacket may be necessary for protection. Saddles at supports
PVC pipe	Leak at fittings	Cracked ends of pipe	Cut off defective ends before makeup at joints.
Pipe clamps/hangers	Loose or missing	Vibration	Tighten—use lock nuts
Pipe steel	Internal pitting	Air (oxygen) in system	Bleed system; check air control systems; check compression tank air level.
Pressure reducing valves	Erratic flow Fails to shut-off	Pilot line clogged Contamination	Clean and flush line Replace or rebuild valve
Refrigeration filter driers	Excess pressure drop	Dirt	Replace
	Excess moisture	Contamination	Replace

(Continued)

TABLE 25.1.1   Piping System Troubleshooting Chart (*Continued*)

Component	Defect	Cause	Repair and maintenance procedure
Refrigerant rupture disc	Leaking refrigerant	Excess temperature or pressure	Replace disc after system repair—comply with refrigerant procedures.
Relief valves	Leaks	Weak diaphragm; dirt on seat valve spring tension	Flush or clean seat
	Fails to open	Corrosion	Replace valve, if necessary-check pipe for cleanliness. Test at system design.
Strainers	Clogged	Dirt in system	Back flush Remove screen—clean and replace Verify mesh size of screen system cleaning
Thermometers	Separation	Excessive vibration or temperature	Proper location of device
	False reading	Poor heat transfer	Use compound in well to transmit temperature
		Wrong scale device	Replace with device with scale of media range
Unions	Leak	Dirt Misaligning pipe ends	Clean joining surfaces Align pipe ends—tighten
	Flow turbulence	Installation direction	Install union in proper flow direction
Valves	Will not turn	Stem corroded	Clean and lubricate stem (Do not paint stems)
	Will not shut-off	Dirt in seats	Clean seats, discs, gates; Replace as required
Water hammer	Lack of flow	Air in system	Install manual air vents at high points to vent system
	Noise	Condensate in steam flow	Install drip leg and steam trap to collect condensate
Vibration eliminators	Leaking failure	Pipe weight or alignment	Support pipe independently; Align pipe

any dirt accumulation. Relief valves should be manually activated on steam and water systems to verify proper operation.

The loss of media from hydronic systems may result in internal scaling and oxygen corrosion of pipe. This condition will cause clogged strainers, valve seat defects, flow restriction in heat transfer units, and low system efficiency. The system should be cleaned, flushed, and pressure-tested after repair. Hydronic systems should be filled and have air vented through manual vents to obtain full circulation through all components. The media should be analyzed to determine the need for chemical treatment.

As previously stated, piping systems when properly installed should require very limited maintenance. Visual inspections may reveal the need for repairs; however, only skilled mechanics qualified for the specific piping system and understanding the media in that system should attempt any repair procedures.

The piping components of HVAC systems have a much longer life cycle than heat transfer components; therefore, repair is more common than replacement. A highly corrosive atmosphere or internal contamination could result in pipe pitting, scaling, and eventually failing if not maintained. Periodic (annual) inspection and chemical treatment of media should provide good operating conditions for the life of the system.

BIBLIOGRAPHY

1. "Guideline for Quality Piping Installation" 2003, Mechanical Contracting Foundation, Rockville, MD.

25.2   WATER AND STEAM PIPING*

Nils R. Grimm, P.E.

Section Manager, Mechanical,
Sverdrup Corporation, New York, New York

25.2.1   INTRODUCTION

Once the designer has calculated the required flows in gallons per minute (cubic meters per second or liters per second) for chilled-water, condenser water, process water, and hot-water systems or pounds per hour (kilograms per hour) for steam systems and tons or Btu per hour (watts per hour) for refrigeration, calculation of the size of each piping system can proceed.

25.2.2   HYDRONIC SYSTEMS

With respect to hydronic systems (chilled water, condenser water, process water, hot water, etc.), the designer has the option of using the manual method or one of the computer programs.

Whether the piping system is designed manually or by computer, the effects of high altitude must be accounted for in the design if the system will be installed at elevations of 2600 ft (760 m) or higher. Appropriate correction factors and the effects of altitudes 2500 ft (760 m) and higher are discussed in Appendix A of this book.

The following is a guide for design water velocity ranges in piping systems that will not result in excessive pumping heads or noise.

Boiler feed	8 to 15 ft/s (2.44 to 4.57 m/s)
Chilled water, condenser water, hot water, process water, makeup water, etc.	4 to 10 ft/s (1.22 to 3.05 m/s)
Drain lines	4 to 7 ft/s (1.22 to 2.13 m/s)
Pump suction	4 to 6 ft/s (1.22 to 1.83 m/s)
Pump discharge	8 to 12 ft/s (2.44 to 3.66 m/s)

Where noise is a concern, such as in pipes located within a pipe shaft adjacent to a private office or other quiet areas, velocities within the pipe should not exceed 4 ft/s (1.22 m/s) unless acoustical treatment is provided.

Flow velocities in PVC pipe should be limited to 5 ft/s (1.5 m/s) unless special care is taken in the design and operation of valves and pumps. This is necessary to prevent pressure surges (water hammer) that could be damaging to pipe.

Erosion should also be considered in the design of hydronic piping systems, especially when soft material such as copper and plastic is used. Erosion can result from particles suspended in the water combined with high velocity. To assist the designer, Table 25.2.1

---

*Reviewed by Robert O. Couch, Perma-Pipe Corp., Niles, IL.

TABLE 25.2.1 Maximum Water Velocities to Minimize Erosion

Annual operating (h)	Maximum water (ft/s)	Velocity (m/s)
1500	11	3.35
2000	10.5	3.20
3000	10	3.05
4000	9	2.74
6000	8	2.44
8000	7	2.13

shows maximum water velocities that are suggested to minimize erosion, especially in soft piping materials.

Pipe size depends on the required amount of flow, the permissible pressure drop and the desired velocity of the fluid. This may be manually calculated by various methods given in Refs. 1 to 5. An acceptable method of evaluating water flow is the Hazen-Williams formula:

$$f = 0.2083 \times \left(\frac{100}{C} \right)^{1.852} \times \frac{Q^{1.852}}{id^{4.8655}} \qquad (25.2.1)$$

where f = friction head loss in ft of water per 100 ft of pipe (Divide by 2.31 to obtain pounds per square inch)

C = constant for inside pipe roughness (See Table 25.2.2 below)
Q = flow in U.S. gal/m
id = inside diameter of pipe, in

Water velocity in f/s may be calculated as follows:

$$V = 0.408709 \times \frac{Q}{id^2} \qquad (25.2.2)$$

TABLE 25.2.2 Typical Values to Use for the Hazen-Williams Coefficient

Pipe material	C
PVC, FRP, PE	150
Very to extremely smooth metal pipes	130–140
Smooth wooden or masonry pipe	120
Vitrified clay	110
Old cast iron or old steel pipe	100
Brick	90
Corrugated metal	60

where V = velocity in f/s
$\quad Q$ = flow in U.S. gal/m
$\quad$ id = inside diameter of pipe, in in.

If the computer method is chosen to size the hydraulic piping systems, the designer must select a software program from the several that are available. Two of the most widely used are Trane's CDS Water Piping Design program and Carrier's E20-II Piping Data program. In addition to determining the pipe sizes, both programs print a complete bill of materials (quantity takeoff by pipe size, length, fittings, and insulation).

Whichever program is used, the specific program input and operating instructions must be strictly followed. It is common to trace erroneous or misleading computer output data to mistakes in inputting design data. *It cannot be overstressed that in order to get meaningful output data, input data must be correctly entered and checked after entry before the program is run.* It is also a good, if not mandatory, policy to independently check the computer results the first time you run a new or modified program, to ensure that the results are valid.

If the computer program used does not correct the computer output for the effects of altitude when the elevation of the project is equal to or greater than 2500 ft (760 m) above sea level, the computer output must be manually corrected by using the appropriate correction factors listed in Appendix A of this book.

The following describe the programs available to the designer using Trane's CDS Water Piping Design program for sizing hydronic systems.

Water Piping Design (DSC-IBM-123).   This pipe-sizing program is for open and closed systems, new and existing systems, and any fluid by inputting the viscosity and specific gravity. The user inputs the piping layout in simple line-segment form with the gallons per minute of the coil and pressure drops or with the gallons per minute for every section of pipe. The program sizes the piping and identifies the critical path, and then it can be used to balance the piping so that the loops have equal pressure drops.

The output includes

- Complete bill of materials (including pipe sizes and linear length required, fittings, insulation, and tees).
- Piping system costs for material only or for material of labor.
- Total gallons of fluid required.

The following summary describes the program available to the designer using Carrier's E20-II Water Piping Design for sizing hydronic systems.

Water Piping Design (Version 1.0).   This program provides the following:

- Enables the designer to look at the balancing required for each piping section, thereby permitting selective reduction of piping sizes or addition of balancing valves.
- Calculates pressure drop and material takeoff for copper, steel, or plastic pipe.
- Sizes all sections and displays balancing required for all circuits.
- Sizes closed or open systems.
- Corrects pressure drop for water temperature and/or ethylene glycol.
- Calculates gallons per minute of total system.
- Calculates total material required, including fittings.

- Ability to store for record or later changes up to 200 piping sections.
- Ability to change any item and immediately rerun.
- Allows sizing of all normally used piping materials.
- Allows balancing of system in a minimum amount of time.
- Allows easy sizing of expansion tanks and determination of necessary gallons per minute of glycol for brine applications.
- Estimates piping takeoff fitting by pipe size, quantities (linear feet, fittings, valves, etc.).

25.2.3 STEAM SYSTEMS

There are few computer programs available for sizing complex networks of steam piping. Most design is done manually although simple computer programming of the various formulas such as the Fritzsche and Unwin formulas will save a considerable amount of time. Unwin's formula that appears to be the preferred method of district heating engineers is as follows:

$$P = \frac{0.0001306 \times W^2 \times L\left(1 + \frac{3.6}{d}\right)}{y\, d^5} \tag{25.2.3}$$

where P = pressure drop, psi
$\quad W$ = pounds of steam, lb/m
$\quad L$ = length of pipe, lt
$\quad d$ = inside diameter of pipe, in
$\quad y$ = Average density of steam, lb/ft^3

It is advisable to use values for the specific volume corresponding to the average pressure if the drop exceeds 10 to 15 percent of the initial absolute pressure.

Figure 25.2.1 gives a graphical solution to Unwin's formula.

The effects of high altitude must be accounted for in the design when the system will be installed at elevations of 2500 ft (760 m) or higher. Appropriate correction factors and the effects of altitudes 2500 ft (760 m) and higher are discussed in Appendix A.

Table 25.2.3 gives reasonable velocities for steam lines based on average practice. The lower velocities should be used for smaller pipes and the higher velocities for pipes larger than 12 in (30 cm).

Steam piping systems may also be sized by following one of the accepted procedures found in standard design handbook sources such as Refs. 2, 3, 5.

TABLE 25.2.3 Typical Values of Steam Line Velocity

Condition of steam	Psi	Bar	Ft/min	m/s
Saturated	0–15	0–1.03	4000–6000	20.32–30.48
Saturated	50 and up	3.43 and up	6000–10000	30.48–50.08
Superheated	200 and up	13.73 and up	7000–20000	35.56–101.60

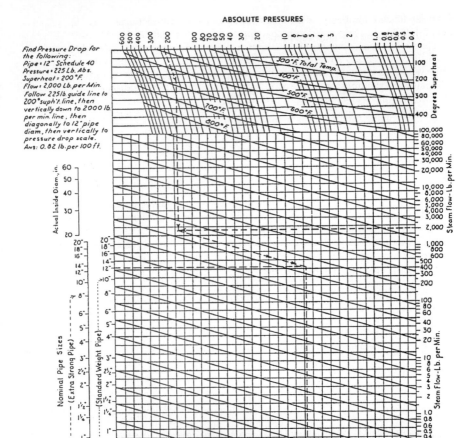

ABSOLUTE PRESSURES

Find Pressure Drop for
the following:
Pipe = 12" Schedule 40
Pressure = 225 Lb. Abs.
Superheat = 200°F.
Flow = 2,000 Lb. per Min.
Follow 225 lb. guide line to
200° suph't. line, then
vertically down to 2000 lb.
per min. line, then
diagonally to 12" pipe
diam., then vertically to
pressure drop scale.
Ans: 0.82 lb. per 100 ft.

FIGURE 25.2.1   Graphical solution to Unwin's formula. (*Courtesy of Perma-Pipe, Inc.*)

25.2.4   *REFRIGERANT SYSTEMS*

Here the designer has the option of using the annual method or at least one computer program.

Whether the piping system is designed manually or by computer, the effects of high altitude must be accounted for in the design when the system will be installed at elevations of 2500 ft (760 m) or higher. Appropriate correction factors and the effects of altitudes 2500 ft (760 m) and higher are discussed in Appendix A.

Liquid line sizing is considerably less critical than the sizing of suction or hot gas lines, since liquid refrigerant and oil mix readily. There is no oil movement (separation) problem in designing liquid lines. It is good practice to limit the pressure drop in liquid lines to an equivalent 2°F (1°C). It is also good practice to limit the liquid velocity to 360 ft/min (1.83 m/s).

The suction line is the most critical line to size. The gas velocity within this line must be sufficiently high to move oil to the compressor in horizontal runs and vertical risers with upward gas flow. At the same time, the pressure drop must be minimal to prevent penalizing the compressor capacity and increasing the required horsepower. It is good practice, where possible, to limit the pressure drop in the suction line to an equivalent temperature penalty of approximately 2°F (1°C). In addition to the temperature (pressure drop) constraints, the following minimum gas velocities are required to move the refrigerant oil.

Horizontal suction lines   500 ft/min (2.54 m/s) minimum

Vertical upflow suction lines   1000 ft/min (5.08 m/s) minimum

The velocity in upflow rises must be checked at minimum load; if it falls below 1000 ft/min (5.08 m/s), double risers are required. To avoid excess noise, the suction line velocity should be below 4000 ft/min (20.32 m/s).

The discharge (hot-gas) line has the same minimum and maximum velocity criteria as suction lines; however, the pressure drop is not as critical. It is good practice to limit the pressure drop in the discharge (hot-gas) line to an equivalent temperature penalty of approximately 2 to 4°F (1 to 2°C).

If the manual method is used to size the project, refrigerant piping systems should be calculated by following one of the accepted procedures found in standard design handbook sources such as Refs. 3, 6, and 7.

If the computer method is used to size the project hydraulic piping systems, the designer must choose a program from among the several available. Two of the most widely used are Trane's CDS Water Piping Design program and Carrier's E20-II Piping Data program. In addition to determining the pipe sizes, both programs print a complete bill of materials (quantity takeoff by pipe size, length, fittings, and insulation). Whichever program is used, it is mandatory that the specific program's input and operating instructions be strictly followed. It is common to trace erroneous or misleading computer output data to mistakes in inputting design data into the computer. *In order to get meaningful output data, input data must be correctly entered and checked after entry before the program is run.* It is also a good, if not mandatory, policy to independently check the computer results the first time you run a new or modified program, to ensure that the results are valid.

If the computer program used does not correct the computer output for the effects of altitude when the elevation of the project is equal to or greater than 2500 ft (760 m) above sea level, the computer output must be manually corrected by using the appropriate correction factors, listed in Appendix A.

DX Piping Design (Version 1.0).   Described in the following summary, this program is available to the designer using Carrier's E20-II DX Piping Design to size the refrigerant systems.

• This program will determine the minimum piping size to deliver the refrigerant between compressor, condenser, and evaporators while ensuring return at maximum unloading.

• This program is able to size piping systems using ammonia and Refrigerants 12, 22, 500, 503, 717.

• This program is capable of calculating low-temperature as well as comfort cooling applications.

• This program determines when double risers are needed, sizes the riser, and calculates the pressure drop.

- This program will include accessories in the liquid line and automatically calculates the subcooling required.
- This program permits entering, for all fittings and accessories, pressure drops in degrees Fahrenheit or pounds per square inch.
- This program will size copper or steel piping.
- This program can select pipe size based on the specific pressure drop.
- This program will calculate the actual pressure drop in degrees Fahrenheit and pounds per square inch for selected size.
- This program will estimate piping takeoff, listing by pipe size the quantities of linear feet, fittings, valves, etc.

REFERENCES

1. Cameron hydraulic data published by Ingersoll Road Company, Woodcliff Lake, NJ.
2. "Flow of Fluids through Valves, Fittings, and Pipe," Technical Paper 410, Crane Company, New York.
3. *1993 ASHRAE Handbook, Fundamentals*, ASHRAE, Atlanta, GA, 1985, chap. 33.
4. Carrier Corp., *Handbook of Air Conditioning System Design*, McGraw-Hill, New York, 1965, part 3, chaps. 1, 2.
5. Ibid., part 3, chaps. 1, 4.
6. Ibid., part 3, chaps. 1, 3.
7. *Trane Reciprocating Refrigeration Manual*, Trane Company, La Crosse, WI, 1989.

25.3 OIL AND GAS PIPING

Cleaver-Brooks, Division of Aqua-Chem, Inc.
Milwaukee, Wisconsin

25.3.1 INTRODUCTION

The fuel oil piping system consists of two lines. The suction line is from the storage tank to the fuel oil pump inlet. On small burners the fuel oil pump is an integral part of the burner. The discharge line is from the fuel oil pump outlet to the burner. On systems that have a return line from the burner to the storage tank, this return line is considered part of the discharge piping when the piping losses are calculated.

25.3.2 OIL PIPING

25.3.2.1 Suction

Suction requirements are a function of:

1. Vertical lift from tank to pump
2. Pressure drop through valves, fittings, and strainers
3. Friction loss due to oil flow through the suction pipe. This loss varies with:

   a. Pumping temperature of the oil, which determines viscosity
   b. Total quantity of oil being pumped
   c. Total length of suction line
   d. Diameter of suction line

To determine the actual suction requirements, two assumptions must be made, based on the oil being pumped. First, the maximum suction pressure on the system should be as follows:

No. 2 oil	12 in Hg (305 mmHg)
No. 4 oil	12 in Hg (305 mmHg)
Nos. 5 and 6 oil	17 in Hg (432 mmHg)

Second, the lowest temperature likely to be encountered with a buried tank is 40°F (5°C). At this temperature the viscosity of the oil would be:

No. 2 oil	68 SSU* (12.5 cSt)
No. 4 oil	1000 SSU (21.6 cSt)

In the case of Nos. 5 and 6 oil, the supply temperature of the oil should correspond to a maximum allowable viscosity of 4000 SSU (863 cSt). This viscosity corresponds to a supply temperature of 110 to 225°F (43 to 105°C) for commercial grades of Nos. 5 and 6 oils.

Then, using Fig. 25.3.1 and entering at 4000 SSU and going horizontally to the No. 5 fuel range, the maximum corresponding temperature is about 70°F (21°C). Likewise, the maximum corresponding temperature for No. 6 fuel is about 115°F (46°C).

The suction pressure limits noted above also allow for the following:

1. The possibility of encountering lower supply temperatures than indicated above, which would result in higher viscosities
2. Some fouling of suction strainers
3. In the case of heavy oil (Nos. 5 and 6), pump wear, which must be considered with heavy oils (See Figs. 25.3.3 to 25.3.6 for suction pressure curves.)

Strainers.   It is a good practice to install suction-side strainers on all oil systems to remove foreign material that could damage the pump. The pressure drop associated with the strainer must be included in the overall suction pressure requirements.

Strainers are available as simplex or duplex units. Duplex strainers allow the ability to inspect and clean one side of the strainer without shutting down the flow of oil.

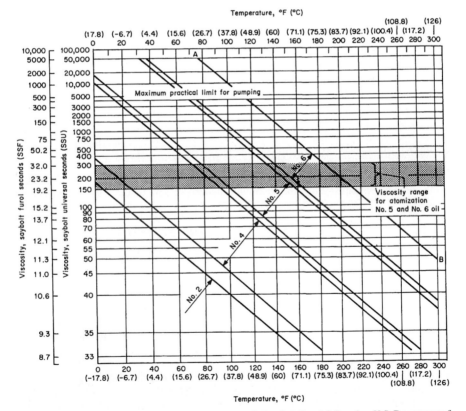

FIGURE 25.3.1   Viscosity-temperature curves for fuel oil Nos. 2, 4, 5, and 6. Based on U.S. Department of Commerce's Commercial Standard CS12-48. (*Courtesy of Cleaver-Brooks.*)

25.3.2.2 Discharge

Pumps. Pumps for fuel oil must be chosen based on several design criteria; viscosity of fuel oil, flow requirements, discharge pressure required, and fluid pumping temperature.

Viscosity. Charts for commercial grades of fuel oil are shown in Fig. 25.3.1. The pump must be designed for the viscosity associated with the lowest expected pumping temperatures.

Flow. Fuel oil pumps should be selected for approximately twice the required flow at the burner. The additional flow will allow for pressure regulation, so that constant pressure can be supplied at the burner.

Pressure. The supply pressure of the pump is based on the required regulated pressure at the burner.

A system utilizing a variable orifice for flow control typically requires from 30 to 60 psig (207 to 414 kN/m^2). The metering orifice type of system can be used on all grades of fuel oil. Burners utilizing an oil metering pump usually limit the supply pressure to prevent seal failure. As with metering orifices, there is no limitation on the grade of fuel oil used.

Temperature. The temperature of the oil must be considered, to ensure that the seals and gaskets supplied can withstand the fluid temperature.

Pumping. The major difference between calculating hydronic and fuel oil piping systems is that the actual specific gravity of the oil being pumped must be accounted for.

The design pump head is equal to the suction lift, dynamic piping loss (including fittings, valving, etc.), and required supply pressure at the burner (if applicable).

Figure 25.3.2 should be used to determine the equivalent length of straight pipe that results in the same pressure drop as the corresponding pipe fitting or valve.

Figures 25.3.3 to 25.3.8 should be used to determine the appropriate dynamic piping losses with respect to type of oil being pumped, flow rate, and pipe size. The total equivalent length of straight pipe for fittings and valving, from Fig. 25.3.8 must be added to the total length of horizontal and vertical piping before multiplying by the appropriate piping loss factor.

The pressure loss for each strainer generally must be calculated separately and added to the total.

To obtain the suction lift in inches (millimeters) of mercury (Hg) from the bottom of the suction pipe (in the tank) to the boiler connection or pump suction centerline, multiply this vertical distance in feet (meters) by 0.88155 inHg/ft of water (73.428 mmHg/m of water) by the specific gravity of the oil being pumped.

For No. 2 oil with a specific gravity of 0.85 at maximum 40 SSU and 100°F (37.8°C);

$$\text{Suction lift} = Cd \qquad (25.3.1)$$

where the suction lift is in inHg (J), C is in inHg/ft (mmHg/m), and d is in ft (m).

Heaters. Heaters are used to increase fuel oil temperatures, to provide the viscosity to atomize properly. Oil temperatures corresponding to a viscosity of 100 SSU [2×1.6 centistokes (cSt)] or less are recommended.

Heating can be accomplished by using hot water, steam, electricity, or a combination of these. Most packaged boilers have heaters that utilize electric elements for initial warmup and then transfer to either hot water or steam when the boiler has reached sufficient temperature and pressure. The heater sizing should be based on the supply pump design flow rate and temperature.

Electric heaters are commonly used to preheat heavy fuel oils on low-temperature hot-water boilers or on startup of a high-temperature hot-water or steam boiler.

The watt density of an electric heater should not exceed 5 W/in^2 (0.007 W/mm^2) because of dangers with vapor lock and coking on the heater surface. When steam is used as the heating medium for heavy oils, the steam pressure used should have a saturation temperature at least equal to the desired oil outlet temperature.

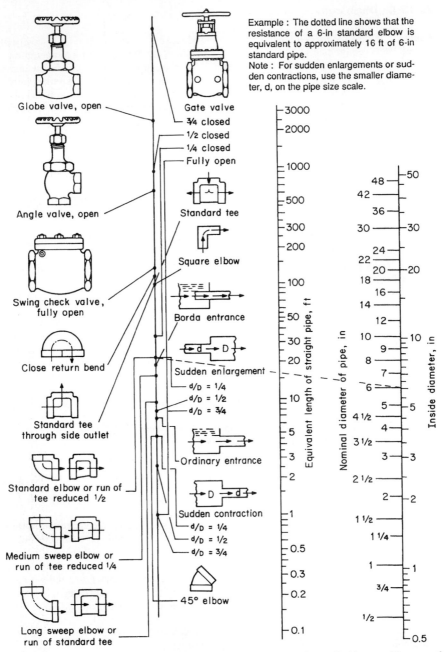

Example : The dotted line shows that the resistance of a 6-in standard elbow is equivalent to approximately 16 ft of 6-in standard pipe.
Note : For sudden enlargements or sudden contractions, use the smaller diameter, d, on the pipe size scale.

FIGURE 25.3.2   Friction losses in pipe fittings. The chart may be used for any liquid or gas. (*Courtesy of Cleaver-Brooks.*)

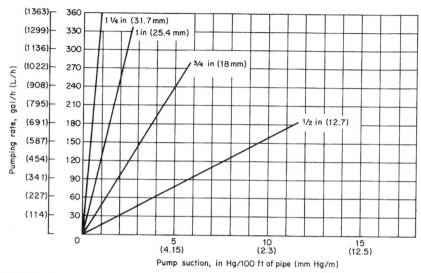

FIGURE 25.3.3 Pump suction curves for No. 2 fuel oil. Curves are based on a pumping temperature of 40°F (4.4°C), or 68 SSU. (*Courtesy of Cleaver-Brooks.*)

The flow of steam is controlled by using a solenoid valve that responds to a signal from the oil heater thermostat.

Some steam heaters include electric heating elements to allow firing of oil on a cold startup. When sufficient steam pressure is available, the electric heater is automatically de-energized.

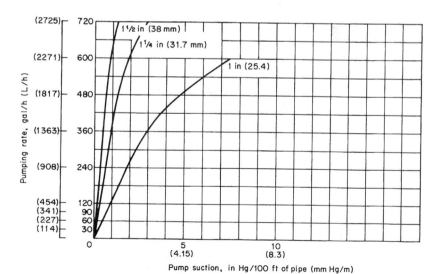

FIGURE 25.3.4 Pump suction curves for No. 2 fuel oil. Curves are based on a pumping temperature of 40°F (4.4°C), or 68 SSU. (*Courtesy of Cleaver-Brooks.*)

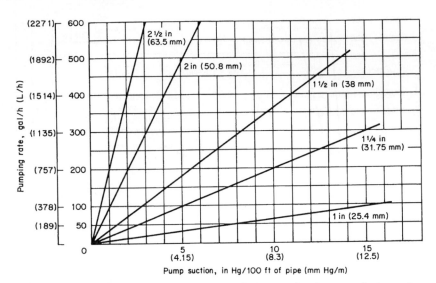

FIGURE 25.3.5 Pump suction curves for No. 4 fuel oil. Curves are based on a pumping temperature of 40°F (4.4°C), or 1000 SSU. (*Courtesy of Cleaver-Brooks.*)

Steam from the boiler is regulated to the desired pressure for sufficient heating. If the boiler pressure exceeds the steam heater pressure by 15 lb/in² (1 bar) or more, superheated steam will be produced by the throttling process. Steam heater lines should be left uninsulated to allow the steam to desuperheat prior to entering the heater. It is common practice to discharge the steam condensate leaving the oil heater to the sewer, to eliminate the

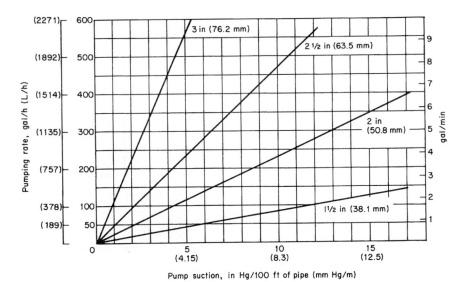

FIGURE 25.3.6 Pump suction curves for Nos. 5 and 6 fuel oils. Curves are based on a pumping limit of 4000 SSU. (*Courtesy of Cleaver-Brooks.*)

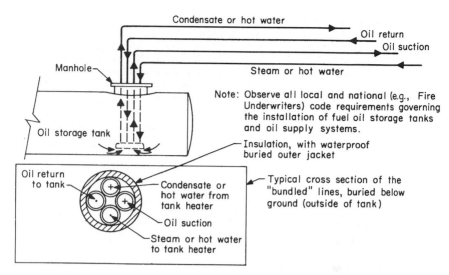

Note: The temperature of the oil suction line should not exceed 130°F (54.4°C). Higher temperatures could cause vapor binding of the oil pump, which would decrease oil flow.

FIGURE 25.3.7   Tank heaters. (*Courtesy of Cleaver-Brooks.*)

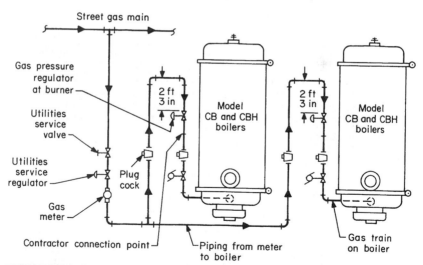

FIGURE 25.3.8   Gas piping to boiler. The figure illustrates the basic gas valve arrangement on boilers and shows the contractor's connection point for a typical installation. Actual requirements may vary depending on local codes and local gas company requirements, which should be investigated prior to both the preparation of specifications and construction. (*Courtesy of Cleaver-Brooks.*)

possibility of contaminating the steam system in the event of an oil leak. The heat from the condensate is usually reclaimed prior to dumping it.

Excessive steam temperatures can also cause coking in the heater.

Hot-water oil heaters are essentially water-to-oil heat exchangers used to pre-heat oil. However, since the source of heat energy is boiled water circulated by the pump through the heater, any system leak could cause boiler water contamination. Therefore, safety-type heater systems are recommended for this service. Such an exchanger is frequently a double-exchange device using an intermediate fluid.

In cases where the oil must be heated to a temperature in excess of the hot-water supply temperature, supplemental heat must be provided by an electric heater. Tank heaters are commonly an insulated bundle of four pipes submerged in the oil tank. See Fig. 25.3.7. Tank pre-heating is required anytime the viscosity of the oil to be pumped equals 4000 SSU or greater.

Valves

Pressure Relief Valves.   These are installed in the discharge line from the supply pump, to protect the pump and system from over-pressure. Pressure relief valves are also commonly installed on oil heaters to relieve pressure so that oil may circulate even though the burner does not call for oil.

Pressure Regulators.   These reduce system pressure and maintain a desired pressure at the burner.

Oil Shutoff.   There are two commonly used styles of oil shutoff valves for burner service: electric coil and motorized. Electric coil solenoid valves are used on most small industrial and commercial burners. These valves are normally closed valves, and they control the flow of oil fuel to the burner. Two such valves for fuel shutoff are used on commercial and industrial boilers.

The second type of oil shutoff valve is a motorized valve that has a spring return to close. Motorized valves can be equipped with a proof-of-closure switch which ensures that the valve is in the closed position or prevents the burner from igniting if it is not. This type of switch is necessary to meet certain insurance requirements.

Manual Gas Shutoff Valves.   Manual gas shutoff valves are typically a lubricated plug type of valve with a 90° rotation to open or close. The valve and handle should be situated such that when the valve is open, the handle points in the direction of flow.

The number of valves and their locations are based on insurance requirements. Typically, manual valves are installed upstream of the gas pressure regulator, directly downstream of the gas pressure regulator, and downstream of the last automatic shutoff valve.

Automatic Gas Shutoff Valves.   Three types of automatic gas shutoff valves are used on burners: solenoid valves, diaphragm valves, and motorized valves.

Of the three automatic valves, the solenoid is the simplest and generally the least expensive. A controller opens the valve by running an electric current through a magnetic coil. The coil, acting as a magnet, pulls up the valve disk and allows the gas or oil to flow. Solenoid action provides fast opening and closing times, usually less than 1 s.

Diaphragm valves are frequently used on small to medium boilers. These valves have a slow opening and fast closing time. They are simple, dependable, and inexpensive. They are full-port valves and operate with little pressure loss.

Motorized shutoff valves are used for large gas burners that require large quantities of gas and relatively high gas pressures. There are two parts to a motorized valve: the valve and a fluid power actuator. A limit switch stops the pump motor when the valve is fully open. The valve is closed by spring pressure. The valve position (open or closed) is visible through windows on the front and side of the actuator. Motorized valves often contain an override switch which is actuated when the valve reaches the fully closed position. This proof-of-closure switch is needed to meet several different insurance company requirements.

Vent Valves.   Vent valves are normally open solenoid valves that are wired in series and are located between two automatic shutoff valves in the main gas line or, in some cases, the pilot line. The vent valve vents to the atmosphere all gas contained in the line between the two valves.

Flow Control Valves

1. Butterfly valves are the most commonly used device for controlling the quantity of fuel gas flow to the burner. The pressure drop associated with a fully open butterfly valve is very low. Butterfly valves can be used for control of air flow and, with special shaft seals, can be used for all grades of fuel gas. Linkage arms are connected to the shaft of the valve and driven directly from the burner-modulating motor.

2. Modulating gas shutoff valves can be supplied with positioning motors that can operate on the on/off or high/low/off principle. In the case of the high/low/off shutoff valves, the air damper is controlled by the valve modulating motor. This allows the valve position to dictate the amount of combustion air necessary for the gas input rate.

3. Pneumatic control valves are often butterfly valves that are driven by a pneumatic actuator. The signal to the pneumatic actuator is proportional to the combustion air flow and positions the valve to deliver the appropriate amount of gas. Often additional signals such as steam flow and combustion air flow are used to determine the signal to the valve and its corresponding position.

Gas Strainer.   It may be advisable to use a strainer to protect the regulators and other control equipment against any dirt or chips that might come through with the gas.

Gas Compressors or Boosters.   If the local gas utility cannot provide sufficient gas pressure to meet the requirements of the boiler, a gas compressor or booster should be used. *Caution:* The use of a gas compressor or booster must be cleared with the local gas utility prior to installation.

25.3.3   GAS PIPING

Figure 25.3.8 illustrates the basic arrangement for piping gas to boilers from street gas mains for a typical installation.

25.3.3.1   Line-Sizing Criteria

The first step in designing a gas piping system is to properly size components and piping to ensure that sufficient pressure is available to meet the demand at the burner. The boiler manufacturer should be consulted to determine the pressure required.

The gas service piping installed in the building must be designed, and components selected, to provide the required fuel gas flow to the boiler at the manufacturer's recommended pressure. The utility supplying gas to the facility will provide the designer with information on the maximum available gas pressure for the site. The gas piping design must be appropriate for the specific site conditions.

The gas train pressure requirements can be expressed as

$$P_s = P_R + P_C + P_P + P_F + P_B + P_{fp}$$

(25.3.2)

where P_s = supply pressure available
  P_R = pressure drop across gas pressure regulator
  P_C = pressure drop across gas train components
  P_P = pressure drop associated with straight runs of pipe
  P_F = pressure drop associated with elbows, tees, or other fittings
  P_B = pressure drop across burner orifice or annulus
  P_{fp} = boiler furnace pressure

Pressure drop calculations for regulators and valves are normally based on the C_v factor or coefficient of value capacity of air or in equivalent feet or diameters of pipe length.

The resistance coefficient k can be used to express the pressure drop as a number of lost velocity heads

$$k = \frac{PV^2}{2g} \qquad (25.3.3)$$

Depending on the information available, the following equations can be used to determine the pressure drop through valves or across regulators:

$$k = f \cdot \frac{L}{D} \qquad (25.3.4)$$

$$P = f \cdot \frac{L}{D} \cdot \frac{PV^2}{2g} \qquad (25.3.5)$$

$$k = \frac{891 d^4}{C_v^2} \qquad (25.3.6)$$

$$H_v = 0.000228 V^2 \text{ inWG} \qquad \text{for air} \qquad (25.3.7)$$

$$P = \frac{k}{144} H_v \qquad (25.3.8)$$

$$C_v = 0.0223 (\text{ft}^3/\text{h}) @ 1\text{-inWG drop}) G \quad \text{for 0- to 2-psig gases} \qquad (25.3.9)$$

where k = resistance coefficient
  f = Darcy friction factor
  L = length of pipe or equivalent length of pipe for fitting, ft
  D = diameter of pipe, ft
  P = pressure drop or differential, lb/in^2
  V = velocity, ft/s
  C_v = valve conductance based on H_2O @ 1 lb/in^2 drop
  g = acceleration of gravity
  H_v = velocity head
  G = gas gravity relative to air = $P/0.0765$
  p = density of flowing fluid, lb/ft^3

Note: Metric units must be converted to English units before Eqs. (25.3.2) to (25.3.9) can be applied.

To determine the losses associated with straight runs of pipe (P_p) and pipe fittings (P_f), Eq. (25.3.2) can be used. Values for equivalent length of pipe or equivalent pipe diameter are listed in Fig. 25.3.4. The pressure drop for the burner orifice or annulus (P_B) can be calculated by using Eq. (25.3.5) and making the appropriate gas density corrections. The furnace pressure P_{fp} is a function of the furnace geometry, size, and firing rate. This pressure is often zero or slightly negative, but for some types of boilers and furnaces it can run as high as 15 in water column (inWC) (381 mm) positive.

25.3.3.2 Gas Train Components

Pressure Regulators. Pressure regulators or pressure-reducing regulators are used to reduce the supply pressure to the level required for proper burner operation. The regulated, or downstream, pressure should be sufficient to overcome line losses and deliver the proper pressure at the burner. Pressure regulators commonly used on burners come in two types: self-operated and pilot-operated.

In a self-operated regulator, the downstream, or regulated, pressure acts on one side of a diaphragm, while a preset spring is balanced against the backside of the diaphragm. The valve will remain open until the downstream pressure is sufficient to act against the spring.

Regulators for larger pipe sizes are normally the pilot-operated type. This class of equipment provides accurate pressure control over a wide range of flows and is sometimes selected even in smaller sizes where improved flow control is desired.

A gas pressure regulator must be installed in the gas piping to each boiler. The following items should be considered when a regulator is chosen.

1. *Pressure rating.* The regulator must have a pressure rating at least equivalent to that in the distribution system.

2. *Capacity.* The capacity required can be determined by multiplying the maximum burning rate by 1.15. This 15 percent over-capacity rating of the regulator provides for proper regulation.

3. *Spring adjustment.* The spring should be suitable for a range of adjustment from 50 percent under the desired regulated pressure to 50 percent over.

4. *Sharp lockup.* The regulator should include this feature because it keeps the downstream pressure (between the regulator and the boiler) from climbing when there is no gas flow.

5. *Regulators in parallel.* This type of installation would be used if the required gas volume were very large and if the pressure drop had to be kept to a minimum.

6. *Regulators in series.* This type of installation would be used if the available gas pressure were over 5, 10, or 20 psig (34.5, 68.0, or 137.9 kPa), depending on the regulator characteristics. One regulator would reduce the pressure to 2 to 3 psig (17.8 to 20.7 kPa), and a second regulator would reduce the pressure to the burner requirements.

7. *Regulator location.* A straight run of gasline piping should be used on both sides of the regulator to ensure proper regulator operation. This is particularly important when pilot-operated regulators are used. The regulator can be located close to the gas train connection, but 2 to 3 ft (0.6 to 0.9 m) of straight-run piping should be used on the upstream side of the regulator. *Note:* Consult your local gas pressure regulator representative. She or he will study your application and recommend the proper equipment for your job.

25.4 PUMPS FOR HEATING AND COOLING*

James E. Hope
Director of Technical Services,
ITT Bell & Gossett, Morton Grove, Illinois

25.4.1 INTRODUCTION

Pumps can be classified into two basic types—*centrifugal* and *positive-displacement*. Each operates on a different principle, and their basic design differences influence their areas of application.

The *centrifugal* pump produces a head differential, and thus a flow, by increasing the velocity of the liquid through the machine with a rotating vaned impeller. The conversion of a portion of the input energy to head and velocity energy by the impeller within the pump casing produces the flow and head differential from suction to discharge. It follows that the design of the impeller and casing controls the relationship between head, flow, and input horsepower (wattage) at any given speed. Since the design of the impeller and casing can be widely varied, the centrifugal pump is a very versatile machine for pumping the liquids required in most applications today.

The *positive-displacement* pump operates on an entirely different principle: the alternation of filling a cavity and then displacing a given volume of liquid. Pumps designated as power, steam, gear, and rotary pumps fall into this classification. Power pumps are used primarily for applications requiring small flows at high pressures. Steam-operated reciprocating pumps are economical when low rates at moderate pressures are required, such as for feed pumps in smaller plants or for utilizing waste process steam. The principal advantage of rotary pumps is their ability to deliver a constant volume of liquid against a varying discharge head.

Whether centrifugal or positive-displacement pumps are involved, it is obvious, (but often overlooked) that all the wetted parts of the pump must be of materials that are compatible with the fluid being pumped in terms of corrosion, erosion, and temperature.

25.4.2 CENTRIFUGAL PUMPS

25.4.2.1 General Theory

The application of centrifugal pumps requires a basic knowledge of some design parameters and the general performance characteristics of the wide range of pumps now available. Since the centrifugal pump produces head by accelerating the liquid through the impeller and then converting the velocity into head in the casing, the basic relation is $H = V^2/2g$, where H is the head in ft (m) of water, V is the velocity in ft/s (m/s), and g is the constant for the acceleration of 32.2 ft/s^2 (0.981 m/s^2). ["Feet (meters) of head" is a common pump term used to express foot-pound (Newton-meter) force per pound (kilogram) mass of specific

---

*Updated chapter written by Warren Fraser and Will Smith, Dresser Pump Division, Dresser Industries, Mountainside, NJ.

energy added to the liquid.] By means of this simple power law, the performance of a given pump at any one speed can be predicted at any other speed merely by varying (1) the head produced by the square of the rotational speed, (2) varying flow or capacity directly as the change in the rotational speed, or (3) varying horsepower (wattage) required by the cube of the speed change. These relationships are designated the "affinity laws" and are expressed mathematically as:

$$\frac{H_1}{H_2} = \frac{N_1^2}{N_2^2} \quad \frac{Q_1}{Q_2} = \frac{N_1}{N_2} \quad \frac{P_1}{P_2} = \frac{N_1^3}{N_2^3} \qquad (25.4.1)$$

where H = head of fluid of liquid being pumped, ft (m)
N = rotational speed of pump, r/min
Q = capacity of pump, gal/min (m³/h)
P = power required, hp (kW)

There are, however, two limitations to the range throughout which the *affinity laws* can be applied. The first is a mechanical limitation of the maximum speed at which the pump is designed to operate. This limitation is set by the manufacturer and should not be exceeded.

The second limitation is the head required on the suction side of the pump to prevent cavitation at the impeller inlet. Adequate suction head, or "net position suction head" (NPSH), is mandatory for satisfactory pump operation and should be carefully checked for any application, regardless of pump size or type. The suction-head limitation is usually stated by the pump manufacturer as the "required net positive suction head" (NPSHR). In practice, the "available net positive suction head" (NPSHA) at the pump suction must equal or exceed the NPSHR. If it does not, the pump's output will be reduced and the pump will be damaged mechanically.

The NPSHA at the pump suction can be determined as soon as the piping design has been completed and the absolute pressure of the source of the liquid to be pumped has been established. Since the NPSH is an atmospheric-pressure head, the friction loss through the piping must be expressed in terms of feet (meters) of head as well as in terms of the absolute pressure of the liquid source. To determine the NPSHA, use the relation

NPSHA = static head + atmospheric pressure − friction head loss
− vapor pressure   (25.42)

In an open suction system, the static head is the height in feet (meters) of the liquid's surface above (negative if below) the center of the pump-inlet connection; in a closed system, it is the positive suction system pressure expressed in feet (meters) of liquid pumped. The atmospheric pressure is the local barometric pressure expressed in feet (meters); at sea level it is 34 ft (10.4 m) of water. The friction head loss is the loss due to fluid friction of the liquid in the suction line.

Example 25.4.1 Say that the NPSHA at the pump suction drawing from a cooling tower basin is as follows:

1. The static difference in elevation between the level in the basin and the pump centerline (which is positive for a submerged impeller) = 6 ft (1.82 m).
2. Since the water surface in the basin is open to the atmosphere, the absolute head on the surface is the head corresponding to the absolute atmospheric pressure = 33 ft (10.2 m).
3. The friction loss in the suction piping at, say, 2000 gal/ (7600 L) min = 3 ft (0.91 m).
4. The vapor pressure of the water at the impeller at, say, 68°F (20°C) = 0.8 ft (24 cm).

Solution   From Eq. (25.4.2) we have

$$NPSHA = 6 \text{ ft} + 33 \text{ ft} - 3 \text{ ft} - 0.8 \text{ ft} = 35.2 \text{ ft}$$
$$= (1.82 \text{ m} + 10.2 \text{ m} - 0.91 \text{ m} - 0.24 \text{ m} = 10.8 \text{ m})$$

Thus the pump selected must be capable of pumping 2000 gal/min (7600 L/min) with an NPSHR of 35.2 ft (10.8 m) or less.

A more generalized means to determine the NPSHR for centrifugal pumps has been developed in the form of specific-speed limitations, as published in the *Hydraulic Institute Standards*. The term "specific speed" is used here to designate the relation between capacity, head, and speed for any given design at the point of maximum efficiency.* It is a dimensionless index number and has no relation to the size of the pump. It does have significance, however, in determining the maximum rotational speed permissible for a given capacity and head and the NPSHA in the design stages of a pump system. Specific speed is defined as

$$N_s = \frac{N\sqrt{Q}}{H^{0.75}} \tag{25.4.3}$$

where N_s = specific speed,† gal/ (min·ft) [m^3/ (s·m)]
N = rotational speed of pump, r/min
Q = capacity of pump at maximum efficiency, gal/min (m^3/s)
H = total head of pump at maximum efficiency, ft (m) of liquid

With a given required flow and head and NPSHA, it is thus possible with the Hydraulic Institute's specific-speed limitations to determine the maximum pump rotational speed. This kind of analysis is useful in determining the most economic relation between the initial pump and system costs.

The concept of specific speed is also useful in designating general performance characteristics for the entire range of good centrifugal pump designs. Figure 25.4.1 to 25.4.4 show the variation in head, capacity, input horsepower (wattage), and efficiency as well as the general impeller configuration for selected specific speeds. It is important to consider the relation between head, capacity, and input horsepower (wattage) throughout the expected range of operation, so that an adequately powered driver can be provided and so that the NPSHR does not exceed the NPSHA at any point. With an increase in flow, the NPSHR increases, while, in most installations, the NPSHA decreases because of the increased piping losses. It follows, therefore, that the NPSHA and NPSHR should be compared not at the design point, but at the maximum expected capacity. The divergence of the NPSHR and NPSHA with increasing flows also points out the danger of overestimating the system discharge head against which the pump must operate. Referring the pump characteristics shown in Figs. 25.4.1 to 25.4.4, it can be seen that if the head is lower than estimated, the pump will operate at an increased capacity and require an increased NPSH that was not allowed for in the pump selection.

Optimum efficiencies for centrifugal pumps are obtained in the 2000 to 3000 specific-speed range. Figure 25.4.5 shows the variation in pump efficiency with specific speed and rated capacity as well as typical impeller designs for the range of specific speeds.

---

*The definition of "specific speed" is the speed of a pump for unit head and unit capacity at the point of best efficiency.
†While N_s is dimensionless in concept, the use of inconsistent units for flow and head results in numerical differences between the conventional and metric systems.

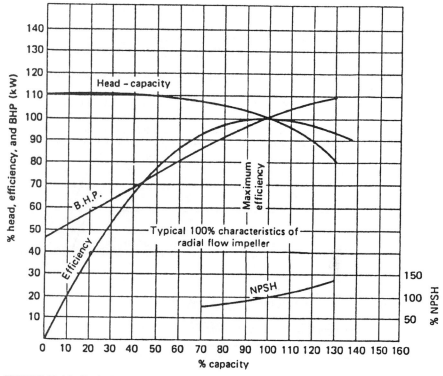

FIGURE 25.4.1 Typical 100 percent characteristics of radial-flow impeller, N_s = 1100 gal · min/ft (21.3 m³ · s/m). (*Courtesy of Dresser Pump Division, Dresser Industries, Inc.*)

When feasible, it is advisable to operate a centrifugal pump at or near the capacity corresponding to the point of maximum efficiency, as shown on the characteristic performance curve. This is not only to save on power cost but also to prevent reduction in the life of pump components, which is caused when operation occurs too far to the right or left of the maximum-efficiency capacity. The impeller vanes are designed to operate at the flows corresponding to those at peak efficiency, and operation at higher or lesser flows means that the vane angles are mismatched to the flow angles; accelerated erosion on the impeller and casing can result, with accompanying noise, vibration, and unstable operation. This is particularly relevant where quiet or continuous operation is required. The recommended range of operation for a typical centrifugal pump is shown in Fig. 25.4.2.

The power requirement for a centrifugal pump can be determined from the relation

$$P = \frac{Q \times H \times S}{C \times E} \qquad (25.4.4)$$

where P = power, BHP (kW)
Q = flow rate, fal/min (m³/h)
H = head of fluid being pumped, ft (m)
S = specific gravity of the fluid

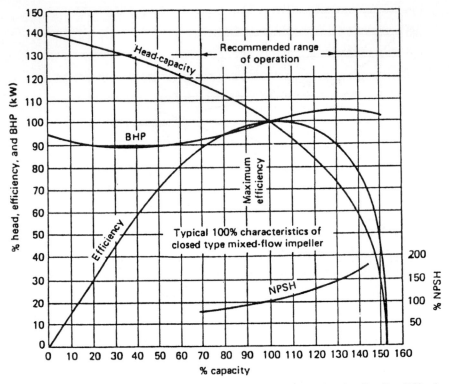

FIGURE 25.4.2 Typical 100 percent characteristics of closed-type mixed-flow impeller. $N_s = 4000$ gal · min/ft (77.4 m³ · s/m). (*Courtesy of Dresser Pump Division, Dresser Industries Inc.*)

C = a constant depending on the system of units used: 3960 for English units (3678 for metric units)

E = pump efficiency, expressed as a decimal

25.4.2.2 Parallel and Series Operation

Two or more pumps can be installed in parallel to increase the capacity of any given system or to match the flow requirements of a variable-capacity system. The increase in flow or the incremental steps in pumping rates will depend on the shape of the head capacity curve, the characteristic of the system curve, and the number of pumps in operation. The system curve is the head against which the pumps must operate and may consist of static head, friction head, or a combination of static and friction head. The intersection of the pump head capacity curve and the system head determines the capacity that will be delivered by the pump(s).

An example of two centrifugal pumps operating in parallel is shown in Fig. 25.4.6. With pump 1 in operation, it will deliver 100 percent capacity and intersect the system curve shown at point A. The same pump operating against a pure static head will intersect at point C and deliver 114 percent capacity. Additional capacity can be produced by operating pump 2 in parallel with pump 1. The two pumps will now intersect the system curve at point B and deliver 153 percent capacity. The two pumps operating against a pure static head will deliver 228 percent capacity and intersect the static head at point D.

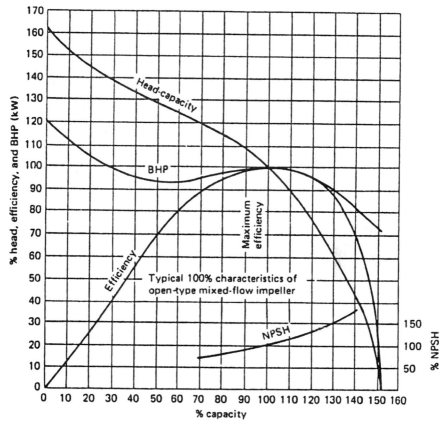

FIGURE 25.4.3 Typical 100 percent characteristics of open-type mixed-flow impeller. N_s = 6250 gal · min/ft (120.9 m³ · s/m). (*Courtesy of Dresser Pump Division, Dresser Industries, Inc.*)

The same two operating in series and pumping against a pure friction curve system is shown in Fig. 25.4.7. In this case, pump 1 operating alone will pump 100 percent capacity and intersect the system curve at point A. When pump 2 operates in series with pump 1, the two pumps will deliver 125 percent capacity and intersect the system curve at point B.

Of course, if the system head becomes as high as the shutoff head for either of the pumps operating in parallel (Fig. 25.4.6), that pump can contribute no capacity to the system. Figure. 25.4.7 illustrates what happens when a second pump is started up in series, as opposed to a second pump being run in parallel as shown in the previous figure.

25.4.2.3 Frame-Mounted and Close-Coupled End-Suction Pumps

Both frame-mounted and close-coupled end-suction centrifugal pumps have wide application for general and specialized services. Since many of these units are used throughout the industry, a high degree of standardization has been achieved along with an interchangeability of parts to reduce the inventory of spare parts and to meet changing pumping requirements.

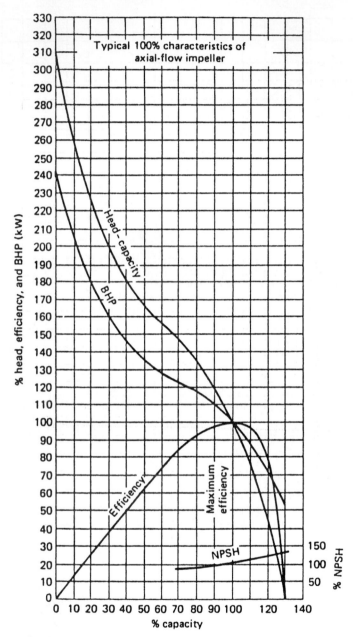

FIGURE 25.4.4　Typical 10 percent characteristics of axial-flow impeller. $N_S =$ 11,000 gal · min/ft (213 m³ · s/m). (*Courtesy of Dresser Pump Division, Dresser Industries, Inc.*)

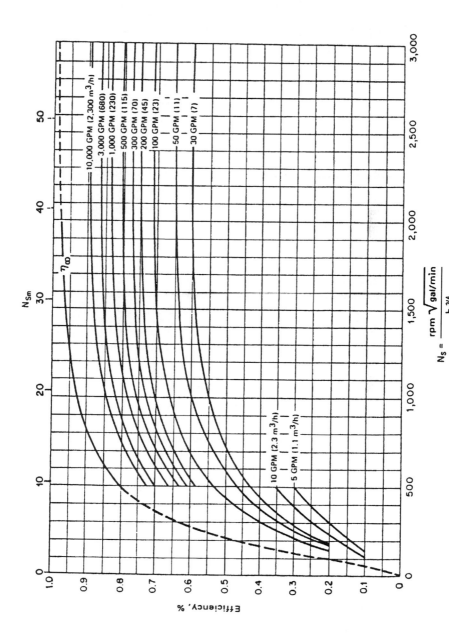

FIGURE 25.4.5 Specific speed vs. efficiency. (*Courtesy of Dresser Pump Division, Dresser Industries.*)

$$N_s = \frac{\text{rpm } \sqrt{\text{gal/min}}}{h^{3/4}}$$

N_{Sm}

Efficiency, %

10,000 GPM (2,300 m³/h)
3,000 GPM (680)
1,000 GPM (230)
500 GPM (115)
300 GPM (70)
200 GPM (45)
100 GPM (23)
50 GPM (11)
30 GPM (7)

10 GPM (2.3 m³/h)
5 GPM (1.1 m³/h)

η_{∞}

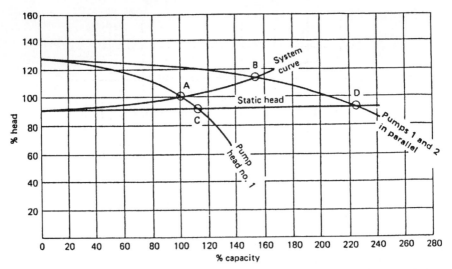

FIGURE 25.4.6   Two pumps in parallel. (*Courtesy of Dresser Pump Division, Dresser Industries, Inc.*)

Figure 25.4.8 shows a typical frame-mounted pump, and Fig. 25.4.9 shows a typical close-coupled pump. As seen in these two figures, the liquid inlets of both types are identical. The frame-mounted pump is used with a separate driver coupled to the pump and mounted on a common baseplate. The close-coupled unit has the liquid end bolted directly onto the motor frame, with the impeller mounted on the motor shaft. The close-coupled unit has the advantage of compactness, and it eliminates the problem of maintaining alignment between pump and driver.

Many HVAC pumps are located in very compact spaces. Some are located above dropped ceilings and are supported by the pipeline in which they are installed. We call these pumps *in-line* or *pipe-mounted* pumps. Usually no baseplate or foundation is required. Figure 25.4.10 illustrates an in-line pump.

Double-Suction Centrifugal Pumps.   The double-suction impeller design is basically two single-suction impellers placed back to back in an integral casing. As shown in Fig. 25.4.11 the incoming flow is divided before entering the impeller, with one-half of the total flow entering each side of the impeller. The inlet velocities and the NPSHR are thereby reduced as compared to a single-suction impeller delivering the same total flow. In many applications, the low NPSHR will often dictate the use of a double-suction pump. The following inherent advantages in double-suction design will require careful consideration in the selection of the type of pump best-suited to a particular set of conditions:

1. Since the impeller is symmetrical about a plane normal to the axis, the hydraulic end thrust is virtually eliminated. This means that smaller bearings are needed to produce the same bearing life as does a single-suction pump of the same rating.

2. In the horizontal split case version (Fig. 25.4.11H) the suction and discharge connections are in the same line, thus simplifying the piping layout just as with in-line pumps.

3. The rotor is readily accessible by removing the top half of the horizontally split volute casing. (Fig. 25.4.11H)

4. Although shown with packing, the vast majority of HVAC pumps will be furnished with mechanical seals.

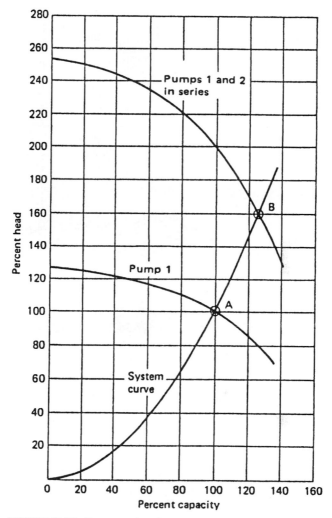

FIGURE 25.4.7   Two pumps in series. (*Courtesy of Dresser Pump Division, Dresser Industries, Inc.*)

5. The *vertical split case* has wide acceptance in the HVAC industry where overhead piping is employed, such as in chilled or hot water distribution.

6. The rotor is readily accessible from either end.

25.4.2.4   Vertical Multistage Pumps

The vertical multistage pump has found wide application throughout the industry primarily because of its compactness, low cost, and vertical arrangement. This type of pump is usually referred to as a "deep well turbine pump," although its application has gone far beyond its

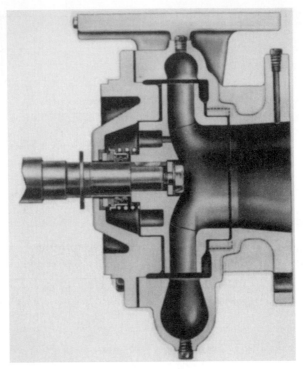

FIGURE 25.4.8   Frame-mounted end-suction pump. (*Courtesy of ITT Bell & Gossett.*)

original use as a well pump. A typical design for use in well service is shown in Fig. 25.4.12. An adaptation for use as a wet-pit general-water-service pump is shown in Fig. 25.4.13.

Figure 25.4.14 shows a further modification for use in closed system and other pressurized suction applications such as pressure boosting.

The vertical multistage pump can be provided with either an open or closed shaft arrangement. The open-shaft design is shown in Fig. 25.4.12 with the intermediate bearings supported in spiders attached at intervals to the column pipe. The bearings are lubricated by the liquid pumped and can be of any appropriate bearing material, although bronze or rubber or a combination of the two is most frequently specified. It is quite evident, however, that the open-shaft design is limited to applications pumping liquids that not only adequately lubricate the bearings but also are noncorrosive and nonabrasive to the bearing material specified.

With the closed-shaft design, the shaft and bearings are enclosed in a cover tube, as shown in Fig. 25.4.13. The bearings are usually of bronze and are threaded externally. The cover tube is threaded internally at both ends to receive the bearings, and at the top of the column a special bearing is provided that can be adjusted to impose an axial load on the cover tube to maintain bearing alignment. The bearings are usually oil-lubricated, with the lubricant supplied from a drop-feed oiler.

There are no thrust bearings in the vertical multistage pump. The thrust of the pump is carried by the special thrust bearing provided in the motor. It is almost universal practice to furnish hollow-shaft motors with deep-well pumps so that the position of the pump rotor can be adjusted to compensate for the accumulation of dimensional tolerances and the stretch of the shaft in deep settings. Another valuable feature of the hollow-shaft motor is

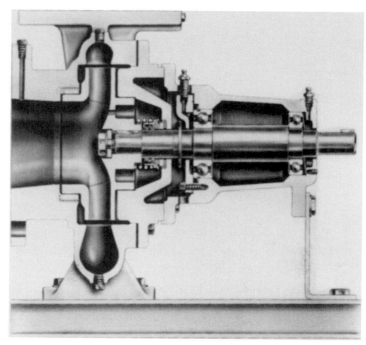

FIGURE 25.4.9   Close-coupled end-suction pump. (*Courtesy of ITT Bell & Grosset.*)

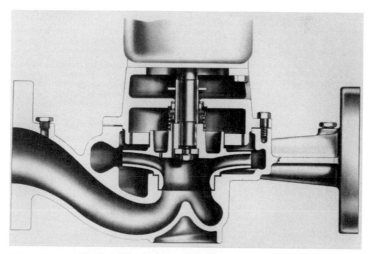

FIGURE 25.4.10   Close-coupled vertical in-line pump. (*Courtesy of ITT Bell & Grosset.*)

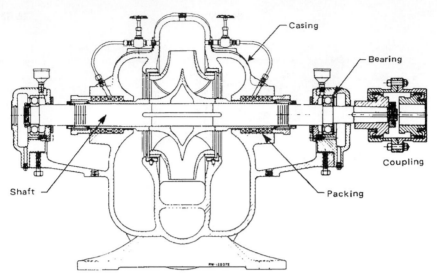

FIGURE 25.4.11H Horizontal split case double-suction pump. (*Courtesy of Dresser Pump Division, Dresser Industries, Inc.*)

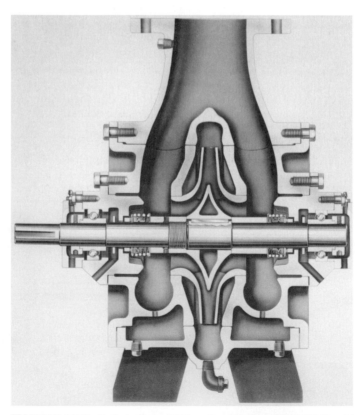

FIGURE 25.4.11V Vertical split case double suction. (*Courtesy of ITT Bell & Gossett.*)

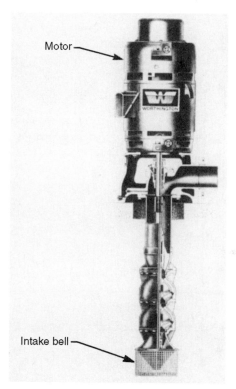

Motor

Intake bell

FIGURE 25.4.12   Deepwell turbine pump. (*Courtesy of Dresser Pump Division, Dresser Industries, Inc.*)

that adaptation of a nonreverse ratchet at the top of the motor that prevents the line shaft of the pump from backing out of the threaded couplings in the event that the motor accidentally operates in a direction of rotation opposite to that for which the pump was designed.

In the case of the wet-pit pump with a short setting (Fig. 25.4.13) or the condensate or process pump (Fig. 25.4.14), a solid-shaft motor with an intermediate flanged coupling between the pump and drive is preferred. The reason for this is that the adjustable feature of the hollow-shaft motor is not required for the shorter-setting arrangements, and a more accurate alignment of pump and motor shaft can be achieved with the solid-shaft motor than with the motor that has a hollow shaft.

25.4.2.5   Submersible Pumps

Submersible pumps are used in many building trades applications ranging from sump pumps to general purpose water pumps, sewage ejectors, and sewage effluent pumps. With submersible pumps the motor driver, as well as the pump case, can be submerged under the liquid being pumped. (See Fig. 25.4.15.)

25.4.2.6   Self-Priming Pumps

Unlike a positive-displacement pump, a centrifugal pump will not start pumping unless the liquid can flow into the pump suction under a positive head. This is an obvious disadvantage

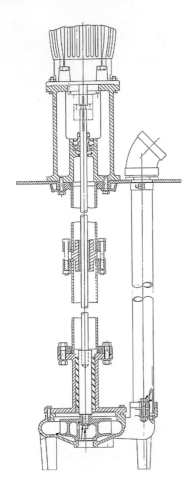

FIGURE 25.4.13 Wet-pit Pump. (*Courtesy of Paco Pumps.*)

for pumps that must operate under a suction lift, and various methods have been developed to circumvent this limitation. There is no problem with the vertical deep-well or wet-well pump, as the impeller is always located beneath the surface of the water level and is self-priming. With the centrifugal pump operating on a lift, a priming device can be used to evacuate the air from the casing and thus raise the water to the impeller eye. An auxiliary vacuum pump or an ejector operating on air pressure is commonly used for this purpose.

When vacuum pumps or ejectors are not available and where starting and stopping is frequent and automatic starting is necessary, self-priming pumps are employed. A typical self-priming pump is shown in Fig. 25.4.16. It is well known that a small centrifugal pump can be primed if sufficient liquid is forced back from the discharge to the suction, and the self-priming design utilizes this device. Priming is accomplished by trapping a small portion of the liquid, each time the pump is shut down, in a reservoir mounted on the pump casing. On the next starting cycle, this trapped liquid recirculates through the impeller, and through this pumping action and the entrainment of air bubbles, the pressure in the suction pipe is reduced to a point sufficient to raise the suction liquid level to the inlet of the impeller. The pump then operates in its normal manner.

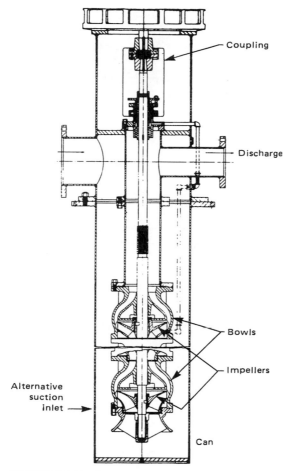

FIGURE 25.4.14 Can-type turbine pump. (*Courtesy of Dresser Pump Division, Dresser Industries, Inc.*)

25.4.2.7 Regenerative Turbine Pumps

The regenerative turbine pump produces head by adding energy to the liquid through a series of straight vanes located only at the periphery of the impeller. As shown in Fig. 25.4.17, the liquid enters the impeller along the axis of the shaft and flows to the periphery of the impeller through clearances on both sides of the impeller, and the repeated entry and reentry of liquid between the vanes produces very small capacities at high heads. Unlike a typical centrifugal pump, the regenerative turbine pump exhibits a very steep head capacity curve and is commonly found in injection pump service.

Operation of regenerative turbine pumps at less than 40 percent of the best efficiency point should be avoided, for the impeller clearances are subjected to accelerated wear, which will sharply decrease pump performance.

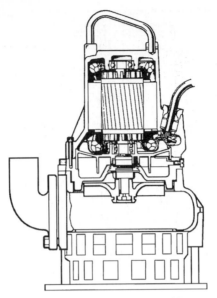

FIGURE 25.4.15   Submersible pump. (*Courtesy of ITT Flygt Corporation.*)

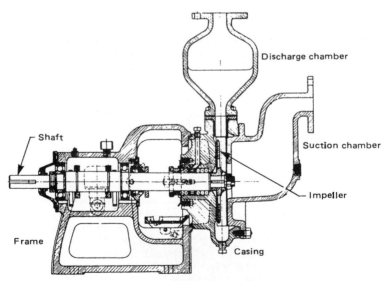

FIGURE 25.4.16   Self-priming pump. (*Courtesy of Dresser, Pump Division, Dresser Industries, Inc.*)

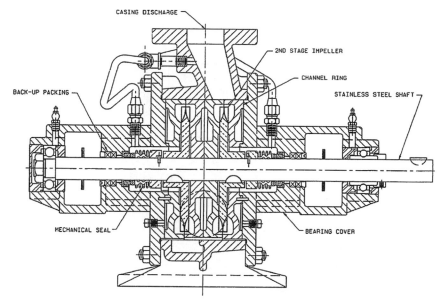

FIGURE 25.4.17 Regenerative turbine pump. (*Courtesy of Dresser, Pump Division, Dresser Industries, Inc.*)

25.4.3 *POSITIVE-DISPLACEMENT PUMPS*

25.4.3.1 Rotary Pumps

One common type of positive displacement pump is the *rotary gear pump,* which is used for handling viscous liquids such as oils, chemicals, glycols, and nonabrasive sludge. A typical gear pump is shown in Fig. 25.4.18. The liquid flows into the suction and is trapped

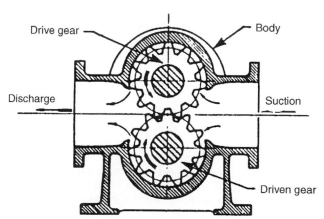

FIGURE 25.4.18 Gear pump. (*Courtesy of Dresser, Pump Division, Dresser Industries, Inc.*)

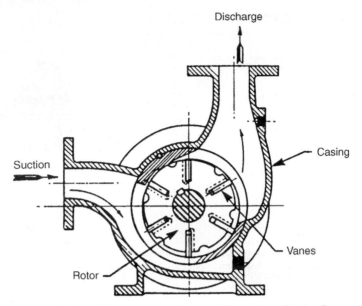

FIGURE 25.4.19   Sliding-vane pump. (*Courtesy of Dresser Pump Division, Dresser Industries, Inc.*)

between the gear teeth and the housing. As the gears turn, the liquid is carried around to the discharge and is squeezed out from between the meshing gear teeth. Spur gears may be used, although double helical gears are usually preferred, as they run more quietly at higher speeds. The output of a gear pump is essentially constant with increasing pressure over a wide range of high efficiency.

Other types of rotary pumps are built with different devices to trap an increment of liquid on the inlet side and then discharge the liquid at a higher pressure at the discharge side. Cams, screws, sliding vanes, and rotating lobes are widely used. A typical sliding-vane pump is shown in Fig. 25.4.19. Sliding vanes, fitted to a slotted rotor, press out against a housing mounted eccentrically to the rotor. As the rotor turns, liquid is trapped between the rotor and the sliding vanes on the inlet side of the pump, and then squeezed out on the discharge side as the sliding vanes are forced back into the rotor by the eccentric housing.

25.4.4   HVAC SYSTEM DESIGNS

Before we discuss system designs it is well to understand the nature of not only the system but also of the system loads. Figure 25.4.20 reflects a typical HVAC load profile for a high-rise office building chilled water system.

The system operates at 60 percent design load 15 times longer than 100% design load. Another design factor that must be understood is the variability in each component of the total system load. We call this variability in load *load diversity*. To best illustrate this concept, consider a typical college campus. Design day loads may collectively total 6000 tons of refrigeration (1 ton of refrigeration, T/R = 12,000 BTUs) but practically you should

% LOAD	% TIME	HRS/DAY
0-10%	0%	0.00
10-20%	0%	0.00
20-30%	3%	0.72
30-40%	5%	1.20
40-50%	10%	2.40
50-60%	25%	6.00
60-70%	25%	6.00
70-80%	15%	3.60
80-90%	12%	2.88
90-100%	5%	1.20

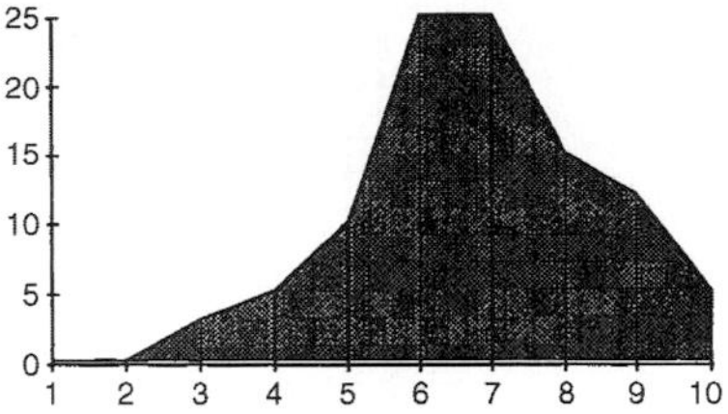

FIGURE 25.4.20 Typical chilled water load profile—office building.

never see a total system load greater than 5000 tons. See Table 25.4.1 to clarify this concept. A portion of the dormitory load comes from the students themselves. The same students are a contributing component of both the classroom and cafeteria design day loads. When the students occupy the dormitories they are not in the classrooms or cafeteria. The loads in both the classrooms and cafeteria therefore must be less than full design day loads. *Design engineers can take advantage of load diversity by designing a 5000 ton chiller plant to very adequately serve a 6000 ton campus.*

The wide variations in load and therefore flow requirements of a typical hydronic system makes a low head rise pump desirable. HVAC pumps commonly have a 12 to 15 percent head rise to shut-off from the best efficiency point.

25.4.4.1 Pump Location

The following applies to all equipment rooms whether intended for heating only, chilled only, or dual chilled and hot water, three- or four-pipe systems.

Distribution pumps must discharge into the piping system with the compression tank connection located at the pump suction. The wrong pump location will decrease system static pressure by an amount equal to the operating pump head. System circulation difficulties

TABLE 25.4.1 Load Diversity: Typical Campus Loads

	Design day load	Actual load
Dormitory	1000 T/R	1000 T/R
Classrooms	1000 T/R	500 T/R
Cafeteria	1000 T/R	500 T/R
Gymnasium	1000 T/R	1000 T/R
Offices	1500 T/R	1500 T/R
Common Spaces	500 T/R	500 T/R
	6000 T/R	5000 T/R

$$\frac{\text{Actual Load} \times 100}{\text{Design Day Load}} = \text{Load Diversity } \% = \frac{500,000}{6000} = 83\%$$

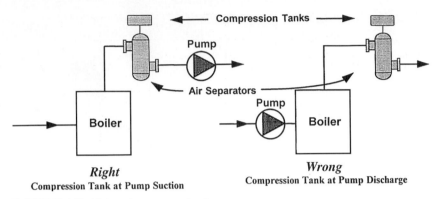

FIGURE 25.4.21   Right and wrong pump location.

and mechanical pump problems may occur by pumping into the compression tank or point of no pressure change.

The properly located pump will increase system static pressure during operation and eliminate trouble potential. See Fig. 25.4.21.

25.4.5 HEATING SYSTEMS

Heat is usually generated at a central point and transferred to one or more points of use. The transfer may be by means of a liquid (usually water or water/glycol mixture), which has its temperature increased at the source (boiler) and then gives up its heat at the point of use such as an air-handling unit (AHU), coil, or radiant panel. It may also be transferred by means of a vapor (usually steam), which changes from a liquid to a vapor at the source, and then gives up its heat at the point of use through condensation. Pumps may be required in both of these methods.

25.4.5.1 Hot Water Systems

A centrifugal pump best meets the requirements of this service. Water is usually used in a closed circuit so that there is no static head that must be overcome. The only resistance to flow is that from friction in the piping and fittings, the boiler, the heating coils or radiators, and the control valves. In selecting the pump, the total flow resistance at the required flow rate should be calculated as accurately as possible, with some thought as to how much variation there might be as a result of inaccuracy of calculations or changes in the circuit because of installation conditions. *It is not good practice to select a pump for a head or capacity considerably higher than that required, as this is likely to result in a higher noise level as well as in increased power consumption.*

The typical heating-only hydronic system will often operate satisfactorily despite design errors because of an inherent safety factor. The safety factor is established by the water flow-heat transfer relationship as shown in Fig. 25.4.22 for a heating terminal selection based on the 200°F supply and a 20°F temperature drop (Δt).

A decrease in terminal unit flow rate to 50 percent of design still allows the order of 90 percent of heat transfer capability. The reason for the relative insensitivity to changing flow rates is that the governing coefficient for heat transfer is the outside or air side

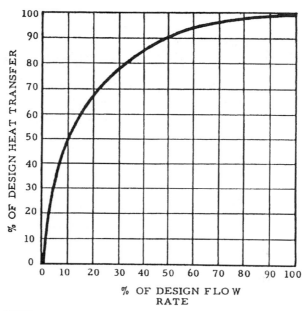

FIGURE 25.4.22 Effect of flow variation on heat transfer for a 20°F design Δt @ 200°F supply temperature.

coefficient; a change in internal or water side coefficient with flow rate does not materially affect the overall heat transfer coefficient. This means:

1. Load handling ability for water to air terminals is basically established by mean air to water temperature difference.

2. A high order of design temperature difference exists between the air being heated and mean water temperature in the coil. A substantial change in mean water temperature is necessary before terminal load ability is measurably changed.

3. A substantial change in mean water temperature (load ability) requires a very substantial change in water flow rate.

A secondary safety factor in terms of application also applies to heating terminals. Unlike chilled water, hot water is not bound to a narrow temperature span in terms of supply water temperature. This means that a reduction in terminal heating capacity as caused by inadequate flow rate can often be overcome by simply raising the supply water temperature and maintaining the same return-water temperature.

The previous comments apply to allowable flow variation for a heating terminal selected for a 20°F (−7°C) temperature and a supply water temperature on the order of 200°F (93°C). Changes in design supply water temperature and design temperature drop will effect permissible flow variation. Given that 90 percent of terminal capacity will be satisfactory for generalized system application, the permissible flow variation can be approximated as in Fig. 25.4.23.

This simple fact remains: Chilled water systems are much less tolerant to flow variation than hot water systems.

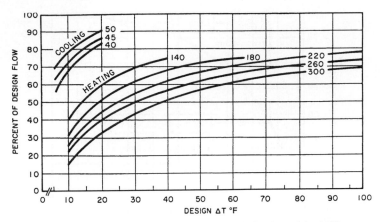

FIGURE 25.4.23   Percent variation of design flow vs. design Δt to maintain 90 percent terminal heat transfer at various supply temperatures °F.

25.4.6   CLOSED SYSTEM DESIGN

Piping systems for water transmission can be considered in two general categories—open systems and closed or sealed systems. Open systems are piping circuits, pumped or gravity circulated, that are open to the atmosphere at some point. Closed systems are designed and installed as hermetically sealed systems and offer several important advantages:

1. When a system is closed, little, if any, makeup water is ever required.

2. With no addition of fresh water there can be no accumulation of oxygen and other corrosive agents. System life is extended indefinitely.

3. Closed systems can be pressurized permitting elevated water temperatures and greater temperature drops. Piping and operating costs can be reduced drastically.

4. Closed systems with positive air control offer improved control, faster temperature response, and quieter system operation.

Obviously, closed hydronic systems are used whenever possible. However, many systems that are intended to be closed, too often end up as open systems. Although a system may be designed as a closed system, it will not be closed unless all components are pressure-tight and leak-proof. Special consideration should be given to pump seals, manual air vents, and tight installations.

The following simplified drawings define terms and configurations of hydronic piping systems.

In the single-pipe system (Fig. 25.4.24), Load 1 receives the full benefit of the hot or chilled water source. Load 2 and all subsequent loads receive supply temperatures degraded by the previous load. An example would be in a chilled water system where Load A has a supply temperature of 45°F (7°C), Load B: 48°F (44°C), Load C: 51°F (46°C), and Load D: 54°F (48°C). Because of the degrading supply temperature, one-pipe systems are seldom used in commercial systems.

Two-pipe systems segregate supply side and return side piping systems. All loads are supplied with unspent design temperature. Two-pipe systems can be either direct return as in Fig. 25.4.25 or reverse return as in Fig. 25.4.26.

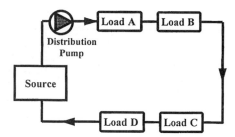

FIGURE 25.4.24   Single-pipe in-line pumping system.

The advantage of reverse return piping is that the system is much easier to balance flows to design conditions as the pressure differential across Load A is approximately the same as across Loads B, C, and D.

The advantage of direct return piping is its inherent lower cost due to less pipe being required. The disadvantage, of course, is the fact that Load A has substantially more differential pressure than Load B; Load B has more than Load C, etc. Most modern systems are two-pipe direct return systems due to the lower costs.

The following two-pipe systems are the most common piping arrangements in the modern commercial HVAC industry. Examples shown reflect chillers as the sources, however, hot water heating systems would be similar except the pump would be located on the outlet side of the boiler, not on the inlet side as depicted with the chillers.

It is desirable to maintain reasonably constant flow across the evaporator section of a centrifugal chiller to prevent freeze-up or unstable performance. Application of a three-way valve to bypass the load makes the total system constant flow. See Fig. 25.4.27.

Application of two-way control valves for each load allows the system to respond to load changes by passing more or less flow through the system. This improves system efficiency but makes the chiller variable flow on the evaporator side. See Fig. 25.4.28.

Addition of a common or decoupling line and a distribution pump is perhaps the most common of all system designs. The chiller pump requires only enough head to circulate design flow through the chiller back through the low pressure drop common line and back to chiller pump suction. The distribution pump would be a true variable volume pump supplying only the capacity required by the two-way control valves. The result is two independent pumping systems having one short pipe connection in common, hence the name common pipe. This is the basic primary/secondary pumping system. See Fig. 25.4.29.

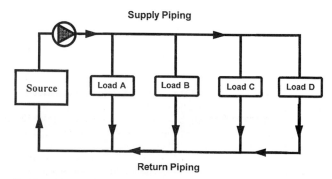

FIGURE 25.4.25   Direct return two-pipe system.

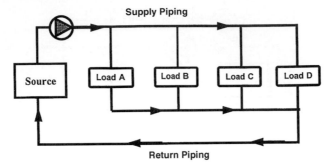

FIGURE 25.4.26   Reverse return two-pipe system.

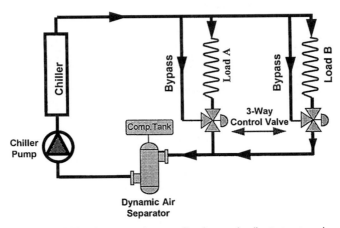

FIGURE 25.4.27   Direct pumped constant flow 3-way valve direct return two-pipe system.

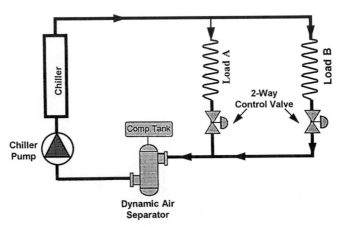

FIGURE 25.4.28   Simple direct pumped/direct return two-pipe system.

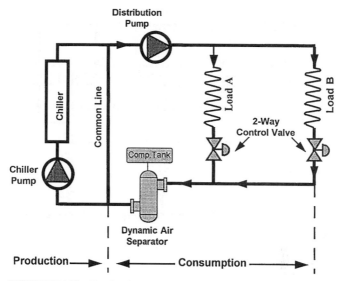

FIGURE 25.4.29 Simple primary/secondary two-pipe system with direct return piping.

Larger campus or airport systems can incorporate a third level of pumping system isolation. Figure 25.4.30 shows a primary/secondary/tertiary pumping system where the chiller circuit represents the primary system. The distribution pump serves the secondary system while the zone pump serves the tertiary system.

Figure 25.4.31 reflects a typical tertiary pumping system but this time with reverse return piping.

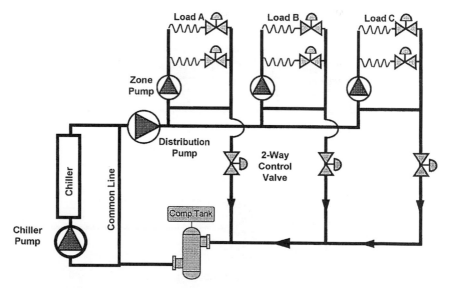

FIGURE 25.4.30 Primary/secondary/tertiary direct return two-pipe system.

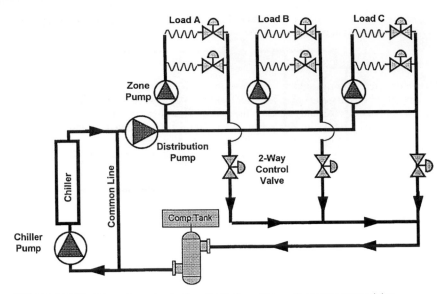

FIGURE 25.4.31 Primary/secondary/tertiary distribution with two-pipe reverse return piping.

The last common distribution system is shown in Fig. 25.4.32—the zone-pumped direct return pumping system. The distribution pumps have been moved out to remote locations such as buildings on a college campus or outer terminals at an airport. Caution must be exercised in zone pumping systems as all distribution pumps are in parallel and are interconnected. A change in one system will cause a corresponding change in all other pumps.

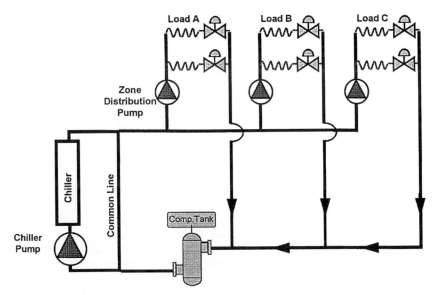

FIGURE 25.4.32 Zone pumped direct return two-pipe system.

The addition of an additional terminal could obsolete all existing pumps in the older terminals due to changes in pump system pressure drop.

25.4.6.1 Mechanical Seals

Mechanical seals are required for all closed system circulating pumps. Pump seals of the packing gland type require constant water leakage to provide seal lubrication. This means fresh water must be constantly added to the system. Since fresh water contains air and other corrosive agents, system life and operation will be seriously affected. In addition, foreign particles of sand and dirt, often found in fresh water supplies, will enter and accumulate within the system. Therefore, water-tight mechanical pump seals should always be specified on closed hydronic systems.

25.4.6.2 Manual Air Vents

Manual air vents should be used where initial venting of high points is necessary to fill the system with water. Automatic air vents, if allowed to operate automatically after the system is placed into operation, are a source of system leakage.

25.4.6.3 Air Control

To assist the design engineer in evaluating the conditions affecting air control within a particular hydronic system, a closer look at the effect of air solubility in water under changing temperatures and pressures is desirable. Air is roughly 80 percent nitrogen and 20 percent oxygen. Within a short time after the initial fill, the air in a properly operating closed system begins to lose its oxygen content through oxidation. Unless fresh water is added, the gas within the system is predominately nitrogen. Therefore, to illustrate the effect of nitrogen (system air) solubility in water under various temperatures and pressures, the chart in Fig. 25.4.33 has been prepared.

The maximum amount of nitrogen that can be held in solution, the left hand scale, is expressed in percentage of water volume, if released from solution and then expanded to atmospheric pressure, zero psig (Pa), while at 32°F (0°C). Obviously if water, at its maximum air solubility level, is heated and/or reduced in pressure, nitrogen will be released from solution. If water temperature and pressure are returned to their original level, and if this free nitrogen is still present, it will be reabsorbed. However, if this free nitrogen can be vented off or separated in some manner, the water will remain in a deaerated condition.

Effective air control for the design engineer of a closed hydronic system means designing a system so that when nitrogen (system air) is released from solution, either by heating or by a reduction in pressure, it will be at only one point. This point of separation should be the connecting point to the compression tank.

It should be emphasized again, however, that this figure illustrates only maximum solubility of nitrogen in water. Total nitrogen in solution can be measured only if removed as a gas and measured at zero psi (Pa), at 32°F (0°C). How then does Fig. 25.4.33 help the design engineer to better understand effective air control for all types of closed hydronic systems?

Figure 25.4.33 stresses the importance of locating air-separating equipment at the point of lowest air solubility, based on both temperature and pressure. This emphasizes the importance of having adequate system pressurization, since low pressure areas can allow nitrogen to be released from solution. This means locating all pumps in the system with respect to the point of no pressure change so that their operating pressures are always added to the pressures existing at each point in the system. Pumps should not be a cause of creating reduced pressure areas at remote points in the system where air can be released from solution. Pumps should always be located so that their pressures add to system pressures at all points.

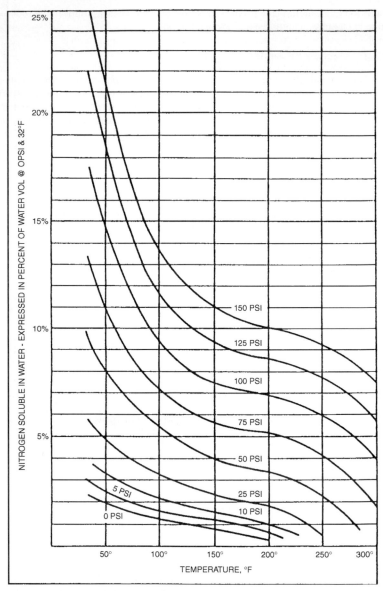

FIGURE 25.4.33   Solubility of nitrogen, in water vs. temperature and pressure. (*Courtesy of ITT Bell & Gossett, from Ref. 1.*)

Ideally, air control is complete when the air separator can be located at a single point, which is the area of highest temperature and lowest pressure. Obviously this is not likely, so the engineer must provide not only effective separating equipment, but design his piping circuitry so that any free air within the system will be carried back to the point where it will be separated and directed to the compression tank. Normally, water velocity above 2 ft/s (61 cm/s) will be adequate to prevent free entrained air from forming air pockets. High points in various piping circuits, in particular, can be collecting points for free air. Proper pipe sizing, particularly on down-feed return piping, is often effective in preventing air traps at high points.

The best location for air-separating equipment, of course, depends upon the particular system. No single rule can apply for all systems. Obviously, when possible, select a point that will have both highest temperature and lowest pressure. When this is not possible, generally the point of highest temperature should receive first consideration as the best point for air separators, even though it is often located at the bottom of the system under full static system pressure. Figure 25.4.33 will show how a large percentage of nitrogen is released from solution when water temperatures rise above 200°F (95°C). Except, possibly, for high-rise buildings, air separation at points of elevated temperatures is generally an effective way to deaerate system water to a point where additional nitrogen release will not take place under pressures at system high points.

Furthermore, since boilers are usually protected by pressure relief valves, air separation at or near the boiler supply then places the point of no pressure change with respect to pump operation near the pressure relief valve. Proper pump location, operating away from the point of no pressure change, will then present a minimum of pump pressure increase to the boiler and its relief valve.

When air-separation equipment is installed at a point of both highest temperature and lowest pressure, of course, it will have its greatest effectiveness.

25.4.6.4 Dynamic Air Separators

A very effective air separator available today is the dynamic air separator pictured in Fig. 25.4.34. Inlet and outlet openings are installed tangentially. Circulation through the separator creates a vortex or whirlpool in the center where entrained air, being lighter, can collect and rise into a compression tank installed above. Instead of relying entirely on low velocity separation, the action of centrifugal force sends heavier air-free water to the outer portion of the tank, allowing the lighter air–water mixture to move into the lower velocity center. An air-collecting screen located in the vortex aids in developing a low-velocity area in the center where air can collect. The dynamic air separator offers the advantage of achieving efficient separation in a much smaller-size tank.

25.4.6.5 Steam Systems

No pumping is required with the smallest and simplest steam systems if there is sufficient level difference between the boiler and condensers (radiators, heating coils, etc.) to provide the required flow. When insufficient head exists between the level of the condensate in the condenser and the boiler to produce the required flow to the boiler, then a pump must be introduced to provide the required head. Since the condensate in the hot well will be at or near its saturation temperature and pressure, the only NSPH available to the pump will be the submergence less the losses in the piping between the hot well and the pump. A pump must be selected that will operate on these low values of NPSH without destructive cavitation.

In many cases, particularly for very large systems, vacuum pumps are used to remove both the condensate and the air from the condensers. This permits smaller piping for the return of condensate and air, more positive removal of condensate from condensers, and, when rather high vacuums [above 20 in (51 cm)] are possible, some control of the temperature at which the steam condenses. The use of vacuum return, particularly with higher vacuums, helps reduce the possibility of exposing freezing heating coils to outside air or to stratified outside and recirculated air. Vacuum return pumps are available for handling air and water. Vacuum, condensate, and boiler-feed pumps with condensate tanks are all available in package form.

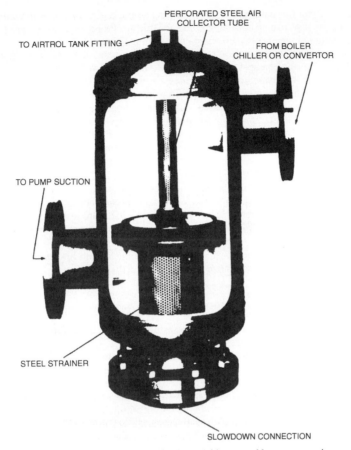

PERFORATED STEEL AIR
COLLECTOR TUBE

TO AIRTROL TANK FITTING

FROM BOILER
CHILLER OR CONVERTOR

TO PUMP SUCTION

STEEL STRAINER

SLOWDOWN CONNECTION

FIGURE 25.4.34   Dynamic air separator with removable system strainer. (*Courtesy of ITT Bell & Gossett.*)

Most condensate pumps are centrifugal. Some are furnished with very low NPSH requirements, less than 2 ft NPSHR. Low NPSH pumps frequently incorporate an impeller inducer as displayed in Fig. 25.4.35. A vacuum pump condensate return system is shown in Fig. 25.4.36.

25.4.6.6 Condenser Water Circulation

Condenser water may be recirculated and cooled by passing through a cooling tower, or it may be pumped from a source such as a well, a lake, or the ocean.

Cooling Tower Water.   Centrifugal pumps are used for the circulation of cooling tower water. The circuit is open at the tower, where the water falls or is sprayed through the air and transfers heat to the air before the water falls to the pan at the base of the tower. A pump then circulates the water through the condenser as shown in Fig. 25.4.37. In this case, the pump must operate against a head equal to the resistance of the condenser and piping plus the static head required to the tower from the water level in the pan.

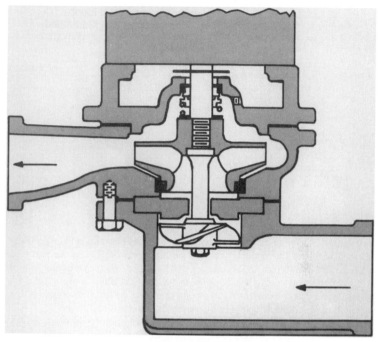

FIGURE 25.4.35   Condensate pump with very low NPSH. (*Courtesy of ITT Bell & Gossett.*)

Figure 25.4.37 shows a somewhat similar circuit except that the condenser level is above the pan water level. The size of the pan of a standard cooling tower is sufficient to hold the water in the tower distribution system so that the pan will not overflow and waste water each time the pump is shut down. This capacity also ensures that the pan will have enough water to provide the required amount above the pan level immediately after startup, without waiting for the makeup that would have been needed if there was any overflow when the pump stopped. When the condenser or much of the piping is above the cooling tower pan's overflow, the amount draining when the pump is stopped would exceed the pan

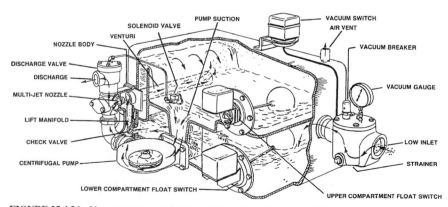

FIGURE 25.4.36   Vacuum pump, condensate return system. (*Courtesy of ITT Bell & Gossett.*)

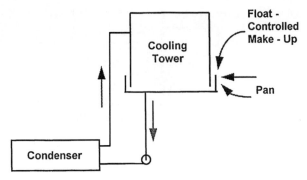

FIGURE 25.4.37   Cooling tower with condenser above pan water level. [*From Igor J. Karassik et al. (eds.),* Pump Handbook, *1st ed., McGraw-Hill, New York,* © *1976. Reproduced with permission.*]

capacity unless means were provided to keep the condenser and lines from draining. In Fig. 25.4.38 it will be noted that the line from the condenser drops below the pan level before rising at the tower. This keeps the condenser from draining by making it impossible for air to enter the system. This is effective for levels of a few feet, but it is useless if the level difference approaches the barometric value. Such large level differences should be avoided, if possible, since they require special arrangements and controls.

When a cooling tower is to be used at low outside temperatures, it is necessary to avoid the circulation of any water outside unless the water temperature is well above freezing. The arrangement shown in Fig. 25.4.39 provides this protection. The inside tank must now provide the volume previously supplied by the pan in addition to the volume of the piping from the tower to the level of the inside tank. Condensers or piping above the new overflow level must be treated as already described and illustrated in Fig. 25.4.38, or additional tank volume must be provided.

The only portion of the inside tank that will be available for the water that drains down after the pump stops is that above the operating level. This operating level, or available head, is fixed by the height of liquid required to avoid cavitation or air entrainment at the inlet to the pump. The size of the pipe at the tank outlet should be determined by the velocity that can be attained from the available head, not by the pressure loss. [A good approximation is $d_p = 0.3\sqrt{\text{gal/min}/h}$, where dp is the pipe diameter in inches (centimeters) and h is the head in feet (meters.)] See Fig. 25.4.40.

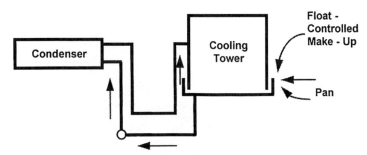

FIGURE 25.4.38   Cooling tower with condenser above pan water level. [*From Igor J. Karassik et al. (eds.),* Pump Handbook, *1st ed., McGraw-Hill, New York,* © *1976. Reproduced with permission.*]

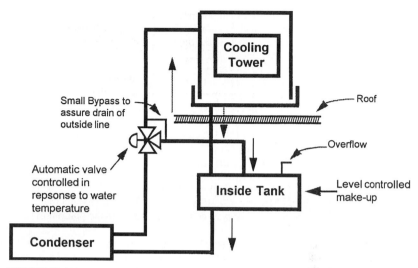

FIGURE 25.4.39   Cooling tower with inside tank for subfreezing operation.

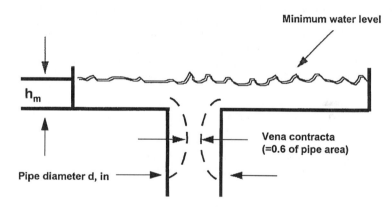

H_m = minimum height of water to prevent air entrainment
        (vortex) at pump section in open cooling tower pans
A = area of pipe = ¼ $d^2/144$, ft^2
A_c = area at vena contracta = 0.6 A, ft^2
Q = required flow in pipe, gal/min
V_c = velocity at vena contracta
      = flow in ft^3/min/A_c ,ft/min
      = $\dfrac{Q(\text{gal/min } 0.1337 \text{ ft}^3/\text{gal})}{0.6(n/4)d^2/144}$
h_m = $V_c/2g$, ft
g = 32.2, ft/s

FIGURE 25.4.40   Calculation of velocity at vena contracta. [*From Igor J. Karassik et al. (eds.),* Pump Handbook, *1st ed., McGraw-Hill, New York, © 1976. Reproduced with permission.*]

Well, Lake, or Ocean Water.   Centrifugal pumps are used for all these services. The level from which the water is pumped is a critical factor. The level of the water in a well will be considerably lower during pumping than when the pump does not operate. In the case of pumping from a lake or from the ocean, the drawdown is usually not significant. When pumping from a pit where the water flows by gravity, there will be a drawdown that will depend on the rate of pumping. With seawater supply there will be tidal variations. A lake supply may have seasonal level differences.

All these factors must be taken into consideration in selecting the level for mounting the pump to ensure that it will be filled with water during start-up. Check or foot valves may be used for this purpose. Also, the head of the water entering the pump at the time of highest flow rates must not be so low that adequate NPSH is not available. To ensure proper pump operating conditions, the pump is frequently mounted below the lowest level expected during zero flow conditions, as well as below the lowest level expected at the greatest flow rate. These conditions may require a vertical turbine-type pump (Fig. 25.4.12). The motor should either be above the highest water level with a vertical shaft between the motor and the pump bowls or be of the submerged type connected directly to the pump bowls.

25.4.7   REFRIGERATION SYSTEMS

For refrigeration systems with temperatures near or below freezing, pumps are often required for brine or refrigerant circulation. The transfer of lubrication oil also frequently requires pumps.

Brine Circulation.   The word "brine," as used in refrigeration, applies to any **salt-based** liquid (1) that does not freeze at the temperatures at which it will be used and (2) that transfers heat by a change solely in its temperature without a change in its physical state. As far as pumping is concerned, brine systems are very similar to systems for circulating chilled water or any liquid in a closed circuit. A centrifugal pump is preferred for this service. It must be constructed of materials suitable for the temperatures and fluids encountered. For salt brines, the pump materials must be compatible with other metals in the system to avoid damage from galvanic corrosion.

Tightness is usually more important in a brine-circulating system than in a chilled water system. This is true not only because of the higher cost of the brine but also because of problems that are caused by the entrance of minute amounts of moisture into the brine at very low temperatures.

Refrigerant Circulation.   For a number of reasons, including pressure and level as well as improvement of heat transfer, refrigerant liquid may be circulated with a pump. Centrifugal pumps with mechanical shaft seals are usually preferred for this purpose

Whereas the fluid is all in liquid form throughout a brine circuit, it is in vapor form during part of its circulation in a **refrigeration circuit**. In the refrigerant-circulating system, most of the heat transfer is by evaporation or condensation or both. Satisfactory performance and longevity of the pump, as well as initial warranty responsibility, often depend on following these instructions.

25.4.7.1   Piping

The suction piping to any pump must be carefully designed to ensure that the flow enters the pump with a minimum of rotation and swirl, and with as uniform a velocity profile as practical. Short-radius elbows, tees, check valves, double elbows, and globe valves should not be used at the suction of a pump. The turbulence created by these fittings is carried directly into the pump and may result in noise and a loss in capacity and efficiency.

The discharge piping and suction should be firmly supported to minimize the load that is transferred to the pump. Excessive piping loads can distort the pump and its alignment to the driver, with resulting damage to the pump and driver bearings. In severe cases, coupling and shaft failures have resulted.

In installations where flexible pipe connections are used, careful consideration must be given to the fact that both the pump and the piping must be anchored rigidly to prevent movement from the piston effect of the flexible connection. Both the piping and the pump are subjected to a load equal to the cross-sectional area of the pipe times the discharge pressure of the pump in pounds per square inch (bars). In some installations where this precaution has not been considered, the resulting piston effect of a flexible pipe connection was sufficient to force the pump off its foundation.

25.4.7.2 Sumps

For wet or dry pit pumps taking suction from an open sump, the severity of the turbulence or swirl in the pump can have a marked effect on pump performance. The flow into the sump should be as straight and uniform as feasible, with as few turns or abrupt changes in cross section as possible upstream from the sump. Many examples have been reported where noise, vibration, loss of capacity, driver overload, and excessive bearing and shaft failures have been corrected by modifying the sump design. The pump manufacturer should be consulted prior to the actual construction of the sump as to the adequacy of its design for the particular pump selected for the intended service.

25.4.7.3 Foundations

A concrete foundation with a securely grouted pump base is the preferred arrangement for either horizontal or vertical pumps. In the case of vertical dry-pit pumps with intermediate line shafting, the cost of installation can be reduced by supporting the line shaft bearings and the driver on structural steel supports. It is essential in this type of construction not only that the structural steel supports be designed for the loads and maximum deflection furnished by the pump manufacturer, but that the natural frequency of the structure be determined prior to installation. The calculated natural frequency should be at least 100 percent higher than the operating frequency of the pump.

Pipe-mounted pumps are designed to be mounted and supported by the piping system itself. In these cases, special attention must be given to the pipe supports, as they must be capable of sustaining the weight of the pump and its drive motor. Larger sizes may require a separate foot stand for support. Check the manufacturer's instructions.

25.4.7.4 Start-up

Prior to start-up of a new HVAC pumping system, all pipe lines should be cleaned of foreign material. Often this is accomplished by pumping a cleaning solution through the system. On certain extremely "dirty" systems it may be advisable to install a temporary pump suction strainer. Careful monitoring of the strainer pressure drop is required in order to avoid damage to the centrifugal pump.

Pump suction line strainers are apparently one of those peculiar "darned if you do—and darned if you don't" propositions. There are two conflicting needs:

• Protection of the system: pumps, valves, condenser, spray nozzles, etc., against dirt and debris.

• The act of placing a fine mesh strainer in the suction piping will make a mockery of the most careful pump suction pressure evaluation. This is because an uncontrollable variable has been introduced; once the strainer gets clogged, cavitation may occur.

The problem can be corrected, however, once it is understood that the centrifugal pump will pass fairly large objects. This means that strainer mesh openings from $^3/_{16}$ in to $^1/_4$ in (48 to 64 mm) can be used if the only function of the strainer is to protect the pump.

Tower pans are usually provided with an exit strainer (at tower inlet to suction piping) of this mesh order. Such tower strainers should be specified since they can be monitored and are easily cleaned without piping drainage.

When tower pan strainers cannot be provided, a large mesh low pressure drop strainer can be placed in the suction line. Such strainers should be strongly specified both as to mesh size [$^3/_{16}$ in min] (48 mm) and pressure drop.

Fine mesh strainers are often needed for protection of the condenser, its valves, and/or spray nozzles. The fine mesh strainer should be placed at the pump discharge, usually between pump discharge and the pump check valve. This location will often simplify the work of the operator in removal and cleaning of the easily clogged basket.

25.4.7.5 End of Curve

Typical HVAC centrifugal pumps have a rising horsepower and rising NPSHR curves with increasing flow. *Operating the pump well beyond its design flow may cause catastrophic failure of the pump due to excessive NPSHR or impeller choking.* When a system's design conditions require operating two or more pumps in parallel, a single pump pumping the same system will attempt to increase capacity to a point beyond its "end of curve." It is therefore always advisable to start up a single pump with its discharge throttled.

REFERENCE

1. D. M. Himmelbau, *Journal of Chemical and Engineering Data*, vol. 5, no. 1, January 1960.

25.5 VALVES

Edward Di Donato
Nordstrom Valves, Inc., Sulphur Springs, Texas

25.5.1 INTRODUCTION

25.5.1.1 Valves for Heating and Cooling Systems

In the majority of piping systems, valves are seldom given the detailed attention they deserve. Valves are usually considered just another piece of plumbing. More importance is given to pumps, compressors, motors, and other more complex equipment. Yet system function depends a good deal on proper valve selection and operation since valves are one of the major controlling elements in any fluid-handling system. The control requirement may be on-off (to isolate system components), throttling (modulation of flow), reduction of fluid pressure, etc. Whatever its function, selection of the proper valve for a given service must take into consideration each operation the valve must perform and the operating conditions under which it must function.

This chapter on valving for HVAC systems is confined to the use of valves for both isolation and balancing applications and offers some considerations in valve selection.

25.5.1.2 Isolation Valves

The location and selection of isolation valves (or stop valves) is an important consideration in plant design. It is usually the practice to locate isolation valves in the inlet and outlet lines to all pumps, condensers, vessels, and long lengths of pipe so that they can be isolated in the event of leaks or to repair equipment.

Although gate valves, globe valves, ball valves, plug valves, and butterfly valves can be used as isolation valves, some knowledge of their operating and sealing characteristics is essential to suitable valve selection.

25.5.2 VALVE SEALING

In a discussion of the phenomenon of sealing in valves, it is interesting to give consideration to the mechanism of closure and how it may affect the sealing function. There are several different valve types, and one can observe significant differences between the sealing actions of the various types.

The fact that there are, indeed, widely differing kinds of valves can be attributed to the relative simplicity of the basic valve functions, which can be produced in a variety of ways. One of these functions is to serve as an intermediate piece of pipe when the valve is in the open position, permitting the contained fluid to pass through with a minimum of resistance. Conversely, in the closed position, the function is to serve as a stopper, or the equivalent of a blind flange, isolating one side of the valve from the other to prevent the flow of the contained fluid.

To judge the merits of any valve, it is usually desirable to evaluate three characteristics:

1. How well does it imitate the piece of pipe in the open position (with respect to flow and pressure loss characteristics)?

2. How well does it imitate the blind flange in the closed position?

3. How easily can it be changed from one position to the other?

25.5.2.1 Flow Characteristics

There are many different types of valves, and not enough space here to cover them all. The ways valves provide the shutoff and sealing function, however, are obviously limited in principle. A closure element must be moved, somehow, from a position in which its effect on flow is minimal to a position in which it completely obstructs the flow. Let's look at a few valve types and see how these movements are accomplished:

In the *conduit-type gate valve* (Fig. 25.5.1), the gate simply slides into and out of the closed position.

There is a degree of similarity between this and the action in a *tapered plug valve*, with the blind side of the plug sliding into and out of position (Fig. 25.5.2). The difference, of course, is that the sliding in this case is achieved by virtue of rotation of the plug in place, whereas in the gate valve there is straight linear motion into and out of position.

The *ball valve* (Fig. 25.5.3) is again similar, rotating as the plug does. In this case, however, the ball shape of the plug simplifies the geometry of the seal between the body and the ball and thus facilitates the use of specially designed sealing elements, as we will see a little later.

Another widely used type of gate valve has a smaller, simpler closure element (Fig. 25.5.4). This uses a wedging action between the gate and the body seats to provide some of the sealing force.

An obvious alternative to sliding the closure to and from its closed position is to move it directly toward and away from the body seat. This is done in *globe-type valves* (Fig. 25.5.5), and we can think of the action as replacing the sliding flange of the other types with a plug or stopper, which must be inserted into the body-seat opening. The conventional *angle valves* (Fig. 25.5.6) and the *Y valves* (Fig. 25.5.7) all use this principle.

To evaluate different types of valves, it is interesting to examine the questions asked earlier. The first was, *how well does the valve imitate the piece of pipe in the open position?* The sliding-closure-element types are generally superior in this respect, as they permit the flow to be straight through when the valve is in the open position. The *wedge gate valve* is not quite as good as the others, because the gap between the seats, when the gate is in the open position, causes some flow turbulence that is not found in the other types.

Actually, the flow capacity of the through-conduit type of valve is somewhat better than is needed in many applications, and in such cases it can be economically attractive to use a *reduced-port valve* (Fig. 25.5.8). In gas or product transmission pipelines, however, valves are usually required to be full-port so that they will permit the passage of pipe scrapers or fluid-separating "pigs."

Conventional globe-type valves (Fig. 25.5.9) cannot provide a straight-through flow path for the fluid and are therefore characterized by somewhat less favorable flow characteristics. Angle-type valves (Fig. 25.5.10) and Y valves (Fig. 25.5.11) are much better. For straight flow lines, Y valves provide the best flow capacity available in a globe-type valve; in addition, their operating characteristics are considered to be superior for some services. Where a piping arrangement requires a 90° elbow, an angle valve provides good flow capacity (Fig. 25.5.12) and can also save on installation costs.

Although not a through-conduit type, *butterfly valves* in the smaller sizes also have favorable flow characteristics (Fig. 25.5.13). Their pressure drops are roughly comparable to those of reduced-port plug valves.

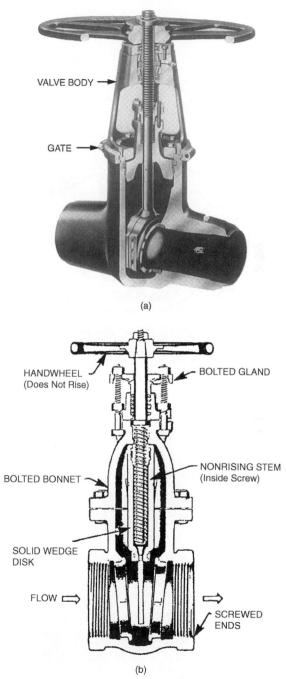

FIGURE 25.5.1 (*a*) Conduit-type gate valve; (*b*) Standard gate valve, (*Courtesy of Nordstrom Valves, Inc., Sulphur Springs, Texas.*)

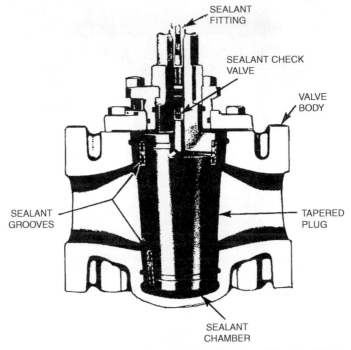

FIGURE 25.5.2   Tapered plug valve. (*Courtesy of Nordstrom Valves, Inc., Sulphur Springs, Texas.*)

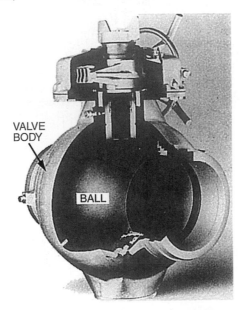

FIGURE 25.5.3   Ball valve. (*Courtesy of Nordstrom Valves, Inc., Sulphur Springs, Texas.*)

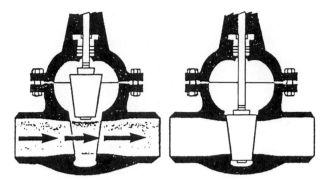

FIGURE 25.5.4 Gate valve, wedge-seal type. (*Courtesy of Nordstrom Valves, Inc., Sulphur Springs, Texas.*)

Valves lacking a straight-through flow path result in a loss of energy and add to pumping power costs. However, in practical terms, the cost of pumping power must be balanced against the high capital cost of valves with very large ports. It has been found, for example, that well-streamlined *plug valves* with port areas 40 to 60 percent of the pipe area often provide an excellent compromise between capital cost and pumping cost.

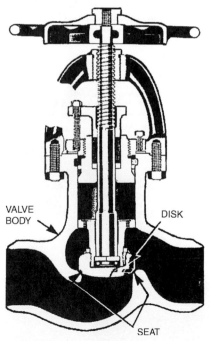

FIGURE 25.5.5 Globe-type valve. (*Courtesy of Nordstrom Valves, Inc., Sulphur Springs, Texas.*)

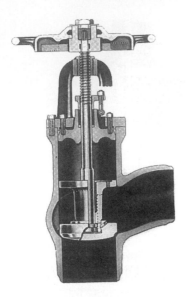

FIGURE **25.5.6** Angle valve. (*Courtesy of Nordstrom Valves, Inc., Sulphur Springs, Texas.*)

Low-pressure butterfly valves use a flat, circular disk mounted on a shaft in the plane of the disk, and the disk rotates within the body by means of a 90° rotation of the shaft (Fig. 25.5.14a).

High-pressure butterfly valves (Fig. 25.5.14b and c), one of the newest types of valve available, have the circular disk plane offset from the shaft so that an uninterrupted surface is available on the periphery for engagement with the body seat. Opening and closing is still accomplished by a 90° rotation of the shaft.

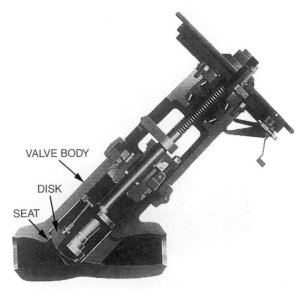

FIGURE 25.5.7   Y valve. (*Courtesy of Nordstrom Valves, Inc., Sulphur Springs, Texas.*)

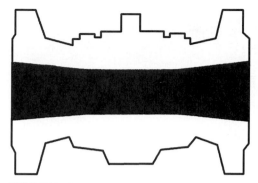

FIGURE 25.5.8   Reduced-port valve. (*Courtesy of Nordstrom Valves, Inc., Sulphur Springs, Texas.*)

25.5.2.2 Closure Tightness

The second basic consideration is, *How well does the valve imitate the blind flange?* Our presumption, of course, is that with the blind flange we have assurance of a reliable closure with no leakage.

In valves, of course, we cannot always assume that there will be no leakage. Even in cases where the valve has no apparent defects, leakage can result from the presence of foreign material, such as sand or metallic corrosion products, being caught between the

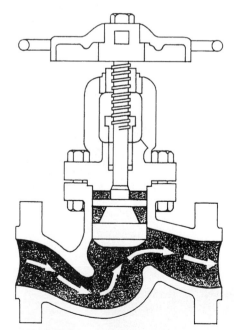

FIGURE 25.5.9 Conventional globe-type valve. (*Courtesy of Nordstrom Valves, Inc., Sulphur Springs, Texas.*)

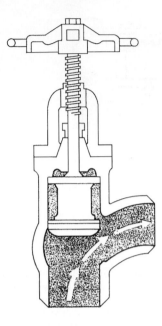

FIGURE 25.5.10   Angle valve. (*Courtesy of Nordstrom Valves, Inc., Sulphur Springs, Texas.*)

sealing elements. There are, in fact, three distinctly different kinds of sealing action utilized in industrial valves:

Metal to Metal. This is the most obvious arrangement of a metallic surface on one closure element contacting a metallic surface on the other closure element. For tight sealing, it is necessary for the surfaces to be held in intimate contact over the entire

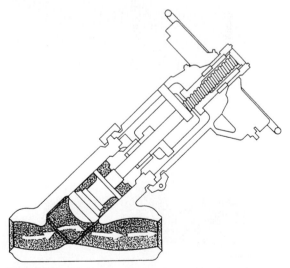

FIGURE 25.5.11   Y valve. (*Courtesy of Nordstrom Valves, Inc., Sulphur Springs, Texas.*)

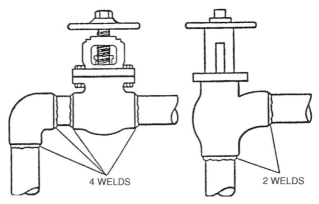

FIGURE 25.5.12   A 90° elbow with an angle valve. (*Courtesy of Nordstrom Valves, Inc., Sulphur Springs, Texas.*)

periphery of the seal. Any discontinuities in this contact, very small or even microscopic, will constitute leakage paths. Also, any imperfections in geometry that would prevent the required continuity of contact will result in leakage. Good metal-to-metal sealing, therefore, is best achieved by providing very smooth surfaces very accurately finished and by providing a relatively large mechanical force pushing the sealing elements against each other.

Sealant material. Recognizing that all surfaces have some degree of imperfection and that even near-perfect surfaces are susceptible to damage, provision is made in some valve types for the introduction of a special sealant material into the sealing contact area (Fig. 25.5.15). This sealant reduces, and in most cases eliminates, leakage by filling the

FIGURE 25.5.13   Butterfly valve. (*Courtesy of Nordstrom Valves, Inc., Sulphur Spring, Texas.*)

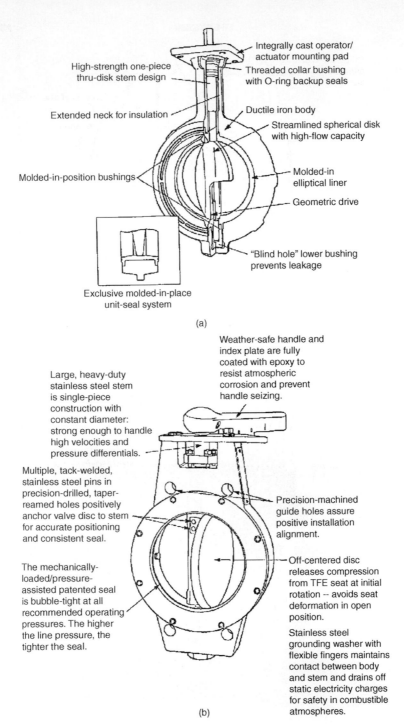

High-strength one-piece
thru-disk stem design

Integrally cast operator/
actuator mounting pad

Threaded collar bushing
with O-ring backup seals

Extended neck for insulation

Ductile iron body

Streamlined spherical disk
with high-flow capacity

Molded-in-position bushings

Molded-in
elliptical liner

Geometric drive

Exclusive molded-in-place
unit-seal system

"Blind hole" lower bushing
prevents leakage

(a)

Large, heavy-duty
stainless steel stem
is single-piece
construction with
constant diameter:
strong enough to handle
high velocities and
pressure differentials.

Weather-safe handle and
index plate are fully
coated with epoxy to
resist atmospheric
corrosion and prevent
handle seizing.

Multiple, tack-welded,
stainless steel pins in
precision-drilled, taper-
reamed holes positively
anchor valve disc to stem
for accurate positioning
and consistent seal.

Precision-machined
guide holes assure
positive installation
alignment.

The mechanically-
loaded/pressure-
assisted patented seal
is bubble-tight at all
recommended operating
pressures. The higher
the line pressure, the
tighter the seal.

Off-centered disc
releases compression
from TFE seat at initial
rotation -- avoids seat
deformation in open
position.

Stainless steel
grounding washer with
flexible fingers maintains
contact between body
and stem and drains off
static electricity charges
for safety in combustible
atmospheres.

(b)

FIGURE 25.5.14 (*a*) Low-pressure butterfly valve. (*b*) and (*c*) Off-center high-pressure butterfly valves. (*Courtesy of Nordstrom Valves, Inc., Sulphur Springs, Texas.*)

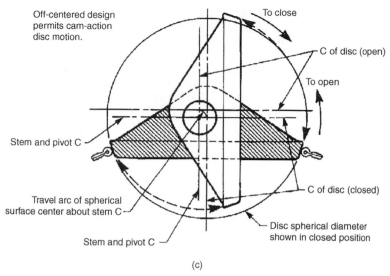

Off-centered design permits cam-action disc motion.

To close

C of disc (open)

To open

Stem and pivot C

C of disc (closed)

Travel arc of spherical surface center about stem C

Disc spherical diameter shown in closed position

Stem and pivot C

(c)

FIGURE 25.5.14 (*Continued*).

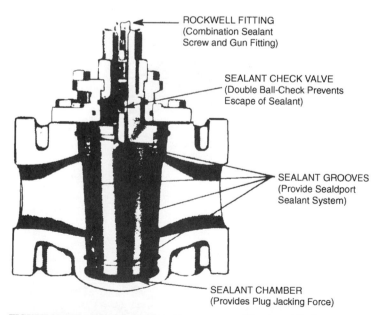

ROCKWELL FITTING
(Combination Sealant
Screw and Gun Fitting)

SEALANT CHECK VALVE
(Double Ball-Check Prevents
Escape of Sealant)

SEALANT GROOVES
(Provide Sealdport
Sealant System)

SEALANT CHAMBER
(Provides Plug Jacking Force)

FIGURE 25.5.15 Sealant material used in contact area. (*Courtesy of Nordstrom Valves, Inc., Sulphur Springs, Texas.*)

scratches and microscopic crevices in the sealing interface. In valve construction using a sliding motion of the closure element relative to the seat, this sealant material provides helpful additional effects of lubrication of the sliding surfaces to prevent wear or damage, and it protects these important surfaces against corrosion.

Nonmetallic seal. This method of improving the reliability of tight valve sealing is the use of a tough, resilient, nonmetallic sealant material in the seat contact area. In all such designs the basic objective is simply to provide a material that is much more deformable than the metal seat sealing face materials, with the intention that the deformable material will be squeezed into the scratches and microscopic crevices in the contacting materials and will thus prevent leakage.

Application. How well, then, do the various types of valves shut off flow in the closed position? Actually, because of the many variables that affect the sealing capability, it is not possible to evaluate any one valve design as "best" in all applications. Certain observations can be made, however, and these will help explain (in at least some cases) why some types of valves are predominantly used in some particular services and not in others.

Metal-to-metal seals without sealant or other nonmetallic backup are widely used in globe-type valves (Fig. 25.5.16). The lifting and reseating movement of the disks in these valves avoids a sliding contact between the sealing surfaces that can cause scratching and

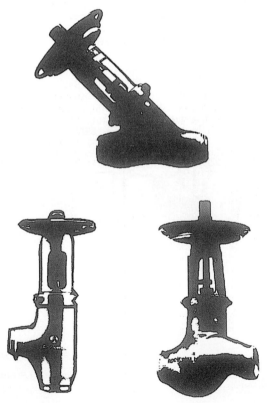

FIGURE 25.5.16   Metal-to-metal seals. (*Courtesy of Nordstrom Valves, Inc., Sulphur Springs, Texas.*)

wear. The mechanism for providing this motion is also capable of applying a strong seating force to the closure element, enhancing the intimacy of mutual contact between the sealing surfaces that minimizes the leakage.

It will also be apparent that with this construction, using appropriate metals in the body, disk, and sealing surfaces, satisfactory valve performance can be achieved over a wide range of temperatures, up to 1112°F (600° C) and higher.

The tapered plug valve is the most universally used example of a valve having metal-to-metal sealing enhanced by the application of a specially compounded sealant between the metal-to-metal sealing surfaces. The plug valve, in contrast to the globe-type valve, does not avoid sliding metal-to-metal contact. In fact, it maximizes this contact by having virtually the entire outside surface of the plug in contact with the corresponding internal surfaces of the body. This extensive surface contact has the desirable effect of spreading the contact load over a large bearing area, thus minimizing the contact stress. With the contact stress kept relatively low, and with the presence of sealant between the surfaces at essentially all times, the likelihood of damage to those surfaces is minimized and there is good assurance of completely tight sealing throughout the life of the valve.

The operating principle of the *tapered plug valve* with an integral sealant system has been applied to conduit-type gate valves (see Fig. 25.5.1) in small to medium sizes and medium to very high pressures. The sliding contact of the closure element (gate) against the body seal, characteristic of most conduit-type gate valves, is mitigated very usefully by the large contact area provided by the extended seat surfaces and, as in the case of the plug valve, by the constant presence of sealant between these parts. These mutually enhancing features—surfaces protected against wear and damage by the presence of sealant, and sealant to provide maximum probability of tight sealing—give this type of valve superior performance in many applications.

The use of a solid, nonmetallic sealant material is also helpful in improving the tight-shutoff reliability of valves. The problem of holding the sealant material in place must be faced; this is complicated, of course, by the fact that nonmetallic materials are generally quite inferior to metals in terms of mechanical strength. An example of what can happen as a result of improper application is shown in Fig. 25.5.17. Here we see sequentially

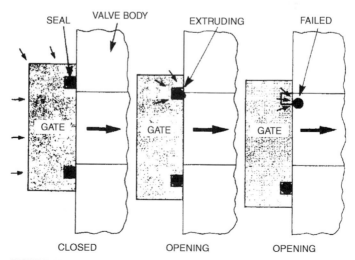

FIGURE 25.5.17 Improper application of sealant. (*Courtesy of Nordstrom Valves, Inc., Sulphur Springs, Texas.*)

how a simple seal can fail if improperly applied. With the elastomer sealing against a gap that opens or closes on its downstream side, the pressure can cause failure of the seal during operation of the valve. Simply changing the location of the elastomer so that the opening or closing gap is on the upstream side, as in Fig. 25.5.18, can eliminate this problem.

The application of this principle is shown in Fig. 25.5.19. Since the valve design provides the shutoff seal on the upstream seat, the sealing action is correct, and the elastomer can function satisfactorily against openings and closings that are under very high pressure.

The need for large full-port conduit-type valves for transmission pipelines has stimulated the development of trunnion-mounted ball valves (see Fig. 25.5.3) with combination sealing methods. In the design illustrated in Fig. 25.5.20, we see the use of all the sealing methods discussed above. The basic relationship of the seat ring to the sphere is intimate metal-to-metal contact between accurately machined and highly finished surfaces. A tough, resilient nylon ring is securely retained in a groove in the seat ring, providing a positive local seal point at the line of contact of the nylon ring with the surface of the sphere. And finally, means are provided for distributing a sealant material to the surface of the sphere through a groove in the seat ring. Conduit gate valves are also offered for large-diameter pipeline service and are usually provided with elastomeric seals (Fig. 25.5.21). Wedge-type gate valves are not generally offered for large pipeline service, but rather for general service applications, including use at elevated temperatures. Consequently, wedge-type gate valves usually have a simple metal-to-metal seating, with no elastomers or provisions for injecting sealant.

25.5.2.3 Ease of Operation

The third basic consideration is: *How easily can the valve be changed from one position to the other?* As will be apparent, there can be a relationship between the relative difficulty of operation and the reliability of tight valve sealing.

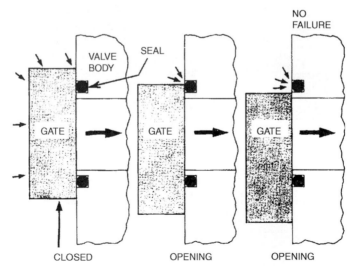

FIGURE 25.5.18 Correcting the problem of Fig. 17. (*Courtesy of Nordstrom Valves, Inc., Sulphur Springs, Texas.*)

FIGURE 25.5.19 Cylindrical plug valve with elastomeric seal. (*Courtesy of Nordstrom Valves, Inc., Sulphur Springs, Texas.*)

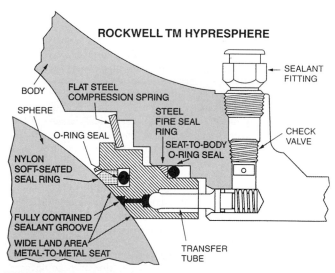

FIGURE 25.5.20 Sealing methods. (*Courtesy of Nordstrom Valves, Inc., Sulphur Springs, Texas.*)

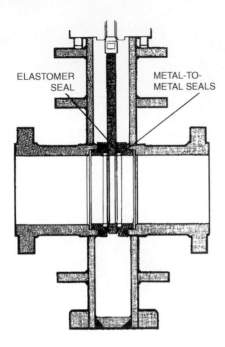

FIGURE 25.5.21 Conduit gate valve with elastomeric seal. (*Courtesy of Nordstrom Valves, Inc., Sulphur Springs, Texas.*)

In all metal-seated valves, for example, our primary concerns are the condition of the sealing surface and the amount of force with which the sealing surfaces are pressed together. In the globe-type valves (including angle and Y designs), the absence of sliding action in their operation results in good preservation of the highly finished surfaces, and the large force provided gives good sealing performance. Perhaps the most troublesome cause of leakage in globe-type valves is damage to the sealing surfaces, which can be caused by foreign particulate materials caught between the disk and body when the valve is closed.

The need for application of a large force in the closure of globe-type valves necessitates a correspondingly large mechanical advantage in the operating mechanism for manual operation. Globe-type valves are usually comparatively slow and difficult to operate.

Gate valves are generally more susceptible to mechanical damage of the sealing surfaces in the opening and closing operations. They are somewhat easier to operate than globe-type valves because the stem is not required to sustain the entire fluid pressure load on the disk, as is the case with globe-type valves. Instead, the fluid pressure load is sustained on gate guides and, ultimately, on the seat. The stem, then, is required only to overcome the sliding friction that results from the movement of the gate sliding on the guides or on the seat. It is apparent, of course, that such sliding under heavy load involves a considerable risk of damage to the sealing surfaces. There is therefore a direct trade-off of valve characteristics. The gate valve is easier to operate than the globe valve, but it is more susceptible to damage during operation.

In conduit-type gate valves, it can be further noted that the fluid pressure load is carried entirely on the seating surfaces, since there is no separation of these surfaces as provided in a wedge gate valve (see Fig. 25.5.21). Now the valve size and the amount of pressure become particularly significant, because with increasing size and increasing pressure the magnitude of the fluid pressure load increases very rapidly (Fig. 25.5.22). It is reasonable

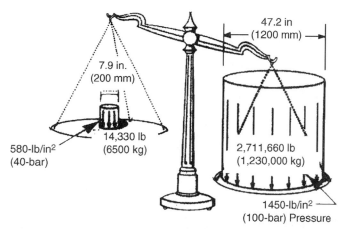

FIGURE 25.5.22 Fluid pressure load. (*Courtesy of Nordstrom Valves, Inc., Sulphur Springs, Texas.*)

to expect, therefore, that as the sizes and pressures increase, gate-type valves will have increasingly serious problems related to the sliding of the gate across the valve seat.

Tapered plug valves (see Fig. 25.5.2) also operate by a sliding movement of the closure element (plug) across the seating surface. As mentioned earlier, the contact area in this case is quite large relative to the effective flow-port area that determines the fluid pressure load. And the presence of the plug-valve sealant between the surfaces serves as protection against wear and damage.

In the case of plug valves, the closure sliding motion is comparable to that in gate valves, but it involves rather considerably less distance because the motion is across the short dimension of a narrow opening.

The operating effort required for ball valves is strongly influenced by a design feature that greatly reduces the force acting between the seat and the sphere. In large ball valves, the ball is pivotally supported on trunnions, which carry the fluid pressure load imposed on the ball. The trunnions are carried in low-friction journal bearings, and since the radius of these bearings (Fig. 25.5.23) is much less than the distance between the center of ball rotation and the seating surface, the sliding distance, and therefore the total energy required to move the valve between the open and closed positions, is considerably less than it is for any of the other valve types discussed.

There must still be contact, and therefore friction, at the seat-sphere contact area. The contact force, however, is controllable by the designer, who may thus strive for the optimum: the minimum contact force that produces good sealing between the seat and the sphere.

25.5.2.4 Valve Integrity in Case of Fire

One additional consideration is of basic interest in the process of evaluating the performance of various types of valve seals. This is the matter of valve integrity when exposed to elevated temperatures resulting from a fire. Since the valve may itself be installed in a line carrying combustible fluid, a gross failure of the valve could literally add fuel to the fire. On the other hand, a valve capable of surviving such exposure could afford effective protection against such a development.

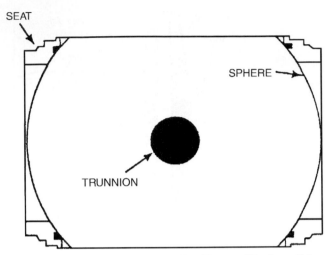

FIGURE 25.5.23   Trunnion in journal bearing. (*Courtesy of Nordstrom Valves, Inc., Sulphur Springs, Texas.*)

It is reasonable to assume that exposure to fire will destroy any nonmetallic sealing materials other than asbestos-containing packing, as is used in stem seals. An important question, then, is: How will the function of the valve be affected by the loss of the elastomeric seal rings or sealants used in the valve? The answers will vary with valve types and designs.

Globe-type valves will remain functional, for they do not rely on heat-sensitive materials.

Wedge gate valves will remain functional. If fire safeness is important, however, consideration should be given to the possibility of pressure buildup in the body of a closed valve (Fig. 25.5.24), since the double seating of this type of valve (as well as of some types of conduit gate valves) can trap fluids in the center section.

Tapered plug valves are highly resistant to fire damage, since the loss of sealant due to high temperature leaves the plug and body intact. Injection of sealant subsequent to cooling has restored fire-damaged valves to useful serviceability on many occasions.

Conduit gate valves may remain substantially functional, depending on individual design details. In general, a closed valve under differential pressure will provide good downstream protection because a loss of nonmetallic sealing materials will simply permit the gate to move slightly downstream and seat against the metal portion of the seat ring. The sealant-sealed conduit-type gate valve mentioned earlier provides excellent fire safety, in that sealant lost by reason of excessive heat can be replaced by injection after cooling.

Ball valves vary considerably in terms of their ability to function during or following exposure to fire. The construction described earlier has good fire safety because of its primary reliance on metal-to-metal ball-to-seat sealing. Some designs place primary reliance on the nonmetallic seat insert and would be expected to leak excessively if the insert were destroyed.

A specific point of concern in trunnion-mounted ball valves is the seal between the seat ring and the body. Since the ball does not press downstream against the seat as the gate in a gate valve does, loss of the elastomeric seal between the seat and body can also constitute a serious failure in such a valve. In the design shown in Fig. 25.5.20, however, the use of a tapered metallic backup ring behind the O-ring provides effective shutoff even when the O-ring is destroyed.

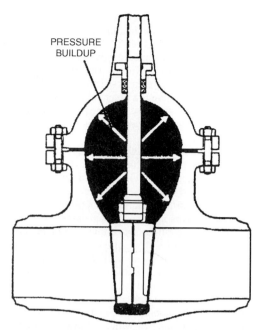

PRESSURE
BUILDUP

FIGURE 25.5.24 Pressure buildup in wedge gate valve.
(*Courtesy of Nordstrom Valves, Inc., Sulphur Springs, Texas.*)

The cylindrical plug valve described earlier (see Fig. 25.5.19) depends on the elastomer for its sealing function and is not considered to be a firesafe construction.

25.5.3 ISOLATION VALVES
AND BALANCING VALVES

Isolation valves spend almost all of their service lives either open or closed, and they experience part-open operation only in those brief periods when they are moved from one position to the other. However, balancing valves are destined to spend most of their service lives partially open, throttled so as to limit the flow in a branch of a system to that which is needed for heating and cooling functions. When performing these duties, valves must imitate variable orifices instead of the straight pipe and blind flanges that isolation valves are called upon to mimic.

Ideally, each branch of a system might be sized with the proper flow capacity to suit its functional needs. Practically, most branches are oversized to various degrees to accommodate increments in standard pipe sizes and other available equipment. Sometimes systems must be sized to accommodate future increases in demand. *Balancing valves* are required in such systems to restore order by introducing the flow resistance that is required to approximate the ideal system. The balancing valves have a function partly comparable to that of modulating control valves; they must dissipate energy. Unlike modulating control valves, which must change position frequently to cope with the changes in system demand, balancing valves are often set to the required position and rarely adjusted after proper balance between system branches is achieved. Sometimes seasonal adjustments or changes to accommodate local variations in heating or cooling loads are necessary, but the objective in many cases is to "set it and forget it."

While the throttling duties of some balancing valves are mild, it must be recognized that energy dissipation is achieved by closing valves partially so as to produce higher-than-usual local flow velocities through orificelike restrictions. In mild cases, there may not be problems for years; in severe cases, there is a potential for dislodging or tearing away soft-seat inserts, for erosion of metal, and for noise damage problems associated with cavitation. Thus balancing-valve applications must be viewed as at least potentially more problematic than isolation-valve applications.

It may be noted that some of the potential problem areas for balancing valves may not be too serious as long as the valve is used only for throttling. Loss of a soft-seat insert or minor metal damage from erosion or cavitation might have little effect on system flow. However, if the same valve should be required to be closed and serve as an isolation valve at some later time, the damage produced by throttling might have a profound effect on tightness. Many prudent engineers include two valves in series in such cases—one for isolation and one for throttling. Where economic considerations make this a problem, valves must be selected very carefully to assure satisfaction of both required functions: isolation and balancing.

While any valve suitable for isolation can also be used for throttling or balancing just by adjusting it to a position intermediate between open and closed, proper selection of valves involves the consideration of several factors. One of these is the flow characteristic of the valve, which describes the relationship of the valve flow coefficient with the amount or percentage of valve opening. An improper flow characteristic may cause an oversensitivity of the system or branch flow to small variations in valve opening. A linear relationship between system flow and valve position is often desirable for easy adjustment; in control-valve terms, this implies that the goal is a "linear installed characteristic." Since this relationship is highly dependent on the amount of fixed resistance that exists in system piping and other equipment in series with a balancing valve, the inherent characteristic of the valve should be such that a linear installed characteristic may be approximated.

Figure 25.5.25 illustrates how a linear installed characteristic may be best approximated by using a valve with an inherent characteristic that is often referred to in control-valve terminology as "equal percentage." Most gate and globe valves have inherent characteristics that are between linear and quick opening. The use of such valves in systems with much fixed resistance produces an installed characteristic that would result in little system-flow change between about 30 and 100 percent opening, so a very sensitive adjustment between closed and 30 percent open would be necessary to balance the system or branch flow. Fortunately, simple quarter-turn ball, plug, and butterfly valves produce inherent characteristics comparable to complex equal-percentage control valves. Properly sized, such quarter-turn valves can provide excellent adjustability in balancing applications.

In present practice, the decision on balancing applications often rests on deciding between a plug valve and a butterfly valve. Usually, both provide inherent flow characteristics that will provide satisfactory adjustment in system or branch flow comparable to that shown in Fig. 25.5.25. There are no simple rules of thumb to guide this decision, but the following guidelines should be considered. (Valve manufacturers should be consulted regarding the specific performance characteristics of specific valve sizes, pressure classes, and actuator types; some characteristics that apply to one size or class may not apply to another.)

25.5.3.1 Pressure and Differential Pressure

Plug valves are offered in a broad range of pressure classes, with different materials and end connections. In most cases, valves are capable of operating at differential pressures

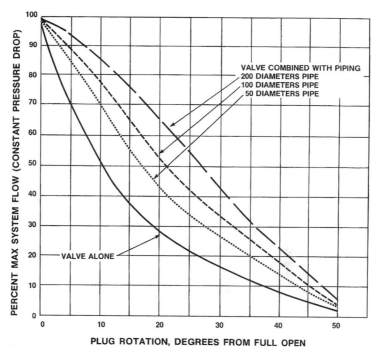

FIGURE 25.5.25 Representative throttling curves showing pipe effects for sizes 1 to 6 Rockwell Nordstrom plug valves. (*Courtesy of Nordstrom Valves, Inc., Sulphur Springs, Texas.*)

equal to the body pressure rating. Some low-pressure butterfly valves are restricted to differential pressures less than their body ratings because of seat, stem, or disk design limitations. Most modern high-performance butterfly valves do not have such limitations, but soft seat inserts may not be suitable for continuous high-differential throttling.

25.5.3.2 Precision of Control

Manual operation is used for most balancing-valve applications. Worm-geared manual actuators are available for most plug and butterfly valves, and they are usually necessary for valves in sizes 6 to 8 in (150 to 200 mm) or larger, depending on the valve type and operating pressure. Such actuators offer excellent capability for precise adjustment. They lock in position under normal conditions. However, severe vibration (which can be produced by cavitating flow) may cause drift even with gears that are normally self-locking. *This is another reason to watch out for cavitation.*

In smaller sizes, the plug ball and butterfly valves may be easily operated with a lever or wrench. It should be recognized that flow through all quarter-turn valves produces a dynamic torque, usually in a direction tending to close a throttled valve. Butterfly valves have relatively low internal friction to resist flow-induced torque; to lever operating valves usually requires "index plates," which require the lever to be locked into a discrete position between open and closed. This limits the precision of the balancing adjustment. Many plug valves have sufficient internal friction to permit them to be adjusted to any position with a removable wrench without requiring a locking device to hold the plug in place. For most moderate throttling and balancing duties, this is an advantage, allowing small plug valves to be adjusted very precisely. Nevertheless, high flows and pressure drops may cause valve drift, particularly if there is vibration (again a reason to watch out for cavitation).

25.5.3.3   Isolation Capability of Balancing Valves

Section 25.5.3 described in detail the requirements for a successful isolation valve. This section describes the requirements for throttling and trimming. Fortunately, the plug and butterfly valves that are often used for balancing are also usually excellent as isolation valves. However, where valves used for balancing are also called upon to serve an isolation function, some secondary features must be considered.

Most butterfly valves (like most ball valves) achieve their isolation capabilities through the use of soft seats made of elastomeric or polymeric materials (such as TFE). When new and intact, most such valves easily provide drop-tight shutoff of closed valves at both low and high differential pressures. In contrast, most plug valves are of metal-to-metal seating construction but provide for sealant injection to achieve drop-tight isolation. Providing that there is no valve damage, these factors indicate an advantage to the butterfly valve, which requires no sealant injection. Still, considering the severity of some of the services encountered in balancing-valve applications, the risks of damage to soft seats must be recognized. Unless a butterfly valve is provided with a backup metal-to-metal seat, damage or destruction of the primary soft seat may produce leakage rates that are unacceptable in sensitive HVAC applications. If excessive leakage due to seat damage should require shutdown of an entire system or interruption of service to a large area, another type of valve (or two types in series) should be considered.

In such cases, the plug valve offers the advantage that sealant injection will usually provide the required shutoff requirements without system shutdown or valve removal for maintenance. Soft-sealed valves can usually be repaired, but this requires system downtime. Even where the damage to a metal plug valve is so severe as to require maintenance, sealant injection will usually permit the maintenance to be deferred until a convenient schedule period.

25.5.3.4   Cavitation Resistance

Cavitation is a two-stage process consisting of (1) the formation of vapor bubbles in a liquid where local static pressure falls below the vapor pressure (boiling point) and (2) the collapse of such bubbles where the pressure rises above the vapor pressure. Even where both the inlet and outlet pressures of a throttled valve are well above the vapor pressure, cavitation may occur as a result of the low pressures existing at high-velocity regions within the valve. Vapor bubbles or cavities produced within the valve will collapse either downstream of the throttling restriction in the valve or in the downstream piping.

Cavitation effects may range from hardly noticeable to violent, depending on many considerations. "Mild" cavitation may cause intense "roar" and "rocks and gravel" noises; in this region, vibration and damage to metal parts can be severe.

A detailed discussion of the cavitation resistance of specific valve types and applications is outside the scope of this chapter. Qualified consultants or valve manufacturers with good experience and test data should be involved in evaluating questionable applications. Readers should recognize that some data published in usually reputable sources may show only where "severe" cavitation occurs. Less severe conditions may produce noise and damage that would be unacceptable in some HVAC applications. The following suggestions are intended only to provide a preliminary evaluation of potentially serious or questionable applications:

Calculate the system cavitation parameter K_{sc} by the equation

$$K_{sc} = \frac{\Delta P}{P_{in} - P_{vap}}$$

where P_{in} is the inlet pressure and P_{vap} is the vapor pressure.

- If $K_{sc} = 0.1$, there should be no problem with any standard quarter-turn valve.
- If $K_{sc} = 0.2$, there should be no problem with any throttled butterfly valve except in the worst cases.
- If $K_{sc} = 0.3$, there should be no problem with any throttled plug valve except in the worst cases.

For application outside the limits given above, the valve manufacturer should be consulted. For a more detailed description of cavitation, flow characteristics, etc., see Ref. 1.

REFERENCE

1. *Flow Manual for Quarter-Turn Valves*, Nordstrom Valves, Sulphur Springs, TX.

CHAPTER 26
DUCTWORK*

26.1 DUCT SIZING

Nils R. Grimm, P.E.
Consulting Engineer, Hastings, New York

26.1.1 INTRODUCTION

The function of a duct system is to provide a means to transmit air from the air-handling equipment (heating, ventilating, or air conditioning). In an exhaust system the duct system provides the means to transmit air from the space or areas to the exhaust fan to the atmosphere.

The primary task of the duct designer is to design duct systems that will fulfill this function in a practical, economical, and energy-conserving manner within the prescribed limits of available space, friction loss, velocity, sound levels, and heat and leakage losses and/or gains.

With the required air volumes in cubic feet per minute (cubic meters per second) determined for each system, the zone and space requirements known from the design load calculation, and the type of air distribution system [such as low-velocity single-zone, variable-air-volume (VAV) or multizone or high-velocity VAV or dual duct] decided upon, the designer can proceed to size the air ducts.

The designer must also choose one of three methods to size the duct systems: the equal-friction, equal-velocity, or static regain method. Of the three, the equal-friction and static regain methods are used most often. The equal-velocity method is used primarily for industrial exhaust systems where a minimum velocity must be maintained to transport particles suspended in the exhaust gases.

Static regain is the most accurate method, minimizes balancing problems, and results in the most economical duct sizes and lowest fan horsepower. It is also the only method that should be used for high-velocity comfort air-conditioning systems.

The equal-friction method is used primarily on small and/or simple projects. If manual calculations are made, this method is simpler and easier than static regain; however, if a computer is used, this advantage disappears.

Typical duct velocities for low-velocity duct systems are shown in Table 26.1.1. For high-velocity systems, typical duct velocities are shown in Table 26.1.2. The velocities

---

*Updated by the editor.

TABLE 26.1.1 Suggested Duct Velocities for Low-Velocity Duct System, ft/min (m/s)

Application	Main ducts		Branch ducts	
	Supply	Return	Supply	Return
Residences	1000 (5.1)	800 (4.1)	600 (3)	600 (3)
Apartments Hotel bedrooms Hospital bedrooms	1500 (7.6)	1300 (6.6)	1200 (6.1)	1000 (5.1)
Private offices Directors rooms Libraries	1800 (9.1)	1400 (7.1)	1400 (7.1)	1200 (6.1)
Theaters Auditoriums	1300 (6.6)	1100 (5.6)	1000 (5.1)	800 (4.1)
General offices Expensive restaurants Expensive stores Banks	2000 (10.2)	1500 (7.6)	1600 (8.1)	1200 (6.1)
Average stores Cafeterias	2000 (10.2)	1500 (7.6)	1600 (8.1)	1200 (6.1)
Industrial	2500 (12.7)	1800 (9.1)	2200 (11.2)	1600 (8.1)

TABLE 26.1.2 Suggested Duct Velocities for High-Velocity Duct System, ft/min (m/s)

Application	Main duct		Branch duct	
	Supply	Return	Supply	Return
Commercial institutions	2500–3800 (12.7–19.3)	1400–1800 (7.1–9.1)	2000–3000 (10.2–15.2)	1200–1600 (6.1–8.0)
Public buildings				
Industrial	2500–4000 (12.7–20.3)	1800–2200 (9.1–11.2)	2200–3200 (11.2–16.3)	1500–1800 (7.6–9.1)

suggested in Tables 26.1.1 and 26.1.2 may have to be adjusted downward to meet the required noise criteria.

Whether the duct system is designed manually or by computer, the effects of high altitude must be accounted for in the design if the system will be installed at elevations of 2500 ft (760 m) or higher. Appropriate correction factors and the effects of altitudes of 2500 ft (760 m) and more are discussed in Appendix A.

26.1.2 MANUAL METHOD

If the manual method is used to size the project duct systems, they should be calculated by following one of the accepted procedures found in standard design handbooks, such as Refs. 1 and 2. A detailed discussion on all handling-system design is shown in Ref. 3.

For industrial dilution, ventilation, and exhaust duct systems, they should be calculated and sized by the procedures set forth in Ref. 4.

26.1.3 COMPUTER METHOD

26.1.3.1 VariTrane Duct Designer

An in depth look... Product Details, October 2001.

Read about:

Introduction	Resizing noncritical paths
Duct configurator	Elevation adjustments
Equal friction	Constant and variable volume systems
Static regain	Existing buildings
Leakage correction	Ductulator
Thermal correction	Fitting loss calculator
Insert feature	Trane VAV box selection
Fitting entry methods	Auto-balance

Introduction.   The VariTrane Duct Designer duct design program was written to help optimize design while obtaining a minimum pressure system. The program is based on engineering data and procedures outlined in the *ASHRAE Fundamentals Handbook*. It includes tested data from the ASHRAE Fitting Database and from United McGill to provide the most accurate modeling possible.

The program consists of three subprograms or applets: *Duct Configurator, Ductulator*, and *Fitting Loss Calculator*. Each of these applets can be used individually or combined to provide a complete duct analysis.

Duct Configurator.   Duct configurator sizes and analyzes supply duct systems. Quickly model the conditions for any duct system and specifications for each section, terminal unit, and diffuser. The section-by-section approach easily incorporates existing ductwork and fittings.

Design Methodologies

Equal Friction.   This design methodology sizes the supply duct system for a constant pressure loss per unit length. This is the most widely used method of sizing lower pressure, lower velocity duct systems. The main disadvantage of this method is that there is no equalization of pressure drops in duct branches unless the system has a symmetrical layout.

Note: Set the optimal design friction rate per length of duct for the entire duct system in Duct Configurator by calculating the friction rate in the first trunk section of the system (the section connected directly to the fan). If the friction rate calculated for the trunk section is undesirable for the entire system, override the friction rate by selecting "No" to "Use calculated friction rate?" Set it directly in the field that appears.

Static Regain.   This design methodology sizes the supply duct system to obtain uniform static pressure at all branches and outlets. Much more complex than equal friction, static regain can be used to design systems of any pressure or velocity. Duct velocities are systematically reduced, which allows the velocity pressure to convert to static pressure, offsetting friction losses in the succeeding section of duct. Systems designed using static regain require little or no balancing. One disadvantage of this methodology is that oversized ducts can occur at the ends of long branches. However, VariTrane Duct Designer lets you limit this problem by specifying a minimum velocity constraint.

Note: Set the minimum velocity constraint directly on the main tab. Choose from one of the three values listed or specify your own velocity constraint.

Fitting Entry Methodologies.   Two new fitting entry methods in Duct Configurator add increased flexibility to the data entry process. They include:

The **Lowest Friction** method reduces entry time by limiting the amount of operations required for fitting selections. When checked, the fitting chosen is based on the fitting type and shape. Fitting types include transition, 45-degree wye, 90-degree tee, bullhead tee, and cross; along with fan connection, entry, and exit, which are available under certain circumstances. When a fitting type is selected, a fitting is automatically chosen based on lowest frictional loss, which eliminates a lengthy search through the available fittings.

The **Quick Pick Fitting** method allows a user to customize the fitting types and available fittings within those types. When this method is selected the user selects a fitting type and shape. From these choices a customized list of fittings is presented based on the user's preference of fittings. The selections for a given fitting type must be set prior to using this method, but after the initial setup, the list of customized fittings will be available for any duct systems modeled in the future.

The **ASHRAE** method is the traditional method of inputting fittings and it provides the full assortment of ASHRAE fittings. When checked, every ASHRAE database fitting from a given fitting type is available for selection. This method offers the most accurate, but also the most tedious, entry method.

These methodologies may be used in conjunction with one another at any time in the entry process. For example, halfway through entering a duct system with the **Lowest Friction** method the user discovers a fitting that is not a close enough match to any of the fitting types. At this point the user can switch to the **ASHRAE** or the **Quick Pick Fittings** method of fitting selection and then back again if necessary.

Trane VAV Box Selections.   VariTrane Duct Designer now offers the ability to make Trane VAV box selections directly from the program through a seamless integration with the TOPSS program (Trane Official Product Selection Software). This eliminates the tedious and time-consuming process of transferring product selection data into Duct Configurator. It also provides additional sizing criteria and increased accuracy in the duct design procedure. The ability to export the VAV box selections to the TOPSS software gives the user the ability to further analyze the box selections.

Modifying an existing duct system layout is easier with the new insert feature. When a selection is inserted between two sections, it is connected to the downstream section by a transition of the same shape and is connected to the upstream section by the fitting that existed prior to the insertion. The transition can then be changed to a junction to create additional takeoffs in the system.

26.1.3.2 Modified Duct System

Features

Auto-Correct for Leakage.   When this feature is checked, Duct Configurator automatically recalculates the airflow required at the supply fan to overcome leakage losses in the system. The amount of leakage is based on the leakage class you select. Leakage class is an

ASHRAE term that denotes the leakage airflow per unit of duct surface area at a specific static pressure. For more information on leakage, see Chapter 32 in the 1997 *ASHRAE Fundamentals Handbook.*

Auto-Correct for Thermal Losses/Gains. During the journey from the supply fan to the conditioned spaces, the air stream may undergo various thermal changes. When this feature is checked, Duct Configurator automatically recalculates the airflow required at the supply fan to adjust for heat loss or gain based on the thermal insulation specified for each section and the temperatures specified inside and outside of the duct sections.

Auto-Balance the System. By checking this advanced feature, Duct Configurator will automatically add dampers in noncritical paths, where required, to balance the pressures in the system. Whenever the static overpressurization exceeds 1.0 inWG in a given section, the application inserts an orifice rather than a damper to balance that section. This is done for acoustic optimization, because orifices are less noisy than dampers when high-pressure drops are required to balance the system.

When this feature is not checked, Duct Configurator identifies the critical path and determines the required duct sizes and supply fan airflow accordingly, but it does not correct the oversized ductwork or reduce over pressurization with dampers in the noncritical paths.

Resizing Noncritical Paths. Would you like to reduce your duct material and installation costs up to 40 percent or more without affecting your annual energy costs? You can by using the new improved Resize Noncritical Paths feature in Duct Configurator. Duct Configurator can optimize the noncritical path sizes prior to adding any dampers or orifices. All noncritical paths sized using either Equal Friction or Static Regain have excess static pressure. By reducing the sizes of the duct sections in the noncritical paths, the majority of the excess static pressure is converted to velocity pressure. The remaining excess static pressure can then be reduced using dampers as described above.

Note: By enabling auto balancing with resizing noncritical paths, the calculation time increases dramatically. This is due to the tremendous amount of iterations that can accumulate for each path. For this reason it is suggested to run this feature after the initial design has been optimized.

Adjust for Elevation Differences. Duct Configurator does system calculations both at the elevation you specify and at sea level. You can specify elevation changes for individual sections of duct if there are large elevation differences from the inlet to the outlet of a given section.

Constant/Variable Volume Systems. Duct Configurator can model both constant and variable volume systems. Constant volume systems are systems in which the *block airflow* equals the *peak airflow*. If these values are not the same, it is said that there is *diversity* in the system. Diversity is the decimal value that describes the ratio of block airflow and peak airflow.

$$\text{Diversity} = \frac{\text{block airflow}}{\text{peak airflow}}$$

If the diversity is something other than 1, the system is a variable volume system. This can be input in one of two ways. Diversity factor or block airflow can be input on the fan tab, but not both. The program then calculates the remaining variable. Peak airflow is always known since it is input on the VAV or diffuser tab for each terminal device. The diversity is then equally distributed throughout the system. For example, if the diversity factor is 0.8 at the fan and 1.0 at each VAV box, the root section diversity factor would be slightly more than 0.8 and the section immediately preceding the VAV box would have a diversity factor slightly less than 1.0.

Note: Modeling downstream of VAV boxes cannot currently be done, therefore VAV boxes and diffusers cannot be used in the same modeled duct system. To account for the losses downstream of a VAV box use the Ductulator in conjunction with the Fitting Loss Calculator to determine pressure loss in each section of the longest run. Then add all of the losses together and enter that value in Duct Configurator.

Existing Building or Design Constraints. Another feature of the Duct Configurator applet is the ability to easily meet design constraints or model existing ductwork. To do this, set maximum of fixed section size constraints.

Note: An unlimited number of take-offs can be connected to a section defined as a plenum.

Spreadsheet. Make design changes and check for entry errors with ease through the spreadsheet view in Duct Configurator. Once in the spreadsheet view, double click on a field category to get further information. Use copy/paste functionality to make global changes!

Ductulator. This applet transforms Trane's popular manual duct sizing and layout calculator into a convenient PC-based tool. Use this applet to quickly size system components and determine the appropriate nominal duct size for equal friction applications.

Fitting Loss Calculator. This applet lets you quickly identify the optimal fittings and sizes for each duct section by comparing their efficiency and cost.

Note: Return and exhaust systems cannot yet be modeled using Duct Configurator. To account for these losses, use the Ductulator in conjunction with the Fitting Loss Calculator to obtain the static pressure losses through the return and exhaust ductwork.

These applets, combined with an integral database of accurate performance data for hundreds of ASHRAE and United McGill fittings, allow you to confidently model new or existing supply duct systems, whether round, rectangular, flat, or oval. Once you complete your design, print reports with detailed information about all aspects of the duct system, including bills of material.

Electronic Ductulator

The electronic version of the Trane Ductulator simplifies the manual process of sizing ductwork based on equal-friction methodology. It computes the friction rate or pressure drop in a particular duct so that you can specify an appropriate nominal size.

The equal-friction-duct sizing methodology attempts to keep the pressure drop, per unit length of duct, relatively constant throughout the system. It is widely used to size relatively low-velocity constant-volume systems and to size the flexible ductwork between each VAV box and diffuser.

Use the entries in the Ductulator window to describe the physical characteristics of the duct and specify two performance parameters (typically airflow and velocity). The program will calculate the friction rate and dimensions. With that information, enter a nominal duct size that approximates the calculated value(s) and let the program determine the new friction rate and velocity.

Ductulator requires the following information to solve the equal-friction-related equations:

- Duct shape
- Material type
- Roughness
- Duct ID
- Length

Plus any two of these variables: airflow, friction rate, and velocity dimensions. For round duct, enter the diameter plus airflow or friction rate or velocity. For rectangular or oval duct, either:

• Enter the height and width plus airflow or friction rate or velocity. These entries can be made in any order.

• Enter one dimension (either height or width) plus two other variables from among airflow, friction rate, and velocity. This method fixes the duct height or width if you enter the value before completing the other entries.

Remember that you can represent fittings and terminations by defining equivalent lengths of straight duct. On completing these calculations, the program automatically displays the Ductulator Report.

REFERENCES

1. *1993 ASHRAE Handbook, Fundamentals*, ASHRAE, Atlanta, GA, 1985, chap. 33, "Duct Design."
2. Carrier Corporation, *Air Conditioning System Design*, McGraw-Hill, New York, 1965, part 2, chaps. 1–3.
3. *Engineering Design Reference Manual for Supply Air Handling Systems*, United McGill Corp., 1996.
4. Committee on Industrial Ventilation, *Industrial Ventilation—A Manual of Recommended Practice*, American Conference of Governmental Industrial Hygienists, Lansing, MI, 1989.

BIBLIOGRAPHY

Publications of the Air Diffusion Council, Cincinnati, OH.
HVAC Design Software Trane Corp. La Crosse, WI. www.trane.com

26.2 DUCT MAINTENANCE

Michael S. Palazzolo
President, Safety King, Inc., Utica, Michigan

Michael J. Palazzolo
General Manager, Safety King, Inc., Utica, Michigan

26.2.1 A PRACTICAL DEFINITION

Ductwork is a system of passageways through which air is delivered to, and/or extracted from, the interior spaces in buildings. The image called to mind by the word "ductwork" is usually the long, enclosed, aluminum passageways, rectangular in cross section, through which heated or cooled air is supplied to the rooms of homes or offices by the furnace or air conditioner fan units. But ductwork is not always long, or enclosed, or aluminum, or rectangular in cross section, and does not always convey heated or cooled air and may not even be connected to a furnace or air conditioning. In practice, ductwork must be understood to include all the air-handling components of the heating, ventilating, and air-conditioning (HVAC) system (Fig. 26.2.1).

The HVAC system of a building includes return air grilles, return air ducts to the Air-Handling Unit, or AHU, (except ceiling plenums, those spaces above a suspended ceiling from which return air may be extracted), interior surfaces of the AHU, air wash systems, filters, filter housings, fan housings and blades, mixing boxes, coil compartments, condensate drain pans, spray eliminators, supply air ducts, humidifiers and dehumidifiers, turning vanes, reheat coils, and supply diffusers. However, the HVAC system of a building sometimes is not considered to include the air heating or cooling equipment (furnaces and air conditioners) themselves.

26.2.2 TASKS OF THE HVAC SYSTEM

The HVAC system of a building has three primary tasks: to deliver air, to return air, and to make up air.

To fulfill the air delivery task, the HVAC system conveys, to occupied building zones or conditioned spaces, air that has been pressurized, heated, cooled, humidified, dehumidified, washed, or filtered. In its return-air role, the HVAC system extracts air from occupied building zones or conditioned spaces in order to pressurize, heat, cool, humidify, dehumidify, wash, or filter it. To accomplish the makeup air task, the HVAC system carries air to occupied building zones or conditioned spaces to replace air being removed by equipment such as range hoods, spray booths, exhaust fans, or clothes dryers.

26.2.3 MAINTENANCE REQUIREMENTS OF THE HVAC SYSTEM

Properly maintaining the HVAC system of a building requires several ongoing activities:

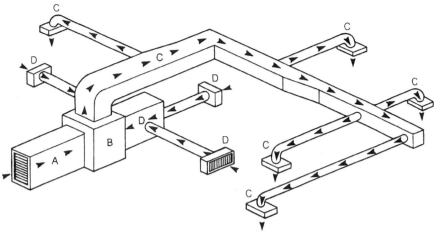

A. Fresh air intake (not always present)
B. Central heating/cooling equipment
C. Supply duct system
D. Return duct system

FIGURE 26.2.1 Typical HVAC system schematic. [*Reproduced by permission of the North American Insulation Manufacturers Association (NAIMI).*]

26.2.3.1 Inspection Checklist, Schedule, and Log

In all but the smallest buildings, a written inspection checklist (in hard-copy or electronic form) should be drawn up (for example, Fig. 26.2.2—four parts). This inspection checklist should be sufficiently detailed to serve unaided as documentation to guide an inspector unfamiliar with both the building and its HVAC system through the entire inspection program. The procedure should be updated at least annually, or more often as changes may dictate, to ensure that it reflects the most recent additions to or alterations in the components of the HVAC system. The inspection procedure, each time it is updated, should be dated or coded in such a way that its versions may be controlled. Some care should be exercised to limit the number of hard copies that are made to guard against inadvertent use of an expired version of the procedure.

Whenever a new version is created, all hard copies of the previous version should be discarded. A copy of the current version should be under the control of the chief building engineer, or of a person in an equivalent position, whose responsibility is to see that orderly and regular inspections are carried out as called for in the inspection procedure. This person should be prepared to produce this copy for examination by the building engineer audit team or auditor should such an audit (or any other similar audit) be common practice in the building.

A schedule for inspections, matching the requirements of the inspection procedure, should be created and maintained in a place accessible to all personnel directly concerned with inspection and maintenance of the HVAC system. The schedule should specify dates and times for inspections to be carried out and identify the engineering personnel responsible for carrying out each one.

A log of inspections should also be created. This log, as it is completed during and following each inspection, becomes the official and sometimes legal record of the maintenance of the building HVAC system. As such, it should be a first priority of building

HVAC Checklist - Short Form

Building Name: _____ Address: _____

Completed by: _____ Date: _____ File Number: _____

MECHANICAL ROOM

■ Clean and dry? _____ Stored refuse or chemicals? _____

■ Describe items in need of attention _____

MAJOR MECHANICAL EQUIPMENT

■ Preventive maintenance (PM) plan in use? _____

Control System

■ Type _____

■ System operation _____

■ Date of last calibration _____

Boiler

■ Rated Btu input _____ Condition _____

■ Combustion air: is there at least one square inch free area per 2,000 Btu input? _____

■ Fuel or combustion odors _____

Cooling Tower

■ Clean? No leaks or overflow? _____ Slime or algae growth? _____

■ Eliminator performance _____   _____

■ Biocide treatment working? (list type of biocide) _____

■ Spill containment plan implemented? _____ Dirt separator working? _____

Chillers

■ Refrigerant leaks? _____

■ Evidence of condensation problems? _____   _____

■ Waste oil and refrigerant properly stored and disposed of? _____

FIGURE 26.2.2   HVAC checklist [part 1].

engineering to ensure the log is completed for each inspection and that the log is maintained in a place accessible to all personnel directly concerned with inspection and maintenance of the HVAC system. The log should show the date each inspection is carried out, the person or persons conducting the inspection, the findings of the inspection, and the action plan (with completion dates shown) for all unscheduled maintenance activity generated as a result of the inspection.

Since all three of these documents, the procedure, the schedule, and the log, are of such singular importance to the maintenance of the building HVAC system, building engineering should consider maintaining all three, if in hard-copy form, in a loose leaf binder kept

HVAC Checklist - Short Form

Building Name: _____ Address: _____

Completed by: _____ Date: _____ File Number: _____

AIR HANDLING UNIT

■ Unit identification _____ Area served _____

Outdoor Air Intake, Mixing Plenum, and Dampers

■ Outdoor air intake location _____

■ Nearby contaminant sources? (describe)_____

■ Bird screen in place and unobstructed? _____

■ Design total cfm _____ outdoor air (O.A.) cfm _____ date last tested and balanced _____

■ Minimum % O.A. (damper setting) _____ Minimum cfm O.A. $\dfrac{\text{(total cfm x minimum \% O.A.)}}{100}$ = _____

■ Current O.A. damper setting (date, time, and HVAC operating mode) _____

■ Damper control sequence (describe) _____

■ Condition of dampers and controls (note date) _____

Fans

■ Control sequence _____

■ Condition (note date) _____

■ Indicated temperatures   supply air _____ mixed air _____ return air _____ outdoor air _____

■ Actual temperatures   supply air _____ mixed air _____ return air _____ outdoor air _____

Coils

■ Heating fluid discharge temperature _____ ΔT _____ cooling fluid discharge temperature _____ ΔT _____

■ Controls (describe) _____

■ Condition (note date) _____

Humidifier

■ Type _____ If biocide is used, note type _____

■ Condition (no overflow, drains trapped, all nozzles working?) _____

■ No slime, visible growth, or mineral deposits? _____

FIGURE 26.2.2   HVAC checklist [part 2].

in a single location. If the documents are in electronic form, consider storing all three in a single folder or directory.

26.2.3.2 Balancing of Air Flow

Proper maintenance of the HVAC system of a building is impossible if proper balance of air flow is not insured in all occupied building zones or conditioned spaces. If the air flow in any part of the HVAC system is allowed to become unbalanced with respect to zone

HVAC Checklist - Short Form

Building Name: _____ Address: _____

Completed by: _____ Date: _____ File Number: _____

DISTRIBUTION SYSTEM

Zone/ Room	System Type	Supply Air ducted/ unducted	cfm	Return Air ducted/ unducted	cfm	Power Exhaust cfm	control	serves (e.g. toilet)

Condition of distribution system and terminal equipment (note locations of problems)

■ Adequate access for maintenance? _____

■ Ducts and coils clean and obstructed? _____

■ Air paths unobstructed?   supply _____ return _____ transfer _____ exhaust _____ make-up _____

■ Note locations of blocked air paths, diffusers, or grilles _____

■ Any unintentional openings into plenums? _____

■ Controls operating properly? _____

■ Air volume correct? _____

■ Drain pans clean? Any visible growth or odors? _____

Filters

Location	Type/Rating	Size	Date Last Changed	Condition (give date)

FIGURE 26.2.2   HVAC checklist [part 3].

requirements, either too much or too little air flow, the occupants of the affected zone are likely to take action that can only result in premature or excessive collection of particulate matter and/or serious damage to the equipment itself. These actions include taping over air supply or cold-air return grilles thought to be related to air flow problems, removal and loss of grilles or diffusers, abnormal thermostat settings causing excessive running of heating or cooling apparatus, overly frequent thermostat setting changes causing excessive start/stop cycling of heating or cooling apparatus, disabling of thermostats thought to be responsible for perceived lack of air flow balance, and blocking off of parts of the building creating new

HVAC Checklist - Short Form

Building Name: _____ Address: _____

Completed by: _____ Date: _____ File Number: _____

OCCUPIED SPACE

Thermostat types ——

Zone/ Room	Thermostat Location	What Does Thermostat Control? (e.g., radiator, AHU-3)	Setpoints		Measured Temperature	Day/ Time
			Summer	Winter		

Humidistat/Dehumidistat types ————————————————————————————————

Zone/ Room	Humidistat/ Dehumidistat Location	What Does It Control?	Setpoints (%RH)	Measured Temperature	Day/ Time

■ Potential problems (note location) _____

■ Thermal comfort or air circulation problems (drafts, obstructed airflow, stagnant air, overcrowding, poor thermostat location)

_____

_____

■ Malfunctioning equipment ———————————————————————————————————

■ Major sources of odors or contaminants (e.g., poor sanitation, incompatible uses of space)

FIGURE 26.2.2   HVAC checklist [part 4].

zones not allowed for in HVAC system design. These actions, taken by occupants in order to secure their own comfort, are usually the result of inadequate attention to air flow balance. All these actions will mean more frequent and costly cleaning and repair of affected HVAC system components.

Two things are needed to avoid these difficulties and ensure proper balance of air flow in all occupied building zones or conditioned spaces: First, systematic monitoring of occupants and/or spaces must be carried out. And second, at least one person per shift on the building engineering staff must be properly and completely trained in the specific air flow balancing techniques and equipment applicable to the HVAC system components of the building.

In smaller buildings, monitoring of occupants and spaces can be a simple matter of walking the various spaces of the building, noting temperature and other air quality characteristics of each space, and asking occupants, if any, to comment on their perceived level of comfort. In large buildings with many occupants, it may be necessary to adopt a more formal process, utilizing regular measurements of, logging of, and corrective action on temperature, humidity, particulate matter, or other air quality measurables connected to HVAC system air flow balance. Formal outside training may be necessary to assure one staff person on each shift capable of adjusting air flow balance throughout the building. This training can sometimes be obtained from the manufacturer of the HVAC system components affecting airflow balance. Some community colleges also provide courses which include training in specific air flow balance techniques and equipment. More frequently, such information is obtained by engineering staff members by careful study of equipment, blueprints, and engineering manuals supplied with HVAC system components at the time of installation. It is critical that this information, once reliably established, be documented and stored. If air flow balance techniques and equipment use are written down or stored electronically, training of new engineering staff members is greatly simplified and tends to produce a more uniform approach to air flow balance over time. Attention paid to air flow balance throughout the building HVAC system will pay for itself many times over in reduced cleaning and repair costs.

26.2.3.3   Replacement or Repair of Components—Choosing a Contractor

Even with best-in-class inspection procedures and consistent attention to balancing of air flow throughout the HVAC system, components still break down or wear out and must be repaired or replaced. Some of these tasks are very simple and can be carried out in-house, but most (particularly major failures) will require the services of one or more outside professional mechanical, heating, cooling, or plumbing contractors. This is particularly true when replacement or repair involves the use of expensive or specialized equipment you do not have in-house. Even when an HVAC system component repair or replacement can be handled in-house, you may often choose to bring in a contractor. In some cases, there may be service contracts in place. In effect, a service contract means that the services of the contractor have already been paid for, so there is every reason to make use of the contractor's services. In other cases, the use of a contractor to address a replacement or repair of an HVAC system component may mean faster results. This could be important, especially in occupied buildings in which the component failure deprives occupants of heated or cooled air in work spaces. It is worth remembering too that the use of a contractor always means that you and your staff can attend to the many vital building maintenance tasks for which you are responsible, rather than putting everything on hold while you address an HVAC system component repair or replacement.

If you decide to employ a contractor, and no service contract is in place, you will need to choose a contractor. This will generally be a matter of consulting historical files or lists to find out which contractors have performed similar services in your building in the past. If you find you must utilize a contractor unknown to you, invest the time it takes to talk with the contractor in advance and satisfy yourself that he or she appears to be knowledgeable and professional. Make sure you see applicable insurance, licenses, and references and do not contract until you have checked references. A few minutes of checking can eliminate expensive "do-it-again" costs and sometimes even higher legal costs in cases of service failure that can only be resolved by litigation.

26.2.4 CLEANING OF HVAC SYSTEM COMPONENTS

The HVAC system components in your building will require regular cleaning. Because of health, safety, and regulatory issues, many buildings now have in place an HVAC system component cleaning schedule. If you do not have such a preventive maintenance cleaning schedule for HVAC system components, create one. Consult your inspection, repair, and cleaning records and use these to estimate appropriate intervals between scheduled cleanings. Your inspection procedures should be sufficiently thorough to reveal special cleaning needs that may arise for various reasons between regularly scheduled cleanings.

The cleaning can be carried out by your staff or an outside contractor. Typically, an organization will not have in-house the specialized vacuum equipment and other gear required to do the job thoroughly and will need to bring in a professional to do the work. One word of caution: be sure that the contractor you have has the insurance and licenses your state requires. Discovering such a lack after an expensive service failure or a lawsuit is too late.

If you must change contractors for some reason or have no service history to guide you, you may have to hire a contractor with whom you have had no experience. Interview the prospective contractor in person or by telephone before you contract. Here are several suggestions to guide the interview.

Ask to see the contractor's insurance certificates, licenses (be sure that you know your state's licensing requirements), and references. Tell the contractor you will call at least one of the listed references and ask a fellow building engineering professional whether he or she would hire this contractor to do the kind of work you need done. Check references and the Better Business Bureau after the interview.

Ask the contractor about other professional credentials he or she may have (formal training, degrees, years of experience, membership in professional associations, etc.) and whether the people actually doing the work have appropriate experience and training.

Inquire into the methods the contractor will use. Ask him or her to describe to you how the work will be done. Ask about equipment that may be used and ductwork access practices, if applicable, employed by the contractor. If the job involves porous and/or non-metallic surfaces, ask what methods are used to clean these surfaces. If the contractor is going to "clean" such surfaces by spraying glue, shellac, or other encapsulant on them, you need to know that in advance. In some cases, such an approach may make sense, but discuss it in advance with the contractor and make sure that you know what "cleaning" you are buying. Generally, use of an encapsulant would only be appropriate after rigorous application of source-removal techniques.

Ask what hazardous waste disposal practices the contractor makes use of should such materials be encountered during HVAC system component cleaning. HVAC system component cleaning occasionally uncovers asbestos, leaky refrigerant lines, and other wastes that pose potential harm to building occupants, cleaning technicians, or the environment.

If the cleaning will involve liquid treatments of HVAC system component surfaces, for deodorizing or sanitizing, find out what substances will be used and ask for material safety data sheet (MSDS) documentation on each one. Sanitizing treatments, to kill mold, mildew, bacteria, etc., are generally administered by self-contained foggers or vaporizers, but may also be delivered via air-driven spray units. These treatments can be a valuable complement to normal cleaning procedures, especially in cases involving building occupants who may have heightened sensitivity to airborne contaminants because of asthma or other conditions. Make sure that the sanitizing treatment product is EPA registered and accepted by the FDA and will dissipate in a known (and reasonable) time.

Price is, of course, an important contracting issue. If the contractor's pricing seems excessively high or low, ask him or her to explain the pricing. Investigate at least one other contractor for comparison.

Here is a list of additional questions you may consider asking a potential HVAC system component cleaning contractor.

1. How long has your company been cleaning HVAC system components?
2. What percentage of your business is dedicated to HVAC system component cleaning?
3. What is your experience in cleaning systems similar to those in my facility?
4. Who will be the on-site supervisor responsible for this project? How many projects of a similar scope has he or she been responsible for?
5. Is your firm a member in good standing of the National Air Duct Cleaners Association (NADCA) and can you provide us with a current membership certificate?
6. Will you use *source removal* techniques in accordance with NADCA Standard ACR 2002 when cleaning my system?
7. Do you have a complete understanding of NADCA Standard ACR 2002 and will you comply with all of its provisions on this job?
8. Do you have a comprehensive in-house safety program with training for employees?
9. Are you knowledgeable about site preparation issues for a project of this scope?
10. Is your equipment in good repair and proper working order? When was it purchased and how long has it been in use?

Once a new contractor has been chosen, and the cleaning work completed, do a post-cleaning inspection and verification. In some special cases, it may be appropriate to hire an independent lab to do particulate and biological sampling in work spaces affected by the HVAC system components that have been cleaned. Some labs will do before-and-after-cleaning sampling to quantify certain kinds of indoor air quality measurables. If you intend to have such sampling done, notify the cleaning contractor before the work is done and give a copy of the results to the contractor, whether favorable or not. In most HVAC system component cleaning, such sampling will not be necessary. A before-and-after visual inspection of the HVAC system components involved will tell you whether the contractor has done the job agreed upon.

26.2.5 PROCEDURES FOR DUCT CLEANING*

There are three generally accepted methods:

- Contact vacuum method
- Air washing method
- Power brushing method

26.2.5.1 Contact Vacuum Method

Principle of Operation. Conventional vacuum cleaning of interior duct surfaces through openings cut into the ducts is satisfactory so long as reasonable care is exercised.

---

*Reproduced by permission of the North American Insulation Manufacturers Association (NAIMA)

The risk of damaging duct surfaces is minimal. Only high efficiency particle arresting (HEPA) vacuuming equipment should be used if vacuum equipment will be discharging into occupied space. Conventional vacuuming equipment may discharge extremely fine particulate matter back into the building air space, rather than collecting it.

This process may leave particulate matter in the duct which may later become airborne and contaminate the occupied space. This may occur because the duct is not under negative pressure during the cleaning operation.

Cleaning Procedure.   See Fig. 26.2.3.

- Direct vacuuming will usually require larger access openings in the ducts in order for cleaning crews to reach into all parts of the duct. Spacing between openings will depend on the type of vacuum equipment used and the distance from each opening it is able to reach.

- The vacuum cleaner head is introduced into the duct at the opening furthest upstream and the machine turned on. Vacuuming proceeds downstream slowly enough to allow the vacuum to pick up all dirt and dust particles. The larger the duct, the longer this will take. Observation of the process is the best way to determine how long it takes before linings are considered sufficiently clean.

- When observation indicates the section of duct has been cleaned sufficiently, the vacuum device is withdrawn from the duct and inserted through the next opening, where the process is repeated.

26.2.5.2   Air Washing Method

Principle of Operation.   A vacuum collection device (preferably a powerful truck-mounted vacuum positioned outside the building) is connected to the downstream end of the section being cleaned through a predetermined opening. It is recommended that the isolated area of the duct system being cleaned be subjected to 1 in (25 mm) (minimum) static pressure to ensure proper transport of loosened material (take care not to collapse the duct).

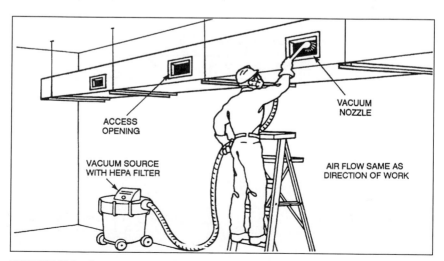

FIGURE 26.2.3   Contact vacuum cleaning.

Compressed air is introduced into the duct through a hose terminating in a "skipper" nozzle. This nozzle is designed so that the compressed air propels it along inside the duct. This dislodges dirt and debris which, becoming airborne, are drawn downstream through the duct and out of the system by the vacuum collection equipment. (Dirt and dust particles must be dislodged from duct surfaces and become airborne before they can be removed from the duct system.) If the vacuum collector discharges to occupied space, HEPA filtration should be used. The compressed air source should be able to produce between 160 and 200 psi (965 and 1100 kPa) air pressure, and have a 20-gal (75 L) receiver tank, for the air washing method to be effective.

This method is most effective in cleaning ductwork no larger than 24 in · 24 in (60 cm · 60 cm) inside dimensions.

Cleaning Procedure.   See Fig. 26.2.4.

* Openings for cleaning purposes often need be no larger than those cut for borescope inspection purposes. These should be drilled through the duct wall and the insulation (if present) at intervals which depend on type of equipment and duct size.

* The vacuum collection equipment is turned on and the proper negative pressure established. The compressed air hose with the skipper nozzle is inserted into the hole farthest upstream.

* The skipper nozzle is allowed to travel downstream slowly enough to allow the skipping motion to dislodge dirt and dust particles. The larger the duct, the more time this will take; observation of the process in each case is the best way to determine how long. When observation suggests the section of duct has been cleaned sufficiently, the compressed air hose is withdrawn from the duct and inserted in the next downstream hole, where the process is repeated. The condition of the cleaned section may be inspected with a borescope inserted in the hole.

26.2.5.3   Power Brushing Method

Principle of Operation.   As with the air-washing system, a vacuum collection device (preferably a powerful, truck-mounted vacuum positioned outside the building) is

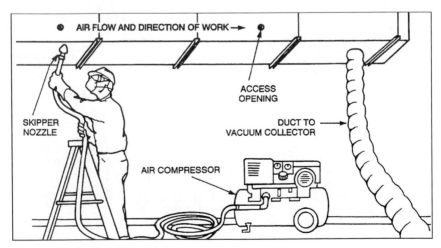

FIGURE 26.2.4   Air washing method.

connected to the downstream end of the section being cleaned through a predetermined opening. Pneumatically or electrically powered rotary brushes are used to dislodge dirt and dust particles that become airborne, are drawn downstream through the duct system, and are evacuated by the vacuum collector. Power brushing works satisfactorily with all types of duct, and all types of fibrous glass surfaces, provided the bristles are not too stiff nor the brush permitted to remain in one place for such a long time that it may damage them. If the vacuum collector discharges to occupied space, HEPA filtration should be used.

Note: It may prove impossible to use the power brushing method to clean fibrous glass ducts with tie rod reinforcement.

Cleaning Procedure. See Fig. 26.2.5.

• Power brushing will usually require larger access openings in the ducts in order for cleaning crews to reach inside and manipulate the equipment. However, fewer openings may be required; some power brushing devices are able to reach up to 20 ft (6 m) in either direction from an opening.

• Once the isolated section of the duct to be cleaned is under negative pressure (as described in the air-washing method), the rotary power brushing device is introduced into the duct at the opening furthest upstream. The brushes are worked downstream slowly to allow them to dislodge dirt and dust particles.

• When working in ducts with fibrous glass airstream surfaces, the brushes should be kept in motion so as not to gouge or dig into the surfaces. Again, the larger the duct, the more time this will take; observation of the process in each case is the best way to determine how long. When observation suggests the section of duct has been cleaned sufficiently, the power brush is withdrawn from the duct and inserted in the next downstream opening, where the process is repeated.

Note: Access to ductwork is key to the success of all three methods of cleaning ductwork. Whether in-house or contracted, optimal results cannot be expected if ductwork is inaccessible above ceilings, under slabs, or in unreachable crawl spaces or attics.

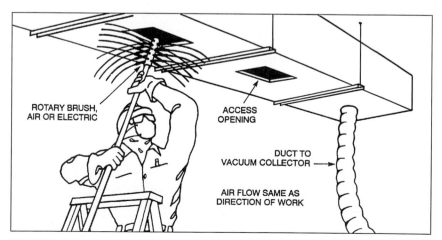

FIGURE 26.2.5 Power brushing method.

RESOURCES

Following is a list of applicable document sources and documents on cleaning of HVAC system components:

Indoor Air Quality Information Clearinghouse
U.S. Environmental Protection Agency
P.O. Box 37133
Washington, DC 20013
1(800) 438-4318/Fax (301) 588-3408
The Inside Story: A guide to Indoor Air Quality.
Building Air Quality: A Guide for Building Owners and Facility Managers.

National Air Duct Cleaners Association (NADCA)
1518 K Street, N.W., Suite 503
Washington, DC 20005
(202) 737-2926/Fax (202) 347-8847
NADCA Standard 1992-01, Mechanical Cleaning of Non-Porous Air Conveyance System Components.
Understanding Microbial Contamination in HVAC Systems Introduction to HVAC System Cleaning Services.

National Air Filtration Association (NAFA)
1518 K Street, N.W., Suite 503
Washington, DC 20005
(202) 628-5328/Fax (202) 638-4833
NAFA Guide to Air Filtration (ISBM 1-884152-00-7).

North American Insulation Manufacturers Association (NAIMA)
44 Canal Plaza, Suite 310
Alexandria, VA 22314
(703) 684-0084/Fax (703) 684-0427
Fibrous Glass Duct Construction Standards, 1993.

Sheet Metal and Air Conditioning Contractors National Association (SMACNA)
4201 Lafayette Center Drive
Chantilly, VA 22021
(703) 803-2980
HVAC Duct Construction Standards-Metal and Flexible.

U.S. Department of Labor
Occupational Safety & Health Administration Room N3651
200 Constitution Avenue, N.W.
Washington, DC 20210
(202) 219-6666
All About OSHA (OSHA 2056).
Control of Hazardous Energy (OSHA 3120) Respiratory Protection (OSHA 3079).
Chemical Hazard Communication (OSHA 3084) Personal Protective Equipment (OSHA 3077).

American Society of Heating, Refrigerating, and Air Conditioning Engineers (ASHRAE)
1791 Tullie Circle, NE
Atlanta, GA 30329
(404) 636-8400

APPENDIX A
ENGINEERING GUIDE FOR ALTITUDE CORRECTIONS

A.1   INTRODUCTION

Altitude affects the operation of air-conditioning equipment in the following areas:

• Psychrometric air properties
• Air density
• Temperature level of stream-heating equipment

A.1.1   Effect on Psychrometric Air Properties

Altitude affects the psychrometric properties of air such as enthalpy, dew-point temperature, and specific humidity. This is best shown visually by the construction of an altitude psychrometric chart using the following relationship:

$$W = \frac{0.622 P''}{P - P''}$$

where W = specific humidity, lb water vapor/lb dry air (kg water vapor/kg of dry air)
P'' = partial pressure of water vapor at the dew-point temperature, psia (kPa absolute)
P = barometric pressure, psia (kPa absolute)

By assuming various values of the dew-point temperature and calculating values of W corresponding to the barometric pressure, the saturation curve may be drawn for air at elevations above sea level. This curve shows that air at elevation has a higher specific humidity than air at sea level for a given dry-bulb temperature and percent relative humidity. The higher specific humidity causes the enthalpy of air to be higher for a given dry-bulb temperature and percent relative humidity or for a given value of dry-bulb and wet-bulb temperature.

Note: The material in this appendix, including tables and illustrations, is taken from Carrier Corporation's *Engineering Guide for Altitude Effects*, 1967, courtesy of Carrier Corporation. Edited for this handbook by Nils R. Grimm.

A.1.2   Effect on Air Flow

The density of air at altitude varies inversely as the absolute pressure from the perfect gas law:

$$p = \frac{P}{RT}$$

where p = density, lb/ft^3 (kg/m^3)
  P = pressure, lb/ft^2 (Pa)
  R = gas constant for air = 53.3 (287.05)
  T = absolute temperature, °F (°R or K)

Therefore, to maintain the same mass flow rate at altitude as at sea level, the air-volume flow rate must be increased at a rate inversely proportional to the air-density ratio. If the air-volume flow rate is held constant, the mass flow rate decreases at a rate directly proportional to the air-density ratio.

A.1.3   Effect on Temperature Level of Equipment Using Steam

Although altitude causes no change in the saturation temperature-pressure relationship, there is a change in the temperature corresponding to a certain gauge pressure. For example, at sea level the saturation temperature of 5 psig (0.34 bar) steam is 227°F (108.5°C), while at 7500 ft (2285.9 m) the saturation temperature is 216.3°F (102.4°C). The effect of altitude on temperature level should be considered in steam unit heaters, heating and ventilating units, and absorption refrigeration machines.

In the range of 0 to 10,000 ft (0 to 3050 m), the effect of altitude on the thermal properties of air (such as viscosity, thermal conductivity, and specific heat) is very small and can be disregarded.

A.2   ADJUSTMENT DATA FOR VARIOUS KINDS OF AIR-CONDITIONING EQUIPMENT

A ready reference index (Table A.1) is provided to allow quick reference to correction factors for various air-conditioning products. *For some products, it is impractical to give the correction factor. In such cases, refer to the applicable section of this appendix referenced in the table for the procedure or recommendations for correcting for altitude effects.*

A.2.1   Open and Hermetic Compressors

Compressor capacity is expressed as a function of saturated discharge temperature and saturated suction temperature. Regardless of compressor type, the capacity is the same at altitude as at sea level for the same saturated suction and discharge temperatures. At altitude, however, there is a change in the gauge pressure corresponding to the saturated discharge and saturated suction temperatures. For example, the gauge pressure at 5000 ft (1500 m) of elevation is higher by 2.6 psi (0.179 bar) than the gauge pressure at sea level (see Table A.2.).

TABLE A.1   Ready Reference Index

Effect of altitude on air-conditioning equipment capacity

	Altitude correction factors			
	Elevation, ft (m)			
Product name	2500 (760)	5000 (1500)	7500 (2300)	10,000 (1005)
Reciprocating compressors	None	None	None	None
Condensing units (water-cooled)	None	None	None	None
Air-cooled condensers	0.95	0.90	0.85	0.80
Evaporative condensers	1.00	1.01	1.02	1.03
Liquid coolers	None	None	None	None
Absorption refrigeration machines	See Sec. A.2.6, p. A.7			
Centrifugal refrigeration machines (open and hermetic type)	None	None	None	None
Reciprocating liquid chilling packages with air-cooled condensers	0.98	0.97	0.95	0.93
Induction-room terminals: Chilled-water Hot water—gravity and forced air	0.93 0.95	0.86 0.90	0.80 0.85	0.74 0.81
Steam heating Electric heating	See Sec. A.2.11, p. A.16 See Sec. A.2.13, p. A.17			
Air terminal units and outlets	See Sec. A.2.14, p. A.17			
Condensing units (air-cooled)	0.98	0.97	0.95	0.93
Central station air-handling units: Chilled-water and direct-expansion Steam and hot water	See Sec. A.2.15, p. A.19 See Secs. A.2.11, p. A.16, and A.2.15, p. A.19			
Chilled-water fan-coil units: Chilled-water Steam and hot water	See Sec. A.2.9, p. A.8 See Secs. A.1.3, p. A.2, and A.2.15, p. A.19			
Direct-expansion fan-coil units: Direct-expansion, total capacity SHF 0.40–0.95 Direct-expansion, total capacity SHF 0.95–1.00 Direct-expansion, sensible capacity SHF 0.40–0.95 Hot water	0.97 0.93 0.92 0.95	0.95 0.86 0.85 0.90	0.93 0.79 0.78 0.85	0.91 0.73 0.71 0.81
Steam	See Sec. A.1.3, p. A.2			
Room fan-coil units: Chilled-water, total capacity SHF 0.40–0.95 Chilled-water, total capacity SHF 0.95–1.00 Chilled-water, sensible capacity SHF 0.40–0.95 Hot water	0.97 0.93 0.92 0.95	0.95 0.86 0.85 0.90	0.93 0.80 0.78 0.85	0.91 0.74 0.71 0.81
Steam	See Sec. A.2.11, p. A.16			

TABLE A.1   Ready Reference Index (*Continued*)

Effect of altitude on air-conditioning equipment capacity

	Altitude correction factors			
	Elevation, ft (m)			
Product name	2500 (760)	5000 (1500)	7500 (2300)	10,000 (1005)
Central station air-handling units with sprayed coil:				
Chilled-water and direct-expansion	See Sec. A.2.17, p. A.20			
Steam and hot water	See Secs. A.2.11, p. A.16, and A.2.15, p. A.19			
Unit Heaters:				
Steam	See Sec. A.2.11, p. A.16			
Gas-fired	See Sec. A.2.15, p. A.19			
Hot water	0.95	0.90	0.85	0.81
Heating and ventilating units:				
Steam	See Sec. A.1.3, p. A.2			
Hot water	0.95	0.90	0.85	0.81
Roof-mounted air-conditioning units:				
Direct-expansion, total capacity SHF 0.40–0.95	0.98	0.96	0.94	0.92
Direct-expansion, total capacity SHF 0.95–1.00	0.96	0.92	0.88	0.84
Direct-expansion, sensible capacity SHF 0.40–0.95	0.92	0.85	0.78	0.71
Gas-fired heating	See Sec. A.2.16, p. A.19			
Package air-conditioning unit with air-cooled condenser:				
Direct-expansion, total capacity SHF 0.40–0.95	0.98	0.96	0.94	0.92
Direct-expansion, total capacity SHF 0.95–1.00	0.96	0.92	0.88	0.84
Direct-expansion, sensible capacity SHF 0.40–0.95	0.92	0.85	0.78	0.71
Package air-conditioning unit with water-cooled condenser:				
Direct-expansion, total capacity SHF 0.40–0.95	0.99	0.985	0.98	0.97
Direct-expansion, total capacity SHF 0.95–1.00	0.96	0.93	0.90	0.87
Direct-expansion, sensible capacity SHF 0.40–0.95	0.93	0.87	0.80	0.74
Package air-conditioning unit with air- or water-cooled condenser:				
Hot water	0.95	0.90	0.85	0.80
Steam	See Secs. A.1.3, p. A.2, and A.2.11, p. A.16			
Gas-fired furnace	See Sec. A.2.16, p. A.19			
Centrifugal fans	See Sec. A.2.18, p. A.23			
Motors	See Sec. A.2.20, p. A.24			
Pumps	See Sec. A.2.19, p. A.24			

Altitude thermal capacity = sea-level thermal capacity x altitude correction factor.

TABLE A.2 Altitude Pressure for Air

Altitude	Barometric pressure		Gauge-pressure correction
ft	inHg	(psia)	psia
0	29.92	(14.70)	
500	29.38	(14.40)	−0.30
1,000	28.86	(14.19)	−0.51
1,500	28.33	(13.91)	−0.79
2,000	27.82	(13.58)	−1.12
2,500	27.32	(13.41)	−1.29
3,000	26.82	(13.20)	−1.50
3,500	26.33	(12.92)	−1.78
4,000	25.84	(12.70)	−2.00
4,500	25.37	(12.44)	−2.26
5,000	24.90	(12.23)	−2.57
5,500	24.43	(12.01)	−2.69
6,000	23.98	(11.78)	−2.92
6,500	23.53	(11.55)	−3.15
7,000	23.09	(11.33)	−3.37
7,500	22.65	(11.10)	−3.60
8,000	22.22	(10.92)	−3.78
8,500	21.80	(10.70)	−4.00
9,000	21.39	(10.50)	−4.20
9,500	20.98	(10.30)	−4.40
10,000	20.58	(10.10)	−4.60
m	mmHg	(bar)	bars
0	759.98	(1.014)	0
152.4	746.25	(0.993)	−0.021
304.8	733.04	(0.979)	−0.035
457.2	719.58	(0.959)	−0.055
609.6	706.63	(0.937)	−0.077
762	693.93	(0.925)	−0.089
914.4	681.23	(0.910)	−0.104
1066.7	668.78	(0.891)	−0.123
1219.1	656.34	(0.876)	−0.138
1371.5	644.4	(0.859)	−0.155
1523.9	632.46	(0.843)	−0.171
1676.3	620.52	(0.828)	−0.186
1828.7	609.09	(0.812)	−0.202
1981.1	597.66	(0.797)	−0.217
2133.5	586.49	(0.781)	−0.233
2285.9	575.31	(0.766)	−0.248
2438.3	564.39	(0.753)	−0.261
2590.7	553.72	(0.738)	−0.276
2743.1	543.31	(0.724)	−0.290
2895.4	532.89	(0.710)	−0.304
3047.9	522.73	(0.697)	−0.317

A.2.2 Water-Cooled Condensing Units

Since air is not used in these condensers, operation at altitude causes no change in capacity.

A.2.3 Air-Cooled Condensers

When an air-cooled condenser is operated at altitude, the actual air volume is maintained constant (for the same air temperature and fan speed) but the air-weight flow decreases. This causes the capacity of the condenser to decrease.

To obtain the capacity of altitude for these condensers, multiply the published values of capacity by the following factors:

Elevation, ft (m)	Altitude capacity factors
2,500 (760)	0.95
5,000 (1500)	0.90
7,500 (2300)	0.85
10,000 (3050)	0.80

A.2.4 Evaporative Condensers

There is a slight increase in the capacity of an evaporative condenser at altitude when operating at the same fan speed as at sea level.

To obtain the capacity at altitude for the above units, multiply the published values of capacity by the following factors:

Elevation, ft (m)	Altitude capacity factors
2,500 (760)	1.00
5,000 (1500)	1.01
7,500 (2300)	1.02
10,000 (3050)	1.03

A.2.5 Liquid Coolers

Since the heat exchange is between water and refrigerant, altitude operation causes no change in capacity.

A.2.6 Absorption Refrigeration Machine

The capacity of the absorption refrigeration machine is a function of the absolute pressure of the steam applied to the machine. Since the capacity of the above unit is expressed as a function of gauge pressure, it is necessary to correct for the reduction in absolute pressure at altitude. The correction of capacity for operation at altitude may be obtained by subtracting from the actual gauge pressure the pressure correction tabulated below to obtain a pseudo gauge pressure. Use this pseudo gauge pressure to obtain absorption-machine capacity:

Elevation, ft (m)	Gauge-pressure correction, psig (bar)
2,500 (760)	−1.3 (−0.0897)
5,000 (1500)	−2.6 (−0.179)
7,500 (2300)	−3.6 (−0.248)
10,000 (3050)	−4.6 (−0.317)

A.2.7   Air-Cooled Condensing Units

Graphical plots of air-cooled condenser capacity at sea level and at various altitudes in combination with compressor capacity show that the decrease in condensing-unit capacity at altitude is small compared to the reduction in capacity of the air-cooled condenser alone. To obtain the capacity at altitude for air-cooled condensing units, multiply the published values of capacity by the following factors:

Elevation, ft (m)	Altitude capacity factors
2,500 (760)	0.98
5,000 (1500)	0.97
7,500 (2300)	0.95
10,000 (3050)	0.93

A.2.8   Liquid Chilling Units with Air-Cooled Condensers

Graphical plots made of air-cooled condensers' capacity at sea level and at various altitudes in combination with condenserless liquid chilling units show that the decrease in capacity at altitude is small compared to the reduction in capacity of the air-cooled condenser alone.

   To obtain the capacity at altitude for the above units, multiply the published values of capacity by the following factors:

Elevation, ft (m)	Altitude capacity factors
2,500 (760)	0.98
5,000 (1500)	0.97
7,500 (2300)	0.95
10,000 (3050)	0.93

A.2.9   Chilled-Water Central Station Air-Handling Units; Chilled-Water Fan-Coil Units

Altitude capacity may be compared with sea-level capacity on the basis of constant air-volume flow (varying-weight flow) or constant air-weight flow (varying-volume flow).

For the above units the comparison is most conveniently made on the basis of constant air-weight flow. This shows that the capacity of the units increases if the cooling coil is dehumidifying. The comparison is based on the same entering-air dry-bulb and wet-bulb temperature, the same water flow, and the same entering-water temperature. Therefore, to obtain the same capacity from the unit operating at altitude with the same air-weight flow, the same entering dry-bulb and wet-bulb temperature, and same entering-water temperature, a smaller amount of water is required.

The following procedure is recommended to determine the gallons per minute (gpm) required at altitude when the sensible heat factor (SHF) is 0.95 or less.

Given: entering-air dry-bulb and wet-bulb temperatures, leaving-air dry-bulb and wet-bulb temperatures, entering-water temperature, and standard cubic feet per minute (cfm).

1. Calculate the load using the formula

$$GTH = 4.45 \times cfm(h_1 - h_2).$$ (A.1)

where GTH = grand total heat, Btu/h
 h_1 = entering-air enthalpy, Btu/lb at altitude entering-air wet-bulb temperature. Use altitude psychrometric charts, Figs. A.1 to A.4.
 h_2 = leaving-air enthalpy, Btu/lb at altitude leaving-air wet-bulb temperature. Use altitude psychrometric charts, Figs. A.1 to A.4.
 cfm = rate of air flow, standard cfm. If cfm is given at altitude, convert to standard cfm in order to use Eq. (A.1). (Standard cfm = altitude cfm × air-density ratio at altitude.)

2. Using the load calculated in step 1, the same standard air cfm, and the entering-air enthalpy at sea level (with the specified entering wet-bulb temperature), calculate the value of the leaving-air enthalpy at sea level as follows:

$$h_{2s} = h_{1s} - \frac{GTH}{4.45 \times cfm}$$ (A.2)

where h_{1s} = entering-air enthalpy at sea level and specified altitude entering wet-bulb temperature, Btu/lb
 h_{2s} = leaving-air enthalpy at sea level, Btu/lb

The saturation temperature which corresponds to the above value of enthalpy is the leaving-air wet-bulb temperature for sea-level operation.

3. From the manufacturer's performance chart, determine the sea-level adp using the entering wet-bulb temperature specified, the assumed coil surface and standard air velocity, and the sea-level leaving wet-bulb temperature determined from step 2.

4. Determine the gallons/minute required at sea level from published aparatus dewpoint (adp) ratings using the adp from step 2, the load determined from step 1, and the design entering-water temperature.

5. Multiply the sea-level gpm determined from step 4 by the following factors:

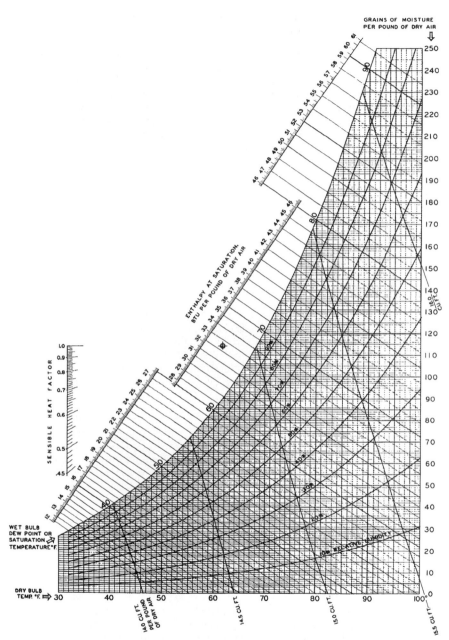

FIGURE A.1 Altitude psychrometric chart, elevation 2500 ft (762 m). Barometric pressure, 27.3 inHg = 13.4 psia (93 kPa); normal temperatures. Metric conversions: C = (°F − 32)/1.8; kJ (0.0004299) = Btu/lb; kg/kg (6997.9) = grains of moisture/lb of dry air; m/kg (16.01846) ft/lb of dry air.

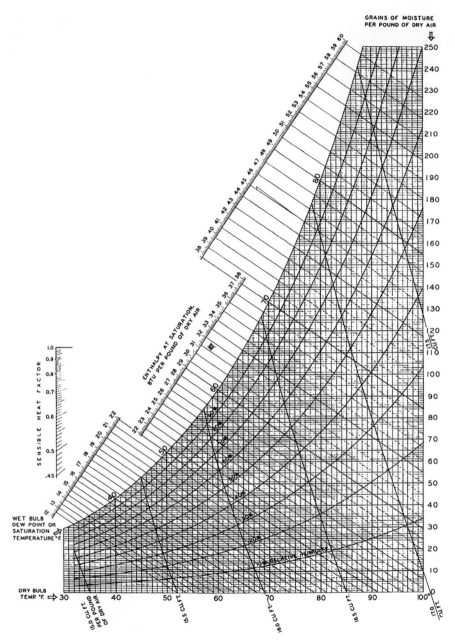

FIGURE A.2   Altitude psychrometric chart, elevation 5000 ft (1523.9 m). Barometric pressure, 24.9 inHg = 12.23 psia (84 kPa); normal temperatures. For metric conversion factors, see Fig. A.1.

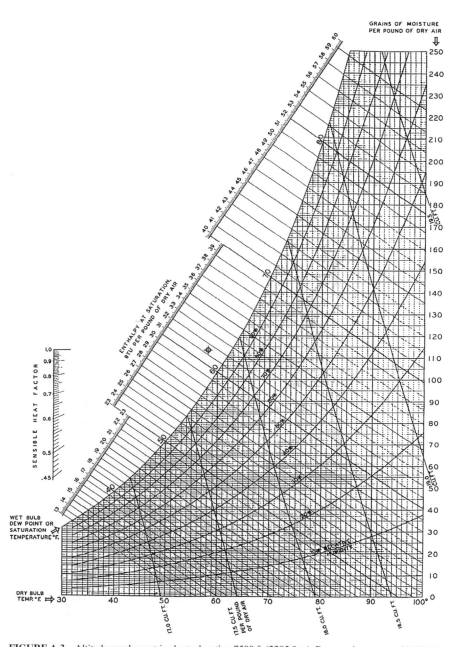

FIGURE A.3 Altitude psychrometric chart, elevation 7500 ft (2285.9 m). Barometric pressure, 22.7 inHg = 11.1 psia (76.5 kPa); normal temperatures. For metric conversion factors, see Fig. A.1.

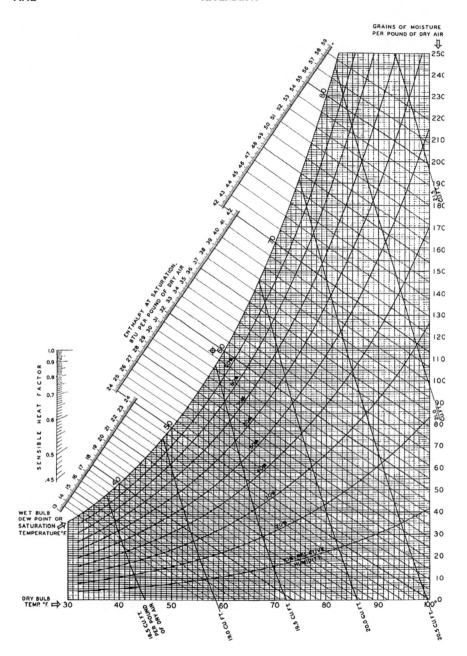

FIGURE A.4   Altitude psychrometric chart, elevation 10,000 ft (3047.9 m). Barometric pressure, 20.6 inHg = 10.1 psia (69.63 kPa); normal temperatures. For metric conversion factors, see Fig. A.1.

Elevation, ft (m)	Altitude sensible heat factor, 0.40–0.95
2,500 (760)	0.97
5,000 (1500)	0.94
7,500 (2300)	0.91
10,000 (3050)	0.88

6. Determine actual leaving dry-bulb temperature at altitude by the formula

$$t_2 = t_s + \text{BF}(t_l - t_s) \qquad \text{(A.3)}$$

where t_s = adp at altitude–saturation temperature corresponding to h_s in the formula

$$h_s = h_1 - \frac{h_1 - h_2}{1 - \text{BF}} \qquad \text{(A.4)}$$

t_1 = entering-air dry-bulb temperature, °F (°C)
t_2 = leaving-air dry-bulb temperature, °F (°C)
BF = bypass factor. Determine from the manufacturer's performance chart using standard air cfm.
h_s = enthalpy at t_s, Btu/lb (W/kg). Use altitude psychrometric charts, Figs. A.1 to A.4.
h_1 = entering-air enthalpy at altitude entering-air wet-bulb temperature, Btu/lb (W/kg). Use altitude psychrometric charts, Figs. A.1 to A.4.
h_2 = leaving-air enthalpy at altitude leaving-air wet-bulb temperature, Btu/lb (W/kg). Use altitude psychrometric charts. Figs. A.1 to A.4.

7. For chilled-water units operating with an SHF of 0.95 or greater, consider the unit to be doing all sensible cooling and to have the same capacity at altitude as at sea level. Follow the procedure for determining the gpm for a dry coil at sea level as given in the manufacturer's published data.

A.2.10   Direct-Expansion Central Station Air-Handling Units

Altitude capacity may be compared with sea-level capacity on the basis of constant air-volume flow (varying-weight flow) or constant air-weight flow (varying-volume flow).

For the above units the comparison is most conveniently made on the basis of constant air-weight flow. This shows that there is an increase in the capacity of these units if the cooling coil is dehumidifying. The comparison is based on the same air-weight flow at sea level and altitude, the same entering dry-bulb and wet-bulb temperature, the same refrigerant, and the same suction temperature. Therefore, to obtain the same capacity from the unit operating at altitude with the same air-weight flow, the same entering dry-bulb and wet-bulb temperature, and the same refrigerant, a higher saturated suction temperature is required at altitude than at sea level.

The following procedure is recommended to determine the required suction temperature at altitude when the sensible heat factor is 0.95 or less.

Given: entering-air dry-bulb and wet-bulb temperatures, leaving-air dry-bulb and wet-bulb temperatures, and standard cfm.

1. Calculate the load using formula (A.1):

$$\text{GTH} = 4.45 \times \text{cfm}(h_1 - h_2)$$

where GTH = grand total heat, Btu/h
      h_1 = entering-air enthalpy, Btu/lb at altitude entering-air wet-bulb temperature.
           Use altitude psychrometric charts, Figs. A.1 to A.4.
      h_2 = leaving-air enthalpy, Btu/lb at altitude leaving-air wet-bulb temperature.
           Use altitude psychrometric charts, Figs. A.1 to A.4.
   cfm = rate of air flow, standard cfm. If cfm is given at altitude, convert to standard
           cfm in order to use Eq. (A.1). (Standard cfm = altitude cfm × air-density
           ratio at altitude.)

2. Using the load calculated in step 1, the same standard air cfm, and the entering-air enthalpy at sea level (with the specified entering wet-bulb temperature), calculate the value of the leaving-air enthalpy at sea level, using formula (A.2), as follows:

$$h_{2s} = h_{1s} - \frac{\text{GTH}}{4.45 \times \text{cfm}}$$

where h_{1s} = entering-air enthalpy at sea level and specified altitude entering wet-bulb temperature, Btu/lb
    h_{2s} = leaving-air enthalpy at sea level, Btu/lb as calculated

The saturation temperature which corresponds to the above value of enthalpy is the leaving-air wet-bulb temperature for sea-level operation.
    English units should be used with Eq. (A.2). To obtain metric equivalents:

$$\frac{(1)*(2)}{W}(3.413) = \text{Btu/h}$$

$$\frac{\text{kJ}}{\text{kg}}(0.0004299) = \text{Btu/lb}$$

$$\frac{\text{m}^3}{\text{s}}(2118.9) = \text{cfm}$$

$$\frac{\text{L}}{\text{s}}(2.1189) = \text{cfm}$$

3. From the manufacturer's performance chart, determine the sea-level adp using the entering wet-bulb temperature specified, the assumed coil surface and standard air velocity, and the sea-level leaving wet-bulb temperature determined from step 2.

4. From the published adp ratings, using the load and adp determined above, obtain the value of the suction temperature required for operation at sea level.

5. To obtain the value of the suction temperature required for operation at altitude, increase the sea-level suction temperature obtained from step 4 by the following number of degrees:

Elevation, ft (m)	Altitude sensible heat factor, 0.40–0.95
2,500 (760)	0.6
5,000 (1500)	1.3
7,500 (2300)	2.0
10,000 (3050)	2.6

6. Determine actual leaving dry-bulb temperature at altitude by formula (A.3):

$$t_2 = t_s + BF(t_1 - t_s)$$

where t_s = adp at altitude-saturation temperature corresponding to h_s in formula (A.4)

$$h_s = h_1 - \frac{h_1 - h_2}{1 - BF}$$

t_1 = entering-air dry-bulb temperature, °F (°C)
t_2 = leaving-air dry-bulb temperature, °F (°C)
BF = bypass factor. Determine from the manufacturer's performance chart using standard air cfm.
h_s = enthalpy at t_s, Btu/lb (W/kg). Use altitude psychrometric charts, Figs. A.1 to A.4.
h_1 = entering-air enthalpy at altitude entering-air wet-bulb temperature, Btu/lb (W/kg). Use altitude psychrometric charts, Figs. A.1 to A.4.
h_2 = leaving-air enthalpy at altitude leaving-air wb, Btu/lb (W/kg). Use altitude psychrometric charts, Figs. A.1 to A.4.

Maximum Standard Air Coil Face Velocities. To be consistent with established limits of standard air face velocity (to avoid problems with water carryover), multiply the maximum allowable air face velocity at sea level by the following factors to establish the maximum standard air velocity at altitude:

Elevation, ft (m)	Maximum standard air face velocity factor	Maximum actual air face velocity factor
2,500 (760)	0.95	1.05
5,000 (1500)	0.91	1.10
7,500 (2300)	0.87	1.15
10,000 (3050)	0.83	1.20

For example, the limiting standard air face velocity is 700 ft/min (3.05 m/s). The maximum allowable standard air face velocity at 7500 ft (300 m) would be $0.87 \times 700 = 609$ ft/min ($0.87 \times 3 = 2.61$ m/s). The maximum allowable actual air face velocity at 7500 ft (300 m) would be $1.15 \times 700 = 805$ ft/min ($1.15 \times 3.05 = 3.507$ m/s).

A.2.11 Induction Units; Central Station Air-Handling Units; Fan-Coil Units; Room Fan-Coil Units; Heating and Ventilating Units; Unit Heaters; Packaged Air-Conditioning Units

Steam Heating. Operation of the above units at altitude with a steam-heating coil requires a correction for the reduction in absolute steam pressure and a correction for the reduced air-weight flow at the same fan speed. The effect of these two corrections is combined in the values shown in the second table below.

To obtain the heating capacity at altitude for the above units, multiply the published values of capacity at the altitude cfm by the following factors:

	Altitude capacity factors			
	Steam pressure, psig (bars)			
Elevation, ft (m)	2 (0.138)	10 (0.69)	50 (3.45)	100 (6.9)
2,500 (760)	0.92 (0.0634)	0.93 (0.0611)	0.94 (0.0648)	0.94 (0.0648)
5,000 (1500)	0.85 (0.0586)	0.87 (0.0600)	0.88 (0.0607)	0.88 (0.0607)
7,500 (2300)	0.78 (0.0538)	0.80 (0.0552)	0.82 (0.0566)	0.83 (0.0572)
10,000 (3050)	0.71 (0.0490)	0.74 (0.0510)	0.77 (0.0551)	0.78 (0.0538)

A.2.12 Induction Units; Fan-Coil Units; Room Fan-Coil Units; Unit Heaters; Packaged Air-Conditioning Units

Hot Water. Operation at altitude affects the capacity of the above units only in the reduction of air-weight flow through the unit. To obtain the heating capacity at altitude for these units, multiply the published value of capacity at the altitude cfm by the following factors:

Elevation, ft (m)	Altitude capacity factors
2,500 (760)	0.95
5,000 (1500)	0.90
7,500 (2300)	0.85
10,000 (3050)	0.81

A.2.13 Induction Units

Electric Heat. Operation at altitude does not affect the capacity of the above units since the capacity is determined only by the power consumption of the electric heating elements. It is necessary, however, to increase the minimum required actual primary air volume over that published in order to maintain the same minimum air-weight flow. Failure to compensate for the reduction in air-weight flow may trip the heater element's thermal overload.

Minimum Required Primary Air; Altitude cfm (m³/s)

Heater wattage	Elevation, ft (m)			
	2500 (760)	5000 (1500)	7500 (2300)	10,000 (3050)
500	39 (18.4)	43 (20.3)	47 (22.2)	51 (24.1)
1000	77 (36.3)	85 (40.1)	93 (43.9)	102 (48.1)
1500	110 (51.9)	121 (57.1)	132 (62.3)	146 (68.9)
2000	148 (69.8)	162 (76.5)	178 (84.0)	197 (93.0)
2500	181 (85.4)	198 (93.4)	218 (102.9)	240 (113.3)
3000	219 (108.3)	241 (113.7)	264 (124.6)	291 (137.3)
3500	252 (118.9)	277 (143.5)	304 (143.5)	335 (158.1)
4000	290 (136.9)	319 (150.6)	350 (165.2)	385 (181.7)
4500	323 (156.4)	354 (167.1)	382 (180.3)	429 (202.5)
5000	362 (170.8)	396 (186.9)	437 (206.2)	489 (226.5)

Note: 1000 L/s = 1 m³/s.

A.2.14   Air Terminal Units and Outlets

When the sea-level cfm has been determined for a terminal unit or outlet, the required unit discharge air volume at altitude is equal to the sea-level (standard) cfm (m³/s) divided by the air-density ratio (see Table A.3). Where volume regulators are part of the terminal, they should be set accordingly.

For example, if a terminal unit or outlet is operating at 5000 ft (1500 m) of elevation and the design air flow required is 100 standard cfm (0.0472 m³/s), the unit or outlet should be set to discharge an actual air volume of

$$\frac{100}{0.832} = 120 \text{ cfm} \quad \frac{0.0472}{0.832} = 0.0567 \text{ m}^3/\text{s}$$

This allows the correct air-weight flow to be supplied to the conditioned space.

Where sound-level guides are available for outlet selection, altitude cfm should be used.

A.2.15   Central Station Air-Handling Units; Chilled-Water Fan-Coil Units; Heating and Ventilating Units

Hot Water.   When operated at constant fan speed, the above units have a reduced capacity at altitude due to the reduced air-weight flow through the units. To obtain the heating capacity at altitude, the following procedure is recommended:

1. Multiply the altitude cfm (m³/s) by the air-density ratio to determine the standard cfm.
2. Divide the standard cfm (m³/s) by the coil face area to determine standard air face velocity.
3. Use this face velocity with the specified gpm to determine the heat transfer index from the published curves and gpm for the coil surface required.

TABLE A.3 Altitude Density for Air

Altitude	Air-density ratio	Density	Specific volume
ft	At 70°F	lb/ft³ at 70°F	ft³/lb at 70°F
0	1.00	0.0750	13.33
500	0.982	0.0737	13.58
1,000	0.964	0.0722	13.86
1,500	0.947	0.0710	14.10
2,000	0.930	0.0697	14.36
2,500	0.913	0.0684	14.61
3,000	0.896	0.0672	14.88
3,500	0.880	0.0660	15.17
4,000	0.864	0.0648	15.45
4,500	0.848	0.0635	15.76
5,000	0.832	0.0625	16.00
5,500	0.817	0.0613	16.32
6,000	0.801	0.0601	16.66
6,500	0.786	0.0590	16.98
7,000	0.772	0.0579	17.28
7,500	0.757	0.0567	17.66
8,000	0.743	0.0557	17.95
8,500	0.729	0.0545	18.38
9,000	0.715	0.0536	18.69
9,500	0.701	0.0526	19.05
10,000	0.687	0.0515	19.45
m	At 21.1°C	kg/m³ at 21.1°C	m³/kg at 21.1°C
0	1.000	1.201	0.8326
152.4	0.982	1.181	0.8467
304.8	0.964	1.157	0.8643
457.2	0.947	1.137	0.8795
609.6	0.930	1.116	0.8961
762	0.913	1.096	0.9124
914.4	0.896	1.076	0.9294
1066.7	0.880	1.067	0.9372
1219.1	0.864	1.038	0.9634
1371.5	0.848	1.017	0.9833
1523.9	0.832	1.001	0.999
1676.9	0.817	0.982	1.0183
1828.7	0.801	0.963	1.0384
1981.1	0.786	0.945	1.0582
2133.5	0.772	0.927	1.0787
2285.9	0.757	0.908	1.1013
2438.3	0.743	0.892	1.1211
2590.7	0.729	0.873	1.1455
2743.1	0.715	0.859	1.1641
2895.4	0.701	0.843	1.1862
3047.9	0.687	0.825	1.2121

4. Calculate the unit capacity to verify that it equals or exceeds the specified capacity from

$$Q = \text{HTI} \times 1000(T_w - T_a) \qquad (A.5)$$

where Q = unit capacity, Btu/h
  HTI = heat transfer index
  T_w = entering-water temperature, °F
  T_a = entering-air temperature, °F

English units should be used with Eq. (A.5). To obtain metric equivalents:

$$W(3.413) = \text{Btu/h}$$
$$°C(1.8) + 32 = °F$$

A.2.16 Unit Heaters and Duct Furnaces; Roof-Mounted Heating Units; Furnaces

Gas-Fired.   Operation at altitude affects the unit capacity because the flue passages allow a constant volume of combustion air to pass through the unit regardless of altitude. At altitude, therefore, the combustion air-weight flow through the unit is reduced, causing a reduction in unit capacity.

The recommendation of the American Gas Institute for derating gas-fired heating units for altitude operation is as follows: "Ratings need not be corrected for elevations up to 2000 ft (610 m). For elevations above 2000 ft (610 m), ratings should be reduced 4 percent for each 1000 ft (305 m) above sea level."

The gas flow rate should be adjusted to the reduced capacity by one of the following methods of control:

1. Reduce manifold pressure by changing the setting of the pressure-regulating valve. This provides only a partial degree of gas-flow control but may be sufficient to provide the proper gas flow when operating at low altitudes without any change of the orifice size.

2. Change to a different-size orifice to provide proper gas flow. This change is usually necessary for operation at higher altitudes. Consult the manufacturer's engineering department for correct orifice size for the application. After changing to the correct orifice size, it may also be necessary to vary the setting of the pressure regulator to obtain the proper gas-flow rate.

For gas-fired duct furnaces, an accessory modulating gas valve is available for installation downstream from the pressure-regulating valve. This varies the gasflow rate to the furnace at the required rate for heating.

A.2.17 Direct-Expansion Fan-Coil Units; Chilled-Water Room Fan-Coil Units; Roof-Mounted Air-Conditioning Unit; Package Air-Conditioning Unit

General.   In using the altitude factors to determine unit capacity for the above units, refer first to the published ratings. If the sensible heat factor (SHF) shown for the design conditions is less than 0.95, the unit is dehumidifying at altitude. In these cases the published values of total capacity and sensible capacity should be multiplied by the factors under the 0.40 to 0.95 sensible heat factor columns shown in various of the unnumbered tables in this appendix. If the sensible heat factor shown for the design conditions is greater than 0.95, the unit still may be dehumidifying at altitude due to the higher specific humidity of altitude air for a given design dry-bulb and wet-bulb temperature. To verify whether or not the unit is dehumidifying at altitude, the adp should be calculated from formula (A.4):

$$h_s = h_1 - \frac{h_1 - h_2}{1 - BF}$$

where h_s = enthalpy at t_s, Btu/lb. Use altitude psychrometric charts, Figs. A.1 to A.4.
    h_1 = entering-air enthalpy at altitude entering-air wet-bulb temperature, Btu/lb. Use altitude psychrometric charts, Figs. A.1 to A.4.
    h_2 = leaving-air enthalpy at altitude leaving-air wet-bulb temperature, Btu/lb. Use altitude psychrometric charts, Figs. A.1 to A.4.
    BF = bypass factor. Determine from published data using standard air cfm.
    t_s = adp at altitude–saturation temperature corresponding to h_s

The actual air-condition line should then be drawn on the altitude psychrometric chart from the entering-air conditions to the adp.

If the slope of this line is less than 0.95 m, multiply the published values of total capacity and sensible capacity by the factors under the 0.40 to 0.95 sensible heat factor columns shown in various of the unnumbered tables of this appendix. If the slope of the conditions line is in the range of 0.95 to 1.00, multiply the published values of capacity by the factors under the 0.95 to 1.00 sensible heat factor columns in tables already noted.

Direct-Expansion Fan-Coil Units. For these units, the comparison of altitude capacity with sea-level capacity is made on the basis of constant air-volume flow (varying-weight flow).

To obtain the total heat capacity at altitude for these units, multiply the published values of total heat capacity at the altitude cfm (m³/s), entering-air wet-bulb temperature, and coil-refrigerant temperature by the following factors:

Elevation, ft (m)	Altitude total capacity factors	
	Altitude SHF 0.40–0.95	Altitude SHF 0.95–1.00
2,500 (760)	0.97	0.93
5,000 (1500)	0.95	0.86
7,500 (2300)	0.93	0.79
10,000 (3050)	0.91	0.73

To obtain the sensible heat capacity at altitude for these units, multiply the published values of sensible heat capacity at the altitude cfm (m³/s), entering-air wet-bulb temperature, and coil-refrigerant temperature by the following factors:

Elevation, ft (m)	Altitude sensible capacity factors for altitude SHF 0.40–0.95
2,500 (760)	0.92
5,000 (1500)	0.85
7,500 (2300)	0.78
10,000 (3050)	0.71

Chilled-Water Room Fan-Coil Units. For these units, the comparison of altitude capacity with sea-level capacity is made on the basis of constant air-volume flow (varying-weight flow).

To obtain the total heat capacity at altitude for these units, multiply the published values of total heat capacity at the entering-air wet-bulb temperature, gpm (m³/s), and entering-water temperature by the following factors:

Elevation, ft (m)	Altitude total capacity factors	
	Altitude SHF 0.40–0.95	Altitude SHF 0.95–1.00
2,500 (760)	0.97	0.93
5,000 (1500)	0.95	0.86
7,500 (2300)	0.93	0.80
10,000 (3050)	0.91	0.74

To obtain the sensible heat capacity at altitude for these units, multiply the published values of sensible heat capacity at the altitude cfm (m³/s), entering-air wetbulb temperature, and coil-refrigerant temperature by the following factors:

Elevation, ft (m)	Altitude sensible capacity factors for altitude SHF 0.40–0.95
2,500 (760)	0.92
5,000 (1500)	0.85
7,500 (2300)	0.78
10,000 (3050)	0.71

Direct-Expansion Roof-Mounted Air-Conditioning Unit; Direct-Expansion Package Air-Conditioning Unit. Since the above units are rated in terms of cfm, the comparison of altitude capacity with sea-level capacity is made on the basis of constant air-volume flow (varying-weight flow).

To obtain the total heat capacity at altitude for these units, multiply the published values of total heat capacity at the altitude cfm (m³/s), entering-air wet-bulb temperature, and condensing temperature (for water-cooled units) or temperature of air entering condenser (for air-cooled units) by the following factors:

Unit with Air-Cooled Condenser

Elevation, ft (m)	Altitude total capacity factors	
	Altitude SHF 0.40–0.95	Altitude SHF 0.95–1.00
2,500 (760)	0.98	0.96
5,000 (1500)	0.96	0.92
7,500 (2300)	0.94	0.88
10,000 (3050)	0.92	0.84

Unit with Water-Cooled Condenser

Elevation, ft (m)	Altitude total capacity factors	
	Altitude SHF 0.40–0.95	Altitude SHF 0.95–1.00
2,500 (760)	0.99	0.96
5,000 (1500)	0.985	0.93
7,500 (2300)	0.98	0.90
10,000 (3050)	0.97	0.87

To obtain the sensible heat capacity at altitude for these units, multiply the published values of sensible heat capacity at the altitude cfm (m³/s), entering-air wetbulb temperature, and condensing temperature by the following factors:

Unit with Air-Cooled Condenser

Elevation, ft (m)	Altitude sensible capacity factors	
	Altitude SHF 0.40–0.95	Altitude SHF 0.95–1.00
2,500 (760)	0.92	0.96
5,000 (1500)	0.85	0.92
7,500 (2300)	0.78	0.88
10,000 (3050)	0.71	0.84

Unit with Water-Cooled Condenser

Elevation, ft (m)	Altitude sensible capacity factors	
	Altitude SHF 0.40–0.95	Altitude SHF 0.95–1.00
2,500 (760)	0.93	0.96
5,000 (1500)	0.87	0.93
7,500 (2300)	0.80	0.90
10,000 (3050)	0.74	0.87

For all fan-coil units having a belt-driven evaporator fan, the fan speed may be increased at altitude to circulate a larger air volume through the evaporator. This increases the unit capacity so that the altitude capacity factors shown are increased. For cases where the evaporator fan speed is increased, capacity values must be obtained from the unit manufacturer for each specific application.

A.2.18 Centrifugal Fans

Altitude affects both the fan static pressure and brake horsepower as stated by the fan law: "For constant capacity and speed, the horsepower and pressure vary directly as the air density (directly as the barometric pressure and inversely as the absolute temperature)."

Fan tables and curves are based on air at standard conditions. For operation at altitude a correction should be applied. With a given capacity [cfm (m³/s)] and total static pressure at altitude operating conditions, the adjustments are made as follows:

1. Determine the air-density ratio (Table A.2).
2. Calculate the equivalent static pressure by dividing the given static pressure by the air-density ratio.
3. Enter the fan tables at the given capacity [cfm (m³/s)] and the equivalent static pressure to obtain fan speed and a pseudo brake horsepower. The speed is correct as determined.
4. Multiply the pseudo brake horsepower by the air-density ratio to find the true brake horsepower at the operating conditions.
5. Determine which class of fan is required by referring to the manufacturer's fan catalog, using the equivalent static pressure and the fan outlet velocity.

In determining the required fan horsepower for equipment where values of horsepower (sea level) are published as a function of external static pressure, the adjusted horsepower must be determined from adjusted values of total static pressure (not external pressure) following the procedure given above.

Refer to Sec. A.4. "System Pressure Loss" for static pressure adjustment. Where unit static pressure drop is not available, estimate or obtain it from the manufacturer.

A.2.19   Pumps

Altitude affects the operation of pumps installed in an open system because it reduces the available net positive suction heat (NPSH). The available NPSH must always be equal to or greater than the required NPSH in order to produce flow through the pump. A lack of sufficient NPSH allows the liquid to flash into vapor inside the pump, often resulting in unstable flow and cavitation.

Refer to Carrier Crop.'s *System Design Manual*, part 8, chap. 1, or Chap. 25.4 for the procedure to calculate the available NPSH.

Pumps installed in a closed system are selected on a standard gpm (m³/s) versus head losses, and the effect of NPSH on flow to the pump suction need not be considered.

A.2.20   Motors

Since the effectiveness of cooling air depends on its density, motor cooling is decreased with altitude. To compensate for this decrease, it is necessary to provide additional margin for the increase in motor temperature. The decrease in cooling effectiveness may be offset by one or more of the following:

1. Elimination of the service factor
2. Provision of a higher-temperature insulation
3. Special motor design
4. Use of a larger-size motor.

Up to 3300 ft (1005 m) of elevation, all motors meet their rated temperature guarantees. Between 3300 and 9900 ft (1005 and 3020 m) of elevation, motors operate satisfactorily if not operated above nameplate rating when applied voltage is constant. If a variation in voltage of the magnitude of ±10 percent exists and the motor is to be operated above 3300 ft (1005 m), consult the motor manufacturer for recommendations. It is recommended that the motor manufacturer be consulted for all applications above 9900 ft (3020 m).

A.3 LOAD CALCULATION

The following adjustments are required to calculate loads at high altitude:

1. The design outside- and room-air moisture content must be adjusted to the new elevation by one of the following methods:
   a. If dry-bulb temperature and percent relative humidity are given, divide the specific humidity at sea level by the air-density ratio.
   b. If dry-bulb and wet-bulb temperatures are given, obtain the specific humidity at altitude from the formula

$$W_1 = W_0 + \frac{P_0 - P_1}{P_1} W_s \qquad (A.6)$$

where W_0 = specific humidity at altitude for specified dry-bulb and wet-bulb temperatures, lb/lb of dry air (kg/kg or dry air)
W_s = specific humidity at sea level and saturated wet-bulb temperature, lb/lb of dry air (kg/kg of dry air)
P_0 = barometric pressure, psia (kPa absolute)
P_1 = altitude pressure, psia (kPa absolute)

Note: The values of specific humidity given in table 1, part 1, chap. 2 of Carrier Corp.'s *System Design Manual* are already corrected for the altitude condition.

2. Solar heat-gain corrections should be made for altitude. Refer to part 1, chap. 3, table 6 of Carrier Corp.'s *System Design Manual.*
3. Because of the increased moisture content of the air, the effective sensible heat factor must be corrected. Refer to part 1, chap. 8, table 66 of Carrier Corp.'s *System Design Manual.*

A.4 SYSTEM PRESSURE LOSS

A.4.1 Air Friction through Air-Conditioning Equipment (Filters, Heating Coils, Cooling Coils, etc.)

To obtain correct values of air friction at altitude and also at temperatures other than 70°F (21.1°C), the following procedure is recommended:

1. Correct the altitude cfm (m³/s) to standard cfm (m³/s) by multiplying the altitude cfm by the air-density ratio (Table A.3).
2. From published data determine the value of air friction based on standard cfm flow.
3. Divide the standard air friction obtained from step 2 above by the air-density ratio to obtain the air friction at altitude and temperature.

A.4.2 Air Friction through Ducts

1. Use the altitude cfm (m^3/s) to obtain a value of air friction for standard air from duct design manuals such as *ASHRAE Handbook of Fundamentals*, Chap. 12, "Duct Design."
2. To correct for nonstandard air density at altitude, multiply the value of air friction obtained from step 1 by the following factors:

Elevation, ft (m)	Temp. corr. factor F_1	Temperature, °F (°C)	Temp. corr. factor F_2
2,000 (610)	0.944	0 (−17.8)	1.120
4,000 (1220)	0.890	50 (10)	1.031
6,000 (1830)	0.838	100 (37.8)	0.957
8,000 (2440)	0.788	150 (65.6)	0.894
10,000 (3050)	0.742	200 (93.5)	0.840
		250 (121.1)	0.792
		300 (143.9)	0.749

Duct air friction (at altitude and temperature) = duct air friction (standard air) $\times F_1 \times F_2$.

BIBLIOGRAPHY

Carrier Corporation: *System Design Manual,* McGraw-Hill, New York, 1960.

APPENDIX B
METRIC CONVERSION TABLES

NOTE: Most of the text in this handbook provides SI metric conversions. However, editorial factors necessitated omissions in some tables and graphs. Listed below are conversions most likely encountered in HVAC.

In the table below the first two digits of each numerical entry represents a power of 10. For example, the entry "−02" expresses 2.54×10^{-2}. Standard abbreviations are used as appropriate.

To convert from	To	Multiply by
AREA		
acre	m^2	+03 4.046
circular mil	m^2	−10 5.067
in^2	m^2	−04 6.4512
ft^2	m^2	−02 9.290
mi^2	m^2	+06 2.589
$yard^2$	m^2	−01 8.361
DENSITY		
lb/in^3	kg/m^2	+04 2.768
lb/ft^3	kg/m^2	+01 1.602
ENERGY		
Btu (mean)	joule	+03 1.056
Btu/lb	J/kg	+03 2.324
ft·lb	joule	+00 1.356
ENERGY/AREA TIME		
$Btu/(ft^2 \cdot s)$	W/m^2	+04 1.135
FORCE		
lb	N	+00 4.448
oz	N	−01 2.780
LENGTH		
ft	m	−01 3.048
in	mm	+01 2.54
mil	m	−05 2.540
mi (U. S. statute)	m	+03 1.609
yd	m	−01 9.144
MASS		
oz	kg	−02 2.835
lb	kg	−01 4.536
ton (short)	kg	+02 9.072

(Continued)

(*Continued*)

To convert from	To	Multiply by
	POWER	
Btu/h	W	−01 2.931
(ft·lb)/hr	W	−04 3.766
hp [550 (ft·lb)/s]	W	+02 7.457
	PRESSURE	
bar	Pa	+05 1.00
ft H_2O	Pa	+03 2.989
inHg	Pa	+03 3.386
lb/in^2 (psi)	Pa	+03 6.895
	SPEED	
ft/s	m/s	−01 3.048F
mi/h	m/s	−01 4.470
	TEMPERATURE	
°F	°C	(°F − 32)/1.8
	VISCOSITY	
cSt	m^2/s	−06 1.00
cP	(Pa·s)	−03 1.00
	VOLUME	
barrel (42 gal)	m^3	−01 1.590
fluid oz (U.S.)	m^3	−05 2.597
ft^3	m^3	−02 2.832
gal (U.S. liquid)	m^3	−03 3.785
gal (U.S. liquid)	L	00 3.785
in^3	m^3	−05 1.639
ppm	mg/L (of H_2O)	00 1.000

APPENDIX C
GLOSSARY OF HVAC TERMS

Robur Corporation
Evansville, Indiana

Absorption   The process by which one material extracts other substances from a mixture of gases or liquids.

Access Door   A door or panel provided in any structure, as in a duct, wall, etc., or in a cooling or heating unit, to permit inspection and adjustment of the inside components.

Air, Ambient   The air surrounding an object.

Air Change   The amount of air required to completely replace the air in a room or building; not to be confused with recirculated air.

Air Circulation   Natural or forced movement of air.

Air Cleaner   A mechanical, electrical, or chemical device (usually a filter) for removing dust, gas, vapor, fumes, smoke, and other impurities from air.

Air Conditioner   A machine that controls the temperature, moisture, cleanliness, and distribution of air.

Air Conditioner, Unitary   An air conditioner consisting of several factory-made components within an insulated casing, normally including an evaporator (cooling) coil, a condenser and compressor combination, and controls and fans for distributing conditioned air; some also contain filters, heaters, and dampers.

Air Conditioning   The process of treating air to control its temperature, moisture, cleanliness, and distribution.

Air-Conditioning System, Year-Round   A system that controls the total environment during all seasons under all comfort requirements.

Air Diffuser   An outlet which discharges supply air in a spreading pattern.

Air Handler   A fan or blower for moving air within a distribution system.

Air-Handling Unit   An air handler, heating element and cooling coil, and other components in a cabinet or casing.

Air, Recirculated   Air drawn from a space and passed through the conditioner and discharged again into the conditioned space.

Air, Ventilation   The quantity of supply air drawn from outdoors needed to maintain the desired amount of oxygen and the quality of air within a designated space.

Anemometer   An instrument for measuring the velocity of air.

Atomize   To break up a liquid into a fine spray, as in oil burner nozzles and atomizing water humidifiers.

Baffle   A surface used to deflect or direct air flow, usually in the form of a plate or wall.

Barometer   A device for measuring atmospheric pressure.

Blow (Throw)   In air distribution, the distance an air stream travels from an outlet to a point where air motion is reduced to a specific flow measured in feet (meters) per minute (terminal velocity).

Blower (Fan)   An air-handling device for moving air in a distribution system.

British Thermal Unit (Btu)   A unit for measuring heat quantity equal approximately to the amount of heat produced by the burning of an ordinary wooden kitchen match. More specifically, it is the amount of heat required to raise the temperature of 1 lb of water $1°F$.

Capacity Reducer   A device, usually placed at the energy source, to reduce the Btuh capacity of the unit; usually used to decrease a furnace's output without decreasing its air-handling capacity; frequently required in well-insulated, tightly constructed buildings.

Capillary Tube   A small-diameter tube (metering device), whose bore and length are designed to permit the passage of a specific amount of liquid refrigerant or other liquid at a specific pressure drop.

Centigrade   A metric scale for measuring temperature. $0°C$ is $+32°F$, and $100°C$ is $+212°F$.

Change of Air   Introduction of new air to a conditioned space, measured by the number of complete enclosed-space changes per unit time, usually per hour.

Charge, Refrigerant   The amount of refrigerant in a system; to put in the refrigerant charge.

Chill   To reduce temperature moderately without freezing.

Chimney Effect   The tendency of air or gas to rise in a duct or other vertical passage when heated.

Coil   A heating or cooling element made of pipe or tubing, with or without extensions or fins.

Coil, Cooling   A commonly used term meaning the evaporator.

Comfort Zone   The range of temperatures, humidities, and air velocities at which the greatest percentage of people feel comfortable.

Compression   In a compression refrigeration system, a process by which the pressure of the refrigerant is increased.

Compression, ratio of   The ratio of absolute pressures after and before compression.

Compressor   A device for increasing the pressure of heat-laden refrigerant vapors thus increasing the heat level (super-heating) within those vapors so that the heat contained can be released into the outside atmosphere.

Condensate   The liquid formed by the condensation of a vapor, e.g., moisture on cold window glass, or dew.

Condensation   The process whereby a vapor is changed to a liquid by removal of heat after its dew point (condensation temperature) is reached.

Condenser   A vessel, or arrangement of pipes or tubing, in which vapor is liquified by removal of heat.

Condensing Unit   A refrigerating machine consisting of one or more power-driven compressors, condensers, liquid receivers, and other components.

Conduction   The process of transferring heat along the elements of a substance, as from a tube to a fin.

Conduit   A tube or pipe used for conveying liquid or gas; also, a tube or pipe in which wire or other pipes may be inserted for protection.

Connecting Rod   A device connecting a piston to a crankshaft used in reciprocating compressors.

Control   A device for regulating a system, or part of a system, in normal operation. Such a device can be either manually operated or automatic.

Convection   The transfer of heat by movement of fluid or air.

Cycle, Refrigeration (Absorption)   The complete cycle of a refrigerating agent through a system whereby the agent absorbs heat, transports it to a point where a second, unrelated, substance extracts the heat from the initial agent which is returned through the system to absorb more heat and repeat the process. Energy source for such a cycle can be one of several different types.

Cycle, Refrigeration (Mechanical)   Complete cycle of a refrigerant from a condensing unit (compressor-condenser) through a system of lines or tubing to the evaporator coil and back into the condensing unit; a cycle in which heat is absorbed in the evaporator and given up in the condenser.

Damper   A device for adjusting the amount of air flowing through an outlet, inlet, or duct.

Damper, Fire   A damper placed in a duct or at an outlet to prevent or retard the emission of smoke and/or heat into a room or space during a fire. Such devices are also designed to prevent or retard the spread of fire throughout the structure, thus confining it to a specific area.

Defrosting   The removal of accumulated ice or frost from a cooling element or coil.

Dehumidification   The reduction of water vapor in air by cooling the air below the dew point; removal of water vapor from air by chemical means, refrigeration, etc.

Dehydrate   To remove water in all forms from other matter.

Dew Point   The temperature at which water vapor turns to liquid.

Draft   A current caused by the movement of air from an area of high pressure to an area of low pressure; usually considered objectionable.

Drier   A device placed in a refrigeration system to attract and collect unwanted moisture which may be in the system.

Dry   To separate or remove liquid or vapor from another substance, such as moisture from a refrigerant.

Duct   A pipe or closed conduit made of sheet metal or other suitable material used for conducting air to and from an air-handling unit.

Entrainment   The induction of room air into an airstream from an outlet. (Also, see Induction.)

Evaporator   That part (heat exchanger) of a cooling system in which refrigerant is vaporized.

Exfiltration   Air flow outward from an enclosed space through a wall, leak, membrane, etc.

Exhaust Opening   An opening through which air is exhausted from a room that is being cooled, heated, or ventilated.

Fan Coil Unit   An air-handling unit especially designed to condition small spaces.

Filter   A device for removing dust particles from air or unwanted elements from liquids.

Filter Drier   A combination of a liquid refrigerant line strainer and dehydrator.

Fluid   Any liquid or gas.

Freezing Point   The temperature at which liquids will solidify or freeze upon removal of heat (e.g., the freezing point of water is +32°F or 0°C).

Furnace   That part of an environmental system that converts gas, oil, electricity, or other fuel into heat for distribution within a structure.

Gage (Gauge)   An instrument for measuring temperatures, pressures, or liquid levels. Also, an arbitrary scale of measurement for sheet metal thicknesses, wire and drill diameters, etc., liquid levels within boiler pipes, tanks, and other enclosures.

Gas, Fuel   An expanded hydrocarbon fluid used to provide heat in environmental systems.

Gas, Inert   A gas that neither undergoes nor causes chemical reaction nor change of form.

Gravity, Specific   Density measured on the basis of the known density of a given substance, usually air or water.

Grille   A covering for an opening through which air passes.

Heat   Energy which can be transferred as a result of temperature differences.

Heat, Latent   Heat energy which changes the form of a substance (e.g., ice to water) without changing its temperature.

Heat Sensible   Heat which changes the temperature of a substance without changing its form.

Heat Exchanger   Usually considered to be that part of a furnace that transfers heat from burning fuel or electricity into air or other medium. Also, it is any device in which heat is absorbed by one fluid from another fluid, such as a condenser, evaporator, or boiler.

Heat Pump   A reverse-cycle refrigeration system designed to perform both heating and cooling operations.

Heat Transmission   Any time-rate flow of heat, usually refers to conduction, convection, and radiation.

High Side   That part of a refrigeration system where refrigerant is under the greatest amount of pressure where heat is rejected (condenser section). This is area of highest temperatures and highest pressures.

Humidifier   A device that adds moisture to warm air being circulated or directed into a space.

Humidistat   A device designed to regulate humidity input by reacting to changes in the moisture content of the air.

Humidity, Relative   The percentage of moisture in the air measured against the amount of moisture the air could hold at a given temperature. For example, since cold air is capable of holding less moisture than warm air, if the temperature drops and the moisture volume remains constant, the relative humidity will increases. Conversely, if the temperature rises and the moisture volume remains constant, the relative humidity will decrease.)

Induction   The entrainment of room air into an air stream from an air outlet.

Infiltration   Air flow inward into a space through walls, leaks around doors and windows or through the building material used in the structure.

Insulation, Thermal   A material having a relatively high resistance to heat flow, used primarily to reduce heat loss and heat gain.

Liquid Line   The tube or piping carrying liquid refrigerant from the condenser or receiver of a refrigeration system to a pressure-reducing device.

Louver   A series of vanes that permit directional adjustment of air flow. They are usually installed on outside grilles or intake openings to impede water entry into ducts, elbows, and to minimize turbulence; they may also include indoor grilles to eliminate light penetration.

Low Side   Those parts of a refrigeration system where heat is absorbed at evaporator pressure; the cooling coil or evaporator with associated components; also commonly referred to as the suction side; this is the area of lowest temperature and lowest pressure.

Main (Trunk Line)   Pipe or duct for distributing fluids such as air, water, or steam to various branch ducts and collecting fluids from various branches.

Makeup Air Unit   Unitary equipment which introduces outside air into a space and conditions it to offset air exhausted from that space.

Manometer   An instrument for measuring pressures by the difference in liquid volume.

Meter   An instrument for measuring rates of flow of energy or fuel over a given period. Also, in the metric system of measurement, it is a distance of 39.37 in.

Motor, Air (Pneumatic Motor)   An air-operated device normally used for opening and closing dampers and valves.

Multizone System   An air-conditioning system designed to serve one or more areas having different heating/cooling/humidification requirements. Such a system can consist of more than one heating/cooling/humidification unit, or can consist of a single unit controlled by dampers, bypasses and thermostats.

Muffler, Noise Reducing   A device for the prevention or reduction of sound or vibration being transmitted from one space to another via air-conditioning distribution systems or equipment.

Oil Separator   A device for separating oil or oil vapor from refrigerant.

Plenum Chamber   An air compartment connecting one or more ducts with the heating or cooling unit. Any enlarged section within an air duct with several duct, grille, or diffuser connections.

Pneumatic   Any device actuated by air pressure is said to be pneumatic.

Power Roof Ventilator   A motor-driven exhaust fan mounted above the roof on a roof curb.

Pressure, Absolute   That pressure which will register on a pressure, gauge, plus atmospheric pressure present, e.g., a gauge pressure of 68.5 pounds per square inch (psi) (5 bars) at sea level would represent an absolute pressure of 83.2 (68.5 plus 14.7 equals 83.2) [6 bars]. In the United States, absolute pressure is expressed as pounds per square inch absolute, or as psia. Elsewhere, in bars or pascals.

Pressure, Atmospheric   Pressure resulting from the weight of the atmosphere which is 14.696 pounds per square inch of surface (1 bar, 100 kPa); or that pressure in the outdoors or within a space that is present as a result of the forces of nature with no artificially induced pressure change.

Pressure Drop   That pressure lost between any two points of a piping or duct system due to friction, leakage, or other reasons.

Pressure, Static   Force (per unit area) expended against the walls of a container such as an air duct. Commonly, force is measured in heating and air-conditioning in pounds per square inch, or inches of water pressure.

Pressure, Total   A combination of both static and velocity pressures.

Pressure, Velocity   Force of air as it moves in an air duct.

Psychrometer   A thermometer-like instrument for measuring wet bulb and dry bulb temperatures simultaneously.

Radiation   Transmission of heat or energy by electromagnetic waves, such as heat transmitted by an electric heater and absorbed by a disjoined substance.

Radiator   A mechanical device that transmits heat by the process of radiation.

Refrigerant   The fluid in a refrigeration cycle that absorbs heat at low temperatures and rejects heat at higher temperatures.

Register   A combination grille and damper assembly covering an air-supply outlet, designed to distribute air into a room.

Return Air   Air recirculated through a return air system to an air-handling unit or furnace.

Saturation   A condition that occurs when air contains all the moisture that it can possibly hold at a given temperature.

Sight Glass   A glass tube used to indicate the liquid condition and level in pipes, tanks, bearings, and similar equipment.

Steam   Water that has turned to vapor as a result of the application of heat at a given pressure, or the reduction of pressure at a given temperature.

Strainer   A device for withholding foreign matter from a flowing liquid or gas.

Sun Effect   Heat transmitted into space through glass and building materials exposed to the sun.

System, Duct   A series of tubular or rectangular sections, elbows, and connectors fabricated as a channel to carry air from one point to another.

Temperature   The heat content of a substance in degrees Fahrenheit or Centigrade.

Temperature Drop   A measurement of the difference in heat between two points of a system, such as at the furnace plenum and at the outlet grille.

Temperature, Dry-Bulb   The temperature of the air at any given location indicated by an accurate thermometer and not influenced by outside interferences such as radiation or water.

Ton of Refrigeration   Removal of 12,000 Btu of heat per hour, of 200 Btu per minute (360 W), from a given area. The heat required to melt one ton of ice (288,000 Btu) in a period of 24 h.

Unit   An assembly for heating, cooling, dehumidifying, and/ or ventilating.

INDEX

ABOUT THE EDITOR

Robert C. Rosaler is an engineering and management consultant with more than 40 years' experience as an engineer/executive at companies large and small. The editor of two previous well-received books on HVAC—*HVAC Maintenance and Operations Handbook* and *HVAC Systems & Components Handbook*, both from McGraw-Hill, he resides in Santa Rosa, California.